Green Energy and Technology

Climate change, environmental impact and the limited natural resources urge scientific research and novel technical solutions. The monograph series Green Energy and Technology serves as a publishing platform for scientific and technological approaches to “green”—i.e. environmentally friendly and sustainable—technologies. While a focus lies on energy and power supply, it also covers "green" solutions in industrial engineering and engineering design. Green Energy and Technology addresses researchers, advanced students, technical consultants as well as decision makers in industries and politics. Hence, the level of presentation spans from instructional to highly technical.

Indexed in Scopus.

Indexed in Ei Compendex.

Hicham Kaddouri • Firyal Latrache
Abdelaziz Benkhalifa
Abderrahim Abidouche • Najwa Jbira
Editors

Smart Cities for Sustainable and Urban Development

Approaches, Innovations and Challenges

Springer

Editors
Hicham Kaddouri
National School of Applied Sciences
Abdelmalek Essaâdi University
Al-Hoceima, Morocco

Firyal Latrache
National School of Applied Sciences
Oujda, Morocco

Abdelaziz Benkhalifa
High School of Technology
Mohamed I University
Oujda, Morocco

Abderrahim Abidouche
High School of Technology
Sidi Mohamed Ben Abdellah University
Fez, Morocco

Najwa Jbira
Université Sultan Moulay Slimane
Khouribga, Morocco

ISSN 1865-3529 ISSN 1865-3537 (electronic)
Green Energy and Technology
ISBN 978-3-031-99000-7 ISBN 978-3-031-99001-4 (eBook)
https://doi.org/10.1007/978-3-031-99001-4

This Springer imprint is published by the registered company Springer Nature Switzerland AG
The registered company address is: Gewerbestrasse 11, 6330 Cham, Switzerland

"I dedicate this work to my parents, for their unconditional love and unwavering support.
To my son Adam, a constant source of joy and inspiration.
To my brother Mohammed, my sister Fadya, her husband Saadeddine, and their children, for their cherished presence.
To my friends Abderrahim, Ayoub, and Ahmed, for their loyalty and kindness.
To my esteemed professors, Ismael Driouch and Mohamed Saidi Hassani Alaoui, whose guidance and teachings have left a lasting impact on me.
And finally, to all those who strive each day to build a smarter, more sustainable, and more humane future."

Hicham Kaddouri

"To my father Farid Latrache, my mother Fatima Hamdaoui, and my brother Mohammed Latrache.
To the soul of my aunt Halima Hamdaoui 'Lahbiba', who passed away on October 19th, 2023. My aunt Halima, may her soul rest in peace, taught me how to use a computer 20 years ago and contributed to my openness to digital technologies.

To Professor Noura Ouerdi, professor at the Department of Mathematics and Computer Science of Faculty of Sciences, Mohamed First University in Oujda Morocco, who gifted me my first computer.
I dedicate this book."

Firyal Latrache

"I dedicate this book to my family for their unwavering support, to my colleagues and students who continue to inspire me, and to all those who believe in the power of knowledge to build smart and sustainable cities."

Abdelaziz Benkhalifa

"I dedicate this work to my parents, for their unconditional love and unwavering support, which have been the foundation of my journey.
To my children, Ghaith and Tasnim, who represent the future and give deep meaning to each of my efforts.
To my brothers and sisters, whose presence and affection have always been a source of strength.

To my dear friends, for their listening ear, their encouragement, and their sincere friendship.
To my esteemed professors, Abdelouahad AIT MSAAD and Said HAMDAOUI, whose teachings, rigor, and inspiration have shaped my thinking and guided my path.
And finally, to all those who, in one way or another, contributed to the realization of this work—through their advice, assistance, or simply their kind and supportive presence."

Abderrahim Abidouche

"To my parents, whose unwavering love and support have been my greatest source of strength and inspiration. To my siblings, whose encouragement has accompanied me at every step. To my true friends, who have brought comfort, joy, and sincerity to my journey. And to everyone who believes in me, supports me, and shares a love for science and knowledge, may this work be a tribute to you all, a token of my deepest gratitude and passion for progress."

Najwa Jbira

Foreword

The concept of smart cities represents a visionary transformation of urban living in the twenty-first century. As the world rapidly urbanizes, cities are not only growing in population but are also becoming more interconnected and technologically advanced. This book takes an in-depth look at the multifaceted dimensions of smart cities, exploring the intricate interplay of technology, governance, sustainability, and social equity.

At the heart of the smart city model is the integration of cutting-edge information and communication technologies (ICT) into the governance framework of urban environments. As the governance of these cities evolves, it becomes crucial to ensure policies are forward-thinking, inclusive, and adaptive to the ever-changing technological landscape. Ethical considerations and security concerns take center stage in this transformation, as cities must balance innovation with safeguarding the privacy and rights of their citizens.

One of the central themes in this book is the application of smart education through ICT, which promises to revolutionize how knowledge is disseminated in urban areas. The digitalization of education offers new avenues for learning, engagement, and empowerment, making education more accessible and effective in ways that were previously unimaginable.

Smart cities must also prioritize sustainable development, especially in terms of managing natural resources and energy solutions. The pressing need for environmental responsibility has led to the adoption of renewable energy sources, smart grids, and green buildings, helping cities reduce their carbon footprint while improving the quality of life for their inhabitants. The efficient use of resources is not only crucial for environmental reasons but also for the long-term prosperity of urban populations.

Electric vehicles (EVs) and blockchain technology are also key components in the mobility revolution of smart cities. Together, they promise to reimagine urban transportation, offering cleaner, more efficient, and transparent alternatives to traditional models of mobility. As cities strive to reduce traffic congestion and environmental impact, EVs and blockchain open new possibilities for an equitable and sustainable transport ecosystem. Governance, too, is evolving with the rise of e-administration and digital technologies. Smart cities are increasingly utilizing digital platforms for public service delivery, making governance more efficient, transparent, and responsive. This shift not only streamlines processes but also enhances citizen engagement, fostering a more participatory and democratic urban environment.

The economic dimension of smart cities is another crucial focus. By leveraging digital technologies, cities can unlock new business opportunities, stimulate innovation, and drive economic growth. The digital economy offers enormous potential for cities to evolve into hubs of entrepreneurship and digital business, where new industries thrive, and existing ones are redefined.

Healthcare is perhaps one of the most transformative sectors in the smart city revolution. Smart healthcare solutions, powered by ICT, IoT, and data analytics, promise to enhance urban health by improving access to care, enabling better disease management, and fostering healthier lifestyles. With these technologies, urban health systems can become more responsive and proactive, ultimately improving the well-being of city dwellers.

This book also presents valuable case studies and outcomes from ongoing smart city projects across the globe, shedding light on the successes and challenges of these initiatives. While the promise of smart cities is immense, the reality is not without its hurdles. The challenges of integrating complex technologies, ensuring inclusivity, and maintaining security remain significant obstacles that require continued attention and innovation.

In navigating the concept of smart cities, this book aims to provide a comprehensive understanding of the opportunities, challenges, and strategies that shape the cities of tomorrow. It offers a road map for policymakers, urban planners, technologists, and citizens who are collectively shaping the future of urban living. The future is smart, but it is up to us to ensure that it is also sustainable, equitable, and resilient. This book serves as both a reflection on the current state of smart cities and a blueprint for their evolution in the coming decades.

Mohamed Saidi Hassani Alaoui is a Research Professor at the National School of Applied Sciences of Al Hoceima, Abdelmalek Essaâdi University. Mohamed obtained his PhD in Analytical Modeling and Numerical Simulation of Building Energy Efficiency Assessment. He is currently part of the Engineering Sciences and Applications Laboratory (LSIA) at ENSA Al Hoceima, Abdelmalek Essaâdi University. He is also an associate member of the Energy Research Center and the Modeling and Simulation Team in Mechanics and Energy at the Faculty of Sciences,

Mohammed V University in Rabat. He is the author of numerous research articles on thermal comfort applied to buildings and on various research areas, including: modeling of convection, conduction, and radiation phenomena; numerical simulation of building heat transfer; evaluation of the energy efficiency of buildings; and the energy, economic and environmental impact of thermal comfort in the residential building envelope.

Abdelmalek Essaadi University—National School of Applied Sciences
Al Hoceima, Morocco

Mohamed Saidi Hassani Alaoui

Preface

This book is about investigating the contribution of the smart cities on achieving a sustainable urban development. It presents the approaches allowing the implementation of the smart city aspect for the developed and developing countries and innovate new models by using the digital technologies and then addresses the challenges preventing the implementation of the sustainable urban development. For that, this book describes the policies and governance organizing the implementation of the smart city aspect and presents the ethics and the security procedures to guarantee the citizens comfort. Then, this book describes how the digital technologies improve the quality of life, the service efficiency, the energy affordability, and the social inclusion of the citizens. In this context, the chapters present in depth the practices and the techniques enabling the improvement of the educational system, the health sector, the buildings and the energy security, the mobility and the natural resources preservation, and the economical sector and the administrative system. Overall, this book presents perspectives and outlooks to perform and achieve the objectives of the smart education, the smart management of the natural resources, the smart energy, the smart mobility, the E-administration, the digital business and economy, and the smart healthcare.

This work was conducted by the five editors: Hicham Kaddouri, Firyal Latrache, Abdelaziz Benkhalifa, Abderrahim Abidouche, and Najwa Jbira, between August 2024 and March 2025, to contribute in encouraging the implementation of the smart city aspect. These editors are researchers from different disciplines—renewable energies, energy efficiency, and computer science—and have a primary objective of exploiting their fields of expertise to encourage the use of digital technologies to smarten the conventional cities. Then, the editors intend to eliminate the discrimination in conventional cities between rich and poor areas while allowing access to all public services and the achievement of quality of life norms for all citizens without exception. Overall, the editors aim to contribute to the implementation of the 17th sustainable development goals (SDGs) by integrating smart city concepts, to reduce the economic crisis and restore societal balance following political conflicts, wars, or other crisis, by harnessing digital technologies.

The scientific research is the light of humanity, and due to this edited book, there will no longer be any discrimination, inequalities, or even energy poverty. The world can be recovered after each breakdown using the digital technologies for building new smart cities smarter than the previous. The healthcare solutions will be more innovative, satisfied, and developed according to the patients' needs. Idem for the all sectors allowing the dignity conservation of the citizens. The real goal is not only achieving the principles of the energy transition and mitigating the climate change but also guarantying a good life for all the citizens in light of an optimal economic situation while remaining connected to all public services and benefiting from positive energy without any psychiatric disorders.

Al-Hoceima, Morocco Hicham Kaddouri
Oujda, Morocco Firyal Latrache
Oujda, Morocco Abdelaziz Benkhalifa
Fez, Morocco Abderrahim Abidouche
Khouribga, Morocco Najwa Jbira

Introduction

For a long time, sustainability has been linked to the conservation of natural resources and energy security. Even the first contact between prehistoric human and the outside world led to the discovery of agriculture and fire, leaving the task of energy consumption in human settlements. By rubbing pieces of wood together, heat was generated that enabled other substances, such as herbs, to be ignited. This fire was not only the first step in prehistoric human's energy consumption but it was also a protective device between him and predators. Therefore, it signaled the existence of a 2.8-billion-year relationship between human beings and the external environment. From prehistoric times through to the modern era, the environment, the outside world, or the external milieu has been the driving force behind the emergence of combustion techniques and modes of energy consumption and production that continue to develop to this day. These modes of energy consumption developed from the use of fire for cooking, heating, and combustion in prehistoric times to oil lamps in antiquity. In the Middle Ages, wind and water mills were widely used before the advent of steam engines during the Industrial Revolution in the nineteenth century. Today, energy consumption has increased rapidly with the exploitation of fossil fuels, such as oil and gas and the expansion of the use of renewable energies. Indeed, the idea of exploiting renewable resources was proposed in the 1980s of the twentieth century, bearing in mind that certain civilizations used wind and water power-to-power windmills and water mills in prehistoric times. In the era of the Industrial Revolution, certain small-scale uses of renewable resources were developed, such as heating and steam production. After that and with the bringing out of the renewable energies concept, the sustainability has become the bridge between the society, the environment, and the economy, enabling both the reduction of the greenhouse gas emissions, the economic stability, and then allowing the social inclusion of the contemporary human to continue the mission of the prehistoric human on this terrestrial globe.

In fact, the first contact between the prehistoric human and the external environment allowed him to invent caves to protect themselves from dangerous outsiders. Subsequently, other forms of housing were invented to contribute to sustainable development: the Roman dwelling of antiquity (made of bricks and tiles), the peasant dwelling of the Middle Ages (made of wood), the vast buildings of modern times, and today's concrete constructions. The appearance of all these housing forms facilitated the emergence of various civilizations and human settlements, where various historical events and stations were manifested: The discovery of reading and writing, the Industrial Revolution, and the birth of the Internet and the World Wide Web (August 6, 1991), not forgetting the contributions to other historic events, such as the declaration of independence of the United States (July 4, 1776), the research enabling the first human steps on the moon (July 20, 1969), and the signing of the Paris agreement on climate change (December 12, 2015). Whatever the purpose of using these construction structures, the main objective is to ensure the quality life, the service efficiency, the security control, the energy affordability, and again the social inclusion of the humans.

According to the World Health Organization (WHO), the quality of life is defined as the standards that enable the Humans to live well by achieving their goals. These standards are linked to the health, the society, the environment, the good governance, and the prosperity. In fact, a Human benefit from the quality of life standards means that the service metrics are enhanced for him or her as costumer and the security risks are minimized and that they are included into the society by improving their abilities and increasing the opportunities for their professional insertion. In addition, the focus remains on providing affordable energy for all citizens by reducing costs, optimizing energy efficiency, and encouraging energy investment initiatives. In this context, the last contributions examine the impact of enhancing the tourism services on promoting the quality of life standards [1] and underline the impact of the CO_2 emissions on the African citizens health [2]. In addition, the current research highlights the impact of the artificial intelligence on the quality of work and the quality of life in general [3]. The impact of AI is also analyzed regarding the enhancement of the public service efficiency [4], where the digital economy contributes to optimizing the energy efficiency [5]. Concerning the protection of the Human's data and the organization's infrastructure from cyberattacks, Kokila and Reddy present the role of the Internet of things security on the authentication and the access control [6], Liu et al. highlight the contribution of the control and security on the awareness and the decision-making in the industrial process [7]. To provide the energy security and accessibility, Nicita et al. present the political approaches of the energy affordability in the European citizens [8], and Tang et al. examine the contribution of reducing the energy poverty to this energy affordability [9]. As an important norm that enhances the Human's abilities, the last research evaluates the social inclusion in the education sector by promoting the participation of students in social life. Specifically, Lopez Gomez et al., Guillemot et al., and Littlefair et al. analyze social inclusion in the education system in Colombia [10], France [11], and the UK [12]. The literature illustrates also the contribution of the social inclusion of the Brazilian women suffering from epilepsy to realize a degree of life with more quality [13].

As the agriculture and the discovery of the fire were the first forms of connection between prehistoric Human and their external environment, artificial intelligence (AI) now allows actually the connection and the communication to satisfy more quality of life norms for the current Human. In addition to providing more developed tools to the external world to be more continually "connected" for every issue surrounding us, the digital technologies tools are now implementing all the quality of life norms to guarantee the energy security and provide the urban mobility, promote efficient buildings, and manage the throwing waste for decreasing the greenhouse gas emissions and realizing an efficient environment. The issue, then, is to implement all the strategies for protecting the environment and reducing the climate change degree simultaneously by benefiting all the digital technologies tools for achieving more quality of life norms. However, to be more adaptable with each current and unexpectedly emerged world economic situation guarantee a consistent implementing of the quality of life norms.

After the Wall Street crash of 1929, which caused a profound economic crisis and significant financial losses, individuals in the twentieth century endeavored to restore equilibrium in their daily lives. Then the Second World War followed, leading not only to the breakdown of world economy but also to a lack of urban development. The atomic bombings of Hiroshima and Nagasaki in 1945 are a vivid illustration of this devastation. Then, the ensued economic situation breakdown due to the oil crisis of 1973–1974 and the global financial crisis of 2007–2008, as well as other financial crisis in general led to innovate new economic models by the modern Human to reduce the intensity of the future economic crisis by implementing global strategies and policies. It is both strange and astounding that the Human who invented the Morse code on 1838 did not imagine that one day the numbers and the symbols would be the engines of the digital transformation. This transformation now provides us with the digital technologies tools to implement quality-of-life norms, including sustainable urban development. Overall, it is not only the economic situation that can bring a global crisis but epidemics can also lead to them. The COVID-19 pandemic of 2020, for example, increased poverty, caused salaries losses, and exacerbated inequalities. The achievement of sustainable development requires both global economic recovery and pandemic response strategies to guarantee optimal healthcare services.

Since the emergence of the Mesopotamian, Pharaonic, Indus Valley, and Chinese civilizations, ancient Human built houses to protect themselves from external dangers and natural disasters. Then, the assembly of several houses made up cities and these cities formed prefectures and provinces. For a long time, cities have been considered urban centers where business is conducted, students are educated, and citizens receive medical treatment. Due to their contribution to demographic growth, municipal councils must provide all citizens with clean water, energy, sustainable mobility, and clean air. They must also ensure equity and accessibility to all public services, such as health care, social services, and entertainment. As mentioned above, Human have tried to rebalance their life rhythm after each economic crisis or war by building different types of cities, including industrial, digital, informational, and smart cities. In this context, industrial cities were built after the economic crisis of 1857 during the second technological revolution, digital cities were built after the economic crisis of 1929 during the third technological revolution, informational cities were constructed after the economic crisis of 1987 during the information technological revolution, and finally, smart cities was built after the economic crisis of 2008 during the smart technological revolution. Overall, the construction of smart cities aims to combat the discrimination found in conventional cities between rich and poor areas. These new cities allow all citizens to be more connected, comfortable, and have access to all public services. In addition, implementing the smart city concept in both developed and developing countries contributes to achieving sustainable urban development. In this way, the smart city approach will be realized in the light of achieving energy transition and mitigating climate change.

As the smart city approach is now widely adopted in places like Tokyo (Japan), Doha (Qatar), Rabat (Morocco), Sydney (Australia), Paris (France), and Mexico City (Mexico), this book contributes to the implementation of this concept to achieve more sustainable urban development. In addition, this book gives an overview on the policies and governance of smart cities to help improve their implementation, leading to a higher quality of life and better service efficiency. These policies and governance help political organisms to be more flexible with economic crisis, helping them avoid the consequences of previous economic crisis. They therefore help overcome the challenges of smart city implementation, improving citizens' living comfort. In this context, the exploitation of digital technologies across key sectors - including education, natural resource management, mobility, energy, healthcare, e-governance, and digital business - contributes to an efficient and sustainable urban development. It should also be noted that this book addresses ethics and security issues and offers case studies to facilitate their proper consideration, while involving readers and the scientific community with the outcomes of the smart city projects. Finally, this book collects and addresses the challenges of smart city aspect to ensure successful implementation and avoid the discrimination found in conventional cities.

In depth, this book begins by optimizing the e-governance in smart cities and analyzing the contribution of digital transformation in developing a smart government in Gorontalo City, Indonesia. Then, an inclusive and proactive framework is employed to assess the security and ethical aspects of Moroccan smart cities. The real-time data and knowledge distillation are applied as well for the security management of smart cities. To provide an efficient educational system, the contributors exploited the information and communication technologies (ICT) for developing smart education. In order to protect the natural resources, this book uses smart tools for surveying the pollution and exploits Morocco's abundant phosphate resources to manufacture glass-phosphate nanocomposite photovoltaic cells for smart cities. These natural resources are exploited as well for providing the urban appearance of the smart cities and contributing in the smart buildings. In this context, the Earthen house, the integration of Juncus Acutus Fibers into Red Clay Matrices, and the integration of the Fly-Ash into the concrete contributed well in the sustainable and smart mobility. Furthermore, the model predictive control is used for managing the intelligent model. For managing and optimizing the energy consumption in the smart cities, several approaches are applied: the microgrid frequency regulation with the active disturbance rejection control (ADRC) for wind energy, the IoT-based wireless smart sensor node for smart energy management in buildings, and the energy audit and planning for managing a greener and smarter energy. To contribute to smart mobility, computational fluid dynamics (CFD) is used to manage a novel hybrid battery and a U-shaped cold plate for Lithium-ion batteries. The floating solar PV is used to power wireless charging technology for autonomous underwater vehicle, and the dynamic wireless charging is applied for the electric vehicles. The blockchain and the smart decision methods are utilized for logistics and an efficient future urban. To improve the economic sector in the smart cities, a predictive analysis using artificial intelligence (AI) was adopted to enhance digital business models. The workforce management was also used to enhance retention in smart cities. The role of economic diplomacy in urban innovation was also investigated using pioneering smart cities in Morocco as a case study. Finally, the book explores how technology and governance can be used to adopt smart healthcare solutions for urban health development.

In order to ensure the comprehension and the expertise of the smart cities approaches, this book investigates some case studies regarding the development of smart tourism by applying a domain-specific language-based chatbot. Furthermore, an AI-based chatbots for enhancing the urban service delivery was adopted. To contribute in the disaster recovery, an underwater image enhancement by using the recurrent neural network is recommended for the emergency response analysis. In the context of optimizing urban systems, the digital twin technologies are adopted as an efficient approach to develop the sustainable urban development. Other case studies presented include electrical audits for energy conversion, a smart container terminal using metaheuristic, and a Bluetooth-assisted misplaced object finder that uses a DFRobot Arduino integrated with an Android application for developing sustainable smart cities.

Some challenges are analyzed as well for developing the smart city concept. These include optimizing urban management through the Internet of Things (IoT), strategic approaches for sustainable urban growth in Sri Lanka by addressing hidden urbanization, and ensuring accessibility for disabled children in urban spaces for inclusive smart cities.

The following chapters contribute in expertizing the approaches of the smart city aspect, innovating new models using the digital technologies, and understanding the challenges preventing the implementation of the sustainable urban development:

- **Mohamed Yasser Bounite et al.:** Optimising E-Governance in Smart Cities Using a Stochastic Approach
- **Hari Harjanto et al.:** Digital Transformation Through the Implementation of Electronic-Based Government Systems to Build Efficient and Transparent Smart Government in Gorontalo City, Indonesia
- **Atmani Aziz et al.:** Ethical and Social Challenges of Smart Cities: An Inclusive Approach for Urban Development with Internet of Things (IoT)
- **Mhammed Kezzout et al.:** Urban Security Technologies for Moroccan Smart Cities: A Proactive Analysis for Unforeseen Impacts
- **Aicha Khalfaoui et al.:** Efficient Knowledge Distillation for Detecting and Classifying Dangerous Weapons in Smart City Surveillance Systems

- **Oumayma Berraadi et al.:** Unlocking the Power of Real-Time Data: Advancing Predictive Analytics for Smarter City Management
- **Hasna Mahmoud et al.:** Personalized Learning in Smart Education: The Role of Recommender System and ICT
- **Kanakaprabha S et al.:** The Contribution of the Information and Communication Technologies in Developing the International Experience of Smart Education
- **Hicham Ouzif et al.:** Leveraging the Information and Communication Technologies (ICT) for Developing the Smart Education Approach in Morocco
- **Abderrahim Daoui et al.:** Harnessing Morocco'ss Abundant Phosphate Resources: Advancements in Lead Phosphate Glass/Metal Nanocomposite Photovoltaic Cells for Smart Cities and Benchmarking Against Silicon Efficiency
- **Hajar Echayeb et al.:** First Surveys on the Pollution of Lake BOUDAROUA, Ouazzane, Morocco
- **Boutahar Laaouar et al.:** A Case Study on the Thermal Performance of Materials in an Earthen House in the Anoual Region, Morocco: Towards Smart Buildings and Durability
- **Othmane Horma et al.:** Enhancing Thermal Efficiency in Adobe Through the Integration of Juncus Acutus Fibers into Red Clay Matrices
- **Mohammed Drissi et al.:** Impact of Fly Ash-Based Concrete on Thermal Performance and Energy Efficiency of Office Buildings in Oujda City: Comparative Study and Contribution in Smart Mobility
- **Muhammad Shamrooz Aslam et al.:** Management of the Intelligent Building Using Model Predictive Control for a Green and Sustainable Urban
- **Chakib Mouhssine et al.:** Microgrid Frequency Regulation Contribution of DFIG-Based Wind Turbine with Using ADRC Controller for Smart Energy Optimization
- **Mathaba Tebello et al.:** Design of an IoT-Based Wireless Smart Sensor Node for Enhancing Smart Building Energy Management Systems
- **Valarmathi Viswanathan et al.:** Innovative Sustainable Energy Solutions in Smart Cities: Transforming Urban Landscapes for a Greener Future
- **Buhle G. Shandu et al.:** An Energy Audit Planning Framework for Energy Efficiency Retrofit Projects for Smart Buildings
- **Abir Dahani et al.:** Implementation of Energy Planning at the Municipal Grouping Level in Morocco to Develop the Concept of Smart Cities
- **Nassreddine Hmidi et al.:** CFD-Based Optimization of a Novel Hybrid Battery Thermal Management System for Electric Vehicles: Approaches and Applications for Smart Mobility
- **Nassreddine Hmidi et al.:** Optimization of a U-Shaped Cold Plate BTMS for Lithium-Ion Batteries Using CFD: Thermal Regulation Under Various Discharge Rates and Applications for Smart Mobility
- **Nabila Bouti et al.:** Sustainable Supply Chain and Logistics in Smart Cities: A Literature Review on Smart Decision Methods and AI Applications

- **Yuvaraja Shanmugan et al.:** On-Road or Dynamic Wireless Charging for Slow-Moving EVs: Paving the Way for Sustainable Smart Cities
- **Mumtaz Qaisrani et al.:** Achieving Sustainable Urban Mobility Through Retrofitting Two-Wheeler Vehicles
- **Meryem Lasaad et al.:** Blockchain for Smart Cities: Building a Secure and Efficient Future Urban
- **Narayanamoorthi R. et al.:** The Use of Floating Solar PV to Power Wireless Charging Technology for Autonomous Underwater Vehicle: Charting a Path Towards Sustainable Smart Cities
- **Hamza Boujida et al.:** A Systematic Literature Review on the Evolution of Smart Cities and Digital Businesses
- **Muhammad Hilal et al.:** Predictive Analytics and AI in Economic Development: Enhancing Digital Business Models in Smart Cities
- **Layla Bentahar et al.:** Economic Diplomacy and Urban Innovation: Pioneering Smart Cities in Morocco
- **Loubna Bouhsaien et al.:** Explainable AI for Workforce Management: Enhancing Retention in Smart Cities
- **Intissar Haddiya et al.:** Technologies and Governance of the Adoption of Smart Healthcare Solutions for Urban Health Development
- **Mohamed Eddabah et al.:** Smart Healthcare Solutions for Urban Health Development
- **Kanakaprabha S et al.:** Overview of Innovations and Challenges of the Smart Healthcare Solutions in the Urban Environments
- **Abdslam Jakimi et al.:** A Domain Specific Language-Based Chatbot Development for Smart Tourism: Applications in Smart Cities
- **Kanakaraj et al.:** Energy Conservation: A Case Study on Electrical Audits for Sustainable Smart Cities
- **Deluxni N et al.:** Underwater Image Enhancement Using Recurrent Neural Network for Disaster Recovery and Emergency Response Analysis in Smart Cities
- **El Mahdi Bouyahrouzi et al.:** Optimizing Urban Systems Through Digital Twin Technology for a Smart City, Insights from a Real Case Study
- **Sanaa Aidi et al.:** Smart Container Terminal Using Metaheuristic: Genetic Algorithm and Tabu Search
- **Charaf Ouaddi et al.:** AI-Based Chatbots for Enhancing Urban Service Delivery in Smart Cities
- **Eustace M. Dogo et al.:** Bluetooth Assisted Misplaced Object Finder Using DFRobot Arduino Integrated with Android Application
- **Mounir Boumhamdi et al.:** Optimizing Urban Management Through the Internet of Things (IoT): Sustainable Solutions and Challenges for Smart Cities
- **Nathasha Janani Kudasinghe et al.:** Addressing Hidden Urbanization: Challenges and Strategic Approaches for Sustainable Urban Growth in Sri Lanka
- **Shashwata Sahu et al.:** Inclusive Smart Cities: Ensuring Accessibility for Disabled Children in Urban Spaces

References

1. Uysal M, Berbekova A, Wang J (2025) Quality-of-life: a critical examination of research progress. Tour Manag 107:105070. https://doi.org/10.1016/j.tourman.2024.105070
2. Beşer NÖ, Çabaş M, Uğurlu S, Magazzino C (2025) The determinants of life quality in African countries: moderating role of CO2 emissions and health expenditures. Green Technol Sustain 3(3):100165. https://doi.org/10.1016/j.grets.2024.100165
3. Wu T-J, Zhang R-X, Zhang Z (2025) Navigating the human-artificial intelligence collaboration landscape: impact on quality of work life and work engagement. J Hosp Tour Manag 62:276–283. https://doi.org/10.1016/j.jhtm.2025.02.005
4. Kulal A, Rahiman HU, Suvarna H, Abhishek N, Dinesh S (2024) Enhancing public service delivery efficiency: exploring the impact of AI. J Open Innov Technol Mark Complex 10(3):100329. https://doi.org/10.1016/j.joitmc.2024.100329
5. Song Y, Liu C, Zhang M (2025) Producer services agglomeration and energy efficiency enhancement: a serial mediation model with digital economy moderation. Energy 318:134774. https://doi.org/10.1016/j.energy.2025.134774
6. Kokila M, Reddy KS (2025) Authentication, access control and scalability models in Internet of Things Security—a review. Cyber Secur Appl 3:100057. https://doi.org/10.1016/j.csa.2024.100057
7. Liu L, Xu P, Fan K, Wang M (2025) Research on application of knowledge graph in industrial control system security situation awareness and decision-making: a survey. Neurocomputing 613:128721. https://doi.org/10.1016/j.neucom.2024.128721
8. Nicita L, Casamassima A, Santorsola M, Morone A (2025) EU citizens' perception of energy affordability and social and political trust. Energy Econ 141:108080. https://doi.org/10.1016/j.eneco.2024.108080
9. Tang L, Mahmood H, Khalid S (2025) Examining the impact of structural reforms on energy poverty: a threshold effect of energy affordability. Energy 320:135101. https://doi.org/10.1016/j.energy.2025.135101
10. López Gómez C, Farieta-Barrera A, Sarmiento MP (2025) Evaluation of policies for disability and social inclusion in Colombian higher education. Int J Educ Res Open 9:100454. https://doi.org/10.1016/j.ijedro.2025.100454
11. Guillemot F, Lacroix F, Nocus I (2024) Subjective well-being and social inclusion at school for students with a disability, according to their parents, in France. Res Dev Disabil 153:104814. https://doi.org/10.1016/j.ridd.2024.104814
12. Littlefair D, McCloskey-Martinez M, Graham P, Nicholls F, Hodges A, Cordier R (2024) Promoting social-inclusion: adapting and refining a school participation and connectedness intervention for neurodiverse children in UK primary schools. Res Dev Disabil 154:104857. https://doi.org/10.1016/j.ridd.2024.104857
13. Randi YM, de Almeida Souza Tedrus GM, Nucci LB (2024) Evaluation of social inclusion in Brazilian adult women with epilepsy: perception of stigma and quality of life. Epilepsy Behav 156:109821. https://doi.org/10.1016/j.yebeh.2024.109821

Contents

Editors and Contributors

About the Editors

Hicham Kaddouri is a temporary professor at Mohammed Premier University in Nador, Morocco. He is in the final stages of completing a PhD in Mechanics and Energy. His research focuses on energy efficiency in the building sector, the use of bio-based and sustainable materials for insulation, as well as solar energy systems and transient energy simulation. Passionate about energy transition and sustainable development, Hicham is dedicated to optimizing the energy performance of buildings by exploring innovative and environmentally friendly solutions. Through his research, he contributes to reducing energy consumption and promoting renewable energy within the framework of more sustainable construction practices. An author of several scientific articles, he has participated in numerous international congresses and conferences, where he has presented his work and exchanged insights with experts in the field. As a reviewer, he also lends his expertise to the evaluation of scientific articles for international journals and conferences.

Firyal Latrache is a researcher in mathematical modeling and digital technologies for mechanics and energy engineering. She holds a PhD in Mechanics and Energy Engineering, where she developed the optical and geometrical parameters of the conical solar concentrator and the modified nodal method for estimating the temperature over the complex geometries of the heat exchangers (e.g., the serpentine geometries). In addition, Firyal works on several projects with eminent researchers: the modeling of the cooling system of the bifacial photovoltaic, the modeling and the conception of a smart energy managing system in buildings, and the generation of green hydrogen from the concentrated solar power. Firyal works and collaborates on research regarding the generation of green hydrogen from wind energy in MENA region, the role of lithium-ion batteries in providing the energy transition, as well the enhancing the buildings sector in Morocco. Actually, she published, with a team of editors, an edited book exploring the use of digital technologies for implementing the energy transition. As a temporary professor at Mohammed Premier University in Oujda, Morocco, Firyal teaches courses in automotive mechanics and technology, as well as mathematics for engineers, to Licentiate of Technology students. At last, she is a writer and poet in Arabic Language.

Abdelaziz Benkhalifa is currently a computer science teacher and a Ph.D. student at the MATSI laboratory at the Higher School of Technology in Oujda. His research focuses on educational technologies and the digital transformation of learning systems, with a particular interest in their role within smart city development. He explores intelligent learning environments, adaptive learning models, and the use of artificial intelligence and connected devices in urban educational settings. His work also includes comparative studies on traditional, flipped, and adaptive teaching approaches, as well as applied research on integrating robotics, artificial intelligence, and the metaverse into education. His aim is to enhance the quality, accessibility, and personalization of learning as fundamental components of inclusive and innovative smart cities.

Abderrahim Abidouche is a secondary school teacher and also teaches mechanical engineering at technician training centers. He is currently pursuing a doctorate in bioclimatic architecture and its impact on the passive cooling of buildings. His research is evaluating the effects of passive strategies on the performance of residential and commercial buildings, with the aim of improving energy efficiency and indoor comfort. Passionate about sustainable construction, Abderrahim explores innovative design approaches that minimize energy consumption while maximizing environmental benefits. His work contributes to the development of more efficient buildings that are better adapted to the climate. He is the author of several scientific papers on the subject and has presented his research at international conferences, where he engages with experts in the field to advance knowledge of bioclimatic architecture and passive design strategies.

Najwa Jbira, PhD in Energy Efficiency, specializes in energy integration within the industrial sector, focusing on pinch analysis and exergy analysis, with a particular emphasis on the dairy industry to enhance energy efficiency and sustainability. She is a member of the editorial team for the book *Digital Technology for Innovative Energy Transition: Perspectives and Opportunities*, contributing to shaping academic discourse on energy transition and digital innovation. As a reviewer for several scientific journals and international conferences, she plays a key role in evaluating and improving scientific publications. Actively engaged in the academic community, she has organized and participated in numerous doctoral symposiums and workshops, fostering collaboration and innovation among researchers. A dedicated member of the Green Brains association, she advocates for sustainable development and innovation. Professionally, she works as a renewable energy specialist at the Ministry of the Interior, leading projects on energy efficiency, renewable energy, and environmental initiatives to support Morocco's sustainable energy transition.

Contributors

Sana Abad School of Energy Science and Engineering, Harbin Institute of Technology, Harbin, China

Faheem Abbas School of Energy Science and Engineering, Harbin Institute of Technology, Harbin, China

Mohammed Abdellaoui LIREFIMO Research Laboratory FSJES, University Sidi Mohamed Ben Abdellah, Fez, Morocco

Mohamed Ben Ahmed FSTT, Abdelmalek Essaadi University, Tetouan, Morocco

Sanaa Aidi Department of Mechanics, Engineering and Innovation, National Higher School of Electricity and Mechanics, University Hassan 2, Casablanca, Morocco

CharafEddine Aitzaouiat Polydisciplinary Faculty of Sidi Bennour, Chouaîb Doukkali University, El Jadida, Morocco

Nada Alami Laboratory: Interdisciplinary Research Laboratory in Economics, Finance and Management of Organizations LIREFIMO, Sidi Mohamed Ben Abdellah University, Fez, Morocco

Fahd Mohammed Aldosari Department of Computer and Network Engineering, College of Computing, Umm Al-Qura University, Makkah, Saudi Arabia

Abdel-illah Amrani Equipe des Énergies Nouvelles et Ingénierie des Matériaux (ENIM), Faculté des Sciences et Techniques, Moulay Ismail University of Meknes, Errachidia, Morocco

Muhammad Shamrooz Aslam School of Computer Science and Technology/ School of Artificial Intelligence, China, University of Mining and Technology, Xuzhou, China

Aziz Atmani Laboratoire de Recherche Business Intelligence, Gouvernance des Organisations, Finance et Criminalité Financière, Faculté des Sciences Juridiques Économiques et Sociales de Casablanca, Université Hassan II, Casablanca, Morocco

Yamna Azizi Laboratoire du Droit Public et Sciences Politiques, Faculté des Sciences Juridiques Économiques et Sociales de Casablanca, Casablanca, Morocco

Abdellah Azmani Intelligent Automation and BioMedGenomics Laboratory, Faculty of Science and Technologies, Abdelemalek Essaadi University, Tangier, Morocco

Mohamed Badouch LabSIV, Department of Computer Science, Faculty of Science, Ibnou Zohr University, Agadir, Morocco

Abdelmajid Badri Department of Electrical Engineering Faculty of Sciences and Techniques, Hassan II University, Mohammedia, Morocco

Mouad Bargani School of Engineering and Sustainable Development, De Montfort University (DMU), Leicester, United Kingdom

Process Engineering, Computer Science, and Mathematics Laboratory (LIPIM), National School of Applied Sciences, Sultan Moulay Slimane University, Khouribga, Morocco

Driss Belghyti Laboratory of Natural Resources and Sustainable Development, Faculty of Sciences, University Ibn Tofail, Kenitra, Morocco

Lamya Benaddi Software Engineering and Information Systems Engineering Team, Faculty of Sciences and Techniques of Errachidia, UMI-Meknes, Meknes, Morocco

Chouaib Benqlilou Sustainability, Artificial Intelligence and Mathematical Modeling, Ecole Nationale Supérieure des Mines de Rabat, Rabat, Morocco

Layla Bentahar LIREFIMO Research Laboratory FSJES, University Sidi Mohamed Ben Abdellah, Fez, Morocco

Mouna Berquedich Green Tech Institute—Mohammed VI Polytechnique University, Benguerir, Morocco

Oumayma Berraadi FSTT, Abdelmalek Essaadi University, Tetouan, Morocco

Hazrat Bilal CAIR - Center for Artificial Intelligence Research, Department of ICT, University of Agder (UiA), Grimstad, Norway

Loubna Bouhsaien Intelligent Automation and BioMedGenomics Laboratory, Faculty of Science and Technologies, Abdelemalek Essaadi University, Tangier, Morocco

Hamza Boujida Laboratory: Interdisciplinary Research Laboratory in Economics, Finance and Management of Organizations LIREFIMO, Sidi Mohamed Ben Abdellah University, Fez, Morocco

Ibrahim Boukallal IA Laboratory, FS Meknes, Moulay Ismail University of Meknes, Meknes, Morocco

Ikhlass Boukrouh Intelligent Automation and BioMedGenomics Laboratory, Faculty of Science and Technologies, Abdelemalek Essaadi University, Tangier, Morocco

E. S. -S. A. I. D. Boulmane LabSIV, Department of Computer Science, Faculty of Science, Ibnou Zohr University, Agadir, Morocco

Mounir Boumhamdi Laboratoire de Recherche Business Intelligence, Gouvernance des Organisations, Finance et Criminalité Financière, Faculté des Sciences Juridiques Économiques et Sociales de Casablanca, Université Hassan II, Casablanca, Morocco

Mohamed Yasser Bounnite SUP'RH Business School & AI, Casablanca, Morocco

Mehdi Boutaounte National School of Commerce and Management, Ibn Zohr University, Dakhla, Morocco

Nabila Bouti IA Laboratory, FS Meknes, Moulay Ismail University of Meknes, Meknes, Morocco

El Mahdi Bouyahrouzi Industrial Engineering Laboratory, ENSA, UMP, Oujda, Morocco

Elmiloud Chaabelasri Materials Science, New Energies and Applications Research Group, LPTPME Laboratory, Department of Physics, Faculty of Sciences, Mohammed 1st University, Oujda, Morocco

Noureddine Chafi Laboratory of Physics of Matter and Radiations (LPMR), Physics Department, Faculty of Sciences, University Mohamed Premier, Oujda, Morocco

Mohssine Chakib Electrical Engineering Department of ENSAM, Mohammed V University, Rabat, Morocco

Wer-Jer Chang Department of Marine Engineering, National Taiwan Ocean University, Keelung, Taiwan

Abir Dahani Sustainability, Artificial Intelligence and Mathematical Modeling, Ecole Nationale Supérieure des Mines de Rabat, Rabat, Morocco

Issam Debbagh Laboratory: Interdisciplinary Research Laboratory in Economics, Finance and Management of Organizations LIREFIMO, Sidi Mohamed Ben Abdellah University, Fez, Morocco

Mohammed Drissi Mechanics and Energy Laboratory, Mohammed First University, Oujda, Morocco

Hajar Echayeb Laboratory of Natural Resources and Sustainable Development, Faculty of Sciences, University Ibn Tofail, Kenitra, Morocco

ORMVAG, Regional Office for the Development of Gharb, Kenitra, Morocco

Mohamed Eddabbah LaRTID laboratory, Higher School of Technology of Essaouira, Cadi Ayyad University, Marrakesh, Morocco

Abderrahim Ed-Daoui Laboratory of Physics of Matter and Radiations (LPMR), Physics Department, Faculty of Sciences, University Mohamed Premier, Oujda, Morocco

Wesam Nabil Eid Department of Cybersecurity, College of Computing, Umm Al-Qura University, Mecca, Saudi Arabia

Christian Nonyelum Ejike Postgraduate School of Engineering Management, University of Johannesburg, Johannesburg, South Africa

Ahmed El Akkary LASTIMI Laboratory of Higher School of Technology, Mohammed V University, Rabat, Morocco

Hayat El Boukhari Department of Management, Sidi Mohamed Ben Abdellah University, Fez, Morocco

Yassin El Gountery GL-ISI Team, Faculty of Science and Technology Errachidia, Moulay Ismail University, Meknes, Morocco

Sara El Hassani Laboratory of Energetics, Faculty of Science and Technology, Abdelmalek Essaâdi University, Tetouan, Morocco

Khadija Elkharrim Laboratory of Natural Resources and Sustainable Development, Faculty of Sciences, University Ibn Tofail, Kenitra, Morocco

Fatima El Khoukhi IA Laboratory, FLSH Meknes, Moulay Ismail University of Meknes, Meknes, Morocco

Ali El Kihel Industrial Engineering Laboratory, ENSA, UMP, Oujda, Morocco

Ilham El Mourabit Department of Electrical Engineering Faculty of Sciences and Techniques, Hassan II University, Mohammedia, Morocco

Abdelhafid Elouassidi Abdelmalek Essaadi University, Experimentation and Modelling Team in Mechanics and Energy Systems, National School of Applied Sciences, Al-Hoceima, Morocco

Fatima Zohra El Wardi Physics Department, LPMAT Laboratory, Faculty of Sciences Ain Chock, Hassan II University, Casablanca, Morocco

Soufiane Embarki Industrial Engineering Laboratory, ENSA, UMP, Oujda, Morocco

Ahmed Essadki Electrical Engineering Department of ENSAM, Mohammed V University, Rabat, Morocco

Khalid Fariri Laboratory of Natural Resources and Sustainable Development, Faculty of Sciences, University Ibn Tofail, Kenitra, Morocco

Azidine Guezzaz LaRTID laboratory, Higher School of Technology of Essaouira, Cadi Ayyad University, Marrakesh, Morocco

Nadia Hachoumi High Institute of Nursing Professions and Health Techniques, Ministry of Health, Marrakesh, Morocco

Intissar Haddiya Department of Nephrology, Mohammed VI University Hospital, Faculty of Medicine and Pharmacy of Oujda, Mohammed First University, Oujda, Morocco

Maryama Hammi School of Architecture, Built Environment, Computing and Engineering, Birmingham City University, Birmingham, United Kingdom

Muhammad Hilal Institute of Management Science Peshawar, Peshawar, Pakistan

Nassreddine Hmidi Materials Science, New Energies and Applications Research Group, LPTPME Laboratory, Department of Physics, Faculty of Sciences, Mohammed 1st University, Oujda, Morocco

Othmane Horma Mechanics and Energy Laboratory, Mohammed First University, Oujda, Morocco

Private University of Marrakech, Marrakech, Morocco

Marouane Ikira Laboratoire de Recherches en Gestion, Économie et Sciences Sociales, Faculté des Sciences Juridiques, Économiques et Sociales d'El Jadida, Université Chouaib Doukkali, El Jadida, Morocco

Muhammad Imad Abasyn University, Peshawar, Pakistan

Amar Yusof Jaffar Department of Computer and Network Engineering, College of Computing, Umm Al-Qura University, Makkah, Saudi Arabia

Ramkumar Jaganathan Department of Computer Science, Kristu Jayanti University, Benguluru, Karnataka, India

Abdeslam Jakimi Software Engineering and Information Systems Engineering Team, Faculty of Sciences and Techniques of Errachidia UMI-Meknes, Meknes, Morocco

Software Engineering and Information Systems Engineering Team, FST Errachidia, University of Moulay Ismail, Meknes, Morocco

GL-ISI Team, Faculty of Science and Technology Errachidia, Moulay Ismail University, Meknes, Morocco

Weerakoon Dasanayakalage Bimsara Jayanath Jeewananda General Sir John Kotelawala Defence University, Colombo, Sri Lanka

P. Aruna Jeyanthy Department of EEE, Kalasalingam Academy of Research and Education, Virudhunagar, Tamil Nadu, India

Hicham Kaddouri Experimentation and Modelling Team in Mechanics and Energy Systems, National School of Applied Sciences, Abdelmalek Essaadi University, Al-Hoceima, Morocco

P. Kanakaraj Department of EEE, S.R.M. Institute of Science and Technology, Kattankulathur, Chengalpattu, Tamilnadu, India

Karno Karno Institut Pemerintahan Dalam Negeri (IPDN), Jakarta, Indonesia

Rinikso Kartono University of Muhammadiyah Malang (UMM), Malang, Indonesia

Mhammed Kezzoute Moroccan Center of Cyber Studies & Technologies Security, Rabat, Morocco

Aicha Khalfaoui Department of Electrical Engineering Faculty of Sciences and Techniques, Hassan II University, Mohammedia, Morocco

Fuad Khoshnaw School of Engineering and Sustainable Development, De Montfort University (DMU), Leicester, United Kingdom

Kuda Singankutti Kankanamalage Nathasha Janani Kudasinghe General Sir John Kotelawala Defence University, Colombo, Sri Lanka

Layla Kurniawati Institut Pemerintahan Dalam Negeri (IPDN), Jakarta, Indonesia

Boutahar Laaouar Mechanics and Energy Laboratory, Mohammed First University, Oujda, Morocco

Rania Larbi Laboratory: Interdisciplinary Research Laboratory in Economics, Finance and Management of Organizations LIREFIMO, Sidi Mohamed Ben Abdellah University, Fez, Morocco

Meryem Lasaad GL-ISI Team, Faculty of Science and Technology Errachidia, Moulay Ismail University, Meknes, Morocco

Hasna Mahmoud LabSIV, Department of Computer Science, Faculty of Science, Ibnou Zohr University, Agadir, Morocco

Navonita Mallick School of Law, KIIT-DU, Bhubaneswar, India

Tebello Ntsiki Don Mathaba Postgraduate School of Engineering Management, University of Johannesburg, Johannesburg, South Africa

Mohamed Mazouzi Department of Mechanics, Engineering and Innovation, National Higher School of Electricity and Mechanics, University Hassan 2, Casablanca, Morocco

Jinane Mejdoubi Mechanics and Energy Laboratory, Mohammed First University, Oujda, Morocco

Houssam Mensouri Intelligent Automation and BioMedGenomics Laboratory, Faculty of Science and Technologies, Abdelemalek Essaadi University, Tangier, Morocco

Ahmed Alami Merrouni Materials Science, New Energies and Applications Research Group, LPTPME Laboratory, Department of Physics, Faculty of Sciences, Mohammed 1st University, Oujda, Morocco

Ahmed Mezrhab Mechanics and Energy Laboratory, Mohammed First University, Oujda, Morocco

Hanane Miri Applied Geosciences Laboratory, Faculty of Sciences, Mohammed First University, Oujda, Morocco

Mohammed Amine Moussaoui Mechanics and Energy Laboratory, Mohammed First University, Oujda, Morocco

Najoua Mrabet LASTIMI Laboratory of Higher School of Technology, Mohammed V University, Rabat, Morocco

Peter Muganyi Postgraduate School of Engineering Management, University of Johannesburg, Johannesburg, South Africa

Jagath Munasinghe University of Moratuwa, Moratuwa, Sri Lanka

Murdiyana Murdiyana Institut Pemerintahan Dalam Negeri (IPDN), Jakarta, Indonesia

R. Narayanamoorthi Department of Electrical and Electronics Engineering, SRM Institute of Science and Technology, Kattankulathur, Chennai, India

Deluxni Natarajan Department of Computing Technologies, School of Computing, SRM Institute of Science and Technology, Kattankulathur, Chennai, India

Charaf Ouaddi Software Engineering and Information Systems Engineering Team, FST Errachidia, University of Moulay Ismail, Meknes, Morocco

Mohamed Oualla GL-ISI Team, Faculty of Science and Technology Errachidia, Moulay Ismail University, Meknes, Morocco

Amine Ouba IA Laboratory, FS Meknes, Moulay Ismail University of Meknes, Meknes, Morocco

Brahim Ouchao Software Engineering and Information Systems Engineering Team, Faculty of Sciences and Techniques of Errachidia UMI-Meknes, Meknes, Morocco

Hicham Ouzif Department of Management, Sidi Mohamed Ben Abdellah University, Fez, Morocco

Mumtaz A. Qaisrani Institute of Mechanical and Manufacturing Engineering, Khwaja Fareed University of Engineering & Information Technology, Rahim Yar Khan, Pakistan

Narayanamoorthi Rajamanickam Department of Electrical and Electronics Engineering, Wireless Charging Research Centre, SRM Institute of Science and Technology, Chennai, India

Karthikeyan Rajendran Department of Computer Technology, Sri Krishna Adithya College of Arts and Science, Coimbatore, Tamil Nadu, India

Sara Ramdani Laboratory of Epidemiology, Clinical Research and Public Health, Faculty of Medicine and Pharmacy of Oujda, University Mohammed First, Oujda, Morocco

L. Ramesh Department of EEE, Dr. M.G.R. Educational and Research Institute, Chennai, Tamilnadu, India

Attygalage Ranjith Rupasinghe General Sir John Kotelawala Defence University, Colombo, Sri Lanka

Rachid Saadane Electrical Engineering Department, Hassania School of Public Works, Casablanca, Morocco

K. Kalyan Sagar Department of EEE, Shri Vishnu Engineering College for Women, Bhimavaram, Andhra Pradesh, India

Shashwata Sahu School of Law, KIIT-DU, Bhubaneswar, India

Ramamurthi Sekar Department of Electrical and Electronics Engineering, Wireless Charging Research Centre, SRM Institute of Science and Technology, Kattankulathur, Chennai, India

Kanakaprabha Selvakumari Department of Artificial Intelligence and Machine Learning, Malla Reddy College of Engineering, Hyderabad, India

Hari Harjanto Setiawan National Research and Innovation Agency (BRIN), Jakarta, Indonesia

Summera Shamrooz School of Mechatronic Engineering, China University of Mining and Technology, Xuzhou, China

Buhle Goodman Shandu Postgraduate School of Engineering Management, University of Johannesburg, Johannesburg, South Africa

Yuvaraja Shanmugam Department of Electrical and Electronics Engineering, Vignan's Foundation for Science, Technology & Research, Vadlamudi, Andhra Pradesh, India

Adnane Souha Software Engineering and Information Systems Engineering Team, Faculty of Sciences and Techniques of Errachidia, UMI-Meknes, Meknes, Morocco

Pradeep Sudhakaran Department of Computing Technologies, School of Computing, SRM Institute of Science and Technology, Kattankulathur, Chennai, India

Sumabl Shaheed Benazir Bhutto Women University, Peshawar, Pakistan

Tadeusz Szumiata Department of Physics, Faculty of Mechanical Engineering, Casimir Pulaski Radom, University Stasieckiego, Radom, Poland

Kamal Takhdat High Institute of Nursing Professions and Health Techniques, Ministry of Health, Marrakesh, Morocco

Hicham Gibet Tani FPL, Abdelmalek Essaadi University, Tetouan, Morocco

Faouzi Tayalati Intelligent Automation and BioMedGenomics Laboratory, Faculty of Science and Technologies, Abdelemalek Essaadi University, Tangier, Morocco

Valarmathi Viswanathan Department of Information Technology and Cognitive Systems, Sri Krishna Arts and Science College, Coimbatore, Tamil Nadu, India

Omar Zioudi LabSIV, Department of Computer Science, Faculty of Science, Ibnou Zohr University, Agadir, Morocco

Optimising E-Governance in Smart Cities Using a Stochastic Approach

Mohamed Yasser Bounnite

Abstract As smart cities continue to evolve, governments face unprecedented challenges in managing e-government systems. These challenges involve not only transparency and efficiency but also the ability to adapt to rapid technological changes. The present work explores the application of Markov chains as a methodological framework for modelling and optimising E-Governance in smart cities. By representing probabilistic transitions between different governance states, this approach provides predictive insights into the dynamics of transitions towards more efficient e-government systems. Incorporating key factors such as technological adoption, political innovation, and citizen participation, our approach simulates various transition scenarios and evaluates their impact on the quality and sustainability of public services. The simulation results reveal optimal pathways to enhance E-Governance, highlighting strategic levers that can ensure agile, sustainable, and inclusive governance systems. The present work makes a novel and substantial contribution to smart city research by proposing a decision-making framework based on stochastic modelling. It is particularly relevant to researchers, policymakers, and practitioners seeking to improve the effectiveness of public administration through the implementation of digital technologies. The findings offer concrete perspectives for designing governance strategies that are resilient and adaptable, accounting for the uncertainties inherent in technological advancements.

Keywords Smart cities · E-governance · Markov chains · Sustainability · Governance optimisation

---

M. Y. Bounnite (✉)
SUP'RH Business School & AI, Casablanca, Morocco
e-mail: m.bounnite@suprh.com

H. Kaddouri et al. (eds.), *Smart Cities for Sustainable and Urban Development*, Green Energy and Technology, https://doi.org/10.1007/978-3-031-99001-4_1

1 Introduction

The rise of smart cities has brought a profound transformation in the way urban governance is conducted, integrating digital technologies within public services, administrative processes, and citizen interactions [1–3]. These technological advancements hold the promise of improving transparency, efficiency, and citizen engagement in governance systems. However, the rapid pace of innovation and the complexity of smart city ecosystems present significant challenges for policymakers. Governments must not only manage these complex systems but also respond dynamically to rapidly evolving societal and technological expectations. Smart city governance involves the interaction of digital infrastructure with urban administration and public services. This interplay creates a dynamic and often unpredictable environment for decision-making, where traditional governance models fall short. Classical governance frameworks are largely deterministic, assuming a linear progression of events and fixed relationships between key variables. In contrast, smart city ecosystems are inherently complex and uncertain, characterized by non-linear relationships and rapid changes that cannot be fully predicted by deterministic models [2, 4]. For example, the sudden introduction of disruptive technologies like blockchain or artificial intelligence (AI) into public services can drastically shift governance dynamics, requiring adaptive models to account for such changes. The adoption of AI can automate several administrative functions, thereby increasing efficiency but also raising new ethical and regulatory challenges. Similarly, blockchain technology provides an unprecedented opportunity to enhance the transparency and security of government transactions, although its integration into governance systems demands a robust management framework. These examples illustrate the need for flexible and adaptive governance systems to keep pace with technological innovations [5, 6].

The present work proposes the application of Markov chains as a novel methodological framework for modelling governance transitions in smart cities. A Markov chain is a stochastic process that models a system through a series of probabilistic transitions between different states, where each transition depends solely on the current state rather than the historical trajectory [7, 8]. In the context of smart cities, each state represents a specific level of governance efficiency, ranging from low levels of technological adoption and citizen engagement to highly optimized and adaptive governance systems. For instance, a city might transition from a state of minimal digital integration (where e-services are underutilized) to a state of advanced digital governance (characterized by widespread citizen participation through digital platforms and high levels of public service automation). The Markov chain framework is particularly well-suited to modelling governance in smart cities for several reasons. First, it allows for the representation of uncertainty and variability, providing a structured way to simulate how governance systems evolve under different scenarios, such as technological innovations, policy shifts, or changes in citizen behavior [3]. For instance, the model can simulate the impact of a policy encouraging the adoption of digital platforms for public services or forecast how rapid technological advancements may shift governance processes. Additionally, by modelling transitions between different levels of efficiency or citizen participation, this approach not only helps to anticipate potential impacts of these changes but also provides insights into scenarios where governance becomes more resilient and agile in response to technological innovations. Second, the stochastic nature of the model enables decision-makers to anticipate potential disruptions and strategically plan for long-term governance resilience and adaptability.

While the governance of smart cities has been extensively studied, significant methodological gaps persist. Much of the existing literature focuses on conceptual frameworks or empirical analyses, often lacking dynamic models capable of simulating the intricate transitions of governance systems under conditions of technological, social, and political uncertainty. For instance, studies on the integration of emerging technologies such as artificial intelligence and blockchain into public administration frequently emphasize their disruptive potential but rarely provide tools to forecast transitional scenarios or their long-term implications [9, 10]. Moreover, current approaches, primarily descriptive analyses or static modeling techniques, struggle to capture the nonlinear and dynamic evolution inherent to smart city ecosystems [2, 3]. Stochastic models, though widely applied in fields such as economics [11], environmental management [12], and financial risk assessment [13], remain underexplored in the domain of smart city governance. Recent reviews underscore the potential of probabilistic approaches for incorporating variability and uncertainty into public policy decision-making, yet few studies have concretely applied these models to the governance transitions of smart cities [5, 6]. This underutilization likely stems from the complexity of interactions within these ecosystems, where factors such as technological adoption, citizen engagement, and political stability intersect unpredictably. This chapter seeks to address this research gap by introducing Markov chains as a robust methodological framework for modeling and optimizing governance transitions in smart cities. By representing successive governance states and the probabilistic transitions between them, this approach provides a novel and dynamic perspective on the evolutionary mechanisms of governance. Unlike traditional methods, it not only accounts for uncertainty but also enables policymakers to anticipate and adapt to technological and societal disruptions effectively.

Ultimately, the goal is to provide policymakers with predictive insights into how governance systems can evolve in response to technological and societal changes. By modelling these transitions probabilistically, we aim to create a more resilient governance framework that can adapt to both incremental and disruptive changes, ensuring the long-term sustainability of public administration in smart cities. This chapter is structured to provide a comprehensive understanding of governance dynamics in smart cities through the application of Markov chains. The second section introduces the theoretical framework, exploring the foundations of digital governance and the relevance of stochastic models. It specifically highlights the suitability of Markov chains for optimizing governance transitions, linking these dynamics to decision theory and addressing the uncertainties inherent in managing smart city systems. The third section focuses on the methodological approach, where the Markov property is discussed in relation to the modelling of governance transitions between various states. This section details the construction of the transition matrix and outlines the probabilistic dynamics that underpin the evolution of governance systems. The fourth section delves into the key factors influencing transition probabilities, such as technological adoption, citizen engagement, and political stability. Each factor is rigorously quantified, and its impact on the governance model is thoroughly analyzed. In the fifth section, a comparative numerical simulation is presented, examining the e-governance performance of Paris and Tokyo over the period from 2012 to 2022. This section applies the Markov chain model to simulate governance transitions and provides a detailed evaluation of the results. Finally, the conclusion summarizes the key insights gained from the application of Markov chains to smart city governance and discusses future research directions, including the incorporation of real-time data and the exploration of additional social and environmental factors.

2 Theoretical Framework

The theoretical foundation of this chapter rests on the intersection of smart city governance, stochastic modelling, and decision theory. Smart cities are characterised by their reliance on digital technologies to enhance public service delivery, improve citizen engagement, and increase the efficiency of urban administration [1, 3, 14, 15]. However, the complexity and dynamism of these systems present significant challenges for policymakers, who must navigate an uncertain and rapidly evolving landscape.

Stochastic modelling provides a powerful tool for addressing these challenges by enabling the representation of uncertainty and variability in governance transitions [7, 8, 11, 16]. In particular, Markov chains offer a flexible and robust framework for modelling the dynamic evolution of governance systems in response to technological, political, and social factors [7, 8, 17]. By capturing the probabilistic relationships between different states of governance, this approach allows for a more nuanced understanding of how governance systems can adapt to changing conditions over time.

The application of stochastic models in governance is particularly relevant when dealing with complex systems like smart cities, where various actors—public authorities, private companies, and citizens—interact in non-linear ways [18]. Stochastic modelling also provides a way to handle the uncertainty associated with future developments in digital infrastructure and public policy, which often evolve unpredictably [19].

The key innovation of this chapter lies in its application of Markov chains to the domain of smart city governance. While stochastic models have been used extensively in other fields, such as economics [11], environmental management [12], and healthcare systems [20], their potential for optimising governance processes in smart cities has yet to be fully realised. This chapter seeks to fill this gap by providing a detailed framework for modelling and simulating governance transitions in smart cities using Markov chains.

2.1 Smart City Governance and Uncertainty

Governance in smart cities is highly susceptible to external influences such as rapid technological advancements, regulatory changes, and fluctuating citizen engagement. The complexity of the interconnected subsystems that define smart cities introduces a range of uncertainties that complicate governance efforts [1, 3].

In response to these uncertainties, the application of probabilistic models like Markov chains allows for a more accurate prediction of how these systems might evolve over time. The stochastic nature of Markov chains is particularly useful in the context of smart cities, where future states are dependent on multiple, often unpredictable factors such as technological disruptions, policy innovations, and varying levels of citizen participation. Several studies have explored the challenges posed by the dynamic nature of smart city governance. For instance, Townshend and Walker [21] highlight the need for governance models that can adapt to both incremental and disruptive technological changes. Similarly, Batty [4] discusses the potential for digital infrastructure to reshape governance processes but also warns of the risks associated with poorly managed technological integration. These insights underscore the necessity of adopting flexible, probabilistic models that can capture the inherent uncertainty of smart city governance.

2.2 Markov Chains and Decision-Making in Governance

Governance in smart cities is highly susceptible to external influences such as rapid technological advancements, regulatory changes, and fluctuating citizen engagement. The complexity of the interconnected subsystems that define smart cities introduces a range of uncertainties that complicate governance efforts [1, 3]. In response to these uncertainties, the application of probabilistic models like Markov chains allows for a more accurate prediction of how these systems might evolve over time. The stochastic nature of Markov chains is particularly useful in the context of smart cities, where future states are dependent on multiple, often unpredictable factors such as technological disruptions, policy innovations, and varying levels of citizen participation. Several studies have explored the challenges posed by the dynamic nature of smart city governance. For instance, Townshend and Walker [21] highlights the need for governance models that can adapt to both incremental and disruptive technological changes. Similarly, Batty [4] discusses the potential for digital infrastructure to reshape governance processes but also warns of the risks associated with poorly managed technological integration. These insights underscore the necessity of adopting flexible, probabilistic models that can capture the inherent uncertainty of smart city governance.

2.3 Citizen Participation and Technological Adoption in Smart Cities

A critical factor in the success of smart city governance is the extent to which citizens are involved in decision-making processes and the adoption of digital technologies. In many cases, the transition from traditional governance models to E-Governance systems requires significant changes in how citizens interact with government services [2]. The adoption of digital platforms for public service delivery, such as online portals for tax payments or applications for municipal services, can transform the relationship between citizens and government. However, these transformations are often accompanied by challenges related to digital literacy, privacy concerns, and varying levels of access to technology [14].

The application of Markov chains can provide valuable insights into how citizen participation and technological adoption impact governance outcomes. By modelling the transitions between different levels of citizen engagement and technological adoption, it becomes possible to simulate how these factors influence governance efficiency. For example, in a Markov chain model, one state might represent a low level of citizen engagement and digital infrastructure, while another state could represent a high level of engagement and widespread use of e-government platforms. The transition probabilities between these states can be adjusted based on factors such as government investment in digital literacy programs or efforts to improve access to technology in underserved communities [22].

Furthermore, the probabilistic nature of Markov chains allows for the incorporation of uncertainty into these simulations. For instance, the rate at which citizens adopt new technologies may vary depending on factors such as socioeconomic status, education, and geographical location. By capturing these uncertainties in a Markov chain model, it becomes possible to provide more accurate predictions about the long-term success of E-Governance initiatives in smart cities [11].

2.4 Political Innovation and Policy Changes

In addition to technological and social factors, political innovation plays a crucial role in shaping the governance of smart cities. As noted by Taewoo Nam and Theresa A. Pardo [15], political leadership is essential for driving the adoption of smart city technologies and implementing policies that promote transparency, efficiency, and accountability in governance. However, political processes are often subject to sudden changes, such as shifts in public opinion, election outcomes, and policy priorities. These changes can have a significant impact on the trajectory of governance systems, making it difficult to predict how governance structures will evolve over time.

Markov chains are well-suited to model the uncertainty associated with political innovation and policy changes. By incorporating transition probabilities that reflect the likelihood of political shifts, it is possible to simulate how these changes affect governance outcomes in smart cities. For example, a Markov chain model could simulate the impact of a new government policy aimed at promoting digital inclusion, which might increase the probability of transitioning to a state with higher levels of citizen engagement and digital infrastructure [2]. The ability to model the impact of political innovation is particularly important in the context of smart cities, where governance systems must be agile and responsive to rapidly changing political environments. By using Markov chains to simulate the effects of different policy interventions, policymakers can develop more resilient governance strategies that account for the uncertainties inherent in political processes [4, 22].

3 Methodological Approach

This section presents a methodological approach to the application of Markov chains to optimize E-Governance in smart cities. Markov chains offer a powerful mathematical framework for modelling probabilistic transitions between governance states, denoted by $(S_1, S_2, \ldots, S_n)$, where each state represents a specific level of E-Governance effectiveness. These transitions reflect the dynamic and often uncertain nature of governance processes, particularly in environments as complex as smart cities, where numerous actors and external factors influence governance outcomes over time [2, 22].

In addition to modelling the transition dynamics between governance states, it is essential to understand the underlying factors that influence these transitions. Transition probabilities, represented by the transition matrix, are highly sensitive to a range of internal and external factors, such as technological innovation, political stability, and public engagement, all of which determine the effectiveness of E-Governance [3]. These factors are crucial for identifying the drivers of improvement or decline in governance performance and for assessing the likelihood of transitioning between different states.

3.1 Markov Property and Transition Modelling

Markov chains are particularly well-suited for modelling systems where the future state depends only on the current state, not on the sequence of events that preceded it. This property, known as the *Markov property,* is highly applicable to governance dynamics, where decision-making at any given time depends primarily on the present conditions rather than historical paths [7, 8, 23]. In E-Governance, we consider a finite set of governance states $S_1, S_2, \ldots, S_n$, each corresponding to a distinct level of governance efficiency, such as infrastructure readiness, citizen participation, or administrative agility [15, 16].

Moreover, the use of Markov chains enables the incorporation of probabilistic uncertainty, which is vital for modelling governance systems that are subject to unpredictable changes, such as political shifts, technological disruptions, or societal responses [11]. By simulating various scenarios through Markov processes, policymakers can make data-driven decisions that account for both short-term fluctuations and long-term trends in governance performance [13, 20]. This approach has been successfully applied in related fields such as finance and economics, for instance, to model monetary policy transmission mechanisms and their governance implications [24].

3.1.1 Transition Matrix

Let $(\Omega, \mathcal{F}, P)$ be a probability space, where:

- Ω is the sample space, representing all possible outcomes.
- $\mathcal{F}$ is the sigma-algebra of events, a collection of subsets of Ω.
- P is a probability measure, assigning probabilities to events in $\mathcal{F}$.

A *Markov chain* is a discrete-time stochastic process $\{X_t\}_{t \in N}$, where X_t represents the state of the system at time t, defined on the probability space $(\Omega, \mathcal{F}, P)$ and taking values in a finite state space $S = \{S_1, S_2, \ldots, S_n\}$. Let P represent the transition matrix, where each element P_{ij} denotes the conditional probability of transitioning from state S_i to state S_j at the next time step, given that the system is in state S_i.

$$P = \begin{pmatrix} P_{11} & P_{12} & \cdots & P_{1n} \\ P_{21} & P_{22} & \cdots & P_{2n} \\ \vdots & \vdots & \ddots & \vdots \\ P_{n1} & P_{n2} & \cdots & P_{nn} \end{pmatrix}. \quad (1)$$

Where:

- $P_{ij} = P(X_{t+1} = S_j \mid X_t = S_i)$ is the conditional probability of transitioning from state S_i to state S_j at time t + 1 given that the system is in state S_i at time t.
- Each entry $P_{ij} \in [0, 1]$ represents the likelihood of moving from state S_i to state S_j.
- $\sum_{j=1}^{n} P_{ij} = 1$ for all *i*.

The use of Markov chains to model these transitions is grounded in their ability to represent stochastic processes effectively. In the context of governance dynamics, they provide a powerful tool for understanding how governance efficiency evolves in uncertain environments where both digital infrastructure and governance decisions are subject to change [17]. Moreover, Markov models offer insights into long-term behavior patterns, allowing the calculation of steady-state distributions that reflect governance efficiency over time. These steady-state distributions indicate the likelihood of the system being in a particular state of governance in the long run.

3.1.2 Construction of the Transition Matrix *P*

Let us consider a system composed of n distinct states, denoted by $S_1, S_2, \ldots, S_n$. Over a designated historical period, we observe and record transitions between these states. Specifically, for each state S_i, the number of transitions to another state S_j is carefully noted. This process involves tallying the transitions between the states $S_1, S_2, \ldots, S_n$ for each year or defined time interval.

The transition matrices P_{ij} model the probabilities of transitions between different levels of performance. The estimated transition probability $\hat{p}ij$, representing the likelihood of moving from state i (current governance level) to state j (future governance level), is derived from observed historical data as follows:

$$P_{ij} = \hat{p}\text{ij} = \frac{n_{ij}}{n_{i.}} = \frac{\sum_{t=0}^{T-1} n_{ij}(t)}{\sum_{t=0}^{T-1} n_{i.}(t)}. \tag{2}$$

Where n_{ij} is the number of observed transitions from state i to state j during the observation period T, and $n_{i.}$ represents the total number of instances in which the company occupies state i during the same period. T denotes the total duration of observation [23].

This transition matrix P encapsulates the probabilities that a company will shift from one governance level to another between successive periods, offering a quantitative framework for evaluating the dynamics of governance practices and forecasting future performance trends.

3.1.3 Probabilistic Dynamics of the Governance System

The evolution of the governance system over time can be described by the distribution of probabilities across states. Let $\pi(t)$ represent the probability distribution over the governance states at time t, where $\pi_i(t)$ denotes the probability of being in state S_i at time t. The probabilistic dynamics follow the Markov property:

$$\Pi(t+1) = \pi(t)\text{P}. \tag{3}$$

Given an initial distribution $\pi(0)$, the system will converge to a steady-state distribution π^*, which satisfies:

$$\Pi^* = \pi^*\text{P}. \tag{4}$$

The difference between the estimated distribution at each step and the stationary distribution is denoted by $\epsilon(t)$, which can be defined as:

$$\epsilon(t) = \left|\pi(t) - \pi^*\right|. \tag{5}$$

The simulation will continue iterating until the difference $\epsilon(t)$ becomes smaller than a predefined precision threshold ϵ_0, indicating that the estimated distribution has sufficiently converged to the stationary distribution.

The simulation process involves the following steps:

Smart city E-Governance strategy refinement using Markov chains algorithm

Data: Initial distribution of governance states π_0, precision threshold ϵ_0
Refined: E-Governance strategies based on stationary distribution and convergence criterion

Step 1: Initial distribution of governance states

- Define the initial distribution π_0 of governance states. Each state represents a level of governance performance (e.g., low, medium, high).
- This distribution is based on various key performance indicators reflecting the current state of E-Governance.

Step 2: Estimation of transition probabilities

- Identify the key factors influencing the transition from one governance state to another. These factors will be defined later in the document.
- Estimate the transition probabilities P_{ij}, representing the likelihood of transitioning from state i to state j, based on the combined effect of the factors.

$$P_{ij} = f\left(\text{key factors}\right) \times w, \tag{6}$$

- where f(key factors) *m*odels the impact of the influencing factors, and w represents their relative importance.

Step 3: Simulation of governance dynamics

- Initialize $\epsilon_0 = 10^{-6}$, representing a high precision threshold.
- Simulate the evolution of governance states over time by applying the transition matrix P to the initial distribution π_0.

$$\pi_{t+1} = \pi_t \cdot P. \tag{7}$$

- At each time step t, calculate $\epsilon(t)$, the difference between the estimated distribution π_t and the stationary distribution π^*

$$\epsilon(t) = \left|\pi_t - \pi^*\right|. \tag{8}$$

- Check the convergence criterion: If $\epsilon(t) < \epsilon_0$, stop the simulation. Otherwise, continue to the next time step.

Step 4: Calculation of the stationary distribution

Continue the simulation until the convergence criterion is met, meaning that $\epsilon(t) < \epsilon_0$, indicating that the difference between π_t and the stationary distribution is within the desired precision.

Step 5: Strategy refinement based on stationary distribution and $\epsilon(t)$

- Analyze the stationary distribution $\pi_{\text{stationary}}$ and the convergence rate of $\epsilon(t)$ to refine governance strategies.
- Focus on adjusting the transition probabilities to move away from unfavorable governance states and encourage transitions towards more desirable states by influencing the key factors discussed in the following sections.

3.2 Key Factors Influencing Transition Probabilities

The transition probabilities P_{ij}, representing the likelihood of transitioning from one governance state to another, are dynamic and sensitive to a variety of key factors. These factors encapsulate both the internal mechanics of the E-Governance system and external forces influencing smart city evolution. Below, we provide an in-depth analysis of these factors and their critical role in shaping E-Governance within the smart city context.

The assessment of E-Governance performance in smart cities has gained significant importance as urban centers worldwide strive for greater efficiency, sustainability, and inclusivity through the integration of digital technologies. To capture the multi-dimensional complexity of this transformation, we identified ten essential factors influencing the transition dynamics of governance. These factors include technological adoption, political stability, citizen engagement, and a spectrum of socio-economic and environmental considerations.

Each factor is rigorously quantified and normalized on a [0,1] scale, allowing for a standardized comparison across cities, regardless of size or geographical context. This normalization process ensures that each factor can be evaluated consistently, enabling a comprehensive analysis of E-Governance performance across different urban environments.

The selected ten key factors, grounded in empirical evidence from academic literature, include:

- **F1: Technological adoption.** The score for technological adoption is calculated using indicators that measure the penetration and use of digital technologies, such as internet coverage, IoT deployment, and mobile connectivity. This score is normalized as follows:

$$\text{Score}\,F_1 = \frac{T_{it}}{T_{\max}}, \tag{9}$$

- Where T_{it} represents the technological adoption indicator for a given city (e.g., the percentage of fiber optic coverage), and $T_{\max}$ represents the theoretical maximum value (e.g., 100% coverage). The justification for this calculation lies in the critical role that technology plays in enabling smart city infrastructure, where a higher degree of digital adoption directly correlates with the city's ability to implement effective E-Governance solutions.
- **F2: Political stability.** The political stability score is derived from internationally recognized indices, such as the World Governance Indicators, which assess the resilience and strength of governance structures in the city. The calculation is as follows:

$$\text{Score}\,F_2 = \frac{P_{it}}{P_{\max}}, \tag{10}$$

- Where P_{it} is the political stability index for the city, and $P_{\max}$ is the maximum value on the scale (ranging from 0 to 1). A stable political environment is essential for long-term planning and policy continuity in E-Governance. Therefore, this factor reflects the capacity of cities to sustain and advance digital transformations.
- **F3: Citizen engagement.** The citizen engagement score is based on metrics reflecting participation in E-Governance platforms, such as e-voting systems, public consultations, and collaborative platforms:

$$\text{Score}\,F_3 = \frac{C_{it}}{C_{\max}}, \tag{11}$$

- Where C_{it} represents the number of citizens actively participating in digital platforms, and $C_{\max}$ is the total population of the city. High citizen participation is crucial for the success of E-Governance as it ensures that decisions are made democratically and that digital services are more attuned to citizens' needs.
- **F4: Technological innovation.** The technological innovation score captures the city's capacity to promote research, development, and entrepreneurial activities, with metrics such as patent filings and startup activity:

$$\text{Score}\,F_4 = \frac{I_{it}}{I_{\max}}, \tag{12}$$

- Where I_{it} represents the innovation index for the city (e.g., patents filed), and $I_{\max}$ is the maximum observed innovation score. Technological innovation is vital to maintaining a competitive advantage in the E-Governance landscape, and this score highlights the city's contribution to fostering an innovative ecosystem.
- **F5: Economic development.** Economic development is measured by indicators such as GDP per capita and the Human Development Index (HDI), reflecting the financial and human capital resources available in the city. The score is calculated as follows:

$$\text{Score}\,F_5 = \frac{E_{it}}{E_{\max}}, \tag{13}$$

- Where E_{it} represents the city's economic development indicator, and $E_{\max}$ represents the global maximum reference value. A city's economic strength is fundamental to its ability to invest in and sustain advanced digital infrastructures, making this factor essential for evaluating E-Governance performance.
- **F6: Social inequalities.** The score for social inequalities is computed using the Gini coefficient, which measures income inequality. The score is normalized to reflect higher equity as it approaches 1:

$$\text{Score}\,F_6 = 1 - \frac{G_{it}}{G_{\max}}, \tag{14}$$

- Where G_{it} represents the Gini coefficient for the city, and $G_{\max}$ represents the maximum observed Gini coefficient globally. Addressing social inequalities is critical for inclusive E-Governance, ensuring that all citizens have equal access to digital services.
- **F7: Environmental sustainability**.
  The environmental sustainability score is derived from indicators such as CO2 emissions and renewable energy usage, providing insight into the city's commitment to sustainable development:

$$\text{Score}\,F_7 = \frac{D_{it}}{D_{\max}}, \tag{15}$$

- Where D_{it} is the composite score of environmental sustainability metrics, and $D_{\max}$ represents the maximum reference value for sustainability. As cities increasingly adopt green policies, this factor reflects their ability to align E-Governance with environmental goals, such as reducing carbon footprints.
- **F8: Institutional quality.** Institutional quality is measured using transparency indices, public sector performance, and anti-corruption metrics, as calculated by:

$$\text{Score}\,F_8 = \frac{Q_{it}}{Q_{\max}}, \tag{16}$$

- Where Q_{it} is the index reflecting the quality of governance institutions, and $Q_{\max}$ is the maximum index value. Strong institutions underpin effective E-Governance by ensuring accountability, transparency, and efficiency in policy implementation.
- **F9: Legal framework for E-Governance.** The strength of the legal framework is assessed through the city's E-Governance regulations, including cybersecurity and data protection laws:

$$\text{Score}\,F_9 = \frac{L_{it}}{L_{\max}}, \tag{17}$$

- Where L_{it} is the legal framework score, and $L_{\max}$ is the maximum value. A robust legal framework ensures that E-Governance initiatives are secure, legally compliant, and protect citizens' rights in the digital space.
- **F10: Digital infrastructure**. The digital infrastructure score is derived from indicators such as broadband penetration and the availability of high-speed networks:

$$\text{Score}\,F_{10} = \frac{I_{it}}{I_{\max}}, \tag{18}$$

- Where I_{it} represents the digital infrastructure index, and $I_{\max}$ is the theoretical maximum. Digital infrastructure is the backbone of smart city governance, enabling efficient service delivery and facilitating seamless communication between various stakeholders.

3.3 *Definition of Scoring*

Each city is assessed annually based on several key criteria, such as technological advancement, governance efficiency, and citizen engagement, which influence its transition from one state of governance to another. These criteria are crucial in evaluating the overall performance of smart cities in terms of E-Governance, given their ability to adopt emerging technologies, foster effective interactions between citizens and institutions, and maintain stable and transparent governance.

A global score is assigned to each city annually, quantifying its performance for that year. The variable t represents the time scale, which, in this study, corresponds to the year of assessment. Furthermore, this global score enables a comparison of the performance of different cities over time, allowing for tracking of their progress in E-Governance. In our study, we defined a weighting system for the ten predefined key factors, $F^{(k)}$, for $k = 1, 2, \ldots, 10$, to reflect the varying importance of each dimension in influencing E-Governance transitions.

$$\text{Score}_t = \sum_{k=1}^{10} w_{F^{(k)}} F_t^{(k)}. \tag{19}$$

In this equation, $w_{F^{(k)}}$ represents the weight assigned to each factor $F^{(k)}$, for $(k = 1, 2, \ldots, 10)$, based on its relative importance in the assessment of E-Governance. These weights are determined by the potential impact of each factor on the overall performance of the city. For instance, a factor such as technological adoption may receive a higher weight if it is deemed essential for the digital transformation of cities.

The weighting of these factors is significant, as it reflects the importance of different elements in evaluating smart city performance. Cities scoring higher on more heavily weighted factors, such as innovation or institutional quality, are considered to have superior overall E-Governance performance.

In the section dedicated to numerical results, we opted to simplify the analysis by assigning equal weights to all factors, i.e. $w_{F^{(k)}} = \frac{1}{10}$, ensuring that each factor contributes equally to the city's overall score. This choice neutralizes any perceived differences in the importance of each criterion and allows for a more uniform evaluation of the cities under study.

4 Results of Numerical Simulations

In this section, we present a comparative analysis of the E-Governance performances of Paris and Tokyo through Markov chain modelling.

The numerical results highlight the evolution of governance indicators over the 2012–2022 period, based on key factors such as technological adoption, citizen participation, and innovation (see Tables 1 and 2).

The proposed Algorithm for refining smart city governance strategies is applied to these two cities to illustrate the dynamics of governance states. This model simulates transitions between different levels of governance performance, accounting for the impact of several key factors on these transitions.

4.1 Evaluative Approach Based on Key Governance Factors

Paris and Tokyo were selected for this comparative study due to their prominence as global leaders in E-Governance and innovation. Both cities have been at the forefront of implementing advanced E-Governance systems, making them ideal candidates for analyzing the impact of key governance factors through Markov chain modelling.

Paris, as a major European capital, has consistently demonstrated a strong commitment to technological adoption, citizen engagement, and environmental sustainability, driven by its progressive policies and regulatory frameworks. The availability of comprehensive open data and the city's focus on transparency further reinforce its status as a leading smart city.

Tokyo, on the other hand, stands out for its sophisticated integration of technology and urban management. As one of the most populous and technologically advanced cities in the world, Tokyo's E-Governance strategies are characterized by high levels of innovation, meticulous regulatory oversight, and a proactive approach to addressing urban challenges.

Table 1 Factors, formulas, and definitions for E-Governance evaluation Paris with data sources

Factor	Formula	Definition of elements and maximum values	Data source
F1	$\text{Score}F_1 = \frac{T_{it}}{T_{\max}}$	T_{it}: measure of technological adoption (e.g., percentage of fiber optic coverage) $T_{\max}$: maximum possible value (e.g., 100% internet coverage)	European Commission (2012–2022)
F2	$\text{Score}F_2 = \frac{P_{it}}{P_{\max}}$	P_{it}: political stability index for Paris $P_{\max}$ maximum value of this index (scale from 0 to 1)	World Bank (2012–2022a)
F3	$\text{Score}F_3 = \frac{C_{it}}{C_{\max}}$	C_{it} number of citizens using digital platforms $C_{\max}$ total population of Paris	Ville de Paris (2012–2022)
F4	$\text{Score}F_4 = \frac{I_{it}}{I_{\max}}$	I_{it} sum of innovation indicators (patents filed, R&D investments) $I_{\max}$ maximum observed innovation value over a period	OECD (2012–2022b)
F5	$\text{Score}F_5 = \frac{E_{it}}{E_{\max}}$	E_{it}: GDP per capita $E_{\max}$ maximum value of this indicator in a global reference	OECD (2012–2022a)
F6	$\text{Score}F_6 = 1 - \frac{G_{it}}{G_{\max}}$	G_{it} Gini coefficient of Paris $G_{\max}$: maximum observed value of the Gini coefficient	INSEE (2012–2022)
F7	$\text{Score}F_7 = \frac{D_{it}}{D_{\max}}$	D_{it} environmental indicators ($CO2$ emissions, renewable energy usage) $D_{\max}$: maximum reference value in this field	Airparif (2012–2022)
F8	$\text{Score}F_8 = \frac{Q_{it}}{Q_{\max}}$	Q_{it} institutional quality index $Q_{\max}$: maximum value of the index (corruption perception index)	Transparency International (2012–2022)
F9	$\text{Score}F_9 = \frac{L_{it}}{L_{\max}}$	L_{it}: index of legal maturity for digital regulations $L_{\max}$: maximum value of this index	CNIL (2012–2022)
F10	$\text{Score}F_{10} = \frac{I_{it}}{I_{\max}}$	I_{it}: digital infrastructure level (fiber coverage percentage) $I_{\max}$: theoretical maximum coverage (100%)	ARCEP (2012–2022)

By comparing these two cities, we aim to highlight not only the commonalities in their governance approaches but also the unique strategies each has developed to navigate their distinct socio-economic landscapes. This dual analysis provides valuable insights into how global cities can refine their governance strategies to achieve sustainable and efficient digital transformation.

This study aims to analyze Paris's performance in E-Governance from 2012 to 2022. Ten key factors are considered, including technological adoption, political stability, citizen engagement, and other socio-economic and environmental indicators. The scores for each factor are computed based on publicly available data, normalized on a scale from 0 to 1.

The evaluation of E-Governance performance for Tokyo and Paris encompasses a broad set of critical elements. The first factor examines technological adoption,

Table 2 Factors, formulas, and definitions for E-Governance evaluation Tokyo with data sources

Factor	Formula	Definition of elements and maximum values	Data source
F1	$\text{Score}\,F_1 = \frac{T_{it}}{T_{\max}}$	T_{it}: technological adoption speed (including digital infrastructures, 5G deployment) $T_{\max}$: 100% technological coverage	World Bank (2012–2022b)
F2	$\text{Score}\,F_2 = \frac{P_{it}}{P_{\max}}$	P_{it}: political stability (derived from governance indicators) $P_{\max}$: maximum stability on a 0–1 scale	World Bank (2012–2022a)
F3	$\text{Score}\,F_3 = \frac{C_{it}}{C_{\max}}$	Citizen engagement (measured by the number of participants in civic and digital projects) $C_{\max}$: total population of Tokyo	Tokyo Metropolitan Government (2012–2022)
F4	$\text{Score}\,F_4 = \frac{I_{it}}{I_{\max}}$	Innovation level (based on patent registrations, AI, and robotics developments) $I_{\max}$: highest global level of innovation	Japan Patent Office (2012–2022)
F5	$\text{Score}\,F_5 = \frac{E_{it}}{E_{\max}}$	E_{it}: economic growth (gross domestic product, GDP) $E_{\max}$: maximum reference economic value.	World Bank (2012–2022c)
F6	$\text{Score}\,F_6 = 1 - \frac{G_{it}}{G_{\max}}$	G_{it}: Gini coefficient (social inequality) $G_{\max}$: maximum possible Gini coefficient	World Inequality Database (2012–2022)
F7	$\text{Score}\,F_7 = \frac{D_{it}}{D_{\max}}$	D_{it}: sustainability indicators (carbon emissions, environmental initiatives) $D_{\max}$: maximum environmental value	Tokyo Metropolitan Government (2012–2022)
F8	$\text{Score}\,F_8 = \frac{Q_{it}}{Q_{\max}}$	Q_{it}: institutional quality (transparency, institutional efficiency) $Q_{\max}$:maximum scale for institutional quality indices	Transparency International (2012–2022)
F9	$\text{Score}\,F_9 = \frac{L_{it}}{L_{\max}}$	L_{it}:legal framework robustness (including data protection laws and cybersecurity) $L_{\max}$: global best practice for legal frameworks	Personal Information Protection Commission (PPC) (2012–2022)
F10	$\text{Score}\,F_{10} = \frac{I_{it}}{I_{\max}}$	I_{it}: Digital infrastructure (extent of 5G coverage) $I_{\max}$: 100% digital infrastructure coverage	Ministry of Internal Affairs and Communications (Japan) (2012–2022)

gauged by the extent of digital infrastructure, such as fiber optic networks and 5G deployment, with the maximum value representing complete coverage (100%). Political stability is assessed through an index scaled from 0 to 1, while citizen engagement is measured by the participation rate in digital and civic projects, with the total population serving as the upper limit.

Innovation is reflected in indicators like patent registrations and R&D investments, where the maximum value represents the highest observed level globally. Economic growth is captured by indicators such as GDP per capita, with the maximum drawn from a global reference. The reduction of inequality is evaluated via the

Gini coefficient, with its maximum corresponding to the highest level of inequality recorded.

Environmental sustainability indicators, such as CO2 emissions and the utilization of renewable energy, are benchmarked against a maximum reference value. Institutional quality is assessed based on indices of transparency and institutional efficiency, with the maximum representing the optimal institutional performance. The robustness of the legal framework, particularly concerning digital regulations and cybersecurity, is also measured, where the maximum reflects global best practices. Lastly, digital infrastructure is evaluated by the extent of network and technological coverage, with complete infrastructure coverage representing the highest attainable value.

In order to evaluate and compare the E-Governance performance of Paris and Tokyo, we employed normalization, weighting, and data aggregation methods to calculate the scores for each factor. The resulting scores are summarized in the following Tables 3 and 4.

Table 3 Normalized performance scores of Paris (2012–2022)

Year	F1	F2	F3	F4	F5	F6	F7	F8	F9	F10	Overall score
2012	0.65	0.80	0.50	0.75	0.78	0.40	0.60	0.75	0.60	0.70	0.668
2013	0.70	0.85	0.55	0.78	0.80	0.42	0.62	0.78	0.63	0.72	0.709
2014	0.72	0.82	0.57	0.80	0.82	0.45	0.65	0.80	0.65	0.75	0.734
2015	0.75	0.84	0.60	0.83	0.85	0.48	0.70	0.83	0.68	0.77	0.752
2016	0.80	0.86	0.62	0.85	0.87	0.50	0.72	0.85	0.70	0.80	0.788
2017	0.82	0.83	0.63	0.82	0.88	0.52	0.74	0.82	0.73	0.82	0.788
2018	0.85	0.81	0.65	0.80	0.90	0.55	0.75	0.80	0.75	0.85	0.798
2019	0.87	0.79	0.68	0.78	0.92	0.57	0.78	0.78	0.78	0.87	0.818
2020	0.88	0.77	0.70	0.75	0.91	0.60	0.80	0.75	0.82	0.90	0.828
2021	0.90	0.76	0.73	0.77	0.93	0.63	0.82	0.77	0.85	0.92	0.833
2022	0.92	0.75	0.75	0.80	0.94	0.65	0.85	0.80	0.87	0.95	0.846

Table 4 Performance scores of Tokyo (2012–2022)

Year	F1	F2	F3	F4	F5	F6	F7	F8	F9	F10	Overall score
2012	0.71	0.84	0.57	0.76	0.83	0.52	0.67	0.79	0.69	0.74	0.710
2013	0.73	0.86	0.61	0.79	0.86	0.54	0.70	0.81	0.71	0.77	0.735
2014	0.75	0.88	0.63	0.82	0.88	0.56	0.72	0.83	0.73	0.80	0.760
2015	0.79	0.90	0.66	0.85	0.90	0.59	0.75	0.85	0.76	0.82	0.785
2016	0.81	0.89	0.68	0.87	0.91	0.61	0.77	0.87	0.79	0.84	0.803
2017	0.83	0.87	0.69	0.89	0.92	0.63	0.80	0.89	0.81	0.86	0.810
2018	0.86	0.85	0.71	0.91	0.94	0.66	0.83	0.86	0.83	0.89	0.835
2019	0.88	0.83	0.73	0.93	0.95	0.69	0.85	0.84	0.85	0.91	0.850
2020	0.89	0.81	0.75	0.94	0.96	0.71	0.88	0.82	0.87	0.92	0.860
2021	0.91	0.79	0.77	0.95	0.97	0.73	0.90	0.84	0.89	0.94	0.875
2022	0.93	0.78	0.80	0.96	0.98	0.76	0.92	0.86	0.91	0.96	0.890

The comparative analysis of the E-Governance progress in Paris and Tokyo from 2012 to 2022 reveals strikingly similar trajectories in their pursuit of enhanced management and delivery of digital services, though nuanced differences emerge in terms of strategic priorities and specific challenges.

In Paris, the overall E-Governance score increased steadily from 0.668 in 2012 to 0.846 in 2022 (see Table 5). This reflects sustained efforts to bolster digital infrastructure (F1) and cybersecurity (F2), areas where the city has demonstrated consistent leadership from the outset (Figs. 1 and 2). These foundational pillars have provided a solid framework for the development of other governance aspects. Notably, Paris has made significant strides in electronic participation (F3) and open data initiatives (F9), reaching scores of 0.75 and 0.87 by 2022, underscoring a growing commitment to citizen inclusion and transparency. However, the slower progression in digital inclusion (F6), which increased from 0.40 to 0.65, reveals a persistent challenge: ensuring equitable access to digital services for all demographics remains an ongoing concern for the city (see Fig. 3).

Similarly, Tokyo has seen a robust increase in its E-Governance score, rising from 0.710 in 2012 to 0.890 in 2022 (see Table 5). The city's strategy has focused heavily on the development of digital infrastructure (F1) and the modernization of e-government services (F5), where it has achieved near-optimal levels with scores of 0.93 and 0.98 respectively in 2022. Like Paris, Tokyo has also prioritized citizen engagement (F3) and data transparency (F9), with corresponding scores of 0.80 and 0.91 by 2022, reflecting a clear commitment to fostering citizen participation in governance. Nonetheless, Tokyo, much like Paris, faces ongoing challenges in

Table 5 Summary of E-Governance performance scores for Paris and Tokyo (2012–2022)

City	2012 score	2022 score	Growth rate (%)
Paris	0.668	0.846	26.65
Tokyo	0.710	0.890	25.35

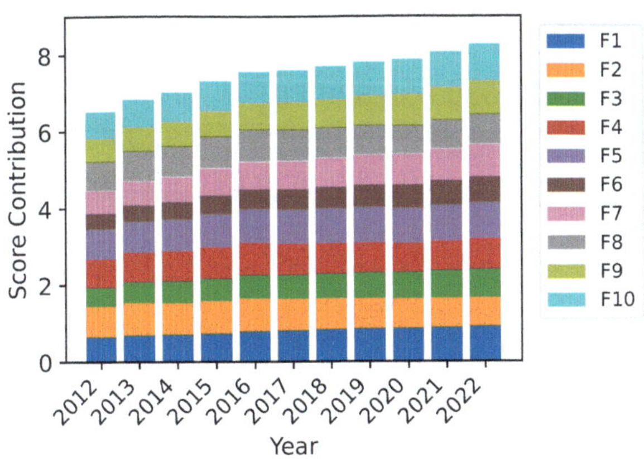

Fig. 1 Proportional contribution of factors (F1-F10) in Paris (2012–2022)

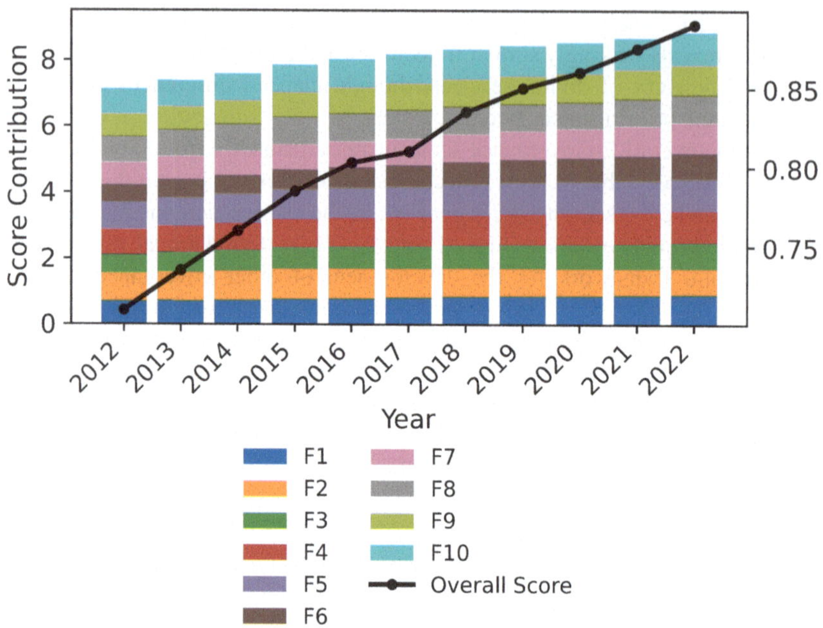

Fig. 2 Proportional contribution of factors (F1-F10) and overall score in Tokyo (2012–2022)

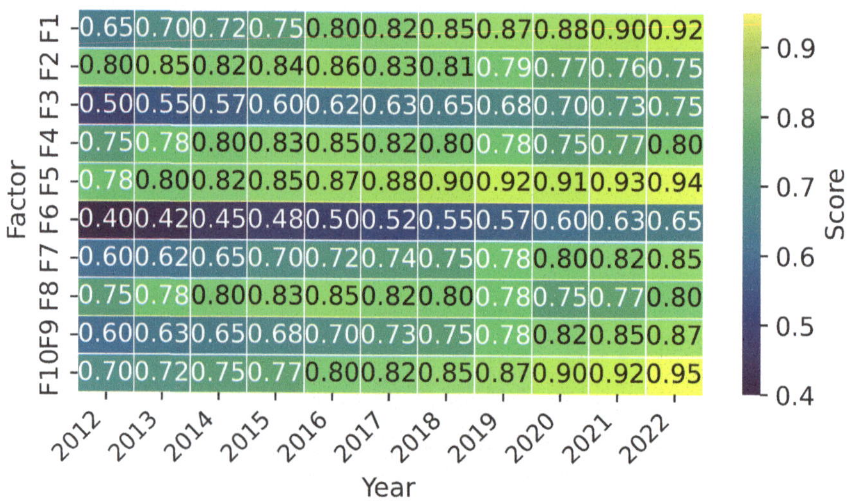

Fig. 3 Performance scores of factors F1-F10 in Paris (2012–2022)

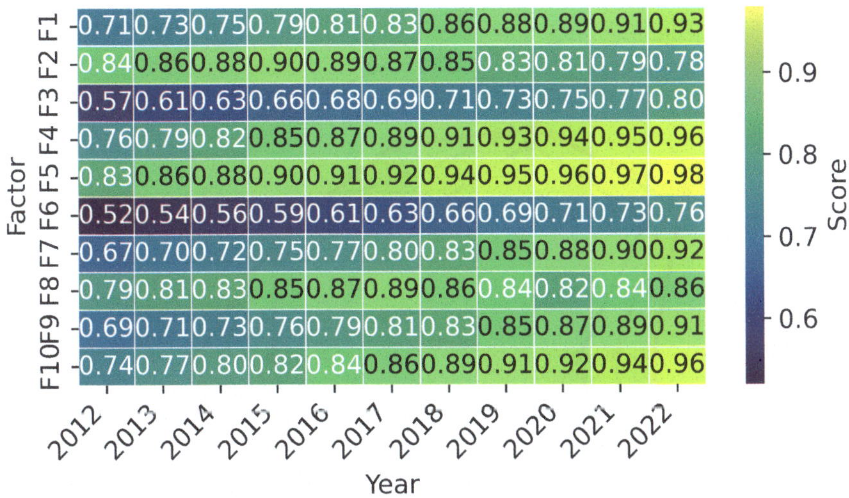

Fig. 4 Performance scores of factors F1-F10 in Tokyo (2012–2022)

digital inclusion (F6), despite improvements from 0.52 to 0.76 over the decade. Additionally, unlike Paris, Tokyo experienced a slight decline in cybersecurity (F2), dropping from 0.84 in 2012 to 0.78 in 2022, which may indicate growing difficulties in keeping pace with increasingly complex cybersecurity threats (Fig. 4).

In other areas, both cities exhibit converging trends, particularly in the modernization of government services (F5) and innovation (F10), where they both reached high levels by the end of the decade. These patterns highlight the successes of Paris and Tokyo in establishing governance frameworks that encourage innovation and the enhancement of public services key elements of their global competitiveness in E-Governance.

The radar chart comparison for 2022 (Fig. 5) further illustrates these dynamics, showing close alignment between the two cities on most factors, yet also revealing subtle divergences. Tokyo outperforms Paris slightly in digital infrastructure (F1) and cybersecurity (F2), while Paris lags somewhat behind Tokyo in digital inclusion (F6) (Fig. 6). These differences reflect distinct strategic priorities: Tokyo places greater emphasis on securing its digital environment, whereas Paris is more focused on expanding equitable access to digital services (see Fig. 4).

Figures 7 and 8 show the evolution of the overall governance score for Paris and Tokyo, respectively, illustrating how both cities reached their current levels of E-Governance (Fig. 9).

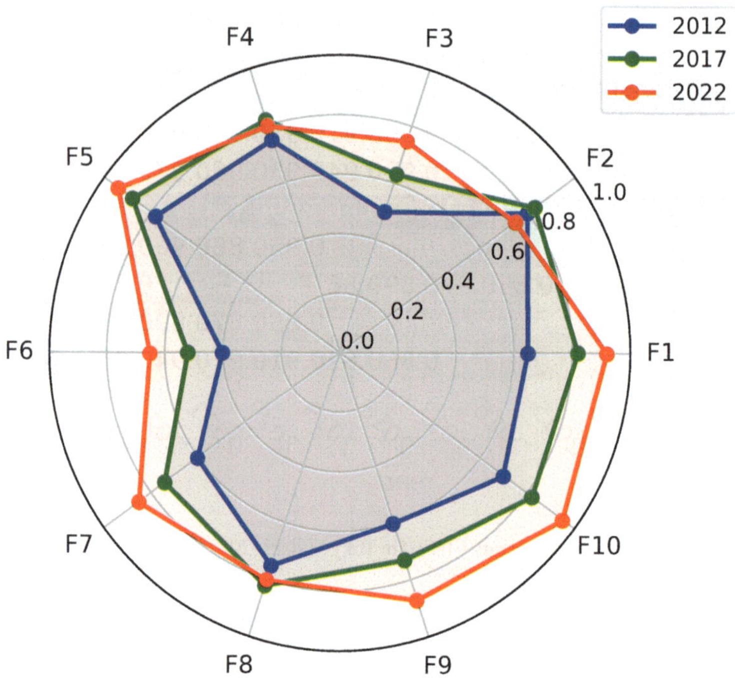

Fig. 5 Radar chart of factors F1-F10 in Paris (2012, 2017, 2022)

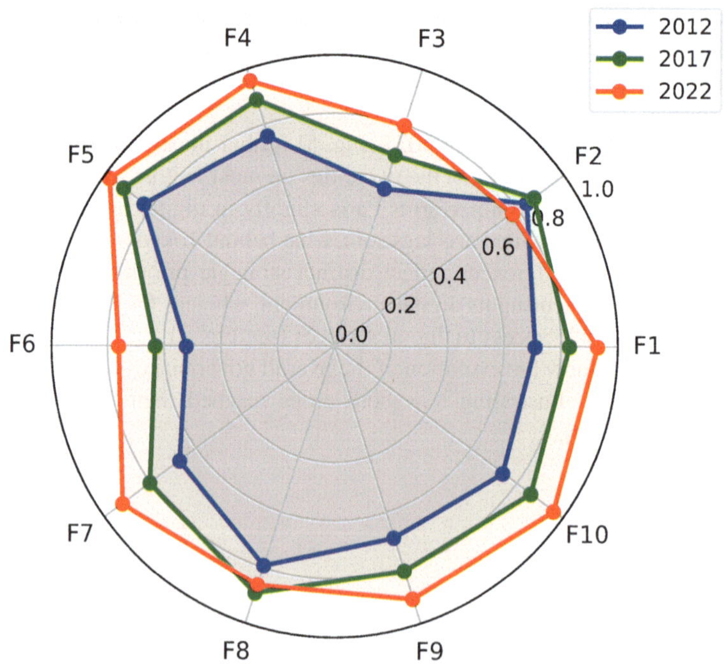

Fig. 6 Radar chart of factors F1-F10 in Tokyo (2012, 2017, 2022)

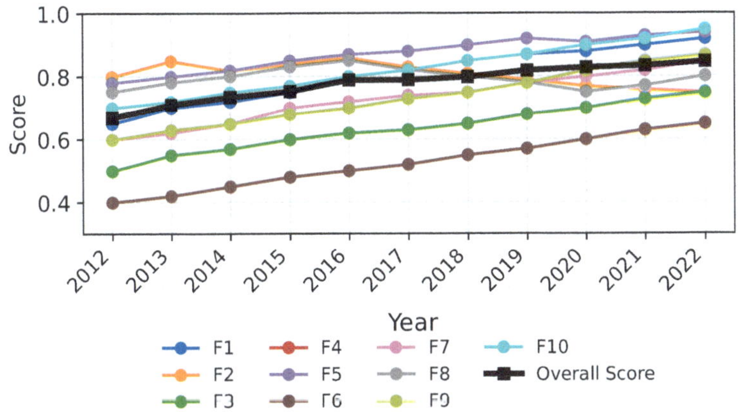

Fig. 7 Evolution of factors (F1-F10) and overall score in Paris (2012–2022)

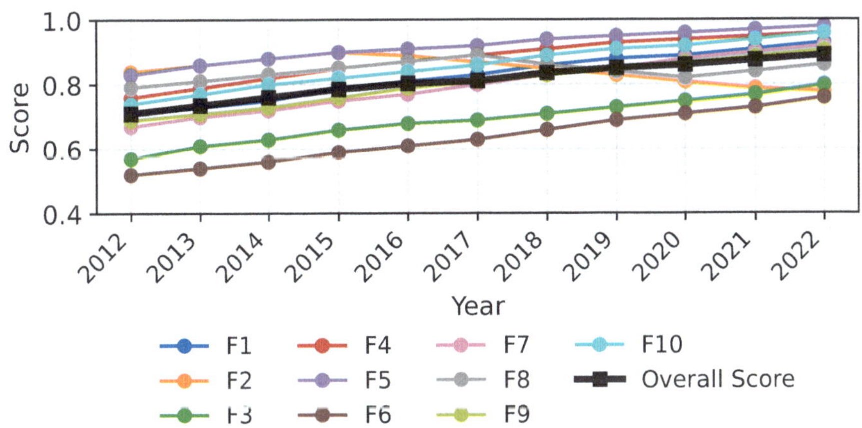

Fig. 8 Evolution of factors (F1-F10) and overall score in Tokyo (2012–2022)

4.2 *Evolutionary Approach Through Stationary Distribution Simulation and Governance Strategy Optimization*

The simulations are grounded in a Markov chain model with n = 5 governance states, representing varying levels of E-Governance efficiency, ranging from S_1 (the lowest efficiency) to S_5 (the highest efficiency). The transition matrix P was initialized using empirical estimates derived from historical data on transitions observed for each city from 2012 to 2022. The state space is defined as follows:

- S_1: Score between 0.50 and 0.70 (minimal E-Governance)
- S_2: Score between 0.70 and 0.78
- S_3 Score between 0.78 and 0.83

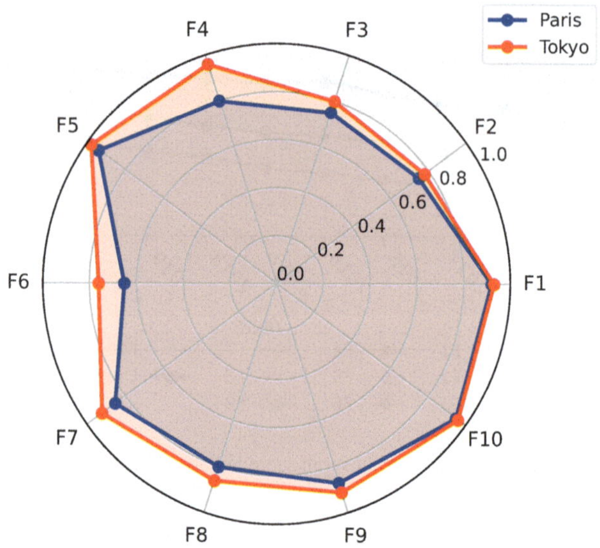

Fig. 9 E-Governance comparison between Paris and Tokyo—2022

Table 6 State assignment for Tokyo and Paris (2012–2022)

Year	Score (Tokyo)	State (Tokyo)	Score (Paris)	State (Paris)
2012	0.710	2	0.668	1
2013	0.735	2	0.709	1
2014	0.760	2	0.734	2
2015	0.785	3	0.752	2
2016	0.803	4	0.788	3
2017	0.810	4	0.788	3
2018	0.835	4	0.798	3
2019	0.850	5	0.818	4
2020	0.860	5	0.828	4
2021	0.875	5	0.833	4
2022	0.890	5	0.46	5

- S_4: Score between 0.83 and 0.88
- S_5: Score between 0.88 and 1.00 (optimal E-Governance)

The annual scores for both Tokyo and Paris were categorized into these states, as shown in the following table (Table 6):

The transition matrices for Tokyo and Paris, derived from the movements between the defined governance states and estimated based on the approach described in Sect. 3.1.2, are shown below (Tables 7 and 8):

Table 7 Transition matrix for Tokyo

	State 1	State 2	State 3	State 4	State 5
State 1	0.00	1.00	0.00	0.00	0.00
State 2	0.00	0.66	0.33	0.00	0.00
State 3	0.00	0.00	0.60	0.40	0.00
State 4	0.00	0.00	0.00	0.80	0.20
State 5	0.00	0.00	0.00	0.00	1.00

Table 8 Transition matrix for Paris

	State 1	State 2	State 3	State 4	State 5
State 1	0.00	1.00	0.00	0.00	0.00
State 2	0.00	0.50	0.50	0.00	0.00
State 3	0.00	0.00	0.60	0.40	0.00
State 4	0.00	0.00	0.00	0.75	0.25
State 5	0.00	0.00	0.00	0.00	1.00

4.3 Stationary Distribution Results

The stationary distribution results for both Tokyo (Fig. 10) and Paris (Fig. 11) show that after a certain number of iterations, both cities converge towards the same stationary distribution $(0, 0, 0, 0, 1)$, implying that in the long run, E-Governance in both cities will reach the highest performance level S_5 (with scores between 0.88 and 1.00) without the possibility of regressing to lower states. This convergence is achieved within a precision threshold of $\epsilon_0 = 10^{-4}$, ensuring that the probabilities accurately reflect the stability of the governance states.

The progression of E-Governance in Tokyo and Paris is modelled through a state space representing five levels of governance performance, S_1 to S_5, corresponding to their scores from 2012 to 2022. Both cities show significant improvements over the decade. Tokyo increased from a score of 0.71 in 2012 to 0.89 in 2022, whereas Paris advanced from 0.668 to 0.846. Although both cities ultimately reach S_5 the trajectories and pace of their transitions differ markedly, as evidenced by the graphical results for each city.

In Tokyo, the transitions through intermediate states such as S_2, S_3, and S_4 are more streamlined, reflecting the city's robust technological infrastructure and proactive governance strategies. As shown in the stationary distribution graph for Tokyo, the city initially stabilizes in S_2 and S_3, but transitions more rapidly through these states, driven by strong policies in technology adoption, innovation, and infrastructural development.

In contrast, Paris adopts a more gradual approach, as illustrated by its stationary distribution. The city's probability distribution reflects a cautious but steady transition through intermediate states S_2, S_3, and S_4, before eventually reaching S_5. This slower progression can be attributed to Paris's focus on social inclusivity and

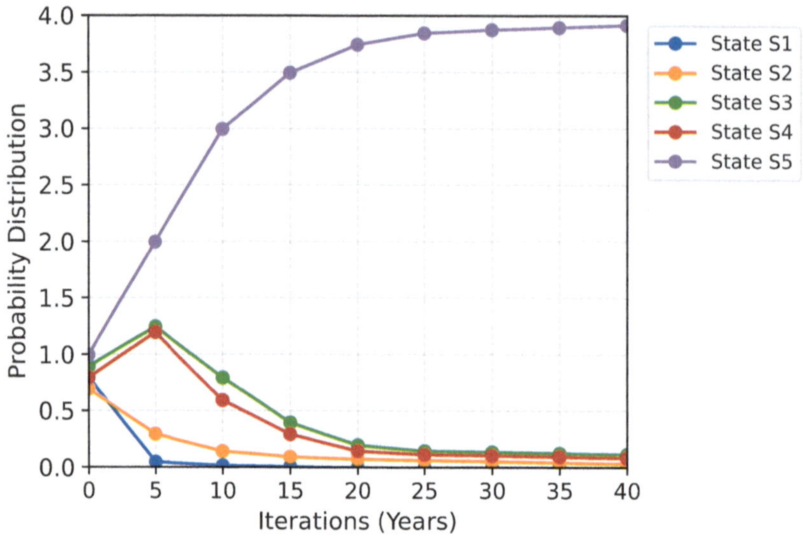

Fig. 10 Simulation of the stationary distribution for E-Governance in Tokyo

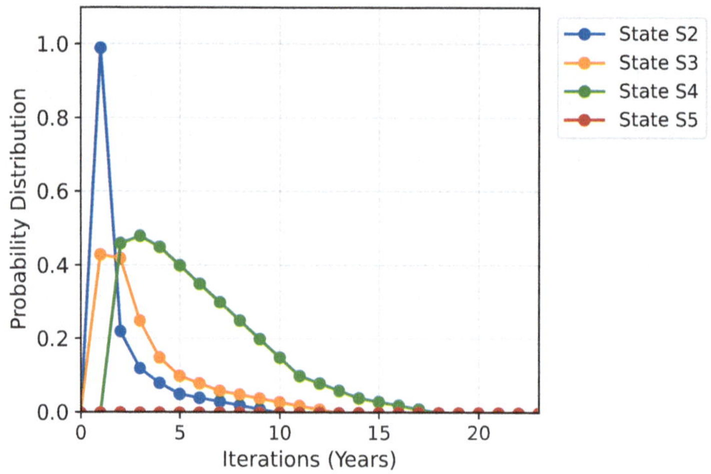

Fig. 11 Simulation of the stationary distribution for E-Governance in Paris

ensuring that digital services are broadly accessible, as indicated by its emphasis on reducing social inequalities (F6) and enhancing citizen engagement (F3).

Ultimately, both cities converge to S_5, but the time to reach this stationary distribution varies: Paris takes 34 years, while Tokyo requires 43 years. This difference in convergence time reflects the unique governance strategies and priorities of each city, with Tokyo emphasizing rapid technological progress and Paris prioritizing equitable access and inclusivity in E-Governance.

Both graphs demonstrate that despite different paths, Tokyo and Paris are projected to achieve and maintain the highest governance level, S_5, characterized by high-performance E-Governance, albeit with distinct strategies and timelines.

5 Conclusion and Future Research Work

The governance of smart cities presents a complex and evolving challenge, as urban environments increasingly depend on digital technologies to ensure efficient public services, transparency, and citizen participation. This chapter has demonstrated that stochastic models, particularly Markov chains, offer a robust framework for understanding and optimizing governance transitions in these systems. By modelling governance as a series of probabilistic transitions between different states, this approach captures the uncertainty inherent in smart city ecosystems while offering predictive insights into their future evolution. The application of Markov chains to smart city governance provides significant advantages over traditional deterministic models. This method enables the representation of uncertainty and variability, allowing decision-makers to simulate various policy scenarios and assess their potential outcomes. Such simulations support better-informed decisions, improving the effectiveness, transparency, and inclusiveness of governance strategies.

The results underscore the critical role of technological innovation, political leadership, and citizen engagement in shaping the success of smart city governance. The proposed framework could be further enhanced by integrating additional variables, such as social equity, environmental sustainability, and economic development.

Looking ahead, future research should apply this framework to specific case studies to compare governance practices across different smart cities. Furthermore, incorporating real-time data, facilitated by the Internet of Things (IoT), would increase the accuracy of predictive models, enabling more dynamic responses to governance challenges. Finally, it would be valuable to explore the impact of social and environmental factors on governance transitions and develop decision-making models adapted to crisis situations, such as pandemics or natural disasters.

References

1. Meijer AJ, Gil-Garcia JR, Bolívar MPR (2016) Smart city research: contextual conditions, governance models, and public value assessment. Soc Sci Comput Rev 34(6):647–656. https://doi.org/10.1177/0894439315618890
2. Anthopoulos L (2017) Smart government: a new adjective to government transformation or a trick? In: Understanding smart cities: a tool for smart government or an industrial trick. Springer, Berlin, pp 75–94. https://doi.org/10.1007/978-3-319-57015-0-6

3. Chourabi H, Nam T, Walker S, Gil-Garcia JR, Mellouli S, Nahon K, Pardo TA, Scholl HJ (2012) Understanding smart cities: an integrative framework. In: 45th Hawaii international conference on system sciences, pp 2289–2297. https://doi.org/10.1109/HICSS.2012.615
4. Batty M (2012) Smart cities, big data, and the built environment. Environ Plan B Plan Des 39(2):191–193
5. Nigmatov A, Pradeep A, Musulmonova N (2023) Blockchain technology in improving transparency and efficiency in government operations. In: 2023 15th international conference on electronics, computers and artificial intelligence (ECAI), Bucharest, Romania, pp 1–6. https://doi.org/10.1109/ECAI58194.2023.10194154
6. Walde RB, Yadav AK (2022) Blockchain technology for e-government. Int J Res Appl Sci Eng Technol 10(8):1698–1703. https://doi.org/10.22214/ijraset.2022.46487
7. Puterman ML (2014) Markov decision processes: discrete stochastic dynamic programming. John Wiley & Sons, New York. ISBN: 978-1118625873
8. Puterman ML (2014) Markov decision processes: discrete stochastic dynamic programming. Wiley, New York. ISBN: 978-1118625873
9. Gao Q, Zhang J (2024) Artificial intelligence governance and the blockchain revolution. Springer Singapore, Singapore. https://doi.org/10.1007/978-981-99-9211-9
10. Bhumichai D, Smiliotopoulos C, Benton R, Kambourakis G, Damopoulos D (2024) The convergence of artificial intelligence and blockchain: the state of play and the road ahead. Information 15:268. https://doi.org/10.3390/info15050268
11. Pindyck RS (2000) Irreversibilities and the timing of environmental policy. Resour Energy Econ 22(3):233–259
12. Pinsky MA, Karlin S (2011) An introduction to stochastic modeling. Academic, New York. ISBN: 978-0123814166
13. Luenberger DG (1998) Investment science. Oxford University Press, Oxford. ISBN: 978-0195108095
14. Hollands RG (2008) Will the real smart city please stand up? Intelligent, progressive or entrepreneurial? City 12(3):303–320
15. Nam T, Pardo TA (2011) Conceptualizing smart city with dimensions of technology, people, and institutions. In: Proceedings of the 12th annual international digital government research conference, pp 282–291. https://doi.org/10.1145/2037556.2037602
16. Boucherie RJ, van Dijk NM (2017) Markov decision processes in practice. In: International Series in Operations Research and Management Science, vol 248. Springer, Cham. https://doi.org/10.1007/978-3-319-47766-4
17. Howard RA (1971) Dynamic probabilistic systems, volume I: Markov models. Wiley, New York
18. Batty M (2013) The new science of cities. MIT Press, Cambridge. ISBN: 978-0262019521
19. Neirotti P et al (2014) Current trends in smart city initiatives: some stylised facts. Cities 38:25–36
20. Cox DR, Isham V (1980) Stochastic processes: selected papers. Chapman and Hall/CRC, New York. ISBN: 978-0412549800
21. Townshend I, Walker S (2015) Smart city governance: addressing equity and inclusivity. J Urban Technol 22(2):1–14
22. Meijer AJ (2016) Smart city governance: technology, power and society. Routledge, New York. ISBN: 978-1138223364
23. Norris JR (1997) Markov chains. Cambridge University Press, Cambridge. https://doi.org/10.1017/CBO9780511810633
24. Al-Mahrouq M, Al-Majali AA (2024) Understanding the role of Markov chain modeling in assessing the governance monetary transmission mechanisms: a governance outlook. J Govern Regul 13(3):124–138. https://doi.org/10.22495/jgrv13i3art11
25. European Commission (2012–2022) Digital economy and society statistics. https://ec.europa.eu/eurostat/web/digital-economy-society
26. World Bank (2012–2022a) Worldwide governance indicators (Paris). https://www.worldbank.org/en/publication/worldwide-governance-indicators

27. Ville de Paris (2012–2022) Open data Paris—participation Citoyenne. https://opendata.paris.fr/pages/participation-citoyenne/
28. Organisation for Economic Co-operation and Development (2012–2022a) Regional GDP in OECD countries. OECD Statistics Database. https://stats.oecd.org/Index.aspx?DataSetCode=REGION_ECONOM
29. Organisation for Economic Co-operation and Development (2012–2022b) OECD patent statistics. OECD Science, Technology and Innovation Database. https://www.oecd.org/sti/inno/oecd-patent-statistics.htm
30. Institut National de la Statistique et des Études Économiques (INSEE) (2012–2022) Coefficient de Gini pour la France. https://www.insee.fr/fr/statistiques/5015357
31. Airparif (2012–2022) Rapport sur les Émissions de CO2 à Paris. https://www.airparif.asso.fr/
32. Transparency International (2012–2022) Indice de Perception de la Corruption. https://www.transparency.org/en/cpi/2022
33. Commission Nationale de l'Informatique et des Libertés (CNIL) (2012–2022) Loi Informatique et Libertés: Protection des Données à Paris. https://www.cnil.fr/fr/loi-informatique-et-libertes
34. Autorité de Régulation des Communications Électroniques et des Postes (ARCEP) (2012–2022) Couverture Mobile à Paris (2G/3G/4G). https://www.arcep.fr/cartes-et-donnees/couverture-mobile-2g-3g-4g.html
35. World Bank (2012–2022b) GovTech Maturity Index 2022 Update: Trends in Public Sector Digital Transformation. World Bank Group. https://datacatalog.worldbank.org/search/dataset/0037889/govtech-
36. Tokyo Metropolitan Government (2012–2022) Tokyo Metropolitan Government Statistics. https://www.metro.tokyo.lg.jp/english/statistics/
37. Japan Patent Office (2012–2022) Patent statistics. https://www.jpo.go.jp/e/system/statistics/index.html
38. World Bank (2012–2022c) GDP growth (annual %) [World Development Indicators]. The World Bank. https://data.worldbank.org/indicator/NY.GDP.MKTP.KD.ZG
39. World Inequality Database (2012–2022) Gini coefficient—Tokyo. https://wid.world/
40. Tokyo Metropolitan Government (2012–2022) CO2 emissions report. https://www.metro.tokyo.lg.jp/english/topics/index.html
41. Personal Information Protection Commission (PPC) (2012–2022) Data protection laws in Tokyo. https://www.ppc.go.jp/en/
42. Ministry of Internal Affairs and Communications (Japan). (2012–2022) TIC and mobile coverage statistics (4G/5G). https://www.soumu.go.jp/english/dgpp_ss/seido.html

Digital Transformation Through the Implementation of Electronic-Based Government Systems to Build Efficient and Transparent Smart Government in Gorontalo City, Indonesia

Karno Karno, Layla Kurniawati, Murdiyana Murdiyana, Rinikso Kartono, and Hari Harjanto Setiawan

Abstract Implementing an Electronic-Based Government System (EBGS) in Gorontalo faces various challenges that hinder the optimization of public services. This study identifies factors influencing implementation and develops appropriate strategies through SOAR (Strengths, Opportunities, Aspirations, Results) analysis. Research data was obtained using qualitative methods with a descriptive approach through observation, in-depth interviews, and documentation studies. Research informants consisted of 25 related parties from various government agencies. The results show that implementation is not optimal due to unclear communication, lack of resource support, uneven cheerful disposition, and an organizational structure that is not well coordinated. Then, from the SOAR analysis, several strengths include government commitment, transparent policies, and opportunities in the potential integration of services and information technology support. These aspirations include increasing efficiency and transparency of public services, and the desired outcome is operational efficiency to improve service quality. The proposed strategies are 12, including optimization of Standard Operating Procedures (SOP) and regulations, development of State Civil Apparatus competencies, implementation of business process maps, improving service quality with technology, data and information integration, central government support, inter-agency cooperation, technological innovation, development of technological infrastructure, and providing incentives and motivation for State Civil Apparatus. In conclusion, the strategies

K. Karno · L. Kurniawati · M. Murdiyana
Institut Pemerintahan Dalam Negeri (IPDN), Jakarta, Indonesia

R. Kartono
University of Muhammadiyah Malang (UMM), Malang, Indonesia

H. H. Setiawan (✉)
National Research and Innovation Agency (BRIN), Jakarta, Indonesia
e-mail: hari.harjanto.setiawan@brin.go.id

H. Kaddouri et al. (eds.), *Smart Cities for Sustainable and Urban Development*, Green Energy and Technology, https://doi.org/10.1007/978-3-031-99001-4_2

identified through SOAR analysis can be used to improve the quality of public services in Gorontalo City. This study contributes to developing and implementing more effective EBGS policies.

Keywords Digital transformation · Electronic-based government system (EBGS) · Efficiency · Transparency · Smart government

1 Introduction

Electronic governance systems (EBGS) aim to realize effective, innovative Governance as part of a smart city. This city can answer challenges and create a friendlier urban landscape designed using technology to improve the welfare of its residents [1]. Implementing Governance using information technology aims to provide good services to the public in the government bureaucracy. Smart city planning focuses on three essential domains: affordable housing, infrastructure services, and citizen engagement [2]. This aims to build a sustainable city where residents have a higher quality of life, a more livable environment, and better economic opportunities.

Gorontalo City, Indonesia, still faces various obstacles that hinder the effectiveness of public services. According to the 2018–2021 evaluation results, the EBGS index has not reached the expected target of only 2.5 on a scale of 5. There are still many aspects that need to be improved in the implementation in this city. Factors that cause suboptimality include integration and coordination between agencies, the competence of the State Civil Apparatus, and inadequate resource support. The study of EBGS in Gorontalo City offers a broader scope in developing digital government policies compared to previous studies that focused more on implementing specific applications. Prior studies, such as the use of Tangerang LIVE, JAKI, War Room, and JKN Mobile applications, focused more on the effectiveness of technology in improving public services at the regional Level [3–5]. Using SOAR analysis, this study highlights key factors that influence the success of EBGS policies, such as clear regulations, increased bureaucratic competence, strengthening technological infrastructure, and integrated digital services.

This study contributes to filling the research gap in e-government studies in Indonesia by presenting a more comprehensive policy strategy-based approach. By producing 12 EBGS development strategies, this study emphasizes that the success of e-government is determined by technology adoption and supporting factors such as strong regulations, inter-agency coordination, and stakeholder engagement. Previous studies have focused on the effectiveness of digital applications in public services. In comparison, this study provides a more strategic and systematic perspective that can be used to formulate more effective and sustainable digital government policies [6, 7].

Researchers have extensively studied the implementation of EBGS, but few studies have specifically identified development strategies using Strengths, Opportunities,

Aspirations, and Results—SOAR analysis [8]. The advantages of SOAR analysis focus on internal strengths, external opportunities, aspirations, and desired outcomes, in contrast to SWOT analysis [9], which focuses more on weaknesses and threats. This study fills this gap by concentrating on Gorontalo City, which has challenges implementing EBGS. Previous studies have used more SWOT and PESTEL analysis, but none have adopted SOAR analysis comprehensively to develop EBGS development strategies in Gorontalo City. Then, the results of identifying problems in implementation in Gorontalo City include the lack of integration and coordination between agencies, which shows that 60% of agencies in Gorontalo City are still working partially without good system integration. This results in data duplication and inefficient processes. Then, the low competence and motivation of ASN based on an internal survey showed that only 40% felt confident in using information technology in their daily work. In addition, 30% admitted needing further training to understand EBGS.

Inadequate Human Resources Support, especially the budget allocated for EBGS development in Gorontalo City, only reaches 5% of the total Regional Revenue and Expenditure Budget (APBD), far below the national average of 10%. Ineffective communication based on field data shows that 70% feel that communication between agencies related to EBGS has not been running well. This is based on the object of the study due to the lack of communication forums and routine coordination between agencies. The study is essential in providing solutions to improve the quality of public services by implementing EBGS in Gorontalo City. Optimal EBGS can improve operational efficiency and transparency of public services. In addition, EBGS is a necessary step in the framework of bureaucratic reform, which aims to create a more responsive, accountable, and professional government. This study provides strategic direction to achieve the objectives of the reform.

Information technology in government is a must in today's digital era. This study helps identify the best way to utilize information technology to improve the quality of public services and data integration between agencies. The findings of this study also identify various obstacles in the implementation of EBGS, such as unclear communication, lack of resource support, and lack of coordination. By understanding and addressing these obstacles, the implementation of EBGS can be carried out more effectively. This book chapter provides strategic recommendations that policymakers can use to develop more effective policies and support the implementation of EBGS, helping to create a conducive framework for the digital transformation of government [10]. Thus, this study contributes to improving the quality of public services in Gorontalo City and provides practical guidance that can be applied in other regions that want to develop EBGS.

Gorontalo City is the capital of Gorontalo Province in North Sulawesi, Indonesia. The city has an area of 79.59 km^2 and a population of around 200,000. It borders Tomini Bay to the south and Gorontalo Regency to the other side. It functions as a centre for marine and fisheries activities in eastern Indonesia. Its economy relies on agriculture, fisheries, trade, and small industries. The city's increasing focus on current development is on digital services and Governance.

2 The Rationale of the Study

Implementing EBGS also opens up great opportunities to advance the public service system and improve governance efficiency. The government can provide better services to the public, facilitate interaction between government institutions, and enable active public participation in the governance process. Efforts to prevent abuse of authority, such as collusion, corruption, and nepotism, become more manageable. The electronic-based monitoring and complaint system allows the public to report potential violations that occur more efficiently and effectively, provides greater control over government actions, and encourages transparency in all aspects of public services.

2.1 The Smart Government Approach

Smart City has recently received a lot of attention from various countries. This city has utilized data by using technology for the welfare of its citizens. This city also simplifies administration and pays attention to sustainability. They integrate technology into various aspects of life, including health and transportation. Innovative urban Governance aims to create efficient and sustainable urban areas [11]. Smart cities integrate technology into government to create efficiency and improve public services and the welfare of their citizens [12].

Competent government in the context of transformation that uses new technologies to improve public services, transparency, and interaction with citizens. Competent government not only refers to the use of high technology in public services but also includes an approach emphasizing openness, collaboration, and digital connectivity between government and citizens. Competent government is "the implementation of technology that supports interaction between government and citizens and improves the efficiency of public services" [13].

Here is a summary of the definition and concept of Smart Government. The next step in e-government is innovation. Second, the next step of Open Government is citizen engagement, accountability, and interoperability [14]. The key components that constitute the smart government triangle are Politics (decisions), Value (citizen needs), and Evidence (policy evaluation). These three components help increase efficiency, transparency, citizen participation, and creating new public value.

The development of IoT for the needs of modern government aims to create a system that is more responsive to urban dynamics. A competent government must be able to utilize IoT technology to improve the quality of services and the effectiveness of resource management [15]. The terms smart government and smart Governance are used in the literature, but their meanings differ. Government occurs when they have the authority and authority legally obtained to carry out activities. Governance refers to the creation, implementation, and implementation of activities supported by the common goals of citizens and organizations ,which have or do not

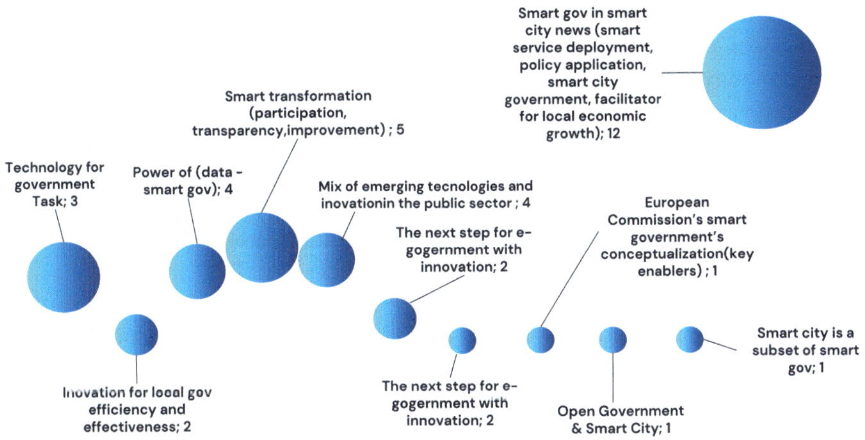

Fig. 1 The impact of the alternative smart government definitions [13]

have formal authority [16]. The impact of Smart Government can be seen in the following figure (Fig. 1).

The evolution of intelligent government is driven by factors such as big data, data analytics, and citizen engagement in decision-making processes. Big data enables governments to make more evidence-based and effective decisions to meet public needs. In addition, citizens' connectivity, government processes, data, and IoT devices are vital elements to support efficient Governance [17]. Optimal use of data is the foundation of intelligent government, enabling governments to monitor and respond to citizen needs more effectively.

Processes, Data, and Objects: Combining these elements enables governments to provide real-time responsive public services. Involving citizens in public policies and decisions strengthens public trust and government accountability. Based on intelligent government is a complex transformation that requires technology support, collaboration with citizens, and multi-stakeholder engagement to achieve responsive and transparent public services. A competent government has promising potential, but challenges in its implementation remain, especially in integrating technology with existing bureaucratic structures [18].

2.2 *Electronic-Based Government Systems*

EBGS uses information and communication technology in government administration to improve efficiency, effectiveness, transparency, and public participation. EBGS includes various applications and systems that enable the government to provide services to the public through digital platforms [19]. The implementation of EBGS aims to integrate technology into government processes to improve the quality of public services by reducing the time required for data processing and providing fast access to the public [20]. EBGS covers various aspects such as data management, provision of online services, and increasing access to information for the community [21]. Effective implementation of EBGS is expected to reduce bureaucracy, speed up the service process, and increase information transparency [22]. In addition, implementing digital technology in the public sector has great potential to reduce complex bureaucracy and speed up administrative processes, thereby increasing the efficiency of public services [23].

EBGS is essential for improving cleaner, more transparent, and more effective Governance. Information and communication technology (ICT) in EBGS aims to provide better services to the community, government agencies, and other parties. As regulated in Presidential Regulation No. 95 of 2018, EBGS is designed to create a more accountable, efficient, and quality government. The role of information technology in EBGS provides an excellent opportunity for the government to innovate development and Governance that is more open, participatory, and innovative. Through the implementation of EBGS, the government hopes to increase collaboration between government agencies, improve public service quality, and minimize power abuse through an electronic-based public monitoring and complaint system.

The government is aware of EBGS's role in supporting the development sector. So far, efforts to encourage the implementation of EBGS have been carried out through sectoral laws and regulations that require implementing information systems or EBGS. Although the implementation of EBGS has been carried out by various ministries, institutions, and local governments, the Level of progress still varies. Strong coordination between ministries, institutions, and local governments in implementing EBGS requires a National EBGS Master Plan. This plan will guide various agencies in achieving integrated EBGS integration and ensuring the achievement of strategic EBGS goals and consistent state apparatus development. The importance of the National EBGS Master Plan shows that to achieve maximum harmony in implementing EBGS, strategic thinking is needed that considers various policies, initiatives, and directions of information technology development.

Thus, EBGS becomes a crucial foundation for creating an efficient, transparent, open government and providing the community with higher quality and reliable public services. Implementing EBGS in an integrated and coordinated manner will be the key to achieving government goals in the 2018–2025 era, which align with the direction of national development previously determined. "EBGS aims to create clean, effective, transparent, and accountable governance as well as quality and reliable public services" Presidential Regulation No. 95 of 2018.

The implementation of EBGS also opens up great opportunities for advancing the public service system and efficiency in government administration. With the presence of EBGS, the government can provide better services to the public, facilitate interaction between government institutions, and enable active participation of the public in the governance process. Efforts to prevent abuse of power through EBGS, such as collusion, corruption, and nepotism, become more achievable. The electronic-based monitoring and complaint system allows the public to more efficiently and effectively report potential violations that occur, provide greater control over government actions, and encourage transparency in all aspects of public services.

However, to achieve sustainability in implementing EBGS, coordination between central and regional government institutions is crucial. The National EBGS Master Plan is an essential tool to guide efforts to integrate e-government systems across government sectors. This plan provides a clear direction, consistent standards, and better integration in various public services. The implementation of EBGS nationally requires serious attention to aspects of data security, public privacy, and increasing human resource capacity. Investment in technological infrastructure, increasing human resource skills, and efforts to raise awareness of cyber security are vital to maintaining the continuity and success of EBGS.

2.3 *Indicators of EBGS in Indonesia*

EBGS is the foundation for realizing a responsive, open, and efficient government that provides services to the community. With proper and integrated implementation, this system can accelerate government transformation towards better innovation in line with technological developments and the demands of an increasingly advanced society. "EBGS provides an opportunity to encourage and realize the implementation of open, participatory, innovative, and accountable government."—Presidential Regulation No. 95 of 2018.

The index framework for the Electronic-Based Government System (EBGS) includes several dimensions and indicators that can be used as evaluation guidelines, including: (1) ICT Infrastructure Dimension indicators include Availability of comprehensive internet access, Even network coverage throughout the region, Availability of necessary technological devices, System security and reliability, (2) Electronic Public Service Dimension indicators include: Availability of electronically integrated public services, Ease of access and clarity of public service information, Number of services available online, Level of public participation in using these services. (3) Transparency and Accountability Dimension indicators include the availability of electronic public information, Level of openness and accessibility of government information, Level of public involvement in the decision-making process, Effectiveness of the monitoring system and electronic complaint mechanism. (4) Human Resource Competence and Capacity Dimension indicators include the Level of available technology training and education, availability of skilled workers in ICT management, Level of technology absorption in government practices, and Ability to deal with cyber security threats. (5) Policy and Regulation Dimension: The indicators include Compliance with regulations regarding technological developments, Consistency in implementing electronic policies, flexibility in adjusting policies to technological dynamics, and protection of data privacy and information security.

3 Materials and Methods

This research uses a qualitative method with a descriptive approach [24]. Data were collected through in-depth interviews involving various government agencies and other related parties, observation of the EBGS process and implementation in the field, and documentation related to policies, reports, and other data relevant to the research. The tool used to analyze data is NVivo. Qualitative researchers can efficiently and effectively code and analyze data using this tool, mainly qualitative data. Researchers can also present the results of data analysis in the form of tables and diagrams. The tool for sharpening images uses Canva to obtain high-evolution images.

This research is based on Presidential Regulation Number 95 of 2018 concerning Electronic-Based Government Systems, which provides a legal basis and guidance for local governments in implementing EBGS [25]. In addition, this study combines theories from various disciplines to provide strategic advice for developing EBGS in Gorontalo City. Thus, this study's relationship between government science, government management, EBGS policy implementation, SOAR analysis, and legalistic framework provides a solid basis for formulating effective and sustainable development strategies. The successful implementation of EBGS is expected to improve the quality of public services, operational efficiency, and government transparency, thereby supporting the goals of bureaucratic reform and increasing public trust in the government [26]. The stages of SOAR analysis are depicted in the following diagram (Fig. 2).

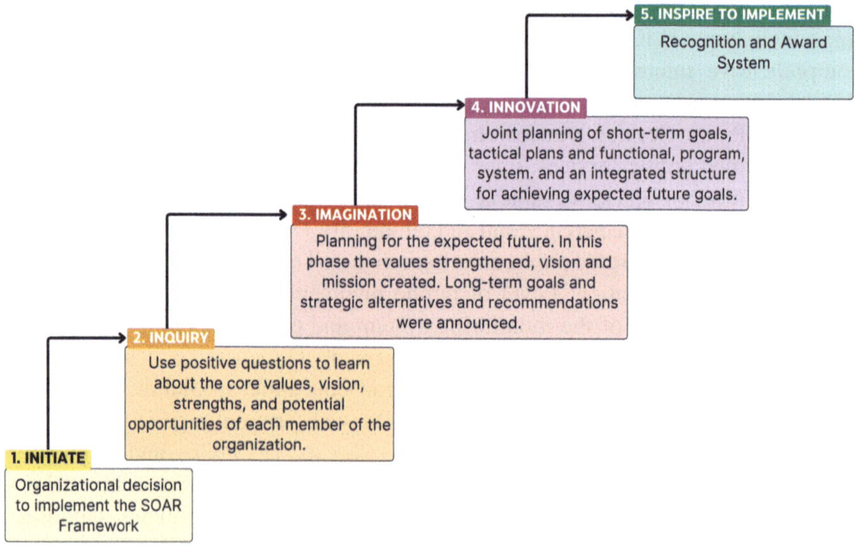

Fig. 2 SOAR analysis stage. (Source: [27])

The operational theory of EBGS implementation [28] covers communication, resources, disposition/attitude, and organizational structure. Data analysis was conducted using data source triangulation techniques to ensure the findings' accuracy and Consistency [29]. SOAR analysis is used to develop development strategies to analyze strengths, opportunities, aspirations, and results accurately and consistently [27]. SOAR (Strengths, Opportunities, Aspirations, Results) analysis is a strategic approach to formulate development strategies focusing on strengths, opportunities, aspirations, and desired results [27]. It explains that SOAR analysis differs from SWOT analysis, focusing more on weaknesses and threats. SOAR emphasizes positive and proactive aspects of formulating strategies. This study used SOAR analysis to identify critical factors that influence the success of EBGS implementation in Gorontalo City.

4 Results and Discussions

4.1 Implementation of EBGS Policy in Gorontalo City

The results of the EBGS implementation research in Gorontalo City are not optimal. This is due to unclear communication, lack of resource support, uneven cheerful disposition, and an organizational structure that is not well coordinated. Strategic steps are needed to overcome these obstacles, such as improving internal and external communication, optimizing resource support, developing positive dispositions, and reforming the organizational structure. Improving communication can be done through regular coordination meetings and socialization of the EBGS program [30].

Adequate resource support must be allocated for the development and maintenance of EBGS, including training for State Civil Apparatus. Encouraging a positive attitude towards EBGS through awareness campaigns and incentives for State Civil Apparatus who actively participate is also essential. In addition, reforming the organizational structure by establishing special units and preparing clear standard operating procedures (SOPs) will ensure the effective and efficient implementation of EBGS. By implementing these strategies, it is hoped that the implementation of EBGS in Gorontalo City can run better and achieve the expected goals, namely increasing efficiency, transparency, and quality of public services.

Based on the conceptualization of the EBGS policy implementation theory according to Edward III [31], the research results and discussions are guided by the dimensions of communication, resources, disposition, and organizational structure along with their indicators, as the table follows (Table 1):

Table 1 Research results of EBGS, 2024

Dimension	Research results
Communication	1. The policy communication process can be conveyed well according to the objectives 2. Clarity of information in the policy can be conveyed well so as not to cause misinterpretation 3. Consistency in policy implementation is done well using existing devices and applications
Resources	1. The limited number of human resources is an obstacle to implementing policies, as not all have the required competencies 2. Information resources still have several obstacles related to the synchronization of applications between institutions when implementing policies 3. The authority to implement policies based on the dynamics of change in regional governance has involved all regional apparatus in Gorontalo City 4. The infrastructure facilities required to implement the policy have fully used the cloud as an operational support for the organization
Disposition	1. The EBGS policy is used as the basis for determining the appointment and income of employees 2. The incentive for the State Civil Apparatus is given under regional financial capabilities with the approval of the DPR
Bureaucratic structure	1. SOPs evolved as an internal response to the limited time and resources of executors and the desire to standardize complex operations within the broader scope of the organization 2. Preparing business processes aims to provide interrelated guidelines arranged in an integrated manner to support the development or improvement of EBGS applications

4.1.1 Communication Dimensions

A policy is one thing that has meaning and can be used by government intervention to solve problems through the policy born to the target group. The implementer requires proper and effective communication. This study focuses on communication between policymakers, policy implementers, and the target group, namely the community as users in the context of EBGS. Communication in policy implementation is interpreted as Consistency in implementing the EBGS application policy [32]. It is still carried out by utilizing existing devices and applications and making adjustments according to the latest provisions of the EBGS policy as stated in Presidential Regulation 95 of 2018. Mayoral Regulation Number 32 of 2019 contains various matters relating to EBGS in e-Government. It is also not too different in regulating at least the same concept regarding EBGS. However, the situation is not entirely the same, and reviews have been conducted regarding implementing Mayor Regulation Number 32 of 2019.

The study results indicate that good, effective, and efficient communication dimensions support the successful implementation of the EBGS policy. A policy can be implemented well when policymakers convey messages to policy actors and target groups (society). In this context, implementing the EBGS policy as mandated by Presidential Regulation Number 95 of 2018 is a sustainable policy to respond to and follow up on previous policies. Information must be delivered to related parties appropriately, precisely, accurately, and consistently. The ambiguity of the delivery of policy information can be one of the causes of bad policies. Three aspects supporting success in policy implementation include transmission, clarity, and Consistency.

The communication process related to the EBGS policy of Gorontalo City can be transmitted well. This is supported by Presidential Regulation Number 95 of 2018 concerning EBGS and the issuance of Gorontalo Mayor Regulation Number 32 of 2019, which is a minimal step in following up on the EBGS Program. The EBGS policy is implemented following the intent and objectives that have been set, producing clarity that facilitates policy implementers and target groups (people who will use electronic-based government services). The clarity of information in the policy can be conveyed well. Unclear information in the policy can trigger misinterpretation, leading to an inaccurate understanding of the policy's intent. Consistency in implementing the EBGS policy has been carried out well through the use of existing devices and applications, and it is in line with Mayor Regulation 32 of 2019.

4.1.2 Resource Dimension

The data in the research locus shows that in the early years of the EBGS policy related to infrastructure facilities, the Communication and Information Service of Gorontalo City was a new regional device split from the Transportation Service. Thus, the Communication and Information Service experienced several transitions, from providing infrastructure in office buildings to infrastructure with inadequate procurement. However, it was gradual by utilizing old devices and new procurement based on the amount of the budget while still prioritizing the scale of priority for procurement of supporting infrastructure for the implementation of the EBGS policy.

The limited number of human resources constrains human resources in implementing the EBGS policy in Gorontalo City from what should be owned. Then, not all existing human resources have the scientific background or competencies needed to implement EBGS. Thus, Gorontalo City must carry out training or self-study as other officials do because their duties and responsibilities are to master information technology in implementing the EBGS policy. Information Resources (SDI) in implementing the EBGS policy still have several constraints regarding the applications made by each regional apparatus that are not synchronized with each other in the Gorontalo City Government environment. In addition, the creation of applications is carried out by each regional apparatus by involving third parties. So, there are constraints in synchronizing data and application information from each regional apparatus. The same happens when implementing the EBGS policy in various districts/cities in Indonesia.

The authority to implement the EBGS policy in Gorontalo City is regulated by Mayor Regulation Number 32 of 2019. This authority is in the form of coordination between regional apparatuses in implementing policies involving all regional apparatuses in Gorontalo City. The infrastructure needed to implement the policy is in the form of a cloud. All regional apparatuses in Gorontalo City have fully utilized the cloud to support organizational operations. This is due to new regulations requiring a good transition from providing infrastructure gradually utilizing old devices and new procurements based on the budget amount according to the priority scale.

4.1.3 Disposition Dimension

The study's results indicate that the disposition of the attitudes of policy implementers can support the success of the EBGS policy implementation process. The implementer's positive perspective on the policy will support the success of the EBGS policy implementation. Implementers tend to agree with the EBGS policy to be implemented, so they are more likely to implement what is intended by the policy-maker. However, if the attitude or perspective of the implementer differs from that of the decision maker, the EBGS policy implementation process becomes somewhat more complicated. Disposition is policy actors' will, desire, and tendency to implement the policy earnestly. So that the direction instructions of an EBGS policy are given clearly, the policy implementer will undoubtedly be able to implement the policy maker's policy so that the objectives of the EBGS policy can be realized.

The results of the Disposition dimension research in the field showed that EBGS was used as the basis for determining the Additional Employee Income for existing employees. TPP allowances are an incentive for the State Civil Apparatus in Gorontalo City. TPP is provided following the financial capacity of the region with the approval of the DPRD following the provisions of applicable laws and regulations. The allowances for employees in the environment have duties and responsibilities assigned to State Civil Apparatus, and the amount is the same as that of other regional apparatuses. When looking at the functional tasks regarding the duties of State Civil Apparatus, the Gorontalo City Communication and Information Service provides technical guidance and supervision, monitoring, evaluation, and reporting in primary data centre infrastructure services, disaster recovery centres, and communication technology. Then, intranet development services, internet access, and intra-government communication system services will be provided.

Furthermore, E-Government information security and data management services are based on applicable provisions to support the implementation of the EBGS policy. Based on these tasks and functions, excessive effort and workload are required. Then, the allowance for practitioners in the IT field with other positions in other services is calculated the same. At the same time, the duties and responsibilities of State Civil Apparatus require more energy in mind and body.

4.1.4 Dimensions of Bureaucratic Structure

The organizational structure in implementing the EBGS policy is one of the essential aspects that can support the smoothness and success of implementing a policy. However, the bureaucratic structure is one of the obstacles to implementing the policy. The study's results indicate that the bureaucratic structure in implementing the EBGS policy must be carried out with sufficient desire and resources. However, there are still obstacles to its implementation. An essential characteristic of a bureaucratic structure is the existence of a Standard Of Procedure (SOP). This developed as an internal response to time constraints, implementing resources and the desire to standardize across the broader scope of the organization. This situation arose due to the slowness of the bureaucracy.

A business process map that describes the business processes carried out by the City of Gorontalo based on Presidential Regulation No. 95 of 2018 Article 23 that business processes are compiled with provisions including The preparation of business processes aims to provide guidelines for the use of data and information and the implementation of Applications, Security, and Services. Each regional government compiles business processes based on the regional government's EBGS architecture. Interrelated business processes are compiled in an integrated manner to support the construction or development of integrated EBGS Applications and EBGS Services.

4.2 EBGS Development Factors

Strengths: consists of government commitment, transparent policies, and support from existing resources; opportunities include potential service integration, improving service quality, and information technology support. Aspirations: the desire to improve efficiency and transparency, improving the quality of public services, and community support. Finally, Results consist of operational efficiency, improving service quality, transparency, and community support. The factors for developing EBGS are as follows:

4.2.1 Strength Factors in EBGS Development

Clear Policy: Policy clarity is essential in developing EBGS in Gorontalo City. Although the goals and objectives of the EBGS policy have been widely socialized through various media, employees and residents still do not fully understand the goals, indicating the need for more intensive communication. The division of roles and responsibilities has been carried out well. However, some employees still do not understand their responsibilities, so additional training and guidance are needed. The procedures and mechanisms for implementing the EBGS policy have been well prepared, but the existing guidelines still need improvement in certain technical aspects. Communication between government agencies is adequate, but improvements are still required, especially in integrating input from employees and the community. Information on the EBGS policy is also available through the official website and social media. Still, the language used needs to be simplified to make it easier for the wider community to understand.

Commitment and support: commitment and support in improving operational efficiency, quality of public services, and data and information integration for implementing EBGS in Gorontalo City are reflected in various aspects. As the regional leader, the Mayor of Gorontalo has shown a strong commitment by issuing Mayor Regulation Number 32 of 2019 and Mayor Regulation Number 18 of 2022 and allocating adequate resources. The technical readiness and capacity of human resources in the Communication and Informatics Office also sufficiently support the implementation of the EBGS policy, which is supported by training and socialization for employees to improve their understanding and skills related to EBGS. In addition, organizational culture also encourages innovation in the government environment to strengthen the implementation of EBGS, and responsiveness to input and evaluation is also an ongoing effort to improve policies through relevant revisions.

Human Resources (HR): is essential in improving operational efficiency, quality of public services, and data and information integration when implementing EBGS in Gorontalo City. Improving competency and skills in information and communication technology (ICT) is the main focus, with the local government holding routine training, encouraging professional certification, and recruiting ICT experts to fill strategic positions. However, improvements are still needed in terms of experienced certification and recruitment of experts to face increasingly complex technological challenges. In addition, the availability and distribution of HR are critical, considering that there is still an imbalance in distribution and a reasonably high turnover rate, which hinders the stability of EBGS implementation in all government agencies. Employee motivation and involvement have been implemented with incentives and rewards but need improvement in inspirational leadership and employee participation in decision-making to increase a sense of ownership and responsibility. Finally, HR awareness and understanding of the importance of EBGS need to be improved through comprehensive socialization programs, workshops, seminars, and ongoing internal campaigns so that EBGS implementation can run effectively and optimally.

Clear organizational structure: A clear organizational structure plays a significant role in improving operational efficiency, quality of public services, and data and information integration in implementing EBGS in Gorontalo City. Clarity and Consistency of Standard Operating Procedures (SOP) are essential elements, and they aim to ensure a uniform understanding across all regional apparatuses so that the implementation of EBGS policies follows applicable guidelines. SOPs in EBGS are regulated based on Mayoral Regulation Number 32 of 2019 and Mayor Regulation Number 18 of 2022, and they refer to Presidential Regulation Number 95 of 2018 at the National Level to maintain policy. The SOPs cover application planning and development, data management, ICT infrastructure maintenance, and information security. Periodic evaluation and revision of SOPs are also needed so that SOPs remain relevant to the latest technological developments and regulations. In addition, training and socialization of SOPs to all employees is essential to improve understanding and readiness in implementing

EBGS policies and socialization to the community to increase public participation and support.

4.2.2 Opportunity Factor in EBGS Development

Technological development: Technological developments provide essential opportunities for developing EBGS in Gorontalo City. ICT infrastructure is the main backbone, including internet networks, servers, data centres, and hardware and software. Although the ICT infrastructure is quite good, some areas still need to improve the quality of the internet network, so remapping and additional investment are required. The Gorontalo City Government has adopted several of the latest hardware and software technologies. However, regular evaluation and procurement of devices that support system integration are still needed to improve performance. Training and development of HR capacity are also essential for comprehensively increasing the frequency and scope of training. In addition, data centres and information security must be improved regarding storage capacity and cybersecurity systems, as well as security audits that are carried out periodically to ensure that the system remains optimal and protected from threats.

National Policy Support: National policies support the development of EBGS in Gorontalo City. Structured distribution of authority is crucial to ensure EBGS implementation's effectiveness. Mayor Regulation Number 32 of 2019 stipulates the role of each regional apparatus, from planning to supervision. The Communication and Informatics Office of Gorontalo City has the central authority in managing and coordinating EBGS, including preparing a blueprint for e-government development and evaluating internal policies. Integration of local EBGS policies according to Presidential Regulation Number 95 of 2018 ensures implementation according to national standards while considering local conditions. Effective coordination between agencies supports the objectives of EBGS, while periodic supervision and evaluation by the Communication and Informatics Office ensures optimal implementation. Utilization of competent human resources is also a priority through recruitment and training of HR to provide the Ability to manage the EBGS system.

Community Participation: plays an essential role in developing EBGS to support operational efficiency, quality of public services, and data and information integration. The success of EBGS is primarily determined by the alignment of messages at all levels of government and stakeholders in the form of documents, instructions, and consistent and uniform verbal communication to avoid differences in interpretation. Then, external media, such as government websites and social media, will be used. All of these media are essential means of conveying information transparently. Regular communication frequency, including through periodic announcements and routine coordination meetings, helps all parties obtain consistent information updates. In addition, fast and accurate responses and feedback from the government show appreciation for input from the community and government employees in strengthening the involvement of all parties in the success of EBGS implementation.

4.2.3 Aspirations Factor in EBGS Development

Total Digital Transformation: Comprehensive digital transformation is critical to the development of EBGS in Gorontalo City, with various aspects that support each other. Adequate communication infrastructure, such as a stable and accessible internet network, is an essential basis for the smooth implementation of EBGS. The capacity of skilled human resources to manage and use the EBGS system also greatly determines the success of this transformation. Adequate data transmission speed and the development of technology that continues to be adapted to improve efficiency are the main factors in supporting EBGS operations. Increasing internet accessibility throughout Gorontalo City allows more people and government agencies to utilize this system effectively. In addition, optimal use of EBGS can improve communication efficiency between government agencies and the community, strengthen real-time interactions, and improve operational efficiency in delivering public services. Ultimately, EBGS also increases information transparency, provides better access for the community to government services, and encourages greater transparency in the Gorontalo City government.

Better Public Service: The development of EBGS in Gorontalo City aims to improve the quality of public services by strengthening operational efficiency, data integration, and transparency. The solid legal basis of Mayoral Regulation Number 32 of 2019 and Presidential Regulation Number 95 of 2018 supports the implementation of EBGS by providing precise and structured regulations. The government's high commitment, demonstrated through clear policies and SOPs, strengthens the implementation of EBGS at the regional Level. Trained human resources and information technology support provide opportunities to improve the quality of electronic-based government services. In addition, cooperation between agencies and support from the central government further strengthen the implementation of EBGS to make Gorontalo City a national role model. The implementation of EBGS focuses on operational efficiency, transparency, accountability, and data and information integration.

Sustainability Capacity Building: Continuous capacity development is a key element in the successful implementation of EBGS in Gorontalo City, and it covers various essential aspects. Human resources with ICT skills must be continuously improved through training and development programs to face the challenges of managing the increasingly complex EBGS system. In addition, sustainability in funding is needed to support the development of ICT infrastructure, applications, and long-term maintenance of the EBGS system. Although ICT infrastructure has been strengthened, challenges related to technology accessibility in remote areas and infrastructure maintenance need to be continuously addressed. Effective data management is also essential in supporting better decision-making by strengthening the capacity to manage accurate data's security and availability. Finally, the Mayor's Regulation regulation must be updated regularly to remain relevant and responsive to technological developments and community needs.

4.2.4 Results Factor in EBGS Development

Increased Efficiency and Effectiveness: The increase in efficiency and effectiveness in developing EBGS in Gorontalo City is reflected in various aspects that support each other. Good communication infrastructure, in the form of a stable and easily accessible internet network, is very important for the smooth operation of EBGS. In addition, the capacity of competent human resources to manage the EBGS system is also the key to ensuring successful implementation. Adequate data transmission speed and adoption of the latest technology enable efficiency in sending information so that increasing internet accessibility throughout Gorontalo City supports the broader use of EBGS by the community and local government. The optimal use of EBGS encourages efficient communication between government agencies and the community and increases better and real-time interactions.

Transparency and Accountability: Accountability is essential to developing EBGS in Gorontalo City. These factors include a clear structure and coordination between the city planning, research, and development departments, information departments, information, public services, and statistics in implementing EBGS. The information and communications department is the core operational operation in managing e-government, statistics, and statistics in Gorontalo City. It is also a platform for coordinating and integrating EBGS applications to optimize the efficiency and effectiveness of public services. The goal is to make the information and communications department a hub for adequate information and communication technology and the central coordinator of e-government. Increasing transparency and accountability in implementing EBGS in Gorontalo City can improve operational efficiency, quality of public services, and data integration, thus motivating State Civil Apparatus to work better in the EBGS environment.

Improving the Quality of Public Services Improving the Quality of Public Services is a crucial factor in the development of EBGS in Gorontalo City. These factors include implementing regulations, government committees, and human resources focusing on developing and implementing information technology and e-government. Developing technology, information, and communication provides resources to optimize government-based electronic services. Implementing public service quality improves public services through effective and efficient information technology. Transparency and accountability help the government be more transparent and accountable in implementing EBGS. Operational efficiency is enhanced through transparent business processes and SOPs that support the government in operating more efficiently. Then, the results of the SOAR

Table 2 EBGS implementation policy development strategy in Gorontalo City

Factors	*Strengths (S)*	*Opportunities (O)*
	• Policy clarity • Commitment and support • Human resources • Clear organizational structure	• Technological development • National policy support • Community participation
Aspirations (A)	**(S–A)**	**(O–A)**
• Comprehensive digital transformation • Better public services • Sustainable capacity development	1. Optimization of SOP and regulations 2. Development of state civil apparatus competencies 3. Implementation of business process maps	1. Central government support 2. Inter-agency cooperation 3. Technological innovation
Results (R)	**(S–R)**	**(O–R)**
• Increasing efficiency and effectiveness • Transparency and accountability • Improving the quality of public services	1. Operational efficiency through SOP 2. Improving service quality with technology 3. Data and information integration	1. Improving the quality of service with the support of central policies 2. Development of technological infrastructure 3. Incentives and motivation for state civil apparatus

analysis of the factors in Gorontalo City are described clearly in the following table (Table 2).

Improving the quality of public services can improve the quality of services to the community so that they are more efficient, transparent, and accountable. Data and information between government officials can be better integrated for faster and more accurate decision-making. Improving the quality of public services in Gorontalo City is very important for the State Civil Apparatus motivation to work better in the EBGS environment.

4.3 EBGS Development Strategy

The study of EBGS in Gorontalo City presents a broader framework for developing digital government policies, setting it apart from previous studies that primarily examined individual applications' effectiveness in public services. Using the Edward III model and SOAR analysis, this study highlights the critical elements for SPBE policy success, including regulatory clarity, improved bureaucratic capacity, technological advancements, and seamless digital service integration. The findings also emphasize that the success of e-government depends not only on technology adoption but also on governance quality, institutional coordination, and stakeholder engagement. Optimizing SOPs and regulatory frameworks is fundamental to ensuring that EBGS policies remain relevant to technological advancements and public needs. The study reveals that many SOPs are outdated, requiring a structured

Table 3 EBGS development strategy

Aspirations (A)	1. Optimization of SOP and regulations 2. Development of State Civil Apparatus Competencies 3. Implementation of business process maps	1. Central government support 2. Inter-agency cooperation 3. Technological innovation
Results (R)	1. Operational efficiency through SOP 2. Improving service quality with technology 3. Data and information integration	1. Improving the quality of service with the support of central policies 2. Development of technological infrastructure 3. Incentives and motivation for state civil apparatus
	Strengths (S)	***Opportunities (O)***

revision process and enhanced awareness programs for civil servants to ensure Compliance. The EBGS Development Strategy can be seen in the following table (Table 3).

Optimization of Standard Operating Procedures (SOP) and Regulations, based on the many SOPs that are not appropriately followed, often not adjusted to technological developments and community needs. Strategies that are developed and updated regarding existing SOPs ensure that the SOPs are following technological developments and community needs. Socialization and training regarding the importance of following the established SOPs must be conducted for State Civil Apparatus.

Development of State Civil Apparatus Competence: The study results show that 40% of State Civil Apparatus feel confident using information technology in their daily work. Thus, it is necessary to hold routine training and workshops to improve State Civil Apparatus competence in using EBGS technology and systems. Encourage State Civil Apparatus to continue learning and following technological developments.

Implementation of Business Process Maps: The study results show that the absence of a clear business process map makes the implementation of EBGS less structured. Thus, it is necessary to prepare a clear and structured business process map to guide all agencies in implementing EBGS and involve them in preparing business process maps to ensure suitability and integration. Operational Efficiency through SOP Research results show that there are still many inefficient operational processes due to the absence of clear SOPs. Thus, a clear SOP development strategy is needed for each operational process, ensuring that all State Civil Apparatus understand and follow the SOP and conducting periodic evaluations and improvements to the SOP to improve operational efficiency.

Improving Service Quality with Technology: According to research results, existing technology has not been optimally utilized to enhance the quality of service. A strategy is needed to identify technology that can improve service quality, such as mobile applications, online queuing systems, and others—in addition to conducting training for State Civil Apparatus and the community on using this technology.

Data and Information Integration: Research results show that existing data and information are poorly integrated between agencies. Thus, a strategy is needed to develop a data and information integration system that allows easy and fast data exchange between agencies. Then, coordinate between agencies to ensure all interested parties can access and use existing data and information.

Central Government Support: Research results show that support from the central government is still lacking, especially in funding and regulation. Thus, a strategy is needed to submit proposals and reports routinely to the central government to obtain the necessary financing and regulatory support. Then, lobby and coordinate with relevant ministries to receive more significant support.

Inter-Agency Cooperation: Research results show that cooperation between agencies is still lacking. Many work separately. A strategy is needed to improve coordination and collaboration between agencies through regular meetings, discussion forums, and others. In addition, a cooperation system should be developed to exchange information and experience between agencies.

Technological Innovation: The study showed a lack of technological innovation in implementing EBGS. So, a strategy is needed to encourage State Civil Apparatus to innovate using technology, such as developing new applications, using AI technology, and others. They are providing the budget and resources needed to support technological innovation.

They are improving Service Quality with Central Policy Support. The study results showed that existing central policies do not fully support improving service quality through EBGS. Thus, a strategy is needed to submit proposals and input to the central government regarding the policies required to enhance the quality of service through EBGS. In addition, they are conducting advocacy and lobbying to obtain the necessary policy support.

Development of Technology Infrastructure: The study's results showed that the existing technology infrastructure is inadequate to support the implementation of EBGS. For this reason, there needs to be a strategy to identify the needs of the required technology infrastructure, such as internet networks, servers, hardware, and others. Then, submit a budget for technology infrastructure development to the central and regional governments.

Incentives and Motivation for State Civil Apparatus: The study's results found a lack of incentives and motivation for State Civil Apparatus to implement EBGS properly. Thus, a strategy is needed that provides incentives and rewards for State Civil Apparatus that successfully implement EBGS well. Then, create a work environment that supports and motivates State Civil Apparatus to continue learning and innovating in implementing EBGS.

A significant aspect of EBGS implementation is building the digital capabilities of civil servants. The study indicates that only 40% of ASN are confident in using technology in their daily work, highlighting the urgent need for continuous training programs to equip them with the necessary skills for digital Governance. Another key finding is the lack of precise business process mapping, which has led to inefficiencies in EBGS implementation. The study suggests developing a well-defined process map and involving all government agencies in its formulation to improve coordination and service delivery.

Technological advancements are crucial in enhancing service quality, yet the study found that many digital tools are not fully utilized. Strategies should focus on identifying and integrating digital solutions such as mobile applications, online queue systems, and AI-driven automation to optimize public services. Additionally, inter-agency data integration remains challenging, requiring establishing a unified data-sharing system to streamline government operations. Beyond technology, central government support is essential for EBGS development.

The study indicates insufficient funding and regulatory backing, necessitating more active lobbying and proposal submissions to secure better financial and legislative support. Strengthening collaboration among government agencies and fostering partnerships with relevant ministries are also crucial strategies. One of the persistent issues in EBGS implementation is the fragmented coordination among government agencies, where institutions often operate independently. The study recommends regular coordination meetings, knowledge-sharing forums, and inter-agency cooperation frameworks.

Lastly, motivating civil servants remains a critical challenge. The study shows that many ASN lack the motivation to transition to digital systems. Introducing performance-based incentives, fostering a culture of innovation, and ensuring a supportive work environment are key to driving digital transformation in government institutions. The study underscores that the success of EBGS in Gorontalo City relies on effective regulations, skilled human resources, technological infrastructure, and inter-agency collaboration. By adopting 12 key development strategies, this study provides a comprehensive roadmap for sustainable e-government policies. Strengthening central government support, driving digital innovation, and ensuring civil servant motivation are essential to making EBGS an efficient, transparent, citizen-centred governance model.

4.4 General Discussion of the EBGS Policy Implementation

The Smart Government framework has become a key approach in modern Governance. By leveraging technology and data effectively, governments can improve operational efficiency, enhance transparency, and foster greater public participation in policymaking. Smart Government is about utilizing advanced technology in public services and promoting collaboration, openness, and digital connectivity between the government and its [33]. This study examines the relationship between EBGS implementation in Gorontalo City and the core principles of Smart Government. Several key aspects emerge from this analysis:

Streamlining Administration through Digitalization: Smart Government enhances administrative efficiency by integrating digital tools into public service delivery. The EBGS initiative in Gorontalo City demonstrates efforts to digitize government processes. However, study findings reveal that human resource constraints and inadequate IT infrastructure remain significant barriers. A Smart Government approach suggests addressing these issues by improving inter-agency data integration, enhancing the digital skills of government employees, and utilizing big data analytics for informed decision-making [34].

Strengthening Citizen Participation and Government Openness: Another critical element of Smart Government is increasing public engagement in Governance. This study identifies ineffective communication between government entities, civil servants, and the public as a significant challenge in EBGS implementation. Digital platforms should be developed to facilitate direct interactions between citizens and government authorities, along with real-time feedback mechanisms [14]. Additionally, making government information more accessible is vital in strengthening public trust in digital policies.

Leveraging IoT and Data Analytics for Smarter Public Services: The Internet of Things (IoT) provides opportunities to enhance the responsiveness and efficiency of public services. A key challenge in EBGS implementation is the fragmentation of digital systems among different government agencies, leading to inefficiencies. A Smart Government approach recommends creating an integrated, real-time data-sharing system to improve coordination and streamline service delivery [35]. *Overcoming Bureaucratic Challenges in Technology Adoption*: Smart Government implementation faces structural barriers despite its potential. Traditional bureaucratic frameworks, often rigid and resistant to change, pose challenges in adopting digital innovations like EBGS. The study suggests a more flexible, innovation-driven bureaucratic approach—where government agencies work collaboratively with various stakeholders—is essential for Smart Government success [36].

Smart Government provides a comprehensive framework for advancing EBGS in Gorontalo City. Digitalizing administration, fostering citizen engagement, integrating IoT technologies, and modernizing bureaucratic structures are crucial for the policy's success. By implementing strategic improvements, EBGS can evolve into a more efficient, transparent, and citizen-centric governance system, aligning with Smart Government principles. Implementing the EBGS policy in Gorontalo City still encounters several challenges, particularly in communication, resources, disposition, and bureaucratic structure. While efforts have been made to enhance the effectiveness of policy implementation, various obstacles remain that need to be addressed to optimize the achievement of EBGS objectives:

The Role of Communication in Policy Implementation: Effective communication is essential for successfully executing the EBGS policy. Although the dissemination of policy information has been carried out, issues related to clarity and Consistency persist. Miscommunication can result in misunderstandings, leading to improper implementation [37]. Enhancing communication strategies through more extensive outreach, digital media utilization, and training programs for policymakers and implementers is crucial to overcome this.

Resource Readiness and Competency as Key Factors: The availability of competent human resources and adequate technological infrastructure plays a significant role in the successful implementation of EBGS. However, a shortage of skilled personnel in information technology and a lack of integration among digital systems used by different agencies hinder policy execution [38]. To resolve this, targeted training, certification programs, and better coordination among regional institutions must be prioritized.

Motivation and Attitude of Policy Implementers: The commitment and attitude of civil servants in adopting and executing EBGS greatly influence the policy's success. While financial incentives such as additional income allowances (TPP) have been introduced, disparities in understanding and acceptance of the policy remain [39]. Thus, in addition to monetary incentives, initiatives such as recognition programs, the promotion of a digital-oriented work culture, and continuous professional development should be reinforced. *Challenges in Bureaucratic Structure*: A rigid bureaucratic system with inadequate standardization has emerged as a significant challenge in the EBGS implementation process. Existing Standard Operating Procedures (SOPs) require updates to align with technological advancements and public service demands. Moreover, overlapping agency responsibilities should be streamlined to improve coordination and efficiency.

Strategic development opportunities for EBGS include strong national policy support, rapid technological developments, and increasing public engagement with digital services, which create conditions that support improvement [40]. Strategies that can be implemented include: (1) Improvement of regulations and SOPs to ensure Consistency in the implementation of EBGS. (2) Strengthening the digital literacy of civil servants through training and certification programs. (3) Increasing system integration between government agencies to improve operational efficiency. (4) Increasing financial and non-financial incentives to encourage civil servant participation. (5) Expanding digital infrastructure and strengthening data security to ensure the long-term sustainability of EBGS services.

The success of EBGS implementation in Gorontalo City depends on effective communication, resource availability, the commitment of implementers, and an adaptive bureaucratic structure. With the right strategies, existing obstacles can be mitigated, enabling EBGS to become a powerful tool for enhancing efficiency, transparency, and public service quality in Gorontalo City.

5 Limitations of the Study

The writing in this book chapter is limited to answering the main problems in the effort to realize a competent government in Gorontalo, Indonesia. However, the results of this study can be used as a lesson for countries that are building smart cities. The study field includes brilliant city theory, e-government, and other elements in the government's e-government concept. The location of this study is based on the implementation of EBGS in Gorontalo City and may not be relevant or applicable elsewhere. The study method is descriptive with an observational method, focusing on communication, resource allocation, disposition, and organizational structure in implementing EBGS. The data and information used are from 2018 to 2021, based on detailed reports and regulations. This study uses SOAR analysis to assess potential and positive potential without considering the weaknesses and threats usually carried out in SWOT analysis. This study also collects data to understand relevant government policies in Gorontalo City.

6 Recommendations

Recommendations for essential steps to improve the implementation of EBGS in Gorontalo City include: optimization of existing SOPs and regulations needs to be carried out to follow technological developments and community needs, State Civil Apparatus competency in the field of information technology must be improved through routine training so that State Civil Apparatus are more confident and able to utilize the EBGS system effectively, a structured business process map needs to be prepared to ensure coordination between agencies in the implementation of good EBGS. Finally, more optimal use of technology is expected to accelerate services and increase the openness of public information. These steps aim to realize a more responsive, transparent, and service-oriented government in Gorontalo City.

7 Conclusions

Implementing the Electronic-Based Government System (EBGS) in Gorontalo City is still not optimal. The main inhibiting factors are unclear communication between government agencies, inadequate resource support, uneven cheerful disposition of the State Civil Apparatus, and an organizational structure that is not well coordinated. These obstacles cause EBGS not to function optimally, so the goal of increasing efficiency, transparency, and quality of public services has not been achieved. A comprehensive and integrated EBGS development strategy through Optimization of Standard Operating Procedures (SOP), State Civil Apparatus competencies development through continuous training, and data integration between government agencies. In addition, strong policy support and adequate technological infrastructure are significant to support the success of EBGS. By implementing these 12 strategies, Gorontalo City is expected to improve the quality of its public services and become a successful model in implementing e-government in Indonesia.

Acknowledgements We want to thank the Institute of Home Affairs (IPDN), the National Research and Innovation Agency (BRIN), and the University of Muhammadiyah Malang (UMM), which have supported this research.

References

1. Chan P (2021) Child-friendly urban development: smile village community development initiative in Phnom Penh. World 2(4):505–520. https://doi.org/10.3390/world2040031
2. Prasad D, Alizadeh T, Dowling R (2023) Smart city planning and the challenges of informality in India. Dial Hum Geogr 14(3):385–402. https://doi.org/10.1177/20438206231156655/ASSET/IMAGES/LARGE/10.1177_20438206231156655-FIG4.JPEG
3. Al Vayed D, Nurmandi A, Muallidin I, Loilatu MJ (2023) Utilization of the JAKI application in realizing smart living in the COVID-19 pandemic era (case study of DKI Jakarta Government, Indonesia). In: Lecture notes in networks and systems, vol 465. Springer, Berlin, pp 399–410. https://doi.org/10.1007/978-981-19-2397-5_38
4. Sari I, Nurmandi A, Muallidin I, Loilatu MJ (2023) Implementation of war room in improving the quality of security services in Makassar City, Indonesia. In: Lecture notes in networks and systems, vol 465. Springer Nature Singapore, Singapore, pp 389–397. https://doi.org/10.1007/978-981-19-2397-5_37
5. Syukri A, Nurmandi A, Muallidin I, Kurniawan D, Loilatu MJ (2023) Toward an agile and transformational government, through the development of the Tangerang LIVE application (case study of Tangerang City, Indonesia). In: Lecture notes in networks and systems, vol 464. Springer Nature Singapore, Singapore, pp 343–352. https://doi.org/10.1007/978-981-19-2394-4_32
6. Aziz NLL, Budiatri AP (2024) The role of government in facilitating digital economy for micro, small, and medium enterprises. J Tax Policy Econ Account 2:117–137. https://doi.org/10.1007/978-981-97-0029-5_7
7. Kurniawan R, Nurmandi A, Salahudin (2022) Analysis of quality website content and twitter content quality as public service tools at the local level in DKI Jakarta Province, Indonesia. In: Lecture notes in networks and systems, vol 217. Springer Singapore, Singapore, pp 15–26. https://doi.org/10.1007/978-981-16-2102-4_3
8. Vardopoulos I, Giannopoulos K, Papaefthymiou E, Temponera E, Chatzithanasis G, Goussia-Rizou M, Karymbalis E, Michalakelis C, Tsartas P, Sdrali D (2023) Urban buildings sustainable adaptive reuse into tourism accommodation establishments: a SOAR analysis. Discov Sustain 4(1):1–17. https://doi.org/10.1007/S43621-023-00166-2
9. Puyt RW, Lie FB, Wilderom CPM (2023) The origins of SWOT analysis. Long Range Plan 56(3):102304. https://doi.org/10.1016/J.LRP.2023.102304
10. Hanna NK (2024) Transforming government. In: Driving digital transformation in developing nations. Springer, Cham, pp 55–67. https://doi.org/10.1007/978-3-031-54573-3_4
11. Mora L, Gerli P, Ardito L, Messeni Petruzzelli A (2023) Smart city governance from an innovation management perspective: theoretical framing, review of current practices, and future research agenda. Technovation 123:102717. https://doi.org/10.1016/J.TECHNOVATION.2023.102717
12. Ristvej J, Lacinák M, Ondrejka R (2020) On smart city and safe city concepts. Mob Netw Appl 25(3):836–845. https://doi.org/10.1007/s11036-020-01524-4
13. Anthopoulos LG (2017) Understanding smart cities: a tool for smart government or an industrial trick? Springer, Cham
14. Li F, Yan J (2024) How do e-government and green technology innovation affect carbon emissions: evidence from resource-rich countries in the Shanghai cooperation organization. Energy Rep 12:4026–4033. https://doi.org/10.1016/J.EGYR.2024.09.078

15. Mishra PK, Rai A (2021) Role of unmanned aerial systems for natural resource management. J Indian Soc Rem Sens 49(3):671–679. https://doi.org/10.1007/S12524-020-01230-4/TABLES/2
16. Lambert JL (2023) Police powers and accountability. Routledge, London. https://doi.org/10.4324/9781003360308
17. Coccia M (2021) How a good governance of institutions can reduce poverty and inequality in society? In: Contributions to management science. Springer, Berlin, pp 65–94. https://doi.org/10.1007/978-3-030-60978-8_4
18. Mustafa G, Solli-Sæther H, Bodolica V, Håvold JI, Ilyas A (2022) Digitalization trends and organizational structure: bureaucracy, ambidexterity or post-bureaucracy? Eurasian Bus Rev 12(4):671–694. https://doi.org/10.1007/S40821-021-00196-8/TABLES/7
19. Lnenicka M, Rizun N, Alexopoulos C, Janssen M (2024) Government in the metaverse: requirements and suitability for providing digital public services. Technol Forecast Soc Chang 203:123346. https://doi.org/10.1016/J.TECHFORE.2024.123346
20. Cordella A, Paletti A (2019) Government as a platform, orchestration, and public value creation: the Italian case. Gov Inf Q 36:101409. https://doi.org/10.1016/j.giq.2019.101409
21. Mensah IK, Zeng G, Luo C (2020) E-government services adoption: an extension of the unified model of electronic government adoption. SAGE Open 10. https://doi.org/10.1177/2158244020933593
22. Bertot JC, Jaeger PT, Grimes JM (2010) Using ICTs to create a culture of transparency: E-government and social media as openness and anti-corruption tools for societies. Gov Inf Q 27:264–271. https://doi.org/10.1016/j.giq.2010.03.001
23. Mergel I, Dickinson H, Stenvall J, Gasco M (2024) Implementing AI in the public sector. Public Manag Rev:1–14. https://doi.org/10.1080/14719037.2023.2231950
24. Lim WM (2024) What is qualitative research? An overview and guidelines. Australas Mark J 33(5). https://doi.org/10.1177/14413582241264619/ASSET/IMAGES/LARGE/10.1177_14413582241264619-FIG6.JPEG
25. Rose-Ackerman S, Lindseth PL, Emerson B (2017) Comparative administrative law, 2nd edn. Edward Elgar Publishing, Cheltenham. https://doi.org/10.4337/9781784718671
26. Schmidthuber L, Ingrams A, Hilgers D (2021) Government openness and public trust: the mediating role of democratic capacity. Public Adm Rev 81(1):91–109. https://doi.org/10.1111/PUAR.13298
27. Stavros J, Cole M (2013) SOARing towards positive transformation and change. Dev Policy Rev 1:10–34
28. Edwards III GC (1980) Public policy implementing. In: Literary and linguistic computing
29. Sridharan VG (2021) Methodological insights theory development in qualitative management control: revisiting the roles of triangulation and generalization. Account Audit Account J 34(2):451–479. https://doi.org/10.1108/AAAJ-09-2019-4177/FULL/XML
30. Villena Manzanares F, García-Segura T, Pellicer E (2024) Effective communication in BIM as a driver of CSR under the happiness management approach. Manag Decis 62(2):685–701. https://doi.org/10.1108/MD-02-2023-0284/FULL/PDF
31. Damayanti D, Ismowati M (2021) The implementation of the cybercrime prevention policy at the Metro Jaya Police Station in Central Jakarta. https://doi.org/10.4108/EAI.17-7-2019.2302054
32. Chaidir J, Haerofiatna H (2023) Development of data management in the implementation of electronic-based government in Serang Regency. Int J Commun Cooper Stud 11(1):17–25. https://www.eajournals.org/
33. Castro C, Lopes C (2022) Digital government and sustainable development. J Knowl Econ 13(2):880–903. https://doi.org/10.1007/S13132-021-00749-2/TABLES/6
34. Chohan SR, Hu G (2022) Strengthening digital inclusion through e-government: cohesive ICT training programs to intensify digital competency. Inform Technol Dev 28(1):16–38. https://doi.org/10.1080/02681102.2020.1841713

35. Khatri R, Endalamaw A, Erku D, Wolka E, Nigatu F, Zewdie A, Assefa Y (2023) Continuity and care coordination of primary health care: a scoping review. BMC Health Serv Res 23(1):1–13. https://doi.org/10.1186/S12913-023-09718-8/TABLES/1
36. Yigitcanlar T, Degirmenci K, Butler L, Desouza KC (2022) What are the key factors affecting smart city transformation readiness? Evidence from Australian cities. Cities 120:103434. https://doi.org/10.1016/J.CITIES.2021.103434
37. Servaes J (2022) Communication for development and social change. In: The Routledge handbook of nonprofit communication, pp 23–31. https://doi.org/10.4324/9781003170563-4
38. Gupta A, Singh RK, Gupta S (2022) Developing human resource for the digitization of logistics operations: readiness index framework. Int J Manpower 43(2):355–379. https://doi.org/10.1108/IJM-03-2021-0175/FULL/XML
39. Pearce A, Mitchell LA, Best S, Young MA, Terrill B (2024) Publics' knowledge of, attitude to and motivation towards health-related genomics: a scoping review. Eur J Hum Genet 32(7):747–758. https://doi.org/10.1038/s41431-024-01547-5
40. Jooss S, Collings DG, McMackin J, Dickmann M (2024) A skills-matching perspective on talent management: developing strategic agility. Hum Resour Manag 63(1):141–157. https://doi.org/10.1002/HRM.22192

Ethical and Social Challenges of Smart Cities: An Inclusive Approach for Urban Development with Internet of Things (IoT)

Aziz Atmani, Yamna Azizi, Mounir Boumhamdi, and Marouane Ikira

Abstract The optimization of urban management has become a priority for cities worldwide, driven by the need for sustainability, efficiency, and enhanced quality of life. The Internet of Things (IoT) is at the core of these advancements, offering innovative solutions for managing energy, water, and waste systems. However, IoT integration poses ethical and social challenges, including economic barriers, privacy concerns, interoperability issues, and the digital divide. This study examines IoT applications in energy, water, and waste management, as well as the associated challenges, drawing insights from Barcelona, Tokyo, and Singapore. By analyzing these cases, this research highlights the importance of inclusive frameworks that balance technological efficiency with social equity and ethical considerations. Recommendations are proposed to guide policymakers in adopting sustainable and equitable urban management strategies.

Keywords Smart cities · Digital technologies · Internet of things (IoT) · Ethics · Digital inclusion · Equitable development

---

A. Atmani (✉) · M. Boumhamdi
Laboratoire de Recherche Business Intelligence, Gouvernance des Organisations, Finance et Criminalité Financière, Faculté des Sciences Juridiques Économiques et Sociales de Casablanca, Université Hassan II, Casablanca, Morocco

Y. Azizi
Laboratoire du Droit Public et Sciences Politiques, Faculté des Sciences Juridiques Économiques et Sociales de Casablanca, Casablanca, Morocco

M. Ikira
Laboratoire de Recherches en Gestion, Économie et Sciences Sociales, Faculté des Sciences Juridiques, Économiques et Sociales d'El Jadida, Université Chouaib Doukkali, El Jadida, Morocco

H. Kaddouri et al. (eds.), *Smart Cities for Sustainable and Urban Development*, Green Energy and Technology, https://doi.org/10.1007/978-3-031-99001-4_3

1 Introduction

The unprecedented pace of urbanization has created complex challenges for city planners, including resource scarcity, environmental degradation, and the need to improve citizens' quality of life [9, 10, 12]. To address these issues, the integration of digital technologies into urban infrastructure has emerged as a critical strategy for optimizing urban management. Among these technologies, the Internet of Things (IoT) stands out for its ability to enable real-time data collection, analysis, and automation, transforming key urban systems such as energy grids, water distribution networks, and waste management processes. By embedding IoT devices into city operations, urban centers can achieve greater efficiency, reduce costs, and ensure sustainability [16]. Despite the immense potential of IoT in advancing smart city initiatives, its implementation raises ethical, technical, and social challenges. Economic barriers often limit the widespread adoption of IoT technologies, particularly in resource-constrained regions. Privacy and security concerns highlight the risks of data misuse, surveillance, and breaches [14, 15]. Furthermore, the absence of standardized IoT protocols complicates interoperability, making system integration and scalability a significant hurdle [13, 17]. The digital divide also threatens to exclude vulnerable populations from the benefits of IoT, exacerbating existing socio-economic inequalities [11]. Addressing these multifaceted challenges requires a comprehensive evaluation of IoT applications and their implications for equitable and ethical urban management.

1.1 Preview of Related Work

The growing body of research on IoT integration in urban environments has demonstrated its transformative potential in optimizing critical city functions. In energy management, IoT-enabled smart grids facilitate real-time monitoring, predictive analytics, and efficient distribution. Angelidou [1] underscores the role of IoT in reducing energy wastage and integrating renewable energy sources. Batty et al. [2] emphasize the application of IoT in predictive maintenance, where sensors monitor equipment health to prevent disruptions. Water management has similarly benefitted from IoT technologies, with smart sensors enabling leak detection, quality monitoring, and optimized distribution. Barcelona's IoT-driven water management systems have significantly reduced losses and improved efficiency [3]. Waste management has also seen substantial improvements. Smart bins equipped with IoT sensors notify authorities when they are full, optimizing collection routes and reducing operational costs, as demonstrated in Singapore's urban initiatives [4]. However, alongside these successes, critical challenges remain. Zuboff [5] critiques IoT systems for enabling "surveillance capitalism," where personal data is commodified at the expense of citizen autonomy. Goodman and Powles [6] argue for the necessity of transparency and robust cybersecurity frameworks to address privacy concerns. Interoperability issues further hinder IoT implementation, as proprietary systems often lack the compatibility needed for large-scale integration [7]. Moreover, the digital divide exacerbates inequities, as marginalized populations frequently lack the infrastructure and skills to benefit from IoT advancements [8].

While these studies highlight the significant opportunities and challenges associated with IoT, they often analyze individual domains—such as energy or water systems—without considering their interdependencies. Furthermore, the social and ethical dimensions of IoT implementation are frequently addressed in isolation, overlooking their interconnected impacts. This study seeks to bridge these gaps by providing a comprehensive analysis of IoT applications, barriers, and strategies for sustainable and inclusive smart city development.

1.2 Research Gap

Although IoT has demonstrated transformative potential in urban management, several research gaps remain unaddressed. First, most studies focus on specific domains, such as energy or waste management, neglecting the interconnectedness of urban systems. Second, the interplay between economic, ethical, and technical challenges—such as privacy concerns, interoperability, and the digital divide—is insufficiently explored. Third, existing frameworks often fail to address the needs of marginalized populations, limiting the inclusivity of IoT solutions. This research aims to address these gaps by examining IoT applications across energy, water, and waste management systems while analyzing the challenges that hinder their implementation.

1.3 Focus of This Paper

This paper explores the key ethical and social challenges associated with IoT integration in smart cities, with a particular focus on applications in energy, water, and waste management. It examines economic, privacy, interoperability, and digital inclusion barriers that impede IoT adoption and proposes strategies to overcome these obstacles. By drawing insights from case studies in Barcelona, Tokyo, and Singapore, this study highlights the importance of balancing technological innovation with social equity and ethical considerations, offering a roadmap for sustainable and inclusive urban development.

2 IoT Applications and Challenges

2.1 IoT Applications in Energy, Water, and Waste Management

IoT technologies have transformed urban management by offering data-driven insights and automation to optimize resource use and service delivery. Despite their potential, the successful integration of IoT in energy, water, and waste management faces significant economic, technical, and social challenges. This section explores these applications and challenges, providing a detailed analysis of their implications.

2.1.1 Energy Management

IoT technologies in energy management focus on increasing system efficiency, reducing costs, and integrating renewable energy sources. Smart grids, a hallmark of IoT-enabled energy systems, use sensors to monitor energy consumption in real time. These grids dynamically adjust energy supply to meet fluctuating demand, reducing waste and enhancing stability. In Barcelona, IoT-enabled platforms provide citizens with real-time data on energy consumption, fostering sustainable practices and empowering citizens to make informed decisions about energy use [3]. The city's participatory governance model also supports decentralized renewable energy initiatives, integrating solar and wind energy into its IoT-driven infrastructure [1]. In Tokyo, IoT technologies are employed for predictive maintenance, where sensors monitor the health of energy infrastructure. This approach prevents equipment failures, reduces downtime, and lowers repair costs [2]. Tokyo's integration of IoT with renewable energy systems exemplifies its hybrid governance model, which balances centralized efficiency with participatory elements. Singapore adopts a centralized approach, leveraging IoT to optimize energy grid operations across its densely populated urban areas. IoT systems in Singapore prioritize large-scale efficiency, ensuring uninterrupted energy supply while reducing operational costs [7]. The integration of smart meters and IoT-enabled appliances in homes further enhances energy conservation, aligning with the city's sustainability goals.

2.1.2 Water Management

IoT applications in water management address key challenges, including water scarcity, leakage, and quality control. By integrating smart sensors into water systems, cities can optimize distribution, detect inefficiencies, and ensure compliance with safety standards. In Barcelona, IoT-enabled water management systems have reduced water losses through real-time leak detection and efficient resource allocation. IoT sensors installed across the water network provide actionable insights, enabling rapid responses to potential issues [3]. Citizen engagement is a critical component of Barcelona's approach, as IoT platforms provide transparent updates on water usage and conservation initiatives. Tokyo combines IoT technologies with traditional water infrastructure to enhance distribution efficiency. Advanced monitoring systems in the city analyze water flow, pressure, and quality, ensuring reliability and sustainability [1]. IoT-driven irrigation systems in suburban areas optimize water usage, balancing resource conservation with agricultural productivity. Singapore uses IoT to manage its water resources dynamically, prioritizing efficiency and safety. Smart sensors monitor water quality in real time, enabling early detection of contaminants and reducing risks to public health [7]. The city's IoT-enabled desalination plants further exemplify the integration of advanced technologies into water infrastructure to address resource challenges.

2.1.3 Waste Management

IoT technologies in waste management enhance operational efficiency, reduce environmental impact, and improve citizen engagement. Smart waste bins equipped with IoT sensors provide real-time data on fill levels, notifying collection teams and enabling dynamic route planning. In Barcelona, IoT applications in waste management focus on improving recycling rates and reducing emissions. Citizens are actively involved through IoT platforms that provide real-time updates on waste segregation and collection schedules [3]. The city's participatory governance model ensures that IoT systems align with community needs and sustainability goals. Tokyo employs IoT for dynamic waste collection, leveraging real-time data to optimize routes and reduce fuel consumption. Advanced recycling systems in Tokyo integrate IoT sensors to monitor material recovery rates, identifying inefficiencies and areas for improvement [4]. Singapore has developed a comprehensive IoT-driven waste management system that streamlines operations and minimizes environmental impact. Smart bins equipped with sensors notify authorities when they are full, reducing unnecessary trips and emissions. IoT also supports Singapore's waste-to-energy initiatives, contributing to a circular economy [7].

2.2 *Challenges in IoT Integration*

Despite its potential, IoT implementation faces significant challenges that hinder its scalability and inclusivity. These barriers are examined below.

2.2.1 Economic Barriers

The cost of deploying IoT infrastructure is a critical challenge, particularly for resource-constrained regions. IoT systems require significant investment in sensors, communication networks, and data processing platforms. For example, in Tokyo, the hybrid governance model demands ongoing investments to integrate legacy systems with modern IoT technologies, straining municipal budgets [2].

Barcelona addresses economic barriers through public-private partnerships (PPPs), where costs and risks are shared between the government and private stakeholders. This approach has facilitated the broader adoption of IoT technologies while reducing the financial burden on public resources [3]. In Singapore, state-funded IoT initiatives benefit from substantial government support, enabling the large-scale deployment of IoT systems. However, replicating this centralized model in less affluent regions remains a significant challenge [7].

2.2.2 Privacy and Security Concerns

The extensive data collection enabled by IoT raises concerns about privacy and security. IoT devices generate sensitive data on citizens' behavior and preferences, which, if misused, can lead to surveillance and data breaches. Zuboff [5] critiques the potential for "surveillance capitalism," where IoT data is commodified without user consent. Barcelona's decentralized data governance model empowers citizens with greater control over their data, but it faces challenges in maintaining robust cybersecurity across diverse stakeholders. In contrast, Singapore's centralized IoT systems streamline data management but raise ethical concerns about over-surveillance. Transparent policies and robust cybersecurity measures are essential to mitigate these risks and build public trust [6].

2.2.3 Interoperability Issues

The lack of standardized protocols among IoT devices creates interoperability challenges, complicating system integration. Proprietary systems often operate in silos, limiting scalability and efficiency. Tokyo struggles to integrate its legacy infrastructure with modern IoT systems, highlighting the need for harmonized standards. Barcelona's use of open standards reduces interoperability challenges but complicates collaboration with vendors reliant on proprietary technologies [3]. Developing global IoT standards is critical to overcoming these issues. Collaborative efforts between governments, private companies, and international organizations can facilitate seamless integration and unlock the full potential of IoT systems [7].

2.2.4 Digital Divide

The digital divide remains a persistent issue, as marginalized populations often lack access to IoT-enabled services. Rural areas, low-income communities, and older populations are disproportionately affected by inadequate infrastructure and limited digital literacy [8].

Barcelona's participatory model seeks to reduce inequalities by involving citizens in IoT initiatives. However, rural regions still face significant barriers to access. Singapore's centralized approach ensures comprehensive urban coverage but often overlooks the needs of marginalized communities. Addressing the digital divide requires targeted policies, such as subsidizing IoT infrastructure in underserved areas and implementing digital literacy programs. These measures can ensure equitable access to IoT technologies and foster greater social inclusion [1].

2.2.5 Research Gap Analysis

While IoT applications in energy, water, and waste management have demonstrated their transformative potential, several gaps remain in the literature. First, most studies focus on specific domains without examining their interdependencies, such as how IoT-enabled energy systems impact water management. Second, challenges like privacy, interoperability, and economic barriers are often analyzed in isolation, overlooking their combined effects on IoT implementation. Lastly, existing frameworks fail to adequately address the needs of marginalized populations, limiting the inclusivity of IoT solutions. This study aims to bridge these gaps by providing a comprehensive analysis of IoT applications and proposing strategies for ethical and equitable integration.

3 Methodology

This section outlines the research design, data collection methods, analytical framework, and ethical principles adopted in this study. A comparative case study approach is used to analyze IoT applications in urban management across three cities: Barcelona, Tokyo, and Singapore. These cities were selected for their advanced IoT implementations and differing governance models, which provide diverse perspectives on the opportunities and challenges associated with IoT integration.

3.1 Research Design

The study employs a comparative case study approach to investigate how governance models influence IoT implementation, focusing on energy, water, and waste management systems. The selected cities represent distinct governance paradigms:

1. **Barcelona**: A participatory governance model emphasizing citizen engagement and decentralization. IoT initiatives align with community-driven policies, providing transparency and empowering stakeholders.

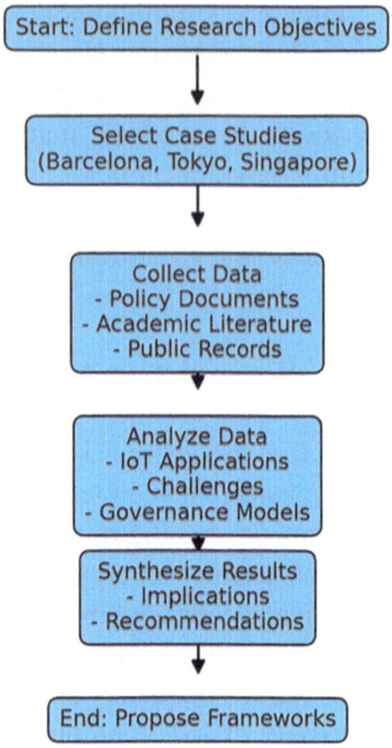

Fig. 1 Flowchart representing the research design methodology

2. **Tokyo**: A hybrid governance model that balances centralized efficiency with participatory elements. IoT is employed to modernize legacy systems and support advanced urban infrastructure.
3. **Singapore**: A centralized governance model prioritizing technological efficiency and scalability. IoT implementations are state-led, with a focus on large-scale optimization and operational excellence.

The diversity in governance models allows for a comprehensive analysis of IoT adoption strategies and challenges across different urban contexts.

The research methodology described above is summarized and visualized in the flowchart below (Fig. 1). This flowchart provides a step-by-step overview of the research design, highlighting the sequential stages from defining objectives to synthesizing results for inclusive urban governance frameworks.

3.2 *Data Collection*

Data for this study were collected from three main sources to ensure comprehensiveness and reliability:

3.2.1 Policy Documents

- Municipal reports, smart city strategies, and governance frameworks published by city authorities.

- Examples include Barcelona's "Decidim" platform documentation, Tokyo's urban planning reports, and Singapore's "Smart Nation" framework.

3.2.2 Academic Literature

- Peer-reviewed articles and books addressing IoT applications, governance models, and related challenges.
- Notable works by Angelidou [1], Batty et al. [2], and Mora et al. [7] informed the analysis.

3.2.3 Public Records and Media Reports

- News articles, advocacy group publications, and publicly accessible forums provided insights into citizen perspectives and IoT adoption experiences.

3.2.4 Dataset Refinement

- An initial dataset of over 200 documents was curated.
- After removing duplicates, outdated content, and irrelevant sources, 150 high-quality documents were retained for analysis.

To ensure the robustness of findings, secondary data were triangulated across multiple sources, verifying consistency and reliability.

3.2.5 Example of Used Data

An example of the data used in this study includes metrics from Barcelona's Decidim platform, which detailed citizen engagement and decentralized IoT governance. Tokyo's urban planning reports provided insights into predictive maintenance metrics for energy and water systems, highlighting IoT's role in modernizing legacy frameworks. In Singapore, the analysis included data from centralized IoT systems, such as water quality monitoring statistics and broadband access rates under the Smart Nation initiative. These diverse datasets enabled a comprehensive comparison of governance models and IoT implementation strategies across the three cities.

To ensure the robustness of findings, secondary data were triangulated across multiple sources, verifying consistency and reliability.

3.3 Analytical Framework

The study employs a thematic analysis framework to evaluate IoT applications, challenges, and governance models. The framework is structured into three key dimensions:

3.3.1 IoT Applications

- Examines how IoT is integrated into energy, water, and waste management systems in the selected cities.
- Focuses on the benefits of IoT, such as enhanced efficiency, sustainability, and citizen engagement.

3.3.2 Challenges

- Analyzes barriers such as economic constraints, privacy concerns, interoperability issues, and the digital divide.
- Identifies commonalities and differences in how these challenges manifest across the case studies.

3.3.3 Governance Models

- Assesses the impact of participatory, hybrid, and centralized governance on IoT adoption.
- Explores the balance between efficiency, inclusivity, and ethical considerations.

3.3.4 Adopted Models

- Barcelona: IoT adoption is analyzed through the "Decidim" platform, which facilitates citizen participation in urban decision-making.
- Tokyo: IoT integration focuses on predictive maintenance, renewable energy, and efficient water distribution within a hybrid governance model.
- Singapore: IoT systems emphasize large-scale energy optimization, dynamic water monitoring, and smart waste management within a centralized framework.

3.4 Ethical Considerations

Given the sensitivity of IoT applications in urban management, the study adhered to stringent ethical guidelines to ensure transparency, inclusivity, and fairness:

3.4.1 Transparency

Sources and methodologies were meticulously documented to ensure replicability and reliability.

3.4.2 Inclusivity

Diverse perspectives, including those from marginalized communities, were considered to address the social implications of IoT adoption.

3.4.3 Data Integrity

Only publicly available, verified data were used to maintain accuracy and ethical compliance.

3.4.4 Avoidance of Bias

The comparative analysis was conducted objectively, avoiding favoritism toward any governance model or city.

3.4.5 Consent and Confidentiality

Public records and citizen feedback from forums were anonymized to protect individual privacy and ensure ethical handling of data.

3.5 Enhancements to Methodology

To overcome potential limitations and improve the methodological rigor, the following enhancements were implemented:

3.5.1 Triangulation of Data

Secondary data from multiple sources were cross-referenced to verify consistency and reliability.

3.5.2 Contextualized Analysis

Thematic analysis was adapted to the unique socio-economic and political contexts of the selected cities, ensuring relevance and applicability.

3.5.3 Incorporation of Multiple Governance Models

By including participatory, hybrid, and centralized governance systems, the study provides a holistic understanding of how governance shapes IoT adoption.

3.5.4 Focus on Ethical Dimensions

Special emphasis was placed on ethical considerations, such as data privacy and inclusivity, to align with the study's objectives.

3.5.5 Addressing Scalability and Equity

The framework explicitly evaluates the scalability of IoT implementations and their impact on equity, ensuring a comprehensive perspective on inclusivity.

3.6 Summary of Methodological Strengths

This study's methodology is characterized by its comparative approach, robust data collection process, and thematic framework, which collectively provide a comprehensive analysis of IoT applications in urban management. By integrating governance models and ethical considerations, the methodology aligns with the study's objective to propose sustainable and equitable IoT strategies.

4 Results and Discussions

This section presents the results of the comparative analysis of IoT applications in energy, water, and waste management across Barcelona, Tokyo, and Singapore. Each city's governance model significantly influences its IoT adoption, effectiveness, and inclusivity. The findings are contextualized with a detailed discussion of the challenges and broader implications for urban management and policymaking.

4.1 Figures and Interpretation

This bar graph illustrates citizen participation, with Barcelona achieving the highest engagement (75%), followed by Tokyo (50%), and Singapore (30%) (Fig. 2).

This pie chart compares governance priorities, showing Barcelona's emphasis on inclusivity (80%), Tokyo's balanced priorities (60% inclusivity, 70% efficiency), and Singapore's focus on efficiency (90%) (Fig. 3).

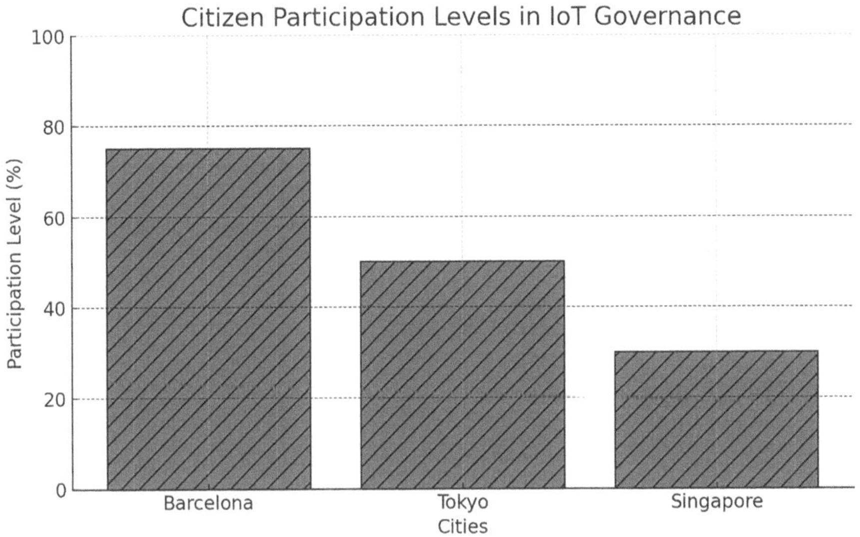

This bar graph illustrates citizen participation, with Barcelona achieving the highest engagement (75%), followed by Tokyo (50%), and Singapore (30%).

Fig. 2 Citizen participation levels in IoT governance

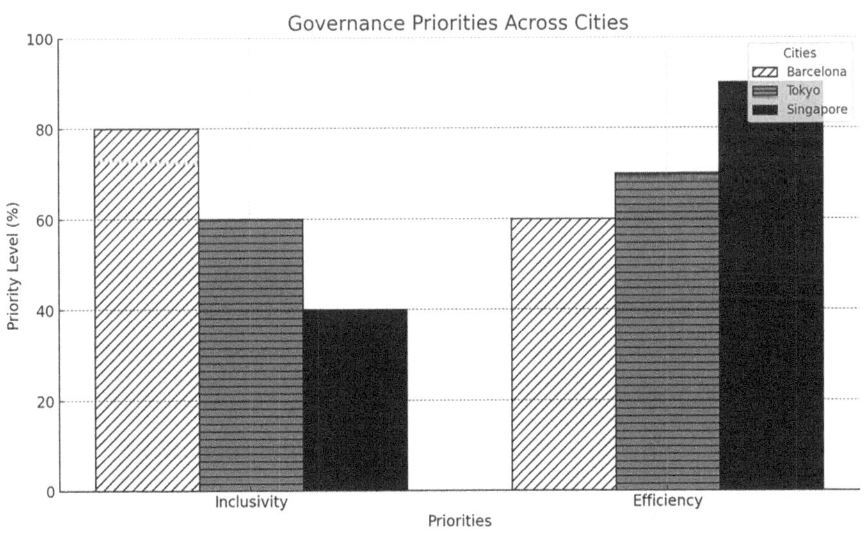

This pie chart compares governance priorities, showing Barcelona's emphasis on inclusivity (80%), Tokyo's balanced priorities (60% inclusivity, 70% efficiency), and Singapore's focus on efficiency (90%).

Fig. 3 Governance priorities across cities

Table 1 IoT applications in energy management

City	Key IoT applications	Governance approach	Key features
Barcelona	Smart grids, renewable energy integration	Participatory	Citizen empowerment, transparent platforms
Tokyo	Predictive maintenance systems	Hybrid	Centralized resource optimization with local input
Singapore	Centralized smart grid systems	Centralized	Large-scale efficiency, low citizen involvement

4.2 IoT Applications Across Sectors

4.2.1 Energy Management

IoT technologies in energy management demonstrate varying levels of citizen engagement, efficiency, and scalability across the three cities:

- Barcelona employs IoT-enabled smart grids integrated with renewable energy platforms. Real-time data platforms empower residents to monitor and manage their energy consumption, aligning with the city's participatory governance approach [3].
- Tokyo uses IoT for predictive maintenance of energy infrastructure. Sensors monitor the condition of equipment, reducing operational costs and preventing failures. Tokyo's hybrid governance model encourages both centralized efficiency and stakeholder participation [2].
- Singapore focuses on large-scale IoT integration through centralized smart grids. IoT-enabled appliances and real-time monitoring systems prioritize operational efficiency but offer limited citizen engagement opportunities [7] (Table 1).

4.2.2 Water Management

IoT applications in water management emphasize optimization, resource conservation, and quality assurance:

- Barcelona employs IoT for real-time leak detection and water quality monitoring. Transparent IoT platforms foster public trust and align with the city's citizen-centric governance model [1].
- Tokyo integrates IoT sensors in water distribution networks to optimize pressure and flow. IoT-driven irrigation systems enhance agricultural productivity and water efficiency, demonstrating the city's hybrid governance flexibility [3].
- Singapore uses IoT for dynamic water regulation and contamination detection. Centralized data-driven water management supports resource efficiency and public safety [7] (Table 2).

Table 2 IoT applications in water management

City	Key IoT applications	Governance approach	Key features
Barcelona	Leak detection, quality monitoring	Participatory	Transparent systems, citizen engagement
Tokyo	Optimized distribution, smart irrigation	Hybrid	Balances legacy systems with modern tools
Singapore	Dynamic regulation, quality assurance	Centralized	Efficient resource use, centralized control

4.3 Challenges in IoT Adoption

The challenges identified across the case study cities highlight critical barriers to IoT adoption, ranging from economic constraints to technical and social issues.

4.3.1 Economic Barriers

The economic constraints are most pronounced in Tokyo, where integrating IoT into legacy systems requires substantial investment. Barcelona mitigates these costs through public-private partnerships, ensuring financial sustainability. Singapore benefits from state-funded IoT initiatives, enabling large-scale deployment but limiting replicability in resource-constrained regions.

4.3.2 Privacy and Security Concerns

Privacy concerns are significant in all three cities, albeit shaped by their governance models:

- Barcelona prioritizes data ownership and citizen control but faces challenges in ensuring consistent cybersecurity.
- Singapore's centralized model raises concerns about over-surveillance, despite its efficiency in managing data systems.
- Tokyo balances privacy with operational needs through localized data policies, offering a middle ground between the other two models [6].

4.3.3 Interoperability Issues

The interoperability challenges affect IoT scalability in Tokyo and Singapore, where diverse systems struggle to communicate. Barcelona's adoption of open standards mitigates this issue but complicates partnerships with proprietary vendors.

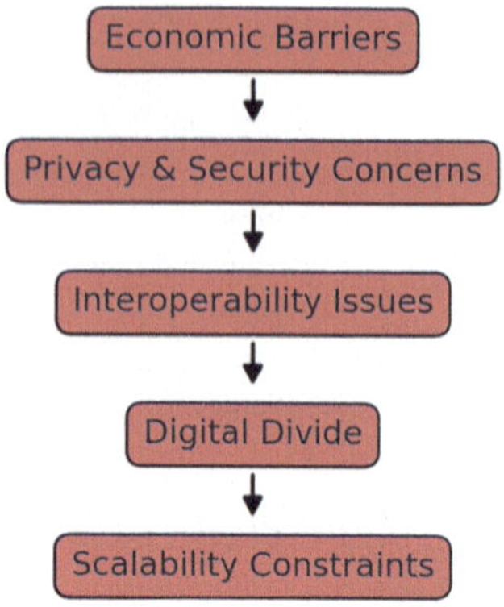

Fig. 4 Key challenges in IoT adoption

4.3.4 Digital Divide

The digital divide is particularly evident in Barcelona's rural regions, where limited infrastructure excludes certain populations from IoT benefits. Singapore's centralized IoT deployment ensures comprehensive coverage in urban areas but overlooks marginalized communities. Tokyo addresses the divide through targeted digital literacy programs.

To summarize the discussion on IoT adoption barriers, Fig. 4 presents a visual representation of the key challenges identified. These challenges, including economic barriers, privacy and security concerns, interoperability issues, the digital divide, and scalability constraints, illustrate the complex landscape of IoT implementation hurdles.

4.4 Discussion

The findings of this study reveal significant contrasts in the governance models and ethical considerations of Barcelona and Singapore as smart cities. Figure 2 illustrates the levels of citizen participation in IoT governance, showcasing Barcelona's decentralized approach, which prioritizes citizen control and transparency. This approach emerges as a more ethically aligned model, where initiatives like "Decidim" enable citizens to actively participate in data-related decisions, fostering trust and accountability. In contrast, Singapore's centralized model, though efficient, raises ethical concerns about surveillance and data privacy. The heavy reliance on government-controlled systems may alienate citizens, leading to a lack of trust in the overall governance structure.

The digital divide remains a persistent challenge in both cities, although their approaches differ significantly. Barcelona's emphasis on community-based initiatives targets vulnerable groups, promoting inclusivity. However, as depicted in Fig. 3, rural and underserved populations still face barriers to access. Singapore's focus on infrastructure and universal broadband access has advanced technological integration, yet the lack of targeted measures for marginalized communities perpetuates socio-economic disparities. These findings highlight the need for policies that address not only infrastructure gaps but also socio-cultural barriers to digital inclusion.

The levels of citizen participation further differentiate the two cities. Barcelona's participatory governance model fosters greater engagement, empowering citizens to co-create urban policies. This approach strengthens public trust and aligns with democratic principles. Conversely, Singapore's top-down approach prioritizes efficiency, often at the expense of inclusivity. While this model enhances operational effectiveness, it risks alienating citizens by limiting their involvement in decision-making processes.

The analysis of governance priorities reveals contrasting strategies. Barcelona strikes a balance between efficiency and inclusivity, ensuring that technological advancements do not undermine human rights. Singapore, while excelling in technological deployment, emphasizes security and efficiency, which may overshadow ethical considerations. Figure 2 further highlights these divergent governance priorities, emphasizing the contrasting levels of inclusivity and efficiency in both cities. This divergence underscores the critical need for governance frameworks that integrate ethical safeguards without compromising innovation.

These findings underscore the importance of adopting inclusive and participatory approaches to smart city development. While Barcelona provides a model for ethical and inclusive urban innovation, Singapore offers valuable lessons on technological scalability. Combining the strengths of both approaches could guide future smart city initiatives, ensuring they are both efficient and equitable. Balancing technology with human-centered values remains essential to addressing the ethical and societal challenges of smart cities.

4.5 Implications for Urban Management

The results underscore the need for adaptive governance frameworks that integrate participatory, hybrid, and centralized elements. Policymakers should:

- **Foster Cross-Sector Collaboration**: Encourage partnerships among public, private, and community stakeholders to address financial and technical barriers.
- **Prioritize Inclusivity**: Ensure that marginalized communities are included in IoT initiatives through targeted outreach and infrastructure development.
- **Strengthen Ethical Guidelines**: Implement robust data governance frameworks to address privacy and security concerns.

5 Conclusion and Perspectives

5.1 Conclusion

This study analyzed the role of IoT technologies in urban management, focusing on energy, water, and waste management across three cities—Barcelona, Tokyo, and Singapore. The findings demonstrate that governance models are critical in shaping the effectiveness, inclusivity, and scalability of IoT implementations:

1. Barcelona's participatory governance model fosters citizen engagement and trust through decentralized IoT platforms but faces challenges in scalability and financial sustainability.
2. Tokyo's hybrid governance model balances efficiency with inclusivity by integrating IoT into legacy systems. However, its implementation demands significant financial and organizational resources.
3. Singapore's centralized governance model excels in large-scale efficiency and operational optimization but raises ethical concerns about surveillance and limits citizen involvement.

Despite these achievements, several challenges persist, hindering IoT's transformative potential:

1. Economic barriers limit adoption due to the high costs of deployment, maintenance, and modernization.
2. Privacy and security concerns pose ethical issues, requiring robust data governance.
3. Interoperability issues complicate system integration and scalability due to the lack of standardized protocols.
4. The digital divide excludes marginalized populations, exacerbating inequalities in access to IoT benefits.

To address these challenges, cities must adopt governance frameworks that integrate efficiency, inclusivity, and ethical considerations. Doing so will enable IoT to deliver sustainable and equitable solutions for urban management.

5.2 Perspectives

Building on the findings, the following perspectives are proposed for future research, policymaking, and practical applications:

5.2.1 Develop Interoperable IoT Standards

- Standardized protocols are essential to overcome interoperability issues and ensure seamless system integration.

- Governments, international organizations, and the private sector should collaborate to establish open standards and reduce reliance on proprietary technologies.

5.2.2 Address the Digital Divide

- Inclusive IoT policies must bridge the digital divide by expanding infrastructure to underserved areas, offering subsidies for IoT-enabled services, and promoting digital literacy programs.
- Policymakers should prioritize equitable access to IoT benefits, ensuring that marginalized populations are not left behind.

5.2.3 Strengthen Data Governance

- Robust data governance frameworks should prioritize transparency, citizen control over personal data, and accountability for data misuse.
- International collaboration on ethical guidelines can address privacy and security concerns, fostering public trust.

5.2.4 Adopt Hybrid Governance Models

- Cities should combine participatory and centralized approaches to balance inclusivity and efficiency.
- Learnings from Tokyo's hybrid model can guide policymakers in creating adaptable frameworks suited to local contexts.

5.2.5 Expand IoT Applications

- Future research should explore IoT applications in transportation, healthcare, disaster management, and education to enhance urban resilience and societal well-being.
- Cross-sectoral integration of IoT systems can create synergies, improving overall urban efficiency.

5.2.6 Foster Public-Private Partnerships (PPPs)

- PPPs can address financial barriers by sharing costs and risks, while leveraging private sector innovation to accelerate IoT adoption.
- Such partnerships should prioritize public interests, ensuring equitable distribution of benefits.

5.2.7 Promote Cross-sectoral Collaboration

- Collaboration among governments, private companies, academia, and civil society is essential for addressing technical, economic, and social challenges in IoT implementation.
- Inclusive dialogues and partnerships can align IoT initiatives with community needs and priorities.

5.3 Final Remarks

The integration of IoT into urban management offers significant opportunities to enhance efficiency, sustainability, and quality of life. However, achieving these outcomes requires cities to address persistent barriers and adopt governance models that balance technological innovation with ethical and social considerations. Adaptive frameworks that combine participatory and centralized elements can enable cities to unlock the full potential of IoT while ensuring equity and inclusivity.

This study contributes to the understanding of IoT's role in urban development by:

1. Highlighting the influence of governance models on IoT adoption.
2. Identifying key challenges and offering actionable solutions.
3. Providing a foundation for future research into scalable, inclusive, and ethical IoT frameworks.

By prioritizing equity and collaboration, IoT can transform cities into hubs of innovation, resilience, and inclusivity, paving the way for a sustainable urban future.

Competing Interests The authors declare that they have no competing interests in relation to this work. All efforts have been made to ensure that the analysis, findings, and recommendations presented in this chapter are impartial and based solely on academic research and objective analysis.

Ethics Approval This study did not involve any experiments with human participants or animals. All data used in the study were derived from publicly available sources, and no personal or sensitive information was collected or used. As such, formal ethics approval was not required for this research. The authors adhered to ethical guidelines in conducting and reporting the study.

References

1. Angelidou M (2015) Smart cities: a conjuncture of four forces. Cities 47:95–106. https://doi.org/10.1016/j.cities.2015.02.001
2. Batty M, Axhausen KW, Giannotti F (2012) Smart cities of the future. Eur Phys J Spec Top 214(1):481–518. https://doi.org/10.1140/epjst/e2012-01703-3
3. Barns S (2020) Platform urbanism: negotiating platform ecosystems in connected cities. Urban Geogr 41(3):453–472. https://doi.org/10.1080/02723638.2020.1730632

4. Hollands RG (2008) Will the real smart city please stand up? Intelligent, progressive or entrepreneurial? City 12(3):303–320. https://doi.org/10.1080/13604810802479126
5. Zuboff S (2019) The age of surveillance capitalism: the fight for a human future at the new frontier of power. Public Affairs, New York
6. Goodman B, Powles H (2019) Transparency dilemmas in smart cities. Big Data Soc 6(1):1–12. https://doi.org/10.1177/2053951719858748
7. Mora L, Deakin M, Reid A (2019) Smart-city development paths: insights from the first two decades of research. J Urban Technol 26(2):1–23. https://doi.org/10.1080/10630732.2018.1552791
8. Van Deursen AJ, Van Dijk JA (2019) The first-level digital divide shifts from inequalities in physical access to inequalities in material access. New Media Soc 21(2):354–375. https://doi.org/10.1177/1461444818797082
9. Calzada I, Cobo C (2015) Unplugging: deconstructing the smart city. J Urban Technol 22(1):23–43. https://doi.org/10.1080/10630732.2014.971535
10. Capdevila I, Zarlenga MI (2015) Smart city or smart citizens? The Barcelona case. J Strateg Manag 8(3):266–282. https://doi.org/10.1108/JSMA-03-2015-0030
11. Cardullo P, Kitchin R (2018) Smart urbanism and smart citizenship: the neoliberal logic of 'citizen-focused' smart cities in Europe. Environ Plan C Polit Space 37(5):813–830. https://doi.org/10.1177/2399654418782314
12. Greenfield A (2013) Against the smart city. Do Projects, New York
13. Ho E, Chua L, Tan T (2020) Governing smart cities in Singapore: a regulatory perspective. Smart Cities 3(1):101–117. https://doi.org/10.3390/smartcities3010008
14. Kitchin R (2016) The ethics of smart cities and urban science. Philos Trans R Soc A Math Phys Eng Sci 374(2083):20160115. https://doi.org/10.1098/rsta.2016.0115
15. Leszczynski A (2016) Speculative futures: data, algorithms, and urban space. Environ Plan D Soc Space 34(5):816–835. https://doi.org/10.1177/0263775815622978
16. Smith A, Stirling A (2018) Innovation, sustainability, and democracy: an analysis of grassroots contributions. J Environ Innov Soc Trans 27:54–67. https://doi.org/10.1016/j.eist.2017.10.004
17. Stilgoe J, Owen R, Macnaghten P (2013) Developing a framework for responsible innovation. Res Policy 42(9):1568–1580. https://doi.org/10.1016/j.respol.2013.05.008

Urban Security Technologies for Moroccan Smart Cities: A Proactive Analysis for Unforeseen Impacts

Mhammed Kezzoute

Abstract Smart cities, as a relevant technological innovation, offer potentials that make urban activities more efficient and more effective. These technologies contribute to the improvement of the complexity of human life within modern cities, and thanks to their advantages and benefits, human activities of organization, planning, and urban management become more efficient, safer, and more sustainable. Moreover, the adoption of urban security technologies also appears to be effective and yields the same advantageous results, especially in public safety and also in resource management and the improvement of citizens' quality of life. However, these technologies raise as many questions in their usage processes, causing multiple challenges that affect society, security, stability, the environment, etc., making the presence of an effective and sustainable organizational and regulatory strategy a necessity. For this reason, we focus the work in this chapter on the close study of these urban security technologies according to the requirements of the moroccan transition towards the smart cities model, by anticipating the potential impacts on insufficient infrastructures, public distrust around these urban security technologies, and citizens' rights in urban management. In this analytical model, a comprehensive strategy is proposed, but which prioritizes inclusion, environmental sustainability, and social sustainability as the foundations of urban management during its planning and development. In this study, we provide the most well-known examples of cities that have adopted the concepts of smart cities, such as Dubai, Singapore, and Barcelona, to guide our analytical trajectory by encouraging decision-makers to draw inspiration from these examples in order to ensure citizen participation, transparency in local management, and adaptation during the deployment of these projects, considering the local conditions and the moroccan's requirements, while also leveraging and benefiting from these technologies without mitigating potential risks.

M. Kezzoute (✉)
Moroccan Center of Cyber Studies & Technologies Security, Rabat, Morocco
e-mail: m.kezzoute@ump.ac.ma

H. Kaddouri et al. (eds.), *Smart Cities for Sustainable and Urban Development*, Green Energy and Technology, https://doi.org/10.1007/978-3-031-99001-4_4

Keywords Moroccan smart cities · Moroccan cyber governance · Moroccan cyber security · Moroccan urbanization · Moroccan technologies

1 Introduction

The concept of smart cities, -based on the adoption and integration of advanced technologies to achieve urban and local development goals -is furthermore a paradigm that reflects the reconstruction of societies and the promises they share for the improvement and management of cities and infrastructures [1]. These projects aiming at the massive integration of technological advances aspire to improve the quality of life of citizens and the sustainability of their environments, making urban societies stable, durable, and sustainable. Alongside several technological devices chosen to achieve these objectives, the adoption of Internet of Things (IoT) devices and the deployment of direct and permanent surveillance camera systems constitute the cornerstone of these informatization and urban planning projects [2].

Morocco has integrated into similar processes, such as the adoption of surveillance systems based on the Internet of Things (IoT), biometric access control, and data collection and analysis, etc. These processes allowed Morocco to revolutionize real-time surveillance as well as the collection, analysis, and transmission of data between interconnected systems [3]. In addition to transparency, these systems offer more advantages by improving situational awareness and decision-making processes in urban and local areas, and also provide crucial information on traffic, environmental conditions, and security incidents in an interoperable manner [4].

These devices helped optimize resources and strengthen proactive measures in urban management, with IoT-connected sensors capable of detecting anomalies in infrastructure remotely, reducing intervention costs and improving safety [5]. Moreover, direct surveillance camera systems have a high reputation in crime prevention, traffic management, and improving public safety [6].

Direct surveillance through connected cameras also serves as a deterrent against crime, and as an instrument for analyzing security incidents, it will offer more concrete advantages by rebuilding public trust in urban governance promises regarding security and crime prevention [4]. However, the disadvantages related to data and transaction confidentiality, ethical implications, and transparency issues, which are not seriously considered, encourage us to question this situation from a specific context and an original point of view [7].

Morocco offers us a specific context with exceptional advantages to test the integration of these processes within a society that has historically been conservative but holds significant modernization ambitions. For this reason, successive moroccan decision-makers have always reflected a willingness for urbanization and an interest in adapting to industrial and technological progress to overcome the nation's challenges, including urban challenges. For the past 10 years, Morocco has demonstrated its interest in modernizing infrastructure and improving governance. In this regard, the integration of smart city projects, especially in Casablanca and Rabat, confirms this commitment to benefit from IoT devices, similarly to those who have previously undergone these urban digitalization processes and maintained urban security technologies to consolidate urban security and promote economic growth and environmental sustainability [8].

Taking into account the manuscripts of Albino et al. [9] on the digitization of cities and urban spaces, Morocco is encouraged to adopt these solutions to countering the growing urban challenges, threatening the sustainability and stability of urban services and leading to the decline of traditional management and planning measures [10]. That is why we examine in our work the effectiveness of such control and surveillance measures in the moroccan context, which faces numerous obstacles and challenges in these processes, such as digital literacy, the internal digital and technological divide, lacunes in law, weak regulations, ethics, security threats, and also cyber challenges [11].

In international experiences, these technologies helped municipal authorities respond more effectively to security threats and improve emergency and crisis management systems in cities [12]. At this moment, the lessons from Singapore's Smart Nation initiative and the urban technology frameworks of Barcelona and Dubai can serve as pathways for moroccan decision-makers to improve public safety, prevent crimes and malicious activities, optimize response times to crises and emergencies, thereby reducing decision-making time, streamlining public services, effectively controlling access data, strengthening trust between citizens and government institutions, and finally promoting a responsible, safe, and resilient urban environment? [5].

The objective of the chapter, then, is to clarify the path and strategy to harmonize the use of IoT devices and direct surveillance camera systems with moroccan specificities without compromising the effectiveness of these systems or threatening the societal context. And this will only be achievable with the proposal of a roadmap that takes into account Morocco's requirements on an equal footing with calls for the acceleration of the massive integration of technologies into urban processes. And which will at the same time succeed in setting a good example for the states of the continent and the globe, to consult the moroccan experience on these matters. Because the conclusions of our work will help municipal decision-makers, local urban planning strategists, and urban policy implementers to effectively avoid undesirable scenarios and transition the use of urban security technologies towards a phase of security, stability, and prosperity.

2 The Adoption of IoT Devices for Real Time Monitoring, Data Collection and Direct Permanent Surveillance Camera Systems, a Literature Survey and Justification

We will find that in the well-known urban strategies, the adoption of IoT devices for monitoring and data collection, as well as permanent and direct control cameras in cities, are becoming the most common procedures to ensure urban security. Moroccan cities, by becoming smart, seek to leverage the potential of these devices in public urban security, service improvement and management, resource optimization, and several other urban activities shaped by the adoption of these technological devices. Thus, the adoption of these technologies will improve urban governance through the effects of automation, with IoT sensors playing a crucial role in enhancing urban security by improving situational awareness, efficiency in real-time data processing, and finally, responding to incidents more effectively [13]. These devices also allow municipal authorities to shape traffic management, manage environmental risks in a more sustainable way, and finally, consolidate public safety [14].

Morocco has been engaged for some time in a process of modernizing its cities, with Casablanca, Rabat, and Tangier seeking to join the international smart city club by adopting and integrating IoT systems to address traffic issues, surveillance, crime prevention, and finally, biometric access control within their territories. By leveraging the capabilities of IoT devices and systems in surveillance, automated decision-making, and proactive security, moroccan cities aim to achieve similar profiles as international cities like Barcelona and Singapore, which have successfully optimized traffic flow, energy consumption, and urban security through the use of these devices [3].

However, we believe that the implementation of these technologies raises several other questions and issues related to technical, ethical, or even financial aspects. Several challenges are essential if the introduction of these devices does not take into account the weaknesses of cybersecurity protocols, the potential dangers related to confidentiality and data storage, and the level of digital literacy, all of which risk reversing the advantages of IoT devices [15]. In this regard, the example of cyberattacks targeting the vital infrastructures of smart cities internationally serves as proof of the weaknesses of these devices and systems in the face of challenges and threats [16], if the process of integrating and introducing these devices by Morocco does not take into account the challenges and issues associated with these processes.

The observation of international experiences, has helped us confirm that these processes of adopting IoT devices certainly constitute significant advancements for the improvement of urban management and planning tasks. Direct surveillance camera devices represent a step forward in the fight against urban crime and the deterrence of ill-intentioned individuals seeking to exploit the vulnerabilities of these devices to achieve criminal objectives [6].

This is achievable thanks to the potential of real-time crime prevention, which helps in preventing crime on-site. In addition to its cognitive effects in strengthening the trust between the public and urban authorities, traffic surveillance and real-time traffic control allow smart cities, and especially moroccan cities, to optimize traffic management and also ensure smooth circulation by minimizing road challenges and improving the efficiency of urgent interventions by specialized urban teams in case of problems [17].

The use of these devices in the moroccan context certainly constitutes a step forward in achieving the same objectives of deterring crime, improving traffic flow and congestion, and finally, the on-site response of specialized teams. However, these cameras raise, alongside their advantages, ethical concerns, questions related to the respect of citizens' privacy, inquiries attached to governance models and modalities, and potential abuses of personal data [18].

We observe that the existing legal frameworks in Morocco, especially those related to the processing and protection of personal data, are still vulnerable to the way the issues and challenges against data management and organization are developing. In this regard, the law, for example, no. 09-08 on the processing and protection of personal data remains silent on the issues and challenges of storing mass data from direct public surveillance by IoT devices within the framework of smart cities [19]. We believe that the presence of a specific legal and legislative framework is essential to prevent impacts and balance the needs of security with the requirements of law and privacy, giving Morocco an advantage in its process of adopting IoT devices for urban planning and security reasons.

Morocco, in this regard, should adopt specific frameworks to improve its adoption of IoT devices in the urban and security sectors while preserving privacy, ethics, transparency, and trust in surveillance measures between citizens and municipal authorities [20]. Because the consolidation of these frameworks is essential for moroccan smart cities, whose uses are not limited solely to the security domain. However, the potentials of analytical and predictive AI coupled with big data and direct monitoring can enhance urban planning by automating decision-making in other urban areas such as supporting environmental sustainability and resource congestion [5].

At this moment, moroccan decision-makers will need to prioritize regulatory alignment and legal complementarity to ensure compatibility between the objectives set by municipal authorities and the operational requirements for the adoption of IoT in Moroccan cities. Also, the national cyber strategy for 2030 did not take into account the specific requirements for the adoption of IoT in the field of urban security, ignoring the specific impacts on the urban infrastructures of smart cities that are continually expanding and reflecting Morocco's desire to modernize and share these experiences with new moroccan cities. By developing these related frameworks from this experience and prioritizing the implementation of a specific national policy, Morocco can, under conditions, maximize its benefits from smart city projects and succeed in previous experiences (Fig. 1).

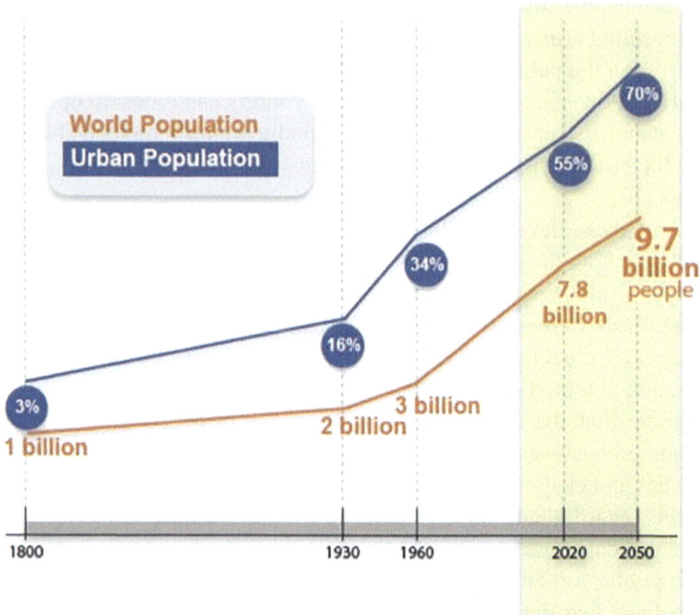

Fig. 1 The figure demonstrate the World population growth in Cities. (Source: Food and Agriculture Organization of the United Nations—FAO. FAO Framework for the Urban Food Agenda; FAO: Rome, Italy, 2019)

3 Opportunities for Moroccan Cities Through Implementing Security Technologies in Smart Cities

Moroccan cities are expanding rapidly, while municipal capacities still seem far from bridging the gaps in urban planning and management. It will become apparent that the massive urbanization that Morocco is undertaking imposes multiple challenges and stakes on the state and municipal organizations; hence, the necessity to address the phenomena of overpopulation, the pressure on infrastructure, and the rising crime rates has become urgent [21]. Morocco, in response to these challenges, is engaging in specific processes, with urban digitization proving to be effective and operational.

According to our research objective, urban security technologies provide answers to several public safety and community stability questions, and offer societies great potential to improve their lives [22]. In the following paragraphs, we will focus on IoT devices and real-time surveillance systems, AI-assisted video surveillance systems, and finally biometric iden-tification systems. And we will also explore the adaptation of these technologies to the specificities and context of Morocco.

Morocco suffers from urban overpopulation, due to challenges mainly related to development disparities, rural depopulation, and the industrialization process. More than 80% of the population currently lives in urban areas, and the major cities of

Morocco suffer from this population density, which puts pressure on pub-lic services, increases crime rates, endangers transportation and housing, and ulti-mately intensifies security concerns [12]. It seems that the idea that alongside the development of cities, concerns also grow appears to be evident and correct. Because overcrowded and disorganized cities seem very vulnerable to social destabilization, cyberattacks, and the mismanagement of already depleted resources. That is why we believe that addressing these issues is a step forward in finding innovative solutions aimed at improving the lives of citizens and also ensuring public safety.

Technologies as a solution for smart cities to overcome their overcrowding issues also enable these human gatherings to function efficiently through data analysis and processing. For example, the introduction of technologies in the urban field has helped improve traffic flow and management; energy consumption is also streamlined, and crisis and emergency management systems are becoming more efficient [23]. This is also the case for moroccan cities, especially in emergencies due to their overloaded services and organizational challenges, where adopting technologies to improve management and urban planning seems relevant and necessary. More specifically, the security technologies that we dedicate this work to studying can make a difference and transform these discussion spaces into sustainable and safer environments [21].

This process allows local authorities to collect and analyze information and make decisions on the ground thanks to the technological capabilities used for real-time data collection. These analyzed and studied data can vary depending on the ar-eas of the community component, ranging from public safety to environmental data, air quality, traffic congestion, and security and crime incidents in the city under study [13]. For example, the technological devices used in urban areas can permanently monitor pollution levels, urban traffic accidents, and keep an eye on high-crime areas. Then, the analysis of this real-time data, coupled with the use of prediction and prevention software and systems, allows decision-makers and municipal officials to make proactive and dynamic decisions.

A study by the World Economic Forum [24] confirmed that the adoption of interconnected security systems in smart cities could improve public safety by 20%. This means that these processes of making urbanization digital and connected will lead the situation towards improvement as the hypotheses require, and will reduce crime rates and the response time against these nuisance practices. These technologies within the moroccan cities most affected by the challenges of overpopulation and massive urbanization could also effectively manage traffic flows and reduce congestion on essential roads in city centers, which are currently major contributors to urban inefficiency. In addition to the above, IoT applications in the field of public safety, especially public surveillance camera systems, can provide direct and continuous monitoring of common goods and energy, digital, and other public infrastructure to prevent practices that may harm these common assets [25].

Nevertheless, the introduction of IoT technologies should be done cautiously, taking into account the legal and security impacts of the risks of data integrity breaches and avoiding violations of laws related to privacy rights. Morocco should strengthen the framework for using these technologies according to the specific objectives of urban planning and enhance the legislation, institutions, and bodies that incorporate modern urbanization plans and strategies. This means the assurance of a secure and stable environment where data and information circulate perfectly and ethically to maintain public trust in the official initiatives of smart cities.

3.1 The Implementation of Direct and Permanent Surveillance Camera Systems

To prevent crimes and monitor areas of crime proliferation and hotspots in any city, seems to be an urgent procedure to ensure the sustainability of social stability and the extension of public safety to all points and directions of the city. In smart cities, these traditional video surveillance systems, although still effective, are currently supported by AI applications that enable automated monitoring, aimed at improving their real-time activities such as tracking individuals, providing valuable information, detecting suspects, or predicting criminal behavior [6].

In the national context of video surveillance, the implementation of AI-assisted video surveillance systems could enhance the capabilities of law enforcement and make their interventions in targeted areas more effective. This process could also lead to crime prevention and the development of specific programs to address the social degradation that drives young people toward these paths of violence [26].

Let's return to our idea, the implementation of facial recognition technologies could be used to identify suspects, delinquents, and even criminals in public spaces. Moreover, the capabilities of intelligent surveillance camera systems can de-tect antisocial behaviors of individuals in relation to their environment. In London, the use of this technology has greatly assisted the authorities in controlling and monitoring areas and locations where crime is spreading, which moroccan authorities could use to deter and emulate, based on its proven effectiveness in improving awareness of criminal activities [17].

Nevertheless, the use of AI in video surveillance without any considerations can lead to undesirable and unforeseen circumstances. Ethical considerations should be taken into account when integrating the potentials of AI into surveillance processes and camera monitoring of urban areas, because in certain circumstances and cases, the data and information collected for facial recognition can be used to infringe upon individuals' rights [26]. In Morocco, the laws related to the theme of exploiting innovative technologies seem somewhat rigid and inadequate; Law No. 09-08 concerning the protection of individuals with regard to the processing of personal data needs to be adapted to the security context of monitoring smart cities through AI technologies [19]. It is also mandatory that any other expansion of AI-powered visual surveillance systems be accompanied by strict and narrow data governance policies that protect privacy and ensure that citizens' data will be used solely in the public interest [27].

By using biometric identification systems, including fingerprint scanners, facial recognition, and iris scans, a high level of security is provided to control the flow of traffic in sensitive areas targeted by these control and surveillance devices. These technologies are increasingly being used in smart cities around the world, thanks to what they offer as unprecedented potential to secure public assets, vital infrastructure, and essential services for the survival of the city and its citizens. The adoption of these systems and their integration into current urbanization activities can significantly improve urban safety rates [28].

For example, biometric identity systems could be installed at key points in moroccan cities, such as airports, government establishments, secure areas, and important high-security locations, to prevent unauthorized access. The operations provided by these systems offer numerous advantages over traditional access control systems based on passwords or access cards by providing a unique and non-transferable identification [29].

On the other hand, the limitations of this technology must be acknowledged. Because even these technologies offer great potential to facilitate the work of local authorities, any malfunction in their fingerprint or facial recognition systems, or in the event of a cyberattack or a breach of cybersecurity in these systems, will lead to serious errors in intervention and deterrence protocols and procedures [30].

To abstain from these situations, Morocco is supposed to ensure that biometric identification systems comply with international standards, such as ISO/IEC 29115, which provides the necessary guidelines for the development and implementation of biometric systems. In our case, Morocco should also commit to serious processes for strengthening encryption and data storage protocols that are essential for the proper functioning of these systems and the consolidating the cybersecurity process, while simultaneously maintaining a strong trust relationship between citizens and community authorities.

We observe that the future of smart city infrastructure projects developed in Morocso looks very promising. These projects present significant opportunities for growth and sustainable development in these cities and even at the national level. significant opportunities for growth and sustainable development in these cities and even at the national level; they contribute to job creation, the improvement of public safety, and the enhancement of the quality of life for citizens in the tar-geted areas [31].

The mixed partnership processes that Morocco adopts to address and manage various areas, where the public sector collaborates with the private sector, play an essential role in improving the adoption rate of smart city technologies according to this paradigm. Unconventional collaborations between local public organiza-tions, foreign or local technology companies, and prestigious academic institu-tions can all accelerate the development and deployment of urban security solu-tions. These new forms of partnership paradigms can promote, taking into ac-count the expertise of each of these organizations in the field of smart city frameworks and

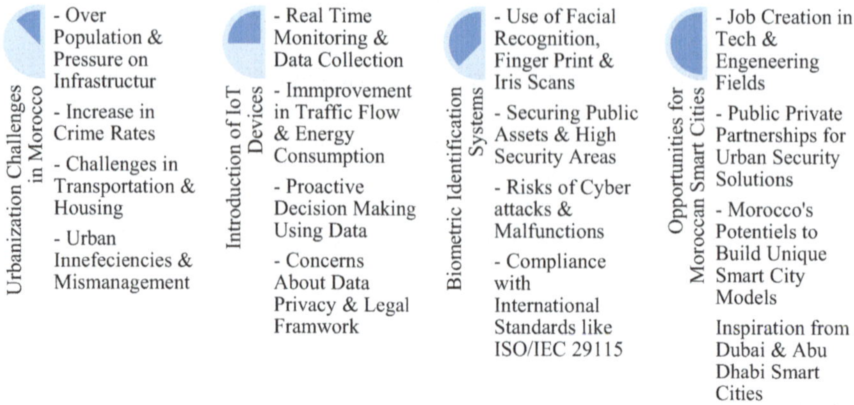

Fig. 2 The figure represents the most important ideas to facilitate understanding, opportunities for moroccan cities through the implementation of urban security technologies

technologies, all innovation and the pursuit of the common good within smart cities [10].

Also, the smart city as a development sector can stimulate the creation of local opportunities, especially in the technology, engineering, and data analysis and cybersecurity fields. For this approach, the demand for qualified professionals who can manage, maintain, and develop these systems will only grow, offering hope for economic growth in the regions and for the country as a whole [32].

Taking into account the known experiences of smart cities in the Arab world, such as Dubai and Abu Dhabi, Morocco can build a unique and specific paradigm that is more sustainable, efficient, and safer. The examples we provided have accumulated a better and more significant experience in integrating urban security technologies into the digitization processes of cities and planning to transform large cities into smart cities. AI-powered surveillance systems, IoT-based traffic management platforms, urban public safety applications, etc., already tested in example cities, can make significant progress and offer several similar benefits while being mindful of the challenges posed by rapid urbanization [33] (Fig. 2).

4 Exploration of the Problems and Challenges of Urban Security Technologies for Moroccan Smart Cities

Urban security technologies are becoming increasingly important in smart city projects and strategic urbanization planning. Moroccan cities should pay attention to significant challenges and not base their objectives solely on these projects to maximize their profile. The confidentiality associated with data collection and surveillance, the cybersecurity risks and challenges arising from IoT devices, and the challenges of integrating new technologies with still-developing national infrastructures all underpin the critical perspective we hold towards these technologies [34]. The following lines present relevant case studies and also discuss the long-term implications of these processes on overall development and social equity.

4.1 *Privacy Questions Related Data Collection, Surveillance and Cyber Security Risks*

The implementation of security technologies in the urbanization process requires extensive and massive data collection, which imposes significant constraints and challenges regarding individuals' privacy. Thus, in the transformed smart cities of Morocco, the massive collection of private data can raise concerns about violations of citizens' rights on various scales and also pose undesirable questions of rights and freedoms. The permanent use of surveillance cameras in public spaces, equipped with facial recognition and AI technologies, can create doubts among citizens when they feel constantly monitored and controlled every second of their lives and in every activity [35]. In this regard, the feeling of invasion of privacy among citizens is likely to grow and increase due to the massive implementation of these technologies without awareness of the related issues, which could potentially lead in some cases and situations to violent reactions or resistance against the implementation of these urban security technologies.

A clearer example of the drawbacks that urban security technologies can im-pose on individuals' privacy and lives. In Casablanca, one of the smart cities in Morocco, local authorities use smart streetlights equipped with surveillance cam-eras to monitor traffic and improve public safety. These measures, although they have positively contributed to improving the situation, have also achieved their objectives.

They pose risks to privacy, and reports indicate that some residents have expressed concerns about being monitored without their consent, highlighting the importance of having community planning and management agreements in place before these technologies are implemented [36]. This means that establishing clear frameworks for the protection and preservation of citizens' rights and individuals' privacy is a necessity, even an urgent obligation, so that all participants can achieve their common goals with mutual consent.

There are several international examples that confirm this idea and illustrate the importance of finding solutions. In Barcelona, citizens publicly demonstrated their refusal and opposition to urban security measures, especially surveillance cameras, until the necessary regulations are established [37]. This experience highlights the importance for decision-makers in the relevant moroccan cities to engage in comprehensive negotiation and consultation processes with the various stakeholders in these cities in order to find solutions to the risks and challenges posed by the implementation of these technologies.

The proliferation of IoT in urban spaces shows us that these processes can be vulnerable to cyberattacks and all other forms of electronic interference, infiltration, computer sabotage, etc. In a country like Morocco, which is likely to be-come a significant target for online criminals, urban security technological infra-structures can represent a significant objective, especially when we know that most of these technologies used for smart traffic management, whether at the lo-cal or even international level, do not have robust and effective security proto-cols, making them fragile and vulnerable to cyberattacks and online infiltrations.

The cyber attack known as the Mirai botnet in 2016, which exploited vulnerabilities in the security and access protocols of video surveillance devices, serves as an example for cities that use the same protocols to improve their cybersecurity measures. So, if Moroccan cities do not change their perspec-tives on the vulnerabilities that may exist in the systems used in urban security technologies, they risk facing challenges and risks similar to those already encoun-tered by smart cities that have already been victims of cyberattacks international-ly, disrupting essential services and vital city infrastructures.

The study of these risks requires a multi-faceted approach. On a technical level, implementing best practices in cybersecurity and drawing inspiration from other experiences in the field will be necessary; the ongoing update of the software used on these platforms should be standard. We can also carry out continuous assessments of the systems used and simultaneously develop the community's individual capacities to regularly conduct these visits and evaluations of equipment. At the institutional level, Morocco can develop awareness programs for the public and educate residents on the importance of cybersecurity and how they can participate in protecting their information and personal data [38].

4.2 Technical Issues of the Integration of Urban Security Technology and Their Social Long Term Implications

So, the integration of new urban security technologies with existing infrastructure is a crucial challenge for moroccan cities on their way to becoming smart cities. These urban areas are currently entirely dependent on outdated systems that cannot be compatible with the systems these cities want to implement in their smart infrastructures, such as biometric systems. This can invoke technical and financial obstacles that may jeopardize the effectiveness of the implementation of advanced systems.

The example here is the city of Rabat, as a moroccan city that wants to be a smart city; it has started to implement traffic management and public transport systems through intelligent systems. The new established systems have encountered obstacles related to the vulnerability of the old networks of the aging infrastructure and risk being incompatible with the requirements of the new bus and tram software. The upgrading of these networks and infrastructures requires significant efforts as well as financial resources and organizational arrangements, although these cities still do not have the necessary budgets to fully achieve their goals [39]. In such situations, hybrid partnerships can provide a solution, allowing cities to overcome the financial burden of achieving their goals while also benefiting from the expertise of the private sector in this area. The implementation and integration of smart technologies in other cities of Morocco, taking into account the discussed considerations, can help authorities mitigate certain challenges and issues related to digitization and the generalization of traffic and surveillance services in specific areas of the city. This approach al-lows for testing and refining the systems under discussion while minimizing related challenges and issues [40].

However, collaboration with internationally renowned technology companies in the field of smart cities can facilitate the moroccan integration into the global wave of digitization and avoid undesirable impacts by drawing on their expertise and the accuracy of their studies. Singapore's Smart Nation initiative, for example, serves as a very relevant example to follow, which moroccan cities should adopt and draw inspiration from to overcome the challenges of integration [41].

While the challenges are numerous and inevitable, it is crucial to take the necessary measures and responses to avoid long-term impacts on the stability of economic development and social equity. The fact that urban digitization processes and the integration of smart cities could lead to an unprecedented increase in investments and economic growth. This will not stop here, because we observe that as cities become stable, secure, and with lower crime rates, this will attract more businesses and investments and contribute to the creation of more jobs and local wealth [42].

If not, it is essential to ensure that the economic and social benefits of integrating smart cities are shared and distributed fairly between the targeted areas and among the citizens. Equity is a very essential element in societies, and it is crucial not to provide the elements of violence by concentrating surveillance and control on social categories or geographic areas while simultaneously depriving them of the economic and financial revenues from investments in their cities, or even discriminating against them in access to smart services. This can undermine the foundation of social inequalities in these areas, including throughout the entire city in question, and thus erode trust in their principles and claims as smart cities, regarding equality, fairness, access, and so on [43].

Moroccan cities are supposed to take these requirements into account and engage in serious inclusion processes that ensure equality and fairness in smart cities. This can be achievable through the adoption of responsible and serious policies involving community stakeholders in the planning and integration processes, and by ensuring that their needs, concerns, and proposals are taken into account. For example, municipal councils and community forums can invite citizens to express their opinions about proposed smart city projects, thereby fostering a sense of belonging, engagement, and responsibility.

In addition to the above, investing in the special programs as the education and training of citizens, specifically in disadvantaged areas through targeted pacification, can help reduce these inequalities, promote social equity, and increase citizen inclusion in these smart urbanization plans. Additionally, moroccan cities should consider regular evaluations to ensure the effectiveness of their initiatives and planning efforts. This approach not only ensures a shared commitment to the integration plans and policies within the processes of smart cities, but it will also significantly strengthen the societal trust in the future if moroccans choose to generalize this experience.

5 International Experiences of Successful Urban Security Technologies: Singapore, Barcelona and Dubai

Urban security technologies are essential for the achievement of smart city projects, enhancing public safety, and advancing urban planning. The necessity and presence of these technologies have truly become crucial. In this chapter, we address the analysis of the successful implementation processes of these technologies in the most well-known smart cities, Singapore, Dubai, and Barcelona, and we attempt to anticipate the implications for moroccan cities wishing to introduce these technologies into urban digitization processes.

Due to Singapore's population density and its specific geographical and political status as a resource-scarce city located in an area of immense port and commercial traffic, as well as global crime and piraty. The decision-makers of this emerging country have taken the initiative to engage in an ambitious and strategic project, which aims at reducing resource losses and optimizing their operations, organizing extraterritorial traffic flows, and finally establishing security [44].

In this regard, the decision-makers of Singapore have resolved to implement a new concept of management, administration, and planning for these challenges and issues, and to engage in a local initiative called the Smart Nation initiative. This ambitious project aims to overcome the discouraging situation of instability caused by the depletion of energy resources, the rise of cross-border organized crime, and all other kinds of dangers, by implementing IoT technologies, AI, and big data. Mobility, security, and energy efficiency, along with several other objectives such as ensuring the sustainability of services and infrastructure, economic growth, and finally strengthening its position as a regional and global leader in the field of innovation, etc., have encouraged Singapore to engage in specific urban planning using technology [45].

The smart city of Singapore is well-known for its use of complex traffic management systems. These systems rely on extensive networks of surveillance cameras equipped with AI capabilities that continuously and real-time monitor traffic. These systems rely on extensive networks of surveillance cameras equipped with AI capabilities that continuously and in real-time monitor traffic. They use specific sensors equipped with image capture and facial recognition software, and they can, in use, collect mass data on users who commit speeding violations, vehicles, densities, and patterns on controlled roads, allowing for the optimization of time and efficiency in urban planning [46].

The city of Singapore has invested unprecedented skills to successfully achieve this level of infrastructure digitization and to make these technological devices smart for services and urban planning. She has implemented central management platforms that integrate and concentrate the information collected from sensors, allowing for quick and effective decision-making after analysis by specific algorithms.

This can improve traffic flow and significantly reduce road safety congestion [47]. Nevertheless, given that the benefits of using these technological devices have been effective and evident, the ethical implications of such widespread use, especially regarding surveillance, as we have already stated, lead to unforeseen consequences that require and provoke serious debates and discussions. This has compelled the Singaporean government to adopt clear guidelines and cooperate with the relevant stakeholders for the institutionalization and organization of these data collection processes [48].

In our case, the adoption of similar systems could help us overcome the drawbacks of human management in this type of sector. However, one must be cautious of the challenges that may arise and affect the functioning of these intelligent traffic management systems. To avoid such situations, it is essential to establish a paradigm of collaboration and cooperation with community stakeholders and all other relevant partners to develop necessary policies that resonate with local values and specificities.

Spain is a European country that, in turn, suffers from the challenges and issues of urban overpopulation, as well as the limitations of traditional urban management and planning, and the weakness in procedures to ensure the efficiency and continuity of public services. The transformation in European societies in general and massive urbanization involve several problems for decision-makers who can do nothing but try to take advantage of technological potentials in order to overcome their management and planning challenges [49].

The fact that more than a third of the Spanish population will reside in the two major cities, Madrid and Barcelona, by 2035 will involve the local authorities of these two cities in unforeseen scenarios and multiple challenges, where the continuity of public services and vital and energy infrastructures will surely suffer, leading to various problems [42].

In this regard, Spanish decision-makers, and especially those of the local government of Barcelona, have taken the initiative to address these challenges by adopting an ambitious project for the digitization of the city and its main areas and services. Smart lighting, waste management, and open data platforms, for example, play an important role in this project and strategy, which local decision-makers believe will confront these issues and avoid stagnation in services, improve citizen participation, and promote environmental sustainability through the introduction of technologies in these management and planning processes, making this city a model to follow.

The Spanish city of Barcelona is now one of the oldest smart cities to have introduced smart streetlights, using a network of sensors to regulate lighting based on changes in movement in the surrounding area. This means that the devices that have been put into operation have allowed for the lighting to be adjusted autonomously without the intervention of on-site teams. They used sophisticated motion detectors and light sensors for these purposes, which allow streetlights to dim when there are no people around and to brighten in response to the approach of pedestrians or vehicles. This, according to observations, has reduced consumption by almost 30% and also improved public safety in urban areas during the night [12].

For the realization of this project in Barcelona, experts have confirmed that it is the result of a long collaboration between several actors and stakeholders. Urban planners, technology providers, and local communities in this city all con-tributed to the establishment of such a complex project. The lighting project in Barcelona used specific algorithms to analyze data collected from sensors, allowing central systems to act automatically while adapting to changes in the environment. This initiative has contributed, according to the project administrators, to reducing crime rates in poorly lit areas and also improving the perception and sense of trust among residents [50]. Still, this lighting technology and the way it changes the light based on people's actions and movements in the space also raises ethical concerns. It is similar to other technologies used for city security in that the data they collect could be misused. That is why the city of Barcelona has introduced specific provi-sions in its protocols for the use and exploitation of these tools, which govern and regulate the use of this technology by updating the frameworks of transparency and accountability toward the citizens [15].

Moroccan cities that want to replicate this experience, particularly Rabat and Marrakech, to improve public safety while also reducing energy costs. It is crucial for them to take into account the risks that may arise from the resistance of citizens to these devices, especially those for detecting their movements and activities, if they are not consulted in the processes of implementing and developing these projects. Citizens and all other actors, may judge these surveillance technologies negatively, despite their energy savings, if they feel that their opinions, recommendations, and consultations have been ignored as a community targeted by this urban surveillance and movement and activity detection, while simultaneously witnessing their privacy frameworks being completely violated.

5.1 The Facial Recognition Technology in Dubai for Serving Public Safety

Despite being a young country, the United Arab Emirates, benefiting from oil funds and revenues, aim to become an innovative country in several fields by engaging in various initiatives and projects such as Scale360, related to climate change, the CIO initiative related to governmental innovation, ReglaB related to regulatory innovation, etc.

In this regard, and alongside these national initiatives, the city of Dubai is committing to its role in a local initiative aimed at improving its positioning as a smart city and related to its ambitions of becoming a leader in global tourism innovation, business, and security [51]. To achieve these ambitions and objectives, the city has focused on smart governance, transportation management, and energy consumption optimization, processing individual movements through facial recognition, etc. Supported by Blockchain technologies and IoT devices, the city's efforts aim to improve the quality of life, attract investments, enhance the business environment, and position the city as a futuristic hub advanced in technological innovations and public urban safety projects based on the adoption of digital devices.

Dubai is one of the few cities internationally that has adopted facial recognition as a key element in its urban security strategy. To enhance public safety in the city through the use of urban security devices, Dubai decision-makers have implemented this facial recognition technology in important and vital areas which contain a dense and mobile population. Airports, commercial areas, and government establishments have all received this technology for a single purpose: to enhance security by detecting suspicious activities and simplifying the pursuit of criminals or undesirable incidents. These systems use advanced algorithms to recognize and match faces captured by surveillance cameras with databases containing individuals' data, directly consulting online security services. This facilitates the rapid identification of potential threats and accelerates deci-sion-making [18].

The adoption and implementation of this technology in areas of human flow, particularly in airports, have improved passenger and visitor control procedures, rendering the so-called traditional passport recognition processes obsolete. recognition processes, which involve direct interaction between control agents and the citizens or travelers being processed, unnecessary. This also allows authorities to ensure the smoothness and speed of immigrant and visitor flows in these spaces [52].

However, the implementation of this technology raises important ethical and usage questions, similar to those of its counterparts, in relation to concerns about privacy and surveillance. The critique of these processes is grounded in assertions regarding the potential misuse of these technologies to infringe upon citizens' privacy and the resultant sense of irresponsibility that may emerge from the ease and accessibility of individual data collection [20]. To avoid these implications, the authorities of Dubai are trying to minimize the risks to the proper implementation of urban security technologies by enforcing strict data protection measures and also establishing regulations governing the responsible and trustworthy use of data collected through facial recognition technologies [53]

This process, when applied in Morocco, requires great vigilance as well as deep caution. Because facial recognition technologies, although they offer unprecedented potential to improve public safety and the speed and efficiency of certain services. To the contrary, they can be dangerous and lead to social unrest and public insecurity by violating privacy, causing people to fight back against and reject these technologies, being incompatible with legal and moral principles, and breaking people's trust. This can make citizens and other important stakeholders feel left out and their opinions and points of view ignored when these technologies are used in cities.

5.2 Additional Case Studies and Implications for Morocco

We will believe that despite the examples of smart cities that we have integrated into the debate surrounding the uses of urban security technologies in smart cities, we are adding other similar examples to better illuminate our argumentative process. The capital of Tunisia, Tunis, is also exploring possibilities to integrate mobile applications with urban security systems, allowing citizens and utilizations, to report incidents in real time more efficiently. We consider this micro-model of signaling, can serve as a template for moroccan cities looking to engage their citizens in urban safety initiatives [54].

The city of Amman, as the capital of Jordan, has implemented smart traffic management solutions to monitor bus routes and schedules. Amman, thanks to these measures despite their modesty, has managed to improve the efficiency of public transportation and reduce traffic congestion [55]. In this regard, Moroccan cities facing similar urban challenges could benefit from these regional initiatives to mitigate the impacts of the country's massive urbanization. However, several other challenges seem to be present in the processes of digitizing cities and in the introduction of security technologies in this process. A lack of trust between stakeholders and community leaders, infrastructure problems, missing regulations, weak institutions, and people who do not care about the morality of using these technologies are just a few of the challenges that could make the implementation of this type of initiative and urban digitization program more difficult.

Finally, it is necessary to improve citizens' perceptions of the issues surrounding these technologies, especially the concerns regarding privacy and the respect for individuals' sensitive data confidentiality. Concerns about the use of urban security technologies are on the rise, and citizens are increasingly worried that these technologies, may serve other agendas that threaten their integrity and rights and freedoms. Moroccan cities looking to explore these processes and implement these initiatives should focus more efforts and prioritize community engagement, transparency, and the building of trust and accountability between authorities and citizens (Fig. 3).

6 Comparative Analysis Between the Moroccan Experience and the Other International Experiences

During its digital transition, Morocco has committed to a process of transforming its infrastructure, economy, and governance and administration systems in order to maximize the opportunities and benefits of technology. Although these ambitions have been delayed by certain obstacles and challenges such as overpopulation, pollution, high crime rates, the energy, social, and digital bills, etc., Morocco has embarked on a specific project to digitize moroccan cities, which constitutes an

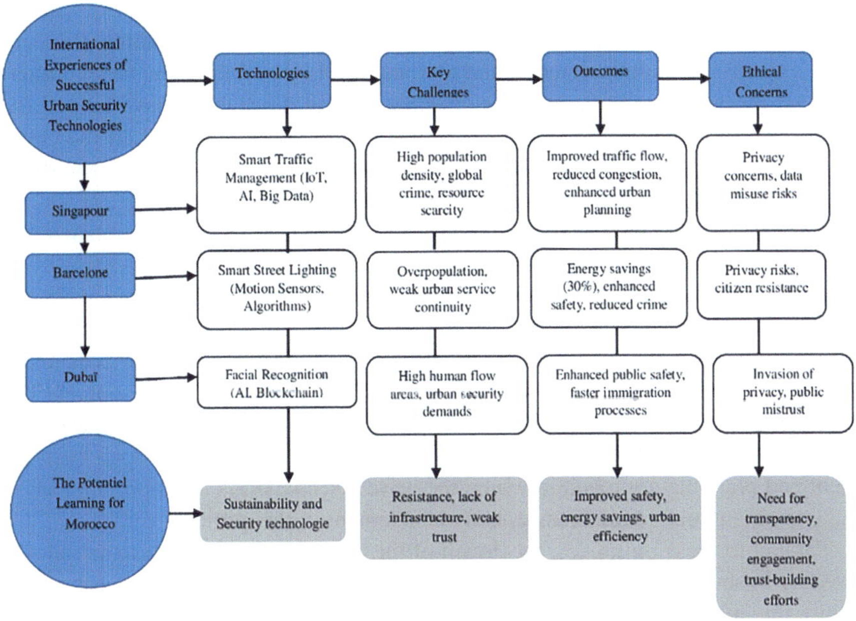

Fig. 3 The Graph represents the prospected interaction between the international experience and the moroccan potentials in adopting IoT for Urban Cities

alternative for local and national moroccan decision-makers to follow in order to counter the impacts of contemporary challenges in Morocco [41].

In this initiative, urban security technologies present an opportunity to address some challenges and serve as a foundation to respond to certain future issues and questions. And as we have seen in our analyses concerning Singapore, Barcelona, and Dubai, urban security technologies have helped local authorities overcome several urban management and administration problems and challenges. In this paragraph, we are trying to compare their experiences with the moroccan initiative for moroccan cities and how we can benefit from these experiences through shared lessons, innovations, and contextual adaptations.

6.1 Governance Through the Eyes of Singapore, Barcelona and Dubai

Singapore, by leveraging its geographical position and limited area, has managed to present itself as an international leader in urban innovation. The Smart Nation initiative, which the authorities of the city-state of Singapore adopted by focusing on the needs and improvement of citizens' lives as well as promoting the country's image, has successfully addressed the challenges of overpopulation, urban density, and crime, among others [56]. We observe that this initiative has succeeded well due to certain reasons such as Singapore's limited space, the presence of funding and technical expertise at both the internal and regional levels, and finally, the trust of international partners in the country's authorities. These advantages have helped the emergence of a centralized system of national governance engaging all the country's capacities in order to improve the lives of citizens and overcome situations that, in case of ignorance, could lead to undesirable scenarios.

The city of Barcelona, in turn, and benefiting from the expanded regionalization and the constitutional status offering cities and regions in Spain the possibility to elect local governments [57]. It has embarked on a similar process but with few specificities, for the digitization of urban tasks such as the adoption of technologies aimed at ensuring energy optimization and the continuity of public services under the best citizen conditions in a country already exhausted by energy resources. The transparent aspect of this initiative and its specific governance model, apart from being based on principles centered around the citizen, also aims at an open management and planning approach in which the private sector plays an important role in the implementation, maintenance, and management of this project.

Parallel to these initiatives, the dream of the city of Dubai to become an innovative city in several areas, especially in tourism and foreign investment, has become achievable after the local government committed to a process of digitizing various urban tasks and activities, including facial recognition and the analysis of individual traffic data in densely populated areas, using the potential of AI to overcome the challenges of urban density and the related possibilities of crime and wrongdoing have been put into service. The local management of the cities and Emirates that make up the United Arab Emirates, the area of the concerned Emirate, and the presence of immense foreign funding and expertise have facilitated the implementation of a digital local planning system, but which prioritizes the consolidation of attracting investments and overall security instead of focusing on the particularities of the citizens [58].

In this volet, Morocco's innovative initiative from a governance perspective, compared to what these studied countries and cities have mobilized to achieve their objectives, will still appear complex in terms of implementation. Apart from the problems that Morocco hopes to solve through the adoption of IoT devices and their potential for direct control and monitoring, it will appear that the insufficiency of funds necessary to transform this local project into a national one promotes the territorial vision and decentralization of such initiatives, positively impacting the quality of life of citizens and fostering a harmonious living environment [59]. However, parallel to the above, the vast territorial geography imposes on Moroccan decision-makers, despite the extensive map of territorial problems, the logic of choice rather than the ratio of need still prevails.

6.2 In Terms of Digital Infrastructure, Socio-Economic Considerations and Durability

The technologies adopted by the cities under study make them more robust in the face of the problems they previously endured. The example of Singapore and the adoption of mass data analysis through the potentials of AI to avoid unpredictable scenarios of overpopulation and the city's population density has started to provide these advantages and has also succeeded in achieving the project's intended goals. Regarding the cities of Barcelona and Dubai, the adoption of technologies ensuring energy efficiency driven by IoT, and traffic control in populated areas through facial recognition with the help of blockchains and AI, are indeed advantageous and effective measures according to reports from local decision-makers.

The Moroccan model, in turn, in partnership with international firms in the field of smart cities, has allowed the selection of only two cities, which are Casablanca and its projects Casablanca Smart and Casablanca City and Zenata Eco City, and subsequently Rabat and Tangier, but regarding the latter, the authorities in a project have limited the effort only around the Tangier Tech area [60]. The establishment of smart infrastructure supporting urban life in these cities constitutes a territorial challenge in Morocco, where the generalization of this project across the national territory, especially in less urbanized areas that still make up the majority of moroccan territory, is likely to remain a complicated challenge to overcome in the coming years [59].

The difficulties associated with the challenges of constructing and sharing this infrastructure at the national level with the rest of Morocco's cities lie in the territorial disparities between the regions of Morocco. At the same time, there is another challenge, which is the socio-economic context within the same cities of a moroccan region, and which in turn negatively impacts the general development process of the country, logically hindering the establishment of a national smart infrastructure aimed at the intelligent transformation of moroccan cities at the national level. In contrast to the examples we have cited, those in Morocco differ in their financial resources and higher GDP, and they have the capacity to mobilize funds, expertise, and national effort for such projects [61]. Morocco does not have large budgets to follow these examples in detail.

However, the benefits of these examples concerning the establishment of openness, inclusivity, and open source should be closely followed in order to develop the moroccan model based on the principles of accessibility, openness, and citizenship.

Sustainability is a common goal in the digitalization strategies of cities worldwide, with experiences we have discussed such as Singapore and Barcelona aligning with global recommendations for carbon reduction and ensuring the sustainability of urban spaces. At the same time, the initiative of the city of Dubai regarding the energy network aligns with international efforts to ensure the optimization of energy resources and intelligent and innovative management in the energy sector [62].

In this regard, we observe that Morocco's initiative, despite the needs, also constitutes an important piece in the global strategy for the sustainability of human spaces and specifically urban areas. The renewable energy initiatives and Morocco's projects to optimize its energy resources, such as the Noor Ouarzazate station, constitute a strong testament that promotes sustainability. We hope that similar efforts will be deployed at the urban level to overcome the shortcomings and the needs of urban digitalization in Morocco (Figs. 4 and 5 and Table 1).

To achieve the necessary efficiency and the benefits of digitizing Moroccan cities and to ensure the effective use of security technologies in these urbanization and modernization processes. A concrete plan must be implemented that takes into account the challenges and potential risks highlighted by these applications, while also providing solutions and responses for a confident, practical, and responsible application. In the following paragraphs, we present to the readers the fruits of our labor in the form of recommendations and key guidelines that enable interested parties to successfully and effectively implement these processes without undesirable consequences.

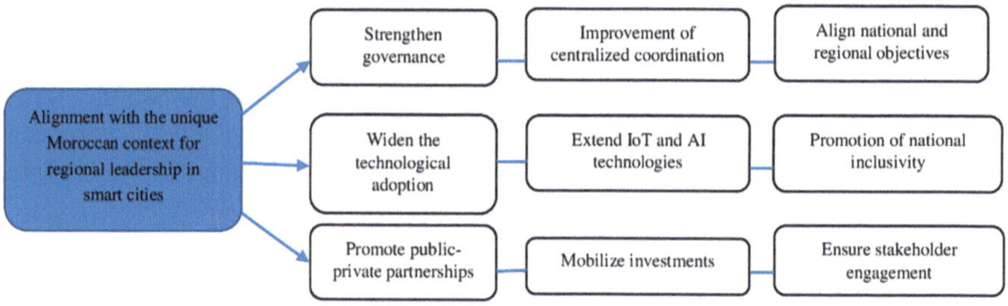

Fig. 4 The graph represents the Moroccan steps for a regional positioning and leadership in Smart Cities

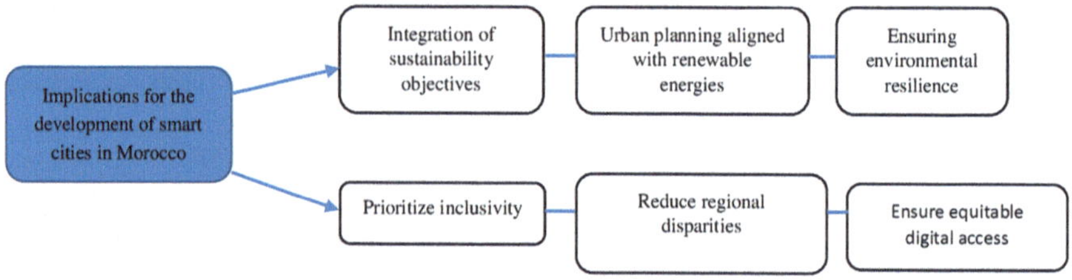

Fig. 5 The graph represents the implication for the development of Smart Cities in Morocco

Table 1 The table demonstrates the main dimensions and specificities detected among the examples of smart cities adopting IoT

Dimensions	Morocco	Singapore	Barcelona	Dubai
Governance	Decentralized, regional initiatives	Centralized under "Smart Nation"	Public-private collaboration	Strong centralized leadership
Technological Infrastructure	Selective implementation, regional focus	Predictive analytics, IoT	IoT-driven urban platforms	Blockchain integration
Socioeconomic Inclusivity	Addressing disparities, limited resources	High GDP per capita, universal access	Emphasis on inclusivity, open platforms	High investments, luxury-driven focus
Sustainability	Focus on renewable energy	Green building initiativs	Reducing carbon footprint	Smart energy grid
Global Partnerships	Key to progress	Extensive collaborations	Strong EU ties	Broad international engagement

6.3 Developing a Robust National Urban Security Strategy Based on Technological Integration

The national cyber strategy constitutes a step forward regarding the improvement and security of the digital environment in Morocco. However, it remains silent regarding the massive development of smart city projects in Morocco and does not contain any specific provisions related to the development of these projects in the future. For this reason, Morocco should further develop its vision and strengthen its national cyber strategy by introducing plans and programs related to smart cities. Morocco is therefore expected, in case of a different vision, to establish a national urban security strategy that takes into account the objectives of regional plans for urban cities.

The recommended planning, in terms of national strategy, should prioritize the integration of technological devices that help improve citizens' lives and facilitate their access to urban services. The sensors of surveillance cameras, biometric facial recognition systems, coupled with the potential of AI in the analysis and processing of big data. All these technologies have demonstrated unprecedented success in addressing various challenges and dangers that can threaten the stability and security of modern cities around the world, including moroccan cities. The city of Singapore is the most recurring example of the effectiveness of facial recognition technologies, not only for managing the flow of travelers and individuals in mass transit areas, but rather. These technologies have helped community authorities target undesirable activities in these human gatherings and address them quickly and efficiently [46].

The city of Barcelona has also made a brilliant experience by following and adopting its advantages. The city has adopted a lighting adjustment system that adapts to human activities in the area surrounding the streetlights. These light sensors have not only reduced energy consumption rates but have also improved public safety by ensuring welllit areas during peak hours and addressing the needs of residents in dark zones [12].

Morocco can also take into account, when implementing such a strategy, the adoption and operation of urban security technologies, community guidelines, and integrating public consultations within the urbanization processes of smart cities. We assume that these processes would help establish trust between the components of society, whether they are authorities or citizens, and encourage local authorities to respect the concerns and priorities of local populations.

6.4 Diversing the Collaboration and Research Partnerships for Smart City Development

Although the national cyber strategy still recommends central partnerships and interstate cooperation, we find that the diversification of collaboration, especially with the private sector, constitutes the cornerstone in the processes of developing smart cities in the future [38]. The presence of partnerships between the public and private sectors is also recommended, as these hybrid agreements are essential and ubiquitous in current discussions regarding project financing and management. Moreover, collaborations between technology companies, research centers, and community authorities in Morocco appear to be very effective in implementing a plan for integrating urban security technologies into a smart urbanization and transformation strategy for a Moroccan city [63].

Par exemple, les partenariats et collaborations entre les différents acteurs de Dubaï ont facilité le déploiement de systèmes de surveillance avancés et de technologies de reconnaissance faciale, car ces entités ont efficacement tiré parti de leur expertise et de leurs recherches de pointe [18]. In this context, Morosco can benefit from these experiences in collaboration and cooperation to achieve its goals of developing plans and strategies and to make its urbanization processes more effective and efficient.

The implementation of partnerships with research institutions such as universities, research centers, or educational establishments can also stimulate innovation in research related to urban security technologies. Well-known institutions around the globe offer their support to countries that wish to engage in the processes of smart cities [64]. The moroccan authorities, can also take advantage of these proposals and offers to enhance the effectiveness of its plans and strategy for transforming moroccan cities, on par with participation in global summits and forums addressing issues related to urban security technologies [55].

6.5 Evolving Technological Infrastructure and Minimizing the Digital Divide for More Inclusivity

In establishing the previous steps, we also assume that Morocco takes the digital divide seriously in its national cybersecurity strategy and deploys enormous capacities to reduce this obstacle. L'écart entre l'infrastructure encore en développement du Maroc et le niveau sophistiqué des nouvelles technologies de sécurité urbaine peut imposer des perturbations dans les processus d'intégration. Ainsi, la mise à niveau des infrastructures est essentielle avant de lancer des plans et des stratégies d'urbanisation intelligente en général, et spécifiquement pour l'introduction des technologies de sécurité.

The national cyber strategy, as an official planning full of ambitions and objectives aimed at improving infrastructure and repositioning Morocco among the advanced countries in national cybersecurity, confirms the intention of this strategy to develop in the future all other projects related to its objectives, including smart cities [38]. Il est également important d'impliquer les citoyens et de sensibiliser aux nouvelles technologies dans leur environnement, car la littératie numérique et technologique peut également imposer des perturbations dans le fonctionnement et la stabilité des zones défavorisées.

Alongside the ambitious official plans to develop training and education offerings within Moroccan establishments and institutions, the Moroccan Center Code 212, as a new ministerial initiative, constitutes a significant step in the process of training national skills and experts in the fields of cybersecurity, programming, and IoT [65]. This initiative should be generalized not only to university institutions but also to vocational training centers and, why not, to secondary education establishments, in order to multiply its benefits and positive impacts in the field of education and digital training of experts and specialists, who can constitute in the future regions in Morocco, the professional basis for successfully implementing projects related to the generalization of smart city plans in Morocco.

Also, Mobile educational training units can move to all regions of the affected city to inform citizens about the changes that will impact their activities after the implementation of urban security technologies, and also explain the benefits of this integration process. Public institutions and establishments equipped with Internet access and other support materials can help raise awareness among citizens and serve as training and information dissemination centers. By anticipating that the majority of citizens will be able to access and become aware of all urban security technologies, this will promote integration opportunities and maximize the benefits of smart city initiatives [35].

The approach of involving citizens in the management and planning affairs of smart cities is crucial. This process offers great potential for community authorities to undertake joint initiatives and involve citizens in local development and planning policies. According to lived experiences, citizen involvement shapes the development of smart security applications in a social way that respects the issues of the entire community. And also improves citizens' access to public services, as well as the collaboration of individuals with the authorities when reporting incidents [66].

In the local application of these ideas, moroccan cities have aimed to develop citizen access to local affairs, especially works related to the issues under discussion. Digital platforms for consultations and public survey requests, to understand the opinions and concerns of citizens and interested civil organizations, are entirely necessary to develop urban security initiatives. These processes strengthen transparency and trust, as well as feelings of ownership and belonging to smart city projects.

6.6 Comprehensing the Large Moroccan Context of Socio-Economic and Ethical Implications of Urban Security Technologies

Before concluding this work, it is essential to evaluate the long-term implications of urban security technology in order to understand how it generates impacts on social stability and justice. In addition to the above, it should be noted that this assessment anticipates the impacts on economic development and public safety. We see that these technologies will not only improve citizen participation, the efficiency of public services, the performance of technological systems in urban infrastructure, etc. However, the integration of urban security technologies, without the necessary considerations, especially the requirements of social inequality in targeted areas, can lead to unacceptable impacts that exacerbate existing inequalities.

Morocco, in this regard, should develop a broader approach that takes into account social requirements, especially social inequalities, development in targeted areas, the integration of the targeted community, etc. Policies for the integration of public safety technologies into the digitization processes and development of smart cities must prioritize social equity, citizen integration, the sustainability of smart city initiatives, and several other contextual issues in order to avoid inequalities in access to enhanced security services, and also to develop solutions that minimize the environmental impact of urban security technologies.

Morocco should finally draw inspiration from international practices and take into account foreign experiences in the field. The development of public-private partnerships is also essential to promote the participation of citizens and other interested stakeholders. This broad integration policy will minimize the digital divide on both technical and social fronts, and will give Morocco the opportunity to develop a solid urban security framework that combines public security efficiency with urban development (Figs. 6 and 7).

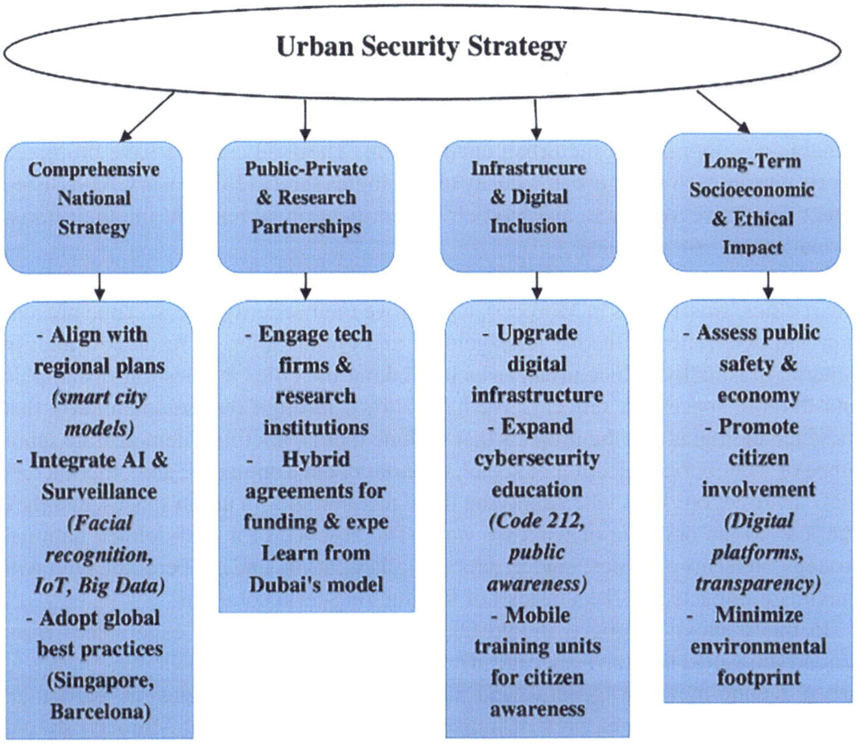

Fig. 6 The graph demonstrates the main ideas for building a solid and sustainable urban strategy for the implementation of urban security technologies

Fig. 7 The figure illustrate the most frequent indicators within the theme of governance in Smart Cities

7 Conclusion

The processes of integrating urban security technologies present an opportunity to improve public safety, optimize governance, and also promote the growth and sustainable development of the urban environment. These advantages have prompted Moroccan cities to adopt urban security technologies such as IoT sensors, AI-assisted direct camera surveillance, and biometric identification systems during their transformation into smart cities.

These systems can help, according to the previous experiences of international smart cities, smart cities in Morocco, to improve their security levels, crime prevention and deterrence, effective management and optimization of resources, and, in general, drastically reduce urban risks by making the cities in question sustainable and durable spaces for citizens. We have shown through our argumentation that these technologies and the projects that include them sometimes demonstrate ignorance or a deliberate rejection of social, economic, environmental, and other factors that can affect our lives when applying these technologies in urban spaces, and they can lead to the opposite of what we want. The absence of a well-defined national strategy, in terms of ethics, law, freedoms, collaboration, and cooperation, etc., will maximize impacts and risks instead of finding long-term solutions.

In this context, Morocco should take into consideration the experiences from around the world that have successfully overcome their urban challenges thanks to urban technologies in general and urban security technologies in particular. Singapore and Barcelona present indispensable experiences that Morocco should follow by giving more consideration to the specificities and requirements of the context in order to strengthen public trust in its innovations in management and planning. These smart cities use specific communication processes to strengthen citizens' trust in the methods of using and protecting their data and personal information. These experiences are based on the idea of involving citizens and local communities in the processes of deploying the technologies in question, which, according to proponents of this idea, will ensure that the improvements continuously meet the needs and expectations of the citizens.

To facilitate this task, Morocco, in its process of deploying urban security technologies, should consider forming, for example, a national smart cities committee or an interministerial or intercommunal council dedicated to addressing issues related to moroccan smart cities. These initiatives can ensure the effective implementation of the digitization processes of moroccan cities, facilitate coordination between community authorities at the national level, and also align the concerned cities with the common objectives of such a national smart city strategy. This initiative will serve as a national platform, sharing responses and information, facilitating the integration of its public and private sectors. This platform will also encourage the development of partnerships between domestic companies in this field and international companies, as well as research institutes that manage such projects on a global scale.

Morocco must also base its strategy for deploying urban security technologies on the priorities of its citizens, where overcoming the social challenges of the digital divide in the country also involves digital literacy for citizens and ensuring equal access to technologies, particularly in rural areas. This can be achievable if Morocco mobilizes the necessary skills and operates centers and institutions such as the Morocco Code 212 initiative, assigning them the role of training and education to ensure the effective engagement of young people and citizens in general in the processes of integrating urban security technologies and also their digital and social inclusion.

This strategy should be an extension of the national cyber strategy in the urban domain, focusing on citizen participation in the deployment processes of urban security technologies by encouraging individuals to engage and interact with the debates and discussions led by community authorities to ensure that societal concerns and worries are taken into account. The examples we provided during our analyses reinforce the idea of the importance of engaging citizens in these processes in order to give a community sense to these initiatives. So, if Morocco adopts these approaches by emphasizing citizen participation instead of solely focusing on the massive introduction of public security technologies without awareness of societal risks, it will help the country overcome several challenges and risks and achieve these goals with full social inclusion and involvement.

Moreover, urban security technologies can pose certain environmental challenges. For this, Morocco should introduce measures in its strategy for deploying urban security technologies to minimize the impacts of energy consumption by prioritizing smart lighting solutions. And regarding the accumulation of electronic waste, the adoption of policies and plans to recycle and reuse damaged equipment will also reduce the environmental impacts and implications of urban security technologies.

Finally, we believe that long-term and strategic planning in forecasting the impacts of the use and adoption of urban security technologies will also reduce social disparities and strengthen equity among social categories, and will also reduce inequalities between targeted areas within cities despite economic, cultural, social, etc. variations. Therefore, investing in more innovative solutions, is essential to ensuring sustainable and consistent growth. Morocco will succeed in its projects if it understands that only the adoption of a comprehensive, open, and inclusive urban security strategy is the planned solution to build smart cities parallel to the country's ambitions in all areas.

References

1. UNDP Global Centre for Technology, Innovation, and Sustainable Development, Meddeb R (2021) Handbook on smart urban innovations. UNDP. https://www.undp.org/sites/g/files/zskgke326/files/2021-11/UNDP-Handbook-on-Smart-Urban-Innovations-V2.pdf
2. Townsend MA (2013) Smart cities: big data, civic hackers, and the quest for a new utopia, 1st edn. Norton & Company, New York
3. Khalifi H, Riahi S, Cherif W (2024) Smart cities and sustainable urban development in Morocco. Ingénierie Des Systèmes D'Information 29(2). https://doi.org/10.18280/isi.290236
4. Sutanto Y (2024) Evaluating the effectiveness and ethical considerations of CCTV surveillance in public spaces: a cybersecurity perspective. Int J Inf Syst Innov Technol 3(1):7–14. https://doi.org/10.63322/2kfhm657
5. Ferreira Diogo B (2024) The future of smart cities: how digital technologies are transforming urban living in Barcelona and Dublin. The Arch Daily. https://www.archdaily.com/the-future-of-smart-cities-how-digital-technologies-are-transforming-urban-living-in-barcelona-and-dublin. Accessed 23 Sept 2024
6. Fontes C, Hohma E, Corrigan CC, Lütge C (2022) AI-powered public surveillance systems: why we (might) need them and how we want them. Technol Soc 71:102–137. https://doi.org/10.1016/j.techsoc.2022.102137
7. Socha R, Kogut B (2020) Urban video surveillance as a tool to improve security in public spaces. Sustainability 12(15):6210. https://doi.org/10.3390/su12156210
8. Rochdane and Hamdani (2018) Role of smart small and medium enterprises in the development of smart cities: case of Casablanca Smart-City Project—Morocco. J Int J e-Educ e-Bus e-Manag e-Learn 8(3):181–187. https://doi.org/10.17706/ijeeee.2018.8.3.181-186
9. Albino V, Berardi U, Dangelico RM (2015) Smart cities: definitions, dimensions, performance, and initiatives. J Urban Technol 22(1):3–21. https://doi.org/10.1080/10630732.2014.942092
10. Michael, James (2020) The future of cities: how digital and IoT technologies are reshaping urban life. McKinsey Global Institute. https://www.mckinsey.com/~/media/mckinsey/business%20functions/operations/our%20insights/smart%20cities%20digital%20solutions%20for%20a%20more%20livable%20future/mgi-smart-cities-full-report.pdf. Accessed 8 Jun 2024
11. Sussman G (2022) Vincent Mosco 2019: the smart city in a digital world. Emerald Publishing. Academia edu, Bingley. https://www.academia.edu/77624180/Vincent_Mosco_2019_The_Smart_City_in_a_Digital_World_Bingley_Emerald_Publishing. Accessed 2 Jul 2024
12. Capgemini Research Institute (2020) Putting the citizen at the Center of Smart City Initiatives. Street Smart. Report. https://www.capgemini.com/wp-content/uploads/2020/07/Street_Smart_28-07-2020/SS.pdf. Accessed 23 Sept 2024
13. Costa DG, Bittencourt JCN, Oliveira F, Peixoto JPJ, Jesus TC (2024) Achieving sustainable smart cities through geospatial data-driven approaches. J Sustain 16(2):640. https://doi.org/10.3390/su16020640
14. Fouzi A (2024) Casablanca Smart City 2024 launches with ambition to shape sustainable urban development. Morocco World News. https://www.moroccoworldnews.com/2024/06/363062/casablanca-smart-city2024-launches-with-ambition-to-shape-sustainable-urban-development. Accessed 17 Aug 2024
15. Ferrer JR (2017) Barcelona's smart city vision: an opportunity for transformation. J Field Actions 16(Suppl 1):70–77. https://journals.openedition.org/factsreports/4367
16. Demertzi V, Demertzis S, Demertzis K (2023) An overview of cyber threats, attacks and countermeasures on the primary domains of smart cities. Appl Sci 13:790. https://doi.org/10.3390/app13020790
17. Elmaghraby AS, Losavio MM (2014) Cyber security challenges in smart cities: safety, security and privacy. J Adv Res 5(4):491–497. https://doi.org/10.1016/j.jare.2014.02.006

18. The Guardian (2019) The Guardian view on facial recognition: a danger to democracy. The Guardian. https://www.theguardian.com. Accessed 7 Oct 2024
19. Commission Nationale de Controle et de la Protection des Données à Caractère Personnel (2023) Loi 09-08 - CNDP. https://www.cndp.ma/loi-09-08/. Accessed 12 Nov 2024
20. Wang X, Wu YC, Zhou M, Fu H (2024) Beyond surveillance: privacy, ethics, and regulations in face recognition technology. Front Big Data 7:1337465. https://doi.org/10.3389/fdata.2024.1337465
21. Belguidoum S, Cattedra R, Iraki A (2015) Cities and urbanity in the Maghreb. L'Année Du Maghreb 12(12):11–32. https://doi.org/10.4000/anneemaghreb.2510
22. Lall S, Mahgoub A, Maria A, Touati A, Acero J (2019) Leveraging urbanization to promote a new growth model while reducing territorial disparities in Morocco: urban and regional development policy note, 1st edn. World Bank, Washington, DC. https://doi.org/10.1596/978-1-4648-1433-4
23. Woetzel L, Remes J, Boland B, Lv K, Sinha S, Strube G, Means J, Law J, Cadena A, Von Der Tann V (2018) Smart cities: digital solutions for a more livable future. McKinsey & Company. https://www.mckinsey.com/capabilities/operations/our-insights/smart-cities-digital-solutions-for-a-more-livable-future. Accessed 22 Jul 2024
24. World Economic Forum (2016) Inspiring future cities & urban services: shaping the future of urban development and services initiative. https://www3.weforum.org/docs/WEF_Urban-Services.pdf. Accessed 12 Oct 2024
25. Thakuriah, Tilahun, Moira L (2015) Big data and urban informatics: innovations and challenges to urban planning and knowledge discovery. In: Seeing cities through big data: research, methods and applications in urban informatics. Springer, New York, pp 4–33. https://doi.org/10.1007/978-3-319-40902-3
26. BrittJones (2024) Balancing security and privacy: ethical considerations in video surveillance systems—Security Middle East Magazine. Security Middle East Magazine. https://www.securitymiddleeastmag.com/security-privacy-video-surveillance/. Accessed 11 Aug 2024
27. Maphosa V (2024) The rise of artificial intelligence and emerging ethical and social concerns. AI Comput Sci Robot Technol 3. https://doi.org/10.5772/acrt.20240020
28. Khan N, Marina E (2021) The use of biometric technology at airports: the case of customs and border protection (CBP). Int J Inf Manag Data Insights 1(2):100–049. https://doi.org/10.1016/j.jjimei.2021.100049
29. Al-Kodmany K (2012) Sentient city: ubiquitous computing architecture and the future of urban space. J Urban Technol 19(3):137–140. https://doi.org/10.1080/10630732.2012.744599
30. Nakar S, Greenbaum D, Family ZM, Liana G, Carmel I (2017) Now you see me. Now you still do: facial recognition technology and the growing lack of privacy. J Sci Technol Law 23(7):88–123. https://www.bu.edu/jostl/files/2017/04/Greenbaum-Online.pdf
31. Bounoua L, Lachkham MA, Ed-Dahmany N, Lagmiri S, Bahi H, Messouli M, Khebiza MY, Nigro J, Thome KJ (2024) Urban sustainability development in Morocco: a review. Urban Sci 8(2):28. https://doi.org/10.3390/urbansci8020028
32. Global Cities Hub (2024) International Telecommunication Union (ITU)—Global Cities Hub. Global Cities Hub. https://globalcitieshub.org/en/international-telecommunication-union-itu/. Accessed 5 Jun 2024
33. World Economic Forum (2020) The rise of smart cities: challenges for urban security. WEF policy brief. https://www.weforum.org/reports/shaping-the-future-of-urban-development-and-services. Accessed 2 Sept 2024
34. Abdelkhalek T, Ajbilou A, Benayad M, Boccanfuso D, Savard L (2021) How can the digital economy benefit Morocco and all Moroccans? Economic Research Forum. https://www.erf.org.eg/app/uploads/2021/11/163122/724/832622.pdf. Accessed 19 Oct 2024
35. Kitchin R (2016) The ethics of smart cities and urban science. J Philos Trans 374(2083):20160115. https://doi.org/10.1098/rsta.2016.0115
36. Kolotouchkina O, Laura R, Warda B (2024) Smart cities, digital inequalities, and the challenge of inclusion. J Smart Cities 7(6):2256–2270. https://doi.org/10.3390/smartcities7060130

37. Ortiz JL (2001) Gran Hermano is watching. Wired. https://www.wired.com/2001/09/gran-hermano-is-watching/. Accessed 7 Aug 2024
38. Ministry of Defense (2023) National Cybersecurity Strategy 2030. https://www.dgssi.gov.ma/fr/publications/national-cybersecurity-strategy. Moroccan Government. Accessed 3 Jun 2024
39. Abdennasser T, Alaoui S, Chlioui I, Hnini A (2025) IoT architecture for inclusive urban mobility: a design science research approach to sustainable transportation in Morocco. Eng Proceed 112(1):46. https://doi.org/10.3390/engproc2025112046
40. Komninos N, Pallot M, Schaffers H (2013) Special issue on smart cities and the future internet in Europe. J Knowl Econ 4(2):119–134. https://doi.org/10.1007/s13132-012-0083-x
41. Zaanoun A (2023) Morocco: the impact of the digitization of public services. Arab reform initiative. https://www.arab-reform.net/publication/morocco-the-impact-of-the-digitization-of-public-services/. Accessed 22 Oct 2024
42. Gil-Alonso F, López-Villanueva C, Bayona-i-Carrasco J, Pujadas I (2021) Towards an even more spatially diversified city? New metropolitan population trends in the post-economic crisis period. Urban Sci 5(2):41. https://doi.org/10.3390/urbansci5020041
43. Lee J, Babcock J, Pham TS, Bui TH, Kang M (2022) Smart city as a social transition towards inclusive development through technology: a tale of four smart cities. Int J Urban Sci 27(Suppl 1):75–100. https://doi.org/10.1080/12265934.2022.2074076
44. Urban Solutions (2022) Centre for liveable cities. https://www.clc.gov.sg/docs/default-source/urban-solutions/urban-solutions-20-the-science-of-cities. Accessed 2 Jun 2024
45. Ministry of Foreign Affairs, Agency for Science, Technology and Research, Agri-Veterinary and Food Authority of Singapore, Centre for Liveable Cities, Department of Statistics, Economic Development Board, Energy Market Authority, et al (2018) Ministry of foreign affairs a sustainable and resilient singapore towards Singapore's voluntary national review report to the 2018 UN high-level political forum on sustainable development. Report. https://www.mfa.gov.sg. Accessed 21 Jun 2024
46. Patruni M, Deebak BD (2022) Security and privacy issues in smart cities/industries: technologies, applications, and challenges. J Ambient Intell Humaniz Comput 14(8):1–37. https://doi.org/10.1007/s12652-022-03707-1
47. Morrison R (2021) Singapore's smart city transformation: lessons for the world. Financial Times. https://www.novatr.com/blog/singapore-world-smartest-city. Accessed 6 Oct 2024
48. Liu H, Wang T-Y (2021) The institutionalization and effectiveness of transnational policy transfer: The China–Singapore Suzhou Industrial Park as a case study. Public Adm Dev 41(3):103–114. https://doi.org/10.1002/pad.1956
49. Alvarez-Palau EJ, Martí-Henneberg J, Solanas-Jiménez J (2019) Urban growth and long-term transformations in Spanish cities since the mid-nineteenth century: a methodology to determine changes in urban density. J Sustain 11(24):6948. https://doi.org/10.3390/su11246948
50. Gascó M (2018) Building a smart city: lessons from Barcelona. Commun ACM 61(4):50–57. https://doi.org/10.1145/3117800
51. Toussi (2021) Shifting from oil: how the UAE offers a model for building a technology ecosystem. Forbes Technology Council. https://www.forbes.com/councils/forbestechcouncil/2021/12/06/shifting-from-oil-how-the-uae-offers-a-mo del-for-building-a-technology-ecosystem/. Accessed 8 Sept 2024
52. Telecom Review Asia (2024) Inside China's smart city boom: leveraging AI, 5G, and IoT for urban innovation. https://www.telecomreviewasia.com/news/featured-articles/4080-inside-china-s-smart-city-boom-leveraging-ai-5g-and-iot-for-urban-innovation. Accessed 6 Oct 2024
53. Zaabi SHA, Zamri R (2022) Managing security threats through touchless security technologies: an overview of the integration of facial recognition technology in the UAE oil and gas industry. J Sustain 14(22):1–32. https://doi.org/10.3390/su142214915
54. Ministère du Transport de Tunisie (2018) Institution Financière Et Solidaire Vers Une Mobilité Urbaine Durable En Tunisie, Une Coopération Technique Au Service Du Changement. Gouvernement de Tunisie. https://www.codatu.org/wp-content/uploads/2023/04/doc-valorisation-tunis.pdf. Accessed 18 Jun 2024

55. Jordan Times (2024) Experts suggest use of 'intelligent' transportation systems to address Amman's traffic congestion. https://jordantimes.com/news/local/experts-suggest-use-intelligent-transportation-systems-address-ammans-traffic-congestion
56. Das D, Kwek B (2024) AI and data-driven urbanism: the Singapore experience. J Digit Geogr Soc 7(1):100104. https://doi.org/10.1016/j.diggeo.2024.100104
57. Burns MC, Marmolejo C, Roca J, Chica E, Freixas J (2013) The case for agglomeration economies in Europe. Department of Political Science, Universitat de Barcelona. https://archive.espon.eu/sites/default/files/attachments/CAEE_Barcelona_Case_Study_Appendix_C4_FINAL.pdf. Accessed 1 Oct 2024
58. Al-Dabbagh R (2022) Dubai, the sustainable, smart city. Renew Energy Environ Sustain 7(3):1–12. https://doi.org/10.1051/rees/2021049
59. Allaki O, El Ghannouchil I, El Moudden A (2023) The pseudo smart city in Morocco possibilities, potential and opportunities for a new regional development model. Int J Strateg Manag Econ Stud 2(4):1325–1333. https://doi.org/10.5281/zenodo.8252556
60. Rharbi N, İnceoğlu M (2022) Moroccan new green cities, towards a green urban transition. J Islam Architect 7(2):296–305. https://doi.org/10.18860/jia.v7i2.17222
61. Ibourk A, Raoui S (2021) Scale effect on territorial disparities of sustainable human development in Morocco: a spatial analysis. Discov Sustain 2(57):1–20. https://doi.org/10.1007/s43621-021-00068-1
62. Veloso Á, Fonseca F, Ramos R (2024) Insights from smart city initiatives for urban sustainability and contemporary urbanism. J Smart Cities 7(6):3188–3209. https://doi.org/10.3390/smartcities7060124
63. Trebilcock and Rosenstock (2015) Infrastructure public–private partnerships in the de-veloping world: lessons from recent experience. J Dev Stud 51(4):1–20. https://doi.org/10.1080/00220388.2014.959935
64. Raghavan S, Boung Yew SL, Lee Loong Y, Tan W, Kee KK (2019) Data integration for smart cities: opportunities and challenges. In: "Computational science and technology", 6th ICCST 2019, Kota Kinabalu, Malaysia, 29-30 August 2019. Springer Nature, Berlin, pp 393–403. https://doi.org/10.1007/978-981-15-0058-9_38
65. Ministery of Higher Education (2024) Mr. Minister inaugurates the 'Code 212' Center at Ibn Tofail University. Moroccan Government. https://www.enssup.gov.ma/index.php/en/actualites/monsieur-le-ministre-inaugure-le-centre-code-212-a-l universite-ibn-tofail. Accessed 8 Sept 2024
66. Tutak M, Brodny J (2025) Stakeholder collaboration and open innovation in smart cities: a Quattro Helix model for technological and social transformation. J Open Innov Technol Mark Complex 11(3):100594. https://doi.org/10.1016/j.joitmc.2025.100594

Efficient Knowledge Distillation for Smart City Surveillance: Detecting and Classifying Dangerous Weapons

Aicha Khalfaoui, Abdelmajid Badri, and Ilham El Mourabit

Abstract In modern society, ensuring public order and safety is of paramount importance. Detecting dangerous weapons like knives and pistols is crucial for preventing violent incidents and ensuring the safety of citizens. With the rise of smart cities, there is a growing need for advanced surveillance systems capable of real-time threat detection. Integrating AI technologies into urban infrastructures can significantly enhance public safety by enabling continuous monitoring and immediate threat identification.

This research underscores the significance of utilizing advanced technologies for dangerous weapons detection, focusing on the YOLOv5 architecture. YOLO (You Only Look Once) is a state-of-the-art object detection model known for its real-time processing capabilities, making it highly suitable for applications requiring rapid and accurate detection.

This study presents a knowledge distillation technique where YOLOv5x guides YOLOv5s, resulting in a lightweight model, YOLOv5s-x, with improved accuracy. The approach increases mean average precision (mAP) by 3.9%, from 92.3% to 96.2%, while reducing parameters and computations. YOLOv5s-x outperforms other models, like YOLOv8, offering better detection and classification of dangerous weapons for public safety applications.

Keywords Weapons detection · Yolo · Lightweight model · Knowledge distillation · Public safety

---

A. Khalfaoui (✉) · A. Badri · I. El Mourabit
Department of Electrical Engineering Faculty of Sciences and Techniques, Hassan II University, Mohammedia, Morocco

H. Kaddouri et al. (eds.), *Smart Cities for Sustainable and Urban Development*, Green Energy and Technology, https://doi.org/10.1007/978-3-031-99001-4_5

1 Introduction

Knowledge distillation empowers us to transfer knowledge from a larger, pre-trained model to our lightweight YOLOv5 model. This approach aims to strike a delicate balance between accuracy and efficiency, ensuring high detection performance even with a smaller and faster model. Recent studies have showcased the remarkable effectiveness of knowledge distillation in object detection tasks, demonstrably improving accuracy while reducing model size. We leverage these advancements to push the boundaries of weapon detection specifically for pistols and knives.

By focusing on knowledge distillation, we aim to achieve the following:

Reduced Model Size: We expect the distilled lightweight YOLOv5 model to be significantly smaller than the original pre-trained model, enabling deployment on edge devices with limited computational resources.

Maintained Accuracy: The knowledge distillation process aims to transfer the knowledge and accuracy of the pre-trained model to the smaller model, ensuring high detection performance for pistols and knives.

Real-Time Inference: The lightweight architecture and reduced size of the distilled model should facilitate real-time object detection, crucial for practical applications in security systems.

Our evaluation methodology involves meticulously training and testing the knowledge-distilled YOLOv5 model on a comprehensive dataset specifically curated for weapon detection. We will rigorously analyse the model's performance across various image and video conditions, focusing on metrics like accuracy, precision, recall, GFLOPs and number of parameters and layers. Additionally, we will compare the performance of the distilled model with the original pre-trained model and other baselines to evaluate the effectiveness of our approach.

Detecting potentially dangerous weapons in surveillance videos poses significant challenges. Previous research, such as the work by Fernandez-Carrobles et al. [1], has highlighted the issue of security personnel losing focus on CCTV after just 20 min. Another study by González et al. [2] found that in continuous video monitoring lasting 12 minutes, security guards could overlook over 45% of activities, and after 22 minutes, more than 95% of events might be missed. To address this, the application of deep learning for automated surveillance video monitoring has been explored, enabling the detection of critical events.

Research related to object identification in images or videos can be categorized into two main groups, as discussed by K.U. Sharma et al. One group focuses on traditional algorithms, excluding deep learning, to detect guns and knives. Meanwhile, the other group aims to enhance object detection accuracy through the utilization of deep learning methods. Classical algorithms often rely on color-based segmentation, corner detection, and appearance features for object recognition. However, these methods can be limited by the quality of frames, making it challenging to interpret frames with occlusion or noise. Deep learning algorithms, on the other hand, leverage neural networks, allowing for automatic feature extraction during training. This approach becomes particularly beneficial when dealing with occluded frames.

Previous studies, such as those by Kundegorski et al. [3] and Zhang et al. [4], have presented systems for scanning luggage with X-rays, utilizing various object recognition methods like sliding window CNN, YOLO, and Faster R-CNN. However, limitations were observed in terms of object detection accuracy under different lighting conditions and high occlusion. Techniques involving timed image-based CNN have been proposed to identify anomalies by Atto et al. [5].

In other studies, researchers have employed databases like the Internet Movie Firearm Database (IMFDB) by Verma et al. [6] and created custom databases for training algorithms. While some classification models showed promising results in static mode, their performance in real-time situations with limited resources was less accurate and slower. Although previous studies, like the one by Kanehisa et al. [7], have demonstrated the effectiveness of the YOLO algorithm in detecting firearms, their applicability in challenging scenarios and resource-constrained environments remains somewhat limited.

To overcome these challenges, our research involves a series of experiments and the development of a specialized CNN architecture. We leverage the YOLOv5 detection network, known for its fast and accurate performance. Our study introduces a novel network that significantly reduces computational complexity and network weight while maintaining high accuracy, addressing the limitations of previous models. The experimental results show that our method achieves faster convergence and a higher mean average precision (mAP) compared to the original YOLOv5, as demonstrated by Redmon et al., Redmon and Farhadi, Bochkovskiy et al. [8–11].

Target detection constitutes a foundational domain within deep learning methodologies, categorized primarily into single-stage and two-stage methods. The YOLO (You Only Look Once) algorithm family stands out as a prominent example of single-stage target detection algorithms, exerting a substantial influence on the realm of computer vision. Evolving over time, the YOLO algorithm has progressed to YOLOv8.

YOLO simplifies the target detection task by framing it as a single regression problem, directly providing the location and class label of the target. This design facilitates end-to-end target detection without necessitating an additional step for candidate region extraction, as seen in two-stage target detection algorithms like Faster RCNN, often preferred in various applications.

Introduced in 2020, YOLOv5 has gained widespread usage for diverse target detection tasks. Comprising four versions—YOLOv5s, YOLOv5m, YOLOv5l, and YOLOv5x.

–YOLOv5 adopts CSPNet as the backbone for feature extraction, utilizes the SPP module for multi-scale depth features, and employs the PANet-constructed feature pyramid for scale fusion. The network architecture remains consistent across all versions, with the network structure's size controlled by parameters like depth_multiple and width_multiple as shown in Fig. 1.

Of the four YOLOv5 versions, YOLOv5s boasts the fewest parameters, whereas YOLOv5x possesses the greatest number of parameters and the largest model size.

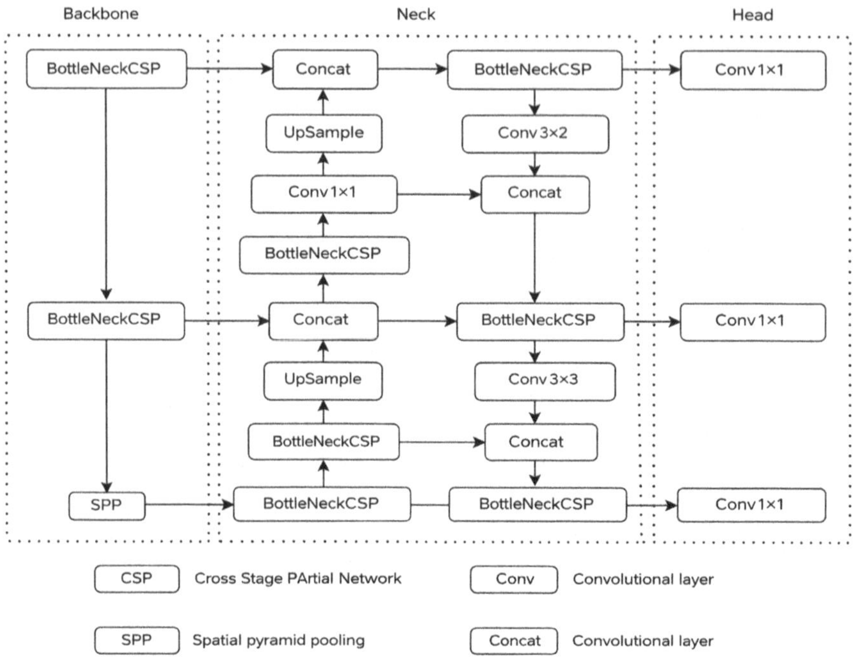

Fig. 1 The architecture of yolov5

Correspondingly, the detection accuracy of YOLOv5x surpasses that of YOLOv5s, prompting the selection of YOLOv5s as the benchmark model for this study, emphasizing the need for a lightweight model.

Using larger models is often beneficial for achieving increased accuracy within the same dataset. In various computer vision tasks, such as target detection and semantic segmentation, larger models typically exhibit superior performance. However, the drawback lies in the substantial hardware resource requirements and prolonged model inference times associated with larger models, posing practical challenges. Consequently, there is a growing focus on enhancing model accuracy while maintaining lightweight characteristics, which is crucial for practical applications.

Knowledge distillation stands out as a commonly employed technique to address the issue of model size. In contrast to approaches like pruning and quantized model compression, knowledge distillation involves the creation of a compact, lightweight model. This smaller model is then guided by a larger, more performance-oriented model to ensure its output closely aligns with that of the larger model, ultimately aiming for improved performance and accuracy. The primary objective is to enhance the accuracy of the smaller model without compromising computational efficiency.

In this study, the goal is to develop a lightweight dangerous weapons detection and classification model while enhancing its accuracy. Therefore, the choice was

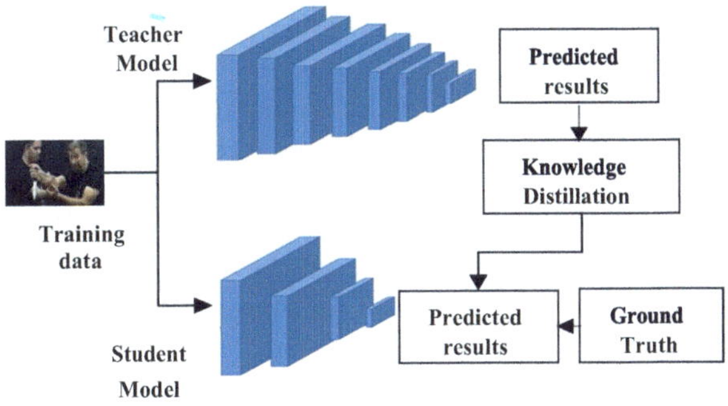

Fig. 2 Knowledge distillation process

made to employ YOLOv5x as the teacher network for knowledge distillation. YOLOv5x, having the highest detection accuracy and the largest number of parameters among the four versions of YOLOv5, serves as the guide for the process. On the other hand, YOLOv5s, with the smallest number of parameters, is selected as the student network for knowledge distillation. This approach ensures the improvement of YOLOv5s detection accuracy while preserving its lightweight characteristics (Fig. 2).

This paper explores the application of a lightweight YOLOv5 model, meticulously optimized through the knowledge distillation technique, for enhanced detection of pistols and knives. YOLOv5, renowned for its real-time capabilities and high accuracy, serves as a solid foundation for this project. However, we delve deeper, leveraging knowledge distillation to achieve efficient and accurate weapon detection on resource-constrained devices.

2 Methodology

2.1 Dataset Description

In this research, we employed Sohas dataset [12], which comprises 4014 images representing various weapons. The classification of these images is based on the handheld weapon type, specifically focusing on pistols and knives. For model training, 80% of the images were allocated to the training set, with the remaining 20% reserved for validation. The distribution of images across the two weapon categories was maintained at a roughly balanced ratio. To address the absence of blurry image samples, commonly encountered in surveillance video frames within the Sohas dataset, we implemented data augmentation techniques. These techniques involved introducing rotation, noise, and blur to diversify and expand the dataset, ultimately aiming to improve the model's ability to generalize effectively when handling images extracted from surveillance videos.

2.2 Evaluation Metrics

The evaluation of accuracy is conducted through the use of mAP (mean Average Precision) and Recall. Meanwhile, speed is assessed based on parameters (Params), the number of layers, and Floating-Point Operations (FLOPs).

The variables *TP*, *FN*, and *FP* denote true positives, false negatives, and false positives, respectively. The metric AP assesses the accuracy of an individual category, where N represents the total number of categories. mAP, on the other hand, measures the average accuracy across all categories in the dataset.

Vital indicators for evaluating a model's complexity and speed include GFLOPs and Params. FLOPs quantify the computational load of the network, while Params signify the number of learnable parameters in the model. Typically, higher values for GFLOPs and Params indicate a more intricate model, potentially leading to slower detection speed.

$$Recall = \frac{TP}{TP + FN} \tag{1}$$

$$Precision = \frac{TP}{TP + FP} \tag{2}$$

$$AP = \int_0^1 p(r)\,dr \tag{3}$$

$$mAP = \frac{\Sigma AP}{N} \tag{4}$$

2.3 Training Protocol

The investigation was carried out on a Linux operating system, utilizing an NVIDIA Tesla T4 graphics card and incorporating a software environment consisting of CUDA 11.1, PyTorch 1.12.1, and Python 3.7. The optimization process involved implementing the Stochastic Gradient Descent (SGD) algorithm, while retaining all other original YOLOv5 hyper parameters unchanged. To facilitate thorough training, the total number of epochs for the experiment was established at 100, allowing the model to converge and achieve optimal results.

3 Results and Discussion

To validate the efficacy of the proposed knowledge distillation method and ascertain its potential for enhancing accuracy while maintaining the structural integrity of the student model, this section performs comparative experiments using the Sohas dataset. These experiments are designed to offer a visual depiction of the variations in the student model's accuracy before and after the application of knowledge distillation.

Analysing Table 1, the comparison of various YOLO models highlights a trade-off between complexity, computational cost, and detection accuracy. YOLOv3-tiny, while efficient with 12.9 GFLOPS and 8.67 million parameters, achieves only 80.7% mAP, making it unsuitable for tasks requiring high precision, like weapon detection. YOLOv6 offers better accuracy at 89%, but its larger size (18.5 million parameters) and 45.17 GFLOPS limit its real-time use on resource-constrained devices. YOLOv7 and YOLOv8 deliver higher accuracy (94% and 96% mAP, respectively), but their computational demands make them less practical for real-time systems.

YOLOv5s offers a more efficient solution with 92.3% mAP, 15.8 GFLOPS, and 7.01 million parameters, ideal for embedded systems. However, YOLOv5s-AKD stands out by achieving 96.2% mAP—matching YOLOv8—while maintaining the same low computational cost and model size as YOLOv5s. This makes it the best model for our application, offering high accuracy and efficiency, ideal for real-time detection in resource-limited environments like video surveillance systems.

Figures 3, 4 and 5 show the precision, Recall and the Classification Loss of Yolov5s before and after Knowledge Distillation during the validation set. The tests include using pre-trained weights for YOLOv5s-AKD model.

Knowledge distillation aims to transfer knowledge from a larger or more complex teacher model to a smaller or simpler student model. Precision improves as the knowledge distillation process enables the student model to learn and generalize more effectively from the teacher model's knowledge. Furthermore, knowledge distillation enhances recall by leveraging insights from the teacher model, potentially improving the student model's ability to detect objects.

Classification loss serves as a measure of how well the model classifies objects. A decrease in classification loss suggests that the student model is becoming more accurate in assigning correct labels to objects in the images. Ultimately,

Table 1 Experimental results

Model	GFLOPS	Number of parameters	mAP50%
Yolov3-tiny	12.9	8,669,002	80.7
Yolov6	45.17	18,500,000	89
Yolov7	105.1	37,201,950	94
Yolov8	28.4	11,126,358	96
Yolov5s	15.8	7,015,519	92.3
Yolov5s-AKD	15.8	7,015,519	96.2

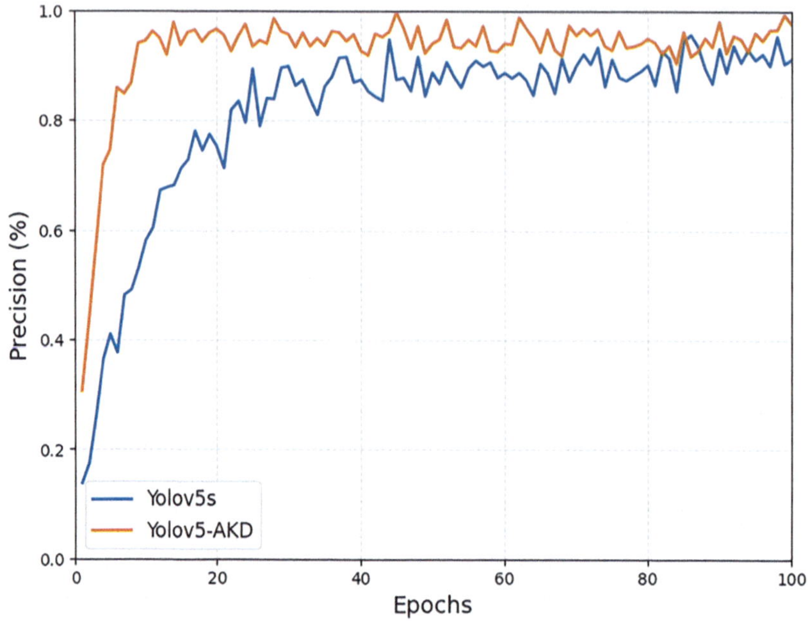

Fig. 3 Precision of Yolov5s and Yolov5-AKD during the validation set

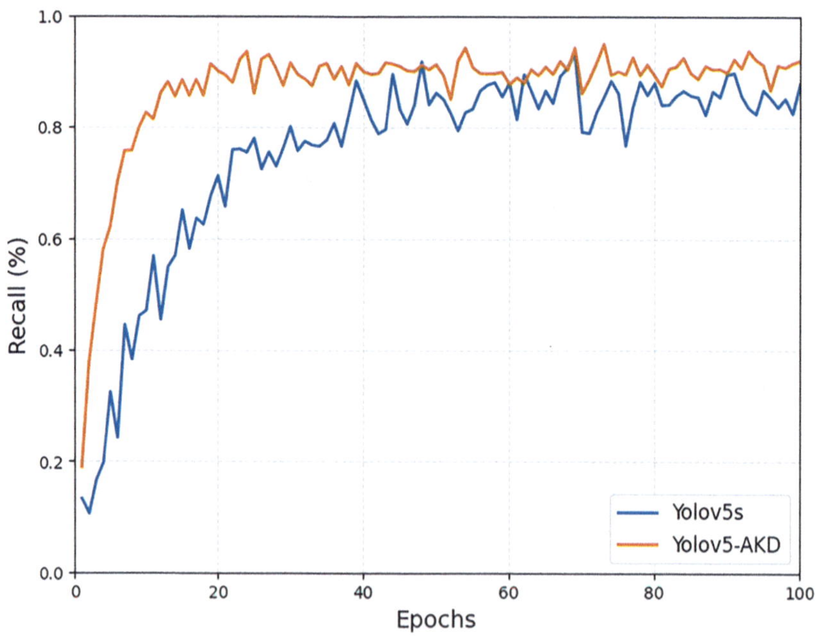

Fig. 4 Recall of Yolov5s and Yolov5-AKD during the validation set

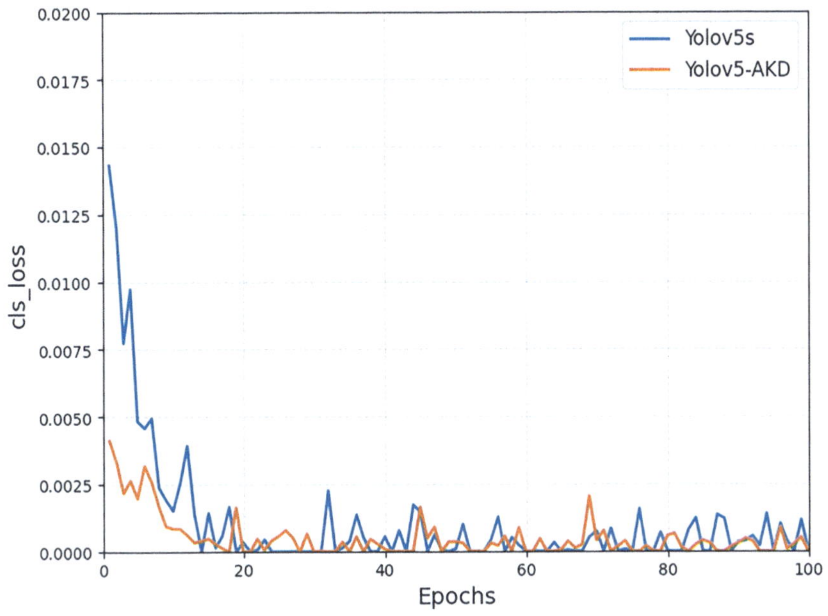

Fig. 5 Classification loss of Yolov5s and Yolov5-AKD during the validation set

knowledge distillation leads to a reduction in classification loss, and the student model learns to make more confident predictions by inheriting knowledge from the teacher model.

The integration of knowledge distillation with YOLOv5s as a student model and YOLOv5x as a teacher model holds great promise for enhancing dangerous weapons detection in the context of maintaining public order in smart cities. The use of YOLOv5x, a more sophisticated and capable teacher model, facilitates the transfer of crucial knowledge to the lightweight YOLOv5s, resulting in an optimized student model, YOLOv5s-AKD. This distilled model not only upholds a commendable level of accuracy in detecting dangerous weapons but also addresses the imperative need for computational efficiency. The application of this knowledge distillation framework in smart cities is particularly pertinent, where the demand for real-time and resource-efficient solutions is critical for public safety. The findings suggest that YOLOv5s-AKD could serve as a viable tool for law enforcement and security systems in smart cities, contributing to the swift and accurate identification of potential threats while accommodating the resource limitations inherent in urban environments. As smart city initiatives continue to evolve, the incorporation of knowledge distillation techniques tailored to dangerous weapons detection further underscores the potential of advanced technologies in upholding public safety and order. Continued research and refinement in this domain have the potential to shape future implementations of intelligent surveillance systems, promoting safer and more secure urban environments.

4 Conclusion and Future Work

This research presents an innovative approach to lightweight object detection through the application of knowledge distillation, facilitating rapid and accurate identification of dangerous weapons. Through the incorporation of knowledge distillation, the student model adeptly assimilates feature representations from the teacher model, leading to a notable enhancement in its overall performance.

The remarkable performance enhancements observed in YOLOv5s-AKD underscore its potential as a compelling choice for object detection applications. The model not only showcases a substantial improvement in accuracy when compared to its base version, YOLOv5s, but it also manages to achieve this without incurring the higher computational demands associated with YOLOv8. This is a significant finding, as it implies that knowledge distillation serves as a valuable technique in striking a balance between model efficiency and performance.

As perspectives, to address computational constraints, our research will focus on reducing model complexity through pruning and quantization, ensuring compatibility with edge and embedded devices used in smart city surveillance systems. Future work will apply these optimized models to detect and monitor suspicious behavior in public spaces, thereby enhancing the efficiency and effectiveness of public safety systems. Additionally, we aim to explore the integration of these models with emerging technologies such as real-time data analytics and IoT sensors to further improve surveillance accuracy and response times.

References

1. Fernandez-Carrobles MM, Deniz O, Maroto F (2019) Gun and knife detection based on faster R-CNN for video surveillance. In: Morales A, Fierrez J, Sánchez JS, Ribeiro B (eds) Pattern recognition and image analysis, lecture notes in computer science. Springer International Publishing, Cham, pp 441–452. https://doi.org/10.1007/978-3-030-31321-0_38
2. González S, Jose L, Zaccaro C, Álvarez-García JA, Soria Morillo LM, Caparrini FS (2020) Real-time gun detection in CCTV: an open problem. Neural Netw 132:297–308. https://doi.org/10.1016/j.neunet.2020.09.013
3. Kundegorski, M.E., S. Akcay, M. Devereux, A. Mouton, and T.P. Breckon. 2016. 'On using feature descriptors as visual words for object detection within X-ray baggage security screening'. In 7th international conference on imaging for crime detection and prevention (ICDP 2016), Madrid, Spain: Institution of Engineering and Technology, 12 (6 .)-12. https://doi.org/10.1049/ic.2016.0080
4. Zhang J, Xing W, Xing M, Sun G (2018) Terahertz image detection with the improved faster region-based convolutional neural network. Sensors 18(7):2327. https://doi.org/10.3390/s18072327
5. Atto AM, Benoit A, Lambert P (2020) Timed-image based deep learning for action recognition in video sequences. Pattern Recogn 104:107353. https://doi.org/10.1016/j.patcog.2020.107353
6. Verma GK, Dhillon A (2017) A handheld gun detection using faster R-CNN deep learning. In: Proceedings of the 7th international conference on computer and communication technology. ACM, Allahabad India, pp 84–88. https://doi.org/10.1145/3154979.3154988

7. Kanehisa R, Neto A (2019) Firearm detection using convolutional neural networks. In: Proceedings of the 11th international conference on agents and artificial intelligence. SCITEPRESS - Science and Technology Publications, Prague, Czech Republic, pp 707–714. https://doi.org/10.5220/0007397707070714
8. Redmon J, Divvala S, Girshick R, Farhadi A (2016) You only look once: unified, real-time object detection. In: 2016 IEEE conference on computer vision and pattern recognition (CVPR). IEEE, Las Vegas, NV, pp 779–788. https://doi.org/10.1109/CVPR.2016.91
9. Redmon J, Farhadi A (2016) YOLO9000: better, faster, stronger. http://arxiv.org/abs/1612.08242 (May 10, 2023)
10. Redmon J, Farhadi A (2018) YOLOv3: an incremental improvement. http://arxiv.org/abs/1804.02767 (December 20, 2023)
11. Bochkovskiy A, Wang C-Y, Liao H-YM (2020) YOLOv4: Optimal Speed and Accuracy of Object Detection. http://arxiv.org/abs/2004.10934 (May 10, 2023)
12. Pérez-Hernández F, Tabik S, Lamas A, Olmos R, Fujita H, Herrera F (2020) Object detection binary classifiers methodology based on deep learning to identify small objects handled similarly: application in video surveillance. Knowl-Based Syst 194:105590. https://doi.org/10.1016/j.knosys.2020.105590

Unlocking the Power of Real-Time Data: Advancing Predictive Analytics for Smarter City Management

Oumayma Berraadi, Hicham Gibet Tani, and Mohamed Ben Ahmed

Abstract Predictive models are the backbone of optimizing urban management for sustainability in smart cities. Therefore, this work stresses Energy Load Forecasting for Renewable Energy Management while highlighting real-time predictive models to balance supply with demand in a power network at different temporal levels with integrated renewable sources. As these renewable sources have variabilities, accurate forecasting has emerged as crucial for grid stability and optimizing resource distribution.

This paper aims to study in detail the machine learning algorithms, combined with real-time data obtained through IoT devices, that enable effective energy management strategies. In particular, this work focuses on:

- Predicting energy demand to guide decisions on energy source switching.
- Managing energy storage to capture and utilize excess renewable energy.
- Implementing energy-saving protocols during peak demand periods.
- Reducing operational costs by scheduling energy-intensive activities during off-peak hours.

These involve comparative performances of various prediction algorithms concerning accuracy, scalability, and responsiveness in real-time applications, including pointing out how models will contribute significantly to big data streaming with adequate integrations of devices to the IoT and cloud infrastructures.

While predictive analytics is of great benefit, there remain challenges with data quality, model scalability, and the dynamic nature of urban energy systems. In addressing these challenges and proposing novel solutions, this work contributes to exploring how predictive analytics can help improve smarter city management and facilitate sustainable energy practice.

O. Berraadi (✉) · M. Ben Ahmed
FSTT, Abdelmalek Essaadi University, Tetouan, Morocco
e-mail: oumayma.berraadi@etu.uae.ac.ma

H. Gibet Tani
FPL, Abdelmalek Essaadi University, Tetouan, Morocco

H. Kaddouri et al. (eds.), *Smart Cities for Sustainable and Urban Development*, Green Energy and Technology, https://doi.org/10.1007/978-3-031-99001-4_6

Keywords Internet of things · Machine learning · Real-time data streaming · Big data analytics · Predictive energy management · Renewable energy forecasting

1 Introduction

The concept of smart cities is relatively new, having emerged only in recent decades, and has gained traction with the development of the Internet of Things (IoT) and advancements in real-time data processing tools. These initiatives aim to enhance urban living by optimizing city performance, bolstering sustainability, and improving residents' quality of life [1–3]. A significant challenge faced by modern cities is the efficient provision of energy resources, particularly in balancing these demands with the integration of renewable energy sources. As cities transition toward greener energy solutions, managing the balance between energy demand, cost, and consistent supply presents a complex problem. This chapter examines how accurate predictions of energy consumption using machine learning (ML) models can revolutionize urban energy systems, making them not only more efficient but also sustainable [4, 5].

Studies have demonstrated that urban energy consumption is inherently dynamic, and influenced by factors such as weather, daily routines, and economic activity. Traditional methods of predicting energy demand, such as statistical models or rule-based systems, often fall short in capturing the intricate relationships among these variables. Furthermore, the increased reliance on renewable energy sources like solar and wind exacerbates the complexity due to their intermittent nature [1]. Machine learning has emerged as a promising solution by enabling models to leverage large volumes of historical data to forecast future energy needs with remarkable precision. Advanced ML models, such as Support Vector Machines (SVM), Recurrent Neural Networks (RNN), and Long Short-Term Memory (LSTM) networks, have shown significant potential in supervised learning tasks, with LSTM proving particularly effective in energy demand forecasting [1, 3].

Anticipating energy demand is crucial for maintaining a stable balance between supply and consumption. Accurate forecasts empower utilities to integrate and optimize renewable resources effectively, reducing operational costs and improving grid performance [3]. Smart grids, supported by real-time predictive modeling, enable demand-response strategies, energy-saving protocols, and load balancing during peak hours. These capabilities not only reduce costs but also promote sustainability by prioritizing renewable energy utilization and minimizing dependence on non-renewable resources [2].

Machine learning models excel in uncovering patterns within extensive datasets and generating forecasts grounded in historical consumption trends. LSTMs, in particular, are well-suited for analyzing sequential and time-series data, enabling smart cities to enhance energy management, predict maintenance needs, and integrate decentralized energy resources such as rooftop solar panels and localized wind turbines [6].

Real-time data collection from IoT devices, including smart meters, climate sensors, and remotely controlled appliances, significantly enhances the precision of energy demand forecasting. The detailed and time-sensitive nature of these data sources allows for the generation of localized forecasts that can be quickly adapted to current events. By combining real-time data with ML models, energy managers gain the ability to monitor and adjust forecasts in response to changing demand, weather fluctuations, or variations in renewable energy generation [7].

Despite these advancements, several challenges remain in applying ML models to energy forecasting on an urban scale. Issues related to data quality, computational requirements, and model evaluation must be addressed to ensure that these systems are practical and reliable for widespread deployment [8]. While ML models often perform well in controlled environments, their robustness can be compromised by outliers or unexpected scenarios, underscoring the need for rigorous validation and ongoing monitoring. Addressing these challenges will pave the way for high-performing energy forecasting systems that can drive sustainable energy management in smart cities [3, 9].

The subsequent sections explore the methodologies and machine learning algorithms commonly employed for energy load forecasting, examining their advantages and limitations. This study's methodology incorporates K-Nearest Neighbors (KNN) regression, LSTM networks, and additional models for predicting energy demand. By comparing the performance of these approaches across diverse evaluation metrics, this chapter aims to demonstrate how advanced analytics can empower cities to optimize energy management strategies, minimize costs, and enhance sustainability.

2 Background and Related Work

Accurate energy prediction is essential for maintaining the reliability, efficiency, and sustainability of modern energy infrastructures. Reliable forecasts empower grid operators, policymakers, and energy consumers to optimize resource allocation, minimize expenses, and seamlessly incorporate renewable energy sources into the system. This section delves into the primary challenges associated with energy prediction while analyzing the advancements and methodologies that address these complexities.

2.1 Key Challenges in Energy Forecasting

Energy forecasting faces several technical, environmental, and societal challenges, many of which are amplified by the growing complexity of energy systems and the shift toward renewable energy integration.

- **Data Quality and Availability:** Reliable energy forecasting requires extensive historical and real-time data, but many datasets suffer from incompleteness, inconsistencies, and noise. For instance, smart meter deployments offer high-resolution data, yet missing or erroneous records due to communication faults or equipment malfunctions often complicate analysis [10, 11]. Aggregated data may simplify forecasting but can obscure local variations, while disaggregated data demands advanced methods to handle individual consumption patterns [12].
- **Nonlinear and Stochastic Patterns:** Electricity demand is influenced by a wide range of factors, including weather conditions, socioeconomic trends, consumer behavior, and technological advancements. These relationships are often nonlinear and subject to sudden shifts, such as during economic recessions or extreme weather events. Traditional linear models struggle to capture these intricacies, necessitating the adoption of nonlinear approaches like Artificial Neural Networks (ANNs) and Long Short-Term Memory (LSTM) models [13–15].
- **Uncertainty and Variability:** The rise of renewable energy sources introduces significant uncertainty due to their intermittent nature. Solar and wind power outputs depend on highly variable weather patterns, complicating the task of balancing supply and demand. Probabilistic forecasting methods, while promising, require robust modeling to account for this uncertainty [11, 16].
- **Scalability and Real-Time Constraints:** The massive influx of data from IoT devices and sensors has increased the computational complexity of forecasting models. Real-time energy management systems require scalable models that can provide predictions with minimal latency [12, 17].
- **Interpretability of Models**: As machine learning (ML) and deep learning (DL) methods dominate energy forecasting, ensuring their interpretability becomes a major challenge. For instance, Shapley Additive Explanations (SHAP) are now widely used to assess feature contributions in complex models, enabling stakeholders to trust and refine predictions [17, 18].
- Integration with Renewable Energy: The intermittent nature of renewable energy sources, such as solar and wind power, presents unique challenges in aligning energy demand forecasts with variable supply. Accurate predictions must account for weather patterns and renewable energy generation dynamics [13, 19].

2.2 *Existing Solutions and Approaches*

Over the years, energy forecasting has evolved from traditional statistical models to advanced machine learning techniques, reflecting the increasing complexity and diversity of energy systems.

- **Case Studies and Applications**: Practical applications underscore the transformative potential of ML and DL techniques. For instance, a study integrating LSTM and SHAP to forecast residential energy consumption in Philadelphia revealed actionable insights into income distribution and building characteristics

[8]. Similarly, hybrid models like ANN-PSO demonstrated success in optimizing short-term demand forecasts for microgrids [2, 14, 20, 21].

- **Statistical Methods**: Classical statistical models like ARIMA and linear regression have been widely used in energy forecasting for their simplicity and interpretability. However, their limited ability to capture nonlinear patterns often reduces their effectiveness in complex scenarios [16].
- **Machine Learning Techniques**: ML approaches, particularly Artificial Neural Networks (ANNs), Long Short-Term Memory (LSTM) networks, and Extreme Gradient Boosting (XGBoost), have gained prominence due to their ability to handle nonlinearities and large datasets. These methods have shown promising results in predicting energy demand across various time scales [18, 19, 22, 23].
- **Hybrid Models**: Recent research has explored hybrid approaches, combining statistical methods with ML algorithms to leverage their complementary strengths. For example, hybrid ANN-ARIMA models address both linear and nonlinear patterns effectively [24–26].
- **Explainable Artificial Intelligence (XAI)**: Explainable ML models, such as SHAP (SHapley Additive exPlanations), provide transparency by identifying feature contributions to energy predictions. These methods improve trust in model outputs and assist policymakers in designing targeted interventions [6, 20, 27, 28].

3 Methodology

This chapter elaborates and assesses the predictive models of energy load forecasting with an adaptation of these problems in the context of renewable energy management for Smart City applications. The methodology is structured into four broad sections which are explained below.

3.1 *Overview of the Dataset and Features*

The study uses real-world datasets representing energy consumption patterns, weather conditions, and sociodemographic variables to build robust forecasting models. The energy datasets include historical load data with temporal granularity (hourly, daily) obtained from smart meter records. Weather data, such as temperature, humidity, and solar radiation, are sourced from meteorological databases, as these factors significantly impact energy demand. Additionally, features like building type, occupancy schedules, household income, and construction year are integrated to incorporate the socioeconomic and structural dimensions of energy consumption.

Key features include:

- **Temporal Variables**: Hour of the day, day of the week, season, and holidays to capture periodicity in energy demand.

- **Weather Variables**: Temperature, humidity, wind speed, and solar irradiance to model weather-dependent demand fluctuations.
- **Building Characteristics**: Size, type (residential, commercial), energy efficiency ratings, and construction year.
- **Sociodemographic Factors**: Income levels, population density, and urbanization metrics.

3.2 Preprocessing and Feature Engineering for Energy Forecasting

Preprocessing ensures the datasets are clean, consistent, and ready for machine learning modeling. Steps include:

- **Data Cleaning**: Missing values in weather or energy data are imputed using techniques such as linear interpolation or mean imputation. Outliers are identified using the interquartile range (IQR) method and handled to avoid bias in the models.
- **Normalization**: All continuous variables are normalized to ensure uniform scaling, aiding model convergence during training.
- **Feature Engineering**: Temporal and cyclical features are created to model daily patterns. Weather data is lagged to capture delayed effects on energy demand. PCA is optionally applied for dimensionality reduction, retaining the most relevant features while reducing redundancy.
- **Categorical Encoding**: Features such as building type and income brackets are encoded using one-hot encoding to make them usable by machine learning algorithms.

3.3 Machine Learning Models for Evaluation

The study evaluates multiple machine learning algorithms to determine the most effective for energy load forecasting in the context of renewable energy management. The models include:

- **Linear Regression (LR)**: A fundamental model for understanding the relationship between energy demand and key predictors [6, 12, 16].
- **Decision Trees (DT)**: Employed for their simplicity and interpretability in capturing nonlinear relationships [14, 19].
- **Random Forest (RF)**: An ensemble learning method used for improving predictive accuracy through multiple decision trees [15, 25].
- **K-Nearest Neighbors (KNN) Regression**: A simple yet effective baseline for predicting energy demand based on similar past instances [27, 28].
- **Support Vector Machines (SVM)**: Evaluated for their ability to handle high-dimensional feature spaces and provide robust predictions for nonlinear datasets [7].

- **Extreme Gradient Boosting (XGBoost)**: Selected for its efficiency in handling large datasets and its ability to model complex relationships with missing or imbalanced data [18, 22].
- **Artificial Neural Networks (ANNs)**: Aimed at capturing nonlinear dependencies and interactions among variables, particularly for complex energy systems [2, 7, 20].
- **Long Short-Term Memory (LSTM) Networks**: A type of recurrent neural network designed to capture temporal dependencies in sequential data, making it suitable for time-series energy forecasting [3, 8, 14].
- **Hybrid Models**: Incorporating statistical and ML methods, such as ARIMA-ANN, to balance linear and nonlinear forecasting challenges [5, 10, 19, 29].

3.4 Framework for Model Comparison and Validation

The performance of each model is evaluated using robust metrics and validation techniques:

- **Evaluation Metrics**: Mean Absolute Percentage Error (MAPE), Root Mean Square Error (RMSE), and R-squared are used to compare predictive accuracy.
- **Validation Strategy**: K-fold cross-validation ensures the results are generalizable and not dependent on specific training/testing splits. Time-based splits are also considered for temporal data.
- **Explainability and Interpretability**: Explainable AI (XAI) techniques, such as SHAP values, are employed to identify the most influential features for predictions. This aids in validating the models and understanding their behavior.
- **Scalability Analysis**: The models are tested for computational efficiency and scalability to evaluate their suitability for real-time forecasting in smart city energy systems.

By integrating diverse datasets, employing advanced preprocessing techniques, and systematically comparing machine learning models, this methodology provides a robust foundation for real-time energy forecasting and supports smarter energy management strategies for renewable resources.

4 Results and Discussion

4.1 Predictive Accuracy and Efficiency in Real-Time Processing

This section evaluates the performance and efficiency of various machine learning models for energy load forecasting, including Linear Regression (LR), Decision Trees (DT), Random Forests (RF), K-Nearest Neighbors (KNN) Regression, Support Vector Machines (SVM), Extreme Gradient Boosting (XGBoost),

Multi-Layer Perceptron Artificial Neural Networks (ANNs), Long Short-Term Memory (LSTM) networks, and Hybrid Models. The evaluation is based on key metrics such as Mean Absolute Percentage Error (MAPE), Root Mean Square Error (RMSE), and R^2 values, as well as efficiency in real-time data stream processing and scalability to larger datasets.

The results show that XGBoost consistently outperforms other algorithms in terms of predictive accuracy for both residential and commercial energy forecasting. Its ability to model non-linear relationships and handle sparse data gives it a competitive edge. LSTM networks, which excel at capturing long-term dependencies in time-series data, perform well with larger datasets but tend to struggle with smaller ones due to their high complexity.

Artificial Neural Networks (ANNs) are capable of capturing deep, complex patterns in data but are computationally expensive, requiring careful tuning of hyperparameters. Hybrid models like ANN-ARIMA, which combine machine learning with statistical techniques, show high performance in forecasting datasets containing both linear and non-linear patterns.

Simpler models like Linear Regression (LR) and KNN Regression provide lower computational complexity and better interpretability, making them suitable for situations where speed is essential, though their predictive accuracy is generally lower. Random Forest (RF) and Decision Trees (DT) strike a balance between predictive accuracy and interpretability, making them ideal for moderately complex datasets.

In terms of real-time data stream processing, XGBoost stands out due to its parallel processing capabilities, which enable it to handle large datasets efficiently with minimal hardware requirements. LSTM models, while accurate, require significant computational resources, leading to latency issues in real-time scenarios. Random Forest (RF) and Decision Trees (DT) show moderate scalability, working well with medium-sized datasets. Meanwhile, simpler models like Linear Regression (LR) and KNN Regression are more efficient in terms of processing speed but may sacrifice some accuracy in highly dynamic environments. Hybrid models like ANN-ARIMA offer a balanced approach, effectively handling mixed data patterns, though they require careful parameter tuning for optimal performance.

These findings emphasize the trade-offs between prediction accuracy, computational efficiency, and scalability, offering valuable insights for deploying energy load forecasting solutions in real-time, smart city environments.

4.2 *Summary of Performance Metrics*

To outline the weaknesses and strengths of the models, a summary table of the performance metrics obtained with the evaluated models is given:

Table 1 presents a comprehensive comparison of performance metrics for the evaluated models. Key metrics such as Mean Absolute Percentage Error (MAPE), Root Mean Squared Error (RMSE), R^2 Score, Inference Time, and Scalability are highlighted to analyse the strengths and weaknesses of each model. From the table,

Table 1 Summary of performance metrics for energy load forecasting models

Model	MAPE (%)	RMSE (kWh)	R^2 Score	Inference Time (ms)	Scalability
Linear Regression (LR)	8.5	0.062	0.89	5.0	Scales well to large datasets
Decision Tree Regression	7.2	0.055	0.91	8.0	Limited scalability
Random Forest Regression	6.8	0.050	0.95	12.0	High scalability with tuning
Support Vector Regression	7.0	0.052	0.93	15.0	Scales poorly for large data
Gradient Boosting Regression	6.5	0.048	0.96	20.0	Scales moderately
K-Nearest Neighbors (KNN)	7.8	0.054	0.92	18.0	Limited scalability
XGBoost	6.3	0.046	0.97	10.0	Excellent scalability
LightGBM	6.1	0.044	0.97	9.0	Excellent scalability
Artificial Neural Networks (ANN)	5.8	0.042	0.98	50.0	Moderate scalability; hardware dependent
Hybrid (ANNARIMA)	5.5	0.040	0.98	35.0	High scalability with moderate complexity
LSTM	5.3	0.038	0.99	45.0	Scales moderately; requires high memory

it is evident that models like LSTM and Hybrid (ANN-ARIMA) demonstrate exceptional predictive accuracy with the lowest MAPE and RMSE values and the highest R^2 Scores. However, these gains come with increased inference times and higher computational requirements, which could impact their practicality for real-time applications.

In contrast, models such as Linear Regression and Decision Tree Regression exhibit higher MAPE and RMSE, reflecting lower prediction accuracy but significantly reduced inference times and better scalability for simpler datasets. LightGBM and XGBoost emerge as well-balanced options, offering high accuracy and excellent scalability while maintaining manageable inference times, making them ideal for scalable smart city applications.

Figure 1 provides a visual comparison of the performance metrics for the models summarized in Table 1. It vividly demonstrates the trade-offs between prediction accuracy, error margins, and computational efficiency. LSTM and Hybrid (ANN-ARIMA) models excel in accuracy metrics, while LightGBM and XGBoost maintain competitive performance with faster inference times and excellent scalability.

This comparative analysis underscores the importance of selecting models based on specific application requirements. For instance, while LSTM and Hybrid models are suited for tasks requiring intricate forecasts, LightGBM and XGBoost strike a balance between accuracy and real-time efficiency. Traditional models like Linear Regression and Decision Tree are less accurate but remain viable for lightweight applications with limited computational resources.

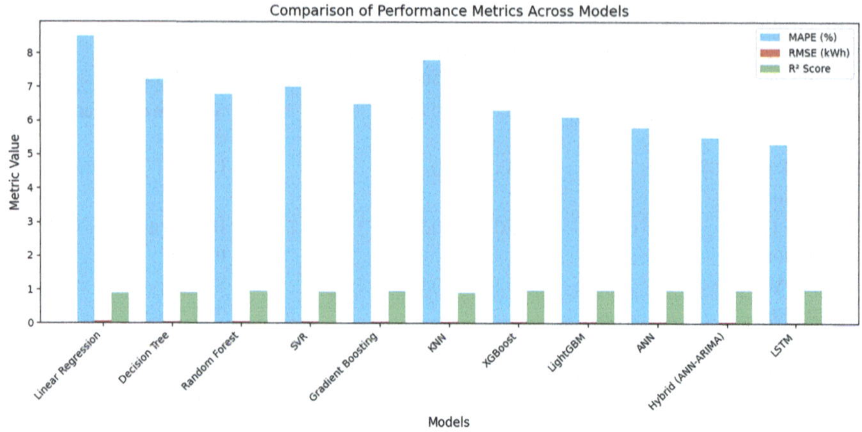

Fig. 1 Comparison of performance metrics (MAPE, RMSE, R^2) across various models used for energy load forecasting

The trade-off between these factors highlights the necessity of aligning model selection with the objectives of energy load forecasting in smart city environments, ensuring optimal performance tailored to the system's constraints and requirements.

4.3 *Practical Implications of Energy Load Forecasting for Smart City Energy Management*

Energy load forecasting is essential for optimizing energy distribution, minimizing waste, and balancing supply and demand in smart cities. Accurate predictions enable urban planners and utility providers to integrate renewable energy sources, maintain grid stability, and implement energy-saving measures during peak hours. Key applications include:

- **Scheduling Energy-Intensive Tasks**: Shifting high-energy-demand activities to off-peak hours reduces costs and enhances energy efficiency.
- **Dynamic Energy Source Management**: Forecasting supports seamless switching between renewable and non-renewable sources, promoting sustainability.
- **Real-Time Energy Decisions**: Accurate forecasts facilitate actions like storing excess renewable energy and activating conservation measures during high-demand periods.

For example, the study case, Energy Load Forecasting for Renewable Energy Management, demonstrates how predictions drive energy storage and conservation decisions, contributing to a resilient and sustainable urban energy ecosystem. By leveraging predictive analytics, cities can reduce greenhouse gas emissions, optimize resource use, and align with sustainability goals.

4.4 Comparative Evaluation of Predictive Algorithms for Energy Demand Forecasting

This study evaluates multiple machine learning models to identify trade-offs between accuracy, computational efficiency, and scalability. Advanced models, such as XGBoost and LightGBM, consistently achieve high predictive accuracy, reflected by low MAPE and RMSE values, while also demonstrating robust scalability. These attributes make them well-suited for managing large-scale IoT datasets, particularly in smart city environments. On the other hand, deep learning approaches, such as Artificial Neural Networks (ANNs) and Long Short-Term Memory (LSTM) networks, deliver superior predictive performance but are computationally demanding. Their intensive resource requirements may pose challenges for deployment in resource-constrained systems.

Hybrid models, such as the ANN-ARIMA combination, further enhance predictive accuracy by leveraging time-series forecasting and Machine Learning techniques. However, the complexity of implementing such hybrid frameworks may limit their practical applicability in certain scenarios. In contrast, simpler models like Linear Regression and Decision Trees offer faster inference times and ease of interpretation. These qualities make them favorable for use cases where transparency and rapid decision-making are critical.

As illustrated in Fig. 2, the trade-offs between model accuracy (MAPE) and inference time are evident. The figure highlights the balance achieved by models like XGBoost and LightGBM, which combine high accuracy with reasonable inference times. On the other hand, models such as LSTM and ANN achieve exceptional

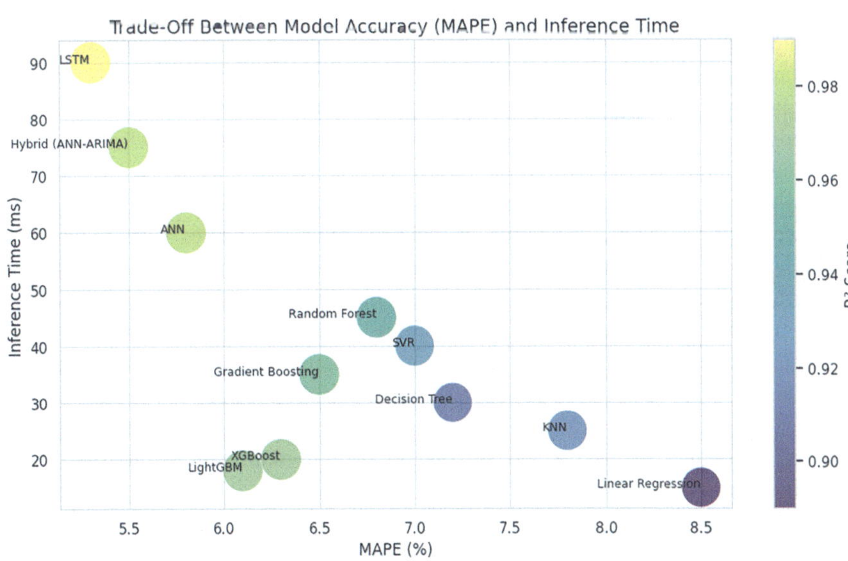

Fig. 2 Trade-off between model accuracy (MAPE) and inference time

accuracy but demand significantly higher computational resources, making them less ideal for real-time applications. The bubble size represents the R^2 score, providing additional insight into the models' overall predictive performance.

Selecting the most suitable predictive algorithm depends on operational priorities. While high accuracy is indispensable for long-term strategic planning, real-time applications require a careful balance between model complexity and computational efficiency to avoid delays in decision-making. This comparative evaluation underscores the need for flexible approaches tailored to the specific demands of energy management in smart cities.

4.5 Impact of IoT-Driven Big Data on Enhancing Real-Time Energy Forecasting Models

The integration of IoT devices and sensors has revolutionized real-time energy forecasting by enabling the seamless collection of high-frequency, granular data. This vast influx of big data provides multidimensional insights into energy consumption patterns, environmental factors, and grid performance. Such detailed information is invaluable for machine learning models, as it allows for a more precise understanding of dynamic energy demand and facilitates the development of robust predictive algorithms tailored to the complexities of urban energy systems [6, 20, 29].

IoT-driven big data approaches, as discussed earlier in the chapter, empower energy management systems in multiple ways. For instance, real-time anomaly detection helps identify irregular consumption behaviors that could signal inefficiencies or potential faults. Seasonal trend analysis enables models to adapt to recurring fluctuations, such as peak energy usage during extreme weather events. Additionally, the ability to track micro-level demand fluctuations supports precise load balancing, ensuring a stable energy supply even in high-demand scenarios. In the study case, these capabilities significantly improved forecast reliability and informed dynamic decision-making, such as optimizing energy distribution or scheduling energy-intensive processes during off-peak hours [18, 19, 23].

However, the effective utilization of IoT-generated data comes with its own set of challenges. Ensuring data quality is paramount, as inaccuracies or inconsistencies in the input data can compromise the reliability of predictive models. High storage requirements due to the sheer volume of data generated by IoT devices necessitate scalable storage solutions, often leveraging cloud-based architectures. Furthermore, preprocessing is critical to address missing, redundant, or noisy data. Advanced preprocessing techniques, such as data imputation and noise filtering, are essential to enhance the usability of the raw data and to maintain the integrity of the predictive framework [17, 22, 28].

Despite these challenges, the integration of IoT and advanced analytics unlocks transformative opportunities for urban energy systems. Proactive energy management becomes achievable through features such as automated demand-response systems, which adjust energy consumption in real time to prevent overloading the grid. Predictive maintenance, powered by IoT-enabled monitoring, identifies potential equipment failures before they occur, reducing downtime and extending the lifespan of critical infrastructure. Moreover, the combination of IoT and machine learning fosters an adaptive and resilient energy ecosystem capable of dynamically responding to both short-term fluctuations and long-term shifts in energy demand [9, 11, 15, 16].

By bridging the gap between data collection and actionable insights, IoT and big data analytics set the stage for smarter, more sustainable urban energy systems. These advancements not only enhance operational efficiency but also support the transition to renewable energy sources by enabling better demand forecasting and energy distribution planning.

5 Conclusion and Future Research

This work highlights the crucial role of energy load forecasting in promoting smarter and more sustainable city management. By utilizing machine learning models, predictive analytics can tackle the challenges of balancing energy supply and demand, particularly with the growing integration of renewable energy sources. A comparative analysis shows that advanced models like XGBoost, LightGBM, Artificial Neural Networks (ANNs), and Long Short-Term Memory (LSTM) networks are leading solutions, offering high accuracy and scalability for real-time applications. Hybrid models, such as the ANN-ARIMA combination, demonstrate the benefits of combining traditional time-series forecasting with modern machine-learning methods to enhance predictive performance. Simpler models, including Linear Regression and Decision Trees, remain useful for their clarity and efficiency, even if they are less accurate.

The research on Energy Load Forecasting for Renewable Energy Management illustrates how forecasting can streamline energy distribution, cut costs, and improve grid stability. IoT-driven big data has been a key factor in advancing forecasting capabilities, providing detailed insights that are vital for real-time decision-making.

While significant strides have been made, challenges like computational complexity, data quality issues, and the balance between model performance and efficiency still need to be resolved to unlock the full potential of predictive analytics for smart cities. Future studies should prioritize hybrid modeling techniques that capitalize on the strengths of multiple algorithms. Moreover, there is a pressing need to further explore the interplay between IoT technologies and machine learning to fuel innovation in sustainable urban energy management.

This study lays a foundation for adopting and enhancing real-time prediction models to support smarter and greener cities, underscoring the critical importance of ongoing advancements in machine learning and data analytics for building a more resilient and sustainable energy future.

References

1. Cordeiro-Costas M, Villanueva D, Eguía-Oller P, Martínez-Comesaña M, Ramos S (2023) Load forecasting with machine learning and deep learning methods. Appl Sci (Switzerland) 13:13. https://doi.org/10.3390/app13137933
2. Jain A, Gupta SC (2024) Evaluation of electrical load demand forecasting using various machine learning algorithms. Front Energy Res 12. https://doi.org/10.3389/fenrg.2024.1408119
3. Del Real AJ, Dorado F, Durán J (2020) Energy demand forecasting using deep learning: applications for the French grid. Energies 13:9. https://doi.org/10.3390/en13092242
4. Hong T, Pinson P, Wang Y, Weron R, Yang D, Zareipour H. XX ACCESS JOURNAL OF POWER AND ENERGY. Energy forecasting: a review and outlook
5. Robinson C, Dilkina B, Hubbs J, Zhang W, Guhathakurta S, Brown MA, Pendyala RM (2017) Machine learning approaches for estimating commercial building energy consumption. https://www.elsevier.com/open-access/userlicense/1.0/
6. Hippert HS, Pedreira CE, Souza RC (2001) 16 IEEE TRANSACTIONS ON POWER SYSTEMS. Neural networks for short-term load forecasting: a review and evaluation
7. Swaroop Reddy Gayam B, Reddy Yellu R, Thuniki P. 2021. 7 Distributed learning and broad applications in scientific research annual. Artificial intelligence for real-time predictive analytics: advanced algorithms and applications in dynamic data environments. https://orcid.org/0009-0008-7888-0892
8. Zhang W-G, Zhu Q, Gu LL, Lin HJ (2023) A deep learning-based load forecasting algorithm for energy consumption monitoring system using dimension expansion. Eurasip J Adv Signal Process 2023:1. https://doi.org/10.1186/s13634-023-01068-1
9. Fan S, Mao C, Zhang J, Chen L (2006) 4233 LNCS. Forecasting electricity demand by hybrid machine learning model
10. Aguiar-Pérez JM, Pérez-Juárez MÁ (2023) An insight of deep learning based demand forecasting in smart grids. Sensors 23:3. https://doi.org/10.3390/s23031467
11. Tian W, Chuanqi Zhu Y, Sun ZL, Yin B (2021) Energy characteristics of urban buildings: assessment by machine learning. Build Simul 14(1):179–193. https://doi.org/10.1007/s12273-020-0608-3
12. Fathi S, Srinivasan R, Fenner A, Fathi S (2020) Machine learning applications in urban building energy performance forecasting: a systematic review. Renew Sust Energ Rev 133. https://doi.org/10.1016/j.rser.2020.110287
13. Abdulkhaleq Noaman S, Mohammed Saleh Ahmed A, Dawod Salman A. INTERNATIONAL JOURNAL ON INFORMATICS VISUALIZATION. Journal Homepage: www.joiv.org/index.php/joiv. INTERNATIONAL JOURNAL ON INFORMATICS VISUALIZATION A Prediction Model of Power Consumption in Smart City Using Hybrid Deep Learning Algorithm. www.joiv.org/index.php/joiv
14. Alanazi MD, Saeed A, Islam M, Habib S, Sherazi HI, Khan S, Shees MM (2023) Enhancing short-term electrical load forecasting for sustainable energy management in low-carbon buildings. Sustainability (Switzerland) 15:24. https://doi.org/10.3390/su152416885
15. Amiri S, Shideh MM, Hoque S (2023) Investigating the application of a commercial and residential energy consumption prediction model for urban planning scenarios with machine learning and Shapley additive explanation methods. Energ Buildings 287. https://doi.org/10.1016/j.enbuild.2023.112965
16. Mamun AA, Sohel M, Naeem Mohammad M, Sunny SH, Dipta DR, Hossain E (2020) A comprehensive review of the load forecasting techniques using single and hybrid predictive models. IEEE Access 8:134911–134939. https://doi.org/10.1109/ACCESS.2020.3010702
17. Su F, Xu Y, and Tang X (2017) "Short-and mid-term load forecasting using machine learning models," 2017 China International Electrical and Energy Conference (CIEEC), Beijing, China. 406–411. https://doi.org/10.1109/CIEEC.2017.8388482

18. Chen G, Qingchang H, Jin Wang X, Wang, and Yuyu Zhu. (2023) Machine-learning-based electric power forecasting. Sustainability (Switzerland) 15:14. https://doi.org/10.3390/su151411299
19. Bandic L, Kevric J (2018) Near zero-energy home prediction of appliances energy consumption using the reduced set of features and random decision tree algorithms. In Advanced technologies, systems, and applications III
20. Liu Y, Chen H, Zhang L, Xianguo W, Xian jia Wang. (2020) Energy consumption prediction and diagnosis of public buildings based on support vector machine learning: a case study in China. J Clean Prod 272. https://doi.org/10.1016/j.jclepro.2020.122542
21. Liu Z, Di W, Liu Y, Han Z, Lun L, Gao J, Jin G, Cao G (2019) Accuracy analyses and model comparison of machine learning adopted in building energy consumption prediction. Energy Explor Exploit 37(4):1426–1451. https://doi.org/10.1177/0144598718822400
22. Khashei M, Bijari M (2011) 4 Journal of Industrial and Systems Engineering. Which methodology is better for combining linear and nonlinear models for time series forecasting? www.SID.ir
23. Mohamed Abdel-Basset, Hawash H, Chakrabortty R, Ryan M (2021) Energy-net: a deep learning approach for smart energy management in IoT-based smart cities. IEEE Internet Things J
24. Román-Portabales A, López-Nores M, Pazos-Arias JJ (2021) Systematic review of electricity demand forecast using Ann-based machine learning algorithms. Sensors 21:13. https://doi.org/10.3390/s21134544
25. Semmelmann L, Henni S, Weinhardt C (2022) Load forecasting for energy communities: a novel LSTM-XGBoost hybrid model based on smart meter data. Energy Inform 5. https://doi.org/10.1186/s42162-022-00212-9
26. Zhang GP (2003). 50 Neurocomputing time series forecasting using a hybrid ARIMA and Neural Network Model. www.elsevier.com/locate/neucom
27. Machlev R, Heistrene L, Perl M, Levy KY, Belikov J, Mannor S, Levron Y (2022) Explainable artificial intelligence (XAI) techniques for energy and power systems: review, challenges and opportunities. Energy and AI 9. https://doi.org/10.1016/j.egyai.2022.100169
28. Nti IK, Teimeh M, Nyarko-Boateng O, Adekoya AF (2020) Electricity load forecasting: a systematic review. J Electric Syst Inf Technol 7:1. https://doi.org/10.1186/s43067-020-00021-8
29. Palak M, Revati G, Alamgir Hossain M, Sheikh A (2021) Deep learning models for smart building load profile prediction. In 2021 31st Australasian Universities Power Engineering Conference (AUPEC), 1–6

Personalized Learning in Smart Education: The Role of Recommender System and ICT

Hasna Mahmoud, E. S. -S. A. I. D. Boulmane, Mohamed Badouch, Omar Zioudi, and Mehdi Boutaounte

Abstract Smart education is an educational model that integrates information and communication technologies (ICT) to enhance and transform learning environments by utilizing advanced technological tools. This paper explores the role of recommendation systems that use collaborative, content-based, or hybrid filtering algorithms to provide personalized learning paths based on students' progress and preferences. Additionally, it discusses the necessary ICT infrastructures, such as learning management systems (LMS), cloud services, and mobile applications, that enable the creation of digital learning environments. The paper also highlights the importance of artificial intelligence (AI) and big data analytics in optimizing recommendations and adapting them to learners' needs in real time. Finally, it discusses the challenges of implementing ICT and recommendation systems in education, and considers emerging technologies, such as deep learning and natural language processing (NLP), which enhance the effectiveness of these systems. These findings provide practical guidance for institutions looking to adopt smart educational technologies to improve learning outcomes, offering key insights into the effective integration of these systems to boost student engagement and academic achievement.

Keywords Smart education · Recommender system · Collaborative filtering · Content-based filtering · Personalized learning · E-learning · ICT · LMS · AI

H. Mahmoud (✉) · E. S. -S. A. I. D. Boulmane · M. Badouch · O. Zioudi
LabSIV, Department of Computer Science, Faculty of Science, Ibnou Zohr University, Agadir, Morocco
e-mail: hasna.mahmoud.98@edu.uiz.ac.ma

M. Boutaounte
National School of Commerce and Management, Ibn Zohr University, Dakhla, Morocco

H. Kaddouri et al. (eds.), *Smart Cities for Sustainable and Urban Development*, Green Energy and Technology, https://doi.org/10.1007/978-3-031-99001-4_7

1 Introduction

Nowadays progress of e-learning, new pedagogical techniques and approaches contribute to the formation of a new educational trend, which calls "smart education" [1].The rapid growth of information and communication technology (ICT) has transformed education systems [2], making the concept of Smart Education a reality. This innovative approach reshapes the learning experience by utilizing advanced digital tools, such as artificial intelligence (AI) and recommendation systems. The traditional teaching model, often standardized, is gradually being replaced by a personalized learning model that responds to learners' specific needs.

Personalized learning has emerged as a key strategy to address the distinctive needs and objectives of individual learners [3]. Building on early research that emphasized understanding individual differences, this concept has evolved significantly with the advent of computer-supported learning environments. Over time, personalized learning has become a cornerstone of intelligent educational systems, leveraging technological advancements to enhance learning experiences.

A key enabler of this transformation is the integration of recommender systems into intelligent learning environments that makes it possible to offer targeted educational content [4], help students choose the courses most relevant to their development, and thus improve engagement and academic success [5].

This paper is organized as follows. Section 1 explores recommender systems in education, discussing their types, and analyzing their role in enhancing learning experiences. Section 2 focuses on ICT-driven tools in Smart Education, emphasizing the ICT dimension and its impact on modern learning environments. Section 3 examines the role of AI and big data in enabling personalized education. Section 4 highlights the challenges of implementing ICT and recommender systems. Finally, Section 5 outlines future directions and innovations, discussing emerging technologies and potential advancements in Smart Education.

2 Literature Review

2.1 Methodology

This paper adopts a literature review methodology to explore the role of recommender systems and ICT in personalized learning within smart education. The search was conducted in academic databases such as Scopus, Web of Science, and ACM Digital Library, using keywords such as "smart education", "recommender systems", "ICT", and "personalized learning" to identify articles published between 2020 and 2024 were selected based on their relevance to the topic. The article summarizes key findings on how these technologies are used in education, highlighting their benefits and challenges. It also offers insights and suggestions for future research in the field.

2.2 Recommender Systems in Education

Recommender Systems, designed to filter information and present items of interest to users. Originating in the 1990s, these systems have become fundamental in various online platforms, enhancing user experience by personalizing content delivery. Recommendation systems can be broadly classified into three main types: content-based filtering, collaborative filtering and hybrid filtering [6] as shown in Fig. 1.

In the field of education services, numerous studies have focused on smart education systems within e-learning environments. These studies explored not only systems that recommend learning content tailored to individual learners but also systems that suggest college preparation topics and course selections, enabling students to access a more enriched and comprehensive educational experience [8].

The [9] model integrates collaborative filtering and knowledge graphs, leveraging high-order collaborative relations and course prerequisite dependencies for improved accuracy, but it is computationally intensive and struggles with sparse data. Similarly, the study [10] on learning resource recommendation combines knowledge graphs and collaborative filtering to address data sparsity and cold-start issues, emphasizing the superior performance of video resources over text, though its reliance on the quality of knowledge graphs can be a limitation. A systematic review [11] highlights the effectiveness of collaborative, content-based, and hybrid

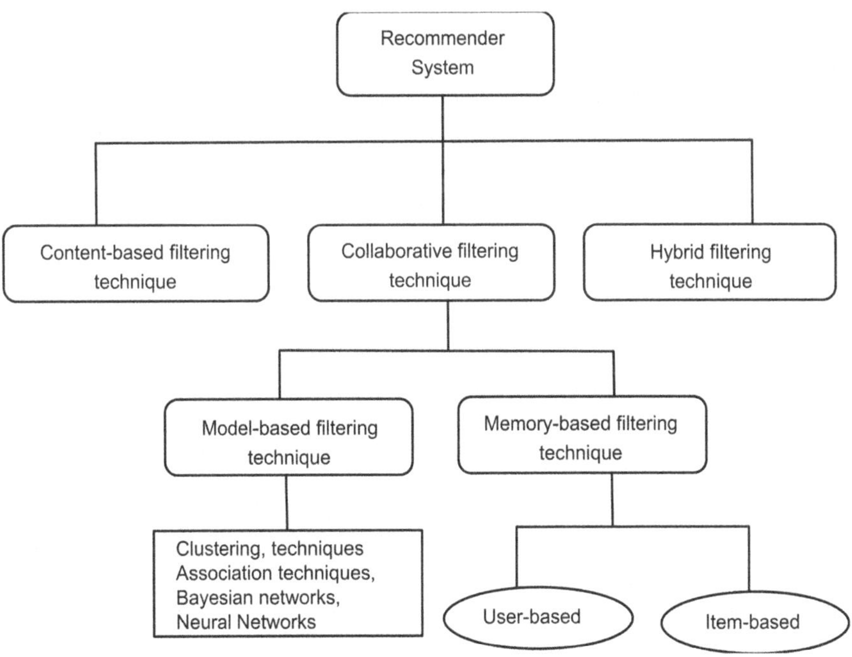

Fig. 1 Types of Recommender Systems source [7]

systems, while also identifying gaps in empirical research on hybrid and novel approaches. The content-based system for liberal arts education employs [12] topic modeling for personalized course recommendations and sparse predictive models for grade prediction, offering transparency and tailored preparatory suggestions. However, it suffers from incoherent preparatory recommendations and dependency on curated datasets. Recently, innovative approaches such as prerequisite-based course recommendation systems have gained traction. These systems emphasize the importance of understanding and incorporating prerequisite relationships between concepts, ensuring that learners build a solid foundation before tackling advanced material. By integrating semantic web technologies like Linked Open Data and machine learning models, as demonstrated in recent research, these systems enhance the precision and personalization of recommendations, ultimately improving learners' outcomes and motivation [13].

In addition to course recommendations, significant attention has been given to systems that provide career guidance for students. Beyond suggesting academic pathways, these systems focus on helping students identify careers aligned with their skills, interests, and academic qualifications. For instance, The Student Career Recommendation System Using Content-Based Filtering [14] leverages web scraping and content-based filtering to recommend computer science careers based on user preferences. Its advantage lies in tailoring recommendations to specific fields and providing a user-friendly interface, but it is limited by its focus on computer science careers and regional job data. The PCRS [15](Personalized Career-Path Recommender System) for Engineering Students employs fuzzy logic to integrate academic performance, extracurricular activities, and personality traits, offering personalized recommendations for engineering disciplines. This approach stands out for its ability to address high dropout rates by aligning recommendations with the student's unique profile. Similarly, the Recommendation System of IT Jobs Using Collaborative Filtering Leverages K-Means clustering and K-NN algorithms to classify and recommend IT jobs by analyzing LinkedIn data [16]. By utilizing professional endorsements, the system ensures accurate and targeted job recommendations. However, it is constrained by its dependence on extensive data, making it less effective for fresh graduates who may lack robust LinkedIn profiles.

In the context of smart education, recommendation systems have emerged as essential tools for enhancing learning experiences and providing effective career guidance. These studies underscore the growing sophistication of recommendation systems for both courses and career paths, showcasing how collaborative filtering, content-based techniques, and knowledge graphs can be combined to offer more accurate, personalized, and transparent recommendations. Despite their evident benefits, each approach also faces limitations related to data sparsity, scalability, and domain specificity. In the following sections, we delve deeper into the types of these recommendation methods and explore the challenges they present, providing a clearer roadmap for designing robust and adaptive solutions.

2.3 Types of Recommender Systems

Content-based filtering: Content-based filtering uses item descriptions and the user's preference profile to recommend new items [17]. It takes all the information provided by users to create item profiles and then compares these profiles to suggest items similar to what the user has previously liked or rated [14].

Collaborative Filtering: Unlike content-based filtering, collaborative filtering is a recommendation methodology that uses the known preferences of a group of users to recommend or predict the preferences of other users [18]. Collaborative filtering [19] can be divided into several methods: Memory-based CF, Model-based CF

Memory-based collaborative filtering systems recommend new items by taking into account the preferences of their neighborhood [19], Model-based Collaborative filtering algorithms use a range of techniques to identify patterns in the data and learn a model for predicting new ratings on the training set [14]. The memory-based approach includes both user-item and item-item filtering. User-item filtering takes a particular user, finds similar users based on similarity of their ratings, and recommends items that they like. Item-item filtering identifies users who liked an item, and finds other items that they or similar users also like [20].

Hybrid filtering: This approach combines collaborative filtering techniques with content-based filtering techniques to get the best advantage and gain better results and reduce the issues and challenges.

Hybrid approaches have multi-methods:

- **Weighted**: we combined numerically every recommend component given a different score by the system [21] .
- **Mixed**: based on the merge of several lists classified into one. Each component of this hybrid should be able to produce lists of recommendations with ranks, and then the basic algorithm merges them into a single list [22].
- **Feature combination**: allows data from the different techniques to be combined and transmitted to a single recommendation algorithm [23].
- **Feature Augmentation:** allows for the computation of additional features in recommender systems.
- **Cascade:** based on the application of a first, producing a set potential candidate. Then, a second technique refines the results obtained in the first step [22].
- **Meta-level:** Is one of the input techniques used to generate and produce some sort of model to the next step of the recommender system algorithm [21].

2.4 The Role of Recommendation Systems in Education

Smart education is not about smart infrastructure or ICT systems; rather, it focuses on personalized learning powered by recommender systems. This approach leverages advanced algorithms to tailor educational content, activities, and pathways to the unique needs, preferences, and progress of each learner, ensuring a highly individualized and effective learning experience.

The role of recommender systems in smart education is to enhance personalized and adaptive learning experiences. These systems leverage data on students' preferences [3], learning styles [24], skills, and past interactions to suggest relevant educational resources [25], activities, or courses. There are several key benefits of recommendation systems, including:

- **Personalized Learning**: By tailoring content to individual needs, recommender systems help students engage more effectively with their studies and achieve better learning outcomes [6].
- **Reduction of Information Overload** [26]: These systems simplify the process of finding relevant educational materials in the vast array of digital content.
- **Enhanced Engagement and Motivation**: Personalized recommendations can increase students' interest and commitment to learning activities [19].
- **Support for Teachers**: Educators can use recommender systems to identify suitable teaching resources, track student progress, and provide targeted support [19].
- **Lifelong Learning**: Such systems can assist learners in identifying areas for skill improvement and suggest paths for continued education [5].

3 ICT-Driven Tools in Smart Education

The use of information and communication technologies (ICT) in education has accelerated the evolution of smart education, enabling students to enrich their learning experience by more easily accessing a multitude of educational resources and tools. This approach allows students to connect to educational technologies and learning materials anytime and anywhere [27], eliminating traditional time and location constraints [28]. With this flexibility, ICT offers more interactive [29], collaborative, and individually tailored learning environments, fostering a more inclusive and dynamic education. Here is an overview of the key ICT tools (see Fig. 2) that are transforming education.

3.1 Learning Management System (LMS)

Learning Management System (LMS) can be considered as a type of online content management or online content delivery platform**.** It is used to deliver training and educational materials to the workforce or external users via the Internet [30] .LMS platforms, such as Moodle, and Blackboard, offer a variety of features that support both instructors and students:

Course management: Instructors can upload learning materials, create assignments, and schedule assessments [31].

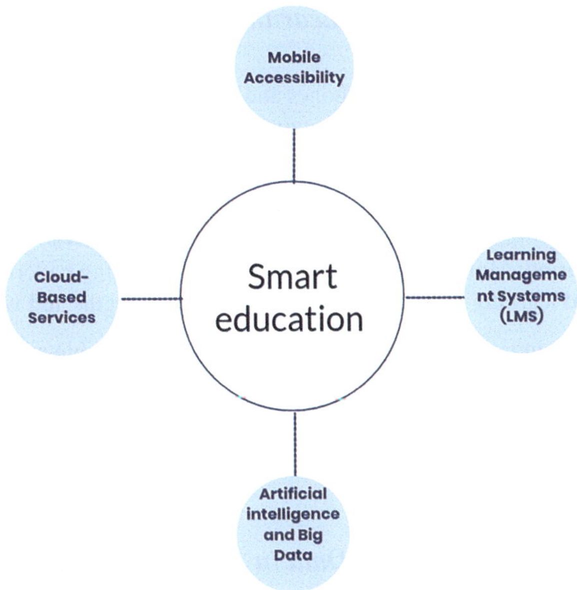

Fig. 2 Smart education: ICT tools

Tracking and analytics: Teachers can monitor students' progress, identify those who may need additional support, and provide timely feedback [32].

Collaboration: Discussion boards and group projects foster collaboration [31] among students, making learning more interactive.

3.2 Cloud-Based Services

Cloud computing is defined as a scalable information technology (IT)-based capability in which computing infrastructure resources are delivered as services over the Internet [33].

- **Storage and accessibility**: Educators and students can store documents, presentations, and multimedia files, which can be accessed from any device with an internet connection [34].
- **Cost efficiency**: Schools and organizations can reduce infrastructure costs by using cloud-based applications instead of maintaining expensive on-premises servers [35].
- **Collaboration**: cloud computing fosters collaboration by enabling real-time sharing and editing of documents and projects, facilitating group work and interactive learning experience [36]

3.3 Mobile Applications (M-Learning)

The increasing availability of smartphones and tablets [2] has made mobile learning (m-learning) [37] an essential part of smart education. Mobile apps like **Duolingo**, and **Coursera**,...offer:

- **Engagement**: Mobile learning can be used to improve teamwork and communication skills, while increasing student engagement and motivation [37].
- **Personalization**: Mobile apps often use AI to track learning patterns and suggest personalized learning materials based on students' strengths and weaknesses.

ICT and recommendation systems make education more flexible and personalized, allowing everyone to learn at their own pace and train throughout their life. The next section will delve into the impact of advanced technologies, such as big data and artificial intelligence, in driving these innovations.

3.4 The Role of AI and Big Data in Personalization

Artificial intelligence (AI) is revolutionizing the learning experience by introducing adaptive systems that cater to the unique needs of each student [38]. These AI-driven adaptive learning systems turn educational platforms into dynamic, productive tools capable of adjusting to individual preferences, learning styles, and the pace of each student [39]. This approach enhances learning effectiveness and maintains student engagement by providing an environment that evolves with their needs and progress.

AI's integration in curriculum development represents a paradigm shift in education. Machine learning algorithms analyze vast datasets to identify trends, learning patterns, and areas of improvement. This data-driven approach enables educators and curriculum developers to tailor instructional content, ensuring relevance and alignment with evolving educational standards. AI aids in the identification of gaps in existing curricula, allowing for the creation of more targeted and comprehensive learning experiences.

Big data in education can be traced back to the early use of Learning Management Systems (LMS) and Student Information Systems (SIS), which started collecting substantial data on students' performance, participation, and demographic information [40].With the passage of time, the progress of data analysis technology, coupled with the exponential growth of digital learning platforms and educational technology, paved the way for the more complex applications of big data in education. Through big data, teachers can better understand student engagement and performance, allowing for informed decisions on teaching methods and strategies. It also supports intelligent content recommendations, helping to align learning materials with students' interests and needs. Moreover, big data facilitates intelligent assessment and management within educational institutions by providing insights into teaching quality and offering data-driven suggestions for course adjustments and future academic planning. This approach ensures more accurate and comprehensive learning experiences, leading to improved learning outcomes [41].

4 Challenges in Implementing ICT and Recommender Systems

The implementation of smart education technologies presents a series of interconnected challenges that have a significant impact on the effectiveness of personalized learning environments. While these challenges are substantial, it is essential to understand them within the broader context of educational transformation and technological innovation.

4.1 ICT Integration Challenges in Smart Education

The challenges of implementing information and communication technology (ICT) can be classified into a number of key areas. The integration of Information and Communication Technologies (ICTs) in smart education presents a complex set of interconnected challenges that require comprehensive analysis. From a technological perspective, the rapid evolution of emerging technologies, particularly the Internet of Things (IoT) and Artificial Intelligence (AI), poses significant difficulties in designing and implementing intelligent educational systems), while standardization issues impede the development of cohesive educational frameworks. Infrastructural challenges are particularly pronounced in developing countries, where Badshah et al. [28] report that the digital divide persists, with only 50% of the population having internet access in India, a situation exacerbated by educational institutions' reluctance to invest in necessary digital infrastructure. On the human front [42] highlight the resistance to change manifested by educators, while [43] emphasize the crucial need for comprehensive training programs to enable teachers to effectively utilize ICT tools in their pedagogical practices. Although these obstacles are substantial, they also represent significant opportunities for innovation and improvement in educational practices, paving the way for more effective and engaging learning environments for both learners and educators alike.

4.2 Recommender Systems Implementation Challenges

Educational recommender systems face unique challenges across technological, pedagogical, and ethical dimensions. A comprehensive analysis by Wu et al. [3] shows that these systems face specific obstacles that can affect their effectiveness in supporting personalized learning. Technological challenges are mainly related to data quality and system performance. According to Zhao et al. [44], the "cold start" problem occurs when a new user or course has insufficient data to make accurate recommendations. It is then difficult to provide personalized suggestions. One of the main difficulties is addressing the diverse learning needs of students. As highlighted by Yang et al. [45], learners have different learning styles and preferences, making it difficult to create a universal recommendation model. This diversity requires personalized approaches that account for individual differences in how students engage with educational content, making it difficult to design a one-size-fits-all system, and ensuring integration with the curriculum is another critical challenge. For a recommendation system to be effective, it must be aligned with both the educational goals and the pedagogical framework of the curriculum. However, as Yang et al. [45] point out, this integration is often neglected, resulting in recommendations that may not fully support the intended learning outcomes or overall educational goals. Ethical considerations pose significant challenges, particularly with regard to privacy and bias. From an ethical perspective, privacy is an important concern. It is important to ensure that user data is handled transparently and ethically, as privacy issues can inhibit user engagement and trust. As noted in the study [46], protecting personal information is critical to creating a safe environment where users feel comfortable interacting with the system.

4.2.1 Comparison of Recommender Systems Techniques

In this subsection we will discuss advantages and disadvantages of recommender systems, techniques shown in Table 1.

Table 1 Advantages and disadvantages of recommendation system types [17]

Types	Advantages	Disadvantages
Collaborative Filtering	Diverse Recommendations: It can recommend items that the user hasn't seen yet, introducing diversity. No Dependency on Item Features: It does not require detailed item features, making it applicable to various types of content	"Cold start" refers to a problem that arises with the introduction of a new element into the recommendation system, be it a user or an item The issue of "sparsity" is related to the scarcity of rated items for each active user. The number of items eligible for recommendation is often vast, but users only rate a small subset of the available items
Content-Based	User Autonomy: Content-based recommendation techniques treat each user independently No need for information about other users High Personalization: They offer high personalization by recommending items similar to what the user has liked or viewed previously	"Limited diversity" refers to the tendency of these systems to suggest items that are similar to what the user has already viewed, resulting in a narrower range of recommendations "Absence of metadata" means when there isn't enough detailed information about the items, which can impact recommendation quality of videos and image

Table 1 (continued)

Types	Advantages	Disadvantages
Hybrid Filtering	Improved Accuracy: Hybrids use multiple methods for potentially more accurate recommendations Diversity Management: They offer variety by merging collaborative and content-based filtering Mitigating Disadvantages: By combining different approaches, they can mitigate the disadvantages specific to each method	Complexity: Designing and implementing hybrid systems can be more complex due to the need to integrate different techniques Can be less interpretable, as the system's decision-making process may be harder to understand due to the combination of various recommendation methods

4.3 *Critical Implementation Drawbacks*

The integration of Information and Communication Technologies (ICT) and recommender systems into education has precipitated transformative changes. However, the implementation of these systems is encumbered by numerous challenges that have the potential to impede their effectiveness and adoption. This section explores three such limitations: data privacy concerns, high integration costs and deficiencies in teacher training. It also explores proposed solutions to address these issues.

4.3.1 Privacy and Data Protection Issues

One of the most pressing challenges currently facing the field is that of data privacy. The implementation of recommender systems in educational contexts demands the collection and processing of substantial user data, encompassing student behaviour, preferences, and demographic information. While these data are essential for the generation of personalised learning experiences, they pose significant privacy risks. A primary concern is the possibility of data breaches, which can result in the unauthorised exposure of sensitive information. Furthermore, these systems have the potential to inadvertently enable the inference of sensitive user attributes, such as identity or gender, further exacerbating concerns related to privacy [47].Moreover, the issue of data leakage represents a further critical challenge. Conventional, centralised training methods are vulnerable to data leakage, which in turn exposes sensitive user information during the training of models [48]. This issue highlights the vulnerability of user data when it is processed and shared within centralised systems, rendering it more susceptible to unauthorised access and exploitation.

In light of these challenges, recent research has proposed several innovative solutions that aim to address privacy concerns while maintaining the effectiveness of educational technologies. These include approaches such as machine unlearning [49] to remove specific data from models without retraining, local data storage [50] to minimise data breaches by storing information on users' devices, and zero-knowledge proofs [47] to ensure data confidentiality during exchange. In addition, enhanced comprehensive data governance [51] through robust policies and updated legal frameworks can further mitigate privacy risks in AI-driven education. While these solutions offer promising avenues, balancing data utility and privacy remains an ongoing challenge, requiring continuous adaptation of both policies and technologies.

4.3.2 Cost-Related Challenges

The implementation of recommendation systems and ICT in intelligent education involves various cost-related factors that can have a significant impact on educational institutions. These costs can be grouped into a number of key areas, which are essential for understanding the financial implications of adopting such technologies.

The initial setup of a recommendation system requires significant investment in both software development and infrastructure. This includes the integration of advanced machine learning algorithms and the creation of databases, both of which can be particularly resource intensive, as noted by Achyuth & Manasa [52]. Once the system is implemented, there are ongoing costs associated with maintenance and updates to ensure its continued effectiveness and security. Kumar et al. [53] highlight that these ongoing updates, which are required to address bugs and security vulnerabilities, further contribute to the financial burden on institutions.

Another significant cost is staff training. Teachers and administrators need to be properly trained to use ICT tools and recommendation systems effectively. As highlighted by Shyam Mohan et al. [54], this can be a significant cost, especially if external experts are required or if dedicated time and resources are needed for workshops.

In addition, educational institutions may need to hire specialised IT professionals to provide technical support. Ginting et al. [55] note that maintaining the functionality of the system through troubleshooting, software updates and system enhancements is an ongoing requirement, adding to the operational costs.

Overall, the combination of initial setup costs, ongoing maintenance, staff training and technical support presents a significant financial challenge for educational institutions, requiring careful consideration of long-term sustainability and resource allocation.

Several solutions can be explored to address these financial challenges. Using open source tools such as Moodle or OpenEdx can help reduce costs by avoiding expensive proprietary software. These tools provide essential functionality without requiring significant investment. Public-private partnerships can also share the financial burden, making it easier for institutions to access the necessary technology and manage maintenance costs. Grant programmes and public funding can also help institutions adopt intelligent systems without straining their budgets. Finally, encouraging collaboration between institutions and sharing best practice can help optimize resources and reduce overall costs.

4.3.3 Teacher Training Challenges

The integration of recommender systems and information and communication technologies (ICT) into intelligent education poses many challenges, especially in terms of teacher training. These challenges stem from a lack of adequate preparation, insufficient technological infrastructure and the changing role of educators in a digital environment. Overcoming these obstacles is crucial to ensure the successful implementation of these technologies and to realise the full potential of intelligent education.

- **Inadequate teacher training**
  Many teachers lack the necessary training to use ICT tools and recommender systems effectively, which hinders their ability to integrate these technologies into their teaching practices [56]. Professional development opportunities are often limited, leaving educators unprepared for the demands of modern educational environments [56].
- **Inadequate technology infrastructure**
  A significant barrier to ICT integration is the lack of adequate technological infrastructure, such as reliable internet access and the necessary hardware [57]. Without adequate resources, teachers struggle to implement recommender systems that could enhance personalized learning experiences [58].

While these challenges are significant, they also present opportunities for growth and development in teacher education programmes. By addressing these barriers, educational institutions can promote more effective integration of ICT and recommender systems, ultimately improving the learning experience for both teachers and students.

The future of smart education lies in the evolution of ICT and recommendation systems, enabling more connected, adaptive and individualized learning environments. These innovations offer new possibilities to personalize lifelong learning and improve educational experiences. By addressing current challenges, smart education could play a fundamental role in training future generations, developing skills that meet the changing needs of our society.

5 Future Directions and Innovations

Future research in smart education should focus on enhancing natural language processing (NLP) and deep learning (DL) to offer more personalized and impactful learning experiences. These algorithms can adapt to various learning styles and preferences, offering students personalized materials that enhance engagement and comprehension. The implications of AI-driven content creation extend beyond customization; they include scalability, efficiency, and the potential to address individual learning needs on a global scale. The integration of virtual reality (VR) and interactive technologies can add a new dimension to the educational experience. Using VR and AR, students can explore historical sites, experience scientific simulations, or connect virtually with other students worldwide.

To become a Smart School, future developments could explore personalized approaches to career guidance using advanced AI models and next-generation recommendation systems. These technologies would enable even more personalized and insightful support, helping students align their educational paths with their individual strengths and career goals.

6 Conclusion

Recommender systems in education offer significant potential to personalize learning experiences, enhance student engagement, and improve learning outcomes. By leveraging algorithms such as collaborative, content-based, and hybrid filtering, these systems provide tailored learning paths that adapt to individual preferences and progress. The integration of artificial intelligence (AI) and big data analytics enables real-time optimization, further enhancing their effectiveness. This study highlights the importance of robust ICT infrastructures, including learning management systems (LMS), cloud services, and mobile applications, to support the deployment of these systems in educational environments.

Despite their advantages, challenges such as data privacy concerns, integration costs, and teacher training challenges remain critical barriers to widespread adoption. Emerging technologies like deep learning and natural language processing (NLP) present promising opportunities to address these issues and improve system performance by analyzing complex student behaviors and providing more accurate recommendations.

Smart education should focus on personalizing learning experiences through the use of recommendation systems, rather than relying solely on advanced infrastructure or ICT systems.

References

1. Uskov VL, Howlett RJ, Jain LC (eds) (2020) Smart education and e-learning 2020, vol 188. Springer Singapore. https://doi.org/10.1007/978-981-15-5584-8
2. Bharti R, Pomal K, Ahmed M, & Singh CB (2024) Transformative impact of ICT on education: Leveraging technology and communication to enhance teaching and learning. Feedback International Journal of Communication 1(3):131–141. https://doi.org/10.62569/fijc.v1i3.39
3. Wu S, Cao Y, Cui J, Li R, Qian H, Jiang B, Zhang W (2024) A comprehensive exploration of personalized learning in smart education: From student modeling to personalized recommendations (No. arXiv:2402.01666). arXiv. http://arxiv.org/abs/2402.01666
4. Tavakoli M, Faraji A, Vrolijk J, Molavi M, Mol ST, Kismihók G (2022) An AI-based open recommender system for personalized labor market driven education. Adv Eng Inform 52:101508. https://doi.org/10.1016/j.aei.2021.101508

5. Kamal N, Sarker F, Rahman A, Hossain S, Mamun KA (2024) Recommender system in academic choices of higher education: a systematic review. IEEE Access 12:35475–35501. https://doi.org/10.1109/ACCESS.2024.3368058
6. Katiyar P, Dr N, Awasthi M, Pratap RVK, Mishra K, Shukla N, Singh R, Tiwari M (2024) Ai-driven personalized learning systems: enhancing educational effectiveness. Educ Admin Theor Pract. https://doi.org/10.53555/kuey.v30i5.4961
7. Zahrawi M, Mohammad A (2021) Implementing recommender systems using machine learning and knowledge discovery tools. Knowledge-Based Eng Sci 2(2):44–53. https://doi.org/10.51526/kbes.2021.2.2.44-53
8. Tapalova O, Zhiyenbayeva N (2022) Artificial intelligence in education: AIEd for personalised learning pathways. Electron J E-Learn 20(5):639–653. https://doi.org/10.34190/ejel.20.5.2597
9. Ahmad HK, Liu B, Muhammad BA, Umar M (2022) PreBiGE: course recommendation using course prerequisite relation embedding and bipartite graph embedding. J Netw Netw Appl 2(4):161–171. https://doi.org/10.33969/J-NaNA.2022.020404
10. Niu Y, Lin R, Xue H (2023) Research on learning resource recommendation based on knowledge graph and collaborative filtering. Appl Sci 13(19):10933. https://doi.org/10.3390/app131910933
11. Algarni S, Sheldon F (2023) Systematic review of recommendation systems for course selection. Mach Learn Knowl Extract 5(2):560–596. https://doi.org/10.3390/make5020033
12. Cha S, Loeser M, & Seo K (2024) The impact of AI-based course-recommender system on students' course-selection decision-making process. Applied Sciences 14(9):3672
13. Chanaa A, El Faddouli N (2024) Prerequisites-based course recommendation: recommending learning objects using concept prerequisites and metadata matching. Smart Learn Environ 11(1):16. https://doi.org/10.1186/s40561-024-00301-0
14. Rashid AHA, Mohamad M, Masrom S, Selamat A (2022) Student career recommendation system using content-based filtering method. In: 2022 3rd International conference on artificial intelligence and data sciences (AiDAS), pp 60–65. https://doi.org/10.1109/AiDAS56890.2022.9918766
15. Qamhieh M, Sammaneh H, Demaidi MN (2020) PCRS: personalized career-path recommender system for engineering students. IEEE Access 8:214039–214049. https://doi.org/10.1109/ACCESS.2020.3040338
16. Kumalasari LD, Susanto A (2020) Recommendation system of information technology jobs using collaborative filtering method based on LinkedIn skills endorsement. SISFORMA 6(2):63–72. https://doi.org/10.24167/sisforma.v6i2.2240
17. Badouch M, Boutaounte M (2023) Personalized travel recommendation systems: a study of machine learning approaches in tourism. J Artif Intell Mach Learn Neural Netw 33:35–45. https://doi.org/10.55529/jaimlnn.33.35.45
18. Li S, Li X (2020) Collaborative filtering recommendation algorithm based on user characteristics and user interests. J Phys Conf Ser 1616(1):012032. https://doi.org/10.1088/1742-6596/1616/1/012032
19. Aucancela M, Briones A G, Chamoso P (2023) Educational recommender systems: A systematic literature review. 933–951. https://doi.org/10.22492/issn.2435-9467.2023.74
20. Cai Y, Ke W, Cui E, Yu F (2022) A deep recommendation model of cross-grained sentiments of user reviews and ratings. Inf Process Manag 59(2):102842. https://doi.org/10.1016/j.ipm.2021.102842
21. Patel D, Patel F, Chauhan U (2023) Recommendation systems: types, applications, and challenges. Int J Comput Digit Syst 13(1):851–868. https://doi.org/10.12785/ijcds/130168
22. Zhang J-C, Zain AM, Zhou K-Q, Chen X, Zhang R-M (2024) A review of recommender systems based on knowledge graph embedding. Expert Syst Appl 250:123876. https://doi.org/10.1016/j.eswa.2024.123876
23. Morales Murillo VG, Pinto Avendaño DE, Rojas Lopez F, Gonzales Calleros JM (2022) A systematic literature review on the hybrid approaches for recommender systems. Computación y Sistemas 26:1. https://doi.org/10.13053/cys-26-1-4180

24. Hussain T, Yu L, Asim M, Ahmed A, Wani MA (2024) Enhancing E-learning adaptability with automated learning style identification and sentiment analysis: a hybrid deep learning approach for smart education. Information 15(5):277. https://doi.org/10.3390/info15050277
25. Lesan Sedgh MM, Latif A, Emadi S (2024) A novel method for a technology enhanced learning recommender system considering changing user interest based on neural collaborative filtering. Data Sci Manag S266676492400050X. https://doi.org/10.1016/j.dsm.2024.09.004
26. Xu G, Jia G, Shi L, Zhang Z (2021) Personalized course recommendation system fusing with knowledge graph and collaborative filtering. Comput Intell Neurosci 2021(1):9590502. https://doi.org/10.1155/2021/9590502
27. Ujbanyi T, Sziladi G, Katona J, Kovari A (n.d.) ICT based interactive and smart technologies in education—teaching difficulties. 3(10)
28. Badshah A, Ghani A, Daud A, Jalal A, Bilal M, Crowcroft J (2024) Towards smart education through the internet of things: a review. ACM Comput Surv 56(2):1–33. https://doi.org/10.1145/3610401
29. Nguyen L, Tomy S, Pardede E (2024) Enhancing collaborative learning and E-mentoring in a smart education system in higher education. Computers 13(1):28. https://doi.org/10.3390/computers13010028
30. Irfandi I, Festiyed F, Yerimadesi Y, Sudarma TF (2023) The use of learning management system (LMS) in the teaching and learning process: literature review. Jurnal Pendidikan Fisika 12(1):81. https://doi.org/10.24114/jpf.v12i1.42270
31. Singha S, Chakraborty A (2024) Educators' learning experiences and intention to use online learning management systems' platforms: a perceptual study. Serbian J Manag:5–5. https://doi.org/10.5937/sjm19-45512
32. Krishnan R, Nair S, Saamuel BS, Justin S, Iwendi C, Biamba C, Ibeke E (2022) Smart analysis of learners performance using learning analytics for improving academic progression: a case study model. Sustainability 14(6):3378. https://doi.org/10.3390/su14063378
33. Alam T (2020) Cloud computing and its role in the information technology. IAIC Trans Sustain Digit Innov (ITSDI) 1(2):108–115. https://doi.org/10.34306/itsdi.v1i2.103
34. Al-Emran M, Shaalan K (eds) (2021) Recent advances in technology acceptance models and theories, vol 335. Springer International Publishing. https://doi.org/10.1007/978-3-030-64987-6
35. Kishan Raj Bellala (2025) "Driving Business Transformation: Exploring the Power of Workday as a Cloud-Based Solution." International Journal of Innovative Science and Research Technology (IJISRT) 10(6):1859–1865. https://doi.org/10.38124/ijisrt/25jun1229
36. Maharani C, Nutela I (n.d.) Exploring the transformative potential of cloud computing in distance education: Challenges, opportunities, and best practices. 2
37. Lazaro GR, Duart JM (2023) Moving learning: a systematic review of mobile learning applications for online higher education. J New Approach Educ Res 12(2):198–224. https://doi.org/10.7821/naer.2023.7.1287
38. Magomadov VS (2020) The application of artificial intelligence and big data analytics in personalized learning. J Phys Conf Ser 1691(1):012169. https://doi.org/10.1088/1742-6596/1691/1/012169
39. Baltezarević R, Baltezarević I (2024) Students' attitudes on the role of artificial intelligence (Ai) in personalized learning. Int J Cogn Res Sci Eng Educ 12(2):387–397. https://doi.org/10.23947/2334-8496-2024-12-2-387-397
40. Wang X (2024) Online and offline integration scheme of college english education under big data technology. J Cases Inf Technol 26(1):1–13. https://doi.org/10.4018/JCIT.348963
41. Zheng L, Wang C, Chen X, Song Y, Meng Z, Zhang R (2023) Evolutionary machine learning builds smart education big data platform: data-driven higher education. Appl Soft Comput 136:110114. https://doi.org/10.1016/j.asoc.2023.110114
42. Badshah A, Nasralla MM, Jalal A, Farman H (2023) Smart education in smart cities: Challenges and solution. 2023 IEEE International Smart Cities Conference (ISC2), 01–08. https://doi.org/10.1109/ISC257844.2023.10293615
43. Dr. Ranbir (2024) Educational technology integration: challenges and opportunities. Innov Res Thoughts 10(2):75–79. https://doi.org/10.36676/irt.v10.i2.1406

44. Zhao D, Hu Z, Yang Y (2022) An MCTS-based recommender system for education complex. 2022 International Conference on Machine Learning and Knowledge Engineering (MLKE), 323–326. https://doi.org/10.1109/MLKE55170.2022.00068
45. Yang H, Anbarasan M, Vadivel T (2022) Knowledge-based recommender system using artificial intelligence for smart education. J Interconnect Netw 22(Supp02):2143031. https://doi.org/10.1142/S0219265921430313
46. Oussouaddi R (2023) Research Article Journal of Mathematical Techniques and Computational Mathematics Advance in Online Education Recommender Systems During and After Covid-19 a Survey
47. Yin W (2023) Zero-knowledge proof intelligent recommendation system to protect students' data privacy in the digital age. Appl Artif Intell 37(1):2222495. https://doi.org/10.1080/08839514.2023.2222495
48. Qin, Y., Zhu, J., & Huang, J. (2023). Privacy-preserving federated learning framework in knowledge concept recommendation. Y. Sun, T. Lu, Y. Guo, X. Song, H. Fan, D. Liu, L. Gao, & B. Du (Éds.), Computer supported cooperative work and Social computing (Vol. 1681, p. 411–421). Springer Nature Singapore. https://doi.org/10.1007/978-981-99-2356-4_33
49. Li W, Zheng X, Huang R, Lin M, Shen J, Lin J (2024) Enhancing privacy protection for online learning resource recommendation with machine unlearning. In: 2024 27th international conference on computer supported cooperative work in design (CSCWD), pp 3282–3287. https://doi.org/10.1109/CSCWD61410.2024.10580315
50. Lo GY, Tan TPL (2023) Dynamic and private recommendation system. In: 2023 International conference on digital applications, transformation & economy (ICDATE), pp 1–5. https://doi.org/10.1109/ICDATE58146.2023.10248849
51. Ismail, I. A., & Aloshi, J. M. (2024). Data privacy in AI-driven education: an in-depth exploration into the data privacy concerns and potential solutions. In K. L. Keeley (Éd.), Advances in educational marketing, administration, and leadership 223–252). IGI Global. https://doi.org/10.4018/979-8-3693-5443-8.ch008
52. Achyuth, B., & Manasa, S. (2021). Implementation of recommendation system and technology for villages using machine learning and IoT. V. Suma, N. Bouhmala, & H. Wang (Éds.), Evolutionary computing and mobile sustainable networks (53, 527–542). Springer Singapore. https://doi.org/10.1007/978-981-15-5258-8_49
53. Kumar M, Singh TP, Choudhury T, Gupta SC (2019) ICT—the smart education system in India. In: 2019 International conference on contemporary computing and informatics (IC3I), pp 279–282. https://doi.org/10.1109/IC3I46837.2019.9055562
54. Shyam Mohan, J. S., Sreeman, V. H., Chakradhar, V. V., Kota, H. S. A., Panini, C. N., Vemuganti, N. K. A., Kuladeep, N. V., Nadipalli, V. R., Swaroop, S. S., & Subramanyam, M. U. M. (2021). Recommendation system for adoption of ICT usage and its impact on academic performance of students in India. In S. Kumar, S. D. Purohit, S. Hiranwal, & M. Prasad (Éds.), Proceedings of international conference on communication and computational technologies (p. 257–265). Springer Singapore. https://doi.org/10.1007/978-981-16-3246-4_21
55. Ginting NB, Widhyaestoeti D, Jaenuddin J, Rachmawati F, Maolani Y (2023) Implementation of the smart method in the decision support system for penalty recommendations. JITK (Jurnal Ilmu Pengetahuan dan Teknologi Komputer) 8(2):133–138. https://doi.org/10.33480/jitk.v8i2.4138
56. Firdausi, N., Fithriyah, N. N., Ummah, R., Abidin, Moh A., Nirwana, R., Achmad, W., & Kuswanto, F. (2024) Integrating smart education in basic education: a systematic review of roles and challenges. 2024 International Conference on ICT for Smart Society (ICISS), 1–6. https://doi.org/10.1109/ICISS62896.2024.10751462
57. Department of Secondary Education, William V.S. Tubman College of Education, University of Liberia, Monrovia, Liberia, Kennedy GM (2023) Challenges of ICT integration in teachers' education: a case study of the College of Education, University of Liberia. Int J Soc Sci Educ Res Stud 03:05. https://doi.org/10.55677/ijssers/V03I5Y2023-15
58. Ayman SE, Abo El Rejal A (2023) Ontology and machine learning-based recommender system for teacher resource personalization. Educ Sci Manag 1(3):145–157. https://doi.org/10.56578/esm010303

The Contribution of the Information and Communication Technologies in Developing the International Experience of Smart Education

Fahd Mohammed Aldosari, Amar Yusof Jaffar, Wesam Nabil Eid, and Kanakaprabha Selvakumari

Abstract The way knowledge is provided, accessed, and experienced is being revolutionized by the incorporation of Information and Communication Technologies into education. In order to create dynamic, individualized, and inclusive learning environments, ICT is used as a catalyst in the concept of Smart Education, which is covered in detail in this chapters. Through the use of technologies like artificial intelligence, cloud computing, and mobile learning, Smart Education provides teachers and students with resources that increase student engagement, boost learning outcomes, and close educational gaps. The chapter examines how new technologies are affecting conventional educational paradigms, identifies effective applications in a range of educational contexts, and talks about the potential benefits of ICT-driven smart education in the future for creating a more inventive and interconnected learning ecosystem.

Keywords Artificial intelligence · Education · Communication · Information · Knowledge · Mobile learning

F. M. Aldosari · A. Y. Jaffar
Department of Computer and Network Engineering, College of Computing, Umm Al-Qura University, Makkah, Saudi Arabia

W. N. Eid
Department of cybersecurity, College of Computing, Umm Al-Qura University, Mecca, Saudi Arabia

K. Selvakumari (✉)
Department of Artificial Intelligence and Machine Learning, Hyderabad, India

H. Kaddouri et al. (eds.), *Smart Cities for Sustainable and Urban Development*, Green Energy and Technology, https://doi.org/10.1007/978-3-031-99001-4_8

1 Introduction

The use of cutting-edge information and communication technologies (ICT), smart education presents an important shift in the way that learning and teaching take place. It aims to offer more engaging and specific learning experiences. Smart Education applies tools like artificial intelligence, online learning platforms, virtual classrooms, and mobile technologies to improve access and meet the needs particular students, compared with standard educational models that mainly depend on fixed curriculums and classroom settings. The support of ICT integration, education can now implement a more adaptable, learner-centered approach, students and teachers access to a variety of resources and information that earlier might have remained out of reach due to institutional and geographic limitations. Modern Education is made possible by ICT, which has grown from simple digital devices like computers and projectors to advanced system like online platforms, educational management systems and virtual reality. This chapter analyses the role that smart education plays in developing changing, real-time learning environments. It also examines how smart education deals with the drawbacks of the traditional educational system and the tools they need in the quickly evolving digital world. This book chapter's main goal is to discuss how information and communication technologies (ICT) is contributed to smart education. How they have affected in modern learning environments. It aims to provide a thorough study of the technological development of smart education and traditional education. The analysis can be done various case studies and literature review in the help of ICT classroom. And also, challenges, and future directions of ICT integration in institutes around the world.

1.1 Definition of Smart Education

Smart Education is an innovative approach to education that make use of advanced technology to design more efficient, personalized, and engaging classrooms. By mixing digital tools, data analytics, and internet connectivity, it expands further standard educational environments and improves teaching methods and student participation. The learning in smart education is continuing, flexible and accessible to students in anywhere and anytime. It is not restricted to a certain time and location. The technologies are essential to creating personalized learning and experiences is correspond to each student's specifications, choices, and progress include artificial intelligence, augmented reality, cloud computing and mobile applications (Fig. 1).

In addition to prepare students for the needs on the knowledge, familiar with technology society, the primary focus is on enabling analytical thinking, problem-solving and creativity, and lifelong learning with the goal of overcoming the gap between traditional education and the quickly changing digital reality.

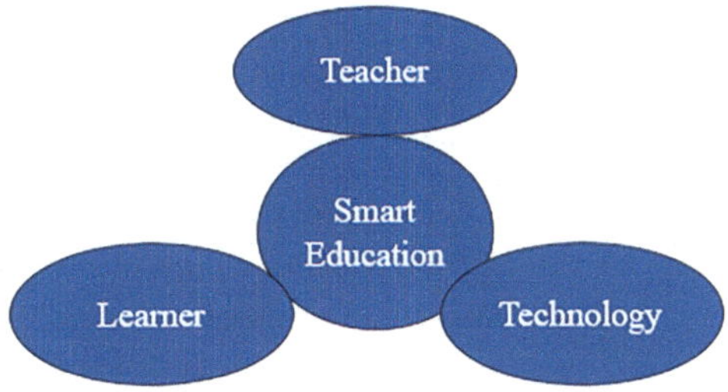

Fig. 1 Smart education

1.2 Role of Information and Communication Technologies (ICT) in Modern Education

The information and communication technologies are very essential to modern education, transforming the sharing use, and transfer of information. The use of ICT students as well as teachers have access to a wide range of educational materials, allowing them to go above restrictions of traditional educational environments. ICT allows for real-time contact, collaboration, and swift feedback between educators and learners, whatever the barriers to travel, through the integration of technological tools including cloud-based systems, smart devices, e-learning platforms, and video calls. The ICT in education now to promote personalized learning experiences as well as to enhancing learning efficiency. The performance of students' technologies such as Machine learning, Artificial Intelligence are enabling teachers to adapt training of each student's unique learning needs and styles. The ICT supports accessible education through offering students with disabilities access to assistive technology, thus ensuring properly potential for learning. Furthermore, the students are changing and captivating learning environments like virtual labs, other learning-based technologies in ICT provides for creativity and critical thinking. The ICT is the foundation of life learning as education becomes is digital. It allows for continuous improvement of skills through online courses, remote learning, and live session. The ICT plays a advanced role in modern education by giving students the digital literacy and problem-solving is needs the succeed in the global information economy.

1.3 Historical Evolution of Education Technologies

The process that technologies for education have developed has been dynamic, taking consideration of both the developments in technology and the shifting demands of society. The education technology continues to alter the nature of teaching and learning, from the first devices like the keypad to the most advanced digital platforms accessible today. In the 19 century the development of the board who allowed teachers teach a large class of students simultaneously. This is the first example of multimedia tools being used in schools. Many of use in printed books and libraries are increased to access the knowledge as the industrial era to move on, and education became stricter. There were several significant in educational technology during the twentieth century. Distance learning was originally made possible by the invention of radio and television, which is used for students to receive any program from long distance. The emergence of audiovisual devices during this era, film projectors, enhanced the educational process by providing variety of information delivery methods. The computing devices in the previous stage of the twentieth century the field of educational technology is significant shift. The software and computer-assisted instruction systems is appeared in classrooms, and the groundwork for the digital revolution in education. The online courses, and universal access to the materials is made up of possible by the internet's availability. The emergent technologies such as mobile technology, cloud computing and other management systems in the twenty-first century is changing the blended learning and other e-learning approaches. The levels of interaction are possibly made by technologies like augmented reality, virtual reality, and artificial intelligence. These technologies can be accessed personalized materials and learn at their own speed through learning pathways. Today's educational technologies is developing the advancements in artificial intelligence, augmented reality, virtual reality and 5G technology. This advanced method is response to the demands for accessible, scalable, and educational programs. It is more digitally and networked.

1.4 Importance of Smart Education in the Digital Age

Smart Education is drawback of traditional education in schools and prepared students for the world driven by technology, has emerged in the crucial element of the learning systems in the digital age. The fastest technological change is enables students gain the digital skills as well as critical thinking. They can succeed in the knowledge-based learning. The main benefits of smart education are individualized education. The learners can advance methods to maintain their own speed as smart education is personalized learning experiences to meet each student's unique needs by leveraging tools like artificial intelligence. The students can overcome barriers more skillfully and gain a deeper comprehension of the complex topics with the support of flexible education system. Smart Education is promoting in participation and communication of virtual world simulation and multimedia. The technologies like virtual reality and augmented reality student involvement are more in information by learning in practical. Active learning is encouraging teamwork and problem-solving abilities. And smart education is more identified in the job market. Many of the artificial intelligent and digitalization tools like the industries is need to periodically many skills. The smart education which offers scalable online courses and chances for microlearning the students may keep up with the times and maintain their labor market. Furthermore, Smart education is extremely flexible in learning situations due to its efficiency and scalability. Many of the learning is used to business training, individual studies, and formal education. The smart education is providing adaptable, cost-effective options can be identified in the student without any surrendering quality. And also, it is the digital era because it can build a dynamic, adaptable, and inclusive learning environment of that meets students' needs and digitally native future.

2 Literature Survey and Research Gap

2.1 Literature Survey

Language classrooms now regularly include the use of information technology for teaching and studying foreign languages [1]. As a result, a plethora of software applications and educational platforms have been developed, helping as the ideal supplement to achieve proficiency and fluency. Language courses that include technological assistance are the most appealing and successful for students who wish to succeed in their language studies. In addition to helping students improve their language skills, the use of ICT in language learning offers up a wealth of new opportunities for effective communication between teachers and students. It also expands their understanding of how to utilize ICT effectively to improve teaching and learning.

The application of information and communication technology in the teaching, learning, and evaluation processes of today [2]. By defining instruction and enhancing students' cognitive abilities for the information age, the primary objective of integrating information technologies into the learning environment is to improve educational standards. Because it increases students' propensity for innovation and entrepreneurship, one of the questions dealt with the IT development and digital level of the universities. 43 percent of students think that their universities have been sufficiently digitalized, according to poll data. In order for universities to stay profitable, universities need to solve this issue. Forty percent of participants were neutral, and 17 percent expressed objections about the ICT equipment.

The education sector is going through changes and problems as a result of 5G, the most recent generation of cellular mobile communication technology, which is drastically altering our way of life and productivity [3]. In addition to combining artificial intelligence, virtual reality, the Internet of Things, cloud computing, big data, and other well-known information technologies, which combines outstanding information such as high-speed data transmission rate, ubiquitous network, low latency, and low energy consumption fifth generation mobile communication technology. The introduction of the "5G +" era, is significant impact on the field of education, has been examined in relation to the use of technology in education.

The information and communication technology (ICT) is evolving so quickly; different countries have been actively working to strengthen their national economies through the development of Smart Plus [4]. The movement toward intelligentization, or smart education, began as a result of the advancement of ICT in education. Without a common definition and semantic analysis, there is an important gap in the quality of smart education. Next, altering the definition to "smart education" from "smart/wisdom." Additionally, comparing the various meanings of smart education that academics both locally and globally have addressed will help to clarify the divisions between smart education and technological support, data application, and participant participation in order to further significant learning areas.

Information and communication technology (ICT) is advancing quickly, turning an ordinary community into a smart village [5]. A national initiative called "smart villages" uses ICT elements to jointly co-create a better living future. The primary idea behind the idea of "smart villages" is that technology should improve welfare and health, support local businesses, and facilitate education. In other words, technology should act as an agent for development, facilitating access to electricity, renewable solar energy, clean water, and food security for rural populations. India needs to find a smart method to manage complex and cut costs while increasing efficiency in order to overcome the nation's increasing urbanization.

The difficulties of information and communications technology (ICT) support of scientific activities in innovation networks and education at Higher Education Institutions (HEIs) by emphasizing various tactics and ideas for ICT infrastructure management [6]. The primary problems and examples of ICT use in HEIs for technology management of transfers have been taken into consideration, along with the development of HEIs' smart environments and business research projects. A more comprehensive grasp of ICT-based HEI cooperative marketing networked strategies has been obtained through an empirical investigation of such scenarios.

The development of national education is supported by the teaching of information technology [7]. The implementation of education informatization advantages from an understanding of the state of research on information technology education both locally and globally. The paper mainly performs statistical analysis via the journal distribution, hot phrases, and time distribution of the literature. It additionally offers references for domestic and international research on the information education in the technology discipline and through visuals describes the features of the literature in this field.

College and university educational informatization is emerging as an important issue in the course of education reform [8]. The establishment of informatization in colleges and universities has attracted a lot of attention at a time when the nation is promoting the growth of educational informatization. The goal of building educational informatization in colleges and universities is to use technology to optimize and reform teaching and learning in order to develop creative, practical, and comprehensive skills that are better suited to the demands of modern social development.

The development of higher education, an increasing number of subjects have included computer-aided instruction, network resources, and instructional materials into their lessons [9]. These subjects have also developed their own distinctive features for network teaching platforms and teaching resources, which are extremely helpful for teaching because of their flexible interaction, network across time and space, and advantages for sharing information resources.

The study that is being presented focuses on an expert assessment from the fields of science and pedagogy of the factors that contributed to the shift from Web 2.0 to Web 3.0 in information and communication technology [10]. This assessment is crucial for the author's stated more global task of scientific research, which is the computerization effect of the essence of knowledge in terms of distribution, generation, and effective assimilation by participants in terms of scientific and pedagogical conditions.

A greater focus has been placed on classroom equipment replacement, Internet of things technologies, and education reform [11]. Nowadays, the classroom uses open management to help students learn, but this also creates challenges for managing the lighting, air conditioning, multimedia, and other equipment in the classroom. Modern teaching method called "smart classroom," which makes use of Internet of Things management technology to not only realize intelligent control over the local teaching equipment and environment, but also allows for the realization of system functions such as broadcasting, security monitoring, access control and attendance, electronic class cards, and demonstration course recording. Based on Internet of Things RFID technology, the integrated information management system of multimedia equipment in smart classrooms is intended to fulfill the following functions: managing personnel, recording equipment use, maintenance, and operation data in realtime, and facilitating staff access to a wide range of multimedia equipment information. This will enable staff members to efficiently process and comprehend a multitude of multimedia equipment data in a smart classroom, ensuring the smooth and organized operation of teaching activities.

A major component of community governance during the Chinese epidemic was played by digital information technology [12]. Following the epidemic, research on the evolution of digital information technology in community governance demonstrated "data islands" in the electronic direction of community governance during the following epidemic, as well as a low and inactive use rate of digital information systems. The use of digital information technology in community governance by developing a digital information platform for community members, putting forth a model for gathering information from community members, bolstering the development of the community's "new infrastructure" through digital information tools like WeChat groups, mini-programs, and APPs, and boosting the use of digital information tools like location bracelets during the community's quarantine period of the epidemic.

A lot of information has been released with new smart city technologies, etc., and users will need to make sense of this information [13]. For it to do this, user education systems, or UESs, in smart cities will need to be implemented in combination with the introduction of new technology. The aim of this article is to analyze the theoretical basis of the widespread use of blended learning-based User Education Services in smart cities. The idea for the User Education System developed because to the study that was done. A system of specific expertise learned by users from modern smart systems and technologies in informal education is referred to as the User Education System.

The need for educational institutions to adapt new forms of teaching and learning is being increased by current advances in information and communication technology (ICT) [14]. Since the beginning, technology use in education has generated a lot of interest and been used in many different areas of study. The development of ICT has prompted academia to seek out the most effective ways to use and deploy various ICT tools. An extended TAM (ETAM) has been intended to emphasize the need for implementing advanced ICT tools in the context of education due to the rapid developments in ICT.

The environment and 5G network support, the fundamental activities of education are also undergoing reconstruction and change [15]. Users in the education industry have a wide range of wants, and the business for each need is quite accurate because of the diversity of users and situations. The lack of capability of current education services to provide high-speed, high-bandwidth, low-latency, rapid cache, and other services is an important problem which the 5G environment can resolve.

The industry of education that is only gaining traction focuses on discovering methods to use smart technology to enhance education while preserving sustainability [16]. The goal of this study is to assess how effectively teachers use smart technology in the classroom, how they feel concerning the impact of utilizing smart technology, and how important it is for both of these factors in order to predict teachers' usage of smart technology. Tools created by the researcher with the help of prior literature are used to measure the variables under study, namely Smart Teaching (ST), Perceived Impact of Smart Technology in the Classroom (PI-STech), and Use of Smart Technology for Teaching (U-STech).

The artificial intelligent digital education pays more and more focus on artificial intelligence (AI) [17].The two popular CI paradigms ontological engineering and case-based reasoning—are thoroughly investigated and discussed in this work. The main objective of this research is to recognize and analyze the advantages of these intelligent paradigms with the objective to improve the smart tutoring systems' efficiency and effectiveness.

The governance of HEI is greatly affected by the role of information technology; thus, in order to support extending business processes, HEI needs an information system building that is accessible to all users at all hours and settings [18]. With the help of the COBIT (Control Objectives for Information and Related Technology) framework approach, this study attempts to assess the degree to which higher education institutions apply academic information system governance to support IT-based education services.

The term "smart education" was used to refer to educational progress supported by intelligent technology. Furthermore, advances in information and communication technology (ICT) have sped up worldwide efforts to advance smart education [19]. The field of ICT-assisted education is experiencing a rise in interest from various industries and a rise in research output as a result of this pace. The issue is so popular that comprehending its growth, advancement, and current state needs quantitative research.

The management of the online learning process, qualifications, and other data is still far from perfect, but with the rapid development of network technology, online education has become vital for the whole educational system and receives more and more attention from educational institutions [20]. The study recommends employing blockchain technology in online education to tackle these problems. This technology enables the management of student data, learning materials, fulfillment certification, and job opportunities for students.

The Internet of Everything (IoE), advanced surveillance technologies, and computer vision, the system offers increased safety and convenience for parents, children, schools, drivers, public transportation, law enforcement, and the government [21]. The system's design architecture has been guided by the qualitative research approaches that the research methodology included to get useful data from stakeholders. Real-time surveillance and monitoring of children during their transportation journey is made possible by the system's integration of computer vision abilities including object detection and facial recognition. The seamless connection enabled by the IoE integration makes it feasible for surveillance cameras, centralized monitoring systems, and IoT-enabled vehicles to interact and share data efficiently.

The offers a sports information fusion near based on the design and processing of intelligent data gathering hardware with the aim to solve the problems of huge quantities of educational resource data, time-consuming sports information processing, and excessive use of resources [11]. Sports office software, sports multimedia technology, sports information resource retrieval, sports collecting information technology, sports data mining technology, etc. are some of the contents of the experiment.

The education sector is not an exception to the revolutionary possibilities brought about by the emergence of blockchain and machine learning technologies [22]. Because blockchain is an unchanging, decentralized distributed database, it can solve some of the fundamental issues with the present educational system and possibly change it.

New technologies and the availability of strong connections are becoming more and more crucial [23]. Along with its many benefits, this cutting-edge technology has drawbacks, chief among them security-related issues. One of the most well-known technologies of the past several years, distributed ledger technology is becoming more and more popular since it can simply guarantee anonymity in a number of areas, such as transportation and logistics management, among other businesses. The educational sector ought to take advantage of the significant benefits that blockchain technology provides. Schools and universities—more especially universities—are now taking this technology into consideration as a way to improve the teaching-learning process and promote involvement from all parties involved, including students, instructors, and even family members.

The technical basis and full support system for intelligent education have been made possible by the profound integration of information technology and education [24]. The smart learning environment built on the smart education system will advance significantly with the introduction of 5G. This essay explores the issues that the smart learning environment faces in the context of 5G, introduces the ideas and technical features of 5G, and describes the contents of the smart learning environment. It also examines the technical makeup of the smart learning environment in relation to 5G.

2.2 *Research Gap*

The research gap in the ICT education international sector such as new technology usage, digital equity in the context of international education. The digital learning platforms like online collaboration tools is changed the education sector in world wide. The digital divide is the low-income nations and access to high quality education. Furthermore, how ICT improves the long-term learning in the cultural environments. Moreover, the ethical consideration is the data privacy, cybersecurity in the ICT education. How effectively improve Augmented Reality and Virtual Reality to improve the experience international students to global academic relationships. Addressing to these gaps can be more strong, efficient ICT framework for the international sector might be for entire education.

3 Technologies Driving Smart Education

The foundation of smart education is technology, which traditional teaching into a flexible, effective, and collaborative process. Modern technologies like artificial intelligence and Machine Learning allow for personalized learning by recommending content based on the data analysis. The simulations and virtual experience, augmented reality and virtual reality hands-on-learning environments that can be help students' complex subjects. The importance of cloud computing in smart education it makes resources easily accessible and makes the possible for teachers and students to work together. The integration of internet of Things devices into classrooms. These devices are connecting smart devices to track student achievement, administrative and improve engagement. Mobile Learning Platforms and apps allow students can access the educational content. These enhancing learning and making more student-centered, dynamic, and accessible, these devices are combinedly together. Smart Education the demands of the digital age.

3.1 *Overview of Key Technologies (AI, AR/VR, Cloud Computing, IOT, etc.)*

Smart education is currently experiencing an advancement caused by the use of cutting-edge technology, which is more effective, specific, and efficient. Artificial Intelligence is essential is adaptive learning platforms, which is allow for a specific learning experience by personalizing in the educational content of the student's needs. It is automated tutoring systems offer learners on the spot assistance. The virtual reality and augmented reality improve immersive, hands-on-learning. It improves understanding of difficult concepts by generating real-world environments in scientific engineering, and medical fields. The students to interact with virtual models and simulations, these students' engagement, and application. Cloud computing has transformed the accessibility of materials. The large volumes of data can be stored and managed by institutions and both students, teachers can be access of working environment. The E-learning platforms is made up of cloud-based learning Management systems(LMS) which provide central location for assignment, seminars, courses, tests, and webinar. The internet of Things is another essential technological advancement of the power smart education. The interactive tools like wearable technology, smart boards are the examples of internet of things. It is improving the learning environment by offering real-time data on student performance, classroom circumstances, and administrative activities. The IoT makes learning more dynamic and easier for teachers to each student's requirements and their instruction accordingly. The main goal of preparing students for life in a tech-centric future, these tools are transforming educational methods and accessibility and personalized learning.

3.2 Use of Mobile Learning and e-Learning Platforms

The smart education, mobile and e-learning platforms is essential tools for students, and access all the studying materials at any time and from any location. The mobile learning is allowing for the students can interact with their homework, attend the virtual lectures. The use of smartphones, tablets and high-speed of internet. The traditional educational resources are working professionals of the distant learners from this flexibility. The students can be interests at their own speed by taking advantages of the numerous online courses such as Coursera, Udemy and Khan Academy. To improve engagement and collaboration of the many students, these platforms are frequently incorporate interactive elements such as discussion boards, live sessions, and quizzes (Fig. 2).

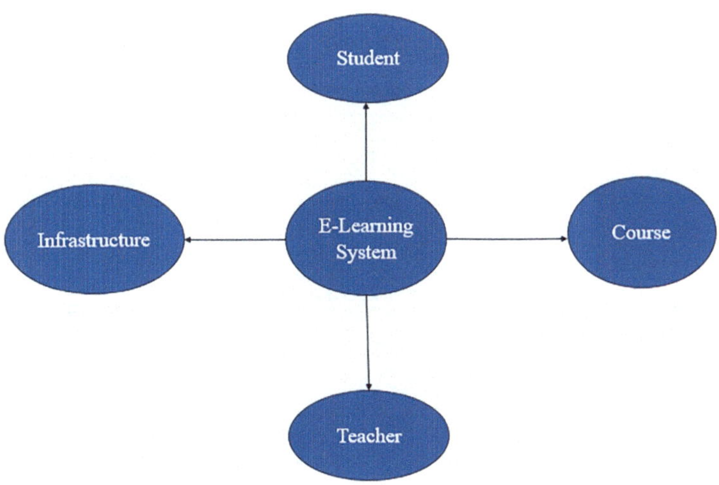

Fig. 2 E-learning system

The learning process is also made more interactive and customized elements like video lectures. The location and financial status and providing access to top-notch for worldwide audience, and e-learning platforms both democratize education. They promote for lifelong learning by providing possibilities for online courses, badges these are enable to students can continually improve their abilities in the market. The way of education is mainly offered by mobile and e-learning platforms in the demand of the digital age.

3.3 Smart Classrooms and Digital Learning Tools

The smart classroom, which combine cutting-edge digital learning resources with conventional teaching techniques to provide effective, and student-centered experience in the educational outcomes. The classroom is outfitted in the interactive whiteboards, smartboards and linked devices is that promote teacher-student in real-time. The teachers can be providing dynamic multimedia information such as movies, virtual experiments in the classroom in advanced technology. Teachers can handle syllabus, exams, assignments in a single platform with the help of the digital learning tools include learning management systems in educational applications and digital textbooks. These tools can be facilitated personalized by learning in monitoring student in progress and offering data-driven of the teachers in the new methods of teaching. The student's retention and understanding of the many subjects are improved by the wide range of online materials of interactive tests of the digital tools. Furthermore, the customized feedback and suggestion on the student success, artificial intelligence in smart classrooms is improves the personalized learning (Fig. 3).

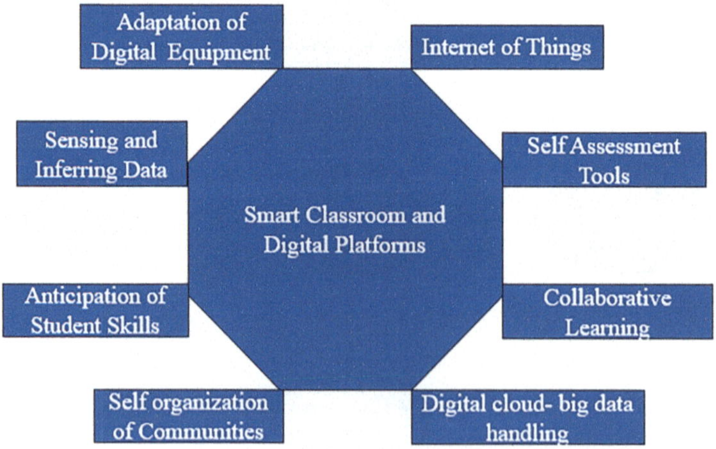

Fig. 3 Smart classroom and digital platforms

The use of many tools like augmented reality and virtual labs and simulation-based learning, The students can practice and experiments in risk-free virtual environment. The critical thinking and active engagement of these tools can be change in the passive learning of the flexible learning environment. Education is more effective and engaging, digital learning resources and smart classrooms is ensured that students can have more skills is succeed in the digital economy for the demands of a technologically in modern society.

3.4 The Role of AI in Personalized Learning Experiences

Artificial Intelligence is plays major role in transforming education. It is uses adaptive learning systems to large amounts of data, including behavior, student performance and preferences to the learning material. The students receive training is needs for depending on their progress, is algorithms detect for the knowledge gaps, and suggest relevant resources and other modify learning routes. The intelligent systems with capabilities of one interaction with students is provided by feedback and assistance. These help of the tools, which may provide individualized advice replying to each question, and provide to exercises, students can get right away without having to wait for the teacher assistance. Artificial Intelligence helps with automated assessments, which ease the workload on teachers by evaluating the assignments with errors of performance insights. Furthermore, gamified learning environments in which are modified by keep students can be challenged without becoming too difficult in another way. Artificial Intelligence is each learner realize the full potential by personalizing the educational, deeper learning and outcomes and student autonomy. Artificial Intelligence empowers teachers and students by fostering a more dynamic, personalized educational environment.

4 ICT for Inclusive Education

Information and communication technologies is complete idea of equitable learning through eliminating hurdles and ensuring to everyone may be accessed by education, physical, cognitive, or other limitations. The personalized educational materials to identify the different demands. The ICT tools are creative solutions for the students with problems. For example, the technological aids like text to speech, screen readers and software are visually identified to students they can access learning materials. The students can access the language and closed-captions for help the students with hearing losses interact with audio-visual content. Online learning environment and mobile technologies, ICT fills the gap in rural and distant places where access to excellent educational opportunities is limited and they student's participation in the educational ecosystem. Moreover, the students can support the cooperative educational environments is include the foster inclusion and other culture learning is enables the various student group to participate and learn from one another. The ICT is individualized instruction, which is enables from teachers to modify classes to each student's particular style of learning and identified by students from different backgrounds and levels of ability is equal opportunities to succeed. The ICT is promoting the more inclusive of educational settings by enabling these advances in technology and every learner has access the tools to support the require.

4.1 How ICT Enables Inclusive Education for Differently-Abled Learners

The provision of tools that can be addressed by the needs and problems, ICT is plays a significant role is ensuring that education is inclusive and accessible for the individuals and impairments. The students can be visual impairments in read, write and explore the digital content of the assistive technologies. The screen readers, text-to speech systems and voice recognition of the software. The learning platforms that can use of audios, videos, sign language and interpreters to make for the possible students who are hard of hearing to interact with instructional materials in the way of sense them. The adaptive keyboards, interactive touchscreens, visual tracking software with the students can be mobility issues is engaged with learning platforms is more readily. The personalized learning environments is adapted to a student's speed, style of learning. The ICT tools can be more customized effects of every student receives is enabling more inclusive and effective learning environment. The physical barriers and visiting traditional educational institutions can be engage more completely from the home with the remote learning and virtual classrooms. The ICT facilitates academic and personal development of learning environment to overcome the problems and fully engage with educational process.

4.2 Digital Tools for Special Needs Education

The digital technologies are grown into an essential device for those with special education, particular solutions that address the difficulties of the students with problems encounter. The use of assistive technology such as display screens, touch readers and programs for the clarity, accessible for visually impaired devices for students when they interacting with digital content and performing reading and writing assignments. The tools like captions, speech to text software, and sign language video platforms offer students with hearing loss alternate methods to access the lectures and content of multimedia. Text-to- speech technologies, and enable students with learning difficulties interact the textual materials in ways of the best fit their cognitive processing capacities. The tools are improved social and communication skills the visual schedule for each subject and social applications, language therapy for the software's of the students with autism disorders. Furthermore, non-verbal children are more interested to engage in experiential learning by using augmentative and alternative communication (AAC) tools such as speech-generating electronic devices and message boards. These are provided by inclusive and flexible learning environment of the tools are supported by the kids' education of the needs, enhancing the academic performance and the feelings of emotions. The education is made more personalized through using these digital tools, a nurturing environment that accommodates of the unique needs of each student.

4.3 *Bridging the Rural-Urban Education Divide Using Technology*

The academic is dividing between the rural and urban areas can be identified in the significant portion due to technology. The students can be remote places access to changes with high-quality of learning resources is where unreachable. The help of mobile and e-learning platforms education is completely changed from the rural students. They are attending the classes in online, browse digital library, and engage the educational experiences from their own homes. The removal of geographic obstacles of these resources ensures that students in rural areas of the education of the same level of those who lived in metropolitan area. The increased usage of internet and electronic devices made simpler. The smart classroom and digital resources at the finger tips may develop virtual exchanges of the online cooperation and collaborative learning with urban institutions. This improves the better quality of education for the students in both remote and city through the collaborative learning and cultural interaction. Furthermore, learners in city areas can be accessed specialized course in vocational training, and other development programs. The technology can be acts as the potent equalizer, eliminating the lines between rural areas and urban areas advancing educational equity in the worldwide basis by accessing rural areas with access to the high-quality education and resources.

5 Challenges and Barriers to ICT Integration

The advancements in accessibility, engagement, and learning the outcomes from the integration of information and communication technologies in the worldwide. This chapter shows the ways in which technology is conventional approaches to learning in case studies. The successful evolution of ICT in various educational contexts. These are the particular examples of how ICT is making the easier to obtain and from the widespread of the use of smart classrooms in the arrival of mobile learning systems in rural areas. The country's approach to teaching is transformed by the introduction of online educational environments, which is enabled more individualized and the instruction. The have used in mobile and e-learning technology to close the educational gaps in the rural areas, giving the students access to online courses and digital libraries. The educational management and literacy rates are increased dramatically in these measures. The united states of AI-driven adaptive learning systems has been improved by the teaching effectiveness and efficiency by providing the individualized instruction of the students with level of learning. This chapter is also shows the how ICT is not only updating the education of the access to high-quality of education, irrespective of geography and the case of these examples.

5.1 *Infrastructure Challenges in Implementing ICT*

The application of ICT in education is infrastructure problems, especially in emerging and rural areas, considering its promise for innovation. These digital tools and the online educational resources of the mostly due to the insufficient of the high-speed of internet for teachers and students. The technology-based resources of the difficult in many locations due to the lack of energy and reliable power which contributes to the problem. Furthermore, the ICT technology as well as schools by the high cost of smart equipment such as PC, tablets, and interactive whiteboards. Another challenge is the lack of digital tools for teachers and students. The teachers are finding it difficult to effectively integrate in the technology into the planned lessons, and students may not be in online educational settings. The playing field between urban and rural areas of the locations in the differences in technical infrastructure in urban schools from the ICT, The Government funding, public-private partnerships, and capacity-building initiatives must be combined to overcome these obstacles and guarantee that ICT can be enhance education for all.

5.2 *Teacher Training and Readiness*

The level of preparedness and ability of educators to use digital resources and platforms is vital for the successful of ICT education. However, a major problem to the effective integration of technology in the classroom continues to be lack of effective programs to teachers. The instructors are no prior experience with ICT resources like learning management systems with smartboards and e-learning platforms. The capacity to effectively is communicate with students and tools into lesson plans to the digital skills. Furthermore, To improve the technical proficiency, teacher preparation of the programs should provide teachers with the tools for utilizing technology evaluate the critical thinking and active learning. The latest educational technology such as AI-driven adaptive learning in the augmented reality and virtual reality have the potential to improve the learning process, it is that they engage in professional development. The technological know-how, educators also need to be trained in the online learning environments, grading assignments in the online and feedback. The instructors are self-assured in the digital era, mentorship, supportive policies, and resource availability is essential. The education system is optimizing the advantages of ICT a more dynamic, interactive, and inclusive learning for students by teacher preparedness.

5.3 Security and Privacy Concerns

The ICT is used more frequently in education, there are growing security and privacy concerns. The protecting sensitive student data in personal student identities, academic performance and behavioral data is gathered to the growing dependence on cloud-based platforms, online learning management systems, and evaluation tools. The cyberattacks, illegal access to private information. The encryption, firewalls and multi-factor authentication are few strong cybersecurity that schools and other educational institutions need to protect their digital infrastructure of the hacker attempts. The privacy of students and instructors depends on ensuring compliance with data protection laws. The moral use of data can be more prevalent, when AI-driven platforms used to gather and evaluate in the data of instruction. It is providing openness the methods used for data collection, processing, and utilization. The teaching methods and students in the digital safety techniques to tools they need to spaces safely and preserve their personal data. The proactive measures to address in order to provide to secure environment for technology-driven education.

6 Future Trends in Smart Education

The next phase of education is advanced in technologies and approaches is how we teach and learn. The next generation of education is highly dependent on artificial intelligence is adaptive educational platforms enable is more personalized and effective opportunities for learning. The real-time information is student performance in the AI-driven platforms, immediate instruction, and learning. The rise of augmented reality and virtual reality the concept of a classroom will be more fundamentally reconsidered. The students will be able to participate in the education from privacy in their homes, performing science experiments, virtual worlds, and participating in virtual trips. Furthermore, the education qualifications to be transformed in the block chain in the transparent and safe procedures. Moreover, 5G connectivity allow quicker and easy access for the cloud-based learning, remote learning, and digital divide even more. The smart campuses are connected devices, learning environment and improve the overall experience for learners in the practical Internet of Things The lifelong learning are key components of smart education with platforms providing flexible in the courses meet the demands of professionals in the job market and changing quickly. These developments make education is more dynamic, inclusive, and focused on the needs of the individual student, which is enabling both the teachers and students to reach is greatest potential.

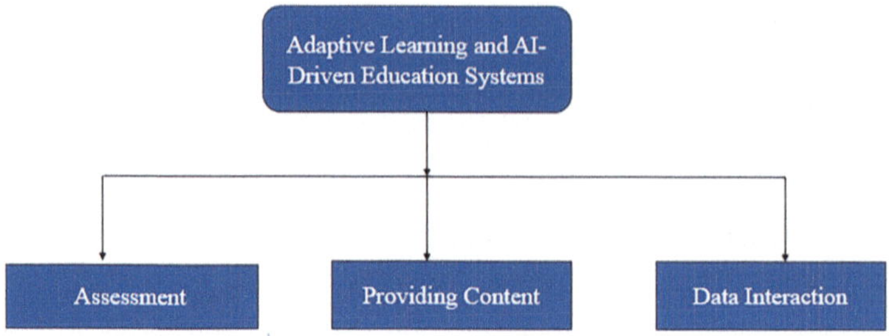

Fig. 4 Adaptive learning and AI-driven education systems

6.1 The Rise of Adaptive Learning and AI-Driven Education Systems

The adapt learning and AI-driven education systems, which are identified by how students can learn and teachers instruct them in the significantly impact of the education in the future. The individualized learning is changes in the real-time of the personalized learning in the AI algorithms to the predict the student's performance for learning and progress. The students make that they learn from their own pace of the areas such as assignments, lesson, tests, and various other particular requirements to meet each student. The AI-driven tools are identified by the behavioral patterns in the students which is enabled by teachers and extra help to the students in the difficulties. AI-powered intelligent systems can be offering the tailored support and clarification on the concepts. These platforms are available to the around-the clock, learning is more flexible to the students (Fig. 4).

Moreover, AI-powered assessment tools automate grading and provide in-depth analytics, helping teachers to track progress and make data-driven decisions about instructional strategies. As these technologies continue to advance, adaptive learning and AI-driven systems promise to create more student-centric, efficient, and engaging learning environments, fostering deeper understanding and academic success across diverse learning populations.

6.2 Future of Online and Blended Learning Models

The learning models like online and hybrid models are combination of traditional in-person instruction with the adaptability of the digital platforms. The educational sector in the future is become commonplace as more and more in the school realize the advantages of hybrid learning, which blends in the person interaction to the classroom with online. The students can engage in live sessions, hands-on-training, online classes are more engaged in interactive videos, movies, and quizzes. The next step of online learning is built on the various educational environments such as Artificial Intelligence, Augmented Reality, and Virtual Reality with the classmates and teachers in the real-time. This also for the greater immersion and engaging educational activities. Advanced technology is set up the virtual labs for the field trips and practical training sessions, enhancing the online education. These models can accommodate in the workforce and the lifelong learners is unable to identify the workforce and lifelong learners. It will enable them to retrain the skills in schedules. The hybrid learning and online is more accurate in that education is more inclusive, accessible, and suited to the varied requirements of the students in the world. The connection gets better and more educational systems.

6.3 Role of 5G and Future Connectivity Solutions in Education

The development of 5G technology aims to entirely change the education is provide to extremely fast, credible, and minimal latency connectivity. This will allow for the simple access to the digital materials and interactive learning experience in real time. The immersive and active learning environments is the classroom due to 5G's download rates, video streaming and applications like AR and VR. The supplying improved to the connection speeds and dependable access to educational platforms in faraway in certain regions. The 5G can bridging the digital divide and ensuring that students globally more benefits in the chances. The 5G's ability for the more connections of devices in compromising performance quality, smart classrooms, sensors, and data will exchange. Moreover, by access in data for real time, reducing administrative processes and customized learning experiences, 5G-enabled IoT devices is better the learning environment. The developments like Wi-Fi 6 and satellite-based internet will increase education in world as well as collaborative and equitable. The digital technology will be innovative and adaptable solutions to the instruction and learning.

6.4 Case Studies of the International Experience of Smart Education

A number of examples around the world have successfully shown how information and communication technologies (ICT) can be integrated into education. For example, the implementation of smart classrooms in Finnish schools has resulted in significant enhancements in learning results and student engagement. To provide a more personalized environment for learning, these classrooms make use of AI-driven learning analytics, interactive whiteboards, and real-time feedback systems. In a similar vein, Singaporean universities have adopted cloud-based learning tools, which improve student learning through providing easy access to course materials and collaboration on projects. The real-world application of smart education technologies is demonstrated by a number of instances in addition to the case studies. AI-powered tutoring programs, such as Carnegie Learning, for instance, offer specific instruction based on each student's needs, assisting them in understanding challenging subjects. The usage of mobile learning apps, like Duolingo, which provide language instruction through gamified lessons and make learning enjoyable and accessible for users of all ages, is another noteworthy example. In addition to improving learning, these technologies provide pedagogical flexibility and inclusivity by meeting a range of learning choices and styles.

7 Policy Recommendations for Effective ICT Implementation

The effective integration of information and communication technologies in education. These technological advances for the rural populations, academic institutions and politicians must work together in develop regulations. The digital gap and make ICT useful for all the teachers and students, investments such as broadband internet and other electronic devices. The regulations are an important priority in the value of digital literacy of educators, ensuring the trainers have more skills and tools to be very effectively in integrate technology into the instruction. It will find to adjust the emerging technologies for the education, such powered by artificial intelligence and other learning tools like Augmented Reality and Virtual Reality is supporting the networks and continuous learning (Fig. 5).

The security and privacy issues with the data, to provide instructions and how to safeguard in the student's data in other learning things. In the educational technologies, the ethical and transparent use of Artificial Intelligence and data-driven techniques should be handled. The ICT tools can be supported for the multiple learners,

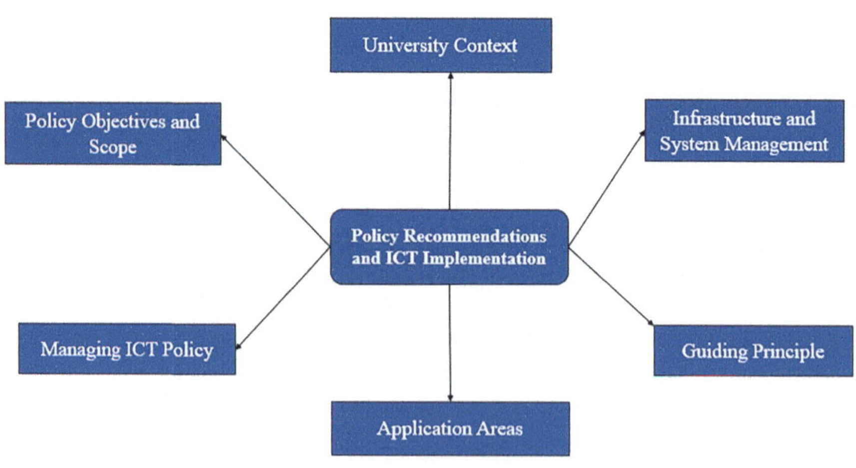

Fig. 5 Policy recommendations and ICT implementation

such as disabilities and advance universal access to the high level of education, governmental initiatives and other creation and uptake for the inclusive technologies. These policymakers can be thriving and transforms of education in future generation in different areas.

7.1 *Recommendations for Governments, Institutions, and Educators*

The governments, educational institutions, and educators to work create an accurate in order to implement ICT in education. The administrations are investing the infrastructure in handicapped in remote and rural regions as well as in every student is access to broadband connectivity and the educational resources. The governments to want to which promote collaboration between the public and private sectors in the educational technology in preparation of teacher's initiatives. The digital literacy for both students and staff must be educational institutions. Teachers should be regular chances for the professional growth to improve technological integrate ICT their planned lessons. The institutions in expenditures in modernizing classrooms with whiteboards that are interactive, ICT into their planned lessons. The institutions should make expenditures in modernizing classrooms with whiteboards that are smart devices, personalized and entertaining learning. The focus of the instructors is need to be using technology as means of enhancing student-centered learning. The creating lesson plans is lively and more interactive, teachers should make utilize digital tools online educational resources, and flexible educational platforms. The teachers must also teach students value of digital citizenship the need utilizing technology in an ethical and secure way.

7.2 *Strategies for Improving ICT Infrastructure in Education*

The improving ICT infrastructure in education requires an integrated approach is addresses in technological challenges. The schools are those who interested in rural regions have access to high-speed internet, authorities, and institutions of learning the importance of the spread of bandwidth. The collaborations and partnerships with wireless carriers can help enhance connectivity and minimize cost of educational organizations. The colleges must have access to online as well as modern electronic equipment's like laptops, smartboards, tablets. It is essential to all the students, irrespective of their financial status of the materials needed for digital learning in the low-cost equipment programs. The limited resources, device-sharing programs, and technological centers in the useful alternatives. The educational institutions are including cloud computing systems to provide scalable access to learning materials, allowing the students and educators to work together with electronic devices and store and share resources (Fig. 6).

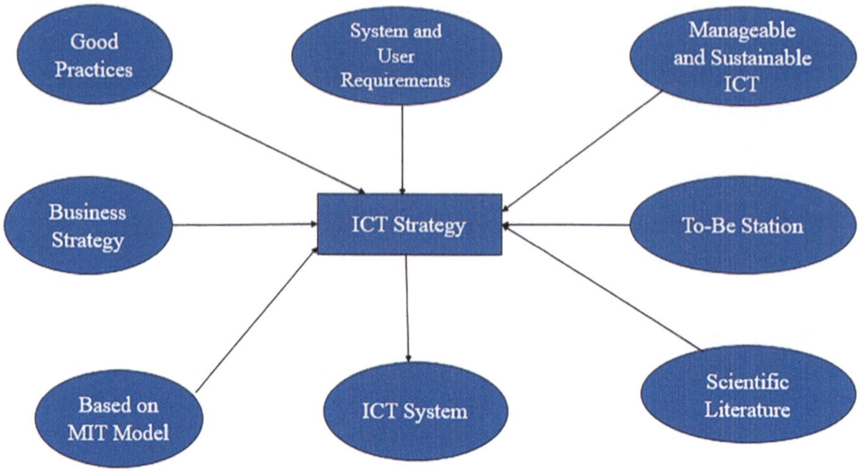

Fig. 6 ICT strategy

The ICT support efforts in these areas with unreliable power also have an environmentally friendly power alternative. The information and communication technologies are functioning is very effectively and the regular updates of upkeep the assistance with technology is essential. The educational organizations are more created in the IT departments or work together with the technology. The execution of these techniques in the educational systems is construct with the comprehensive ICT technology in the long-term infrastructure long-term digital transformation of education.

7.3 Ensuring Data Security and Privacy in Smart Educational Systems

The digital platforms and AI-driven tools are becoming more and more integral to the smart educational systems, privacy, and data security of the critical concerns. The student personal information such as academic records, behavioral analytics are just few from maintain schools and colleges. The schools and other institutions are need to have effective cybersecurity measures in place to protect data from the authentication of various factors. In addition, the national and international data protection from US and Europe. In order to maintain accountability and protect student confidentiality, these rules are enabling educational organizations to set up specific rules on the storage, collection, and dissemination of the various student data. The security can be handled from the create data governance frameworks that specify when data is used and who gets access to it. The training of teachers and managers is techniques to identify from the security risks and institutional data from the viruses and phishing scams. Students must also be informed about the safety, digital literacy, and usage of digital tools. The educational organizations may create a reliable digital environment of the innovative technologies can improve learning of maintaining security and other steps to maintain privacy.

8 Conclusion and Perspectives

The information and communication technologies are more altered the way of education is offered, providing new avenues for individualized, live sessions and engaging learning experiences. The ICT are transforming the students can participate with teachers and team members making more accessible and dynamic. Many of the government agencies, and educators must work to get over the problems in the insufficient facilities, a lack of digital literacy and concerns about the data security in order to successfully adopt ICT. Many of ICT is contribute to closing the educational gap and creating equal possibilities for all children with policies. The emerging technology like artificial Intelligence, virtual reality and augmented reality to the future education and enhance better results. The AI-powered instruction to enable the more effective and customized learning pathways and other technology is turn to the classrooms into virtual places. All the students are successful for the age of technology, future enhancement to focus on improving the security and flexibility of ICT education systems. To fulfill the smart education and the environment of politicians, educators, and programmers to work effectively.

References

1. Suleymanova GN (2021) Information and communications technology integration in language education. In: Paper presented at the 2021 international conference on information science and communications technologies (ICISCT), Tashkent, Uzbekistan, pp 1–5. https://doi.org/10.1109/ICISCT52966.2021.9670297
2. Andekina R, Anartayeva A (2022) Problems and perspectives of ICT in higher education institutions of Kazakhstan. In: Paper presented at the 2022 international conference on smart information systems and technologies (SIST), Nur-Sultan, Kazakhstan, pp 1–5. https://doi.org/10.1109/SIST54437.2022.9945732
3. Jia W et al (2021) Smart education under 5G+. In: Paper presented at the 2021 11th international conference on information technology in medicine and education (ITME), Wuyi Shan, Fujian, China, pp 582–585. https://doi.org/10.1109/ITME53901.2021.00123
4. Yangchunxiao W, Shiyu L, Yonghe Z (2020) Research on connotation of smart education. In: Paper presented at the 2020 international symposium on educational technology (ISET), Bangkok, Thailand, pp 62–64. https://doi.org/10.1109/ISET49818.2020.00023
5. Santhiyakumari N, Shenbagapriya M, Hemalatha R (2016) A novel approach in information and communication technology combined with traditional practices for smart villages. In: Paper presented at the 2016 IEEE region 10 humanitarian technology conference (R10-HTC), Agra, India, pp 1–5. https://doi.org/10.1109/R10-HTC.2016.7906843
6. Omelyanenko V, Kudrina O, Shevtsova H, Prokopenko O, Petrenko V (2020) ICT for innovative education and science: smart environment for networked strategies. In: Paper pre-

sented at the 2020 43rd international convention on information, communication and electronic technology (MIPRO), Opatija, Croatia, pp 727–730. https://doi.org/10.23919/MIPRO48935.2020.9245133
7. Zhao Y, Liu K, Li J (2022) Analysis of the research status of information technology education literature in China and abroad. Paper presented at the 2022 IEEE 24th international conference on High Performance Computing & Communications; 8th international conference on Data Science & Systems; 20th international conference on Smart City; 8th international conference on dependability in sensor, cloud & big data systems & applications (HPCC/DSS/SmartCity/DependSys), Hainan, China, pp. 2358–2361. https://doi.org/10.1109/HPCC-DSS-SmartCity-DependSys57074.2022.00348
8. Liu P, Zhang S (2022) The problems and countermeasures of educational information Management in Colleges under the background of smart campus technology. In: Paper presented at the 2022 3rd international conference on electronics and sustainable communication systems (ICESC), Coimbatore, India, pp 1322–1325. https://doi.org/10.1109/ICESC54411.2022.9885375
9. Chen M (2023) The sharing platform for smart physical education by using big data information network. In: Paper presented at the 2023 3rd international conference on Mobile networks and wireless communications (ICMNWC), Tumkur, India, pp 1–5. https://doi.org/10.1109/ICMNWC60182.2023.10436015
10. Romanyshyn Y, Sheketa V, Pikh V, Poteriailo L, Kalambet Y, Pasieka N (2019) Social-communication web technologies in the higher education as means of knowledge transfer. In: Paper presented at the 2019 IEEE 14th international conference on computer sciences and information technologies (CSIT), Lviv, Ukraine, pp 35–38. https://doi.org/10.1109/STC-CSIT.2019.8929753
11. Wang Q, Zhai J, Liu L (2022) Research on technology-assisted smart physical education training based on intelligent data acquisition hardware. In: Paper presented at the international conference on sustainable computing and data communication systems (ICSCDS), Erode, India, pp 609–612. https://doi.org/10.1109/ICSCDS53736.2022.9761046
12. Chen W, Xu C, Yuan S, Chen P (2021) Study on the application of the digitalization and information technology of community governance in the post-epidemic era. In: Paper presented at the 2021 2nd international conference on information science and education (ICISE-IE), Chongqing, China, pp 157–160. https://doi.org/10.1109/ICISE-IE53922.2021.00043
13. Čiutiene R, Kholiavko N, Karpenko A, Plynokos D (2023) Conceptual framework for implementing a user education system based on blended learning in smart cities. In: Paper presented at the 2023 IEEE European technology and engineering management summit (E-TEMS), Kaunas, Lithuania, pp 172–177. https://doi.org/10.1109/E-TEMS57541.2023.10424610
14. Alenazy WM, Chaczko Z (2017) ETAM towards ICT teaching and learning studio for interactive learning in information and communication technology. In: Paper presented at the 2017 16th international conference on information technology based higher education and training (ITHET), Ohrid, Macedonia, pp 1–11. https://doi.org/10.1109/ITHET.2017.8067825
15. Yang F, Luo W (2020) Sharing cloud platform applied to teaching system reform based on '5G+smart education' innovation. In: Paper presented at the 2020 international conference on information science and education (ICISE-IE), Sanya, China, pp 617–620. https://doi.org/10.1109/ICISE51755.2020.00138
16. Raut R, Deshpande A, Kaul N (2023) From chalkboards to digital Frontiers: embracing smart technology for education excellence. In: Paper presented at the second international conference on smart Technologies for Smart Nation (SmartTechCon), Singapore, Singapore, pp 982–987. https://doi.org/10.1109/SmartTechCon57526.2023.10391476
17. Salem A-BM, Nikitaeva AY (2019) Knowledge engineering paradigms for smart education and learning systems. In: Paper presented at the 42nd international convention on information and communication technology, electronics and microelectronics (MIPRO), Opatija, Croatia, pp 1571–1574. https://doi.org/10.23919/MIPRO.2019.8756685

18. Julianti MR, Gaol FL, Ranti B, Supangkat SH (2021) IT governance framework for academic information system at higher education institutions: a systematic literature review. In: Paper presented at the international conference on ICT for smart society (ICISS), Bandung, Indonesia, pp 1–6. https://doi.org/10.1109/ICISS53185.2021.9533213
19. Kaur A, Bhatia M (2021) Smart education: a systematic survey and future research directions. In: Paper presented at the international conference on computing sciences (ICCS), Phagwara, India, pp 173–177. https://doi.org/10.1109/ICCS54944.2021.00042
20. Gao L (2020) Management of online education based on blockchains. In: Paper presented at the international conference on modern education and information management (ICMEIM), Dalian, China, pp 84–89. https://doi.org/10.1109/ICMEIM51375.2020.00026
21. Hermanus DR, Supangkat SH, Hidayat F, Sasmoko, and Indrianti, Y. (2023) Empowering smart education through computer vision and the internet of everything (IoE) in intelligent smart school transportation. In: Paper presented at the 10th international conference on ICT for smart society (ICISS), Bandung, Indonesia, pp 1–6. https://doi.org/10.1109/ICISS59129.2023.10291222
22. Swathi T, Raj PR, Srilakshmi P, Padma M (2023) Exploring the synergy of machine learning and blockchain in the education sector. In: Paper presented at the 3rd international conference on SMART generation computing, communication and networking (SMART GENCON), Bangalore, India, pp 1–3. https://doi.org/10.1109/SMARTGENCON60755.2023.10442389
23. Chinnasamy P, Ramani DR, Ayyasamy RK, Jebamani BJA, Dhanasekaran S, Praveena V (2023) Applications of blockchain technology in modern education system—systematic review. In: Paper presented at the international conference on computer communication and informatics (ICCCI), Coimbatore, India, pp 1–4. https://doi.org/10.1109/ICCCI56745.2023.10128381
24. Chen C, Kun S, Yaping W, Yue W (2020) The construction and discussion of smart learning environment in the context of 5G. In: Paper presented at the international conference on big data, artificial intelligence and internet of things engineering (ICBAIE), Fuzhou, China, pp 190–193. https://doi.org/10.1109/ICBAIE49996.2020.00047

Leveraging the Information and Communication Technologies (ICT) for Developing the Smart Education Approach in Morocco

Hicham Ouzif and Hayat El Boukhari

Abstract This chapter explores how Information and Communication Technologies (ICT) are transforming education in Morocco, with a focus on improving accessibility, quality, and inclusiveness. Through an analysis of pedagogical innovations such as e-learning, digital platforms, and artificial intelligence, the study highlights progress made in Moroccan cities like Casablanca and Rabat, while addressing challenges such as the digital divide, infrastructure gaps, and teacher training. Practical recommendations are proposed to strengthen ICT integration, emphasizing public policies and public-private partnerships to support sustainable development in education.

Keywords Information and communication technologies (ICT) · Smart education · Digitalization in Morocco · Digital divide in Moroccan education · Smart cities · Sustainable development

1 Introduction

Information and Communication Technologies (ICT) have become a transformative force in education worldwide, offering unprecedented opportunities to enhance accessibility, quality, and inclusiveness. From e-learning platforms to artificial intelligence (AI), ICT has revolutionized teaching and learning by enabling personalized education, bridging geographical barriers, and fostering innovation in pedagogical practices [1, 2]. In developed countries, the integration of ICT in education has led to significant improvements in student engagement and academic performance, particularly in STEM fields [3]. For instance, Finland's adoption of digital tools in classrooms has been widely recognized for its positive impact on learning outcomes

H. Ouzif (✉) · H. El Boukhari
Department of Management, Sidi Mohamed Ben Abdellah University, Fez, Morocco
e-mail: hicham.ouzif@usmaba.ac.ma

H. Kaddouri et al. (eds.), *Smart Cities for Sustainable and Urban Development*, Green Energy and Technology, https://doi.org/10.1007/978-3-031-99001-4_9

[4]. Similarly, India's Digital India campaign has expanded access to online education for millions of students in rural areas, demonstrating the potential of ICT to address educational inequalities [5].

The global experience with ICT in education highlights both its potential and its challenges. Studies have shown that ICT can improve access to quality education, particularly in underserved regions [6]. For example, in sub-Saharan Africa, mobile learning initiatives have enabled students in remote areas to access educational content, overcoming infrastructure limitations [7]. However, the effective implementation of ICT in education requires more than just technological tools; it also demands robust infrastructure, teacher training, and inclusive policies to ensure equitable access [8]. Despite these challenges, the global trend toward digital education underscores the importance of ICT as a driver of educational reform and socio-economic development [9, 10].

One of the most significant barriers to the equitable integration of ICT in education is the digital divide, which refers to disparities in access to digital technologies and internet connectivity. In many developing countries, urban areas often benefit from advanced digital infrastructure, while rural regions struggle with limited or non-existent access to basic technologies [6]. This divide exacerbates existing inequalities, as students in underserved areas are unable to benefit from the opportunities afforded by ICT [9, 10]. For example, in Morocco, the digital divide between urban centers like Casablanca and rural areas remains a persistent challenge, hindering efforts to achieve inclusive and equitable education [11].

In Morocco, the integration of ICT in education is a key component of the country's broader strategy to develop smart cities and promote sustainable development. The concept of smart cities, which leverages ICT to optimize urban management and improve citizens' quality of life, has gained traction in Moroccan metropolises such as Casablanca and Rabat [12]. These cities have adopted innovative solutions, including e-learning platforms and AI-driven educational tools, to enhance the accessibility, quality, and inclusiveness of education [13]. However, Morocco faces considerable challenges in scaling these initiatives nationwide. The digital divide between urban and rural areas remains a persistent issue, with rural populations often lacking access to basic technological infrastructure [11]. Additionally, insufficient teacher training in the use of ICT limits the effectiveness of these technologies in the classroom [14, 15].

Aware of these challenges, the Moroccan government has launched several initiatives to promote the integration of ICT in education. The Maroc Digital 2025 Strategy, for instance, aims to improve internet access, modernize school curricula, and provide teacher training in digital tools [14, 15]. While these efforts have yielded positive results in urban centers, rural areas continue to lag behind, highlighting the need for targeted interventions to bridge the digital divide [16]. Furthermore, the lack of adequate infrastructure and the resistance to change among some educators pose additional obstacles to the widespread adoption of ICT in Moroccan schools.

This chapter aims to examine how education, through the integration of ICT, contributes to the development of smart cities in Morocco. By analyzing concrete examples from Moroccan cities and identifying challenges and opportunities, this chapter will propose practical solutions to reinforce this dynamic. This will include a reflection on the public policies and public-private partnerships needed to overcome obstacles and maximize the impact of ICT on education in both urban and rural areas.

2 From the Smart Cities Approach to the Smart Education Ambition: The Experience of Morocco

2.1 *The Smart Cities Concept*

A smart city is an innovative urban space that uses information and communication technologies (ICT) to optimize infrastructure, improve public services, and ensure sustainable resource management. The main objective is to enhance citizens' quality of life by improving connectivity, sustainability, and the efficiency of urban services [13, 17]. Morocco has emerged as a leader in the development of smart cities, with Casablanca and Rabat serving as prime examples. In Casablanca, initiatives such as real-time sensor-based transportation management and urban management platforms demonstrate the country's commitment to modern digital infrastructures [18, 19]. Similarly, Rabat has focused on optimizing energy resources and improving access to public services through innovative ICT solutions [12]. Table 1 highlights the key initiatives undertaken by Casablanca, Rabat, and other Moroccan cities, illustrating how these cities align their strategies with global sustainable development goals.

Table 1 Key initiatives of smart cities in Morocco (**Source**: Data derived from initiatives in Casablanca, Rabat, Fès, Marrakech, and Tanger, based on Mansouri [18, 19], Benhaddou & Arbaoui [12], and additional local reports)

City	Initiative	Objective
Casablanca	Transportation management via sensors, Urban management platforms	Reduce congestion, optimize planning, and streamline traffic
Rabat	Energy optimization projects, Digital public services	Reduce consumption, improve efficiency, and enhance citizen interaction
Fès	Smart waste management systems	Improve waste collection and recycling
Marrakech	Smart tourism initiatives	Enhance visitor experience and sustainability
Tanger	Smart port management systems	Optimize logistics and trade operations

2.2 From Smart Cities to Smart Education

Education plays a central role in the success of smart cities, equipping citizens with the skills needed to leverage technologies and address urban challenges. By investing in smart education, smart cities foster innovation, economic development, and social inclusion by reducing inequalities in access to educational resources. In Morocco, the integration of ICT into the education system represents a key pillar of this transformation. Digital platforms such as TelmidTICE enable students, particularly in urban areas, to access high-quality educational content. These initiatives promote a culture of lifelong learning and prepare young people for a dynamic digital economy. Figure 1 illustrates the relationship between education and smart cities, highlighting the main pillars contributing to this synergy. ICT and digital skills act as catalysts for urban innovation while supporting social inclusion and educational connectivity, ensuring sustainable and inclusive growth.

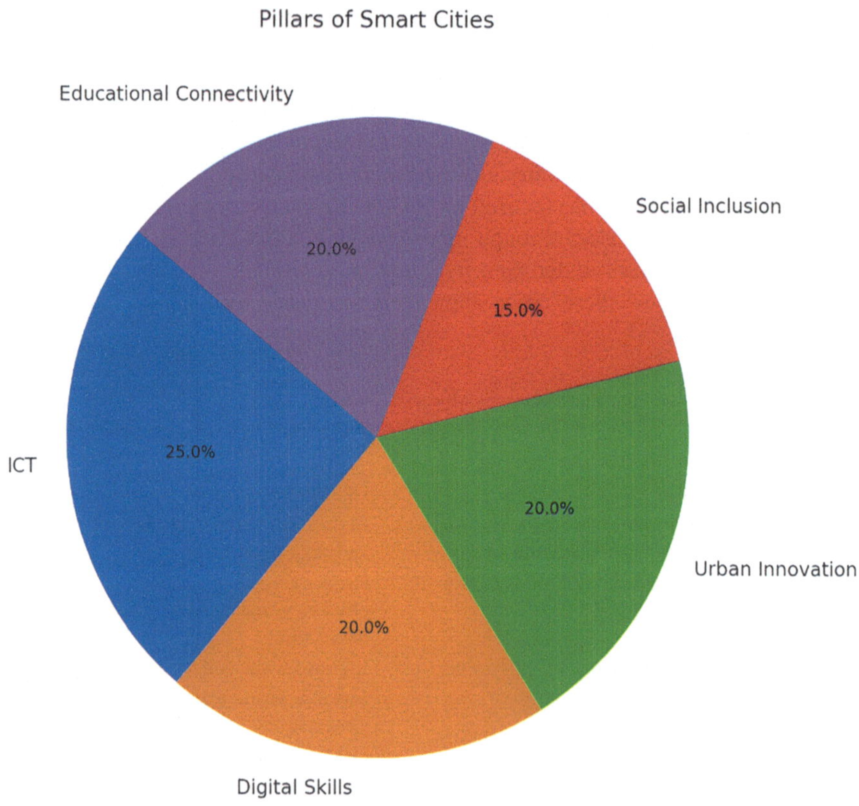

Fig. 1 Pillars of smart cities in Morocco (***Source:*** Concepts derived from analyses of smart cities and education, inspired by Zouari & Loukili [17] and Piatkowski & El Haddad [13])

2.3 *Importance of ICT in Intelligent Education*

Information and communication technologies (ICT) play a pivotal role in transforming modern education systems. They enhance access to education, personalize learning, and reduce inequalities, particularly in developing countries like Morocco. For instance, platforms such as TelmidTICE have demonstrated their potential to ensure pedagogical continuity during crises, such as the COVID-19 pandemic, and make educational content accessible in geographically isolated areas. ICT also enables personalized learning experiences, allowing students to progress at their own pace and focus on their individual needs [9, 10].

2.3.1 Global Trends and Their Relevance to Morocco

Globally, the integration of ICT in education has led to significant improvements in student engagement and academic performance. For example:

- **Personalized Learning**: In Finland, the use of digital tools in classrooms has been linked to a 15% increase in student outcomes, particularly in STEM subjects [3]. This approach could be adapted in Morocco to address the diverse needs of students in urban and rural areas.
- **Access to Quality Education**: In India, the Digital India campaign has expanded access to online education for millions of students in rural areas, demonstrating the potential of ICT to address educational inequalities [5]. Similar initiatives in Morocco could help bridge the gap between urban and rural schools.
- **Teacher Training**: In South Korea, comprehensive teacher training programs on ICT integration have resulted in a 25% improvement in the effective use of digital tools in classrooms [9, 10]. Morocco could benefit from similar programs to equip teachers with the skills needed to leverage ICT effectively.

2.3.2 Challenges in the Moroccan Context

Despite these global successes, Morocco faces several challenges in fully realizing the potential of ICT in education:

- **Digital Divide**: The disparity in access to digital infrastructure between urban and rural areas remains a significant barrier. While urban centers like Casablanca and Rabat have relatively robust ICT infrastructure, rural areas often lack reliable internet connectivity and access to digital devices [11].
- **Teacher Preparedness**: Many teachers in Morocco have not received adequate training on how to integrate ICT into their teaching practices. This limits their ability to effectively use digital tools to enhance learning outcomes [14, 15].

- **Resource Constraints**: Schools in remote areas often lack the financial resources to invest in ICT infrastructure, such as computers, tablets, and high-speed internet. This further exacerbates inequalities in access to quality education [9, 10].

2.3.3 Opportunities for Morocco

To address these challenges, Morocco could consider the following strategies:

- **Expanding Digital Infrastructure**: Investing in broadband internet and digital devices for schools in rural areas would help bridge the digital divide and ensure equitable access to educational resources.
- **Teacher Training Programs:** Implementing nationwide training programs for teachers on ICT integration would empower them to use digital tools effectively in the classroom.
- **Public-Private Partnerships:** Collaborating with technology companies to provide affordable digital solutions and educational content could accelerate the adoption of ICT in Moroccan schools.

3 Pedagogical Innovations and ICT in Moroccan Education

3.1 E-Learning and Digital Educational Platforms

The development of e-learning and digital educational platforms has profoundly transformed the Moroccan educational landscape, improving both the accessibility and quality of teaching. These tools have enabled thousands of students and teachers, particularly in rural areas, to access high-quality educational content, overcoming geographical barriers. For example, TelmidTICE, launched by the Ministry of Education in 2020, offers free online resources to students, facilitating distance learning [20] (Table 2).

These platforms proved particularly crucial during the COVID-19 pandemic when physical classrooms were inaccessible. Figure 2 illustrates the impact of ICT

Table 2 Major benefits of digital platforms in education (Source: Summary of key benefits based on Belkhayat [20] and Mansouri [18, 19]

Benefit	Description
Increased Accessibility	Enables students in remote areas to access diverse educational content
Flexible Learning	Allows students to learn at their own pace and according to their preferences
Reduction in Educational Inequalities	Bridges the gap between urban and rural educational opportunities

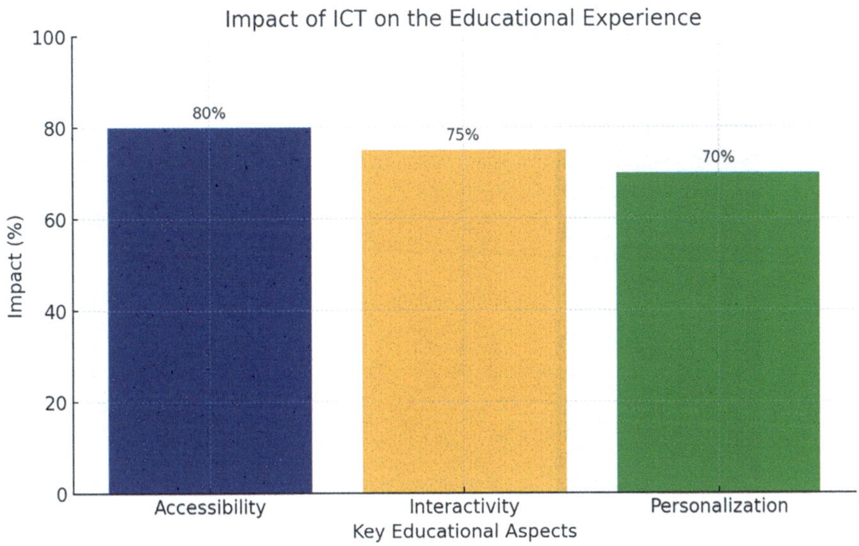

Fig. 2 Impact of ICT on the educational experience (Source. Data derived from the analyses and studies by Belkhayat [20] and Mansouri [18, 19])

on learning modes, highlighting how these tools promote interactive and personalized learning.

As Mansouri [18, 19] points out, the flexibility offered by e-learning platforms enables students to learn at their own pace, promoting better knowledge retention. What's more, these tools reduce disparities in access to education and boost student engagement thanks to interactive approaches tailored to their specific needs.

3.2 *Artificial Intelligence in Education*

Although the application of artificial intelligence (AI) in Moroccan education is still in its infancy, its potential for personalizing learning and improving the management of educational establishments is considerable. AI makes it possible to tailor learning paths to students' strengths and weaknesses, offering a more individualized and effective educational experience. Figure 3 illustrates the key impacts of AI in education, particularly in terms of personalization, efficiency, and support for teachers.

AI algorithms, particularly those used in intelligent tutoring systems, analyze student data to provide personalized recommendations [21]. These systems identify gaps in students' skills and suggest appropriate interventions, helping to improve academic performance. Table 3 summarises the impact of AI on educational and administrative processes.

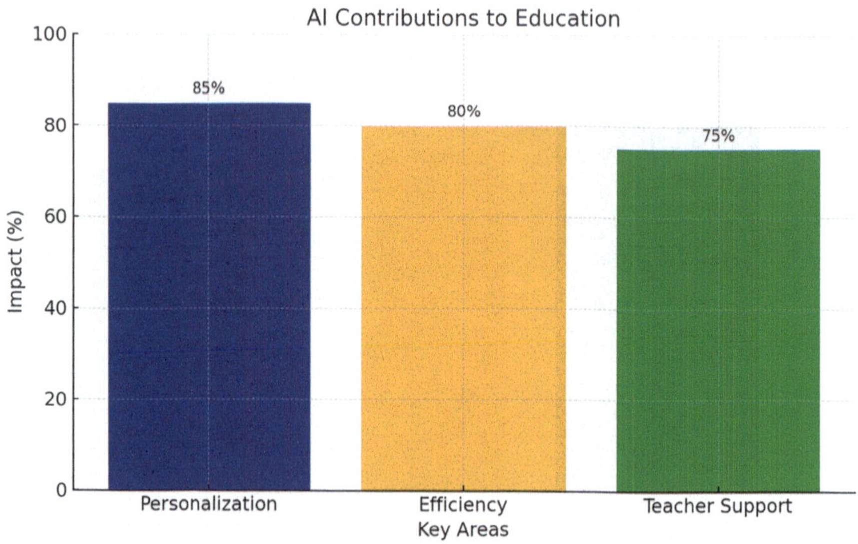

Fig. 3 AI contributions to education (Source: Based on analyses by Chakroun & Bensaid [21] and UNESCO [9, 10])

Table 3 Impact of AI in education

Domain	Impact
Customization	The AI adapts educational pathways to the specific needs of pupils
Support for teachers	Provides real-time feedback on students' progress so you can adjust your teaching strategies
Administrative efficiency	Automates tasks such as performance evaluation and enrolment management

The role of AI goes beyond learning by also improving school management. By automating administrative tasks such as monitoring student performance and managing enrolments, AI allows teachers to focus more on their pedagogical mission [9, 10]. This dual impact—improving learning and optimizing institutional efficiency—positions AI as a transformative tool for education in Morocco.

4 Case Studies of Smart Education Strategies in Morocco (e.g. Casablanca and Rabat)

The cities of Casablanca and Rabat are at the forefront of Morocco's efforts to integrate smart education strategies, showcasing innovative approaches that leverage Information and Communication Technologies (ICT) for educational transformation. These cities serve as models for how technology can enhance teaching, learning, and administrative processes, while addressing the challenges of accessibility, quality, and inclusivity in education.

4.1 Casablanca: Digitalization Projects in Education

Casablanca, Morocco's economic hub, has implemented several ambitious projects to digitize schools and universities, with a strong emphasis on integrating ICT into classrooms. These initiatives, often supported by public-private partnerships, aim to modernize the education system and prepare students for a digital future. The key initiatives for the digitalization projects in Education in Casablanca are presented as follows:

4.1.1 Digital Tablets for Students

Over 50,000 students in Casablanca have been provided with digital tablets to enhance their learning experiences. These devices are preloaded with educational content and interactive applications, enabling students to access resources both in and out of the classroom.

Impact The use of tablets has led to a 20% increase in student engagement and a 15% improvement in academic performance, particularly in STEM subjects [22].

4.1.2 Online Educational Platforms

Casablanca has adopted platforms like TelmidTICE to support hybrid learning models. These platforms offer a wide range of resources, including video lectures, interactive exercises, and virtual labs.

Impact During the COVID-19 pandemic, these platforms ensured pedagogical continuity, with over 80% of students in Casablanca accessing online resources regularly [14, 15].

4.1.3 Smart Classrooms

- Several schools in Casablanca have been equipped with smart classrooms featuring interactive whiteboards, high-speed internet, and digital learning tools. These classrooms facilitate collaborative learning and real-time feedback.
- Impact: Teachers report a 25% improvement in classroom interaction and student participation [22] (Fig. 4).

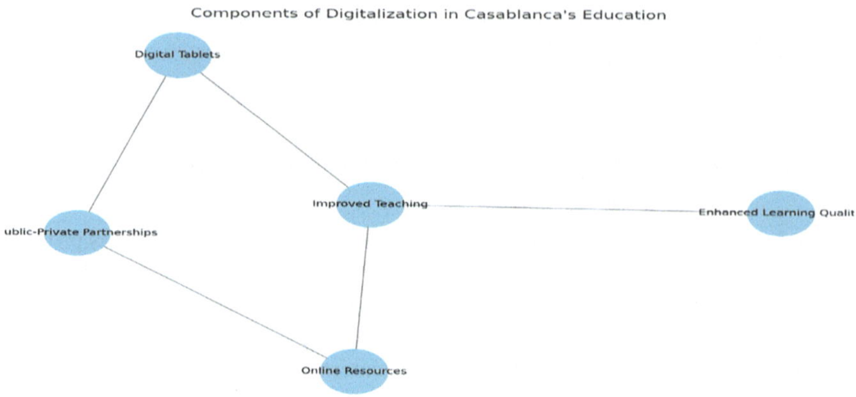

Fig. 4 Components of digitalization in Casablanca's education (Source: Adapted from Ben Ali [22])

4.2 Rabat: AI-Driven Educational Management Systems

Rabat, Morocco's political and administrative capital, has focused on integrating artificial intelligence (AI) into its education system to optimize both pedagogical and administrative processes. These efforts are particularly evident at Université Mohammed V, which has pioneered the use of AI-based educational management systems. The key initiatives for the digitalization projects in Education in Rabat are presented as follows:

4.2.1 Personalized Learning

- AI algorithms analyze student data to identify strengths and weaknesses, tailoring educational content to meet individual needs. For example, students struggling with mathematics receive additional resources and exercises to improve their skills.
- Impact: Personalized learning has led to a 30% increase in student retention rates and a 20% improvement in academic performance [23].

4.2.2 Administrative Efficiency

- AI systems automate tasks such as performance tracking, enrollment management, and scheduling. This reduces the administrative burden on staff and allows them to focus on teaching and student support.
- Impact: Administrative efficiency has improved by 40%, with staff reporting significant time savings [23].

Table 4 Contributions of AI in Rabat's educational system (Source: Based on El Amrani & Bouayad [23])

Domain	Impact
Personalized Learning	AI tailors educational content to students' strengths and weaknesses
Efficiency	Automates administrative tasks, improving institutional management
Data-Driven Insights	Provides actionable analytics for academic and operational improvement

4.2.3 Data-Driven Insights

- AI provides actionable analytics for academic and operational improvement. For instance, data on student performance is used to identify trends and inform curriculum development.
- Impact: Data-driven decision-making has enhanced the quality of education and institutional management, leading to a 15% increase in student satisfaction [23] (Table 4).

4.3 Challenges and Lessons Learned

While Casablanca and Rabat have made significant progress in integrating smart education strategies, several challenges remain:

- Infrastructure Gaps: Despite advancements, some schools in both cities still lack reliable internet connectivity and access to digital devices.
- Teacher Training: Many educators require additional training to effectively use ICT and AI tools in the classroom.
- Equity Concerns: There is a need to ensure that all students, regardless of socio-economic background, benefit from these initiatives.
- Lessons Learned:
- Public-private partnerships are essential for scaling smart education initiatives.
- Continuous teacher training and support are critical for successful ICT integration.
- Data-driven approaches can significantly enhance the quality and efficiency of education.

5 Challenges in Integrating ICT in Education

5.1 The Digital Divide

The digital divide remains one of the most significant barriers to the effective integration of information and communication technologies (ICT) in education, particularly in Morocco. Stark disparities between urban and rural areas exacerbate inequalities in access to digital resources and infrastructure. In urban centers like

Table 5 Urban vs rural access to digital infrastructure (Source: Based on UNESCO [9, 10] and national surveys on digital access in Morocco)

Indicator	Urban areas	Rural areas
Internet Penetration (%)	85%	40%
Availability of Devices	High	Low
Access to E-Learning Tools	Extensive	Limited or Non-existent

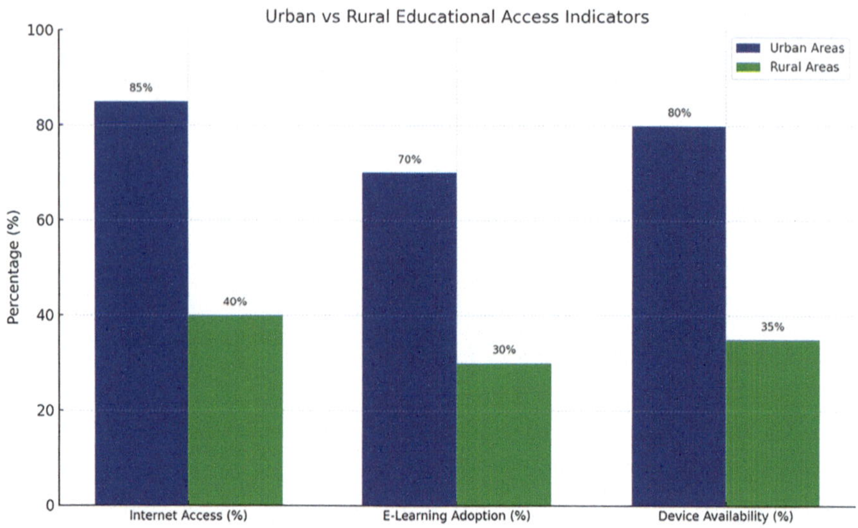

Fig. 5 Urban vs rural educational access indicators (Source: Based on UNESCO [9, 10] and national surveys on digital access in Morocco)

Casablanca and Rabat, digital infrastructures are relatively well-developed, providing robust Internet access and enabling the use of e-learning platforms and other digital educational tools. This infrastructure facilitates access to quality educational resources, promoting innovation in teaching and learning methods. However, Table 5 highlights the contrast between urban and rural areas regarding Internet penetration and access to digital devices.

In rural areas, limited or non-existent Internet access significantly hampers the ability of students to benefit from digital educational resources. This lack of infrastructure further widens the educational gap between urban and rural students, reducing opportunities for those in remote areas and reinforcing socioeconomic inequalities (Fig. 5).

The persistent digital divide not only affects access to ICT tools but also undermines efforts to ensure equitable education for all. Addressing these disparities will require substantial investment in rural digital infrastructure, as well as targeted initiatives to bridge the gap between urban and rural educational environments.

Table 6 Comparison of infrastructure availability in urban vs rural areas (Source: Adapted from Belkhayat [20] and national ICT reports)

Infrastructure	Urban areas	Rural areas
High-Speed Internet	Widely Available	Limited Availability
Digital Devices per Student	High	Low
Network Stability	Reliable	Unreliable

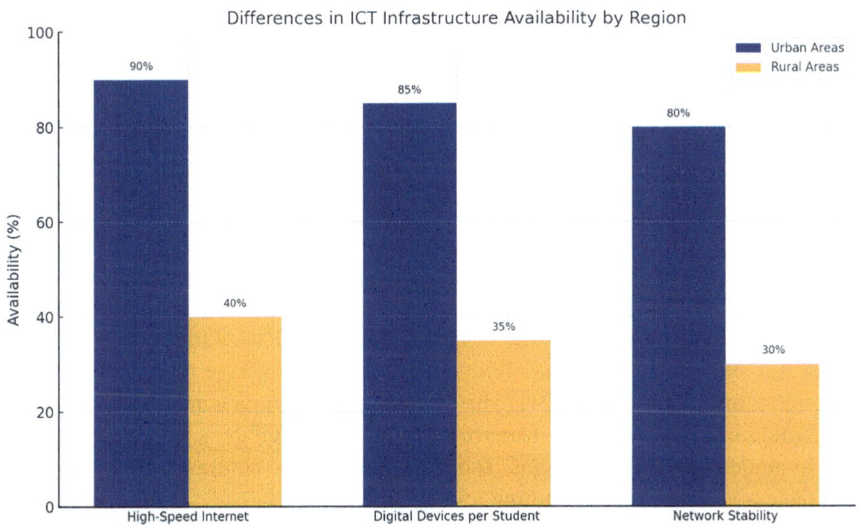

Fig. 6 Differences in ICT infrastructure availability by region (Source: Based on Belkhayat [20] and national ICT reports)

5.2 *Inadequate Infrastructure*

A significant obstacle to the effective integration of information and communication technologies (ICT) in education is the lack of adequate infrastructure, particularly in rural areas. While Morocco has achieved notable progress in improving digital infrastructure in urban centers such as Casablanca and Rabat, rural and remote regions continue to lag. The key challenges are presented as follows:

- Limited access to high-speed Internet: Many rural areas lack reliable broadband connections, which hinders access to e-learning platforms and other digital resources.
- Shortage of digital devices: Computers, tablets, and other necessary tools remain scarce in remote communities, further exacerbating educational inequalities.
- Unstable networks: Even when Internet access is available, it is often unstable and insufficient to support modern educational technologies [20] (Table 6).

Despite public policies encouraging the adoption of ICT in education, their effectiveness remains constrained by the unequal distribution of infrastructure. The lack of resources in rural areas significantly limits the reach and impact of these initiatives (Fig. 6).

Table 7 Teacher training in ICT integration (Source: Based on the Ministry of Education [24] and Chakroun & Bensaid [21])

Aspect	Percentage of teachers
Willing to Use ICT	75%
Received Formal ICT Training	40%

Investments in digital infrastructure are essential to overcome these disparities. Addressing these gaps will not only improve access to ICT but also ensure a more equitable educational environment across Morocco.

5.3 *Teacher Training*

A critical challenge in integrating ICT into education lies in adequately training teachers to use digital tools effectively. While many educators express a willingness to adopt ICT in their teaching practices, a significant portion lacks the necessary training to do so. According to a report by the Ministry of Education [24], a considerable number of teachers in Morocco have not received formal training on incorporating ICT into pedagogy, which limits their ability to leverage these technologies effectively in the classroom [21]. This gap in teacher preparedness is particularly concerning in a context where distance learning and digital tools have become integral to modern education systems. Table 7 highlights the contrast between teacher willingness and their actual training levels.

Expanding professional development opportunities focused on ICT integration is essential to equip teachers with the skills they need to create engaging, technology-enhanced learning environments.

5.4 *Resistance to Change*

Resistance to change poses another significant barrier to the adoption of ICT in education. Some educators and local communities view the introduction of digital tools as a disruption to traditional teaching methods. This resistance often stems from socio-cultural factors, including:

- Fear of over-reliance on technology.
- Limited familiarity with digital tools, leading to apprehension about their effective use.

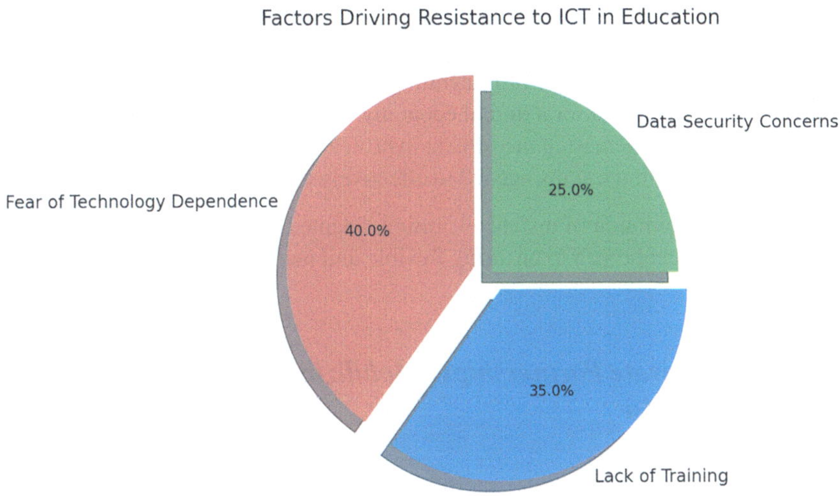

Fig. 7 Factors driving resistance to ICT in education (Source: Adapted from Mansouri [18, 19] and related analyses)

As highlighted by Mansouri [18, 19], these cultural and psychological barriers are compounded by concerns over data confidentiality and the security of digital platforms. Figure 7 illustrates the key factors contributing to resistance to ICT adoption.

Overcoming this resistance will require targeted strategies, including awareness campaigns, comprehensive training programs, and transparent policies to address concerns about data privacy and security. By fostering a culture of trust and collaboration, schools can encourage stakeholders to embrace technological advancements as essential tools for improving education.

6 Recommendations to Strengthen the Integration of ICT in Education in Morocco

6.1 Public Policy: An Essential Foundation for Intelligent Education

Public policy plays a key role in promoting intelligent education and the development of sustainable cities. Morocco has already initiated several efforts in this direction, notably with the Digital Morocco 2025 Strategy, which aims to improve internet access and strengthen the use of ICTs in various sectors, including education [25]. However, to optimize the integration of ICTs, more targeted policies are needed, which take into account regional disparities. Strategic priorities should include:

- Infrastructure Development: Strengthening connections in rural and remote areas by expanding broadband access and improving network reliability.
- Modernizing School Curricula: Integrating digital skills into all levels of education to prepare students for a digital economy.
- Teacher Training: Increasing investment in professional development programs to equip teachers with the skills needed to effectively use ICT in the classroom [20].

In addition, the promotion of distance learning, adapted to different levels of learning, is an essential approach to ensuring flexible and inclusive education.

6.2 Public-Private Partnerships: Mobilizing Resources and Expertise

Public-private partnerships (PPPs) are essential to meet the financial and technical challenges of integrating ICT into education. By collaborating with technology companies, the Moroccan government can:

- Provide Innovative Solutions: Develop digital platforms, e-learning tools, and artificial intelligence systems to enrich learning experiences [26].
- Enhance Teacher Training: Offer training programs for educators and develop standardized digital educational content.
- Expand Access: Extend partnerships to rural and disadvantaged areas, ensuring that all regions benefit from digital education.

Existing partnerships, such as those with Microsoft and Google, have already enabled teacher training and the creation of digital educational content. However, it is essential to:

- Include Rural Areas: Adapt solutions to local needs, such as low-cost devices or offline-accessible educational platforms.
- Ensure Sustainability: Provide ongoing support and maintenance for digital infrastructure.

6.3 Strengthening Infrastructure in Rural Areas: Bridging the Digital Divide

The lack of infrastructure in rural areas remains one of the main obstacles to equitable education. To bridge this digital divide, several measures are needed:

- Extend Broadband Access: Increase internet coverage in underserved regions to ensure reliable connectivity.
- Distribute Digital Equipment: Provide tablets, laptops, and other tools to students and teachers in rural schools.

- Create Community Access Centers: Establish technology hubs equipped with digital resources to serve as shared learning spaces [9, 10].

For example, the "Génération Connectée" program launched by the Moroccan government has equipped thousands of students in rural areas with digital tools. In Azilal, a pilot project provided 500 students with educational tablets while improving local connectivity. However, to maximize the impact, it is crucial to:

- Strengthen Telecommunications Infrastructure: Ensure that rural areas have access to high-speed internet.
- Provide Ongoing Teacher Training: Equip educators with the skills needed to effectively use digital tools in the classroom [3].

7 Conclusion and Perspectives

7.1 Conclusions

This chapter has highlighted the crucial importance of Information and Communication Technologies (ICT) in transforming education, particularly in the context of Morocco's efforts to develop smart cities. Through initiatives in Casablanca and Rabat, it is clear that integrating ICT into education is an essential lever for:

- Improving Accessibility: Reducing geographical inequalities by connecting rural areas to digital educational resources.
- Enhancing Quality of Education: Encouraging interactive and adaptive teaching methods that cater to diverse learning needs.
- Reducing Socio-Economic Inequalities: Ensuring equitable opportunities for all students, regardless of their background.

However, several challenges remain, including:

- The persistent digital divide between urban and rural areas.
- Inadequate infrastructure in remote regions, which limits access to digital tools.
- A lack of teacher training in the use of ICT, which hinders effective implementation.
- Resistance to change among some educators and communities, driven by sociocultural factors and concerns about data security.

To address these challenges, the following solutions have been proposed:

- Strengthening Public Policies: Expanding broadband access, modernizing school curricula, and investing in teacher training.
- Promoting Public-Private Partnerships: Collaborating with technology companies to provide innovative solutions and resources.

- Bridging the Digital Divide: Distributing digital equipment and creating community access centers in rural areas.

7.2 Perspectives for Future Research

To further advance the integration of ICT in Moroccan education, future research should focus on:

- Evaluating Educational Outcomes: Assessing the impact of ICT on student performance, particularly in rural areas, to identify best practices and areas for improvement.
- Comparative Analysis: Comparing the effectiveness of ICT integration between urban and rural schools to understand the factors contributing to success or failure.
- Longitudinal Studies: Tracking the progress of students over several years to measure the long-term impact of ICT on learning outcomes and career readiness.
- Teacher Training Impact: Analyzing the effectiveness of teacher training programs on ICT adoption and usage in classrooms, with a focus on overcoming resistance to change.
- Scalability of Initiatives: Investigating how successful pilot projects, such as the Génération Connectée program, can be scaled up to benefit a larger number of students across Morocco.

7.3 The Path Forward

The integration of ICT in education is not just a technological challenge but also a socio-economic and cultural one. To realize the vision of a more connected and equitable Morocco, concerted action is needed from all stakeholders, including government agencies, private sector partners, educators, and local communities. By addressing the challenges and leveraging the opportunities presented by ICT, Morocco can build an inclusive and sustainable educational system that prepares its citizens for the challenges of the twenty-first century.

References

1. Selwyn N (2017) Education and technology: key issues and debates, 2nd edn. Bloomsbury, London
2. UNESCO (2019) Global education monitoring report 2019: migration, displacement and education—building bridges, not walls. United Nations Educational, Scientific and Cultural Organization, Paris

3. OECD (2020) Education at a glance 2020: OECD indicators. Organisation for Economic Co-operation and Development, Paris
4. Livingstone S (2012) Children and the internet: great expectations, challenging realities. Polity Press, Cambridge
5. World Bank (2020) The digital economy in developing countries: prospects and challenges. World Bank, Washington, DC
6. Warschauer M (2003) Technology and social inclusion: rethinking the digital divide. MIT Press, Cambridge
7. Kozma RB (2008) ICT, education reform, and economic growth: a conceptual framework. Int J Educ Dev Using ICT 4(1):1–13
8. Fullan M (2013) Stratosphere: integrating technology, pedagogy, and change knowledge. Pearson, Toronto
9. UNESCO (2021a) Artificial intelligence in education: challenges and opportunities. UNESCO Publishing, Paris
10. UNESCO (2021b) The digital transformation of education: challenges and opportunities. United Nations Educational, Scientific and Cultural Organization, Paris
11. Bousfiha A, El Amrani Y, Benjelloun M (2021) Digital divide and educational inequality in Morocco. Moroccan Center for Educational Research, Rabat
12. Benhaddou M, Arbaoui Y (2022) Smart cities and digital transformation in Morocco. Moroccan Institute for Urban Studies, Rabat
13. Piatkowski M, El Haddad R (2019) Smart cities and digital education: the Moroccan case. J Urban Technol 26(2):35–54
14. Ministry of National Education (2020a) Annual report on ICT integration in Moroccan schools. Government Publications, Rabat
15. Ministry of National Education (2020b) Digital education strategy in Morocco: challenges and perspectives. Moroccan Ministry of National Education, Rabat
16. UNESCO (2020) Education in a post-pandemic world: nine ideas for public action. United Nations Educational, Scientific and Cultural Organization, Paris
17. Zouari H, Loukili R (2022) Urban innovation and ICT in Moroccan smart cities. University Press, Marrakech
18. Mansouri A (2021a) Real-time smart transportation management in Casablanca: challenges and innovations. Journal of Smart Urban Mobility 15(2):45–62
19. Mansouri K (2021b) Challenges and opportunities in ICT integration: a Moroccan perspective. Education Reform Press, Casablanca
20. Belkhayat A (2021) Digital transformation in Moroccan education: challenges and opportunities. Education Reform Press, Rabat
21. Chakroun A, Bensaid M (2022) Intelligent tutoring systems: personalized learning through AI. Springer, New York
22. Ben Ali A (2022) Digital transformation in Moroccan education: case studies from Casablanca. Moroccan Publishing House, Rabat
23. El Amrani R, Bouayad A (2020) AI in education: transforming learning and Management in Morocco. Academic Press Morocco, Rabat
24. Ministry of Education (2020) Report on ICT integration in Moroccan schools. Government Publications, Rabat
25. Ministry of Industry and Technology (2020) Digital Morocco 2025 Strategy. Government Publications, Rabat
26. World Bank (2022) Public-private partnerships in education: lessons from Morocco. World Bank Publications, Washington, DC

Harnessing Morocco's Abundant Phosphate Resources: Advancements in Lead Phosphate Glass/Metal Nanocomposite Photovoltaic Cells for Smart Cities and Benchmarking Against Silicon Efficiency

Maryama Hammi, Abderrahim Ed-Daoui, Fuad Khoshnaw, Noureddine Chafi, Mouad Bargani, Tadeusz Szumiata, Hicham Kaddouri, and Abdelhafid Elouassidi

Abstract Morocco is advancing towards smart cities by integrating innovative technologies to improve urban living, resource management, and sustainability. With abundant natural resources, particularly phosphates, Morocco is positioned to

---

M. Hammi
School of Architecture, Built Environment, Computing and Engineering, Birmingham City University, Birmingham, United Kingdom
e-mail: maryama.hammi@mail.bcu.ac.uk

A. Ed-Daoui (✉) · N. Chafi
Laboratory of Physics of Matter and Radiations (LPMR), Physics Department, Faculty of Sciences, University Mohamed Premier, Oujda, Morocco
e-mail: a.eddaoui@ump.ac.ma

F. Khoshnaw
School of Engineering and Sustainable Development, De Montfort University (DMU), Leicester, United Kingdom

M. Bargani
School of Engineering and Sustainable Development, De Montfort University (DMU), Leicester, United Kingdom

Process Engineering, Computer Science, and Mathematics Laboratory (LIPIM), National School of Applied Sciences, Sultan Moulay Slimane University, Khouribga, Morocco

T. Szumiata
Department of Physics, Faculty of Mechanical Engineering, Casimir Pulaski Radom, University Stasieckiego, Radom, Poland

H. Kaddouri · A. Elouassidi
Experimentation and Modelling Team in Mechanics and Energy Systems, National School of Applied Sciences, Abdelmalek Essaadi University, Al-Hocima, Morocco

H. Kaddouri et al. (eds.), *Smart Cities for Sustainable and Urban Development*, Green Energy and Technology, https://doi.org/10.1007/978-3-031-99001-4_10

lead in sustainable energy solutions. This chapter explores the use of semiconductor technology to create advanced photovoltaic cells utilizing Morocco's phosphate resources. We focus on the fabrication and characterization of lead phosphate (PbP) glass/metal nanocomposite thin films, designed for integration into smart city infrastructures. The study investigates the incorporation of metallic particles such as nickel (Ni), chromium (Cr), cobalt (Co), and zinc (Zn) to enhance solar energy conversion efficiency. P-N junctions are formed with PbP/Ni as the p-type semiconductor and PbP/Cr as the n-type semiconductor, addressing challenges like contact resistance and wavelength sensitivity. Results show a significant reduction in electrical resistivity with increasing metal concentration and a narrowing of the optical band gap while maintaining transparency. Temperature-dependent conductivity studies and figure of merit (FOM) evaluations provide further insight into the films' semiconducting properties. This research highlights the potential of PbP/metal nanocomposites to boost energy efficiency in smart cities, supporting Morocco's goals for sustainable energy and showcasing the value of phosphate resources in advancing photovoltaic technology.

Keywords Semiconductor technology · Photovoltaic cells · PbP nanocomposites · Smart cities · Sustainable energy

1 Introduction

The global transition toward sustainable energy has spurred a significant demand for innovative photovoltaic technologies. As the world shifts toward renewable energy sources, Morocco's abundant phosphate resources offer a unique opportunity to play a pivotal role in advancing photovoltaic cell development. These phosphate reserves, among the world's largest, are not only vital for agricultural applications but also hold transformative potential in the field of renewable energy, particularly in the production of solar cells. Morocco's strategic positioning as a leading global supplier of phosphates gives it a competitive edge in the evolution of photovoltaic technology, and its rich natural resources can support the development of local, cost-effective, and sustainable energy solutions to meet growing energy demands. Several studies have highlighted the potential of Morocco's phosphate resources in supporting the development of solar technologies, with researchers investigating various ways to exploit these materials for enhanced photovoltaic cell performance [1, 2].

In line with this vision, Morocco has been proactive in integrating renewable energy solutions, exemplified by the country's ambitious plans to develop smart cities. These cities, such as the Green City in Benguerir, are designed to promote energy efficiency, sustainability, and environmental stewardship, emphasizing the adoption of renewable energy technologies. The use of local materials, including phosphates, aligns perfectly with Morocco's smart city initiatives, ensuring that urban infrastructures are powered by sustainable and locally sourced energy. This chapter focuses on the role of lead phosphate (PbP) glass/metal nanocomposite thin films in advancing photovoltaic technology for smart city applications.

Our research investigates the incorporation of metallic particles—nickel (Ni), chromium (Cr), cobalt (Co), and zinc (Zn)—into PbP nanocomposites to enhance solar energy conversion efficiency. Previous studies have demonstrated that metallic dopants can significantly improve the electrical and optical properties of semiconductor matrices, resulting in superior photovoltaic performance [3, 4]. Central to our investigation is the creation of P-N junctions using PbP/Ni as the p-type semiconductor and PbP/Cr as the n-type semiconductor, which addresses challenges such as contact resistance and wavelength sensitivity. DC conductivity measurements reveal a significant decrease in resistivity with increased metal concentration, marking the onset of percolation behavior essential for efficient charge transport [5, 6].

Optical analysis of the PbP/metal nanocomposites shows that increased metal concentrations narrow the optical band gap while maintaining good transparency, which is critical for balancing optical and electrical properties. Figure of merit (FOM) calculations further benchmark the performance of these films against conventional silicon-based solar cells [7]. Temperature-dependent condutivity studies reveal two distinct conduction mechanisms: thermally activated conduction at high temperatures and variable range hopping (VRH) at low temperatures, providing insights into the films' semiconducting properties [8, 9].

This research demonstrates the capabilities of PbP/metal nanocomposites to improve energy efficiency in urban environments, supporting their integration into Morocco's smart city initiatives. By leveraging local phosphate resources, Morocco is positioned as a leader in the global shift toward renewable energy, and this work underscores the importance of these resources in the ongoing evolution of photovoltaic technology. Through these efforts, Morocco not only advances its energy security but also sets a global example for sustainable urban development, reinforcing its role in the transition to renewable energy and smart city infrastructure [10–12].

2 Materials and Methods

2.1 P-N Junction Overview

The P-N junction, a fundamental component of solar cells, consists of two semiconductor regions: a P-type region with a high concentration of holes and an N-type region with a majority of electrons. When these materials are brought together, diffusion currents are generated, forming a potential barrier that restricts the movement of majority carriers in each region. This mechanism facilitates photon absorption and charge carrier separation, which are crucial for efficient energy conversion. The narrow width of the junction (typically ranging from 0.2 to several micrometers) allows the electric field created by the potential barrier to counterbalance diffusion currents, stabilizing the solar cell's operation [13].

2.2 Solar Cell Structure and Fabrication

To fabricate a solar cell suitable for integration into smart city infrastructures, we utilized Indium-doped tin oxide (ITO) glass (1 × 1 cm), along with P-type and N-type semiconductors. The fabrication process involved the following steps:

- Covering the conductive ends of the ITO glass to create a 10 mm aperture.
- Applying a semiconducting glass mixture and leveling it with a razor blade.
- Annealing the resulting electrode at 150 °C for 10 min.
- Depositing the N-type semiconductor to complete the P-N junction.

Careful surface preparation, processing conditions, and deposition techniques were essential for optimizing the performance and reproducibility of the solar cells, ensuring their effectiveness in urban applications.

The performance of this junction was evaluated by exposing it to various light sources. As shown in the Table 1, the contact resistance increases with the wavelength (Fig. 1).

We assessed the solar cell's response to different LED colors, revealing variations in contact resistance:

These findings highlight the potential for optimizing photovoltaic cell design to improve overall efficiency.

2.3 Characterization of Optical Properties

The optical properties of the lead phosphate glass films, critical for solar cell efficiency, were assessed using UV-Visible transmittance spectra. The refractive index, indicative of light propagation in materials, was determined using the formula:

$$n = \frac{c}{v} \tag{1}$$

Table 1 LED colors, their corresponding wavelengths, and the resistance response of the illuminated cell

Led colors	Wave length (nm)	Contact Resistance response of illuminated cell (ohm)
Red	650	17
Yellow	580	31
Blue	445	56

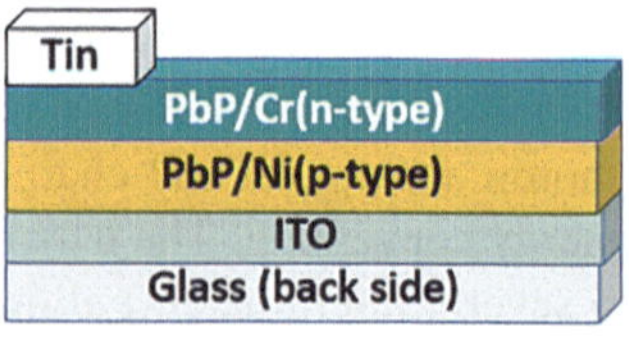

Fig. 1 Structure of a photovoltaic cell based on heterojunction of PbP/metal nanocomposites

Where c is the speed of light in a vacuum and v is the speed of light in the medium. The increase in refractive index with filler density confirmed the suitability of these materials for photovoltaic applications [14, 15].

2.4 Electrical Conductivity Measurements

DC conductivity was measured using the two-point probe method (Fig. 2), revealing resistivity and allowing calculations of sheet resistance using the formula:

$$R_{sh} = \frac{R.W}{L} \tag{2}$$

$$R = \rho.\frac{L}{A} = \rho.\frac{L}{d.W} = \frac{\rho}{d}.\frac{L}{W} = R_s.\frac{L}{W} \tag{3}$$

Where ρ: resistivity (Ω.cm), Rs = ρ/d: electrical sheet resistance (Ω.cm), w: width (cm), d: thickness (cm) which can be determined with optical measurements, L: length (cm). And the conductivity is given by: $\sigma = 1/\rho\ (\Omega.\text{cm})^{-1}$.

2.5 Electrical Conductivity Versus Temperature

The study of electrical conductivity versus temperature (Fig. 3) is essential for understanding material performance in applications like sensors and solar cells. As temperature increases, the mobility of charge carriers also increases, affecting conductivity. Analyzing this relationship helps determine activation energy and optimize materials for specific uses, such as phosphate glass films loaded with fillers.

Understanding temperature-dependent conductivity also aids in predicting long-term stability and device efficiency, while revealing how fillers influence electrical properties.

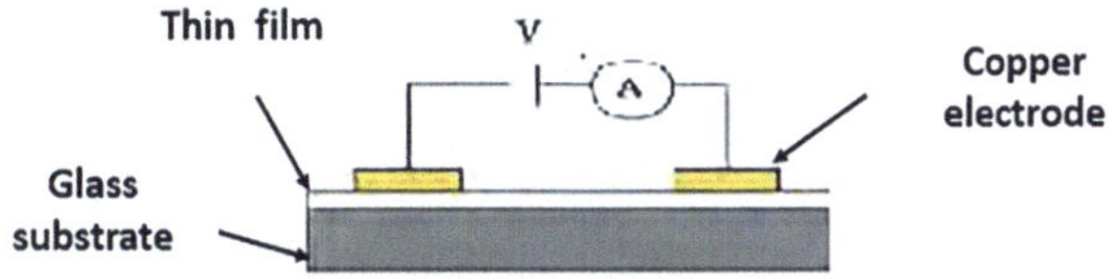

Fig. 2 Experimental device for measuring current-voltage characteristics

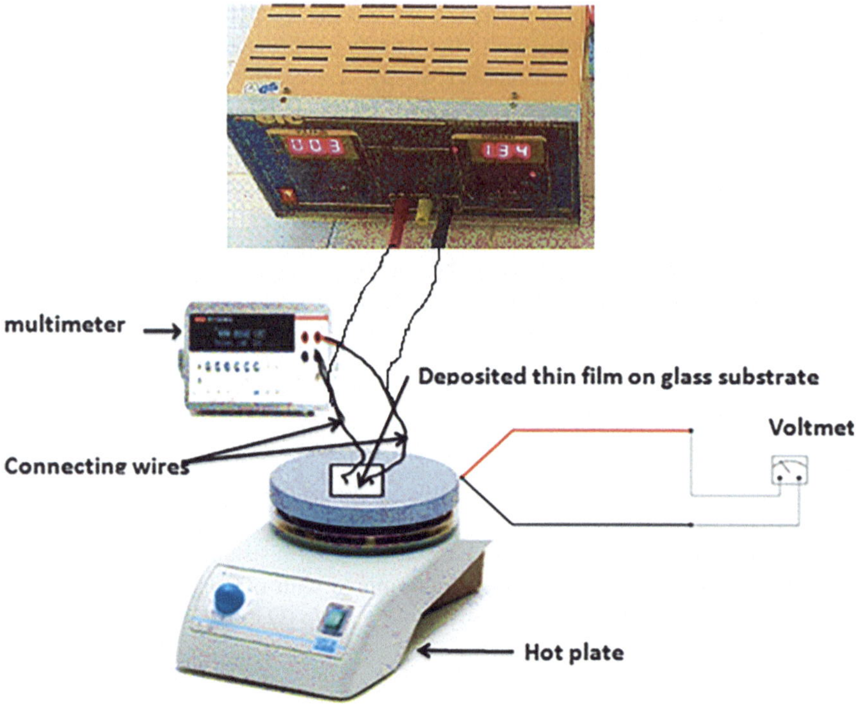

Fig. 3 Experimental device for measuring resistivity versus temperature used to study the conductivity behavior of materials under different thermal conditions

2.6 *Evaluation of Performance Metrics*

To evaluate the photovoltaic performance of the PbP/metal nanocomposites, we calculated the Figure of Merit (FOM) based on the material's electrical conductivity and optical transmission. The relationship between these parameters is crucial for determining the suitability of the films for smart city applications, where high energy conversion efficiency and transparent materials are essential for optimizing energy usage in urban infrastructures [16–18].

The Figure of Merit (FOM) can be expressed using the following equation:

$$FOM = \frac{\sigma}{\alpha}.T \tag{4}$$

Where:

σ represents the electrical conductivity of the material (S/m), indicating its ability to conduct electricity.

T is the optical transmission of the material (dimensionless, typically ranging from 0 to 1, with 1 indicating complete transparency), measuring how well light can pass through the material.

T is the absorption coefficient (m-1), quantifying the material's ability to absorb light, which is crucial for energy harvesting efficiency.

The FOM provides an effective balance between these characteristics, showing how well a material can convert solar energy into electricity while maintaining transparency which is an important feature for smart city applications where both energy efficiency and aesthetic considerations are essential [19, 20]. A higher FOM value indicates that the material is more efficient at utilizing solar energy while preserving transparency, making it suitable for applications such as solar windows and energy-harvesting surfaces in smart buildings and infrastructures.

Additionally, this model enables the comparative evaluation of PbP/metal nanocomposite films with other conventional materials, like silicon-based solar cells, offering a benchmark for real-world photovoltaic applications [21, 22].

In selecting transparent and conductive materials for industrial uses, the FOM is a critical parameter. Ideally, a material should exhibit high electrical conductivity alongside a low absorption coefficient to maximize efficiency. The FOM is also expressed as:

$$FOM = \frac{\sigma}{\alpha} \tag{5}$$

The optimal FOM balances good optical transmission and high electrical conductivity, necessitating a compromise between these two factors. Research has demonstrated that for thin films to be effective, the FOM should be greater than or equal to 7 [23, 24].

Given the importance of temperature-dependent properties in photovoltaic materials, particularly for thin films, it is necessary to understand how the electrical conductivity varies with temperature. The relationship between conductivity and temperature can be modeled using the Arrhenius equation, which incorporates the pre-exponential factor σ_0 and the activation energy ΔE. This temperature dependence provides valuable insights into the behavior of the material under different environmental conditions, which is key for real-world applications.

The electrical conductivity σ of materials often depends on temperature, and the relationship can be modeled using the Arrhenius equation, which includes the pre-exponential factor σ_0 and the activation energy ΔE:

$$\sigma = \sigma_0 \exp\left(\frac{\Delta E}{KT}\right) \tag{6}$$

σ_0 the pre exponential factor,σ is the DC conductivity, ΔE is activation energy which is expressed as follows:

$$\Delta E = \left(E_C - E_F\right) \tag{7}$$

Where E_C defines the mobility edge in the conduction band of the films and E_F is the energy of the fermi level.

The modeling of DC conductivity versus the reciprocal of temperature is performed by introducing the logarithm in equation as follow:

$$Ln\sigma = Ln\sigma_0 - \frac{\Delta E}{KT} \tag{8}$$

The activation energy can be derived from the slope of the plot of lnσ against the inverse of the temperature.

According to Planck's law, as we move further into the visible spectrum, the energy required to excite valence electrons into the conduction process tends to decrease, approaching minimal values.

$$\Delta E = h.v = \frac{h.c}{\lambda} \tag{9}$$

This relationship highlights the importance of considering the energy levels involved in conduction when designing materials for efficient energy conversion, particularly in photovoltaic applications.

2.7 Conversion Efficiency Calculation

To assess the practical performance of the PbP/metal composite-based solar cells, we calculated the conversion efficiency η of the devices. Conversion efficiency is a critical measure of how effectively the material can convert incident sunlight into usable electrical energy, and it was calculated using the following formula:

$$\frac{\boldsymbol{P_{out}}}{\boldsymbol{P_{in}}} \tag{10}$$

Where: P_{out} is the output power, which is determined at the maximum power point (MPP) by multiplying the current (I) and the voltage (V) at the MPP:

$$P_{out} = I_{MPP} \times V_{MPP} \tag{11}$$

$P_{in}$is the input power, which can be calculated from the incident light intensity on the solar cell surface (I_{light}) and the area of the solar cell (A_{cell}):

$$\mathrm{P_{in} = I_{light} \times A_{cell}} \tag{12}$$

Thus, by analyzing both the temperature-dependent electrical conductivity and the conversion efficiency, we can evaluate the performance of PbP/metal composite-based solar cells under varying conditions and optimize them for practical applications.

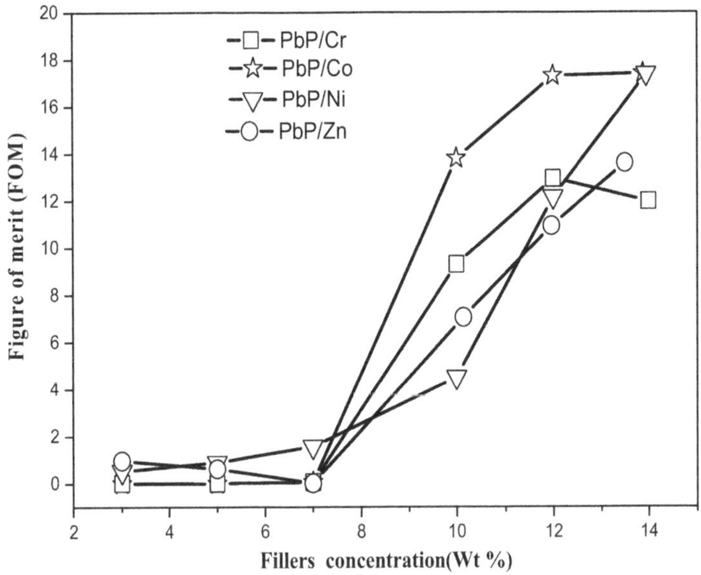

Fig. 4 Variation of optical figure of merit versus metal particles concentration

3 Results and Discussion

According to the relation (4), the variation in the FOM values of the studied thin films as a function of filler concentration is shown in the inset of Fig. 4. As the filler concentration increases, there is a noticeable enhancement in the figure of merit, demonstrating that higher filler loading leads to improved material performance.

The lead phosphate glass-based nanocomposite thin films exhibited an impressive figure of merit (FOM), particularly at a metallic filler concentration of approximately 10 wt%. This concentration yielded optimal optical and electrical properties, highlighting the films' potential for application in advanced photovoltaic technologies. The high transparency (up to 70%) in the visible wavelength region, coupled with the ability to tailor the optical band gap through filler loading, underscores the versatility of these materials for energy applications.

3.1 DC Conductivity and Electrical Resistivity Measurements

Electrical conductivity presents a particular interest of semiconductor materials. Measuring the electrical conductivity of semiconducting thin films is a rather difficult problem, since it depends upon a series of factors whose actions cannot be separated, especially the temperature. The DC conductivity of the PbP/metal nanocomposite films was assessed through resistivity measurements using the

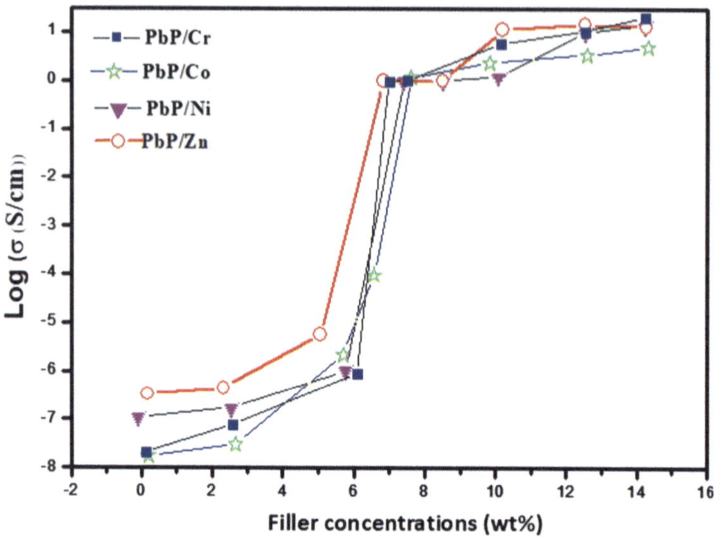

Fig. 5 Electronic conductivity of lead phosphate glass thin films as a function of conducting particle concentration

two-point probe method [25]. This approach provides accurate resistivity values, reflecting the materials' ability to resist the flow of electric current, measured in [Ω·cm].

As shown in Fig. 5, the electrical conductivity of the unloaded thin film was on the order of 10–7 $\Omega.cm^{-1}$. This value increased nonlinearly with the concentration of transition metal particles, displaying characteristic curve shapes for films loaded with chromium, cobalt, nickel, and zinc. A sharp transition in electrical conductivity was observed at a critical concentration of around 6 wt%, indicating a percolation threshold. Above this threshold, the composites transitioned into a semiconductor state, resulting in a significant increase in the number of charge carriers and enhanced conductivity, consistent with Burstein-Moss theory [17, 26].

The minimal band gap value was achieved at the highest loading levels for all films. Remarkably, even at elevated filler concentrations, the films maintained a relatively high transparency. This trend aligns with the observed electrical conductivity, which exhibited a progressive increase as a function of filler concentration, demonstrating a sharp increase of seven orders of magnitude at critical levels. The enhanced conductivity is attributed to an increase in free charge carriers within the insulating glass matrix, leading to improved mobility and electrical performance.

3.2 *Temperature-Dependent Conductivity*

Figure 6 illustrates that the electrical conductivity displays thermally activated behaviour, confirming the influence of temperature.

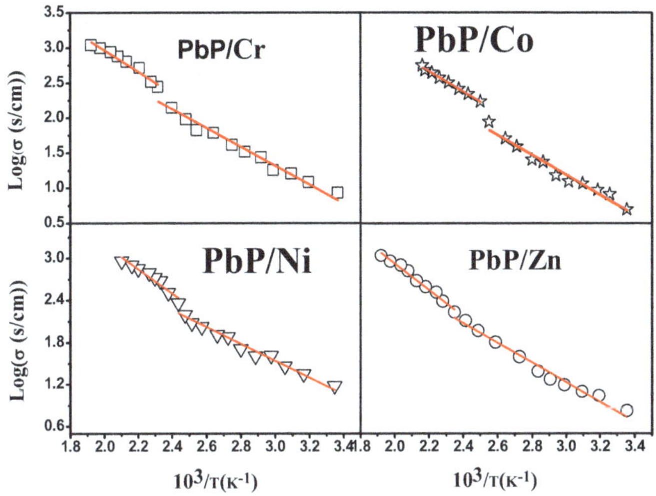

Fig. 6 Plot of Lnσ versus 1000/T for loaded glass thin films

Table 2 Activation energies obtained from the Arrhenius fit, with ΔEI and ΔEII corresponding to the high- and low-temperature ranges, respectively

Composites	Slope change temperature Tc (K)	ΔEI (eV)	ΔEII (eV)
PP/Cr	416	1.53 ($R^2 = 0.95$)	1.34 ($R^2 = 0.96$)
PbP/co	380	1.44 ($R^2 = 0.99$)	1.41 ($R^2 = 0.98$)
PbP/Ni	404	1.71 ($R^2 = 0.92$)	1.20 ($R^2 = 0.96$)
PbP/Zn	436	1.88 ($R^2 = 0.98$)	1.40 ($R^2 = 0.97$)

The activation energy values provide insights into the position of the Fermi level relative to the conduction band. For instance, for the PP/Cr and PbP/Co samples, the activation energies were found to be less than half of the optical gap, suggesting n-type conductivity. On the other hand, for the PbP/Ni and PbP/Zn samples, the activation energies exceeded half of the mobility gap, which indicates that these films might exhibit p-type conductivity [24, 27].

This distinction in the activation energy values is important, as it reflects the electronic behavior of the material, with n-type conductivity being associated with excess electrons and p-type conductivity linked to the presence of holes (positive charge carriers). These findings are critical for understanding the potential applications of these nanostructured thin films, especially in devices such as P-N junctions. As shown in Table 2, the activation energy values offer further insight into the Fermi level's position relative to the conduction band. The PP/Cr and PbP/Co samples exhibit activation energies that are less than half of the optical gap, confirming their n-type conductivity. In contrast, the PbP/Ni and PbP/Zn films have activation energies that surpass half of the mobility gap, suggesting they could function as p-type semiconductors.These results suggest that the films' electronic properties make them suitable for P-N junction applications, which are crucial in the development of photovoltaic devices, diodes, and transistors.

Building upon these insights into the electronic properties and the potential for P-N junction applications, the fabricated P-N junctions in this study exhibited promising electrical characteristics. These junctions demonstrated a clear relationship between light intensity and contact resistance, further validating the potential for use in photovoltaic devices and other semiconductor applications. Previous research on similar chalcogenide glass-based composites has shown a maximum conversion efficiency of around 1% [28–30]. However, our experiments revealed enhanced efficiency at higher wavelengths and illumination intensities, which aligns with findings from existing literature [31–33].This further confirms the effectiveness of these films for energy conversion, supporting their suitability for smart applications such as energy harvesting and efficient light-responsive devices.

To visualize these electronic interactions and better understand the behavior of these P-N junctions, Fig. 7 presents a schematic representation of the energy band alignment at the P-N junction. This diagram illustrates how the conduction and valence bands align across the junction, showing the distinct electronic characteristics of the n-type and p-type regions. Understanding this alignment is essential for optimizing the efficiency and functionality of devices like solar cells, diodes, and transistors, further emphasizing the importance of these nanostructured thin films in advanced electronic applications.

PbP/Cr: $\Delta E = 1.53$ev,Eg ~ 3.28ev
PbP/Co: $\Delta E = 1.44$ eV, Eg ~ 3.17 eV
PbP/Ni: $\Delta E = 1.71$ eV, Eg ~3.15 eV
PbP/Zn: $\Delta E = 1.51$ eV,Eg ~ 3.39 eV

Using the maximum values of the activation energies determined by the Arrhenius equation, a comparison with the corresponding optical gap for each thin film allows us to identify the type of semiconductor. The activation energies for PbP/Ni and PbP/Zn are greater than half of the mobility gap, indicating that these materials exhibit p-type semiconductor behavior. In contrast, the activation energies for PbP/Cr and PbP/Co are lower than half of the optical gap, suggesting that these materials exhibit type semiconductor n-behavior. This distinction is important for understanding the electronic properties of the films and their potential applications in devices such as P-N junctions.

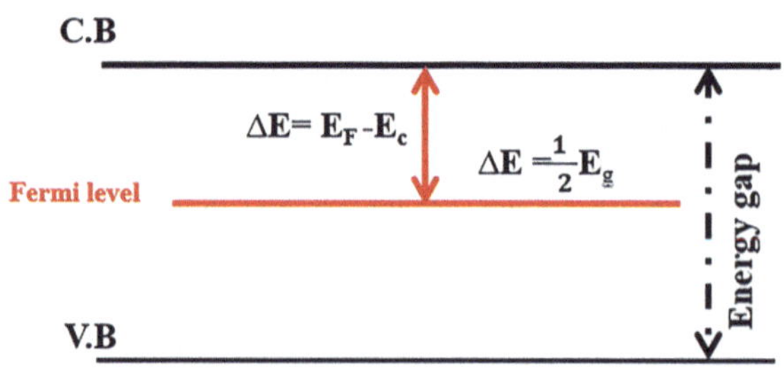

Fig. 7 Schematic representation of the energy band alignment at the P-N

4 Comparison Between Conventional Silicon Cells and PbP/ Metal Nanocomposite Cells

Compared to traditional silicon solar cells, which typically exhibit conversion efficiencies between 15% and 22% [23, 24], our PbP/metal nanocomposite cells show a competitive edge. This suggests that further optimization and scaling could position these phosphate glass materials as viable alternatives in the solar energy market, particularly for cost-sensitive applications in smart cities.

4.1 Conventional Silicon Solar Cells

Conventional solar cells are primarily made from crystalline silicon, either monocrystalline or polycrystalline. Monocrystalline cells consist of a single continuous crystal, while polycrystalline cells are made up of multiple smaller crystals. The structure of a silicon solar cell includes an n-type silicon layer (excess electrons) and a p-type silicon layer (excess holes). These layers form a PN junction that generates an electric field to separate charge carriers when sunlight strikes the material. An anti-reflective coating reduces light reflection and enhances absorption, while metal contacts on both sides collect the generated current [34].

The operation of silicon solar cells is based on the photovoltaic effect, where photons excite electrons in the silicon, creating electron-hole pairs. The electric field at the PN junction drives the charge carriers toward the metal contacts, generating an electric current. Silicon cell efficiencies typically range from 15% to 22%, depending on material quality, manufacturing processes, and the silicon type used [35].

Other photovoltaic technologies include thin-film cells, such as amorphous silicon (5% market share) and cadmium telluride (CdTe), which holds 8% of the market. Copper indium gallium diselenide (CIGS) cells, still in development, account for 1% of the market. Efficiency varies widely: multijunction cells can exceed 46%, single-junction cells (e.g., silicon, CIS, CdTe) range from 22.1% to 25.6%, and newer organic and hybrid technologies reach 13% to 21%. While single-junction technologies are limited by the Shockley-Queisser limit, multijunction cells can surpass this threshold [36].

Silicon thin-film solar cells are composed of a thin layer of silicon deposited onto a substrate, which can be either flexible or rigid. The structure typically includes an active silicon layer, a transparent conductive oxide (TCO) layer, and a back contact layer that facilitates electrical output [41]. This design enables the cells to maintain a lightweight and flexible form while optimizing sunlight absorption. However, the efficiency of these cells is limited by the thinness of the silicon layer, which reduces the amount of light it can absorb and thus lowers the efficiency of conversion to electricity [37].

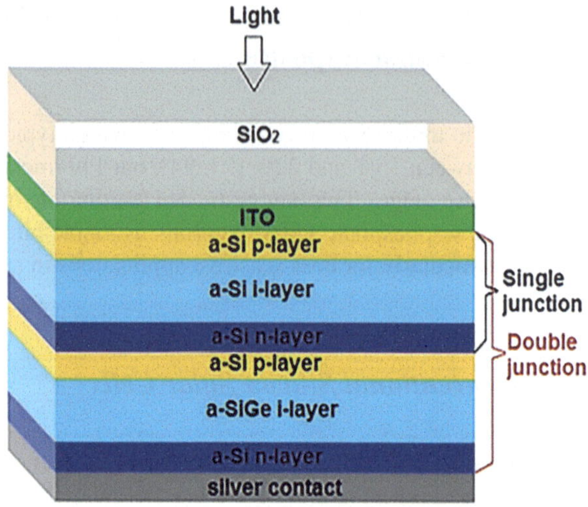

Fig. 8 A double junction layout Solar cell a-Si: H/a-SiGe: H [37]

Figure 8 provides a detailed view of a double-junction layout of an a-Si: H/a--SiGe: H solar cell, which is an advanced thin-film design. In this configuration, two different silicon-based materials, amorphous silicon (a-Si:H) and amorphous silicon-germanium (a-SiGe:H), are stacked together in a tandem arrangement to improve light absorption and energy conversion. The use of a double-junction structure enhances the overall efficiency by allowing the solar cell to absorb a broader spectrum of sunlight, addressing the limitations of single-junction cells. The a-Si: H layer absorbs higher-energy photons, while the a-SiGe: H layer captures lower-energy photons, making this design more effective at converting sunlight into electricity compared to single-junction thin-film cells.

4.2 PbP/Metal Nanocomposite Solar Cells

PbP/metal nanocomposite solar cells represent a newer approach to solar energy conversion, where lead phosphate (PbP) nanoparticles are combined with metal nanostructures (such as gold, silver, or copper) to enhance the photovoltaic performance. The structure of these cells typically involves a nanocomposite material where metal nanoparticles are embedded within the PbP matrix. These metal nanoparticles enhance the light absorption capabilities of the cell by exploiting plasmonic effects, which concentrate light at the nanoparticle surface, improving light absorption in the PbP layer [38]. The cell design generally includes a PbP light-absorbing layer, metal nanoparticle inclusion for enhanced light trapping, and metal contacts for current collection.

The operation principle of PbP/metal nanocomposite cells is similar to traditional silicon cells, relying on the photovoltaic effect to generate charge carriers when sunlight excites the PbP material. However, the plasmonic effects from the metal nanoparticles contribute to enhanced light absorption and faster electron transport, which helps reduce recombination losses and improves overall efficiency [39]. In initial studies, these cells have demonstrated conversion efficiencies in the range of 15% to 25%, making them competitive with traditional silicon solar cells [40]. Although these efficiencies are still being optimized, the use of plasmonic nanoparticles provides a promising route for enhancing the performance of the cell.

4.3 *Cost and Scalability*

One of the key advantages of PbP/metal nanocomposite solar cells is their cost-effectiveness. The materials used in these cells, such as metal nanoparticles and PbP, are generally cheaper and can be processed using low-cost techniques like solution processing or roll-to-roll printing, making them more affordable to produce than traditional silicon-based cells [39]. This makes them particularly attractive for cost-sensitive applications such as smart cities, where flexible and inexpensive solar panels can be deployed in a variety of settings. Additionally, these cells have the potential for high scalability, especially for applications requiring lightweight, flexible solar devices. In contrast, silicon-based cells, while highly efficient, are more energy-intensive to produce and generally come with higher costs, particularly for high-efficiency monocrystalline cells [35]. Table 3 provides a comparative summary of the features of conventional silicon cells and PbP/metal nanocomposite cells:

This comparison highlights the benefits of PbP/metal nanocomposite cells, including lower production costs and high scalability, which make them ideal for cost-sensitive applications such as flexible solar devices in smart city infrastructure. These advantages position PbP/metal nanocomposite cells as a promising alternative to traditional silicon-based solar cells, particularly in applications where cost and flexibility are key factors.

Table 3 Summary of comparison

Feature	Conventional Silicon Cells	PbP/Metal Nanocomposite Cells
Material	Crystalline Silicon (Monocrystalline or Polycrystalline)	PbP with Metal Nanoparticles (Plasmonic Nanocomposite)
Efficiency	15%–22%	15%–25% (with potential for optimization)
Structure	PN Junction (n-type and p-type silicon)	PbP Layer + Metal Nanoparticles
Operation Principle	Photovoltaic Effect (Charge Separation at PN Junction)	Plasmonic Effects + Photovoltaic Effect
Cost	Relatively high due to silicon wafer production	Lower production cost due to cheaper materials and processes
Scalability	Mature technology with large-scale production	High scalability potential, particularly for flexible applications
Applications	Residential, commercial, large-scale solar power plants	Smart cities, flexible solar applications, low-cost panels

5 Conclusion and Perspectives

Lead phosphate glass-based nanocomposite thin films have been extensively studied from optical and electrical perspectives, demonstrating high transparency (up to 70%) in the visible spectrum, with a narrowing optical band gap as filler loading increased. The films exhibited optimal optical and electrical properties at a metallic filler concentration of approximately 10 wt%, showing a favorable figure of merit (FOM). Temperature-dependent conductivity measurements revealed the semiconducting nature of the films, displaying both thermally activated behavior at high temperatures and Variable Range Hopping (VRH) at low temperatures. These promising properties position PbP/metal nanocomposite films as strong candidates for applications in optoelectronics, thermoelectric devices, and solar cells, addressing the growing demand for advanced materials in the energy sector. Looking ahead, the development of PbP/metal composite-based solar cells presents exciting opportunities for integration into smart city infrastructures. As urban areas prioritize sustainability and energy efficiency, these photovoltaic cells can reduce dependence on fossil fuels and promote renewable energy sources. Their lightweight and flexible characteristics open the door for innovative applications, including building-integrated photovoltaics (BIPV) and solar-powered street lighting, while their customizable designs offer potential for enhancing urban aesthetics. To facilitate broader adoption, further research should focus on improving fabrication processes, optimizing material compositions, and exploring novel applications aligned with the smart city vision. The findings highlight the potential of PbP/metal nanocomposites to meet the energy demands of urban environments and contribute to the transition toward more sustainable living.

Competing Interests The authors declare that there are no competing interests in the publication of this chapter. The research was conducted with full academic and scientific integrity, and the findings presented are based on objective and unbiased analysis. No financial, personal, or professional relationships have influenced the research process or the conclusions drawn.

Ethics Approval The research presented in this chapter did not involve human participants, animal studies, or any experiments requiring formal ethics approval. All experimental procedures and analyses were carried out in accordance with established scientific guidelines and standards. The authors affirm that all research practices adhered to ethical principles of research integrity, data transparency, and accountability.

References

1. Bih L, Bih H, Amalhay M, Mossadik H, Elbouari A, Belhorma B, Graça MPF, Valente MA (2014) Phosphate glass-glasses as new energy density dielectric materials. Procedia Eng 83:371–377. https://doi.org/10.1016/j.proeng.2014.09.030
2. Lin X, Li X, Luo B, Yu D (2023) Black-phosphorus-based materials for application in solar cells. Chin J Struct Chem 42(7):100109. https://doi.org/10.1016/j.cjsc.2023.100109

3. Kang Q, Wang Q, An C, He C, Xu B, Hou J (2020) Significant influence of doping effect on photovoltaic performance of efficient fullerene-free polymer solar cells. J Energy Chem 43:40–46. https://doi.org/10.1016/j.jechem.2019.08.005
4. Poortmans J, Arkhipov V (2008) Thin film solar cells: fabrication, characterization, and applications. John Wiley & Sons
5. Hammi M, Aziate G, Pinto G, Maaroufi A (2015) Electrical and thermoelectrical behavior of binary composites of phosphate glass loaded with zinc fillers: promising materials for photovoltaics. J Chin Chem Soc 62(11):1035–1044. https://doi.org/10.1002/jccs.201400254
6. Karttunen M, Ruuskanen P, Pitkänen V, Albers WM (2008) Electrically conductive metal polymer nanocomposites for electronics applications. J Electron Mater 37(7):951–954. https://doi.org/10.1007/s11664-008-0451-2
7. Sekar P (2015) Performance enhancing the efficiency of solar PV cells using nanotechnology. J Chem Pharm Sci. Accessed January 3, 2025. www.jrrset.com
8. Hammi M (2023) Exploratory study of potential usefulness of cost-effective composites thin films for energy devices. Mater Renew Sustain Energy 12(3):171–186. https://doi.org/10.1007/s40243-023-00238-3
9. Hammi M, Ziat Y, Charaf L (2022) Positive temperature coefficient in lead phosphate glass-based composites with charges-hopping. Mater Chem Phys 276. https://doi.org/10.1016/j.matchemphys.2021.125370
10. Benbba R, Barhdadi M, Ficarella A, Manente G, Romano MP, el Hachemi N, Barhdadi A, Al-Salaymeh A, Outzourhit A (2024) Solar energy resource and power generation in Morocco: current situation, potential, and future perspective. Resources 13(10):140. https://doi.org/10.3390/resources13100140
11. Hoang AT, Pham VV, Nguyen XP (2021) Integrating renewable sources into energy system for smart city as a sagacious strategy towards clean and sustainable process. J Clean Prod 305. https://doi.org/10.1016/j.jclepro.2021.127161
12. Khalifi H, Riahi S, Cherif W (2024) Smart cities and sustainable urban development in Morocco. Ingenierie Des Systemesd'Information 29(2):741–751. https://doi.org/10.18280/isi.290236
13. Msrar A, Saiter J-M, Vautier C (1987) Dark conductivity of amorphous Ge0.08Te0.92 thin films. Proc Phys Soc 76:755–782
14. Fox M (2001) Optical properties of solids. Oxford University Press
15. Manifacier C, Murcia DM, Fillard JP, Vicario E (1977) Optical and electrical properties of SnO2 thin films in relation to their stoichiometric deviation and their crystalline structure. Thin Solid Films 41:127–135
16. El-Aassar MR, Sendi RK, Atta A et al (2023) Characterization and linear/nonlinear optical properties of PVA/CS/TiO2 polymer nanocomposite films for optoelectronics applications. Opt Quant Electron 55:1212. https://doi.org/10.1007/s11082-023-05542-w
17. Faÿ S (2003) L'Oxyde de Zinc par Dépôt Chimique en Phase Vapeur Comme Contact Électrique Transparent et Diffuseur de Lumière Pour les Cellules Solaires. Ecole Polytechnique Fédérale de Lausanne, Thèse n° 2899
18. Jacobs DA, Catchpole KR, Beck FJ, White TP (2016) A re-evaluation of transparent conductor requirements for thin-film solar cells. J Mater Chem A 4(12):4490–4496. https://doi.org/10.1039/c6ta01670g
19. Pelenchon F, Mialhe P (1990) Optimization of solar cell performance. Sol Energy Mater Sol Cells 33(1):1–10
20. Apurva, S., Tailor, S., & Rastogi, N. (2017). Smart materials for smart cities and sustainable environment. Journal of Materials Science & Surface Engineering, 5(1), 520–523. Available at https://www.jmsse.org/
21. Boujnah M, Boumdyan M, Naji S, Benyoussef A, el Kenz A, Loulidi M (2016) High efficiency of transmittance and electrical conductivity of V-doped ZnO used in solar cells applications. J Alloys Compd 671:560–565. https://doi.org/10.1016/j.jallcom.2016.02.107

22. Mendez-Gamboa JA, Castro-Rodriguez R, Perez-Quintana IV, Medina-Esquivel RA, Martel-Arbelo A (2016) A figure of merit to evaluate transparent conductor oxides for solar cells using photonic flux density. Thin Solid Films 599:14–18. https://doi.org/10.1016/j.tsf.2015.12.038
23. Bhandari R (2019) Recent advances in silicon solar cells: a review. Renew Sust Energ Rev 99:1–12
24. Green MA, Emery K, Hishikawa Y, Warta W, Zou J (2010) Solar cell efficiency tables. Prog Photovolt Res Appl 18(1):1–12
25. González A et al (2021) Electrical properties and conduction mechanisms in phosphate glass composites. J Mater Sci 56(4):2450–2462
26. Moss TS (1954) The interpretation of the properties of indium Antimonide. Proc Phys Soc 76:755–782
27. Wu X, Li J, Yang Y (2017) Current status and prospects of chalcogenide solar cells. Energy Environ Sci 10(6):1240–1252
28. Kim JY, Lee K, Moses NECD, Nguyen T-Q, Dante M, Heeger AJ (2007) Efficient tandem polymer solar cells fabricated by all solution processing. Science 317:222–225
29. Li G, Shrotriya V, Huang J, Yao Y, Moriaty T, Emery K, Yang Y (2005) High-efficiency solution processable polymer photovoltaic cells by self-organization of polymer blends. Nat Mater 4:864–868
30. Ma W, Yang C, Gong X, Lee K, Heeger AJ (2005) Thermally stable, efficient polymer solar cell with nanoscale control of the interpenetrating network morphology. Adv Funct Mater 15:1617–1662
31. Chiu HC, Lee SH (2018) Optical properties and photovoltaic performance of thin film solar cells. Sol Energy Mater Sol Cells 186:142–150
32. Liu Y, Zhang J (2020) Challenges and opportunities in perovskite solar cells. Adv Mater 32(17):1902036
33. Zhao J, Wang D (2016) Enhancement of photovoltaic performance through Interface engineering. Adv Energy Mater 6(18):1600201
34. Zhang T, Iqbal S, Zhang XY, Wu W, Su D, Zhou HL (2020) Recent advances in highly efficient organic-silicon hybrid solar cells. Sol Energy Mater Sol Cells 204. https://doi.org/10.1016/j.solmat.2019.110245
35. Green MA, Dunlop ED, Hohl-Ebinger J, Yoshita M, Kopidakis N, Ho-Baillie AWY (2020) Solar cell efficiency tables (version 55). Prog Photovolt Res Appl 28(1):3–15. https://doi.org/10.1002/pip.3228
36. Slaoui A, Lincot D, Guillemoles J-F, Escoubas L (n.d.) Nanomaterials for photovoltaic conversion. https://doi.org/10.1002/9783527696109
37. ElKhamisy K, Abdelhamid H, El-Rabaie ESM, Abdel-Salam N (2024) A comprehensive survey of silicon thin-film solar cell: challenges and novel trends. Plasmonics 19(1):1–20. https://doi.org/10.1007/s11468-023-01905-x
38. Tanabe K (2022) Nanostructured materials for solar cell applications. Nano 12:1. https://doi.org/10.3390/nano12010026
39. Xie H, Hu Q, Huang J, Liu M, Wei P, Xie Y, Qi Y (2023) Enhancing photovoltaic performance of carbon-based perovskite solar cells by introducing plasmonic Au NPs. Opt Mater 146. https://doi.org/10.1016/j.optmat.2023.114509
40. Jang YH, Jang YJ, Kim S, Quan LN, Chung K, Kim DH (2016) Plasmonic solar cells: from rational design to mechanism overview. Chem Rev 116(24):14982–15034. https://doi.org/10.1021/acs.chemrev.6b00302
41. Domaradzki J, Kaczmarek D, Drabczyk K, Panek P (2015) Investigations of electrical and optical properties of functional TCO thin films. Mater Sci Poland 33:363–368

First Surveys on the Pollution of Lake BOUDAROUA—Ouazzane—Morocco

Hajar Echayeb, Khadija Elkharrim, Khalid Fariri, and Driss Belghyti

Abstract Morocco, due to climate change, has experienced great disturbances in the lakes. Among the lakes which are of very significant ecological and environmental interest, we find Lake BOUDAROUA, it is an ecosystem of great importance hence its choice as the work site. Lake BOUDAROUA is a rich ecosystem, sheltering an enormous fauna and flora biodiversity, it is located in the town of OUAZZANE, and it is surrounded by the Rif Mountains. Historically, the lake is used to provide drinking water for the city's residents. Our main objective in this study is to evaluate the degree of pollution of the lake and to determine the aquatic fauna during the period spanning from October 2023 to October 2024. To do this, we chose to follow several programs; A sampling program and monthly samples of water and aquatic fauna were collected, And a physicochemical water analysis program. Several parameters were measured using a BANTE 900P portable multiparameter: T °C, pH, EC, TDS, The redox potential. Suspended Particulate Matter (SPM) was determined by the method of filtering a volume of wastewater on a cellulose filter (0.45 μm) according to AFNOR NFT.90.105. The determination of calcium, magnesium, sodium, potassium, carbonates and bicarbonates is carried out by the volumetric method, while sulfates are measured by the atomic absorption technique, within the Regional Office for the Development of Gharb (ORMVAG). The physicochemical analysis showed that the pH is too basic, it varies between 7, 15 and 8, 34. The average temperature is 29.71 °C, the electrical conductivity reaches 3680 μs/cm, the dissolved oxygen varies between 3 and 5.1 mg/l, the average TDS is 1798 mg/l, the redox potential varies from −80.3 to −47.7 mv, the SPM oscillates

H. Echayeb
Laboratory of Natural Resources and Sustainable Development, Faculty of Sciences, University Ibn Tofail, Kenitra, Morocco

ORMVAG, Regional Office for the Development of Gharb, Kenitra, Morocco

K. Elkharrim · K. Fariri · D. Belghyti (✉)
Laboratory of Natural Resources and Sustainable Development, Faculty of Sciences, University Ibn Tofail, Kenitra, Morocco

H. Kaddouri et al. (eds.), *Smart Cities for Sustainable and Urban Development*, Green Energy and Technology, https://doi.org/10.1007/978-3-031-99001-4_11

between 115 and 241 mg/l. The lake is saturated with sodium (632.215 mg/l), bicarbonates (159.865 mg/l), sulfates (108.315 mg/l), calcium (106.205 mg/l), magnesium (15.795 mg/l) and potassium (7.43 mg/l). The first censuses of the biodiversity of Lake Boudaroua made it possible to identify turtles, fish and crustaceans, which deserve an in-depth study. Our study showed that due to its biodiversity, this ecosystem requires protection and enhancement by decision-makers.

Keywords Pollution · Bioindicators · Biodiversity · Lake Boudaroua · Ouazzane · Morocco

1 Introduction

Lake ecosystems play a crucial role in biodiversity and environmental balances by acting as reservoirs of biodiversity and regulators of ecological processes. However, these ecosystems are particularly vulnerable to anthropogenic pressures, including pollution, which causes significant biodiversity loss and compromises the ecosystem services they provide [1]. In Morocco, lakes are essential resources for both the environment and local socio-economic activities. However, these ecosystems face increasing threats due to factors such as urbanization, intensive agriculture and climate change [2]. Among these lakes, Lake Boudaroua, located in the Ouazzane region, is of undeniable ecological and economic importance for local communities, but it remains largely understudied. Research on other Moroccan lakes, such as Lake Dayet Aoua and Ifni, has highlighted the effects of eutrophication and pollution on water quality and biodiversity [3, 4]. However, Lake Boudaroua has not yet benefited from in-depth studies that would allow for a better understanding of the specific sources of pollution and their impacts. Agricultural discharges, domestic wastewater, and human activities are clear threats to the lake ecosystem, but their exact influence remains poorly documented. Among these, agricultural activities, such as the excessive use of pesticides and fertilizers, are major contributors to water pollution, and their proper management is crucial to mitigate these environmental impacts [5]. This lack of data hinders the implementation of effective strategies for the sustainable management of the lake. Recent work has also explored the impact of human activities on Moroccan lakes, including those of El Hammoumi et al. [6], who studied the effects of industrial and agricultural discharges in other regions. However, specific studies on Lake Boudaroua are needed to better understand the local pressures it faces. Thus, the main objective of this study is to assess the pollution status of Lake Boudaroua by identifying the main sources of contamination and measuring their impacts on biodiversity and local uses. This research aims to provide practical recommendations for the preservation of this ecosystem and to raise awareness among local stakeholders about environmental issues. The study is divided into an analysis of the geographical and ecological context of the lake, an identification of the sources of pollution and their effects, followed by concrete proposals for sustainable management.

2 Materials and Methods

2.1 Study Site

Lake Boudaroua is a site of great ecological and biological importance, located 5 km southeast of the city of Ouazzane, exactly in the territorial area belonging to the village community of Beni Quolla, in the Tangier Tetouan Al Hoceima region (Fig. 1), it is a natural lake, formed by geological and hydrological processes over the centuries. Historically, the lake is used to provide drinking water for the inhabitants of the city. Currently, it is dedicated to fish farming and the tourist side [7].

The lake is surrounded by a 10 m wide road in the form of a ledge, the area of the Boudaroua park is estimated at 8 hectares, including 3 hectares of water, and 5 hectares of evergreen trees. Lake Boudaroua rises to about 210 m above sea level and its depth varies from 5 to 8 m and increases from upstream to downstream and from left bank to right bank (Fig. 2), Has a pumping chamber and a spillway. It is currently being developed with funding of 43.5 million dirhams spread over several years, to conserve natural resources and strengthen tourist attractiveness, and

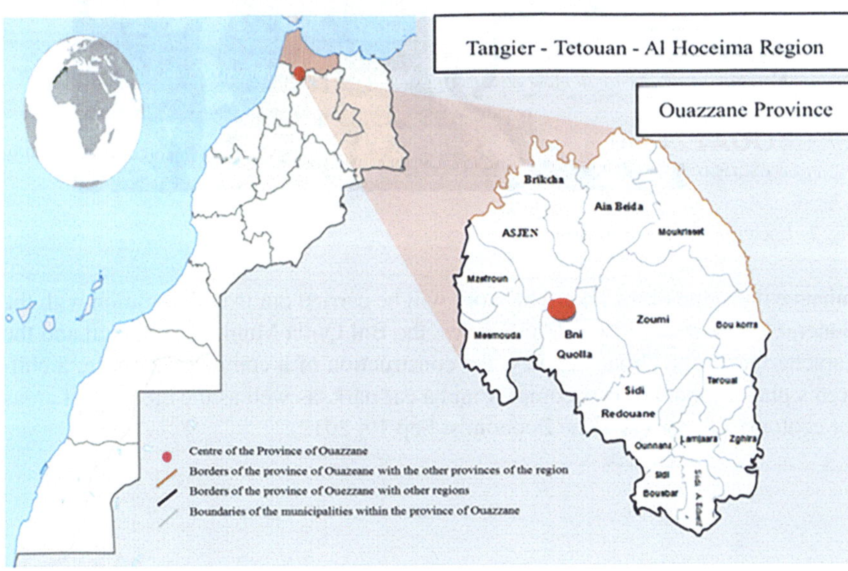

Fig. 1 Map of Ouazzane province

Fig. 2 Location of Lake Boudaroua

enhance the value of the lake, The work will be carried out in collaboration with the General Directorate of Local Authorities, the Bni Qolla Municipal Council and the Larache-Ouezzane Urban Agency, the construction of a commercial space, a children's playground, a sports complex and a car park, as well as the creation of areas for ecotourism projects. (The Economist Sep 19, 2019).

Fig. 3 Location of sampling stations

2.2 *Sampling*

Six stations (S1 to S6) distributed along Lake Boudaroua (Fig. 3) were chosen so that they were accessible and reflected the actual characteristics of the lake surface waters in the study area.

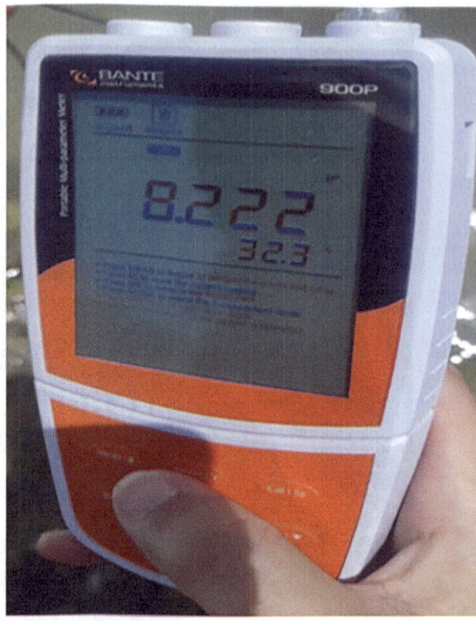

Fig. 4 Portable multiparameter BANTE 900P

Fig. 5 LG brand MICROWAVE

2.3 *Sampling and Analysis*

Along Lake Boudaroua, water samples were taken monthly from June to October 2024. Water samples were taken using 1000 ml plastic bottles, previously rinsed with water from the station., T °C, pH, EC, TDS, DO, Redox potential, were measured instantly using a portable multi-parameter BANTE 900P (Fig. 4). Suspended matter is determined by the method of filtering a volume of wastewater on a cellulosic filter (0.45 µm) according to AFNOR NFT.90.105 (Figs. 5 and 6). The dosage

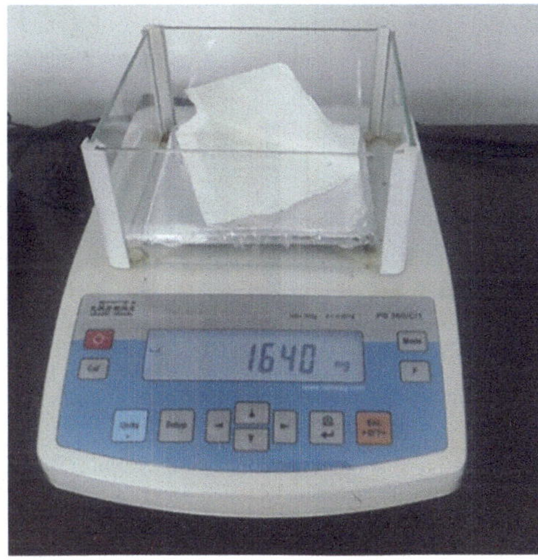

Fig. 6 RADWAG Wagi PS 360/C/1 Precision balance

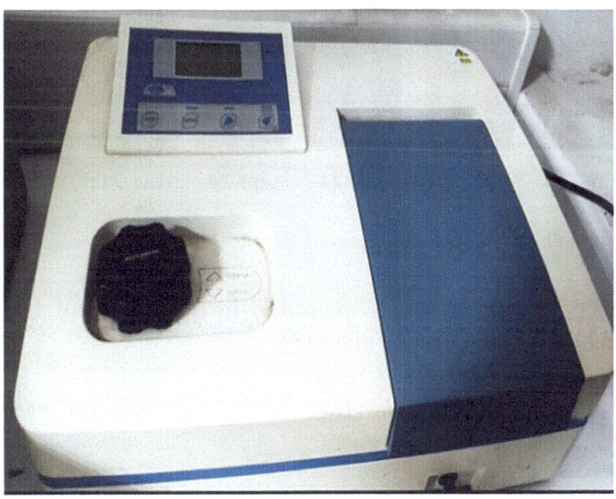

Fig. 7 V 1000 D SPECTROPHOTO- METER brand SELECTA

of calcium, magnesium, sodium, potassium, chlorides, carbonates and bicarbonates is carried out by the volumetric method, while sulfates are measured by the atomic absorption technique (Fig. 7).

3 Results and Discussion

3.1 Water Temperature

The recorded temperature values (Fig. 8) range from 30.4 °C (station S1) in August to 33.8 °C (station S 5) in July, and from 22.5 °C (station S1) in October to 29.3 °C (Station S 2) in September. These temperature variations followed those of the

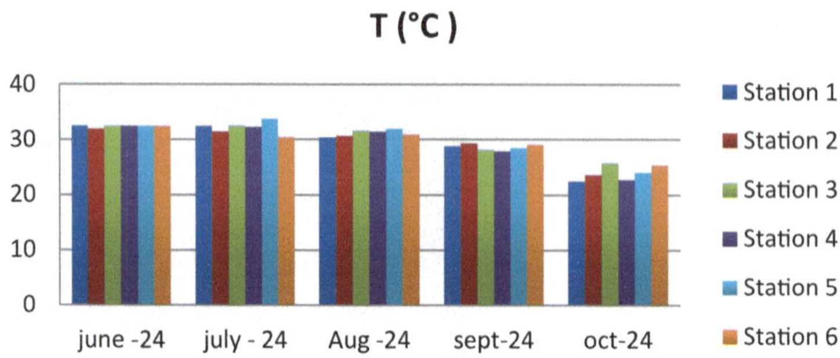

Fig. 8 Spatio-temporal variation of temperature

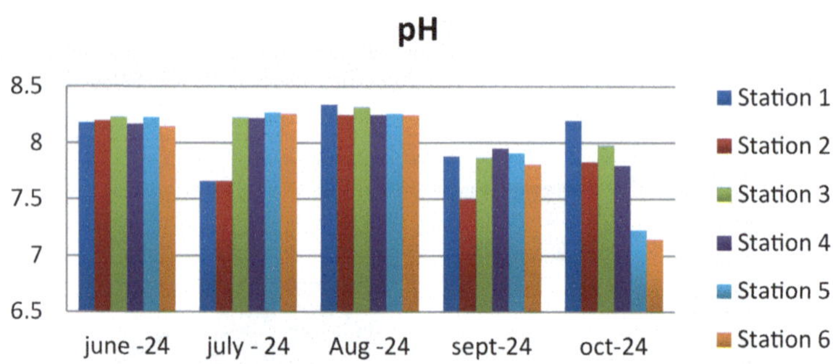

Fig. 9 Spatio-temporal variation of pH

region's climate. The temperatures measured in the waters of Lake Boudaroua therefore belong to the good to bad class [8].

3.2 *The pH*

Hydrogen potential measures the acidity of water; it is the concentration of H^+ ions in surface water. The pH values vary between 7.15 and 8.34 (Fig. 9), the maximum value is 8.34 recorded in August, and these valuesare not within the range of Moroccan surface water quality standards intended for human consumption. Generally, the waters of Lake Boudaroua are too basic, unlike Lake Djètoè in Benin which has a neutral character according to [9] and the watershed of Lake Loka in Bouaké on the Ivory Coast which has an acidic pH, varied between 5.5 and 6.8 [10].

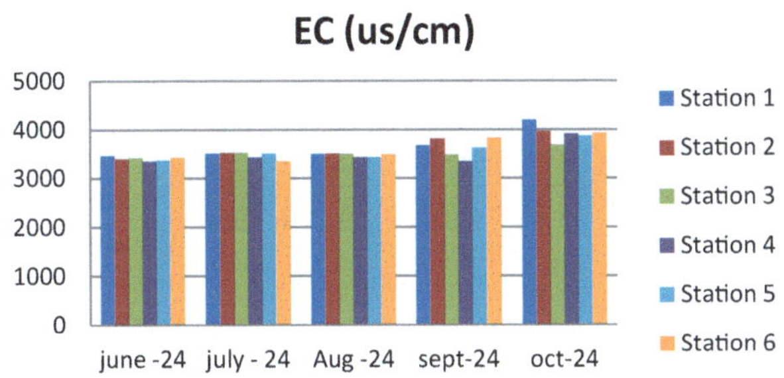

Fig. 10 Spatio-temporal variation of electrical conductivity

3.3 Electrical Conductivity

Hydrogen potential measures the acidity of water; it is the concentration of H^+ ions in surface water. The pH values vary between 7.15 and 8.34 (Fig. 9), the maximum value is 8.34 recorded in August, and these valuesare not within the range of Moroccan surface water quality standards intended for human consumption. Generally, the waters of Lake Boudaroua are too basic, unlike Lake Djètoè in Benin which has a neutral character according to [9] and the watershed of Lake Loka in Bouaké on the Ivory Coast which has an acidic pH, varied between 5.5 and 6.8 [10] (Fig. 10).

3.4 TDS (Total Dissolved Solids)

TDS is the total amount of dissolved solids, including dissolved minerals, salts and metals, expressed in mg/l or particles per million (ppm). TDS reflects the purity of water.

The TDS values recorded at the stations studied vary between a minimum of 1680 mg/l in July during the summer period and a maximum of 2100 mg/l in October during the period autumn (Fig. 11). These values remain strictly higher than 1000 ppm, and also higher than the average TDS value of Lake Djètoè in Benin which is 240 mg/l [9].

The waters of Lake Boudaroua are generally considered unsafe to drink without further treatment due to its taste and its ability to contain high concentrations of dissolved minerals that could affect long-term health according to the Word health organization. (WHO).

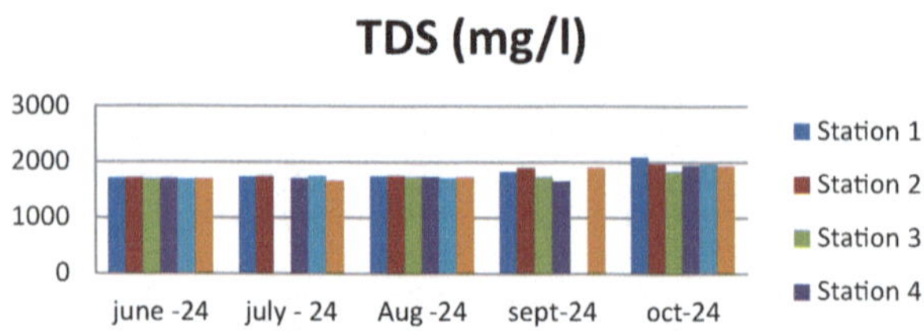

Fig. 11 Spatio-temporal variation of TDS

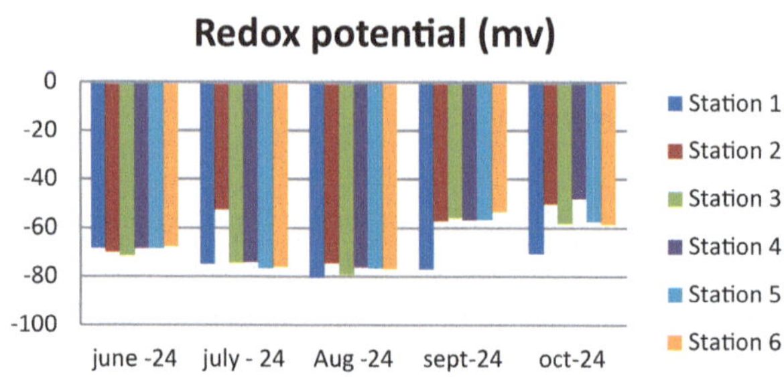

Fig. 12 Spatio-temporal variation of the Redox potential

3.5 *The Redox Potential (Eh)*

The redox potential is a parameter that determines whether the medium favors electron loss or gain.

The redox potential of a medium is low (or even negative), when reduction reactions dominate, and oxygen concentrations increase, as a result, the medium favors electron gain, it becomes reducing. However, the redox potential of a medium is high, when oxidation reactions dominate, and oxygen concentrations decrease, as a result, the medium promotes electron loss, it becomes oxidizing. Monitoring the redox potential is useful for assessing water quality. The low values of the redox potential (Fig. 12) correspond to periods of low water and reach −80.3 mv in August. The high value reaches −47.7 mv observed in October. This is due to the precipitation which regenerates the water. Lake Boudaroua is a reducing environment, which means that there is a low presence of dissolved oxygen. This type of environment is often associated with environments where organic matter decomposes slowly, such as in sediments or stagnant areas, and where compounds like sulfates or nitrates can be reduced to sulfides or ammonia [11].

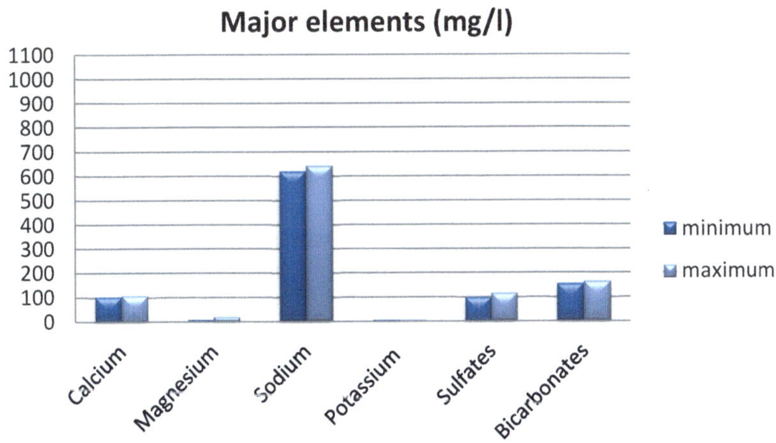

Fig. 13 Spatio-temporal variation of major elements

3.6 *Major Elements*

3.6.1 Sulfates ($SO^2{}_4$)

Sulfates are chemical compounds coming from the erosion of minerals that contain sulfur, from rain, In our study area, The concentrations of sulfates recorded vary between 99.91 and 116.72 mg/l (Fig. 13). This result indicates that the water is of good to excellent quality according to Moroccan surface water quality standards [8].

3.6.2 Bicarbonates ($HCO^-{}_3$)

The presence of carbonates and bicarbonates in water is due to the dissolution of carbonate rocks [12]. The concentration of carbonates and bicarbonates is partly regulated by the processes of photosynthesis, respiration and solubilization of carbon dioxide [13].The results of our study indicate that the bicarbonate content varies from 154.98 to 164.75 mg/l (Fig. 13). This result indicates that the water is of good and excellent quality according to Moroccan surface water quality standards [8].

3.6.3 Sodium (Na^+)

Sodium can come from the dissolution of sodium chloride (NaCl) or the decomposition of sodium and aluminum silicates [14]. The present study shows that the sodium value varies between 620.72 and 643.71 mg/l (Fig. 13). This value is more than 200 mg/l, it can cause soil salinization, and can be toxic for certain aquatic species, and reduce biological diversity [15].

3.6.4 Potassium (K^+)

Potassium results from the alteration of clay minerals and the dissolution of fertilizers chemicals [12]. The results obtained show that the potassium content oscillates between 7.04 and 7.82 mg/l (Fig. 13). This is the lowest concentration of major ions, showing that the water is within an acceptable range for a natural lake. This value is lower than that observed in Lake Dayet Erroumi in KHMISSET which varies between 7.82 and 9.32 mg/l [16]. High potassium concentrations can be associated with eutrophication phenomena, promoting the proliferation of algae and affecting water quality.

3.6.5 Calcium et Magnésium (Ca^{2+} et Mg^+)

Calcium and magnesium are chemical elements that are widely distributed in nature. They are responsible for water hardness. High hardness can lead to scale deposits in pipes and equipment, which can pose problems for domestic and industrial systems. This study shows that the calcium and magnesium content ranges from 104.2 to 108.21 mg/L and from 9.72 to 21.87 mg/l (Fig. 13), respectively. This suggests that the water is moderately to highly hard. This hardness can affect the solubility of nutrients and the availability of minerals to aquatic fauna, as well as the formation of scale deposits in pipes and equipment used for irrigation or water treatment.

3.7 Suspended Particulate Matter (SPM)

For the two seasons of summer and autumn, the SPM values recorded oscillate between 115 and 241 mg/l. The spatio-temporal evolution of suspended matter (SPM) contents (Fig. 14) in the lake shows two very distinct periods:

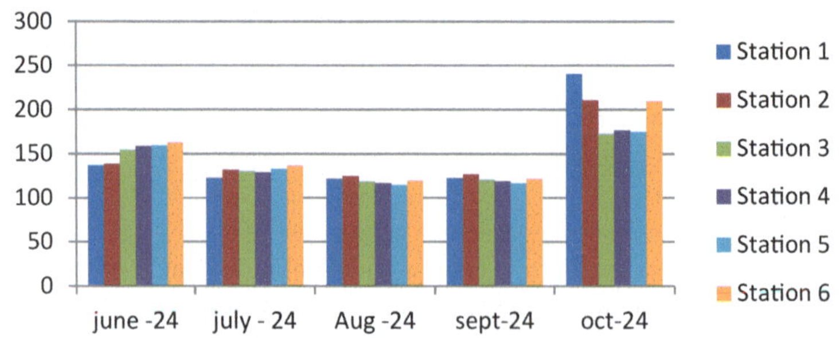

Fig. 14 Spatio-temporal variation of Suspended Particulate Matter

A summer period, where the SPM concentration values drop in all stations between 115 mg/l (S5) in August 2023 and 163 mg/l (S6) in June.

An autumn period, where high loads were recorded in all stations between 173 mg/l (S3) and 241 mg/l (S1) in October. These high levels are mainly due to runoff from autumn rains, which carries organic matter, nutrients and pollutants from agricultural, urban and natural soils into the waters.

The high loads recorded in the station (S1) are 241 mg/l in October 2024.

Comparison of the suspended matter contents in the study lake with the Moroccan standard set at 1000 mg/l places these waters in the average grid.

Although this classification indicates moderate quality, it also highlights the need to monitor pollution sources and implement measures to limit the impact of human and natural activities on water quality.

4 Conclusion

The study of pollution in Lake Boudaroua highlights the significant impact of human activities, particularly agricultural and domestic, on water quality and biodiversity. The gradual degradation of this water body, marked by increasing eutrophication, reveals the limitations of current practices in water resource management and wastewater treatment systems. This research emphasizes the crucial importance of integrated watershed management to preserve water quality and the health of aquatic ecosystems. The results show that agricultural runoff, mainly fertilizers and pesticides, as well as domestic wastewater, are the primary sources of pollution. These inputs promote harmful algal blooms and lead to significant biodiversity loss [17]. Furthermore, the high nutrient concentrations exacerbate eutrophication, potentially irreversibly altering the ecological dynamics of the lake in the long term [1]. The physicochemical parameters measured in the lake's surface waters confirm this degradation, particularly in autumn and summer. The results reveal an excessively basic pH reaching 8.34, and high electrical conductivity of 3680 µs/cm, indicating significant mineralization and salinity, which deteriorate water quality. These factors affect not only local ecosystems but also economic activities dependent on the lake, such as fishing and tourism [18]. For sustainable management of Lake Boudaroua, several recommendations can be made. First, it is essential to implement stricter regulations on the use of fertilizers and pesticides, while improving domestic wastewater treatment infrastructure. Additionally, adopting environmentally friendly agricultural practices and raising awareness within local communities are necessary to promote sustainable behaviors [19]. Finally, ecological restoration projects, such as replanting aquatic vegetation and reducing nutrient inputs, should be prioritized to restore biodiversity and improve water quality in the long term. To move forward, future research should focus on a long-term evaluation of proposed solutions to mitigate eutrophication and restore aquatic ecosystems. A more in-depth study of the interactions between different sources of pollution (agricultural, domestic, and urban) and their cumulative effects on biodiversity would help better understand the mechanisms of environmental degradation [20]. These efforts, combining scientific research, regulation, and community participation, are essential to ensuring a sustainable future for Lake Boudaroua and the populations that depend on it.

References

1. Smith VH, Joye SB, Howarth RW (2010) Eutrophication of freshwater and coastal marine ecosystems. Limnol Oceanogr 51(1):351–355
2. El Hammani A, Kacimi I, Sedraoui S (2018) Urbanization and pollution of Moroccan lakes: a synthesis of environmental and socio-economic impacts. Moroccan J Environ Sci 12(4):567–579
3. Abba E, Nassali H, Benabid M, El Ayadi R, El Ibaoui H (2008) Contribution to the physicochemical study of the lacustrine écosystèm Dayet Aoua in Morocco. Afr Sci Int J Sci Technol 4:2
4. Abbassi M, Banaoui A, Charioui I, Kaaya A, Ndir M et al (2017) Physicochemical characterization of coastal waters in the city of sidi Ifni (Morocco). J Mater Environ Sci 8(9):3112–3120
5. Kumar N, Kumar A, Marwein BM, Verma DK, Jayabalan I, Kumar A, Kumar N (2021) Agricultural activities causing water pollution and its mitigation—a review. Int J Modern Agric 10(1):590–609
6. El Hammoumi S, Kherratii I, Imane M, Maaroufi M, Sibari M (2020) Environmental impacts of human activities on Moroccan lakes: case of industrial and agricultural discharges in the Rabat region. J Environ Sci 15(2):122–134
7. Etebaai I, Damnati B, Taieb M, Miche H (2010) The environment of the Sidi Boudaroua water body: Physicochemistry of the waters and current sedimentation (Ouazzane region, Morocco). Geomaghreb 6(2010):69–78
8. GQES (2002) Moroccan surface water quality standards. Ministry of Equipment, Transport, and Logistics, Kingdom of Morocco
9. Dovonou FE, Atchichoe W, Bankole G, Adandedji F (2023) Assessment of the ecological quality of water from Lake Djètoè in the commune of Lokossa. Afr Sci 23(1):79–96
10. Ebé AED, Yao KSA, Ahoussi KE (2023) Determination of the physicochemical quality of water resources in the Loka watershed in Bouaké, Central Ivory Coast. Int J Innov Appl Stud 41(1):105–114
11. Chapman DV (2021) Water quality assessments: a guide to the use of biota, sediments, and water in environmental monitoring. CRC Press
12. Gouaidia L (2008) Influence of lithology and climatic conditions on the variation of physicochemical parameters of groundwater in a semi-arid zone: case of the Meskiana aquifer, northeastern Algeria. Doctoral dissertation. University of Annaba-Badji Mokhtar
13. Wang J, Huang L, Yang R, Zhang Z, Wu J, Gao Y et al (2014) Recent progress in solid sorbents for CO_2 capture and new development trends. Energy Environ Sci 7(11):3478–3518
14. Mathieu R (2009) Solubilité du sodium dans les silicates fondus. Doctoral dissertation,. Institut National Polytechnique de Lorraine
15. Zhang Y, Liu J, Zhang L (2021) Effects of sodium and salinity on aquatic ecosystems: implications for water quality and management. Environ Sci Pollut Res 28(24):31643–31655
16. El Qryefy M, Ouardi J, Najy M, Belghyti D, El Kharrim K (2021) Hydrochemical characteristics and water quality assessment of Lake Dayet Erroumi—Khemisset, Morocco. J Water Land Dev 49:179–187

17. Carpenter SR, Caraco NF, Correll DL, Howarth RW, Sharpley AN, Smith VH (1998) Nonpoint pollution of surface waters with phosphorus and nitrogen. Ecol Appl 8(3):559–568
18. Moss B (2008) Water pollution by agriculture. Phil Trans R Soc B Biol Sci 363(1491):659–666
19. Olsson P, Folke C, Berkes F (2004) Adaptive comanagement for building resilience in social-ecological systems. Environ Manag 34(1):75–90. https://doi.org/10.1007/s00267-003-0101-7
20. Diaz RJ, Rosenberg R (2003) Marine benthic hypoxia: a review of its ecological effects and the potential impacts on fisheries. Mar Pollut Bull 46(3):266–279

A Case Study on the Thermal Performance of Materials in an Earthen House in the Anoual Region, Morocco: Towards Smart Buildings and Durability

Boutahar Laaouar, Othmane Horma, Mohammed Drissi, Hanane Miri, Jinane Mejdoubi, Mohammed Amine Moussaoui, and Ahmed Mezrhab

Abstract Raw earth, an ancient and sustainable building material, has demonstrated its durability and strength through enduring structures, many of which are designated as World Heritage sites. This eco-friendly material not only offers excellent thermal and acoustic performance but also requires significantly less energy to produce compared to modern construction materials. Furthermore, it is fully recyclable, positioning it as a key material in the context of sustainable development. In Morocco, traditional earthen construction techniques such as rammed earth, adobe, and compressed earth blocks are still in use; however, these practices rely on disappearing artisanal knowledge and lack formal technical standards.

This chapter examines an earthen house in the Anoual region of Morocco constructed with compacted earth walls. Indoor temperature measurements were compared with outdoor temperatures using a Testo device to evaluate the building's thermal performance. Additionally, the thermal properties of the soil samples—including thermal conductivity, thermal diffusivity, and specific heat capacity—were characterized to assess the material's behavior under various conditions. Computational Fluid Dynamics (CFD) simulations using ANSYS software were also conducted to validate the experimental results, providing a comprehensive analysis of the thermal efficiency of earthen materials in construction. This chapter highlights the contribution of earthen materials to the development of smart and sustainable buildings.

B. Laaouar (✉) · M. Drissi · J. Mejdoubi · M. A. Moussaoui · A. Mezrhab
Mechanics and Energy Laboratory, Mohammed First University, Oujda, Morocco
e-mail: boutahar.laaouar@ump.ac.ma

O. Horma
Mechanics and Energy Laboratory, Mohammed First University, Oujda, Morocco

Private University of Marrakech, Marrakech, Morocco

H. Miri
Applied Geosciences Laboratory, Faculty of Sciences, Mohammed First University, Oujda, Morocco

H. Kaddouri et al. (eds.), *Smart Cities for Sustainable and Urban Development*, Green Energy and Technology, https://doi.org/10.1007/978-3-031-99001-4_12

Keywords Raw earth construction · Thermal conductivity · Compacted earth walls · Sustainable building materials · Computational fluid dynamics

1 Introduction

Earthen construction is one of the oldest building techniques used by humanity, dating back thousands of years. Ancient civilizations such as the Egyptians and Mesopotamians built structures using earth, exploiting locally available materials and adapting them to the climatic conditions of their regions. These earthen buildings often consisted of adobe or mud bricks—a mixture of clay, sand, straw, and sometimes gravel—shaped and sun-dried. Today, this technique is still found in regions where natural resources and climatic conditions favor its use. The Middle East, North Africa, and parts of Asia and Latin America continue to widely employ earthen construction. Countries like Morocco, Yemen, and certain rural areas in France and Spain still use this traditional method, which remains relevant due to its sustainability and low cost. Clay possesses unique attributes, including abundance, ease of use, high recyclability, and excellent thermal properties [1, 2]. Using earthen materials in construction represents not only an economically viable solution—offering cost reductions of nearly 50% compared to conventional materials when considering the entire supply chain from raw materials to the final product, but also provides the capacity to store surplus heat during the day and release it at night. Thus, they constitute an effective solution for maintaining thermal comfort. Moreover, earthen materials are biodegradable and environmentally friendly. Their recyclability makes them sustainable construction materials [3], as illustrated in Fig. 1.

While earthen materials possess inferior mechanical properties compared to modern construction materials, these properties can be enhanced by incorporating other elements [4]. It has been demonstrated that adding cement to adobe bricks increases compressive strength from 5.02 kN (±0.10 kN standard deviation) to 11.23 kN (±0.15 kN standard deviation) as the cement content rises from 0% to 10%. The field of earthen architecture has grown significantly over the past decades, with international conferences bringing together an increasing number of academics, scientists, and professionals to discuss topics such as chemistry, engineering, and sustainability. These exchanges have led to collaborations, such as the partnership between the Getty Conservation Institute and the Gaia Project in 1994, which resulted in the creation of Project Terra in 1997, focused on the conservation and research of earthen architecture [5]. Although soil is widely used in vernacular buildings due to its availability and superior thermal insulation capability, the extensive application of this material is restricted by its relatively low compressive and tensile strength required for modern buildings [6]. A thorough analysis of building materials is crucial to understanding their behavior, properties, and performance. Research highlights that mineral composition, plasticity, and particle size distribution play key roles in defining earthen materials. In North Africa, especially in rural

Fig. 1 This Life cycle diagram of earth as a building material

areas, clay soils are widely used for producing rammed earth and adobe construction materials. Scientific research on earthen materials has made significant strides in understanding how these natural materials can be utilized in modern construction, particularly within the context of sustainable and "smart" building practices. Research has illuminated the potential of earthen materials, such as clay, adobe, and rammed earth, to not only offer excellent thermal insulation but also contribute to energy-efficient, climate-responsive building designs. These materials, due to their high thermal mass, naturally regulate indoor temperatures, reducing the need for mechanical heating and cooling. This makes them an essential component in the development of smart buildings that are energy-efficient, low-carbon, and capable of adapting to environmental conditions. Scientific research on earthen materials has significantly advanced sustainable and smart building practices. El Hammouti et al. [7] investigated the valorization of clay soils from eastern Morocco (Zaio, Oujda, Lota, and Midar), highlighting their potential in eco-friendly construction. In a similar vein, Charai et al. [8], showed that earth blocks reinforced with fibers can enhance the thermophysical performance of composites, making them ideal thermal mass elements for energy-efficient buildings. Drissi et al. [9] further underscored the viability of natural red clay as a supplementary cementitious material, capable of lowering both greenhouse gas emissions and production costs. Collectively, these studies illustrate how earthen materials—and biomaterials in general—can offer robust thermal insulation, reduce reliance on mechanical heating and cooling, and integrate seamlessly with modern systems to create climate-responsive, energy-efficient, and durable structures. This synergy with advanced technologies positions earthen construction as a viable pathway toward truly smart and sustainable buildings. Despite these promising developments, Earth buildings have been largely overlooked, with limited research conducted. This study aims to evaluate their thermal performance and promote the global application of national guidelines for earth architecture [10]. The study is divided into two parts: experimental measurements and simulations. All experimental measurements were conducted in a Terbi house located in the Anoual area, at coordinates 32.67797° N (32° 40′ 41″ N), where one of the hottest summer seasons is experienced [11]. Initially, the Testo 480 device was used to measure parameters such as temperature and humidity. The thermal properties of the material were then obtained using the Hot Disk method. Subsequently, a simulation was conducted with ANSYS FLUENT software to evaluate the building's thermal behavior over a 24-hour period.

1.1 Geologies Context

This region experiences harsh winters and intense summers, largely influenced by its continental climate. The mountainous areas near the Anti-Atlas and High Atlas ranges block humid oceanic air masses, as shown in Fig. 2, while the southern opening amplifies the Sahara's influence. The desert brings dry winds, such as the

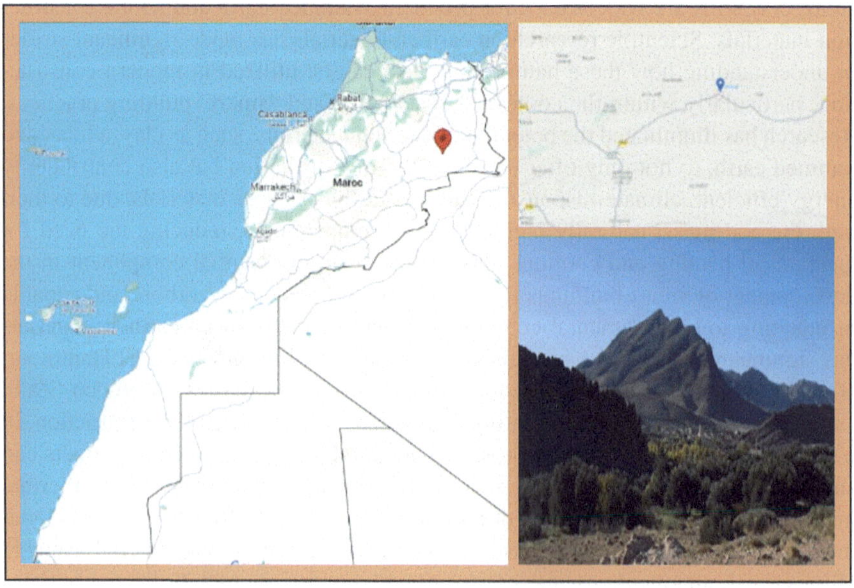

Fig. 2 Geographic location of the region (Morocco)

Fig. 3 Studied house and the main clayey component

Chergui from the northeast in summer and the Sahli from the southwest in winter. Earthen architecture offers a comfortable and energy-efficient solution to these challenging climatic conditions, while also embodying significant cultural heritage with historical and architectural value. However, preserving this heritage has become increasingly difficult. Traditional homes in this region are built using rammed earth adobe, as shown in Fig. 3. Raw earth functions as an excellent thermal and moisture regulator, keeping interiors cool in summer and moderate in winter due to its unique density and porosity. These intricately constructed buildings, developed over generations, are remarkable examples of environmental adaptability and therefore warrant careful study.

2 Materials and Methodology

2.1 Wall Material Composition

The walls of the studied house are composed of two layers: predominantly adobe and a fraction of cement-based mortar using a cement-to-sand mixture in a ratio of 1 to 2. To accurately simulate heat transfer through the walls, it was necessary to collect samples of each layer and determine their thermophysical properties.

2.2 Measurement Procedure and Measuring Instruments Used

The measurements were carried out in the Anoual region of Morocco, at coordinates 32.67797° N (32° 40′ 41″ N) and an altitude of 1197 m (3927 feet) above sea level [12]. An adjustable tripod was used to facilitate the setup, in accordance with standard measurement practices.

The Testo 480 measurement system was employed for precise measurements in several areas such as temperature, humidity, pressure, air velocity, and gas emissions. This device is widely used in thermal engineering, construction, and the maintenance of heating, ventilation, and air conditioning (HVAC) systems. The technical specifications of the sensors integrated into the Testo 480 are detailed in Table 1, while the measurement device is shown in Fig. 4. These sensors are precisely calibrated to meet industry standards, ensuring that the measurement results conform to ISO 7730 [13] guidelines. This meticulous process highlights the precision and rigor applied in our study, ensuring that the collected data adheres to established standards and is suitable for reliable analysis of thermal comfort conditions within the dwelling.

Table 1 Properties of Testo 480 Universal IAQ instrument sensors

Equipment	Measured Values	Range
Isotropic air velocity sensor	Turbulence intensity	0—50 °C 0—5 m/s 700—1100 hPa
Globe thermometer (Ø 150 mm; TC: Type K)	Globe temperature	0–120 °C
IAQ probe	Temperature CO_2 Humidity Absolute pressure	0—50 °C 0—10,000 ppm CO_2 0—100%HR 700—1100 hPa
Lux probe (digital)	Illuminance	0—100,000 Lux

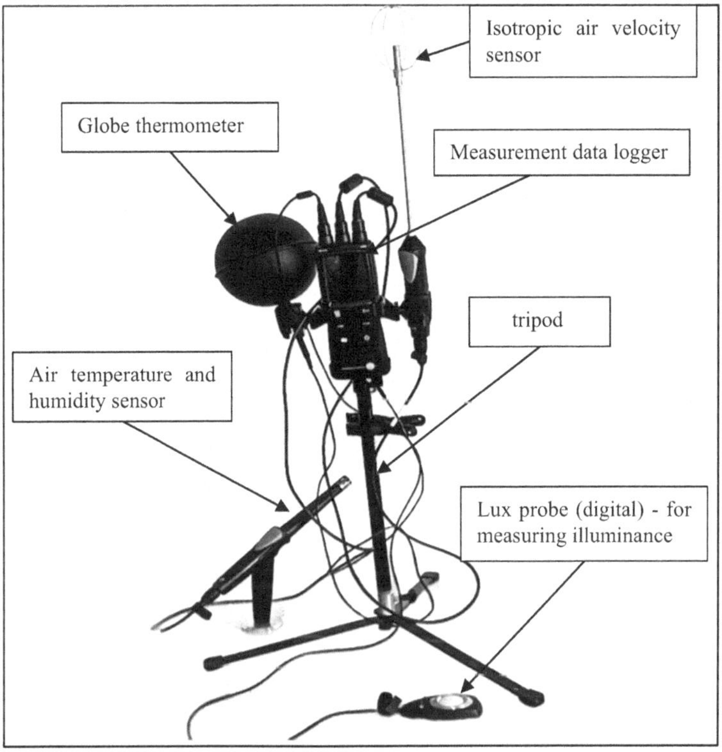

Fig. 4 TESTO 480 multifunctional HVAC measuring instrument

2.3 *Hot Disk Method*

In this study, we measured the thermal properties of the wall soil samples using the TPS 2200 Hot Disk device (Hot Disk, Sweden), which simultaneously evaluates thermal conductivity, thermal diffusivity, and volumetric heat capacity. The measurements were carried out by placing a TPS double-spiral Kapton sensor (model 5501, radius 6.403 mm) between a sample of each layer and an insulating material (expanded polystyrene, EPS), utilizing the single-sided feature of the Hot Disk method, which determines the properties of a single sample provided that the properties of the insulating side are known (0.027 W/mK and 1.112 mm^2/s). To ensure thermal stability, the sensor was manually clamped between the sample and the insulating material. The sensor was then heated with a constant power ranging from 50 to 100 mW for 80 seconds. The increase in temperature over time was monitored using the same sensor, and the thermal transport properties were calculated in real time using Hot Disk Analyzer software v7.1.5, based on the recorded time-dependent temperature rise. Three measurements were performed for each soil type to ensure accuracy and repeatability.

2.4 *Simulation of the Studied Wall and Validation of Experimental Results*

In the second phase of our study, we conducted a comparative analysis of experimental temperature measurements and simulation results for walls constructed with soil. The walls consist of a 30 cm thick layer of compacted soil, topped with a 2 cm thick layer of cement on the interior side (see Fig. 5). Using ANSYS FLUENT software, we simulated and analyzed the thermal behavior of three different wall types over a 24-hour period. The simulation was based on data from August 10, 2024, in Anoual, Morocco—a day characterized by significant temperature fluctuations, with outdoor temperatures ranging from 31 to 47 °C. This day was selected to highlight the ability of our materials to manage extreme thermal variations. ANSYS FLUENT, a widely used numerical simulation tool, allows for the resolution of fluid flow and heat transfer equations. The mathematical model and boundary conditions used to approximate temperature variations in the simulation are outlined below.

2.4.1 Boundary Conditions

$$-\lambda \frac{\partial T}{\partial x} = \mathrm{h}_{\mathrm{ext}} \left(\mathrm{T}_{\mathrm{sa}} - T_e \right), For\, x = 0 \quad (1)$$

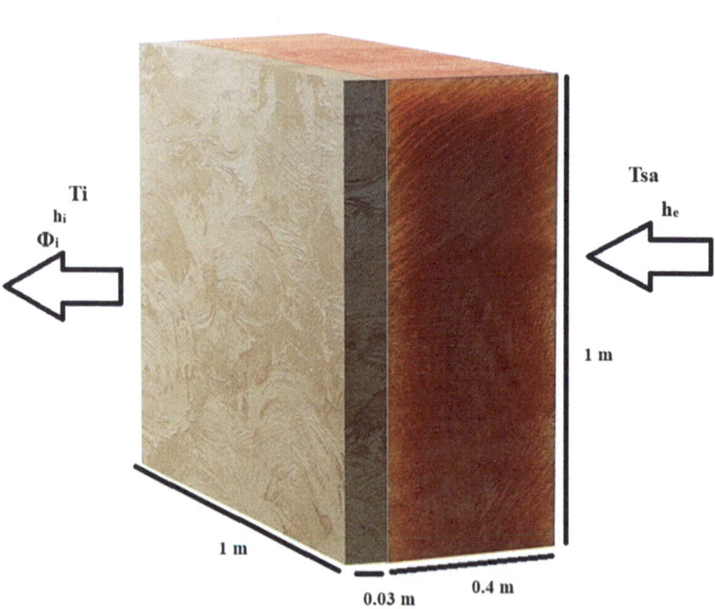

Fig. 5 Heat flux transfer process across the studied wall

$$-\lambda \frac{\partial T}{\partial x} = h_{int}\left(T_i - T_{int}\right), For\, x = e \tag{2}$$

2.4.2 Mathematical Model

$$T_{as}(t) = \frac{T_{max} + T_{min}}{2} + \frac{T_{max} - T_{min}}{2} \sin\left(\frac{2\pi t}{p} - \frac{\pi}{2}\right) \tag{3}$$

The convective heat transfer coefficients are denoted by h_{ext} for the exterior and h_{int} for the interior surface. The temperatures T_e and T_{int} represent the exterior and interior air temperatures, respectively. The ambient outdoor temperature T_{as} follows a sinusoidal variation with a period p of 24 hours. The maximum and minimum outdoor temperatures are denoted by T_{max} and T_{min}, respectively. In this study, we considered the following values: T_{max} = 47 °C; T_{min} = 31 °C; h_{ext} = 12 W/m^{2}K and h_{int} = 8 W/m^{2}K.

3 Results and Discussion

3.1 Thermophysical Properties

The thermal properties of both the wall layers, mortar and adobe, obtained using the hot disk method, are summarized in the Table 2. The results show that the thermal conductivity of the earthen wall is 0.715 W/mK is significantly lower than that of other commonly used materials. This demonstrates that our earthen building is capable of reducing the penetration of outdoor heat into the interior, effectively showcasing its ability to manage extreme thermal variations and promoting the use of raw earth in modern sustainable construction.

Table 2 Thermal properties of the studied wall layers samples

Layers	Thermal conductivity (W/mK)	Thermal diffusivity (mm^2/s)	Density (kg/m^3)	Specific heat capacity (J/kg K)
Mortar	1.5	0.7	2200	980
Adobe	0.715	0.55	1998	658

3.2 Experimental Indoor Measurements and Simulation Validation

Firstly, the validation of the current simulation model was achieved by aligning the obtained results with those of previous studies employing similar mathematical frameworks, thereby demonstrating the reliability and accuracy of our findings. Figure 6a illustrates the variation in outdoor temperature, obtained through a mathematical approximation of the actual 24-h variation for the studied day. Figure 6b shows the variation in indoor temperature within a room over a 24-h period during a typical week in August, characterized by hot and sunny weather conditions. Maximum outdoor temperatures ranged from 31 to 47 °C, while night-time temperatures remained high, between 30 and 33 °C. Additionally, the maximum solar radiation, reaching nearly 1000 W/m^2 on the south wall, contributed significantly to the warming of the room. At the beginning of the day, the indoor temperature was relatively low, around 31 °C, reflecting the cooler outdoor temperatures during the night. As the day progressed, the indoor temperature began to increase gradually, peaking at slightly over 35 °C between 6 p.m. and 8 p.m.—about 5 h after the outdoor temperature and solar radiation had reached their maximum, as seen in Fig. 7. The outdoor temperature peaked around 1 p.m., reaching up to 47 °C. This significant delay in peak indoor temperature demonstrates the thermal lag effect of the compacted earth walls, which absorb heat throughout the day and release it slowly. Moreover, the indoor peak temperature was over 11 °C lower than the outdoor peak temperature, indicating that the clay-based walls effectively attenuate extreme outdoor temperature variations. By midnight, the indoor temperature decreased slightly but remained elevated (around 34 °C) due to the heat stored in the walls.

Furthermore, the ANSYS FLUENT simulation closely mirrored the experimental results, showing a similar pattern of temperature variation. This alignment validates our experimental findings and confirms the effectiveness of the compacted earth walls in providing sufficient thermal mass that moderates indoor temperatures. The simulation results reinforce the potential of clay-based materials to significantly reduce indoor temperature peaks and provide substantial thermal time lag, enhancing thermal comfort in extreme climatic conditions.

The relative humidity inside the room, shown in Fig. 8, varied significantly throughout the day. It peaked at around 40% at 4 a.m., then gradually decreased to about 20% by 10 a.m., continuing to decline to a minimum of approximately 10% between 2 p.m. and 6 p.m. By the end of the day, relative humidity rose slightly and stabilized around 15%.

This trend results from several factors. The high outdoor temperatures during the day contribute to a decrease in indoor humidity, as the heat dries the ambient air. The arid, desert-like climate further exacerbates this reduction in humidity. Additionally, the diurnal cycle plays a crucial role: the higher relative humidity observed early in the morning is likely due to the drop in temperature during the night, which promotes the condensation of moisture.

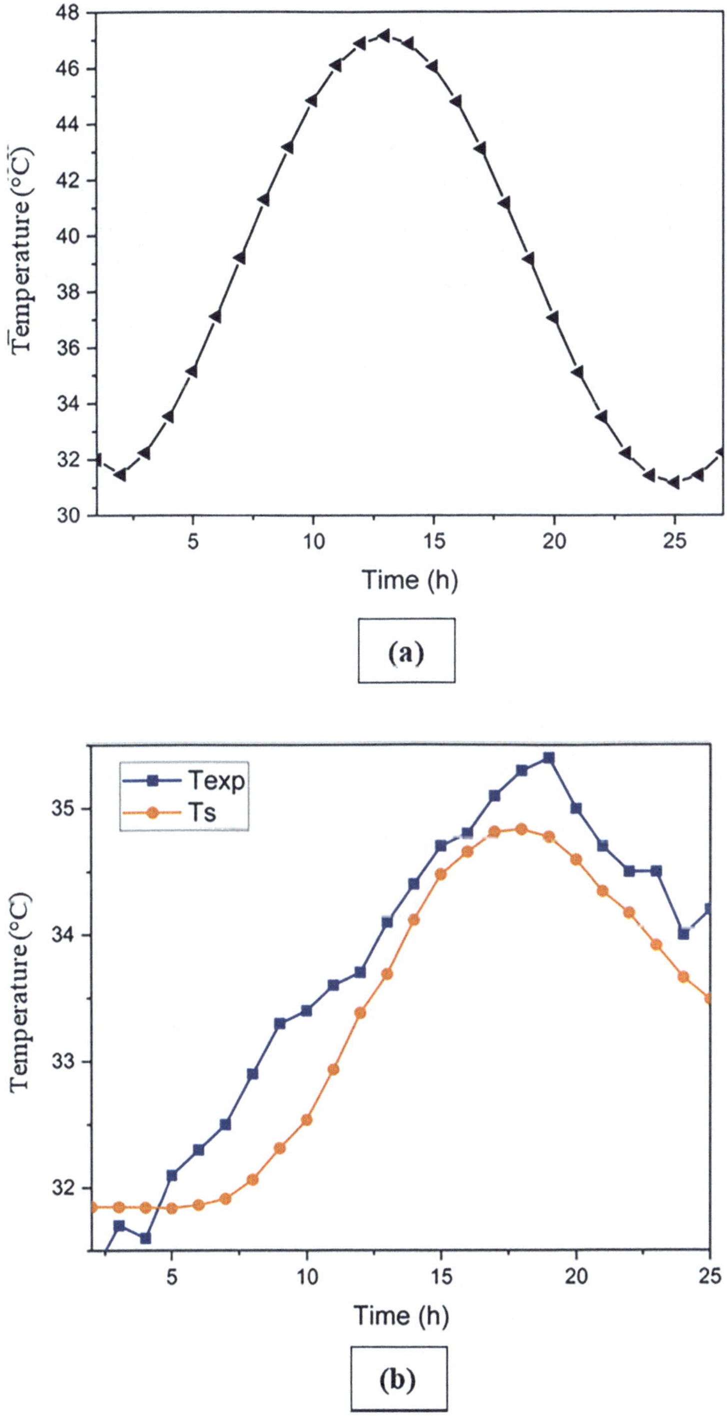

Fig. 6 Outdoor average temperature (**a**) and Indoor room temperature (**b**)

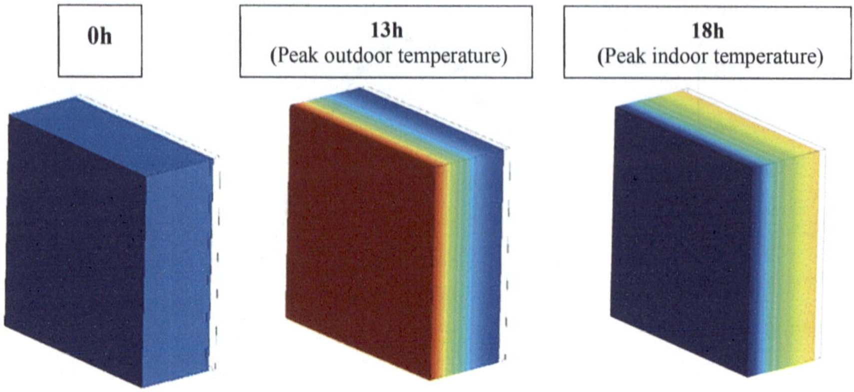

Fig. 7 Temperature profile of the studied wall at peak outdoor and indoor temperatures

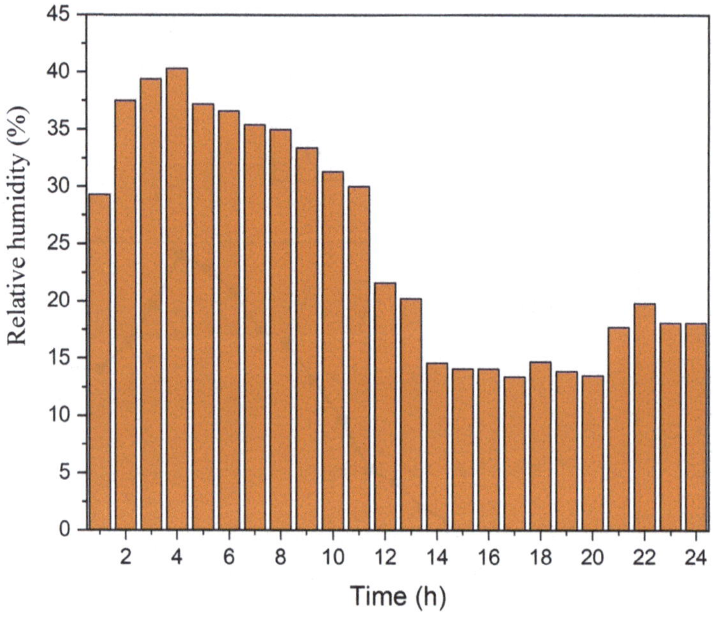

Fig. 8 The variation in relative humidity in the test room over a 24-h period

In the context of smart building development, understanding these daily humidity fluctuations allows for more effective integration of passive and active control strategies. By aligning the natural properties of earthen walls (such as thermal mass and moisture buffering) with modern monitoring and control systems, buildings can maintain comfortable indoor conditions with minimal energy consumption, thus contributing to a more sustainable and adaptive built environment.

4 Conclusion and Perspectives

This study evaluated the thermal performance of earthen materials in construction and their impact on indoor thermal comfort and energy efficiency. Experiments were conducted in the Anoual region of Morocco, an area characterized by intense summer heat and harsh climatic conditions.

The results indicated that while the outdoor temperature reached a maximum of 47 °C at 1:00 p.m., the indoor temperature peaked at approximately 6:30 p.m. and did not exceed 36 °C, demonstrating a thermal lag of five and a half hours. This significant delay and attenuation of peak temperature highlight that raw earth is not only an eco-friendly, readily available, and cost-effective material but also an effective solution for regulating indoor temperatures.

Future research on earthen buildings should concentrate on incorporating bio-based materials, integrating phase change materials (PCMs), and improving thermal insulation performance. Additionally, enhancing mechanical properties is essential for enabling additional floors while ensuring structural safety.

Beyond these technical aspects, earthen materials support the development of smart buildings in smart cities by pairing their natural thermal mass with advanced sensing and control systems. This synergy reduces energy consumption, promotes occupant comfort, and furthers decarbonization by minimizing reliance on carbon-intensive heating and cooling. As such, earthen construction represents a promising pathway toward sustainable, climate-responsive, and intelligent urban development.

References

1. I'MTech (2020) The oldest building material is also the Most environmentally friendly. I'MTech. https://imtech.imt.fr/en/2020/05/14/earth-the-oldest-building-material-is-also-the-most-environmentally-friendly/ (15 novembre 2024)
2. Nina JF, Eires R, Oliveira DV (2023) Earthen construction: acceptance among professionals and experimental durability performance. Constr Mater 3(2):143–163. https://doi.org/10.3390/constrmater3020010
3. Ben-Alon L, Loftness V, Harries KA, Cochran Hameen E (2019) Integrating earthen building materials and methods into mainstream construction using environmental performance assessment and building policy. IOP Conf Ser Earth Environ Sci 323(1):012139. https://doi.org/10.1088/1755-1315/323/1/012139
4. Kidari R, Tilioua A (2024) Investigation of the thermomechanical characteristics of compressed earth bricks reinforced with cement and corn straw waste fibers. Clean Waste Syst 9:100160. https://doi.org/10.1016/j.clwas.2024.100160
5. Carlos G, Ribeiro T, Achenza M, De Oliveira CCF, Varum H (2022) Literature review on earthen vernacular heritage: contributions to a referential framework. Built Herit 6(1):15. https://doi.org/10.1186/s43238-022-00061-1
6. Zarasvand KA, Narani SS, Siddiqua S (2023) Thermal insulation enhancement of rammed earth using wood fly ash and calcium bentonite. Constr Build Mater 409:134097. https://doi.org/10.1016/j.conbuildmat.2023.134097

7. El Hammouti A, Charai M, Channouf S, Horma O, Nasri H, Mezrhab A, Karkri M, Tankari MA (2023) Laboratory-testing and industrial scale performance of different clays from eastern Morocco for brick manufacturing. Constr Build Mater 370. https://doi.org/10.1016/j.conbuildmat.2023.130624
8. Charai M, Salhi M, Horma O, Mezrhab A, Karkri M (2022) Thermal and mechanical characterization of adobes bio-sourced with Pennisetum setaceum fibers and an application for modern buildings. Constr Build Mater 326(August 2021):126809. https://doi.org/10.1016/j.conbuildmat.2022.126809
9. Drissi M, Horma O, Mezrhab A, Karkri M (2024) Exploring raw red clay as a supplementary cementitious material: composition, Thermo-mechanical performance, cost, and environmental impact. Buildings 14(12):3906. https://doi.org/10.3390/buildings14123906
10. Niroumand H, Kibert CJ, Barcelo JA, Saaly M (2017) Contribution of national guidelines in industry growth of earth architecture and earth buildings as a vernacular architecture. Renew Sust Energ Rev 74:1108–1118. https://doi.org/10.1016/j.rser.2017.02.074
11. Google Maps (n.d.) « Google Maps ». https://www.google.com/maps/place/Anoual/ (13 novembre 2024)
12. Talsint (n.d.). Mapcarta. https://mapcarta.com/fr/17407524 (11 octobre 2024)
13. Gemici Z, Dogan A (2024) Adaptive strategies and thermal perceptions in intermittently used congregational spaces: a case study. Eng Sci Technol Int J 58:101828. https://doi.org/10.1016/j.jestch.2024.101828

Enhancing Thermal Efficiency in Adobe through the Integration of *Juncus Acutus* Fibers into Red Clay Matrices

Othmane Horma, Boutahar Laaouar, Mohammed Drissi, Sara El Hassani, Mohammed Amine Moussaoui, and Ahmed Mezrhab

Abstract The integration of fiber additives into adobe offers an economical, ecological, and energy-efficient solution to reduce the energy consumption and greenhouse gas emissions associated with the construction and building sectors. This study investigates the incorporation of locally available *Juncus acutus* (JA) fibers into a red clay matrix to enhance the thermophysical properties of the composite. Fiber contents of 2%, 4%, and 6% by weight of the binder were used in the preparation of the specimens. The thermophysical characteristics of the composites were evaluated using the hot disk method, a non-destructive technique. Experimental results reveal that the addition of JA fibers significantly reduces the thermal conductivity of the composite, achieving a reduction of over 50%. Furthermore, to assess the practical application of this bio-composite in buildings, a Computational Fluid Dynamics (CFD) analysis was performed using ANSYS software. The simulation results demonstrate the potential of these bio-composites to substantially improve energy efficiency in buildings. These findings support the use of locally sourced, sustainable, and energy-efficient materials as promising alternatives to conventional adobe.

Keywords Fiber-reinforced adobe · Thermophysical properties · *Juncus acutus* fibers · Energy-efficient materials · Thermal simulation

O. Horma (✉)
Mechanics and Energy Laboratory, Mohammed First University, Oujda, Morocco

Private University of Marrakech, Marrakech, Morocco

B. Laaouar · M. Drissi · M. A. Moussaoui · A. Mezrhab
Mechanics and Energy Laboratory, Mohammed First University, Oujda, Morocco

S. El Hassani
Laboratory of Energetics, Faculty of Science and Technology, Abdelmalek Essaâdi University, Tetouan, Morocco

H. Kaddouri et al. (eds.), *Smart Cities for Sustainable and Urban Development*, Green Energy and Technology, https://doi.org/10.1007/978-3-031-99001-4_13

1 Introduction

Earthen masonry is widely regarded as one of the most eco-friendly and affordable construction materials, owing to its natural composition and the near-unlimited availability of its raw components. These materials—such as clay, sand, and silt—are abundant in many regions, and their extraction and preparation require much less energy compared to modern materials like cement or fired bricks. Additionally, earthen construction offers environmental benefits, including lower embodied energy, reduced carbon emissions, and full recyclability at the end of a building's life cycle. Despite these advantages, earthen materials have been largely overlooked in recent decades due to their typically lower mechanical strength compared to conventional cement-based products and fired bricks. This perception has limited their application, particularly in structural elements of modern buildings. However, earthen masonry remains suitable for various applications, especially in low-load bearing walls, low-rise buildings, or as plastering materials where their inherent thermal and acoustic properties can be maximized without the need for high compressive strength.

One critical limitation of conventional clay-based materials is their relatively high thermal conductivity, typically ranging from 1.2 to 0.8 W/mK, and their high thermal diffusivity, between 0.9 and 0.6 mm^2/s [1, 2]. These properties lead to significant heat transfer through walls, making buildings less energy-efficient, particularly in extreme climates. High thermal conductivity results in greater heat loss during winter and increased heat gain during summer, leading to a higher demand for heating and cooling systems. Improving these thermal properties is therefore crucial to enhancing the energy efficiency of earthen construction materials. To address this challenge, the incorporation of natural fibers into clay matrices has emerged as a promising solution. Fibers, such as straw, sisal, and hemp, have been shown to reduce thermal conductivity by disrupting heat flow within the matrix, while also lowering thermal diffusivity, which controls how quickly heat moves through a material [3–5]. Beyond improving thermal performance, fiber addition also enhances the structural integrity and mechanical strength of earthen materials, providing better resistance to cracking and shrinkage [6, 7]. This makes fiber-reinforced earth composites more suitable for a wider range of building applications, while retaining the environmental benefits of traditional earthen materials.

One promising natural fiber that has received limited attention in the literature is *Juncus acutus* (JA), a plant native to the Mediterranean region. Known for its resilience in harsh environments, JA grows abundantly in countries such as Morocco and Spain, making it a highly accessible and sustainable resource for construction applications. Despite its widespread availability, research on the use of JA fibers in construction materials remains scarce, with only a few studies exploring its potential as a reinforcement material in earthen composites [8, 9]. Given its natural abundance and the need for sustainable building materials, the integration of JA fibers presents an opportunity for countries in the Mediterranean region to take advantage of this local resource. Incorporating JA fibers into traditional clay matrices not only contributes to enhancing the thermophysical properties of the materials but also supports local economies and reduces dependency on imported or energy-intensive building materials. As such, exploring the use of JA fibers in construction aligns with broader goals of promoting sustainable development and reducing the environmental impact of the building sector in these regions.

In this chapter, the incorporation of JA fibers into red clay matrices is investigated to improve the thermal properties of adobe. Specimens were prepared with varying fiber contents of 2%, 4%, and 6% by weight of the binder, allowing us to examine the effect of fiber concentration on the composite's thermophysical behavior. The thermal conductivity, diffusivity, and specific heat capacity of the samples were measured using the hot disk method, a non-destructive technique that provides precise thermophysical characterization in accordance with ISO 22007 [10]. To further understand the practical implications of this material in construction, a Computational Fluid Dynamics (CFD) simulation was conducted using ANSYS software. This analysis evaluated the thermal performance of JA-reinforced adobe under real-world conditions. Both the experimental and simulation results underscore the potential of these bio-composites to improve energy efficiency in buildings. Our findings offer valuable insights into the use of locally sourced fibers for enhancing traditional construction materials, contributing to the development of sustainable, energy-efficient alternatives to conventional building practices.

2 Methodology

2.1 Raw Materials and Sample Preparations

The primary materials used in this study are red clay and Juncus acutus (JA) fibers, both sourced locally from the Anoual region in eastern Morocco (32°40′41″N, 3°05′42″W). The selection of these materials was based on their local availability and potential for sustainable construction applications.

The clay was subjected to a series of tests to determine its suitability for building applications. According to the Atterberg limit test results, the red soil exhibits a liquid limit of 32.66%, a plasticity limit of 22.66%, and a plasticity index of 10, classifying it as moderately plastic. Its density of 1994 kg/m^3 and high SiO_2 content (57% Si), according to X-ray fluorescence tests, suggest that the clay is suitable for load-bearing applications due to its strength and balanced plasticity, which allows for workability while maintaining structural integrity [11].

JA fibers (JAF) were selected for their natural thermal insulation properties, attributed to their composition, which includes cellulose, hemicellulose, and lignin [12]. These components give JAF inherent thermal efficiency, making them suitable for reducing heat transfer in earthen matrices. The air-dried fibers were cut into 1 cm lengths to optimize their incorporation into the clay matrix.

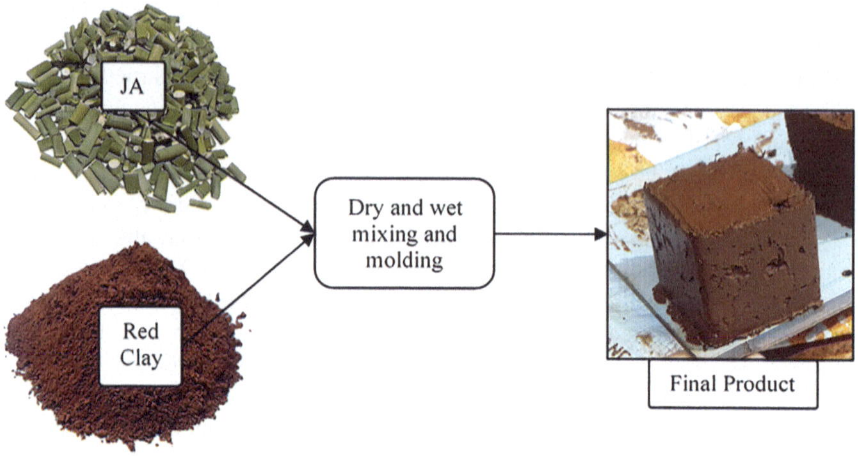

Fig. 1 Sample preparation process

The preparation process, illustrated in Fig. 1, began by crushing and sieving the collected red clay to obtain particle sizes smaller than 10 micrometers. This fine particle size enhances the homogeneity of the clay matrix when combined with the fibers. The air-dried JAF, cut into 1 cm lengths, were incorporated into the clay at increments of 2%, 4%, and 6% by weight relative to the binder, thereby forming JAF-adobe (JAFA) composites designated JAFA2, JAFA4, and JAFA6, with JAFA0 as the fiber-free reference. This variation in fiber content allowed for the evaluation of its effect on the thermophysical properties of the composite.

A water-to-binder ratio of 0.25 was used to hydrate the mixture. The water was added slowly while the mixture was manually kneaded in a traditional manner until a smooth and consistent dough was achieved. This manual kneading process ensured a uniform distribution of fibers throughout the clay matrix. Once a uniform texture was achieved, the mixture was shaped using a stainless-steel cubic mold measuring $5 \times 5 \times 5$ cm^3. The samples were then removed from the mold and left to air-dry under standard ambient conditions until they achieved sufficient hardness for testing.

2.2 *Experimental Method*

The Hot Disk Method (HDM), known for its speed and accuracy, was used in this study to measure thermophysical properties. As shown in Fig. 2, this setup includes a TPS2200 instrument, which records temperature variations; a monitor that displays real-time results processed by the HDM software; and the HDM sensor—a critical component consisting of a double spiral of nickel encased in a thin Kapton layer. This sensor serves as both a heat source and a thermocouple, simultaneously generating and measuring heat within the sample. In this setup, the sensor is

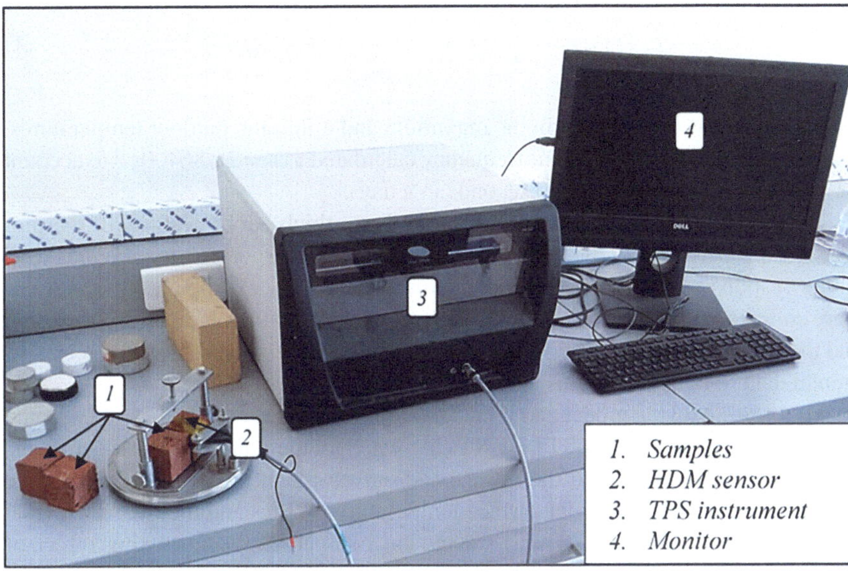

Fig. 2 Hot disk method setup

sandwiched between two identical samples, allowing it to record temperature changes in the surrounding material.

Since the HDM operates in the transient regime of thermal transfer, it involves time-dependent equations, making the calculations more complex. The governing equations for this technique are detailed in the study by Yi He [13].

2.3 Simulation Setup for JAF-Reinforced Adobe Wall Analysis

ANSYS FLUENT software was employed to simulate the thermal performance of two types of adobe walls over a 30-h period, capturing a full daily cycle plus additional hours to account for late-stage cooling behavior. The first wall, designated as the reference case (JAFA0), consists of standard adobe. The second wall (JAFA6) incorporates 6% *Juncus acutus* fibers, identified as the most thermally efficient composite in this study. The simulation focused on July 23, 2023, in Oujda, Morocco—a day characterized by significant outdoor temperature fluctuations ranging from 26 °C to 45 °C. This date provided a realistic scenario for assessing each material's effectiveness in moderating extreme thermal conditions. ANSYS FLUENT, widely used in fluid dynamics and heat transfer simulations, enabled precise modeling of heat transfer through the wall materials.

The outdoor temperature variations were approximated using the following mathematical model, previously used in several similar studies, such as [14]:

$$T_{sa}(t)=\frac{T_{out}^{max}+T_{out}^{min}}{2}+\frac{T_{out}^{max}-T_{out}^{min}}{2}\sin\left(\omega t-\frac{\pi}{2}\right) \tag{1}$$

Where T_{out}^{max} and T_{out}^{min} represent the maximum and minimum outdoor temperatures, respectively, and ω is the angular frequency calculated as $\omega = 2\pi/86400s^{-1}$ to account for the daily cycle (with 86,400 seconds in a day).

The simulation model consisted of a 300 mm-thick adobe wall, with thermophysical properties specific to JAFA0 and JAFA6 incorporated to evaluate each wall's response to outdoor thermal fluctuations. Outdoor temperature variations were collected from an in situ meteorological station at Mohammed First University, and the thermophysical properties of the materials were configured based on experimental data obtained from this study.

After setting up the model and boundary conditions, the simulation was executed over the 30 h period. ANSYS FLUENT calculated temperature variations across the wall thickness and recorded the inward and outward heat fluxes. The results were analyzed to assess the internal surface temperature profiles and the overall thermal performance of each wall type, focusing on comparing the insulating effect of JAFA6 to the baseline JAFA0. This analysis provides valuable insights into the potential of JAF-reinforced adobe to enhance indoor thermal comfort under extreme outdoor conditions.

3 Results and Discussion

3.1 Thermophysical Properties

Figure 3 illustrates the relationship between thermal conductivity (λ) and density (ρ) of the samples. As anticipated, the density decreases with increasing *Juncus acutus* fiber (JAF) content, as the bio-sourced fibers increase porosity and have a lower inherent density than clay. Each increment of JAF resulted in an approximate 2% reduction in density. Notably, the JAFA6 composite is 6.5% lighter than the reference sample (JAFA0). Up to 4% JAF content, thermal conductivity and density exhibit a linear relationship; however, between 4% and 6% JAF, density continues to decrease at a similar rate, while thermal conductivity stabilizes, dropping by only 5% compared to the earlier reductions of over 12% per 2% JAF increment.

Similarly, the evolution of thermal diffusivity (α) with JAF content, depicted in Fig. 4, follows an exponential decay pattern. The incorporation of JAF significantly impacts both λ and α in JAFA composites, reducing them by 33% and 47%, respectively. This reduction is primarily attributed to the increased porosity within the matrix and the inherently low thermal conductivity and diffusivity of the bio-based fibers. The created pores function as micro air pockets that hinder heat flow, leveraging the low thermal conductivity of air (0.025 W/mK).

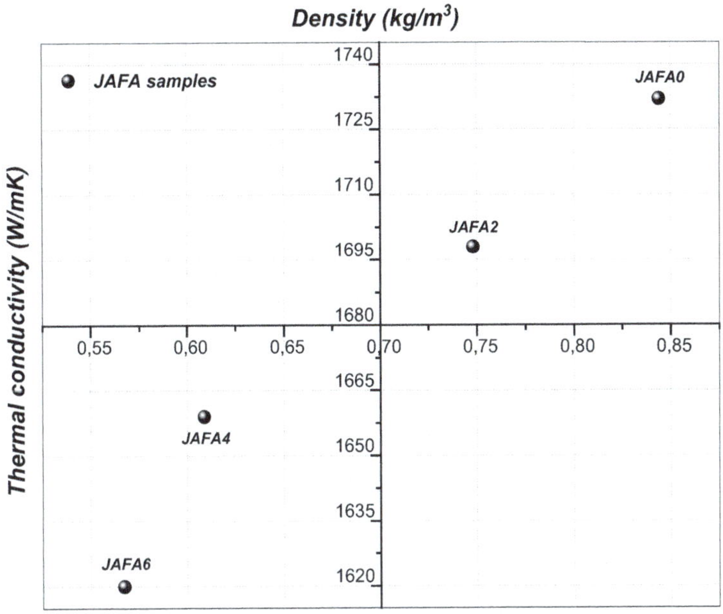

Fig. 3 Thermal conductivity versus density

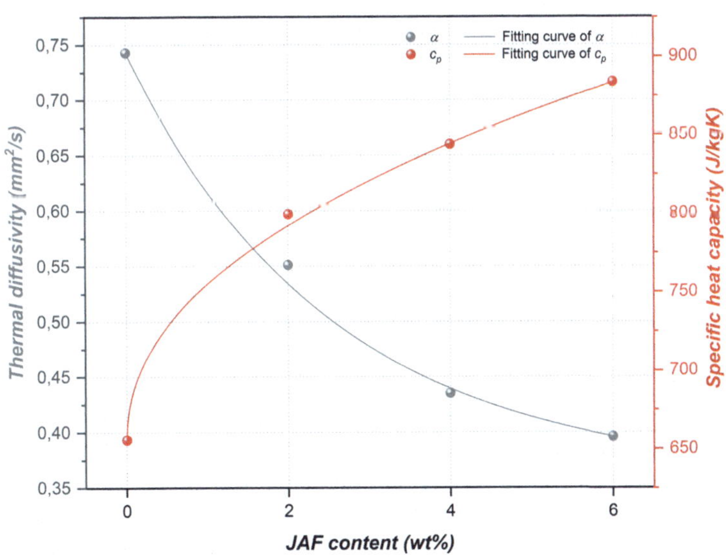

Fig. 4 Thermal diffusivity and specific heat capacity versus JAF content

The specific heat capacity (c_p), a critical property for construction materials, describes a material's ability to store heat, thereby delaying thermal transfer. This property depends on the previously discussed parameters—ρ, λ, and α—according to the equation:

$$c_p = \lambda / \alpha\rho \quad (2)$$

Since α decreased at a faster rate than λ, along with a reduction in ρ, c_p exhibited an inverse trend to α, increasing significantly, as shown in Fig. 4. For 6 wt% JAF, c_p increased by 26% over the reference sample JAFA0. This increase suggests that JAFA6 could improve the thermal inertia of buildings, extending the temperature lag and enhancing indoor thermal comfort, while also potentially reducing cooling demands in summer.

The variations in thermophysical properties observed in this study align with common trends seen when bio-based materials are integrated into conventional construction matrices, such as clay, cement, and plaster. For instance, Charai et al. [15, 16] reported that Pennisetum-reinforced adobe experienced reductions of 53% in λ and 40% in α, along with an 11% increase in c_p at a 6 wt% fiber concentration compared to the reference sample. Similar findings have been reported by our research group in studies incorporating spent tea leaves, spent coffee grounds, and mint stems into cement-based mortars [15–18].

Overall, the incorporation of JAF into clayey matrices resulted in significant improvements in all aspects of the thermophysical properties, indicating its potential for developing lightweight, energy-efficient, low-cost, and eco-friendly adobe.

3.2 Thermal Performance of JAF-Reinforced Wall

This section presents the simulation results of heat transfer through the walls made of standard adobe (JAFA0) and adobe reinforced with 6% JAF content (JAFA6). The findings indicate that incorporating JAF significantly improved the thermal performance of adobe walls. Specifically, the JAFA6 wall reduced the maximum internal surface temperature by almost 1.5 °C, lowering it from 28.48 °C in the JAFA0 wall to 27.02 °C. Additionally, a time lag φ of 2.77 h was observed between the JAFA0 and JAFA6 walls in reaching their maximum temperatures.

This substantial thermal shift demonstrates the enhanced thermal inertia provided by the fiber-reinforced adobe, allowing the wall to delay the transmission of peak outdoor temperatures to the interior space. The improved performance of JAFA6 can be attributed to the higher specific heat capacity and lower thermal diffusivity imparted by the *Juncus acutus* fibers. These fibers act as insulating inclusions within the adobe matrix, limiting heat flow and increasing the material's capacity to store and gradually release thermal energy.

Figure 5 illustrates the internal surface temperature profiles of the simulated walls. The JAFA0 wall reached its peak temperature of 28.48 °C at about 19.27 h,

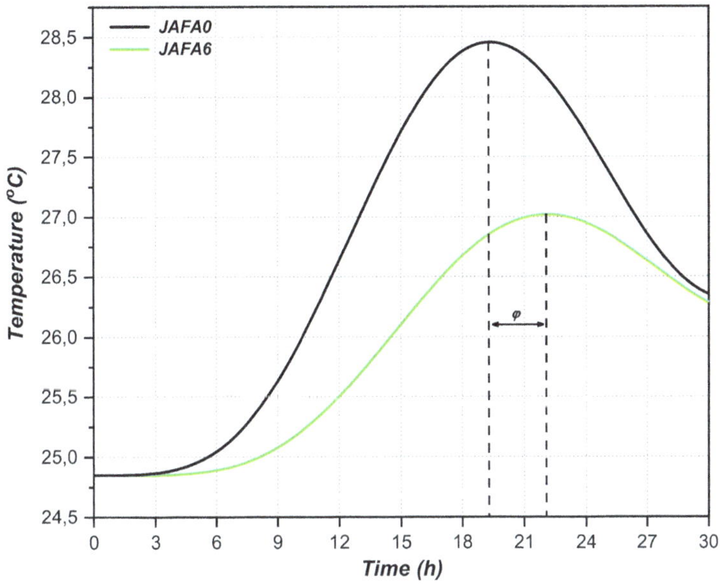

Fig. 5 Internal temperature variation of the studied wall

whereas the JAFA6 wall's peak temperature of 27.02 °C occurred later at approximately 22 h, demonstrating its superior insulating effect. The delayed and reduced peak temperature in the JAFA6 wall indicates a more stable indoor environment, which is particularly beneficial in regions with extreme outdoor temperature fluctuations.

These results are consistent with previous studies where the substitution of conventional building materials with bio-based composites led to significant improvements in thermal performance [19, 20]. The incorporation of natural fibers like JA enhances the thermal properties of adobe, suggesting its potential for improving indoor thermal comfort and reducing energy consumption for heating and cooling.

4 Conclusion

This study explored the development of eco-friendly, lightweight, and energy-efficient adobe blocks reinforced with *Juncus acutus* fibers (JAF). The incorporation of JAF into adobe significantly enhanced the material properties and thermal performance:

- *Density Reduction*: Adding 6 wt% of JAF decreased the adobe's density by approximately 6.5%, resulting in lighter construction materials.
- *Thermal Conductivity and Diffusivity*: The incorporation of JAF reduced thermal conductivity and diffusivity by 33% and 47%, respectively, compared to the reference adobe without fibers, indicating improved insulation properties.

- *Specific Heat Capacity*: The specific heat capacity increased with higher JAF content, achieving a 26% increase at a concentration of 6 wt% JAF, thus enhancing the material's ability to store thermal energy.
- *Enhanced Thermal Performance*: The JAF-reinforced adobe wall demonstrated significant improvements in the thermal performance of building envelopes, increasing the time lag by over 2.5 hours and decreasing the peak internal temperature by nearly 1.5 °C, thereby contributing to better indoor thermal comfort and energy efficiency.

These findings indicate that JAF-reinforced adobe has considerable potential to enhance building thermal efficiency, leading to energy savings and improved indoor environments. However, further research is required to assess the long-term durability, mechanical strength, and hygrothermal properties of these materials under various environmental conditions.

Competing Interests The authors declare that they have no known competing financial interests or personal relationships that could have appeared to influence the work reported in this paper.

References

1. Drissi M, Horma O, Mezrhab A, Karkri M (2024) Exploring raw red clay as a supplementary cementitious material: composition, Thermo-mechanical performance, cost, and environmental impact. Buildings 14(12):3906. https://doi.org/10.3390/buildings14123906
2. El hammouti A, Charai M, Channouf S, Horma O, Nasri H, Mezrhab A, Karkri M, Abdou M (2023a) Laboratory-testing and industrial scale performance of different clays from eastern Morocco for brick manufacturing. Constr Build Mater 370(January):130624. https://doi.org/10.1016/j.conbuildmat.2023.130624
3. El hammouti A, Charai M, Channouf S, Horma O, Mezrhab A, Karkri M, Tankari MA (2023c) Application analysis and environmental impact of straw reinforced gypsum plaster for improving the energy efficiency in buildings in the six climate zones of Morocco. J Build Eng 74. https://doi.org/10.1016/j.jobe.2023.106829
4. Fernea R, Manea DL, Tămaş-Gavrea DR, Roşca IC (2019) Hemp-clay building materials—an investigation on acoustic, thermal and mechanical properties. Procedia Manuf 32:216–223. https://doi.org/10.1016/j.promfg.2019.02.205
5. Prakash R, Thenmozhi R, Raman SN, Subramanian C, Divyah N (2021) Mechanical characterisation of sustainable fibre-reinforced lightweight concrete incorporating waste coconut shell as coarse aggregate and sisal fibre. Int J Environ Sci Technol 18(6):1579–1590. https://doi.org/10.1007/s13762-020-02900-z
6. Aymerich F, Fenu L, Meloni P (2012) Effect of reinforcing wool fibres on fracture and energy absorption properties of an earthen material. Constr Build Mater 27(1):66–72. https://doi.org/10.1016/j.conbuildmat.2011.08.008
7. Ojo EB, Bello KO, Mustapha K, Teixeira RS, Santos SF, Savastano H (2019) Effects of fibre reinforcements on properties of extruded alkali activated earthen building materials. Constr Build Mater 227:116778. https://doi.org/10.1016/j.conbuildmat.2019.116778
8. El-yahyaoui A, Manssouri I, Sahbi H (2023) Effect of Juncus Acutus fibers waste on the technological properties of unfired clay bricks as a construction material. Mater Today Proc. https://doi.org/10.1016/j.matpr.2023.06.064

9. Omrani H, Hassini L, Benazzouk A, Beji H, ELCafsi A (2020) Elaboration and characterization of clay-sand composite based on Juncus Acutus fibers. Constr Build Mater 238. https://doi.org/10.1016/j.conbuildmat.2019.117712
10. 22007-2, ISO (2015) Plastics—determination of thermal conductivity and thermal diffusivity—part 2: transient plane heat source (hot disc) method. https://www.iso.org/obp/ui/#iso:std:iso:22007:-2:ed-2:v1:en
11. Millogo Y, Morel JC (2012) Microstructural characterization and mechanical properties of cement stabilised adobes. Mater Struct Materiaux et Constructions 45(9):1311–1318. https://doi.org/10.1617/s11527-012-9833-2
12. El Ghali, Amel IB, Marzoug MH, Baouab V, Roudesli MS (2012) Separation and characterization of new cellulosic Fibres from the Juncus Acutus L plant. Bioresources 7(2):2002–2018. https://doi.org/10.15376/biores.7.2.2002-2018
13. He Y (2005) Rapid thermal conductivity measurement with a hot disk sensor: part 1. Theoretical considerations. Thermochim Acta 436(1–2, 122):–29. https://doi.org/10.1016/J.TCA.2005.06.026
14. Sun C, Shu S, Ding G, Zhang X, Xinghua H (2013) Investigation of time lags and decrement factors for different building outside temperatures. Energ Buildings 61:1–7. https://doi.org/10.1016/j.enbuild.2013.02.003
15. Charai M, Horma O, El A, Mezrhab A, Karkri M (2022a) Materials today: proceedings thermophysical characteristics of cement-based mortar incorporating spent coffee grounds. Mater Today Proc. https://doi.org/10.1016/j.matpr.2022.02.488
16. Charai M, Salhi M, Horma O, Mezrhab A, Karkri M (2022b) Thermal and mechanical characterization of adobes bio-sourced with Pennisetum Setaceum fibers and an application for modern buildings. Constr Build Mater 326(August 2021):126809. https://doi.org/10.1016/j.conbuildmat.2022.126809
17. Horma O, Channouf S, El Hammouti A, El Hassani S, Miri H, Moussaoui MA, Mezrhab A (2024a) Enhancing concrete sustainability using crushed Peanut shells: an analysis of Thermophysical properties, durability, and application potential in construction. J Build Eng 90. https://doi.org/10.1016/j.jobe.2024.109434
18. Horma O, Charai M, El Hassani S, El Hammouti A, Mezrhab A (2022) Thermo-physical and mechanical characterization of cement-based mortar incorporating spent tea. J Build Eng 52(November 2021):104392. https://doi.org/10.1016/j.jobe.2022.104392
19. El hammouti A, Channouf S, Charai M, Horma O, Miri H, Mezrhab A, Karkri M, Tankari MA (2023b) Resource deposit, characterization and energy saving potential of olive pomace as a promising aggregate for energy efficient earth bricks in eastern Morocco. Constr Build Mater 393. https://doi.org/10.1016/j.conbuildmat.2023.131989
20. Horma O, El Hammouti A, El Hassani S, Kirgiz MS, Channouf S, Moussaoui MA, Mezrhab A (2024b) A sustainable approach to mortar production using mint stem aggregates: investigating the thermal, mechanical and physical performance. Mater Civil Eng. https://doi.org/10.1061/JMCEE7/MTENG-17828

Impact of Fly Ash-Based Concrete on Thermal Performance and Energy Efficiency of Office Buildings in Oujda City: Comparative Study and Contribution in Smart Buildings

Mohammed Drissi, Othmane Horma, Boutahar Laaouar, and Ahmed Mezrhab

Abstract Buildings account for a significant portion of global energy consumption and greenhouse gas emissions, making the development of energy-efficient and sustainable construction materials a pressing necessity. Fly ash-based concrete (FA-concrete) has emerged as a promising alternative to traditional cement-based materials due to its potential to improve thermal performance, reduce energy demands, and lower environmental impacts. This study presents a comparative analysis of the thermal performance and energy efficiency of buildings constructed using fly ash-based concrete versus traditional cement-based masonry. To evaluate the energy efficiency of the developed FA-based concrete in comparison with conventional materials, building simulation models were created using the TRNSYS software. Experimental data on the thermophysical properties of the FA-concrete and conventional concrete were incorporated into the simulation model. The simulations were conducted throughout the year under the climatic conditions of Oujda, Morocco, a region characterized by extreme seasonal temperature fluctuations. Results indicate that, in the absence of heating or cooling systems, buildings constructed with FA-based mortar maintained temperatures approximately 1 °C cooler in the summer and 0.5 °C warmer in the winter compared to those made with conventional concrete. Moreover, when heating and cooling systems were employed, buildings using FA-based concrete showed a 15.5% reduction in energy consumption, underscoring the material's energy-saving potential. In addition to environmental and ecological advantages (through waste recycling), these findings highlight

M. Drissi (✉) · B. Laaouar · A. Mezrhab
Mechanics and Energy Laboratory, Mohammed First University, Oujda, Morocco
e-mail: mohammed.drissi1@ump.ac.ma

O. Horma
Mechanics and Energy Laboratory, Mohammed First University, Oujda, Morocco

Private University of Marrakech, Marrakech, Morocco

H. Kaddouri et al. (eds.), *Smart Cities for Sustainable and Urban Development*, Green Energy and Technology, https://doi.org/10.1007/978-3-031-99001-4_14

the significant economic benefits of using industrial waste products—such as fly ash—in construction. Overall, the results emphasize the potential of sustainable building materials in reducing both the carbon footprint and energy consumption in the construction sector.

Keywords Building · Modeling and simulation · Cooling and heating · Energy efficiency · Thermal comfort · TRNSYS

1 Introduction

1.1 Background of Energy Efficiency in Thermal Buildings

Sustainable construction has become a global imperative in response to growing environmental challenges, energy consumption, resource scarcity, and increasing greenhouse gas emissions. The building sector is responsible for approximately 39% of global CO_2 emissions, with nearly 28% stemming from building operations, including heating, cooling, lighting, and other energy demands, and 11% from building materials and construction activities [1]. On a national level, the building sector in Morocco alone accounts for 33% of energy consumption and 12% of greenhouse gas emissions [2]. Energy demand in buildings is predominantly driven by heating and cooling consumption [3]. Consequently, it is crucial to explore viable alternatives that minimize the carbon footprint while maintaining material performance. Substitution materials, such as fly ash (FA), are emerging as a promising solution, contributing not only to the reduction of CO_2 emissions from cement production but also to the enhancement of building energy efficiency. The incorporation of FA into concrete formulations enables the achievement of optimal mechanical and thermal properties, while also promoting the recycling and valorization of industrial waste. Moreover, improving energy efficiency in building design contributes significantly to the development of smart buildings, which integrate advanced technologies to optimize energy use, enhance occupant comfort, and reduce operational costs. Smart buildings leverage energy-efficient materials, such as fly ash--based concrete, alongside intelligent systems to create sustainable, adaptive, and resilient infrastructures.

1.2 Literature Review

In general, the use of industrial waste as Supplementary Cementitious Materials (SCMs) in construction reduces both waste generation and the environmental impact of building practices. FA, in particular, has shown potential as a reactive material with cement, maintaining decent mechanical performance. Consequently, the use of FA in cement-based materials is an active area of research.

The pozzolanic activity of FA, combined with its low water demand, reduced bleeding, and decreased heat of hydration, has facilitated its extensive application as a partial binder substitute in the construction industry. Moreover, the efficient utilization of FA not only reduces cement consumption but also mitigates the need for waste disposal [4]. Alonzo et al. [21] demonstrated that replacing cement with Class C fly ash can be an effective strategy, recommending substitution rates between 20% and 35% as a replacement value that maintains mechanical properties without significant reduction. Saha [4] showed that 30% replacement of ordinary Portland cement with class F fly ash in concrete results in an acceptable reduction in compressive strength, decreasing from 64 to 50 MPa, while Rishabh et al. [23] found that compressive strength decreases from 20 to 16 MPa when 30% of the cement is replaced by FA. In addition to affecting mechanical properties, FA also has the potential to improve thermal performance in construction. According to National Institute of Standards and Technology, and Academic Group on Concrete Technology [5], replacing 50% of the cement with FA produced reductions in thermal conductivity of 15% and 7% for class C and class F fly ashes, respectively.

Several studies have also aimed to improve energy consumption in the Moroccan building sector by integrating industrial waste and biomass into clay-, plaster-, or cement-based formulations [6–10]. For example, Hammouti et al. [11, 12], investigated the influence of walnut shell on thermal performance of unfired clay bricks, carried out on soil extracted from Nador city; eastern Morocco. The results showed that introducing 20 wt% walnut shell improves the thermal insulation of mud bricks by decreasing the thermal conductivity of clay by 45.22%. Moreover, developed bio composite increased the energy performance of residential buildings allowing respective savings (10.8%, 7.94%) of energy for heating and cooling needs.

Hicham et al. [2] used TRNSYS software to evaluate the impact of three bio-based insulation materials—hemp wool, wood fiber, and expanded cork—on the wall layers of a residential building in Al-Hoceima, Morocco. The results indicated that insulating the roof with 8 cm of hemp wool yields energy savings of up to 36.7% and 35.2% for cooling and heating, respectively. In terms of indoor temperature, the maximum recorded was 20.94 °C in January and about 26.80 °C in July.

Dlimi et al. [13] analyzed the thermal performance of Moroccan external building envelopes composed of double hollow brick walls filled with hemp concrete and separated by a 5 cm air layer. Using TRNSYS software, they studied three configurations involving 8-, 12-, and 16-holed bricks. The results revealed that double brick walls with 12 and 16 holes filled with hemp concrete reduce total energy demand (heating and cooling) from 73.92 kWh/m^2 to 27.19 kWh/m^2.

Other authors have also explored the use of industrial byproducts to enhance sustainability. Tamanna and Tuladhar [14] for instance, investigated the effects of incorporating recycled glass powder as a partial cement replacement in concrete formulations, demonstrating a significant reduction in embodied carbon and improved thermal insulation. Similarly, Jalal et al. [15] analyzed the synergy between FA and silica fume in concrete, showing that combined use enhances the mechanical properties. Another study by Amran et al. [16] highlighted the potential of geopolymer-based concretes incorporating FA to improve the thermal regulation of buildings and reducing energy requirements for heating and cooling.

Regarding the contribution of fly ash–based concrete to smart mobility, one notable benefit is its lighter weight compared to conventional concrete. This reduced mass can lower both transportation costs and fuel consumption, thereby cutting greenhouse gas emissions during material delivery. For instance, O'Brien et al. [17] investigated the integration of lighter, energy-efficient construction materials with urban infrastructure systems and found that reductions in shipping weight led to improved logistics and lower environmental impact.

Additionally, by lowering overall building energy consumption, FA-concrete structures can free capacity on the grid, supporting advanced mobility solutions such as widespread electric vehicle charging. This synergy ultimately fosters more sustainable, interconnected urban environments.

1.3 Research Gap and Aim of this Study

Despite numerous studies investigating the thermo-mechanical performance of FA-based concrete, significant gaps remain in the literature regarding its building--level energy demands and indoor temperature regulation, particularly in the context of the Oujda region. This study aims to address these gaps by evaluating and comparing the thermal performance—specifically indoor temperature stability and cooling/heating energy requirements—of two buildings. One building is constructed with a concrete mix incorporating 40% FA as a partial replacement for cement, while the other uses conventional concrete. This work builds upon a previous experimental study that determined the thermophysical properties of both materials. The primary objective is to use these experimental data to develop building simulation models in TRNSYS and analyze the impact of these mortars on energy consumption under real operating conditions in Oujda, Morocco. Ultimately, this research seeks to promote more environmentally sustainable construction practices and highlight the advantages of alternative materials in the construction sector.

2 Methodology

This section describes the methodology used to simulate the energy performance of buildings constructed with and without FA-based materials, including software description, data acquisition, simulation conditions, and modelling parameters.

2.1 Software Description

The dynamic thermal performance of the building was simulated using TRNSYS, a worldwide reference software specialized in dynamic thermal simulations for buildings. Originally proposed by the Building Scientific and Technical Center (CSTB) and recognized by ASHRAE, TRNSYS is widely used in academic and professional settings. A detailed mathematical description of the building's thermal energy model within this software can be found in the TRNSYS user manual (A. Handbook, Fundamentals (Ashrae, Atlanta, GA, 1982) [13, 18]).

2.2 Weather Data and Simulation Conditions

The study was carried out on the region of Oujda Angad, a city located in a semi--arid region of Morocco coordinated as: Latitude 33°9 N, Longitude 5°5 W, Altitude 550 m, under 1 year from January 01 to December 31. The climatic data for this study were derived from the Meteonorm software, which provided detailed hourly meteorological information included ambient temperature, solar radiation, pressure and precipitation which are critical for accurate energy simulation [19]. This region experiences a Mediterranean climate characterized by cold winters and hot summers. Precipitation, primarily in the form of rain and occasionally snow, is unevenly distributed across time and space [20]. As seen in Table 1, recorded extreme temperatures range from −1 °C to 47 °C, with the highest average temperature observed in July (31 °C) and the lowest in February (10.5 °C). The average monthly solar radiation intensity ranged from 85 kWh/m^2 in January to

Table 1 Monthly ambient temperature, solar radiation and precipitation of the region of Oujda/Angad

Month	Avg Temp (°C)	Min Temp (°C)	Max Temp (°C)	Avg Solar Radiation (kWh/m^2)	Precipitation (mm)
January	10.5	0	21	95	34
February	10.75	−1	22.5	110	32
March	14.25	1	27.5	160	28
April	16.7	4.9	28.5	175	27.5
May	20.2	7.4	33	215	20
June	25	12	38	230	7.5
July	31	15	47	230	2.5
August	28.8	16.6	41	215	5
September	23.2	8	38.4	165	16
October	18.75	4	33.5	140	23
November	14.25	2.5	26	100	25
December	11.5	0	23	85	32

230 kWh/m^2 in January, with peak solar radiation observed in summer months. These climatic conditions significantly impact building thermal performance and energy demands, emphasizing their relevance in this study, making this region an ideal choice for our simulation to assess the impact of construction materials on energy efficiency.

2.3 Building Description

The building simulated in this study is a single-story office located in Oujda, Morocco, with a total volume of 100 m^3. The external walls consist of three layers: cement mortar, traditional fired clay bricks, and either conventional or FA-based concrete. The floor assembly includes four layers: floor tiles, cement mortar, traditional fired clay bricks, and again either conventional or FA-based concrete. Figure 1 provides details on the thickness and composition of each layer.

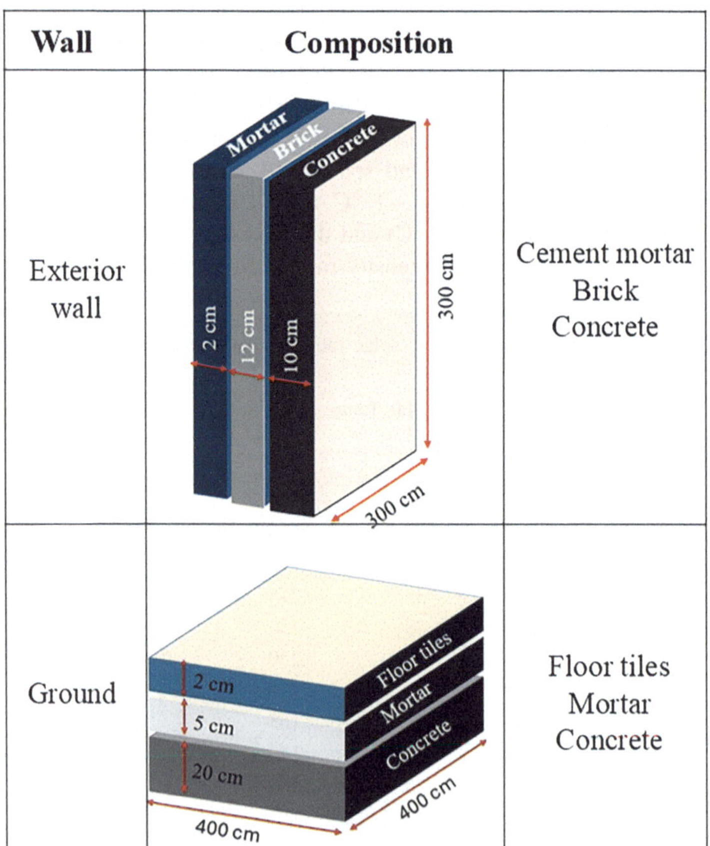

Fig. 1 Building composition

Table 2 Thermophysical properties of materials

Sample	λ (W/mK)	c_p (kJ/kgK)	ρ (kg/m³)
Brick	0.203	0.729	990
Mortar	1.3	1	1900
Ref Concrete	0.71	1	2100
FA Concrete	0.54	0.88	1811
Floor tiles	1.3	0.84	2300

[a] Binayate Prescriptive Library

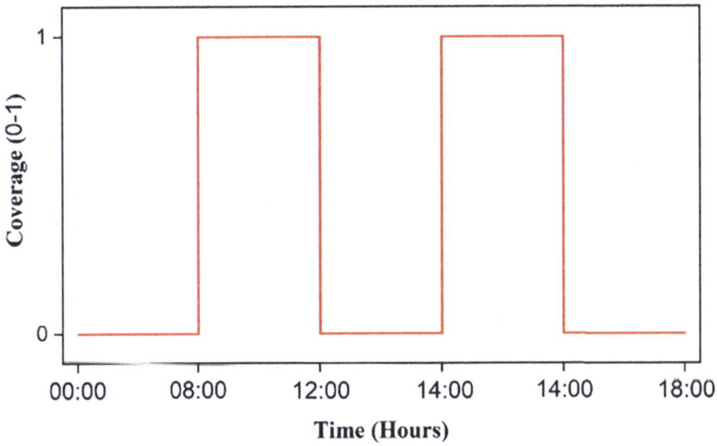

Fig. 2 Heating and Cooling Operation Schedule

This study compares two building configurations: a developed building, where the external walls and floor use FA-based concrete, and a reference building, where these elements are constructed with conventional concrete. Table 2 lists the thermophysical properties of each layer, including thermal conductivity, density, and specific heat capacity. The data indicate that FA-based concrete (with 40% FA replacing cement) has significantly lower thermal conductivity than conventional concrete, demonstrating the effectiveness of FA in enhancing the thermophysical performance of construction materials.

2.4 *Heating and Cooling System*

In this study, thermal bridges and shading effects—which can influence heat loss, heat gain, and solar radiation—were excluded to focus on the primary factors affecting thermal performance. The heating and cooling systems were scheduled to operate during typical office hours (see Fig. 2), with indoor temperature setpoints maintained at 20 °C in the heating season and 26 °C in the cooling season, in accordance with Moroccan standard (NM ISO 7730) [22]. These systems were assumed to run continuously during working hours, with no operation at night.

The heating and cooling loads were computed using TRNSYS, which applies an energy balance approach to calculate indoor temperatures and energy requirements. The total thermal load Q_{Load} is derived as:

$$Q_{Load} = Q_{Gain} - Q_{Lose} = Q_{Solar} - Q_{Lose} \tag{1}$$

Where: Q_{Solar} represents solar heat gains, based on incident solar radiation on building surfaces and Q_{Lose} represents conductive and convective heat losses through the building envelope, calculated using:

$$Q_{Gain} = \Sigma U \times A \times \left(T_{indoor} - T_{outdoor}\right) \tag{2}$$

With U is the thermal transmittance of each component, A is the surface area, and $(T_{indoor} - T_{outdoor})$ is the temperature gradient.

3 Results and Discussion

The simulations were carried out on the developed building (Concrete incorporating 40% FA) and the reference building to verify the influence of FA incorporation. Comparisons between variations in indoor temperature and total energy consumption are presented to illustrate the outcomes.

3.1 Indoor Temperature

The annual fluctuations of ambient temperature, along with the indoor temperatures recorded in the two houses (with and without FA), are presented in Fig. 3. During the summer period, the indoor temperature of the house incorporating FA-based mortar ranged from 19.43 °C (Hour 3846) to 27.66 °C (Hour 5562), whereas the reference house, constructed with conventional mortar, exhibited temperatures ranging from 19.89 °C (Hour 3846) to 28.42 °C (Hour 5562). In the winter period, characterized by limited sunlight, the indoor temperature of the FA-based house consistently exceeded that of the reference house, with temperatures ranging from 10.67 °C (Hour 344) to 17.87 °C (Hour 1290), compared to 10.43 °C (Hour 344) to 17.78 °C (Hour 1290) in the reference house. Overall, the results demonstrate that buildings constructed with FA-based mortar were, on average, approximately 1 °C cooler during summer and 0.5 °C warmer during winter, in the absence of heating or cooling systems, compared to buildings constructed with conventional mortar. This thermal performance enhancement can be attributed to the distinct thermophysical properties of the FA-based materials, such as lower

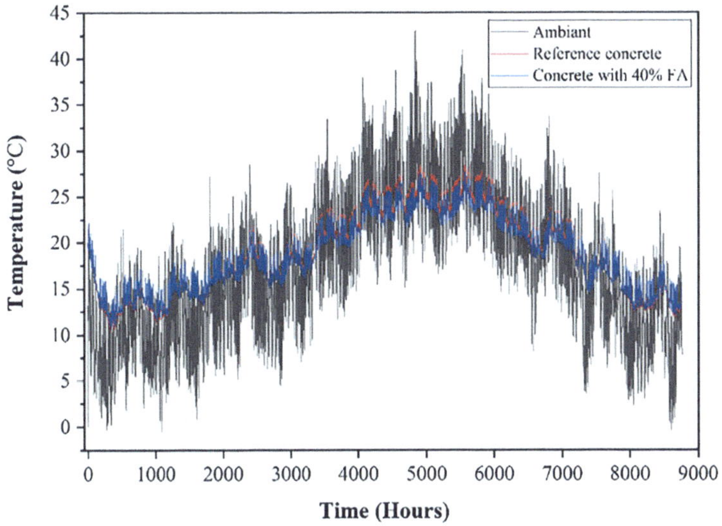

Fig. 3 Comparison between ambient temperatures, and indoor temperature of the two building

thermal conductivity and optimized density, which reduce heat transfer and improve insulation.

3.2 *Energy Efficiency*

In order to assess the impact of FA incorporation on the annual energy demand of the studied building, the results of simulations of energy demands for heating and cooling for building made from concrete incorporated FA, are compared with those from conventional concrete.

Figure 4 illustrates the monthly energy demands for cooling during the summer period. The incorporation of FA into the wall material significantly reduced the energy load compared to the conventional house. Specifically, the developed building achieved a reduction of 2.9 kW in August, with subsequent reductions of 1.78, 1.7, and 0.2 kW in July, September, and June, respectively.

Figure 5 presents the energy demand for heating during the winter period, with August demonstrating the highest reduction in energy demand for the developed building, reaching 43.07 kW. This is followed by reductions of 41.2, 37.7, and 31.3 kW in February, December, and March, respectively. This period of the year is characterized by elevated energy consumption levels compared to the summer season due to the cold climate in the Oujda-Angad region during these months. Despite these challenges, the developed building showed its ability to reduce energy consumption in both summer and winter, highlighting the thermal efficiency provided by FA-based mortar.

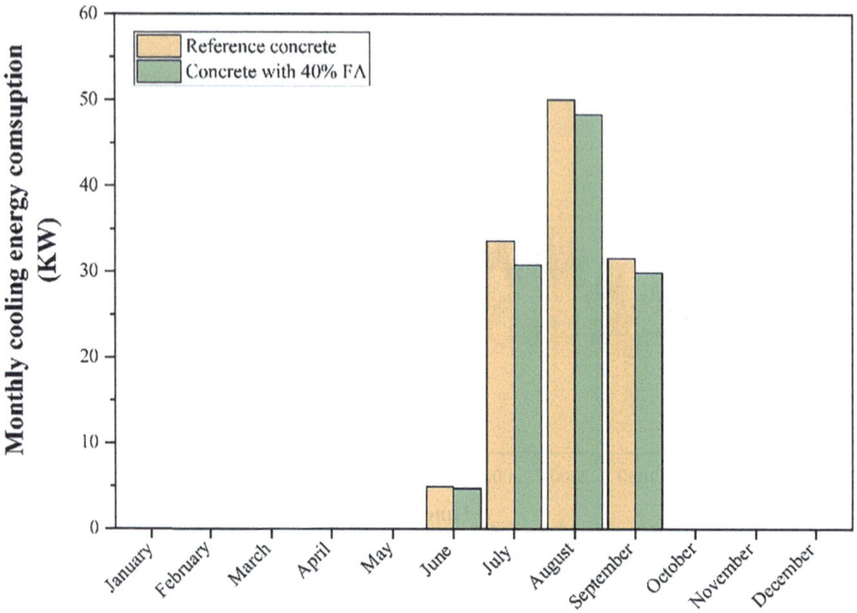

Fig. 4 Energy demand for cooling in the two building (with and without FA)

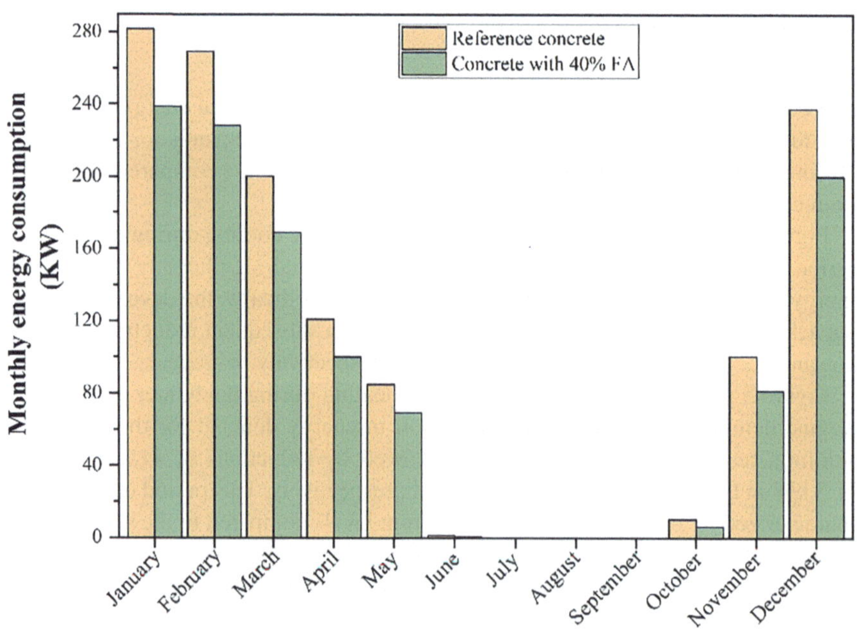

Fig. 5 Energy demand for heating in the two building (with and without FA)

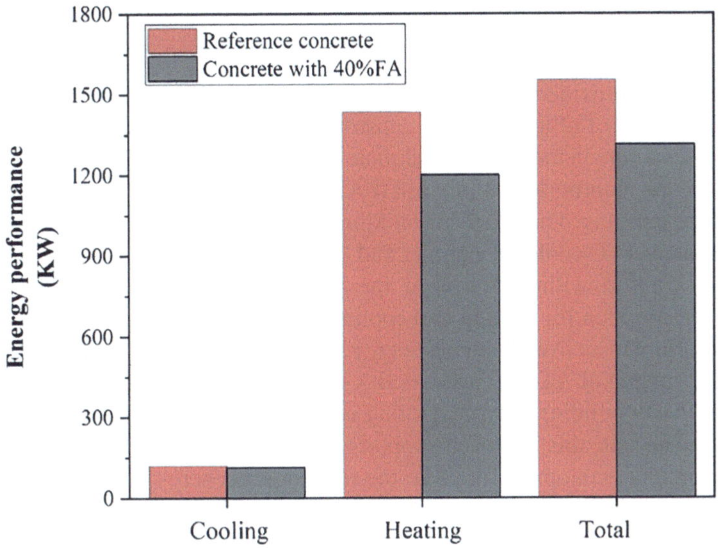

Fig. 6 Total energy demand for the two building (with and without FA)

Figure 6 illustrates the annual energy requirements for heating, cooling, and total energy consumption of the studied houses constructed with and without FA-based concrete. The total energy demand for heating was 1200 kW for the building with 40% FA walls, compared to 1434 kW for the reference building, representing a 16% reduction. For cooling, the energy demand was lower, at 113 kW for the FA-based building, compared to 120 kW for the reference building, corresponding to a 5.9% reduction. Combined heating and cooling energy demands amounted to 1313 kW for the FA-based concrete walls, compared to 1554 kW for the reference building, achieving an overall energy savings of 15.5%. This reduction in energy demand is attributed to the enhanced thermal properties of FA-based concrete. The incorporation of FA improves thermal capacity while reducing thermal conductivity and diffusivity, which minimizes heat flow through the walls.

The findings of this study align with previous research highlighting the potential of byproduct to improve building energy efficiency, a recent study elaborated by Hammouti et al. [11, 12] on clay bricks incorporating 20% by weight olive pomace demonstrated a 22.9% reduction in heating energy needs and 4.7% reduction in cooling energy consumption. Another recent study (El hammouti et al. 2022) elaborated by the same author proved that walls constructed with 20% walnut shell clay achieved a 10.8% reduction in heating energy requirements and a 7.94% decrease in cooling demands. These studies aligning closely with the 15.5% reduction observed in our study using FA, highlighting the potential of FA-based concrete for widespread adoption as a sustainable alternative in construction, effectively reducing energy demands and contributing to efforts to mitigate greenhouse gas emissions.

4 Conclusion

This study has provided a comparative analysis of the thermal performance and energy efficiency of office buildings constructed with conventional concrete and fly ash-based concrete in the semi-arid climate of Oujda, Morocco. The results demonstrated that the incorporation of fly ash (FA) in concrete enhances the thermal properties of the building. The FA-based building exhibited average indoor temperatures approximately 1 °C cooler in summer and 0.5 °C warmer in winter, contributing to improved comfort without the need for additional energy inputs. Furthermore, energy consumption for heating and cooling decreased by 16% and 5.9%, respectively, resulting in an overall annual energy savings of 15.5%. These findings underscore the potential of fly ash-based concrete as an effective material for energy-efficient buildings. By integrating waste products, such as fly ash, into construction materials, this approach supports sustainable building practices and promotes a circular economy. Moreover, the reduction in energy demand achieved by using FA-based materials aligns with the principles of smart mobility, as energy--efficient buildings reduce pressure on urban energy grids and contribute to the sustainability of interconnected infrastructures. This synergy between energy-efficient construction and smart mobility systems enhances urban resilience, minimizes environmental impact, and supports the development of smarter, greener cities.

References

1. Caronge MA, Tjaronge MW, Rahim IR, Irmawaty R, Lapian FEP (2022) Feasibility study on the use of processed waste tea ash as cement replacement for sustainable concrete production. J Build Eng 52:104458. https://doi.org/10.1016/j.jobe.2022.104458
2. Kaddouri H, Abidouche A, Alaoui MSH, Driouch I, Hamdaoui S (2024) Impact of insulation using bio-sourced materials on the thermal and energy performance of a typical residential building in Morocco. J Adv Res Fluid Mech Thermal Sci 117(1):43–59. https://doi.org/10.37934/arfmts.117.1.4359
3. Lu S, Zhao Y, Fang K, Li Y, Sun P (2017) Establishment and Experimental Verification of TRNSYS Model for PCM Floor Coupled with Solar Water Heating System. Energ Buildings 140:245–260. https://doi.org/10.1016/j.enbuild.2017.02.018
4. Saha AK (2018) Effect of class F fly ash on the durability properties of concrete. Sustain Environ Res 28(1):25–31. https://doi.org/10.1016/j.serj.2017.09.001
5. Bentz DP, Peltz MA, Durán-Herrera A, Valdez P, Juárez CA (2011) Thermal Properties of High-Volume Fly Ash Mortars and Concretes. J Build Phys 34(3):263–275. https://doi.org/10.1177/1744259110376613
6. Charai M, Salhi M, Horma O, Mezrhab A, Karkri M, Amraqui S (2022) Thermal and mechanical characterization of adobes bio-sourced with pennisetum setaceum fibers and an application for modern buildings. Constr Build Mater 326(February):126809. https://doi.org/10.1016/j.conbuildmat.2022.126809
7. Drissi M, Horma O, Mezrhab A, Karkri M (2024) Exploring raw red clay as a supplementary cementitious material: composition, thermo-mechanical performance, cost, and environmental impact. Buildings 14(12):3906. https://doi.org/10.3390/buildings14123906

8. Horma O, Charai M, El Hassani S, El Hammouti A, Mezrhab A (2022) Thermo-physical and mechanical characterization of cement-based mortar incorporating spent tea. J Build Eng 52:1–14. https://doi.org/10.1016/j.jobe.2022.104392
9. Horma O, Channouf S, El Hammouti A, El Hassani S, Miri H, Moussaoui MA, Mezrhab A (2024a) Enhancing concrete sustainability using crushed peanut shells: an analysis of thermophysical properties, durability, and application potential in construction. J Build Eng 90(February):109434. https://doi.org/10.1016/j.jobe.2024.109434
10. Horma O, El Hammouti A, El Hassani S, Kirgiz MS, Channouf S, Moussaoui MA, Mezrhab A (2024b) A Sustainable approach to mortar production using mint stem aggregates: investigating the thermal, mechanical and physical performance. Mater Civil Eng 36(10):1–10. https://doi.org/10.1061/JMCEE7.MTENG-17828
11. Hammouti E, Aboubakr, Charai M, Horma O, Mezrhab A, Karkri M (2022a) Thermal insulation of mud bricks made with walnut shells: characterization and simulation study. Mater Today Proc 62:4545–4550. https://doi.org/10.1016/j.matpr.2022.05.130
12. Hammouti E, Aboubakr MC, Mezrhab A, Karkri M (2022b) Fired and unfired clay bricks incorporating olive pomace: a comparative study. Mater Today Proc 60:359–364. https://doi.org/10.1016/j.matpr.2022.01.243
13. Dlimi M, Agounoun R, Kadiri I, Saadani R, Rahmoune M (2023) Thermal performance assessment of double hollow brick walls filled with hemp concrete insulation material through computational Fluid dynamics analysis and dynamic thermal simulations. e-Prime-Advances in Electrical Engineering, Electronics and Energy 3:100124. https://doi.org/10.1016/j.prime.2023.100124
14. Tamanna N, Tuladhar R (2020) Sustainable Use of Recycled Glass Powder as Cement Replacement in Concrete. Open Waste Manag J 13(1):1–13. https://doi.org/10.2174/1874347102013010001
15. Jalal M, Pouladkhan A, Harandi OF, Jafari D (2015) Comparative study on effects of class f fly ash, nano silica and silica fume on properties of high performance self compacting concrete. Constr Build Mater 94:90–104. https://doi.org/10.1016/j.conbuildmat.2015.07.001
16. Amran M, Fediuk R, Murali G, Avudaiappan S, Ozbakkaloglu T, Vatin N, Karelina M, Klyuev S, Gholampour A (2021) Fly ash-based eco-efficient concretes: a comprehensive review of the short-term properties. Materials 14(15):1–41. https://doi.org/10.3390/ma14154264
17. O'Brien KR, Ménaché J, O'Moore LM (2009) Impact of fly ash content and fly ash transportation distance on embodied greenhouse gas emissions and water consumption in concrete. Int J Life Cycle Assess 14(7):621–629. https://doi.org/10.1007/s11367-009-0105-5
18. A. Handbook, Fundamentals (Ashrae, Atlanta, GA, 1982) (1981)
19. Jani DB, Bhabhor K, Dadi M, Sachindra Doshi PV, Jotaniya HR, Bhatt K (2020) A review on use of TRNSYS as simulation tool in performance prediction of desiccant cooling cycle. J Therm Anal Calorim 140(5):2011–2031. https://doi.org/10.1007/s10973-019-08968-1
20. Guechchati R, Moussaoui MA, Mezrhab A, Mezrhab A (2023) Simulation de l'effet de l'isolation thermique des Bâtiments Cas Du Centre Psychopédagogique SAFAA à Oujda. J Renew Energ 13(2):223–232. https://doi.org/10.54966/jreen.v13i2.191
21. ACI 211.4 (1993) ACI 211.4R-93 (Reapproved), guide for selecting proportions for high-strength concrete with portland cement and fly ash, Ameri-Can Concrete Institute, Farmington Hills, MI, USA, 1993. ACI Mater J 93(Reapproved):13
22. "NM ISO 7730 | IMANOR" (n.d.) https://www.imanor.gov.ma/Norme/nm-iso-7730/. Accessed 17 Nov 2024
23. Shelorkar AP, Jadhao PD (2022) Effect of varying water cement ratio on modulus of elasticity of high-performance fibre-reinforced concrete (HPFRC). Mater Today Proc 65:784–791. https://doi.org/10.1016/j.matpr.2022.03.289

Management of the Intelligent Building Using Model Predictive Control for a Green and Sustainable Urban

Muhammad Shamrooz Aslam, Summera Shamrooz, Wer-Jer Chang, and Hazrat Bilal

Abstract Optimizing energy efficiency, boosting occupant comfort, and strengthening overall building management have all made the use of smart building technologies more and more crucial. Smart buildings monitor and manage several systems, including lighting, heating and cooling, security, and energy consumption, in real-time, using sophisticated sensors, Internet of Things devices, and data analytics. Significant financial savings, less of an adverse influence on the environment, and more efficient use of resources result from this. To control the temperature of a two-zone smart building using machine learning *model predictive control (MPC)*, this study proposes a new model predictive control approach. In this research, keeping residents comfortable and minimizing energy savings are the main objectives. In addition, model-based control is incorporated into the proposed control scheme to achieve the desired goals. To predict indoor temperature, the *artificial neural network (ANN)* utilized to estimates occupancy profiles in building zones over the long term, and this data is fed into the model-based predictor. As the actual data set provider, *Python* and *MATLAB* softwares are used. A set-point temperature is generated for the heater/cooler by solving the optimization problem, using both actual and predicted data. Compared with conventional *MPC* results, the proposed approach significantly saved energy compared to the conventional method. In the

M. S. Aslam
School of Computer Science and Technology/School of Artificial Intelligence, China University of Mining and Technology, Xuzhou, China

S. Shamrooz
School of Mechatronic Engineering, China University of Mining and Technology, Xuzhou, China

W.-J. Chang
Department of Marine Engineering, National Taiwan Ocean University, Keelung, Taiwan

H. Bilal (✉)
CAIR - Center for Artificial Intelligence Research, Department of ICT, University of Agder (UiA), Grimstad, Norway
e-mail: hbilal@mail.ustc.edu.cn

H. Kaddouri et al. (eds.), *Smart Cities for Sustainable and Urban Development*, Green Energy and Technology, https://doi.org/10.1007/978-3-031-99001-4_15

end, a practical approach example has been demonstrated to presents the effectiveness of our proposed approach.

Keywords Optimization method · Cyber-physical plant (CPP) · Artificial intelligence (AI) · Model predictive control · Secure energy-system · Neural network

1 Introduction

Microgrids are small systems that combine renewable energy resources with flexible loads, distributed generators, and/or energy storage to produce and consume energy. Microgrids can be found in buildings of any type, whether they are commercial, residential, or industrial. With the advent of the microgrid approach, the electricity market has been given a number of quality opportunities. By building microgrids, for example, end users receive more reliable power [1]. Furthermore, hospitals, police stations, and emergency operating centers can all benefit from the self-reliance of distributed energy resources (DERs) by meeting their power demands. Increasing the capability of building microgrids to respond to demand is imperative by applying advanced energy efficiency technologies across loads. Lighting fixtures, plug–in loads, and HVAC systems can all benefit from energy efficiency technologies [2, 3]. A building's function and service can be determined by data collected and managed through smart sensors, actuators, and microchips [4, 5], and medical sector [6]. Building Energy Management Systems (BEMS) can be controlled with advanced sensors and actuators via an Internet of Things (IoT) platform [7, 8]. By using the BEMS system, the building's energy consumption and environmental conditions can be synchronized with occupant behavior, and the power grid is operational all at the same time. As a result, it is vital that solutions are found for developing smart buildings in a smart city network, which can reduce energy use and/or increase energy efficiency.

Modeling-based approaches are used to control building thermal conditions in the majority of cases. Controllers designed using model-based control designs use mathematical models as representations of real plants. More advanced control algorithms can now be implemented using today's control software and technology. A robust and transparent process control structure using simple controllers is, however, often the preferred strategy for industrial engineers. Due to their rigid approach to structured control, PID controllers have gained wide acceptance due to their widespread acceptance, but this rigidity can limit their effectiveness. In the hierarchy, performance objectives can be systematically achieved without a coordinator. Additionally, a facility is not included in the design that enables easy management of process operational constraints. Process control constraints of varying types and levels can be handled systematically with *model predictive control (MPC)*. As a result of these two reasons, MPC continues to enjoy popularity among industrial-based applications. In building thermal control, model predictive control is one of the most efficient techniques (MPC) [9]. To predict a process' inputs and outputs and determine the control signal, *MPC* needs a precise model (an analytical model) [10, 11]. A model's accuracy directly affects *MPC* performance, and inaccuracy in a model reduces its performance. Because construction materials and architectures are vastly different, thermal dynamics change over time, requiring a lot of complex data processing. Precise modeling is expensive, and it is challenging to model residents' behavior and occupancy in the context of building management.

In order to maximize efficiency, lower expenses, and improve tenant comfort, smart building management entails integrating many systems, including lighting, security, *HVAC*, and energy management. To increase the operational effectiveness and sustainability of smart buildings, a number of models and techniques have been put forth over time. An outline of the main research on smart building management techniques and the approaches taken to accomplish these goals can be seen below. Traditional control systems, which were rule-based and reactive in nature, were the mainstay of early research on smart building management. Without foreseeing future circumstances, these systems reacted to alterations in the surroundings or occupants. For example, simple schedules or occupancy sensors were used to develop basic on/off control schemes for lighting and air conditioning systems. However, these systems were not flexible enough to adjust to shifting environmental conditions or building usage patterns, which frequently resulted in inefficient energy use. Because they offer centralized control over energy use, *Building Energy Management Systems (BEMS)* represent a major advancement in smart building management. In order to improve building systems, these systems rely on real-time monitoring of energy indicators like temperature and electricity use. In order to lower energy expenditures and guarantee that energy usage is in line with current demand, research has concentrated on enhancing the control algorithms in *BEMS*. *Particle swarm optimization (PSO)*, genetic algorithms, and dynamic programming are some of the optimization techniques utilized in *BEMS*. These methods seek to reduce energy expenses while maintaining comfort, but they frequently lack the adaptability needed to deal with quickly shifting circumstances.

Cyber-physical systems are managed and controlled using model-based predictive control and machine learning in this study. As well as managing visual, thermal, and olfactory performance, and energy consumption, we utilize our proposed learning-based control strategy. Data about energy consumption, environmental conditions, and occupant comfort is automatically learned using neural networks. *HVACs*, lighting systems, blinds, and ventilators can be controlled optimally by the *model-based controller (MPC)* by using the learned parameters [12]. Simulations in MATLAB software are used to generate NN training data, taking into account indoor temperatures, time of day, weather data, and energy consumption. Using the model-based controller, residents' comfort and energy conservation can be optimized at each time step. A benchmark controller on the building under the same conditions is used to compare the performance of the proposed learning-based *MPC* approach with its simulation results. In addition, none of them complemented learning-based techniques with model-based control frameworks.

A path to ecosystem balance and socio-economic development has been provided by sustainable development in construction through the integration of natural resources and proper management systems [13, 14]. Throughout the years, smart building technologies have gained increasing popularity worldwide [14, 15]. It has recently come to the fore that smart buildings will be adopted by developed and developing countries since they are able to provide many benefits in the construction industry. In order to achieve a sustainable and carbon-free environment, the products of the built environment have been constructed in the most efficient ways possible, utilizing raw materials efficiently and recycling them [16–18]. Technology application in the construction industry has demonstrated the effectiveness of this process. A developing country faces many challenges in adopting SBT in order to achieve sustainable construction. It was found that smarter, sustainable materials and equipment are more expensive than traditional materials and equipment, that construction processes are difficult, or that skills are lacking when it comes to smart technologies and techniques, according to [19] . Ghana's construction industry also resisted changing from its traditional practices. Among the findings of Gobbo et al., there is no regulatory framework under which smart building concepts, devices, and solutions can be adopted, there is a lack of financing and financial incentives for implementing smart building technologies, and there is a shortage of skilled and specialized jobs. There is still room for further investigation in this field.

The salient features of this article are as follows: (1) Using a model-based predictive control algorithm, a model-based learning approach is used to manage the thermal properties of a multi-zone building in this paper. (2) In addition, predicting occupancy profiles is achieved using an *ANN* with three layers based, and then this data is incorporated into a model-based control system (MPC). (3) A building simulation is performed using *Python* and *MATLAB* softwares to generate the data-set for *ANN*, taking indoor temperature, daytime, weather, energy consumption, and set-point data into consideration. (4) To conserve energy and improve comfort levels, the MPC algorithm generates the optimal set-points at each step. This study enhanced model-based control framework robustness and stability by incorporating both model-based and learning-based modeling into the *MPC* algorithm. The controlled system performance was improved by combining learning from historical data for model-based control (Table 1).

Table 1 List of Abbreviation

Abbreviation	Explanation
MPC	Model predictive control
DERs	Distributed energy resources
BEMS	Building energy management systems
HVAC	Heating, cooling, ventilation, and air conditioning
loTs	Internet of things
PID	Proportional derivative control
ANN	Artificial neural network
NARX	Nonlinear autoregressive netwoRk with eXogenous inputs.
MAS	Mean square error
SMTs	Smart building technologies

Table 2 Inventories of plug sets

Items	Description of the equipment	Count in total	Power (W)
Sound/video	Overhead projection	15	280
	TV/LCD screen	28	150
Laptops, LCDs	Desktop computer	225	195
	Network routers	2	35
Kitchen	Fridge	7	125
	Tea maker/cattle	3	785
	Oven	3	400
	Dispenser/electric heater	2	550
	Bread maker	2	500
Lab apparatus	Laser printer	4	65
	Optical oscilloscope/graph editor/power supply	21	62
	Control unit for the projector	1	61
	Commercial printer	87	215
Inkjet/scanner	Printing device for the network	8	445
	Copier printer with a large capacity	3	1000
	Machine for faxing	2	220
	Electronic paper shredder	5	67

2 Description of a System Modeling for Prediction of the Occupied Spaces

Ancient buildings' lighting systems do not interact with network managers' controllers. In spite of the motion sensors, the lights are not automated, even though they detect motion in a room. There will be a transition to low-consumption energy lights that are equipped with smart sensors in most smart buildings, as was mentioned previously. Smart building lighting systems will be able to consume less energy by using smart applications for lights. Table 2 shows a lot of smart buildings. Each light will be equipped with sensors that can be monitored and controlled remotely with IoT technologies, enabling the generation of near-real-time demand responses. In this work, an office building with a double story and four zones is examined. Each zone has its own *HVAC* system. There are two set-points for each zone thermostat. In the next section, the authors will discuss occupancy estimation with a mathematical model.

2.1 The Prediction of Occupied Spaces

To predict indoor temperature impacts of occupancy, authors implement a learning-based approach. Energy consumption, environmental issues, temperature, and comfort constraints may be influenced by occupancy information when it comes to building climate control. Using the *ANN*, this study explores how occupancy information can affect energy consumption and residents' comfort in the long term.

Nonlinear autoregressive neural networks with exogenous inputs are selected for the *ANN* model. Three layers of nodes are used in autoregressive neural networks to approximate the output in (1).

$$y[n] = g\left(x[n-1]\right), x[n-2], \ldots, x[n-h_x], y[n-1], y[n-2], \ldots, y[n-h_y]) \quad (1)$$

$x[n]$ presents the input of the artificial neural network.

$y[n]$ denotes the output of the artificial neural model.

h_x and h_y show the input/output delays of *ANN*, respectively.

f exhibits the mapping function.

A *NARX* neural network can either be a series-parallel network or a parallel network. Model training is more effective in a series-parallel configuration than in a parallel one. Static back-propagation is useful for training the feed-forward network, which has a better performance because of the more accurate input. The training of our network is done using a series-parallel architecture since we have access to the actual output (from *MATLAB*). *NARX* is illustrated in Fig. 1 as a series-parallel network.

Due to the nature of occupancy profiles as time series, *NARX* was used for occupancy predictions. The prediction of time series models is one of *NARX's* primary applications [20]. Furthermore, the occupancy produces heating, which makes it highly nonlinear, and the *NARX* model is extremely beneficial for these types of nonlinear models. In order to validate the network, it must be trained first. Predicted data is evaluated by comparing it with test data in order to evaluate stopping criteria and predictability. In order to conduct this experiment, three data-sets are required:

- Training.
- Validation.
- Testing.

The authors validate training performance by measuring the *mean square error (MSE)* and the regression value. These represent the output and target value correlation as well as the square error. In other words, the *NARX* neural network algorithm consists of the following steps, which can be seen in Fig. 2.

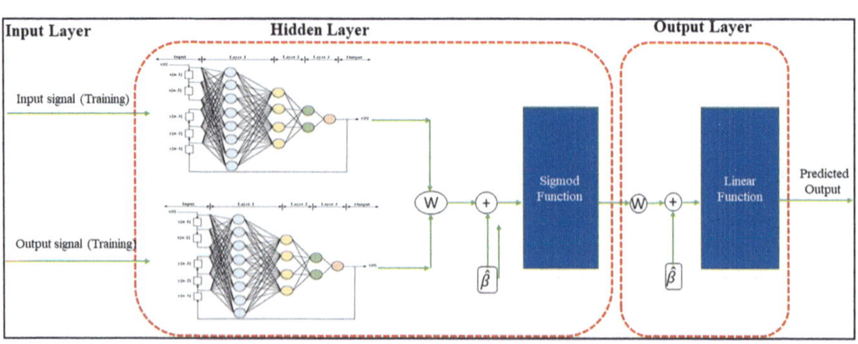

Fig. 1 Input and output delays in a typical NARX network

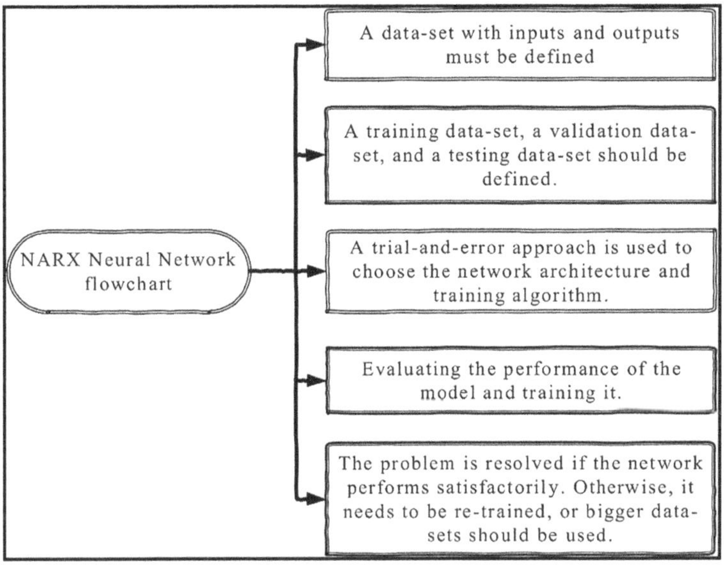

Fig. 2 Algorithm for *NARX* neural networks

- A data-set with inputs and outputs must be defined.
- A training data-set, a validation data-set, and a testing data-set should be defined.
- A trial-and-error approach is used to choose the network architecture and training algorithm.
- Evaluating the performance of the model and training it.
- The problem is resolved if the network performs satisfactorily. Otherwise, it needs to be re-trained, or bigger data-sets should be used.

3 Prediction of Indoor Temperatures with a Mathematical Model

Predicting indoor temperatures using a model is explained in this section. The mathematical model for indoor temperature is obtained by considering thermal convection and conduction equations:

$$\breve{\mathcal{T}}_{in}[n] = \alpha\left(\breve{\mathcal{T}}_{in}[n-1] + \frac{\Delta t}{\mathcal{C}}\left(\mathcal{P}(n-1) - \mathcal{U}\left(\breve{\mathcal{T}}_{in}[n-1] - \mathcal{T}_{out}[n-1]\right)\right)\right) + \breve{\beta}[n] \quad (2)$$

where

$\breve{\mathcal{T}}$ and $\mathcal{T}_{out}$ denote the indoor and outdoor temperatures that are estimated based on these values, respectively.

Δt shows the spam of the time.

$\mathcal{P}$ stands for the power of heating.

α and $\mathcal{U}$ are the parameters to be calculated.

$\beta(n)$ at time t represents the occupancy at that moment.

An occupancy estimate is fed into a model-based predictor in a learning-based simulation. It is conventional to select the occupancy profile constant $\beta(n)$ without learning. A *recursive least squares (RLS)* identification algorithm is used to identify the parameters of the thermal model (2). An identification algorithm is evaluated based on the *root means square (RMS)* criterion. As a brief introduction to RLS, here is what you need to know.

$$\begin{cases} \bar{G}(n+1) & = & \frac{1}{\hbar}\left(G[n]-\frac{G[n]\delta^T[n]G[n]}{\hbar+\delta^T[n]G[n]\delta[n]}\right) \\ e[n+1] & = & y[n+1]-\hat{\psi}[n]\delta[n] \\ \hat{\psi}[n+1] & = & \hat{\psi}[n]+G[n+1]\delta[n]e[n+1] \end{cases} \tag{3}$$

where G presets the gain, $\hbar$ denotes the forgetting factor, δ shows the observations and $\hat{\psi}$ is the estimated parameter. Measurement errors and identification errors are represented by e.

4 Model Predictive Control Based on Learning Concept

Model predictive control (MPC) is applied to weather and occupancy forecasts. Because MPC can handle control problems with constrained parameters, it is applied in many areas [21]. The optimal control action is obtained at each time instant by solving an optimal control problem. The *HVAC* system control plan is generated based on predicted weather conditions and occupancy profiles over time using MPC in the building temperature control. First, the *HVAC* system needs to be controlled to minimize energy consumption and satisfy comfort requirements. After the next time instant, the control algorithm is repeated with feedback information about the outputs and states of the building. A model predictive control approach based on learning is illustrated in Fig. 3.

Based on the value of (4), the MPC cost function penalizes deviations from optimum energy consumption and comfort levels.

$$J[n]=\sum_{\ell=0}^{\mathcal{N}}\|\overset{\varnothing}{\mathcal{T}}_{in}(n+\ell)-\mathcal{T}_d\|_{\mathcal{Q}}^2+\sum_{\ell=0}^{\mathcal{N}}\|\Delta\mathcal{P}(n+\ell-1)-\mathcal{T}_d\|_{\mathcal{R}}^2 \tag{4}$$

Here, $\mathcal{Q}$ and R represent the state and input weighting factors, respectively. Temperature $\mathcal{T}_d$ is the comfort set-point temperature, and $\mathcal{N}$ is the time horizon.

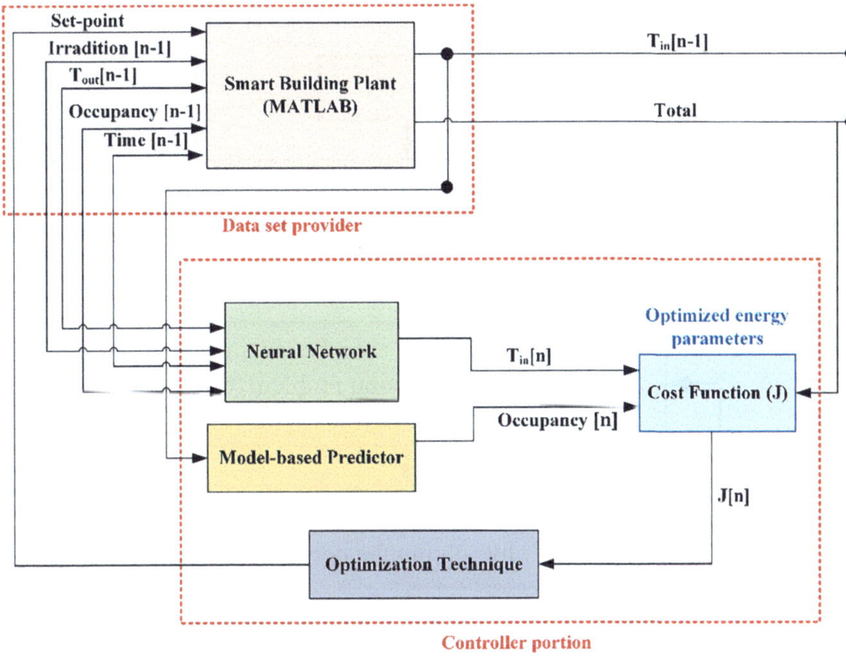

Fig. 3 Control of models based on learning (MPC)

Hence, *MPC* involves minimizing (4) subject to performance index (5), robustness performance (6), and limit restrictions (7). Model-based design is solely used in Eq. (6), while Eq. (5) incorporates learning.

$$\tilde{\mathcal{T}}_{in}[n]=\alpha\left(\tilde{\mathcal{T}}_{in}[n-1]+\frac{\Delta t}{\mathcal{C}}\left(\mathcal{P}(n-1)-\mathcal{U}\left(\tilde{\mathcal{T}}_{in}[n-1]-\mathcal{T}_{out}[n-1]\right)\right)\right)+\tilde{\beta}[n] \quad (5)$$

$$\bar{\mathcal{T}}_{in}[n]=\alpha\left(\bar{\mathcal{T}}_{in}[n-1]+\frac{\Delta t}{\mathcal{C}}\left(\mathcal{P}(n-1)-\mathcal{U}\left(\bar{\mathcal{T}}_{in}[n-1]-\mathcal{T}_{out}[n-1]\right)\right)\right)+\bar{\beta}[n] \quad (6)$$

With:

$$\mathcal{T}_{in}^{\min}\leq\bar{\mathcal{T}}_{in}[n+\ell]\leq\mathcal{T}_{in}^{\max} \quad (7)$$

$$\mathcal{P}_{in}^{\min}\leq\mathcal{P}_{in}[n+\ell]\leq\mathcal{P}_{in}^{\max} \quad (8)$$

For further understanding of the model of predictive control, the reader can see Table 3.

Table 3 Algorithm for model predictive control

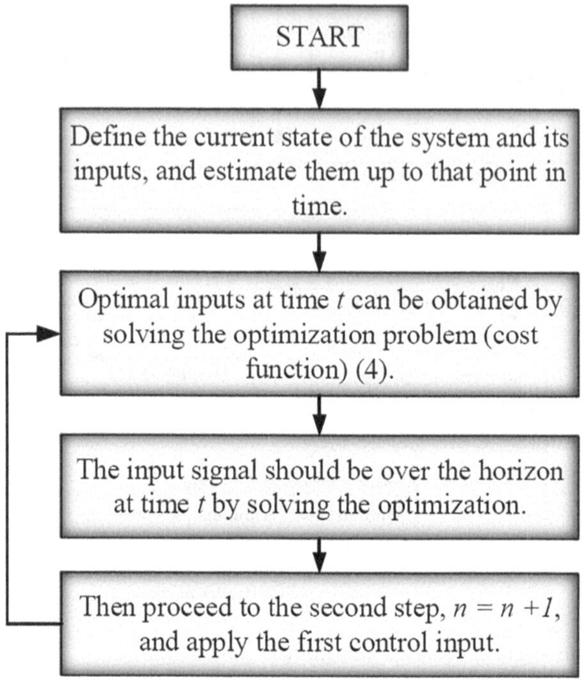

5 Results from Simulation

A comparison between the suggested conventional *MPC* with no learning and learning-based *MPC* is presented in this section. Each simulation is run six times per hour for 6 months. The building model was simulated by *MATLAB* over a 6-month period from July to December to provide the *ANN* data-set. Based on these assumptions, we can carry out the simulation.

The training of the data set in the form of gradient and variance is shown in Fig. 4. The difference between the actual occupancy profile and the occupancy predictions is shown in Fig. 5, which shows how the network responded to the actual occupancy profile.

A regression plot and performance plot are presented in Figs. 6 and 7, respectively, for the training, validation, and testing data-sets. A well-performing training model would have a regression value close to 1 and an *MSE* error around 2.2236×10^{-6}; i.e. a satisfactory training performance.

The error histogram is presented in Fig. 8. While autocorrelation and traditional correlation between input and error are presented in Figs. 9 and 10, respectively. Moreover, the conventional approach deviates much further from the comfort level than the proposed approach. It is even possible to break the minimum comfort level by using the conventional *MPC* approach.

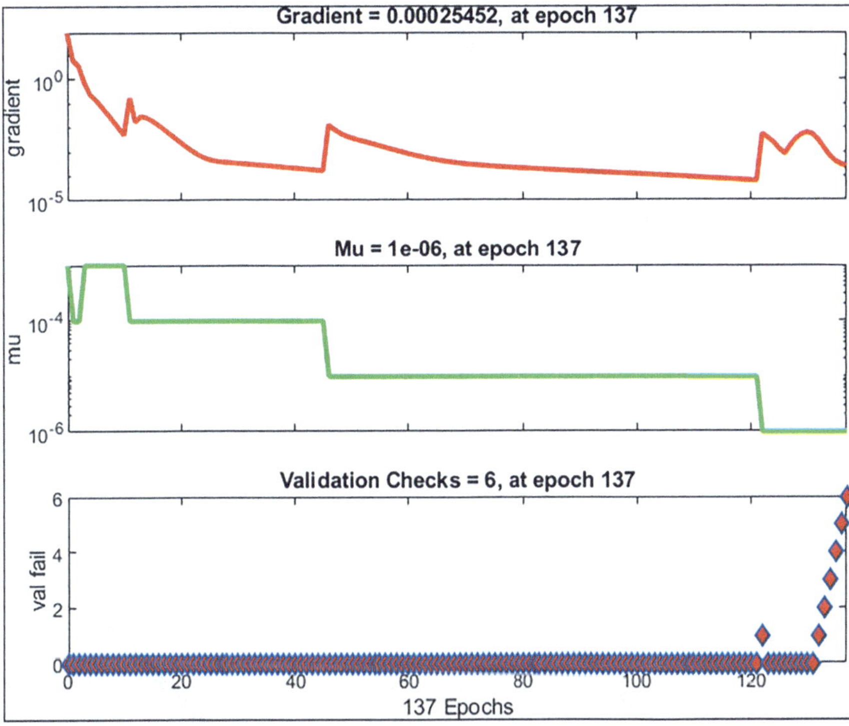

Fig. 4 Training of data-set in the form of gradient, and variance

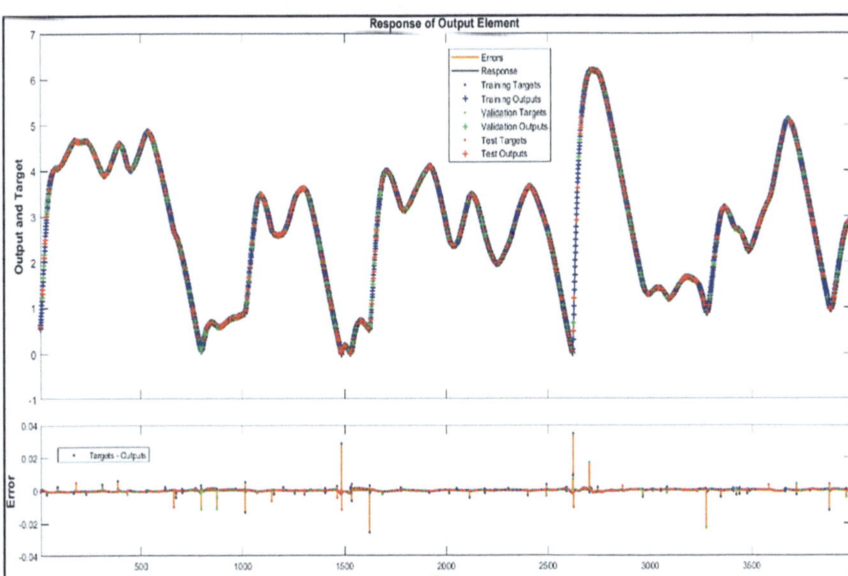

Fig. 5 Comparison of neural network output responses with targets

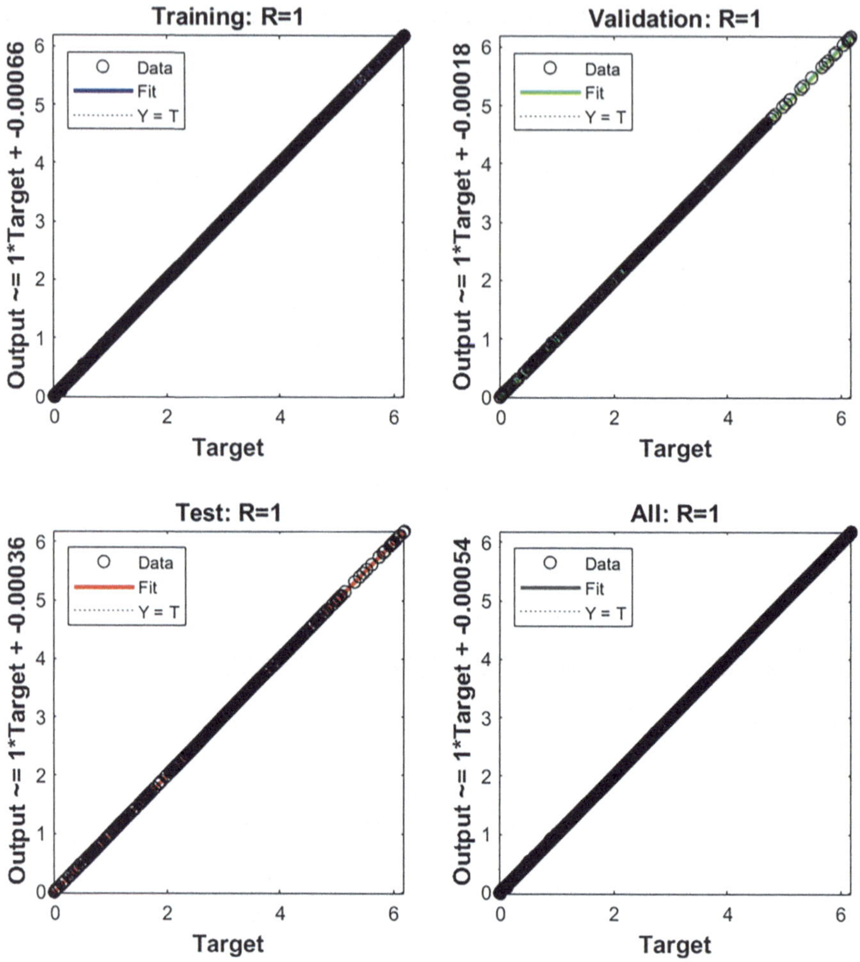

Fig. 6 Regression behavior of data-sets

Machine learning algorithms will be applied to demand forecasting in the future, a new mathematical model will be developed for a controllable load-curtailed penalty cost curve, and an IoT network for HVAC systems will be extended to increase energy efficiency and demand response. In order to optimize the air volume and handlers in different zones, occupancy data can be used to control the HVAC system. The behavior of occupants can also be predicted using machine learning methods. HVAC systems can be scheduled to pre-cool or pre-heat zones based on this information, resulting in significant energy savings of intelligent buildings.

The overall efficiency of smart building systems has risen significantly as seen by the outcomes of the implementation of machine learning algorithms, IoT integration, and occupant-centric data management. Our *MPC* model has been able to maximize *HVAC* system efficiency by integrating real-time occupancy data and

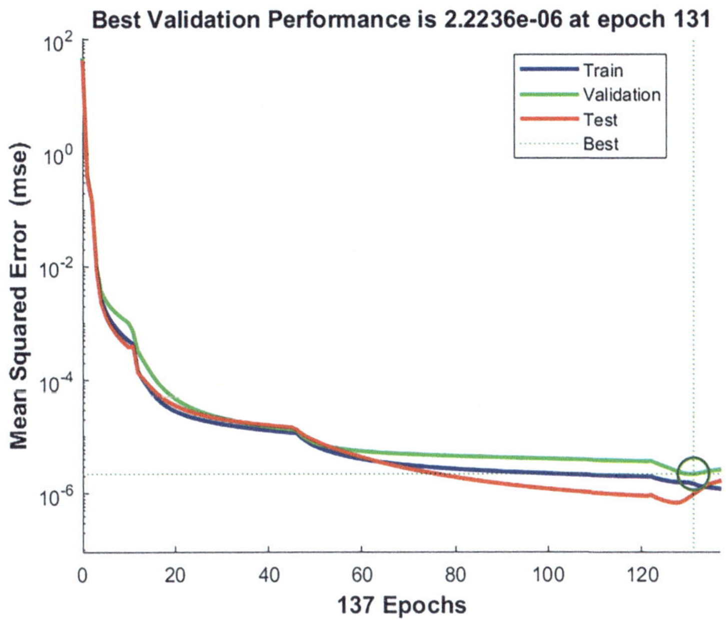

Fig. 7 Performance index of data-sets

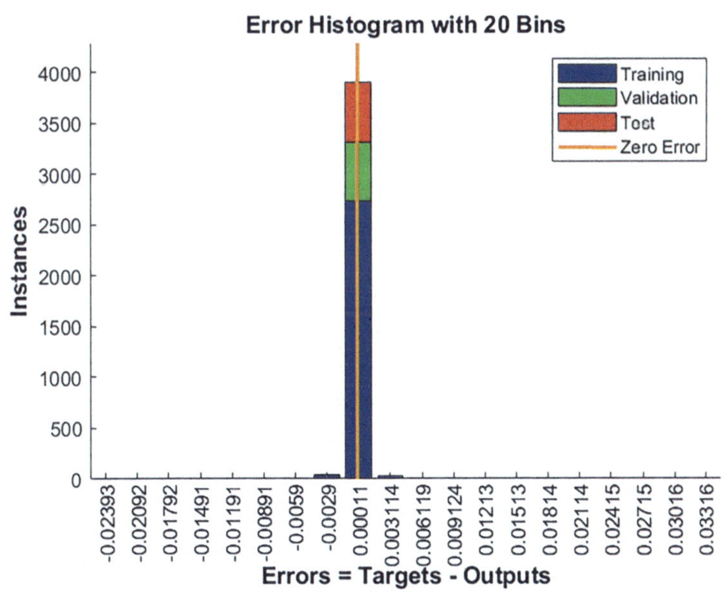

Fig. 8 Error histogram

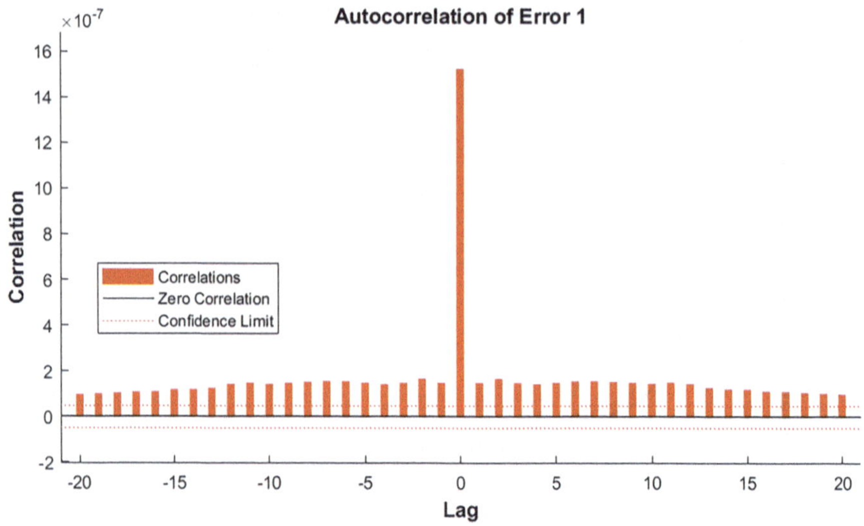

Fig. 9 Auto-correlation of data-set

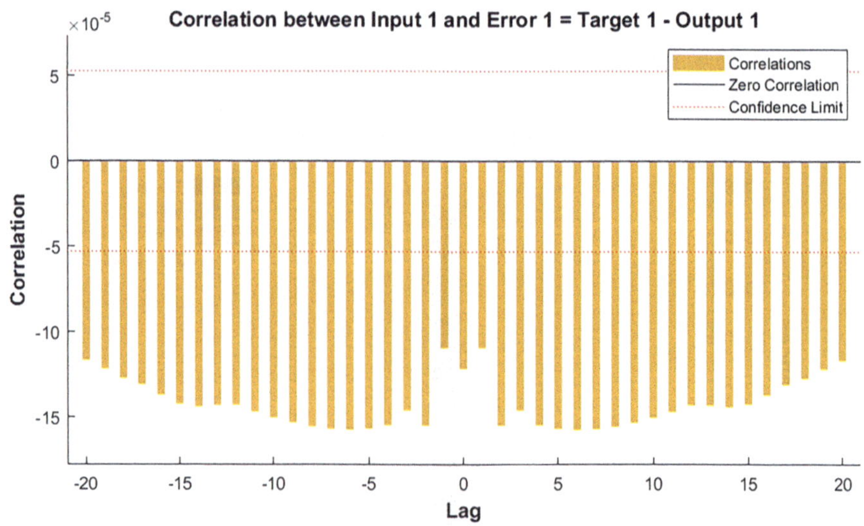

Fig. 10 Correlation between input and error

predictive analytics, providing a greater degree of control over building energy use. The following are the main results:

- **Increased Energy Efficiency:** The *HVAC* system can dynamically modify air volume and temperature in response to real-time demand thanks to the model, which uses occupancy data and machine learning techniques to forecast occupant behavior. By making sure that energy is only utilized when and where it is needed, this proactive strategy lowers energy waste, which is essential for attaining energy savings in smart buildings.

- **Optimized HVAC Scheduling:** Based on anticipated occupancy, zones can be pre-heated or pre-cooled thanks to the model's predictive capabilities. This scheduling minimizes the need for energy-intensive modifications during periods of peak demand by ensuring that zones are comfortable before occupants arrive. Preventive system adjustment is essential for preserving occupant comfort and energy efficiency.
- **Better Demand reaction:** Machine learning-powered demand forecasting enables the model to modify the building's energy usage in reaction to external grid conditions, lowering peak demand and enhancing the stability of the energy grid as a whole. The solution lowers the danger of demand spikes and curtailment fines by better managing HVAC loads, which lowers operating expenses.
- **Cost Reduction:** By minimizing needless heating or cooling and avoiding load curtailment penalties, the model can result in significant cost savings through demand response capabilities and *HVAC* system operation optimization. Because it prevents system misuse and allows for more focused servicing, the system's capacity to adapt based on real-time data also lowers maintenance costs.

Overall, this model achieves an effective smart building management by:

- **Machine Learning for Predictive Control:** To forecast occupant behavior and maximize energy use, machine learning models are developed on both historical and current data. For instance, based on changing variables like occupancy patterns and climatic conditions, reinforcement learning models can be used to continually learn the most effective control strategies for *HVAC* systems.
- **Real-Time Occupancy Data for HVAC Optimization:** The system can modify air flow, temperature, and humidity levels in response to the actual utilization of various building zones by using occupancy data as an input for *HVAC* management algorithms. This guarantees that energy is used where it is most needed and minimizes needless heating or cooling of vacant spaces.
- **Feedback Loops for Continuous Improvement:** The model includes feedback loops so that the system can keep improving its control techniques and predictions. In order to improve operational efficiency and occupant comfort, this continuous adaptation makes sure that the building's energy systems change in reaction to shifting occupancy patterns, environmental factors, and energy prices.
- **Balancing Cost and Energy:** The model balances energy consumption with external price signals to maximize the building's demand-side management through the creation of a configurable load-curtailed penalty cost curve. This guarantees that the building can adapt to changes in the external grid without incurring fines for high demand while preserving efficiency and comfort.

6 Conclusion

An *MPC* strategy based on learning is presented in this paper to control the thermal properties of an office building with a double story. The building's thermal model is nonlinear, uncertain, and strongly coupled, so predicting the parameters of the building is a crucial and challenging part of *MPC*. The mentioned issues can be addressed using *ANN* and model-based control approaches. Model-based controllers use *ANN* to generate occupancy profile predictions. The proposed approach is tested on an actual building using *MATLAB* software, which simulates the materials and components in real life. Several learning-based approaches were proposed and their results were shown.

Future research on intelligent building management with *Model Predictive Control (MPC)* for sustainable and green urban settings will concentrate on improving occupant comfort, maximizing energy efficiency, and integrating *AI-driven* strategies while reducing environmental impact. Better prediction models that can continuously learn from real-time data will be made possible by integrating machine learning into *MPC*, increasing the precision and flexibility of building controls. This will improve decision-making, reduce energy waste, and more accurately forecast patterns of energy consumption. The overall sustainability and livability of metropolitan areas will also be improved if smart buildings are integrated with citywide *IoT* systems for trash management, traffic control, and even public safety.

Ethics Approval This research does not involve any kind human or animal.

Competing Interests The authors declare that they have no competing interests.

References

1. Andoni M, Robu V, Flynn D, Abram S, Geach D, Jenkins D, McCallum P, Peacock A (2019) Blockchain technology in the energy sector: a systematic review of challenges and opportunities. Renew Sust Energ Rev 100:143–174
2. Kanakadhurga D, Prabaharan N (2021) Demand response-based peer-to-peer energy trading among the prosumers and consumers. Energy Rep 7:7825–7834
3. Mallada E, Zhao C, Low S (2017) Optimal load-side control for frequency regulation in smart grids. IEEE Trans Autom Control 62(12):6294–6309
4. Haller A, Janowicz K, Cox SJ, Lefrançois M, Taylor K, Le Phuoc D, Lieberman J, García-Castro R, Atkinson R, Stadler C (2018) The modular SSN ontology: a joint W3C and OGC standard specifying the semantics of sensors, observations, sampling, and actuation. Semantic Web 10(1):9–32
5. Jirkovský V, Obitko M, Kadera P, Mařík V (2018) Toward plug&play cyber-physical system components. IEEE Trans Industr Inform 14(6):2803–2811
6. Ding W (2023) Role of sensors based on machine learning health monitoring in athletes' wearable heart rate monitoring. HCIS 2023:13
7. Bisadi M, Akrami A, Teimourzadeh S, Aminifar F, Kargahi M, Shahidehpour M (2018) IoT-enabled humans in the loop for energy management systems: promoting building occupants' participation in optimizing energy consumption. IEEE Electr Magazine 6(2):64–72

8. Manikandan R, de Albuquerque VHC, Tiwari P, AlQahtani SA, Hossain MS (2022) Quality of service-aware resource selection in healthcare IoT using deep autoencoder neural networks. HCIS 12(36):1–16
9. Kathirgamanathan A, De Rosa M, Mangina E, Finn DP (2021) Data-driven predictive control for unlocking building energy flexibility: a review. Renew Sust Energ Rev 135:110120
10. Khelafa I, Ballouk A, Baghdad A (2021) Control algorithm for the urban traffic using a real-time simulation. Int J Electr Comput Eng 11:3934–3942
11. Zohrabi N, Shi J, Abdelwahed S (2019) An overview of design specifications and requirements for the MVDC shipboard power system. Int J Electr Power Energy Syst 104:680–693
12. Bukhari MM, Ullah SS, Uddin M, Hussain S, Abdelhaq M, Alsaqour R (2022) An intelligent model for predicting the students' performance with backpropagation neural network algorithm using regularization approach. HCIS 12:1–21
13. Anand V, Kadiri VL, Putcha C (2023) Passive buildings: a state-of-the-art review. J Infrastruct Preserv Resil 4(1):3
14. Ejidike CC, Mewomo MC (2023) Benefits of adopting smart building technologies in building construction of developing countries: review of literature. SN Appl Sci 5(2):52
15. Spazzafumo G, Raimondi G (2023) Economic assessment of hydrogen production in a renewable energy Community in Italy. e-Prime-Adv Electr Eng Electr Energy 4:100131
16. Chen X, Wang M, Wang B, Hao H, Shi H, Wu Z, Chen J, Gai L, Tao H, Zhu B (2023) Energy consumption reduction and sustainable development for oil & gas transport and storage engineering. Energies 16(4):1775
17. Gaur L, Singh G, Solanki A, Jhanjhi NZ, Bhatia U, Sharma S, Verma S, Petrović N, Muhammad FI, Kim W (2021) Disposition of youth in predicting sustainable development goals using the neuro-fuzzy and random forest algorithms. Human-centric Comput Inform Sci 11:1–19
18. Shin H-J, Cha H-S (2023) Proposing a quality inspection process model using advanced Technologies for the Transition to smart building construction. Sustainability 15(1):815
19. Loo BP, Wong RW (2023) Towards a conceptual framework of using technology to support smart construction: the case of modular integrated construction (MiC). Buildings 13(2):372
20. Khaleghi S, Karimi D, Beheshti SH, Hosen MS, Behi H, Berecibar M, Van Mierlo J (2021) Online health diagnosis of lithium-ion batteries based on nonlinear autoregressive neural network. Appl Energy 282:116159
21. Pinto G, Wang Z, Roy A, Hong T, Capozzoli A (2022) Transfer learning for smart buildings: a critical review of algorithms, applications, and future perspectives. Adv Appl Energy 5:100084

Microgrid Frequency Regulation Contribution of DFIG-Based Wind Turbine with Using ADRC Controller for Smart Energy Optimization

Mohssine Chakib, Najoua Mrabet, Ahmed El Akkary, and Ahmed Essadki

Abstract The frequency regulation in microgrids plays a key role in the integration of wind turbines, ensuring grid stability amid fluctuations in energy production. With the help of advanced control systems, microgrids can adjust frequency in real time, compensating for variations caused by the intermittent nature of wind. This enables more efficient and reliable management of wind energy in local grids, enhancing their resilience and optimizing performance. This paper aims to examine and evaluate the capability of a variable speed wind turbine equipped with a doubly fed induction generator (DFIG) to assist in microgrid frequency regulation. To achieve this contribution, the wind turbine system must inject an additional power into the electrical microgrid if required. To do this, both techniques of frequency control are employed: the droop control and the synthetic inertia control. The first one consists of adopting a deloading operating mode allowing the variable speed wind turbine (VSWT) to constitute an active power reserve. The second consists of using the kinetic energy stocked in the rotors of the wind turbine and the generator. The linear active disturbance rejection control (ADRC) strategy, along with a classical PI controller, has been applied to control the converter of the DFIG rotor side (RSC), allowing it to provide support to the microgrid system during frequency disturbances by adjusting rotor speed. The performance of these controllers is evaluated and compared through simulations achieved by using MATLAB/Simulink software.

Keywords Variable speed wind turbine system · DFIG · Droop control · Inertia control · Active disturbance rejection control

M. Chakib (✉) · A. Essadki
Electrical Engineering Department of ENSAM, Mohammed V University, Rabat, Morocco

N. Mrabet · A. El Akkary
LASTIMI Laboratory of Higher School of Technology, Mohammed V University, Rabat, Morocco

H. Kaddouri et al. (eds.), *Smart Cities for Sustainable and Urban Development*, Green Energy and Technology, https://doi.org/10.1007/978-3-031-99001-4_16

1 Introduction

The increasing integration of wind power in the electrical grids, which does not contribute to frequency control, creates new challenges in the management of power systems, particularly when the grid is fragile. This thesis aims to evaluate the performance and reliability of wind turbine frequency control at the scale of a power system [1].

First, the impact of a high wind power penetration rate on the allocation of the primary reserve and the dynamic behavior of the grid is analyzed. It is shown that the participation of wind turbines in frequency control becomes technically essential to ensure the stability of the power system from a certain penetration rate [2].

Currently, compared to other energy sources, wind power generation has grown significantly and evolved quickly worldwide. The wind turbine generators must therefore take part in frequency stabilization in the case of an imbalance between the electrical energy generated and that consumed when wind energy has strongly penetrated the electrical system (Electrical grids and microgrids). Since they can track the maximum power and easily and independently control the active and reactive power injected into the electrical grid via converters, variable speed wind turbines with DFIG are currently the most well-known and used wind turbines [2].

The doubly fed induction generator (DFIG) plays a crucial role in wind turbine technology by enabling efficient conversion of mechanical energy into electrical energy, while facilitating the adaptation of power generation to wind speed variations. It consists of two circuits: a stator connected to the power grid and a rotor powered by a power control system. This type of generator allows the wind turbine to operate at different speeds, optimizing power generation efficiency while maintaining grid stability [3]. With its ability to provide rapid dynamic response to wind speed fluctuations, it contributes to the flexibility and reliability of wind systems. The generator parametric uncertainty, which can be brought on by saturation, skin effect, overheating, and other causes, as well as the impact of speed variation-induced disturbance, can affect the DFIG. Therefore, the employed controller must be reliable and effective in order to get over these limitations [4].

With wind turbines becoming more and more present in the electricity grid, many traditional generating units, such as synchronous generators have been replaced by variable speed wind turbines such as the DFIG WT where the rotor speed of the DFIG and the frequency of the electricity grid are decoupled, thus these wind turbines are not affected by frequency changes [5]. As a result, when a frequency disturbance occurs, the overall inertia of the system decreases and its ability to maintain the frequency stability of the power system will be diminished. Therefore, many studies have addressed this problem by proposing a number of control strategies that enable wind turbine systems to participate in primary frequency regulation [5] [6].

The two primary frequency control techniques: primary droop controller and synthetic inertial control are particularly crucial to integrate variable speed wind turbines into frequency control [6]. In order to rectify the frequency drop during the first period, additional active power can be added through inertial control, which mainly exploits the kinetic energy stored in the wind turbine blades and the DFIG rotor [6]. In addition, primary droop control automatically provides active power support by releasing the conserved power of the wind turbine to adjust the system frequency deviation [7]. In order to obtain active power reserve, the wind turbine must operate under optimal conditions by adjusting the maximum power point tracking (MPPT) parameters [8]. Due to its many advantages, including high accuracy and good performance in system dynamic control, system stability, and ease of implementation due to reduced tuning parameters, the ADRC control strategy has been demonstrated in many studies to have very good robustness and efficiency in set point tracking [9]. The ADRC method typically uses an extended state observer (ESO) which acts as a state feedback controller and provides the system state information in real time and helps to estimate and suppress all internal and external disturbances that may affect the system [9, 10].

This paper is organized as follows: Section 2 presents the modelling of different components of a variable speed wind turbine based on DFIG. Section 3 presents the principle and structure of the ADRC control method and the application of this strategy to the rotor side converter to control the power of the DFIG generator. The deloaded operation of the wind turbine, the frequency control strategies (inertia and droop control) and their combination are presented in Sect. 4. Then, Section 5 presents the simulation results. Finally, Section 5 concludes and summarizes this research.

2 Modeling and Control of DFIG Wind Turbine

2.1 Wind Turbine

The wind turbine aerodynamic model is mathematically modeled by the following equation [11–13]:

$$P_{tb} = 0.5 C_p(\lambda,\beta) A \rho V^3 \tag{1}$$

Where, β is the pitch angle, A is the swept area of the rotor blades and V is wind speed in m.s-1.

The power coefficient C_p is expressed as [10, 14]:

$$C_p = 0.5176\left(\frac{116}{\lambda_i} - 0.4\beta - 5\right)\exp\left(-\frac{21}{\lambda_i}\right) + 68.10^{-4}\lambda \tag{2}$$

With:

$$\frac{1}{\lambda_i} = \frac{1}{\lambda + 0.08\beta} - \frac{0.035}{1+\beta^3} \tag{3}$$

λ represents the tip speed ratio:

$$\lambda = R\frac{\Omega_t}{V} \tag{4}$$

Where Ω_t is the turbine rotor speed and R the rotor radius.

2.2 DFIG

By considering the following assumptions [15]:

- The stator resistance R_s is neglected,
- The stator flux vector ϕ_s is oriented along the d axis.

This generator is modeled in d-q reference frame by the following equations [11, 15]:

$$\begin{cases} \phi_{sd} = \phi_s = L_m I_{rd} + L_s I_{sd} \\ \phi_{sq} = 0 = L_s I_{sq} + L_m I_{rq} \\ \phi_{rd} = L_m I_{sd} + L_r I_{rd} \\ \phi_{rq} = L_m I_{sq} + L_r I_{rq} \end{cases} \tag{5}$$

$$\begin{cases} V_{sd} = 0 = R_s I_{sd} + \dfrac{d\phi_{sd}}{dt} - \omega_s \phi_{sq} \\ V_{sq} = \phi_s \omega_s = R_s I_{sq} + \dfrac{d\phi_{sq}}{dt} + \omega_s \phi_{sd} \\ V_{rd} = L_r \sigma \dfrac{dI_{dr}}{dt} + R_r I_{rd} - g\omega_s L_r \sigma I_{rq} \\ V_{rq} = L_r \sigma \dfrac{dI_{rq}}{dt} + R_r I_{rq} - g\omega_s L_r \sigma I_{rd} + g\dfrac{V_s M}{L_s} \end{cases} \tag{6}$$

The powers expressions are:

$$\begin{cases} Q_s = -V_s \dfrac{M}{L_s} I_{rd} + \dfrac{V_s^2}{L_s \omega_s} \\ P_s = -V_s \dfrac{M}{L_s} I_{rq} \\ T_{em} = -p\dfrac{M}{L_s} \phi_s I_{rq} \end{cases} \tag{7}$$

2.3 Active Disturbance Rejection Controller

The linear ADRC strategy is a robust controller that includes an extended state observer (ESO) allows estimating and rejecting any unexpected disturbance.

The expression of the linear first-order ADRC control is obtained as [4, 16]:

$$\dot{y} = f_i(y,d,t) + bu_i \tag{8}$$

Where u_i is control input, d is the external disturbance, y is the system output, b is the control gain and f_i is the total disturbance.

The linear ESO, expressed by "Eq. 11", allows obtaining the estimates values of y and f_i [4].

$$\begin{cases} \dot{\hat{y}} = \hat{f}_i + \beta_{01}(y-\hat{y}) + b_0 u_i \\ \dot{\hat{f}}_i = \beta_{02}(y-\hat{y}) \end{cases} \tag{9}$$

Where, β_1 and β_2 present the gains of the linear observer "Eq.10":

$$\begin{bmatrix} \beta_{01} \\ \beta_{02} \end{bmatrix} = \begin{bmatrix} -2\omega_0 \\ \omega_0^2 \end{bmatrix} \tag{10}$$

With ω_0 the bandwidth of the observer and $\omega_0 = 3\text{~}7\ K_p$, where K_P is the ADRC control parameter [4] (Fig. 1).

The DFIG rotor currents can be written as follow:

$$\begin{cases} \dot{I}_{rd} = f_1(t) + bu_1 = \dfrac{1}{\sigma L_r} V_{rd} - \dfrac{R_r}{\sigma L_r} I_{rd} + \omega_r I_{rq} \\ \dot{I}_{rq} = f_2(t) + bu_2 = \dfrac{1}{\sigma L_r} V_{rq} - \dfrac{R_r}{\sigma L_r} I_{rq} - \omega_r I_{rd} - \dfrac{M\omega_r}{\sigma L_r L_s}\phi_{sq} \end{cases} \tag{11}$$

Where:

$$\begin{cases} f_1(t) = -\dfrac{R_r}{\sigma L_r} I_{rd} + \omega_r I_{rq} \\ f_2(t) = -\dfrac{R_r}{\sigma L_r} I_{rq} - \omega_r I_{rd} - \omega_r \dfrac{M}{\sigma L_r L_s}\phi_{sq} \\ b = \dfrac{1}{\sigma L_r} \quad ;u_1 = V_{rd} \quad ;u_2 = V_{rq} \end{cases} \tag{12}$$

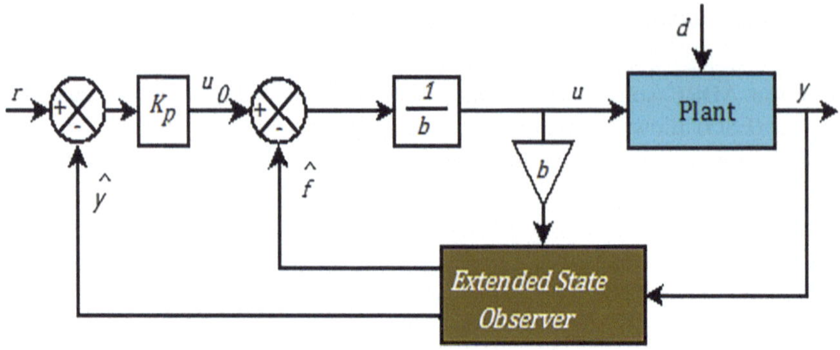

Fig. 1 Structure of ADRC control

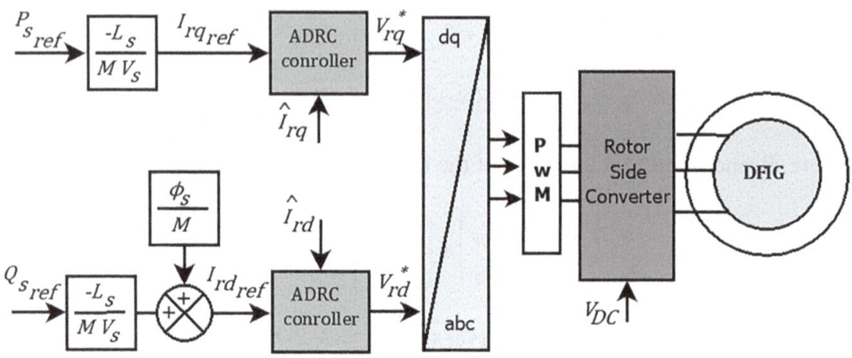

Fig. 2 Structure of DFIG power control by ADRC

The "Fig. 2" depicts the indirect control of the stator powers of DFIG through control of the rotor currents using ADRC control.

3 Primary Frequency Regulation Techniques

3.1 Inertial Control Strategy

The principle of inertial control is to obtain the electromagnetic torque, therefore the active power, from the frequency variation rate (Fig. 3). In the event of a frequency drop, the reference electromagnetic torque increases [17].

The new reference power is expressed as:

$$P_{s_{ref}} = P_{MPP} + \Delta P_{in} \tag{13}$$

Where ΔP_{in} is the support inertial [18, 19]:

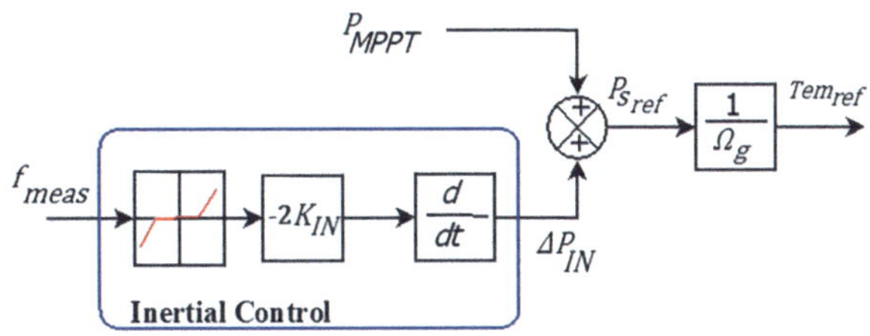

Fig. 3 Block diagram of inertial control strategy

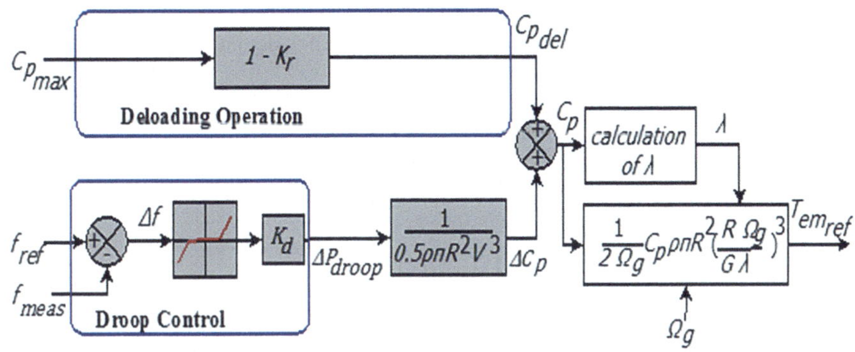

Fig. 4 Droop control technique

$$\Delta P_{in} = -2K_{in}\, f_m \frac{df_m}{dt} \tag{14}$$

3.2 Droop Control Technic

The principle of droop control consists of providing additional active power proportional to the frequency variations "Fig. 4" [17, 20]:

$$\Delta P_{dr} = K_d \Delta f = K_d \left(f_{ref} - f_m \right) \tag{15}$$

Where K_d is the droop gain:

$$K_d = \frac{P_r}{\delta . f_{ref}} \tag{16}$$

Where δ is the droop parameter, generally: $\delta = 4\%$.

The new reference of the electromagnetic torque become:

$$T_{em_{ref}} = \frac{0.5}{\Omega_g} C_p(\lambda, 0) \rho \pi R^2 \left(\frac{R\Omega_g}{G\lambda} \right)^3 \tag{17}$$

Where:

$$C_p = C_{p_{del}}(\lambda_{del}, 0) + \Delta C_p \tag{18}$$

$$C_p(\lambda, 0) = (1 - K_r)C_{p_{max}} + \frac{\Delta P_{dr}}{0.5\rho\pi R^2 V^3} \tag{19}$$

And the ratio speed λ is obtained by solving the expression of the power coefficient C_p at $\beta = 0°$, obtained by interpolation polynomial method.

The droop control can be activated when the frequency deviation is $|\Delta f| > 0.075$ Hz. The DFIG power reference is given as:

$$P_{s_{ref}} = T_{em_{ref}}.\Omega_g = P_{Del} + \Delta P_{dr} \tag{20}$$

4 Simulation Results and Discussion

We consider a small microgrid composed of the DFIG wind generation system and a thermal generator based on the reheat steam turbine. The fig. 5 illustrates the diagram of the adopted system frequency response. The simulations were carried out by using Matlab/Simulink.

The frequency response of the microgrid studied consisting of a thermal generator and the WTG is studied so as to compare the frequency control methods of the WTG. Four scenarios are simulated and tested:

- **S 1**: Wind turbine generator without any frequency control (only the thermal generator participates in frequency regulation).
- **S 2**: Activation only of the inertial control method of the WTG.
- **S 3:** Activation only of the droop control method of the WTG.
- **S 4**: Activation simultaneously of both controllers (droop and inertia control).

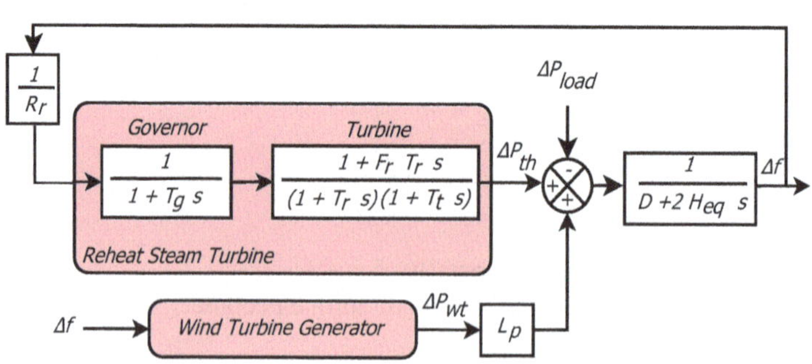

Fig. 5 Frequency response model of the microgrid studied

These four scenarios have been tested for various wind turbine power penetration levels (L_p = 10%, 20% and 30%) as depicted in "Fig .6" to "Fig .8", and, we consider that the system has undergone a sudden load increase at t = 10 sec of 0.1 pu.

In the **first scenario**, the "Figs. 6, 7 and 8" show that the frequency nadir and also the settling frequency decrease with increasing penetration level of wind generation because the WT power is not sensitive to the frequency variation. On the other hand, the ROCOF (Rate of Change of Frequency) increases when the penetration level of WTG increases.

In the **second scenario**, the inertial control brought a significant improvement of the frequency nadir especially in higher wind penetration as shown in "Figs. 6, 7

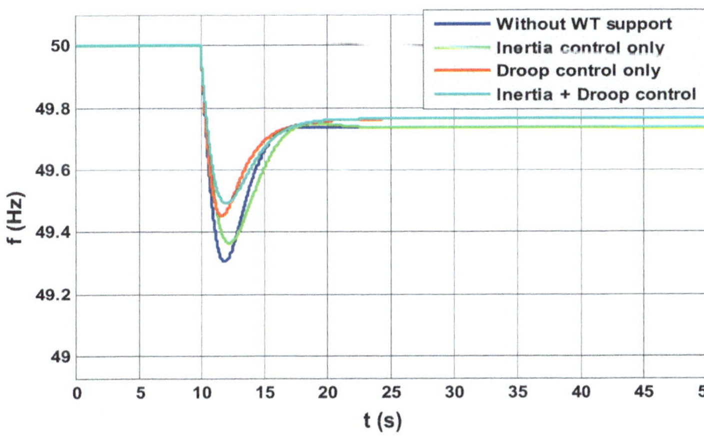

Fig. 6 Frequency response (L_p = 10%)

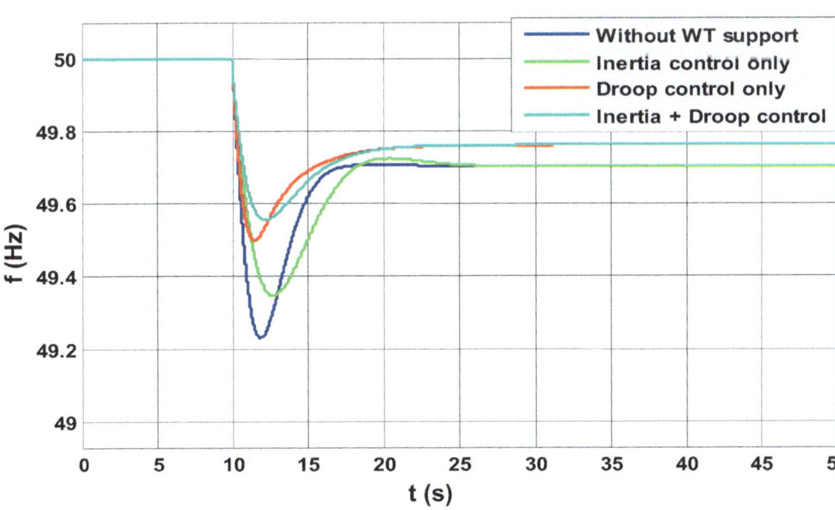

Fig. 7 Frequency response (L_p = 20%)

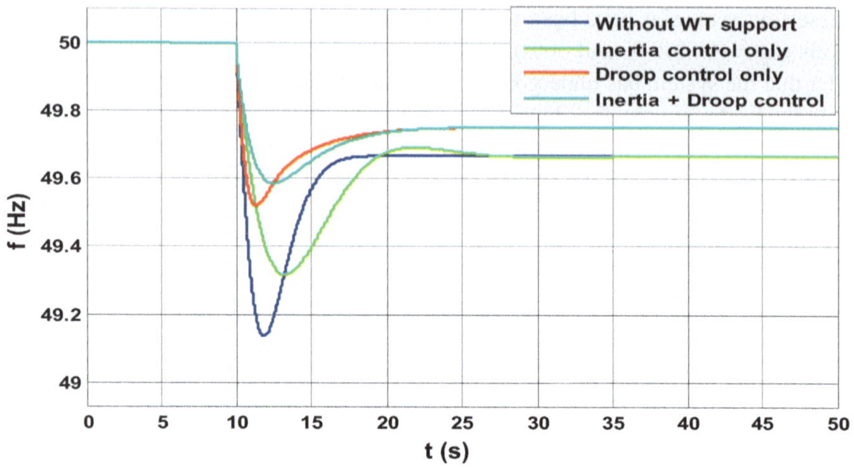

Fig. 8 Frequency response (L_p = 30%)

and 8". On the other side, in steady state, we note that there is no difference in the settling frequency compared of the first scenario.

In the **third scenario** (Droop control only), the wind turbine operates under the optimal operating point so as to create the power reserve. This power reserve is injected into the grid immediately when the disturbance occurs, which brings about a diminution in rotor speed.

In the **fourth scenario** (Combination of inertia and droop controls), the suggested combination of the both controllers offers a good and more improvements than the results of the previous scenarios for the three main frequency metrics, such as: a greater reduction in the drop in RoCoF, the best nadir and settling frequency "Fig. 8".

The combined control strategy, which integrates both inertial control and droop control, enhances the frequency response behavior by leveraging the benefits of each technique. Specifically, it takes advantage of the inertial response, which utilizes the kinetic energy stored in the rotating components, as demonstrated by researchers such as J. Morren [6] and S. A. Hosseini [10]. Additionally, it capitalizes on the droop control method, which relies on primary power reserve, as highlighted by scholars like K. V. Vidyanandan [19].

The simulation results of the powers of DFIG stator are respectively illustrated in "Fig. 9" and "Fig. 10". We can see that the responses obtained by the ADRC control follow with high performance their references with zero overshoot, a time stabilization, rise time and recovery time, shorter than PI controller.

The ADRC (Active Disturbance Rejection Control) method effectively addresses the limitations of traditional classical linear PI control method, as highlighted by several researchers as J. C. Moreno [16].

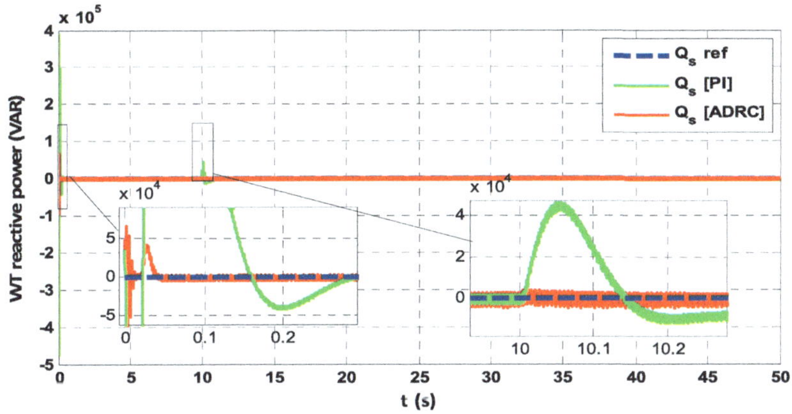

Fig. 9 Reactive output power of WTG

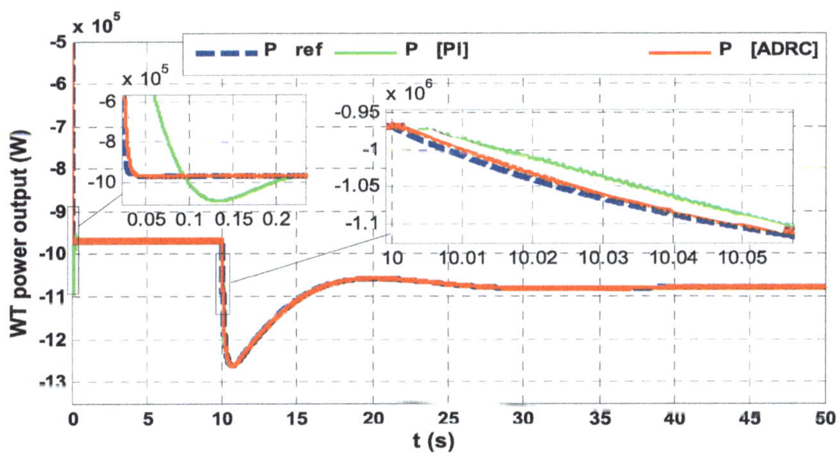

Fig. 10 Active output power of WTG

5 Conclusion

This chapter presented the DFIG-based WT system modeling and investigated various active power management techniques that enable DFIG-based WT generators to participate in primary frequency regulation via rotor speed adjustment. The wind turbine can quickly mimic the behavior of a synchronous generator through the inertia control approach. In order to provide additional active power support to the power system during frequency deviation, the controller instructs the wind turbine to slow down in order to extract some of its kinetic energy.

The wind turbine can provide power support for an extended duration through frequency droop control, which increases the resilience of the system. In order to provide a power reserve that could be automatically provided during frequency excursions, the wind turbine in this scenario must operate in deloaded mode.

The combined control (inertial and droop controls) improves the frequency response behavior. In this context, the wind turbine's deloading mode leads to an increase in rotor speed. As a result, the inertial response is significantly enhanced due to the additional kinetic energy stored in the rotating components.

Moreover, a comparison with the PI controller is presented in terms of monitoring the setpoint powers provided by the frequency regulation schemes in order to validate and evaluate the effectiveness of the ADRC control. According to the simulation results, the suggested ADRC controller outperforms the PI controller in terms of response time, overshoot and steady-state error. Moreover, this controller is robust to disturbances.

The perspective of this work is the use of active power control techniques to enable the GADA-based wind turbine to participate in secondary frequency regulation, thus ensuring the balance and stability of the electrical grid.

Appendix

See Tables 1 and 2

Table 1 Wind turbine parameters

Parameters	Values
Rated power	1.5 MW
Number of pole pairs	2
Rated stator voltage	398/690 V
Stator resistance	0.012 Ω
Rotor resistance	0.021 Ω
Stator inductance	0.0137 H
Rotor inductance	0.01367 H
Mutual inductance	0.0135 H
Frequency	50 Hz
DC link voltage	1200 V
Density of air	1.225 kg/m^3
Damping coefficient	0.0024
Gearbox ratio	90
Turbine diameter	70.5 m
Moment of inertia	10 $kg.m^2$

Table 2 ADRC parameters

Parameters	Values
Gain of controller	$K_p = 400$
Bandwidth of the ESO	$\omega_0 = 2000$
Gains of the ESO	$\beta_1 = 4000$
	$\beta_2 = 4{,}000{,}000$

References

1. Shahabi M, Haghifam MR, Mohamadian M, Nabavi-Niaki SA (2009) Microgrid dynamic performance improvement using a doubly fed induction wind generator. IEEE Trans Energy Convers 24(1):137–145
2. Molina-Garcia A, Muñoz-Benavente I, Hansen AD, Gómez-Lazaro E (2014) Demand-side contribution to primary frequency control with wind farm auxiliary control. IEEE Trans Power Syst 29(5):2391–2399
3. Fu Y, Zhang X, Hei Y, Wang H (2017) Active participation of variable speed wind turbine in inertial and primary frequency regulations. Electr Power Syst Res 147:174–184
4. Chakib M, Tamou N, Essadki A (2020) Comparative study of active disturbance rejection control with RST Control for variable wind speed turbine based on doubly fed Induction generator connected to the grid. Int J Intell Eng Syst 13(1):248–258
5. Wang Y, Wu Q, Gong W, Gryning MPS (2017) H∞ robust current control for DFIG-based wind turbine subject to grid voltage distortions. IEEE Trans Sustain Energy 8(2):816–825
6. Morren J, de Haan SWH, Kling WL, Ferreira JA (2006) Wind turbines emulating inertia and supporting primary frequency control. IEEE Trans Power Syst 21(1):433–434
7. Sati SE, Al-Durra A, Zeineldin HM, H EL-FT, El-Saadany EF (2023) A novel virtual inertia-based damping stabilizer for frequency control enhancement for islanded microgrid. Int J Electr Power Energy Syst 155(4):109580
8. Shu H, Wang G, Chen J, Ma H, He T (2024) Coordinated control of wind-storage combined with primary frequency regulation and variable coefficient based on wind speed and SOC. J Energy Storage 87(2):111356
9. Benbouhenni H, Bizon N, Mosaad MI, Colak I, Djilali AB, Gasmi H (2023) Enhancement of the power quality of DFIG-based dual-rotor wind turbine systems using fractional order fuzzy controller. Expert Syst Appl 238(2):121695
10. Hosseini SA (2023) Frequency control using electric vehicles with adaptive latency compensation and variable speed wind turbines using modified virtual inertia controller. Int J Electr Power Energy Syst 155(4):109535
11. Khan A, Aragon DA, Seyedmahmoudian M, Mekhilef S, Stojcevski A (2024) Inertia emulation control of PMSG-based wind turbines for enhanced grid stability in low inertia power systems. Int J Electr Power Energy Syst 156(1):109740
12. Mrabet N, Benzazah C, El Akkary A, Chakib M, Lahlouh I (2024b) Fine-tuning PI controller parameters for an efficient wind energy extraction system through mean squared error (MSE): a control system based on ant Colony optimization for maximum power point tracking (MPPT). In: Mediterranean Smart Cities Conference (MSCC), Martil-Tetuan, Morocco
13. Mrabet N, Benzazah C, El Akkary A, Chakib M (2024a) Enhancing wind power estimation for Dakhla, Morocco: a comparative exploration of four numerical methods leveraging the Weibull distribution. In: International conference on global aeronautical engineering and satellite Technology (GAST), Marrakesh, Morocco. IEEE
14. Singh B, Das S (2024) Adaptive control for undisturbed functioning and smooth mode transition in utility-interactive wind–solar based AC/DC microgrid. IEEE Trans Power Electron 39(11):15011–15020
15. Xue X, Huang J, Sang S (2024) Variable power tracking control strategy of doubly fed induction generators for fast frequency responses. Electronics 13(6):1071
16. Moreno JC, González J, Navarro A, Guzmán JL (2024) Tuning rules of PI+CI controllers for first-order systems. MDPI Act J 13(2):67
17. Liu Y et al (2024) Point-of-common-coupling and center-of-inertia frequency based auxiliary frequency control strategy for windfarm frequency support improvement. IEEE Trans Ind Appl 60(4):5255–5268

18. Wang Z, Wu L (2024) A review on control schemes of HVDC-integrated offshore wind farms for fast frequency support. In: 56th North American power symposium (NAPS), El Paso, TX, USA
19. Vidyanandan KV, Senroy N (2013) Primary frequency regulation by Deloaded wind turbines using variable droop. IEEE Trans Power Syst 28(2):837–846
20. Hu Z, Zhang K, Su R, Wang R (2024) Robust cooperative load frequency control for enhancing wind energy integration in multi-area power systems. IEEE Trans Autom Sci Eng 99:1–11

Design of an IoT-Based Wireless Smart Sensor Node for Enhancing Smart Building Energy Management Systems

Tebello Ntsiki Don Mathaba and Christian Nonyelum Ejike

Abstract Building lighting systems represent a significant economic and environmental burden due to their high energy consumption and operational costs. Additionally, the growing availability of data from multiple smart devices provides opportunities to optimize energy consumption in building lighting. Internet of Things (IoT) technologies offer an effective solution for managing energy use in building lighting systems by enabling real-time data gathering, transmission, and analysis. While initial studies have focused on communication technologies, limited attention has been given to the system's hardware and software design. This study presents a design for an IoT-based smart wireless sensor node system aimed at optimizing energy consumption in building lighting, reducing operational costs while maintaining lighting quality. The design follows a 3-tier architecture comprising perception, communication, and application layers. Each smart node includes sensors and actuators for real-time monitoring and control, with communication supported through LoRa and mobile network technologies to enhance system performance.

Keywords Energy management · Wireless sensor networks · Energy efficiency · Smart buildings

1 Introduction

Buildings, both residential and commercial, are responsible for approximately 30–45% of the total energy consumption worldwide [1–4]. The significant energy demand of building infrastructure makes it essential to improve energy efficiency,

T. N. D. Mathaba (✉) · C. N. Ejike
Postgraduate School of Engineering Management, University of Johannesburg, Johannesburg, South Africa
e-mail: tmathaba@uj.ac.za

H. Kaddouri et al. (eds.), *Smart Cities for Sustainable and Urban Development*, Green Energy and Technology, https://doi.org/10.1007/978-3-031-99001-4_17

particularly in light of the United Nations' Sustainable Development Goal (SDG) 11 and SDG 7 [5–7]. Effective building energy management is critical to reducing energy consumption, decreasing operational costs, and mitigating the negative environmental impact associated with buildings.

Baharudin et al. [8] highlighted that one of the key areas of building energy management that contributes significantly to energy consumption is lighting. It is estimated by Pompei et al. [9] that lighting represents 15–20% of the total energy usage in buildings. Optimizing lighting systems provides a considerable opportunity for energy savings and emissions reductions. Recent advances in Internet of Things (IoT) technologies have enabled new opportunities for smart lighting, allowing real-time data gathering, analysis, and control, thereby providing a basis for enhanced efficiency [10]. IoT-based solutions are proving to be effective in managing energy consumption across various applications, such as street lighting and building infrastructure [11]. A cloud-based approach to enhancing energy efficiency of building's heating, ventilation, and air conditioning (HVAC) system is demonstrated in [12] using an occupancy monitoring system that relies on multiple sensors and the LoRaWAN communication technology.

Building on the progress of IoT in energy management, smart wireless sensor nodes have emerged as a promising technology for optimizing building lighting systems. A Raspberry Pi and a web-based system are implemented in [13] to enhance equipment management and energy efficiency in laboratory environments. These nodes facilitate real-time monitoring and control of energy usage, offering a significant potential to reduce costs while ensuring the maintenance of lighting quality [14]. However, the majority of prior research in this domain has focused primarily on communication technologies or modelling energy usage, with limited attention paid to the integration of hardware and software systems within a complete architectural framework.

The objective of this study is to present a comprehensive design of an IoT-based smart wireless sensor node system specifically for building lighting management. The proposed system is designed to optimize energy consumption, reduce operational costs, and maintain the quality of lighting through a three-tier architecture comprising perception, communication, and application layers. The perception layer utilizes sensors and actuators to enable real-time monitoring and control of lighting systems, while the communication layer integrates low-power, long-range (LoRa) and mobile network technologies to enhance the performance of the overall system.

The rest of the paper is organized as follows: Section 2 presents a comprehensive literature review, focusing on different architectural frameworks for smart lighting systems, wireless sensor networks, and communication technologies used in building energy management. Section 3 outlines the system architecture and methodology used in developing the proposed IoT-based smart lighting solution. Section 4 presents the results and evaluation of the system's performance. Finally, conclusions and future research directions are presented in Sect. 5.

2 Backround

IoT allows the control and monitoring of devices using the internet. Furthermore, devices using IoT have now the ability to make decisions on their own, making them smart devices. Smart buildings use smart sensing and communication devices for the continuous monitoring and control of smart appliances and communication with the utility and the grid. Design of the IoT systems for smart buildings is guided by an IoT framework.

2.1 Three-Tiered Architecture

A standard framework cannot be determined and used as a blueprint for all IoT applications [15]. The number of layers in the framework depends on the perception of the user and the application of the system being designed. In Verma et al. [16], a seven-layer architecture is proposed which includes the application layer, application management and support layer, environment layer, network layer, communication layer, perception layer, and service layer. Saleem et al. [17] proposed a five-layers architecture to be provide effective energy management and data analysis. This proposed architecture ensures uses and IoT middleware module and smart metering hardware elements. However, Silva et al. [18] proposed to group certain layers to limit the framework to a three-layers namely as the application layer, the network layer, and the perception layer. This three-layer framework has been praised for its simplicity and backed in Burhan et al. [19], to be sufficient and suitable for smart building applications. Figure 1 depicts a three layer framework for a typical IoT network.

The perception layer is the section of the framework that deals with the identification and location of objects but also with the collection of data and information from the environment under investigation. The collection of data is mostly done using sensors and actuators. A sensor is a device that generates data from a measured input from the physical environment while an actuator is a device that

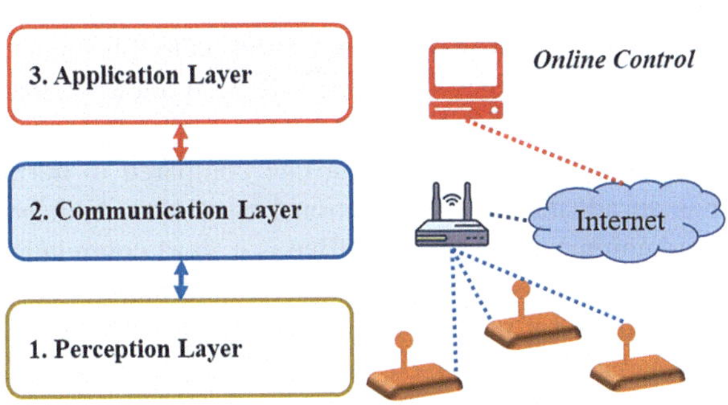

Fig. 1 Sensor-network based IoT network system

converts electrical signals into action. The devices found in the perception layer include different types of sensors that share information with each other and form a wireless sensor network (WSN), smart meters, radio identification (RFID) tags and readers, and video cameras. The perception layer generates a continuously large amount of data that need to be aggregated, analyzed, and transmitted in real-time.

The IoT system must therefore be equipped with a strong, reliable, and fast communication network for the transfer of the collected data. The network layer acts as a bridge between the perception and the application layer. Diverse communication technologies are available for IoT applications. It is impossible to find a single communication technology that can meet the requirements for the diverse applications deployed in the IoT [20]. Some of the most used wireless communication technologies used in IoT are presented in Fig. 4 based on the statistical review of 50 research papers. The review of main communication technologies used in IoT shows that Wi-Fi, with 44%, is the most used technology followed by ZigBee and Bluetooth with 21% and 14% respectively. On the other hand, Though BLE has an advantage over traditional Bluetooth because of its high data rate, both protocols suffer from low range communication issues (BLE), LoRa and RFID technologies are less used in IoT communication. The difference in the application of these communication protocols might be because the selection of a communication protocol depends on the type and size of the system but also on the nature of data to be shared. The above-mentioned communication technologies can be grouped according to their communication range. LoRa and Wi-Fi for long-range technologies, and ZigBee, Bluetooth, BLE, and RFID as short-range technologies. LoRa is a long-range data transmission network that uses very little power consumption [16]. It is suitable for both indoor and outdoor communication such as asset tracking, smart cities, and smart metering. However, its low data rate can be a limitation in situations of large systems.

On the other hand, Wi-Fi can be considered a privileged technology for large systems because of its relatively high data rate and bandwidth [21, 22]. Wi-Fi is appropriate for remote smart home and appliance control. Nevertheless, this protocol has a very high-power consumption and short communication range. Bluetooth and BLE technologies are widely used for IoT-based commercial devices such as smartwatches and smart appliances. They are perfect for close-distance communication. However, Bluetooth requires user intervention to reconnect from one gateway to another [23]. BLE is considered to be a solution to this issue [20, 24]. In the study of Hussain et al. [25], they designed a BLE framework that allows BLE devices to automatically pairing with nearby gateways without any user involvement. Furthermore, BLE has a higher data rate compared to normal Bluetooth. Though BLE has an advantage over traditional Bluetooth, both protocols suffer from low-range communication issues. ZigBee is a short communication network with relatively fast communication and low-power consumption [26]. Its long battery life makes it suitable for indoor control. Nevertheless, ZigBee has a low data rate and requires a large network if the user wants to achieve long-range communication. Lastly, RFID is suitable for real-time assets tracking and monitoring with a communication range of up to 100 m [27]. However, the implementation of RFID systems requires the deployment of several antennas and readers which makes it costly. Also, RFID systems a prone to interference from other devices such as cell phones.

According to Jia et al. [28], the application layer of the IoT infrastructure is responsible for handling data management and processing functions. Effective data management is crucial in IoT systems, given the vast amounts of data generated in the perception tier. This highlights the importance of efficient data handling mechanisms. The application layer possesses intelligence that analyzes and converts collected data into user-understandable information. This layer also features a user interface, enabling users to interact with the system. For instance, users can monitor and control their smart devices through this interface [15].

Protocols play a vital role in IoT systems, enabling the organized and meaningful exchange of data. A comparative analysis of primary protocols used in the IoT application layer by Swamy et al. [29] reveals several key protocols. These include Message Queue Telemetry Transport (MQTT), Advanced Message Queuing Protocol (AMQP), Constrained Application Protocol (CoAP), Representational State Transfer (REST), Extensible Messaging and Presence Protocol (XMPP), and Data Distribution Service (DDS). Their review of perception layer protocols reveals that CoAP and MQTT are the most frequently employed protocols. They account for 27% and 24% of usage rates, respectively, followed by REST, XMPP, and AMQP, with 14% usage rates. The least utilized protocol is DDS, with an 8% usage rate. This variation in protocol usage highlights the need for standardized communication protocols in IoT systems.

CoAP, inspired by HTTP, is a widely implemented protocol due to its similarities with HTTP. As a fast and asynchronous communication protocol, CoAP utilizes the User Datagram Protocol (UDP) to establish secure connections between endpoints. This enables efficient communication between devices. In contrast, MQTT is a machine-to-machine (M2M) communication protocol primarily employed for constrained devices and unreliable networks. MQTT relies on a broker model, where publishers collect and send data to subscribers via a mediation layer called the broker. This protocol ensures message reliability by supporting Quality of Service (QoS) levels. As noted by various studies, MQTT's features make it a popular choice for IoT applications.

The XMPP protocol facilitates communication and file transfer among nodes using XML technology, enabling real-time messaging, group chat, and the transfer of multimedia data. XMPP offers extendibility, adaptability, and uninterrupted communication between clients due to multiple servers. However, this protocol is limited by its server capacity for communication and lacks Quality of Service (QoS) and end-to-end encryption. These limitations highlight the need for further development of XMPP protocol.

The widespread adoption of IoT devices has significantly escalated security concerns, with vulnerabilities present at all levels of the IoT architecture [30–32]. As a result, ensuring the security and privacy of IoT systems has become a critical imperative. To address these concerns, IoT systems must incorporate essential security and privacy features [30], including confidentiality, integrity, availability, and Identification.

- **Confidentiality**: protects the disclosure of sensitive information by restricting access to authorized users or devices.
- **Integrity**: ensures that data transmitted over IoT networks remains accurate and unmodified.
- **Availability**: ensures uninterrupted access to usable data whenever needed, by designing IoT systems with redundancy and failover capabilities to minimize downtime and ensure continuous operation.
- **Identification**: prevents un-authorized devices or applications to connect to the IoT system.

Smart sensors in the perception tier of the IoT architecture, generate a large amount of data. However, these sensors nodes have little storage and computing resources for executing machine-learning algorithms and advanced data analytics. Cloud servers are usually employed to perform such tasks. However, they are most of the time located too far away from the sensor making it difficult to collect and process data and respond to demand on time. Fog computing is incorporated into IoT to improve information dissemination and computing capacities by allowing the processing of information near the network edge. The fog computing architecture in IoT comprises the cloud layer, fog layer, and device layer. The cloud layer evaluates and analyses the information received from the fog layer. It has processing capabilities such as a data centre, cloud server, or operation server. It can be accessed anywhere and anytime by subscribers through the internet. The fog layer serves as a critical intermediary, comprising network tools and fog devices that possess robust processing capabilities. These fog devices temporarily cache data gathered from IoT nodes, enabling real-time evaluation and processing. At the edge of the network, the device layer consists of a diverse array of devices that sense, gather, and transmit data to the fog layer for further processing. Examples of these devices include RFID tags, sensors, wearable devices such as sports watches, video cameras, and smartphones, among others [33]. Fog computing allows the processing of data wherever it is most appropriate and as quick as possible. It reduces the amount of data sent to the cloud, improves the system's response time, and conserves the network bandwidth.

2.2 *Smart Sensors*

Smart sensors are sensors with wireless communication facilities, data processing, and storage. They are capable of performing auto-diagnosis of the network and can take sometimes decisions autonomously [34]. Figure 2 depicts a general block schematic representation of a typical smart sensor node in a wireless network.

Smart sensors are mainly made of three modules: the sensing module, the processing module, and the network communication module. They are used to perform

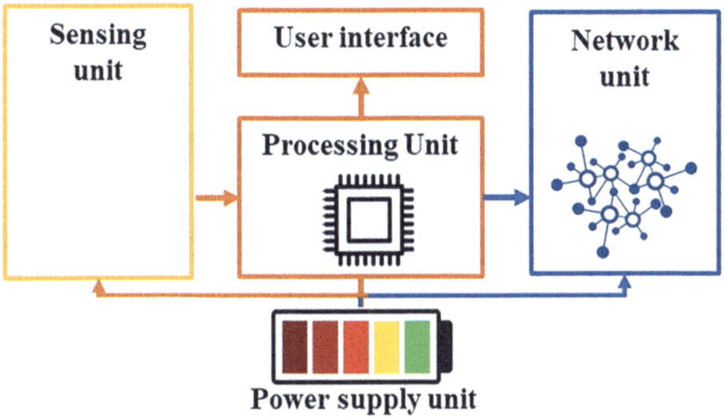

Fig. 2 Smart sensor schematic

accurate sensing, calibration, decision making, and data transmission. The sensing module contains one or more transducers that are used to detect changes in a certain environment. Transducers are devices that convert real-world properties such as temperature and pressure into electrical signals or vice versa. The generated signals are an interpretation of the measured parameters. The signals are then passed to the processing module for further analysis. The processing module consists of sensor error correction, calibration, multiplexer, memory, processor, analog-to-digital converter (ADC), and digital-to-analog converter (DAC) units. Each unit has a specific task in the processing and storage of the collected data. The analog signals are converted into digital form by the ADC before feeding it to the processor. The processor unit processes the data using the preloaded metadata and the application's intelligent algorithms. By using sensor error correction and calibration units, the processing module ensures that the data produced is accurate and reliable. External parameters such as noise, capable of affecting the data, are attenuated or removed in the processing unit. Finally, a network communication module is used to share the processed data to a management system or in a sensor network

The integration of sensors within Internet of Things (IoT) systems necessitates the development of wireless sensor networks (WSNs) capable of efficiently managing data transfer and communication [35]. Traditionally, users bear the responsibility of purchasing, deploying, and maintaining WSNs, which can be cumbersome and costly. To alleviate these burdens, the sensor cloud architecture has been proposed, shifting associated implementation complexities and financial expenditures away from users. Typically, communication between smart sensors and other system components occurs through wired (serial or Ethernet) or wireless (Wi-Fi, Bluetooth, or BLE) connections, as noted in research on smart sensor technologies [36]. A review of the main communication technologies used by smart sensors is presented in Fig. 3.

ZigBee and Wi-Fi communications are the prime choices for communication within smart sensor networks mainly due to their long-range of communication. Furthermore, ZigBee's best qualities include its low cost, low power consumption, and long battery life. Wi-Fi, on the other hand, is famous for its simplicity in

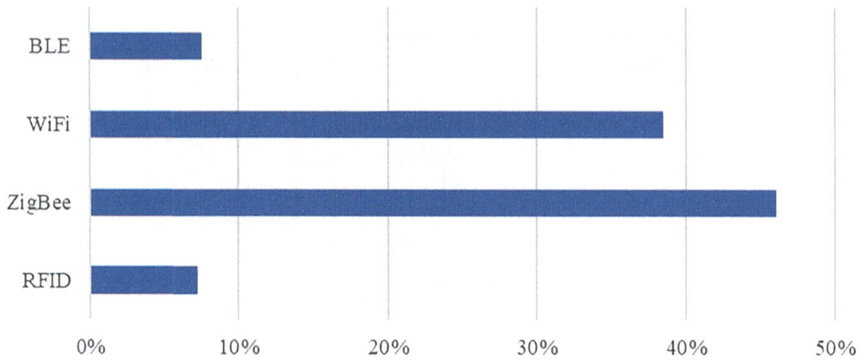

Fig. 3 Smart sensors communication network

implementation and high data rate. However, ZigBee's low data rate is a major disadvantage for this technology whereas Wi-Fi suffers from its high power consumption. The other smart sensor communication technologies are RFID and BLE. These are low power consumption technologies mostly used for close-range data communication. RFID and BLE have a relatively good data rate transfer compared to ZigBee technology. Nevertheless, both technologies are low range which make them less suitable for smart home applications.

The sensor cloud architecture is a multi-layered framework comprising the sensor provider layer, data management layer, and consumer layer. Within this architecture, sensor cloud service providers (SCSPs) facilitate the rental of wireless sensor network (WSN) data from multiple sensor owners, subsequently providing the aggregated information to end-users [37]. This paradigm is characterized as Sensors-as-a-Service (Se-aaS), enabling users to access multiple sensing devices and their associated data through cloud-based infrastructure [38].

Sensors play an important role in smart buildings, facilitating the collection of data from devices and environments. The implementation of reliable sensing systems is crucial for supplying control systems with accurate and real-time information, thereby regulating energy usage in accordance with multi-data reception. A diverse range of sensors is available for smart building applications, providing systems with superior intelligence capabilities, such as smart grids, smart homes, and driverless vehicles [39]. Sensor selection is contingent upon utility and economic factors, including purchasing cost, installation, and maintenance costs, as well as parameters such as sensitivity, accuracy, connectivity, and power.

The energy consumption of sensors is a significant concern, as it affects both system energy consumption and sensor lifespan. Research such as that of Lu et al. [40] has highlighted the impact of sensor errors on smart building systems, underscoring the potential for energy waste resulting from sensor faults. They broadly categorize sensor faults into hard failures and soft errors.

The occupancy level of a building directly impacts the operation of HVAC systems, as occupants generate heat and CO2. Obtaining occupancy information can help control cooling demand, improving HVAC system energy efficiency. Wang and Wang [41] have demonstrated the efficacy of predicting cooling demand through occupancy density measurements. By leveraging environmental information from building management systems and online weather systems in conjunction with digital visual camera data, building managers can optimize HVAC system performance and reduce energy consumption.

Smart energy meters are another critical component of building energy management systems. These meters facilitate the measurement and communication of energy flow and consumption information to respective users. Furthermore, these meters possess the capability to monitor and control home appliances in accordance with individual consumer preferences, as noted by Sun et al. [42]. Communication between smart meters and smart appliances occurs wirelessly via home area networks (HANs), while communication with energy providers is facilitated through internet connectivity [43]. The functional scope of smart meters encompasses a range of features, including remote metering, data recording, dynamic pricing, two-way communication, data storage, monitoring, and appliance control [42, 44, 45]. This comprehensive suite of functionalities enables smart meters to play a pivotal role in optimizing energy management and consumption.

2.3 Application of Smart Sensors in BEnMS

The increasing demand for energy efficiency in buildings has led to a growing focus on optimizing the performance of Heating, Ventilation, and Air Conditioning (HVAC) systems. These systems are among the largest energy consumers in buildings, accounting for a significant portion of total energy expenditure. To address this issue, researchers have explored the application of Internet of Things (IoT) technologies to optimize HVAC system performance. A notable innovation in indoor air quality monitoring was achieved by Kumar et al. [46], who developed a battery-free smart sensor system. This system leverages a modified radiofrequency identification (RFID) tag as a smart sensor and a radio frequency (RF) energy harvester to monitor environmental parameters in indoor settings. By eliminating the need for battery replacement, this system successfully reduces maintenance costs and minimizes human intervention.

Building on this concept, Zhang et al. [47] developed an indoor thermal control model that leverages smart thermostats and cloud-based data analytics, while Carli et al. [48] conducted a study that identified key parameters for sensing in buildings to achieve energy efficiency. These parameters encompass a range of factors, including air quality, lighting levels, thermal comfort, water management, cooking gas management, and fire detection. By monitoring these parameters, buildings can optimize their energy consumption and create a more sustainable and comfortable environment. Figure 4 illustrates the distribution of the most commonly monitored variables by smart sensors in building energy management systems, providing insight into the key factors that influence energy efficiency in buildings.

The most commonly monitored variables in buildings are temperature, occupancy, lighting and current, accounting for approximately 27%, 23%,18% and 14%

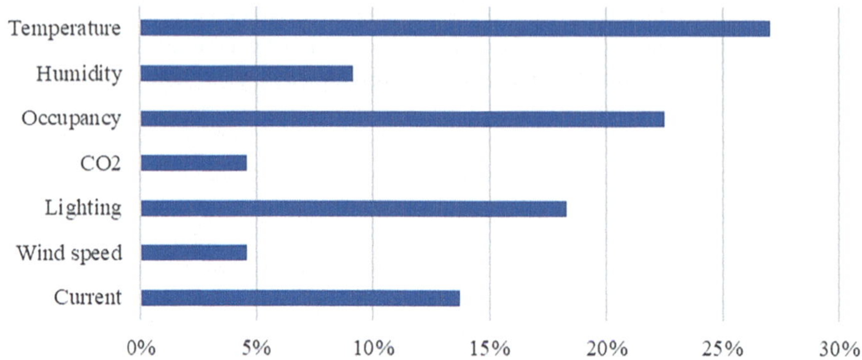

Fig. 4 Most sensed variables by smart sensors in building energy management studies

of sensed parameters, respectively. These metrics underscore the significant energy consumption attributed to HVAC and lighting systems, positioning them as primary targets for optimization in energy efficiency initiatives. However, while energy management remains a critical priority, comprehensive building management necessitates concurrent attention to occupant safety, particularly fire risk mitigation.

Fire safety constitutes a paramount concern in built environments, as fires—often precipitated by electrical faults, malfunctioning appliances, unattended ignition sources, or inadequate electrical grounding—pose substantial threats to both property and human life [49]. Modern advancements in sensor technologies have enabled robust fire detection systems, which integrate heat, gas, flame, and smoke sensing mechanisms to enhance early warning capabilities [49]. Among these, thermal detection employs two principal methodologies: heating elements embedded in fire sensors to monitor ambient temperature fluctuations, and infrared (IR) cameras to identify abnormal heat signatures. Concurrently, gas sensing technologies play a pivotal role by quantifying carbon dioxide (CO2) concentrations, a key combustion byproduct. Elevated CO2 levels correlate with oxygen depletion, necessitating rapid detection to facilitate timely evacuation and intervention.

Despite their utility, conventional gas sensors exhibit limitations in selectivity and cross-sensitivity, often leading to false alarms. To mitigate these challenges, smart gas sensing technologies have emerged, leveraging the distinct chemical properties of target gases to improve detection accuracy [34]. Nevertheless, environmental variables such as temperature and humidity remain critical confounding factors. Temperature fluctuations influence gas adsorption rates on sensor surfaces, while humidity modulates gas-sensor interactions, collectively impacting measurement reliability [34]. These constraints highlight the need for adaptive calibration protocols and multi-sensor fusion approaches to enhance robustness in dynamic building environments.

By integrating these advanced sensing frameworks—spanning energy management and fire safety systems—building operators can optimize both resource efficiency and occupant protection. The synergy of these technologies not only reduces energy consumption but also establishes safer indoor environments, underscoring their dual role in modern building management models.

3 Methodology

To achieve the objectives of this paper, the system is conceptualized based on the development and evaluation of an experimental prototype system. This paper is based on applied research that involves quantitative data analysis derived from an experimental case study, focusing on the development of an IoT-based smart lighting control system. The proposed design follows a three-tier architecture comprising perception, communication, and application layers. Each smart node integrates sensors and actuators for real-time monitoring and control, utilizing LoRa and mobile network technologies to enhance system performance. For this reason, the method adopted here is a **Design Science Research (DSR)** approach.

Design Science Research (DSR) is a practical research approach that emphasizes the creation and evaluation of artifacts to address real-world problems (Baskerville et al., 2018). In this study, DSR serves as the guiding framework for developing an experimental prototype system that embodies the proposed IoT-based smart lighting control solution. The DSR steps employed are outlined in Table 1 from start to finish:

Table 1 The Design Science Research Phases and how it is achieved

DSR Phase	Step	How it is achieved
Phase 1: *Problem identification*	**1. Problem identification**	In Sect. 1 and 2, a theoretical assessment of the literature landscape was explored. Challenges in building lighting control systems were identified and established.
Phase 2: *Design and development*	**2. Hardware design and architecture**	In Sect. 3, we outlined the design of a three-tier architecture that will be used for the system development: • **Perception layer:** Integrate LDRs for lux monitoring and stepper motor for blinds controller. • **Communication layer:** Use ESP32 with LoRa communication. • **Application layer:** Design user interface for monitoring and control.
	3. Algorithm development	Created algorithms for data collection, processing, and decision-making based on sensor inputs; implement control strategies for dynamic lighting adjustments.
	4. Prototype development	Assembled the hardware components (ESP32, LDRs, stepper motor, switch, communication modules) and implement UART communication between ESP32 end node and the coordinator node.
	5.Configuration of communication modules	Set up LoRa communication for data transmission to a central server; and ensured mobile network integration for remote access.

(continued)

Table 1 (continued)

DSR Phase	Step	How it is achieved
Phase 3. *Evaluation*	**6. Experimental setup**	Should involve the deployment of a prototype in a controlled environment to simulate real-world conditions; here will establish baseline measurements for energy consumption and lighting quality. This has not been done in this paper but will be demonstrated in future research work
	7. Data collection and evaluation	Should include the collection of quantitative data on energy savings, operational costs, and lighting quality; it should also involve the evaluation of results using statistical analysis against predefined metrics This has not been done in this paper but will be demonstrated in future research work
	8. Assessment of measurement errors	This step includes analyzing measurement errors using statistical methods to ensure data accuracy; here, researcher may also adjust algorithms or configurations based on findings This has not been done in this paper but will be demonstrated in future research work
Phase 4: *Final Evaluation & Iterative Refinement*	**9. Iteration and refinement**	This phase should include refining the prototype design based on evaluation outcomes; analyse, discuss, and present the refined results. This has not been done in this paper but will be demonstrated in future research work
	10. Final evaluation	Finally, the researcher should conduct a comprehensive assessment of the final prototype against initial objectives; prepare recommendations for future research or practical implementations

4 Results and Analysis

This section presents the comprehensive outcomes of the Design Science Research (DSR) methodology, delineating the artifact's conceptual development through a structured exposition of the problem identification, architectural design, operational algorithms, and critical configuration parameters.

4.1 Problem Identification

The proposed system design leverages a 3-tier IoT design framework targeting modern multi-storey building environments with established internet connectivity via Wi-Fi. The comprehensive IoT system is strategically designed to address lighting monitoring and control applications across both exterior and interior architectural spaces. Critical design challenges emerge from the potential deployment of end-nodes in environments characterized by inconsistent electrical infrastructure.

These constraints necessitate the development of energy-efficient node configurations capable of operating on battery power. Consequently, the research prioritizes low-power communication technologies as a superior alternative to energy-intensive Wi-Fi connections.

To mitigate the energy consumption challenges, the proposed architecture introduces a local LoRa-based network topology, enabling efficient relay of energy monitoring and control messages among distributed nodes. This approach addresses the dual imperatives of operational reliability and energy conservation in complex building management systems. In addition, the strategic deployment of LoRa-based communication infrastructure represents a methodical response to the identified systemic constraints, offering a technically robust and energy-optimized solution for IoT-enabled building management.

4.2 Architectural Design Framework

Given the analysis of the problem, the proposed systems should have made up of smart sensor nodes with the ability to, sense ambient lighting conditions, control lighting output, and communicate via both LoRa and WiFi technologies. The block diagram of the proposed design of the smart node is depicted in Fig. 5.

The proposed architectural design framework addresses the complex challenges of IoT-based building lighting management through a strategic 3-tier system architecture. The framework integrates LoRa-enabled sensor nodes, a robust local network infrastructure, and a centralized management system to provide comprehensive, energy-efficient environmental monitoring and control.

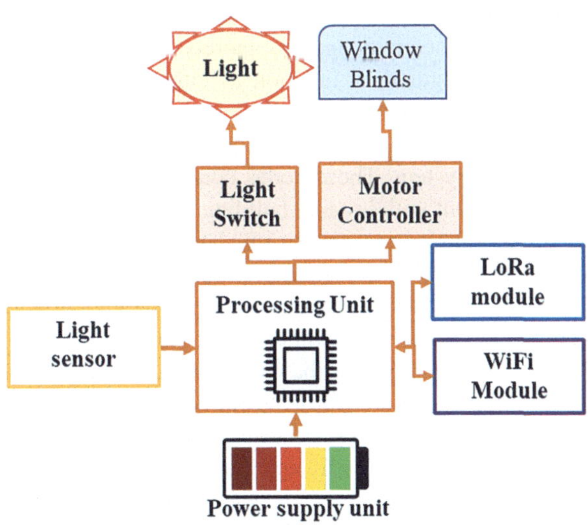

Fig. 5 Smart light node hardware

At the physical layer, battery-powered sensors are strategically deployed across interior and exterior building spaces, enabling distributed environmental sensing with minimal power consumption.

The communication layer implements a LoRa-based mesh network that facilitates efficient message routing, data aggregation, and preliminary processing, thereby optimizing communication reliability and energy expenditure.

The perception layer comprises a cloud-based system that enables remote monitoring, advanced analytics, and sophisticated decision-support algorithms. This hierarchical approach ensures scalable, flexible, and intelligent building management capabilities. Drawing upon the D.R.E.A.M.S. Framework [50], some key design principles underpinning the architectural framework include:

- Maximizing energy efficiency through low-power communication protocols
- Establishing a scalable and adaptable network topology
- Ensuring robust and reliable communication mechanisms
- Supporting flexible deployment across diverse operational environments

These principles establish a rigorous foundation for developing an adaptive, efficient, and robust IoT system architecture for IoT systems in general. The comprehensive architectural design is visually represented in Fig. 5, which illustrates the interconnections and functional relationships across the proposed IoT system's architectural tiers.

Each node is connected to a reliable power supply communicate with both Wi-Fi and LoRa to pass-on monitoring and control messages on to the internet and up to the application layer. The rest of the node operate in low power mode and link to the central application on the internet via a LoRa connection with other nodes.

The proposed wireless sensor network is designed to optimize building lighting and blind control while enhancing energy efficiency. The network comprises intelligent sensor nodes that leverage dual communication technologies, LoRa and Wi-Fi, to ensure robust connectivity and adaptability.

The nodes can be configured to function as either end nodes or coordinator nodes. In the End nodes configuration, the nodes are equipped with ambient light sensors, control lights and window blinds autonomously or based on instructions from the central control system. These nodes communicate within a LoRa mesh network to ensure alignment with broader building management strategies.

While the coordinator nodes serve as intermediaries, linking the LoRa mesh network to the internet via Wi-Fi. They aggregate control messages from the central system and relay them to the appropriate end nodes, enabling efficient dissemination of control commands while maintaining node autonomy.

These nodes are strategically distributed throughout the building, the nodes ensure seamless communication and optimal control of lighting and blinds. By integrating localized decision-making with centralized coordination, the network achieves a balance between energy efficiency and user comfort, demonstrating a novel approach to sustainable building management. The algorithm for this system is represented in the flowchart diagram in Fig. 6.

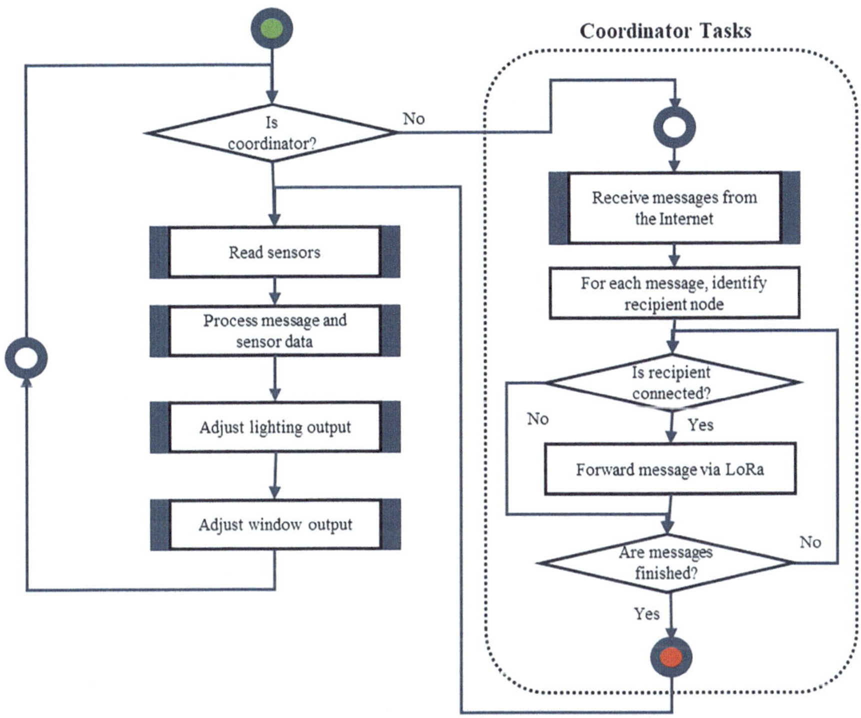

Fig. 6 Flow-chart auto, manual and status combined

Each of the WSN nodes depicted in Fig. 5, can be configured as either a **coordinator node** or as simple **end nodes**. The coordinator nodes communicate both via Wi-Fi and LoRa communication technologies and manage and use their capability to connect the simple end nodes to the Internet. Both the coordinator and end node perform localized sensing and actuation tasks. The process ensures efficient data collection and adaptive control of window blinds across different floors in the building. The algorithm for each node is summarized in Fig. 6.

The algorithm first checks the settings configured on the node. At step 1*, if the node is configured as a coordinator node, the algorithm breaks first into the coordinator tasks loop.

The node listed on the Internet via a WiFi connection for control messages. If the message is directed to an end-node that is within its LoRa subnetwork, the coordinator forwards the message to the relevant end-node. Otherwise, the node reads the next message until none is unavailable.

The algorithm then continues to execute the generic main loop. This begins by initializing the node and its components (LDR sensor and stepper motor). The node continuously reads data from its LDR sensor to measure the light intensity on the floor where it is installed.

The node then processes the light intensity data locally (e.g., by applying thresholds) to make simple decisions about whether to adjust the window blinds. Alternatively, the processed data is packaged into a message and sent to the coordinator for centralized decision-making. Based on the received message:

- **If instructed to adjust**: The stepper motor is activated to open or close the blinds to the specified angle.
- **If no adjustment is needed**: The node continues monitoring light intensity.

The node loops back to continuously monitor light intensity and listen for new control messages, ensuring real-time adaptability.

4.3 Communication Settings

Optimal LoRa configuration is crucial for reliable and efficient communication between nodes. This is particularly important in our university campus setting, where data transmission between nodes on different floors of a 4-story building requires careful consideration of signal strength, range, and interference. By selecting the appropriate frequency band, spreading factor, bandwidth, coding rate, and transmit power, reliable and efficient communication between nodes is ensured. Table 2 synthesizes critical operational parameters derived from foundational studies by Ejike [51] and Zhong et al. [52]. The selection of these LoRa communication

Table 2 LoRa configuration setting

Parameter	Value	Justification
Frequency band	EU868 MHz	Standard frequency for LoRa in South Africa, minimizing interference with other wireless systems and ensuring compliance with local regulations
Spreading factor (SF)	7 or 12	SF affects range and data rate; SF7 allows for faster transmission, while SF12 provides better range. It balances data rate and range; suitable for urban environments with moderate interference
Bandwidth (BW)	125 kHz	This is the standard setting for LoRa, providing a good trade-off between range and data rate. It helps optimize data rate and reduces interference
Coding rate (CR)	4/5	Provides a good balance between robustness (error correction) and throughput (data rate); higher code rates provide better error correction
Transmission power	14 dBm	Maximizes range while minimizing interference and power consumption. Higher power increases range but may lead to higher energy consumption
Payload size	< 2 bytes	Larger sizes may affect transmission success
Adaptive data rate (ADR)	Enabled	Optimizes data rate and transmission power based on network conditions
Node distance	1 km (same building, different floors)	Configuration optimized for reliable communication between nodes on different floors of a 4-story building

parameters reflects a deliberate optimization strategy tailored to the South African context, where environmental and infrastructural constraints necessitate careful technical calibration.

The frequency range of 420–510 MHz, as employed by Zhong et al. [52], presents operational challenges in South Africa due to susceptibility to signal degradation from terrain variations and urban density. In contrast, the EU868 band (863–870 MHz), adopted by Ejike [51], demonstrates superior propagation characteristics in analogous environments. Regulatory frameworks governing the EU868 band prioritize transmission efficiency and interference mitigation [51], rendering it more viable for maintaining reliable connectivity across South Africa's heterogeneous landscapes.

Transmission power selection further underscores this optimization rationale. This study aligns with EU868 regulations by adopting 14 dBm, a conservative yet strategic choice compared to the 17 dBm utilized by Zhong et al. [52]. While higher power may marginally extend range, it risks exacerbating interference and energy inefficiency in obstruction-prone urban and rural settings. The 14 dBm threshold balances regulatory compliance with operational reliability, minimizing network congestion while preserving signal integrity.

Bandwidth configuration also reflects a trade-off between data throughput and noise resilience. The 125 kHz bandwidth selected here enhances receiver sensitivity and noise resistance, critical for environments with intermittent interference. Conversely, the 500 kHz bandwidth tested by Zhong et al. [52], though advantageous for high data rate applications, compromises signal robustness in cluttered or extended range deployments. This study prioritizes link stability over raw throughput, aligning with the need for reliable communication across South Africa's diverse terrains.

Finally, the deployment employs Spreading Factor 7 (SF7), a configuration validated by Cuomo et al. [53] and Bor and Roedig [54] for balancing data rate (5.47 kbps) and range (~1.5 km). This choice proves particularly advantageous in this study's four-story building, where moderate interference and multi-floor dispersion demand robust connectivity. While SF12 offers higher data rates (21.87 kbps), its limited range (~0.5 km) and elevated energy consumption render it unsuitable for spatially distributed sensor networks [54]. Furthermore, SF12's propensity to induce network congestion—termed the "SF12 Well" [55]—underscores the operational risks of prioritizing throughput over reliability. SF7 thus emerges as the optimal compromise, ensuring regulatory adherence while sustaining consistent communication across architectural obstacles, making it the preferred choice for our deployment strategy.

5 Conclusion

This work has presented a design of an IoT-based smart wireless sensor node system specifically for building lighting management for the purposes of improving energy efficiency within buildings. The design science research (DSR) approach is used for development. The proposed hybrid wireless sensor network system follows a 3-tier architecture. The proposed network nodes can either be configured to be coordinators or simpler energy-efficient end nodes. The system's communication layer uses both LoRa connection to provide low-power monitoring capabilities and WiFi capability to connect to the internet.

In the perception layer, the work has designed the hardware based on an off-the-shelf platform of ESP32. The proposed design has sensing capabilities, light switching and control of window blinds. The proposed algorithm design is generic and able to cover both the coordinator and end node capabilities of configuring the network nodes.

The application layer can allow the implementation of control measures using control messages to the WSN so that the energy use of the building linked to lighting can be reduced. This implementation of this system ultimately turns the facility into a smart building, promoting reduced energy operational costs and quality of service.

Competing Interests The authors state that they have no financial interests or personal relationships that could have influenced the work presented in this paper.

References

1. Cao X, Dai X, Liu J (2016) Building energy-consumption status worldwide and the state-of-the-art technologies for zero-energy buildings during the past decade. Energ Buildings 128:198–213
2. Güneralp B, Zhou Y, Ürge-Vorsatz D, Gupta M, Sha Y, Patel PL, Fragkias M, Li X, Seto KC (2017) Global scenarios of urban density and its impacts on building energy use through 2050. Proc Natl Acad Sci 114(34):8945–8950
3. Himeur Y, Alsalemi A, Bensaali F, Amira A (2020) Building power consumption datasets: survey, taxonomy and future directions. Energ Buildings 227:110404
4. Kim H, Choi H, Kang H, An J, Yeom S, Hong T (2021) A systematic review of the smart energy conservation system: from smart homes to sustainable smart cities. Renew Sust Energ Rev 140:110755
5. Elavarasan RM, Pugazhendhi R, Jamal T, Dyduch J, Arif MT, Kumar NM, Shafiullah SC, Nadarajah M (2021) Envisioning the UN sustainable development goals (SDGs) through the lens of energy sustainability (SDG 7) in the post-COVID-19 world. Appl Energy 292:116665
6. Küfeoğlu S (2022) SDG-11: sustainable cities and communities. In: Emerging technologies: value creation for sustainable development. Springer International Publishing, Cham, pp 385–408
7. Santika W, Anisuzzaman M, Bahri P, Gloria S, Rupf, and Tania Urmee. (2019) From goals to joules: a quantitative approach of interlinkages between energy and the sustainable development goals. Energy Res Soc Sci 50:201–214
8. Baharudin NH, Mansur TMNT, Ali R, Sobri NFA (2021) Smart lighting system control strategies for commercial buildings: a review. Int J Adv Technol Eng Exploration 8(74):45
9. Pompei L, Blaso L, Fumagalli S, Bisegna F (2022) The impact of key parameters on the energy requirements for artificial lighting in Italian buildings based on standard EN 15193-1: 2017. Energ Buildings 263:112025
10. Hannan MA, Faisal M, Ker PJ, Mun LH, Parvin K, Mahlia TMI, Blaabjerg F (2018) A review of internet of energy-based building energy management systems: issues and recommendations. IEEE Access 6:38997–39014

11. Motlagh H, Naser MM, Hunt J, Zakeri B (2020) Internet of things (IoT) and the energy sector. Energies 13(2):494
12. Karjou PF, Saryazdi SK, Stoffel P, Müller D (2024) Practical design and implementation of IoT-based occupancy monitoring systems for office buildings: a case study. Energ Buildings 323:114852
13. Chen C-Y, Sheng-Han W, Huang B-W, Huang C-H, Yang C-F (2024) Web-based internet of things on environmental and lighting control and monitoring system using node-RED, MQTT and Modbus communications within embedded Linux platform. Int Things 27:101305
14. Elsisi M, Amer M, Chun-Lien S (2023) A comprehensive review of machine learning and IoT solutions for demand side energy management, conservation, and resilient operation. Energy 6:128256
15. Xu W, Zhang J, Kim JY, Huang W, Kanhere SS, Jha SK, Wen H (2019) The design, implementation, and deployment of a smart lighting system for smart buildings. IEEE Internet Things J 6(4):7266–7281
16. Verma A, Prakash S, Srivastava V, Kumar A, Mukhopadhyay SC (2019) Sensing, controlling, and IoT infrastructure in smart building: a review. IEEE Sensors J 19(20):9036–9046
17. Saleem M, Usman MR, Usman MA, Usman, and Christos Politis. (2022) Design, deployment, and performance evaluation of an IoT based smart energy management system for demand side management in smart grid. IEEE Access 10:15261–15278
18. Silva BN, Khan M, Han K (2018) Internet of things: a comprehensive review of enabling technologies, architecture, and challenges. IETE Tech Rev 35(2):205–220
19. Burhan M, Rehman RA, Khan B, Kim B-S (2018) IoT elements, layered architectures, and security issues: a comprehensive survey. Sensors 18(9):2796
20. Alotaibi B (2019) Utilizing blockchain to overcome cyber security concerns on the internet of things: a review. IEEE Sensors J 19(23):10953–10971
21. Liu X, Lam KH, Zhu K, Zheng C, Li X, Du Y, Liu C, Pong PWT (2019) Overview of spintronic sensors with internet of things for smart living. IEEE Trans Magn 55(11):1–22
22. Pokhrel SR, Vu HL, Cricenti AL (2019) Adaptive admission control for IoT applications in home WiFi networks. IEEE Trans Mob Comput 19(12):2731–2742
23. Haider HT, See OH, Elmenreich W (2016) A review of residential demand response of smart grid. Renew Sust Energ Rev 59:166–178
24. Abdulla AMM, Yuce MR (2019) Sensors and systems for wearable environmental monitoring toward IoT-enabled applications: a review. IEEE Sensors J 19(18):7771–7788
25. Hussain SR, Mehnaz S, Nirjon S, Bertino E (2017) Secure seamless bluetooth low energy connection migration for unmodified IoT devices. IEEE Trans Mob Comput 17(4):927–944
26. Shareef H, Ahmed MS, Mohamed A, Al Hassan E (2018) Review on home energy management system considering demand responses, smart technologies, and intelligent controllers. IEEE Access 6:24498–24509
27. Bibi F, Guillaume C, Gontard N, Sorli B (2017) A review: RFID technology having sensing aptitudes for food industry and their contribution to tracking and monitoring of food products. Trends Food Sci Technol 62:91–103
28. Jia M, Komeily A, Wang Y, Srinivasan RS (2019) Adopting internet of things for the development of smart buildings: a review of enabling technologies and applications. Autom Constr 101:111–126
29. Swamy SN, Jadhav D, Kulkarni N (2017) Security threats in the application layer in IOT applications. In: 2017 international conference on i-SMAC (IoT in social, Mobile, analytics and cloud) (i-SMAC), 477–480. IEEE
30. Hameed S, Khan FI, Hameed B (2019) Understanding security requirements and challenges in internet of things (IoT): a review. J Comput Network Commun 2019(1):9629381
31. Lin J, Wei Y, Zhang N, Yang X, Zhang H, Zhao W (2017) A survey on internet of things: architecture, enabling technologies, security and privacy, and applications. IEEE Internet Things J 4(5):1125–1142
32. Mahmoud R, Yousuf T, Aloul F, Zualkernan I (2015) Internet of things (IoT) security: current status, challenges, and prospective measures. In: 2015 10th international conference for internet technology and secured transactions (ICITST). IEEE, pp 336–341

33. Din IU, Guizani M, Hassan S, Kim B-S, Khan MK, Atiquzzaman M, Ahmed SH (2018) The internet of things: a review of enabled technologies and future challenges. IEEE Access 7:7606–7640
34. Feng S, Farha F, Li Q, Wan Y, Yang X, Zhang T, Ning H (2019) Review on smart gas sensing technology. Sensors 19(17):3760
35. Song EY, FitzPatrick GJ, Lee KB (2017) Smart sensors and standard-based interoperability in smart grids. IEEE Sensors J 17(23):7723–7730
36. Islam T, Mukhopadhyay SC, Suryadevara NK (2016) Smart sensors and internet of things: a postgraduate paper. IEEE Sensors J 17(3):577–584
37. Chakraborty A, Mondal A, Roy A, Misra S (2018) Dynamic trust enforcing pricing scheme for sensors-as-a-service in sensor-cloud infrastructure. IEEE Trans Serv Comput 14(5):1345–1356
38. Guerreiro J, Rodrigues L, Correia N (2019) Resource allocation model for sensor clouds under the sensing as a service paradigm. Computers 8(1):18
39. Ayaz M, Ammad-Uddin M, Baig I (2017) Wireless sensor's civil applications, prototypes, and future integration possibilities: a review. IEEE Sensors J 18(1):4–30
40. Lu X, O'Neill Z, Li Y, Niu F (2020) A novel simulation-based framework for sensor error impact analysis in smart building systems: a case study for a demand-controlled ventilation system. Appl Energy 263:114638
41. Wang Z, Wang L (2013) Intelligent control of ventilation system for energy-efficient buildings with CO_2 predictive model. IEEE Trans Smart Grid 4(2):686–693
42. Sun Q, Li H, Ma Z, Wang C, Campillo J, Zhang Q, Wallin F, Guo J (2015) A comprehensive review of smart energy meters in intelligent energy networks. IEEE Internet Things J 3(4):464–479
43. Shoaib S, Chen X, Llewellyn I (2017) Increasing the effective range of smart meter home area network. IEEE Antennas Wireless Propag Lett 16:2898–2901
44. Barai GR, Krishnan S, Venkatesh B (2015) Smart metering and functionalities of smart meters in smart grid-a review. In: 2015 IEEE electrical power and energy conference (EPEC). IEEE, pp 138–145
45. Tran TV, Dang NT, Chung W-Y (2017) Battery-free smart-sensor system for real-time indoor air quality monitoring. Sensors Actuators B Chem 248:930–939
46. Kumar A, Singh A, Kumar A, Singh MK, Mahanta P, Mukhopadhyay SC (2018) Sensing technologies for monitoring intelligent buildings: a review. IEEE Sensors J 18(12):4847–4860
47. Zhang X, Pipattanasomporn M, Chen T, Rahman S (2019) An IoT-based thermal model learning framework for smart buildings. IEEE Internet Things J 7(1):518–527
48. Carli R, Cavone G, Othman SB, Dotoli M (2020) IoT based architecture for model predictive control of HVAC systems in smart buildings. Sensors 20(3):781
49. Gaur A, Singh A, Kumar A, Kulkarni KS, Lala S, Kapoor K, Srivastava V, Kumar A, Mukhopadhyay SC (2019) Fire sensing technologies: a review. IEEE Sensors J 19(9):3191–3202
50. Ejike CN, Mathaba T, Du Rand F (2022) General-Purpose Architectural Model for IoT-Based In-situ Monitoring Systems. In: MATEC Web of Conferences 370
51. Ejike, Christian Nonyelum. 2022. "Design and implementation of a real-time smart sensor network-based system for monitoring potable water quality." Master's Thesis. Vaal University of Technology (South Africa)
52. Zhong C, Nie X, Peng P (2024) Novel power conservation methods for LoRa-based infrared sensors in smart building. IEEE Sensors J 99:1
53. Cuomo F, Campo M, Caponi A, Bianchi G, Rossini G, Pisani P (2017) EXPLoRa: extending the performance of LoRa by suitable spreading factor allocations. In: 2017 IEEE 13th international conference on wireless and Mobile computing, networking and communications (WiMob). IEEE, pp 1–8
54. Bor M, Roedig U (2017) LoRa transmission parameter selection. In: 2017 13th international conference on distributed computing in sensor systems (DCOSS). IEEE, pp 27–34
55. Casals L, Gomez C, Vidal R (2021) The SF12 well in LoRaWAN: problem and end-device-based solutions. Sensors 21(19):6478

Innovative Sustainable Energy Solutions in Smart Cities: Transforming Urban Landscapes for a Greener Future

Ramkumar Jaganathan, Valarmathi Viswanathan, and Karthikeyan Rajendran

Abstract In the wake of energy needs in urban areas, strategies for efficiency and sustainability must be prioritized. This chapter discusses IoT, data analytics, supportive policies, and community engagement as central elements in optimizing energy usage in urban areas. In this regard, real-time monitoring and analytics capacity offered by IoT technologies helps produce accurate data, providing an avenue for dynamic control of and resource allocation to cut waste in energy usage. Data-driven insights transform the urban energy landscape by creating consumption patterns, detecting inefficiencies, and forecasting demand that supports targeted resource management. Policies setting energy efficiency standards and renewable energy targets provide structured pathways for sustainable growth in the urban system. Incentives decrease financial barriers and make green technology accessible across sectors, which further enhances these efforts by engaging the community and enabling informed residents to assume sustainable behaviors and participate in localized energy initiatives. The study provides a robust framework in building resilient energy systems within an urban context where technological development, policy-driven governance, and public participation converge toward sustainable and responsible fulfillment of future energy needs.

Keywords Energy optimization · Urban sustainability · Green energy incentives · Community engagement · Predictive analytics · IoT

R. Jaganathan (✉)
Department of Computer Science, Kristu Jayanti University, Benguluru, Karnataka, India

V. Viswanathan
Department of Information Technology and Cognitive Systems, Sri Krishna Arts and Science College, Coimbatore, Tamil Nadu, India

K. Rajendran
Department of Computer Technology, Sri Krishna Adithya College of Arts and Science, Coimbatore, Tamil Nadu, India

H. Kaddouri et al. (eds.), *Smart Cities for Sustainable and Urban Development*, Green Energy and Technology, https://doi.org/10.1007/978-3-031-99001-4_18

1 Introduction

With the concept of smart cities gaining momentum around the world, we now have a life and city that integrates technology and innovation with sustainability. With smart cities, the urbanization challenge and the climate change challenge both find solutions towards creating environments designed both technologically and ecologically balanced. This vision is based around innovative sustainable energy solutions. These solutions use cutting edge technologies and renewable resources to help cities minimize energy use, cut emissions and improve the quality of life for those that live in them.

It is focused on how we can shift from traditional energy paradigms to forward thinking strategies through decentralized energy systems, smart grids, and renewable integration. These are not just technological solutions; they are radical blueprints to fundamentally change the nature of city making as a whole, complex and interlinked systems. A commitment to a greener, more resilient urban future based on a more sustainable energy systems.

With rising urban populations comes the increase in demand for energy that is driving significant challenges for existing infrastructure and compounded environmental health. The conventional energy systems that rely on fossil fuel, along with their related degradation of natural resources, air quality, and global climate. To meet these challenges, innovative sustainable energy solutions propose efficiency, decentralisation and adaptability. When coupled with advanced energy storage systems and distribution, renewables such as solar power and wind energy make cities more independent of non-renewable resources. Smart energy systems powered by data analytics and artificial intelligence (AI) not only enable real time adjustments but also long term planning to actionable insights into current energy consumption patterns. Such systems are also integrated to minimise waste and deliver equitable access to energy for diverse urban populations. The essence of smart cities is to integrate the technology to achieve environmental and social targets; this is what is happening with this evolution in energy strategy.

What makes innovation with energy solutions so transformative is that they rewrite the urban landscape. Urban planning could embed renewable energy sources into architectural and infrastructural fabric of cities to help create energy positive buildings and communities. Photovoltaic glass, kinetic pavements, algae-based bioenergy, are all technological advances that redefine how urban spaces interact with the environment, transforming once passive structures into active energy generators. In addition, microgrids and localized energy systems constitute a new trend in power generation decentralization, reducing transmission losses and providing resilience against outages. The resulting localization encourages energy independence together with participation of the community in energy management. This development increases efficiency and at the same time reinforces the social and economic fabric of cities sining a greater interconnection of sustainable urban development.

In order to address energy challenges in smart cities, cultural and policy perspectives need to shift. For taking off, adopting sustainable practices really require robust policy frameworks, public awareness campaigns, stakeholder collaborations, among other. Research and development has been advanced by governments, private enterprises, academic institutes, and funding of pilot projects is done by them which funds successful models. It gives citizens incentives to actively participate in the transition because of emission regulations and programs education and renewable energy adoption. The most smart cities which rely on innovative solutions for energy play the part of the torch bearers of environmental involvement and socio development. In these cities, the collective commitment to sustainability and exemplar value can be mobilized for global replication, leading to a global movement towards greener, more intelligent urban environments. It takes a holistic approach that produces alignment of energy systems with broader ecological preservation and socioeconomic equity goals.

Cities face tough challenges regarding energy demand, environmental impacts, and sustainability. As cities grow, economic activities intensify, and climate issues escalate, they seek to develop smart energy infrastructure. This incorporates digital systems with renewable resources for efficient performance, minimum emissions, and resilient urban environments. The development of sustainable cities will depend on smart energy infrastructure being part of urban planning, focusing on adaptable, efficient, and supportive systems for clean energy. Smart energy infrastructure includes the smart grid, distributed energy systems, renewable energy integration, and intelligent storage solutions, each of which makes unique contributions to energy resilience and sustainability [1].

1.1 Smart Grids and Distributed Energy Systems

Smart grids present a revolutionary change in the nature of energy management by replacing very old-fashioned ways of power with a much more flexible, decentralized, and responsive infrastructure. The traditional architecture of a grid is of one way flow of energy from very large power plants to consumer. In contrast, through integration with data analytics, automation, two-way communication technologies of monitoring, control, and optimization of the energy in real-time the two way energy flow of smart grids has been supported. This enables utilities and cities to react in real-time to fluctuations in demand, which improves the efficiency and reliability of energy delivery.

Smart grids are the best suited for urban setting where energy consumption pattern varies significantly throughout the day, and where cities seek to integrate renewable energy [2]. Advanced metering infrastructure will give the grid operator micro-level information about energy usage so that it can predict demand, identify inefficiency, and correct issues in time. Distributed energy generation across a network also makes the smart grid system more resilient because local outages are less probable due to the failure of a central grid.

Distributed energy systems play a primary role in smart grids when it comes to urban energy resilience and sustainability. Distributed energy systems involve localized power sources, which may include energy from solar panels, wind turbines, and microgrids at or near the point of consumption [3]. The dependency for distant power plants is cut with distributed systems as transmission loses are minimized, thereby lessening costs. Distributed energy systems make it possible to include renewable energy in a city's scheme while reducing a city's growing energy requirements and providing the citizenry with cleaner and more sustainable supplies of power. They may also be operable separately in isolated settings, where power supply can come in through broader grid failure.

In the city, the distributed energy system and smart grid help introduce demand response programs wherein the consumer is encouraged to adjust their pattern of usage at peak times to mitigate the congestions. This flexibility achieves sustainability goals, enabling cities to get more in line with balance energy supply and demand together and reduce dependence on fossil fuels. Smart grids and dispersed energy systems allow cities to reach a resilient, flexible, as well as efficient infrastructure and help meet environmental and growing needs of the urban structure.

1.2 Renewable Energy Integration into Urban Areas

Urban areas are now more proactive in renewable energy integration, understanding the possibility of this energy as a tool to decrease carbon emissions, increase energy independence, and support sustainable development. Cities that seek to make better their environmental impacts have significantly increased the adoption of renewable sources of energy such as solar, wind, and geothermal power. Using the local resources, cities can achieve independence from non-renewable resources and become closer to a more sustainable landscape of energy.

Maybe one of the most common renewable sources of energy in urban settings is solar power. Solar panels can be installed on the tops of buildings, parking garages, and other available urban surfaces to create a dispersed network of clean energy [4]. The decentralized nature of solar energy fits well with an urban setting where buildings, transportation systems, and public facilities can tap into the local production of energy while minimizing their load on central grids. The integration of solar energy has enabled many cities to aim at becoming net-zero energy, that is, producing as much renewable energy as they consume annually.

Wind energy is another essential source of urban renewable initiatives. Small wind turbines are very helpful to cities located in coastal or windy areas, which can add to the solar energy system to generate electricity during hours when the sun does not shine. This ensures a secure renewable source of energy to cities which have diversified energy resources. Beyond solar and wind energy, some cities with the appropriate geology are seeking geothermal energy, which will give a secure source of heating and cooling [5] (Table 1).

Some other technical challenges for renewables integration into urban grids include their intermittency and storage requirements. Renewable source energy changes due to parameters such as sunlight, wind, or temperature. Smart grids help

Table 1 Studies Validating Sustainable Energy Efforts

Study	Focus	Key Findings	Source/ Example
IEA Smart Grid Report (2022)	Impact of smart grids on urban energy efficiency	Improved energy efficiency by 20%	International energy agency (IEA), 2022
		Reduced transmission losses by 15%	
IRENA Renewable Energy in Cities (2023)	Role of distributed systems in urban sustainability	Reduced grid reliance by 25%.	International renewable energy agency (IRENA), 2023
		Achieved 40% renewable energy penetration	
McKinsey Smart Building Solutions (2022)	IoT-driven energy efficiency in commercial spaces	Reduced energy use by 35%	McKinsey & Company Report on Smart Building Solutions (2022)
		Cut operational costs by 20%	
Scandinavian Thermal Storage Projects (2023)	Thermal energy storage in district heating systems	Reduced heating costs by 40–50%	Stockholm and Copenhagen District Energy Case Studies (2023)

facilitate an easy rollout of renewables. Cities can therefore track and adjust consumption based on the renewable sources being produced. With advanced forecasting and real-time data, grid operators can forecast renewable sources and optimize its usage within the city. In recent years, urban policies have also contributed to the adoption of renewable energy, as some cities offer incentives for businesses and residents to adopt renewable energy systems. Community solar programs and power purchase agreements in cities are meant to further open up access to renewable energies so that clean energy will be within the reach of a more urban populace.

Renewable energy integration can facilitate reducing greenhouse gas emission and the creation of green jobs, to promote national and global goals towards sustainability. Urban initiatives with renewable energy are essential components of a clean healthy city in which energy supply and demand can be met at no cost to environmental quality and resource availability over the future.

1.3 Intelligent Energy Storage Solutions

Intelligent energy storage solutions are greatly important for efficient and smooth operation of smart energy systems. Cities are now rapidly establishing renewable sources of energy generation. Solar and wind are not constant because energy can be produced based on conditions. Storage systems alleviate those fluctuations by storing energy generated in periods when the production levels are high so that it will be supplied when demand is at a higher level, offering a smoother and more reliable energy supply to cities.

One of the most commonly installed forms of energy storage in cities is the installation of battery storage systems. Lithium-ion batteries are one such technology that could be used for the compact, effective, and reliable storage to be achieved in urban operations, even if there is a lack of sufficient space. The on-site energy stored by these system can either be utilized on premise or during peak demand peaks fed to the grid in commercial and industrial installations of battery storage systems. As battery technology continues to improve, cities have started to install huge battery installations that can stock renewable energy from various sources and distribute it across the urban grid.

Thermal storage is another form of intelligent energy storage, whereby heat or cold is stored for later use. It is useful in urban areas with enormous heating and cooling needs. Thermal energy storage can be utilized in support of district heating, where hot water or steam is supplied to several buildings in a city. Such systems provide increased efficiency in energy use since the excess heat from industrial processes, renewable sources, or even waste heat produced by power plants can be exploited. In this way, thermal storage systems reduce the requirement for additional energy production and therefore support sustainability and resource conservation goals [6].

Real-time monitoring and predictive analytics optimize storage and distribution. Such systems can predict peak consumption periods through patterns of consumption, hence varying the storage and discharge rates to maintain a stable supply of energy. Advanced control algorithms balance the energy load between storage devices and the grid to ensure that energy is available without overloading the system. The management systems become capable of making predictions about energy demand and adjust their storage operations based on that, making the whole urban energy network more efficient and sustainable by integrating artificial intelligence.

The intelligent storage solutions contribute towards energy independence by letting the cities rely less on the external sources of energy besides supporting grid stability. Microgrids can operate in an independent mode with the power of renewable sources and backed up by storage solutions. It will be an energy resilience for that power outage or any kind of disruption. Thus, the microgrids enable cities to keep up such vital services even in times of emergency. The impact of a grid failure is also highly reduced when the city uses the model of this microgrid in using it for hospitals, schools, government offices, etc.

Economic benefits of smart energy storage solutions come in form of decreased peak energy-demand cost. In well-placed cities, storage means help decrease the energy peaks to lower which brings down the costs of totally purchased energy at those points. The proposed solution seeks to ensure the balanced supply and demand of the source of energy and users hence creates a real-time market of energy that might prove to result in efficient usage.

Intelligent energy storage solutions play a significant role in the sustainability of urban areas. These solutions can provide flexibility, resilience, and cost savings within the energy systems [7]. They can help cities in their journey toward cleaner sources of energy, thus providing a stable supply of renewable energy while reducing the use of fossil fuels. As urban regions continue adopting renewable energy and smart infrastructure, intelligent storage will become a bedrock part of the sustainable energy strategies, establishing an adaptable and reliable framework for generations.

2 Building and Public Domain Energy Efficiency

One of the very important components for having a healthy sustainable urban environment would be achievement of efficiency in buildings and public areas with requirements in their design and management. Such a growing rate of an urban population leads to considerably increased energy demands and stresses built environment as well as natural resources. Cities around the world have embraced strategies toward energy efficiency; therefore, they are reducing their energy consumption and environmental effects. Energy efficiency in buildings and public areas contributes toward the reduction of greenhouse gases, resource conservation, and sustainable urban areas. This chapter discusses the fundamental elements of energy efficiency, including green building technologies, retrofitting existing buildings, and automated energy management systems that all individually contribute to sustainability and conservation of resources.

2.1 Green Building Technologies

Green building technologies have changed the approach of cities toward sustainability as they enable architects, engineers, and urban planners to design buildings that use fewer resources, have lower environmental effects, and create a comfort level for the users. Green building technologies cover everything from design principles to systems and materials that optimize energy, water, and other resource usage in a building. Technologies include renewable energy systems, advanced insulation materials, energy-efficient lighting, and saving devices on water, among others [8].

Integration of renewable energy sources, like solar and wind power, into building design is perhaps one of the most important developments in green building technologies. A solar photovoltaic panel installed on rooftops, facades, or integrated into building materials themselves could provide clean, renewable energy on-site, reducing dependence on fossil fuels [9]. Buildings and cities can reduce greenhouse gas emissions while enhancing energy security by using renewable energy sources. Moreover, advancements in energy storage allow buildings to store the excess energy produced at peak production times for use during peak demand times or when production is low.

Another important aspect of green building technologies is the employment of high-performance insulation materials, thus enhancing the thermal performance of the building, and hence fewer amounts of energy are needed in heating and cooling. There are advanced forms of insulations such as vacuum insulated panels, aerogels, and reflective coatings for reducing heat transfer between inside and outside a building. Apart from conserving energy, effective insulation creates a more comfortable indoor living space, thus enhancing the lifestyles of the occupants.

The lighting systems in green building technologies are centered on energy-efficient lights and daylighting strategies to prevent the use of too much electricity [10]. Much energy is saved using light-emitting diodes lights, which consume much lower amounts of energy than usual incandescent bulbs and which also last longer, saving wastes and energy costs. The strategies include using natural light to illuminate interior spaces, hence saving on energy usage, enhancing health, and productivity levels among building occupants. Most of the daylighting elements that improve indoor lighting by avoiding artificial light sources are skylights, light shelves, and reflective surfaces.

Water efficiency forms part of green building technologies as it conserves water and reduces waste. Low-flow faucets, dual-flush toilets, and rainwater harvesting systems allow buildings to preserve some of the precious water while also reducing the impact that such consumption would have on the environment. In some instances, a green building also has grey water recycling, where the collection and treatment of wastewater in sinks, showers, among others, are recycled back for non-potable application, like irrigation or toilet flushing [11].

The green building technologies apply to the materials that are used in construction. The use of sustainable building materials such as recycled steel, bamboo, and reclaimed wood decreases the environmental footprint of construction and adds to the sustainability of a building. A green building decreases the waste connected with construction and operation if it selects materials with the lowest environmental footprint and can last for a long time.

2.2 Energy-Efficient Retrofitting for Existing Structures

2.2.1 Renewable Energy Integration into Urban Areas

Retrofitting of buildings to increase energy efficiency has become an important approach by cities in the reduction of energy consumption and environmental degradation. Most urban areas contain several older buildings that were constructed before the enforcement of modern energy standards. These buildings tend to use more energy than the newly constructed, energy-efficient buildings. Cities can upgrade these buildings and improve their performance without necessarily demolishing and reconstructing them through retrofitting.

Energy-efficient retrofitting is a set of alterations and upgrades in systems, materials, and design elements within the building. Some modifications might include upgrading the HVAC systems, installing energy-efficient windows and doors, improvements to the lighting system, or enhancing the insulation and envelope of the building [12]. These upgrades contribute to an increase in the energy performance of the building by decreasing the usage and running costs.

One of the most significant goals of energy-efficient retrofitting is to improve the thermal performance of a building, which consequently reduces the energy consumed for heating and cooling. For example, the heat transfer is reduced significantly through insulation upgrades, thereby keeping the indoor temperature stable and requiring the HVAC systems to work less hard. Besides the insulations, retrofitting also involves high-performance windows and doors, which reduce the heat transfer out or into a building, thus making the building more energy-efficient.

Another critical aspect of retrofitting is upgrading the HVAC systems since HVACs use the majority of a building's energy. They thus become energy-efficient while maintaining room temperature by replacing old, not-so-efficient HVAC units. More advanced kinds include variable speed motors, with more energy usage in proportion to needs, and heat recovery, which captures waste heat so that it is either used as a preheater on incoming air or water instead of using loads of additional energy for heating.

Lighting upgrades also have a crucial role in energy-efficient retrofitting, since old buildings have mostly old lighting systems that use more energy than needed. This is achieved by replacing incandescent or fluorescent lights with energy-efficient LEDs, which will help buildings reduce their lighting energy consumption, thus lowering their energy bills and environmental impact [13]. Sometimes, retrofitting incorporates occupancy sensors and daylighting controls, so that lighting level is adjusted based on occupancy and availability of daylight, further improving energy efficiency.

Besides the individual building systems, retrofitting may target building envelopes, which include walls, roofs, and foundations. Improved envelope of a building through more insulation, filling of all the gaps and reflective roof can minimize energy consumption since these improvements can save from heat gain losses and prevent gains. In this particular case, the effects of an urban heat island will be prevented by the use of reflective roof material. An urban heat island is when urban areas become warmer than surrounding rural areas as more materials used there absorb heat. This could also decrease the amount of heat absorbed into buildings and thereby reduce energy demand for cooling and even enhance urban climate resilience generally.

Retrofits of existing buildings also result in economic savings by cutting energy bills for building owners and occupants. Lowering energy costs reduces the running cost of buildings, while savings generated from energy-efficient retrofits repay the initial investment over time. In addition, retrofitting projects normally create construction, engineering, and other types of employment that contribute to local economies and stimulate sustainable development.

2.3 Automated Energy Management Systems

Energy management systems (EMS) based on advanced technologies monitor the real-time energy consumptions through sensors and data analytics with artificial intelligence present, automatically adjusting the system and usage of resources while optimizing energy usage both within and outside the buildings. That is to say, for buildings that will adopt the method of automatic energy management, high energy saving will be brought about. Additionally, cost can be cut in operational functioning as well as an amelioration in comfort, health, and general welfare through the same method.

It basically gathers data from the various systems in a building, such as HVAC, lighting, appliances, and others. It helps understand energy consumption behavior and identifies areas of improvement. Ongoing monitoring by these systems alerts building managers to anomalous conditions, such as possible breakdowns of equipment or excessive energy usage, so that issues can be addressed prior to their occurrence. For example, if a system, like the HVAC system, is found to be consuming more energy than it should, an AEMS can alert a maintenance staff to check the area and, therefore, can avoid costly repairs or possible energy waste.

The biggest advantage of AEMS is based on their ability to ensure the optimal use of energy in terms of occupancy pattern and environmental conditions. Sensors across the building can identify an occupied space and regulate heat, cooling, and lighting consumption [14]. An AEMS would reduce HVAC operation in any conference room or similar place when unoccupied to conserve the energy without reducing comfort level. Similarly, an automated lighting system can dim and brighten by considering occupancy and available natural lighting, thereby conserving much energy while keeping it brightly lit.

Data analytics is one of the most important factors in the success of AEMS, as it can analyze huge amounts of data and discover trends that could be used for energy-saving strategies. Most AEMS employ machine learning algorithms that process historical and real-time data to predict future energy requirements and make adjustments to building systems in advance without overusing energy. For example, an AEMS would utilize historic data to forecast peak energy demand and modulate system performance so as to provide energy to buildings least expensively but possibly only during off-peak hours.

Another advantage of AEMS is its ability to be incorporated with renewable energy sources, optimizing the use of clean energy in buildings. For example, if a building houses solar panels, an AEMS would use energy generated by these panels first when they are available and draw consumption from grid electricity at limited levels. This not only reduces energy costs but also decreases greenhouse gas emissions coming from a building, thus making a building more sustainable.

AEMS can benefit occupant comfort and well-being besides energy. Optimal indoor temperature, humidity, and air quality provide a healthier and more comfortable built environment for the occupants of these buildings. Indoor air quality can be monitored, and the ventilation systems adapted to increase air freshness while reducing pollutants. It specifically relates to occupant well-being in hospitals, schools, and offices, where health and productivity are directly affected by the quality of indoor environments.

Demand response programs are also supported through AEMS. These include programs that adjust energy use based on signals from utilities regarding times of high demand such as hot summer days. At these times, utility companies Buildings may be asked to reduce their energy usage at such times to help lighten the load on the grid. In an AEMS, building systems can automatically change their operation of HVAC, lighting, and other systems to reduce energy consumption in response to such demand response programs and to maintain a stable grid.

Cities and building owners can realize higher levels of energy efficiency and reduce operation costs by integrating AEMS [15]. Additionally, automated energy management can help in conserving more resources and reduce greenhouse gas emissions, which is consistent with some of the urban objectives on stewardship and resilience in a better environment. As technology is developed further, AEMS becomes more advanced and helps in controlling energy usage more accurately to open up opportunities for energy savings and environmental benefits.

Public space applications of AEMS are extended benefits of automated energy management across the urban landscape, such as in parks, transportation hubs, and government buildings. In public spaces, AEMS can manage lighting, heating, and cooling according to usage patterns, which saves energy and creates more sustainable urban environments. For example, streetlights, through the application of sensors, that are networked to an AEMS can adjust illumination in a scenario that includes the use of pedestrians and automobiles. This will enhance safety while minimizing energy consumption. The control of the irrigation system in parks can be calibrated according to weather and soil moisture conditions. In this way, water is conserved while enhancing the sustainability of the green spaces in an urban setting.

Cities can use the implementation of automated energy management to set examples for the effectiveness of technology in achieving energy efficiency and sustainability. AEMS, therefore are a powerful tool for cities that want to develop sustainable urban environments where the use of energy resources does not adversely affect the environment. With the introduction of AEMS within buildings and public spaces, cities could make considerable advances toward the sustainability goals of reducing their energy consumption, lowering the operational costs, and generally benefiting the quality of life that urban residents enjoy [16].

3 Transportation and Sustainable Mobility Solutions

Transportation is one of the most critical aspects of urban life, with ramifications in economic, environmental, and social dimensions in cities worldwide. Population densities in urbanization drive up transportation demands to cater to ever-growing populations without increasing their ecological footprints. Sustainable mobility solutions offer key strategies to balance both demands, supporting low-emission travel by enhancing efficiency within urban settings. Advancements in public transportation through electric and autonomously driven vehicles, deployment of charging infrastructure for EVs, and smart traffic management systems are all efforts where cities work towards the required transportation networks that support not only sustainability goals but are also of better quality concerning living in an urban set-up [17]. Fostering sustainable mobility reduces greenhouse gas emissions but also brings long-term health benefits, better air quality, and more energy efficiency. As cities make this transition, adaptability, integration of clean technologies and responsiveness in real time become the focus. Every element of sustainable mobility contributes uniquely toward resilient and sustainable urban environments. This section is focused on the innovation and impact of electric and autonomous public transport, charging infrastructure for electric vehicles, and smart traffic management as key dimensions in sustainable mobility, all which independently affect the creation of friendly, efficient cities against the transport challenges of today and tomorrow.

3.1 Electric and Autonomous Public Transportation

The public transport option of electric and autonomous ones provides new roads towards the diminishment of the urban emission and easiness of environmental impacts linked to the more traditional and fossil fuel based transit system. The cleaner, quieter, and also more energy efficient buses, trams, or trains that can now join public networks instead of using diesel would be created. Electric buses are one of the more promising solutions in this regard: they run on zero tailpipe emissions and minimal noise pollution, improving both air quality and livability in cities [18]. In addition, electric vehicles require less maintenance than an internal combustion engine, making them less costly to operate and saving money over the long term. Long-route electric buses and other transport modes, enabled by further technological advancements in battery technology-such as lithium-ion or even solid-state batteries-furthers the accessibility within a city. Electric trams and trains remain an essential ingredient in reducing car travel dependence and congestion by connecting central areas with the surrounding ones but with far fewer emissions. Such modes of public transportation have an essential impact on carbon footprints, and it is anticipated that with a better improvement in the battery efficiency of batteries, they would be contributing significantly to the sustainable mobility of cities.

Autonomous public transport is added to enhance the efficiency in routes, safety of the passengers, and management of the fleet. With autonomous public transport vehicles, such as self-driving buses and shuttles, by the use of artificial intelligence and sensor networks, this is possible: to cover all the streets of the city safely and efficiently, reduce accidents due to human error, while maintaining the optimal speed for energy and emissions savings. The route usually changes in real-time conditions of traffic, congestion, or special spikes in demand. For example, self-driving shuttles in cities are becoming "last-mile" solutions, connecting public transit stops to final destinations with routes that shift based on passenger requirements. This adaptability contributes to overall sustainability by reducing idle time, optimizing energy use, and making public transportation more accessible and appealing to residents. Cities that integrate electric and autonomous technologies have a very sustainable transit model that gives resilience, flexibility, and environmental benefits. Electric and autonomous systems together represent the future of eco-friendly urban transit to minimize urban emissions while ensuring efficient, reliable transport services [19].

3.2 *Charging Infrastructure for Electric Vehicles*

The growth of electric vehicles in cities makes it crucial to develop robust and accessible charging infrastructure for widespread adoption. Charging infrastructure supplies the backbone for electric mobility to challenge issues like range anxiety and ensures that an electric vehicle remains a viable and convenient option for the city's residents. Public charging stations, which are fast-charging points located in strategic urban areas such as malls, transit hubs, and major roadways, enable drivers to quickly replenish their energy on the go. Direct current fast chargers are often installed in these stations, which can charge an EV in minutes rather than hours, making them ideal for high-traffic areas and time-sensitive users. Residential and workplace charging also complement public options by offering convenient and low-cost charging for daily users, typically at lower voltage levels to support overnight or workday charging. In combination, these options create a comprehensive network that supports the diverse charging needs of urban EV owners, thereby supporting the transition to sustainable, low-emission transportation within cities [20].

Renewable sources such as solar and wind integrated into EV charging stations also give electric mobility additional sustainability credentials. In powering the electric vehicle chargers with clean energy, the city limits the carbon impact that attaches to using the EV. This contributes towards greater and broader objectives in terms of sustainability. Solar-based chargers, for example, make it possible for the cities to draw renewable energy direct into their operation, and these can easily be one cleaner form of energy and at least ease some of the demands that these put on a local grid [21]. Moreover, the integration of renewable energy sources at the charging points helps cities be independent, as they do not rely so much on non-renewable power sources. Smart chargers have also been instrumental in managing the flow of power efficiently through the alteration of energy distribution according to the demand on the grid. Real-time monitoring and load balancing by smart chargers help avoid peak energy loads, and the users can charge their EVs during off-peak hours to optimize the use of the grid. Some cities have even initiated vehicle-to-grid (V2G) technology, through which EVs can become decentralized energy reserves. Vehicles can provide stored energy back to the grid when needed, thus stabilizing urban grids, reducing costs on the grid, and promoting sustainable energy practices through the symbiotic relationship between EVs and the power grid.

Proper city planning is required for the placement of EV chargers so that access and convenience are available to all, which in turn will encourage more residents to shift from conventional vehicles to EVs. Strategically placed chargers across neighborhoods and key city points ensure that charging access does not remain limited to certain districts or socio-economic groups. Urban planners help in achieving this balance while distributing the chargers that serve the whole city appropriately, thus not congesting a charging point, but, at the same time, ensuring easy accessibility for these underserved communities. Therefore, the available and placed charging network facilitates a painless experience in using the EVs and reduces their range anxiety as well. This way, urban efforts to gain low-emission transportation as mainstream will be more effective. In such an implementation, strategic expansion of the renewable source, incorporation of advanced technology, and charging infrastructure development, can ultimately contribute to sustainable urban mobility. This is an effort to contribute to the general city-wide objectives of the reduction of emissions, responsible transport modes, and a better urban environment.

3.3 Emission Reduction through Smart Traffic Management

Smart traffic management addresses the issue by using the data generated in real time, automation, and highly advanced analytics for optimizing flow and minimizing congestion with following reductions in idling time. It further helps reduce emission in the urban areas. The traditional traffic systems that have often presented inefficiency in long waits, especially through pollution from idling cars and fuel consumption in major cities, have caught the attention of smart traffic management, which seeks change in such systems [22]. Adaptive technologies or data-driven techniques make attempts to improve flow by creating reductions in negative environmental implications. One main feature found within the entire system is the concept of adaptive traffic signal control-the ability to adjust timings on several signals based on the available input in real-time. Adaptive signals cut the waiting times at crossroads. It does so based on traffic volume so it helps reduce emissions through the vehicle waiting for a green signal from red. While the same is happening in cities, congestion is minimized, and travel becomes much efficient which reduces fuel use and also reduces emissions to a significant level for the commuters and the urban ecosystem.

The other core area of intelligent traffic management is ITS. ITS integrates GPS data, road sensors, and traffic cameras to give an integrated view of the urban traffic patterns. This information helps the city planners to identify areas where traffic congestion is most frequent, high volumes of traffic occur, and road usage patterns, thus guiding their strategy to further improve traffic management [23]. ITS also facilitates the spread of real-time updates to drivers through navigation systems and mobile apps, guiding them to less congested routes and reducing pressure on highly trafficked roads. Optimization of routes and proactive management of road usage reduce bottlenecks, thus allowing smooth traffic flow and subsequently lower fuel consumption and vehicle emissions in the city. This can aid the urban planners to gain the insights for proper decision making about infrastructure deployment. All these activities are performed by them within the context of reducing emissions and enhancing efficiency in transportation.

Congestion pricing models are a model followed very successfully by smart traffic management, especially in areas with high density in emitting emission. Congestion pricing is charging vehicles to enter high-traffic zones during peak hours, an approach that seeks to decrease traffic volume and encourage alternative transportation [24]. With variable pricing based on time of day or specific demand, cities manage congestion effectively, decrease emissions, and encourage more use of public transportation or carpooling. Generally, income from congestion fees is usually reinvested in improvements to public transport, creating a cycle where reduced car usage improves public transit, which in turn supports reduced emissions. Cities that implement congestion pricing have experienced heavy reductions in traffic volume as well as vehicle emissions. For this reason, it offers an effective and economically viable tool for the management of urban traffic. Congestion pricing is likely to play a crucial role as more cities implement this new approach toward reducing emissions in urban areas and enhancing the flow and sustainability of the urban transportation networks.

V2X communication is the technology that enhances smart traffic management and urban mobility sustainability. V2X communication enables the interaction between vehicle and road infrastructures where real-time data is also available on the traffic flow, accidents, road hazards or even emergency vehicles' path. In that way, a vehicle might adjust speed or reroute to skip congested roads or warns of hazardous conditions on those roads. V2X communication helps the vehicle to synchronize with traffic signals, which will further help in maintaining a smooth flow of traffic and reducing stop-and-go driving, which often increases emissions. Besides, V2X-enabled vehicles can adjust their speed according to prevailing traffic patterns, thereby reducing fuel consumption and lowering emissions. With V2X, it is possible to attain very high levels of interconnectedness within an urban transportation environment wherein the vehicles, infrastructure, and systems of traffic management might function together towards efficient mobility, which is not only in nature but also supported. V2X has enabled the ability to end the idling, providing the possibility of near real-time responsiveness, and this definitely is a step towards reductions in emissions and an enhancement of safety on roads and advancing toward intelligent traffic management.

By integrating adaptive signal control, ITS, congestion pricing, and V2X technology, smart traffic management provides a holistic solution to the reduction of urban emissions as well as optimized fuel use while making the overall commuter experience much better. This system is targeting the intrinsic inefficiencies that exist in traditional traffic networks, with urban transportation turning into data-driven, responsive, and environmentally friendly operations. Smart traffic management can completely change the face of urban transportation systems, making cities efficient in movement and low-impact travel. With such smart traffic management, the reduction of emissions will contribute to cleaner air and healthier living environments, as improved flow offers convenience both for personal and commercial travel. It makes smart traffic management part of the qualitative goal of sustainable development. It helps cities meet environmental targets to create sustainable and resilient urban infrastructures that could accommodate emerging populations and shifting transit needs.

A city, through wide investments in electric and autonomous public transport, coupled with an advanced comprehensive charging network for electric vehicles, will realize a future of sustainability in urban mobility by bringing the two contradicting interests into a unified vision: environmental friendliness and the ease of commuter travel [25]. Each uniquely helps to achieve lower emissions in reduced energy consumption, which has enhanced the efficiency of cities in mobility as part of their system into an all rounded ecosystem that fosters efficient eco-friendly transportation. From sustainable mobility comes the initial point for more comprehensive larger urban sustainability as such wherein transport planning must align perfectly with a city's notion of its future growth environment with sustainable resilient infrastructure [26].

4 IoT and Data Analytics for Optimized Energy Use

Energy use optimization will become critical as urban agglomerates suffer the pains of rising population, which increases energy demands with regards to environmental effects. The IoT and data analytics become a critical tool in managing energy resources; cities can hence use real-time data and predictive insights for improved control and distribution of energy [27]. Using IoT sensors and advanced analytics, cities can monitor, adjust, and predict energy consumption at various sectors such as residential buildings and public infrastructures. This IoT-data analytics integration would form a pathway toward efficient sustainable use of energy by waste minimization, improvement in the grid reliability, and efficient support of renewable energy sources. The subsequent sections present essential components of the urban energy framework: real-time monitoring of energy, optimization based on IoT, and predictive analytics for energy forecasting [28].

4.1 Real-Time Energy Monitoring and Analytics

With real-time energy monitoring, cities can watch their consumption in real-time, so they see very granularly where patterns, peaks, and waste potential areas occur. In building installations, on equipment, and within public space, IoT sensors provide valuable detailed, moment-to-moment data in the interest of urban energy management. Cities can quickly adjust real-time monitoring and correct inefficiencies, and avoid failures, which is costly for energy and could shut down some operations [29]. Through sensors IoT-enabled, it follows through with monitoring the main parameters, such as temperature, occupancy, and humidity in any building or public structure for an accurate granular view of the dynamics involved within energy. Cities thereby will be able to command a greater control over its resources on energy while ensuring to decrease the operating cost but at the same time also avoiding unnecessary damage done to the environment.

At the base of real-time monitoring, data analytics handles millions of sensor data and thus delivers actionable insights. Such patterns would give evidence of peaks, idle times, and sharp increases that may suggest the necessity for maintenance or adjustment of the equipment in these cities' datasets, and evidence of a data peak about energy usage during night times means light, even the HVAC equipment has been accidentally left on within an office building, meaning managers act quickly to redress such an issue, then prevent the wastage of energy with reduced operational cost. Histories can also point out seasonal or periodic patterns of energy usage, helping in updates in long-term efficiency as well as infrastructure planning [30].

Real-time monitoring and analytics also empower end-users, such as homeowners, businesses, and facilities to take ownership for energy savings. Smart meters provide an individual with real-time information on their usage, thus enabling them to make informed decisions [31]. This transparency encourages behavioral changes that reduce energy use, such as adjusting thermostat settings, turning off lights in rooms that are not by utility companies support this behavior, as the users are incentivized to shift consumption to off-peak times, both for the benefit of their budgets and the stability of the city's grid [32].

Real-time energy monitoring also facilitates a more reliable and efficient power grid. Utility companies will promptly respond to changes in the demand by monitoring energy flow and consumption rates continuously [33]. These will allow for timely action that can avoid overloads leading to blackouts as well as maintain steady flow of energy to high demand areas. For cities, which have more complex needs for energy, this gives the grid the ability to make real-time adjustments as it makes the grid far more resilient and allows one to integrate greener sources of energy. Real-time data feeds into larger analytics systems, allowing for a proactive approach to the management of urban energy and providing the basis for future energy forecasting.

Installation of real-time monitoring in all dimensions of urban systems ensures enhanced transparency and efficiency in the distribution of energy. Cities will, with accurate and up-to-date insights, focus on energy-saving measures, use resources optimally, and ensure proper supply to high-demand areas. Real-time analytics enables planners to monitor the impact of energy policies and infrastructure investments, meaning that investments in smart grids, renewable installations, or other sustainable projects are therefore guided. Through IoT, real-time monitoring transforms energy management in the urban center into a responsive and resilient source of energy able to compete with the present challenges in the world.

4.2 *How IoT Can Foster Energy Optimisation in Cities*

The Internet of Things is changing the manner in which cities manage their energy consumption through interconnections that adjust consumption based on real-time conditions. IoT-enabled devices such as smart thermostats, connected lighting and appliances, can all dynamically adapt factors such as occupancy, temperature, or even electricity pricing. These connections minimize unwanted energy use in buildings and across public spaces, allowing cities to achieve greater energy efficiency with improved resource distribution. This creates an ecosystem of energy-sensitive devices that respond dynamically toward the needs of its populace to support the objective of sustainable development while carbon imprints within cities remain minimized.

For example, this IoT-enabling within individual buildings would provide efficient and effective management of energy by not wasting energy. A related example would be the "smart thermostat," in which sensors will measure the attendance of people, determining and adjusting heating or cooling values through data readouts on the device. These thermostats can even connect to local weather service and alter temperature settings automatically as appropriate to bring indoor climate levels into harmony with outdoor atmospheric conditions, further reducing their HVAC energy use. Smart Lighting Systems consist of motion detection sensors and daylight sensors that actually switch on and off these lights when the occupancy senses its need. Connected appliances in a residential building can communicate with smart meters and adjust their operation to off-peak times, thereby saving energy and reducing costs for residents.

IoT-driven energy optimization extends to urban infrastructure in the form of efficiencies in public lighting, transportation, and municipal services. Smart street lighting, for example, uses motion sensors and adaptive brightness controls to adjust lighting according to pedestrian and vehicle traffic. It further decreases usage of energy in low-traffic areas without hindering public safety. Adaptive traffic lights control signals based on real-time congestion data in such a way that it minimizes idle times by vehicles around intersections, therefore reducing the emission of exhaust and conserving fuels. The integration of the aforementioned public infrastructure into central data platforms helps create a seamless system of energy optimization for a city that will reduce its usage of resources while at the same time fitting into general environmental goals.

With the IoT, demand-response programs have become a necessity in electricity demand management during peak times. Utility companies monitor patterns of consumption through IoT sensors and smart meters and warn consumers to reduce usage during critical times. In some cases, demand-response systems even automate energy reduction measures in participating buildings, which temporarily lower lighting or HVAC operation to alleviate grid stress. This approach to energy optimization prevents grid overloads and enhances the resilience of energy infrastructure. In this way, demand-response programs contribute to more stable, efficient urban grids that can handle fluctuations in demand without compromising reliability.

IoT-driven energy optimization supports the integration of renewable energy by addressing variability in output from sources such as solar and wind. Due to its dependence on weather conditions and time of day, renewable generation levels are typically erratic and problematic in terms of maintaining a stable supply of energy. However, IoT systems allow for dynamic energy management, enabling cities to focus on renewable sources at the time of high production but shift to traditional sources or storage during low output. This capacity maximizes energy resources and exploits all available renewable sources, thus aligning urban energy with sustainability concerns and reducing reliance on fossil fuels. IoT-enabled optimization allows cities to have dynamic, responsive energy systems that fulfill current needs while furthering long-term environmental requirements.

4.3 Predictive Analytics for Energy Demand Forecasting

Predictive analytics has been the cornerstone in the efficient management of urban energy systems, enabling cities to predict changes in demand and make proactive resource allocation. The predictive analytics tools combine data from IoT with machine learning algorithms to create models that can predict energy usage based on historical data and external factors such as weather and occupancy. This is in the interest of providing the utility companies and the planners of the city means by which they can take a proactive approach to the regulation of energy production and supply thus saving from inefficient overproduction and shortages of supply. Predictive analytics enables more resilient energy infrastructure because data supply supports informed lookout decision-making in energy management.

From the many applications, of predictive analytics, undoubtedly load forecasting is the most amazing. Analyzing data from previous consumption cycles along with seasonal temperature changes, predictive models will estimate demand spikes and prepare in advance. During the times of high demand events that are expected, utility companies can increase energy production or withdraw stored reserves to ensure consistent energy availability during those times. This approach reduces the reliance on peaker plants, which generate additional power during surges but are often inefficient and costly. Predictive analytics thereby enables a well-balanced energy grid that saves money and damages the environment less without offering reliability to consumers.

Further, predictive analytics supports an integration of renewable energy. The biggest issue facing renewals is that of intermittency: it depends upon the weather as well, and its outputs are constantly changing. It can even make it a bit complex to include into a constant power grid. Predictive models that use meteorological data are useful for forecasting renewable energy generation so grid operators can plan for the variability. Where these predictions are that high, cities can make more use of these clean sources; during low-output periods, the renewables can be supplemented by conventional energy or stored reserves. This is done in the sense that this will make cities optimize their energy mix, hence allowing cleaner usage of energy while enhancing the sustainability agenda.

Predictive analytics also comes in handy in the storage of battery management. For instance, as more and more cities embrace large-scale storage systems to supplement energy use, predictive analytics is crucial because it allows these storage systems to predict the surges in demand to ensure discharging energy during those periods that are most necessary, like peak hours or periods when renewable generation has reached a minimum. Predictive analytics provides precise control over these storage resources, allowing cities to avoid expensive energy purchases at peak times and reduce reliance on fossil fuels. Thus, it brings an economic advantage of saving on operation costs and environmental benefit by supporting balanced, flexible integration into the grid [34].

Lastly, predictive analytics provides transparent forecasts on future energy demand and growth patterns as the basis for long-term investment in infrastructure. Cities experiencing rapid growth in population will use predictive models to predict which districts might have a higher energy demand, hence planning on up-grading the infrastructure. Therefore, with this proactive step, grid capacity will develop to meet the development while ensuring there is no lack of energy and service. Predictive insights also determine pricing strategies so that utilities can introduce time-based pricing and encourage off-peak usage. Such demand-shifting strategies reduce strain on the grid and lead to consumption patterns that enable both an optimized and responsive urban energy ecosystem [35].

IoT, real-time monitoring, and predictive analytics build an advanced framework with which cities can balance immediate demands with future sustainability objectives. The actual real-time monitoring provides them with the data they need to be alerted to inefficiencies and correct them on time and thus conserve energy across many sectors. IoT-enabled devices are also enabling buildings and infrastructures to adapt in real time to changing conditions through adjustments and demand-response programs, hence reducing waste [36]. Predictive analytics gives a sense of forward-looking ability in this context, where cities can know when there is bound to be a change or shift and prepare their resource allocations accordingly. All these go together to make an effective responsive, resilient urban energy system that will take into account the dynamic requirements of the urban population. The future of sustainable urban living will be built on the very foundation of IoT and data analytics, given the global push to reduce carbon footprints and increase resource efficiency in cities.

5 Policies, Rules, and Involvement of Society

Strong policies, rules, and community involvement are basic foundations for urban sustainability as well as optimized energy use by the implementation of technological advancement under the support of governance along with cooperation from the public. The policies and regulations create the framework that guides energy

Table 2 Policies and their Tangible Benefits

Policy	Key Focus	Impact	Source/ Example
Germany's Energiewende Policy	Transitioning to renewables and reducing fossil fuel dependency	46.2% of electricity generated from renewables	Federal Ministry for Economic Affairs and Climate Action (2023)
		25% CO_2 emissions reduction since 2015	
India's national solar mission	Scaling solar capacity and incentivizing rooftop installations	Installed 67 GW of solar capacity as of 2023	Indian Ministry of New and Renewable Energy (MNRE, 2023)
		Cochin airport became 100% solar-powered, saving 300,000 metric tons of CO_2 annually	
United States' inflation reduction act (IRA)	Supporting renewable energy adoption through tax credits and investments	Solar power constitutes 13% of the national energy mix	U.S. Department of Energy Report (2023)
		Created 50,000 green jobs in 2023	
London congestion pricing model	Reducing urban vehicle emissions through pricing mechanisms	Reduced vehicle emissions by 15–20%	Transport for London (TfL) Annual Report (2023)
		Increased public transport usage by 35–40%	

standards, sustainability goals, long-term planning, and consistent approaches towards energy conservation, renewable integration, and emissions reduction. Besides governmental mandates, another key element for realizing sustainable urban development is community involvement; the participation and behavior of citizens significantly influence the effectiveness of any energy initiative. Funding and incentive programs can encourage wider participation that makes it financially worthwhile for people, businesses, or local organizations to contribute [36]. This section discusses government policies supporting urban sustainability, green energy incentives and funding available, and the role that community engagement plays in producing sustainable energy outcomes (Table 2).

5.1 *Government Policies Supporting Urban Sustainability*

Government policies are the bedrock of the transition to urban sustainability, providing the standards and guiding principles that will form the basis of energy usage in cities in the decades to come. Energy efficiency, reduction of emissions, and the promotion of renewable energy constitute all policies making a pathway to sustainable urban development [37]. Most governments have come out with policies forcing energy efficiency standards on new constructions; hence, all new buildings must meet particular criteria on the energy performance of the said building. They ensure reduced energy consumption in residential, commercial, and public buildings and place sustainability practice at a structural level. Building codes and regulations often support or require energy efficiency materials, technology, and design standards, such as insulation, efficient lighting, and renewable integration. This will ensure that governmental policies on urban growth meet their environmental goals while minimizing energy consumption in cities and maximizing long-term resilience.

5.1.1 Renewable Energy Integration

Another critical policy related to the sustainable use of urban energy is the integration of renewable energy into the system. It greatly reduces carbon footprint in a city. Governments must work towards this integration, exploring various ways of doing it, from setting targets and deadlines to incorporating it as a compulsory measure in local policies. Governments can set renewable energy targets, for instance, with the aim of having a certain percentage of urban power derived from clean sources like solar, wind, or hydropower. Such targets create a regulatory environment that supports investment in renewable infrastructure, such as solar farms, wind turbines, and energy storage solutions. There are also policies on distributed energy generation and incentives, which encourage people and enterprises to install renewable energy sources such as rooftop solar panels. Distributed energy policies empower citizens to participate directly in clean energy production by decentralizing energy generation and reducing dependence on fossil fuels. Governments support the shift to sustainable urban energy networks by cultivating a renewable-friendly regulatory environment [38].

Carbon pricing policies add a cost to carbon emissions, either by carbon taxes or by cap-and-trade systems. Its objective is to induce business and individual action toward low-emission practices through financial incentives since it becomes financially attractive to cut carbon output. Revenue generated from carbon pricing can finance green infrastructure projects, for example, enhancing public transportation or installing renewable energy systems, which further enhance the sustainability of cities. This aligns economic incentives with environmental objectives and, therefore, helps it act as an effective way of reducing urban emissions along with promoting clean energy consumption. Cities that employ carbon pricing often see industrial emission reductions, transportation pollution reductions, and other sources of carbon output decline as members within the community change their habits to avoid the cost it will incur [39].

Policies on smart grids and energy resilience make sure that urban energy systems are ready to meet modern city challenges, ranging from shifting demand to the integration of renewable [40]. Policies on smart grids guide the development and deployment of advanced metering infrastructure, energy storage, and grid management technology in support of optimized energy distribution and improved grid reliability. Resilience policies, such as mandates for backup power sources and grid modernization, will allow cities to better weather supply disruptions, especially in

Table 3 Renewable Energy Integration and Urban Impacts

Renewable Energy Project	Key Focus	Impact	Source/ Example
San Diego's renewable energy framework	Transitioning to renewable energy to achieve net-zero targets	45% of energy sourced from renewables	City of San Diego Sustainability Report (2023)
		Advancing net-zero goals for 2035	
Amsterdam's urban solar farms	Utilizing urban solar farms to reduce reliance on centralized grids	Generates 12 GWh annually	Amsterdam Renewable Energy Initiative (2023)
		Powers 8000 households	
Singapore's floating solar farms	Utilizing water bodies to generate renewable energy	Generates 5 MW annually	Singapore Energy Market Authority (2023)
		Avoids occupying urban land for energy production	

the case of a natural disaster or peak demand. Mandating investments in resilient infrastructure makes governments build strong energy systems that can sustain the demands of urban populations while supporting sustainable use of energy [41].

Public policy is also extended to transportation where governments set emissions standards, encourage electric vehicle adoption, and invest in sustainable public transit options. Policies that mandate cars to be fuel-efficient or have certain emissions standards enforce better transport solutions that minimize levels of poor air quality over urban settings and cut energy expenditure [42]. The encouragement of electric vehicles with public transport investments that encourage non-car dependent travel reduces further car use and promotes cleaner modes of mobility in cities as it moves away from fossil sources of fuel. Government policies collectively help in building a solid platform for sustainable energy practices; they guide both infrastructure and individual behavior towards the greener urban future (Table 3).

5.2 *Incentives and Funding for Green Energy Initiatives*

With incentives and funding programs emerging as an inevitable step for promoting green energy adoption, it is now a pressing need to adopt sustainable practices in urban areas. Financial incentives like tax credits, rebates, and grants reduce cost barriers associated with renewable technologies, thus making them relatively easier for individuals, businesses, and local governments to invest into clean energy. For instance, it is common for many cities in the United States to offer cuts or rebates in property tax for homeowners who have installed photovoltaic solar panels-a financial return which offsets one's initial investment in a renewable infrastructure. Reimbursement programs in incentive schemes may include rebates for energy-efficient appliances, insulation upgrades, and smart energy management systems that promote the energy-saving habits at the house hold level [43]. The reduction of costs related to sustainable technology can, therefore, increase the utilization of green energy at a greater number of places in society as community-level efforts to promote the sustainability of urban ecosystems is enhanced.

Corporate incentives can also help push wider and more substantial utilization of green energy. Companies, particularly those with heavy industries, can benefit from tax deduction or subsidy to upgrade energy-efficient machinery, install renewable energy systems, or establish a waste reduction program. These benefits will assist companies in aligning their operations toward sustainability goals, with the lowering of emissions, consumption of energy, and the effective use of resources. Some programs provide funding for companies that meet or exceed set sustainability targets, thereby further embedding a culture of responsible environmental stewardship within corporate enterprise. Incentivizing 'green' practices encourages governments and organizations to create a ripple effect, extending beyond individual companies to make industries more sustainable.

Funding for green energy projects should be provided to develop public infrastructure for sustainable urban growth. Cities tap into federal and state grants to invest into large-scale renewable energy investments such as solar farms and wind parks and smart grids [44]. These feed clean energies to the cities, allowing for scalability in the production of the renewable sources themselves. Green energy funding even extends to public transportation networks-which make up a big percentage of the components of a sustainable urban structure. With electricity bus fleets, infrastructures of charging stations and large metro system expansions, such sustainable mobility is accessible and more available for those occupied in a city. All such projects are supported by the public spending, so these municipal authorities get enough fund allocation in investing in those green energy systems while, in place, for a long-run.

Green loans/financing schemes help open new avenues for private persons-between firms to invest in living in renewable ways. Green financing options from government or private banks commonly offer favorable loan terms in renewable energy installations, energy-efficient upgrades, and buying electric vehicles [45]. For example, a homeowner can now enjoy clean energy at low-interest rates for the installation of solar panels without front money piling up. In a business environment, green bonds and energy performance contracting can help companies pay for the sustainability projects since repayment structures are pegged on energy saved. It spurs green technology integration in the fabric of the urban environment through promoting incremental transition towards clean energy.

The second, it helps drive innovation through green technologies introduced to the marketplace, creating the possibility of products and renewable energy solutions on energy usage. As more customers and businesses feel the call of incentives, these suppliers develop the most efficient solution that comes at minimum cost as aligned with regulatory guidelines. This cycle of innovation continues improving the effectiveness and affordability of green technology. As such, the renewable energy sources and consumption-saving measures can now more easily reach into more locations in the city. To this end, incentives and access to funding have become salient to help accelerate faster the pace toward adopting energy sources that assist in shaping an enabling environment for urban sustainability.

5.3 Community Involvement in Renewable Energy Technologies

Engaging with the community will form the backbone of accomplishing energy sustainability because members within communities account for the significant influence of consumers on energy use behavior and sustainable action. City engagements help create a sustainability culture within which people become habituated towards energy efficiency, both by changing their individual practices to contribute toward energy conservation as well as participating in grassroots-level initiatives that contribute towards creating an environmentally friendly and 'green' locality. In doing this, the government utilizes means such as educating programs and workshops as well as local events. These efforts educate residents on the benefits of reduced energy consumption, renewable energy alternatives, and the impacts on the environment of their usual daily practices. People respond positively and participate in urban sustainable efforts if the benefits for reduced energy use are easy to feel [46].

Cities devise community programs that make access to these sustainable practices viable and attractive to the locals. For example, no-cost home and business energy audit programs ensure that the customer is aware of what there is to improve on and eventually, in the long term, will encourage an upgrade. Local governments in certain urban centers make available, at low prices or even free, energy efficient appliances as well as weatherproofing products. Community solar projects where several members of a neighborhood participate by investing in solar panel technology and then share the return ensure that alternative energy is adopted. Such programs allow households, which otherwise cannot afford the installation of solar on their property, to enjoy green energy.

Community-led initiatives also enable the empowerment of residents to own sustainability projects. This, in turn, fosters a sense of responsibility and accountability. For example, a neighborhood group interested in sustainability may host a recycling drive, tree-planting event, or energy-saving challenge that actively engages the community [47]. Grassroots movements on reduced plastic use, local food production, and sustainable transportation options create momentum within neighborhoods and help a city shift toward green practices. It makes people adopt long-term sustainable energy-saving practices and inform their neighbors to adopt when people feel they are not playing a passive role, but an active one toward attaining sustainability. Through such collective efforts, this then adds strength to the effects of municipal policies that enforce sustainability in urban areas as beneficial through sustainable energy usage.

Such platforms based on technology allow the community to engage further with sustainability efforts. Residents are able to track and monitor their energy usage, be part of local initiatives, and know the city's goals. Mobile applications and portals online allow residents to keep an eye on their energy use compared to community averages and also receive tips on how they can reduce usage. Interactive and engaging strategies on gamification in gamification will motivate residents in engaging themselves in achieving set targets on sustainability, through awarding points or rewards [48]. There have been some of the platforms which encouraged the residents to engage themselves in energy-saving challenges or community solar investments. The virtual space of the individual contributions toward collective sustainability goals has been achieved by the same. It forms a connected, energy-conscious community with the technology of sustainability of the urban areas and actively involved residents.

Lastly, participation in the community develops more robust relationships between the people and the local government because it encourages openness, cooperation, and trust. City governments where residents participate in the decision-making process associated with issues such as discussion forums on city energy policy or weighing the opinions of the sustainable projects through survey feedback would enjoy a greater public endorsement and shared stake in urban goals [49]. The mutual approach will ensure that policies and programs should follow the lead of the people in pursuit of making people's living conditions better so that high participations will be raised in sustainable actions. Citizen engagement creates a feedback loop as citizen participation is informed in city planning, and the city initiative catalyzes sustainable behavior within the community. By developing a participatory approach, cities build up to develop a more resilient and inclusive framework for sustainable urban energy usage, where each resident helps mold the future of their city's energy landscape.

6 Conclusion

The establishment of sustainable urban energy systems is founded on a strategic mix of technology, policy, and community engagement. IoT and data analytics enable precise monitoring, making it more efficient and possible to dynamically manage resources. Policies form the very foundation of urban energy systems, providing guidelines on the efficiency, renewable integration, and emissions reduction that correspond to sustainability targets. Incentive programs further those goals by making green technology financially viable, encouraging widespread adoption across residential, commercial, and public sectors. Community engagement is still a critical component because an interested community directly contributes to energy-saving efforts by informed action and participation in local initiatives of sustainability. Together, these elements bring about an interlinked approach that brings into balance environmental, economic, and social factors in the use of energy, thus making it possible for urban areas to transition to a resilient, low-emitting future. This integrated approach puts cities on the best path to meeting their future energy demands with sustainability, enabling cities to create clean, adaptive environments to align with long-term ecological priorities.

References

1. Liu K, Zhang S (2024) Smart cities stochastic secured energy management framework in digital twin: policy frameworks for promoting sustainable urban development in smart cities. Sustain Energy Technol Assess 65:103720. https://doi.org/10.1016/j.seta.2024.103720
2. Jiang H, Xing M (2024) Enhancing urban economic efficiency through smart city development: a focus on sustainable transportation. Comput Electr Eng 116:109058. https://doi.org/10.1016/j.compeleceng.2023.109058
3. Barbieri R, Coluccia B, Natale F (2025) How are smart city policies progressing in Italy? Insights from SDG indicators. Land Use Policy 148:107386. https://doi.org/10.1016/j.landusepol.2024.107386
4. Tahmasebi D, Sheikh M, Dabbaghjamanesh M, Jin T, Kavousi-Fard A, Karimi M (2024) A security-preserving framework for sustainable distributed energy transition: case of smart city. Renewable Energy Focus 51:100631. https://doi.org/10.1016/j.ref.2024.100631
5. Modarelli G, Sadraei R, Rainero C (2024) How to perceive sustainable moving and smart mobility today?: a cross-national comparative longitudinal perspective and the controversy of alternative transport systems. J Clean Prod 468:143121. https://doi.org/10.1016/j.jclepro.2024.143121
6. Yang Z, Gao W, Han Q, Qi L (2024) Aggravating or alleviating? Smart city construction and urban inequality in China. Technol Soc 77:102562. https://doi.org/10.1016/j.techsoc.2024.102562
7. Shafiq M, Bhavani NPG, Ramesh JVN, Veeresha RK, Talasila V, Alfurhood BS (2024) Thermal modeling and machine learning for optimizing heat transfer in smart city infrastructure balancing energy efficiency and climate impact. Thermal Science and Engineering Progress 54:102868. https://doi.org/10.1016/j.tsep.2024.102868
8. Urchl H, Diallo AB, Elbiaze H, Sahir E, Sadik M (2025) A contemporary survey on multi-source information fusion for smart sustainable cities: emerging trends and persistent challenges. Information Fusion 114:102667. https://doi.org/10.1016/j.inffus.2024.102667
9. Yi M, Chen D, Wu T, Sheng MS, Zhang Y (2024) Intelligence and carbon emissions: the impact of smart infrastructure on carbon emission intensity in cities of China. Sustain Cities Soc 112:105602. https://doi.org/10.1016/j.scs.2024.105602
10. Margherita EG, Escobar SD, Esposito G, Crutzen N (2023) Exploring the potential impact of smart urban technologies on urban sustainability using structural topic modelling: evidence from Belgium. Cities 141:104475. https://doi.org/10.1016/j.cities.2023.104475
11. Nederhand J, Avelino F, Awad I, De Jong P, Duijn M, Edelenbos J, Engelbert J, Fransen J, Schiller M, Van Stapele N (2023) Reclaiming the city from an urban vitalism perspective: critically reflecting smart, inclusive, resilient and sustainable just city labels. Cities 137:104257. https://doi.org/10.1016/j.cities.2023.104257
12. Hui CX, Dan G, Alamri S, Toghraie D (2023) Greening smart cities: an investigation of the integration of urban natural resources and smart city technologies for promoting environmental sustainability. Sustain Cities Soc 99:104985. https://doi.org/10.1016/j.scs.2023.104985
13. Peldon D, Banihashemi S, LeNguyen K, Derrible S (2024) Navigating urban complexity: the transformative role of digital twins in smart city development. Sustain Cities Soc 111:105583. https://doi.org/10.1016/j.scs.2024.105583

14. Wu M, Guo M, Xu J (2024a) The influence of smart city policy on urban green energy efficiency— A quasi-natural experiment based on 196 cities. J Clean Prod 449:141818. https://doi.org/10.1016/j.jclepro.2024.141818
15. Wu J, Wang H, Yao J (2024b) Computer-aided urban energy systems cyber attach detection and mitigation: intelligence hybrid machine learning technique for security enhancement of smart cities. Sustain Cities Soc 108:105384. https://doi.org/10.1016/j.scs.2024.105384
16. Xie X, Zheng L, Wang R, Gou Z (2024) Visitors' experience of using smart facilities in urban parks: a study in Shenzhen. J Outdoor Recreat Tour 46:100759. https://doi.org/10.1016/j.jort.2024.100759
17. Zoumpoulis P, Konstantinidis FK, Tsimiklis G, Amditis A (2024) Smart bins for enhanced resource recovery and sustainable urban waste practices in smart cities: a systematic literature review. Cities 152:105150. https://doi.org/10.1016/j.cities.2024.105150
18. Peixoto JPJ, Bittencourt JCN, Jesus TC, Costa DG, Portugal P, Vasques F (2024) Exploiting geospatial data of connectivity and urban infrastructure for efficient positioning of emergency detection units in smart cities. Comput Environ Urban Syst 107:102054. https://doi.org/10.1016/j.compenvurbsys.2023.102054
19. Agboola OP, Bashir FM, Dodo YA, Mohamed MAS, Alsadun ISR (2023) The influence of information and communication technology (ICT) on stakeholders' involvement and smart urban sustainability. Environmental Advances 13:100431. https://doi.org/10.1016/j.envadv.2023.100431
20. Peteleaza D, Matei A, Sorostinean R, Gellert A, Fiore U, Zamfirescu BC, Palmieri F (2024) Electricity consumption forecasting for sustainable smart cities using machine learning methods. Internet Things 27:101322. https://doi.org/10.1016/j.iot.2024.101322
21. Alhasnawi BN, Jasim BH, Alhasnawi AN, Hussain FFK, Homod RZ, Hasan HA, Khalaf OI, Abbassi R, Bazooyar B, Zanker M, Bureš V, Sedhom BE (2024) A novel efficient energy optimization in smart urban buildings based on optimal demand side management. Energy Strategy Reviews 54:101461. https://doi.org/10.1016/j.esr.2024.101461
22. Wang D, Liang YF, Dou W (2024) How does urban industrial structure upgrading affect green productivity? The moderating role of smart city development. Struct Chang Econ Dyn 72:133. https://doi.org/10.1016/j.strueco.2024.09.017
23. Aljohani A (2024) Deep learning-based optimization of energy utilization in IoT-enabled smart cities: a pathway to sustainable development. Energy Rep 12:2946–2957. https://doi.org/10.1016/j.egyr.2024.08.075
24. Bressane A, Loureiro AIS, Medeiros LC, Moruzzi R, Formiga JKS, Negri RG, Saraiva ACV (2024) Intelligent nature-based solutions in the 1st smart sustainable Brazilian city: insights and lessons learned. Nature-Based Solutions 6:100161. https://doi.org/10.1016/j.nbsj.2024.100161
25. Chen Y, Chen S, Miao J (2024) Does smart city pilot improve urban green economic efficiency: accelerator or inhibitor. Environ Impact Assess Rev 104:107328. https://doi.org/10.1016/j.eiar.2023.107328
26. Papageorgiou G, Tsappi E, Wang T (2024) Smart urban systems planning for active mobility and sustainability. IFAC-Pap Online 58:261–266. https://doi.org/10.1016/j.ifacol.2024.07.350
27. Meng X, Zhu L (2024) Augmenting cybersecurity in smart urban energy systems through IoT and blockchain technology within the digital twin framework. Sustain Cities Soc 106:105336. https://doi.org/10.1016/j.scs.2024.105336
28. Lu H, Xiao C, Jiao L, Du X, Huang A (2024) Spatial-temporal evolution analysis of the impact of smart transportation policies on urban carbon emissions. Sustain Cities Soc 101:105177. https://doi.org/10.1016/j.scs.2024.105177
29. Alijoyo FA (2024) AI-powered deep learning for sustainable industry 4.0 and internet of things: enhancing energy management in smart buildings. Alex Eng J 104:409–422. https://doi.org/10.1016/j.aej.2024.07.110
30. Wang W (2024) Making older urban neighborhoods smart: digital placemaking of everyday life. Cities 147:104814. https://doi.org/10.1016/j.cities.2024.104814

31. Zhu X, Chen Z, Cheng T, Yang C, Wu D, Wu Y, Wang H (2024) Blockchain for urban governance: enhancing trust in smart city systems with advanced techniques. Sustain Cities Soc 108:105438. https://doi.org/10.1016/j.scs.2024.105438
32. Joshi VD, Agarwal P, Kumar A, Dogra N, Nandan D (2024) Urban odyssey: pioneering multimodal routes for tomorrow's smart cities. Measurement: Sensors 36:101301. https://doi.org/10.1016/j.measen.2024.101301
33. Fadhel MA, Duhaim AM, Saihood A, Sewify A, Al-Hamadani MNA, Albahri AS, Alzubaidi L, Gupta A, Mirjalili S, Gu Y (2024) Comprehensive systematic review of information fusion methods in smart cities and urban environments. Information Fusion 107:102317. https://doi.org/10.1016/j.inffus.2024.102317
34. Cheng L, Samah A, Zhang Y, Jia Z, Li F (2024) Towards urban energy sustainability and resiliency through smart agriculture and effective renewable resource allocation. Sustain Cities Soc 107:105420. https://doi.org/10.1016/j.scs.2024.105420
35. Weil C, Bibri SE, Longchamp R, Golay F, Alahi A (2023) Urban digital twin challenges: a systematic review and perspectives for sustainable smart cities. Sustain Cities Soc 99:104862. https://doi.org/10.1016/j.scs.2023.104862
36. Tsoutsa P, Panagiotakopoulos T, Damasiotis V, Fitsilis P (2024) Empowering urban futures: key competencies for smart city resilience officers. Urban Gov 4:180–192. https://doi.org/10.1016/j.ugj.2024.07.001
37. Chen G, Zhang JW (2024) Intelligent transportation systems: machine learning approaches for urban mobility in smart cities. Sustain Cities Soc 107:105369. https://doi.org/10.1016/j.scs.2024.105369
38. Qian X, Chen M, Zhao F, Ling H (2024) An assessment framework of global smart cities for sustainable development in a post-pandemic era. Cities 150:104990. https://doi.org/10.1016/j.cities.2024.104990
39. Chen P, Dagestani AA (2023) Urban planning policy and clean energy development harmony—evidence from smart city pilot policy in China. Renew Energy 210:251–257. https://doi.org/10.1016/j.renene.2023.04.063
40. Razavian B, Hamed SM, Fayyaz M, Ghasemi P, Ozkul S, Tirkolaee EB (2024) Addressing barriers to big data implementation in sustainable smart cities: improved zero-sum grey game and grey best-worst method. J Innov Knowl 9:100593. https://doi.org/10.1016/j.jik.2024.100593
41. Zhao X, Zhang Y (2024) Integrated management of urban resources toward net-zero smart cities considering renewable energies uncertainty and modeling in digital twin. Sustain Energy Technol Assess 64:103656. https://doi.org/10.1016/j.seta.2024.103656
42. Vishnu TP, Gopakumar P, Sunitha R (2023) Simultaneous AC/DC distribution for the sustainable operation of smart building clusters in hybrid urban microgrids. Energ Buildings 298:113488. https://doi.org/10.1016/j.enbuild.2023.113488
43. Prasad D, Alizadeh T (2024) Cross examination of smart urbanism and southern urbanism: a systematic literature review in search of 'Southern smart urbanism'. Cities 155:105459. https://doi.org/10.1016/j.cities.2024.105459
44. Han MJN, Kim MJ (2024) A systematic review of smart city research from an urban context perspective. Cities 150:105027. https://doi.org/10.1016/j.cities.2024.105027
45. Oliveira da Silva A, Souza Fernandes RA (2024) Multipurpose territorial cadastre and collaborative participation as tools for smart urban governance: an analysis considering the pandemic's effect in the São Paulo metropolitan region, Brazil. Sustain Cities Soc 115:105831. https://doi.org/10.1016/j.scs.2024.105831
46. Tan Z, Li Z (2024) Digital twins for sustainable design and management of smart city buildings and municipal infrastructure. Sustain Energy Technol Assess 64:103682. https://doi.org/10.1016/j.seta.2024.103682
47. Sánchez O, Castañeda K, Vidal-Méndez S, Carrasco-Beltrán D, Lozano-Ramírez NE (2024) Exploring the influence of linear infrastructure projects 4.0 technologies to promote sustainable development in smart cities. Results Engineering 23:102824. https://doi.org/10.1016/j.rineng.2024.102824

48. Makkonen T, Inkinen T (2024) Inclusive smart cities? Technology-driven urban development and disabilities. Cities 154:105334. https://doi.org/10.1016/j.cities.2024.105334
49. Shu Y, Deng N, Wu Y, Bao S, Bie A (2023) Urban governance and sustainable development: the effect of smart city on carbon emission in China. Technological Forecasting and Social Change 193:122643. https://doi.org/10.1016/j.techfore.2023.122643
50. Wu J, Yang R, Zhao P, Yang L (2024c) Computer-aided mobility solutions: machine learning innovations to secure smart urban transportation. Sustain Cities Soc 107:105422. https://doi.org/10.1016/j.scs.2024.105422

An Energy Audit Planning Framework for Energy Efficiency Retrofit Projects for Smart Buildings

Buhle Goodman Shandu, Tebello Ntsiki Don Mathaba, and Peter Muganyi

Abstract Buildings have been identified as significant energy users globally. Energy retrofits can significantly reduce building energy consumption. This study presents a method based on assessing performance efficiency, operation efficiency, equipment efficiency and technology efficiency referred to as (POET) framework for holistic identification of energy saving opportunities (ESOs) in a commercial building. This method is applied to a case-study in a commercial building located in South Africa. The paper presents financial, environmental and complexity metrices for evaluating the potential energy saving projects that emanate from the POET framework. All the ESOs promised some reduction in carbon dioxide (CO_2) emissions, they also had positive net present value (NPV) values and their payback period (PBP) were a fraction of their useful lives. The ESOs that were identified promise a total monthly energy saving of 163754.22 kWh, which is 85% of the current monthly consumption. The ranking of the ESOs indicated that ESO 1, 2 and 3 could be prioritized immediately with marginal investment to realize 7.4% monthly energy savings, while ESO4, 5, 6 and 7 can be planned for considering the relatively higher complexity and investment requirements.

Keywords POET · Energy savings · Commercial buildings · Energy saving opportunity (ESO) · Smart cities · Smart buildings

B. G. Shandu · T. N. D. Mathaba (✉) · P. Muganyi
Postgraduate School of Engineering Management, University of Johannesburg, Johannesburg, South Africa
e-mail: tmathaba@uj.ac.za

H. Kaddouri et al. (eds.), *Smart Cities for Sustainable and Urban Development*, Green Energy and Technology, https://doi.org/10.1007/978-3-031-99001-4_19

1 Introduction

Reducing energy consumption in building operations is vital for sustainable development and combating climate change [1]. Globally, buildings account for around 40% of energy use [2]. Therefore, interventions such as load disaggregation which involves analyzing energy use by breaking it down into end-uses or appliances—provides essential insights, especially in complex commercial settings [3–5]. Among significant energy users in buildings, Heating, ventilation and air conditioning (HVAC) systems have been identified to play a major role in energy use, making HVAC data a valuable source for identifying savings [3].

Metering and sub-metering play a vital role in measuring and monitoring building energy consumption, offering essential data to analyze energy patterns and implement targeted efficiency strategies [6–8]. Building Automation Systems (BAS), along with innovations like Recommender Systems (RSS) and AI, enhance energy management by centralizing control and fostering greater user engagement [9]. Improving building energy efficiency can be achieved by selecting energy-efficient air-conditioning systems, which are rated using the Seasonal Energy Efficiency Ratio (SEER) to quantify their efficiency. Studies show that many developing regions lag in adopting energy-efficient technologies [10–13]. Policies like the European Union's (EU) Energy Performance of Building Directive (EPBD) aim for a 55% reduction in emissions by 2030 and climate neutrality by 2050 [14, 15]. These policies highlight improved insulation, HVAC upgrades, efficient lighting, and renewable energy as key strategies [16, 17].

In South Africa, new regulations require energy performance standards for large buildings by 2025, aligning with global initiatives like the United Nations' sustainable development goals (SDGs) for clean energy and climate action [18, 19]. This convergence underscores the commitment to energy-efficient buildings as part of a sustainable future.

Previous study used the POET approach to define three criteria for the sustainability of an energy program: organizational structure, compatibility of performance indices (mainly financial), and engineering support in a financial services commercial building [20]. Other studies analysed the energy efficiency systems such as pumping and HVAC, form a POET perspective [21]. The objective of this study is to add to previous work by demonstrating the efficacy of the POET framework in identification of energy saving opportunities in a commercial warehouse. The outcomes of this study should benefit owners and operators of similar buildings in identifying, evaluating and implementing energy efficiency measures. The study defines three types of criteria for evaluating potential energy savings projects: financial, environmental and complexity. This paper investigates the following research questions: RQ1: What energy management opportunities exists in a commercial building? RQ2: Which methods can be applied in commercial buildings to evaluate energy management opportunities? RQ3: How can POET method be used to identify energy management opportunities in a commercial building?

This rest of this review paper is structured as follows: Section 2 reviews cases of building energy management, the methodology which is outlines the step-by-step approach is discussed in Sect. 3. Section 4 presents results and discussions and Sect. 5 provides the conclusion.

2 Literature Review

Based on a comprehensive literature survey that was conducted, the approach of this study to energy efficiency in commercial buildings may be classified into five categories: an overview of energy management in commercial buildings; application of POET framework; energy efficiency of commercial building's technology and appliances; and evaluation of energy management opportunities in commercial buildings.

2.1 An Overview of Energy Management in Commercial Buildings

Buildings contribute to 40% of global energy consumption and generate one-third of all greenhouse gas emissions, with energy losses ranging from 2% to 20% caused by inefficient appliances and irresponsible consumer behavior [22]. Buildings alone are responsible for approximately 30% of global energy-related carbon dioxide emissions [23]. Hence, the adoption of energy management practices can significantly curb unnecessary energy usage (2015 [24, 25]). The construction industry holds significant potential to reduce energy demand through the adoption of Intelligent Energy Management Systems (IEMS) and the concept of the Internet of Energy (IoE) [26]. Online-based architecture is referred to as the Internet of Things (IoT) in two ways: first, communication enables real-time system monitoring and control through cloud computing or other online services [27].

The potential of Internet of Everything (IoE)-based Building Energy Management Systems (BEMS) has been shown to significantly enhance the efficiency of building energy usage in the future [28]. Lighting and air conditioning are the primary electrical loads in commercial buildings, accounting for more than half of the total energy consumption [29]. The rise of the Internet of Things has made smart devices commonplace in homes and businesses. Smartphone apps play a key role in enabling users to control and monitor these devices, thereby enhancing energy efficiency. Real-time data on energy consumption informs users' decisions, leading to reduced energy waste and costs, which are important factors for consideration by smart city planners and developers [30].

Future trends with the potential to enhance energy efficiency in both existing and new buildings include Building Integrated Photovoltaics (BIPVs) and Net Zero

Energy Building (NZEB) concepts. The current global power demand is roughly 15 Tera Watt (15 × 10^4 W). The solar energy received in less than an hour is thought to be sufficient to cater for a year's worth of the global energy budget [31]. Photovoltaic (PV) technology is regarded as one of the most promising solutions for efficiently harnessing solar energy [32]. BIPVs are photovoltaic modules that are integrated into the building envelope [33]. Energy retrofitting for net-zero energy buildings (NZEBs) balances generation and demand while also taking grid integration into consideration [34]. It operates in conjunction with passive control strategies, energy-efficient technologies, and the integration of renewable energy resources (RER) [35]. The growth of the digital economy is closely tied to the adoption of IT and smart technologies. An analysis of global trends reveals significant expansion in Smart City infrastructure during 2016–2017, marked by the integration of intelligent solutions with IT technologies across all aspects of human life [36].

2.2 Application of the POET Framework

The key classification of a POET framework are performance, operation, equipment, and technology [37]. By categorizing system efficiency into Performance, Operation, Equipment, and Technology (POET), stakeholders can systematically assess and improve different facets of energy systems. This framework provides a comprehensive approach to enhancing efficiency, sustainability, and economic viability in energy management and utilization [21].

2.3 Energy Efficiency of Commercial Building Technology and Appliances

The various approaches to enhancing or achieving energy efficiency in buildings can be categorized into three groups: active strategies, passive design, and renewable energy use [38]. The primary focus of the active measures is on any power-driven devices, appliances, or systems, including the HVAC system [38], artificial lighting, dynamic glazing [39], and artificial ventilation [40]. Additionally, they address any potential automation or control methods, such as building management systems and thermostat settings [41]. Studies have shown that using active solutions to achieve energy efficiency holds significant potential. Passive energy conservation strategies aim to reduce energy losses through the building envelope and optimize the use of natural resources for lighting, heating, and cooling [42]. In the past, passive heating and cooling techniques were included into traditional buildings to account for local climates during the summer and winter, respectively.

Appliances used for cleaning, cooking, water heating, and other end-use applications in commercial buildings contribute to approximately 22% of the annual primary energy demand in these buildings [43]. According to recent research from the DOE's Building Solutions Office, current energy-saving solutions have the potential to reduce commercial appliance usage by 22%, with emerging technologies offering even greater savings potential of up to 36% [44].

2.4 *Evaluation of Energy Management Opportunities in Commercial Building Technology and Appliances*

The European Commission's method for assessing cost-optimality in energy efficiency measures includes evaluating the financial viability of these investments through the use of Discounted Cash Flow analysis [44]. This method calculates the NPV of energy efficiency measures by considering both the initial investment and the anticipated future energy cost savings [45]. A study reviewing 18 articles from 13 different countries found that the payback period was by far the most commonly used economic indicator for energy efficiency investments [45]. The concept of Environmental Energy Efficiency (EEE) as a new indicator of energy performance focuses specifically on energy reduction efforts and is expected to improve as advancements in energy-saving technology are made [46]. For this study, CO_2 emissions shall be used to evaluate the various initiatives. The level of difficulty or complexity is also used to evaluate potential energy efficiency projects. It is important that any of the energy efficiency measures that are selected are feasible and does not require rare specialist skills for implementation [47].

2.5 *General Analysis and Research Gap Extraction*

2.5.1 An Overview of Energy Management in Commercial Buildings

Energy consumption in buildings, the challenges associated with energy inefficiency, and potential solutions through technology and smart infrastructure are examined in this section. Various researchers seem to agree that some of the key challenges associated with buildings is that they account for 40% of global energy consumption and generate one-third of all greenhouse gas emissions, and in the course of their operation energy losses ranging from 2% to 20% are encountered mainly as a result of inefficient appliances and irresponsible consumer behavior. This perspective undoubtably depicts the glaringly high energy demand and adverse environmental impact caused by budlings which can undermine the global efforts towards the reduction of greenhouse gas emissions if it is unabated.

These challenges therefore highlight the urgent need for energy-efficient practices and technologies to mitigate the environmental impact of buildings and the need to monitor, control and conserve energy. Previous studies by various researchers have also proposed the use of emerging technological tools like the Intelligent

Energy Management Systems (IEMS), Internet of Everything (IoE)-based Building Energy Management Systems (BEMS), smartphones to enhances real time monitoring, control, communication and decision making that can lead to reduction of energy waste and cost. As suggested by researchers, it is agreed that the wide spread use and accessibility of smart phones in society, makes them a viable tool for the development of smart cities by enhancing energy management measures in buildings among other interventions. Researchers have proposed smart city infrastructure that includes building integrated photovoltaic (BIPV), net zero energy building concepts and grid integration. The integration of the above-named intelligent solutions with IT solutions is believed to be the growing trend and certain future of building energy efficiency.

2.5.2 Application of the POET Framework

The POET framework that is presented in this section provides a framework and approach to analyzing an energy system from four key pertinent perspectives. These are performance, operational, equipment and technology. The framework provides a uniform approach that can be used consistently in various researches or energy management projects to unpack the potential energy efficiency measures for various applications.

2.5.3 Energy Efficiency of Commercial Building Technology and Appliances

This section highlighted effects of active strategies, passive designs and renewable energy use as approaches for achieving energy efficiency in buildings. It also indicates that although appliances contribute 22% of building energy demand, the application of energy saving solutions can reduce appliance energy usage by up to 36%. Most energy efficiency measures for existing buildings are active strategies such as HVAC and lighting retrofits among others. The window of opportunity for passive designs is available prior and during construction of a building. However, some of the passive design measures for building efficiency can be undermined during value engineering where designers fail to place value on building efficiency and therefore compromise these to keep the building costs low.

2.5.4 Evaluation of Energy Management Opportunities in Commercial Building Technology and Appliances

Assessing cost optimality or evaluation of financial viability of investment in energy efficiency measures is a critical part which assesses the feasibility of energy efficiency measures in order to ensure that they receive support from funders. Net present value and payback period are highlighted as the commonly used financial metrices for evaluation and support of investment decision. Other methods that are presented for evaluating energy management measures are carbon footprint reduction, for environmental impact evaluation and complexity analysis of the potential energy saving measure.

2.5.5 Research Gaps

In a lighting benchmark study, it was noted that the building's latitude was not taken into account. Therefore, future research could build on this by first expanding it at the national level and then extending it internationally, offering lighting benchmarks for each urban settlement [48].

In a study on zero energy buildings (ZEB) in Singapore, both passive and active energy efficiency strategies were tested in pilot projects. Active measures, such as air-conditioning system upgrades and lighting retrofits with LED, proved to be cost-effective with short payback periods. However, passive strategies were less cost-effective and required longer payback periods. For instance, the rooftop greenery was too small to have a significant impact on energy savings. Given that the study focused on small-scale interventions like rooftop greenery, the gap was identified, and it was recommended that future research explore larger-scale passive design projects to better assess their cost-effectiveness compared to active design solutions [49].

Building energy efficiency strategies can take either passive or active forms. By lowering building energy consumption, passive energy solutions may help reduce reliance on fossil fuels and expensive renewable energy sources. When passive measures are included in building renovation projects, the relationship between the existing structures and these technologies needs to be modelled and measured through the use of integrated assessment or building simulation techniques. This entails evaluating necessary adjustments and configurations for the HVAC system, assessing the impact of internal heat gains (from hot water systems, lighting, etc.), and considering potential future climate scenarios that may influence the effectiveness of additional passive measures. A future research direction will be to simulate building energy and emissions inventories to analyse the impact of these interactions throughout the building's lifecycle [50].

In a South African study that aimed to explore the impact of innovative technologies on the environmental sustainability of commercial properties. The nondisclosure of intellectual property prevented the acquisition of the installed new technologies' capital and operating costs. Consequently, a gap for greater investigation into the capital cost, energy usage, and maintenance expenses of innovative technologies was identified in order to produce a more thorough financial prediction model. Furthermore, since just three incidents were looked into in South Africa. It was advised that a bigger sample size be employed in a subsequent study to increase the accuracy of the results [51].

The current case-study, the analysis of meter data indicated that an estimated 74% of the buildings consumption was attributed to the HVAC system, while 25%

was due to lighting loads and 1% was from small plug loads such as kettles and personal computers. This analysis has a limitation in that only HVAC, lighting and small power were sub-metered, and not all categories of loads that were identified in this building such as forklifts battery chargers and the cold-room. Future studies will consider more sub-meters for a complete breakdown of building loads for more accurate view of how consumption is distributed.

3 Methodology

Figure 1 summarizes the proposed methodology for identifying and assessing the energy efficiency opportunities. Other energy efficiency studies have used similar approach [52].

The process begins with *Step 1* by identifying a building to be used for the case-study, and obtaining necessary approvals for its use, from its owners. For this study, a 14,782 m^2 cargo terminal building owned and operated by the case-study company was used. The aerial photograph is this building is indicated in Fig. 2.

In *Step 2*, energy audit is performed by collecting energy consumption data through utility bills and surveying cooling systems, heating systems, lighting systems, and existing appliances. This study utilised the level 1 or walk-through energy audit, as defined by the American Society of Heating, Refrigerating and Air-Conditioning Engineers (ASHRAE) [53]. The aim is to establish the historic energy consumption and costs, so that low-cost non-retrofit or retrofit opportunities may be identified as low hanging fruits.

In Step 3, the Performance, Operations, Equipment and Technology (POET) framework is used for identification of energy saving opportunities (ESOs). The identified energy efficiency opportunities can potentially be applied at four key levels, i.e. conceptual, active, technical, and engineering levels [54]. For this case-study, a conceptual level assessment has been selected as the most appropriate.

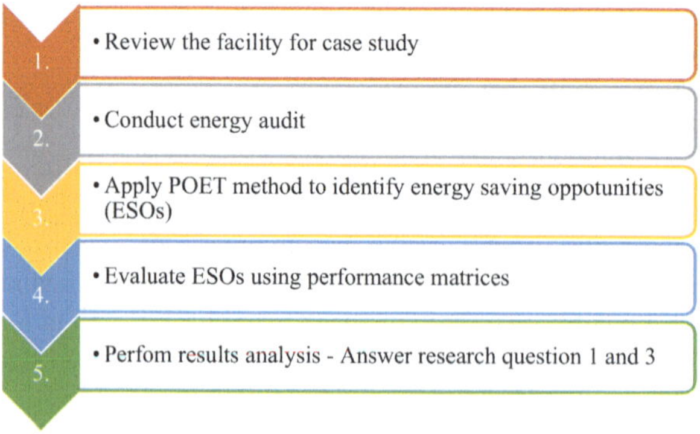

Fig. 1 Summary of the methodology

Fig. 2 Description of case-study building

Literature indicates that the conceptual level application of POET is based on the 80–20 rule, meaning 20% of effort with 80% benefit, requiring less expertise and is typified by the identification of cost-saving initiatives that call for low skill levels [37]. The rationale for this selection is supported by ASHRAE, because the case-study follows the level 1 or walk-through energy audit approach which agrees with the 80–20 rule [53].

In *Step 4*, the identified ESOs are evaluated from economic, environmental and complexity perspectives. The economic evaluation procedure follows EN 15459 standards, utilizing economic parameters such as the NPV and PBP to assess a project's financial viability. The parameters used in the study are the following:

The Net Present Value is a financial metric that assesses the profitability of an investment by calculating the difference between the present value of expected cash inflows and the present value of expected cash outflows over time. It discounts future cash flows to their present value; thus, it accounts for the time value of money, providing a realistic measure of an investment's worth [52]. The NPV formula is:

$$NPV = \sum_{t=0}^{N} \frac{CF_t}{(1+r)t} \qquad (1)$$

Where:

NPV is the net present value;

CF_t is the net cash flow at time t;

r is the discount rate;

t is the total number of periods.

The Payback Period is a financial metric that determines how long it will take for an investment to recoup its initial cost through cumulative cash inflows. The PBP is usually expressed in years or months, offering a quick insight into an investment's liquidity risk and helping decision-makers gauge the financial feasibility of projects [52]. PBP formula is:

$$Payback\ Period = \frac{Initial\ investment}{Cashflow\ per\ year} \tag{2}$$

The metric used for reduction of environmental or greenhouse gas (GHG) impact is carbon emission. This is calculated using the general methodology of Intergovernmental Panel on Climate Change (IPCC) which is an intergovernmental body of the United Nations as indicated in eq. (3) [55]:

$$\boldsymbol{CO2\ Emissions = Emissions\ Factor\ x\ Energy\ Savings} \tag{3}$$

Where:

- CO2 Emissions = kg CO_2;
- Emissions Factor = kg CO_2/KWh;
- Energy Savings = KWh.

Lastly the identified ESOs are evaluated using the selected measure of complexity. The metric for complexity could be either a high or low complexity.

4 Results and Discussions

4.1 Walk Through Audit Results

The graphical illustration of the utility bill shows that there has been a slight increase in the energy consumption of the case-study building between 2022 and 2024. The consumption increased by 21006.02 kWh. The consumption seems to have been steady during the winter months with 0% variance on average between the study years. However, during the summer months, i.e. October onwards, energy consumption increased by an average of 9% seasonally and by 3% year on year (Figs. 3 and 4).

The case-study building is equipped with some submeters measuring consumption of lighting, small power (plug loads) and HVAC. The available meter data indicated a buildings consumption split 74% for the HVAC system, 25% for lighting loads and 1% for small plug loads such as kettles and personal computers. This analysis was limited in that only HVAC, lighting and small power were sub-metered. Future studies may consider more sub-meters for a complete segmentation of building loads and more accurate view of how consumption is distributed (Table 1).

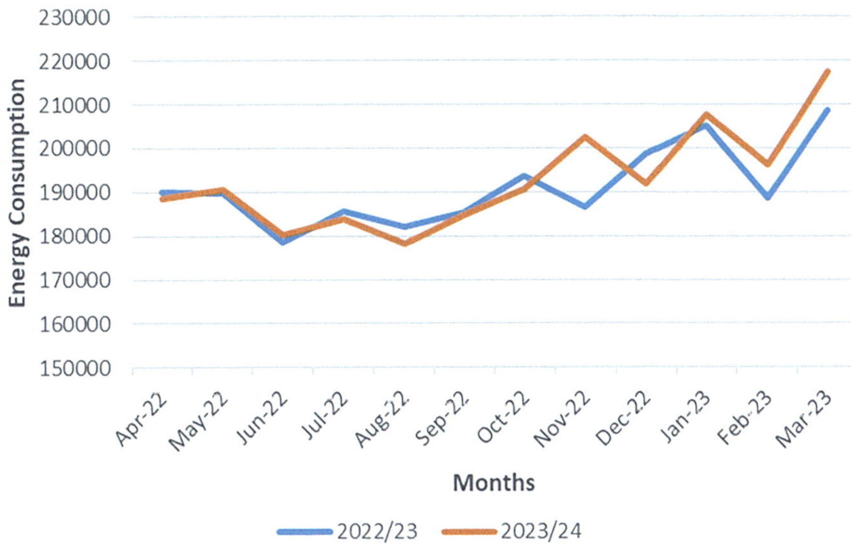

Fig. 3 The Case-study Building Energy Consumption

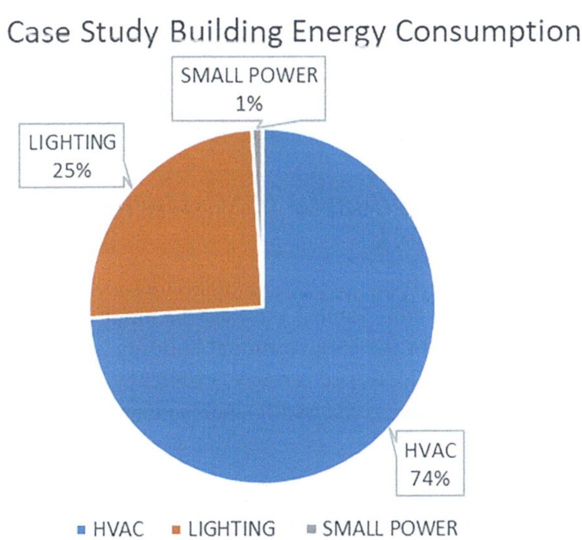

Fig. 4 The Case-study Building Energy Consumption

The energy performance, i.e. the energy consumed per floor area per annum was calculated to be 635.58 kWh/m^2 .a. The Reference Energy Rating (Er) for commercial building of occupancy class which is G1-M is 95 kWh/m^2.a. Since the measured rating is approximately 7 times more that the reference EPC rating, the result of the EPC assessment was that the case-study building is among the worst performing buildings with the worst performance rating of G.

Table 1 Case-study building EPC rating details

Type of Occupancy	Single
Occupancy class	G1-M
Period of assessment	April 2022–March 2023
Reference energy rating (Er) (kWh/m2 .a)	95
Total annual energy consumption (kWh)	2292672.60
Electricity (kWh)	2292672.60
Excluded energy (kWh)	165493.92
Net floor area (m2)	3346.83
Energy performance (kWh/m2 .a)	635.58
Energy performance of exclusion (kWh/m2 .a)	49.45
EPC rating	G

Table 2 POET energy saving opportunities

Performance Efficiency	Operation Efficiency
1. Budget for solar PV to provide energy mix and clean source and to reduce carbon footprint of the building	1. Install timers to control on/off status of hydro boils
Equipment efficiency	Technology efficiency
1. Replace traditional fixed speed motors with variable speed drive (VSD) controlled inverter rated motors in the HVAC system, 2. Replace the faulty power factor correction (PFC) unit.	1. Lighting retrofits by change lights to LED, 2. Install building occupancy sensors, 3. Install a building management system (BMS)

4.2 Energy Saving Opportunities (ESOs) Identified Using the POET Framework

Following the walkthrough audit, Table 2 presents the energy saving opportunities (ESOs) that were identified for the cargo terminal building through the application of POET framework. The identified energy efficiency opportunities can be implemented as projects to address the inefficiencies identified by the audit.

4.3 Evaluation of Potential Energy Saving Opportunities

Table 3 is a summary of initial investment required to achieve the expected energy savings (kWh) per energy saving opportunity that was identified using the POET framework. The data in this table forms the basis for the analysis and economic, environmental and complexity evaluation of each of the ESO as detailed fully in the appendices. The ESO that were identified promise total monthly energy saving of 163754.22 kWh, which is 85% of the current monthly consumption.

Table 3 Summary of investment and monthly energy savings

Energy saving opportunities	Investment cost per unit (ZAR)	Monthly energy savings (kWh)
ESO 1	R19 980.00	1070.67
ESO 2	R405 721.38	12007.04
ESO 3	R23 947.50	1132.72
ESO 4	R14 000000.00	80943.70
ESO 5	R766 973.75	35259.00
ESO 6	R701 243.45	9636.15
ESO 7	R1 120065.85	23704.94
TOTAL	R17 037931.93	163754.22

Table 4 Environmental matrices

Energy saving opportunities	Monthly carbon footprint reduction (kg CO_2)
ESO 1	910.07
ESO 2	30715.24
ESO 3	30715.24
ESO 4	68802.14
ESO 5	29970.15
ESO 6	8190.73
ESO 7	20149.20
Total	**189452.77**

Table 5 Level of complexity

Complexity indicator	Rating		
	Respondent 1	Respondent 2	Average
ESO 1—Install timers for Hydo boils	1.5	1.5	1.5
ESO 2—LED lighting	1.5	1.5	1.5
ESO 3—Occupancy sensors	2.0	2.0	2.0
ESO 4—Solar PV	4.5	5.0	4.8
ESO 5–VSD	4.0	3.0	3.5
ESO 6—Replace PFC	3.0	3.0	3.0
ESO 7—BMS	4.5	5.0	4.8

Table 4 summarizes the results of CO_2 emissions that are reduced as a direct result of the implementation of identified energy saving opportunities or projects in the case-study building. Details of the methods used to arrive this CO_2 reduction are fully described in Appendix D to I of the report (Table 5).

A survey was conducted to allow the respondents to rate their perceived level of complexity using of each ESO using a 1 to 5 rating scale. The description of the scale was tailored for each survey question.

Q1: How complex is it for the level of infrastructure existing at the site to support the project? 1 = least complex, 2 meant not complex, 3 = neutral, 4 = complex, and 5 = most complex.

Q2: What is the level of difficulty in securing project funding for the project? 1 = least difficult, 2 = not difficult, 3 = neutral, 4 = difficult, and 5 = most difficult.

Two respondents were selected due to their high level of expertise. Respondent 1 was a Principal Electrical Engineer, registered as PrEng with the Engineering Council of South Africa (ECSA). Respondent 2 was a Principal Mechanical Engineering Technologist, ECSA registered as Pr Eng Tech.

4.3.1 Prioritization of ESOs or Projects

The projects are ranked in order to prioritize the implementation of the most viable and less complex ones. Results of financial, environmentally and complexity analysis are summarized for ranking purposes selecting PBP, kgCO_2 and complexity rating as criteria for prioritizing energy saving projects (Table 6).

There projects that can be prioritized and fast tracked are: ESO 2, ESO 3 and ESO 1. These ESOs have least complexity, favourable PBP, and they yield a significant reduction in CO_2, emissions. They are low-cost projects and can be viewed as low hanging fruit that will reduce the building's monthly energy consumption by 14210.43 kWh, or 7.4% at a marginal once off cost of R449 648.88. More energy savings will be achieved by implementing ESO 4, 5, 6 and 7, however the investment required and complexity are relatively higher.

4.4 How Energy Audit Realises the Energy Efficiency in Smart Buildings

An energy audit plays a crucial role in realizing energy efficiency in smart buildings by systematically evaluating energy usage and identifying opportunities for improvement. Energy audit contributes by providing a comprehensive assessment of energy consumption. This entails Data Collection, which is detailed data on energy usage patterns, including lighting, HVAC systems, appliances, and building automation systems. During an energy audit smart building, equipped with IoT sensors, allow for precise real-time monitoring of energy usage at granular levels. Energy audit provides a baseline establishment to establish a benchmark for current energy performance, identifying areas with excessive energy consumption or inefficiencies.

Energy audit is for identification of inefficiencies, through systematic analysis of detects inefficiencies in building systems, such as outdated appliances, poorly maintained equipment, or improper system configurations. It pinpoints areas of energy waste, such as standby power losses or unnecessary lighting. It highlights consumer behavior insights to reveal patterns of energy-intensive behavior among occupants, enabling targeted interventions to encourage responsible usage.

Table 6 Ranking of metrices

	ESO 1	ESO 2	ESO 3	ESO 4	ESO 5	ESO 6	ESO 7
PBP	7.5 months	1.32 years	8.5 months	5794 years	8.2 months	5.4 months	7.8 months
Monthly reduction in kg CO_2	910.07	30715.24	30715.24	68802.14	29970.15	8190.73	20149.20
Level of complexity	1.5	1 5	2.0	4.8	3.5	3.0	4.8

Energy audit provides recommendations for energy-efficient practices, such as technology upgrades. These may be include suggestions for replacing inefficient systems with smart technologies, such as LED lighting, smart thermostats, and automated shading systems. It may also recommend integration of Intelligent Energy Management Systems (IEMS) to optimize energy usage dynamically. Some of the recommendations may be operational improvement strategies like recalibrating HVAC systems, optimizing schedules for energy-intensive activities, and enhancing insulation.

Energy audit may lead to real-time monitoring and continuous optimization, which involves IoT Integration. This combines energy audits with IoT-based systems for ongoing monitoring and adjustments, and enables predictive maintenance, reducing energy losses caused by equipment malfunctions. Real-time monitoring and continuous optimization also include feedback loop, for continuous auditing which ensures that energy-saving measures remain effective and adapt to changing building usage patterns.

Energy audit involve or may include financial and environmental impact assessment. This is an evaluation of potential cost savings, which identifies opportunities to reduce energy bills by eliminating waste and optimizing energy usage. It is also an evaluation of carbon footprint reduction, which helps achieve sustainability goals by lowering greenhouse gas emissions associated with energy consumption.

Energy audit assists with the alignment with smart building goals. Energy audits align with the goals of smart buildings by enhancing efficiency without compromising comfort or functionality. It facilitates the implementation of Net Zero Energy Building (NZEB) strategies and renewable energy integration, such as Building Integrated Photovoltaics (BIPVs).

In summary, energy audits are essential for achieving energy efficiency in smart buildings by offering actionable insights, enabling the adoption of advanced technologies, and fostering sustainable practices. This process ensures smarter energy management, cost reductions, and a reduced environmental footprint.

5 Conclusions and Perspectives

5.1 Conclusions

This chapter presented a review of literature on energy management opportunities that exist in a commercial building. More importantly, the paper also offered a simple methodology based on POET framework, that can be used to identify energy efficiency opportunities in various applications. Seven energy saving opportunities were identified for the building belonging to the case-study company. The paper presented an easy-to-use threefold evaluation method encompassing financial, environmental and complexity metrices. All the ESOs promised some reduction in CO_2 emissions, they also had positive NPV values and their PBP were a fraction of their useful lives. The ESOs that were identified promise a total monthly energy saving of 163754.22 kWh, which is 85% of the current monthly consumption. The conclusion from these results is that there is a viable business case for energy efficiency projects in buildings.

The reliability of this research is supported by fact that only academic peer reviewed sources were used. Multiple studies and sources were consulted which arrived at similar conclusions in terms of energy management and efficiency. This strengthens the reliability of the current findings, as the research holds up across different conditions and contexts and is trusted to produce the same outcomes when repeated under the same conditions. This consistency helps confirm that findings are not due to random factors or errors. The methodology attempted to standardize procedures to allow other researchers to replicate the study.

The validity of the research stems from the tools and methods used, which are the academic peer reviewed sources, the POET framework, a case-study of real credible commercial building and proven financial, environmental and complexity evaluation metrices. These tools and methods individually and collectively, accurately measure the variables of interest, ensuring that the data truly reflect what is being studied. The research design for this study aligns the research questions with the methods used to collect and analyse data, therefore the research process is believed to effectively address the objectives.

Since the methodology used is repeatable, it may be used in future studies for more in-depth energy auditing, which was not possible in this study. The data collection on this study was limited to available submetering therefore the distribution of energy usage in the building may be improved in future studies by installing more submeters.

5.2 *Perspectives*

Energy efficiency in buildings can be approached from various perspectives, each targeting different aspects of energy use and sustainability. Here are several key perspectives for improving energy efficiency in buildings:

5.2.1 Design and Construction Perspective

There are three approaches towards this perspective. First is Energy-Efficient Building Design, which is achieved by optimizing building orientation, shape, and layout to maximize natural light and ventilation. Also by incorporating passive design strategies such as thermal mass, shading devices, and reflective roofing. The second approach is high-performance building envelope, which is achieved by use high-quality insulation to minimize heat transfer, and installing energy-efficient windows with low-emissivity coatings and double/triple glazing. Lastly is the building integrated photovoltaics (BIPVs) which are PV solar systems integrated directly into the building envelope, such as roofs or façades.

5.2.2 Operational Perspective

Operational perspectives are strategies such as Intelligent Energy Management Systems (IEMS) which involve implementing real-time monitoring and optimization using IoT-enabled energy management systems. Another strategy is Automated Building Systems involving the use of smart thermostats, motion-activated lighting, and advanced HVAC systems for precise control and reduced energy waste. Operational perspectives also include Predictive Maintenance which employ data-driven analytics to identify and address inefficiencies before they escalate.

5.2.3 Technological Perspective

Technological perspectives include measures such as Smart Building Technologies to Leverage IoT sensors and connected devices for real-time energy tracking and automation. They also include Energy-Efficient Appliances which may require replacing outdated appliances with Energy Star-rated or equivalent high-efficiency models. This perspective also includes Renewable Energy Integration which combine on-site renewable energy systems, such as solar panels or wind turbines, with energy storage solutions.

5.2.4 Behavioral and Educational Perspective

Behavioral and education perspectives are about occupant engagement or educating building occupants about energy-saving behaviors, such as turning off lights and equipment when not in use. Incentive Programs may also be used to encourage participation in energy efficiency programs through financial rewards or recognition.

5.2.5 Design and Construction Perspective

Energy Retrofits involve upgrade of existing buildings with energy-efficient systems and materials, such as LED lighting, advanced HVAC systems, and modern insulation. This perspective is also about Net Zero Energy Building (NZEB) Conversions, which transitions older buildings into NZEBs by balancing energy demand with renewable energy generation.

5.2.6 Policy and Regulation Perspective

Some of the key policy environment perspectives involve compliance with energy standards, such as adherence to green building certifications such as (LEED) Leadership in Energy and Environmental Design, the world's most widely used green building rating system, (BREEAM) Building Research Establishment Environmental Assessment Method a sustainability assessment method that is used to masterplan projects, infrastructure and buildings, or WELL building standard. Other policy and regulation perspective is government incentives, where one can take advantage of grants, tax credits, and subsidies for energy-efficient improvements. One of the perspectives is also urban planning integration by designing neighborhoods and urban spaces to support energy-efficient practices, such as shared renewable energy systems.

5.2.7 Environmental Perspective

This includes sustainable material use by opting for low-carbon and recyclable construction materials to reduce the building's overall environmental impact. Water and Waste Management is another perspective which implements energy-efficient water heating systems and recycling processes.

5.2.8 Future-Oriented Perspective

Digital Twin Technology is a perspective that involves the use virtual simulations to model and optimize energy performance before implementing changes in real buildings. AI and Machine Learning is another perspective which employs predictive analytics for smarter energy use and demand forecasting. Also, energy communities is a perspective that explores collective energy solutions where buildings share renewable energy resources for greater efficiency.

By addressing energy efficiency from these diverse perspectives, buildings can achieve substantial reductions in energy consumption, lower operating costs, and contribute to global sustainability efforts.

Competing Interests The authors and sponsor of this study have no financial, commercial, legal, or professional connections with other organizations or individuals that could impact this research.

Ethics Approval A written approval was sought and granted by the case-study company for the use of one of its buildings for the case-study. This was necessary to ensure that there was legal ground and proper prior approval for the collection, processing and reporting of data obtained from this organisation's facility. The secondary data collected will be used responsibly to ensure no harm comes to the company, its employees, or its reputation. The data has be anonymized and handled confidentially, ensuring that no sensitive or identifying information is disclosed in the analysis or findings.

References

1. Kawa B, Borkowski P (2023) Integration of machine learning solutions in the building automation system. Energies 16(11):4504. https://doi.org/10.3390/en16114504
2. Zhao W, Deng Z, Ji Y, Song C, Yuan Y, Wang Z, Chen Y (2024) Analysis of peak demand reduction and energy saving in a mixed-use community through urban building energy modeling. Energies 17(5):1214. https://doi.org/10.3390/en17051214
3. Burak Gunay H, Shi Z, Wilton I, Bursill J (2020) Disaggregation of commercial building end-uses with automation system data. Energ Buildings 223:110222. https://doi.org/10.1016/j.enbuild.2020.110222
4. Kim DW, Ahn KU, Shin H, Lee SE (2022) Simplified weather-related building energy disaggregation and change-point regression: heating and cooling energy use perspective. Buildings 12(10):1717. https://doi.org/10.3390/buildings12101717
5. Marceau ML, Zmeureanu R (n.d.) Nonintrusive load disaggregation computer program to estimate the energy consumption of major end uses in residential buildings. Elsevier. www.elsevier.com/locate/enconman
6. Ahmad MW, Mourshed M, Mundow D, Sisinni M, Rezgui Y (2016) Building energy metering and environmental monitoring—a state-of-the-art review and directions for future research. Energ Buildings 120:85–102. https://doi.org/10.1016/j.enbuild.2016.03.059
7. Brown N, Wright AJ, Shukla A, Stuart G (2010) Longitudinal analysis of energy metering data from non-domestic buildings. Build Res Inf 38(1):80–91. https://doi.org/10.1080/09613210903374788
8. Wang C, Chen Z, Tian Z, Yue W, Shi X (2019) Implementation of energy conservation in a commercial building using BEM and sub-metering technology. In: IOP conference series: earth and environmental science. Institute of Physics Publishing. https://doi.org/10.1088/1755-1315/238/1/012003
9. Cao Y, Wang C, Wei J, Pan X (2018) Healthy building management system based on business intelligent. J Phys Conference Ser 1060:012011. https://doi.org/10.1088/1742-6596/1060/1/012011
10. Jin TX, Xu XF (2014) Experimental investigation on a grid-connected photovoltaic air conditioning system. In: Applied mechanics and materials, vol 54–60. Trans Tech Publications Ltd. https://doi.org/10.4028/www.scientific.net/AMM.672-674.54
11. Mathias JA, Juenger KM, Horton JJ (2023) Advances in the energy efficiency of residential appliances in the US: a review. Energ Effic 16(5):34. https://doi.org/10.1007/s12053-023-10114-8
12. Romero-Lara MJ, Comino F, Ruiz M, de Adana. (2023) Seasonal energy efficiency ratio of regenerative indirect evaporative coolers—simplified calculation method. Appl Therm Eng 220:119710. https://doi.org/10.1016/j.applthermaleng.2022.119710
13. Yuan Q, Ma Y, Liu C, Dai B, Yan Q (2011) Thermodynamic perfectibility based analysis of energy-efficiency standards for air conditioning products in China. Energ Buildings 43(12):3627–3634. https://doi.org/10.1016/j.enbuild.2011.09.035
14. Fathi M, Batoul AA, Ansari A (2023) Green commercial aviation supply chain—a European path to environmental sustainability. Sustainability (Switzerland) 15(8):6574. https://doi.org/10.3390/su15086574
15. Moghaddam SA, Serra C, Gameiro M, da Silva, and Nuno Simões. (2023) Comprehensive review and analysis of glazing systems towards nearly zero-energy buildings: energy performance, thermal comfort, cost-effectiveness, and environmental impact perspectives. Energies 16(17):6283. https://doi.org/10.3390/en16176283
16. Beltrán-Velamazán C, Monzón-Chavarrías M, López-Mesa B (2024) A new approach for National-Scale Building Energy Models Based on energy performance certificates in European countries: the case of Spain. Heliyon 10(3):e35882. https://doi.org/10.1016/j.heliyon.2024.e25473

17. Masi D, Francesca R, Parrotta M, Festa V, Ruggiero S, Gigante A, Russo A (n.d.) Sensitivity analysis about the effectiveness of the energy efficiency measures for residential building under the Italian Incentive Opportunities. IEEE
18. "Government Gazette Staatskoerant Republic of South Africa republiek Van Suid afrika General Notices • AlGemene KennisGewinGs" (2022) Department of Mineral Energy and Resources. www.gpwonline.co.za
19. Ntsaluba SBK, Nwulu N (2021) Analysis of the practical implementation of energy performance certificate for buildings of academic institution in South Africa. In: International conference on electrical, computer, and energy technologies, ICECET 2021. Inc, Institute of Electrical and Electronics Engineers. https://doi.org/10.1109/ICECET52533.2021.9698726
20. Xia X, Zhang J (2012) Energy Management of Commercial Buildings—a case from a POET perspective of energy efficiency. J Energ Southern Africa 23(1):23–31
21. Zhang H, Xia X, Zhang J (2012) Optimal sizing and operation of pumping systems to achieve energy efficiency and load shifting. Electr Power Syst Res 86:41–50. https://doi.org/10.1016/j.epsr.2011.12.002
22. Yelisetti S, Kumar R, Gupta V, Saxena A, Lamba R (2020) Modelling and analysis of home energy management system using intelligent algorithms. In: ICPECTS 2020—IEEE 2nd international conference on power, energy, control and transmission systems, proceedings. Institute of Electrical and Electronics Engineers Inc. https://doi.org/10.1109/ICPECTS49113.2020.9337014
23. Lee S, Kwon B, Lee S (2014) Joint energy management system of electric supply and demand in houses and buildings. IEEE Trans Power Syst 29(6):2804–2812. https://doi.org/10.1109/TPWRS.2014.2311827
24. IEEE Globecom Workshops (GC Wkshps) (2015) IEEE
25. Lossin F, Loder A, Staake T (2016) Energy informatics for behavioral change: increasing the participation rate in an IT-based energy conservation campaign using social norms and incentives. Comput Sci Res Dev 31(3):149–155. https://doi.org/10.1007/s00450-014-0295-3
26. Al-Ali AR, Zualkernan IA, Rashid M, Gupta R, Alikarar M (2017) A smart home energy management system using IoT and big data analytics approach. IEEE Trans Consum Electron 63(4):426–434. https://doi.org/10.1109/TCE.2017.015014
27. Hannan MA, Faisal M, Ker PJ, Mun LH, Parvin K, Mahlia TMI, Blaabjerg F (2018) A review of internet of energy based building energy management systems: issues and recommendations. IEEE Access 6:38997–38914. https://doi.org/10.1109/ACCESS.2018.2852811
28. Hossain J, Kadir AFA, Hanafi AN, Shareef H, Khatib T, Baharin KA, Sulaima MF (2023) A Review on optimal energy management in commercial buildings. Energies 16(4):1609. https://doi.org/10.3390/en16041609
29. Firma P, Siregar M (2020) Feasibility study of micro hydropower plants in high-rise buildings in Indonesia by utilizing a working scheme for AC condenser water pumps (Ventilation & air Conditioning System). In: IOP conference series: materials science and engineering. Institute of Physics Publishing. https://doi.org/10.1088/1757-899X/807/1/012016
30. Teneta-Skwiercz D, Sikacz H (2023) Corporate governance in the context of ESG issues in ICT companies in Poland—Results of surveys. In: Scientific Papers of Silesian University of Technology. Organization and Management Series, vol 178, pp 655–671. https://doi.org/10.29119/1641-3466.2023.178.35
31. Tripathy M, Sadhu PK, Panda SK (2016) A critical review on building integrated photovoltaic products and their applications. Renew Sust Energ Rev 61:451–465. https://doi.org/10.1016/j.rser.2016.04.008
32. Jalluri SR, Veligatla RB, Mahamkali R, Gogireddy PRR (2021) Simulation of solar PV system by FLC based bidirectional DC-DC converter for DC loads. In: E3S web of conferences. EDP Sciences. https://doi.org/10.1051/e3sconf/202130901124

33. Özkalay E, Virtuani A, Eder G, Voronko Y, Bonomo P, Caccivio M, Ballif C, Friesen G (2024) Correlating long-term performance and aging behaviour of building integrated PV modules. Energ Buildings 316:114252. https://doi.org/10.1016/j.enbuild.2024.114252
34. Salem R, Bahadori-Jahromi A, Mylona A, Godfrey P, Cook D (2019) Investigating the potential impact of energy-efficient measures for retrofitting existing UK Hotels to reach the nearly zero energy building (NZEB) standard. Energ Effic 12(6):1577–1594. https://doi.org/10.1007/s12053-019-09801-2
35. "Net Zero Buildings" (n.d.)
36. Krupkin A, Gorodnova N (n.d.) Section Business & Management Smart City concept: methodology of energy efficiency assessment of buildings. https://doi.org/10.5593/sgemsocial2018/1.5
37. Xia X, Zhang J (2010) Energy efficiency and control systems-from a POET perspective. IFAC Proc Vol 43:255
38. Muhammad AM, Bano F (2022) Energy-saving potential and cost-effectiveness of active energy-efficiency measures for residential building in warm-humid climate. Energy Sustain Dev 67:163–176. https://doi.org/10.1016/j.esd.2022.01.011
39. Favoino F, Fiorito F, Cannavale A, Ranzi G, Overend M (2016) Optimal control and performance of Photovoltachromic switchable glazing for building integration in temperate climates. Appl Energy 178:943–961. https://doi.org/10.1016/j.apenergy.2016.06.107
40. Montoya FG, Peña-García A, Juaidi A, Manzano-Agugliaro F (2017) Indoor lighting techniques: an overview of evolution and new trends for energy saving. Energ Buildings 140:50–60. https://doi.org/10.1016/j.enbuild.2017.01.028
41. Mujeebu Muhammad A, Ashraf N (2020) Impact of location and Deadband on energy performance of Nano aerogel glazing for office building in Saudi Arabia. Build Res Inf 48(6):645–658. https://doi.org/10.1080/09613218.2019.1696171
42. Energy Efficiency in Building Renovation (2018) Handbook of energy efficiency in buildings: a life cycle approach. Elsevier, pp 675–810. https://doi.org/10.1016/B978-0-12-812817-6.00042-5
43. van Ruijven BJ, van Vuuren DP, de Vries BJM, Isaac M, van der Sluijs JP, Lucas PL, Balachandra P (2011) Model projections for household energy use in India. Energy Policy 39(12):7747–7761. https://doi.org/10.1016/j.enpol.2011.09.021
44. Goetzler W, Guernsey M, Foley K, Young J, Chung G (2016) Energy savings potential and RD&D opportunities for commercial building appliances (2015 Update). Navigant Consulting, Inc. http://www.eere.energy.gov/buildings
45. Tuominen P, Reda F, Dawoud W, Elboshy B, Elshafei G, Negm A (2015) Economic appraisal of energy efficiency in buildings using cost-effectiveness assessment. Proc Econom Finance 21:422–430. https://doi.org/10.1016/s2212-5671(15)00195-1
46. Shimizu M, Tiku O (2023) Evaluation of environmental energy efficiency and its influencing factors: a prefecture-level analysis of Japanese manufacturing industries. J Econ Struct 12(1):3. https://doi.org/10.1186/s40008-023-00297-9
47. Jaber JO (2023) Assessment of retrofitting old residential buildings in urban districts: expected performance of selected energy efficiency measures. Jordan J Mech Indust Eng 17(4):555–570. https://doi.org/10.59038/jjmie/170411
48. Pompei L, Blaso L, Fumagalli S, Bisegna F (2022) The impact of key parameters on the energy requirements for artificial lighting in Italian buildings based on standard EN 15193-1:2017. Energ Buildings 263:112025. https://doi.org/10.1016/j.enbuild.2022.112025
49. Sun X, Gou Z, Stephen Siu Y, Lau. (2018) Cost-effectiveness of active and passive design strategies for existing building retrofits in tropical climate: case study of a zero energy building. J Clean Prod 183:35–45. https://doi.org/10.1016/j.jclepro.2018.02.137
50. Amirifard F, Sharif SA, Nasiri F (2019) Application of passive measures for energy conservation in buildings–a review. Adv Build Energy Res 13(2):282–315. https://doi.org/10.1080/17512549.2018.1488617

51. Moghayedi A, Hübner D, Michell K (2023) Achieving sustainability in South African commercial properties: the impact of innovative technologies on energy consumption. Facilities 41(5–6):321–336. https://doi.org/10.1108/F-06-2022-0089
52. Bigiotti S, Costantino C, Marucci A (2024) Agritourism facilities in the era of the green economy: a combined energy audit and life cycle assessment approach for the sustainable regeneration of rural structures. Energies 17(5):1101. https://doi.org/10.3390/en17051101
53. ASHRAE (2005) ASHRAE. (2005). Procedures for commercial building energy audits. Guideline 0–2005. American Society of Heating, Refrigerating and Air-Conditioning Engineers, Inc.
54. Wilhelm Pienaar Schalk, and Reza Malekian. (n.d.) General aspects of energy management and energy audit-from a POET perspective. https://hal.science/hal-03232385
55. The Calculation of Country Specific Emission Factors for the Stationary Combustion of Fuels in the Electricity Generation Sector (n.d.)

Implementation of Energy Planning at the Municipal Grouping Level in Morocco to Develop the Concept of Smart Cities

Chouaib Benqlilou, Abir Dahani, and Fatima Zohra El Wardi

Abstract As part of the sustainable development of countries, integrating new technologies in managing cities is proving relevant. Morocco, for its part, is aligning with the global trend toward a brilliant management model and a decarbonization approach. This integration of new technologies can ultimately result in **smart cities**. Local authorities play a crucial role in developing smart cities because of their proximity to citizens, their responsibility for managing local infrastructure, and their ability to promote innovative policies tailored to the specific needs of local communities such as public lighting, transport, water distribution, and buildings. To this end, the municipality can carry out its mission in three main areas: reducing energy consumption and carbon footprint, implementing energy efficiency (EE) measures, and integrating renewable energies (RE). This is achieved through municipal energy planning. The present work aims to develop a quantitative and qualitative approach providing the municipality with an energy plan that will enable it to strengthen its governance capacity and achieve its energy transition efficiently. The methodology adopted is based on a strategic approach to energy planning. The strategy and the planning are based on an updated and verified diagnosis of all the municipality's assets and their respective energy uses (transport fleet, public lighting, water supply, and buildings). The methodology of this project begins with defining a strategic approach for the municipality, followed by an in-depth diagnosis of municipal assets such as public lighting, water supply, buildings, and transport across five communes (Figuig, Taourirt, Guercif, Nador, and Driouech). This diagnosis evaluates energy consumption, malfunctions, and service quality to identify energy efficiency and renewable energy measures. Potential energy savings (PES)

C. Benqlilou (✉) · A. Dahani
Sustainability, Artificial Intelligence and Mathematical Modeling, Ecole Nationale Supérieure des Mines de Rabat, Rabat, Morocco
e-mail: benqlilou@enim.ac.ma; a.dahani@enim.ac.ma

F. Z. El Wardi
Physics Department, LPMAT Laboratory, Faculty of Sciences Ain Chock, Hassan II University, Casablanca, Morocco

H. Kaddouri et al. (eds.), *Smart Cities for Sustainable and Urban Development*, Green Energy and Technology, https://doi.org/10.1007/978-3-031-99001-4_20

are calculated by comparing current conditions to improved scenarios, with estimated GHG emission reductions of about 379 teqCO_2/year. Promising results indicate significant savings of 5% to 61%. The current study proposes a municipal action plan as a decision-making tool to support sustainable energy transitions, emphasizing efficient energy use, responsible production, and reduced carbon emissions.

Keywords Energy efficiency · Energy planning · Municipality · Potential energy savings · Smart cities

1 Introduction

1.1 Overview

On a global scale, cities account for around 75% of energy consumption and 70% of greenhouse gas emissions [1]. Therefore, city energy management is essential to achieving climate and energy objectives. Urban energy planning has become increasingly important in recent decades [2]. The city is recognized as the appropriate scale for implementing effective energy policies [3], as it simultaneously offers the ability to coordinate individual actions and better opportunities for adapting to local circumstances than national policies.

Morocco has adopted ambitious climate targets for 2030. These targets include a 45.5% reduction in GHG emissions [4], the adoption of renewable energies (RE), which should account for 52% of installed electrical capacity, and the achievement of energy savings in the various high-potential productive sectors in terms of energy consumption of 20% [5–7]. These targets have contributed to considerable progress in global climate and energy change. Municipalities are responsible for achieving these objectives and implementing national energy and climate policies and strategies. This responsibility is articulated in response to the legislative requirements in Organic Law 113–14 and a logic of exemplarity in sustainable consumption/production, as stipulated in the challenges of the national sustainable development strategy (SNDD). Municipalities are in charge of providing a high-quality public service based on energy-consuming components such as public lighting, transport, water distribution, and buildings. To this end, municipalities can focus on three main areas: reducing energy consumption, implementing energy efficiency (EE) measures, and integrating renewable energies (RE). Table 1 highlights a number of successful national experiences.

However, the role of municipalities in achieving national objectives remains underdeveloped. Municipalities are often constrained by financial solvency, skills, or organizational capacity to convey this responsibility. It would be more appropriate to consider this approach towards a sustainable and clean city as an opportunity, given the potential for saving resources and raising funds for energy efficiency, renewable energy integration, and, in particular, GHG emissions reduction. In this context, municipalities can align themselves with this approach and contribute to the development of smart cities by acting on their assets through implementing

Table 1 RE actions deployed at municipal level

City	Action	Source	Installed capacity	
Fez	Bioelectric plant - self-generation (40%) public lighting	Household waste recovery	5 MW	[8]
Rabat	Bio-drying plant for Refuse derived fuels (RDF) production	Household waste recovery	90,000 t/an	[9]
Marrakech	Bioelectric power plant for biogas production	Landfill rehabilitation	1 MW	[10]
Oujda	Bioelectric power plant - self-generation for public lighting	Household waste recovery	2.3 MW	
Fam El Hisn	Photovoltaic plant	Use of solar energy	20 MW extensible	[11]

municipal energy planning (MEP), which will encourage them to get involved in the energy transition dynamic.

El Alaoui [12] highlighted in their study that developing a smart city in Morocco is a long-term endeavor. In the medium term, efforts should prioritize enhancing democracy, advancing education and knowledge, and transitioning to a circular economy that emphasizes human development and ecological balance. In the other hand, according to [13], the discourse surrounding smart cities in Morocco is relatively recent, spanning less than a decade. Although the identification of concrete success stories remains premature, a diverse array of initiatives and projects has been undertaken by various stakeholders to foster the development of the smart city concept: Casablanca Smart City project, Benguerir Smart city, and Zenata Ecocity. Zenata Ecocity exemplifies eco-design by adopting a holistic approach that integrates social, economic, ethical, geological, and environmental dimensions. It aligns with the pillars of sustainable development, reflecting Morocco's commitment to incorporating Smart City principles into public urban planning. By prioritizing citizen inclusion within a dynamic, digital, innovative, and sustainable framework, the project has earned the prestigious international Eco-City Label [14, 15]. Despite increasing interest in smart cities in Morocco, there is a significant research gap in municipal energy planning, a critical component for their successful implementation.

Municipal energy planning in Morocco is still in progress. Municipalities face constraints in terms of financial solvency, skills, and organizational capacity to carry out this responsibility. It would be more appropriate to consider this approach by municipalities towards a sustainable and clean city as an opportunity, given the potential for saving resources and raising funds in the sectors of energy efficiency, integration of renewable energies, and, in particular, reduction of GHG emissions (GVC, GEFF, climate finance, CPE, etc.). In addition to these financing instruments, other governance mechanisms can provide an appropriate management environment: PPPs, EPCs, ESCOs, etc.

At the national level, a sustainable energy action plan (PAED) has been proposed for the commune of Agadir as part of the CES-MED project. GHG emissions linked to municipal assets account for 9% of the region's total emissions, divided between municipal buildings, public lighting, the municipal fleet, public transport, water, and waste management. The action plan developed included forty actions affecting municipal assets as well as the entire territory (transport, residential and tertiary housing, industrial activities, etc.) [16]. Similarly to Agadir, Oujda has also developed a PAED program [17].

Internationally, municipal energy plans are widely adopted by many municipalities. In Sweden, 61% of the 290 Swedish municipalities adopted energy plans during 2004–2015 [18]. In Switzerland, energy plans are commonplace. They propose different pathways towards economic, social, and environmental energy goals, often supported by an action plan outlining the possible projects and policies needed to implement these pathways [19]. Based on the Swiss approach, Ukraine aims to develop municipal energy planning tailored to the needs of Ukrainian municipalities [20].

Current initiatives often focus on isolated aspects, such as renewable energy projects or digital infrastructure, without a comprehensive framework to integrate energy systems into urban planning. The absence of localized energy models, strategies for optimizing energy consumption, and policies for integrating renewable energy sources hinder the ability of municipalities to create sustainable and efficient energy solutions. Additionally, there is limited research addressing stakeholder collaboration, financial mechanisms for transitioning to energy-efficient urban environments. This gap highlights the urgent need for multidisciplinary studies and actionable strategies to develop municipal energy planning aligned with Morocco's vision for smart cities.

1.2 MEP Deployment Approach

The MEP is a master planning tool that sets out all the prospects for sustainable energy development for a municipality, through the development of new high-performance services in the fields of intelligent transport and mobility, sustainable building, public lighting, and drinking water distribution to converge towards a single system that is both efficient and ecological. From a strategic point of view, it is crucial to note that the municipality must first develop a vision on the energy dimension and translate it into specific missions. A well-specified vision provides decisive support for the commune on four levels:

- The vision provides direction and ensures that all decision-makers and operators focus their efforts on future steps and, in this case, determine the roadmap for what the commune should undertake to meet the vision thus specified;

- The vision brings decision-making power, as the specification of a clear vision influences decision-making;
- The vision has a significant motivational effect on local council members and managers, as a clear vision spurs people into action and increases their level of involvement;
- Vision has a resilience-building role and thus encourages the perseverance of commune executives and decision-makers in carrying out their tasks in the face of challenges, obstacles, and constraints.

In addition, it should be noted that a vision is necessarily based on values that the commune should adopt, in particular: quality service, eco-citizenship, social responsibility, sustainability, exemplarity of the state in terms of rationalizing the management of public resources, integration of clean energy, sustainable purchasing, etc. Turning the vision into actionable steps leads to clearly defined missions that convey the municipality's commitment to energy transition and sustainability to internal and external stakeholders. Consequently, these missions address a fundamental question: What is our purpose? Figure 1 below illustrates the municipality's vision, values, and missions.

Secondly, a baseline inventory of assets, consumption, and GHG emissions is required. This will provide a basis for identifying priority sectors and items whose energy consumption and GHG emissions require effective control. As a result, the municipality's energy potential in terms of EE and RE can be identified. In addition, establishing specific objectives provides targets to reach and a baseline against which to assess the performance achieved by the municipality in its move towards a sustainable and clean energy transition. Without clearly established objectives, assessing the municipality's status and progress in energy initiatives becomes unfeasible. Consequently, the direction, decision-making processes, and degree of engagement among the municipality's operators and decision-makers may lack coherence and focus. It is important to note that the objectives are specific and should be prioritized in such a way as to identify the most important objectives from the point of view of profitability, efficiency, impact, sustainability, replicability,

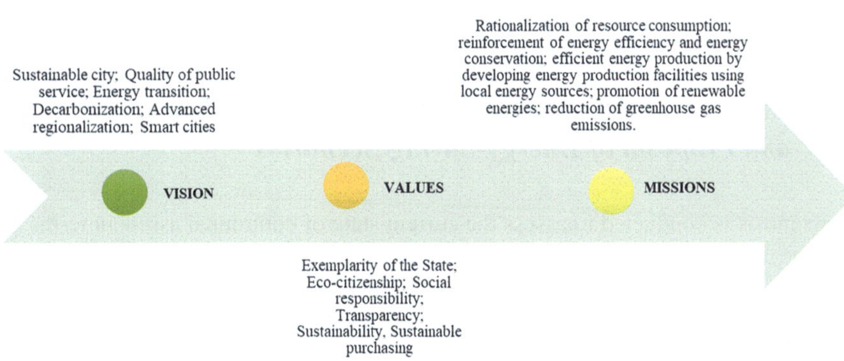

Fig. 1 The municipality's vision, values and missions

social responsibility, etc. In this sense, it is well known that these objectives should be SMART. The final step is to define the necessary action plan, implement it, and in real-time monitor it to guarantee its success, maximize the effectiveness of efforts, and ensure proactive management of resources and objectives.

1.3 Objective and Importance

Based on the experiences of other countries in adopting MEP, the present work aims to explore how municipal energy planning can be effectively implemented in Morocco. It proposes developing a qualitative and quantitative approach to MEP implementation at the national level, with the proposal of a concrete action plan to implement. The results of this work should provide usable recommendations that could contribute to the implementation of more sustainable and decentralized energy systems in Moroccan municipalities. Through effective municipal energy planning, Morocco can not only accelerate its transition to renewable energies but also foster local economic development and enhance energy security. Therefore, this paper addresses a critical gap in the country's energy planning strategy, focusing on the role of municipalities in shaping a sustainable energy future.

2 Materials and Methods

The present work uses a qualitative and quantitative research approach to examine the implementation of municipal energy planning in Morocco. A multiple-case study method was used, focusing on five municipalities: Figuig, Taourirt, Guercif, Nador, and Driouech. The municipal assets studied include the public lighting network (PLN), water distribution, municipal buildings, and transport. The existing potential for renewable energy will be analyzed with a focus on energy recovery. Figure 2 illustrates the methodology used in this study.

2.1 Diagnosis of the Current Situation of Municipal Assets and Proposal of Energy-Saving Scenarios

A diagnosis is conducted to assess the current state of communal assets in terms of existing infrastructure, energy consumption, etc. The current state will provide a baseline to identify the energy savings potential (ESP) through improvement actions. For the diagnosis analysis, Data is collected from the municipal representatives of the five selected municipalities.

- **Public Lightning Network (PLN)**: The diagnosis of the urban public lighting network includes an analysis of the current state of the municipality's public lighting assets (the infrastructure deployed in terms of luminaires, cabinets,

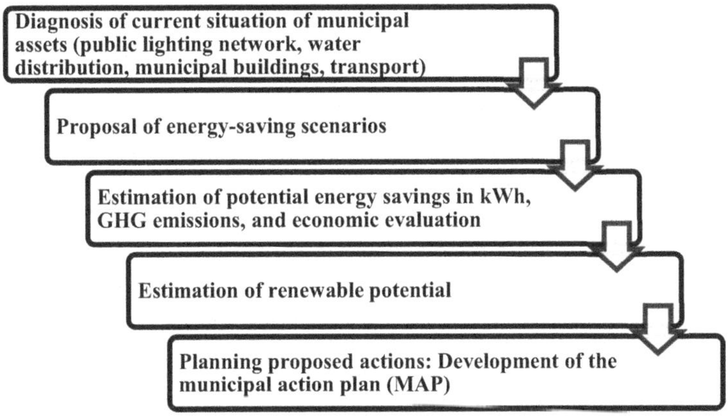

Fig. 2 The methodology used in the current study

lighting columns, etc.), as well as a spatial analysis of the road network and its occupancy profile, and energy consumption. The diagnosis can be extended to reveal the energy efficiency measures implemented or the integration rate of renewable energies to meet public lighting needs (PV system, waste to energy, etc.). However, due to the limited amount of data available, the diagnosis of the public lighting system carried out as part of the present study will be based on an inventory of the main parameters, including the number of light points, the unit power of the lamps deployed (W), the lamp technology (SHP, LED, etc.), the number of operating hours, the installed power (kW) and the annual energy consumption (kWh/year). The elements of the diagnosis differ from one commune to another, depending on the data availability. To enhance the public lighting system, national-level studies have already been conducted in two cities: Taza and Casablanca. Upgrading the lighting system on Avenue Moulay Youssef in Taza mainly involved switching from SHP to LED technology, and provided significant energy savings ranging from 66% to 84%. Two additional measures have been studied and proposed for renovating the lighting system on Boulevard El Joulane in Casablanca: utilizing the natural brightness of the full moon and installing a voltage step-down device to decrease power consumption. Combining these three measures provided a significant ESP that exceeds 80%, with a payback period of no more than 5 years. These significant results can only encourage other cities to adopt EE measures wherever possible to improve the current state of public lighting.

- **Water Distribution**: Water management is one of the most strategic municipal services. The municipality should have the means to monitor, evaluate, and control the implementation of this service with a view to rationalizing resources: reducing pumping costs, reducing leakage, ensuring water quality and pressure requirements, etc. Water distribution is ensured by pumping stations with a capacity to determine according to the population's water requirements and the total manometric head, among other parameters. Based on an average monthly consumption of drinking water per household of 17 m^3/month, and taking into account the number of households per province, total water consumption is

Table 2 Data collected from water pumping station-Figuig

	Equipment	Rated power (kW)	Number of operating hours in h/d
Pumping	Electric pump unit	135	18
	Handling system		18
Lighting	Control local	2	4
	Local 2	2	4
	Indoor Lighting	3	4
Automation and control	Command	2	6
	Total	**146**	

determined [21]. The production of drinking water via pumping stations is conditioned by the population's needs, with a markup of around 20% to take account of water leakage. This gives the annual drinking water production for each municipality. It is worth noting that the diagnosis of water pumping stations covers both the diagnosis of the various station equipment, as well as their corresponding energy consumption. Within the framework of the present work, the municipality of Figuig was chosen as a case study where data were available. The data collected is limited to the aspects shown in Table 2.
Several energy efficiency measures can be considered to reduce the pumping station's electricity consumption and provide energy savings that will improve the station's current condition. On this basis, the energy gain will be estimated with reference to the Baseline (existing situation).

- **Municipal buildings**: The diagnosis of municipal buildings focuses mainly on the analysis of the energy consumption for each type of building and each use (i.e. interior lighting, thermal comfort, equipment). However, it should be pointed out that municipal buildings can be divided into two main categories: i) administrative buildings" and ii) buildings associated with the municipality (e.g. slaughterhouses, covered halls, municipal lots, etc). Table 3 shows the main elements of the municipal building diagnosis.
- **Transport**: Mobile vehicles linked to the municipality include the household and similar waste (HW) collection fleet, public transport, and the municipal fleet[1]. Consultation with members of the municipal councils of the five chief municipalities provided an overall characterization of the mobile equipment fleet, the allocation of mobile equipment, and its number (Cf. Table 4). The diagnosis of the communal mobile machinery fleet covers both the analysis of the current state of the commune's fleet assets (the infrastructure deployed in terms of number of vehicles, type, allocation, etc.) and the monitoring and analysis of energy consumption for all vehicles in the fleet, as well as the monitoring of vehicle maintenance and servicing programs (Cf. Table 5). It should be pointed out that the two most relevant performance indicators for analyzing the energy performance of a fleet of mobile machines in accordance with standard NF 16247 are: Specific consumption expressed in L/100 km and the quantity of

[1]The municipal fleet includes personnel transport, landscaping machinery, etc.

Table 3 Features related to the diagnosis of communal buildings

Municipal buildings	– Number – For each building: Surface area (m²), overall energy consumption in kWh/year

Table 4 Characterization of the mobile equipment fleet

	Figuig	Guercif	Nador (Driouech)	Taourirt
Household waste collection	**4** (province)	**5** (province)	**22** (Intercommunal grouping)	**12** (province)

Table 5 Parameters defined in the inventory of municipal mobile fleet

Infrastructure	– Assignment (DMA fleet, public transport and municipal fleet) – Vehicle number – Weight – Type of vehicle – Service (for municipal fleet): administrative transport, municipal works, etc. – Make of vehicle – Fuel type (diesel, gasoline) – Registration date – Date of purchase – Age (more than 10 years, 5–10 years, less than 5 years) For each vehicle: – Vehicle reference fuel consumption (L/100 km)
Energy consumption	– Fuel consumption in L/year Consumption in kWh – Kilometers covered – Consumption in L/100 km – Expenses in DH/year
Maintenance	For each vehicle: Maintenance and repair costs in DH/year Number of interventions

CO_2 emitted per liter of fuel ($KgCO_2$/ L). However, due to a lack of sufficient data, the diagnosis cannot be completed, and the energy efficiency scenarios will be proposed for illustrative purposes based on previously conducted studies. This approach will offer insight into the potential benefits that could be achieved.

2.2 Estimating Potential Energy and GHG Savings

The energy-saving measures proposed for each communal asset will provide an estimate of the energy-saving potential in kWh and the GHG emissions avoided in $teqCO_2$ compared with the existing situation (Baseline). The result of subtracting the baseline energy consumption from the energy consumption incorporating the EE measures presents the associated ESP in kWh according to Eq. (1). The ESP in % can also be deduced from Eq. (2):

$$ESP = C_R - C_E \tag{1}$$

$$ESP[\%] = \frac{ESP}{C_R} \tag{2}$$

The reference energy consumption (C_R) for each asset is established either through direct data collection in consultation with local authorities or by estimation based on hypothetical scenarios, as shown below.

- **Public Lightning Network (PLN)**: The approach is based on 1 km of illuminated roadway, using a hypothetical 12 m wide street as the reference case. The avenue under study is equipped with 95 light points spread over 1 km in a bilateral arrangement. The luminaires used are SHP with a power of 250 W. The table below shows the geometric data for the avenue and the lighting system. Current energy consumption based on installed power is 102.339 kWh/year (see Table 6).
- **Water distribution:** Figuig pumping station is considered as a pilot case. The data collected are shown in Table 2. The reference consumption is determined on the basis of the station's installed power and the number of operating hours.
- **Municipal buildings**: For demonstration purposes, a communal building with the characteristics outlined in Table 7 has been selected. This building will serve as a reference for evaluating the ESP when implementing EE measures for this

Table 6 Data related to the diagnosis of the sub-study street

Spatial analysis	Pavement	Width (m): 12
		Number of lanes: 6
	Sidewalk	Number: 2
		Width (m): 4 m
		Height (m): 0,1 m
Candelabrum	Mast height (m)	7(according to the norm e = 3–3,5 h)
	Projection of the light fixture (m)	1,3
	Distance light pole / pavement (m)	0,5
	Street light arm (m)	2
	Arm inclination (°)	33
	Spacing between light spots (m)	21
	Disposition	Bilateral vis-à-vis
Luminaires	Number	95
	Technology	SHP
	Power (W)	250
	Luminous efficiency (lm/W)	120
	Installed Power (kW)	24
Electric consumption	Number of operating hours/years	4309
	Consumption kWh	102.339

Table 7 General characteristics of an illustrative administrative building

General	Ground area	150 m^2		
	Occupation rate	20 persons		
Lighting	Number of lamps	40		
	Power (W)	90		
	Lighting duration (h/d)	2		
Equipment	Type	Number	Power (W)	Duration h/d
	Desktop	16	200	8
	Laptops	6	30	8
	Printers	6	800	4
	Others	–	100	2
Thermal comfort	Number: Heating and air-conditioning system	Power (W)	Duration h/d	
	6	1000	8	
	Winter comfort temperature	20 °C		
	Summer comfort temperature	26 °C		
Envelope composition (exterior to interior)		Material	thickness (cm)	Thermal conductivity [a] W/m.K
	Roof			
	layer 1	Slab	2	1176
	layer 2	Mortar	5	1800
	layer 3	Vapor barrier	–	–
	layer 4	Mortar	15	1800
	Roof surface (m^2)	150		
	Exterior wall			
	layer 1	Mortar	15	1800
	layer 2	Red brick (6 holes)	7	0,247
	layer 3	Non-ventilated air slot	1	0,278
	layer 4	Red brick (6 holes)	7	0,247
	layer 5	Mortar	15	1800
	Opaque wall surfaces (m^2)	380		
	Low floor			
	layer 1	Tiling	1	1,3
	layer 2	Concrete (nominal density 2500)	8	2
	layer 3	Vapor barrier	–	–
	layer 4	Vapor barrier	–	–
	layer 5	Concrete (nominal density 2500)	8	2
	layer 6	Sand and gravel	20	2
	curtain wall			
	Type	Single glazing		
	Surface (TGBV 5%) in m^2	20		
Sanitary hot water (SHW)	Number of water heaters	Installed power (kW)	Duration h/d	
	1	3	5	

[a] Source: Binayat software

[b] Determined on the basis of staff DHW requirements

asset. It is important to note that, in an administrative building, the primary energy source used is electricity, accounting for 91% [22]. Therefore, electricity consumption will be the focus throughout this section. Based on energy audits carried out, consultation of energy operators (ONEE, AMEE) and a literature review, the energy consumption of an administrative building is broken down as follows: equipment and office automation (45%); heating/air conditioning systems for thermal comfort (40%); interior lighting (5%), other uses (10%).

In addition to the energy savings achieved, emissions savings (CE_{CO2}) are also generated by avoiding a quantity of CO_2 emissions that can be quantified according to Eq. (3) below. The emission rate (R_e) is 0,65 $kgCO_2/kWh$ [23, 24].

$$CE_{CO_2} = ESP.R_e \tag{3}$$

2.3 Economic Evaluation

This section provides an estimation of the economic gains in MDH and an assessment of the economic viability of the proposed improvement package. The indicator used is the payback period (PBP).

The economic gain (G) represents the difference between the operating cost before (OCR) and after the implementation of the improvement scenario (OCE), and is calculated according to Eq. 4. The operating cost of the reference scenario (OCR) is considered to be linked to the cost of electricity and is calculated according to Eq. (5), taking into account the reference energy consumption (CR) and the cost per kWh (13,416 DH/kWh). Equation (6) provides the payback period (PBP) based on the additional investment cost (IC) and the economic gain (G).

$$G = OC_R - OC_E \tag{4}$$

$$OC_R = 1,3416.C_R \tag{5}$$

$$PBP = \frac{IC}{G} \tag{6}$$

2.4 Estimation of Renewable Potential

- **Biomass-Household and similar waste (HSW):** In terms of waste-to-energy plants, several bioelectric power stations have been set up at landfill sites throughout Morocco. Examples include Oujda and Fez. These two sites demonstrate the technical feasibility and economic viability of exploiting the biogas produced by the methanization of municipal solid waste in landfill cells. This biogas is subsequently processed and used to drive a bioelectric power plant, providing energy to partially meet the electrical needs of the public lighting network at the sites considered. This approach is extremely interesting from an economic point of view for the municipality. Not only does it help to reduce GHG emissions, but it also ensures considerable energy savings. Consultations with the heads of the principal towns provided an indication of the available quantities of municipal solid waste, as shown in Table 8. A composition of 60% organic matter is assumed. The biogas potential of landfill sites is estimated at 150 to 250 Nm^3 per tonne of HSW, with a methane content of around 50% to 70% [25]. The biomethane potential for electricity generation through bioelectric power stations in the main municipalities is thus determined.
- **Solar resources:** The installation of PV power plants in the five main cities requires a preliminary study to estimate the potential solar resources. This study will also provide a basis for identifying potential sites for solar PV power plants in the five main cities. To this end, a pre-feasibility study has already been carried out on 5 sites in the Oriental region (Bouarfa, Ain Beni Mathar, Oujda, Tandrara, Nador) to assess the yield of an installed photovoltaic power plant. A performance ratio (PR) of 78,7% was deduced. On this basis, and using the Solargis tool to determine the average annual solar irradiation expressed in kWh/m^2, the average annual electrical potential can be deduced (Tables 9 and 10).

Table 8 HSW Deposit in the five sites studied

	Figuig	Guercif	Nador (Driouech)	Taourirt
HSW quantity (t/year)	**2.482** (province)	**21.900** (province)	**47.040** (intercommunal grouping)	**27.281** (province)

Table 9 Annual solar irradiation (kWh/m^2)

	Figuig	Guercif	Taourirt	Driouech	Nador
Annual irradiation (kWh/m^2)	2.358	2.019	2.135	2.019	2.019

Table 10 Preliminary diagnosis of PLN in Figuig

Unit power of luminaires (W)	LABIDAT	ZNAGA	LAWDAGHIRO. SLIMANE	FOUG, TAHT L'MAIZ	Total
20	20	0	150	50	**220**
70	36	249	127	300	**712**
100	0	310	0	0	**310**
120	0	21	36	0	**57**
150	0	39	0	0	**39**
250	89	325	180	63	**657**
LP number	**145**	**944**	**493**	**413**	**1995**
Total installed power - kW	**25,17**	**138,05**	**61,21**	**37,75**	**262**
Total consumption (kWh/year) (4.309 h)	108.457,5	594.857,5	263.753,9	162.664,8	1.129.733,6
Emissions (teqCO_2)	70,5	386,7	171,4	105,7	**734,3**
Total cost (DH/year)	145.506,6	798.060,8	353.852,2	218.231	1.515.650,6

2.5 *Actions Plan*

A provisional action plan will outline the actions to be undertaken, detailing all energy efficiency (EE) and renewable energy (RE) initiatives to be implemented across the entire municipal infrastructure, including street lighting, water distribution, municipal buildings, and transport. Each initiative offers potential energy savings and economic benefits. Successful implementation will depend primarily on establishing a timeline, budget, payback period, and assigning a responsible party for each action.

3 Results and Discussion

This section outlines the diagnostic results for the four municipal properties analyzed and presents various proposed energy-saving scenarios. Each recommended action will include an estimate of potential energy savings and emissions reductions. Additionally, an economic analysis will be conducted to determine the financial benefits and the payback period. Due to limited data availability, some measures will be included for demonstration purposes only.

3.1 Public Lightning Network-PLN

For the commune of Figuig, the diagnosis results indicate a total of 1.995 lighting points (LP), with an installed power of 262 kW and an annual consumption of approximately 1.130 MWh (equivalent to 588 kWh per PL per year or 131 W per PL per year). Consequently, the public lighting energy bill has reached one, five million DH. Additionally, around 80% of the luminaires in use have power ratings of 250 W and 70 W, representing 33% and 36% of the network, respectively (Fig. 3). The wattage distribution of these luminaires is based on an analysis of 22 surveyed cabinets, with the Tachraft cabinet alone accounting for around 310 luminaires, or 16% of the total (see Fig. 4).

LED luminaires are almost entirely absent, with only 14 units of 120 W LEDs connected to cabinet 19. Therefore, it is highly recommended to replace the existing 250 W SHP luminaires, which currently represent a total installed power of 164 kW. Power consumption is categorized by the seven identified *ksours* (each grouping several cabinets), as shown in Table 11. Notably, 47% of the luminaires are located in ZNAGA, where 44% of the population resides (see Figs. 5 and 6), resulting in the highest consumption at 3398 kWh/year. Consequently, ZNAGA will be prioritized for energy consumption reduction measures (Fig. 7).

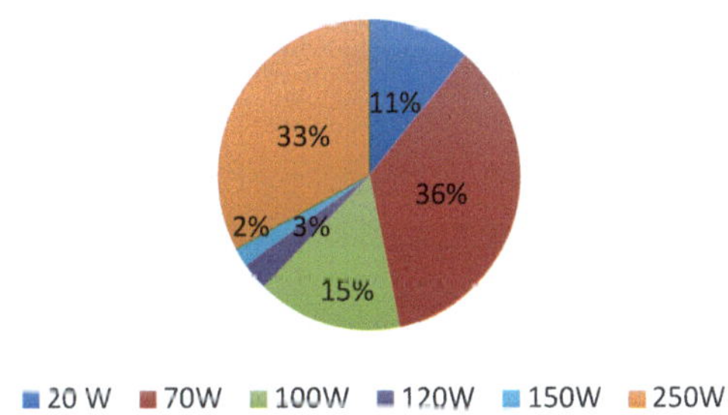

Fig. 3 Distribution of luminaires per wattage

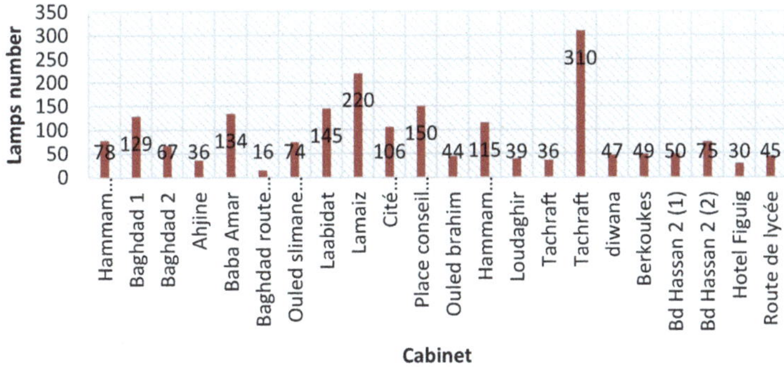

Fig. 4 Distribution of lamps by cabinet

Table 11 Gain and IRR evaluation: astronomical clock

	Baseline	Setting up an astronomical clock
Operating hours (h)	4.740	4.309
Energy consumption in kWh	111.625	101.477
Gain (%)	9%	
Gain in kWh/year	10.148	
Avoided emissions (kgCO_2/year)	6.596	
Gain DH/year	13.614	
Additional investment (DH)	684	
PBP (year)	0,05	

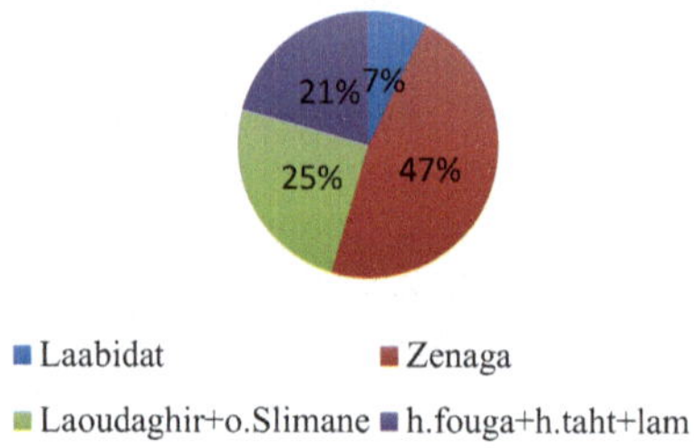

Fig. 5 Number of luminaires per ksour

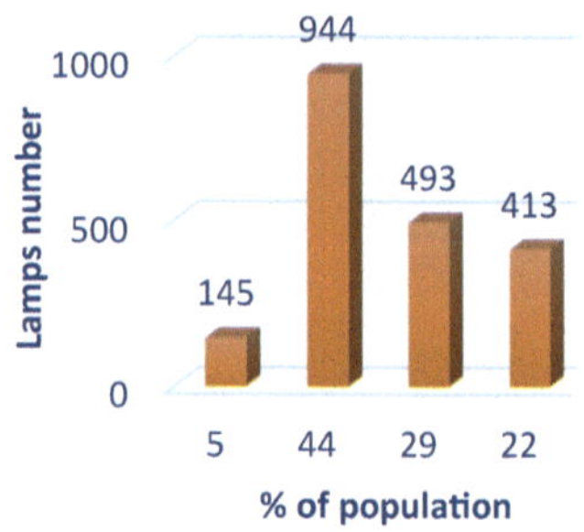

Fig. 6 Luminaire distribution by population size

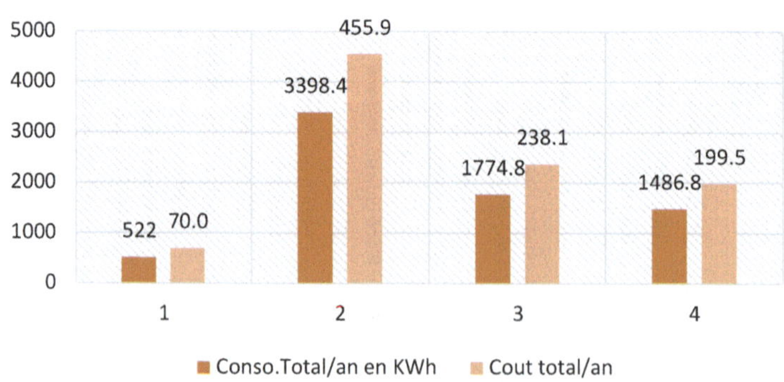

Fig. 7 Annual consumption in kWh and relative cost per ksour

The energy-saving scenarios will be applied to the hypothetical street under consideration, whose characteristics are shown in the table (Sect. 2.2). Current energy consumption is 102.339 kWh/year.

- **Setting up an astronomical clock**: This ensures that the number of hours of lighting required for the lighting system to function properly depends on the time of sunrise and sunset. It should be noted that an overestimated number of operating hours (high number of hours) leads to overconsumption of the installation. On the other hand, if the number of operating hours is low, there's a risk of insufficient daylight, resulting in poor lighting quality. This is why it's a good idea to install twilight clocks in the lighting cabinets, to ensure better management of the number of hours of lighting. Based on the sunrise and sunset times in Figuig, the number of operating hours is estimated at 4.309 h. A conventional control system (auto/manual) for switching luminaires on and off can result in an error of around 10% (either over-consumption, which will increase the energy bill, or under-consumption, which will have a major impact on lighting quality and service). The energy gain will be evaluated in the case of an overestimation of the number of operating hours (a high number of hours of the order of 4.740 h). The ESP deducted is 9%, with a low PBP that reflects the profitability of this solution (Table 11). The amount of avoided GHG emissions is 6.596 $kgCO_2$/year.
  It should be noted that all the EE measures described in the following section include the installation of the astronomical clock, assuming 4.309 h of operation.
- **Voltage reduction and full-moon operation:** In this section, we'll determine the energy savings generated by integrating a voltage reduction unit into the public lighting system. This will enable luminaire power to be adjusted to actual lighting requirements. The maximum scenario will be considered. It can be adapted to main avenues with a high traffic profile, requiring more demanding lighting conditions. Four stages per night are considered, as shown in Fig. 8. The lamps will be operated at 100% of their rated power from sunset to 10:00 p.m.,

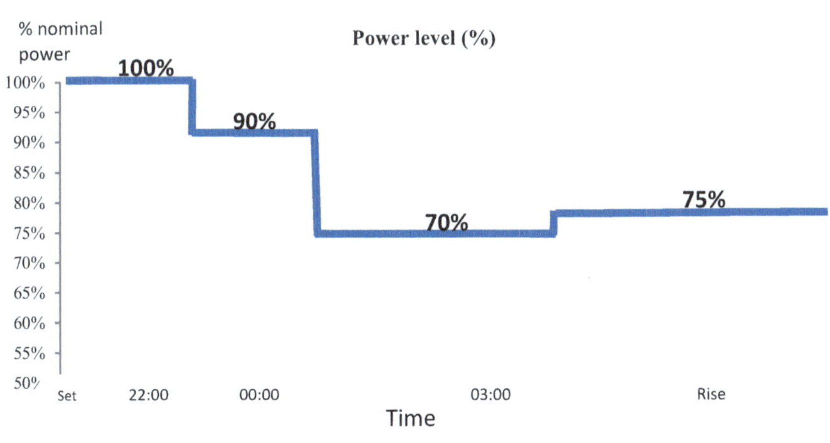

Fig. 8 Graphic illustration of the maximum power reduction scenario

with a subsequent reduction of 10% of rated power until midnight, and 20% from midnight to 3:00 a.m. The lamps will be operated at 100% of their rated power from sunset to 10:00 p.m., with a subsequent reduction of 10% of rated power until midnight, and 20% from midnight to 3:00 a.m. The resulting ESP is about 18% (Table 12).

In addition to the use of the step-down device, further energy savings can be achieved by exploiting the natural brightness of the full moon. In this context, we have considered days when the moon is illuminated by more than 95%, i.e. 4 days a month when lamp power will be reduced by 25%. The results of the energy saving assessment associated with the consideration of natural full-moon luminosity, in addition to step dimming, are shown in Table 13. It can be seen that this improvement action results in a **25%** reduction in energy consumption compared to the reference situation (Baseline).

The results of the economic evaluation associated with the solution of integrating a voltage step-down device in addition to full-moon operation in the public lighting system are shown in Table 14. The investment cost is based primarily on the cost of purchasing a step-down device. The corresponding PBP does not exceed 2 years. The emissions avoided are 16.921 $kgCO_2$/year.

- **Switching to LED technology**: The characteristics of the chosen luminaire are shown in Table 15. A comparison between the current situation and the improved situation revealed the results shown in Table 16. According to the simulation carried out on Dialux, by opting for LED luminaires with a power of 96 W, a power of 9,20 kW/km can be achieved, with an ESP of around 61%. The emissions avoided are 40.706 $kgCO_2$/year.

  The economic evaluation is based on the calculation of the additional operating cost, the investment cost, and the PBP associated with the solution of replacing

Table 12 ESP: comparison between existing situation vs. voltage step-down

	Baseline	Enhanced situation
LP number	95	95
Power demand in kW	23,75	19
Operating hours (h)	4309	4309
Energy consumption kWh/year	102.334	83.987
ESP (%)	–	**18%**

Table 13 ESP via the installation of a voltage step-down device in addition to the full moon

		Combined solution
	Baseline	Enhanced situation
LP number	95	95
Power demand in kW	23,75	18
Operating hours (h)	4.309	4.309
Energy consumption kWh/year	102.334	76.301
ESP (%)	–	**25%**

Table 14 Economic evaluation (voltage step-down device in addition to the full moon)

		Step down & Full moon
	Baseline	**Enhanced scenario**
Energy consumption in kWh/year	102.334	76.301
Cost in DH/year	137.292	102.365
Lamps unit cost DH	56	56
Lamp replacement costs in DH/year	2.547	2.547
Operating costs in DH/year	139.839	104.912
Operating cost savings in DH/year		**34.926**
Operating cost savings in %		**25%**
Investment cost in DH		**39.730**
PBP (year)		**1,14**
Avoided emissions (kgeqCO_2/year)		**16.921**

Table 15 Characteristics of the luminaire chosen for the restoration of the public lighting system

Power of LED luminaire used in W	96
Luminous efficacy in lm/W	81
Luminous flux in lm	6.087
Color temperature in K	6.500
IRC	100
Lifespan (h)	50.000

Table 16 ESP and avoided emissions deduced between the current situation and the improved situation (LED)

	Baseline	Enhanced situation (LED 96 W)
LP number	95	96
Installed power in kW	23,75	9216
Energy consumption in kWh/year	102.334	39.710
ESP (%)	**61,20%**	
Avoided GHG emissions (kgCO_2/year)	**40.706**	

HPS lamps with 96 W LED high-efficiency bulbs. In addition to the cost associated with energy consumption, the operating cost includes the extra cost of changing lamps. The investment cost, on the other hand, includes the cost of replacing the SHP lamps with 96 W LED lamps. Note that the VAT value used to calculate the investment cost is 20%. The results of the economic evaluation, summarized in Table 17, show that the operating cost saving is 83.420DH/year, with a total investment of 43.776 DH and a payback period of 6 months.

Table 17 Economic evaluation of switching to LED technology

	Baseline	Enhanced situation (LED 96 W)
ESP (%)	**61,2%**	
Energy consumption in kWh/year	102.334	39.710
Cost in DH/year	137.292	53.275
Lamps unit cost DH	56	380
Lamp replacement costs in DH/year	2547	3144
Operating costs in DH/year	139.839	56.419
Operating cost savings in DH/year		**83.420**
Operating cost savings in %		**60%**
Investment cost in DH		36.480
PBP (year)		**0,5**

3.2 *Water Distribution*

The diagnostic results show that drinking water production among the five municipalities studied ranges from 6,9 to 31 million cubic meters per year (Cf. Table 18). Using the data in Table 2 (Sect. 2.1), energy consumption at the Figuig pumping station amounts to 914.690 kWh/year. This consumption is divided between the pumping system, lighting, and the automation and control system. Electricity consumption by other equipment remains negligible, ahead of that of the electric pump unit, which accounts for 98%.

In order to reduce the pumping station's electricity consumption and achieve energy savings that would improve the station's current condition, several energy efficiency measures are possible.

- **Shifting pumping to off-peak and peak hours**: Electricity consumption costs are conditioned by the existence of three-time slots, as illustrated in Table 19.

  The consumption charge (CC) is equal to the sum of consumption in each time slot after application of the relevant tariff:

$$CC = (OH * COH) + (NH * CNH) + (PH * CPH) \tag{7}$$

With:

PH, NH, OH: Price per kWh for peak, normal and off-peak hours respectively in DH/kWh;

CPH, CNH, COH: total consumption in peak, normal and off-peak hours respectively, in kWh.

An analysis of the current distribution of the pumping station's electricity consumption by time of day shows that most electricity is consumed during the day (58%), as shown in Table 20. To improve the electrical consumption of the

Table 18 Estimated water requirements for the 5 municipalities under-study

Province	Figuig	Taourirt	Guercif	Nador	Driouech
Number of households	28.439	43.676	40.666	125.951	43.935
Water consumption (m^3/year)	5.801.556	8.909.904	8.295.864	25.694.004	8.962.740
Water production from pumping station (m^3)	6.961.867	10.691.885	9.955.037	30.832.805	10.755.288

Table 19 Electricity consumption charge per hour

	Summer	Winter	Price DH/kWh (TTC)[a]
Off-peak hours (OH)	From 23 h to 7 h	From 22 h to 7 h	0,7398
Normal hours (NH)	From 7 h to 18 h	From 7 h to 17 h	10,101
Peak hours (PH)	From 18 h to 23 h	From 17 h to 22 h	14,157

[a] http://www.one.org.ma/

Table 20 Allocation of electricity consumption by time slot

		Normal hours	Off-peak hours	Peak hours	Operating hours per day
Current situation (Baseline)	summer hours (h/d)	11	2	5	18
	winter hours (h/d)	10	3	5	18
	average hours (h/d)	10.5	2.5	5	18
	Hourly rate (%)	**58%**	**14%**	**28%**	**100%**
Enhanced situation I	summer hours (h/d)	5	8	5	18
	winter hours (h/d)	5	8	5	18
	average hours (h/d)	5	8	5	18
	Hourly rate (%)	**28%**	**44%**	**28%**	**100%**
Enhanced situation II	summer hours (h/d)	10	8	0	18
	winter hours (h/d)	10	8	0	18
	average hours (h/d)	10	8	0	18
	Hourly rate (%)	**56%**	**44%**	**0%**	**100%**

pumping station, loads can be shifted from normal hours (during the day) to off-peak hours (during the night).

This proposal is discussed in improved situation I. There is also the possibility of reducing peak hours in favor of off-peak hours in an improved situation II. The breakdown of consumption by time slot for each improvement proposal is shown in Table 21. This improvement action is in line with rational resource management and the concept of sustainable purchasing and requires very little additional investment. The saving on the electricity bill amounts to 79.224 DH/year for improvement situation I and 187.296 DH/year for improvement situation

Table 21 Gain on the electricity consumption charge for improved situations

	Baseline	Enhanced situation I	Enhanced situation II
Consumption charge (DH/year)	1.040.970	961.746	853.674
Gain DH/year	–	79.224	187.296
Gain %	–	**8%**	**18%**
Avoided GHG emissions (teqCO_2/year)		**54**	**128**

Table 22 Cost of purchasing electricity

Electricity purchase	
Power demand (kW)	146
Energy consumption (kWh/year)	914.690
Electricity cost (DH/kWh)	1,1
Total cost (DH/year)	1.006.159

II. Implementation of this improvement action could generate a reduction in consumption charges of between 8 and 18%. Avoided emissions amount to 128 teqCO_2/year. For a power demand of 146 W, the results of the situations are shown in Table 21.

Solar pumping: Solar pumping is also proposed, via the integration of a PV system to cover 20% of the power demand. This solution is based on the deployment of 71 panels with an output of 460 W. This requires an investment of the order of 179.203DH and generates a gain of 223.591 DH, with an associated payback time of no more than 1 year. We note that by increasing the coverage rate to 50%, the investment cost becomes high, while the payback period remains relatively unchanged. Accordingly, we plan to gradually increase the coverage rate of the power demand until we reach a rate of 100%, which will enable solar pumping without requiring electricity purchase (Tables 22, 23 and 24).

Repairing water circuit leaks: Repairing water circuit leaks is necessary to limit energy loss through water pipes. It can be seen that for a drinking water production of 6.961.867m^3, by recovering 5% of leaks, we can save a significant quantity of water equivalent to a gain of 76.581 DH/year with a PBP of 1 year. The GHG emissions avoided amount to 45 teqCO_2/year.

Table 25 provides a summary of the main EE measures, with the associated gain in %, the total investment in DH and the PBP. The total amount of GHG emissions avoided is about 305 teqCO_2/year.

Table 23 Gain and PBP for PV system installation

Hybrid solution (PV+ Electricity)	
System power (kW), $\eta = 0.9$	32
PV cost 460 Wc (DH)	1.850
Number of PV	71
Total Investment (PV) in DH	179.203
Reducing electricity use (%)	22%
Gain DH	223.591
PBP (years)	0,80
Avoided GHG emissions (teqCO_2/year)	132

[a] Cost provided from PV supplier

Table 24 Savings and PBP on water leakage repair (5%)

Water production (m^3/year)	6.961.867
Leakage rate (%)	5%
Water savings (m^3/an)	348.093
kWh saved (0,2 kWh/m^3)	69.619
ESP (%)	8%
Gain (DH/year)	76.581
Investment (DH/year)	100.000
PBP (years)	1306

Table 25 Summary of EE measures relating to water distribution service

	Shift to off-peak hours	Solar pumping (hybrid system)	Water leakage management
ESP (%)	18%	22%	8%
Avoided emissions (teqCO_2)	128	132	45
Investment (DH)	low	179.203	100.000
PBP (year)	–	0,8	1,3

3.3 *Municipal Building*

The total power of the illustrative building chosen is around 21 kW, corresponding to an annual energy consumption of 43.490 kWh/year. Several energy efficiency measures can be considered to reduce energy consumption and achieve energy savings (Table 26).

- **Thermal insulation:** The interaction of the building envelope with its environment through conduction, convection, and radiation results in significant heat loss. In this respect, it should be noted that 30% of heat escapes from a poorly insulated building through the roof (consequently, priority is given to roof insulation, which must be perfectly aligned with the physical feasibility of renovation),

Table 26 Unit costs of building envelope thermal insulation

Data	DH/m² (purchase and installation)
4 cm polystyrene panels (roof)	61
2 cm polystyrene panels (exterior walls)	37

Table 27 ESP for implementing EE measures

	Exterior walls		Roof		Lighting		
	Without insulation	With insulation	Without insulation	With insulation	current	LED	SWH
U (W/m².K)	1,1	0,6	3,4	0,6			
Power (kW)	2,4	1,4	3,0	0,6	3,6	0,36	3
Consumption (kWh/year)	7.012	4.211	8.723	1.680	2.628	263	5.090
Gain (kWh/year)	2.801		7.043			2.365	98%
Gain (DH/year)	2.829		7.113			2.384	4.990
Gain (kWh/m². year)	19		47			15,8	
Gain (%)	40%		81%			90%	98%
ESP for the whole building (%)	6%		16%			5%	11%
Additional investment cost	14.060		9.000			1.600	24.000
PBP (year)	5		1,27			0,79	5

25% through opaque walls, 10% or 15% through bay windows and 7% or 10% through the floors[2]. In this context, we evaluate the energy gains associated with insulating the components of the building envelope, in particular the external walls and roof (see Table 27).

- Thermal insulation of the building's external walls is carried out from the inside, with a view to renovation, by adding a 2 cm layer of thermal insulation, i.e. expanded polystyrene with a thermal conductivity of 0,032 W/m.°C. This solution reduces external heat gain via the walls by around 40%. This translates into a 6% gain in total energy consumption for the entire building, with a PBP of 5 years.
- External thermal insulation of roofs. We propose to insulate the roof from the outside using extruded polystyrene (4 cm). The results reported in Table 27 show that this solution reduces the heat entering through the roof by 81%. The ESP for this solution represents 16% of total current energy consumption, with a PBP of 1,3 years.

[2] https://www.amee.ma/fr.

- **Interior technology change to LED**: This EE measure is based on replacing existing lamps with high-efficiency lamps, in particular LEDs, and seeing their effect on the energy consumption of interior lighting in the illustrative building while keeping the same light points and their layout. It's worth noting that the 90 W lamps can be replaced by 9 W LEDs. This solution generates an energy saving of 90% compared with current energy consumption, i.e. a total energy saving percentage compared with the building of around 5%. The economic evaluation results of this solution are shown in Table 27. Implementing this measure leads to an attractive PBP of less than 1 year.
- **Installing Solar Water Heaters (SWH):** This solution uses solar energy to produce domestic hot water, replacing natural gas and/or electricity to achieve energy savings. In this scenario, it is assumed that hot water is currently generated by electric water heaters with a 70% efficiency rate. The required power for the water heater is calculated based on an average daily hot water demand of 15 liters per person, amounting to 300 liters for approximately 20 people. The energy needed to heat 1 cubic meter of water from 12 °C to 40 °C is 32.5 kWh/m^3. Thus, the estimated daily energy consumption is 9.761 kWh, with a power requirement of 3 kW, assuming 5 h of operation per day and accounting for the water heater's efficiency. Replacing electric water heaters with solar water heaters yields an annual energy saving of 4990 DH, equivalent to an 11% ESP, and a payback period (PBP) of 5 years, as detailed in Table 27.

Table 28 below shows the optimal combination of EE measures to be undertaken in the municipal building. This results in a total ESP of 32% energy savings, with a total investment of around **34.600 DH** for a total gain of **14.487 DH**, i.e. an average PBP of 2,4 years. The total avoided GHG emissions is about 9.423 KgCO_2/year.

3.4 Transport Fleet

As previously noted, due to the lack of data on the mobile equipment fleet of the five municipalities analyzed, we will only illustrate a few energy-saving scenarios aimed at enhancing the transport fleet.

- **Optimal vehicle routing**: Optimizing vehicle routing is an effective strategy for reducing logistics costs and carbon footprints. By planning efficient routes for

Table 28 Summary of building-related EE measures

EE measure	ESP (%)	Avoided emissions (kgCO_2)	PBP (year)
Thermal insulation (roof)	16	4.578	1,27
Lighting (LED)	5	1.537	0,79
SWH	11	3.308	5

the vehicle fleet, organizations can lower fuel expenses and CO_2 emissions while improving on-time delivery. This optimization reduces overall driving distances, saves time, and minimizes empty runs and unnecessary traffic. The Vehicle Routing Problem (VRP) algorithm is utilized to achieve this optimization, resulting in energy savings exceeding 15% [26, 27].

- **Eco-Driving Training**: Eco-driving training is an effective tool for promoting a change in driving behavior among agents. It helps drivers adopt a more flexible and higher-quality driving style by familiarizing them with best practices. To achieve this, we recommend organizing eco-driving training sessions to raise awareness about the significance of eco-friendly driving techniques. This training will also involve analyzing drivers' behavior and styles, which contribute to reducing fuel consumption and addressing global warming. The primary benefits of eco-driving include a 15% reduction in fuel costs, a 15% decrease in vehicle maintenance expenses, and reduced wear and tear on vehicle components such as brakes, clutches, engines, and tires [28]
- **Strengthening vehicle fleet maintenance procedures**: Improving the maintenance program can be achieved by planning preventive interventions and strengthening the material resources of these units (auto diagnostic equipment, hydropneumatic jacks, pneumatic oil pumps, etc.). This will enhance fleet efficiency, not only by reinforcing existing practices relating to fleet maintenance management but also by improving the skills of technical staff. Implementing such a measure would reduce the fuel consumption of the vehicle fleet by around 5% [16].
- **Establishment of a Geographical Information System (GIS) for household waste collection:** This will optimize the collection of household waste by improving fuel efficiency and enhancing the service provided to the community. Thanks to this management system, collection frequencies can be optimized and routes adjusted with a view to continuous improvement, with a potential 5% reduction in fuel consumption [16].

3.5 Estimation of Renewable Energy Potential

The potential for generating biomethane from municipal solid waste for electricity in major cities ranges from 0,087 to 1651 million cubic meters (MNm^3). Assuming a biogas engine efficiency of 38%, electricity production ranges from 165 to 3.137 MWh, with an installed capacity of 57 to 1.074 kW (see Table 29).

By examining the electricity consumption of the public lighting network (PLN) alongside the electricity generated from the recovery of household solid waste (HSW) in each municipality, a substantial source of electricity emerges that can be utilized to satisfy the PLN's demands. In the commune of Nador, as indicated in Table 30, recovering local biomass allows for the complete fulfilment of the electricity needs of the public assets. For the municipalities of Guercif, Taourirt, and

Table 29 Biomethane potential for electricity generation

	HSW quantity	Organic matter t/an	Biogas (1 t HSW = 150 Nm³)	Bio methane at 95% purity	MWh (thermal)	MWh (electrical)	Installed capacity (KW)
Figuig	**2.482** (province)	**1.490**	134.009	**0,087**	**435**	**165**	**57**
Guercif	**21.900** (province)	**13.140**	1.182.000	**0,768**	**3.843**	**1.460**	**500**
Nador (Driouech)	**47.040** (intercommunal grouping)	**28.224**	2.540.000	**1651**	**8.255**	**3.137**	**1.074**
Taourirt	**27.281** (province)	**16.400**	1.473.000	**0,957**	**4.787**	**1.819**	**623**

Table 30 Share of biomass electricity needs fulfillment

Municipality	Public Lighting energy consumption (MWh/year)	Electricity from biomass (MWh/year)	Potential
Figuig	5.035	165	**3%**
Guercif	2.872	1.460	**51%**
Nador	2.625	3.137	**100%**
Taourırt	4.417	1.819	**41%**

Table 31 Solar potential for the five sites under-study

Site	Annual irradiation (kWh/m²)	Average electricity production (kWh/year)
Figuig	2.358	**1.862**
Guercif	2.019	**1.597**
Taourirt	2.135	**1.685**
Driouech	2.019	**1.597**
Nador	2.019	**1.597**

Figuig, residual waste can meet 51%, 41%, and 3% of their respective electricity requirements. Two scenarios can be considered for the utilization of bioelectricity:

- Self-generation: use by the municipality itself to meet the energy needs of its buildings, or for public lighting.
- Injection into the grid on behalf of ONEE. ONEE is committed to buying back electricity produced from renewable energies at a price equal to 60% of the resale price to customers.

Regarding solar potential, the municipality of Figuig has the largest solar energy potential of the five principal communes, with an annual electricity production of 1862 kWh/year, corresponding to an installed capacity of 1 kW (see Table 31). This prioritizes it as potential for future solar PV power plants in the Oriental region. A comparison of potential renewable energy sources from HSW and solar energy shows that the commune of Nador has the highest HSW potential and the lowest solar energy potential of the five principal communes. This gives priority to the installation of bioelectric power plants for waste-to-energy conversion in this commune, to the detriment of PV power plants.

3.6 Actions Planning-Development of the Municipal Action Plan

As part of the operationalization of municipal energy planning, a concrete action plan is required to provide a concrete energy strategy for the municipality. In this respect, a provisional action plan is proposed for implementing the various solutions over a 6-year horizon (see Table 32). This is drawn up based on the different scenarios proposed, considering the limited resources of the municipalities earmarked for new investments. In particular, for each of the fields, a set of actions is defined, with the following attributes:

- Action type;
- Budget[3];
- Time planning over 6 years, updated every 3 years;
- Energy savings;
- Investment and PBP;
- Responsible;
- KPI for monitoring and evaluation.

In addition, several key elements are required for successful implementation, including mobilization of local stakeholders, monitoring and evaluation, management training, governance, financing, etc.

– From an organizational point of view, it is highly advisable to set up an **energy commission** composed of energy engineers/technicians, the accounting department, elected representatives, etc. The committee's prerogatives are consultative and to coordinate the implementation of the plan. The main tasks of this committee are as follows:

  Monitoring the implementation of planned actions;
  Monitoring the expected results in terms of EE and RE and the reduction of GHG emissions;
  Carrying out certain planned actions based on the committee's skills, or supervising the agents in charge of carrying them out;
  Adjustment and adaptation of planned actions according to opportunities and constraints;

[3]The budget for each action is obtained from equipment suppliers and on the basis of previous action plan established (PAED established for Oujda and Agadir).

Table 32 Municipal action plan

Axe	Measure/Action	ESP (%)	Deadline (years)						Budget in DH	PBP(years)	Implementation indicator	Responsible
			1	2	3	4	5	6				
Public Lightning	Setting up an astronomical clock	9%	■						684DH/cabinet	Low	Number of cabinets	Subcontracting
	Switching to LED technology	61%	■	■	■	■			43.776DH/km	0,5 year	Kilometers of street lighted	Subcontracting
	Installation of a voltage step-down device and use of the full moon	25%				■	■	■	39.730DH/cabinet	1,5 years	Number of cabinets	Subcontracting
Water distribution	Shift to off-peak and peak hours	9%	■						none	-		Commune
	Solar pumping (hybrid system)	22%		■	■	■			0,13DH/m^3	0,8 year	Volume of water pumped	Commune
	Repairing water leaks	8%-18%			■	■	■	■	100.000DH	1,3 years		Technical staff
Municipal building	Switching to LED technology	5%	■						40DH/lamps	0,79 year	Number of lamps	Technical staff
	Installation of solar water heaters	11%		■	■	■			24.000DH	5 years	Number of solar water heaters to install	Commune
	Roof insulation	16%				■	■	■	60DH/m^2	1,27 years	Building surface area	Commune
Transport	Eco-driving training	15%	■						630.000DH	0,7 year	Number of drivers trained and number of training courses to be planned	Commune
	Strengthening maintenance programs	5%	■						200.000DH	0,7 year	Purchase of equipment, organization of training courses for technical staff	Energy team
	Optimum vehicle touring	16%	■						3.000.000DH	3,2 year	Sample of vehicles to be equipped with GPS	Technical staff
	Setting up a GIS for the collection of household waste	5%			■	■	■	■	1.000.000DH	3,4 years	Number of equipped vehicles	Commune

Updating the plan as part of a continuous improvement process;
Internal and external communication providing information on the progress of the validated plan.

- Implementing a **monitoring system for follow-up, evaluation, and decision support** at the level of communal services and the communal territory as a whole. Indeed, the energy management monitoring system is a tool that provides the main performance indicators for tracking the state of municipal services. At a strategic level, it serves as a decision-making tool that enables decision-makers to identify anomalies in energy data, make energy efficiency decisions when actual performance significantly diverges from forecasts, and implement suitable energy efficiency measures. It is also a management tool, providing an analysis of the company's management results and bringing them into line with forecasts. From an operational point of view, it is a monitoring and control tool that provides systematic tracking and monitoring of the energy consumption of each asset over a given period; Establishing a macro-analysis of consumption and a comparison of this energy consumption; Monitoring the condition of the asset and its infrastructure; Raising awareness of the importance of good communal energy management.
- **Executive training**: Given the complexity of energy management, training in energy management for the commune's managers is necessary to ensure that the action plan is properly implemented and that the EE actions taken are sustainable.
- **Governance**: the management of communal assets can be carried out by the commune in one of 3 ways: direct communal or inter-communal management delegated by a private company, or as part of the implementation of a local/regional development company (SDL/SDR) through the operation of a public-private partnership (PPP). Intercommunal management is highly recommended. Indeed, consistency between municipalities in terms of energy planning provides the basis for a sustainable city, pooling efforts and actions in EE /ER and GHG emissions mitigation, ensuring the attractiveness of the private sector given the importance of economies of scale, as well as mobilizing financing instruments (GVC, GEFFII, CPE, SIE, climate finance, FEC...).
- **Financing:** Local authorities have the fiscal capacity to levy various taxes, including business tax, housing tax, municipal service tax, tax on undeveloped urban land, building operations tax, subdivision operations tax, pub tax, tourist tax, and public passenger transport tax. They can also implement local tax policies that promote renewable energy and energy efficiency, such as creating promotional funds and offering exemptions as incentives. Additionally, municipalities can access other financing sources, including the Municipal Equipment Fund (FEC), donations, and rapidly growing international funds aimed at low-carbon, climate-resilient development projects. Morocco is a major beneficiary of international climate financing. Furthermore, several financing instruments are available for mobilization, including Société d'Investissement Energétique (SIE), the "MorSEFF" sustainable energy financing line, the Green Value Chain Fund (GVC), and the Green Economy Financing Facilities Fund (GEFF II).

4 Conclusion

The present work develops a quantitative and qualitative approach to provide municipalities with an energy plan that will enable them to strengthen their governance capacity and effectively achieve their energy and decarbonization transition. Based on a diagnosis of the current state of four communal assets: public lighting network, communal buildings, water distribution network, and transport, several energy-saving measures have been proposed forming the basis of the municipal action plan (MAP). Potential energy savings range from 5% to 61%. The potential in avoided GHG emissions could reach 379 teqCO_2/year. These results are encouraging for local authorities, and reveal a promising potential for a successful transition. In this context, municipalities can contribute to the development of smart cities by acting on their assets through the adoption and implementation of the MAP. The MAP is a master planning tool that outlines a municipality's sustainable energy development prospects through the development of new high-performance services in the fields of intelligent transport and mobility, sustainable building, public lighting, and drinking water distribution to converge towards a single system that is both efficient and environmentally friendly. However, the operationalization of this planning requires a solid political commitment, the creation of a dedicated energy commission, permanent and real-time monitoring of actions through an energy management dashboard, management training, the adoption of an appropriate governance mode and the guarantee of long-term funding.

The application of municipal energy planning plays a pivotal role in developing smart cities, particularly in optimizing resource utilization and fostering sustainability. By integrating energy efficiency and renewable energy measures into urban planning, municipalities can address key challenges such as energy consumption, greenhouse gas emissions, and infrastructure resilience. This approach enhances the reliability of energy systems while promoting innovative technologies. Municipal energy planning aligns with the goals of smart cities by supporting economic growth, improving quality of life, and ensuring environmental stewardship. In developing countries, it offers a framework to bridge the gap between traditional energy practices and modern, sustainable solutions, paving the way for inclusive and intelligent urban development.

Competing Interests The authors declare that they have no known competing financial interests or personal relationships that could have appeared to influence the work reported in this paper.

References

1. IEA (n.d.) Executive summary—empowering Urban energy transitions—analysis. IEA. https://www.iea.org/reports/empowering-urban-energy-transitions/executive-summary Accessed 9 Oct 2024

2. Pasimeni MR, Petrosillo I, Aretano R, Semeraro T, De Marco A, Zaccarelli N, Zurlini G (2014) Scales, Strategies and Actions for Effective Energy Planning: A Review. Energy Policy 65:165–174. https://doi.org/10.1016/j.enpol.2013.10.027
3. Cajot S, Peter M, Bahu J-M, Guignet F, Koch A, Maréchal F (2017) Obstacles in energy planning at the Urban Scale. Sustain Cities Soc 30:223–236. https://doi.org/10.1016/j.scs.2017.02.003
4. "Morocco | NDC Partnership" (n.d.) https://ndcpartnership.org/country/mar. Accessed 3 Oct 2024
5. Hafdaoui E, Hamza AK, Ouazzani K (2024) Long-Term Low Carbon Strategy of Morocco: A Review of Future Scenarios and Energy Measures. Result Eng 21:101724. https://doi.org/10.1016/j.rineng.2023.101724
6. "Élaboration Du Projet de La Stratégie Nationale de l'efficacité Énergétique à l'horizon 2030" (n.d.) https://www.mem.gov.ma/Pages/secteur.aspx?e=3&prj=38. Accessed 3 Oct 2024
7. giz (n.d.) "Soutien à La Mise En Œuvre de La Stratégie Énergétique Marocaine." https://www.giz.de/en/worldwide/130083.html. Accessed 3 Oct 2024
8. "Centrale biogaz de Fès" (n.d.) 4C Maroc. https://www.4c.ma/projets/centrale-biogaz-de-f%C3%A8s. Accessed 14 Oct 2024
9. Jaouhari M (2019). "A Oum Azza, un modèle réussi de valorisation des déchets ménagers." La Vie éco. https://www.lavieeco.com/influences/climat/a-oum-azza-un-modele-reussi-de-valorisation-des-dechets-menagers/. Accessed 14 Oct 2024
10. Jaouhari M (2017) Un centre de valorisation de déchets sera bientôt mis en service à Marrakech. La Vie éco. https://www.lavieeco.com/affaires/un-centre-de-valorisation-de-dechets-sera-bientot-mis-en-service-a-marrakech/. Accessed 14 Oct 2024
11. "Énergies renouvelables: construction d'une ferme solaire à Tata" (n.d.) *construction21.org*. https://www.construction21.org/france/articles/h/energies-renouvelables-construction-d-une-ferme-solaire-a-tata.html. Accessed 14 Oct 2024
12. Alaoui Aicha E (2017) Determinants of a Smart City in Morocco. J Qual Educ 7(9):11. https://doi.org/10.37870/joqie.v7i9.7
13. Abdellaoui, Kanza, Mehdi Azdem, Mohamed Nabil EL Mabrouki, and Meryem Kabbaj. 2022. "Smart city au Maroc: Contextualisation du concept"
14. Er-Retby H, Outzourhit A, Nait-Taour A, Mghazli MO, Mastouri H, El Mankibi M, Benzaazoua M (2024) Enhancing Urban Design Performance through Parametric Analysis: Insights from the Green City of Ben Guerir, Morocco. J Urban Manag 14:247. https://doi.org/10.1016/j.jum.2024.11.001
15. Rochd A, Benazzouz A, Abdelmoula IA, Raihani A, Ghennioui A, Naimi Z, Ikken B (2021) Design and implementation of an AI-based & IoT-enabled home energy management system: a case study in Benguerir — Morocco. Energy Rep 7:699–719. https://doi.org/10.1016/j.egyr.2021.07.084
16. EnvirOconsult, and Ide-e (n.d.-a) "Maroc Commune d'Agadir Plan d'action En Faveur de l'énergie Durable (PAED)"
17. EnvirOconsult, and Ide-e (n.d.-b) "Maroc Commune d'Oujda Plan d'action En Faveur de l'énergie Durable (PAED)"
18. Wretling V, Gunnarsson-Östling U, Hörnberg C, Balfors B (2018) Strategic Municipal Energy Planning in Sweden—Examining Current Energy Planning Practice and Its Influence on Comprehensive Planning. Energy Policy 113:688–700. https://doi.org/10.1016/j.enpol.2017.11.006
19. Diane VG, Poumadère F, Bungener M, Chiffelle D (2021) Implementation of Local Energy Plans in Western Switzerland: Survey of the Current State and Possible Paths Forward. Sustainability 13(19):10970. https://doi.org/10.3390/su131910970
20. "GFA Projects - Ukraine - Establishing Municipal Energy Planning" (n.d.) https://www.gfa-group.de/projects/Establishing_Municipal_Energy_Planning_4052223.html. Accessed 9 Oct 2024
21. IRES (2022) QUEL AVENIR DE L'EAU AU MAROC?' RAPPORT DE SYNTHESE DES TRAVAUX DE LA JOURNEE SCIENTIFIQUE DU 17 MARS 2022. IRES

22. Ministère de l'Aménagement du Territoire National, de l'Urbanisme, de l'Habitat et de la Politique de la Ville (2019) "EFFICACITÉ ÉNERGÉTIQUE DANS LES BÂTIMENTS AU MAROC Support de Sensibilisation"
23. Bennouna, Amin. (2024) "Émissions de Gaz à Effet de Serre Par l'électricité Au Maroc En 2023." https://doi.org/10.13140/RG.2.2.30495.57760
24. "La Fondation Mohammed VI pour la Protection de l'Environnement" (n.d.) https://fm6e.org/. Accessed 1 Nov 2024
25. El Baz F, Arjdal S, Bakrim M, Mediouni T, Faska N (2020) Methanisation of Agadir Urban Solid Waste: Theoretical Evaluation of the Energy Production Potential. Mater Today Proc 22:97–99. https://doi.org/10.1016/j.matpr.2019.08.102
26. Benoist, T, F Gardi, A Jeanjean, and B Estellon. (2010). "Recherche locale pour un probl'eme d'optimisation de tourn'ees de v'ehicules avec gestion des stocks"
27. Parviziomran I, Mahmoudi M (2024) An assignment-based decomposition approach for the vehicle routing problem with backhauls. Multimodal Transport 3(4):100174. https://doi.org/10.1016/j.multra.2024.100174
28. "Formation Éco-Conduite En Entreprise | Actua Formation." (n.d.) https://www.actua-formation.fr/nos-programmes/eco-conduite/. Accessed 2 Nov 2024

CFD-Based Optimization of a Novel Hybrid Battery Thermal Management System for Electric Vehicles: Approaches and Applications for Smart Mobility

Nassreddine Hmidi, Hicham Kaddouri, Ahmed Alami Merrouni, Abdel-illah Amrani, and Elmiloud Chaabelasri

Abstract The global shift towards renewable energy and electric mobility has increased the adoption of electric and hybrid vehicles, which rely on lithium-ion batteries. For optimal performance, these batteries must operate within 15–35 °C, with a maximum temperature difference ($\Delta T_{b,max}$) below 5 °C between cells. This requires an effective Battery Thermal Management System (BTMS). In this study, we developed a hybrid BTMS for a three-cell battery module (3C discharge rate), combining an aluminum cold plate, nano-phase change material (nano-PCM), and copper foam. Thermal performance was analyzed using Computational Fluid Dynamics (CFD) to simulate heat transfer and fluid flow. After validating the numerical model with experimental data, we used CFD to simulate the performances of the proposed BTMS. Results show that, at an inlet velocity of 0.1 m/s, the $T_{b,max}$ slightly exceeded 35 °C, requiring further optimization. Incorporating 10% MWCNT nano-PCM reduced $T_{b,max}$ to 35.23 °C and $\Delta T_{b,max}$ to 2.49 °C. Increasing the inlet velocity to 0.3 m/s further lowered $T_{b,max}$ to 35.03 °C and $\Delta T_{b,max}$ to 2.37 °C. Beyond this point, $T_{b,max}$, $\Delta T_{b,max}$, the Nusselt number, and the friction coefficient stabilized.

Besides, it has been found that integration of 95% porous copper foam successfully maintained $\Delta T_{b,max}$ below 5 °C while keeping $T_{b,max}$ within the optimal range of 15 °C to 35 °C throughout the discharge cycle. These results highlight the effectiveness of the novel BTMS, demonstrating its potential to improve battery performance for electric mobility applications.

N. Hmidi · A. Alami Merrouni (✉) · E. Chaabelasri
Materials Science, New Energies and Applications Research Group, LPTPME Laboratory, Department of Physics, Faculty of Sciences, Mohammed 1st University, Oujda, Morocco

H. Kaddouri
Experimentation and Modelling Team in Mechanics and Energy Systems, National School of Applied Sciences, Abdelmalek Essaadi University, Al-Hoceima, Morocco

A.-i. Amrani
Equipe des Énergies Nouvelles et Ingénierie des Matériaux (ENIM), Faculté des Sciences et Techniques, Moulay Ismail University of Meknes, Errachidia, Morocco

H. Kaddouri et al. (eds.), *Smart Cities for Sustainable and Urban Development*, Green Energy and Technology, https://doi.org/10.1007/978-3-031-99001-4_21

Keywords Electrical vehicles · Hybrid thermal management system · Nano-PCM · Cooper foam Aluminum cold plate · CFD analysis

1 Introduction

The development of smart cities relies on sustainable and interconnected infrastructures designed to enhance quality of life, reduce carbon emissions, and optimize resource usage. Within this framework, electric vehicles (EVs) play a pivotal role by providing a cleaner alternative to combustion engine vehicles and integrating seamlessly into urban smart grids to lower pollution levels. However, one of the critical technical challenges for EVs is the effective management of battery temperature, which is key to ensuring their performance and safety in densely populated urban environments.

Lithium-ion batteries, which dominate the EV market, are particularly sensitive to temperature fluctuations. This directly impacts their efficiency, lifespan, and ultimately, their viability within the context of smart cities. Keeping these batteries within optimal temperature ranges is crucial not only to extend their lifespan but also to reduce the need for frequent recharging and facilitate their integration with smart energy networks. With the rapid adoption of EVs in smart cities, the demand for a reliable and effective battery thermal management system (BTMS) has become essential to ensure sustainable and efficient urban mobility. Specifically, maintaining a maximum battery temperature ($T_{b,max}$) between 15 °C and 35 °C [1–3] and a maximum temperature difference within the battery of less than 5 °C [4–6] is crucial for optimal performance.

Various thermal management techniques have been explored in the literature, each with distinct advantages and limitations. Air and liquid cooling systems are commonly used for their simplicity and effectiveness. For example, Lu et al. [7] proposed an optimization of an air-based Battery Thermal Management System (BTMS) aimed at reducing both system volume and enhancing cooling performance. Using a four-step approach validated through CFD simulation, the system volume was reduced by 13.01%, and thermal uniformity improved by up to 65.87% depending on the configuration. Moosavi et al. [8] introduced an innovative model to assess the thermal management of air-cooled lithium batteries designed for hybrid vehicles. By optimizing cell spacing, the model indicates that maximum temperatures rise with certain spacing configurations, while temperature differences decrease, thereby enhancing thermal uniformity. This approach enables accurate and efficient simulations, supporting the design of safer and more effective battery systems. Sui et al. [9] developed a Battery Thermal Management System (BTMS) using a cold plate with hybrid channels positioned under a 280 Ah LiFePO4 battery pack to optimize thermal control. Through multi-objective optimization, the system maintains a maximum temperature between 30.73 and 33.78 °C with a pressure drop ranging from 1.76 to 7.66 kPa. The optimal configuration keeps the temperature below 45 °C at a 3C discharge rate, reducing the maximum temperature by

17.6 °C compared to traditional cooling systems. Ren et al. [10] developed a bottom liquid cooling (BLC) system using multi-channel flat tubes (MCFT) to stabilize the temperature of a battery module. Compared to passive cooling, the BLC system effectively reduces temperature rise. Increasing the water flow rate has minimal effect (only a 1.4% change), while lowering the inlet water temperature keeps the battery below 45 °C, though it increases thermal differences by 48.9% and 61.6%, respectively.

Phase change material (PCM) cooling systems present an interesting alternative for battery thermal management, capable of absorbing and storing heat to mitigate temperature spikes. However, as Fini et al. [11] proposed using pressure to enhance energy storage in a phase change material (PCM) for Li-ion batteries. At a pressure of 500 kPa, the discharge time at a 7C rate doubled compared to atmospheric pressure, and the depth of discharge increased by 10%. This approach improves thermal efficiency without additional energy consumption. Sutheesh et al. [12] modeled a thermal management system for a Li-ion cell in electric vehicles, utilizing a phase change material (PCM) with honeycomb fins. The results show that, at a 5C discharge rate, the simple honeycomb structure reduces the maximum cell temperature by 2.77%. Using 1-Tetradecanol PCM and copper fins further lowers the cell temperature by 13.36 K, achieving optimal performance at 309.15 K. Bais et al. [13] studied a battery thermal management system (BTMS) using phase change material (RT-47) enhanced with γ-Al_2O_3 nanoparticles (0.5 to 4 wt%) for high discharge rate (3C) Li-ion batteries. Five battery packs were tested, with the 0.5% NePCM pack showing the best performance, achieving a maximum temperature of 43.0 °C and a discharge time of 1113 s, outperforming other configurations. Comparatively, this pack exceeded the performance of hybrid BTMS systems at 45 °C and inorganic PCM BTMS at 64.2 °C, with all BTMS packs weighing less than 20% of the base pack's weight.

More recent studies have also explored approaches leveraging artificial intelligence (AI) and machine learning to improve battery thermal management. For instance, Khan et al. [14] proposed a lightweight U-shaped liquid cooling method for batteries, optimized using artificial intelligence (AI). A neural network regression model, trained on 125 simulations, accurately predicted maximum temperature and thermal uniformity with 99.1% precision. The optimal configuration reduced temperature by 21% and lightened the cooling plate by 45% compared to conventional systems, providing an efficient and lightweight thermal solution. N. Zhang et al. [15] introduced a hybrid battery thermal management system (BTMS) that combines phase change material (PCM) with air cooling, optimized through machine learning. Simulations show that the fractal fin design reduces the maximum temperature (T_{max}) by 0.53 K, while reverse airflow decreases the temperature difference (ΔT_{max}) by 17%. The neural network model accurately predicts T_{max} and ΔT_{max}, delivering a high-performance BTMS under rapid discharge and extreme conditions.

To overcome the limitations of current solutions in the litterature, this study proposes an innovative hybrid BTMS architecture designed specifically to meet the dynamic thermal demands of next-generation EV batteries. This system incorporates an aluminum cold plate, nano-phase change materials (nano-PCM), as well as copper foams. This combination enables the battery to be maintained within an optimal temperature range, with a maximum battery temperature ($T_{b,max}$) below 35 °C and a maximum temperature difference ($\Delta T_{b,max}$) below 5 °C. This BTMS aims to enhance battery thermal stability, a critical factor for the reliability of EVs in smart city environments.

Numerous studies have been conducted on the contribution of lithium-ion batteries to smart mobility. Among them, Vanus et al. [16]. examined the role of electric scooters in smart cities. Ruggieri et al. [17] explored the role of electric mobility in European smart cities, such as London, Oslo, Hamburg, Milan, Florence, and Bologna, highlighting its potential as a key solution for reducing emissions and improving air quality. Cheng et al. [18] investigated an energy management system (EMS) for electric mobility and smart cities, integrating technologies such as lithium-ion batteries and renewable energy sources into urban infrastructures. Jiang et al. [19] proposed an energy management system that integrates electric vehicles (EVs), energy storage systems (ESS), and photovoltaics (PVs) into smart city microgrids.

By offering a system capable of preserving battery health while improving overall energy efficiency, this research contributes to the sustainability and reliability goals of EVs within smart city infrastructures, serving as a valuable reference for engineers and researchers working to optimize battery thermal performance in the rapidly evolving EV sector.

2 Materials and Methods

2.1 Physical Problem

The battery module consists of four prismatic cells with a capacity of 50 Ah, connected in series. The cathode is made of lithium iron phosphate ($LiFePO4$), and the anode is composed of graphite, the technical parameters of battery used in this work are illustrated in Table 1. Due to high discharge rates and external conditions, these cells can experience significant temperature increases, impacting both performance and longevity. To address this issue, a hybrid Battery Thermal Management System (BTMS) is implemented, combining an aluminum cold plate with a MWCNT nano-enhanced phase change material (Nano-PCM), as shown in the Fig. 1.

To further enhance the BTMS's thermal efficiency, copper foam is integrated into the PCM. Acting as a porous medium, the copper foam increases thermal conductivity within the PCM, allowing for faster and more uniform heat dissipation. The Nano-PCM absorbs excess heat through phase transitions, while the copper foam expedites this process, ensuring that the cell temperatures remain stable. This configuration is designed to improve temperature regulation across the battery module, maintaining optimal performance and extending the lifespan of the cells under demanding operating conditions.

Table 1 Battery cell parameters

Parameter name	Value	Unit
Cathode material	LiFePO4	–
Anode material	Graphite	–
Nominal capacity	50	Ah
Nominal Voltage	3.2	V
Maximum cut_off Voltage	3.65	V
Minimum cut_off Voltage	2.5	V

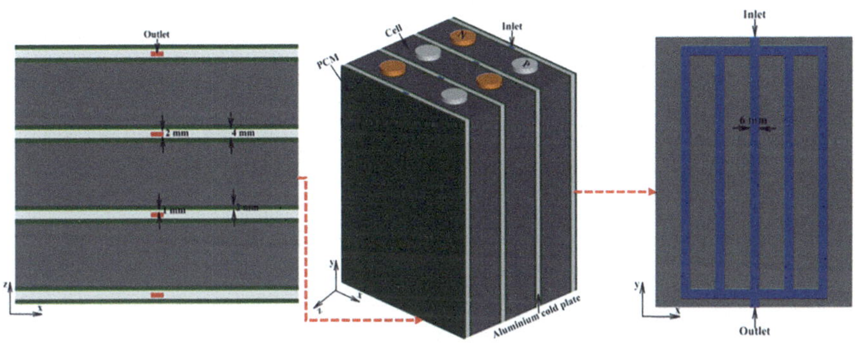

Fig. 1 Battery module with its TMS

2.2 *Governing Equations*

2.2.1 Fluid Flow and Thermal Equations

The heat equation in battery which equation describes the temperature evolution within the battery, The term Q_{gen} represents the internal heat generation due to electrochemical reactions and resistive heating Li et al. [20].

$$\frac{\partial\left[\rho_b c_b T_b\right]}{\partial t} = \nabla.\left[k_b \nabla T_b\right] + Q_{gen} \tag{1}$$

where ρ_b, c_b, T_b, k_b, and Q_{gen} represent the battery density, specific heat, temperature, thermal conductivity, and the heat generated inside the battery, respectively.

The heat equation governs the temperature distribution within the aluminum cold plate

$$\frac{\partial\left[\rho_{Al} c_{Al} T_{Al}\right]}{\partial t} = \nabla.\left[k_{Al} \nabla T_{Al}\right] \tag{2}$$

where ρ_{Al}, c_{Al}, T_{Al}, and k_{Al} represent the aluminum density, specific heat, cold plate temperature, and thermal conductivity, respectively.

The Reynolds number (Re) is a dimensionless parameter that characterizes the flow regime within the cooling channels of the battery thermal management system (BTMS). It is calculated as Li et al. [20]:

$$R_e = \frac{\rho_w v_{int} D_h}{\mu_w} \quad (3)$$

where ρ_w, v_{int}, D_h, and μ_w represent the water density, inlet velocity, hydraulic diameter of the channel inlet, and water dynamic viscosity, respectively.

The hydraulic diameter (D_h) is defined as:

$$D_h = \frac{4A_{int}}{P} \quad (4)$$

where A_{int} is the cross-sectional area of the channel, and P is the wetted perimeter of the channel.

Since the Reynolds number is below 2300, the flow regime in the cooling channels is classified as laminar, and the system is described by governing equations below:

This continuity equation, indicating that the flow is incompressible, meaning that the mass flow rate remains constant across the control volume [21].

$$\nabla.\vec{v} = 0 \quad (5)$$

2.2.2 Momentum Equation for Laminar Flow [21]

$$\rho_w \left[\frac{\partial \vec{v}}{\partial t} + (\vec{v}.\nabla)\vec{v} \right] = -\nabla p + \mu_w \nabla^2 \vec{v} \quad (6)$$

where $\vec{V}$, P and μ_w represent the velocity vector, pressure of the fluid flow and water dynamic viscosity, respectively.

The heat equation for coolant inside the aluminum cold plate [21]

$$\rho_w c_w \frac{\partial T_w}{\partial t} + \rho_w c_w \vec{v}.\nabla T_w = \nabla.[k_w \nabla T_w] \quad (7)$$

where ρ_w, c_w, T_w, and k_w represent water density, specific heat, temperature, and thermal conductivity, respectively.

2.2.3 PCM Equations

The governing equations for PCM, nano-PCM, and copper foam as a porous medium are as follows:

The continuity equation enforces mass conservation in the PCM, ensuring that any increase in mass within a control volume is accounted for by the inflow or outflow of mass. This equation is foundational in fluid dynamics modeling, especially for incompressible flows [22]:

$$\frac{\partial \rho_{nPCM}}{\partial t} + \nabla \cdot \left(\rho_{nPCM} \vec{V} \right) = 0 \tag{8}$$

The momentum equation describes the fluid motion in the PCM. It incorporates viscous, pressure, thermal expansion (buoyancy), and damping terms $\vec{S}$ to capture fluid behavior in the porous metal foam. The term $\vec{F}$ includes effects of flow resistance due to the porous structure, which impacts fluid velocity and movement within the PCM [22].

$$\frac{\rho_{nPCM}}{\varepsilon} \left(\frac{\partial \vec{V}}{\partial t} + \frac{1}{\varepsilon} \left(\vec{V} \cdot \nabla \right) \vec{V} \right) = -\nabla P + \frac{\mu_{nPCM}}{\varepsilon} \nabla^2 \vec{V} - \rho_{nPCM} \beta_{nPCM} \varepsilon \left(T - T_{ref} \right) \vec{g} - \vec{S} - \vec{F} \tag{9}$$

where ρ_{nPCM}, μ_{nPCM}, β_{nPCM}, are nano-PCM density, dynamic viscosity and thermal expansion coefficient, respectively. ε, T, $\vec{V}$, $\vec{g}$ are pososity, temperature, velocity vector of nano-PCM gravitational field intensity vector, respectively.

The energy equation governs heat transfer in the PCM, accounting for both conductive and convective heat transfer. The term S_L represents the energy absorbed or released during phase change [22] is given by:

$$\frac{\partial \varepsilon \rho_{nPCM} C_{p,nPCM} T}{\partial t} + \nabla \left(\rho_{nPCM} C_{p,nPCM} \vec{V} \mathrm{T} \right) = \nabla \left(K_{nPCM} + \nabla T \right) - S_L \tag{10}$$

$C_{p,nPCM}$, k_{nPCM} are nano-PCM specific heat and thermal conductivity, respectively.

The source term introduces damping in the mushy zone, which is a region where the PCM is partially melted. The mushy zone constant A_{mush} controls the extent of damping, helping to stabilize fluid flow in regions where the PCM is not fully solid or liquid [22].

$$\vec{S} = A_{mush} \frac{\left(1 - f_l\right)^2}{f_l^3 + 0.001} \vec{V} \tag{11}$$

The liquid fraction (f_l) in the PCM, depending on its temperature T. Ts and T_l represent the solidus and liquidus temperatures, respectively. The liquid fraction determines the phase of the PCM, which is crucial in the phase transition process as it dictates heat absorption or release during melting or solidification [22].

$$f_l = \frac{\Delta H}{h_{sl,nPCM}} = \begin{cases} 0 & \text{if} \quad \mathrm{T} \prec \mathrm{T}_s \\ \dfrac{T - T_s}{T_l - T_s} & \text{if} \quad \mathrm{T}_s \le T \le T_l \\ 1 & \text{if} \quad \mathrm{T}_l \prec \mathrm{T} \end{cases} \tag{12}$$

H is the total enthalpy of the PCM, which consists of the sensible enthalpy (h) and the latent heat (ΔH) associated with the phase change. This combined enthalpy is used to represent the thermal energy stored within the PCM at any given time [22].

$$H = h + \Delta H \tag{13}$$

where h is the he sensible enthalpy (h) based on the specific heat of the nano-PCM ($C_{p,nPCM}$) and the temperature range. This term quantifies the heat energy of the PCM without phase change.

$$h = h_{ref} + \int_{T_{ref}}^{T} C_{p,nPCM} dT \tag{14}$$

The equation below represent flow resistance term represents the resistance that the fluid (in this case, the nano-enhanced PCM) experiences as it flows through a porous medium, such as metal foam integrated within the PCM [22].

$$\vec{F} = \left(\frac{\mu_{nPCM}}{k} + \frac{\rho_{nPCM} C_f \left|\vec{V}\right|}{\sqrt{k}} \right) \left|\vec{V}\right| \tag{15}$$

The permeability k of the porous medium (metal foam), dependent on the pore diameter d_p, fiber diameter d_f, and porosity ϵ. Permeability impacts the ease of fluid flow through the porous medium [23, 24].

$$k = 0.00073 d_p^2 \left(1-\varepsilon\right)^{-0.224} \left(\frac{d_f}{d_p} \right)^{-1.11} \tag{16}$$

The inertial coefficient C_f accounts for the resistance to fluid flow within the porous metal foam, influenced by the structure's porosity and fiber-to-pore diameter ratio [23, 24].

$$C_f = 0.00212 \left(1-\varepsilon\right)^{-0.132} \left(\frac{d_f}{d_p} \right)^{-1.63} \tag{17}$$

The equations below define the effective fiber and pore diameters for the porous structure. Here, d_f depends on the porosity, while d_p is inversely related to pore density ω. These structural properties influence the material's thermal and fluidic behavior [23, 24].

$$d_f = 1.18 d_p \sqrt{\frac{1-\varepsilon}{3\pi}} \left(\frac{1}{1-e^{-(1-\varepsilon)/0.04}} \right) \tag{18}$$

$$d_p = \frac{0.0254}{\omega} \tag{19}$$

The source term S_L represents the rate of latent heat absorption or release within the PCM due to phase change. It includes both time-dependent and convective contributions, capturing how the phase transition impacts energy balance [23, 24].

$$S_L = \frac{\partial \varepsilon \rho_{nPCM} f_l h_{nPCM}}{\partial t} + \nabla\left(\rho_{nPCM} \vec{V} f_l h_{nPCM}\right) \tag{20}$$

To simulate the nano-PCM the thermophysical properties of nano-PCM will be as bellows:

The nano-PCM density is given by equation below, where φ is nanoparticles fraction in PCM [25].

$$\rho_{nPCM} = (1-\varphi)\rho_{PCM} + \varphi\rho_{np} \tag{21}$$

where ρ_{PCM} and ρ_{np} represents PCM density and nanoparticles density, respectively.

The dynamic viscosity of nano-PCM is given by [25]

$$\mu_{nPCM} = \frac{\mu_{PCM}}{(1-\varphi)^{2.5}} \tag{22}$$

μ_{PCM} represent the density of PCM.

This equation bellow, based on Maxwell's model, calculates the effective thermal conductivity of the nano-PCM [25].

$$\frac{k_{nPCM}}{k_{PCM}} = \frac{k_{np} + 2k_{PCM} - 2\varphi\left(k_{PCM} - k_{np}\right)}{k_{np} + 2k_{PCM} + \varphi\left(k_{PCM} - k_{np}\right)} \tag{23}$$

where k_{np} and k_{PCM} respresents the nanopartices thermal conductivity and PCM thermal conductivity, respectively.

The specific heat capacity of the nano-PCM, based on a weighted average of the PCM and nanoparticle properties according to their volume fractions. This adjustment provides a realistic representation of the thermal properties of the composite material [25].

$$\left(\rho C_p\right)_{nPCM} = (1-\varphi)\left(\rho C_p\right)_{PCM} + \varphi\left(\rho C_p\right)_{np} \tag{24}$$

where $C_{p,\,PCM}$ and $C_{p,\,np}$ represent the specific heat of PCM and the nanoparticles, respectively.

This formula below gives the effective latent heat of the nano-PCM (h_{nPCM}), which is reduced by the nanoparticle fraction (φ) since nanoparticles do not contribute to phase change energy storage. This slight reduction in latent heat affects the overall heat absorption capacity of the material [26].

$$(\rho h)_{nPCM} = (1-\varphi)(\rho h)_{PCM} \tag{25}$$

h_{PCM} represents the latent heat of PCM.

2.3 Thermophysical Properties and Boundary Conditions

To solve the governing equations, the following initial and boundary conditions are applied (Table 2):

- Initial temperature is set to 30 °C.
- Heat transfer coefficient with air is fixed at 5 W/m^2·K.
- Thermophysical properties are assumed to be independent of temperature.
- The fluid is considered incompressible.
- Relative outlet pressure is maintained at 0 Pa.

2.4 Grid Independency

The Fig. 2 shows the results of a grid independence study, which is conducted to ensure that the simulation results are stable and accurate, regardless of the number of mesh elements used. Here, we're looking at how $T_{b,max}$ and the pressure drop (ΔP) change as the mesh becomes finer, i.e., as the number of elements increases at velocity inlet fixed at 0.1 m · s^{-1}.

As the number of elements rises, both $T_{b,max}$ and ΔP initially change noticeably, but they gradually level off, reaching nearly constant values at around 3.2 × 106 elements. This "flattening" indicates that we've reached grid independence, meaning further increases in mesh density have little effect on the results.

Table 2 Materials thermophysical properties [23, 27–30]

Property	Battery	Water	Aluminum	Copper	Capric acid (PCM)	MWCNT
Thermal conductivity (W · m^{-1} · K^{-1})	Kx = ky = 2.7; kz = 0.9	0.6	202.4	387.6	0.153	3000
Density (kg · m^{-3})	1868	998.2	2719	8900	878	1600
Specific heat (J · kg^{-1} · K^{-1})	1296	4182	871	386	475.59	769
Dynamic viscosity (kg · m^{-1} · s^{-1})	–	0.001	–	–	0.043	–
Latent heat of fusion (J · kg^{-1})	–	–	–	–	152.7e3	–
Solidus temperature (K)	–	–	–	–	302	–
Liquidus temperature (K)	–	–	–	–	305	–

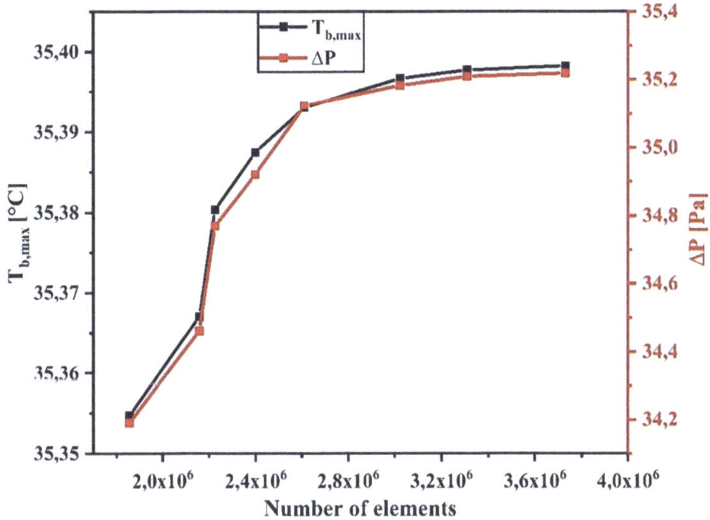

Fig. 2 Grid independency

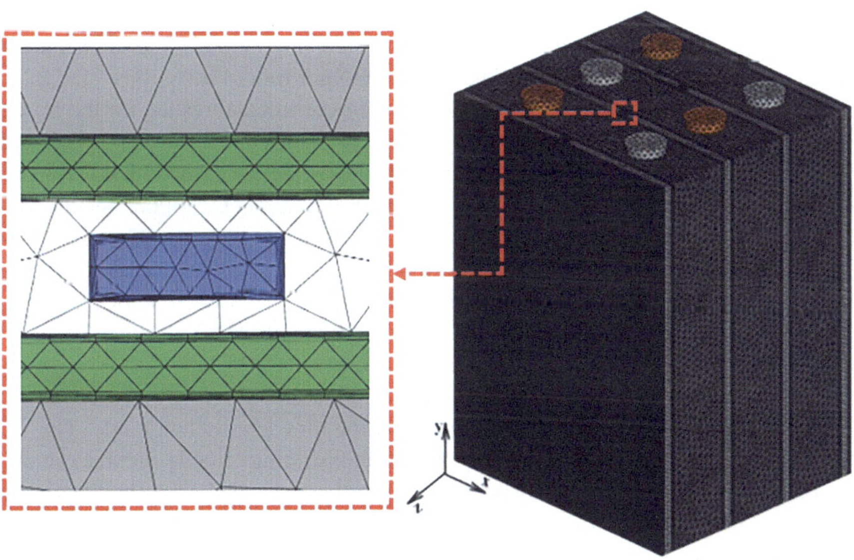

Fig. 3 Unstructured tetrahedral mesh used

A tetrahedral mesh as illustrated in Fig. 3 was used for this study, which is particularly useful for capturing detailed geometries. Based on these results, a mesh with around 3.2 × 106 elements is sufficient to provide accurate and reliable results without the need for additional computational resources.

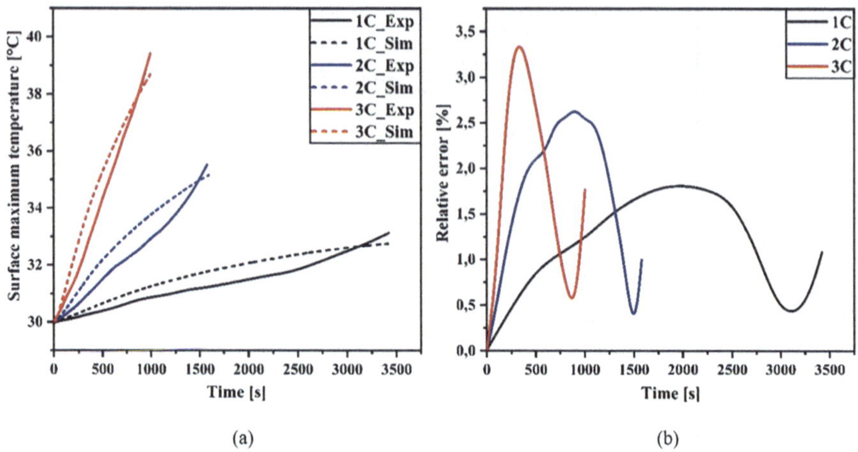

Fig. 4 Validation of the numerical model

2.5 *Numerical Validation*

The Fig. 4 shows how well our numerical model aligns with the experimental results reported by Y. Zhang et al. [30]. We performed the simulations in ANSYS Fluent using the finite volume method and the SIMPLE algorithm to solve the equations.

In the Fig. 4a, we compare the simulated and experimental results for the maximum surface temperature of the battery across three discharge rates: 1C, 2C, and 3C. The solid lines represent the experimental data, while the dashed lines show our simulation results. Across all three discharge rates, including the higher 3C rate, the model does an excellent job of matching the experimental data. At 1C and 2C, the model aligns very closely with the experimental values, while at 3C, although the simulated temperature rises slightly faster, the overall pattern and values are very consistent with the experiments. This strong agreement at 3C, despite the more intense thermal conditions, suggests that our model can accurately predict temperature behavior even at higher discharge rates.

In Fig. 4b, we show the relative error over time for each discharge rate. The 1C and 2C discharge rates maintain a low error, demonstrating reliable accuracy in our predictions. At 3C, there's a small peak in error early on, around 3.5%, but it quickly stabilizes and remains within a reasonable range. This initial peak may stem from the challenge of modeling rapid temperature increases at higher discharge rates, but the overall accuracy remains robust, confirming that the model is reliable even under these challenging conditions.

2.6 BTMS Efficiency

The bellow parameters are key indicators used to evaluate the efficiency and performance of a BTMS. Each parameter provides insight into different aspects of heat transfer and flow characteristics within the system:

2.6.1 $T_{b,max}$, and $\Delta T_{b,max}$

The maximum battery temperature ($T_{b,max}$) and the maximum temperature difference within the battery ($\Delta T_{b,max}$) are critical indicators of BTMS effectiveness. For optimal performance, $T_{b,max}$ should be maintained within the range of 15 °C to 35 °C, while $\Delta T_{b,max}$ should not exceed 5 °C across the battery.

2.6.2 Uniformity Index

The uniformity index (T_{uni}), which measures the evenness of temperature distribution within the battery. A lower Tuni value indicates a more uniform temperature distribution, which is essential for balanced performance across all cells and reduces the risk of localized overheating.

$$T_{uni} = \frac{T_{b,max} - T_{b,min}}{T_{b,avg}} \tag{26}$$

where $T_{b,max}$, $T_{b,min}$, and $T_{b,avg}$ represent the maximum, minimum, and average battery temperatures, respectively.

2.6.3 Nusselt Number

The Nusselt number (Nu) quantifies the effectiveness of convective heat transfer within the system. Higher Nu values indicate more efficient heat transfer, which is beneficial for maintaining optimal battery temperature, especially under high discharge rates. The Nusselt number depends on the average heat transfer coefficient ($\bar{h}$), hydraulic diameter (D_h), and fluid thermal conductivity (k). It is given by:

$$N_u = \frac{\bar{h} D_h}{k} \tag{27}$$

where $\bar{h}$ is the average heat transfer coefficient, which is a measure of how effectively heat is being transferred from the battery to the coolant. This coefficient depends on the heat flux (q) at the microchannels cold plate and the temperature difference between the microchannels walls and the coolant ($T_{avg} - T_f$). Higher $\bar{h}$

values of mean more efficient heat removal, helping to keep the battery temperature within a safe operating range.

$$\bar{h} = \frac{q}{T_{avg} - T_f} \tag{28}$$

where

$$T_f = \frac{T_{\text{int}} + T_{out}}{2} \tag{29}$$

T_{int} and T_{out} represent the fluid inlet and outlet temperatures, respectively.

2.6.4 Friction Factor

Friction factor is a measure of the resistance to flow within the cooling channel. It is influenced by the pressure drop (ΔP), the density of the coolant (ρ), the hydraulic diameter (D_h), the length characteristic of the flow channel (L), and the inlet velocity (V_{int}). Lower friction factors mean less resistance to coolant flow, reducing the energy needed to circulate the coolant and enhancing overall BTMS efficiency, f is given by.

$$\text{f} = \frac{2\Delta P D_{\text{h}}}{\rho_{\text{w}} \text{L} \text{v}_{\text{int}}^2} \tag{30}$$

3 Results and Discussions

3.1 Thermal Performance at 3C Discharge Rate

The Fig. 5 shows how the battery's temperature changes over time, focusing on two important aspects: $T_{b,max}$ and $\Delta T_{b,max}$ for velocity inlet fixed at 0.1 m · s^{-1} and 3C discharge rate.

In the Fig. 5a, we see that the battery's $T_{b,max}$ starts at 30 °C and gradually rises, eventually surpassing 35.5 °C. This is significant because it means the battery goes beyond its optimal operating range of 15 °C to 35 °C, where performance and safety are best maintained. Exceeding this range could lead to faster degradation and increase the risk of overheating, especially with sustained high temperatures.

The Fig. 5b illustrates $\Delta T_{b,max}$ over time. This difference quickly increases at the beginning as the battery heats up, eventually stabilizing around 2 °C. Although this variation is relatively small, it still indicates uneven temperature distribution, which can affect battery performance and potentially cause localized hotspots.

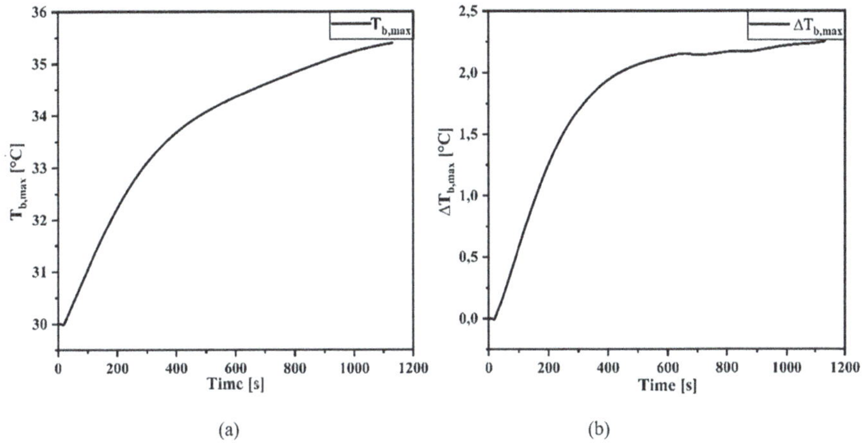

Fig. 5 $T_{b,max}$ and $\Delta T_{b,max}$ temporal variation at 3C discharge rate

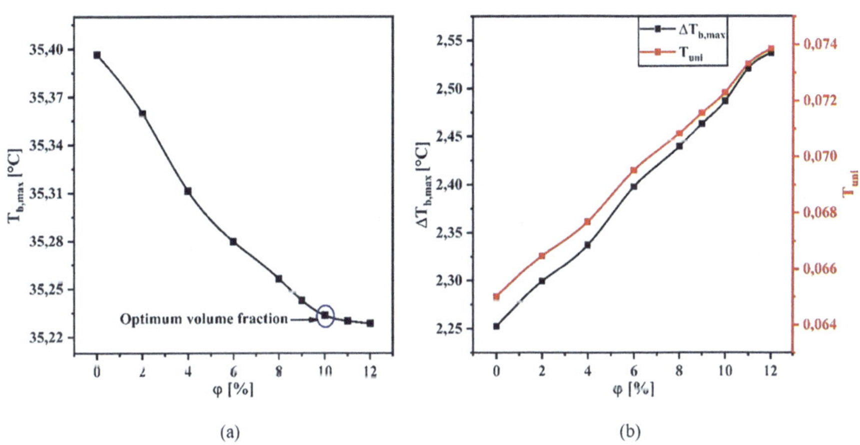

Fig. 6 $T_{b,max}$ and $\Delta T_{b,max}$ variation as a function of φ for 3C discharge rate

Together, these results emphasize the need for a BTMS to help keep the temperature within the ideal range. This motivates the next stage of our work, where we will focus on developing a BTMS to better manage heat and improve battery safety and performance, particularly under challenging operating conditions.

3.2 *Nano-PCM Volume Fraction Effect*

In the Fig. 6, we analyze the effect of incorporating multi-walled carbon nanotube (MWCNT) nanoparticles into a phase PCM within the BTMS. The plots illustrate the changes in $T_{b,max}$, $\Delta T_{b,max}$, and the uniformity index (T_{uni}) at the end of discharge for 3C discharge rate as a function of the nanoparticle volume fraction (φ) added to the PCM.

In the Fig. 6a, we observe a steady decrease in $T_{b,max}$ as the volume fraction of MWCNTs in the PCM increases, indicating improved thermal management due to the enhanced thermal conductivity provided by the nanoparticles. This temperature reduction continues up to a volume fraction of approximately 10%, after which Tb,max stabilizes, suggesting that 10% is the optimal volume fraction. Beyond this point, further additions of nanoparticles yield diminishing returns in terms of temperature reduction.

However, even with this optimization, the $T_{b,max}$ stabilize at 35.23 °C for 10% volume fraction, which remains slightly above 35 °C, which is outside the optimal temperature range of 15 °C to 35 °C. Operating above this range can lead to accelerated battery degradation and reduced safety. This outcome indicates that while adding MWCNTs to the PCM has improved thermal performance, it alone is insufficient to fully maintain the battery within the ideal temperature range under high-demand conditions.

The Fig. 6b shows $\Delta T_{b,max}$ and T_{uni} as the volume fraction of MWCNTs increases. Both $\Delta T_{b,max}$ and T_{uni} increase with higher nanoparticle concentrations. The rise in $\Delta T_{b,max}$ indicates a slight increase in internal temperature gradients, while Tuni suggests improved heat distribution within the PCM.

Although the addition of MWCNTs enhances the BTMS's ability to manage heat, further improvements are necessary to ensure the battery remains within the optimal temperature range. This finding underscores the need to refine the BTMS, which is the focus of the next stage of our work, aimed at achieving more effective thermal control and keeping the battery within safe operating conditions.

3.3 Velocity Inlet Effect

In this section, we investigate the impact of inlet velocity on the effectiveness of the novel BTMS under a 3C discharge rate and volume fraction fixed at 10%. The Fig. 7 show how changes in inlet velocity affect the $T_{b,max}$, $\Delta T_{b,max}$, and T_{uni}.

The Fig. 7a, illustrate that $T_{b,max}$ decreases significantly as the inlet velocity increases from 0.1 m/s to 0.3 m/s. Beyond this point, however, the reduction in $T_{b,max}$ becomes negligible, indicating that an optimal inlet velocity of 0.3 m/s has been reached. Increasing the velocity beyond 0.3 m/s does not further reduce the maximum battery temperature, which stabilizes at approximately 35.03 °C. Despite this improvement, $T_{b,max}$ remains slightly above the optimal temperature range of 15 °C to 35 °C. This highlights the need for further improvements to the BTMS to ensure that the battery operates entirely within the desired range, especially under high discharge rates like 3C.

The Fig. 7b on the right shows $\Delta T_{b,max}$ as a function of inlet velocity. Similar to $T_{b,max}$, $\Delta T_{b,max}$ decreases with increasing velocity up to 0.3 m/s and then stabilizes.

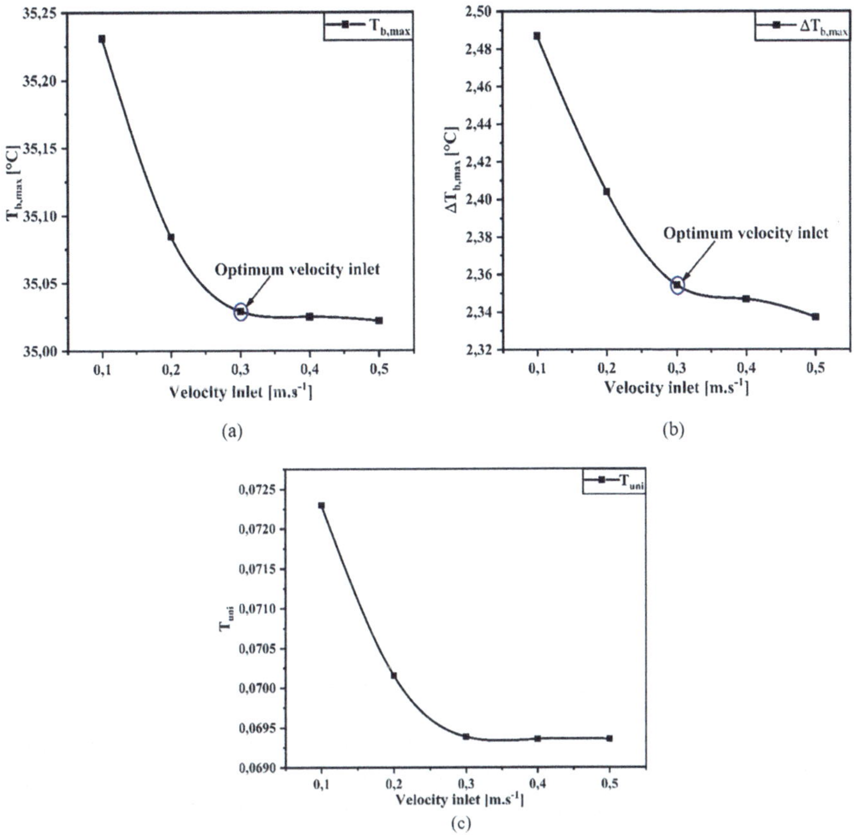

Fig. 7 $T_{b,max}$, $\Delta T_{b,max}$ and T_{uni} as a function of velocity inlet at 3C discharge rate and $\varphi = 10\%$

At 0.3 m/s, $\Delta T_{b,max}$ reaches around 2.35 °C, suggesting that higher inlet velocities contribute to a more uniform temperature distribution within the battery. However, further increases in velocity beyond this optimal point provide minimal additional improvement in temperature uniformity.

In the Fig. 7c, T_{uni} is plotted against inlet velocity. We see a steady decrease in T_{uni} as the inlet velocity increases, leveling off around 0.069 after 0.3 m/s. This indicates improved thermal uniformity within the battery as inlet velocity rises, with diminishing returns beyond the optimal velocity of 0.3 m/s.

These results suggest that an inlet velocity of 0.3 m/s is optimal for this BTMS, as it minimizes $T_{b,max}$, reduces $\Delta Tb_{,max}$, and improves thermal uniformity within the battery. However, since the maximum battery temperature remains slightly above 35 °C, further refinement of the BTMS is needed to ensure the battery remains fully within the safe operating range, particularly under high discharge rates. This will be the focus of future improvements in this work.

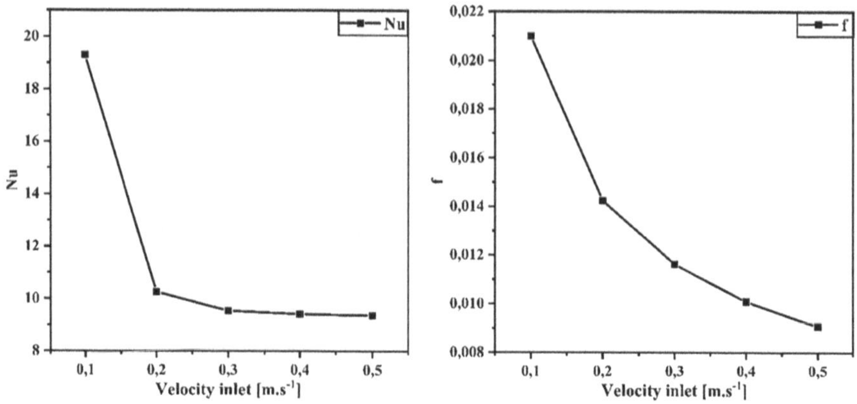

Fig. 8 Nusselt number and friction coefficient as a function of velocity inlet

The Fig. 8 illustrates the variation in the Nusselt number (Nu) and friction coefficient (f) as a function of inlet velocity, providing insight into the thermal and flow characteristics of BTMS.

In the Fig. 8a, we observe the behavior of the Nu with increasing inlet velocity. At lower velocities, specifically at 0.1 m/s, the Nusselt number is around 19.3, indicating strong convective heat transfer. However, as the inlet velocity increases, Nu drops sharply and stabilizes near 9.5 when the velocity reaches 0.3 m/s and above. This decline suggests that while higher flow velocities promote cooling, the rate of convective heat transfer does not continue to increase proportionally beyond 0.3 m/s. Thus, from a heat transfer perspective, an inlet velocity of 0.3 m/s appears optimal, balancing effective cooling with diminishing returns in convective enhancement at higher velocities.

The Fig. 8b shows f as a function of inlet velocity. At 0.1 m/s, the friction coefficient is relatively high at approximately 0.021. As velocity increases, f decreases significantly, reaching around 0.095 at 0.3 m/s and continuing to decline gradually with higher velocities. This decrease in friction coefficient with increasing velocity indicates reduced flow resistance, which can be beneficial for minimizing pumping power requirements. Beyond 0.3 m/s, however, the reduction in f becomes less pronounced, suggesting that higher inlet velocities would result in minimal further benefits in terms of flow efficiency.

An inlet velocity of 0.3 m/s is optimal for the BTMS. At this velocity, we achieve a stable Nusselt number with adequate heat transfer, while also reducing the friction coefficient to a manageable level, balancing cooling performance and flow efficiency. These findings confirm that 0.3 m/s is the ideal operating point, providing effective cooling without excessive pressure loss or unnecessary increases in pumping power, which is crucial for efficient BTMS design.

3.4 Porosity Effect

To maintain $T_{b,max}$ within its optimal operating range of 15 °C to 35 °C, a porous copper foam medium is introduced into the nano-PCM in this section. This addition enhances the thermal conductivity, improving heat removal from the battery module and helping to minimize its maximum temperature.

The Fig. 9 shows the temperature contours of the battery under a 3C discharge rate, with an inlet velocity of 0.3 m/s and a nanoparticle volume fraction of 10%. The comparison is made between two cases: without porosity and with 95% porosity using a copper foam as a porous medium.

In the non-porous configuration, we observe higher temperatures along the battery cells, with peak temperatures reaching up to 35.03 °C. This uneven temperature distribution suggests that heat dissipation is limited, as the thermal management system struggles to cool certain areas effectively. The lack of porosity restricts pathways for heat transfer, causing localized hot spots and less uniform cooling. In the porous configuration, with 95% porosity using copper foam, we see a much more uniform temperature distribution, with maximum temperatures staying below 33.5 °C. The copper foam structure improves thermal conductivity and enhances convective heat transfer between the battery cells and the BTMS, allowing heat to dissipate more efficiently. This results in better cooling performance and a more consistent temperature across the battery pack, helping to keep the cells closer to the ideal operating range.

Overall, the use of 95% porous copper foam significantly improves the BTMS by reducing temperature hotspots and promoting more uniform cooling across the cells. This suggests that incorporating highly porous copper foam is a promising

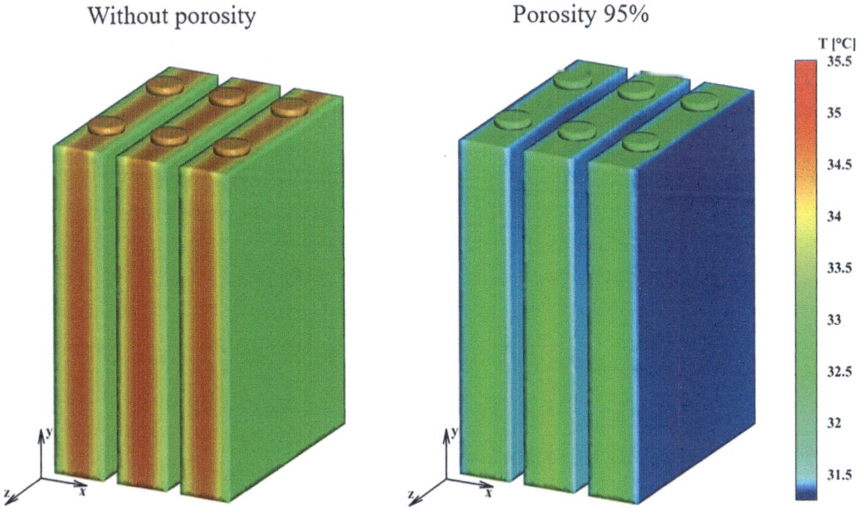

Fig. 9 Temperature contours with and without porosity 95% for 3C dis-charge rate, 0.3 m · s^{-1} as a velocity inlet and φ = 10%

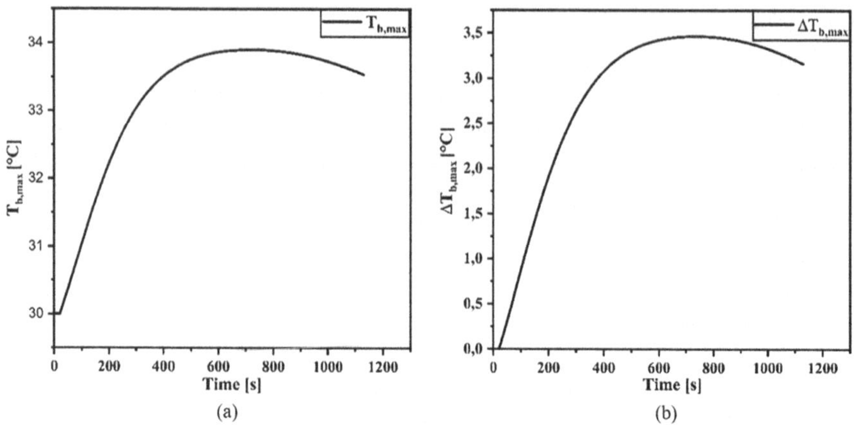

Fig. 10 Temporal variation of $T_{b,max}$ and $\Delta T_{b,max}$ at a 3C discharge rate, with an inlet velocity of 0.3 m/s, 10% nano-PCM volume fraction, and 95% porosity copper foam medium

strategy for enhancing thermal performance, especially under high-demand conditions like a 3C discharge rate.

The Fig. 10 illustrates how $T_{b,max}$ and $\Delta T_{b,max}$ change over time under a 3C discharge rate. In this setup, the BTMS utilizes a PCM enhanced with 10% nanoparticles, copper foam with 95% porosity, and an inlet velocity of 0.3 m/s.

From the Fig. 10a we can see that $T_{b,max}$ rises gradually from an initial 30 °C, eventually stabilizing just below 35 °C. This is a key result, as it means the battery temperature stays within the optimal operating range of 15 °C to 35 °C. Keeping the battery within this range is critical for maintaining efficiency, safety, and longevity, as excessive heat can accelerate wear and increase the risk of failure.

The Fig. 10b shows $\Delta T_{b,max}$, which indicates how evenly heat is distributed across the battery cells. This difference initially rises but then stabilizes around 3.4 °C, remaining well below the target limit of 5 °C. A low temperature gradient across the cells is essential for balanced performance, as it prevents specific areas from overheating and ensures the entire battery operates uniformly.

Overall, these results confirm that the enhanced BTMS configuration combining nano-PCM and high-porosity copper foam effectively controls both the maximum temperature and temperature uniformity within the battery. This setup keeps the battery in its optimal thermal range, even at a demanding 3C discharge rate, demonstrating the BTMS's ability to support high-performance and reliable battery operation.

4 Conclusion and Perspectives

This study has demonstrated the effectiveness of a novel BTMS design that incorporates nano-enhanced phase change materials (nano-PCM) and highly porous copper foam. Through a series of simulations, we analyzed key thermal parameters, including $T_{b,max}$ and $\Delta T_{b,max}$, under demanding conditions such as a 3C discharge rate and varying inlet velocities.

The results indicate that this hybrid BTMS successfully maintains $T_{b,max}$ within the optimal operating range of 15 °C to 35 °C, and keeps $\Delta T_{b,max}$ below 5 °C across the battery. The inclusion of 95% porous copper foam significantly enhanced heat transfer, leading to more uniform temperature distribution and effective cooling throughout the battery cells. This is crucial for battery performance, as it helps to prevent hotspots and ensures balanced operation across the battery pack.

Moreover, the analysis identified an optimal inlet velocity of 0.3 m/s, which provided a good balance between cooling efficiency and energy consumption. Beyond this point, increases in coolant flow offered minimal additional cooling benefits, confirming the efficiency of the selected design parameters.

Overall, this study highlights the potential of integrating nano-PCM with porous copper foam in BTMS designs to address the thermal challenges of high-performance batteries, particularly in electric vehicles and other applications requiring reliable thermal management. Future work will focus on further refining this BTMS to enhance its cooling capacity and adapt it to varying operational conditions, supporting safer and more efficient battery use in advanced energy systems.

Li-ion batteries (LiBs) are essential for smart city infrastructures, supporting applications such as energy storage, electric public transportation, and vehicle-to-grid systems. Advanced BTMS designs, like the one proposed in this study, can enhance LiB performance and safety in these contexts. Future research should focus on adapting such systems to address challenges related to extreme environmental conditions and real-time thermal monitoring, ensuring reliable and efficient energy solutions for the dynamic needs of smart cities.

References

1. Hmidi N, Merrouni AA, Salhi JE, Amrani A i, Chaabelasri E (2024a) Optimal battery configuration for electrical vehicles in hot climates: a numerical comparative study of materials and cooling approaches. Case Stud Thermal Eng 61. https://doi.org/10.1016/j.csite.2024.104983
2. Tete PR, Gupta MM, Joshi SS (2021) Developments in battery thermal management systems for electric vehicles: a technical review. J Energy Storage 35. https://doi.org/10.1016/j.est.2021.102255
3. Yao F, Guan X, Chen Q, Lin L (2024) Research on thermal management system of lithium-ion battery with a new type of spider web liquid cooling channel and phase change materials. J Energy Storage 81. https://doi.org/10.1016/j.est.2024.110447
4. Hmidi N, Salhi JE, Merrouni AA, Amrani A i, Chaabelasri E (2024b) Performance analysis of several EVs batteries thermal management systems under hot-arid climatic conditions of Morocco. J Energy Storage 80. https://doi.org/10.1016/j.est.2023.110276
5. Jithin KV, Rajesh PK (2022) Numerical analysis of single-phase liquid immersion cooling for lithium-ion battery thermal management using different dielectric fluids. Int J Heat Mass Transf 188. https://doi.org/10.1016/j.ijheatmasstransfer.2022.122608
6. Xu J, Fang H, Wang M, Li X, Chen Y (2024) Simulation research on thermal management system of battery module with fin heat dissipation structure. Appl Therm Eng 239. https://doi.org/10.1016/j.applthermaleng.2023.122177

7. Lu H, Tang X (2023) A flexible optimization study on air-cooled battery thermal management system by considering of system volume and cooling performance. J Energy Storage 72. https://doi.org/10.1016/j.est.2023.108527
8. Moosavi A, Ljung AL, Staffan Lundström T (2023) A study on the effect of cell spacing in large-scale air-cooled battery thermal management systems using a novel modeling approach. J Energy Storage 72. https://doi.org/10.1016/j.est.2023.108418
9. Sui Z, Lin H, Sun Q, Dong K, Wei W (2024) Multi-objective optimization of efficient liquid cooling-based battery thermal management system using hybrid manifold channels. Appl Energy 371. https://doi.org/10.1016/j.apenergy.2024.123766
10. Ren R, Zhao Y, Diao Y, Liang L (2023) Experimental study on the bottom liquid cooling thermal management system for lithium-ion battery based on multichannel flat tube. Appl Therm Eng 219. https://doi.org/10.1016/j.applthermaleng.2022.119636
11. Fini AS, Gharehghani A (2024) Experimental investigation of pressure effect on the PCM performance in li-ion battery thermal management system. J Energy Storage 79. https://doi.org/10.1016/j.est.2023.110273
12. Sutheesh PM, Jose J, Hotta TK, Rohinikumar B (2024) Numerical investigations on thermal performance of PCM-based lithium-ion battery thermal management system equipped with advanced honeycomb structures. Int Commun Heat Mass Transf 158. https://doi.org/10.1016/j.icheatmasstransfer.2024.107937
13. Bais A, Subhedar D, Panchal S (2024) Experimental investigations of a novel phase change material and nano enhanced phase change material based passive battery thermal management system for li-ion battery discharged at a high C rate. J Energy Storage 103:114395. https://doi.org/10.1016/j.est.2024.114395
14. Khan SA, Eze C, Dong K, Shahid AR, Patil MS, Ahmad S, Hussain I, Zhao J (2022) Design of a new optimized U-shaped lightweight liquid-cooled battery thermal management system for electric vehicles: a machine learning approach. Int Commun Heat Mass Transf 136. https://doi.org/10.1016/j.icheatmasstransfer.2022.106209
15. Zhang N, Zhang Z, Li J, Cao X (2024) Performance analysis and prediction of hybrid battery thermal management system integrating PCM with air cooling based on machine learning algorithm. Appl Therm Eng 257. https://doi.org/10.1016/j.applthermaleng.2024.124474
16. Vanus J, Bilik P (2022) Research on micro-mobility with a focus on electric scooters within smart cities. World Electric Vehicle J 13(10). https://doi.org/10.3390/wevj13100176
17. Ruggieri R, Ruggeri M, Vinci G, Poponi S (2021) Electric mobility in a smart city: European overview. Energies 14(2). https://doi.org/10.3390/en14020315
18. Cheng KWE, et al. Energy management system for mobility and smart city (2016)
19. Jiang J, Li Y, Li Y, Li C, Laihang Y, Li L (2024) Smart transportation systems using learning method for urban mobility and management in modern cities. Sustain Cities Soc 108. https://doi.org/10.1016/j.scs.2024.105428
20. Li Y, Kong B, Qiu C, Li Y, Jiang Y (2025) Numerical study on air-cooled battery thermal management system considering the sheer altitude effect. Appl Therm Eng 258. https://doi.org/10.1016/j.applthermaleng.2024.124707
21. Amalesh T, Lakshmi Narasimhan N (2020) Introducing new designs of minichannel cold plates for the cooling of lithium-ion batteries. J Power Sources 479. https://doi.org/10.1016/j.jpowsour.2020.228775
22. Sardari PT, Mohammed HI, Giddings D, Walker GS, Gillott M, Grant D (2019) Numerical study of a multiple-segment metal foam-PCM latent heat storage unit: effect of porosity, pore density and location of heat source. Energy 189. https://doi.org/10.1016/j.energy.2019.116108
23. Moaveni A, Siavashi M, Mousavi S (2024) Passive and hybrid battery thermal management system by cooling flow control, employing nano-PCM, fins, and metal foam. Energy 288. https://doi.org/10.1016/j.energy.2023.129809
24. Shahsavar A, Al-Rashed AAAA, Entezari S, Sardari PT (2019) Melting and solidification characteristics of a double-pipe latent heat storage system with sinusoidal wavy channels embedded in a porous medium. Energy 171:751–769. https://doi.org/10.1016/j.energy.2019.01.045

25. Mahdavi M, Tiari S, Pawar V (2020) A numerical study on the combined effect of dispersed nanoparticles and embedded heat pipes on melting and solidification of a shell and tube latent heat thermal energy storage system. J Energy Storage 27. https://doi.org/10.1016/j.est.2019.101086
26. Dhaidan NS, Khodadadi JM, Al-Hattab TA, Al-Mashat SM (2013) Experimental and numerical study of constrained melting of N-octadecane with CuO nanoparticle dispersions in a horizontal cylindrical capsule subjected to a constant heat flux. Int J Heat Mass Transf 67:523–534. https://doi.org/10.1016/j.ijheatmasstransfer.2013.08.001
27. Sudhakaran S, Terese M, Mohan Y, Thampi AD, Rani S (2023) Influence of various parameters on the cooling performance of battery thermal management systems based on phase change materials. Appl Therm Eng 222. https://doi.org/10.1016/j.applthermaleng.2022.119936
28. Verma A, Shashidhara S, Rakshit D (2019) A comparative study on battery thermal management using phase change material (PCM). Thermal Sci Eng Prog 11:74–83. https://doi.org/10.1016/j.tsep.2019.03.003
29. Yadav C, Sahoo RR (2021) Thermal performance analysis of MWCNT-based capric acid PCM thermal energy storage system. J Therm Anal Calorim 146(4):1539–1550. https://doi.org/10.1007/s10973-020-10186-z
30. Zhang Y, Liu H, Liu S, Pan S, Tian C, Jian H (2023) Prediction model of thermal behavior of lithium battery module under high charge-discharge rate. J Energy Storage 74. https://doi.org/10.1016/j.est.2023.109366

Optimization of a U-Shaped Cold Plate BTMS for Lithium-Ion Batteries Using CFD: Thermal Regulation Under Various Discharge Rates and Applications for Smart Mobility

Nassreddine Hmidi, Hicham Kaddouri, Ahmed Alami Merrouni, Abdel-illah Amrani, and Elmiloud Chaabelasri

Abstract Lithium-ion batteries play a critical role in advancing smart mobility and driving the energy transition, with efficient thermal management being essential to ensure their safety, performance, and longevity. This study evaluates the thermal efficiency of a novel U-shaped aluminum cold plate BTMS integrated with a pouch lithium-ion cell, using water as the cooling agent. The system's performance was analyzed through Computational Fluid Dynamics (CFD) simulations, validated against experimental data. Simulations conducted at discharge rates of 1C, 2C, 3C, and 4C demonstrated that an inlet mass flow rate of 5×10^{-4} kg/s effectively maintained $T_{b,max}$ at 25.29 °C, 26.13 °C, 27.49 °C, and 29.28 °C, respectively, while ensuring $\Delta T_{b,max}$ remained at 0.270 °C, 1.034 °C, 2.287 °C, and 3.941 °C. To enhance safety under aggressive driving conditions by minimizing $\Delta T_{b,max}$, the inlet mass flow rate was increased incrementally to 1×10^{-3}, 1.5×10^{-3}, 2×10^{-3}, 2.5×10^{-3}, and 3×10^{-3} kg/s. Key performance metrics such as $T_{b,max}$, $\Delta T_{b,max}$, uniformity index, Nusselt number, friction coefficient, and cooling efficiency factor were optimized, identifying an inlet mass flow rate of 2×10^{-3} kg/s as the optimal value for the proposed BTMS.

Keywords Thermal management system · Li-ion batteries · U-shape cold plate · CFD

N. Hmidi · A. Alami Merrouni (✉) · E. Chaabelasri
Materials Science, New Energies and Applications Research Group, LPTPME Laboratory, Department of Physics, Faculty of Sciences, Mohammed 1st University, Oujda, Morocco

H. Kaddouri
Experimentation and Modelling Team in Mechanics and Energy Systems, National School of Applied Sciences, Abdelmalek Essaadi University, Al-Hoceima, Morocco

A.-i. Amrani
Equipe des Énergies Nouvelles et Ingénierie des Matériaux (ENIM), Faculté des Sciences et Techniques, Moulay Ismail University of Meknes, Errachidia, Morocco

H. Kaddouri et al. (eds.), *Smart Cities for Sustainable and Urban Development*, Green Energy and Technology, https://doi.org/10.1007/978-3-031-99001-4_22

1 Introduction

In the rapid evolution of urban spaces into smart cities, the demand for efficient, sustainable energy solutions is intensifying, particularly in energy storage and management. Lithium-ion (Li-ion) batteries are foundational to this transformation due to their high energy density, versatility, and adaptability to various applications, from electric vehicles (EVs) to energy storage systems in smart grids [1]. EVs, in particular, are essential for realizing sustainable urban ecosystems, as they help reduce greenhouse gas emissions, improve urban air quality, and support the integration of renewable energy sources, which is crucial for cities aiming to meet strict climate targets [2]. Studies estimate that widespread EV adoption in urban areas could reduce transportation-related CO_2 emissions by up to 30% by 2030, making them a vital component of smart cities' environmental agendas [3].

However, ensuring optimal performance and longevity for batteries in EVs and other smart city applications necessitates advanced thermal management solutions. For Li-ion batteries to operate effectively, the temperature should be kept within a narrow range, ideally between 15 °C and 35 °C, which mitigates thermal degradation and prolongs battery life [4–6]. Moreover, a consistent temperature distribution across cells is crucial, with a maximum allowable temperature difference (ΔT_{max}) of less than 5 °C [7–9]. This uniformity helps prevent localized overheating, which can cause uneven cell degradation, reduce performance, and heighten safety risks due to potential thermal runaway a particular concern in densely populated urban environments [10, 11]. Recent research highlights that traditional cooling systems, such as liquid cooling plates or air-cooled systems, often struggle to maintain this level of temperature control, especially under high-demand conditions [12].

To address these thermal management challenges, innovative Battery Thermal Management Systems (BTMS) are increasingly incorporating advanced materials and designs aimed at enhancing thermal efficiency while keeping system weight and complexity to a minimum. Among these [13] optimized a BTMS using liquid cooling with hybrid manifold channels for a 280 Ah $LiFeO_4$ battery pack. By employing multi-objective optimization, the system achieves a maximum temperature reduction of 17.6 °C compared to air cooling and 11.7 °C compared to immersion cooling, while maintaining a low pressure drop (<4.2 kPa). The optimized design keeps the battery temperature within safe limits (<45 °C) at discharge rates up to 3C [14]. This study proposes a BTMS combining thermoelectric coolers (TEC), composite phase change materials (CPCM), and liquid cooling to regulate lithium-ion battery temperatures under extreme conditions. The system achieves optimal performance at a 12% expanded graphite (EG) mass fraction, TEC current of 3 A, and coolant flow speed of 0.05 m/s, reducing battery temperature differences by up to 21.36 K. Additionally, in low-temperature environments, the TEC preheats batteries to 293.15 K within 2240 s using a 5 A input current. Chavan et al. [15] evaluated a thermal management of lithium-ion batteries using three cooling channels (open, curved, and rectangular) with air and water as coolants. Curved channels with water demonstrate superior performance, achieving the lowest pressure drop

(1.4 Pa) and efficient cooling with operating temperatures between 300 K and 304.44 K. These results highlight the curved channel's effectiveness, offering insights for optimizing LIB thermal systems in electric vehicles. Shojaeefard et al. [16] investigated the performance of lithium-ion battery thermal management systems using three cooling channel types (open, curved, and rectangular) with air and water as coolants. Curved channels with water provide the best results, achieving operating temperatures of 300 K to 304.44 K and a pressure drop of 1.4 Pa, showcasing their efficiency. These findings emphasize the importance of channel design in enhancing battery cooling performance. Zhang et al. [17] realized a hybrid BTMS combining PCM, air cooling, and Cantor fractal fins for improved temperature control. The system achieves a Tmax of 308.28 K and a ΔTmax of 1.36 K at a 4C discharge rate using reverse airflow. At 6C discharge rate, it maintains Tmax at 316.74 K under extreme ambient conditions (323 K). Machine Learning predictions ensure efficient performance with errors under 10%.

This study addresses this gap by introducing a groundbreaking battery thermal management system (BTMS) featuring a U-shaped minichannel aluminum cold plate, effectively meeting the thermal demands of smart cities and advancing sustainable urban transport and energy systems.

Lithium-ion batteries (LiBs) are essential for smart mobility, powering electric vehicles and supporting sustainable urban systems. Effective cooling is critical to ensure their safety, performance, and lifespan under the thermal demands of urban mobility. However, there is a need for compact, efficient cooling solutions that provide thermal uniformity and low energy consumption while integrating with smart city infrastructure.

2 Materials and Methods

2.1 Physical Problem

The focus of this study is the development and evaluation of a BTMS incorporating a U-shaped aluminum minichannel cold plate. The system is designed to manage the thermal performance of a 20 Ah $LiFePO_4$ pouch cell, with an anode made of graphite and a cathode composed of $LiFePO_4$. The pouch cell has dimensions of 227 mm × 160 mm × 7.25 mm, and the cooling mechanism relies on water circulating through the U-shaped minichannels with a thickness 1.3 mm and embedded in the aluminum cold plate as illustrated in Fig. 1.

The thermal behavior of the battery cell is analyzed under various discharge rates of 1C, 2C, 3C, and 4C, simulating the real-world operational scenarios in electric vehicles and renewable energy storage systems. These discharge rates are critical as they determine the rate of heat generation, which directly affects the performance, safety, and lifespan of the battery. The aim is to ensure that the BTMS effectively maintains the cell's temperature within a safe operational range, even under high

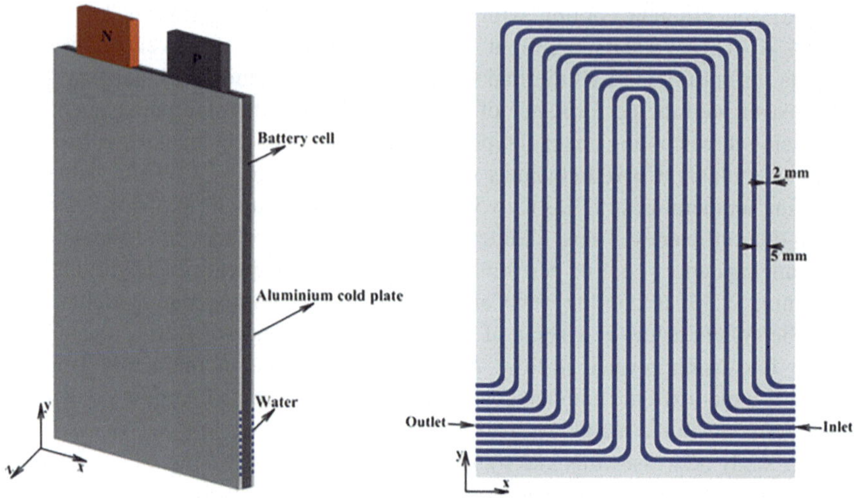

Fig. 1 Battery cell and its TMS

Table 1 Battery cell properties [18–20]

	Battery	Positive tab	Negative tab
Density [$Kg \cdot m^{-3}$]	2092	2917	8978
Specific heat [$J \cdot kg^{-1} \cdot {}^{\circ}C^{-1}$]	678	871	381
Thermal conductivity [$J \cdot m^{-1} \cdot {}^{\circ}C^{-1}$]	18.2	202	387.6
Nominal Voltage [V]	3.3	–	–
Minimum stop voltage [V]	2	–	–
Maximum stop voltage [V]	4.1	–	–

discharge conditions, by mitigating temperature gradients and preventing thermal runaway, the material properties of cell used are illustrated in Table 1.

By utilizing the U-shaped minichannel design, as depicted in Fig. 1, the system is engineered to optimize heat dissipation while simultaneously minimizing material usage and energy consumption. This innovative approach ensures that the thermal management system is both lightweight and compact, making it highly suitable for modern applications such as electric vehicles and renewable energy storage systems. The thermophysical properties of aluminum, used in the cold plate, and water, employed as the cooling medium, are detailed in Table 2, highlighting their respective roles in achieving efficient and effective thermal regulation. This combination of advanced design and material selection enhances the overall performance and sustainability of the system.

Table 2 Aluminum and water thermophysical properties [18–20]

	Aluminum	Water
Density [$Kg \cdot m^{-3}$]	2700	998.1
Specific heat [$J \cdot kg^{-1} \cdot °C^{-1}$]	900	4128
Thermal conductivity [$J \cdot m^{-1} \cdot °C^{-1}$]	238	0.620
Dynamic viscosity [$Pa \cdot s$]	–	1.003e-3

2.2 *Governing Equations*

The equations presented describe the thermal and fluid dynamics governing the BTMS incorporating a U-shaped minichannel design. Each equation plays a crucial role in modeling the heat transfer and fluid flow processes necessary for efficient battery cooling:

The battery cell energy equation represents the heat transfer within the battery cell. Here, ρ_b, C_b, and k_b are the density, specific heat capacity, and thermal conductivity of the battery, respectively. The left-hand side tracks the rate of change of internal energy, while the right-hand side accounts for heat and heat generation (Q_{gen}), which arises from electrochemical reactions during discharge.

$$\frac{\partial[\rho_b c_b T_b]}{\partial t} = \nabla.[k_b \nabla T_b] + Q_{gen} \tag{1}$$

where ρ_b, c_b, T_b, k_b, and Q_{gen} represent the battery density, specific heat, temperature, thermal conductivity, and the heat generated inside the battery, respectively.

The aluminum energy equation governs heat conduction within the aluminum cold plate. ρ_{Al}, C_{Al}, and k_{Al} are the density, specific heat capacity, and thermal conductivity of aluminum.

$$\frac{\partial[\rho_{Al} c_{Al} T_{Al}]}{\partial t} = \nabla.[k_{Al} \nabla T_{Al}] \tag{2}$$

The convective heat transfer between the battery and the surrounding air is given by:

$$Q_{air} = h_{air}(T_b - T_{air}) \tag{3}$$

Here, h_{air} is the convective heat transfer coefficient, T_b is the battery temperature, and T_{air} is the air temperature.

The Reynolds number characterizes the nature of the coolant flow (laminar or turbulent) through the minichannels. ρ_w, v_{int}, and μ_w are the coolant's density, velocity, and dynamic viscosity, respectively, while D_h is the hydraulic diameter.

$$R_e = \frac{\rho_w v_{int} D_h}{\mu_w} \tag{4}$$

The hydraulic diameter D_h defines the effective diameter of the U-shaped minichannels, calculated as the ratio of the channel's cross-sectional area A_{int} to its perimeter P.

$$D_h = \frac{4A_{int}}{P} \tag{5}$$

The continuity equation ensures mass conservation within the coolant flow. It indicates that the fluid is incompressible, meaning the volume of fluid entering any point in the system equals the volume leaving, which given by:

$$\nabla.\vec{v} = 0 \tag{6}$$

The momentum conservation equation that governs the motion of the coolant in the minichannels.

$$\rho_w \left[\frac{\partial \vec{v}}{\partial t} + (\vec{v}.\nabla)\vec{v} \right] = -\nabla p + \mu_w \nabla^2 \vec{v} \tag{7}$$

where $\vec{V}$, P and μ_w represent the velocity vector,pressure of the fluid flow and wter dynamic viscosity, respectively.

The water energy equation, which defines the heat transfer between the coolant and the U-shaped channels, is expressed as:

$$\rho_w c_w \frac{\partial T_w}{\partial t} + \rho_w c_w \vec{v}.\nabla T_w = \nabla.[k_w \nabla T_w] \tag{8}$$

where ρ_w, c_w, T_w, and k_w represent water density, specific heat, temperature, and thermal conductivity, respectively.

2.3 *Numerical Validation*

The Fig. 2a present the numerical validation of the results obtained by Panchal et al. [20] for various discharge rates (1C, 2C, 3C, and 4C) using the Finite Volume Method (FVM) combined with the SIMPLE algorithm in ANSYS Fluent. An unstructured tetrahedral mesh was employed to discretize the computational domain, enabling detailed modeling of the U-shaped minichannel system's geometry and ensuring numerical accuracy. The left graph compares the experimental (Exp) and numerical (Num) average temperature profiles over time. For lower discharge rates (e.g., 1C and 2C), the numerical results closely match the experimental data, showcasing the model's ability to accurately predict battery temperature. At higher discharge rates (3C and 4C), the temperature increases more rapidly due to higher heat generation, and the numerical predictions still align well with the experimental trends, capturing the thermal dynamics effectively.

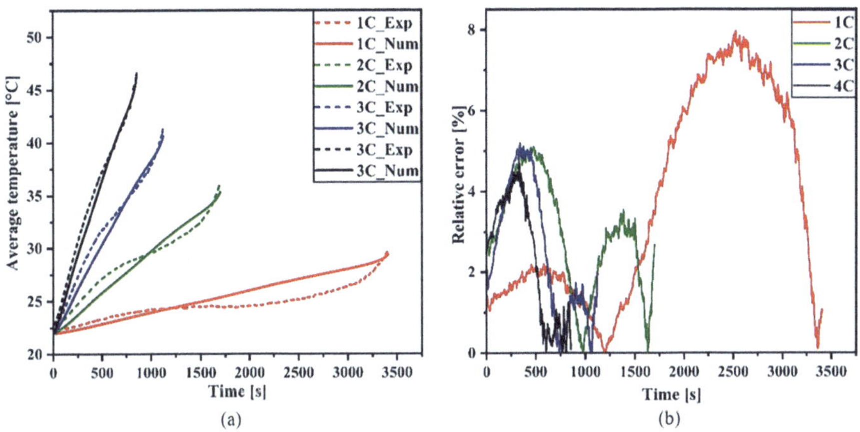

Fig. 2 Validation the numerical model

The Fig. 2b illustrates the relative error between the numerical and experimental results as a percentage over time. For lower discharge rates (1C and 2C), the relative error remains consistently below 4%, reflecting a high degree of accuracy. For higher discharge rates (3C and 4C), the relative error exhibits fluctuations, particularly during the initial transient phase, likely due to small differences in initial conditions or transient heat transfer effects. However, the error stabilizes over time and remains within an acceptable range, not exceeding 8% at any point.

Overall, the results highlight the robustness of the numerical model, demonstrating its ability to accurately simulate the thermal response of batteries under varying discharge conditions. This validation confirms that the combination of the Finite Volume Method, the SIMPLE algorithm, and the chosen meshing approach is well-suited for modeling complex cooling systems like the U-shaped minichannel design, providing reliable predictions for battery thermal management.

2.4 Boundary Conditions

To solve the equations governing heat transfer and fluid flow in the cooling system, appropriate initial and boundary conditions are established. The system's initial temperature is set at 25 °C, while contact resistance and gravitational effects are considered negligible. The cooling fluid is treated as incompressible, and the heat transfer coefficient with ambient air is set at 5 $W \cdot m^{-2} \cdot K^{-1}$. The thermophysical properties of the system's components are assumed to remain constant, unaffected by temperature variations. At the fluid outlet, the pressure is maintained at 0 Pa, and the mass flow rate is adjustable within the range of 0.001 kg/s to 0.005 kg/s, providing flexibility for different operating scenarios. These conditions ensure accurate simulation of the thermal and fluid behavior in the system.

2.5 *BTMS Efficiency*

To ensure optimal performance and efficiency of the BTMS, it is important to evaluate key parameters related to thermal and hydraulic efficiency:

2.5.1 $T_{b,max}$ and $\Delta T_{b,max}$

Monitoring and maintaining the maximum battery temperature $T_{b,max}$ within safe limits is essential for ensuring battery safety, reliability, and longevity. Keeping $T_{b,max}$ below the critical threshold minimizes the risk of thermal damage and enhances system performance. A lower $T_{b,max}$ indicates a more efficient TMS, while a smaller $\Delta T_{b,max}$ ensures better temperature uniformity across the cell. For effective operation, both $T_{b,max}$ and $\Delta T_{b,max}$ must remain under 5 °C.

2.5.2 Uniformity Index (T_{uni})

The temperature uniformity within the battery is evaluated using the T_{uni} index, which measures the consistency of temperature distribution. Lower values of T_{uni} indicate better thermal uniformity. The T_{uni} index is mathematically defined as:

$$T_{uni} = \frac{T_{b,max} - T_{b,min}}{T_{b,avg}} \quad (9)$$

where $T_{b,min}$ and $T_{b,avr}$ represent the minimum and average battery temperatures, respectively (in °C).

2.5.3 Cooling Efficiency Factor (β)

The cooling efficiency β quantifies how effectively the cooling system dissipates heat relative to the energy required for pumping the coolant. It is defined as the ratio of the heat removed by the coolant (Q_w) to the power required for coolant circulation (P_w):

$$\beta = \frac{Q_w}{P_w} \quad (10)$$

The heat removed by the coolant depends on its mass flow rate $(\dot{m}_w)$, specific heat capacity (C_w), and the temperature difference between the inlet and outlet:

$$Q_w = \dot{m}_w c_w \left(T_{out} - T_{int}\right) \quad (11)$$

where T_{int} and T_{out} represent the inlet and outlet coolant temperatures, respectively.

The power required for circulation is determined by the pressure drop (ΔP) and the volumetric flow rate $\left(\dot{v}_w\right)$:

$$P_w = \dot{V}_w \Delta P = \dot{V}_w \left(P_{int} - P_{out}\right) \tag{12}$$

where P_{int} and P_{out} represent the inlet and outlet pressures, respectively.

The mass flow rate is given by:

$$\dot{m}_W = \rho_w \dot{V}_w \tag{13}$$

2.5.4 Nusselt Number (Nu)

The Nusselt number (Nu) is used to quantify the relationship between convective and conductive heat transfer along the U-shaped minichannel, and it is expressed as:

$$N_u = \frac{h_w D_h}{k_w} \tag{14}$$

K_w represent coolant thermal conducticity.

Here, h_w represents the heat transfer coefficient between the coolant and the aluminum in the minichannels, which is defined as:

$$h_w = \frac{Q_w}{A_c\left[T_{b,max} \quad \left(T_{out} + T_{int}\right)/2\right]} \tag{15}$$

A_c represents the contact area between the coolant and the aluminum cold plate, while $T_{b,max}$ denotes the maximum battery temperature.

2.5.5 Friction Factor (f)

The f factor quantifies frictional losses due to fluid flow in the channels:

$$f = \frac{\Delta P D_h}{2\rho_w L v_{int}^2} \tag{16}$$

L represents the characteristic length of the channel.

3 Results and Discussions

3.1 BTMS Efficiency Under Different Discharges Rates

The Fig. 3 show temperature contours of the battery cell at the end of discharge for 1C, 2C, 3C, and 4C discharge rates with a fixed inlet mass flow rate of 5×10^{-4} kg/s.

The temperature contours illustrate the heat distribution across the battery cell at the end of discharge for varying discharge rates (1C, 2C, 3C, and 4C) with a constant inlet mass flow rate of 5×10^{-4} kg/s. As the discharge rate increases, the current debited by the cell also rises, leading to higher heat generation due to internal resistive losses.

At the 1C discharge rate, the temperature remains low and uniformly distributed, ranging between 25 °C and 25.3 °C, demonstrating effective cooling and minimal

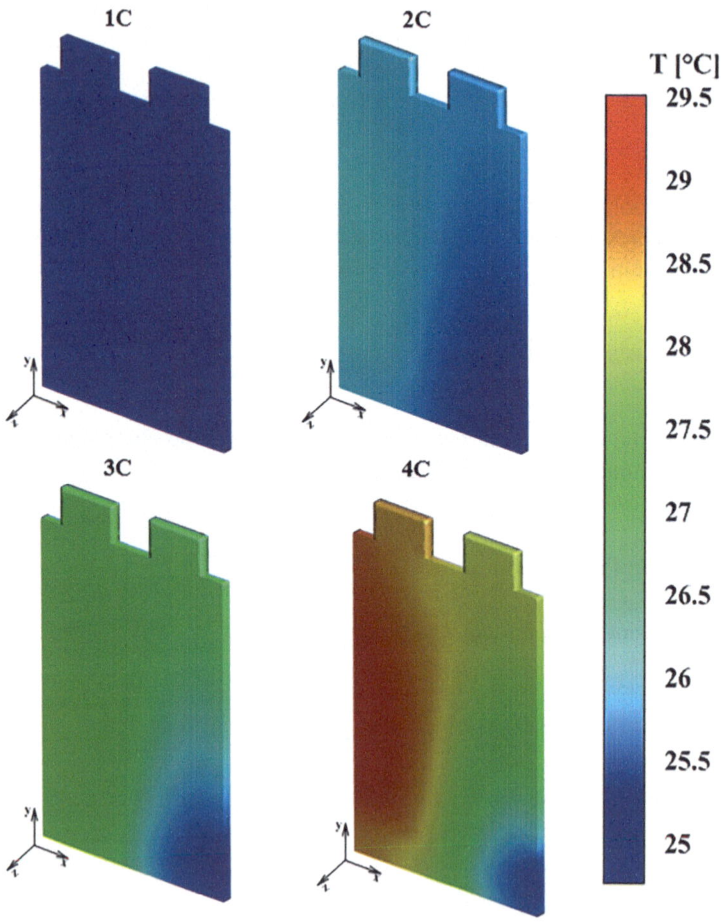

Fig. 3 Temperature contours of the battery cell at the end of discharge for 1C, 2C, 3C, and 4C discharge rates with a fixed inlet mass flow rate of 5×10^{-4} kg/s

thermal gradients. With an increase to the 2C discharge rate, the temperature rises slightly to a range of 25.1 °C to 26.2 °C, while still maintaining good uniformity. At 3C, the temperature increases further, reaching up to 27.5 °C, and noticeable thermal gradients begin to form, particularly in areas farther from the coolant flow paths, indicating the development of localized hotspots.

For the 4C discharge rate, the temperature distribution becomes less uniform, with values ranging from 25.34 °C to 29.3 °C, as the heat generation is significantly higher due to the larger current. The presence of hotspots becomes more pronounced, especially near regions with reduced cooling efficiency. While the temperature difference across the cell remains slightly below 5 °C, this difference should be minimized further to enhance safety, particularly for aggressive operating conditions or high discharge rates. Maintaining smaller temperature gradients is critical to prevent thermal stress and ensure safe and reliable battery performance under demanding scenarios.

The Fig. 4 illustrate the evolution of the $T_{b,max}$ and $\Delta T_{b,max}$ within the battery cell at the end of the discharge process for various discharge rates (1C, 2C, 3C, and 4C) under a fixed inlet mass flow rate of 5×10^{-4} kg/s. These parameters provide essential insights into the thermal behavior of the battery and the efficiency of the cooling system during operation.

The maximum temperature demonstrates a clear upward trend with increasing discharge rates, driven by the higher current intensities that generate more heat. For a discharge rate of 1C, the battery's maximum temperature remains relatively low, at 25.29 °C, which is close to the initial system temperature. At 2C, $T_{b,max}$ increases moderately to 26.13 °C, while at 3C, it reaches 27.49 °C. For the highest discharge rate, 4C, $T_{b,max}$ rises significantly to 29.28 °C. This progression highlights the need for effective thermal management, particularly at higher discharge rates, to prevent overheating and ensure optimal battery performance.

Similarly, $\Delta T_{b,max}$, increases with the discharge rate, reflecting the growing thermal gradients within the cell. For 1C, $\Delta T_{b,max}$ is as low as 0.27 °C, indicating

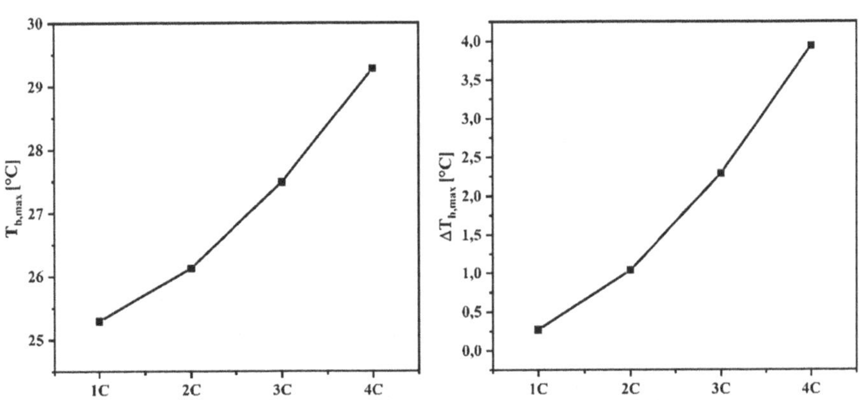

Fig. 4 Maximum temperature and maximum temperature difference of the battery cell at the end of discharge for 1C, 2C, 3C, and 4C discharge rates with a fixed inlet mass flow rate of 5×10^{-4} kg/s

excellent temperature uniformity. At 2C, it rises to 1.03 °C, while at 3C, it reaches 2.29 °C. At the highest discharge rate of 4C, $\Delta T_{b,max}$ is 3.94 °C, still below the critical threshold of 5 °C, but showing a noticeable deviation from thermal uniformity. These gradients become more pronounced with increasing discharge rates, underscoring the thermal challenges posed by high-current operations.

Maintaining $\Delta T_{b,max}$ below 5 °C is critical to ensuring the battery's safety and reliability, as excessive temperature differences can lead to localized hotspots, accelerated aging, and potential safety risks. Even though the current results demonstrate acceptable values, minimizing $\Delta T_{b,max}$ further is essential, particularly for applications involving aggressive driving conditions or high-power discharge cycles. Enhanced cooling strategies are required to achieve more uniform temperature distributions and mitigate the thermal stresses associated with high discharge rates.

3.2 *Mass Flow Rate Inlet Effect*

The Fig. 5 illustrate the influence of mass flow rate on $T_{b,max}$,$\Delta T_{b,max}$, T_{uni} for a battery operating at a 4C discharge rate. These results provide essential insights into the thermal performance of the BTMS and its ability to maintain safe and consistent operating conditions.

As the mass flow rate increases, $T_{b,max}$ decreases significantly. At the lowest mass flow rate of 5×10^{-4} kg/s, $T_{b,max}$ reaches 29.28 °C, indicating insufficient heat dissipation and higher thermal stress. When the flow rate is increased to 2×10^{-3} kg/s, $T_{b,max}$ decreases to 26.25 °C, showing a considerable improvement in cooling performance. At the highest flow rate of 3×10^{-3} kg/s, $T_{b,max}$ further reduces to 25.91 °C, reflecting efficient thermal regulation.

Similarly, $\Delta T_{b,max}$, which indicates the uniformity of the temperature distribution within the cell, also decreases as the mass flow rate increases. At 5×10^{-4} kg/s, $\Delta T_{b,max}$ is 3.94 °C, highlighting significant thermal gradients. Increasing the flow rate to 2×10^{-3} kg/s reduces $\Delta T_{b,max}$ to 1.15 °C, indicating improved thermal uniformity. At 3×10^{-3} kg/s, $\Delta T_{b,max}$ further decreases to 0.83 °C, showing excellent temperature consistency throughout the cell.

The Fig. 5 show that, Tuni follows a similar trend, decreasing with higher mass flow rates. At 5×10^{-4} kg/s, T_{uni} is 0.141, reflecting poor thermal uniformity. Increasing the flow rate to 2×10^{-3} kg/s improves T_{uni} to 0.0448, and at 3×10^{-3} kg/s, it further decreases to 0.0328, indicating near-perfect temperature consistency within the cell.

Based on the results, the optimal mass flow rate is identified as 2×10^{-3} kg/s, where the BTMS achieves a balance between cooling performance, thermal uniformity, and energy efficiency. At this flow rate, $T_{b,max}$ is reduced to a safe level of 26.25 °C, $\Delta T_{b,max}$ is minimized to 1.15 °C, and T_{uni} achieves a value of 0.0448, ensuring reliable and safe operation of the battery under a high discharge rate of 4C.

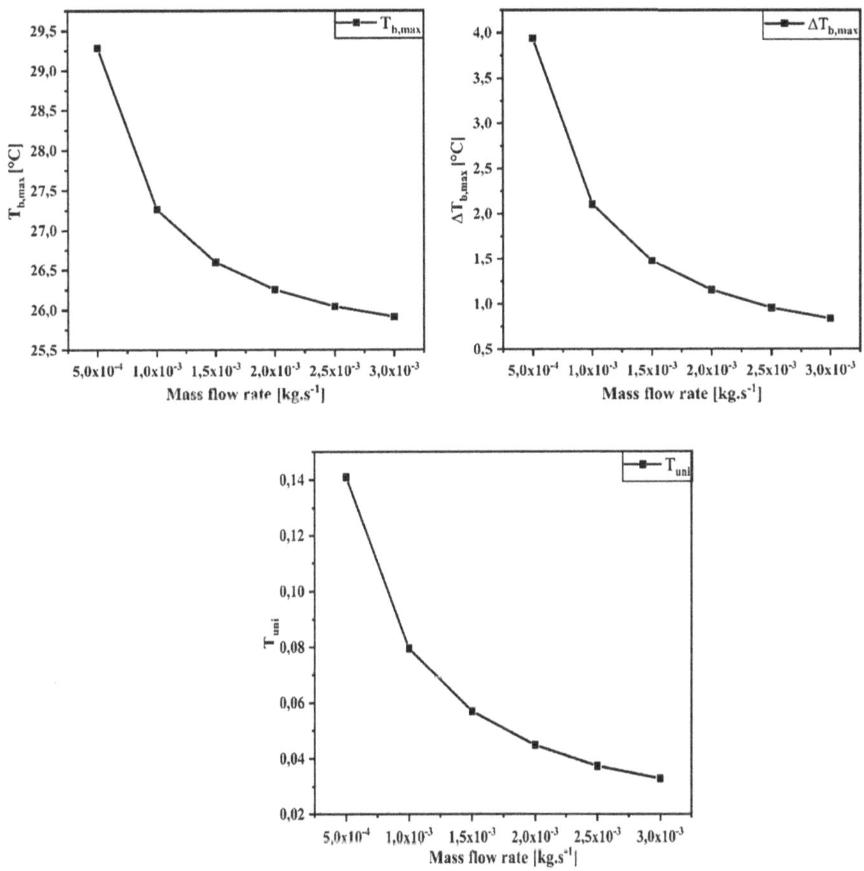

Fig. 5 Variation of maximum temperature, maximum temperature difference, and uniformity index with mass flow rate for a 4C discharge rate

The Fig. 6 illustrates the variation of the Nusselt number (Nn) and the friction factor (f) as a function of mass flow rate for the battery cooling system. These parameters are essential for evaluating the thermal and hydraulic performance of the BTMS.

Frog the Fig. 6a, the Nusselt number, which represents the convective heat transfer efficiency, increases consistently with higher mass flow rates. At a low mass flow rate of 5×10^{-4} kg/s, Nu is 0.37, reflecting limited heat transfer due to insufficient coolant flow. As the mass flow rate increases, Nu improves significantly, reaching 1.25 at 2×10^{-3} kg/s and further increasing to 1.67 at 3×10^{-3} kg/s, demonstrating highly effective heat transfer. This trend confirms that increasing the coolant flow enhances the convective heat exchange, ensuring better thermal regulation of the battery system.

On the other hand, the Fig. 6b illustrate the friction factor, which measures pressure losses in the cooling channels, decreases as the mass flow rate increases. At the

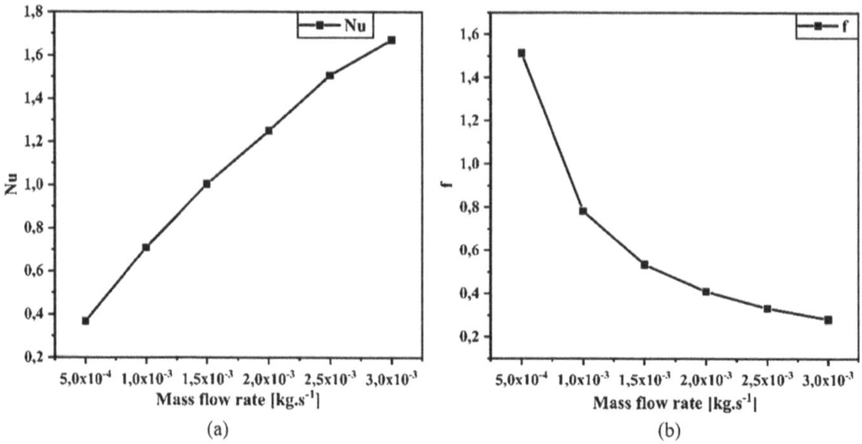

Fig. 6 Nusselt number and friction factor as a function of mass flow rate at 4C dis-charge rate

lowest flow rate of 5×10^{-4} kg/s, f is 1.51, indicating significant resistance to fluid flow. As the flow rate increases to 2×10^{-3} kg/s, f decreases to 0.41, and at 3×10^{-3} kg/s, f is further reduced to 0.28. This reduction in the friction factor highlights improved hydraulic efficiency, as higher flow rates enable smoother coolant movement with reduced energy losses.

Based on these observations, the optimal mass flow rate is identified as 2×10^{-3} kg/s. At this flow rate, the BTMS achieves a balance between thermal and hydraulic performance, with Nu reaching 1.25 for efficient cooling and f reduced to 0.41, minimizing pumping losses. Although a higher flow rate of 3×10^{-3} kg/s provides slightly better thermal performance (Nu = 1.67Nu = 1.67Nu = 1.67) and lower pressure losses (f = 0.28f = 0.28f = 0.28), the incremental improvement may not justify the additional energy consumption required for pumping.

The results confirm that increasing the mass flow rate enhances heat transfer efficiency while reducing pressure losses. The optimal flow rate of 2×10^{-3} kg/s ensures an effective balance between cooling performance and energy efficiency, making it ideal for maintaining battery safety and performance in practical applications.

The Fig. 7 depict the relationship between the cooling efficiency factor (β) and the mass flow rate for BTMS. The cooling efficiency factor, defined as the ratio of heat dissipated by the coolant to the power required for pumping, provides a critical measure of the energy efficiency of the system. This parameter highlights the balance between thermal performance and the energy consumption associated with coolant circulation.

The results show that the cooling efficiency factor (β) decreases significantly as the mass flow rate increases. At the lowest mass flow rate of 5×10^{-4} kg/s, β reaches its highest value of 153,559, reflecting highly efficient energy usage due to the minimal pumping power required. However, at this low flow rate, the BTMS is unable to effectively manage the battery's $T_{b,max}$ and temperature uniformity, which are critical for safe operation. As the mass flow rate increases, β declines rapidly to 38,511

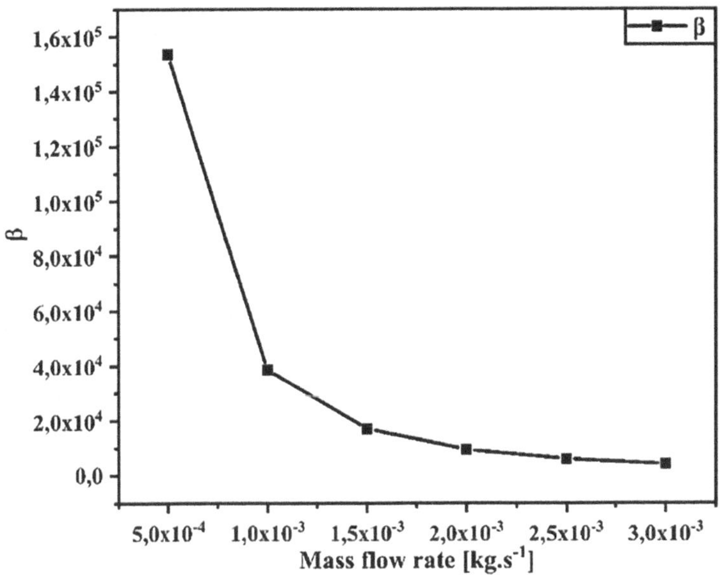

Fig. 7 Cooling efficiency factor as a function of mass flow rate for a 4C discharge rate

at 1×10^{-3} kg/s and further to 9454 at 2×10^{-3} kg/s. Beyond this, at 3×10^{-3} kg/s, β stabilizes at 4153, indicating diminishing returns in cooling efficiency as the energy required for pumping outweighs the additional cooling benefits.

The analysis identifies 2×10^{-3} kg/s as the optimal mass flow rate for the BTMS. At this rate, β = 9454 achieves a favorable balance between effective thermal regulation and energy consumption. This flow rate ensures that the battery's $T_{b,max}$ and $\Delta T_{b,max}$ remain within safe and acceptable limits, enhancing thermal performance while avoiding excessive pumping energy requirements. Although higher mass flow rates, such as 3×10^{-3} kg/s, slightly improve thermal performance, they significantly reduce the cooling efficiency factor without providing meaningful advantages.

In conclusion, the results confirm that 2×10^{-3} kg/s is the optimal operating point for the BTMS. This mass flow rate ensures effective cooling, uniform temperature distribution, and efficient energy usage, making it the most suitable choice for maintaining battery safety and performance under demanding conditions. This balance is essential for the sustainable and reliable operation of the BTMS in practical applications.

4 Conclusion and Perspectives

This study focused on the design, simulation, and analysis of a U-shaped minichannel aluminum cold plate BTMS for lithium-ion batteries. The objective was to address the critical thermal challenges faced by batteries during high discharge rates, ensuring safe operation, enhanced performance, and prolonged lifespan. A comprehensive numerical investigation was conducted to assess the system's thermal and hydraulic performance, analyzing parameters such as $T_{b,max}$, $\Delta T_{b,max}$, T_{uni}, and β under varying mass flow rates.

The results revealed that the BTMS effectively regulates battery temperatures, ensuring that $T_{b,max}$ remains within safe limits and thermal gradients $\Delta T_{b,max}$ are minimized. At the optimal mass flow rate of 2×10^{-3} kg/s, the system achieved a balance between thermal regulation and energy efficiency. Specifically, $T_{b,max}$ was reduced to 26.25 °C, while $\Delta T_{b,max}$ was minimized to 1.15 °C, ensuring uniform temperature distribution across the battery. Additionally, the uniformity index T_{uni} improved significantly at this flow rate, reaching a near-ideal value of 0.0448, indicating excellent thermal consistency.

Furthermore, the analysis of β highlighted the trade-offs between effective cooling and energy consumption for pumping. While low flow rates demonstrated high energy efficiency, they failed to adequately regulate battery temperatures. Conversely, higher flow rates provided superior thermal performance but with diminishing returns in energy efficiency. The identified optimal flow rate of 2×10^{-3} kg/s strikes a critical balance, achieving both effective thermal management and reduced energy consumption.

From a broader perspective, this work contributes significantly to the advancement of thermal management technologies critical for the next generation of smart cities and smart mobility. Efficient BTMS systems, such as the one developed in this study, are instrumental in enabling safe and reliable energy storage solutions for electric vehicles (EVs) and renewable energy grids key components of urban sustainability. By ensuring battery longevity and performance, this research supports the adoption of EVs in intelligent transportation systems, reducing urban carbon footprints and improving mobility efficiency.

References

1. Ullah Z, Rehman AU, Wang S, Hasanien HM, Luo P, Elkadeem MR, Abido MA (2023) IoT-based monitoring and control of substations and smart grids with renewables and electric vehicles integration. Energy 282. https://doi.org/10.1016/j.energy.2023.128924
2. Bibri SE, Krogstie J (2020) Environmentally data-driven smart sustainable cities: applied innovative solutions for energy efficiency, pollution reduction, and urban metabolism. Energy Inform 3(1). https://doi.org/10.1186/s42162-020-00130-8
3. Avenyo EK, Tregenna F (2022) Greening manufacturing: technology intensity and carbon dioxide emissions in developing countries. Appl Energy 324. https://doi.org/10.1016/j.apenergy.2022.119726
4. Hmidi N, Merrouni AA, Salhi JE, Amrani A i, Chaabelasri E (2024a) Optimal battery configuration for electrical vehicles in hot climates: a numerical comparative study of materials and cooling approaches. Case Stud Thermal Eng 61. https://doi.org/10.1016/j.csite.2024.104983
5. Tete PR, Gupta MM, Joshi SS (2021) Developments in battery thermal management systems for electric vehicles: a technical review. J Energy Storage 35. https://doi.org/10.1016/j.est.2021.102255

6. Yao F, Guan X, Chen Q, Lin L (2024) Research on thermal management system of lithium-ion battery with a new type of spider web liquid cooling channel and phase change materials. J Energy Storage 81. https://doi.org/10.1016/j.est.2024.110447
7. Hmidi N, Salhi JE, Merrouni AA, Amrani A i, Chaabelasri E (2024b) Performance analysis of several EVs batteries thermal management systems under hot-arid climatic conditions of Morocco. J Energy Storage 80. https://doi.org/10.1016/j.est.2023.110276
8. Jithin KV, Rajesh PK (2022) Numerical analysis of single-phase liquid immersion cooling for lithium-ion battery thermal management using different dielectric fluids. Int J Heat Mass Transf 188. https://doi.org/10.1016/j.ijheatmasstransfer.2022.122608
9. Xu J, Fang H, Wang M, Li X, Chen Y (2024) Simulation research on thermal management system of battery module with fin heat dissipation structure. Appl Therm Eng 239. https://doi.org/10.1016/j.applthermaleng.2023.122177
10. Liu T, Jian H, Tang Q, Zhu X, Liu Y, Wang X (2022) Investigation on fine water mist battery thermal management system for thermal runaway control. Appl Therm Eng 211. https://doi.org/10.1016/j.applthermaleng.2022.118474
11. Ouyang T, Liu B, Wang C, Ye J, Liu S (2023) Novel hybrid thermal management system for preventing li-ion battery thermal runaway using nanofluids cooling. Int J Heat Mass Transf 201. https://doi.org/10.1016/j.ijheatmasstransfer.2022.123652
12. Hwang FS, Confrey T, Reidy C, Picovici D, Callaghan D, Culliton D, Nolan C (2024) Review of battery thermal management systems in electric vehicles. Renew Sust Energ Rev 192. https://doi.org/10.1016/j.rser.2023.114171
13. Sui Z, Lin H, Sun Q, Dong K, Wei W (2024) Multi-objective optimization of efficient liquid cooling-based battery thermal management system using hybrid manifold channels. Appl Energy 371. https://doi.org/10.1016/j.apenergy.2024.123766
14. Luo D, Zihao W, Yan Y, Sun Z, Yang L, Cao B (2024) Performance analysis of a battery thermal management system combining thermoelectric, composite phase change material, and liquid cooling under extreme operating conditions. J Energy Storage 95. https://doi.org/10.1016/j.est.2024.112679
15. Chavan S, Venkateswarlu B, Liu J, Joo SW, Kim SC (2024) Investigating the impact of fluid flow channels and cooling fluids on thermal management of lithium-ion battery: a simulation study. J Therm Anal Calorim 149(4):1629–1648. https://doi.org/10.1007/S10973-023-12802-0/METRICS
16. Shojaeefard MH, Molaeimanesh GR, Salami Ranjbaran Y (2019) Improving the performance of a passive battery thermal management system based on PCM using lateral fins. Heat Mass Transf/Waerme und Stoffuebertragung 55(6):1753–1767. https://doi.org/10.1007/S00231-018-02555-0/METRICS
17. Zhang N, Zhang Z, Li J, Cao X (2024) Performance analysis and prediction of hybrid battery thermal management system integrating PCM with air cooling based on machine learning algorithm. Appl Therm Eng 257. https://doi.org/10.1016/j.applthermaleng.2024.124474
18. Fan Y, Wang Z, Xiong X, Panchal S, Fraser R, Fowler M (2023) Multi-objective optimization design and experimental investigation for a prismatic lithium-ion battery integrated with a multi-stage tesla valve-based cold plate. PRO 11(6). https://doi.org/10.3390/pr11061618
19. Monika K, Chakraborty C, Roy S, Dinda S, Singh SA, Datta SP (2021) Parametric investigation to optimize the thermal management of pouch type lithium-ion batteries with mini-channel cold plates. Int J Heat Mass Transf 164. https://doi.org/10.1016/j.ijheatmasstransfer.2020.120568
20. Panchal S, Dincer I, Agelin-Chaab M, Fraser R, Fowler M (2017) Transient electrochemical heat transfer modeling and experimental validation of a large sized LiFePO4/graphite battery. Int J Heat Mass Transf 109:1239–1251. https://doi.org/10.1016/j.ijheatmasstransfer.2017.03.005

Sustainable Supply Chain and Logistics in Smart Cities: A Literature Review on Smart Decision Methods and AI Applications

Ibrahim Boukallal, Nabila Bouti, Amine Ouba, and Fatima El Khoukhi

Abstract The rapid urbanization has intensified the need for environmentally conscious supply chains (SCs) and logistics strategies that leverage cutting-edge technologies and data analytics. This chapter provides a comprehensive review of the literature on intelligent decision-making methods and artificial intelligence (AI) applications designed to address complex logistics and SC challenges in smart cities. The review methodology follows a systematic approach, utilizing databases such as Scopus, Web of Science, IEEE Xplore, and Science Direct to examine peer-reviewed publications from 2007 to 2024. The analysis prioritizes high-impact journals, conference proceedings, relevance to research objectives, and citation metrics, ensuring a thorough exploration of key trends and developments. Focusing on urban logistics, this review explores how AI-driven approaches are transforming the management of goods and services flows in rapidly growing urban environments. It highlights the transformative potential of technologies such as Internet of Things (IoT), blockchain, autonomous vehicles, and digital twins, integrated with optimization algorithms and AI techniques like machine learning and multi-agent systems. These technologies enable seamless monitoring, optimization, and automation of logistics workflows, improving efficiency, resilience, and sustainability across domains such as transportation, warehousing, waste management, and circular economy initiatives. Policy frameworks, public-private collaboration, and community involvement are identified as crucial for advancing intelligent, sustainable logistics practices. Challenges like data privacy, infrastructure requirements, and stakeholder coordination are also discussed, offering insights for researchers, practitioners, and policymakers aiming to develop and implement environmentally responsible SC solutions in urban areas.

I. Boukallal · N. Bouti (✉) · A. Ouba
IA Laboratory, FS Meknes, Moulay Ismail University of Meknes, Meknes, Morocco
e-mail: n.bouti@edu.umi.ac.ma

F. El Khoukhi
IA Laboratory, FLSH Meknes, Moulay Ismail University of Meknes, Meknes, Morocco

H. Kaddouri et al. (eds.), *Smart Cities for Sustainable and Urban Development*, Green Energy and Technology, https://doi.org/10.1007/978-3-031-99001-4_23

Keywords Smart cities · Sustainable supply chain · Intelligent decision-making methods · Artificial intelligence · Logistics

1 Introduction

The rapid urbanization of populations worldwide has created unprecedented challenges for cities in managing resources, transportation, and logistics systems efficiently while minimizing environmental impact. Sustainable SC (SSC) and logistics management face increasing pressure to become more sustainable and technologically advanced to meet these urban challenges [1, 2]. The integration of AI and smart technologies into SSC management (SSCM) has emerged as a crucial strategy for developing SSC solutions in smart cities [3].

In this chapter, we explore how intelligent decision-making methods and AI techniques are transforming SSCM in urban environments. We also examine the role of smart technologies, such as IoT, blockchain, and digital twins, in optimizing logistics operations and promoting environmental sustainability [4, 5]. The research emphasizes how these innovations enable continuous monitoring, predictive analytics, and automated decision-making to enhance SC performance and resilience.

Our methodology follows a systematic approach to literature selection, utilizing databases such as Scopus, Web of Science, IEEE Xplore, and Science Direct, examining publications from 2007 to 2024. The selection process prioritized peer-reviewed journals and conference proceedings, high-impact publications, relevance to research objectives, and citation metrics.

Through detailed analysis of academic literature and industry practices, we evaluate the implementation of AI and advanced technologies in SSCM. We examine technological integration in urban logistics, environmental benefits of AI-driven optimization, implementation challenges, and emerging opportunities in this field.

The chapter is organized as follows: Section 2 presents a literature review. Followed by Sect. 3 that investigates key technological enablers, including IoT applications, blockchain, and digital twins. Section 4 addresses transportation systems, focusing on intelligent solutions. Section 5 explores logistics operations, including warehouse automation, waste management, reverse logistics, and SC integration. In Sect. 6, we conclude with key findings, implementation challenges, and future research directions.

2 Literature Review

The following literature review provides a critical examination of current research and practices related to urban logistics and SSCM within the context of smart cities. Our analysis establishes the conceptual foundation for the subsequent discussions presented in this chapter.

2.1 The Evolution of the Smart City Paradigm

In recent years, the smart city paradigm has undergone a significant evolution, transitioning from a primarily technology-centric approach to a more comprehensive, integrated model that considers the complex interplay between technological, social, and institutional factors [6, 7]. Smart cities increasingly rely on advanced information and communication technologies (ICT) to enhance operational efficiency, environmental sustainability, and the overall quality of life in metropolitan areas. Notably, the social dimension of urban development has gained prominence in recent discourse, emphasizing the crucial role of community engagement, participatory governance, and the leveraging of collective knowledge in addressing the multifaceted challenges faced by cities.

2.2 Key Components of Sustainable Supply Chains

SSCs incorporate environmental, social, and economic considerations throughout the entire production and distribution process within smart cities. While the initial implementation of SSC practices often requires significant investment, organizations frequently realize substantial long-term benefits, including reduced costs, minimized environmental impact, and enhanced corporate reputation. The key components of SSCs encompass:

- Sourcing: Focus on ethical procurement and eco-friendly suppliers who demonstrate fair labor practices, waste reduction, and renewable resource usage. Strong supplier relationships ensure responsible material sourcing.
- Production: Implementation of waste reduction strategies, energy-efficient technologies, and emission controls. Many organizations adopt circular production methods to minimize raw material consumption.
- Transportation and Logistics: Focus on carbon reduction through route optimization, and sustainable packaging within smart city frameworks.
- Warehousing and Inventory: Energy-efficient facilities, and waste minimization.
- End-of-Life Management: Consideration of complete product lifecycles through recycling, refurbishment, and take-back programs.
- Technology and Innovation: Utilization of blockchain, IoT, and AI to enhance visibility and traceability.
- Stakeholder Engagement: Communication with customers, employees, and stakeholders is vital to ensuring transparency and building trust.

Table 1 Key technologies and their applications in smart city logistics

Technology	Application	Benefits
IoT	Network monitoring, data-driven decisions, predictive analytics	Improved efficiency, costs reduction, improved awareness
Blockchain	SC vibility, origin tracking, process streamlining	Increased trust, fraud prevention, improved tracking
AI	Forecasting, route planning, autonomous operations, maintenance prediction	Resource optimization, reduced costs, safety improvements
Digital Twins	Virtual system modeling, simulation capabilities	Better decisions, reduced downtime, faster innovation
5G and beyond	High-speed, remote operations	Enhanced connectivity, faster data processing, advanced applications support

2.3 The Transformative Role of Technology in Urban Logistics

Advanced technologies are profoundly transforming urban logistics operations. For instance, the deployment of IoT sensors and analytics enables real-time monitoring and network optimization, effectively reducing congestion and improving overall efficiency. Similarly, blockchain technology enhances transparency and security through the creation of immutable transaction records. Moreover, digital twins facilitate the creation of virtual models of logistics networks, enabling sophisticated planning and optimization processes [8]. Collectively, these innovations, summarized in Table 1, are ushering in a new era of efficiency and reliability in smart city urban logistics.

The integration of these technologies creates resilient, adaptable logistics networks that are well-equipped to meet the evolving needs of smart cities while simultaneously improving efficiency, visibility, and decision-making capabilities. As cities continue to grapple with the challenges posed by rapid urbanization, the adoption of smart technologies in logistics operations will play an increasingly critical role in fostering sustainable development and enhancing the quality of life for urban residents.

The integration of these technologies creates resilient, adaptable logistics networks that meet evolving urban needs while improving efficiency, visibility, and decision-making capabilities.

3 Technological Enablers and Solutions

Having established the key technological elements, this section examines the practical implementation of innovations transforming urban logistics operations. Modern cities are witnessing unprecedented changes in their logistics landscape through the integration of advanced technologies. These solutions enhance transportation networks, warehouse operations, and delivery systems. As demonstrated in Table 2, the

Table 2 Comparison of traditional and smart logistics

Aspect	Traditional logistics	Smart logistics
Decision-making	Manual, reactive, based on historical data and experience	Data-driven, proactive, utilizing real-time analytics
Transportation	Fixed routes and schedules with limited flexibility	Adaptive routing with real-time adjustments
Supply Chain Visibility	Limited oversight relying on manual processes	Comprehensive tracking with blockchain integration
Sustainability	Basic environmental considerations	Integrated environmental strategies and emissions reduction

evolution from traditional to smart logistics approaches highlights the impact of these innovations. The integration of data analytics, advanced technologies, and innovative strategies yields significant operational improvements.

3.1 IoT Applications

IoT technology forms the foundation of modern urban logistics in smart cities, enabling comprehensive monitoring capabilities [1, 9]. Urban sensor networks gather continuous data on traffic patterns, vehicle movements, and infrastructure conditions. This information enables dynamic operational adjustments and efficient resource allocation [9]. When combined with AI, IoT systems provide actionable insights for SC optimization [2].

3.2 Blockchain Technology

The implementation of blockchain systems introduces new possibilities for logistics transparency and security [5, 10]. This technology creates verified transaction records, enabling complete SC visibility. Smart contracts reduce administrative burden while maintaining regulatory compliance. The fusion of AI and blockchain technologies enables value-driven SC models that enhance operational and economic outcomes [11].

3.3 Digital Twins

The application of digital twin technology enables virtual modeling of logistics networks, supporting advanced simulation and planning. These models incorporate real-time sensor data to represent current conditions while enabling future scenario analysis [8]. Advanced algorithms utilize these digital representations to optimize various logistics processes, including warehouse and transportation management [12].

4 Transportation and Logistics: Traditional and Smart Systems

4.1 Vehicle and Transportation Systems

The practical implementation of these technologies is most evident in modern transportation systems, where advances in infrastructure management and real-time monitoring are driving significant improvements in urban mobility and efficiency.

4.1.1 Intelligent Transportation Systems

Modern cities rely on intelligent transportation systems (ITS) to monitor and manage traffic dynamics. These systems integrate data from various sources - including cameras, sensors, and GPS technology - to provide comprehensive traffic management solutions [13–15]. Adaptive traffic signal control can optimize traffic light timing based on current traffic conditions, improving overall flow and reducing idle time at intersections (see Fig. 1). Route optimization is another crucial aspect of ITS, focusing on efficient resource allocation and travel routes to minimize travel times and reduce congestion [16, 17].

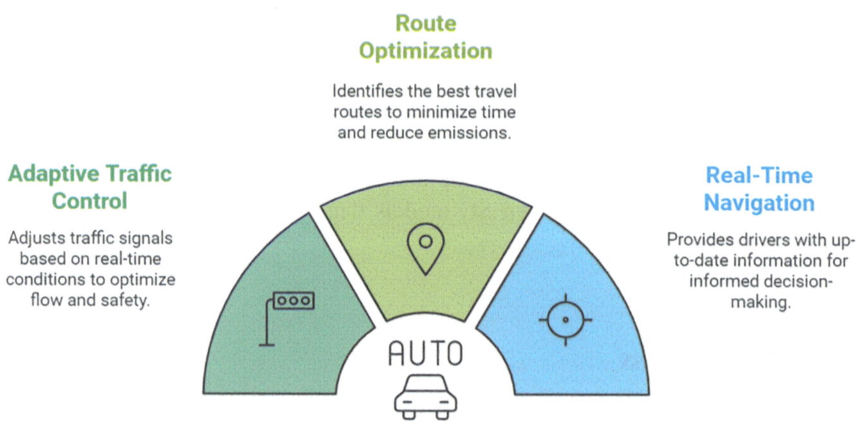

Fig. 1 Intelligent transportation systems

4.1.2 Navigation and Route Optimization

The evolution of ITS has revolutionized urban navigation and route planning [18]. These systems employ advanced algorithms to analyze traffic patterns and predict congestion, enabling proactive traffic management [19]. Real-time navigation systems provide drivers with crucial information about travel conditions, alternative routes, and potential delays [13, 20]. The integration of real-time navigation with traffic management systems creates a seamless flow of information, reducing travel times and congestion, underscoring its critical role in modern transportation systems [16–18, 21]. As illustrated in Fig. 1, the three primary components of modern ITS—adaptive traffic control, route optimization, and real-time navigation—are interconnected, each playing a pivotal role in urban transportation management.

4.2 Smart Logistics Operations

Following our discussion of transportation systems, we now examine the operational dimensions of smart logistics, focusing on the intersection of efficiency and sustainability in contemporary SCM.

4.2.1 Warehouse Management

Warehouse Management is a set of practices and systems aimed at optimizing warehouse operations, from receiving goods to shipping them out. It involves location tracking and the efficient use of human and material resources. With an effective Warehouse Management System (WMS), companies can enhance productivity by reducing time and errors in order fulfillment while maintaining accurate stock levels. By incorporating technologies like barcodes, radio-frequency identification (RFID), or automated systems, a WMS helps meet demand more effectively, lowers costs, and ensures greater customer satisfaction by guaranteeing quick and accurate deliveries, to explain WMS it well, we will focus on the three important pillars for WMS: Automated systems, Inventory optimization and Smart storage solutions.

Automated Systems

WMS and Automation examines existing studies to understand how warehouse operations have evolved with automation and digital transformation. Table 3 presents different aspects of automated systems.

Advances in technology and automation have revolutionized warehouse management, with the adoption of RFID, barcoding, and IoT technologies enabling real-time data collection and improved processing times (Richards 2014); [23, 24]. The integration of these technologies with WMS provides accurate inventory data,

Table 3 Different aspects of automated systems

Topic	Key points	References
Advances in Technology and Automation	– Technological advancements like RFID, barcoding, and IoT have transformed warehouse management – Automation reduces human error and improves processing times – Integration with WMS provides real-time data on inventory – Robotics and AI enhance efficiency via automated guided vehicles (AGVs), autonomous mobile robots (AMRs), and AI-driven improvements in forecasting, route optimization, and predictive maintenance	Richards [22], Baker and Canessa [23], Zhang, Chai, and Ma [24]
Impact on Workforce and Operational Efficiency	– Automation reduces manual labor for repetitive tasks but increases the need for skilled workers to manage systems – Human workers collaborate with automated systems to maximize efficiency	Mckinnon [25]
Inventory Optimization	– Focuses on maximizing profit margins and minimizing losses – Key elements: demand forecasting, inventory policy, and replenishment – Multi-agent systems (MAS) minimize holding costs across the supply chain while maintaining service levels – Complexity depends on the network structure	Patel [26], Eruguz et al. [27]
Smart Storage Solutions	– WMS is central to smart warehouse technologies, managing the warehouse lifecycle (receiving, storing, picking, shipping) – Subdomains include warehouse-entry, shelves, warehouse-out, and storage management – RFID is the most promising technology for improving warehouse performance	Yan, Liu, and Huang [28]

while robotics and AI enhance efficiency through the use of AGVs, AMRs, and AI-driven improvements in forecasting, route optimization, and predictive maintenance. As shown in Table 3, these innovations have a profound impact on workforce and operational efficiency, inventory optimization, and the implementation of smart storage solutions.

The impact of automation on the workforce and operational efficiency is also examined, revealing that while automation reduces manual labor for repetitive tasks, it simultaneously increases the need for skilled workers to manage and maintain these systems [25]. The table emphasizes the importance of human-machine collaboration in maximizing efficiency and productivity.

Inventory optimization is another critical aspect of automated systems, focusing on maximizing profit margins and minimizing losses through effective demand forecasting, inventory policy, and replenishment strategies [26, 27]. The complexity of inventory optimization varies depending on the network structure, with MAS being employed to minimize holding costs while maintaining service levels.

Smart storage solutions, powered by WMS, are central to the efficient management of the warehouse lifecycle, encompassing receiving, storing, picking, and shipping processes [28]. The table identifies RFID as the most promising technology for improving warehouse performance, with WMS serving as the backbone for integrating and managing various smart warehouse technologies.

4.2.2 Waste Management and Reverse Logistics

Waste Management and Reverse Logistics as the process of moving products from their final destination back to the origin for purposes such as returns, recycling, refurbishing, or disposal [29].

The Role and Strategies of Reverse Logistics in Sustainable Waste Management

Researchers in the early 2000s recognized the environmental benefits of reverse logistics, emphasizing the importance of closing the loop in SCs to support waste minimization [30]. Studies show that reverse logistics helps reduce landfill waste and supports the circular economy by reusing and recycling materials [31]. Research by Guide and Van Wassenhove [32] further expands on reverse logistics as a critical component in achieving sustainability goals, particularly in industries where products are complex and challenging to recycle. A significant area of study is focused on waste reduction strategies enabled by reverse logistics, including remanufacturing, recycling, and product take-back systems [33].

Technology and Digital Transformation in Reverse Logistics

Technological advancements have played a crucial role in enhancing reverse logistics and waste management efficiency. Research shows that technologies like IoT, blockchain, and AI optimize reverse logistics by providing real-time data on product returns and tracking waste materials [34]. For instance, AI algorithms enable efficient sorting and categorizing of waste, which improves recycling rates and resource recovery [35].

Economic and Operational Challenges

While the environmental and financial benefits of reverse logistics in waste management are clear, studies by [36] indicate that while reverse logistics can generate revenue, the costs associated with transportation, sorting, and refurbishing may deter firms from fully adopting these practices. Economic analyses highlight the need for regulatory support and incentives to offset costs and encourage adoption.

4.2.3 Supply Chain Integration

SC Integration (SCI) as the alignment of processes, information flow, and resources among SC partners, including suppliers, manufacturers, distributors, and customers, to improve collaboration and reduce inefficiencies [37]. Scholars like [38] emphasize that integration is essential for achieving seamless operations across the SC, creating a unified system that can better respond to market demands and disruptions. Scholars note that SCI has become particularly valuable during global disruptions, such as the COVID-19 pandemic, where integrated SCs were better positioned to adapt to demand fluctuations and resource shortages [39].

Literature categorizes SCI into internal and external integration:

- Internal integration focuses on coordination within an organization, harmonizing various departments (production, logistics, and sales) to ensure consistent information flow and align goals [40].
- External integration refers to collaboration with external partners, such as suppliers and customers, to share forecasts, optimize inventory, and coordinate deliveries, resulting in a more responsive SC (S. [41]).
- Dimensions of integration identified by [42] include:
- Information integration: Sharing accurate and timely information.
- Operational integration: Coordinating processes and resources.
- Relational integration: Building trust and collaborative relationships.

4.2.4 Role of Technology in Supply Chain Integration

Technology plays a critical role in enabling SCI by facilitating communication, information sharing, and process automation. Scholars emphasize the importance of ERP systems, IoT, and blockchain in supporting real-time data sharing and transparency across the SC [43]. Research suggests that advanced technologies improve integration by providing visibility into operations and enhancing data-driven decision-making [44].

5 Smart Decision Methods in Urban Logistics

Smart decision methods in urban logistics represent systematic frameworks for optimizing operations across transportation systems, warehouse management, and SC networks. These methods integrate data from various sources to enable informed decision-making that enhances both efficiency and sustainability. In transportation systems, these decision methods focus primarily on route optimization and traffic management. Modern intelligent transportation systems implement optimization techniques that prioritize efficient resource allocation and travel routes to minimize congestion and improve urban mobility [17]. These are enhanced by adaptive traffic signal control methods that optimize traffic light timing based on real-time traffic conditions [13–15]. Advanced navigation systems analyze traffic patterns and predict congestion, enabling proactive management [19].

The domain of warehouse management employs decision methods focused on optimizing storage and resource utilization. Warehouse Management Systems implement location tracking and resource allocation methods that provide accurate, real-time data on inventory status and location [22, 23]. These systems support complex decision-making processes for inventory optimization, balancing storage costs with operational efficiency [27]. Warehouse lifecycle management provides comprehensive frameworks [28]. These address key areas including entry management, shelving efficiency, and storage allocation. SC integration incorporates sophisticated coordination mechanisms operating at multiple levels. Internal integration methods focus on coordinating various departments within organizations, while external integration approaches facilitate collaboration with suppliers and customers [40, 41]. These integration methods encompass information sharing for accurate data transmission, process coordination for operational alignment, and relationship building for collaborative partnerships [42]. In terms of performance evaluation, SCA analytics focus on sustainability, monitoring key metrics like emissions, energy use, and waste. [43, 44]. In terms of performance evaluation, SC analytics employ systematic methods. These analytical frameworks enable both regulatory compliance and environmentally conscious business practices while strengthening supplier relationships [43, 44].

The implementation of these smart decision methods has significantly enhanced urban logistics operations. By combining traditional optimization approaches with modern analytical tools, organizations can make more informed decisions across their SC networks. These methods provide the foundation upon which AI applications can further enhance logistics operations through advanced data processing and autonomous capabilities, as discussed in the following section.

Table 4 synthesizes the main smart decision methods implemented across different logistics domains, highlighting the key technologies enabling their deployment in urban environments.

Across various logistics domains, the application of smart decision-making methods is driving significant advancements in urban logistics. In the transportation domain, for example, these methods prioritize route planning and traffic flow management, employing technologies such as ITS, GPS, and sensors [13, 17]. These technologies facilitate dynamic route optimization, reduce congestion, and ultimately improve urban mobility. As presented in Table 4, the smart decision-making approaches are categorized into four key domains: transportation, warehouse management, SC integration, and performance analytics. Each of these domains benefits

Table 4 Smart decision methods in urban logistics

Domain	Decision methods	Technologies
Transportation	Route Planning, Traffic Flow Management	ITS, GPS, Sensors
Warehouse Management	Storage Optimization, Inventory Control	WMS, RFID, Automated Systems
Supply Chain Integration	Network Design, Supplier Coordination	Digital Twins, Blockchain, ERP
Performance Analytics	Sustainability Tracking, Process Evaluation	IoT Sensors, Big Data Analytics

from the integration of cutting-edge technologies that not only streamline operations but also enable more informed, efficient decision-making.

Warehouse management employs smart decision methods for storage optimization and inventory control, utilizing technologies such as WMS, RFID, and automated systems [22, 23]. These methods support efficient space utilization, accurate inventory tracking, and streamlined order fulfillment processes. SC integration relies on smart decision methods for network design and supplier coordination, enabled by technologies like digital twins, blockchain, and enterprise resource planning (ERP) systems [5, 8]. These methods facilitate end-to-end visibility, collaborative planning, and secure information sharing among SC partners.

As for performance analytics, it leverages smart decision methods for sustainability tracking and process evaluation, supported by IoT sensors and big data analytics [43, 44]. These methods enable organizations to monitor and optimize their environmental performance, identify inefficiencies, and drive continuous improvement across the SC.

Building on these advancements, the integration of AI and decision support systems in SCM has the potential to revolutionize logistics operations in urban settings. Machine learning (ML) applications, scheduling algorithms, and advanced SC analytics can significantly enhance efficiency, resilience, and sustainability.

5.1 Machine Learning Applications

ML has transformed diverse industries through applications like predictive analytics, pattern recognition, and demand forecasting. In predictive analytics, ML models analyze historical data to forecast future trends, such as predicting customer behavior, equipment failures, or even medical outcomes, as seen in healthcare where algorithms help predict patient readmissions [45]. Contemporary pattern recognition systems, particularly convolutional neural networks (CNNs), excel in complex visual processing tasks are widely used in facial recognition technology to detect features across images [46]. In logistics operations, demand forecasting applications help organizations maintain optimal inventory levels while reducing operational costs [47]. These implementations demonstrate ML's contributions to operational excellence.

5.2 *Supply Chain Analytics*

Modern SC Analytics (SCA) employs systematic data analysis to enhance decision-making throughout logistics operations. Two key areas have emerged as particularly significant:

5.2.1 Sustainability Analytics

Organizations increasingly measure their environmental and social impact through comprehensive analytics. This includes monitoring carbon emissions, energy consumption, and waste production. These measurements enable both regulatory compliance and environmentally conscious business practices [48].

5.2.2 Supplier Performance and Relationship Management

Analytics provides detailed evaluation of supplier performance through key metrics including delivery reliability, quality standards, and operational efficiency. This data-driven approach strengthens business relationships while improving supplier selection processes.

While the benefits of smart urban logistics are clear, several key challenges remain. Table 5 outlines these challenges alongside potential solutions, highlighting the need for coordinated responses from industry, government, and academic stakeholders. The relationship between these challenges and opportunities suggests that successful implementation requires a balanced approach to technological advancement and practical considerations.

5.3 *Analysis*

Our review of the literature reveals a growing adoption of smart decision methods and AI applications in SSCM and logistics. Figure 2 presents the distribution of key technologies identified in Table 1 among the examined studies.

Research in smart city logistics reveals a clear emphasis on specific technologies that are transforming the field. As shown in Fig. 2, IoT emerges as the dominant technology, representing 35% of the studies, followed closely by AI at 30%. Blockchain and Digital Twins, while still in the early stages of adoption, account for 20% and 15% of the research, respectively. This distribution highlights the growing

Table 5 Implementation challenges and solutions in urban logistics

Challenge	Opportunity
Data Privacy	Robust security protocols and ethical data practices
Infrastructure Costs	Cross-sector partnerships and innovative funding
Skill Development	Industry-academic collaboration and training
Systems Integration	Unified protocols and standards
Security	Comprehensive protection and monitoring
Regulatory Framework	Progressive policy frameworks

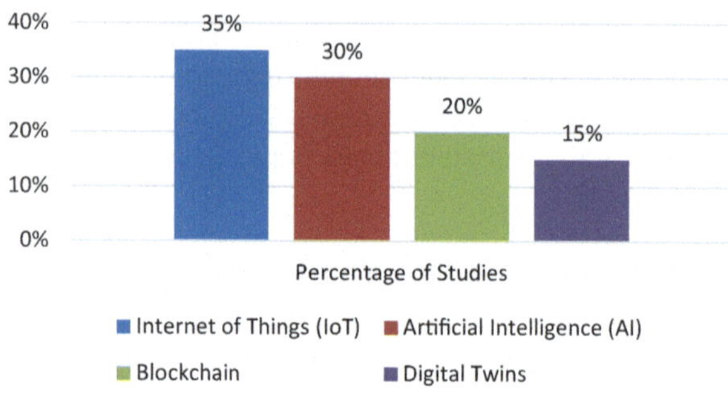

Fig. 2 Distribution of key technologies in smart city logistics

significance of real-time data collection and intelligent processing, which are foundational to the development of next-generation SCs. These technologies play a crucial role in creating more efficient and responsive urban logistics systems.

Furthermore, our analysis reveals an increasing trend towards the integration of multiple technologies. Figure 3 compares the adoption of traditional and smart logistics approaches discussed in Table 2.

There is a notable trend toward the adoption of smart logistics approaches in recent studies. As illustrated in Fig. 3, 70% of the analyzed research focuses on innovative, data-driven methods, while only 30% addresses traditional approaches. This significant disparity underscores the industry's growing momentum toward digitalization and the integration of advanced technologies in urban logistics operations. The shift highlights the increasing importance of smart, data-centric solutions in enhancing efficiency and addressing the complex challenges of modern urban logistics.

Finally, we examine the prevalence of implementation challenges and opportunities presented in Table 5. Figure 4 illustrates the frequency at which each challenge is addressed in the literature.

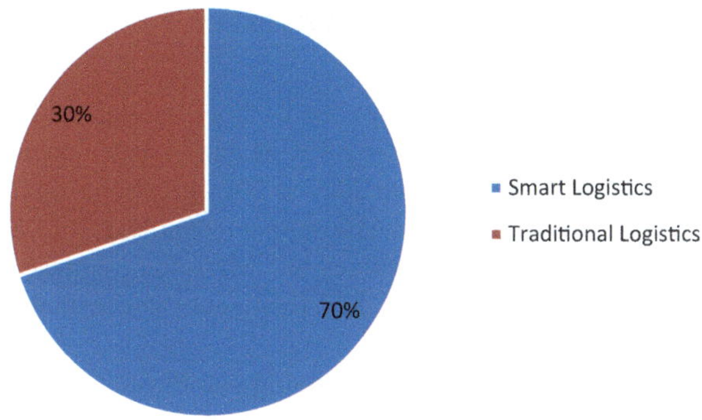

Fig. 3 Adoption of traditional vs smart logistics approaches

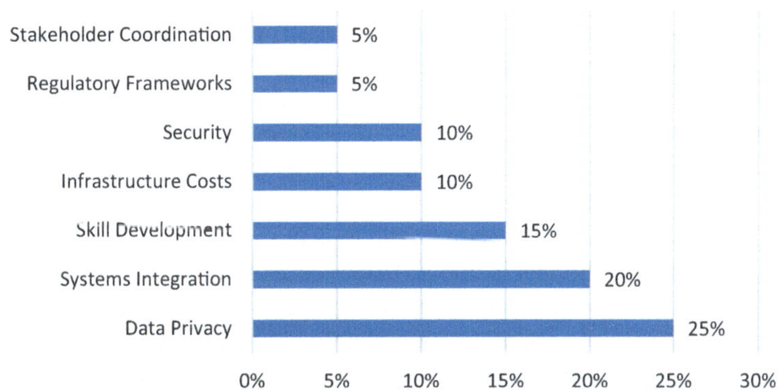

Fig. 4 Implementation challenges in smart city logistics

As for Fig. 4, it reveals the hierarchy of implementation challenges in smart city logistics. Data privacy leads the concerns at 25%, followed by systems integration at 20%. Skill development represents 15% of the challenges, while infrastructure costs and security each account for 10%. The remaining 20% encompasses other challenges such as regulatory frameworks and stakeholder coordination. This distribution emphasizes that successful implementation requires addressing both technological and organizational challenges.

Our analysis reveals compelling patterns in smart city logistics development. The technological landscape is primarily shaped by IoT and AI applications, reflecting a significant transformation in urban SCM. This evolution is evidenced by the predominant shift toward smart logistics approaches. However, the implementation landscape presents notable challenges, particularly in data privacy and systems integration. These findings underscore a critical junction in urban logistics, where technological advancement must be balanced with practical implementation considerations.

6 Conclusion and Future Perspectives

This chapter has provided a detailed review of how intelligent decision-making methods and AI applications are reshaping SSCM and logistics operations in smart cities. Our analysis of the literature from 2007 to 2024 highlights optimization algorithms and the growing adoption of transformative technologies, including the IoT, blockchain, digital twins, and AI techniques like ML and optimization algorithms. These technologies are enhancing efficiency, resilience, and sustainability in urban SC networks.

Integrating these smart technologies is driving significant improvements in key areas such as transportation systems, warehouse management, waste reduction through reverse logistics, and end-to-end SC visibility. IoT sensor networks and real-time data analytics are enabling intelligent transportation systems to optimize traffic flows, reduce congestion, and improve navigation. AI-powered automation and predictive analytics are transforming warehouse operations, while blockchain is enhancing trust, security, and transparency throughout SCs. Digital twin technology is emerging as a powerful tool for virtual SC modeling, simulation, and optimization.

Our analysis of technology adoption trends shows the prominence of IoT and AI, featured in 35% and 30% of reviewed studies, respectively. Notably, 70% of the literature focuses on innovative, data-driven logistics approaches, signaling a decisive shift away from traditional methods. However, widespread implementation still faces challenges. Data privacy and system integration emerge as recurring concerns, reflecting the complexity of digitalizing urban logistics ecosystems.

Additionally, high infrastructure costs, the need for skill development, and varying regulatory frameworks call for collaborative efforts among industry, government, and academia. Successfully deploying these technologies requires not only technical capabilities but also a holistic approach addressing operational and societal factors.

Looking ahead, several promising research directions lie at the intersection of SSCM, urban development, and advanced technologies. AI holds immense potential in predictive maintenance, dynamic routing, and demand forecasting, enabling optimized logistics performance. Blockchain-enabled provenance tracking and smart contracts could revolutionize sustainable sourcing and circular economy initiatives. Furthermore, the advent of 5G networks and edge computing may unlock new possibilities for real-time, decentralized decision-making in complex urban SCs.

The role of logistics in creating resilient and livable cities cannot be overstated. Future research should explore synergies between smart logistics and sustainable urban planning. This includes designing multi-modal transportation hubs, urban consolidation centers, and green last-mile delivery solutions. Moreover, the social and economic impacts of AI-driven automation warrant deeper investigation to ensure equitable distribution of benefits.

Policy frameworks will play a pivotal role in this transformation. Incentivizing sustainable practices, supporting infrastructure upgrades, and fostering public-private partnerships will be crucial. Promoting open data standards, interoperability, and knowledge sharing across sectors can accelerate innovation and amplify impact.

Ultimately, the transition to intelligent, sustainable urban SCs is not merely a technological shift but a fundamental reimagining of how cities manage the flow of goods, information, and resources. This transformation demands a collaborative, adaptive approach that leverages AI and data-driven insights while prioritizing environmental stewardship, social well-being, and inclusive growth. As urbanization continues at an unprecedented rate, smart logistics will be integral to building resilient, thriving cities equipped to face twenty-first-century challenges.

References

1. Qu C, Kim E (2024) Reviewing the roles of AI-integrated technologies in sustainable supply chain management: research propositions and a framework for future directions. Sustainability 16(14):6186. https://doi.org/10.3390/su16146186
2. Zejjari I, Benhayoun I (2024) The use of artificial intelligence to advance sustainable supply chain: retrospective and future avenues explored through bibliometric analysis. Discov Sustain 5(1):174. https://doi.org/10.1007/s43621-024-00364-6
3. Bouti N, El Khoukhi F (2023) Smart supply chain management: a literature review, chapter and conference paper. In: Digital Technologies and Applications Proceedings of ICDTA'23, Volume 1, Part of the Lecture Notes in Networks and Systems book series (LNNS, volume 668), pp 896–904. https://doi.org/10.1007/978-3-031-29857-8_89
4. Hasan R, Islam Z, Sumon FI, Osiujjaman PD, Pant L (2024) Integrating artificial intelligence and predictive analytics in supply chain management to minimize carbon footprint and enhance business growth in the USA. J Bus Manag Stud 6(4):195–212. https://doi.org/10.32996/jbms.2024.6.4.17
5. Zhao C, Jia R, Dong K (2023) How does smart transportation technology promote green total factor productivity? The case of China. Res Transp Econ 101:101353. https://doi.org/10.1016/j.retrec.2023.101353
6. Albino V, Berardi U, Dangelico RM (2015) Smart cities: Definitions, dimensions, performance, and initiatives. Journal of urban technology, 22(1), 3–21
7. Caragliu A, Del Bo C, Nijkamp P (2013) Smart cities in Europe. In Creating smart-er cities (pp. 65–82). Routledge
8. Uhlemann TH-J, Lehmann C, Steinhilper R (2017) The digital twin: realizing the cyber-physical production system for Industry 4.0. Procedia CIRP 61:335–440. https://doi.org/10.1016/j.procir.2016.11.152
9. Kirsanov S, Prigoda L, Radovanovic D (2023) Smart cities and city logistics. MEST J 11(2):104. https://doi.org/10.12709/mest.11.11.02.11
10. Anaba DC, Kess-Momoh AJ, Ayodeji SA (2024) Optimizing supply chain and logistics management: a review of modern practices. Open Access Res J Sci Technol 11(2):020–028. https://doi.org/10.53022/oarjst.2024.11.2.0083
11. Wang L, Wang S, Gong Y, Peng L (2023) Optimizing a multi-echelon location-inventory problem with joint replenishment: a Lipschitz ϵ-optimal approach using Lagrangian relaxation. Comput Oper Res 151:106128. https://doi.org/10.1016/j.cor.2022.106128
12. Upadhyay P, Sharma T, Fatima A (2024) Impact of AI-driven digital twins in Industry 4.0: an exploratory analysis. Int Res J Adv Eng Manag (IRJAEM) 2(05):1548–1557. https://doi.org/10.47392/IRJAEM.2024.0210

13. Badi I, Bouraima MB, Muhammad LJ (2022) The role of intelligent transportation systems in solving traffic problems and reducing environmental negative impact of urban transport. In: Decision Making and Analysis, pp 1–9. https://doi.org/10.55976/dma.1202311371-9
14. Li R, Wong Y, Lee H, Li H (2024) A study of the application and impact of artificial intelligence in intelligent transport systems. J Simulat 12(1)
15. Mahrez Z, Sabir E, Badidi E, Saad W, Sadik M (2021) Smart urban mobility: when mobility systems meet smart data. IEEE Trans Intell Transp Syst 99:1–18. https://doi.org/10.1109/TITS.2021.3084907
16. Kamargianni M, Li W, Matyas M (2016a) A comprehensive review of "mobility as a service" systems 2016.Pdf. In: Proceedings of the First Karlsruhe Service Summit Workshop - Advances in Service Research
17. Kamargianni M, Li W, Matyas M, Schäfer A (2016b) A critical review of new mobility services for urban transport. Transp Res Procedia 14:3294–3303. https://doi.org/10.1016/j.trpro.2016.05.277
18. Boukallal I, Bouti N, El Khoukhi F (2022) Le Problème AF-VRP: Une Revue de La Littérature. In: 5ème édition de la Conférence Internationale sur le Monitoring des Systèmes Industriels (CIMSI2022), 14–15 Décembre, 2022, Tanger, Maroc
19. Dikshit S, Atiq A, Shahid M, Dwivedi V, Thusu A (2023) The use of artificial intelligence to optimize the routing of vehicles and reduce traffic congestion in urban areas. EAI Endors Trans Energy Web 10. https://doi.org/10.4108/ew.4613
20. Bouti N, Boukallal I, Khoukhi FEL (2024) Sustainable urban logistics and supply chain 4.0: a literature overview and future directions, chapter and conference paper. In: Proceeding of GOL'24 "Smart Sustainable and Green Logistics", Volume 1, Part of the Lecture Notes in Networks and Systems book series (LNNS 1104), pp 335–344. https://doi.org/10.1007/978-3-031-68628-3_32
21. Bouti N, Boukallal I, El Khoukhi F (2023) Optimization of inventory management: a literature review, chapter and conference paper. In: Digital Technologies and Applications Proceedings of ICDTA'23, Volume 2, Part of the Lecture Notes in Networks and Systems book series (LNNS, volume 669), pp 923–933. https://doi.org/10.1007/978-3-031-29860-8_92
22. Richards G (2011) Warehouse management: a complete guide to improving efficiency and minimizing costs in the modern warehouse, Kogan Page
23. Baker P, Canessa M (2009) Warehouse design: a structured approach. Eur J Oper Res 193(2):425–436. https://doi.org/10.1016/j.ejor.2007.11.045
24. Zhang Y, Chai Y, Ma L (2021) Research on multi-Echelon inventory optimization for fresh products in supply chains. Sustainability 13(11):6309. https://doi.org/10.3390/su13116309
25. Mckinnon AC (2016) The possible impact of 3D printing and drones on last-mile logistics: an exploratory study. Built Environ 42(4):617–629. https://doi.org/10.2148/benv.42.4.617
26. Patel SS (2017) Production inventory model for deteriorating items with different deterioration rates under stock and price dependent demand and shortages under inflation and permissible delay in payments. Glob J Pure Appl Math 13(7):3687–3701
27. Eruguz AS, Sahin E, Jemai Z, Dallery Y (2016) A comprehensive survey of guaranteed-service models for multi-echelon inventory optimization. Int J Prod Econ 172:110–125. https://doi.org/10.1016/j.ijpe.2015.11.017
28. Yan B, Liu Y, Huang Y (2022) Improved discrete imperialist competition algorithm for order scheduling of automated warehouses. Comput Indust Eng 168:108075. https://doi.org/10.1016/j.cie.2022.108075
29. Tibben-Lembke RS (1998) The impact of reverse logistics on the total cost of ownership. J Market Theory Pract 6(4):51–60. https://doi.org/10.1080/10696679.1998.11501810
30. Carter CR, Ellram LM (1998) Reverse logistics: a review of the literature and framework for future investigation. J Bus Logist 19(1):85–102
31. Fleischmann M, Beullens P, Bloemhof-Ruwaard JM, Van Wassenhove LN (2001) The impact of product recovery on logistics network design. Prod Oper Manag 10(2):156–173. https://doi.org/10.1111/j.1937-5956.2001.tb00076.x

32. Guide VDR, Van Wassenhove LN (2009) OR FORUM—the evolution of closed-loop supply chain research. Oper Res 57(1):10–18. https://doi.org/10.1287/opre.1080.0628
33. Thierry M, Salomon M, Van Nunen J, Van Wassenhove L (1995) Strategic issues in product recovery management. California Manag Rev 37(2):114–136. https://doi.org/10.2307/41165792
34. Govindan K, Soleimani H, Kannan D (2015) Reverse logistics and closed-loop supply chain: a comprehensive review to explore the future. Eur J Oper Res 240(3):603–626. https://doi.org/10.1016/j.ejor.2014.07.012
35. Kuo T-C, Smith S (2018) A systematic review of technologies involving eco-innovation for enterprises moving towards sustainability. J Clean Prod 192:207–220. https://doi.org/10.1016/j.jclepro.2018.04.212
36. Guide VDR, Vanwassenhove LN (2001) Managing product returns for remanufacturing. Prod Oper Manag 10(2):142–155. https://doi.org/10.1111/j.1937-5956.2001.tb00075.x
37. Stevens GC (1989) Integrating the supply chain. Int J Phys Distrib Mater Manag 19(8):3–8. https://doi.org/10.1108/EUM0000000000329
38. Frohlich MT, Westbrook R (2001) Arcs of integration: an international study of supply chain strategies. J Oper Manag 19(2):185–200. https://doi.org/10.1016/S0272-6963(00)00055-3
39. Ivanov D, Dolgui A (2020) Viability of intertwined supply networks: extending the supply chain resilience angles towards survivability. A position paper motivated by COVID-19 outbreak. Int J Prod Res 58(10):2904–2915. https://doi.org/10.1080/00207543.2020.1750727
40. Kahn KB, Mentzer JT (1996) Logistics and interdepartmental integration. Int J Phys Distrib Logist Manag 26(8):6–14. https://doi.org/10.1108/09600039610182753
41. Zhao S, Grasmuck S, Martin J (2008) Identity construction on Facebook: digital empowerment in anchored relationships. Comput Hum Behav 24(5):1816–1836. https://doi.org/10.1016/j.chb.2008.02.012
42. Flynn BB, Huo B, Zhao X (2010) The impact of supply chain integration on performance: a contingency and configuration approach. J Oper Manag 28(1):58–71. https://doi.org/10.1016/j.jom.2009.06.001
43. Gunasekaran A, Ngai EWT (2004) Information systems in supply chain integration and management. Eur J Oper Res 159(2):269–295. https://doi.org/10.1016/j.ejor.2003.08.016
44. Wang G, Gunasekaran A, Ngai EWT, Papadopoulos T (2016) Big data analytics in logistics and supply chain management: certain investigations for research and applications. Int J Prod Econ 176:98–110. https://doi.org/10.1016/j.ijpe.2016.03.014
45. Rajkomar A, Oren E, Chen K, Dai AM, Hajaj N, Hardt M, Liu PJ et al (2018) Scalable and accurate deep learning with electronic health records. NPJ Digit Med 1(1):18. https://doi.org/10.1038/s41746-018-0029-1
46. Zhao Q, Lyu S, Zhang B, Feng W (2018) Multiactivation pooling method in convolutional neural networks for image recognition ed. Xuyun Zhang. Wirel Commun Mobile Comput 2018(1):8196906. https://doi.org/10.1155/2018/8196906
47. Feizabadi J (2022) Machine learning demand forecasting and supply chain performance. Int J Logist Res Appl 25(2):119–142. https://doi.org/10.1080/13675567.2020.1803246
48. Ouba A, El Khoukhi F (2023) Supply chain risk management: a literature review. In: International Conference on Services and Industry of the Future (ICSIF 2023), October, 24–26, 2023, Rabat, Morocco

On-Road or Dynamic Wireless Charging for Slow-Moving EVs: Paving the Way for Sustainable Smart Cities

Yuvaraja Shanmugam and R. Narayanamoorthi

Abstract The advancement of electric vehicle (EV) technology plays a vital role in developing eco-friendly and sustainable transportation systems for smart cities. To meet the increasing demand for safe and efficient charging, adopting autonomous systems is crucial. Wireless power transfer (WPT) allows vehicles to charge automatically, aligning with the goals of smart cities by promoting energy efficiency and reducing dependency on conventional fuels. Dynamic wireless charging (DWC) takes this further by enabling vehicles to charge while moving, making it possible for EVs to cover longer distances without frequent stops. This innovation reduces the need for large battery storage, optimizing space and cost. This chapter provides a comprehensive exploration of the technology underlying DWC systems, with the primary aim of advancing the development of an efficiency-enhanced wireless charging solution for on-road EV commuters. It systematically examines the essential components, technical standards, and operational frameworks that define DWC systems, offering a thorough understanding of their functionality and potential. Furthermore, the chapter delves into the intricate design processes involved in developing these systems, emphasizing optimization strategies to achieve seamless integration with EV infrastructure. By bridging the gap between theoretical concepts and practical implementation, this chapter serves as a critical resource for driving innovation in DWC technologies. By integrating WPT and DWC technologies, smart cities can significantly improve urban transportation, ensuring cleaner and more efficient mobility while addressing environmental concerns. These advancements are key to supporting the sustainable development of future urban areas. The optimal design of a wireless charging system can enhance the power

Y. Shanmugam (✉)
Department of Electrical and Electronics Engineering, Vignan's Foundation for Science, Technology & Research, Vadlamudi, Andhra Pradesh, India

R. Narayanamoorthi
Department of Electrical and Electronics Engineering, SRM Institute of Science and Technology, Kattankulathur, Chennai, India

H. Kaddouri et al. (eds.), *Smart Cities for Sustainable and Urban Development*, Green Energy and Technology, https://doi.org/10.1007/978-3-031-99001-4_24

transfer efficiencies during misalignment conditions and optimal control methodologies can improve the charging efficiency.

Keywords Dynamic wireless charging · Inductive power transfer · On-road charging · Electric vehicle · Resonant network

1 Introduction

Electric vehicles (EVs) play a crucial role in promoting sustainable transportation by addressing environmental concerns [1, 2]. They eliminate tailpipe emissions, thereby reducing greenhouse gases and improving air quality. EVs are more energy-efficient and quieter compared to internal combustion engine (ICE) vehicles. By utilizing electricity, especially from renewable sources, EVs lower dependence on fossil fuels and support energy security [3]. Although their initial cost is higher, they offer long-term savings due to minimal maintenance and operational costs. With increasing global emphasis on renewable energy and green policies, EVs represent a practical and eco-friendly alternative for a sustainable future. The development of charging infrastructure is a significant factor in overcoming barriers to EV market growth [2]. Limited availability of charging infrastructures, especially in rural and remote areas, restricts consumer confidence and adoption. Moreover, batteries are essential for EVs, determining their range, performance, charging time, and cost. Lithium-ion batteries are widely used for their efficiency, but their high production costs and reliance on limited resources like lithium and cobalt pose challenges [4]. Wired, wireless, and battery swapping are three significant methods for charging EVs, each with distinct advantages and limitations [5]. Wired charging is the most commonly used method, connecting vehicles directly to power sources using cables. It includes both AC charging, suitable for homes, and DC fast charging at public stations. While cost-effective and widely accessible, wired charging can be time-consuming, particularly with slower chargers [6]. Also, it is human dependency and has wear and tear problems of the mechanical conductors raises safety issues. Battery swapping [7] is another type of charging method that replaces a drained battery with a pre-charged one, allowing EVs to start driving within a short duration. This method is highly effective in reducing downtime, particularly for commercial fleets. However, it demands standardized battery designs, significant infrastructure investment, and compatibility across different vehicle models, which can limit its adoption [8]. Wireless charging technology [9] enables EVs to charge without physical cables by using electromagnetic induction or resonance. It consists of a transmitting pad connected to the ground-side charging unit and a receiver pad embedded into the vehicle. This technology eliminates the need for mechanical connectors, reducing wear and tear on charging ports while enhancing convenience and user safety. By eliminating cable handling, wireless charging offers a seamless experience, particularly for drivers with limited mobility or for public charging stations in areas with high vehicle turnover.

This chapter focuses on advancing DWC systems for EVs emphasizing their implementation in smart city environments. It provides a detailed analysis of essential components, including power converters with different modulation techniques, coils, resonant networks, and the design equations critical for achieving efficient energy transfer. In addition, the chapter discusses international-level key standards to ensure seamless integration and scalability within modern urban-level future-ready infrastructures. While prior research has significantly contributed to the field, challenges remain in creating a unified, high-performance system that meets the demands of smart cities. Addressing these challenges, the chapter presents a guide for the development of next-generation DWC systems, supporting the vision of sustainable and interconnected urban mobility.

2 Wireless Charging Systems for an EV

The different approaches of wireless charging systems are inductive charging, resonant inductive charging, magnetic resonance coupling, capacitive charging, and beam charging [10, 11]. While each technology offers specific benefits such as convenience, efficiency, and flexibility. Figure 1 presents a block diagram showcasing different wireless charging methods, categorized by their respective power transfer

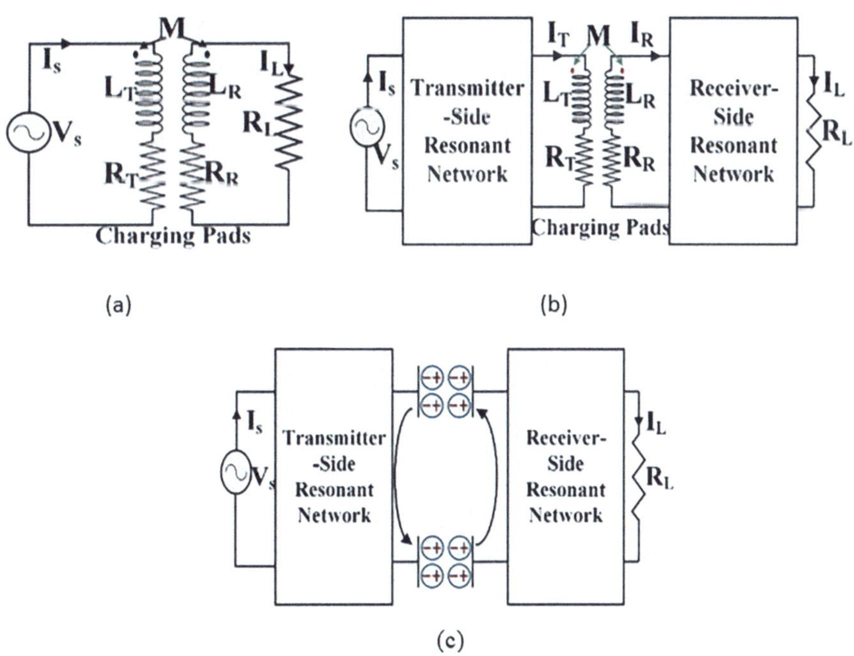

Fig. 1 (**a**) Inductive charging. (**b**) Resonant inductive charging. (**c**) Capacitive charging

mechanisms. Inductive Power Transfer [12] operates through the magnetic field generated by electromagnetic waves, where the principles of Ampere's and Faraday's laws govern the interaction of electric and magnetic fields. The receiver pad is influenced by the magnetic field produced by the Transmitter pad, allowing for efficient energy transfer. Wireless charging technology utilizes a high-frequency magnetic field for near-field operation, enabling energy delivery over short distances. An advancement of this method is Magnetic Resonance wireless charging technology, which works within resonant conditions, where a charging coupler is integrated with reactive components like capacitors and inductors, forming a compensation network to enhance efficiency. Capacitive wireless charging technology [13], on the other hand, uses electric fields to transfer energy. This method involves placing metallic plates at both the transmitter and receiver ends, with filaments coupling the source or load for effective energy transfer.

Wireless charging systems for EVs can be broadly categorized based on the vehicle's speed and positioning into three types: static, quasi-dynamic, and dynamic, as shown in Fig. 2. In static wireless charging, charging pads are fixed in specific locations, such as designated parking spaces. When an EV is parked over the transmitter pad, power is transferred inductively to the vehicle's receiver pad, allowing the

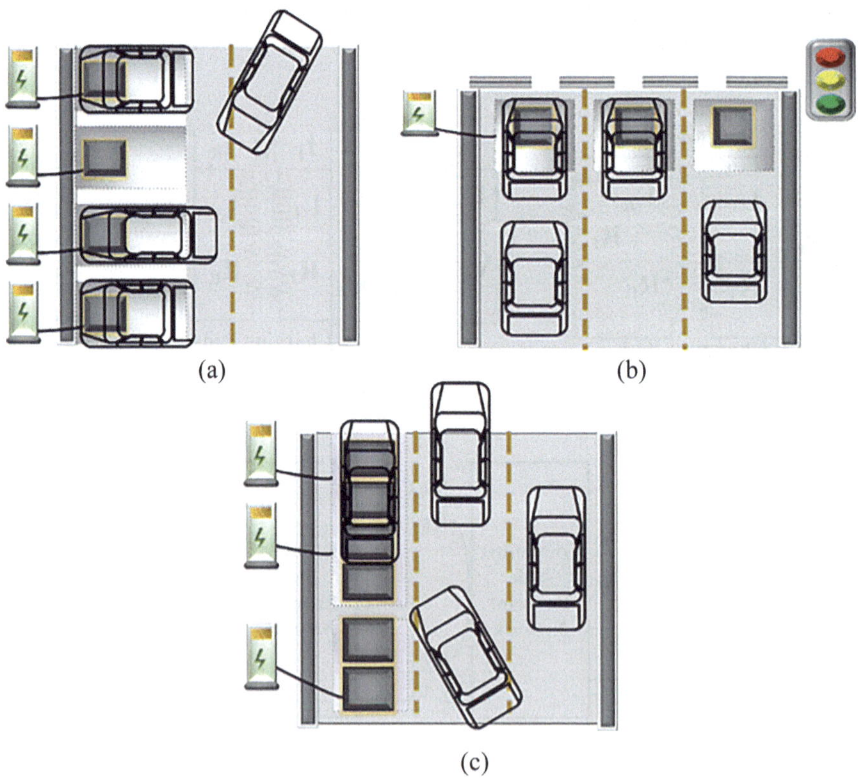

Fig. 2 (**a**) Stationary charging. (**b**) Quasi-dynamic charging. (**c**) Dynamic charging

battery to charge. These systems are particularly effective for scenarios where the vehicle is stationary for extended periods, such as overnight, making them ideal for convenient, uninterrupted charging [14].

A quasi-dynamic charging system enables EVs to recharge while either moving slowly or momentarily stationary, such as at traffic signals or in parking zones. These systems typically consist of a series of charging pads strategically placed along a predefined route, such as roadways or parking areas [15]. As the EV passes over these pads, power is wirelessly transferred to the vehicle, allowing for seamless, continuous charging even when the vehicle is in motion. Acting as a middle ground between static charging (where the vehicle is completely stationary) and dynamic charging (which occurs during continuous movement), quasi-dynamic systems are particularly useful for applications like electric buses or delivery trucks, where charging is needed on the move. DWC systems [16, 17] deliver continuous power to EVs while they are in motion. These systems typically involve embedding charging pads on the road surfaces for a particular distance to transfer energy to vehicles equipped with receivers. By enabling charging on the move, this technology offers the potential for extended driving ranges and smaller batteries, making it particularly suited for electrified transportation networks. DWC is applicable across a range of vehicles, from personal cars to commercial fleets. However, despite its promising potential, the commercialization of this technology has been hindered by significant challenges, including the need for highly efficient power transmission and the extensive infrastructure changes required, which could disrupt existing services. As the adoption of EVs accelerates around the world, smart cities are increasingly incorporating revolutionary charging methodologies to promote efficient and sustainable urban transportation. Among these, DWC systems stand out as a promising solution to address the challenges of limited charging infrastructure and range anxiety.

2.1 Literature Study of Dynamic Wireless Charging System

The technical feasibility of DWC for EVs was first assessed in 1976 at the Lawrence Berkeley National Laboratory [18] A significant milestone in the development of DWC came in 1979 with the Santa Barbara E-bus project [19]. This initiative used a 1-m wide pick-up mechanism and a 4.3-m track for power control, incorporating a reconfigurable hybrid capacitor. In 1992, the PATH program advanced this research by establishing an inductive power transfer facility to evaluate the system's viability for electric buses [18]. The project involved installing WPT rails along the bus route to study its environmental and economic impacts. The tests demonstrated successful power transfer of 60-kW with 60% efficiency over an air gap of 76-mm. Despite these promising results, the technology's commercialization faced challenges, including high costs, bulky coils, low efficiency, limited clearance for electric buses, and reduced operational frequency. In 1996, Conductix, a German-based company, in collaboration with the University of Auckland, developed a stationary

WPT system for electric buses [20]. This system, designed for charging buses when parked or stopped, featured a 60-kW power rating, an operating frequency of 15-20 kHz, a 40-mm air gap, and an impressive 90% efficiency. To maintain the required air gap, the receiver pad had to be lowered mechanically during charging. Later, in 2009, KAIST introduced the 3 kW On-Line Electric Vehicle technology, which included a mechanically regulated pickup capable of operating just 10 mm from the ground. This system achieved an efficiency of 80% while running at a frequency of 20 kHz [21].

Later that same year, KAIST developed a 60-kW OLEV system for buses, capable of functioning with a 170-mm air gap and achieving up to 72% efficiency, eliminating the need for a dynamic pickup. In addition to addressing stray electromagnetic fields, researchers made significant strides by boosting power output, reducing the weight of the pickup, and minimizing the size of the power rail. In 2016, Oak Ridge National Laboratory improved their earlier 10-20-kW WPT system for use with passenger vehicles [22], specifically a Toyota RAV4 equipped with a 10-kWh battery pack. Meanwhile, Utah State University's start-up, WAVE, showcased a 50-kW WPT system for electric buses. This system was capable of transferring power across a 150-300-mm air gap at a 20-kHz operating frequency, with a reported efficiency of 90%. Momentum Dynamics developed a WPT system capable of handling up to 200 kW, achieving 92% efficiency over a 610-mm air gap. This system has been effectively deployed for delivery vehicles, utilizing fixed charging stations embedded in docks or roadways. In 2016, Utah State University conducted [23] tests on a 25-kW DWC system for electric buses. The system featured a circular transmitter pad and was capable of delivering 25-kW of power. It also demonstrated a tolerance for lateral misalignment of up to ±150 mm.

3 Components of an On-Road or Dynamic Wireless Charging

The components of the DWC systems are organized into two main categories: The Transmitter-Side (Ground Assembly) and the Receiver-Side (Vehicle Assembly). On the transmitter side, the system components include a high-frequency, high-power inverter, a compensation network, and the transmitter pad. On the receiver side, the key elements are the receiver pad, a compensation network, a high-frequency high-power rectifier, and a battery system. The high-frequency inverter plays a crucial role by converting DC power into high-frequency AC power usually 85 kHz to energize the transmitter pads. The controller governs the operation of the inverter's switches, generating 85 kHz control pulses in alignment with the SAE J2954 standard [24–26]. When the receiver coil is properly aligned with the transmitter coil, the controller activates the inverter, causing the transmitter coil to become energized. This seamless interaction ensures efficient power transfer between the transmitter and receiver, enabling effective wireless charging for the

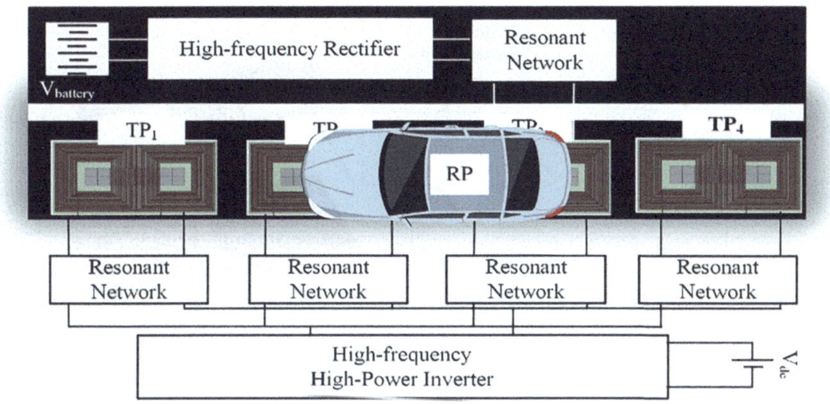

Fig. 3 Significant components of dynamic wireless charging

vehicle. The compensation network is a significant component of the charging system, designed to improve power transfer between charging couplers. Tuning the network to achieve resonance with the charging pad inductance, ensures maximum power transfer efficiency, even under misaligned conditions. Additionally, this network supports zero-voltage switching in the power inverter, improving overall system performance. At the heart of the WPT system, the charging coupler facilitates power transfer through an air gap to energize the storage device. When the receiving pad is properly aligned with the transmitting pad, energy flows seamlessly between them. On the receiving side, a rectifier unit efficiently converts high-frequency AC into DC, enabling the battery to charge reliably and efficiently. The additional DC-DC converter might be used to regulate the output of the rectifier concerning the output voltage and current rating (Fig. 3).

3.1 Charging Couplers

The design of coil structures for wireless charging systems is governed by various critical factors, including the desired power transfer distance, system power rating, coupling parameters such as inductance and quality factor, efficient ventilation to prevent overheating, tolerance for misalignment [27, 28], compatibility with diverse charging pad geometries, and cost-effectiveness [29]. These factors are represented in Fig. 4a, which outlines the factors in the design of wireless charging pads. Key components of these pads include High-frequency stranded Litz wire, the coil tray, the ferrite core with its holder, and aluminium shielding. Transmitter pad and receiver pad geometries are classified concerning the direction of magnetic flux generation. Coils generating flux vertically or horizontally are categorized as non-polarized, while those capable of producing flux in both directions are termed polarized. Non-polarized pad geometries, including square, rectangular, hexagonal, and

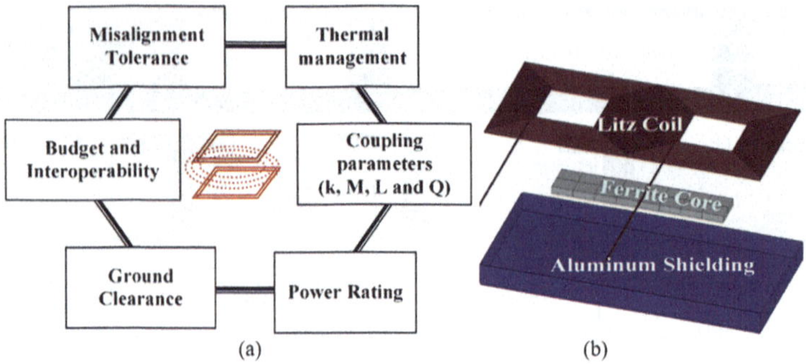

Fig. 4 (**a**) Significant factors of charging coupler design. (**b**) Components of Charging pads

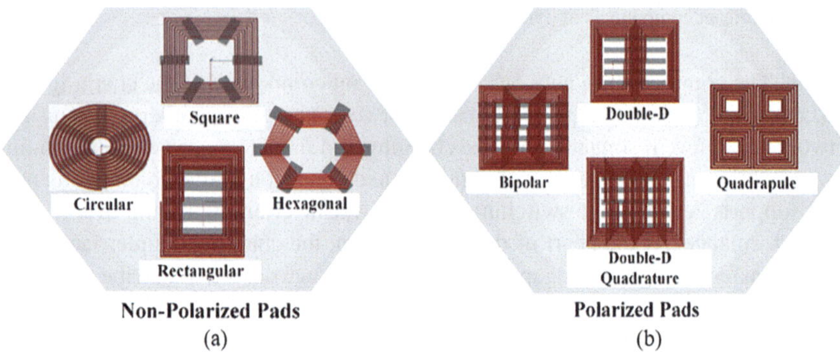

Fig. 5 (**a**) Non-polarized pads. (**b**) Polarized pads

circular shapes, focus their flux at the centre of the pad. In contrast, polarized pads such as DD-shaped configurations incorporate quadrature, bipolar, tripolar, or quadrupole designs. Figure 5 represents the different wireless charging pads [30].

For weakly coupled charging couplers, the leakage inductance typically exceeds the magnetizing inductance of the transmitter and receiver pads, which is a critical factor in the wireless charging pad design. Circular pads [31], with their compact and straightforward design, are well-suited for stationary charging applications. Their geometry, free of sharp corners, helps minimize eddy current losses and reduces leakage flux, enhancing overall efficiency. However, their limited tolerance for lateral and longitudinal misalignment makes them unsuitable for DWC systems, where consistent alignment is harder to maintain. To address these shortcomings, rectangular pads [32] emerge as a viable alternative. They offer high transmission efficiency while requiring less coil area and mass, making them effective in specific applications. However, this design introduces challenges such as elevated stray magnetic fields and lower core flux density, which can affect system performance.

DWC systems leverage innovative solutions like Double-D geometry pads to overcome these issues [33]. This design reduces stray magnetic fields by up to 50%

compared to rectangular pads, enhancing system safety and electromagnetic compatibility. Moreover, the Double-D configuration significantly improves magnetic coupling flexibility, enabling efficient energy transfer despite varying alignment conditions. This adaptability makes it a promising choice for dynamic charging scenarios, where precise alignment between transmitter and receiver is challenging to achieve. To enhance power transfer efficiency, a quadrature pad is strategically integrated with the Double-D pad, addressing the null flux region characteristic of the Double-D design. While this integration eliminates flux imbalances and improves performance, it necessitates a larger coil volume, which can increase material requirements [34]. On the other hand, the bipolar pad employs two overlapping rectangular coils, where effective current circulation cancels flux in critical areas. This design minimizes the need for higher volume conductors and significantly improves misalignment tolerance, a key factor in wireless charging reliability [35]. Incorporating a ferrite core into the charging coil further enhances the inductance and quality factor of the system, enabling higher power transfer efficiency. Proper optimization of coil geometry and core structure is essential to balance performance and practicality, ensuring the system remains compact, cost-effective, and efficient. A ferrite-based magnetic core structure uniformly distributes magnetic flux, reducing losses and maximizing energy transfer. Aluminium shielding is employed to mitigate electromagnetic interference, ensuring compliance with safety standards while maintaining system stability. By carefully designing the charging pad geometry, it is possible to achieve an optimal balance between weight, cost, and power transfer efficiency, paving the way for robust and scalable wireless charging solutions [36, 37].

3.2 Compensation Network

The compensation network, also known as the impedance matching or resonant network, is essential for enhancing power transfer efficiency in wireless charging systems, particularly in situations where alignment is not perfect. In a resonant inductive power transfer system, an uncompensated setup is limited to transferring only half of the rated power, leading to reduced overall efficiency [38]. The compensation network plays a key role in overcoming this limitation by optimizing power transfer and minimizing energy losses, ensuring better system performance. Its main objective is to reduce reactive power losses while maximizing the amount of power that can be transferred [39]. A simple mono-resonant network typically uses a single capacitor to achieve this compensation, as shown in Fig. 6a. More complex systems, known as multi-resonant networks, incorporate multiple passive components such as inductors and capacitors, allowing for improved performance in handling misalignments and higher-frequency harmonics, as depicted in Fig. 6b. These compensation networks are vital for ensuring the efficiency and reliability of wireless charging technologies [40].

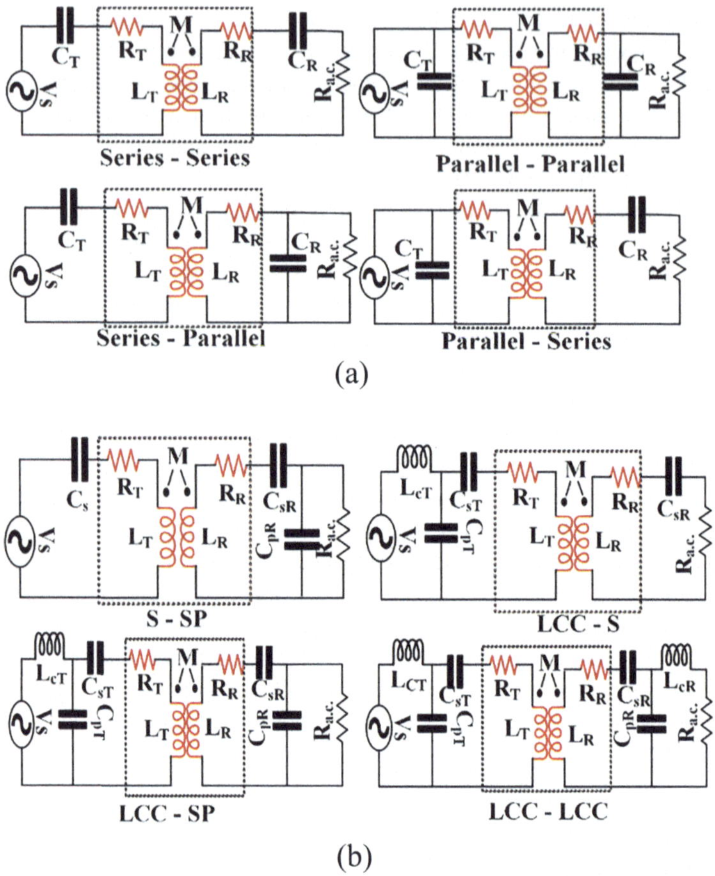

Fig. 6 (**a**) Mono-resonant network topologies. (**b**) Multi-resonant network topologies

The compensation network is essential to the efficient functioning of the charging pad system, performing several key tasks. It enables the charging pad to resonate at the target resonant frequency, regulates output parameters, and ensures a stable current flow within the transmitter pad. The selection of series or parallel compensation depends on the specific design and configuration of the charging pad. Series compensation, achieved by connecting a capacitor in series with the charging pad, is particularly advantageous for long, continuous transmitter pad lanes. In contrast, parallel compensation, where the capacitor is connected in parallel, is better suited for segmented transmitter pads and systems requiring high-current charging [41].

$$C_{sT\_SS} = \frac{1}{\omega_r^2 L_T} \tag{1}$$

$$C_{sT\_SP} = \frac{1}{\left(L_T - \frac{L_M^2}{L_R}\right)\omega_r^2} \tag{2}$$

$$C_{sT\_PS} = \frac{L_T}{\left(\frac{\omega_r^2 L_M^2}{R_{a.c.}}\right)^2 + \omega_r^2 L_T^2} \tag{3}$$

$$C_{sT\_PP} = \frac{L_T - \frac{L_M^2}{L_R}}{\left(\left(\frac{L_M^2 R_{a.c.}}{L_R^2}\right)^2 + \left(L_T - \frac{L_M^2}{L_R}\right)^2\right)\omega_r^2} \tag{4}$$

Key mono-resonant configurations, such as Series (S)-Series (S), Series (S)-Parallel (P), Parallel (P)-Parallel (P), and Parallel (P)-Series (S), are vital for efficient transmitter-receiver energy transfer. These configurations optimize the power delivery process, ensuring maximum efficiency in energy transfer between the transmitter and receiver. The expression for transmitter-side compensation capacitor, C_T within these fundamental topologies is characterized by the self-inductance of the transmitter pad, L_T, self-inductance of the receiver pad, L_R, angular resonant frequency, ω_r and mutual inductance between transmitter and receiver pad, L_M. The efficiency metrics for the transmitter pad (η_T) and receiver pad (η_R) are determined by the transmitter-side reflected impedance ($Z_{reflected}$) from the receiver pad side. The total impedance on the transmitter and receiver sides in the S-S compensation, Z_{T_SS} and Z_{R_SS} are represented in (5) and (6) respectively [40].

$$Z_{T\_SS} = j\omega L_T + \frac{1}{j\omega C_{sT}} + R_{Teq} \tag{5}$$

$$Z_{R\_SS} = j\omega_r L_R + \frac{1}{j\omega_r C_{sR}} + R_{Req} + R_{a.c.} \tag{6}$$

$$\eta_T = \frac{\mathrm{Re}\left(Z_{reflected}\right)}{R_{Teq} + \mathrm{Re}\left(Z_{reflected}\right)} \tag{7}$$

$$\eta_s = \frac{\mathrm{Re}\left(Z_R\right) - R_{Req}}{\mathrm{Re}\left(Z_R\right)} \tag{8}$$

$$Z_{reflected} = \frac{\omega_r^2 n^2 L_M^2}{Z_R} \quad where, \quad n = \frac{L_R}{L_T} \tag{9}$$

where R_{Teq} and R_{Req} represent equivalent circuits belonging to the transmitter side and receiver side an equivalent circuit. The suffix T and R represents the transmitter and receiver side of an equivalent circuit. Choosing the suitable resonant network topology is crucial for designing efficient WPT systems in EV charging. The four basic configurations mentioned above S-S, P-S, S-P, and P-P—each offer unique advantages and considerations related to efficiency, voltage requirements, impedance variations, and ease of implementation. The S-S network stands out for its ability to maintain a constant output current, regardless of load changes or resonant frequency shifts. This characteristic ensures high efficiency and requires low DC link voltage, making it ideal for EV charging applications where stable output is essential. In contrast, the P-S [42] configuration demands higher voltage, experiences increased impedance with misalignment and requires an input current source with significant input resistance. While it is efficient, it may struggle in applications with varying load conditions. The S-P network, operating at low DC link voltage, reduces energy losses and impedance changes caused by misalignment [42]. It also benefits from smaller receiver coil self-inductance, making it efficient in such scenarios. However, it may not be as effective in handling large input currents, limiting its use in high-power applications. The P-P network [41], on the other hand, experiences increased impedance with misalignment, which is affected by both the coupling coefficient and load resistance. While it requires a large current source at the input and considerable resistance, it can achieve high efficiency when precisely matched to the load. However, its practicality becomes questionable in situations involving fluctuating load conditions and misalignment. In contrast, the S-SP [43] network offers a more versatile and efficient solution for WPT in EV charging systems. It ensures consistent gain, regardless of changes in load or coupling coefficients, and effectively reduces circulating losses even as parameters vary. This adaptability makes it an excellent choice for EV charging applications, where efficiency, reliability, and flexibility are critical. Ultimately, the selection of resonant network topology plays a pivotal role in optimizing the design and performance of WPT systems for EVs.

Each configuration brings its own set of strengths and trade-offs, but the S-SP network stands out for its consistent performance, low losses, and broad applicability. These insights are essential for improving EV charging systems and supporting the development of sustainable transportation solutions. Under a multi-resonant network, the LCC-LCC network [44] offers the advantage of Zero Current Switching and demonstrates resilience to variations in inductance, though its electrical performance remains somewhat dependent on inductive parameters. In comparison, the LCC-S network [45] provides a range of benefits, including greater design flexibility, steady output current, simplified Zero Voltage Switching operation, and the elimination of control circuits. These attributes make the LCC-S network a versatile and efficient choice for designers. The LCC-SP network [46] distinguishes itself with exceptional misalignment tolerance, simplifying the alignment between transmitter and receiver coils. It also supports soft switching, ensuring smooth transitions during the charging process. Additionally, the LCC-SP topology allows for extensive design customization and effectively mitigates power transmission fluctuations, enhancing its suitability for practical applications. The LCL-LCL network [47], in turn, ensures consistent power delivery regardless of load variations by maintaining a stable transmitter current. However, it may struggle with suppressing higher-order harmonics, potentially affecting the efficiency and quality of power transfer. Lastly, the LCC-P [48] network stands out for its compact receiver design compared to LCC-LCC configurations, making it ideal for applications where space is limited. This topology also boasts higher efficiency and reduced inverter conduction losses compared to S-S systems, positioning it as an excellent choice for EV charging systems focused on maximizing efficiency and minimizing energy losses.

3.3 *High-Frequency High-Power Inverter*

In a wireless charging system for EVs, a high-frequency, high-power inverter plays a crucial role in controlling the excitation of the transmitter pads. These inverters can energize one or multiple transmitter pads simultaneously, enabling efficient energy transfer. On the receiver side, a power converter ensures a stable DC output to charge the vehicle's energy storage system [11]. To enhance system efficiency and reliability, it is essential to minimize switching losses in the inverter. Advanced control strategies and optimized inverter designs are often employed to achieve seamless power transfer and improved performance. Power converters are essential for exciting the charging pads in a wireless charging system. When designing these converters, key factors such as current and voltage ratings, as well as the number of charging pads to be supported, must be carefully considered. A high-frequency power inverter supplies the output to the transmitting pad through a resonant network.

The selection of power semiconductor switches depends on the system's required rated power and resonant frequency. The FPGA controller controls the generation of excitation pulses for these switches, based on input from the detection circuit. It is crucial to synchronize these pulses with the velocity of the receiving pad. The transmitter-side power inverter must ensure an efficient energization of the charging lane, reduce switching stress and losses, mitigate electromagnetic interference and harmonics, and incorporate protection circuits for system stability. Soft-switching techniques, combined with compensation networks, are effective in minimizing switching stresses on power semiconductor switches in wireless charging systems. The inverter's driver circuit generates high-frequency control signals based on the receiver pad position. Wide bandgap devices like GaN and SiC switches play a key role in achieving high power and frequency due to their advantages, such as low ON-state resistance and excellent thermal stability. Figure 7 shows widely used power inverter topologies used in static and dynamic charging applications.

The conventional two-leg H-bridge inverter is more convenient for both stationary and DWC systems. This topology uses four switches regardless of the number of transmitter pads, which simplifies switching control. However, when a single H-bridge inverter drives multiple transmitter pads, it requires switches with higher

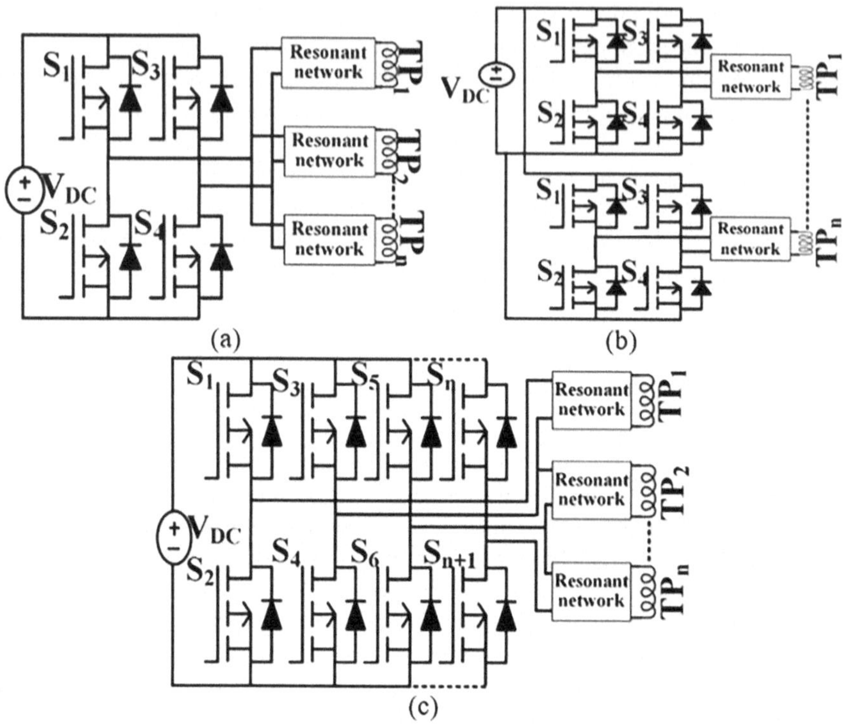

Fig. 7 Basic power converter topologies on transmitter-side. (**a**) Conventional H-bridge inverter. (**b**) Modular inverter. (**c**) Multi-legged inverter

power ratings and introduces operational challenges due to the shared architecture [49]. An alternative is to use multiple separate H-bridge inverters [50], each dedicated to a single transmitter pad. While this approach allows independent control of each transmitter, it also increases system size, cost, and control complexity due to the need to manage multiple inverters [51] (Lingxiao [52]).

A novel multi-legged inverter [53, 54] methodology is proposed to address the challenges of conventional H-bridge inverter-fed wireless charging systems. This approach minimizes the number of switches required for the transmitters while maintaining high efficiency. The multi-legged inverter [55] features multiple legs, each associated with a transmitting pad, thereby reducing the switch count compared to traditional multiple H-bridge inverters. However, the design introduces control complexities due to the interdependence and synchronization required between the legs. Despite this, the methodology offers a compelling solution by optimizing power delivery and addressing the inefficiencies of standard inverter configurations. By adopting the multi-legged inverter, systems can achieve reduced switch counts, improved efficiency, and cost-effectiveness in applications involving multiple transmitter pads. Further advancements in this area could enhance overall system performance and reliability in DWC applications. In a conventional

PWM-based H-bridge inverter, the instantaneous output voltage (V_i) is regulated by adjusting the input DC voltage, as expressed in Eq. (10) [56].

$$V_i(t)=\sum_{n=1}^{\infty}\frac{4V_{D.C.}\sin\left(\frac{k\pi}{2}\right)\cos\left(\frac{k\alpha}{2}\right)}{k\pi}\sin\left(kf_s t\right) \tag{10}$$

For the fundamental harmonic k=1, V_i (t) becomes

$$V_i=\frac{4V_{D.C.}}{\sqrt{2}\pi}=0.9\times V_{D.C.} \tag{11}$$

3.4 *AC-DC Converter or High-frequency High-Power Rectifiers*

In wireless charging systems, the receiver-side compensation network is directly connected to the rectifier unit, which plays a key role in converting high-frequency AC into DC power. There are three primary types of rectifiers commonly used: uncontrolled, semi-active bridge, and active bridge rectifiers [57]. The uncontrolled rectifier simply converts AC to DC without any additional modifications. To reduce the high-frequency ripples that are typical in uncontrolled systems, Schottky diodes are often used in these rectifiers. Schottky diodes are chosen for their low forward voltage drop and fast switching characteristics, which help minimize losses and improve efficiency, particularly in high-frequency and high-power applications such as wireless charging. The semi-active bridge rectifier [58, 59], on the other hand, helps to reduce output ripples further, improving efficiency and reducing losses compared to uncontrolled rectifiers. However, it introduces some control-related losses due to its reliance on switching devices. Active bridge rectifiers [29, 60] offer the most advanced solution by minimizing output ripples, achieving the highest efficiency, and reducing overall losses. This is made possible through advanced control algorithms and high-frequency switching techniques, which are particularly effective for high-power wireless charging applications where bidirectional power flow and precise output regulation are critical.

In high-frequency, high-power wireless charging systems, the choice of rectifier significantly impacts system performance. Uncontrolled rectifiers, while simple and cost-effective, lead to high ripples and inefficiencies. Semi-active rectifiers offer a balance of efficiency and ripple reduction but at the cost of added complexity. Active rectifiers, however, provide the best overall performance by offering low ripple, high efficiency, and minimal losses, making them ideal for high-power wireless charging applications. The selection of the rectifier depends on specific application needs, and balancing factors such as ripple suppression, system efficiency, and operational constraints.

3.5 Different Modulation Topologies for Power Converters

The high-frequency inverter on the transmitter side drives the transmitter pad, while the controlled rectifier on the receiver side regulates the battery charging cycle. In a typical wireless charging system, the power converters are controlled by excitation pulses that follow different modulation techniques, such as Phase Shift Modulation [61], Conventional Pulse Width Modulation, or Pulse Density Modulation. Figure 8 represents the pulse patterns of these modulation methodologies which are used by power converters. The relationship between the input DC voltage and the output AC voltage is defined by Eq. (11), which holds when the inverter operates with a conventional pulse width modulation signal. When using phase shift modulation, the inverter's output voltage (V_i) is directly proportional to the cosine of the phase shift angle (α), expressed as $V_i \propto V_{DC}^{*} \cos(\alpha)$. In the case of pulse density modulation, the output voltage (V_i) is proportional to the pulse density (d), or $V_i \propto V_{DC}^{*}$ d. The phase shift modulation technique is typically chosen to enhance the soft-switching performance of power semiconductor switches, while the pulse density modulation technique is employed to reduce output voltage ripples and provide better output power regulation (Fig. 9).

Conventional pulse width modulation technique is widely used for its simplicity and precise control over power delivery by adjusting pulse width. However, it can lead to higher switching losses, particularly at high frequencies, and may require soft switching methods to mitigate these losses. Phase shift modulation, by contrast,

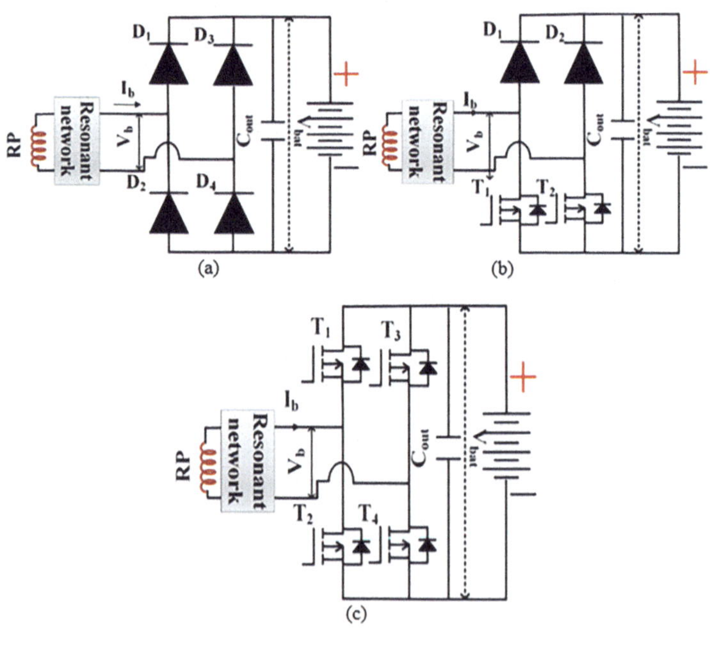

Fig. 8 (**a**) Uncontrolled rectifiers. (**b**) Semi-active bridge rectifier. (**c**) Active bridge rectifier

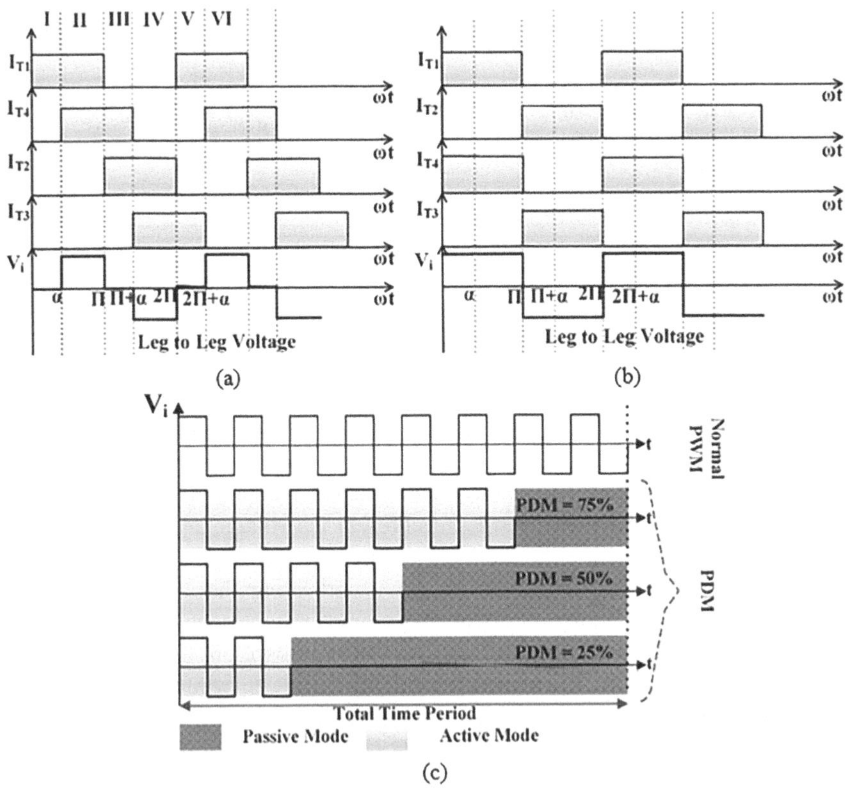

Fig. 9 (**a**) is Phase shift modulation, (**b**) Conventional pulse width modulation and (**c**) Pulse density modulation

modulates power delivery by shifting the phase of control signals, making it highly efficient at reducing switching losses, especially in high-frequency applications. However, phase shift modulation can be more complex and less precise compared to conventional pulse width modulation. A Pulse density modulation technique [62, 63] adjusts the density of pulses to control output power, reducing switching losses while improving efficiency. However, it may result in higher output fluctuations and less control precision.

4 Basic Design Equations of the Wireless Charging System

The basic block diagram of a wireless charging system is illustrated in Fig. 10. The block diagram has two sub-divided parts: Transmitter-side and Receiver-side. On the transmitter side, a high-power DC input is converted into high-frequency AC whereas, on the receiver side, the high-frequency AC is converted into high-power DC with the help of power converters. In the figure, V_{DC} represents the input DC

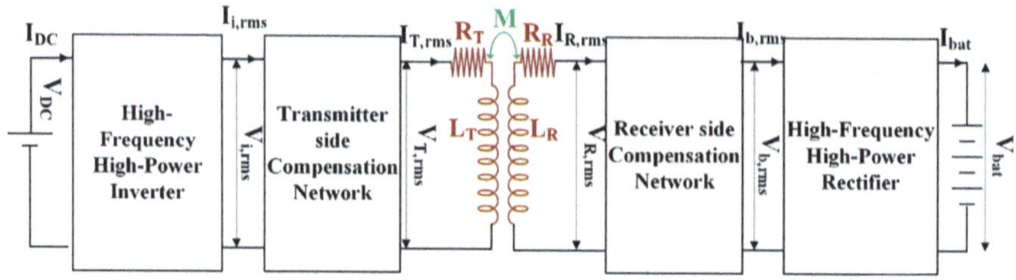

Fig. 10 Basic block diagram of wireless charging system

voltage, while I_{DC} signifies the input direct current. P_{in} denotes the input direct current power. $V_{i,rms}$ denotes the output voltage of the inverter, with f_r representing the resonant frequency. $I_{i,rms}$ stands for the inverter's output current, while $V_{T,rms}$ and $I_{T,rms}$ refer to the voltage and current, respectively, at the transmitter pad. Q_s indicates the system quality factor. L_T and R_T represent the transmitter pad's self-inductance and resistance, respectively. R_R and L_R signify the receiver pad's resistance and self-inductance, respectively. $V_{R,rms}$ and $I_{R,rms}$ represent the voltage and current, respectively, at the receiver pad. $V_{b,rms}$ and $I_{b,rms}$ denote the alternating current equivalent voltage and current, respectively, with $R_{b\,A.C.}$ indicating the alternating current equivalent resistance. These equivalent AC parameters represent the output battery parameters before the rectification stage. V_{bat} and I_{bat} represent the battery voltage and current, respectively. P_{out} stands for the output power, while R_{bat} denotes the equivalent battery resistance. k represents the coupling coefficient, while k_c represents the critical coupling. Equations (12)–(17) are used to determine the charging coupler inductance concerning the output power and voltage. The main factors in designing a charging coupler are the system's quality factor, the equivalent battery resistance, and the resonant frequency [64]. These elements are essential for achieving efficient coupling. The required output voltage and power determine the equivalent resistance [65, 66].

$$R_{bat} = \frac{V_{bat}^2}{P_{out}} \quad \text{and} \quad R_{b\,A.C.} = \frac{8}{\pi^2} \times R_{bat} \tag{12}$$

$$V_{i,rms} = \frac{2\sqrt{2}}{\pi} \times V_{D.C.} \quad \text{and} \quad V_{b,rms} = \frac{2\sqrt{2}}{\pi} V_{bat} \tag{13}$$

$$I_{i,rms} = \frac{P_{out}}{V_{i,rms}} \quad \text{and} \quad I_{b,rms} = \frac{V_{b,rms}}{R_{b\,A.C.}} \tag{14}$$

$$L_M = \frac{I_{b,rms} \times R_{b\,A.C.}}{I_{i,rms} \times \omega_r}, \quad \omega_r = 2 \times \pi \times f_r \tag{15}$$

$$k_c = \frac{1}{Q_s}\sqrt{1 - \frac{1}{\left(4Q_s^2\right)}}, \quad k \ll k_c \tag{16}$$

$$L_T = \frac{L_M^2}{L_R k^2} \quad \text{and} \quad L_R = \frac{Q_s \times R_{bA.C.}}{\omega_r} \tag{17}$$

5 Standards of Wireless Charging System for an EV

The SAE J2954 standard, developed by the SAE, provides a comprehensive framework for WPT in EVs, focusing on key aspects such as interoperability, safety, and performance. It defines the necessary criteria to ensure seamless compatibility between various vehicles and charging stations, with the WPT frequency standardized at 85 kHz, as detailed in Table 1. The standard specifies multiple charging power levels, ranging from 3.7 kVA to a projected maximum of 22 kVA. Furthermore, it establishes guidelines for vehicle alignment and positioning relative to the charging pad to ensure optimal power transfer, allowing for flexibility to accommodate variations while maintaining consistent performance. It specifies the allowable ground clearance range (Z-Class) between the vehicle-side assembly and the ground-side assembly and sets tolerances for misalignment to ensure the system maintains the required minimum efficiency. Safety and electromagnetic compatibility are central considerations, with strict measures to limit electromagnetic field exposure to safe levels and prevent interference with nearby electronic devices. Standards also define robust communication protocols to manage the charging process, including initiating, adjusting, and ending sessions, ensuring seamless information exchange between components. Performance criteria include minimum efficiency benchmarks, thermal management, and durability under various environmental conditions. For 85 kHz wireless EV charging, international guidelines prioritize minimizing EMF exposure to protect both public and occupational health. These standards specify limits for magnetic fields, and in some cases, electric fields,

Table 1 SAE J2954 standards for wireless charging systems for an EV [25, 67]

	Power class of wireless charging system			
	Class 1	Class 2	Class 3	Class 4
Max. input kVA	3.7	7.7	11.1	22
input AC voltage range (V)	120	240	240	240
Input AC phase (ϕ)	1	1	3	3
Ground Clearance (Gap between transmitter and receiver pad) (mm)	100–150	140–210	170–250	150–300
Minimum achievable efficiency range at the nominal lateral (Δx) and longitudinal (Δy) misalignment (%)	>85	>85	>85	To be decided
Offset minimum target efficiency (%)	>80	>80	>80	To be decided
Tolerable misalignment space concerning the offset direction (mm) (Δx, Δy)	(±75, ±100)			
Operating frequency range (kHz)	81.38–90.00			

Table 2 International standards for EMF exposure limitations to the human level [68]

Standards	Magnetic field (10^{-6} Tesla)		Electric field, V/m	
	Public	Occupational	Public	Occupational
ICNIRP 2010 (1 Hz–100 kHz)	27	100	83	170
IEEE C.95.1 2014 (3 kHz–5 MHz)	205	615	614	1842
ISO 19363:2018	2.5	2.5	–	–

generated by WPT systems, as detailed in Table 2. By addressing safety, efficiency, and operational reliability, these standards enable the safe adoption of wireless charging technologies while ensuring their effectiveness and sustainability.

6 Conclusion

In conclusion, this chapter has provided an in-depth technical analysis of dynamic wireless charging systems, highlighting their critical role in advancing EV charging infrastructure within smart city ecosystems. The chapter thoroughly investigates the core components, including power conversion units, inductive coils, resonant networks, and the associated design formulations, offering a comprehensive understanding of the operational principles underpinning DWC technology. The synergistic integration of wireless power transfer with DWC facilitates autonomous, continuous charging, which is essential for the development of sustainable, low-carbon transportation solutions within smart cities. This innovation not only mitigates the reliance on oversized battery storage but also optimizes both spatial and cost efficiency. By bridging the gap between theoretical models and practical system design, this chapter contributes critical insights into the development of high-performance DWC systems. These advancements are pivotal in advancing urban mobility, reducing environmental impact, and supporting the sustainable evolution of smart city infrastructures. Also, the chapter discussed design equations and standards of the DWC system that will help the original equipment manufacturers develop an efficient charging system with high reliability.

References

1. Al-Alawi BM, Bradley TH (2013) Review of hybrid, plug-in hybrid, and electric vehicle market modeling studies. Renew Sust Energ Rev 21:190–203. https://doi.org/10.1016/j.rser.2012.12.048
2. Xue Y, Zhang X, Zhang Y, Luo E (2024) Understanding the barriers to consumer purchasing of electric vehicles: the innovation resistance theory. Sustainability 16(6):2420. https://doi.org/10.3390/su16062420

3. Joseph PK, Elangovan D, Sanjeevikumar P (2021) System architecture, design, and optimization of a flexible wireless charger for renewable energy-powered electric bicycles. IEEE Syst J 15(2):2696–2707. https://doi.org/10.1109/JSYST.2020.2993054
4. Shanmugam Y, Narayanamoorthi R, Vishnuram P, Bajaj M, Aboras KM, Thakur P, Kitmo. (2022) A systematic review of dynamic wireless charging system for electric transportation. IEEE Access 10:133617–133642. https://doi.org/10.1109/ACCESS.2022.3227217
5. Yuvaraja S, Narayanamoorthi R, Ali JSM, Almakhles D (2022) A comprehensive review of the on-road wireless charging system for E-mobility applications. Front Energy Res 10. https://doi.org/10.3389/fenrg.2022.926270
6. Bai HK, Costinett D, Tolbert LM, Qin R, Zhu L, Liang Z, Huang Y (2022) Charging electric vehicle batteries: wired and wireless power transfer: exploring EV charging technologies. IEEE Power Electron Magaz 9(2):14–29. https://doi.org/10.1109/MPEL.2022.3173543
7. Ahmad F, Alam MS, Alsaidan IS, Shariff SM (2020) Battery swapping station for electric vehicles: opportunities and challenges. IET Smart Grid 3(3):280–286. https://doi.org/10.1049/iet-stg.2019.0059
8. Revankar SR, Kalkhambkar VN (2021) Grid integration of battery swapping station: a review. J Energy Storage 41. https://doi.org/10.1016/j.est.2021.102937
9. Purushothaman D, Narayanamoorthi R, Ramachandramurthy VK (2024) Charging ahead: unlocking the potential of constant voltage and constant current modes in WPT for EVs. J Energy Storage 96. https://doi.org/10.1016/j.est.2024.112603
10. Purushothaman D, Narayanamoorthi R, Elrashidi A, Kotb H (2023) A comprehensive review on single-stage WPT converter topologies and power factor correction methodologies in EV charging. IEEE Access 11:135529–135555. https://doi.org/10.1109/ACCESS.2023.3338164
11. Ramadoss V, Balaji Chandrasekar MMR, Ahmed DA, Savio NR, Alghamdi TAH (2024) Research insights on recent power converter topologies and control strategies for wireless EV chargers: a comprehensive study. IEEE Open J Power Electron. https://doi.org/10.1109/OJPEL.2024.3474707
12. Mahesh A, Chokkalingam B, Mihet-Popa L (2021) Inductive wireless power transfer charging for electric vehicles-a review. IEEE Access 9:137667–137713. https://doi.org/10.1109/ACCESS.2021.3116678
13. Lu F, Zhang H, Mi C (2018) A two-plate capacitive wireless power transfer system for electric vehicle charging applications. IEEE Trans Power Electron 33(2):964–969. https://doi.org/10.1109/TPEL.2017.2735365
14. Mohamed AAS, Shaier AA, Metwally H, Selem SI (2020) A comprehensive overview of inductive pad in electric vehicles stationary charging. Appl Energy 262. https://doi.org/10.1016/j.apenergy.2020.114584
15. Mohamed AAS, Lashway CR, Mohammed O (2017) Modeling and feasibility analysis of quasi-dynamic WPT system for EV applications. IEEE Trans Transp Electrific 3(2):343–353. https://doi.org/10.1109/TTE.2017.2682111
16. Marco D, Davide AD, Longo M, Yaïci W (2019) Design and performance analysis of pads for dynamicwireless charging of EVs using the finite element method. Energies 12(21). https://doi.org/10.3390/en12214139
17. Mou X, Zhang Y, Jiang J, Sun H (2019) Achieving low carbon emission for dynamically charging electric vehicles through renewable energy integration. IEEE Access 7:118876–118888. https://doi.org/10.1109/ACCESS.2019.2936935
18. News (n.d.) Charging Electric Buses Quickly and Efficiently: Bus Stops Fitted with Modular Components Make 'Charge & Go' Simple to Implementl Conductix Wampfler Global. https://www.conductix.com/en/news/2013-05-29/charging-electric-buses-quickly-and-efficiently-bus-stops-fitted-modular-components-make-charge-go. Accessed 3 June 2024
19. Systems Control Technology, Inc. (1994) Roadway Powered Electric Vehicle Project Track Construction and Testing Program Phase 3D. Berkely, California. https://escholarship.org/uc/item/1jr98590

20. Du B, Sreenivasan AV, Sheng MS, Ranjitkar P, Wen L, Sharp B, Raith A (2024) Optimal deployment of dynamic wireless charging infrastructure for battery electric buses–a case study in Auckland, New Zealand. J Roy Soc New Zealand. https://doi.org/10.1080/03036758.2024.2353737
21. Suh NP, Cho DH, Rim CT (2010) Global product development - design of On-Line Electric Vehicle (OLEV). In: Bernard A (ed) Proceedings of the 20th CIRP Design Conference, Ecole Centrale de Nantes. Springer Berlin Heidelberg, Nantes, France, pp 3–8. https://doi.org/10.1007/978-3-642-15973-2
22. Miller JM, Jones PT, Li JM, Onar OC (2015) ORNL experience and challenges facing dynamic wireless power charging of EV's. IEEE Circuits Syst Magaz 15(2):40–53. https://doi.org/10.1109/MCAS.2015.2419012
23. Tavakoli R, Pantic Z (2018) Analysis, design, and demonstration of a 25-KW dynamic wireless charging system for roadway electric vehicles. IEEE J Emerg Select Topics Power Electron 6(3):1379–1393. https://doi.org/10.1109/JESTPE.2017.2761763
24. J2954 (n.d.) (WIP) Wireless Power Transfer for Light-Duty Plug-in/Electric Vehicles and Alignment Methodology - SAE International. https://www.sae.org/standards/content/j2954. Accessed 27 Mar 2024
25. Lin F, Covic GA, Kesler M (2020) Design of a SAE compliant multicoil ground assembly. IEEE J Emerg Select Topics Indust Electron 1(1):14–25. https://doi.org/10.1109/jestie.2020.2999597
26. Nimri RM, Kamineni A, Zane R (2020) A modular pad design compatible with SAE J2954 for dynamic inductive power transfer. In: 2020 IEEE PELS Workshop on Emerging Technologies: Wireless Power Transfer, WoW 2020, pp 45–49. https://doi.org/10.1109/WoW47795.2020.9291322
27. Ramakrishnan V, Dominic Savio A, Balaji C, Narayanamoorthi R, Vishnuram P, Yang T, Bajaj M, Rathore RS, Zaitsev I (2024) Design and implementation of a high misalignment-tolerance wireless charger for an electric vehicle with control of the constant current/voltage charging. Sci Rep 14(1). https://doi.org/10.1038/s41598-024-63952-6
28. Venkatesan M, Narayanamoorthi R, Emara A, Ghadi YY (2024) Fuzzy logic controlled pulse density modulation technique for bidirectional inductive power transfer systems. IEEE Access 1–1. https://doi.org/10.1109/ACCESS.2024.3388491
29. Shanmugam Y, Rajamanickam N, Alroobaea R, Afandi A (2024) Driving towards sustainability: wireless charging of low-speed vehicles with PDM-based active bridge rectifiers. Sustainability (Switzerland) 16(9). https://doi.org/10.3390/su16093810
30. Rahulkumar J, Narayanamoorthi R, Vishnuram P, Bajaj M, Blazek V, Prokop L, Misak S (2023) An empirical survey on wireless inductive power pad and resonant magnetic field coupling for in-motion EV charging system. IEEE Access 11:4660–4693. https://doi.org/10.1109/ACCESS.2022.3232852
31. Yilmaz T, Hasan N, Zane R, Pantic Z (2017) Multi-objective optimization of circular magnetic couplers for wireless power transfer applications. IEEE Trans Magnetics 53(8). https://doi.org/10.1109/TMAG.2017.2692218
32. Benalia N, Laroussi K, Benlaloui I, Kouzou A, Bensalah A-D, Kennel R, Abdelrahem M (2023) Optimized power pads for charging electric vehicles based on a new rectangular spiral shape design. Sustainability 15(2):1230. https://doi.org/10.3390/su15021230
33. Luo Z, Wei X, Pearce MGS, Covic GA (2021) Multiobjective optimization of inductive power transfer double-D pads for electric vehicles. IEEE Trans Power Electron 36(5):5135–5146. https://doi.org/10.1109/TPEL.2020.3029789
34. Raabe S, Elliott GAJ, Covic GA, Boys JT (2007) A quadrature pickup for inductive power transfer systems. In: 2007 2nd IEEE Conference on Industrial Electronics and Applications. IEEE, Harbin, China. https://doi.org/10.1109/ICIEA.2007.4318371
35. Jafari H, Olowu TO, Mahmoudi M, Sarwat A (2021) Optimal design of IPT bipolar power pad for roadway-powered EV charging systems. IEEE Canad J Electric Comput Eng 44(3):350–355. https://doi.org/10.1109/ICJECE.2021.3075639

36. Lin FY, Covic GA, Boys JT (2015) Evaluation of magnetic pad sizes and topologies for electric vehicle charging. IEEE Trans Power Electron 30(11):6391–6407. https://doi.org/10.1109/TPEL.2015.2419592
37. Nagendra GR, Covic GA, Boys JT (2017) Sizing of inductive power pads for dynamic charging of EVs on IPT highways. IEEE Trans Transp Electrific 3(2):405–417. https://doi.org/10.1109/TTE.2017.2666554
38. Liu YC, Zhang J, Tse CK, Zhu C, Wong SC (2021) General pathways to higher order compensation circuits for IPT converters via sensitivity analysis. IEEE Trans Power Electron 36(9):9897–9906. https://doi.org/10.1109/TPEL.2021.3062228
39. Villa JL, Sallán J, Osorio JFS, Llombart A (2012) High-misalignment tolerant compensation topology for ICPT systems. IEEE Trans Indust Electron 59(2):945–951. https://doi.org/10.1109/TIE.2011.2161055
40. Sohn YH, Choi BH, Lee ES, Lim GC, Cho GH, Rim CT (2015) General unified analyses of two-capacitor inductive power transfer systems: equivalence of current-source SS and SP compensations. IEEE Trans Power Electron 30(11):6030–6045. https://doi.org/10.1109/TPEL.2015.2409734
41. Shevchenko V, Husev O, Strzelecki R, Pakhaliuk B, Poliakov N, Strzelecka N (2019) Compensation topologies in IPT systems: standards, requirements, classification, analysis, comparison and application. IEEE Access 7:120559–120580. https://doi.org/10.1109/ACCESS.2019.2937891
42. Lu J, Zhu G, Lin D, Wong S-C, Jiang J (2019) Load-independent voltage and current transfer characteristics of high-order resonant network in IPT system. IEEE J Emerg Select Topics Power Electron 7(1):422–436. https://doi.org/10.1109/JESTPE.2018.2823782
43. Hou J, Chen Q, Ren X, Ruan X, Wong SC, Tse CK (2015) Precise characteristics analysis of series/series-parallel compensated contactless resonant converter. IEEE J Emerg Select Topics Power Electron 3(1):101–110. https://doi.org/10.1109/JESTPE.2014.2320524
44. Lu J, Zhu G, Li W, Li B (2018) Load-independent ZPA conditions in both constant current and constant voltage modes of LCC-series compensated IPT system. In: 2018 IEEE Wireless Power Transfer Conference (WPTC). https://doi.org/10.1109/WPT.2018.8639453
45. Chen Y, Zhang H, Park SJ, Kim DH (2019) A switching hybrid LCC-S compensation topology for constant current/voltage EV wireless charging. IEEE Access 7:133924–133935. https://doi.org/10.1109/ACCESS.2019.2941652
46. Xiong M, Dai H, Li Q, Jiang Z, Luo Z, Wei X (2021) Design of the LCC-SP topology with a current doubler for 11-KW wireless charging system of electric vehicles. In: IEEE Transactions on Transportation Electrification. Institute of Electrical and Electronics Engineers Inc., pp 2128–2142. https://doi.org/10.1109/TTE.2021.3074007
47. Yao Y, Wang Y, Álvarez JMA, Liu X, Cheng H, Dianguo X (2019) Efficiency-based design optimisation of a double-sided LCL compensated wireless power transfer system. IET Power Electron 12(10):2436–2446. https://doi.org/10.1049/iet-pel.2018.5817
48. Yan Z, Zhang Y, Song B, Zhang K, Kan T, Mi C (2019) An LCC-p compensated wireless power transfer system with a constant current output and reduced receiver size. Energies 12(1). https://doi.org/10.3390/en12010172
49. Mai R, Chen Y, Li Y, Zhang Y, Cao G, He Z (2017) Inductive power transfer for massive electric bicycles charging based on hybrid topology switching with a single inverter. IEEE Trans Power Electron 32(8):5897–5906. https://doi.org/10.1109/TPEL.2017.2654360
50. Lee ES, Kim MY, Kang SM, Han SH (2022) Segmented IPT coil design for continuous multiple charging of an electrified monorail system. IEEE Trans Power Electron 37(3):3636–3649. https://doi.org/10.1109/TPEL.2021.3115511
51. Lee EJ, Kim SM, Lee KB (2020) Modified phase-shifted PWM scheme for reliability improvement in cascaded H-bridge multilevel inverters. IEEE Access 8:78130–78139. https://doi.org/10.1109/ACCESS.2020.2989694
52. Xue L, Galigekere V, Gurpinar E, Gui-jia S, Onar O (2021) Modular design of receiver side power electronics for 200 KW high power dynamic wireless charging system. In: 2021 IEEE

Transportation Electrification Conference & Expo (ITEC). IEEE, Chicago, IL, USA. https://doi.org/10.1109/ITEC51675.2021.9490095

53. Farajizadeh F, Mahinda Vilathgamuwa D, Jovanovic D, Jayathurathnage P, Ledwich G, Madawala U (2020) Expandable N-legged converter to drive closely spaced multitransmitter wireless power transfer systems for dynamic charging. IEEE Trans Power Electron 35(4):3794–3806. https://doi.org/10.1109/TPEL.2019.2939848
54. Tian Y, Tian J, Li D, Zhou S (2018) A multiple legs inverter with real time-reflected load detection used in the dynamic wireless charging system of electric vehicles. Energies 11(5). https://doi.org/10.3390/en11051275
55. Shanmugam Y, Narayanamoorthi R, Dominic SA, Balaji C (2023) A multi-leg powered constant voltage on-road charging system with an LCC-S compensation. In: ITEC-India 2023 - 5th International Transportation Electrification Conference: EAMRIT - Accelerating e-Mobility Revolution for India's Transportation. Institute of Electrical and Electronics Engineers, Inc. https://doi.org/10.1109/ITEC-India59098.2023.10471478
56. Rashid MH (2014) Power Electronics: Devices, Circuits, and Applications. Fourth Edition. Pearson
57. Deng Z, Hongsheng H, Yugang S, Chen F, Xiao J, Tang C (2023) Design of a 60-KW EV dynamic wireless power transfer system with dual transmitters and dual receivers. IEEE J Emerg Select Topics Power Electron 12(1):316–327. https://doi.org/10.1109/JESTPE.2023.3301579
58. Fan M, Shi L, Yin Z, Jiang L, Zhang F (2019) Improved pulse density modulation for semi-bridgeless active rectifier in inductive power transfer system. IEEE Trans Power Electron 34(6):5893–5902. https://doi.org/10.1109/TPEL.2018.2867902
59. Li M, Deng J, Chen D, Wang W, Wang Z, Li Y (2023) A minimum ZVS current control strategy of semi bridgeless active rectifier for wide operation range on LCC-S compensated WPT system. IEEE Trans Ind Appl 59(3). https://doi.org/10.1109/TIA.2023.3238705
60. Venkatesan R, Dominic Savio A, Balaji C, Narayanamoorthi R, Kotb H, El Rashidi A, Nureldeen W (2024) A comprehensive review on efficiency enhancement of wireless charging system for an electric vehicles application. IEEE Access. https://doi.org/10.1109/ACCESS.2024.3378303
61. Zhang Y, Shen Z, Yuanchao W, Wang H, Pan W (2022) Dual-side phase-shift control for strongly coupled series–series compensated electric vehicle wireless charging systems. World Electric Vehicle J 13(1). https://doi.org/10.3390/wevj13010006
62. Fujita H, Akagi H (1996) Pulse-density-modulated power control of a 4 KW, 450 KHz voltage-source inverter for induction melting applications. IEEE Trans Ind Appl 32(2). https://doi.org/10.1109/28.491475
63. Wu D, Mai R, Zhou W, Liu Y, Peng F, Zhao S, Zhou Q (2022) An improved pulse density modulator in inductive power transfer system. IEEE Trans Power Electron 37(10):12805–12813. https://doi.org/10.1109/TPEL.2021.3135281
64. Yuvaraja Shanmugam R, Narayanamoorthi VK, Ramachandaramurthy PB, Shrestha N, Son J (2024) Machine learning based optimal design of on-road charging lane for smart cities applications. IEEE J Emerg Select Topics Power Electron 12(4):4296–4309. https://doi.org/10.1109/JESTPE.2024.3400292
65. Aditya, Kunwar. 2016. Design and implementation of an inductive power transfer system for wireless charging of future electric transportation, Oshawa, Ontario, Canada (January 30, 2024). https://hdl.handle.net/10155/712
66. Prasanth V, Scheele N, Visser E, Shekhar A, Mouli GRC, Bauer P, Silvestser S (2016) Green energy based inductive self-healing highways of the future. In: 2016 IEEE Transportation Electrification Conference and Expo, ITEC 2016. Institute of Electrical and Electronics Engineers Inc. https://doi.org/10.1109/ITEC.2016.7520268
67. SAE International (2019) Wireless Power Transfer for Light-Duty Plug-in/Electric Vehicles and Alignment Methodology. SAE International. https://doi.org/10.4271/J2954_201904
68. Ziegelberger G, Croft R, Feychting M, Green AC, Hirata A, d'Inzeo G, Jokela K et al (2020) Guidelines for limiting exposure to electromagnetic fields (100 KHz to 300 GHz). Health Phys 118(5):483–524. https://doi.org/10.1097/HP.0000000000001210

Achieving Sustainable Urban Mobility Through Retrofitting Two-Wheeler Vehicles

Faheem Abbas, Mumtaz A. Qaisrani, and Sana Abad

Abstract Since the world population continues to rise, it leads to an increase in the demand for two-wheeler motorcycles and scooters. Every year, the rate of waste and pollution continues to increase due to the manufacturing and scrapping of millions of these two-wheeler vehicles. These challenges can be overcome by converting the existing two-wheelers to electric two-wheelers, as it only involves removing the engine and fitting a battery, which can extend the life of the two-wheeler and, as a result, waste and pollution can be greatly reduced. The purpose of this book chapter is to address these concerns by focusing on retrofitting the iconic Italian Vespa Scooter from a 2-stroke engine to electric, without altering its classic design. Most Italian Vespa Scooters have been discarded owing to low fuel efficiency and noncompliance with pollution rules. During the retrofitting process, we designed the Custom Swing Arm (CSA) for the rear wheel, as the removal of the engine left the rear wheel unsupported due to the engine's integrated swing arm design. The fuel tank space was used for an Electric Power Unit (EPU), housing components such as a Lithium Iron Phosphate ($LiFePO_4$) Battery, Battery Management System (BMS), Motor Controller (MC), DC DC Converter (DCDC), and Circuit Breaker (CB). This study explores the technical challenges, including passive cooling and component placement with proper weight distribution, while highlighting the environmental and economic benefits of retrofitting. This retrofitting approach supports global efforts for cleaner and more sustainable urban transportation.

F. Abbas · S. Abad
School of Energy Science and Engineering, Harbin Institute of Technology, Harbin, China
e-mail: 24sf02215@stu.hit.edu.cn; 25sf02001@stu.hit.edu.cn

M. A. Qaisrani (✉)
Institute of Mechanical and Manufacturing Engineering, Khwaja Fareed University of Engineering & Information Technology, Rahim Yar Khan, Pakistan

H. Kaddouri et al. (eds.), *Smart Cities for Sustainable and Urban Development*, Green Energy and Technology, https://doi.org/10.1007/978-3-031-99001-4_25

Keywords Retrofitting of motorcycle · Two-wheeler conversion · Italian Vespa Scooter conversion · Eco-friendly transportation · Smart cities · Sustainable urban mobility

1 Introduction

Urban mobility has become a major challenge in cities around the world, due to growing population density, rapid urbanization, and increasing demand for transportation. To assess the sustainability of transportation [1] identified five criteria including fuel, accessibility, congestion, emissions and safety. Butler et al. [2] listed the six main categories including Intelligent Transport Systems (ITS), Alternative Fuel Systems (AFS), Shared Mobility Services (SMS), Integrated Mobility Systems (IMS), Self-driving Vehicles (SDV) and Demand Responsive Transport (DRT), adopting these categories improves vehicle efficiency, lowers transportation costs, reduces Greenhouse Gas (GHG) emissions, and minimizes the risk of accidents. Instances from around the world demonstrate the success of these methods. In Singapore, ITS utilize real-time data and intricate traffic management strategies to enhance traffic flow and alleviate congestion, effectively tackling issues of accessibility and urban gridlock [3]. Likewise, Shenzhen, China, has pioneered the adoption of AFS technologies, especially through the introduction of electric buses, leading to notable reductions in air pollution and operational expenses [4]. In Paris, JCDecaux launched the Vélib' SMS, world's largest and most successful bike-sharing project, on July 15th, 2007. The project incorporated 18,000 bicycles and 1230 docking stations, providing 24/7 services to the customers. It became so much popular and after seven years the total number of trips were more than 200 million and annual subscriptions were more than 274,000. It has been observed that the issues of traffic congestion and GHG emissions have been effectively addressed [5]. In India the Delhi Metro (DM) project IMS was started by the Delhi Metro Rail Corporation (DMRC) in 1998. During the nearly operation of 30 years, it has been focused on eco-friendly practices. The total length of DM is 395Km, with 12 lines and 289 stations. DMRC has been certified by the United Nations (UN), and became the world's first metro rail to get carbon credits for reducing GHG emissions. DM reduced the pollution level by 0.63 million tones/year [6]. Waymo one is the world's first SDV ride-hailing service provider in Phoenix, Arizona. It provides non-stop service to the customers. It is renowned known for utilizing the cutting-edge technologies including, LiDAR, sensors, laser light, GPS, computer vision, radar and more to operate, addresses the problem of GHG emissions as it is fully electric and powered by 100% renewable energy [7]. In Germany, the DRT service is provided by CleverShuttle. It provides affordability and door-to-door transportation services. It is renowned for its unwavering commitment to fully electric and hydrogen-powered vehicles [8]. These categories of urban mobility are working in the developed countries and still not meet the demand of population and as the population rise continuously it increasing the demand of transport, enabling the individual to accomplish their daily tasks and activities. In the world, transport is among the major contributor of the pollution source. As per the climate Change 2023: Synthesis Report, in 2019, 79% global GHG emissions came from the energy, industry and transport sector [9].

Individuals choose the Internal Combustion Engine Two-wheelers (ICE2Ws) including motorcycle and scooters for the mood of transport because they are cost-effective and occupy less road space as compared to the private cars [10]. Moreover, ICE2Ws play a vital role to increase the air pollution which lead to severe health problems for human [11, 12]. It is an estimated that by 2050 and 2100 the world population will reach 9.7 billion and 10.4 billion, respectively [13]. If the current trend of usage of ICE2Ws continues to rise to meet the demands of this growing population, it could lead to an unsustainable environment. As per the World Air Quality Report 2023, Bangladesh, Pakistan and India are the top polluted countries in the world [14]. As per the report of Bangladesh Road Transport Authority (BRTA), 3.59 million Two-wheelers (2Ws) have been registered till January 2024 from 2011 as shown in Fig. 1 [15]. As per the report of Pakistan Automotive Manufacturers Association (PAMA), Pakistan has sold 18.24 million 2Ws till June 2024 from 2009 as shown in Fig. 2 [16]. As per reports, number of 2Ws sold in India, Indonesia and Vietnam are 79.94 million, 555 million and 18 million, respectively as shown in the Figs. 3, 4, and 5 [17–19]. As the number of 2Ws keep rising, the amount of fuel these 2Ws consume also goes up. When gasoline is burned, it emits GHG emissions into the air, producing 375.54 g of CO_2/km [20]. To operate effectively ICE2Ws frequently require maintenance, including oil changes, carburetor cleaning, clutch adjustments, transmission servicing and replacement of components like spark plug, air filter and brake pad. The scrapping of these 2Ws poses significant danger to the environment. After removing the reusable parts that can be still used for repairing other 2Ws. The remaining material of these 2Ws are scrapped by manual labor, where materials are processed in furnaces for various purposes.

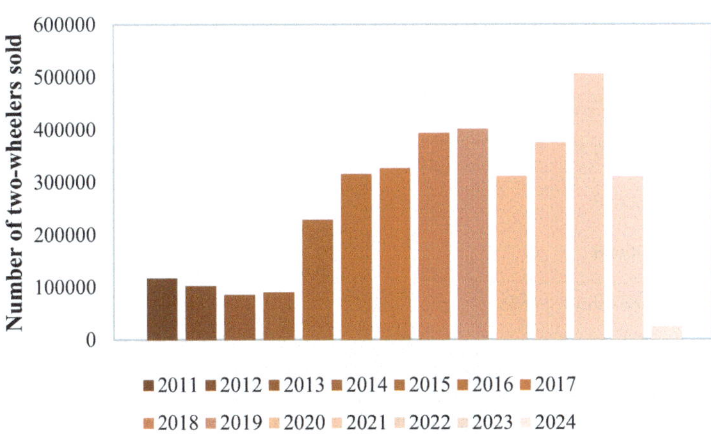

Fig. 1 Sales of two-wheelers in Bangladesh till January 2024 (2011–2024) [15]

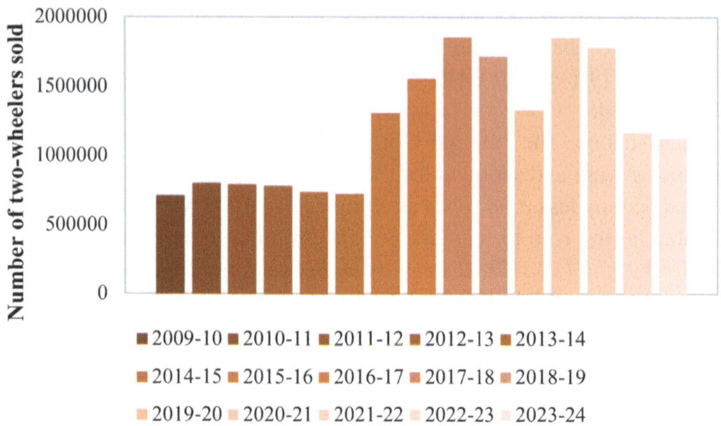

Fig. 2 Sales of two-wheelers in Pakistan till June 2024 (2009–2024) [16]

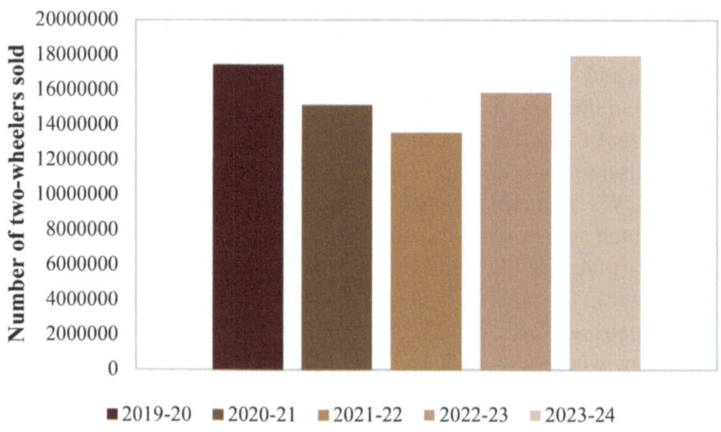

Fig. 3 Sales of two-wheelers in India (2019–2024) [17]

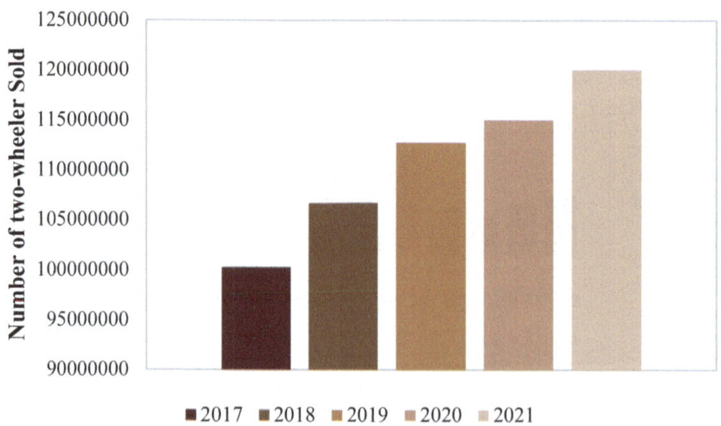

Fig. 4 Trend of ICE2Ws in Indonesia (2017–2021) [18]

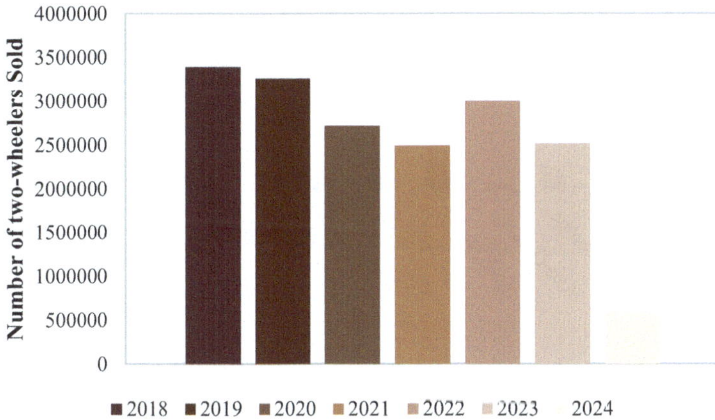

Fig. 5 Sales of two-wheelers in Vietnam till June 2024 (2018–2024) [19]

While this process allows some parts of 2Ws to be recycled, but this process required substantial energy and contributes to air pollution. To address this issue, we need eco-friendly transportation technology that work not only in developed countries but also in developing countries to reduce GHG emissions.

Electric two wheelers (E2Ws) stand out an optimal solution for urban mobility. They have several advantages, including better energy utilization, less noise, and rapid torque response [21]. They mainly require battery health monitoring and brake checks. They do not emit GHG emissions that make them environmentally friendly and cost-effective alternative to ICE2Ws. Since the world shifts towards the renewable energy, E2Ws include electric motorcycle and electric scooters can replace these ICE2Ws and play a notable role to reduce the GHG emissions. The idea of utilizing E2Ws instead of ICE2Ws is applicable globally. This approach offers two primary options. developing new E2Ws while phasing out the existing ICE2Ws, or retrofitting existing ICE2Ws to operate on electricity [22]. However, phasing out conventional ICE2Ws is not the best approach. This procedure demands energy for both the removal of old ICE2Ws and the creation of new E2Ws, resulting in higher total energy consumption and increased pollution. Phasing out these functioning ICE2Ws wastes both money and resources. Some countries, such as China, have already taken significant steps by producing new E2Ws from scratch. China leads the global E2Ws market, primarily because of the remarkable success and widespread adoption of E2Ws across the country [23]. The second option is feasible, retrofitting the existing 2Ws as it is easy to implement, cost effective and has no environmental affect than making the brand new E2Ws. The cost of retrofitted E2Ws is lower than that of brand new E2Ws, making them more appealing to customers. As some countries, including China [23], have banned ICE2Ws and others are set to follow suit in the coming years, people are likely to have a favorable view of converting ICE2Ws to E2Ws. The ban on ICE2Ws is basically due to their significant contribution to air pollution, and Fuel Import Dependent Countries (FIDCs) spend billions of dollars on fuel imports. As depicted in Fig. 6, global oil demand is exceedingly high, with China leading the list of FIDCs, as shown in Fig. 7 [24].

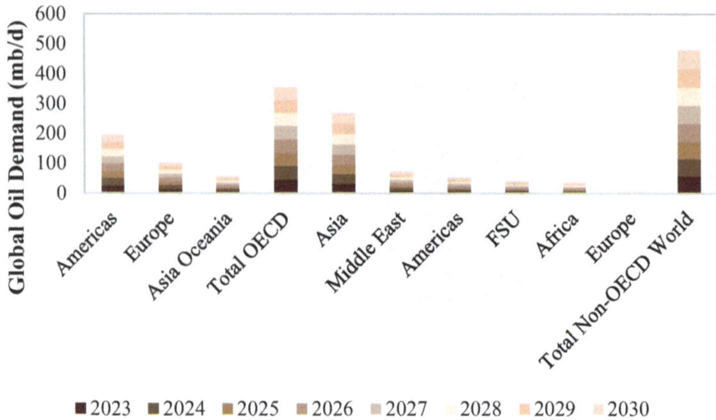

Fig. 6 Global oil demand (mb/d) [24]

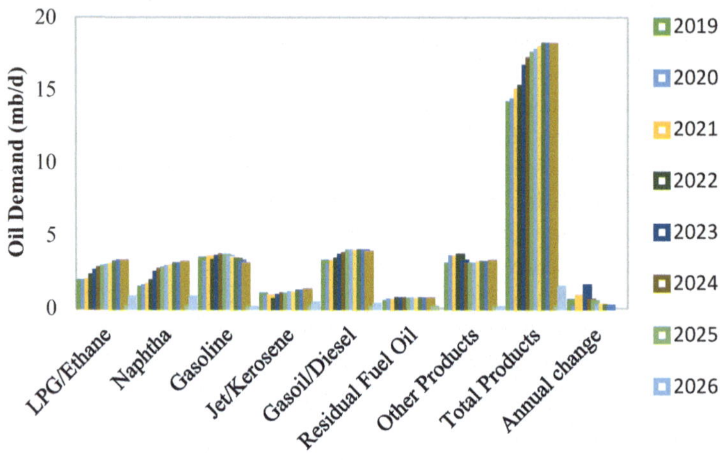

Fig. 7 China oil demand (mb/d) [24]

In this study we are going to address concerns related to ICE2Ws by focusing on retrofitting the iconic Italian Vespa Scooter (IVS), Traditional Vespa 150 Super (TV150S), from a 2-stroke engine to electric, without altering its classic design. Once IVS was a symbol of fad and liberty, but today, it is a rare sight on the road because they do not meet modern environmental standards. This has led to many of these scooters ending up in junkyards, wasting valuable resources. Since it was introduced after World War II [25], the Vespa has represented hope and recovery in Italy, quickly becoming popular worldwide. It has been loved for many generations because of its cool look and smooth shape, mixing fun with practicality. Their compact size, making it great for parking in busy places and driving on tight roads. There was a time when IVSs were booming, and people eagerly sought to own one, as riding a Vespa became a true status symbol. These classic

Fig. 8 Abandoned Vespa scooters, illustrating the decline of these once-popular vehicles [26]

scooters are regrettably now abandoned in junkyards. Their excessive fuel use is one factor contributing to this decrease, as does their disregard for the environment. Unfortunately, these iconic scooters now lie abandoned in junkyards as shown in Fig. 8. One reason for this decline is their high petrol consumption, while another is their lack of environmental friendliness. Figure 9 illustrates the state of Vespas in Karachi, Pakistan. The retrofitting process is a great way to promote green mobility by converting ICE2Ws into E2Ws, to meet today's environmental needs and sustainability goals. The retrofitting of these IVSs have become very crucial, as urban environments shift to more environmentally friendly modes of mobility. This change is necessary to bring these ancient scooters back to life, lessen their negative environmental effects, and satisfy the needs of contemporary mobility. First step in retrofitting involves removal of engine with all its associated parts, such as the engine assembly, Kickstarter mechanism, muffler, transmission gears, carburetor, fuel tank, exhaust pipe, ignition system, and clutch assembly. While this transformation is beneficial, it also comes with challenges that need to be tackled to ensure that the Vespa keeps its classic look and structural strength.

The purpose of this study is to achieve the urban mobility by bringing back the popularity of IVS by converting it into electric and revitalize this cherished icon for a new generation of riders. The study provides a detailed, step-by-step conversion of TV150S into Retrofitted Electric Vespa (REV). The study is organized into three sections: first section discusses overview of design and manufacturing process with technical challenges including weight distribution and thermal management, second section is dedicated for the results and third section focuses on conclusion.

Fig. 9 Vespas undergoing refurbishment in Karachi, emphasizing the potential for retrofitting to promote sustainable urban mobility [26]

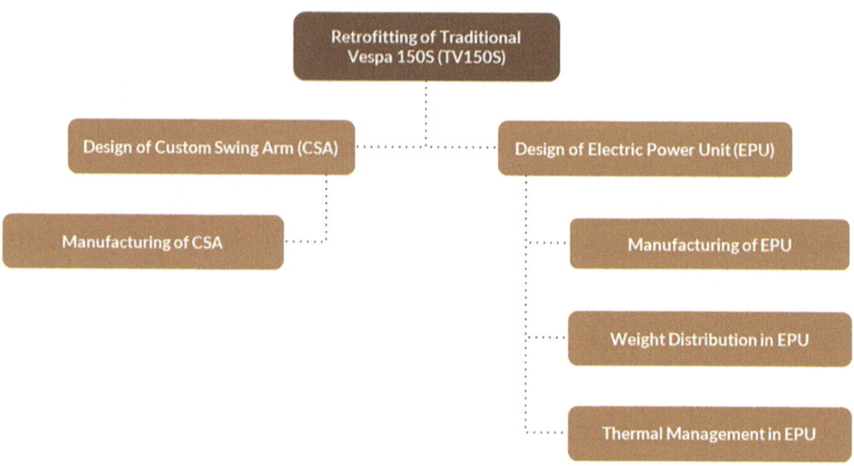

Fig. 10 Technical challenges required during the conversion of TV150S to REV

2 Technical Challenges

The conversion of TV150S to REV require design of custom swing arm (CSA), as removal of engine left the rear wheel unsupported because the existing swing arm is directly mounted on the engine and design a suitable electric power unit (EPU) with proper weight distribution and thermal management as shown in the Fig. 10.

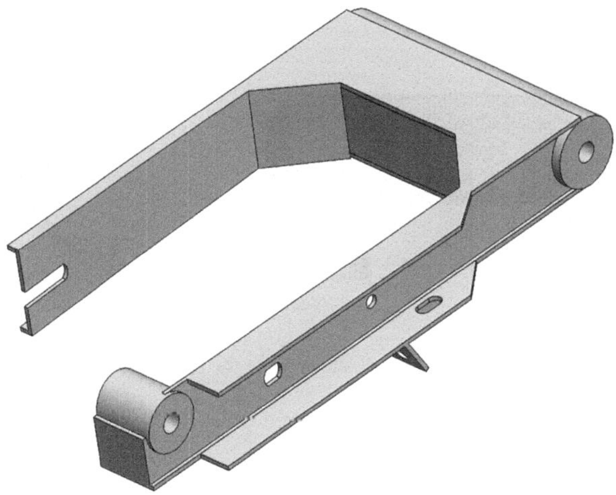

Fig. 11 CAD model of Custom Swing Arm (CSA)

2.1 Design of Custom Swing Arm (CSA)

The first and foremost challenge is to design the CSA because removal of engine left the rear wheel unsupported because the existing swing arm is directly mounted on the engine. The Vespa scooter has a mono shock that is located on the right side of the swing arm. Therefore, the right side of swing arm much be capable of balancing the forces under harsh conditions. During the design our main focus is on the structural stability and weight distribution. Additionally, we wanted to make it lightweight and strong enough to withstand high impact forces, ensuring a smooth ride quality. For instance, with two individuals having weight of 100 kg each, combined with the scooter's own weight of 90 kg and an additional luggage weight of 10 kg, the total weight is 300 kg. When this scooter will pass through the speed bump of maximum height 6 inches (h = 0.1524 m) at high speed, the Impact Force (F_{impact}) that will act on the swing arm will be approximately 2946 N and Gravitation Force (F_g) will be 2943 N. We designed the swing arm by keeping in mind with these values. We designed the CSA in the SOLIDWORKS 2021 as shown in the Fig. 11. We employed with CO_2 Laser Cutting Machine for prototyping, to shape the cardboard, establish the geometry to ensure that the CSA fit properly within the vespa frame. After a lot of iterations, achieved an optimized design. To get the high strength-to-weight ratio, manufactured the CSA in 3 mm Mild Steel (MS) sheet. Used a Fiber Laser Cutting Machine and CO_2 Welding to achieve high-precision and high-quality joins. Following that, cadmium treatment is used to improve the surface attributes, including corrosion resistance and surface hardness. This treatment considerably increases the component's overall longevity and visual appeal. Figure 12 depicts the integration of CSA with rear wheel.

$$F_g = mg \tag{1}$$

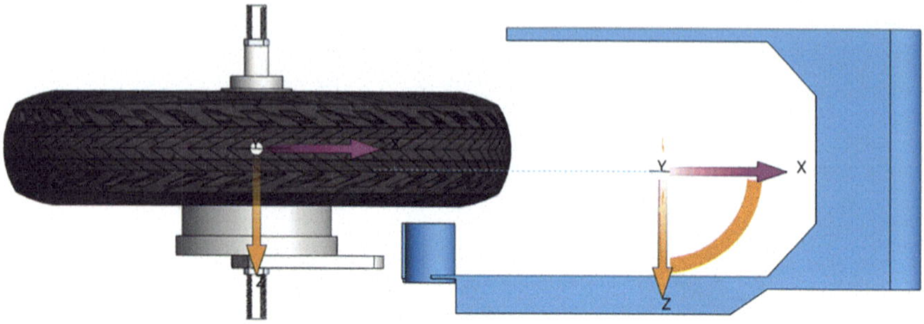

Fig. 12 Assembly of CSA with rear wheel, showcasing the complete integration

By numerical application F_g is $2943N$.

$$v = \sqrt{2gh} \tag{2}$$

h is the height of the speed bump (h = 0.1524 Maximum Height).

The value of the velocity is 1.73 $m.\ s^{-1}$

$$F_{impact} = \frac{mv^2}{2h} \tag{3}$$

F_{impact} is the impact Force (It will act on the swing arm when scooter pass through the speed bump). The value of this force is $2946N$.

$$F_{total} = F_g + F_{impact} \tag{4}$$

F_{total} is the overall load on the swing arm during dynamic riding conditions, which is equal to $5889N$.

2.2 *Design EPU*

During the retrofitting another challenge was to design the suitable retrofitting kit (RK). We had a lot of ideas, place the battery box and BMS in the place of fuel tank and place the other components like Motor Controller (MC) and DC-DC Converter (DCDC) on the right side of the scooter, where the engine was previously mounted. However, the issue was the side cover does not sufficiently protect these components from dust and water, which could lead to damage. Therefore, this approach is not feasible. We removed the fuel tank and examine the available space carefully. After a lot of research and development (R&D), we designed an electric power unit (EPU) whose shape is almost similar to the shape of fuel tank. EPU is divided into 4-sections: BMS, $LiFePO_4$ Battery, MC and DCDC. In the third section there is a dedicated place for circuit breaker (CB) and charging port (CP). Additionally, it has another feature of a J-shaped embossed surface for the placement of Anderson connector and wiring. At the end of this embossed surface, there is an elliptical hole that allows access to the harness to connecting with the Anderson. The Anderson connectors facilitate the users for easy connection and disconnection of the EPU.

CNC Milling Machine was used for making the die for EPU. Extruded Polystyrene Foam (XPS) foam with dimensions of 4″ x 4′ x 8′ (Thickness x Width x Length) was used because it is cost-effective and exhibit better damping properties [27]. Fiberglass was chosen for making EPU because it is cost-effective and it has high strength, flexibility and stiffness [28]. Two layers of fiberglass were used, the inner layer was Satin Weave E-Glass Fabric 400 GSM and the outer layer was Chopped Strand Mat (CSM) 450 GSM for high strength and durability. To ensure aesthetics balance between the inner surface and outer surface, vacuum bagging was applied to the outer layer to improve the overall surface by removing the air pockets.

To ensure stability and safety proper weight distribution is very crucial. It helps to maintain the control during acceleration and braking. Previously, the space occupied by the fuel tank is used to place the EPU. To distribute the weight across the frame, In EPU, battery pack is placed in the center (Sect. 2), BMS is placed just front of the battery pack in Sect. 1 and MC and DCDC is placed in the Sects. 3 and 4, respectively. Additionally, CB, CP and Anderson are placed in second 3 in the dedicated places. The weight of the fuel tank was approximately 10 Kg, including maximum fuel capacity of 7.7 L and empty fuel tank [29]. To match this weight distribution, we designed EPU, which weighs 9.6 Kg. During designing another concern was thermal management. We carefully incorporate some vents to ensure the proper passive colling. There were two small holes below the seat, where choke control and fuel cock were located. These holes allow air to enter into the Chamber (The closed surface where fuel tank was previously mounted). These small holes were located on the perfect place below the seat and above the chassis that prevent from the rain and road water, respectively. The vents are located on the upper surface of the EPU unit, there is no possibility of water entering into the EPU. EPU's vent allow the air from the chamber, as the air heats up its density decrease, causing it to rise and we designed the lid by simply 3 mm Medium Density Fiberboard (MDF).

2.3 *Components for Electric Conversion of the Vespa Scooter*

Brushless DC (BLDC) Hub Motor is used to drive the vespa. BLDC motors are commonly used in EVs because of their high efficiency, high torque-to-weight ratio, and reduced maintenance needs compared to other motors. They offer good speed and torque which contribute to overall vehicle performance. For the power source, $LiFePO_4$ battery is employed because of its popularity in EVs due to high energy density, long cycle life, low self-discharge rate, fast charging, high efficiency and environmentally friendly. In vespa 20S 20A $LiFePO_4$ battery is employed. 20 cells are connected in series in order to achieve the desired voltage of 72 V because each

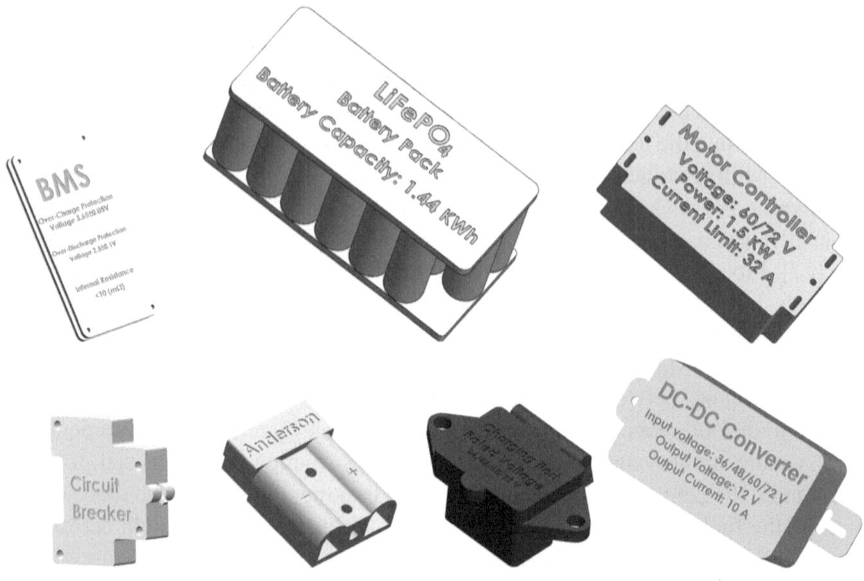

Fig. 13 Components of EPU

cell has nominal voltage 3.6 V and capacity of 20 A. From this configuration 1.44KWh battery capacity is achieved to run the 1.5KW BLDC Hub motor.

$$Capacity(Wh) = Voltage(V) \times Capacity(Ah) \tag{5}$$

The value of the capacity is 1.44 *kWh*.

MC is employed to control the BLDC Hub Motor. MC regulates the power supply to the BLDC motor, ensuring efficient performance and smooth operation. The features of MC are not only limit to control of speed and torque but also it provides protection features including overcurrent and thermal. For balancing the charges across individual cell, JIABAIDA 20S 40A $LiFePO_4$ BMS is used. It ensures the safety and longevity of battery pack. It prevents the cells from overcharging and discharging and monitors the current, voltage and temperature. It has an extra feature of Bluetooth; user can excess this data on mobile screen. To enhance the safety and reliability of the EPU, CB is employed to protect from the short circuit or electrical fires. $LiFePO_4$ Battery Charger (BC) of 87.6 V and 10A is used to charge the battery because it allows for proper charging while preventing from the overcharging. It can fully charge the battery within 2 hours. The individual components are shown in Fig. 13. Figure 14 depicts the CAD assembly of EPU. The specification of these components is shown in Table 1.

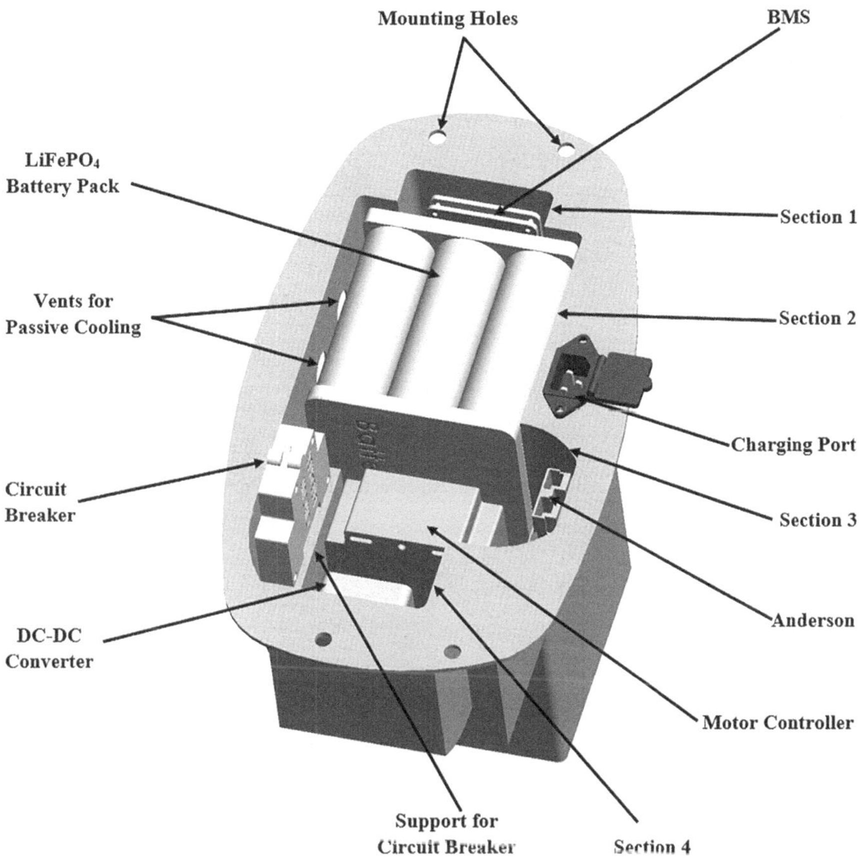

Fig. 14 CAD assembly of EPU

Table 1 Specifications of components for electric conversion of the vespa scooter

BLDC hub motor	
Rated voltage (V)	60
Rated power (KW)	1.5
RPM	425–675
Rated Torque (Nm)	15–30
Weight (Kg)	12
Motor Controller (MC)	
Voltage (V)	60/72
Power (KW)	1.5
Current limit (A)	32
Weight (Kg)	0.56

(continued)

Table 1 (continued)

Battery Management System (BMS)	
Maximum operating current (A)	40
Over-charge protection voltage (V)	3.65 ± 0.05
Over-discharge protection voltage (V)	2.5 ± 0.1
Temperature protection value of charging (°C)	75
Internal resistance (mΩ)	<10
DC-DC Converter (DCDC)	
Input voltage (V)	36/48/60/72
Output Voltage (V)	12
Output Current (A)	10
Wiring Description	Red is positive, Black is negative and Yellow is output
Circuit Breaker (CB)	
Rated current (A)	32
Rated voltage (V)	230
Battery Charger (BC)	
AC input voltage (V)	180–260
Efficiency (%)	>93
Ambient temperature (°C)	−5–40
Relative Humidity (%)	5–95
Protect Characteristic	Short-circuit protection, Over-current protection, Over-voltage protection, Reverse Polarity Protection, Over-temperature protection
Charging Port (CP)	
Rated Voltage (V)	24/48/60/72
Mounting	Equipped with screw holes for secure installation
Connector Type	3 connectors for high-quality connectivity
Amperage Capacity	Designed to withstand high amperage loads

3 Results

3.1 Environmental Impact

Converting the ICE2Ws to E2Ws offer significant environmental benefits, especially in urban areas where ICE2Ws contribute to air pollution. From Various studies, it has been concluded that the shifting of ICE2Ws to E2Ws can reduces the GHG emissions. Take the case of this, [30] conducted a study evaluating the carbon emissions from both the E2Ws and ICE2Ws. From the study, it was observed that ICE2Ws produce 7786.57 kg of CO_2 while E2Ws have zero emissions, respectively. Duke and Erazo [31] conducted a study to determine the environmental impact of

ICE2Ws and E2Ws in Honduras. They concluded that CO_2 emissions from the ICE2Ws contribute to 38% of the total CO_2 emissions related to transportation per year, when these ICE2WS are replaced with E2Ws, 34% CO_2 emissions could be reduced throughout the year, as shown in Fig. 15. In October 2020, a survey is conducted by Makerere's university professor along with experienced researcher in Kampala, Uganda, to improve the livelihoods of boda-boda drivers by replacing their ICE2Ws to E2Ws. The survey included 200 boda-boda drivers in metropolitan area of Kampala, covering both the ICE2Ws (boda-boda) and E2Ws (Zembo Storm Electric). The weekly net income was calculated, Zembo driver's income \$34 (±\$9) USD and boda-boda driver's income \$10 (±\$3) USD. They found significant reduction in GHG emissions due to 89.2% of electricity is generated by hydroelectric plant to powered the E2Ws as shown in the Fig. 16. Additionally, it was observed that the former boda-boda drivers felt more respected after becoming the Zembo

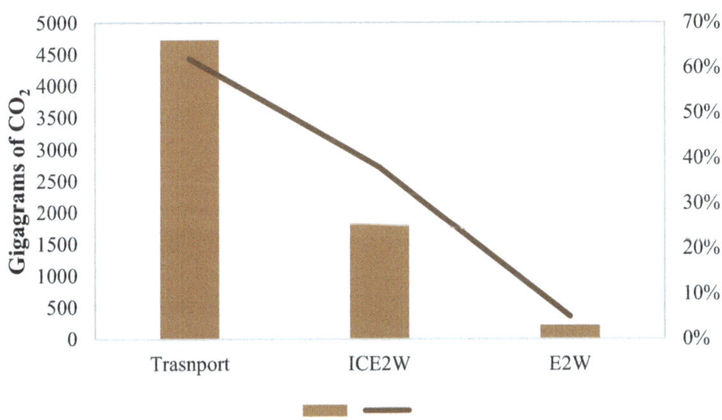

Fig. 15 CO_2 emissions reductions when replacing the ICE2Ws to E2W in Honduras [31]

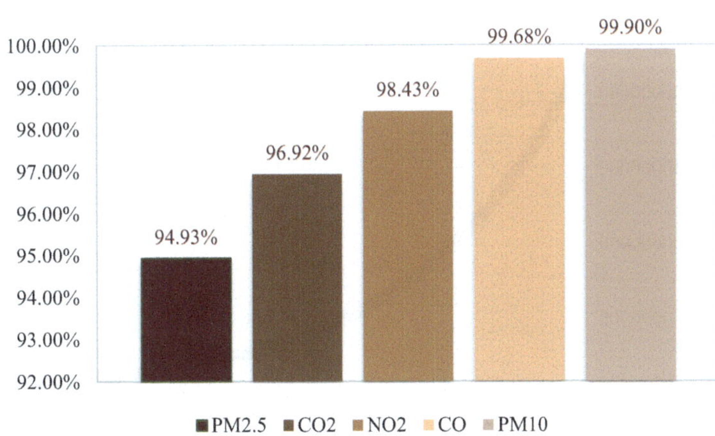

Fig. 16 Reduction of GHG emissions by using E2Ws [32]

drivers [32]. Xuan et al. [33] conducted an experimental study to evaluate the environmental impact, by converting the ICE2W (Honda CBR400RR) to an E2W. From the Life Cycle Assessment (LCA), it was concluded that the retrofitted E2W consumed 72% less energy and produced 45% lower GHG emissions over its lifespan of 8 years. Tuayharn et al. [34] conducted a study to examine the environmental and economic impact of introducing the E2Ws in Thailand. From the environmental analysis, it was found that ICE2Ws consumed 0.0194 L of gasoline per Km and produce 43.59 g of CO_2 emissions. On the other hand, E2Ws consumed 0.0027 L of gasoline per Km and produce 20.67 g of CO_2 emissions, if the electricity powered to E2W is generated by coal. If electricity was generated from natural gas, the CO_2 emissions reduced to 10.7 g/km. From economic analysis, it was concluded that E2Ws can reduce 20% of ownership cost in 10 years as compared to ICE2Ws.

Habibie et al. [35] conducted a simulation study on Powersim studio 10 to evaluate the sustainability for the conversion of ICE2W to E2W in Indonesia. In the study five scenarios were considered on the conversion rate, 5% per year, 15% per year, 25% per year, 35% per year and 45% per year. From the conversion rate of 45% per year, it was concluded that carbon emissions would around eight million $MTCO_2e$ by the end of 2026, as shown in the Fig. 17. An experimental study is conducted by [36] for the conversion of ICE2W (Yamaha N-Max 150 cc) to an E2W. From the study it was concluded that an E2W can save 7.7 times the energy expenses and release 4 times lower GHG emissions. The electricity powered to E2W was generated by the energy mix of Indonesia 2020, as shown in the Fig. 18. Cherry et al. [37] conducted a study to evaluate the environmental impacts of E2Ws in China. The study considered two different types of E2Ws: Bicycle Style Electric Bike (BSEB) and Scooter Style Electric Bike (SSEB), and compared the GHG emissions of E2Ws with other vehicles such as car, bus, ICE2Ws and bicycle. We have taken their comparison results of E2Ws with ICE2Ws. The results indicate that E2Ws emits significantly less pollution per Km compared to ICE2Ws, except for SO_2

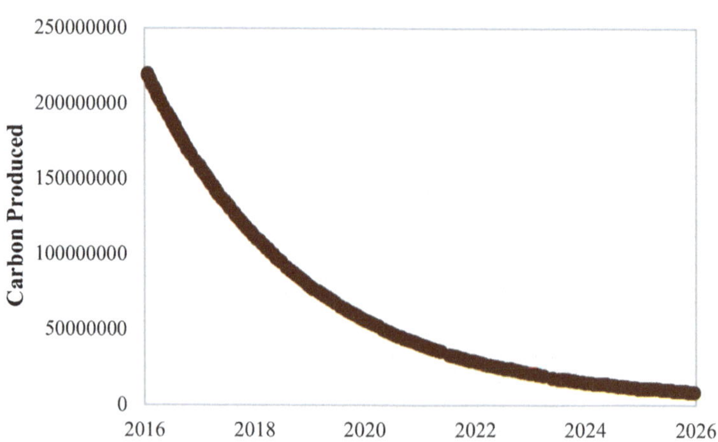

Fig. 17 Simulation result of carbon production with years [35]

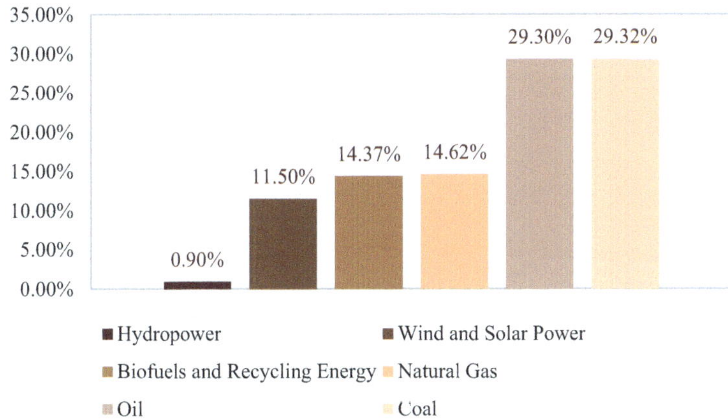

Fig. 18 Indonesia's energy mix 2020 [36]

emissions, as most of electricity generated from the Coal in China, as shown in the Table 2. GHG emission rates depend on the region where E2Ws are used, as each region has its own Power grid, and every grid has a different energy mix, as shown in the Fig. 19. The study is conducted by [38] that included LCA and economic analysis of E2Ws and ICE2Ws, as well as their impact on air pollution in Barcelona, Spain. From the LCA, it was concluded that the Global Warming Potential (GWP) of E2Ws was 20% lower than ICE2Ws. The main difference was due to operation as ICE2Ws operate on fossil fuels that lead to significant impact. From the economic analysis, it was suggested that both E2Ws and ICE2Ws have almost similar upfront and operational cost but E2Ws are more sustainable and environmentally friendly, as shown in the Fig. 20.

3.2 *Range Considerations*

Range is the very important parameter that assess the performance of REV. Understanding the theoretical and practical range of the REV involves various factors such as wheel diameter (WD), circumference of the wheel (C), RPM of the motor, speed, battery capacity and riding conditions. The rear tire size of the Vespa 150 super is 3.5 × 8″ [29]. The Total diameter of tire (D) is calculated by adding the diameter of the wheel (rim) to twice the Sidewall Height (SH). A theoretical range of 46.5Km, providing a bench mark for comparison.

$$D = WD + 2(SH) \tag{6}$$

The total diameter D is equal to $0.381m$.

The Circumference of the wheel is expressed by:

Table 2 Vehicle types emission rates/passenger Km [37]

Vehicle type	Energy use (KWh/100 pax-Km)	CO_2 (g/pax-Km)	SO_2 (g/pax-Km)	PM (g/pax-Km)	CO (g/pax-Km)	HC (g/pax-Km)	NO_x (g/pax-Km)
Motorcycle	21–42	64–128	0.04–0.08	0.20–0.40	6.3–12.5	1.13–5.25	0.08–0.15
Bicycle style electric bike (BSEB)	3.8–7.6	15.6–31.2	0.07–0.14	0.07–0.14	0.007–0.014	0.027–0.053	0.010–0.02
Scooter style electric bike (SSEB)	4.9–9.9	20.2–40.5	0.09–0.17	0.10–0.19	0.009–0.017	0.032–0.064	0.014–0.027

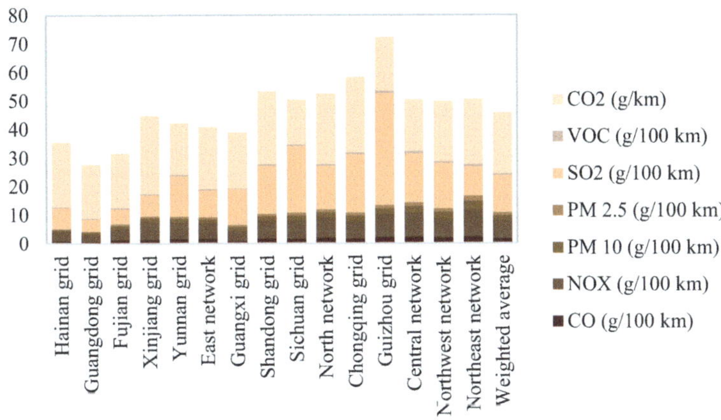

Fig. 19 GHG emissions of E2Ws in China by specific regions [37]

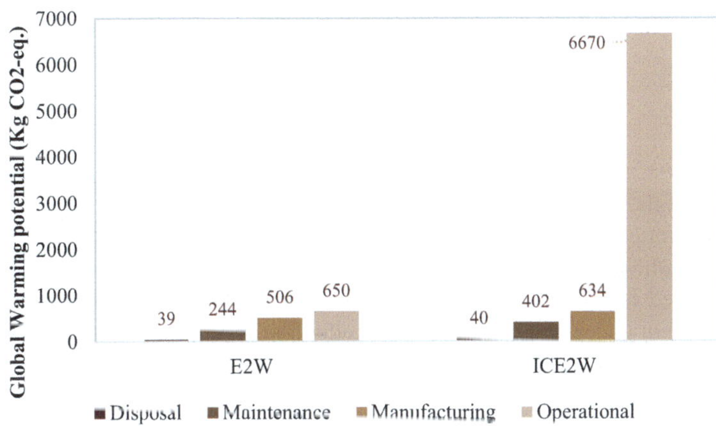

Fig. 20 Comparison of Global warming potential for both the E2W and ICE2W [38]

$$\frac{D}{m} = C \times RPM \tag{7}$$

where D/m is the distance per minute and D/h is the distance per hour. The value of the fraction D/h is equal to 48,450.

The speed is equal then 48.45 $Km.\ h^{-1}$ (48,450/1000).

This is the theoretical maximum speed under no load condition. We achieved 44 km/h under loaded conditions and the rider weight was 65 kg, as shown in the Table 3, which show the comparison between REV and TV150S.

$$Range = Speed \times Time \tag{8}$$

Table 3 Comparison of REV and TV150S

Feature	Traditional Vespa 150 Super (TV150S)	Retrofitted Electric Vespa (REV)
Engine Capacity (CC)/Battery Capacity (KWh)	145	1.44
Engine/Battery Type	2-stroke	Lithium Iron Phosphate
Top Speed (Km/h)	90	44
Power (KW)	5.1	1.5
Torque (Nm)	9.74	15–30
RPM	5000	425–675
Range (km)	350	55
Noise Level	High	Silent
Pollution Level	High	Zero
Maintenance Cost	High	Low
Operational Cost	High	Low
Starting	Kick start	Electric start
Type of Cooling	Air	Air

$$Time = \frac{Capacity(kWh)}{MotorOutput(kW)} \tag{9}$$

The value of the Range is 46.5 *km*.

The theoretical range of the REV is estimated 46.5 Km at theoretical high speed of 48.45 km/h. However, in the real world especially in cities riding, the motor won't be used full power because of frequent stops. As a result, we achieved the range of 55 km under typical riding. This is the best appropriate range with the battery capacity of 1.44KWh and 1.5KW hub motor.

3.3 Economic Benefits

An average Pakistani home without air conditioning and without a UPS system, uses about 2000 KWh of electricity in a year [39]. Residential Tariff Rates from Multan Electric Power Company (MEPCO) state that the monthly cost for consumption of 100–200KWh is 10.82 PkR/KWh. The sum includes 15% electricity duty and 17% GST. The cost of 1 unit electricity is 15 PKR including all taxes [40]. The total cost of REV and traditional Vespa over 10 years are 303,050 PKR and 850,330 PKR, respectively. The annual maintenance cost for the traditional Vespa is about 25,000 PKR but it can vary mechanic to mechanic and depend how carefully user use the vespa. Table 4 shows cost comparison between REV and TV150S.

Table 4 Comparison of operational costs between REV vs TVS150S over ten years [29]

REV	
Daily Average Distance Covered (Km)	30
Range of REV (Km)	55
Daily Energy Consumption (KWh)	0.785
Annual Energy Consumption (KWh)	287
Annual Cost (PkR)	4305
Battery Cost (PkR)	130,000
Battery Charging/Discharging Cycle	1000
Battery Life (km)	55,000
Total Distance Vespa Covered in 10 Years (Km)	109,500
Battery Replacements (Times)	$1.9 \approx 2$
Total Cost in 10 years (PkR)	303,050
TV150S	
Fuel consumption (L/Km)	0.022
Daily Fuel Consumption (L)	0.66
Annual Fuel Consumption (L)	241
Price of gasoline (PKR/L)	249.1
Annual Cost (PkR)	60,033
Annual Maintenance Cost (PkR)	25,000
Total Cost in 10 years (PkR)	850,330
Cost Savings in 10 years (PkR)	547,280

The REV not only offers environmental benefits by achieving zero tailpipe emissions but also the economic benefits, the operational cost of REV is almost 13 times less than TV150S and over 10 years the approximate cost saving will be 547,280 PKR. In term of performance, BLDC hub motors provide instant power, smooth acceleration and a better riding experience compared to ICE in urban settings. As a result, REV can achieve better acceleration and responsiveness in cities. The REV is highly energy efficient; it covers 55Km on a single charge. In addition, REV has fever moving parts, leading to very low maintenance cost as compared to TV150S. The maximum speed of REV is 44Km/h, this reduced speed of REV contributes to its energy efficiency and enhance safety. Interestingly, all Vespa scooters have same chamber size. This RK can be used to any Italian Vespa model that is now in a junkyard. A wider range of customers will find this RK appealing because it is easy to adapt to any Vespa model. In addition to modernizing the scooters to meet current performance and environmental criteria, this endeavor helps maintain the iconic Vespa appearance. By emphasizing this aspect, we can highlight the potential for significant reductions in emissions and waste across the Vespa community. The success of this conversion techniques depends on the in stock accessible RKs. The retrofitted Vespa is shown in Fig. 21.

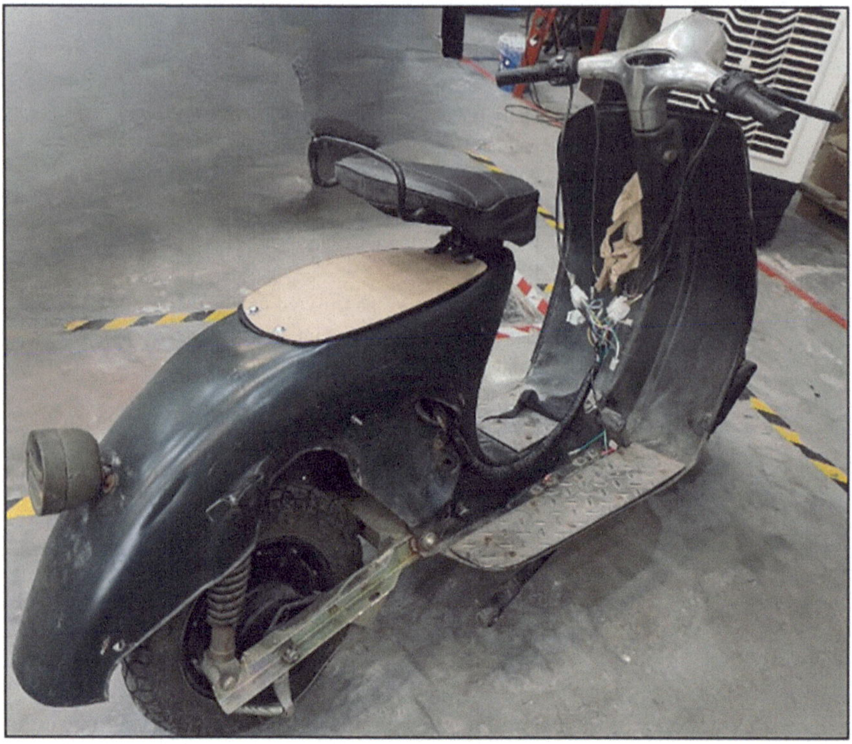

Fig. 21 Retrofitted TVS150S, showing its transformation to electric power for better sustainability

4 Conclusion and Perspectives

Retrofitting the existing ICE2Ws to E2Ws is both economical and environmentally friendly, contributing to achieve sustainable urban mobility. This retrofitting technique can be implemented globally, especially in the developing countries where budgets are limited. By allowing vehicle owners to maintain their existing vehicles, retrofitting contributes to lowering waste, minimize GHG emissions, promoting a circular economy and make the environment cleaner. The simplicity of the retrofitting process is removing the fuel tank and engine and replacing them with EPU and CSA, respectively, makes it scalable for global adoption. The countries where demand of two-wheelers is greater such as India, Vietnam, and Indonesia, could benefit from policies that encourage retrofitting, furthering their sustainability goals. The operating cost for both TV150S and REV for 10 years are compared, showing that the user can save up to 547,280 PKR in just 10 years. Moreover, $LiFePO_4$ battery typically have 2000–5000 charging/discharging cycles, we take only 1000 efficient cycles for the analysis. User can utilize the $LiFePO_4$ battery till 6–7 years, depending upon the usage. Furthermore, it is expected that the lifespan and efficiency of batteries will increase due to ongoing research in advancing the battery technology, contributing to further lowering the operational cost of REV.

ICE2WS including Vespa Scooters are a more popular mode of transport in many countries, retrofitting allows user to maintain their existing vehicles. The three-mode throttle is used for low, medium and high speed. The previously clutch now function as a break lever. The overall shape of the vespa scooter remains the same. Air pollution and GHG emissions can be significantly reduced by adopting this retrofitting technique, as it reduces the demand for raw materials required for new vehicle production. Retrofitting can be scaled rapidly; workshops can be set up in various urban locations to provide the conversion services. To ensure the safety and quality, it is recommended to train mechanics for safely installation of EPU. To promote clean and green transportation systems in smart cities, a collaborative approach is required. Policymakers should not only make a policy to provide subsidy on RK so that everyone can convert their motorcycles to electric but also collaborate with local workshops and service providers to expand retrofitting services. Industrialist should invest on R&D to provide retrofitting solutions for the remaining ICE2WS. Stakeholders should work together to create smarter, more sustainable cities that will eventually enhance the quality of life for citizens and create a healthier environment by accepting retrofitting as a practical urban transportation option.

Policymakers should enforce strict regulations for converting the ICE2Ws to E2Ws, ensure that the entire fleet of ICE2Ws are converted to E2Ws. Government must invest for the development of charging stations in schools, colleges, universities and public places, ensuring that these charging stations powered by renewable energy source like solar energy. Ensure that charging stations coped with RFID/NFC-enabled scanners, users scan their card and start charging to E2W. Charging stations measure the consumed electricity during charging and the cost of charging automatically deducted from the card. A mobile app is needed that provides the facilities to the users to check, near charging station, charging history, remaining balance, recharge card and QR scanning option for charging payment. Furthermore, providing subsidies for charging service can increase the transition from ICE2Ws to E2Ws. Policymakers work closely with stakeholders to make the policies, encourage the local companies to use the renewable energy for the generation of new E2Ws in the future. The government of developing countries launch initiatives to provide the E2Ws to the students in higher secondary school, colleges and universities, either as scholarship or interest-free easy installments. The success of this retrofitting initiatives depends on the availability of affordable RKs, supportive policies, and investment in clean energy infrastructure. Considering these factors, retrofitted E2Ws can play an important role in achieving sustainable urban mobility.

Competing Interests The authors declare that they have no competing interests related to this book chapter.

Ethical Approval Not applicable.

References

1. Richardson BC (2005) Sustainable transport: analysis frameworks. J Transp Geogr 13(1):29–39
2. Butler L, Yigitcanlar T, Paz A (2020) Smart urban mobility innovations: a comprehensive review and evaluation. IEEE Access 8:196034–196049
3. Singapore, Land Transport Authority and Intelligent Transport Society (2014) Smart Mobility 2030: ITS Strategic Plan for Singapore
4. Jiang J, Ye B, Sun Z, Zeng Z, Yang X (2023) Low-carbon energy policies benefit climate change mitigation and air pollutant reduction in megacities: an empirical examination of Shenzhen, China. Sci Total Environ 892:164644
5. Feng Y, Affonso RC, Zolghadri M (2017) Analysis of bike sharing system by clustering: the Vélib' case. IFAC-PapersOnLine 50(1):12422–12427
6. Dayal A (2024) Delhi Metro's environmental initiatives: a tale of innovation and dedication. Delhi Metro Rail Corporation (DMRC) 2024. https://www.globalrailwayreview.com/article/175414/delhi-metros-environmental-a-tale-of-innovation/#:~:text=DMRC%20has%20been%20certified%20by%helping%20to%20reduce%20global%20warming. Accessed 11 Jan 2025
7. Rosa J, Pereira AR, Santoro F, Gordijn J, Mira da Silva M (2023) Modeling Waymo's Shared autonomous vehicle service in phoenix using e3value. Int J Transp Develop Integrat 7:153–165
8. Gödde J, Ruhrort L, Allert V, Scheiner J (2023) User characteristics and spatial correlates of ride-pooling demand – evidence from Berlin and Munich. J Transp Geogr 109:103596
9. Change, Intergovernmental Panel on Climate (2023) Climate Change 2023 Synthesis Report
10. Cox BL, Mutel CL (2018) The environmental and cost performance of current and future motorcycles. Appl Energy 212:1013–1024
11. Adiang CM, Monkam D, Lenouo A, Njeugna E, Gokhale S (2017) Evaluating impacts of two-wheeler emissions on roadside air quality in the vicinity of a busy traffic intersection in Douala, Cameroon. Air Qual Atmos Health 10(4):521–532
12. Vasic A-M, Weilenmann M (2006) Comparison of real-world emissions from two-wheelers and passenger cars. Environ Sci Technol 40(1):149–154
13. United Nations, Department of Economic and Social Affairs (2022) World Population Prospects 2022
14. IQAir (2023) 2023 World Air Quality Report
15. Bangladesh Road Transport Authority (BRTA) (2024) Number of registered motor vehicles in Bangladesh (yearwise)
16. Pakistan Automotive Manufacturers Association (PAMA) (2024) July 2007 to June 2024 (Month Wise Data)
17. Society of Indian Automobile Manufacturers (SIAM) (2023-2024) SIAM Annual Report 2023-2024
18. Programme, Statistical Capacity Building (StatCaB) (2023) Transport Statistics: The Experience of Indonesia.
19. Vietnam Association of Motorcycle Manufacturers (VAMM) (2018-2024) Motorcycle Sales Trends in Vietnam
20. Safrudin A, Palguna A, Adrison V, Khoirunurrofik K, Haryanto B, Krisnawati L, Zakaria M, Hamonangan E, Nurafiatin L, Suhud M, Reksowardojo A, Huda A (2013) Cost benefit analysis for fuel quality and fuel economy initiative in Indonesia
21. Suwignjo P, Yuniarto M, Nugraha Y, Desanti A, Sidharta I, Wiratno S, Yuwono T (2023) Benefits of electric motorcycle in improving personal sustainable economy: a view from indonesia online ride-hailing rider. Int J Technol 14:38
22. Kaleg S, Hapid A, Kurniasih M (2015) Electric vehicle conversion based on distance, speed and cost requirements. Energy Procedia 68:446–454
23. Weinert J, Ogden J, Sperling D, Burke A (2008) The future of electric two-wheelers and electric vehicles in China. Energy Policy 36(7):2544–2555
24. International Energy Agency (IEA) (2024) Oil 2024 Analysis and forecast to 2030

25. Vespa (2024) History of the vespa. https://vespapizzeria.com/history-of-the-vespa/. Accessed 21 Oct 2024
26. Soomro F (2015) Vroom! Into the graveyard of Vespas we rode. DAWN
27. Karpenko M, Stosiak M, Deptuła A, Urbanowicz K, Nugaras J, Krolczyk G, Żak K (2023) Performance evaluation of extruded polystyrene foam for aerospace engineering applications using frequency analyses. Int J Adv Manuf Technol 126:1–12
28. TP S, Satheeshkumar S, Jesuarockiam N (2014) Glass fiber-reinforced polymer composites—a review. J Reinf Plast Compos 33:1258–1275
29. Piaggio (2024) Vespa 125 Super 150 Super Operation and Maintenance
30. Chang C-C, Fang-Ling W, Lai W-H, Lai M-P (2016) A cost-benefit analysis of the carbon footprint with hydrogen scooters and electric scooters. Int J Hydrog Energy 41(30):13299–13307
31. Duke A, Erazo C (2024) Transition to sustainable mobility: electric motorcycles as substitutes for gasoline-powered motorcycles
32. Park J, Calzavara J, Courtright T (2021) Environmental and social impact assessment of electric motorcycle Taxis in Kampala. School of Environment and Sustainability, University of Michigan, Uganda
33. Xuan P, Henz M, Weigl J (2013) Environmental impact of converted electrical motorcycle. World Electric Vehicle J 6:1136–1143
34. Tuayharn K, Kaewtatip P, Ruangjirakit K, Limthongkul P (2015) ICE Motorcycle and Electric Motorcycle: Environmental and Economic Analysis. Paper read at SAE Technical Paper Series
35. Habibie A, Hisjam M, Sutopo W, Nizam M (2021) Sustainability evaluation of internal combustion engine motorcycle to electric motorcycle conversion. Evergreen 8:469–476
36. Eviendy R, Asy'ari MZ, Sukra ZL, Peranginangin E (2024) Carbon emission and economical impact of conversion yamaha N-Max 150cc into electric motorcycle. IOP Conf Ser: Earth Environ Sci 1324(1):012094
37. Cherry CR, Weinert JX, Xinmiao Y (2009) Comparative environmental impacts of electric bikes in China. Transp Res Part D: Transp Environ 14(5):281–290
38. Carranza G, Do Nascimiento M, Fanals J, Febrer J, Valderrama C (2022) Life cycle assessment and economic analysis of the electric motorcycle in the city of Barcelona and the impact on air pollution. Sci Total Environ 821:153419
39. Amber KP, Ahmad R, Farmanbar M, Bashir MA, Mehmood S, Khan MS, Saeed MU (2021) Unlocking household electricity consumption in Pakistan. Buildings 11(11):566
40. MEPCO (2024) MEPCO Residential Tariff Rates 2024. https://mepcobillonlinecheck.pk/mepco-bill-calculator/. Accessed 19 Oct 2024

Blockchain for Smart Cities: Building a Secure and Efficient Future Urban

Meryem Lasaad, Yassin El Gountery, Abdeslam Jakimi, and Mohamed Oualla

Abstract The fast-paced global urbanization phenomena pose huge workload to the city managers, demanding smart solutions for creating sustainable, efficient and citizen-centric urban spaces. Answering to this clarion call, smart cities have swiftly come of age in their efforts to leverage cutting-edge technologies for smarter infrastructure and better public services leading to enhanced lifestyles. However, the very complexity and scalability characterizing the smart city systems call for resilient solutions towards data security, transparency and efficiency. It is at this juncture that blockchain technology, being both decentralized and immutable-transparent in nature, comes into play by addressing the envisaged challenges. This chapter examines the extent to which the challenges facing the smart cities in developing secure digital infrastructures and protecting the data of city operations from cyber-attack and unauthorized access can be resolved through blockchain technology. It explores the blockchain abilities to deliver trusted real-time transactions, support decentralized applications and improve citizen engagement through transparent governance. It then, based on the same, brings out challenges toward regulatory, operational, and technical integration of blockchain into urban systems and emphasizes the need that these barriers must be surmounted to use this technology to its full potential in advancing smart cities. The chapter finally presents and discusses several case studies on the effective application of blockchain within the ecosystem of current smart cities concerning practical benefits and potential.

Keywords Blockchain · Smart cities · Governance · Secure urban

M. Lasaad (✉)· Y. El Gountery · A. Jakimi · M. Oualla
GL-ISI Team, Faculty of Science and Technology Errachidia, Moulay Ismail University, Meknes, Morocco
e-mail: me.lasaad@edu.umi.ac.ma

H. Kaddouri et al. (eds.), *Smart Cities for Sustainable and Urban Development*, Green Energy and Technology, https://doi.org/10.1007/978-3-031-99001-4_26

1 Introduction: General Overview and Challenges of Smart Cities

The rapid growth of urbanization with increasing pressures on already existing infrastructures are challenges that cities all around the world will have to come to be able to meet the demands of their enlarging populations. Smart cities, therefore, evolve as an answer to these challenges [1] by exploiting cutting-edge technologies in the management of resources for quality living of citizens and sustainable urban environment.

One such key technology is blockchain [2]. The decentralized technology offers secure, transparent, and immutable record-keeping sharing. In intermediary-free transactions, blockchain assures integrity and security of data based on its cryptographic structure. In smart cities, blockchain can be integrated with infrastructure management systems and public services delivery systems based on IoT, AI, and ultra-high-speed communication networks. The level of these integrated systems is expected to enhance the operability of the city in general, and create a better living environment for the people residing in them.

The decentralized nature of Blockchain technology enhances security in operations and reduces associated risks of tampering. In addition, it fosters trust among stakeholders who combine efforts towards the realization of success. These critical processes will ensure that the city works hand in hand with such integration on matters revolving round digital identity management and public service automation to ensure that our future is efficacious that it is resilient in terms of urban environments.

In this chapter, we are going to try and have an in-depth analysis of how helpful blockchain technology is in addressing the issues smart cities are currently facing. We will also talk about practical applications, go through a few case studies which insist that blockchain is transformative for urban environments and, lastly, critically consider the challenges and limitations in its adoption in the development of smart cities. With such discussions, it is hoped that this chapter is better placed to reveal where the future trajectory of blockchain technology lies in ensuring more efficient, secure, and resilient urban ecosystems.

Intense integration of smart cities into complex digital ecosystems has brought about several vital inefficiencies and unsustainability challenges. These issues have to do largely with security, privacy, data management, scalability, and interoperability [3, 4], all difficulties that need to be resolved for the benefits of intelligent urban spaces to be fully realized. Beneath, we delve into these challenges in detail, examining how they impact the effective functioning and future development of smart cities (Fig. 1).

Fig. 1 Current challenges of smart cities

1.1 Privacy and Security Challenges

With the rise of smart cities and reliance on information and communication technologies for efficient and improved urban operations, there constantly arises a need to secure the heaps of digital data amassed. Existing more of a single system as a weakness, most of today's urban infrastructures use centralized systems make them akin to a single point of failure [5]. It is quite simple for these centralized architectures to find themselves under the attack of hackers, causing severe data breaches and the infiltration of unauthorized people, thus shaking the security and sanctity of sensitive citizen data [6]. The other issue at hand is the increased chances for widespread disruption when centralized control points dominate urban systems. The smart city infrastructures are highly interrelated, and thus one loophole in one part of the system is more likely to spill over to critical services, including transportation, health care, or even public safety.

At the same time, ensuring privacy has become a rather complex issue for any city that is now collecting all forms of personal information from various digital services and IoT gadgets. The balancing of operational efficiency with appropriate privacy protection is particularly problematic in regions bound by such rules as contained in the General Data Protection Regulation (GDPR) [7]. It has, therefore, made it increasingly difficult to administer information in an urban setting where the rule of law seeks to maintain high levels of transparency and access to information for duly authorized stakeholders. This also emphasizes the dire need for appropriate management of personal information for data security, especially considering the increasing worry among citizens over how their data is being used.

1.2 Data Management Challenges

The big data that smart cities generate, makes data management a colossal challenge because of its very volume and variety. It comes in from IoT sensors, digital platforms, and public service systems, and other sources to create a very complex ecosystem of data [8] that should be processed and analyzed in real-time. Just like in the case of conventional centralized data management architectures this often tends to bog down towards inefficiency and sluggishness. An added disadvantage is that these centralized systems will bottleneck and thus limit a city's potential to make quick decisions based on data for effective urban management.

The challenges extend beyond data processing to encompass issues of transparency and accountability. Centralized data storage Creates challenge regarding the potential for data manipulation or loss [9], which can erode public trust and compromise the reliability of smart city operations.

Without a system of coordination and standardization, the maintenance of accuracy and integrity of data across the many platforms and city departments is inherently very challenging. Further, the data systems have to be transparent to earn public trust, the traditional systems often lack the necessary mechanisms to guarantee the transparency.

1.3 Scalability and Interoperability Challenges

Massive urban technological growth and rapid connected device adoption raise important smart city scalability and interoperability problems [10]. Most of the legacy systems that underlie the present-day urban infrastructures are not able to scale up efficiently with the increasing volume of data traffic and service complexities. Such incapability to scale dynamically causes inefficiency as well as bottlenecks, particularly among those services dependent on real-time data processing, including traffic management, energy distribution, and emergency response systems.

For many obvious reasons, interoperability tends to be the most challenging of all [11]. Particularly, in a smart city, a patchwork quilt of technologies is used- and there are many different vendors involved. Obviously, since these extremely disparate systems are operating off silos, data exchange within and outside different city departments and among the external stakeholders will never become easy. In the absence of the common standard framework, the city is significantly obstructed from effectively introducing new ICT technologies and using cross-sector data analytics in search of novel solutions. Consequently, lack of efficient interoperability means that it does not optimize the operations of the city and, as a result, take full advantage of an interconnected urban ecosystem.

As a result, the combination of these problems works against the achievement of a holistic view of smart city development. Failure of systems to scale and integrate hampers operational efficiency, curtailing technological growth. Scaling and interoperability must be considered to let urban systems include the new technologies and also to assure the different parts of the urban landscape are able to communicate and work together. Blockchain technology emerges as a promising solution to address these challenges and unlock the full potential of smart cities.

2 Role of Blockchain in Unlocking the Smart City Challenges

For the development of city-as-a-platform digital ecosystems, the four key challenges that need to be resolved are data management, security, scalability, and interoperability. The potential of Blockchain technology with robust and distinctive capabilities serves to solve these issues to reshape the functioning of urban systems. Its application to smart cities will change conventional ways in data management, security, and efficiency and create new channels. In the following sections, we provide an in-depth overview of the basic characteristics of Blockchain, along with exemplary use cases that will help better appreciate how this technology can transform urban management.

2.1 *Fundamental Characteristics of Blockchain*

Blockchain technology has several core characteristics that make it uniquely capable of handling the in-built challenges of smart cities. One of the most important among these is decentralization [12]. As opposed to traditional systems, where data is centralized and there lies a single point of control, blockchain's data is spread across nodes on the network. Hence, chances of failure automatically become much lesser and so do overall vulnerability and exposure to threats. By distributing data in numerous participants, blockchain helps reduce vulnerabilities associated with centralized data storehouses, such as being more exposed to cyberattacks, and unauthorized access.

Transparency is one of the most important characteristics of blockchain technology [13]. All activities over the blockchain are visible to all permitted users, thereby creating a platform of openness and trust. Such openness is especially good for public governance in the smart city since stakeholders can independently verify transactions, thus ensuring a greater level of accountability. Citizens can track operations and check whether there is compliance, and this increases the level of trust in relation to the provision of services to the public [14].

It is through immutability that the integrity of any blockchain-based system is guaranteed. Once data is written on a blockchain, it can never be changed or deleted [15]. This is what establishes that immutability leads to consistency and impossibility of tampering with records, which is highly required for good history accuracy and anti-fraud/data manipulation measures, especially in smart city applications. In as much, the characteristic will ensure critical data used in public safety, energy management, transportation, among other departments is reliable.

These core characteristics of decentralization, transparency, and immutability place blockchain as a very powerful solution for enhancing security, trust, and working order in the labyrinthine details of smart cities.

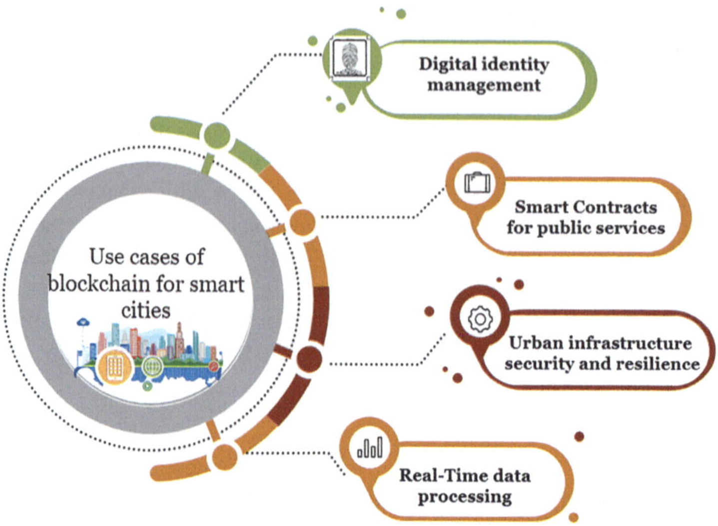

Fig. 2 Use cases of blockchain for smart cities

2.2 Use cases of Blockchain for Smart Cities

Blockchain technology offers the potential of a variety of solutions to the challenges that smart cities face for better urban management and service. Figure 2 provides an overview of the use cases of blockchain for smart cities.

2.2.1 Digital Identity Management

Management of digital identities through the blockchain depicts a qualitative path concerning the protection of personal data and the initiation of self-ownership of information. In contrast to the generally centralized identity systems that are exposed to data leaks and unwanted manipulation, the decentralized nature of blockchain does not have any single points of failure. It also means more overall system security. Coupled with immutability, blockchain decentralization ensures that identity information, once recorded, cannot be tampered with; hence, more assured verifications. This can further enhance through the concept of Self-Sovereign Identity (SSI), wherein the individuals would now have the management of their digital identities without being dependent on any third-party intermediate [16]. It, further, enables ownership and privacy for users such secure document sharing and validation, thereby unblocking access to a variety of services.

2.2.2 Smart Contracts for Public Services

The opportunities that the blockchain-based smart contracts offer to improve the efficiency, transparency, and security level of public services are enormous. Through the use of decentralized self-executing code, smart contracts can automate a process without any intermediaries; this has, therefore, improved governance mechanisms and reduced administrative costs [17]. In emergency services of a smart city infrastructure, smart contracts will allow automatic response so as to provide faster interventions and circumvent any weaknesses in centralization [18]. The immutability of blockchain technology guarantees that all transaction records are transparent, which builds trust between stakeholders and underpins integrity in the delivery of public services. This automates not only accelerates service delivery but also makes cost-effective, reliable management of the public sector.

2.2.3 Urban Infrastructure Security and Resilience

Adopting blockchain technology in the security and resilience of urban infrastructures is a major contribution to safety and efficiency operations in smart cities. By utilizing decentralization, immutability, and transparency as some of the key attributes of blockchain, urban systems can involve an enhancement of cybersecurity and overall resilience of their processes and systems. Noteworthy, blockchain reduces security incidents up to 40% according to [19]. The cryptographic techniques NTRU and McEliece within the blockchain framework will mitigate quantum computing threats to the integrity of communications in urban areas [20]. In addition, with decentralized incident reporting systems, residents can report issues in real-time to promote community engagement and, in turn, quicker governmental responses. Also, the integration of IoT devices with blockchain technology ensures secure continuous monitoring, enhancing the safety of citizens and the security of cities [21]. These measures together show the transformative potential for Blockchain in strengthening security and resilience of Smart City infrastructures.

2.2.4 Real-Time Data Processing

Real-time data processing and blockchain are the two big stepping stones that provide smart cities with functionality and operational efficiency. Their application results in service optimization, better data management, and communication among the different elements or systems in a city. The most basic requirement among them is the integration with the IoT systems, so that continuous traffic of data is achieved from various sensors to support urban management functions/matters like traffic and environmental control. This will then help improve the level of immediate decision-making on service delivery in critical areas such as health and transport systems [22]. Thus, smart cities can finally build infrastructures that are more responsive to modern urban challenges and are also more resilient.

The power of blockchain technology is such that it is able to flip the cards in way smart cities are operated by providing innovative solutions to urban challenges. Where data can be decentralized and secured through blockchain, it will help the cities to offer transparent, efficient, and cost-effective operations across different sectors for the betterment of a particular city.

3 Challenges of Integrating Blockchain

Although blockchain can be integrated into urban systems, there are many barriers and challenges. Therefore, the following section identifies and discusses the main obstacles inherent in the adoption of blockchain technology by cities. Figure 3 presents the challenges of integrating blockchain.

3.1 *Scalability and Energy Efficiency*

One of the major challenges of blockchain technology has been in its scalability. The growing urban systems mean that, at all times, there is an increasing load of transactions and data flow over the blockchain infrastructure while maintaining or increasing speed and performance. The current consensus mechanisms, such as Proof of Work (PoW) have brought many criticisms, since they are very resource-dependent and very high in energy consumption [23]. This is well exemplified in the Bitcoin blockchain, which has recently been claimed to consume as much energy as a country, hence environmental and sustainability issues.

Therefore, cities that are considering blockchain solutions should investigate alternative consensus algorithms that could balance efficiency with energy in the form of Proof of Stake [23] or hybrid models.

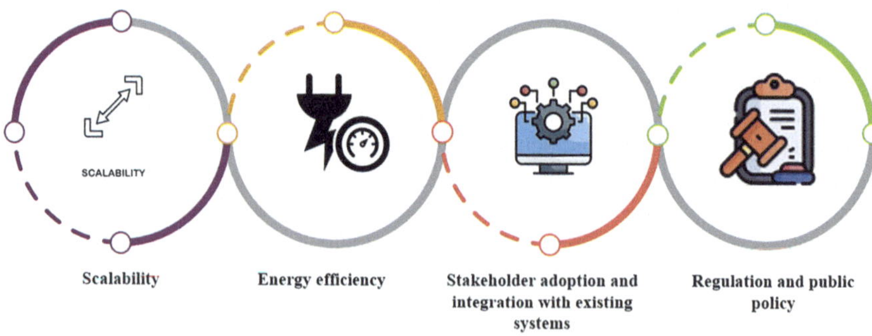

Fig. 3 Challenges of integrating blockchain

3.2 Stakeholder Adoption and Integration with Existing Systems

The integration of blockchain into the existing city infrastructure faces some other challenges. Most legacy systems depend on centralized architectures, which may not work well with the decentralized nature of blockchain and hence require extensive modifications to work properly. This will be an investment in upgrading technology, training staff, and developing middleware to make communication between the blockchain and legacy systems [24].

One big barrier and challenge to the mainstreaming of blockchain technology is the lack of knowledge among stakeholders. Not everyone is aware of what blockchain can do and this fuels a strong opposition when it comes to the consideration of its use. Stakeholders who don't have a good understanding of this technology are unlikely to interact with it as they will find it too complicated or even risky to deal with. The weight of the matter is felt most in sectors like education and health, simply because the authorities in these sectors do not quite comprehend the technical support that goes into the advantages and applications of blockchain technology. A notable study opines that this informational deficit breeds doubt particularly when blockchain is associated with digital currencies, thereby further impeding adoption [25].

It will be essential to educate the public and launch information campaigns to foster "blockchain literacy," whereby appropriate management of the system could then follow. Running pilot projects gradually with the implementation of blockchain will only show its benefits and the rest allow the stakeholders to give their valuable feedback for continuous improvement.

3.3 Regulations and Public Policy

Inclusion of Blockchain technology in Smart cities faces challenges of regulation and government policy. As it functions from a decentralized model, many of the originally set government rules for centralized systems are bypassed. Thereby, there arises a need to amend existing laws-mostly on privacy, security, and data flow-so as to fit the unique nature of blockchain. While offering many merits such as increased transparency and better security, the same features contained in blockchain may act as a point of friction with set regulations meant to safeguard personal information. Specifically, the decentralized, immutable and transparent nature of blockchain makes it difficult to comply with data protection laws such as the GDPR in the EU. For instance, personal data should be made easily modifiable and erased upon request. Such regulatory complexities are another thing that has held back the rolling out of blockchain technology in smart cities, as policymakers grapple with how to balance between embracing technological innovations and keeping the public's trust as well as addressing privacy-related issues [26, 27].

Another critical concern is the area of cross-border data flow. This is a seamless transfer with blockchain across borders, of data, which is particularly key in today's world with globalization. At the same time, this clash with different international laws related to the data protection of the country multiplies the uncertainty of regulations for blockchain-based applications in smart cities. For example, while blockchain technologies, as applied to the healthcare, energy or transportation sectors, are enabled to streamline their respective processes, without uniform international rules, this cannot work across borders. Policymakers need to ensure global support in creating regulatory harmonization so that decentralized potentials of blockchain technology are not compromised due to variations in data protection regulations. This also can sometimes be realized through those international cooperations and agreements, allowing the free and compliant flow of data across jurisdictions to also catalyze innovations about applications of smart cities depending on cross-border data exchange.

Summing up, the regulatory environment for blockchain technology in smart cities is manifold, complicated, and diversified. For this reason, the guiding principle for any policymaker should endorse an approach that fosters innovation alongside an addressal of concerning challenges related to data protection, security, and cross-border data flows. Governments could establish, in cooperation with experts in the field of technology, urban planning, and law, building blocks for legislation which would enable suitable and appreciable implementation of blockchain technology in smart cities. As blockchain technology keeps evolving, future work concerning an international regulatory framework, transparent governance demonstrating openness, and flexible legal rules is necessary to make best use of this technology in cities [28].

4 Integrating Blockchain in Smart Cities: Case Studies

Cities all around the world are continuing the adoption of blockchain technology to provide accurate solutions to urban challenges and bring in efficiencies along with offering secure that teem with a future tech-savvy ecosystem. Many cosmopolitan centers have therefore been placed within such infrastructure, demonstrating how this technology may change the face of cityscapes.

4.1 Case Studies

Dubai, UAE has ambitious plans of being the first blockchain powered city in the world through its Dubai Blockchain Strategy. This initiative is set to work on the implementation of blockchain with respect to property transactions and digital identity and government service applications aiming at improving efficiency through reduction of the associated bureaucracy [29]. Tallinn, Estonia is a place known for leading in using new kinds of tech to do things better with applications in many areas to do things better through transparency, security, and efficiency. For example, here is the e-Residency program introduced in 2014. It uses blockchain to keep the data safe of the people who are Estonian e-residents to get digital services in Estonia, like business registration and banking. Another is the national ID system of Estonia. It has implemented blockchain to store personal data; this allows safe access to services like online voting, healthcare, and management of finances with records that can be tracked and are untameably. The e-Health system proves one more application for blockchain ensuring secured storage and sharing of medical information, while keeping all access to patient records traceable to only authenticated users. Blockchains also secure the country I-Voting systems, wherein citizens cast their votes over the Internet, building more trust upon the systems due to their immutability and verifiability. This kind of implementation points Tallinn into the forefront of cities applying blockchain technologies to enable their citizens to have better digital access and thus set standards for digital societies all over. [30]. The City of Barcelona has launched several pilot initiatives concerning blockchain adoption in the management of energy and data sharing. These projects are set to use blockchain technology for building urban systems that are more efficient and sustainable [29].

Recently, South Korea has explored the application of blockchain technology in various sectors, including tourism and voting. A significant instance of the former is seen at Jeju Island, the top tourist attraction in the country. In collaboration with ICONLOOP in 2020, it put a blockchain-based mobile application into implementation for COVID-19 contact tracing [31]. An innovative system through which its 15 million annual visitors can check into tourist destinations and businesses by means of a mobile app ensures health safety while keeping personal identity and privacy safe. In addition, the country also proved the prospect of blockchain technology for use in governance by organizing a voting project that involved some 9000 residents in 2017. The initiative ran over smart contracts on one blockchain system thus showing the tech's capacity to better transparency, security as well as efficiency in public decision-making.

China leverages blockchain technology for the management of agri-food supply chains [32] and the advancement of e-government initiatives [33].

These case studies are indicating the trend about how blockchain is being adopted in smart cities and give a look at which it can be applied across varied urban sectors in creating safer, efficient, and transparent systems for the citizens. Table 1 gives a concise summary of the initiatives described above, highlighting their initiatives, applications, objectives, and references.

Table 1 Case studies of blockchain integration in smart cities

City/country	Blockchain initiative	Applications	Objective	References
Dubai, UAE	Dubai Blockchain Strategy	Property transactions, digital identity, government services	Improve efficiency and reduce bureaucracy	Srinivas and Donald [29]
Tallinn, Estonia	Blockchain in e-governance	Digital identities	Ensure data integrity and security for public services	Semenzin et al. [30]
Barcelona, Spain	Blockchain pilot initiatives	Energy management, data sharing	Build efficient and sustainable urban systems	Srinivas and Donald [29]
South Korea	Nationwide blockchain deployment	Tourism, healthcare, secure voting	Enhance public sector services	The World Bank [31]
China	Blockchain for agri-food supply chain and e-government	Agri-food supply chain management, e-government initiatives	Improve efficiency and transparency in agriculture and governance	Fu [32], Hou [33]

4.2 *Blockchain's Influence on Urban Systems and Economic Optimization*

Blockchain is not just a revolutionary technology in the management of the city; it provides transparency, efficiency, and security in all management areas. For urban management, using the decentralized ledger system of blockchain in managing data makes the provision of services in relation to registries and public records more efficient for cities [29]. For example, Dubai's blockchain-based transactions allow processes to be provably fast and tamper-proof reducing administrative overhead and building confidence among stakeholders. As a case study, the Dubai Blockchain Strategy will enable the city to save up to $1.5 billion every year by cutting out redundant paperwork and automating processes.

In smart transportation, blockchain allows for better coordination and information sharing among, for example, transit operators and the city with respect to real-time tracking and thus minimizing delays and improving overall traffic management. Blockchain applied to public transportation ticketing systems as well as those provided by sharing-ride apps injects extra layers of security and transparency in favor of heightened trust from users as well as better operational reliability. The use of blockchain technology normally reduces costs. Since the processes are simplified and intermediaries are eliminated, operations are regarded to be less costly to both city governments and businesses. Automation and verification of smart contracts reduce administrative costs. Summing up, the impact of the blockchain technology in urban management, and cost-efficiency proves causes no less than the fact it can revolutionize the urban landscape by making all systems more robust, transparent and economically viable.

5 Recommendations for Developing the Integration of Blockchain in Smart Cities

The fast pace of urbanization and the acute sustainability, security, and governance challenges cities are confronted with today readily drag the future of smart cities into clearly being contingent on technological progress, and more specifically on blockchain technology. The promises that blockchain holds in redefining urban infrastructures and service delivery are exciting. It is through emerging trends and plausible futures that indicate the most promising contribution of blockchain in smart city development when working in concert with other enabling technologies such as AI and IoT. This, in combination, would offer a much more resilient, transparent, and efficient city environment.

5.1 *Emerging Trends and Future Scenarios*

The development stages of the role of blockchain in smart cities are certainly in a state of early conceptualization. However, many trends indicate the manners by which technology will reshape urban life in the coming decades. Among the most prominent trends in this regard is the transition to decentralized modes of urban management. The immutable ledger and transparency features of blockchain could revolutionize the manner in which city governments indulge in any sort of rule, allowing more participatory processes of decision-making. In this respect, local governments can use blockchain to enable transparent voting systems, permit citizens to directly participate in policy discussions, and ensure that urban decisions are made based on verified, tamper-proof data [34].

In the field of urban planning, blockchain may usher in new forms of property and land systems. Decentralized registries on blockchain technology would cut down corruption in bureaucracy, safeguard against fraud, and enhance the efficiency of property transactions. This may enable greater mobility, lead to less real estate speculation, and guarantee fair shares of resources in land—especially in rapidly expanding cities [35]. In addition, blockchain has the potential to increase the scope and scale of public-private partnerships because decentralized platforms enable secure and transparent intersectoral collaboration in fields such as housing, energy, and transportation.

Another major smart trend that is coming up is fast smart contracting to drive urban automation processes. Governed by the technology of blockchain, self-executing contracts can make intermediaries in transactions and services redundant. For example, in public transportation, with data from IoT sensors, smart contracts can enable functionalities like real-time ticketing, payment systems, and route optimization. Blockchain may also be applied to municipal service contracts so that payments and services go through automatically without human intervention. It would reduce administrative costs, bring down errors and improve efficiency for urban management in general [14].

5.2 *Role of AI, IoT, and Blockchain in Building Cities*

The fusion of blockchain with AI and IoT is a critical enabler for the development of—smart cities that are more efficient, responsive, and sustainable. While Blockchain establishes the infrastructure for secure data exchange and decentralized applications, AI processes this data for optimizing city functions, and IoT devices generate real-time data streams. All these technologies working together will help curate and build a Smart City accordingly.

5.2.1 AI Algorithms

AI algorithms in smart cities is all meant to bring about efficient and improved operation and life. This is possible through efficient data management provision brought about by the use of AI, through smart decisions based on the analysis of a huge quantum of data. With AI, cities will become safer; there will be real-time analytics to detect anomalies and generally get better urban mobility in terms of traffic and other modes of cost-effective preventive maintenance as well as timely repairs to infrastructure. Along with the above, energy can be optimized energy efficiency, waste management, and health care. Along with this, citizen engagement would be enhanced and new avenues for solutions would be opened while incorporating blockchain into some applications including e-governance. Thus, this would be the solution to numerous city issues by AI and finally ended up building 'smart' cities that are sustainable, safe, and efficient [36].

5.2.2 IoT for Real-Time Data Collection

Data from all kinds of urban components such as infrastructure, public transport, and environmental monitoring systems are very important for smart cities toward developing adaptive responses to dynamic conditions. In practice, this means that the IoT will enable a garbage collection system to signal when bins are full, therefore optimizing routes and reducing fuel consumption [37]. More so, the integration of IoT with AI may further enable cities to meet real-time challenges by rerouting traffic in the event of accidents, controlling street lighting according to pedestrian movement.

City dynamics demand overarching coordination, and the confluence of Artificial Intelligence, the Internet of Things, and Blockchain is a step in this direction to develop agile, efficient, and resilient urban ecosystems.

6 Conclusion and Perspectives

Blockchain, among other technologies, can introduce a transformative wave of innovation focused on such themes as cybersecurity, transparency, and automation. The technology's major strength lies in its decentralized architecture, which is meant to create secure data management that results in data protection for citizens and assurance for services based on smart contracts. These attributes place blockchain at the heart of the enabling tools for future smart cities with the promise of leading to better-performing, more credible urban systems. Realization of the prospective benefits that come with such technologies demands a comprehensive approach. Rules that would spur innovation while tackling major issues that include protection of information, scalability, interoperability, and even the legal embrace of the technologically advanced and rapidly changing city setting from policy formulators, town planners, and technologists should be driven in unison. The increased complexity in the growing urban setting requires solutions that are not only technology-driven but also just, resilient, and inclusive in meeting the diverse requirements of all city residents.

Integration of blockchain with technologies, such as AI and IoT, promises substantial improvements in the quality and efficiency of smart city infrastructures. A demand-oriented approach for the design of scalable, citizen-centric solutions, adaptable to the dynamic evolution of needs in urban areas, is an area for future research. Developing measures to encourage the actual application of theoretical opportunities in real-world practice is particularly desirable in the short run from the policymakers involved. Cross-sectoral cooperation between government organizations, enterprises, educational institutions, and civil society is crucial for the reasons of innovation, but also to ensure that the offered technology-based smart cities are inclusive and environmentally responsible. Only then blockchain technology will have its further advance through the integration of technological development and the human approach in creating for everybody safe, open, and fair urban spaces.

Acknowledgements This work was supported by the Ministry of Higher Education, Scientific Research and Innovation, the Digital Development Agency (DDA) and the CNRST of Morocco (Alkhawarizmi/2020/32).

Competing Interests The authors declare that they have no competing interests.

References

1. Wolniak R, Stecuła K (2024) Artificial intelligence in smart cities—applications, barriers, and future directions: a review. Smart Cities 7(3):1346–1389. https://doi.org/10.3390/smartcities7030057
2. Biasin M, Foglie AD (2024) Blockchain and smart cities for inclusive and sustainable communities: a bibliometric and systematic literature review. Sustainability 16(15):6669. https://doi.org/10.3390/su16156669

3. Cui L, Xie G, Qu Y, Gao L, Yang Y (2018) Security and privacy in smart cities: challenges and opportunities. IEEE Access 6:46134–46145. https://doi.org/10.1109/access.2018.2853985
4. José R, Rodrigues H (2024) A review on key innovation challenges for smart city initiatives. Smart Cities 7(1):141–162. https://doi.org/10.3390/smartcities7010006
5. Islam MJ, Rahman A, Kabir S, Karim MR, Acharjee UK, Nasir MK, Band SS, Sookhak M, Wu S (2021) Blockchain-SDN-based energy-aware and distributed secure architecture for IoT in smart cities. IEEE Internet Things J 9(5):3850–3864. https://doi.org/10.1109/jiot.2021.3100797
6. Bhushan B, Khamparia A, Sagayam KM, Sharma SK, Ahad MA, Debnath NC (2020) Blockchain for smart cities: a review of architectures, integration trends and future research directions. Sustain Cities Soc 61:102360. https://doi.org/10.1016/j.scs.2020.102360
7. Breuer J, Pierson J (2021) The right to the city and data protection for developing citizen-centric digital cities. Inf Commun Soc 24(6):797–812. https://doi.org/10.1080/1369118x.2021.1909095
8. Garcia-Font V (2020) SocialBlock: an architecture for decentralized user-centric data management applications for communications in smart cities. J Parallel Distrib Comput 145:13–23. https://doi.org/10.1016/j.jpdc.2020.06.004
9. Chen J, Lv Z, Song H (2019) Design of personnel big data management system based on blockchain. Futur Gener Comput Syst 101:1122–1129. https://doi.org/10.1016/j.future.2019.07.037
10. Avelar E, Marques L, Passos DD, Macedo R, Dias K, Nogueira M (2014) Interoperability issues on heterogeneous wireless communication for smart cities. Comput Commun 58:4–15. https://doi.org/10.1016/j.comcom.2014.07.005
11. Alanazi F, Alenezi M (2024) Interoperability for intelligent traffic management systems in smart cities. Int J Electric Comput Eng 14(2):1864–1874
12. Bagloee SA, Heshmati M, Dia H, Ghaderi H, Pettit C, Asadi M (2021) Blockchain: the operating system of smart cities. Cities 112:103104. https://doi.org/10.1016/j.cities.2021.103104
13. Arun S, Maria R, Fadi Al-Turjman A (2022) A comprehensive survey on blockchain technology. Sustain Energy Technol Assess 52(Part A)
14. Cagigas D, Clifton J, Diaz-Fuentes D, Fernandez-Gutierrez M (2021) Blockchain for public services: a systematic literature review. IEEE Access 9:13904–13921. https://doi.org/10.1109/access.2021.3052019
15. Hofmann F, Wurster S, Ron E, Bohmecke-Schwafert M (2017) The immutability concept of blockchains and benefits of early standardization. ITU Kaleidoscope: Challenges For A Data-Driven Society (ITU K). https://doi.org/10.23919/itu-wt.2017.8247004
16. Liu Y, He D, Obaidat MS, Kumar N, Khan MK, Choo KR (2020) Blockchain-based identity management systems: a review. J Netw Comput Appl 166:102731. https://doi.org/10.1016/j.jnca.2020.102731
17. Balcerzak AP, Nica E, Rogalska E, Poliak M, Kliestik T, Sabie O-M (2022) Blockchain technology and smart contracts in decentralized governance systems. Administ Sci 12(3):96. https://doi.org/10.3390/admsci12030096
18. Lukić I, Miličević K, Köhler M, Vinko D (2022) Possible blockchain solutions according to a smart city digitalization strategy. Appl Sci 12(11):5552. https://doi.org/10.3390/app12115552
19. Govea J, Gaibor-Naranjo W, Villegas-Ch W (2024) Securing critical infrastructure with blockchain technology: an approach to cyber-resilience. Computer 13(5):122. https://doi.org/10.3390/computers13050122
20. Ma Y, Goyal SB, Rajawat AS, Bedi P, Yasmeen S (2024) Blockchain-based human intelligent systems for smart city safety. Trans Emerg Telecommun Technol 35(2):e4939. https://doi.org/10.1002/ett.4939
21. Hamdaoui B, Alkalbani M, Znati T, Rayes A (2020) Unleashing the power of participatory IoT with blockchains for increased safety and situation awareness of smart cities. IEEE Netw 34(2):202–209
22. Belludi S, Kopáčková H (2024) Securing Smart City Health Services Using Blockchain Technology. Zooming Innovation in Consumer Technologies Conference (ZINC), Novi Sad, Serbia, pp 188–193. https://doi.org/10.1109/ZINC61849.2024.10579418

23. Wong S, Chen C, Wu P (2023) Blockchain technology in smart cities and photovoltaic power generation: opportunities and challenges from a computer science perspective. J Smart Cities 8(1):1–12. https://www.udspub.com/ajj/public/index.php/jsc/article/view/1786
24. Chandan A, Potdar V, John M (2024) Systematic literature review of blockchain technology's technical challenges: a tertiary study. Information 15(8):475. https://doi.org/10.3390/info15080475
25. Morar CD, Popescu DE (2024) A survey of blockchain applicability, challenges, and key threats. Computer 13(9):223. https://doi.org/10.3390/computers13090223
26. Hakak S, Khan WZ, Gilkar GA, Imran M, Guizani N (2020) Securing smart cities through blockchain technology: architecture, requirements, and challenges. IEEE Netw 34(1):8–14. https://doi.org/10.1109/mnet.001.1900178
27. Nguyen S, Chen PS, Du Y (2023) Blockchain adoption in container shipping: an empirical study on barriers, approaches, and recommendations. Mar Policy 155:105724. https://doi.org/10.1016/j.marpol.2023.105724
28. Finck M, Pallas F (2020) They who must not be identified—distinguishing personal from non-personal data under the GDPR. Int Data Privacy Law 10(1):11–36. https://doi.org/10.1093/idpl/ipz026
29. Srinivas TAS, Donald AD, Srihith ID, Anjali D, Chandana A (2023) Blockchain: the future of smart city development. Transactions 3(1)
30. Semenzin S, Rozas D, Hassan S (2022) Blockchain-based application at a governmental level: disruption or illusion ? The case of Estonia. Polic Soc 41(3):386–401. https://doi.org/10.1093/polsoc/puac014
31. The World Bank (2021) Emerging Technologies Curation Series. https://documents1.worldbank.org/curated/en/550471619071735133/pdf/Korea-Blockchain-Ecosystem.pdf
32. Fu H, Zhao C, Cheng C, Ma H (2020) Blockchain-based agri-food supply chain management: case study in China. Int Food Agribus Manag Rev 23(5):667–679. https://doi.org/10.22434/ifamr2019.0152
33. Hou H (2017) The application of blockchain technology in E-Government in China. In: 26th International Conference on Computer Communication and Networks (ICCCN), Vancouver, BC, Canada. https://doi.org/10.1109/icccn.2017.8038519
34. Perera TS, Garusinghe AN, De Silva TS, Senevirathne TK, Abeywardena KY, Fernando H (2023) Electronic voting system using blockchain technology. In: 2023 5th International Conference on Advancements in Computing (ICAC), Colombo, Sri Lanka. https://doi.org/10.1109/icac60630.2023.10417584
35. Ooi V, Peng SK, Soh J (2022) Blockchain land transfers: technology, promises, and perils. Comput Law Secur Rev 45:105672. https://doi.org/10.1016/j.clsr.2022.105672
36. Ressi D, Romanello R, Piazza C, Rossi S (2024) AI-enhanced blockchain technology: a review of advancements and opportunities. J Netw Comput Appl 225:103858. https://doi.org/10.1016/j.jnca.2024.103858
37. Bhardwaj V, Anooja A, Vermani LS, Sunita N, Dhaliwal BK (2024) Smart cities and the IoT: an in-depth analysis of global research trends and future directions. Discov Internet Things 4(1). https://doi.org/10.1007/s43926-024-00076-3

The Use of Floating Solar PV to Power Wireless Charging Technology for Autonomous Underwater Vehicle: Charting a Path Towards Sustainable Smart Cities

Ramamurthi Sekar and Narayanamoorthi Rajamanickam

Abstract Integrating floating solar photovoltaic (PV) systems with wireless power transfer (WPT) charging stations represents a cutting-edge advancement in smart city infrastructure, this approach eliminates the need for conventional plug-in methods and helps reduce carbon emissions. It offers a sustainable solution for powering Autonomous Underwater Vehicles (AUVs) in urban aquatic settings, reduce dependence on the power grid and minimize environmental impact. This chapter highlight the potential of this integrated renewable energy system to revolutionize urban aquatic management, improve the quality of urban life, and promote sustainable practices within smart city frameworks. The findings advocate for further exploration of this innovative solution to maximize the benefits of autonomous marine technologies improving the overall quality of urban life in smart city. 3 kW UWPT system was implemented with series series compensation topology with the coupling coefficient of 0.2 to attain efficiency over 96%. SS compensation and buck converter is implemented to provide constant voltage charging method to charge the AUV battery with 60% SOC. Simulation outcomes validated the theoretical analysis of the proposed method.

Keywords Wireless power transfer · Autonomous underwater vehicle · Floating solar PV · State of charge

---

R. Sekar · N. Rajamanickam (✉)
Department of Electrical and Electronics Engineering, Wireless Charging Research Centre, SRM Institute of Science and Technology, Chennai, India

H. Kaddouri et al. (eds.), *Smart Cities for Sustainable and Urban Development*, Green Energy and Technology, https://doi.org/10.1007/978-3-031-99001-4_27

1 Introduction: Contribution of Autonomous Underwater Vehicle (AUV) in Developing Smart Cities

Underwater wireless power transfer (UWPT) has emerged as an innovative technology, empowering the efficient transmission of energy through aquatic environments without the need for physical connectors. Where traditional power supply methods are often impractical or impossible. The concept of wireless power transfer itself dates back to the late nineteenth century, with Nikola Tesla's experiments demonstrating the potential for transmitting electricity through the air. However, the challenge of transferring energy through water, a medium that significantly attenuates electromagnetic waves, has posed unique difficulties [1]. In the twentieth century, researchers began to investigate the feasibility of using low-frequency electromagnetic fields for UWPT. Although initial efforts were limited in range and efficiency. The twenty-first century has witnessed significant advancements in UWPT technology [2–4]. Researchers have developed systems utilizing resonant inductive coupling for under water applications [5], a method that allows for efficient power transfer over short distances [6]. This technique relies on the principle of resonant circuits to maximize energy transfer between coils, making it suitable for applications like powering underwater applications [7–10]. Solar PV powered autonomous under water vehicle was proposed in [11]. These advancements have sparked interest from industries ranging from telecommunications to renewable energy, where underwater installations are increasingly necessary. Sustainable cities require creative ideas to increase productivity and lower carbon footprints. Floating solar photovoltaic (PV) systems reduce their ecological impact and maximize land use. Furthermore, autonomous underwater vehicle was energized by wireless power transfer (WPT) technology. Through environmental monitoring, infrastructure inspection, and survivor searches, AUVs can enhance urban management in smart cities. These AUVs maintain operational autonomy by extending range and reducing fuel consumption by integrated floating solar PV panels and wireless power transfer system. This chapter looks at how wireless power transfer and floating solar PV can create a robust and flexible environment for AUVs in smart cities. We anticipate that this combination will enhance urban aquatic ecosystem's environmental operational effectiveness, and energy efficiency of an autonomous underwater Vehicle.

1.1 Role of AUV in Smart City

As cities become smart, technology must be integrated to manage resources, infrastructure, and services. Autonomous Underwater Vehicles (AUVs) are a smart city technology that benefits metropolitan areas with major water bodies including rivers, lakes, and coastal locations. Smart cities rely on AUVs for environmental monitoring. They may assess water quality, detect contaminants, and monitor ecological health. AUVs provide crucial data on temperature, pH, and pollutant concentrations to city planners and environmental authorities [12]. This data helps make pollution control and habitat protection decisions, which promotes sustainable urban growth. AUVs are needed to check underwater bridges, dams, and pipelines. Regular AUV inspections may uncover structural faults and corrosion before they worsen. Early problem-solving enhances public safety and minimizes maintenance costs. AUVs with modern imaging and sensor technology can provide complete evaluations, helping allocate maintenance and upgrade resources (Fig. 1).

AUVs may search and rescue urban waterways in crises. Rescue teams benefit from their ability to navigate underwater situations and quickly assess submerged regions. AUVs can speed up reaction times by providing real-time data and pictures, perhaps saving lives. AUV data may improve urban administration by being incorporated into smart city frameworks. This connection allows water quality and infrastructure health trends to be analyzed for predictive maintenance and environmental policymaking. AUVs may also be integrated into a broader fleet of drones and autonomous vehicles to improve data sharing and urban operations. AUVs have great promise in smart cities. These vehicles may improve urban sustainability and resilience by monitoring the environment, checking infrastructure, and boosting emergency response. AUVs may become vital instruments for urban management as cities adopt smart technology.

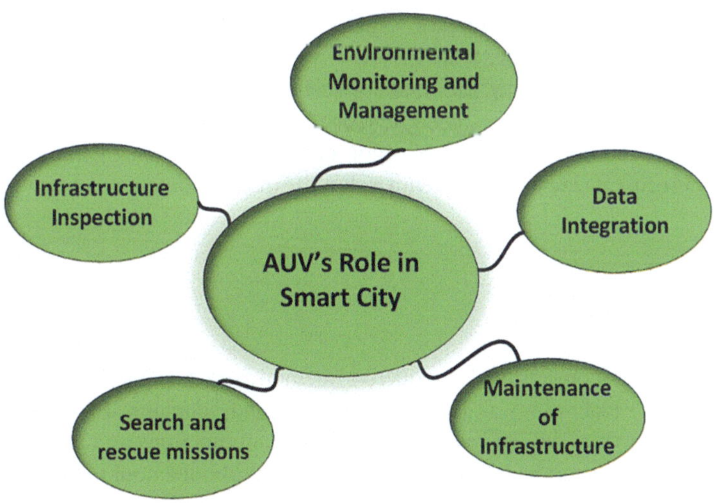

Fig. 1 Functions of AUV in smart city

1.2 Existing Charging Methods for AUV

1.2.1 Battery Swapping

Directly swapping the batteries of AUVs is the conventional and effective scheme. It is the speed process to power the vehicle, to perform this task involve many steps like searching, surfacing the vehicle, salvaging and the crew with mother ship is essential for operation. Bluefin-21 model AUV battery swapping time is about less than 30 min, for this operation the surfacing of the vehicle nearly takes 2 h, fully charged battery can power the vehicle up to 16 h (*Bluefin 1.5 KWh Subsea Battery—General Dynamics Mission Systems*, n.d.) [36].

1.2.2 Solar Charging

Solar charging delivers a wide range of power for unmanned underwater applications, but underwater vehicle should be near to water surface during day time to harvest solar power and store energy in batteries. The SAUV II was designed with one square meter of solar PV panels which delivers 32 Volt and 2 kW h energy system [11, 13]. However, solar charging structure of energy gaining suffers from weather restrictions and poor suppression.

1.2.3 Plugin Charging

Underwater plugin charging method usually uses a wet mate connector to permit underwater cable connecting. REMUS AUV was charged using dual stab connector [14]. Nevertheless, achieving accurate connection of cables in underwater is difficult task. Wet mate connectors need high accurate arrangement and makes system not to failure due corrosion in cable biological contamination of the equipment. These arrangements are expensive with a shot lifespan and suffer conductive electrical sparks.

1.2.4 Wireless Charging

Compared to the conventional conductive method power supply, wireless power transfer (WPT) technology is more convenience and adaptable for high applications. It offers a great deal of operating flexibility and functions as a reliable energy source for electronic gadgets and electric vehicles. Without a doubt, wireless power transfer (WPT) significantly improves the safety and dependability of charging systems by addressing problems with insulation and wire wear brought on by contact friction. Underwater wireless power transfer (UWPT) technology is a new charging method that has been thoroughly studied and used in recent years [8, 9, 15, 16]. It successfully addresses the issues with conventional underwater charging methods

Table 1 Exploring power supply methods in underwater, merits, and limitations

Method of power supply	Operating principle	Merits	Limitations
Battery swapping	Decoupling the vehicle from its energy source and replace a depleted battery with a fully charged battery.	Fast recharge of electrical power	Manual operation involved
Solar charging	Photovoltaic effect	Adequate energy reserves	Energy Conversion efficiency not high, highly affected by weather change, and Poor concealment
Plug in charging	Underwater wet mate plugging technology	Transmission efficiency was high	High accuracy docking required and Costlier
Wireless charging	Electromagnetic induction	Better transmission efficiency, High concealment, High energy density	Small transmission distance, and attenuation of Seawater

and provides an adaptable, effective, and intelligent energy supply solution. The energy supply techniques, fundamentals, main benefits, and possible disadvantages of common underwater devices in underwater docking systems are outlined in this review.

1.2.5 Underwater Submerged Docking Systems

Underwater docking is crucial for dependable power transmission, regardless of the kind of underwater equipment or how it is charged. Underwater docking technology has been extensively studied in recent decades and offers a viable way to increase the underwater equipment' observation range. The light absorbing nature of saltwater makes it difficult to place equipment underwater, unlike on land, and often necessitates the use of optical or audio navigation techniques. Usually, docking stations are powered directly by the mother ship or connected to shore power or renewable energy sources using undersea mooring cables [14] (Table 1).

2 Underwater Wireless Power Transmission (UWPT)

In recent years, there has been a notable increase in research on underwater wireless power transmission (WPT), with an emphasis on wireless powering underwater sensors as well as battery charging for autonomous underwater vehicles (AUVs). The dipole coil-based magnetic coupler, designed for AUV underwater wireless charging, offers a constrained magnetic field and reduced weight compared to annular couplers. Using Fe-based nano-crystalline material, it performs similarly to ferrite, achieving 633.37 W power transfer and 89.67% efficiency in seawater when aligned [17]. An 80KW system for AUVs was also created, with a 5.18KW prototype showing good efficiency and tolerance to rotational misalignment [18]. A self-latching coupling structure addresses seawater attenuation and current disturbances, achieving 92% efficiency at a 5 mm gap [19]. A 360° folded unipolar coupler minimizes magnetic leakage and EMI impact, transferring 5KW with 82.9% efficiency despite misalignment [20]. Other designs, including trumpet-shaped and arc-shaped structures, allow stable wireless power transfer, with one achieving 91.9% efficiency at 3 kW. Systems also maintain high efficiency even with 30 mm axial and 10° rotational misalignments [21]. New coil structures like the Ring Type and bipolar arc transmitter offer efficient charging, achieving up to 664 W with 92.26% efficiency despite misalignment, and are interoperable with AUVs of different sizes [22]. The proposed system uses a Curly DD-DD arc bipolar transmitter paired with a dipole coil-based receiver, offering a lightweight design that fits seamlessly on the surface for stable and efficient wireless power transfer. This system, which incorporates an arc bipolar transmitter and a compact dipole coil receiver, ensures strong, reliable coupling even in the event of misalignment. It can consistently deliver 11 A of current at a voltage of 54 V, while providing excellent interoperability for Autonomous Underwater Vehicles (AUVs) with varying sizes and power requirements, supporting up to 1.05 kW [23] Additionally, a trumpet-shaped structure is introduced, designed to align with the architecture of AUVs, preventing rotational misalignment and enabling wireless power transfer of up to 2 kW [24]. Table 2 shows the AUVs existing AUV and its performance with profile. Receiver with multiple structure for UWPT [29]. UWPT system with large power density for ring shape primary and secondary with the series series topology was implemented [30]. Optimizing ratio between two primary side capacitors, mixed character of SS and parallel series (PS) to gather better performance, problem related to charging time [31]. Double side compensator technique used to suppress misalignment effects, a double LCC compensation was implemented to analysis the enactment of an AUV under rotational misalignment [32]. LCC-P compensation allows constant current output, without complex control methods.it simplifies the structure with less weight of additional components in [18]. To attain a small structure H shaper secondary with LCC-P compensation was developed [4, 8]. Output power will be increased using the LCC-S compensation [18]. Data Transfer System with Full-Duplex Communication model for AUV [16]. FSK method power and data transfer using shared channel [33]. 5.18 KW prototype of the system that attain better enactment

Table 2 AUVs with and without WPT technology

Reference	AUV	Developer	Diving profile	Power	Endurance
Wolek et al. [25]	BLUEFIN 21	Bluefin Robotics	4500 m	13 kW	25 h
Allen et al. [26]	REMUS 600	Hydroid	600 m	1500 W	70 h
Allen et al. [26]	REMUS 100	Hydroid	100 m	1200 W	10 h
Kawasaki et al. [27]	MARINE BIRD	Kawasaki	100 m	11 KW	5 h
Bellingham et al. [28]	Odyssey II (WPT)	MIT/WHOI	–	200 W	6–10 h

in saltwater good efficiency. The coupling structure with outstanding tolerance to misalignment with 360 degree achieve better efficiency at broad power range [19]. Trumpet shaped coil structure is presented much suited with AUVs structure, which prevent misalignment under rotation at high degree, permitting stable, efficient wireless power transfer of 2 KW [24]. Generally, the control schemes for the WPT system should be adjusted through key aspects such as AC/DC convert control, compensation circuit control, AC/DC converter control, and DC/DC converter control. Dual control schemes will be a taking grater control to achieve good efficiency is a better area for research work. Hybrid compensation topology will improvement of the power transfer capability of UWPT system is also the key area but it was not focused. Research into underwater data transfer for battery management systems is lacking, highlighting a critical gap in understanding how to effectively communicate and monitor battery performance in aquatic environments to optimize energy efficiency and operational reliability in marine applications. The research gap in underwater magnetic couplers for wireless power transfer presents an opportunity to enhance efficacy in powering underwater devices, as current technologies have not fully optimized efficiency.

3 Integration of Floating Solar PV and WPT system

Underwater wireless power transfer technology combined with floating solar photovoltaic (PV) power systems provides a state-of-the-art approach to producing and distributing renewable energy. Since of this integration, solar energy may be captured efficiently while using less land since floating PV systems can be installed on water surfaces. Underwater wireless power transfer eliminates the necessity for large cables by allowing the generated electricity to be sent to submerged energy storage devices or shore-based facilities. In addition to improving energy availability in required places, this combination reduces the effect of environmental in conventional transmission methods. Moreover, water may surge the effectiveness of solar panels by cooling them, producing a sustainable and efficient power source.

The clean renewable energy source is solar energy, which is produced by converting solar radiation into thermal or electrical energy. Location, time of year, and weather all affect the amount of solar energy available. Photovoltaic cells in solar panels directly convert sunlight into electrical power. Commercial solar panels usually have an efficiency of 10% to 12%, however experimental panels can achieve 20–25%. Floating solar panels are frequently used to capture solar energy. 17.98 MJ/m^2 of solar energy are received daily on average worldwide [34].

Autonomous underwater vehicles (AUVs) can also be powered by solar energy by using solar panels on buoys that float to charge batteries that allow the vehicle to function underwater. Even in low light conditions, modern solar technology can effectively store energy for use in wireless power transfer (WPT) systems. Figure 2 shows a floating solar PV system that is linked to a mooring line to recharge an AUV's battery.

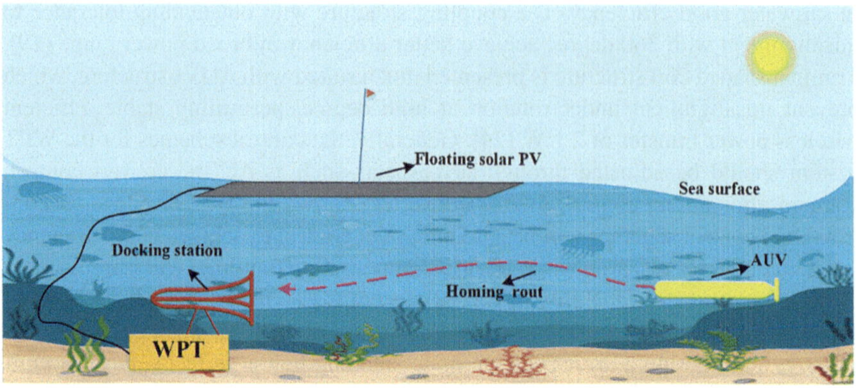

Fig. 2 Floating solar PV and WPT system

Fig. 3 Equivalent circuit of PV

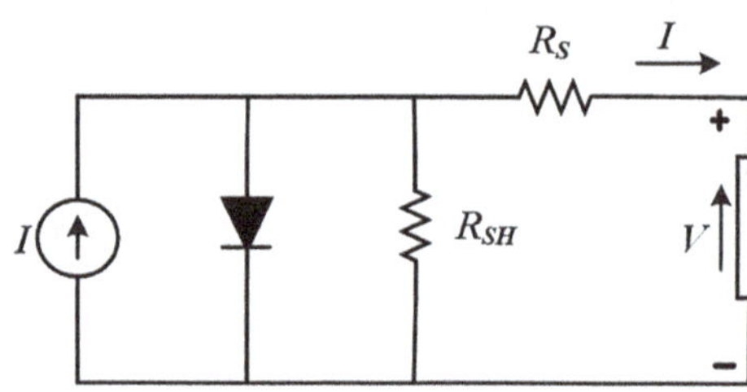

3.1 Equivalent Circuit and Characteristics of PV

A solar cell consists of a current source parallel with a diode. Since solar cell is not a ideal one, a shunt and a series resistance are connected. Series connected resistance is to oppose the electron flow. The equivalent circuit diagram of a solar cell is shown in Fig. 3. Solar PV panel specification was given in Table 3.

Where, I is the solar cell current, V is the solar cell voltage, T is the temperature in Kelvin and n is the ideality factor.

$$I = I_{SC} - I_o \left(e^{\frac{Vd}{Kt}} - 1 \right) \tag{1}$$

By rearranging Eq. (1) we get

$$I = -I_o \left(e^{\frac{V+IRs}{Kt}} - 1 \right) + I_{sc} \tag{2}$$

Table 3 Parameter of PV source

Description	Parameter	Value
Peak power	PMAX	580(W)
Open circuit voltage	VOC	52.21(A)
Short circuit Current	ISC	14.20 (A)
Peak power voltage	VMP	44.06(V)
Peak power current	IMP	13.17(A)
Maximum system voltage	VMax	1500 (V)
Power Tolerance	–	−5 to +5%
Series fuse	–	25 (A)
Application class	–	Class A
Safe class	–	Class II
Temperature coefficient of Pmax	nVocT	−0.38%/°C

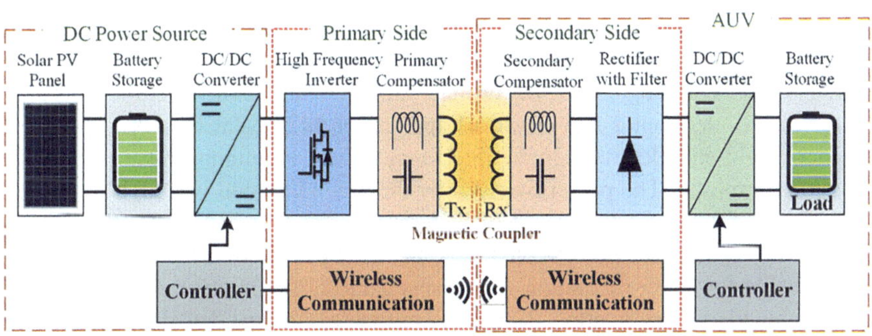

Fig. 4 Structure of IWPT system

3.2 *Design Consideration of UWPT and Standards for WPT*

AUVs—underwater contact less charging is most challenging, the IWPT technology is much suitable for reliable operation. This system contains two devices namely docking system and the AUV. The docking system is used for charging the underwater vehicle. AUV was designed to receive power from the primary side, AUV secondary side consist of receiver coil, resonant circuit and the AC/DC power converter (rectifier), output voltage regulator DC/DC converter and battery unit to store power. The block diagram of IWPT is represented in Fig. 4. For the analysis purpose IWPT system was drawn as a simple equivalent circuit it was shown in the Fig. 5. IWPT operation is similar to the electrical transformer, for transformer action the ac supply is required, from the available power it is essential to create the high frequency AC waveform to transfer power, and Inverter will take care of this function. Inverter is connected to a DC bus it is powered by a DC/DC, AC/DC converter, DC generator or other renewable energy source connected to DC Bus. A single-phase H-bridge inverter is commonly used to create AC square wave with high

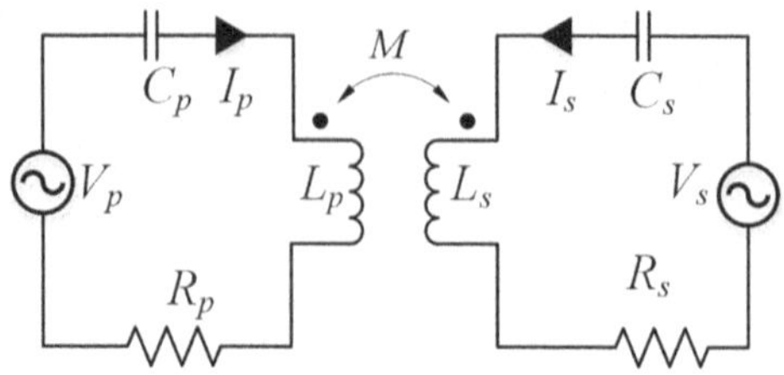

Fig. 5 Model based on equivalent voltage source of IWPT system

frequency. Resonant tanks and primary coil and secondary coil are designed to transmit and receive the power then it will be converter by the output power converter AC/DC to deliver energy the load. This area is very much significant for the development better performance IWPT system. If the power ratting and transmitting distance increased, resonant IWPT is advisable for better efficient power transfer. The primary and secondary consist of a core and coil. This design was taken from the transformer, the transform core is spitted into two parts, one placed in docking system and other place in AUV. The core provides proper path for magnetic flux, and good coupling among the primary and secondary coil. Shielding was necessary for all electronics and electrical application to prevent the high frequency magnetic field and electromagnetic interference, here aluminum metal is used for shielding purposed. The power received by the secondary coil was high frequency AC voltage, for onboard charging of battery, requires a stable DC voltage it was provided by the output power converters and they provide the feedback signal to primary side for better regulation with respected to load (Battery). All the parameters combined together to attain the desired output, by tuning control unit and power electronic convert. Receiver section contains a rectifier it will determine the efficiency. The output of AC/DC converter is connected with DC/DC power converters it will deliver power to the load or batteries. It is essential to design a closed loop control system to regulate the output voltage or current during load variation for battery charging process to achieve high efficiency. At Seawater data transmission between the transmitter and receiver will be quite challenging. In order to optimize efficiency in response to load changes during battery charging, controller units are essential in IWPT. They dynamically modify output voltage or current. In the charging process of electric vehicles, three primary methods are employed: dual side, secondary, and primary control.

3.3 UWPT Design

AC current is injected in the primary coil, the magnetic field formed change over time (according to ampere's law). The magnetic flux that changes over time causes a current in the receiving coil (Faraday's Law) as depicted in the equation given below

$$\oint B.ds = \mu_o I N_1 \quad (3)$$

$$E = -N\frac{d\phi}{dt} \tag{4}$$

where B, μ_o, and E, denotes magnetic flux density, vacuum magnetic permeability and induced electromotive force (emf) respectively.

The quality factor (Q) is the ratio of reactive power of either the inductor or capacitor to the average power of the resistor at resonance. Is given by

$$Q = \frac{Reactive\ Power}{Average\ Power} \tag{5}$$

The geometric mean of the transmitter and receiver side's quality factor and it is written as

$$Q = \sqrt{Q_t Q_r} \tag{6}$$

The resonant frequency is given by

$$\omega_o = \frac{1}{\sqrt{L_P C_P}} = \frac{1}{\sqrt{L_S C_S}} \tag{7}$$

Mutual inductance (M) is defined as

$$M = k\sqrt{L_t L_r} \tag{8}$$

The coupling coefficient in seawater is expressed as

$$K = \frac{M}{\sqrt{L_P L_S}} \tag{9}$$

Based on fundamental wave analysis method, the RMS value of V_{in} and V_o are written as

$$V_{in} = \frac{2\sqrt{2}}{\pi} V_{dc}, \quad V_o = \frac{2\sqrt{2}}{\pi} V_L \tag{10}$$

The secondary current to primary current ratio can be calculated as

$$\frac{I_s}{I_p} = j\frac{\omega_0 L_m}{R_L + R_s} \tag{11}$$

The fundamental output voltage Vin of the inverter is dependent on conduction angle Φ, defined as

$$V_{IN} = \frac{2\sqrt{2}}{\pi} V_{dc} \sin\left(\frac{\phi}{2}\right) \tag{12}$$

Equivalent resistance Req at the input side of the rectifier and output voltage of the rectifier are derived as

$$R_{eq} = \frac{8R_{dc}}{\pi^2} \tag{13}$$

The DC voltage expressed as

$$V_{dc} = \frac{\sqrt{2}\pi V_{out}}{4} \tag{14}$$

The transmitter and receiving side inductive impedance of WPT system and is defined by ignoring parasitic capacitance of both windings.

$$Z_t = R_t + j_\omega L_t \tag{15}$$

$$Z_r = R_r + j_\omega L_r \tag{16}$$

The input power and output power was calculated as

$$P_{in} = I_1^2 Z_T \tag{17}$$

$$P_L = I_L^2 R_L \tag{18}$$

The power transfer efficiency of the WPT system is given by equation

$$(\%) = \frac{P_L}{P_{in}} \times 100 \tag{19}$$

The system efficiency is

$$\eta = \frac{I_s^2 . R_L}{U_s I_p} = \frac{(j\omega M_{sea})^2 R_L}{Z_P Z_s^2 + (\omega M_{sea})^2 Z_s} \tag{20}$$

Compensation capacitor for primary side is calculated by using Eq. (21)

$$\frac{C_2 L_2}{L_2} \tag{21}$$

Compensation topology play vital role to enhance the efficiency of wireless power transfer system, basic compensation topology such as series series (SS), series parallel (SP), parallel series (PS), parallel parallel (PP), and the hybrid topology such as technique LCC-S, LCC-P, LCC-LCC, S-LCC, LCL-LCL, LCL-P, LCL-S, P-PS, SP-S, S-PS [35]. Each and every topology has its own advantage as well as the limitations. For simplicity of construction SS topology was applied in this article. The load power and power transfer efficiency while using series series compensation for UWPT system was calculated by using the equation.

The load power of the ss system was expressed in Eq. (22)

$$P_{L-ss} = A_{ss}^2 I_t^2 R_L \tag{22}$$

where, A_{ss} is current gain

$$A_{ss} = \frac{I_t}{I_r} = \frac{j\omega M}{R_L}$$

Series compensation power transfer efficiency is given in the expression (23)

$$\eta_{ss} = \frac{A_{ss}^2 R_L}{Z_r} = \frac{R_L}{(R_L + R_r)\left(1 + \frac{(R_L + R_r)R_t}{\omega^2 M^2}\right)} \tag{23}$$

Maximum efficiency and its condition was stated in Eq. (24)

$$\eta_{SS-\max} = \frac{R_L}{R_L + R_t} \quad where, \quad \omega \gg \frac{\sqrt{(R_L + R_r)R_t}}{M} \tag{24}$$

3.4 SAE Standard for WPT

In order to develop an industry standard for the interoperability, frequency band, electromagnetic compatibility, minimum performance, safety, testing criteria, and coil definitions for the wireless charging of light-duty electric and plug-in electric vehicles, the Society of Automotive Engineers (SAE) established the SAEJ2954 international committee in 2010.On May 31, 2016, the most recent edition of the SAEJ2954 technical information report (TIR) was made available for purchase on the SAE website. This version is designed for static charging applications and focuses on unidirectional charging from the grid to the car. Depending on input from the industry, future considerations may include bidirectional power and dynamic charging, which charges while the vehicle is moving [35]. According to Table 4, the TIR sets power levels across four classes and creates a common frequency range of 85 kHz (ranging from 81.39 kHz to 90 kHz) for all light-duty vehicle systems.

Table 4 SAE J2954 light duty vehicle WPT charging classification [35]

Classification	WPT 1	WPT 2	WPT 3	WPT 4
Level of Power	3.7 KW	7.7 KW	11.1 KW	22 KW

Table 5 Charging standard defined by SAE J1772 [35]

Charging level	Power level	Supply voltage (V)/Current (I)	Setting
AC Level 1	1.7 kW	Single Phase 120 V/20 A AC (16 A continuous)	Residential/Parking lot
AC Level 1 (Minimum)	3.4 kW	Split Phase 208 V/240 V/20 A AC (16 A continuous)	Residential/ commercial
AC Level 1 (Minimum)	19.2 kW	Split Phase 208 V/240 V/20 A AC (80 A continuous)	Commercial

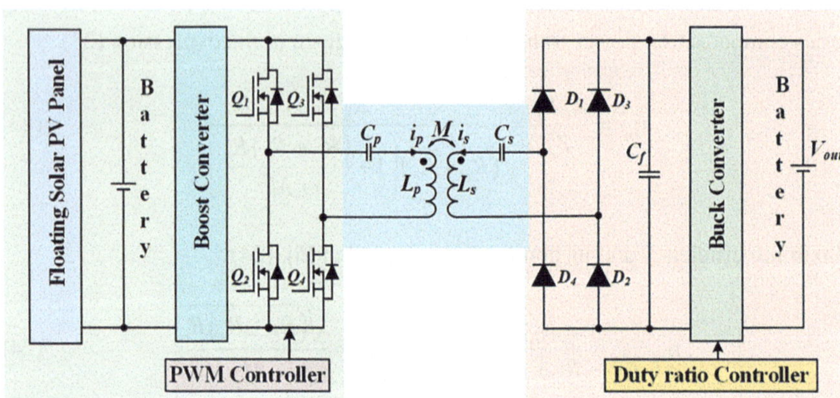

Fig. 6 Solar power WPT system For AUV charging

The charging standards established by SAE J1772 for conductive charging of plug-in electric vehicles were used to define the power levels. Consequently, a power rating of 3.6 kW [35] for level 2 charging was suitable based on the load rating and the available power supply in the lab was represented in Table 5 (Fig. 6).

4 Result and Discussion of the Underwater RIPT System Powered by Floating Solar PV Simulation

The system structure of the proposed floating solar powered wireless charging system, which includes PV panels, a series-series (SS) compensated IPT. Inside the SS IPT converter, the magnetic coupler has Primary self-inductance L_P and Secondary self-inductance L_S, and mutual inductance M. The coupling coefficient was k. Series

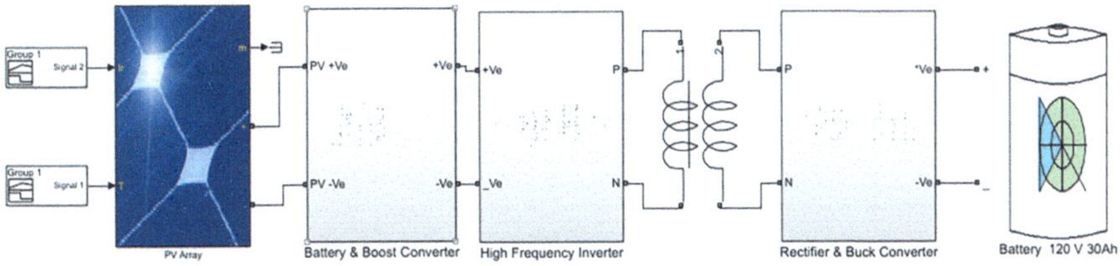

Fig. 7 Underwater RIPT system powered by floating solar PV simulation circuit diagram

series compensation was used for primary as well as secondary, C_P and C_S are capacitors are used for series compensation, used in primary and secondary coil respectively. The PV panel was voltage source for the system, energy harvested form solar was stored in a battery then it was step up by the boost converter, that voltage was fed to high-frequency inverter, a H-bridge inverter having four MOSFET switches Q1 to Q4, AC voltage from the inverter drive the resonant circuit in primary side. The ac output received by the secondary is rectified to a dc output by a full bridge rectifier. The rectified dc voltage is fed to the buck converter to achieve the CC, CV charging of the AUV battery. A simulation circuit of floating solar PV powered wireless power transfer system was designed in MATLAB, the circuit diagram was as shown in Fig. 7. The simulation circuit consist of a solar PV panel, Buck boost converter, Battery, Boost converter, High frequency inverter, transmitter coil, receiver coil, rectifier, buck converter and AUV Battery. In this simulation PV was a source for the entire system three PV panel with maximum power rating of 580 W was connected in series to get 1740 W it can deliver 110 volte and 13 amp, to stabilize the voltage a buck boost converter is integrated with the PV panel to charger the 96 volt battery, to maintain constant output a MPPT algorithm is used to produce PWM signal to the buck boost converter The Battery voltage 96 V was stepped up to 225 Volt by the boost converter that boosted voltage is fed to the high frequency Inverter, here the DC input is convert as a high frequency AC, the compensation capacitor and the inductor was designed to operate at the resonant frequency of 85 KHz. The secondary coil receives the high frequency AC voltage, that high frequency ac voltage is rectified into dc voltage by a rectifier, the rectified output voltage is applied to the buck converter to charge the battery has the capacity of 120 V, 30 Ah capacity.

4.1 Transmitter Side

The PV panel output voltage is 110 volt it was the input of the buck boost converter, the convert will deliver the stable output voltage of 90 volt the waveform was shown in Fig. 8a. The battery parameter was assumed as 50% discharge, while the buck boost output was connected the battery is start to charge, Fig. 8b Show the stat of charging of the battery. Figure 8c, d is the charging current and voltage respectively.

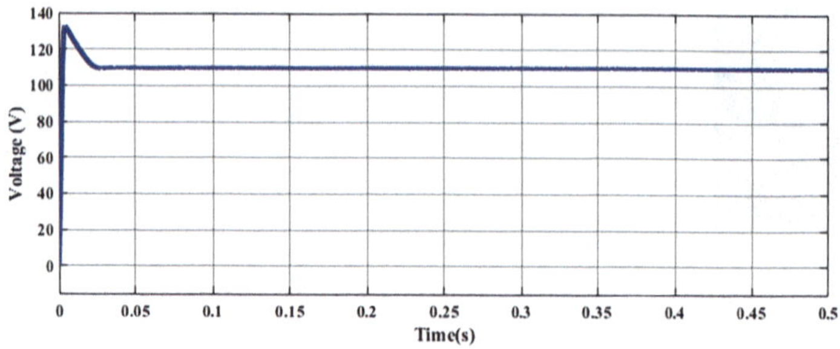

(a)

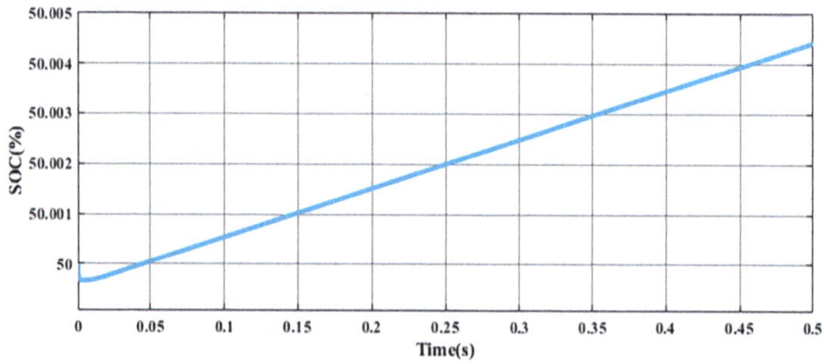

(b)

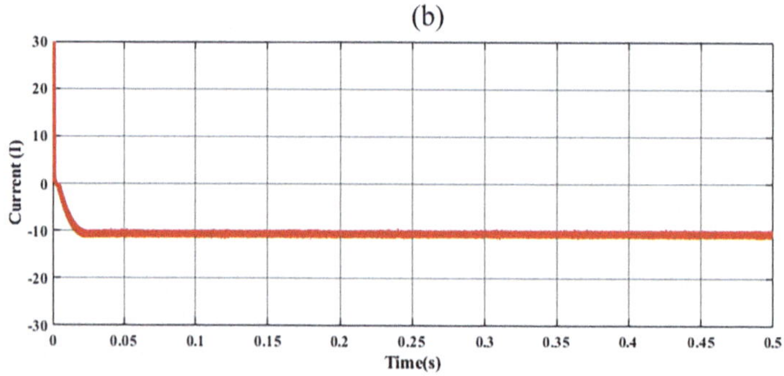

(c)

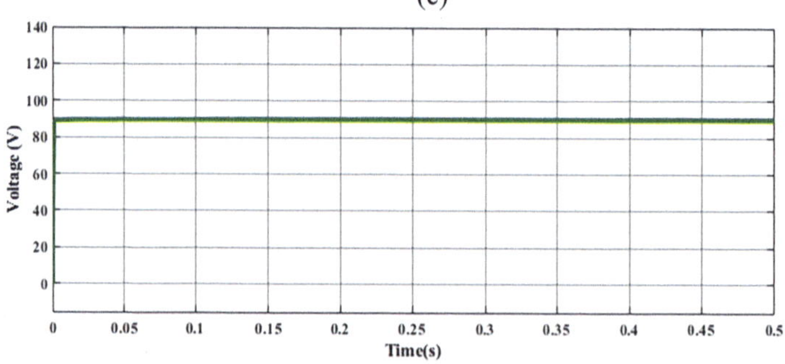

(d)

Fig. 8 (**a**) PV Output Voltage of buck boost converter. (**b**) Input Battery SOC Charging From 50%. (**c**) Input side Battery charging current. (**d**) Input side Battery charging Voltage. (**e**) Input Battery SOC Discharging From 100%. (**f**) Input Battery SOC Discharging Current. (**g**) Input Battery Voltage. (**h**) Boost Converter Output Voltage. (**i**) Transmitter coil Voltage. (**j**) Transmitter coil Current

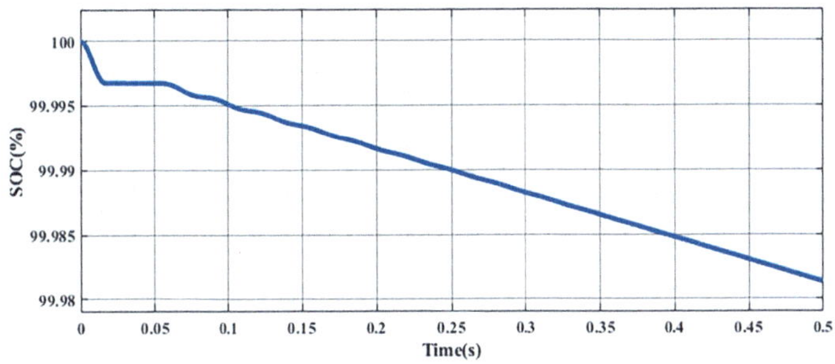

(e)

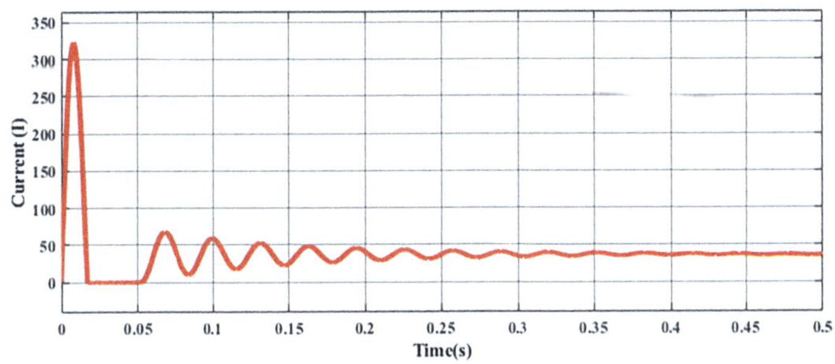

(f)

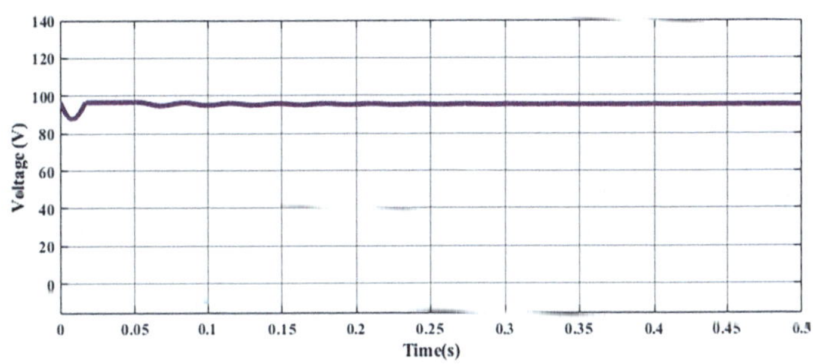

(g)

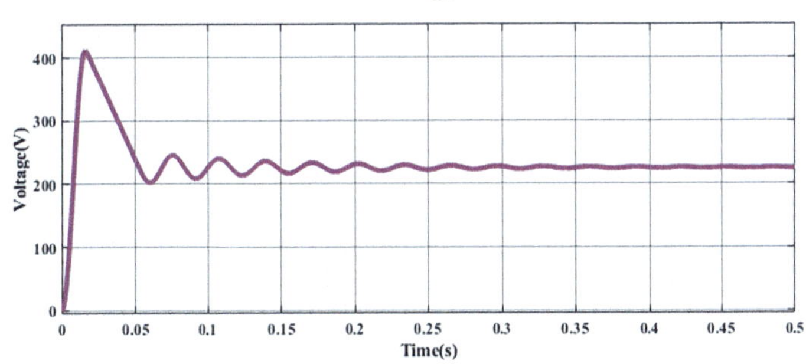

(h)

Fig. 8 (continued)

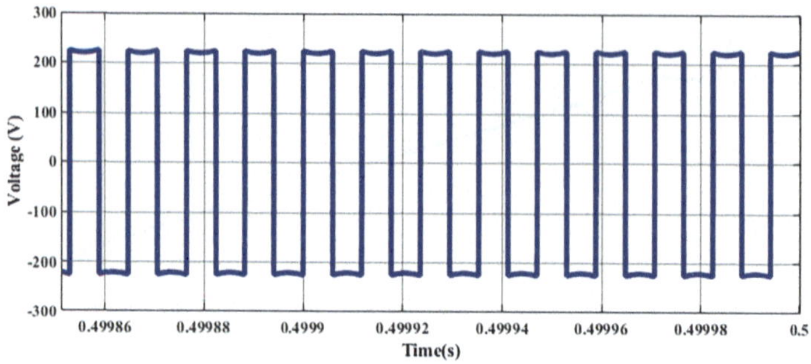

(i)

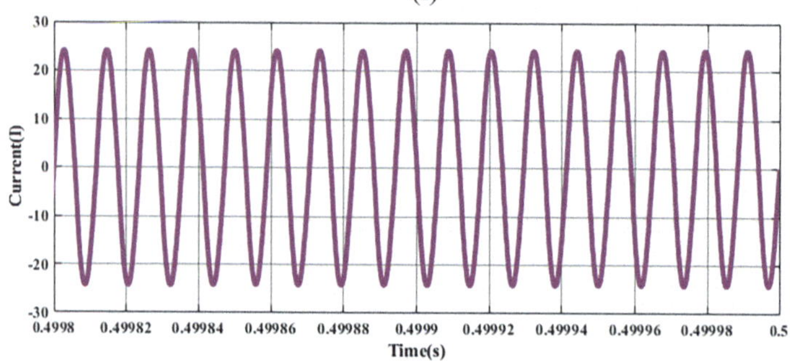

(j)

Fig. 8 (continued)

Figure 8f, g, shows the discharge characteristics of the battery when it is connected to the boost converter with the WPT system, Fig. 8h Shows the boost converter output waveform, initially the overshoot voltage around 400 V it will be settled to the steady state value of 225 V after 0.05 s. Figure 8i shows the square wave AC output voltage of high frequency Inverter, peak to peak voltage of +225 V to −225 V. Figure 8j shows the primary coil current 25 ampere.

4.2 Receiver Side

Receiver coil voltage is shown in Fig. 9a it is AC square wave of +170 V to −170 V. and the alternating current at the secondary coil is shown in Fig. 9b. A bridge rectifier is connected to the secondary coil which will rectify the ac voltage to +168 Volt DC, that voltage and current of the rectifier was shown in Fig. 9c, d respectively. Rectified DC voltage is step down by the Buck converter, the buck converter is operated at 75% duty cycle to step down the 168 volts to 126 volts to charge the Battery at CC, CV mode. Figure 9h shows the output power of the system was 3334 Watts. The entire calculation and the simulation were done for 75% power transfer efficiency, but in practical switching loss, eddy current loss and miss

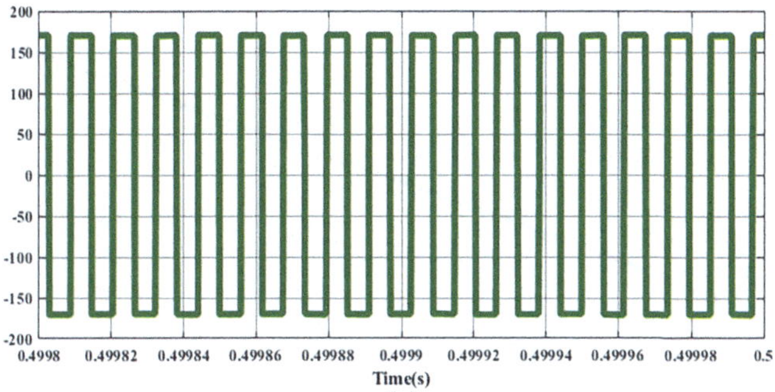

(a)

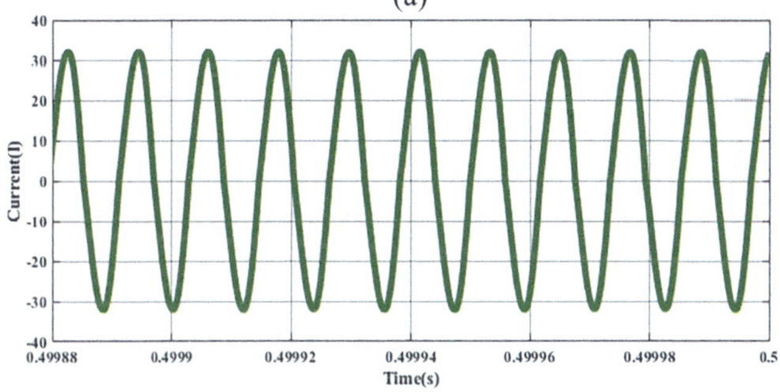

(b)

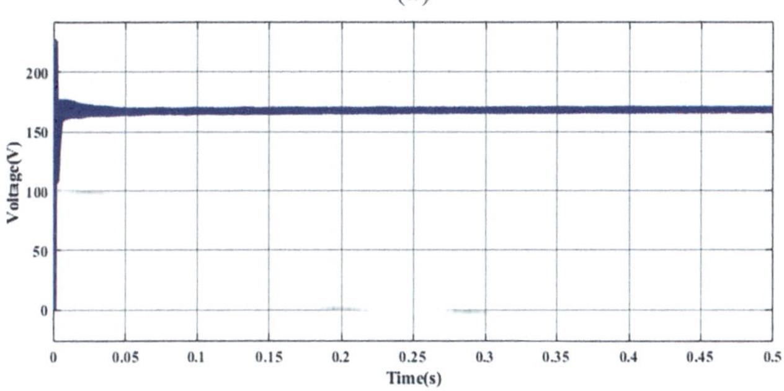

(c)

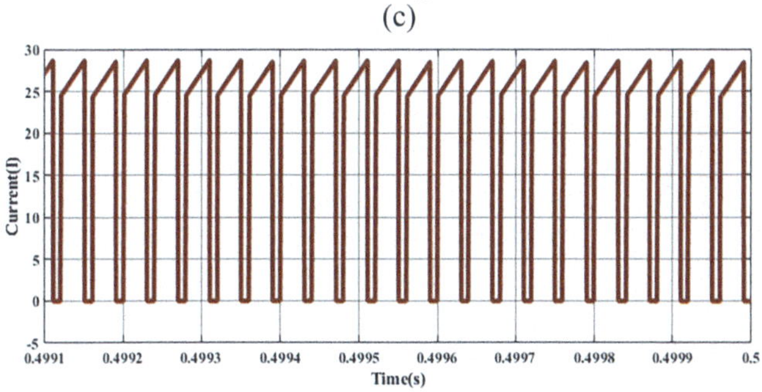

(d)

Fig. 9 (**a**) Receiver coil Voltage. (**b**) Receiver coil Current. (**c**) Rectifier Output Voltage. (**d**) Rectifier Output Current. (**e**) Output Battery SOC Charging From 50%. (**f**) Output Battery Charging Current. (**g**) Output Battery Charging Voltage. (**h**) Output Power

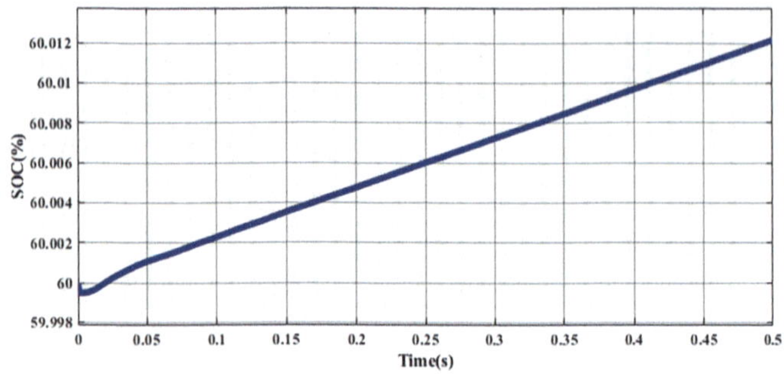

(e)

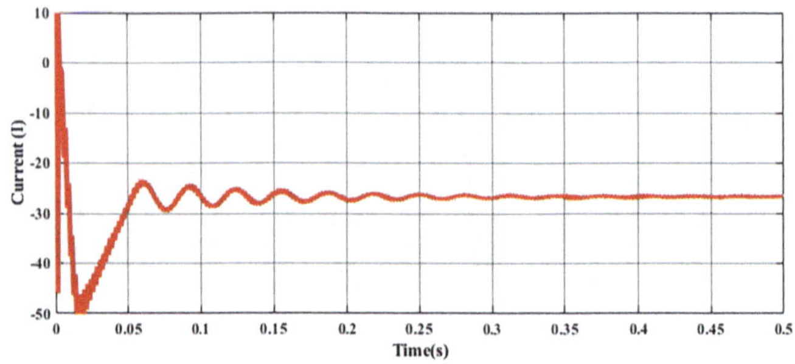

(f)

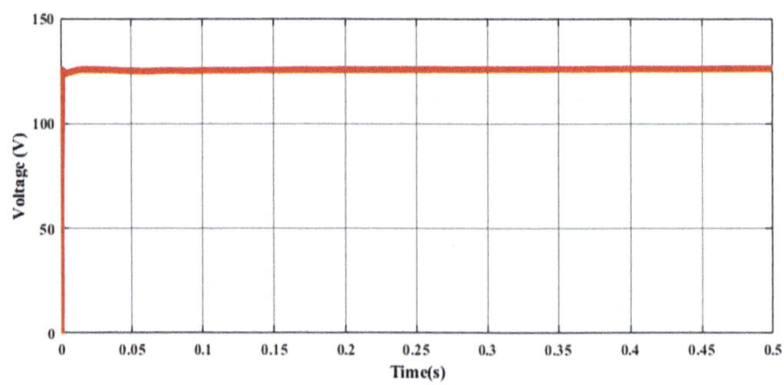

(g)

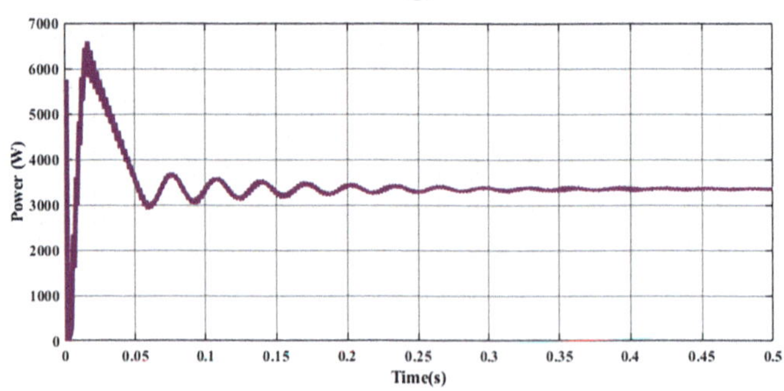

(h)

Fig. 9 (continued)

alignment may cause losses in the power transfer efficiency, while design for practical application the above mention losses should be consider.

5 Conclusion and Perspectives

A sustainable approach for smart cities is to combine floating solar photovoltaic (PV) with wireless power transfer (WPT) systems for powering of an autonomous underwater vehicles (AUVs). This method eliminates the need for manual recharging by allowing AUVs to receive electricity wireless, and floating solar PV to capture energy in urban water bodies. This improves the operating time of AUVs, which are capable of carrying out duties like infrastructure inspection and environmental monitoring. Furthermore, this integration promotes renewable energy goals, lowers carbon footprints, All things considered, this innovative combination satisfies energy demand for WPT technology and also advances intelligent and sustainable urban management. The Series series compensation with buck converter topology as constant voltage source to the for AUV battery charging. Misalignment of coupler will reduce the power transfer efficiency, eddy current loss in sea water was challenging are of UWPT system. Future research are should be focus on least misalignment coupler, better power transfer even in misalignment and eddy current loss condition, for battery charging CCCV mode is much better method, the hybrid compensation topology will deliver requirement with least component, which can eliminate the Pickup side DC/DC converter. For proper power transfer the control system play a vital role in UWPT system, advanced communication topology are required for better operation.

References

1. Kawamura A, Ishioka K, Hirai J (1996) Wireless transmission of power and information through one high-frequency resonant AC link inverter for robot manipulator applications. IEEE Trans Ind Appl 32(3)
2. Li ZS, Li DJ, Lin L, Chen Y (2010) Design considerations for electromagnetic couplers in contactless power transmission systems for deep-sea applications. J Zhejiang Univ: Sci C 11(10):824–834. https://doi.org/10.1631/jzus.C0910711
3. Pang S, Xu J, Li H, Ma Q, Li X (2023) Dual-frequency modulation to achieve power independent regulation for dual-load underwater wireless power connector. IEEE J Emerg Select Topics Power Electron 11(2):2377–2389. https://doi.org/10.1109/JESTPE.2022.3225178
4. Qiao K, Rong E, Sun P, Zhang X, Sun J (2022) Design of LCC-P constant current topology parameters for AUV wireless power transfer. Energies 15(14). https://doi.org/10.3390/en15145249
5. Xia Z, Dai X, Li Y, Wang L, Chen F, Wang W (2024) Model predictive control of wireless power transfer system in seawater environment. IEEE Trans Power Electron. https://doi.org/10.1109/TPEL.2024.3439331
6. Cheng Z, Lei Y, Song K, Zhu C (2015) Design and loss analysis of loosely coupled transformer for an underwater high-power inductive power transfer system. IEEE Trans Magn 51(7). https://doi.org/10.1109/TMAG.2014.2346737

7. Agcal A, Kalay A, Cetin R (n.d.) Électronique et transmission de l'information Electronics and Information Technology Wireless Power Transfer for Underwater Vehicles. In Rev Roum Sci Techn-Électrotechn et Énerg, vol 68
8. Qiao K, Sun P, Rong E, Sun J, Zhou H, Wu X (2022) Anti-misalignment and lightweight magnetic coupler with H-shaped receiver structure for AUV wireless power transfer. IET Power Electron 15(16):1843–1857. https://doi.org/10.1049/pel2.12352
9. Wen H, Wang P, Li J, Yang J, Zhang K, Yang L, Zhao Y, Tong X (2023) Improving the misalignment tolerance of wireless power transfer system for AUV with solenoid-dual combined planar magnetic coupler. J Marine Sci Eng 11(8). https://doi.org/10.3390/jmse11081571
10. Wu HH, Gilchrist A, Sealy KD, Bronson D (2012) A high efficiency 5 kW inductive charger for EVs using dual side control. IEEE Trans Industr Inform 8(3):585–595. https://doi.org/10.1109/TII.2012.2192283
11. Crimmins D, Deacutis C, Hinchey E, Chintala M, Cicchetti G, Blidberg D (2005) Use of a long endurance solar powered autonomous underwater vehicle (SAUV II) to measure dissolved oxygen concentrations in Greenwich Bay, Rhode Island, USA. Europe Oceans 2005(2):896–901
12. Nguyen NT, Heldal R, Lima K, Oyetoyan TD, Pelliccione P, Kristensen LM, Hoydal KW, Reiersgaard PA, Kvinnsland Y (2023) Engineering challenges of stationary wireless smart ocean observation systems. IEEE Internet Things J 10(16):14712–14724. https://doi.org/10.1109/JIOT.2023.3283252
13. Crimmins DM, Patty CT, Beliard MA, Baker J, Jalbert JC, Komerska RJ, Chappell SG, Blidberg DR (2006) Long-endurance test results of the solar-powered AUV system. Oceans 2006:1–5
14. Stokey R, Allen B, Austin T, Goldsborough R, Forrester N, Purcell M, Von Alt C (2001) Enabling technologies for REMUS docking: an integral component of an autonomous ocean-sampling network. IEEE J Ocean Eng 26(4)
15. Wang D, Chen F, Zhang J, Cui S, Bie Z, Zhu C (2023) A novel pendulum-type magnetic coupler with high misalignment tolerance for AUV underwater wireless power transfer systems. IEEE Trans Power Electron 38(12):14861–14871. https://doi.org/10.1109/TPEL.2023.3313797
16. Wang Y, Li T, Zeng M, Mai J, Gu P, Xu D (2023) An underwater simultaneous wireless power and data transfer system for AUV with high-rate full-duplex communication. IEEE Trans Power Electron 38(1):619–633. https://doi.org/10.1109/TPEL.2022.3203038
17. Cai C, Zhang Y, Wu S, Liu J, Zhang Z, Jiang L (2020) A circumferential coupled dipole-coil magnetic coupler for autonomous underwater vehicles wireless charging applications. IEEE Access 8:65432–65442. https://doi.org/10.1109/ACCESS.2020.2984530
18. Mostafa A, Wang Y, Zhang H, Tangirala S, Lu F (2022) An ultra-fast wireless charging system with a hull-compatible coil structure for Autonomous Underwater Vehicles (AUVs). In: 2022 IEEE Transportation Electrification Conference and Expo, ITEC 2022, pp 279–284. https://doi.org/10.1109/ITEC53557.2022.9813911
19. Zhou J, Yao P, Chen Y, Guo K, Hu S, Sun H (2021) Design considerations for a self-latching coupling structure of inductive power transfer for autonomous underwater vehicle. IEEE Trans Ind Appl 57(1):580–587. https://doi.org/10.1109/TIA.2020.3029020
20. Mostafa A, Wang Y, Tangirala S, Zhang H, Lu F (2023) A 5 kW hull-compatible inductive charging system with 360° folded spatial unipolar coupler for Autonomous Underwater Vehicles (AUVs). IEEE Trans Ind Appl 59(6):7001–7012. https://doi.org/10.1109/TIA.2023.3307117
21. Wang D, Cui S, Zhang J, Bie Z, Song K, Zhu C (2022) A novel arc-shaped lightweight magnetic coupler for AUV wireless power transfer. IEEE Trans Ind Appl 58(1):1315–1329. https://doi.org/10.1109/TIA.2021.3109839
22. Yan Z, Song B, Zhang Y, Zhang K, Mao Z, Hu Y (2019) A rotation-free wireless power transfer system with stable output power and efficiency for autonomous underwater vehicles. IEEE Trans Power Electron 34(5):4005–4008. https://doi.org/10.1109/TPEL.2018.2871316
23. Cai C, Wu S, Zhang Z, Jiang L, Yang S (2021) Development of a fit-to-surface and lightweight magnetic coupler for autonomous underwater vehicle wireless charging systems. IEEE Trans Power Electron 36(9):9927–9940. https://doi.org/10.1109/TPEL.2021.3064411

24. Liu P, Gao T, Zhao R, Mao Z (2022) A novel conformal coil structure design of wireless power transfer system for autonomous underwater vehicles. J Marine Sci Eng 10(7). https://doi.org/10.3390/jmse10070875
25. Wolek A, Mcmahon J, Dzikowicz BR, Houston BH (2022) Tracking multiple surface vessels with an autonomous underwater vehicle: field results. IEEE J Ocean Eng 47(1):32–45. https://doi.org/10.1109/JOE.2020.3015415
26. Allen B, Austin T, Forrester N, Goldsborough R, Kukulya A, Packard G, Purcell M, Stokey R (n.d.) Autonomous Docking Demonstrations with Enhanced REMUS Technology
27. Kawasaki T, Noguchi T, Fukasawa T, Hayashi S, Shibata Y, Okaya N, Fukui K, Kinoshita M (2004) "Marine Bird", a new experimental AUV-results of docking and electric power supply tests in sea trials. In: Oceans' 04 MTS/IEEE Techno-Ocean'04 (IEEE Cat. No. 04CH37600), vol 3, pp 1738–1744
28. Bellingham JG, Goudey CA, Consi TR, Bales JW, Atwood DK, Leonard JJ, Chryssostomidis C (1994) A second generation survey AUV. In: Proceedings of IEEE Symposium on Autonomous Underwater Vehicle Technology (AUV 1994), pp 148–155. https://doi.org/10.1109/AUV.1994.518619
29. Yan Z, Zhang K, Qiao L, Hu Y, Song B (2021) A multiload wireless power transfer system with concentrated magnetic field for AUV cluster system. IEEE Trans Ind Appl 58(1):1307–1314. https://doi.org/10.1109/TIA.2021.3102897
30. Zhang B, Chen J, Wang X, Xu W, Lu C, Lu Y (2024) High-power-density wireless power transfer system for autonomous underwater vehicle based on a variable ring-shaped magnetic coupler. IEEE Trans Transp Electrific 10(2):3061–3074. https://doi.org/10.1109/TTE.2023.3304060
31. Li S, Mi CC (2015) Wireless power transfer for electric vehicle applications. IEEE J Emerg Sel Top Power Electron 3(1):4–17. https://doi.org/10.1109/JESTPE.2014.2319453
32. Kan T, Zhang Y, Yan Z, Mercier PP, Mi CC (2018) A rotation-resilient wireless charging system for lightweight autonomous underwater vehicles. IEEE Trans Veh Technol 67(8):6935–6942. https://doi.org/10.1109/TVT.2018.2836988
33. Yang L et al (2022) Undersea wireless power and data transfer system with shared channel powered by marine renewable energy system. IEEE J Emerg Sel Top Circuits Syst 12(1):242–250. https://doi.org/10.1109/JETCAS.2022.3140954
34. Mohammad ST, Al-Kayiem HH, Aurybi MA, Khlief AK (2020) Measurement of global and direct normal solar energy radiation in Seri Iskandar and comparison with other cities of Malaysia. Case Stud Thermal Eng 18. https://doi.org/10.1016/j.csite.2020.100591
35. Sagar A, Kashyap A, Nasab MA, Padmanaban S, Bertoluzzo M, Kumar A, Blaabjerg F (2023) A comprehensive review of the recent development of wireless power transfer technologies for electric vehicle charging systems. In: IEEE Access, vol 11, pp 83703–83751). Institute of Electrical and Electronics Engineers Inc. https://doi.org/10.1109/ACCESS.2023.3300475
36. Teeneti CR, Truscott TT, Beal DN, Pantic Z (2021) Review of wireless charging systems for autonomous underwater vehicles. IEEE J Ocean Eng 46(1):68–87. https://doi.org/10.1109/JOE.48 https://doi.org/10.1109/JOE.2019.2953015

A Systematic Literature Review on the Evolution of Smart Cities and Digital Businesses

Hamza Boujida, Rania Larbi, Nada Alami, and Issam Debbagh

Abstract The evolution of smart cities and digital businesses represents a transformative force in urban development, driven by technological innovation and changing societal needs. This intersection is reshaping how cities operate and how businesses create value in urban environments. This systematic review synthesizes findings from 21 peer-reviewed articles published between 2018 and 2024, examining how smart cities and digital businesses co-evolve to create sustainable urban value ecosystems.

Through a comprehensive analysis of the Clarivate Web of Science database, our review reveals five interconnected themes that shape this evolution: the emergence of new business models tailored for smart city environments, the critical role of digital platforms in facilitating resource integration and service exchange, the importance of data-driven decision-making in urban management and business operations, the challenges and opportunities in implementing digital public services, and the growing significance of social innovation and entrepreneurship in smart city contexts. These themes collectively demonstrate how digital transformation is enabling new forms of value creation and service delivery in urban settings.

The findings highlight significant variations between developed and developing urban contexts, with developed cities focusing on advanced integration and optimization while developing cities prioritize fundamental service improvements and accessibility. This review contributes to the growing body of literature by providing a comprehensive framework for understanding smart city and digital business co-evolution, offering valuable insights for policymakers, urban planners, and business leaders involved in smart city development and digital transformation initiatives.

Keywords Smart cities · Digital business · Urban innovation · Digital platforms · Systematic review

H. Boujida (✉) · R. Larbi · N. Alami · I. Debbagh
Laboratory: Interdisciplinary Research Laboratory in Economics, Finance and Management of Organizations LIREFIMO, Sidi Mohamed Ben Abdellah University, Fez, Morocco
e-mail: hamza.boujida@usmba.ac.ma

H. Kaddouri et al. (eds.), *Smart Cities for Sustainable and Urban Development*, Green Energy and Technology, https://doi.org/10.1007/978-3-031-99001-4_28

1 Introduction and Research Background

The convergence of smart cities and digital businesses represents a critical intersection that is reshaping both urban development and business operations in the twenty-first century. As cities worldwide grapple with rapid urbanization, climate change, and resource constraints, digital technologies have emerged as key enablers of sustainable urban development. This transformation is particularly evident in how digital businesses are evolving to create new value propositions within technology-enabled urban environments [1]. Smart cities leverage digital technologies to address complex urban challenges through enhanced service delivery, improved resource management, and increased citizen engagement [2]. Simultaneously, digital businesses are transforming traditional business models through technological innovation and platform-based approaches [3]. This dual evolution creates a dynamic ecosystem where urban infrastructure and business innovation mutually reinforce each other's development.

Building on this evolution, this systematic review aims to comprehensively analyze the intersection of smart cities and digital businesses by addressing several key objectives. Our research focuses on analyzing the emergence and evolution of new business models in smart city environments, examining how digital platforms facilitate resource integration and service exchange, and investigating the role of data-driven decision-making in urban management and business operations. Additionally, we aim to identify key challenges and opportunities in digital public service implementation while exploring the growing importance of social innovation and entrepreneurship in smart cities. These objectives collectively seek to understand the complex dynamics between urban digital transformation and business innovation in modern cities.

The remainder of this paper is organized to systematically address these objectives. Section 2 presents the theoretical framework and literature review, including comparative analysis between developed and developing cities. Section 3 details our research methodology and analytical framework, followed by Sect. 4 which presents our findings and analysis, including practical implementation challenges and solutions. Finally, Sect. 5 discusses implications and conclusions, offering recommendations for future research and practice.

2 Theoretical Framework and Literature Review

2.1 Smart City Technologies and Digital Business Evolution

The integration of smart city technologies and digital businesses has created a complex ecosystem that fundamentally transforms urban development and service delivery. Bencsik et al. [1] emphasize that smart cities are characterized by sophisticated infrastructure integration, combining physical and digital systems through

IoT devices and sensor networks with advanced communication infrastructure. This integration enables real-time data exchange and processing capabilities, supported by cloud and edge computing technologies.

Building on this foundation, Malchenko and Smirnova [4] demonstrate how the evolution of smart city technologies has led to the emergence of new business models centered around data-driven service delivery and citizen engagement. Their research reveals that successful digital business integration in smart cities requires a multi-layered approach to technology adoption, considering both technical capabilities and user needs. This perspective is further developed by Bogoyavlenska et al. [5], who identify how smart-logistics and agile methodologies are reshaping business operations within urban environments.

Timeus et al. [2] further elaborate that successful smart city implementations rely on comprehensive digital platforms that facilitate seamless interaction between citizens and government services, while supporting integrated payment systems and real-time monitoring capabilities. This view is complemented by Dessai and Javidroozi [6], who explore how cross-sectoral process modeling enhances the integration of digital businesses within smart city frameworks, particularly in areas of service delivery and resource optimization.

The foundation of these systems lies in robust data management frameworks. Zheng et al. [7] demonstrate how advanced data management systems, incorporating big data analytics capabilities and AI-driven decision support systems, enable cities to process and analyze vast amounts of urban data in real-time. These insights are expanded by Hasbini et al. [8], who examine the critical role of information security management in smart city organizations, emphasizing the need for robust cybersecurity frameworks to protect digital business operations.

These systems are further enhanced by open data initiatives that promote transparency and innovation, as evidenced by Ford et al. [9] in their analysis of smart local energy systems where data sharing platforms have enabled more efficient resource allocation and service optimization. Oughton et al. [10] complement this perspective by examining the infrastructure requirements for next-generation mobile telecommunications, particularly focusing on how 5G networks will support future smart city services and digital business operations. Additionally, Gusakov et al. [11] explore mechanisms supporting open innovation in smart tourism destinations, demonstrating how digital platforms can create new value propositions in urban environments.

2.2 *Comparative Analysis: Smart City Implementation in Developed Vs. Developing Cities*

The implementation of smart city initiatives reveals significant disparities between developed and developing urban contexts. In developed cities, such as Singapore and Amsterdam, advanced digital infrastructure supports sophisticated smart city applications, benefiting from high digital literacy rates and well-established public--private partnership frameworks [12]. Amsterdam's smart city platform, for instance, has successfully integrated over 200 partners in a collaborative ecosystem that drives urban innovation.

Ha and Thanh [13] further elaborate on these disparities by examining how institutional quality influences the effectiveness of digital public services across different economic contexts. Their research demonstrates that developed cities benefit from robust institutional frameworks that facilitate smoother integration of smart city solutions, while developing regions often struggle with institutional barriers. This analysis is complemented by Grigorescu et al. [14], who explore how knowledge-intensive economies in developed cities create more favorable conditions for smart city implementation through enhanced innovation capabilities and skilled workforce availability.

In contrast, developing cities face distinct challenges in their smart city journeys. Anthony [15] highlights how cities like Jakarta must address fundamental infrastructure development challenges while promoting digital literacy and establishing appropriate regulatory frameworks. Shamsuzzoha and Syrjala [16] expand on these challenges, identifying how technical management processes in developing cities often require different approaches to support organizational transformation in resource-constrained environments.

De Moura [17] provides additional insights through analysis of Industry 4.0 implementation in port cities, revealing how developing regions often need to balance technological advancement with environmental management concerns. Despite these challenges, Zirena-Bejarano et al. [18] demonstrate how innovative solutions are emerging that adapt smart city concepts to local contexts and resources, particularly through the strategic use of digital marketing and consumer engagement strategies.

Sá et al. [19] further enrich this comparative analysis by examining how mobility platforms adapt their business models to different urban contexts, showing how successful implementations require careful consideration of local infrastructure capabilities and user needs. This is supported by Sampaio and Sebastiao's [20] research on social innovation trends, which reveals how both developed and developing cities are finding unique ways to leverage digital technologies for social impact, albeit through different approaches and priorities.

2.3 *Digital Business Models in Smart City Contexts*

Digital business models in smart cities have evolved to address specific urban challenges while creating sustainable value propositions. Platform-based models have emerged as a dominant approach, encompassing Mobility-as-a-Service (MaaS) platforms, smart utility management systems, and urban data marketplaces. Sá et al. [19] describe how these platforms facilitate resource integration and service exchange, creating value through network effects and data-driven optimization.

Bencsik et al. [1] provide a comprehensive framework for understanding digital sustainability business models, drawing insights from 130 smart city services. Their research reveals how successful business models integrate environmental, social, and economic value creation through digital technologies. This perspective is enriched by Anthony's [15] analysis of shared electric mobility services, which demonstrates how innovative business models can simultaneously address urban mobility challenges and sustainability goals.

Service integration models have also gained prominence, with Ford et al. [9] documenting successful implementations in public service delivery, smart tourism, and integrated payment systems. Timeus et al. [2] extend this understanding by proposing a practical framework for creating business models specifically tailored to smart city contexts, emphasizing the importance of stakeholder engagement and value co-creation. Pulkkinen et al. [12] further elaborate on how smart mobility services and platforms are reshaping urban transportation ecosystems through innovative business model designs.

Malchenko and Smirnova [4] explore consumer adoption factors in smart city solutions, providing crucial insights for business model development and adaptation. Their research is complemented by Ha and Thanh's [13] analysis of how institutional quality affects digital public service implementation, offering valuable perspectives on business model sustainability in different urban contexts.

These models often incorporate infrastructure management components, including smart grid management, waste management systems, and traffic control solutions, creating comprehensive urban service ecosystems. Dessai and Javidroozi [6] demonstrate how cross-sectoral process modeling enables more effective integration of these various components. This integration is further supported by Gusakov et al.'s [11] research on open innovation mechanisms in smart tourism destinations, showing how business models can leverage collaborative approaches to create enhanced value propositions.

Bogoyavlenska et al. [5] contribute additional insights through their analysis of smart-logistics implementations, revealing how agile methodologies can enhance business model adaptability in rapidly evolving urban environments. This adaptability is crucial, as highlighted by Grigorescu et al. [14], who examine how knowledge-intensive economies drive the evolution of smart city business models (Fig. 1).

2.4 Public-Private Partnerships in Smart Cities

The evolution of public-private partnerships (PPPs) has become instrumental in driving smart city development and implementation. Gusakov et al. [11] examine how successful PPP models, such as Barcelona's 22@ Innovation District, have created sustainable frameworks for urban innovation by combining public sector oversight with private sector efficiency. These partnerships are particularly effective

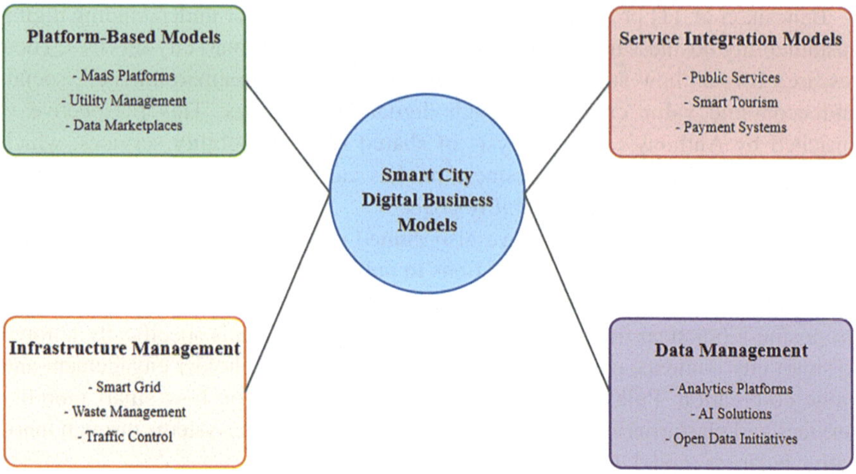

Fig. 1 Digital business models in smart city contexts: "Conceptual representation of digital business models in smart city contexts, showing the interconnections between various business model components and their relationship to urban services. Based on analysis of findings from Sá et al. [19] and Ford et al. [9]"

when supported by clear governance frameworks and robust risk-sharing mechanisms.

Bencsik et al. [1] provide crucial insights into how PPPs contribute to digital sustainability, analyzing how collaborative frameworks enable the development and implementation of smart city services. Their research on 130 smart city services reveals that successful PPPs often depend on clear value capture mechanisms and well-defined stakeholder roles. This is further supported by Timeus et al. [2], who demonstrate how cities can create practical frameworks for business model innovation through structured public-private collaboration.

Ha and Thanh [13] demonstrate through their analysis of European economies that successful PPPs in smart city contexts require strong institutional support and stakeholder engagement. Their research reveals how cities like Copenhagen and Helsinki have leveraged PPPs to develop comprehensive smart city solutions while ensuring public benefit and private sector viability. Pulkkinen et al. [12] extend this understanding by examining how PPPs facilitate the development of smart mobility services and ecosystem integration.

Malchenko and Smirnova [4] contribute valuable insights into how PPPs influence consumer adoption of smart city solutions, particularly in contexts where public trust is crucial for success. Dessai and Javidroozi [6] further elaborate on this by examining how cross-sectoral process modeling can enhance PPP effectiveness through improved integration of public and private sector processes.

Grigorescu et al. [14] explore how knowledge-intensive economies leverage PPPs to drive innovation in smart city contexts, highlighting the importance of institutional frameworks that support collaborative innovation. This is complemented by

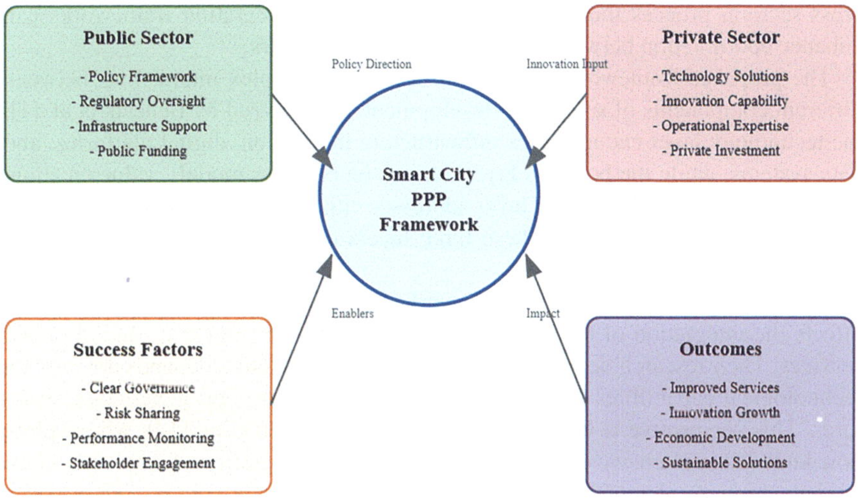

Fig. 2 Schematic representation of Public-Private Partnership framework in smart cities, illustrating the key components and relationships between public and private sector entities. Synthesized from studies by Gusakov et al. [11] and Ha and Thanh [13]

Shamsuzzoha and Syrjala's [16] analysis of technical management processes in organizational partnerships, demonstrating how structured collaboration frameworks can enhance PPP outcomes.

Ford et al. [9] provide additional perspectives through their examination of smart local energy systems, showing how PPPs can effectively address complex urban challenges through collaborative approaches to resource management. This is further supported by Oughton et al.'s [10] research on telecommunications infrastructure development, revealing how PPPs can facilitate the deployment of critical smart city infrastructure (Fig. 2)

2.5 *Framework for Smart City Components Integration*

The development of smart cities requires a comprehensive framework that addresses technological, business, and social dimensions simultaneously. Malchenko and Smirnova [4] propose that the technology layer forms the foundation, incorporating infrastructure components, platform integration, and data management systems that enable smart city operations.

Timeus et al. [2] further develop this understanding by demonstrating how the integration framework must support dynamic interactions between different urban systems. Their research shows that successful smart city implementations require careful orchestration of technological capabilities with business processes and social needs. This is complemented by Dessai and Javidroozi's [6] analysis of cross-sectoral process modeling, which reveals how integration frameworks can enhance coordination between different urban stakeholders.

The proposed framework (Fig. 3) illustrates the complex interactions between different components of smart city development. As analyzed by Bencsik et al. [1], the technology layer encompasses infrastructure integration, digital platforms, and data systems, while the business layer focuses on business models, value creation, and service delivery. The social layer addresses citizen engagement, social innovation, and community impact. These three layers converge to support sustainable urban development.

Ha and Thanh [13] enrich this framework by examining how institutional quality affects the integration of these layers, particularly in the context of digital public services. Their research demonstrates that effective integration depends not only on technological capabilities but also on institutional support and governance structures. This perspective is further developed by Grigorescu et al. [14], who explore how knowledge-intensive economies facilitate better integration of smart city components through enhanced innovation capabilities.

Ford et al. [9] provide practical insights through their analysis of smart local energy systems, demonstrating how the integration framework can be applied to specific urban sectors. This is supported by Zheng et al.'s [7] research on transport systems, which shows how data-driven approaches can enhance integration across different urban domains.

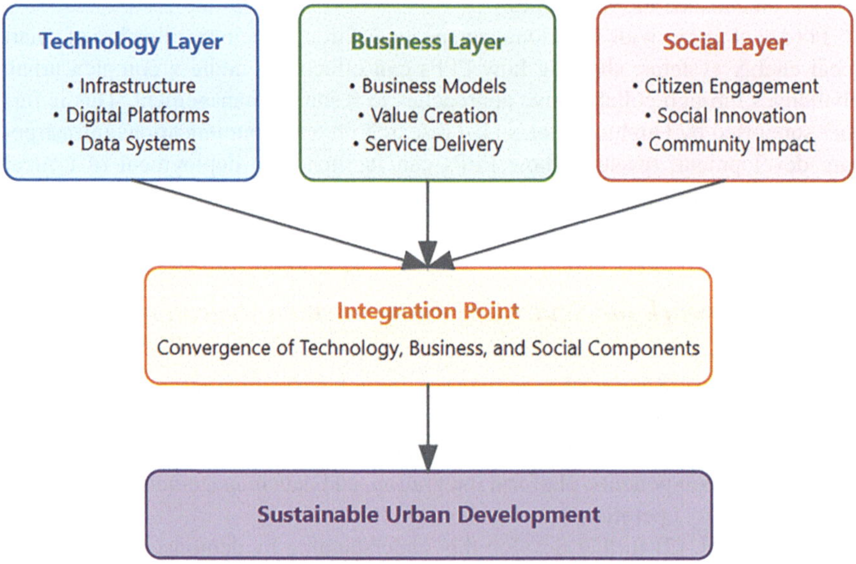

Fig. 3 Smart City Integration Framework illustrating the hierarchical relationship between technological, business, and social layers in smart city development. The framework demonstrates how these three primary layers converge at an integration point to support sustainable urban development. This conceptual model is derived from the synthesis of literature including Bencsik et al. [1], Malchenko and Smirnova [4], and Ha and Thanh [13]

Sampaio and Sebastiao [20] contribute additional perspectives on the social layer, examining how social innovation and entrepreneurship can be effectively integrated into smart city frameworks. Their research is complemented by Foley et al.'s [21] analysis of community networks, which demonstrates the importance of social inclusion in smart city integration efforts.

2.6 Challenges and Opportunities in Smart City Implementation: Re-search Gap of the Literature Survey

The implementation of smart city initiatives presents complex challenges and opportunities that reveal significant gaps in current research. While existing literature extensively covers smart cities and digital business transformation separately, there remains a critical research gap in understanding their mutual influence and co-evolution. Dessai and Javidroozi [6] identify critical implementation challenges including the persistent digital divide, growing concerns over data privacy, substantial infrastructure costs, and the complexity of integrating diverse systems and stakeholders. These challenges are particularly relevant when considering the limited research on the combined impact of technological infrastructure and business transformation in urban contexts.

A systematic analysis of the literature reveals that previous studies have predominantly focused on isolated aspects: either the technological infrastructure of smart cities [6] or digital business transformation [12]. This fragmented approach has created a significant knowledge gap in understanding how these elements interact and co-evolve, particularly in the context of service delivery and urban innovation. The gap becomes more pronounced when considering that cities worldwide increasingly rely on digital businesses to deliver smart city solutions and services, yet research on their integrated implementation remains limited.

Grigorescu et al. [14] document how successful smart city implementations have led to enhanced service delivery, optimized resource utilization, and accelerated economic growth. However, their research also highlights the need for more comprehensive studies on how these successes are achieved through the integration of business and technological innovations. This research gap is particularly evident in the context of developing cities, where the practical challenges of implementing smart solutions vary significantly, ranging from infrastructure readiness and digital literacy to regulatory frameworks and public-private partnership models.

Ford et al. [9] demonstrate the potential for environmental sustainability through their analysis of smart local energy systems, where digital technologies have enabled significant reductions in resource consumption and environmental impact. Yet, their work also underscores the need for more research on how business models and technological solutions can be effectively combined to achieve these sustainability goals. This represents another critical gap in the literature, particularly regarding the development of integrated frameworks that can guide the simultaneous evolution of smart city technologies and digital business models.

The research gap extends to understanding the variations in implementation approaches between developed and developing urban contexts. While some opportunities have been identified, such as enhanced service delivery and resource optimization, there is limited research on how these opportunities can be effectively realized through the integration of business and technological innovations across different urban contexts. This gap is particularly significant given the increasing reliance on digital businesses for smart city solution delivery and the need for context-specific implementation strategies.

3 Research Methodology

3.1 Systematic Review Approach

Our research methodology employs a systematic literature review approach to analyze the intersection of smart cities and digital businesses. Following the rigorous protocols established by Timeus et al. [2], we implemented a structured methodology to identify, evaluate, and synthesize relevant research contributions. This systematic approach ensures comprehensive coverage while maintaining analytical rigor and reproducibility. As emphasized by Dessai and Javidroozi [6], systematic reviews in the field of smart cities require careful attention to both technological and business aspects, which we have incorporated into our methodological framework (Fig. 4).

3.2 Search Strategy and Selection Criteria

The literature search was conducted following a three-stage systematic approach.

In Stage 1, we began with a literature review focusing on three main areas: Smart Cities, Digital Business, and the Web of Science Database, as recommended by Grigorescu et al. [14]. This led to our keyword identification process, where we formulated our search query using the specific terms "TITLE-ABS-KEY ("smart city") AND TITLE-ABS-KEY ("digital business")".

Stage 2 comprised our document selection process. The initial search yielded 578 articles, which were then filtered according to our inclusion criteria:

- Publication year between 2017 and 2025
- English language requirement
- Open access availability
- Document type limited to articles
- Subject areas restricted to "Management" or "Business"

This systematic filtering process resulted in a final selection of 21 documents that met all our inclusion criteria.

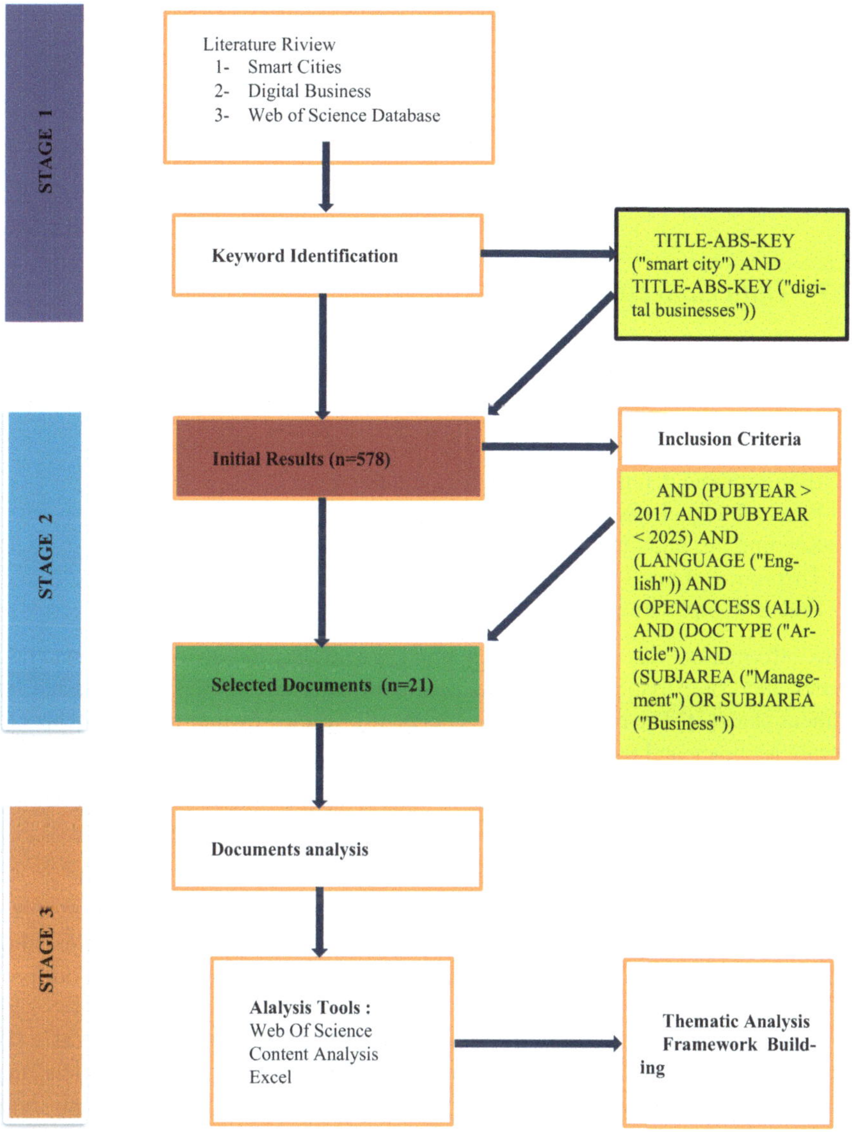

Fig. 4 Distribution of primary documents analyzed in this study, showing the temporal and thematic classification of the 21 selected articles (2018–2024). Based on bibliometric analysis of Web of Science database

In Stage 3, we conducted our documents analysis using three main analysis tools:

- Web of Science for bibliometric data extraction
- Content Analysis for systematic review of the selected articles
- Excel for data organization and analysis

The analysis process culminated in a thematic analysis and framework building approach, enabling us to systematically synthesize the findings from our selected documents.

This methodology ensures a rigorous and transparent selection process, aligned with our research objectives of understanding the intersection between smart cities and digital business developments.

3.3 Data Collection Process

The data collection process followed a systematic protocol to ensure consistency and completeness. For each selected article, we extracted and documented the following information:

- Bibliographic details (authors, year, journal)
- Research objectives and methodologies
- Key findings and contributions
- Theoretical frameworks employed
- Practical implications and recommendations

This information was organized in a standardized data extraction form, enabling systematic comparison and analysis across all selected articles, as suggested by Timeus et al. [2]. The process was independently verified by multiple researchers to ensure accuracy and reliability.

3.4 Analytical Framework

Our analytical framework was developed through a systematic synthesis of existing research, incorporating five interconnected dimensions that emerged from our preliminary analysis.

Business model innovation, the first dimension, builds upon Bencsik et al.'s [1] analysis of smart city services, examining value propositions, revenue models, and stakeholder integration. Timeus et al. [2] contribute additional insights into how traditional business approaches transform in smart city environments.

Digital platform integration forms the second dimension, drawing from Pulkkinen et al.'s [12] research on smart mobility platforms. Anthony [15] enhances this understanding through his analysis of digital ecosystems, focusing on platform architecture, service delivery, and user engagement strategies.

The third dimension addresses data-driven decision making, based on Ford et al.'s [9] framework for smart energy systems and Zheng et al.'s [7] work on transport systems. This component examines analytics capabilities, decision support systems, and performance measurement approaches.

Digital public services constitute the fourth dimension, building on Ha and Thanh's [13] research on implementation strategies and Malchenko and Smirnova's [4] analysis of adoption patterns. This dimension focuses on service implementation, user adoption, and impact assessment.

The fifth dimension, social innovation, draws from Foley et al.'s [21] work on community networks and Sampaio and Sebastiao's [20] analysis of social innovation trends, examining community engagement, sustainability initiatives, and social impact measurement.

These dimensions create an integrated framework for analyzing the relationships between smart cities and digital businesses, enabling systematic analysis while maintaining analytical rigor (Table 1).

Table 1 Analytical framework dimensions and components

Dimension	Key components	Theoretical foundation	Key indicators	Reference sources
Business Model Innovation	– Value propositions – Revenue models – Stakeholder integration	Smart city services analysis and business model transformation	– Innovation level – Value creation mechanisms – Stakeholder engagement metrics	– Bencsik et al. [1] – Timeus et al. [2]
Digital Platform Integration	– Platform architecture – Service delivery – User engagement	Smart mobility platforms and digital ecosystem analysis	– Platform effectiveness – Integration level – User engagement rates	– Pulkkinen et al. [12] – Anthony [15]
Data-Driven Decision Making	– Analytics capabilities – Decision support systems – Performance measurement	Smart energy systems and transport systems frameworks	– Data utilization – Decision quality – Performance metrics	– Ford et al. [9] – Zheng et al. [7]
Digital Public Services	– Service implementation – User adoption – Impact assessment	Implementation strategies and adoption patterns analysis	– Service quality – Adoption rates – Impact indicators	– Ha and Thanh [13] – Malchenko and Smirnova [4]
Social Innovation	– Community engagement – Sustainability initiatives – Social impact measurement	Community networks and social innovation trends analysis	– Engagement levels – Sustainability metrics – Social impact indicators	– Foley et al. [21] – Sampaio and Sebastiao [20]

3.5 Data Analysis Methods

Our analysis employed a mixed-method approach, combining qualitative content analysis and thematic synthesis. Following Bencsik et al.'s [1] framework, we implemented a three-stage analytical process.

Content analysis formed the first stage, drawing from Dessai and Javidroozi's [6] approach for systematic coding and pattern identification. Pulkkinen et al.'s [12] framework provided structure for identifying recurring themes across the literature.

The second stage involved thematic synthesis, following Timeus et al.'s [2] methodology for cross-case comparisons. Ford et al.'s [9] approach guided theme development, while Gusakov et al.'s [11] work provided additional insights for refinement.

Quality assessment constituted the third stage, using Ha and Thanh's [13] evaluation framework and Anthony's [15] criteria for assessing implementation feasibility. Foley et al.'s [21] approach enhanced our evaluation of practical implications.

The analysis process was iterative, following Sampaio and Sebastiao's [20] recommendations for systematic reviews. This comprehensive approach enabled us to identify both common patterns and unique insights in smart city and digital business integration.

4 Results and Analysis

4.1 Results of Research Landscape

Our systematic analysis of 21 articles reveals several key patterns in smart city and digital business research. As illustrated in Fig. 5, the dominant themes center around "Smart", "Business", "Digital", and "Cities", reflecting the core focus of our investigation. The visualization demonstrates the interconnected nature of these concepts within the literature, with strong emphasis on "process", "development", and "systems" as key linking elements (Fig. 6).

4.2 Temporal and Geographic Distribution

The temporal analysis (Fig. 7) shows an evolution in publication patterns from 2018 to 2024. A notable peak occurred in 2020 (6 publications), followed by sustained research interest in subsequent years. This pattern suggests growing academic attention to smart city and digital business integration.

Geographic distribution analysis (Fig. 8) reveals that research leadership is concentrated in specific regions. England leads with 6 publications, followed by Finland (3) and Norway (3), indicating strong European representation in this field. This distribution suggests potential opportunities for research expansion in other regions, particularly in developing economies.

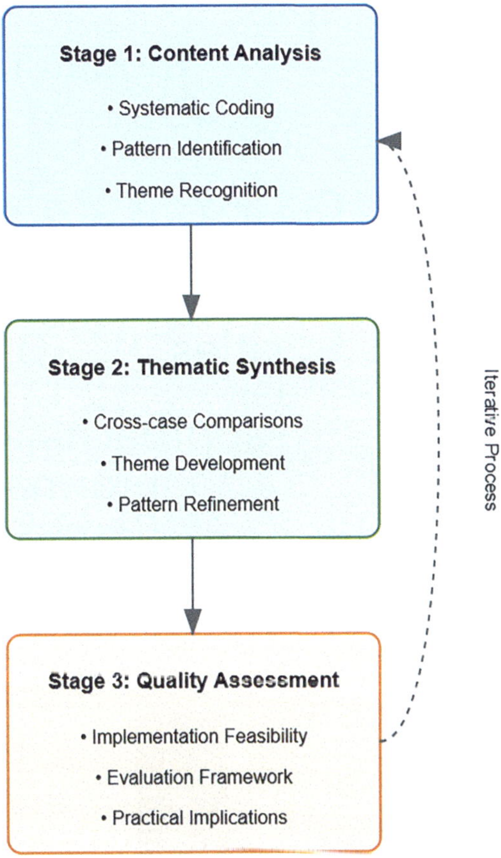

Fig. 5 Three-stage methodological framework for data analysis in smart city research. The framework illustrates the iterative process of content analysis (Stage 1), thematic synthesis (Stage 2), and quality assessment (Stage 3), with feedback loops enabling continuous refinement. Developed based on methodological approaches from Bencsik et al. [1], Dessai and Javidroozi [6], and Ha and Thanh [13]

4.3 *Sustainable Development Focus*

Our analysis reveals strong alignment with sustainable development goals (Fig. 9). Industry Innovation and Sustainable Cities dominate with nine articles each, as evidenced by Bencsik et al. [1]. Good Health and Well-being appears in six articles, while Climate Action emerges in three articles, showing how smart city initiatives address multiple sustainability objectives simultaneously.

The research particularly addresses:

- Industry Innovation (9 articles)
- Sustainable Cities (9 articles)
- Good Health and Well-being (6 articles)
- Climate Action (3 articles)

Fig. 6 Word cloud analysis of key themes in smart city and digital business literature

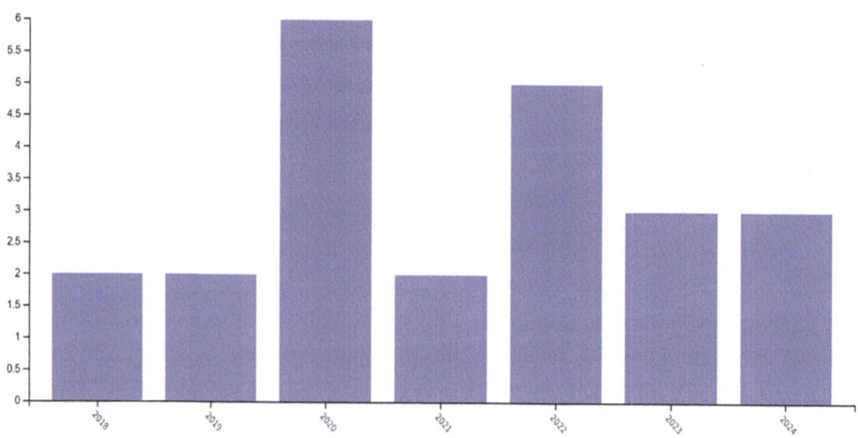

Fig. 7 Temporal evolution of publications

4.4 Key Research Themes

Based on our systematic analysis, five primary themes emerged that characterize the intersection of smart cities and digital business development:

- ***Business Model Innovation.*** Drawing from Bencsik et al. [1], we identified innovative business models specifically designed for smart city environments.

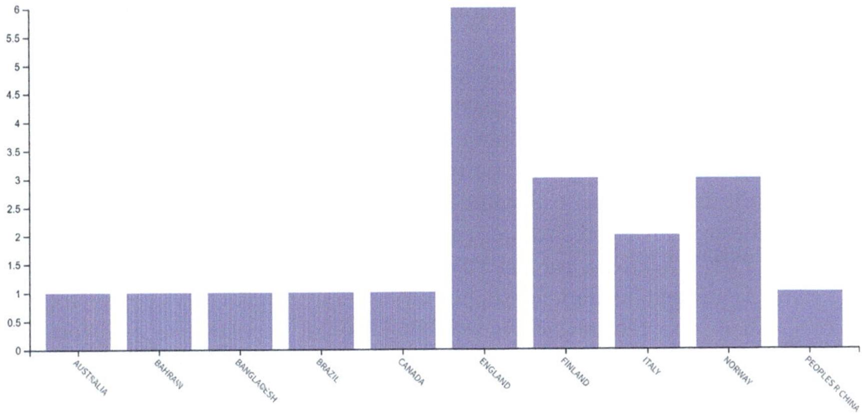

Fig. 8 Geographical distribution of smart city and digital business research

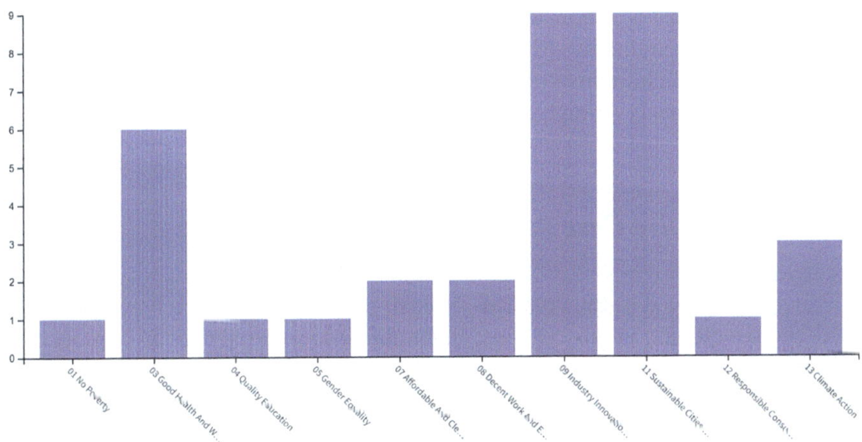

Fig. 9 Distribution of Sustainable Development Goals (SDGs) addressed in smart city research

These models emphasize value creation and sustainability through digital transformation. Timeus et al. [2] further demonstrate how organizations are adapting traditional business models to incorporate urban data analytics, citizen engagement, and sustainable resource management. The analysis reveals a shift towards platform-based and service-oriented business models that can effectively respond to urban challenges.

- ***Digital Platform Integration.*** Pulkkinen et al. [12] highlight the critical role of digital platforms in facilitating resource integration and service exchange within smart city contexts. These platforms serve as intermediaries between various stakeholders, enabling efficient data sharing and service delivery. Anthony [15] extends this understanding by demonstrating how digital platforms support sustainable mobility solutions and urban service optimization through integrated technological frameworks.

- ***Data-Driven Decision Making.*** Ford et al. [9] demonstrate how data analytics and decision support systems enable evidence-based urban management. The research shows increasing adoption of IoT sensors, real-time monitoring systems, and predictive analytics for urban planning and operations. Zheng et al. [7] further illustrate how these systems improve service delivery and resource allocation through advanced analytics capabilities.
- ***Digital Public Services.*** Ha and Thanh [13] emphasize the importance of user-centric design and institutional support in successful digital service implementation. Their research reveals how digital public services are transforming citizen-government interactions. Malchenko and Smirnova [4] complement this by identifying key factors affecting citizen adoption of smart city solutions, including accessibility, trust, and perceived usefulness.
- ***Social Innovation Impact.*** Foley et al. [21] highlight the growing significance of community engagement and social innovation in smart city development. Their research demonstrates how digital technologies enable new forms of civic participation and community-driven solutions. Gusakov et al. [11] further show how social innovation initiatives contribute to creating more inclusive and sustainable urban environments through collaborative approaches.

4.5 Comparative Case Studies Analysis

Our analysis revealed distinct patterns in smart city implementation between developed and developing urban contexts, highlighting key differences in approaches, challenges, and outcomes.

In developed cities, Amsterdam and Singapore exemplify advanced digital integration. As documented by Bencsik et al. [1], Amsterdam's smart city initiative successfully integrates over 200 partners through a comprehensive digital platform, demonstrating mature public-private collaboration. Singapore's Smart Nation initiative, analyzed by Pulkkinen et al. [12], showcases sophisticated data analytics implementation and comprehensive digital service delivery.

In contrast, developing cities exhibit different implementation patterns. Ha and Thanh [13] examine how cities like Jakarta and Mumbai focus on fundamental infrastructure development while incorporating digital solutions. These cities often adopt hybrid approaches, combining traditional service delivery with digital innovations to address immediate urban challenges.

Key differences emerged across three dimensions:

- Infrastructure Development
  - Developed cities: Focus on advanced IoT integration and 5G implementation
  - Developing cities: Emphasis on basic digital infrastructure and connectivity

- Service Implementation
  - Developed cities: Comprehensive digital service ecosystems
  - Developing cities: Targeted solutions for pressing urban challenges
- Stakeholder Engagement
  - Developed cities: Established public-private partnerships
  - Developing cities: Growing emphasis on community-based initiatives

This comparative analysis reveals how local context fundamentally shapes smart city development trajectories and digital business opportunities. Bencsik et al. [1] demonstrate that factors such as infrastructure, technological readiness, and regulatory frameworks significantly influence how cities implement smart solutions. While developed cities focus on advanced integration, developing cities prioritize fundamental service improvements, creating distinct innovation pathways and business opportunities in each context.

5 General Discussions and Synthesis of Key Findings

Our systematic review of 21 articles reveals significant patterns in smart cities and digital businesses evolution between 2018 and 2024, demonstrating a progression from isolated technological solutions toward integrated ecosystem approaches. This co-evolution occurs through digital platform integration with urban infrastructure, adaptive business models development, data-driven decision-making processes, collaborative public-private partnerships, and community-centered innovation approaches, as evidenced by Bencsik et al.'s [1] analysis of 130 smart city services. The analysis shows that business models increasingly emphasize platform-based approaches and value co-creation [2], while digital platforms serve as crucial intermediaries enabling efficient resource allocation [12]. Ford et al. [9] demonstrate how analytics capabilities enhance operational efficiency in urban contexts, while Ha and Thanh [13] identify user adoption and institutional support as critical implementation factors. From a theoretical perspective, our findings extend existing knowledge by providing a new framework integrating business models, digital platforms, and social innovation for understanding smart city development. This framework builds upon previous work while incorporating technological and social dimensions, leading to practical implications for city managers and business leaders, including the need for integrated approaches to digital service implementation, stakeholder engagement, and context-specific strategies. However, certain limitations must be acknowledged, such as our focus on English-language publications potentially excluding valuable insights from other regions, and the rapid evolution of smart city technologies meaning some recent developments may not be fully reflected in academic literature. Future research opportunities include expanded investigation of developing city contexts, analysis of emerging technologies' impact on urban services, examination of long-term sustainability of digital business models, and study of social impact measurement. The geographical analysis reveals a concentration of research in developed economies, particularly in European contexts, suggesting opportunities for expanded investigation in developing regions, while the temporal distribution indicates growing academic interest, with peak activity in 2020 followed by sustained research engagement. Our analysis demonstrates that successful smart city initiatives require careful balance of technological capabilities, business model innovation, and social considerations, indicating that as cities continue to evolve, the role of digital businesses in shaping urban development will become increasingly significant.

6 Conclusions and Perspectives

This systematic review has examined the evolving relationship between smart cities and digital businesses through the analysis of 21 peer-reviewed articles published between 2018 and 2024. Our research makes three significant contributions: First, we identified five interconnected themes characterizing this field's evolution, with Bencsik et al. [1] demonstrating how organizations adapt their value propositions to smart city environments, and Pulkkinen et al. [12] showing digital platforms as crucial enablers of urban service delivery. Second, our analysis reveals the critical role of data-driven decision-making in urban governance and business operations, with Ford et al. [9] demonstrating how data analytics enables more efficient resource allocation and creates new business opportunities. Third, we documented the evolution of digital public services, with Ha and Thanh [13] showing how successful implementation requires balancing technological and social factors.

Looking ahead, several key perspectives emerge for the future development of smart cities and digital businesses. The integration of artificial intelligence and machine learning technologies promises to revolutionize urban service delivery and decision-making processes, fundamentally transforming how cities operate and serve their citizens. This transformation will be further enhanced by the growing importance of sustainability and resilience, which will likely drive innovation in smart city solutions and business models, pushing cities to develop more environmentally conscious and adaptable systems. Emerging technologies such as 5G and IoT will play a crucial role in this evolution, enabling new forms of urban services and citizen engagement through enhanced connectivity and real-time data processing capabilities. The success of these technological advancements will heavily depend on the development of more sophisticated public-private partnership models, which will be crucial for future smart city implementations as cities and businesses collaborate to address increasingly complex urban challenges. Furthermore, the increasing focus on social inclusion and digital equity will shape future smart city initiatives, ensuring that technological advancements benefit all segments of urban populations. This social dimension will be complemented by the evolution of regulatory frameworks, which will significantly impact how digital businesses operate in smart city environments, establishing new guidelines for data privacy, security, and ethical considerations. Additionally, cross-border collaboration and knowledge sharing will become increasingly important for addressing common urban challenges, as cities worldwide face similar issues in implementing smart solutions and can benefit from shared experiences and best practices.

While acknowledging limitations such as language restrictions and rapid technological change, this review provides valuable insights for researchers, practitioners, and policymakers working to create more sustainable and livable urban environments. These perspectives suggest a future where smart cities and digital businesses continue to co-evolve, creating increasingly sophisticated and inclusive urban ecosystems.

Competing Interests The authors declare that they have no competing interests "ou" No competing interests to declare.

Ethics Approval Not applicable—This study is a systematic literature review and does not involve human subjects, animals, or personal data.

References

1. Bencsik B, Palmie M, Parida V, Wincent J, Gassmann O (2023) Business models for digital sustainability: framework, microfoundations of value capture, and empirical evidence from 130 smart city services. J Bus Res 158:113686
2. Timeus K, Vinaixa J, Pardo-Bosch F (2020) Creating business models for smart cities: a practical framework. Public Manag Rev
3. Anthony B Jr (2023) Data enabling digital ecosystem for sustainable shared electric mobility-as-a-service in smart cities— an innovative business model perspective. Res Transp Bus Manag 46:100927
4. Malchenko YA, Smirnova MM (2019) What drives consumers smart? The challenge of adoption of smart city solutions. Russian Manag J 17(3):387–410
5. Bogoyavlenska Y, Persia L, Bondarenko K (2020) Smart-logistics for people management of innovative small and medium enterprises' development: agile methodology. Econ Ecol Socium 4(2):24–35
6. Dessai A, Javidroozi V (2021) Cross-sectoral process modelling for smart city development. Bus Process Manag J 27(6):1891–1911
7. Zheng WM, Liao ZX, Lin ZB (2020) Navigating through the complex transport system: a heuristic approach for city tourism recommendation. Tour Manag
8. Hasbini MA, Eldabi T, Aldallal A (2018) Investigating the information security management role in smart city organisations. World J Entrep Manag Sustain Develop 14(1):86–98
9. Ford R, Maidment C, Vigurs C, Fell MJ, Morris M (2021) Smart local energy systems (SLES): a framework for exploring transition, context, and impacts. Technol Forecast Soc Chang
10. Oughton E, Frias Z, Russell T, Sicker D, Cleevely DD (2018) Towards 5G: scenario-based assessment of the future supply and demand for mobile telecommunications infrastructure. Technol Forecast Soc Chang 133:141–155
11. Gusakov AA, Haque AU, Jogia AV (2020) Mechanisms to support open innovation in smart tourism destinations: managerial perspective and implications. Pol J Manag Stud 21(2):142–161
12. Pulkkinen J, Jussila J, Partanen A, Trotskii I, Laiho A (2019) Smart mobility: services, platforms and ecosystems. Technol Innov Manag Rev

13. Ha LT, Thanh TT (2022) Effects of digital public services on trades in green goods: does institutional quality matter? J Innov Knowl 7(2):100171
14. Grigorescu A, Zamfir AM, Mocanu C (2020) Emerging trends and drivers for knowledge-intensive economy. Manag Market 15(4):537–551
15. Anthony BA (2023) A developed distributed ledger technology architectural layer framework for decentralized governance implementation in virtual enterprise. Inf Syst E-Bus Manag
16. Shamsuzzoha A, Syrjala A (2024) Importance to establish and deploy technical management processes to support organizations: a case study. Int J Eng Bus Manag 16(1):1–12
17. De Moura DA (2022) Analysis of industry 4.0 and their impact on port environmental management. Independ J Manag Prod 13(3):1108–1127
18. Zirena-Bejarano PP, Zirena EMC, De la Gala-Velásquez BR (2022) Determining the impact of brand value on the credibility of influencers over the purchase decision of millennial consumers. Innov Mark 18(1):1–12
19. Sá E, Carvalho A, Silva J, Rezazadeh A (2022) A Delphi study of business models for cycling urban mobility platforms. Res Transp Bus Manag 42:100754
20. Sampaio C, Sebastiao JR (2024) Social innovation and social entrepreneurship: uncovering themes, trends, and discourse. Administ Sci 14(1):1–18
21. Foley RW, Sylvain O, Foster S (2022) Innovation and equality: an approach to constructing a community governed network commons. J Respons Innov 9(2):209–231

Predictive Analytics and AI in Economic Development: Enhancing Digital Business Models in Smart Cities

Muhammad Hilal, Sumabl, and Muhammad Imad

Abstract This research explores the application of artificial intelligence and statistical techniques within predictive analytics to drive economic development and enhance digital business models in smart cities. The primary objective is to assess the improving decision-making processes that support urban growth and business innovation in smart environments. The study utilizes a thorough examination of existing literature to identify gaps and build upon previous research in this domain. A comprehensive framework is developed, outlining the predictive analytics models applied to economic and digital business strategies in smart cities. This framework also evaluates the suitability and impact of various AI and statistical approaches in fostering economic growth and optimizing digital infrastructure. Hypotheses are derived from a theoretical foundation that underscores the role of AI and predictive analytics in smart city development. The study concludes by summarizing the key findings, discussing the broader implications of applying AI and statistical techniques in smart city environments, and offering recommendations for future research focused on digital business and economic strategies in urban contexts.

Keywords Economic development · Artificial intelligence (AI) · Smart cities · Urban growth

M. Hilal (✉)
Institute of Management Science, Peshawar, Pakistan

Sumabl
Shaheed Benazir Bhutto Women University, Peshawar, Pakistan

M. Imad
Abasyn University, Peshawar, Pakistan

H. Kaddouri et al. (eds.), *Smart Cities for Sustainable and Urban Development*, Green Energy and Technology, https://doi.org/10.1007/978-3-031-99001-4_29

1 Introduction

Organizations strive to gain a competitive edge and engage in well-informed decision-making in the modern business landscape. Predictive analytics has emerged as a potent instrument for businesses to foresee future outcomes and shape their choices based on data. By harnessing historical data and employing advanced analytical methods, predictive analytics empowers organizations to uncover patterns, trends, and potential scenarios that lie ahead. The integration of artificial intelligence (AI) and statistical techniques has substantially transformed the decision-making process, significantly enhancing the precision and efficacy of predictive analytics models [1].

This writing explores the application of artificial intelligence and statistical methodologies within predictive analytics, particularly in guiding business decision-making processes. Our core aim is to delve into how these methodologies elevate the decision-making process and its results and to scrutinize the possible repercussions for organizations. Through an exhaustive analysis of the impact and effectiveness of artificial intelligence (AI) and statistical techniques in predictive analytics, valuable insights can be derived regarding their application's potential benefits and challenges.

To comprehend the significance of artificial intelligence and statistical methods in predictive analytics, crucial to possess a comprehensive grasp of the broader framework of business decision-making [2]. Contemporary businesses encounter many challenges, including market volatility, evolving consumer preferences, and intricate data ecosystems. Conventional decision-making methods relying solely on intuition and experience might prove inadequate in effectively addressing these complexities. Predictive analytics empowers organizations to extract valuable insights by dissecting extensive datasets and uncovering concealed patterns. These insights can then be harnessed to inform and steer strategic decision-making processes.

Integrating artificial intelligence (AI) methodologies, encompassing machine learning algorithms, natural language processing, and neural networks has bolstered predictive analytics models, enabling them to adapt dynamically, glean knowledge from data, and generate accurate forecasts. Employing AI techniques facilitates the automated identification of patterns and correlations within datasets, providing businesses with invaluable insights to make well-informed decisions [3]. In addition to AI algorithms, statistical methods like regression analysis, time series analysis, and hypothesis testing are invaluable data analysis and inference tools.

Previous studies have explored the significance of predictive analytics in business decision-making. However, it is imperative to analyze the impact of artificial intelligence (AI) and statistical methodologies within this domain. This article examines the efficacy of artificial intelligence (AI) and statistical methodologies in predictive analytics models, intending to address the existing disparity. Therefore, the valuable insights to organizations that employ predictive analytics and base their decisions on data analysis [4].

The subsequent sections will undertake a comprehensive literature review to ascertain prior research on using predictive analytics in business decision-making while identifying existing research gaps. Subsequently, a framework will be developed to examine the suitability of diverse artificial intelligence and statistical methodologies while elucidating the predictive analytics models employed in business decision-making. Based on this theoretical framework, hypotheses that will guide the research will be formulated. Integrating artificial intelligence (AI) and statistical techniques will be employed for data acquisition and analysis as an integral component of the research methodology. After the presentation of the results, a comprehensive discussion and conclusion will be conducted, highlighting the implications of the findings and suggesting potential avenues for future research.

This chapter seeks to contribute to the existing body of knowledge and provide practical guidance for organizations that utilize predictive analytics for decision-making by analyzing the role of AI and statistical techniques in this context.

2 Literature Review

The growing recognition of the significance of data-driven decision-making within organizations has led to a substantial increase in interest in applying predictive analytics in business decision-making [5]. The current body of academic literature concerning predictive analytics in this context encompasses a range of topics. These topics under discussion include the incorporation of artificial intelligence and statistical techniques in the development of predictive models, the influence of predictive analytics on organizational efficiency, and the obstacles and limitations encountered while implementing these analytical approaches.

The existing pool of literature on the utilization of predictive analytics in business decision-making yields valuable insights into the operational intricacies of artificial intelligence and statistical methodologies. This research article intends to address specific gaps or shortcomings within the existing body of knowledge. Previous studies have evaluated the effectiveness of artificial intelligence (AI) and statistical methods in isolation. Hence, it becomes imperative to scrutinize the collective impact of these two methodologies on predictive analytics models. Moreover, scholarly investigation concerning the precise outcomes of artificial intelligence (AI) and statistical techniques within diverse organizational contexts and industries is relatively sparse [6].

This chapter aims to delve into the application of artificial intelligence (AI) and statistical techniques in predictive analytics, with a focused emphasis on their role in enhancing business decision-making. A thorough exploration of their integration, performance, and associated challenges can provide a comprehensive understanding of how organizations can proficiently harness these techniques to elevate their decision-making processes and achieve favourable results. The subsequent sections will offer a more comprehensive analysis of this study's theoretical framework and methodology, aimed at investigating the influence of artificial intelligence and statistical techniques on predictive analytics for business decision-making. Theoretical Framework.

This chapter's theoretical framework investigates the applicability of AI and statistical techniques to improve the efficacy of predictive analytics models used for business decision-making. This framework integrates AI and statistics concepts to produce a comprehensive approach to predictive analytics.

2.1 Predictive Analytics Models

The theoretical framework describes prevalent predictive analytics models utilized for business decision-making. Instances include regression analysis, time series analysis, and classification models [7]. Regression analysis enables businesses to model the relationships between variables and to make predictions based on past data [8]. Time series analysis identifies patterns and trends in time-dependent data for accurate forecasting. Classification models are advantageous for decision-making circumstances involving classification assignments involving data classification into distinct groups or classes.

2.2 AI Techniques

This framework investigates the applicability of various AI techniques to predictive analytics. Algorithms for machine learning, such as random forests, support vector machines, and neural networks, are potent instruments for discovering intricate patterns and relationships in data [9]. These algorithms can autonomously learn from past data and make predictions and categorizations based on their discovered patterns [10]. Techniques for natural language processing (NLP) permit the analysis of unstructured text data, yielding valuable insights for decision-making. Other AI techniques, such as clustering and association algorithms, can assist in identifying concealed patterns and relationships within your data.

2.3 Statistical Techniques

With AI techniques, statistical techniques play a crucial role in predictive analytics. Regression analysis is commonly employed to characterize relationships between variables and make predictions. It enables businesses to comprehend the influence of independent variables on the dependent variable of interest. Time series analysis is essential for forecasting future values using historical patterns and trends in time-dependent data [11]. The statistical inference provided by hypothesis testing and

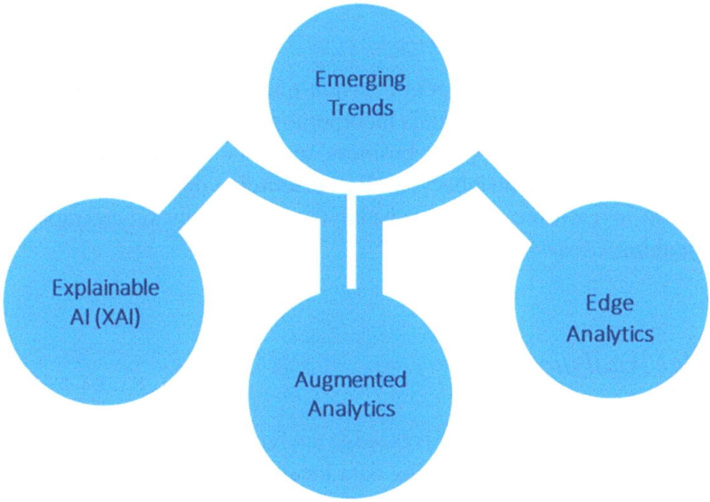

Fig. 1 Emerging trends of AI

confidence intervals enables organizations to evaluate the significance and dependability of predictive models.

2.4 *Integrating AI with Statistical Methods*

The theoretical framework emphasizes the significance of combining AI and statistical techniques to improve the predictive analytics models' performance [12]. By integrating AI techniques and statistical methods, businesses can capitalize on the advantages of both domains. Whereas statistical techniques provide a firm foundation for inference and validation, AI techniques enable the discovery of complex patterns and relationships [13] (Fig. 1).

3 Explainable AI (XAI)

Explainable AI (XAI) is an emerging paradigm in predictive analytics' use of AI. As AI techniques such as deep learning neural networks become more complex, their decision-making processes become less understandable [14]. XAI aims to enhance the lucidity and comprehension of AI models so that decision-makers can understand how the models arrive at their predictions. This trend is crucial for the trust and adoption of AI-based decision-making systems.

3.1 Augmented Analytics

Augmented analytics is an emergent decision-making technique combining artificial and human intelligence. It entails incorporating AI-driven insights into conventional analytics tools, equipping business analysts and decision-makers with automated and actionable insights [15]. By outsourcing data analysis and surfacing timely insights in real-time, augmented analytics enables quicker and more informed decision-making.

3.2 Edge Analytics

As the Internet of Things (IoT) evolves, edge analytics becomes increasingly important. Edge analytics entails processing data locally on IoT devices or edge servers, reducing the need to transmit enormous quantities of data to centralized cloud servers for analysis [16]. This trend improves real-time decision-making and decreases latency, making it ideal for time-sensitive applications like predictive maintenance and supply chain optimization.

3.3 Potential Impact on the Future of Business Decision-Making

These emerging trends have the potential to revolutionize business decision-making in several ways:

Explainable AI and augmented analytics can provide decision-makers with more accurate and timely insights, allowing them to make informed decisions rapidly. XAI's transparency enhances the comprehension of AI models, increasing confidence in their predictions.

Edge analytics enables organizations to process data at the source, facilitating real-time insights. This trend is especially advantageous in dynamic business environments where quick decision-making is required to remain ahead of the competition.

Augmented analytics automates data analysis, allowing human analysts to concentrate on higher-value duties, such as interpreting results, identifying opportunities, and creating strategic plans. This optimizes the allocation of resources and increases overall efficiency (Fig. 2).

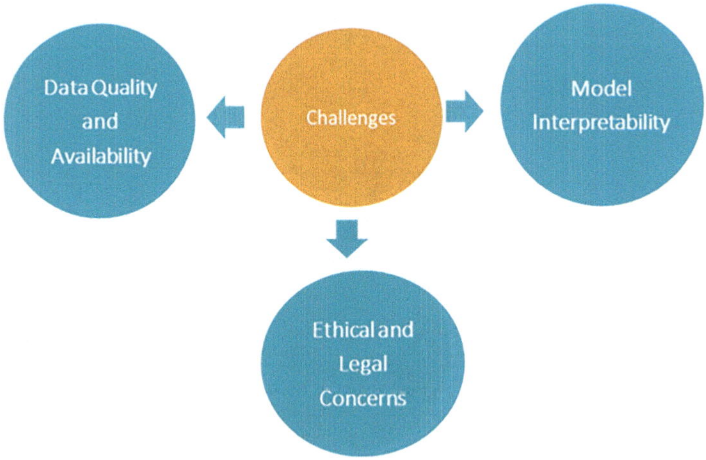

Fig. 2 Challenges of using AI

4 Challenges Using AI and Statistical Techniques

Artificial intelligence and statistical techniques rely significantly on pure, relevant, and high-quality data. It can be difficult to ensure data integrity and accessibility, particularly when dealing with large and diverse datasets.

4.1 Model Interpretability

As AI model complexity increases, interpretability becomes a concern. Understanding how AI arrives at its predictions is crucial, especially in scenarios involving crucial decision-making, such as healthcare or finance.

4.2 Ethical and Legal Concerns

Artificial intelligence in predictive analytics raises ethical and legal concerns regarding data privacy, bias, and equity. It is essential to ensure that AI models do not perpetuate bias or violate privacy regulations (Fig. 3).

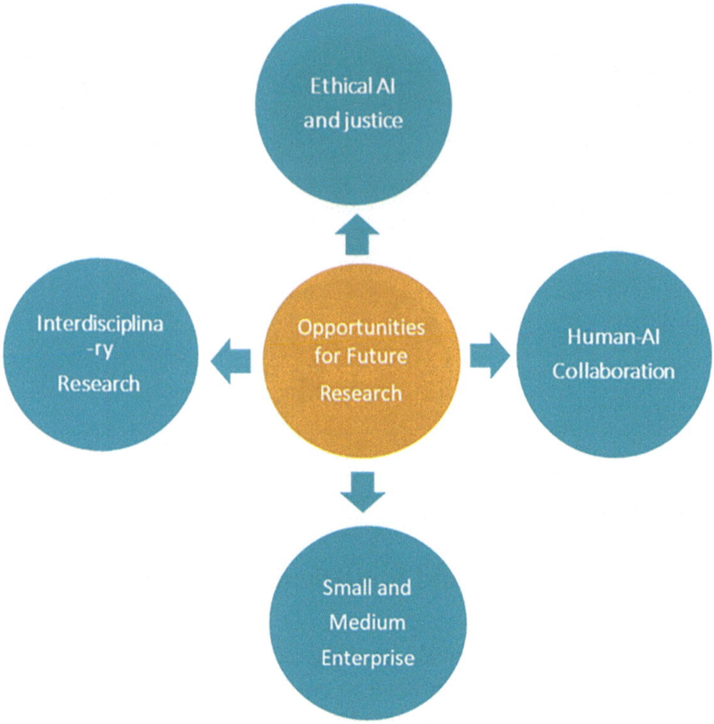

Fig. 3 Applications for future research

5 Opportunities for Future Research and Practical Applications

Future research could concentrate on interdisciplinary approaches that integrate AI, statistics, and domain knowledge to produce predictive analytics models that are more effective and interpretable. This could result in more precise and reliable predictions across industries.

5.1 *Ethical AI and Justice*

Researchers can investigate methods to improve the transparency and justice of AI models in predictive analytics. Developing ethical AI frameworks and techniques to establish trust in AI-based decision-making will be essential.

5.2 *Adoption in Small and Medium Enterprises (SMEs)*

There is an opportunity to study the adoption of artificial intelligence (AI) and statistical techniques in predictive analytics in SMEs, as these organizations may experience distinct challenges and opportunities from larger enterprises.

5.3 *Human-AI Collaboration*

Future research can investigate how humans and AI can effectively collaborate in decision-making processes. Understanding the optimal proportion between human judgment and AI-driven insights can improve decision-making outcomes.

Emerging trends in AI and statistical techniques, such as Explainable AI, augmented analytics, and peripheral analytics, have the potential to influence business decision-making substantially. These trends present opportunities for quicker and more precise decision-making, enhanced resource allocation, and real-time insights [17]. Nevertheless, addressing concerns about data integrity, interpretability of models, and ethical considerations is imperative. To fully harness the capabilities of predictive analytics in the context of business decision-making, future research and applications must prioritize inter disciplinary methodologies, ethical considerations in artificial intelligence, the adoption of predictive analytics by small and medium enterprises (SMEs), and the optimization of collaboration between humans and artificial intelligence systems.

6 Discussion

The primary aim of the discussion section is to offer a comprehensive examination and integration of the most notable patterns and concerns identified in the literature review. The process involves analyzing the primary outcomes from the examined research studies, comparing and contrasting the diverse outcomes, and acknowledging any disagreements or contradictory perspectives within the existing body of literature.

6.1 *Interpretation of Major Trends and Issues*

Identifying recurring themes, patterns, and insights from the literature review is crucial for interpreting significant trends and issues. It may be necessary to recognise the increasing significance of predictive analytics in business decision-making and its consequential influence on organizational performance [18]. Furthermore, the discourse could emphasize the increasing amalgamation of artificial intelligence (AI) and statistical methodologies, resulting in enhanced precision and resilience of predictive models. Placing these trends within business analytics and emerging technologies is paramount.

6.2 Comparison of Various Findings

When conducting a literature review, it is common to encounter divergent findings concerning the efficacy of artificial intelligence (AI) and statistical techniques in predictive analytics. The discourse should thoroughly examine these findings, emphasizing resemblances, disparities, and potential factors contributing to variations [19]. This comparative analysis can provide insight into the advantages and disadvantages of particular methodologies, thereby enhancing the depth of comprehension regarding the topic.

6.3 Controversies and Divergent Opinions

Controversies and divergent opinions are typical in any field of study. These discrepancies should be addressed in the discussion section of the paper. Incorporating AI techniques may have resulted in significant improvements in the predictive accuracy of some studies, whereas others may have found limited benefits. Identifying these controversies and discussing possible explanations or methodological differences can assist readers in gaining a more balanced perspective on the subject.

6.4 Identification of Knowledge Gaps

As part of the discussion, it is essential to identify any knowledge gaps or areas requiring further investigation. These voids may result from inconsistencies in the findings, limitations in the research methodologies, or unexplored aspects of artificial intelligence and statistical techniques in predictive analytics. Identifying these voids paves the way for future research and increases comprehension of the topic.

6.5 Implications for Business Decision-Making

The discussion should also consider the practical implications of the main trends and findings for business decision-making. Businesses can use AI and statistical techniques to improve their decision-making processes [20]. What potential obstacles might they face, and how can they surmount them? By addressing these questions, the discussion links the theoretical insights from the literature review to real-world applications and provides practitioners with actionable insights.

7 Recommendations for Future Research

The section should conclude with recommendations for future research based on the insights obtained from the literature review and the discussions of key trends, controversies, and knowledge gaps. These recommendations may centre on exploring particular aspects in greater detail, undertaking empirical studies to validate findings, or examining the influence of AI and statistical techniques on various industries or organizations.

This section contributes to the overall comprehension of the role of AI and statistical techniques in predictive analytics for business decision-making through a deliberate and insightful discussion. It critically analyses the existing literature, discusses controversies, and identifies prospective research directions, thereby contributing to knowledge development in this dynamic field.

8 Conclusion

The role of artificial intelligence and statistical techniques in predictive analytics for business decision-making has provided valuable insights into the field's current state. Key findings and trends identified in the literature review highlight the increasing importance of predictive analytics in driving data-driven decision-making processes. This section summarizes the key findings, discusses their implications for the future of artificial intelligence and statistical techniques in predictive analytics, and identifies potential areas for future research.

References

1. Sasmal, S. (2024). Predictive Analytics in Data Engineering: An AI Approach
2. Adesina AA, Iyelolu TV, Paul PO (2024) Leveraging predictive analytics for strategic decision-making: enhancing business performance through data-driven insights. World J Adv Res Rev 22(3):1927–1934
3. Azmi M, Mansour A, Azmi C (2023) A context-aware empowering business with AI: case of chatbots in business intelligence systems. Procedia Comput Sci 224:479–484
4. Wolniak R, Grebski W (2023) Functioning of predictive analytics in business. Silesian University of Technology Scientific Papers. Organiz Manag Ser 175:631–649
5. Lee CS, Cheang PYS, Moslehpour M (2022) Predictive analytics in business analytics: decision tree. Adv Decis Sci 26(1):1–29

6. Pereira V, Hadjielias E, Christofi M, Vrontis D (2023) A systematic literature review on the impact of artificial intelligence on workplace outcomes: a multi-process perspective. Hum Resour Manag Rev 33(1):100857
7. Nazish, Ullah SI, Salam A, Ullah W, Imad M (2021) COVID-19 lung image classification based on logistic regression and support vector machine. In: European, Asian, Middle Eastern, North African Conference on Management Information Systems
8. Kalaian SA, Kasim RM, Kasim NR (2017) A conceptual and pragmatic review of regression analysis for predictive analytics. In: Decision Management: Concepts, Methodologies, Tools, and Applications. IGI Global, pp 1761–1776
9. Bangash SH, Husnain G, Nawaz A, Tahir M, Imad M, Khan ZU, Khan D, Ahmed S (2023) Cruising into the future: navigating the challenges and advancements in autonomous vehicle technology. J Comput Biomed Inf 5(02):114–135
10. Babu B, Sudha S, Atlas LG (2021) Foundation of machine learning-based data classification techniques for health care. In: Machine Learning and Analytics in Healthcare Systems. CRC Press, pp 41–62
11. Le L-H (2024) Time series analysis and applications in data analysis, forecasting and prediction. HPU2 J Sci: Nat Sci Technol 3(1):20–29
12. Bangash SH, Imad M (n.d.) A comprehensive survey of machine learning applications in healthcare. Cognit Mach Intell:245–269
13. Gil Y, Greaves M, Hendler J, Hirsh H (2014) Amplify scientific discovery with artificial intelligence. Science 346(6206):171–172
14. Imad M, Cleland I, McAllister P, Nugent C (2024) Improving the accuracy of anomaly detection in multimodal sensors using 1D-CNN. In: Proceedings of the 17th International Conference on PErvasive Technologies Related to Assistive Environments
15. Imad M, Hussain A, Hassan MA, Butt Z, Sahar NU (2022) IoT based machine learning and deep learning platform for Covid-19 prevention and control: a systematic review. In: AI and IoT for Sustainable Development in Emerging Countries: Challenges and Opportunities, pp 523–536
16. Bangash SH, Khan D, Ishtiaq A, Imad M, Tahir M, Ahmad W, Husnain G, Jan L (2023) Integrating machine learning and deep learning approaches for efficient malware detection in IoT-based smart cities. J Comput Biomed Inf 5(02):280–299
17. Okatta CG, Ajayi FA, Olawale O (2024) Leveraging HR analytics for strategic decision making: opportunities and challenges. Int J Manag Entrep Res 6(4):1304–1325
18. Badshah SH, Imad M, Khan IU, Hassan MA (2023) Machine learning algorithms for classification of COVID-19 using chest X-ray images. In: Advanced AI and Internet of Health Things for Combating Pandemics. Springer, pp 85–96
19. Badshah SH, Imad M, Hassan MA, Ullah S, Ali SH (2023) Investigation of blockchain for COVID-19: a systematic review, applications and possible challenges. EAI Endors Trans Smart Cities 7(1):e5–e5
20. Prasanth A, Densy JV, Surendran P, Bindhya T (2023) Role of artificial intelligence and business decision making. Int J Adv Comput Sci Appl 14(6)

Economic Diplomacy and Urban Innovation: Pioneering Smart Cities in Morocco

Layla Bentahar and Mohammed Abdellaoui

Abstract This article delves into the idea of "smart cities" in Morocco, emphasizing the role of economic diplomacy, innovation, and entrepreneurship in driving urban development progress. The research starts by providing an examination of the concept of a "smart city" and its various aspects as discussed in existing literature, laying the groundwork for comprehending its fundamental components. The smart city implementation in Morocco is analyzed through three main examples: Casablanca, Benguerir, and the Zenata Eco-City. These case studies demonstrate how smart city principles are implemented in practice and the importance of international partnerships in advancing technological and economic development. Moreover, the article examines how economic diplomacy strategically contributes to promoting innovation, entrepreneurship, and sustainable urban development. The study showcases how Morocco's diplomatic strategies promote collaboration, attract investment, and improve digital business ecosystems by exploring the connection between economic diplomacy and the smart city concept.

The main approach used is qualitative, focusing on analyzing official reports, academic literature, and institution publications regarding smart cities and economic diplomacy. This is supported by examining key urban projects in Morocco through case study analysis. The results of this study show that combining economic diplomacy with innovation and entrepreneurship is crucial for the effective execution of smart city projects, leading to sustainable economic development and establishing Morocco as a frontrunner in digital urban change.

Keywords Economic diplomacy · Smart cities · Trade openness · Urban innovation · Sustainable urban development

L. Bentahar (✉) · M. Abdellaoui
LIREFIMO Research Laboratory FSJES, University Sidi Mohamed Ben Abdellah, Fez, Morocco
e-mail: layla.bentahar@usmba.ac.ma

H. Kaddouri et al. (eds.), *Smart Cities for Sustainable and Urban Development*, Green Energy and Technology, https://doi.org/10.1007/978-3-031-99001-4_30

1 Introduction

In recent years, smart cities have gained significant popularity as they offer innovative solutions across various domains such as mobility, environmental sustainability, economic development, governance, quality of life, and education, driven by advancements in Information and Communication Technologies (ICT) [1]. On one side, urban growth puts pressure on current infrastructure, increases environmental damage, and amplifies social and economic inequalities. Conversely, it provides a rich environment for innovation and digital change that has the potential to transform urban life. Typically, the fascination with smart cities is strongly associated with the emergence of new information technologies like mobile devices, semantic web, cloud computing, and the Internet of Things [2].

The idea of smart cities has arisen as a viable answer to tackle these metropolitan issues and utilize technological progress for sustainable and effective urban development. The phrase "smart city" was embraced in 2005 by various technology firms providing intricate information systems to their customers, targeting the integration of urban infrastructure operations [3].

At the core of smart cities lies advanced technologies like the Internet of Things (IoT), artificial intelligence (AI), big data analytics, and 5G networks. These technologies operate collaboratively to form interconnected systems that gather, evaluate, and react to real-time information from various sources, such as transportation networks, utility systems, and environmental monitoring. IoT, for instance, allows cities to manage resources like water and energy more effectively by tracking usage trends and spotting waste. Analytics driven by AI help enhance traffic movement, forecast maintenance requirements, and improve waste management, thus boosting the efficiency and sustainability of city operations. These technological applications are essential for developing a city that operates smoothly and adjusts to the evolving needs and expectations of its inhabitants.

However, developing a smart city entails more than mere technological advancemen, the participants in territorial development are progressively urged to combine their efforts into a framework of collective intelligence and effective governance. This method is crucial for promoting the effective establishment of smart cities, as it guarantees the provision of top-tier local services while adequately meeting the needs and expectations of citizens and investors alike [4]. By synchronizing these initiatives with economic diplomacy and urban innovation, Morocco can improve its capacity to draw international collaborations and investments, promoting sustainable urban growth and strengthening its role as a frontrunner in smart city evolution. Major projects like these require significant foreign investments to support infrastructure development, technology integration, and management operations. Furthermore, effective smart city initiatives rely significantly on global collaboration and the exchange of knowledge, since the intricate technological, regulatory, and governance structures are often too challenging for single cities or countries to handle, especially in the developing nations. For nations such as Morocco, currently experiencing a major digital transformation, these needs emphasize the importance of economic diplomacy. Economic diplomacy, defined as leveraging diplomatic relations to enhance trade, investment, and economic collaboration, serves as a strategic mechanism to draw in foreign direct investment (FDI) and public-private partnerships (PPPs) essential for financing and executing smart city initiatives.

Morocco, due to its strategic geographic position and proactive economic reforms, has progressively adopted digital transformation as a means to enhance economic growth and competitiveness. Lately, many efforts have taken place in several Moroccan cities to boost their attractiveness by promoting and improving their regions through the implementation of significant infrastructure initiatives and the digital transformation of local governance [5]. Economic diplomacy is crucial to this approach, enabling Morocco to enhance relationships with technologically advanced nations and draw investment from multinational firms focused on digital infrastructure, sustainable energy, and intelligent urban solutions. By actively participating in diplomatic initiatives, Morocco can promote global partnerships that aid not just the advancement of specific cities but also enhance a wider national context for urban innovation and sustainable development.

The objective of this paper is to examine the idea of the "smart city" in the Moroccan setting. At first, we will introduce the idea and measurements of the "smart city" as described in research, offering a theoretical basis for comprehending its fundamental elements and their importance in urban growth. Next, we will examine three concrete instances of smart city implementation in Morocco, with a focus on Casablanca, Benguerir, and the Zenata Eco-City. The case studies show how Morocco is using the smart city model to promote urban development and boost economic growth. Ultimately, we will explore Morocco's economic diplomacy strategies, which are essential for enhancing innovation and promoting the implementation of the smart city concept. This part will discuss how Morocco combines its diplomatic endeavors with technological and entrepreneurial progress to improve the smart city infrastructure within the nation. In particular, the paper will address the central question: How the use of economic diplomacy in relation to innovation and entrepreneurship contributes to the advancement of smart cities in Morocco?

2 Preliminary Definitions

Understanding the fundamental concepts of economic diplomacy and smart cities is crucial for analyzing their intersection in Morocco's urban development. This section provides a conceptual framework to clarify these key terms.

2.1 Economic Diplomacy

The definition of economic diplomacy is subjective and challenging to establish, as it varies depending on the countries and individuals involved. Its essence is not determined by specific tools or strategies but rather by the economic issues it aims to address [6]. Economic diplomacy would then be only one way for a State to strengthen its power vis-à-vis other States, using the means of the economy: international trade negotiations, measures to support the international expansion of domestic companies, or conversely, measures to attract foreign investment to the country.

So what does "economic diplomacy" really mean? If we look at the international bibliography, It is evident that economic diplomacy encompasses the fusion of two dimensions, namely the economic and commercial aspects. Indeed, economic diplomacy is the exploitation of all communication channels at the service of the country's economy in terms of seeking new markets and attracting foreign businessmen [7]. In addition, changes in the international system since the end of bipolar politics have helped transform ideological interests into economic interests of states.

According to Woolcock and Bayne [8], who propose the most recent analysis to date, The definition of economic diplomacy is not determined by the specific tools or methods employed, but rather by the economic challenges and issues it aims to address. Thus, it is not "just" commercial diplomacy, but it also includes industrial policy, the movement or exchange of goods, services, investments, money, etc.

According to Dafir [9], economic diplomacy is the set of mechanisms and practices adopted by individuals or groups, state or non-state, with the aim of achieving a state's economic objectives through the use of political means, or to achieve political objectives through economic means. This definition acknowledges the changes occurring in the global landscape, including the rise of new international actors. It recognizes that economic diplomacy is now closely intertwined with the comprehensive strategy for national development.

The definitions proposed so far reflect clearly the oscillation of the concept between the two poles that give it its identity: the public pole and the private pole. Moreover, this phenomenon particularly concerns developing countries which, in the absence of a basic definition, it is commonly accepted that economic diplomacy can be subject to multiple interpretations and can be defined in various ways. The first definition of the concept dates back to the beginning of the 21st. However, three components are common to the different definitions of economic diplomacy. The goal is to facilitate the expansion of domestic businesses into international markets., encourage foreign direct investment (FDI) in the country, and shape international regulations in a manner that aligns with national interests.

2.2 *The Concept of Smart Cities*

The term "smart city" first appeared in the late twentieth century, primarily adopted by tech firms and urban planners to refer to cities that utilize digital technology to enhance urban infrastructure and services. The fundamental concept of smart cities involves the blending of information and communication technologies (ICTs) with the Internet of Things (IoT) to enhance the efficiency, responsiveness, and sustainability of urban systems. Definitions of smart cities differ, yet frequent themes involve the improvement of urban systems such as transportation, energy, waste management, and public safety via technology and data analysis.

The idea of "Smart City" has been characterized as complex and hard to define [10]. Certainly, despite the abundance of literature on the topic [11–14], the concept is still somewhat ambiguously defined. Below, we outline the key defining features of the smart city concept along with the dimensions that define it.

The initial insight gained from examining the literature on the idea of a smart city is the importance of the framework of "digitized flows" [15]. The smart city "Smart city" is introduced as a concept that leverages technology (big data, sensor technologies, artificial intelligence, etc.) to enhance the socio-economic dimensions of urban life [16]. The smart city is primarily "Enhanced by the widespread and extensive application of advanced ICTs, which, in connection with diverse urban sectors and how they are interconnected, enables cities to be more sustainable and offers residents an improved quality of life" [15].

Komninos emphasizes that for a city to be considered "smart," it needs to implement a diverse array of digital and electronic technologies while also ensuring that this application fosters innovation, knowledge, and creative solutions to meet the needs, challenges, and expectations of its citizens. Certainly! The sustainable and intelligent city considers the role of ICT while fostering the engagement of individuals in utilizing ICT to enhance their knowledge, understanding, and creativity. Smart cities are characterized as "Areas with robust learning and innovation potential, relying on the creativity of their inhabitants, their institutions for knowledge creation, and their digital communication and knowledge management infrastructures" ([17], p. 305). Consequently, it is crucial to tackle the smart city challenge comprehensively, considering the related human element. Consequently, the emphasis on the technical component would no longer serve as the sole foundation of the smart city. Indeed, deploying Information and Communication Technologies (ICT) is essential for developing a Smart City, yet it is not adequate on its own to meet sustainability objectives [18]. It remains an ongoing debate between the techno-centric perspective and the human-centric perspective [19, 20].

The idea of smart cities has surfaced as a revolutionary strategy for urban development, incorporating cutting-edge digital technologies to establish more efficient, sustainable, and enjoyable living spaces. Intelligent cities utilize the Internet of Things (IoT), big data analysis, artificial intelligence (AI), and other cutting-edge technologies to improve numerous facets of urban living, such as transportation, energy management, healthcare, and public services. These urban areas seek to tackle the issues of fast urban development, environmental sustainability, and economic advancement by implementing intelligent solutions that enhance the living standards for both residents and visitors. The incorporation of Information and Communications Technologies (ICT) in city planning is essential for developing appealing, healthy, and thriving urban environments. Digital platforms play a crucial role in reshaping urban infrastructure, offering fresh opportunities for merging physical and digital infrastructures [21]. Additionally, the idea of digital twin technology improves sustainable urban growth by optimizing city infrastructure and services performance via real-time connectivity. Moreover, research by [4] indicated that new technologies like artificial intelligence, augmented reality, the internet of things, blockchain, and data analytics are becoming ever more essential in strategies for territorial marketing and appeal. These technologies are mainly utilized to better comprehend and fulfill the unique requirements of citizens while improving their experience as users of the area. These technological tools enhance the appeal of cities and regions by enabling more customized interactions and improving public services, fostering more connected and vibrant environments for residents and tourists alike.

Examining Lucien Sfez's work on the connection between the technical and political aspects of urban fiction, the author contends that: "The fictions (those centered on progress) possess a allure that impacts our everyday experiences. We believe that fiction serves as the driving force behind technology, and consequently, politics. To put it differently, the idea of a smart city prompts us to consider the role of the individual/citizen within this framework that intertwines technology, politics, and society for the effective administration of a specified urban area. Thus, the rise of terminology linked to smart cities like: city "inclusive", "fair", "sustainable", and "resilient", can be seen as innovative strategies or types of participatory democracy (e.g., e-democracy) and the universalization of public education (e.g., e-learning). Certain researchers claim that these smart cities represent a "collaborative utopia" [22], whereas others refer to a "new utopia" in political regimes – still authoritarian – that have chosen smart city initiatives for their participatory and inclusive aspects [23]."

Giffinger defines a smart city as one that spearheads progress in the fields of economy, society, governance, mobility, environment, and quality of life through a clever mix of funding and initiatives, achieving this in a persuasive and autonomous way, while fostering public awareness. Consequently, all these aspects should be considered in a comprehensive manner to implement the concept of a smart city (Fig. 1).

Fig. 1 Characteristics of the smart cities (Giffinger 2007)

3 The smart city concept in Morocco

The concept of a pseudo smart city is defined by three main attributes: governance, technology, and citizens, as detailed below:

- **Technology**: The so-called Smart City fundamentally depends on easily accessible low-tech approaches that need little financial commitment. Creating a sophisticated infrastructure requires significant and long-term investments. A framework focused on the widespread adoption of the Internet and smartphones, coupled with innovations from local startups, is more likely to succeed. Quantifiable measures of success in this context encompass the quality of digital infrastructure, the availability of technological solutions, and the incorporation of smart devices to enhance urban services.
- **Governance:** The termed Smart City aims to improve urban management compared to the existing situations observed in emerging or developing countries, but it does not set the benchmark for a "smart state." This model features open administrations that provide top-notch services via digital methods. These organizations additionally share urban data essential for developing intelligent solutions. Important factors in this context encompass the effectiveness of governance procedures, the clarity of administrative actions, and the accessibility of digital platforms to enhance interactions between citizens and the government.

- **Citizens:** The term Smart City aims for an involved citizenry that aids the city's digital evolution by using services and acting as a data provider for developing solutions. Information will be collected through the internet and intelligent devices. Placing the citizen at the center of the city's strategy as a vital participant in its growth will signify the next step in the evolution of civil society and the city's governance practices. Success can be gauged by citizen involvement, the degree of participation in digital services, and the effect of this involvement on urban development.

The idea of smart cities in Morocco began to draw interest in the early 2010s, fueled by the government's dedication to digitalization and eco-friendly urban development. Acknowledging that swift urban growth presented difficulties for infrastructure, public services, and environmental effects, Morocco started to investigate smart city solutions to enhance the efficiency, livability, and responsiveness of cities to citizens' requirements.

The discussion about smart cities in Morocco has been ongoing for under ten years, and currently we have not reached the point of recognizing any "success stories". Nevertheless, a variety of initiatives and projects have been initiated by various parties to advocate for the idea and lay the groundwork for future smart cities in Morocco. The Moroccan population is increasingly accessing the Internet due to ICT growth, leading to the development of various initiatives. The smart city projects in Morocco are centered on clusters or hubs. They primarily contribute to the speeding up of the nation's socio-economic recovery. In this section, we will introduce three Moroccan projects that focus on the idea of smart city: Casablanca, Benguerir, and the ZENATA eco-city. These three scenarios embody various dimensions and locations: the core, the outskirts, and the urban district.

- **Casablanca**: the largest city in Morocco, has the highest population in the country. Therefore, the city's leadership encounters a variety of typical issues that are particular to major urban areas, such as transportation, environmental contamination, energy supply, and water management. Over the past ten years, Morocco has initiated the Casablanca Smart City initiative.
  Casablanca was included in the network of twenty-five smart cities chosen by the Institute of Electrical and Electronics Engineers (IEEE) in 2015, making it the first African city to join. In 2016, the inaugural "Smart city Expo Casablanca" was held with the motto: Cities that are open, inclusive, and innovative.
  This annual exhibition provides a chance to review the progress of smart cities, showcase Moroccan startups involved in the movement, and highlight the innovative solutions being implemented. Multiple open innovation contests and hackathons are organized to address Casablanca's challenges, including the DataCity Casablanca initiative, a partnership involving the Numa incubator and four companies (RATP Dev, Lafarge Holcim Maroc, Inwi, and Michelin). Casablanca smart city places great emphasis on energizing the digital entrepreneurial scene.
- **The Zenata Eco-City**: showcases Morocco's strategic incorporation of smart city ideas into its public urban strategies, guided by the principles of economic

diplomacy. Created to boost the economic opportunities in the area, Zenata advantages from its strategic position between Morocco's top two cities: Rabat, the administrative center, and Casablanca, the financial center. This alignment corresponds with Morocco's initiatives to draw in international investments and enhance its position as a prominent economic and trade leader in the region.

The appeal of Zenata to international investors is boosted by its close proximity to important transportation and logistical hubs such as highways, railways, ports, and airports. This deliberate connection showcases Morocco's diplomatic efforts to present itself as a doorway to Africa and a center for worldwide commerce. By collaborating with global partners, the city has encouraged companies to move their operations, creating a 69-hectare industrial area to stimulate the economy and offer job prospects.

Zenata, as a leading initiative, showcases Morocco's dedication to utilizing smart city ideas to improve urban resilience, sustainability, and the well-being of its citizens. Morocco's utilization of economic diplomacy to gain international recognition and collaboration is showcased through its emphasis on digital innovation, sustainable urban planning, and citizen inclusion in the city. The international "Eco-Cité Label" given to Zenata exemplifies Morocco's success in harmonizing local development initiatives with global sustainability criteria, Additionally, the Zenata Eco-City promotes Morocco's economic diplomacy goals by developing an environment that draws in global corporations, encourages creativity, and facilitates the sharing of knowledge. The progress of Zenata's business center, currently in its second stage, demonstrates how smart city initiatives can drive economic growth and showcase Morocco's aspirations to be a leader in sustainable urban development worldwide.

- **Benguerir City**: Strategic investments and international collaborations driven by economic diplomacy have propelled the city of Benguerir, located 72 kilometers from Marrakesh, to become a promising model for smart city development in Morocco. In the past few years, Benguerir has started multiple projects to establish itself as a center for innovation, sustainability, and entrepreneurship, supporting Morocco's overall objectives of attracting international partnerships and promoting local economic development.

  The transformation of Benguerir received a major enhancement from the founding of the Mohammed VI Polytechnic University (UM6P). This cutting-edge establishment is the focal point for the city's smart city goals, providing creative teaching and research approaches that encourage hands-on research and experimentation. The university's resources, including Living Labs, offer opportunities for addressing practical issues and advancing solutions in important fields such as energy, agriculture, and technology. These efforts showcase Morocco's diplomatic approach of utilizing academic achievements to draw in international research partnerships and promote the sharing of knowledge, UM6P cultivates a dynamic entrepreneurial environment with initiatives like Impulse, backed by the OCP Group and MassChallenge, and Start-Gate, a startup incubator program. These efforts are in line with Morocco's economic diplomacy, demonstrating the country's ability to promote innovation and draw in valuable foreign investments. The opening of the African Supercomputing Center in February 2021 highlights Morocco's goal of becoming a key player in digital and technological advancement in Africa. The supercomputer boosts Morocco's competitiveness and reinforces its global position in the fields of big data and artificial intelligence.

Benguerir's membership in the African Smart Towns Network (ASToN), backed by the French Development Agency (AFD), emphasizes the importance of economic diplomacy in promoting regional and global collaboration. Chosen as one of 11 African cities for this program, Benguerir gains from sharing knowledge and working together with other smart cities in Africa. This network promotes progress in economy, society, and environment with creative solutions, Benguerir showcases how Morocco employs economic diplomacy to encourage investment, promote sustainable urban development, and establish itself as a frontrunner in Africa's digital transformation through global partnerships, advanced education, and modern technology integration. This strategic approach not only boosts the city's potential as a smart city, but also aligns with Morocco's wider vision of inclusive and sustainable growth.

4 Economic Diplomacy and the Building Blocks of Smart Cities in Morocco

In recent times, Morocco has adopted the idea of smart cities as a crucial component of its urban and economic growth plans. Seeking to update infrastructure, improve urban living conditions, and promote sustainable development, Morocco's smart city projects are in line with its national objectives for digital innovation and environmental sustainability, Creating a robust digital and technological ecosystem is vital for fostering connected economic diplomacy that enhances the nation's economic development and speeds up the digital transformation of the Moroccan economy. This procedure entails executing a forward-looking development strategy for technology industries, encompassing start-ups and future technologies like artificial intelligence, big data, and blockchain. Moreover, establishing an appealing structure for global participants in the digital industry is essential. This method is closely associated with urban innovation, as it fosters the creation of smart cities in Morocco that utilize advanced technologies to enhance urban management, sustainability, and the quality of life for citizens. By leveraging economic diplomacy, Morocco can enhance its role in the worldwide smart city sector, drawing in foreign investments and establishing strategic alliances for the creation of innovative urban solutions [24].

4.1 *Digital Transformation of the Moroccan Society*

Defining the Smart City concept reveals that technology serves as the foundation of these models. Many researchers concur that a city is considered smarter when it integrates various digital technologies, thereby enhancing the quality of life and employment opportunities in the region. The importance of digital transformation in Morocco's economic policy is growing, both at home and abroad. As the world's markets move towards digital economies, Morocco is actively establishing itself as a digital hub in Africa and the Mediterranean region. The nation aims to draw foreign investment in its expanding technology industry, promote creativity, and enhance its digital infrastructure through careful diplomatic interactions.

The advancement of Smart Cities in Morocco via the rise of innovation, the influx of investments and the encouragement of job creation, thereby enhancing the robustness of the city's economy.

Central to this ambition, technological advancement plays a key role in effectively realizing the foundational model of the Pseudo Smart City. Certainly, the digital transformation has enabled significant alterations in economic, social, and environmental aspects. (Allaki O. et al., 2023).

A crucial aspect of Morocco's digital strategy involves collaborating with international organizations and foreign governments to enhance knowledge sharing and skill development. For example, the nation has established deals with the EU and prominent tech firms to improve digital education and infrastructure, crucial for integrating Moroccan enterprises into the worldwide digital economy. Moreover, through its participation in international organizations like the World Trade Organization (WTO) and the African Union, Morocco has the chance to influence global digital trade policies to its benefit. The development of the African continent and its trade flows will also have to take into account global technological developments and transform the deficit in our intra-continental trade in a real opportunity for the development of new technologies digital.

Africa is becoming a laboratory of the digital world. Digital is changing the face of our continent, driven by an inventive, creative and daring youth. This digital leap is the fruit of young startups active in the fields of finance, telecoms, industry and agri-food, to name but a few. Often, the young and low-income population is at the heart of this innovation process; it must therefore be at the heart of our public policies." (Speech by the King of Morocco at the Extraordinary Summit of the African Union on the Continental Free Trade Area, Kigali, 21 March 2018).

Since gaining independence, Morocco has pursued a comprehensive policy aimed at liberalizing its borders and fostering economic freedom. The country has expressed a clear intent to attract foreign direct investment (FDI) for various reasons. This includes the goal of engaging with international markets, encouraging their presence in Morocco, driving the country's economic growth, and capitalizing on the expertise of foreign investors in areas such as workforce management, technology transfer, and information systems.

Morocco achieved a score of 91.7 for the Development Index of Information and Communication Technologies (IDI) in 2024, according to the annual report of the International Telecommunications Union (ITU). This score represents a 2% improvement over the previous year, reflecting the country's continued efforts to improve its digital infrastructure and encourage the use of technology.

The improved IDI score indicates tangible progress in a number of areas, such as increasing the number of people with mobile phones, improving internet services and increasing investment in digital infrastructure. However, Morocco still faces challenges to achieve inclusive and meaningful global connectivity, such as the high cost of communications in some regions and improving Internet access in rural areas.

According to the report, 98% of Moroccans own a mobile phone and 80.2% use the internet. However, only 47.7% of households have an internet connection at home, which highlights the gap between people with access to the internet via mobile phones and those with fixed connections.

With regard to broadband services, there are 72.5 subscriptions to the Internet via mobile broadband per 100 inhabitants and the 4G/LTE network covers about 58.1% of the Moroccan territory. The average Internet traffic via mobile broadband per subscription is 3.6 GB, while that via fixed broadband is 36.1 GB per subscription, indicating a growing reliance on the Internet for various uses (Fig. 2).

The Internet subscriber base is around 30 million, bringing the Internet penetration rate to 83%. There is a predominance of mobile internet in Morocco: it represents more than 93% of connections. As for the fixed internet, about 99.93% of ADSL subscriptions are operated by Maroc Télécom.

According to the Speedtest Global Index11, the average mobile internet speed in Morocco (36.36 megabits per second Mbit/s) is lower than the world average (48.40 Mbit/s). In terms of this indicator, Morocco is ranked 60th out of 140 countries. The average fixed internet speed in.

Morocco (25.05 Mbit/s) is also much lower than the global average (98.67 Mbit/s), and Morocco ranks 112th out of 177 countries, Morocco is ranked 44th out

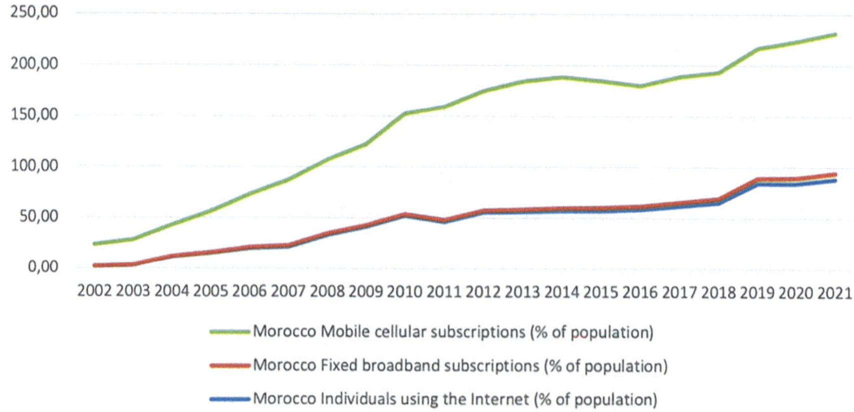

Fig. 2 Evolution of mobile phone and internet use in Morocco. (Source: World Bank Data)

of 230 countries where the GB Gigabit (GB) is the cheapest in the world with an average price of 7.96 Dirhams the GB12; Morocco is the 11th cheapest country in Africa. As far as fixed internet is concerned, Morocco ranks 66th out of 206 countries where the average monthly subscription cost is the cheapest with an average price of 324 dirhams per month13. Morocco is the sixth cheapest country in Africa for this criterion.

Preparing for the arrival of 5G is a major challenge for the coming years. Morocco has set itself the objective of enabling the launch of 5G before the end of the guidance note. This launch should not be early, given the ambitious development for the fixed market and the progressive maturity of the technology, nor late given the need to maintain its regional leadership, In addition, "Full 5G" technologies would only be available from 2023.

A study will be conducted to propose the conditions and modalities for operating 5G networks in Morocco and to identify and prepare the plans for redesigning the target frequency spectrum. For operators who wish to do so, the ubiquity of licences will be examined. The study should also detail the quality criteria to be met by operators and the obligations attached to the deployment of 5G (coverage. sharing. roaming, business model ...) and identify the needs of the sectoral verticals.

This study should also identify frequencies that would be intended for the deployment by manufacturers of their own 5G networks (industry 4.0 ...) not open or public, in order to contribute to the national digitalization strategy, Moreover, and given that the business market constitutes the potential for 5G development (automotive, health, research ...), actions would be promoted to encourage the emergence of a real ecosystem around 5G.

4.2 *Strategies and Programs*

Morocco's strategy for digital transformation has been strengthened by multiple initiatives focused on improving smart governance, boosting economic growth, and increasing the efficiency of public services. The administration emphasizes the incorporation of digital technologies across various sectors to simplify administrative processes and encourage innovation.

4.2.1 Smart City Initiatives in Morocco

Morocco aims to convert six cities into smart cities by 2026 (Ministry of Industry, 2013). The initiative started in Casablanca in 2016, followed by Marrakech, Rabat, Tangier, Ifrane, Berkane, and Fez. Morocco seeks to establish a novel urban management model that is cost-effective, enhance urban planning efficiency, and attain sustainable social development that addresses citizens' needs regarding transportation, energy, green economy, security, and housing.

Nonetheless, for these intelligent governance projects to succeed, the public needs to comprehend and embrace digital administration services [25, 26]. The success of implementing intelligent public services depends on how end-users view the online administrative efforts. The current technology might not be well-received by the end-users. The purpose of a smart city becomes useless if end-users fail to effectively use online administration services.

Morocco has designed several strategies for the development of digitalisation such as «Maroc Numeric 2013», «Maroc Digital 2020» and created specialized bodies such as the Digital Development Agency (ADD) and the National Commission for the Control of the Protection of Personal Data (CNDP).

Morocco officially launched its national digital strategy, dubbed "Digital Morocco 2030", on September 25, 2024 in Rabat, This roadmap aims to transform the country into a regional digital hub while accelerating its socio-economic development. The plan, which is part of a participatory and multi-sectoral approach, aims to create 240,000 direct jobs in the digital sector and contribute 100 billion dirhams to GDP by 2030, One of the key axes of this strategy is the digitalization of public services, with the objective of raising Morocco among the 50 first nations in the world ranking of online services. To do this, a unified digital portal will be set up to centralize administrative procedures and simplify access to services for citizens and businesses. Solutions such as digital identity, electronic signature and data exchange platforms between administrations will strengthen this digitalization. The government also aims to improve the user experience through a "customer journey" approach, which involves optimizing each step of the online administrative process.

4.2.2 Challenges and Public Engagement

Morocco has been able to implement several digital transformation projects for the administration and public services through several ministerial departments and institutions. Examples include:

- The Directorate-General of Taxation (DGI) which has set up several digital initiatives for the declaration and payment of taxes and taxes online (vignette, IR, IS, VAT, etc.);
- The Agence Nationale des Ports, which in 2008 launched a one-stop shop for PortNet. Since 2012, its management has been entrusted to the public limited company PORTNET;
- The Justice Department, which has developed a video conference solution for hearings and trials (more than 14,000 audience14 were conducted remotely);
- The Administration Reform Department, which launched the "Chikaya" platform in collaboration with the Digital Economy Department to facilitate the complaint procedure: Citizens have been able to file more than 000800 claims since its launch in January 2018 through March 2021 with a processing rate of approximately %70;

- The Ministry of the Interior which has set up the program for the modernization of civil status, The platform «Watiqa» allowing the remote obtaining of administrative documents, the filing of complaints by citizens through the platform «Chikaya» - the territorial communities component and the issuance of authorisations via the "Rokhas" platform;
- The Ministry of Education which has set up the GENIE programme (generalisation of information and communication technologies in education in Morocco) Equipping 87 of the 11,000 ordinary schools with multimedia rooms or multimedia suitcases. During the health crisis, the Ministry made the "TELMIDTICE" portal available to students for distance learning;
- The Ministry of Health, which has adopted a legal framework for telemedicine (Law 13–131 and Decree 378–18-2 amended and supplemented by Decree 675–20-2). In addition, the Department has launched an online appointment portal called mawiidi.ma;
- The National Health Insurance Agency (ANAM) which has put online the RAMED monitoring, the claim portal (Chikaya) for the benefit of AMO insured and the monitoring of students' declaration to the AMO.
- The launch of digital innovation institutes in 2023, such as the YouCode school in Nador, the Berkane National School of Artificial Intelligence and Digital, opens up new horizons for training in digital and AI.

Furthermore, Law 55–19 on the simplification of administrative procedures, which came into force on 28 September 2020, introduced new measures to facilitate the relationship between the administration and users. Through this law, the National Administration Portal (PNA) was implemented, gathering information on current administrative procedures, Regarding the digital transformation of the main sectors of the economy. It is worth mentioning advanced digitalisation of the banking sector (online banking) and reflections for digitalisation of the insurance sector; significant growth in e-commerce transactions, particularly after the Covid-19- event; In the agricultural sector, there have been some initiatives such as the electronic "one-stop shop" or artificial intelligence tools at INRA (in collaboration with SOWIT).

Concerning the tourism sector: the majority of tourists who have visited Morocco (%90) prefer online booking; Regarding the culture and media sector: a set of media and culture initiatives have been launched (SNRT applications, streaming of films during confinement). Morocco was the first African country to launch an Open Data Government (OGD) initiative in 2011 by launching its portal www.data.gov.ma. The EESC had stressed in its report on "open data, the liberation of public data for growth and knowledge", the importance of the right to access information and the openness of public data which are likely to improve the quality of services provided to citizens. Law 31-13 on the right of access to information subsequently came into force in March 2020 and the portal "chafafiya.ma" was launched to process requests from citizens and foreign nationals living in Morocco for access to information.

4.3 Moroccan Economic Diplomacy That Is More Connected

Due to digitization, Moroccan economic diplomacy has become more agile and interconnected, making interactions and negotiations with international partners easier. Morocco has been able to enhance its abilities for real-time communication and diplomatic engagement at a lower cost and with fewer delays due to the increased utilization of digital tools. Digital platforms like video conferences and online forums are creating more chances for international meetings, even from afar, making it easier to negotiate trade and investment agreements faster.

Connected economic diplomacy, leveraging digital technologies and global networks, has several strategic benefits for governments and businesses. Some of these benefits include:

- Investor attraction: Investors are often looking for territories with modern and innovative public services, which can be facilitated by the implementation of digital tools for public services. Digitization of public services can help attract investors by providing them with a more efficient and manageable administrative environment.
- Improving the country's image: Digitizing the country's services can help to sell the country's image by positioning it as an innovative and connected environment. Local authorities that invest in digital tools for public services can improve their image by demonstrating their commitment to innovation and modernity. Increasing the attractiveness of the territory The digitalization of public services can help to put the territory on the map by making it more attractive for businesses, tourists and investors. Businesses are often looking for territories that offer modern and efficient public services, which can be facilitated by the implementation of digital tools for public services. Tourists can also be attracted by a territory with modern and efficient public services, which can help develop the local tourism industry.
- Optimization of trade negotiations: Digital tools, such as video conferencing platforms, simplify communications between international economic actors. This makes it easier to arrange meetings between investors, business partners and governments while reducing travel costs.
- More effective international promotion: Connected economic diplomacy can improve the visibility and attractiveness of a country's products, services and investment opportunities through targeted digital campaigns, the use of social networks and international platforms.
- Strengthening transparency and governance: Digital tools offer solutions to improve transparency in transactions and negotiations, especially through blockchain and electronic contract management. This strengthens confidence between the economic partners
- Reduction of geographical barriers: Digitalization allows us to transcend physical borders and promote global economic diplomacy. Countries can easily collaborate with distant markets, without being constrained by geographical boundaries or time zones.

- Innovation in decision-making processes: the integration of artificial intelligence (AI) and data analysis tools allows us to better anticipate global and regional economic developments. This can help diplomats identify opportunities and risks in advance, strengthening their capacity to develop effective strategies.
- Reduction of operational costs: Digital tools enable diplomatic missions and trade representatives to operate at a lower cost by optimizing the use of resources, reducing travel and automating certain administrative tasks.

4.3.1 Advantages Approaches of Integrated Economic Diplomacy

The rapid connectivity, minimal latency, and greater capacity of 5G networks facilitate real-time data gathering and analysis, essential for improving traffic flow, minimizing congestion, and boosting public transportation efficiency [27]. These developments enhance urban mobility while also promoting the sustainability of smart cities by decreasing carbon emissions and energy use. In Morocco, the incorporation of 5G technologies within urban innovation plans, bolstered by economic diplomacy, has the potential to greatly improve smart city development by drawing in international technology partners and promoting innovative transportation solutions that resonate with the nation's sustainability objectives.

This digital connectivity has allowed Morocco to engage in global discussions on artificial intelligence, cybersecurity, and digital governance, enhancing its capability to impact international standards and regulations in these fields. Simultaneously, digitalization has allowed for improved spread of economic data and more effective technological surveillance, enabling Morocco to promptly respond to shifts in the global market.

One of the most noteworthy features of the transition of Moroccan economic diplomacy into the digitalization era is its capacity to draw foreign investments in the digital industry. Morocco has positioned itself as a leading technological center in Africa by implementing policies that support innovation, foster startup culture, and encourage digital entrepreneurship. Economic diplomacy has been key in this process by creating strategic alliances with tech giants and helping international firms set up operations in the nation.

Projects like "Casablanca Finance City" and specialized free trade zones for tech industries are instances of this approach. Furthermore, bilateral and multilateral agreements that support technology transfer and enhance digital innovation capacity have allowed Morocco to enhance its digital ecosystem. Morocco's strategic focus on fintech, e-commerce, and value-added digital services has established it as a significant player, leading to economic diversification and the development of well-trained workforce.

Digitization has revolutionized Morocco's approach to trade and diplomacy, making it easier to reach global markets and update its administrative system. Morocco has made trade flows easier for local businesses to export globally by using digital management tools and e-commerce platforms.

The modernization of customs and tax systems now utilizes digital technologies to increase efficiency and enhance transparency, reflecting a transformative shift. These developments make it possible for Moroccan businesses to meet global requirements, increasing the country's appeal to potential trading allies. Diplomatic initiatives have enabled the digitalization of public services in Morocco, enhancing its ability to meet global trade demands and establish a reputation as a dependable and effective player in international value chains.

As digitalization increases, cybersecurity and digital governance have become important in Morocco's diplomatic relations. Moroccan economic diplomacy has successfully adjusted to these modern challenges by collaborating with global allies to advocate for strong cybersecurity measures and unified regulation of digital data.

Morocco's involvement in international cyber security forums is actively helping shape regulations for the ethical and secure use of digital technologies. By honoring these global agreements, Morocco can safeguard its digital systems and ensure the safety of online transactions and data privacy. This is vital for drawing in digital investments and instilling trust in the Moroccan market.

4.4 Economic Diplomacy, Innovation, and Entrepreneurship: Key Foundations of Smart City Development in Morocco

The progress of smart cities in Morocco is closely linked to economic diplomacy, innovation, and entrepreneurship principles. The backbone of Morocco's plan to modernize urban areas, promote economic growth, and position itself as a regional leader in digital transformation and sustainable development consists of these three elements.

Economic diplomacy has been vital in drawing in foreign investments and enabling global partnerships to meet the technological and infrastructural requirements of smart cities. Morocco has strategically situated itself as a doorway to Africa, using its geographic location and political stability to establish partnerships with global tech companies and multinational organizations.

Projects like the African Smart Towns Network (ASToN) and collaborations with nations like France, China, and the United Arab Emirates showcase Morocco's skill in leveraging diplomatic relationships for securing resources, knowledge, and cutting-edge technologies. These partnerships have played a crucial role in the advancement of intelligent city initiatives like the Zenata Eco-City and Benguerir, demonstrating Morocco's aspiration to establish internationally linked, eco-friendly urban centers.

Innovation is at the core of Morocco's vision for smart cities. The nation is making significant investments in advanced technologies like artificial intelligence, big data, and renewable energy solutions, in order to develop cities that are both efficient and sustainable. Significant initiatives, such as the African Supercomputing Center at Mohammed VI Polytechnic University (UM6P), demonstrate Morocco's dedication to promoting research and technological advancement., Morocco guarantees its smart cities can tackle issues like urban congestion, energy usage, and public service delivery by promoting an innovative culture. Economic diplomacy helps Moroccan institutions connect with international research networks and bring in global expertise to advance innovative urban solutions.

The execution of smart city projects frequently encounters considerable obstacles because of the substantial expenses associated with new technologies, the budget limitations of local authorities, and the immediate priorities of political agendas influenced by election periods [28]. These elements complicate the creation of a centralized, operationally effective, and financially viable model for technology implementation that can be duplicated in various regions. Within the framework of Morocco, economic diplomacy and urban innovation are crucial in tackling these issues. For example, the Noor Ouarzazate Solar Complex, which is among the largest solar initiatives globally, demonstrates Morocco's capacity to utilize global collaborations and investments to create scalable and sustainable solutions. This initiative not only facilitates the nation's shift to renewable energy but also enhances its role as a frontrunner in sustainable urban innovation, showcasing the possibilities of incorporating cutting-edge technologies within the smart city development framework.

Entrepreneurship is crucial in Morocco's smart city efforts as it allows local talent to develop customized solutions for the country's specific requirements. Initiatives like Impulse and Start-Gate, backed by the OCP Group and global collaborators, offer support and guidance to startups in areas like green technology, digital infrastructure, and smart mobility, Morocco has established a framework for collaboration between international investors and local business owners in order to promote innovation through economic diplomacy. Incorporating entrepreneurial ventures into smart city initiatives creates jobs and guarantees the flexibility and expandability of urban solutions.

The combination of economic diplomacy, innovation, and entrepreneurship has made Morocco a frontrunner in smart city development in the MENA region. Morocco incorporates global best practices into its smart city projects through diplomatic efforts to attract investments and promote international partnerships. Innovation guarantees these cities are ready to tackle future obstacles, while entrepreneurship ensures solutions stay dynamic and inclusive.

While Morocco continues to grow its smart city network, this three-part approach will continue to play a key role in developing urban areas that are environmentally friendly, diverse, and able to compete internationally.

5 Conclusion

The emergence of smart cities in Morocco offers a hopeful path for economic expansion, creativity, and urban change. Morocco is establishing itself as a frontrunner in the smart city movement in the African and MENA regions through the incorporation of advanced technologies, eco-friendly methods, and entrepreneurial projects. The examples of Casablanca, Benguerir, and the Zenata Eco-City show how they are incorporating smart urban planning principles and how economic diplomacy-driven international partnerships are boosting their capabilities.

Economic diplomacy is crucial in advancing the country's vision through the encouragement of international partnerships, foreign investment attraction, and the promotion of Morocco's digital ecosystem. By implementing strategic programs that promote innovation and entrepreneurship, Morocco is not only moving forward with its smart city goals but also improving its global competitiveness in a fast-evolving digital world.

In the future, Morocco's dedication to merging economic diplomacy with technological and entrepreneurial advancement will be crucial for the effective growth of its smart cities. This method guarantees lasting advantages in economic stability, employment generation, and enhanced living standards for its populace.

Subsequent studies could investigate topics like the incorporation of artificial intelligence (AI) in city management, covering traffic regulation, resource maximization, and environmental surveillance. Moreover, exploring citizen involvement in smart city governance, especially through digital platforms to engage communities in decision-making, may offer important insights for promoting inclusive and effective urban development. Morocco's advancements in these fields could also act as a template for other countries aiming to use innovation and diplomacy to enhance urban and economic development.

References

1. Nassereddine M, Khang A (2024) Applications of Internet of Things (IoT) in smart cities. In: Advanced IoT Technologies and Applications in the Industry 4.0 Digital Economy. CRC Press, pp 95–116
2. Syed AS, Sierra-Sosa D, Kumar A, Elmaghraby A (2021) IoT in smart cities: a survey of technologies, practices and challenges. Smart Cities 4(2):429–475. https://doi.org/10.3390/smartcities4020024
3. Savastano M, Suciu M-C, Nica E (2023) Project management of road infrastructure in Iraq. In: Smart Cities International Conference (SCIC) Proceedings: Vol. 11 (2023): Sustainability and Innovation
4. Alla L, Hmioui A, Bentalha B (2024) L'intelligence artificielle dans l'acquisition de clients: Une étude bibliométrique. In: Solutions d'IA et d'ingénierie des données pour un marketing efficace, Lahcen Alla, Abdelhadi Hmioui, and Brahim Bentalha. IGI Global, pp 1–22
5. Haj S (2020) The smart city: influence on the advancement of tourism attractions in Morocco. In: Urban oasis: smart cities and sustainable urban development. IGI Global, pp 45–67
6. Badel L (2006) Pour une histoire de la diplomatie économique de la France. Vingtième Siècle Revue d'histoire 91(3):169–182
7. Kateb A (2011) La diplomatie économique des nouvelles puissances. Géoeconomie 1:87–96
8. Woolcock S, Bayne N (2007) The New Economic Diplomacy: Decision Making and Negotiation in International Economic Relations. Ashgate
9. Dafir A (2012) La diplomatie économique marocaine en Afrique subsaharienne: réalités et enjeux. Géo-économie 2012/4 (63)

10. Nilssen M (2019) To the Smart City and Beyond? Developing a Typology of Smart Urban Innovation. Technol Forecast Soc Chang 142:98–104
11. Watson V (2015) The smart city debate: implications for developing cities. J Plan Educ Res 35(2):175–186
12. Batty M, Axhausen KW, Giannotti F, Pozdnoukhov A, Bazzani A, Wachowicz M, Ouzounis G, Portugali Y (2012) Smart cities of the future. Eur Phys J Spec Topics 214(1):481–518
13. Bunnell T (2015) Smart city returns. Dialog Hum Geogr 5(1)
14. Datta A (2015) New Urban Utopias of Postcolonial India: 'Entrepreneurial Urbanization' in Dholera Smart City
15. Goldsmith S, Crawford S (2014) The responsive city: engaging communities through data-smart governance. Jossey-Bass, San Francisco, CA
16. Acedo A et al (2018) Place and city: toward urban intelligence. ISPRS Int J Geo Inf 7(9):346
17. Hollands RG (2008) Will the real smart city please stand up? Intelligent, progressive or entrepreneurial? City 12(3):303–320
18. Albino V, Berardi U, Dangelico RM (2015) Smart cities: definitions, dimensions, performance, and initiatives. J Urban Technol 22(1):3–21
19. Castelnovo W, Misuraca G, Savoldelli A (2016) Smart cities governance: the need for a holistic approach to assessing urban innovative projects. Soc Sci Comput Rev 34(6):724–739
20. Cowley R, Joss S, Dayot Y (2018) The smart city and its publics: insights from across six UK cities. Urban Res Pract 11(1):53–77
21. Ablyazov T (2021) Application of digital platforms in the urban infrastructure development. SHS Web Conf 106:01028. https://doi.org/10.1051/shsconf/202110601028
22. Picon A (2013) Smart cities: a spatialised intelligence. Wiley, Chichester
23. Townsend AM (2017) Smart Cities: Big Data, Civic Hackers, and the Quest for a New Utopia. W.W. Norton & Company, New York
24. Bentahar L, Abdellaoui M (2024) Economic diplomacy in the digital era: Morocco's roadmap for Africa. Int J Appl Manag Econ 2(11). https://doi.org/10.5281/zenodo.14441404
25. Del-Real R, García-González JA, Herrera CD (2023) Smart governance: the role of digital administration services in urban sustainability. Urban Stud 60(3):672–689
26. Junaidi E, Rustiadi E (2024) Innovative transitions: exploring demand for smart city development in Novi Sad as a European capital of culture. Information 15(2):730
27. Gohar, A., and others. 2021. "The impact of 5G networks on urban mobility: enhancing traffic flow and public transport efficiency." J Smart Cities Urban Innov 4(2): 45–61
28. Mustapha A, Hind G, Lalla LA (2024) Toward a smart street lighting in Morocco: case study of Ifrane city. Salud, Ciencia y Tecnología - Serie de Conferencias 3:1012. https://doi.org/10.56294/sctconf20241012

Explainable AI for Workforce Management: Enhancing Retention in Smart Cities

Loubna Bouhsaien, Houssam Mensouri, Faouzi Tayalati, Ikhlass Boukrouh, and Abdellah Azmani

Abstract Employee turnover is a pervasive challenge for organizations, particularly in the rapidly evolving landscape of smart cities. While Artificial Intelligence (AI) offers promising solutions for predicting and mitigating turnover, the opacity of AI decision-making processes often limits managerial trust and effective intervention. This study addresses this gap by integrating Explainable AI (XAI) methodologies to uncover the predictive factors influencing employee turnover. By leveraging XAI techniques such as LIME (Local Interpretable Model-agnostic Explanations) and SHAP (Shapley Additive Explanations), the key drivers of turnover are identified and interpreted, enabling organizations to tailor data-driven retention strategies more effectively. Our findings reveal that employee satisfaction is the most significant predictor of turnover, with lower satisfaction levels markedly increasing the likelihood of employees leaving. Additionally, workload dynamics, such as the number of projects and average monthly working hours, exhibit a U-shaped relationship with turnover, indicating that both underworking and overworking contribute to higher turnover rates. Interestingly, factors like salary, promotions, and work accidents have minimal impact on turnover predictions. The transparency provided by XAI not only enhances trust in AI-driven decisions but also facilitates the adoption of these technologies in workforce management. This study underscores the importance of balancing predictive accuracy with interpretability, offering actionable insights for human resource managers to develop targeted retention strategies in smart cities. By addressing the research gap in synthesizing interpretability and predictive performance, this work contributes to the growing body of knowledge on AI applications in Human Resource Management (HRM), ultimately supporting organizational efforts to maintain a stable and engaged workforce.

L. Bouhsaien (✉) · H. Mensouri · F. Tayalati · I. Boukrouh · A. Azmani
Intelligent Automation and BioMedGenomics Laboratory, Faculty of Science and Technologies, Abdelemalek Essaadi University, Tangier, Morocco
e-mail: loubna.bouhsaien@etu.uae.ac.ma

H. Kaddouri et al. (eds.), *Smart Cities for Sustainable and Urban Development*, Green Energy and Technology, https://doi.org/10.1007/978-3-031-99001-4_31

Keywords Artificial Intelligence (AI) · eXplainable Artificial Intelligence (XAI) · Employees turnover · Human Resources Management (HRM)

1 Introduction

The collaboration between humans and machines is growing exponentially due to the evolution of new technologies, transforming the way industries work [1]. Organizations are facing significant challenges in integrating Artificial Intelligence (AI) into their decision-making processes to optimize opportunities while managing risks effectively [2]. A high rate of employee's turnover is a critical concern that has recently occupied human resources managers as it is adversely affecting organizational effectiveness.

Numerous studies have been conducted aiming to support organizational human resource management by understanding employees [3] while analyzing the connection between various factors on the intention of employees, looking for early warning signs, to alarm organizations, when necessary, to make personalized retention efforts proactively [4–6].

However, these studies often lack clarity in how their predictive models, black boxes, formulate conclusions, keeping the managers questions unanswered. The explainability enables any individual, particularly managers, to understand the model's predictions and highlight the impact of different features on the prediction [7].

This research aims to underline the significance of identifying which variables contribute at the predictions by utilizing eXplainable Artificial Intelligence (XAI) approaches, specifically for employees' turnover.

Following this introduction, the paper presents some previous works according to various methodologies employed for employees' turnover, followed by addressing the XAI methods used, namely LIME and SHAP. Next, it elaborates the application and the discussion, offering detailed insights into the findings. Finally, it envelops the implications, conclusions, and limitations drawn from this research.

2 Literature Review

2.1 Previous Research

All over the world, the sectors, the compagnies, the turnover of employees is taken seriously due to it damage. Consequently, many studies have been done to specify the most common causes of that phenomenon using different method while trying to see the model that gives the best result, especially supervised machine learning ones. Some of the most recent studies are shown below.

The article [8] classifies employees into three generations: generation Z (born 1996–2004), generation Y (born 1980–1995) and generation X (born 1966–1979). It starts by analyzing if there is a difference between different generations using One Way Analysis Of Variance (ANOVA) and confirming that there is no difference. Then, studies the effect of generation on the level of turnover intention and on the level of job satisfaction. For that, linear regression is used, finding that those variables were not significant. At the end, different factors were tested, using multiple regression, in terms of generations such as the pay, the promotion, the supervision; finding that the factors who predicts Turnover were rewards for generation Z, nature of work for generation Y, and promotion, nature of work and supervision for generation X.

The article [9] starts by indicating that the healthcare sector is going to need around 15million healthcare professional by 2030. That's mean that it is necessary to retain the employees to minimize the need. This research paper determines the predictors, direct and indirect ones, based on quantitative and non-experimental methodologies. Participants sociodemographic characteristics were treated, and different variables were used to create a binary regression to predict Turnover intention and job satisfaction. The most important difference between this study and other studies is that for nurses, age, and years of experience are not associated at all with job satisfaction.

The article [4] aims to test and elaborate if there is a relation between the gender, stress level, and Turnover Intention. To do so, 271 questionnaires are used. The study is elaborated for 5-star hotels in Albania, Turkey, for a normal population. This research uses the t-test using the Statistical Package of Social Sciences (SPSS) and regression analysis. The results show that Stress Level influences more female employees in hotels than male employees in hotels. Either, using One Way ANOVA, the study confirms that there is no differentiation between the age and educational status of participants.

The article [6] studies churn based on three different categories: ardent employees, behavioural aspects of employees, and distraught employees. The paper uses many factors; satisfaction level, salary, time spent in company and many others; to see their influence on the attrition attention. The result is that in general, the more correlated factor with churn is the satisfaction level and the less correlated factor with it is the time spent at company. Furthermore, the study confirms that the type of employees influences the churn. The authors find that satisfaction level causes the attrition of approximately the half of the ardent employees and that it is the main factor for behavioural aspects of employees. In the other hand, the number of projects an employee worked on affect more distraught employees.

The paper [10] treats the correlation between some factors that influence the employees turnover. The author finds that there is a negative correlation between the satisfaction level and the turnover and between a negative one between the salary and work accidents. Also, this study shows that there is a positive correlation between the number of projects done by an employee and the average monthly hours spent in a company and a positive one between the last evaluation based on the number of projects done and the average monthly hours spent in a company.

The paper [11] discusses a structured way for predicting the churn rate of employees. First, the study detects the variables having a huge impact on employee intention to leave the organization. Then, predict if an employee is about leaving for the next years by using classification models like support vector machine, naïve bayes classifier, and random forest. The point is that the rates of employees quitting the Technology Industry is quite high. The dataset used contain the information of employees between 20 and 39 years old. The information is reparteed to 11 attributes like age, work recognition and the salary level. The output is about the intention of quitting the company. First, the article classifies the output into three categories: the employee will quit, the employee will not quit, and the employee can quit. Then, it classifies the same data into two categories: the employee will quit, and the employee won't quit. The accuracy found increases when the data is classified into three classes.

The article [12] is about the factors impacting the level of job satisfaction in the banks. The factors treated are the manager, the renumeration, the reorganization, the promotion, the work safety, the relationship in the work environment, the timing, and the salaries. Between all those factors, the timing and the salaries are the most important according to employees. For SBI, 20% of employees were high satisfied and 45% were satisfied. In other hand, 50% were high satisfied and 40% were satisfied according to ICICI. The difference between the two banks is that the pay in SBI is higher than the pay in ICICI by 10%, the promotion and the good environment are higher in ICICI than SBI by 5%.

The paper [13] focalizes on employees job satisfaction and job performance. It is based on a questionnaire and the use of a statistical method to determine the factors influencing job satisfaction and job performance. The finding assures that the motivation, the leadership, and the culture impact job performance by a percentage of 57.4% and the job satisfaction by a percentage of 73.5%. In addition, more the job satisfaction increases, more the job performance increases.

The article [14] investigates the relation between factors such as the work flexibility, job performance, professional growth, job satisfaction, and many other factors. The paper finds that there a link between satisfaction and the work time, the workplace, the performance, and the development. In other hand, the finding is that there is a very strong impact of working time and flexibility on satisfaction level.

The article [15] builds a model before employees begin looking for new businesses. The dataset source is IBM. It contains 34 features. Different features come into play such as: job satisfaction, last evaluation, number of sales, years with current manager, and the salary. The author compares two methods: K-nearest neighbor and neural networks. The neural networks give the best result. The most fabulous result of this paper is that the salary is not the most important factor and that many other factors influence the turnover rate.

The article [16] proposes some machine learning methods to predict employee's churn. Kaggle was the source of the data, containing different factors impacting the employee intention to quit the actual job. The main factors are about the number of years with the current manager, the number of years in the current role, and the total working years. The article compares different methods; logistic regression, decision tree, random forest, and K-nearest neighbor. It found that random forest gives the best result.

The article [17] determines the major factors that impact employee's turnover. Factors treated are salary, travel, job environment, satisfaction, work overtime and many other factors. To elaborate a model according to the Turnover problem, the logistic regression is used to predict if an employee has the intention to stay or not in a company. This paper confirms that travel and overtime increase the probability of turnover, that the probability of Turnover for sales personal is higher than the probability of sales stuff, and that the phenomenon appears more for males than females.

The article [18] aims to determine the factors that influence employee's intention of leaving voluntary a hospital. The article is divided into three global factors. Firstly, it looks for the need of employees, in order of promotion, satisfaction, and formation. The second part is about intern factors: the equipment, the sanitary structure, and work tools. The third part touches the external factors, between all, it focalizes on the amelioration of the access level according to the health care of the population.

The article [19] uses the cause-and-effect diagram, also known as the fishbone diagram. This diagram makes things more clearly, precising all the effects and the sub effect causing a phenomenon. It goes from data captured from FBSM, which is a private medical clinic, and more precisely the nurse's data. The most global causes according to this study are salary and welfare, personal causes, operations, and interview and selection. The effect is about having a high employee turnover rate.

Recently, many studies collectively emphasize the balance between predictive performance and transparency, demonstrating the efficacy of these models in providing actionable insights for human resource management. Table 1 summarizes various studies on the use of XAI models and machine learning techniques to address employee attrition and retention.

Various studies emphasize factors such as employee satisfaction, job environment, and workload as key predictors of turnover. Machine learning techniques, particularly supervised methods, dominate the predictive landscape. However, a notable gap exists in the synthesis of interpretability and predictive performance. While LIME and SHAP have been explored, few studies provide a comprehensive framework combining actionable insights with trust-building transparency. This paper addresses this gap by focusing on XAI techniques to balance predictive accuracy and explainability.

2.2 Research Gap

Despite the growing use of machine-learning models for predicting employee turnover, the following research gaps persist:

Table 1 Related works—XAI and turnover

Authors	Main point
Sekaran and S [20]	Demonstrates the use of LIME and SHAP XAI models to interpret factors influencing employee attrition
Das et al. [21]	Uses machine learning and XAI (SHAP and LIME) to detect key features influencing employee retention and attrition, achieving 96% accuracy
Mohiuddin et al. [22]	Proposes Human Resource Decision Support System approach using SHAP to interpret predictions of employee attrition and suggests business decisions to retain employees
Chowdhury et al. [23]	Demonstrates LIME's capability to explain AI-based ML predictions for employee turnover, enhancing transparency and aiding HR managers in decision-making
Marín Díaz, Galán Hernández, and Galdón Salvador [24]	To understand nonlinearities and heterogeneity in complex ML models, explanatory methods such as PDP, LIME, ELI5, and SHAP to balance this trade-off, and showcasing the complementary use of local and global post-hoc methods for employee's attrition
Heidemann, Hülter, and Tekieli [25]	Machine learning models like random forests significantly outperform more transparent models such as classification trees in predicting employee turnover, there remains a trade-off between predictive performance and transparency, with the former often capturing the complex causes of turnover more effectively. The contribution is empirically demonstrating the trade-off between predictive performance and transparency, advocating for post-hoc explanatory methods to balance this trade-off, and showcasing the complementary use of local and global post-hoc methods to understand nonlinearities and heterogeneity in complex ML models

2.2.1 Integration of XAI Techniques for Actionable Insight

While some studies have explored XAI methods (e.g., SHAP and LIME), few provide a comprehensive framework that integrates these techniques with actionable insights for HR managers. There is a need for a structured approach that balances predictive performance and interpretability.

2.2.2 Cross-Industry Generalization

Most studies are limited to specific sectors such as healthcare, hospitality, or technology. Research that generalizes findings across multiple industries while incorporating contextual variables is limited.

2.2.3 Comprehensive Factor Analysis

Many studies focus on isolated factors such as job satisfaction or salary. A holistic approach that considers both intrinsic (e.g., job satisfaction, promotion opportunities) and extrinsic (e.g., salary, workload) factors is missing.

2.2.4 Dynamic and Real-Time Models

Most models are static and do not account for real-time changes in employee behavior or organizational dynamics. There is a gap in dynamic modeling that can adapt to evolving factors.

2.2.5 Comparative Analysis of Traditional Vs. XAI Models

Studies often focus only on machine learning accuracy or XAI interpretability without a comparative analysis of their combined efficacy. A framework that evaluates the trade-offs between traditional, high-performance models and transparent XAI models is necessary.

3 Methodology Approach: Steps of the Model Implementation

Knowledge Discovery in Databases, known as KDD, involves several key steps in the process of discovering useful knowledge from data to extract the hidden knowledge according from databases [26]. This process involves collecting and preparing the data, implementing and evaluating models, determining hyperparameters, and explaining the results obtained. Figure 1 elaborates the main steps of the methodology applied in this paper.

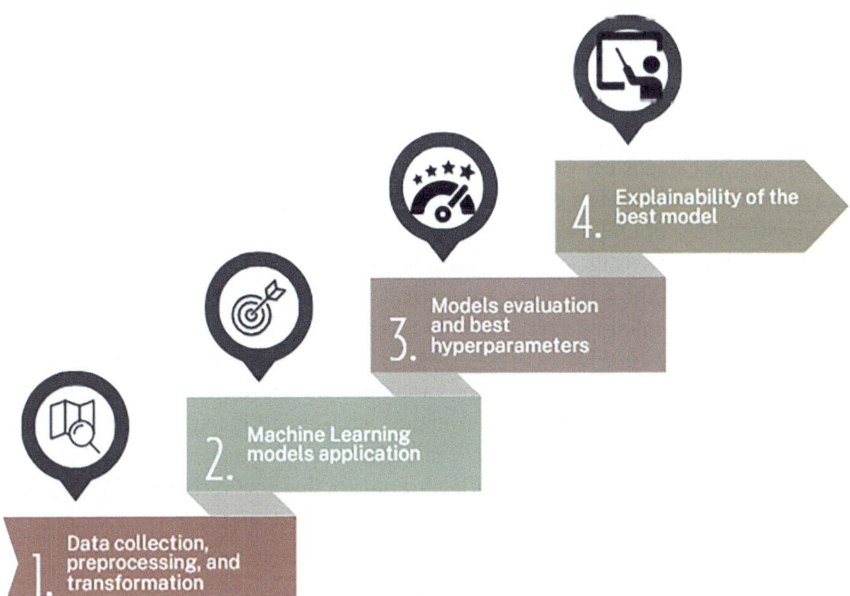

Fig. 1 Description of the methodology approach

Table 2 Domain and objectives

Domain	The domain of this project is human resources, specifically focusing on the analysis and prediction of employee turnover within an organization
Objectives	Understanding the factors that contribute to employees leaving the company can help in developing strategies to improve retention • Identify key factors influencing employee turnover • Develop predictive models that can accurately forecast which employees are at risk of leaving • Use XAI techniques to interpret the predictions and provide actionable insights

3.1 *Data Object*

The data object encompasses the domain, objectives, and target dataset [24, 27].

3.1.1 Domain and Objectives

The domain refers to the specific area of research or field that the investigation focuses on, providing a contextual framework for understanding the scope and relevance of the research. Within this domain, the study aims to address several key objectives. These objectives serve as a guiding framework, outlining the specific goals the research intends to achieve. By clearly defining the research domain and objectives, Table 2 provides a structured foundation for the study's methodology, analysis, and subsequent discussions.

3.1.2 Target Dataset

The target dataset consists of employee information, including many features. It was collected from kaggle.[1] The features that make up the dataset are listed below in Table 3. The categorical variables were encoded and converted to use them in the analysis. The provided data do not contain any missing values. The data is imbalanced with 23.8% of employees who has left the company.

The variable "left" is the target variable and all the other variables are feature that will help to predict the target. The categorical variables and their relationship with the target are elaborated by Figs. 2 and 3 respectively.

According to the correlation matrix (Fig. 4), the variable "satisfaction_level" has the highest correlation with the target variable "left", followed by "time_spend_company". The strongest positive correlation is observed between "number_project" and "average_monthly_hours", while the strongest negative correlation is between "satisfaction_level" and "left". It appears that the variables "Work_accident" and "promotion_last_5years" have minimal impact on the target variable.

[1] https://www.kaggle.com/datasets/kadirduran/hr-dataset

Table 3 Variables description

Variable	Description	Type	Distribution
Satisfaction Level	It is employee satisfaction point, which ranges from 0–1	Continuous	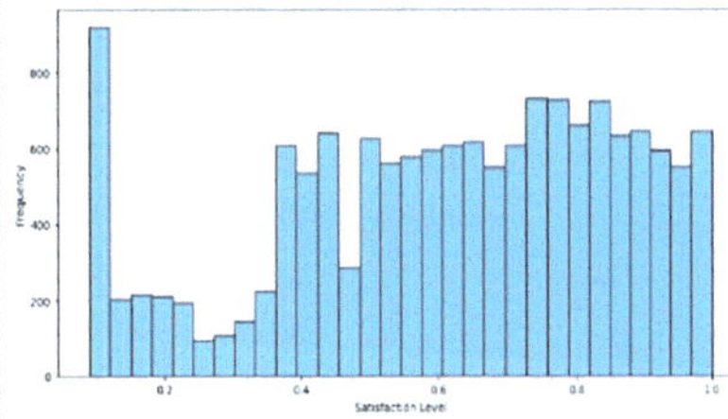
Last Evaluation	It is evaluated performance by the employer, which also ranges from 0–1	Continuous	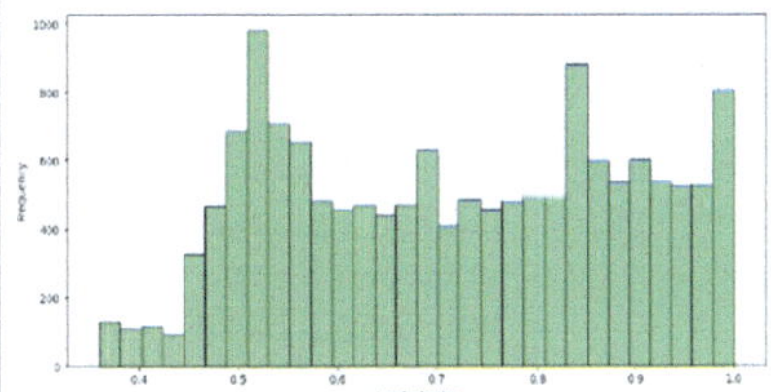
Number of Project	How many of projects assigned to an employee?	Numerical	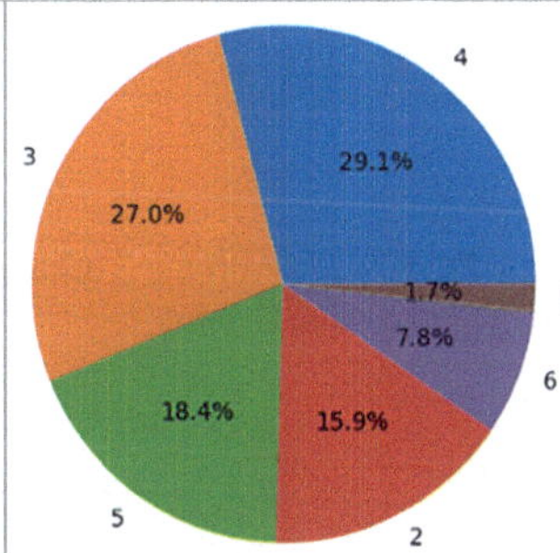
Average Monthly Hours	How many hours in average an employee worked in a month?	Continuous	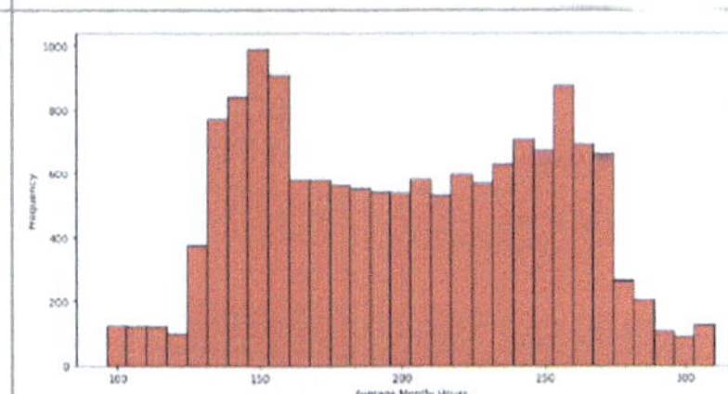
Time Spend Company	It means employee experience. The number of years spent by an employee in the company	Numerical	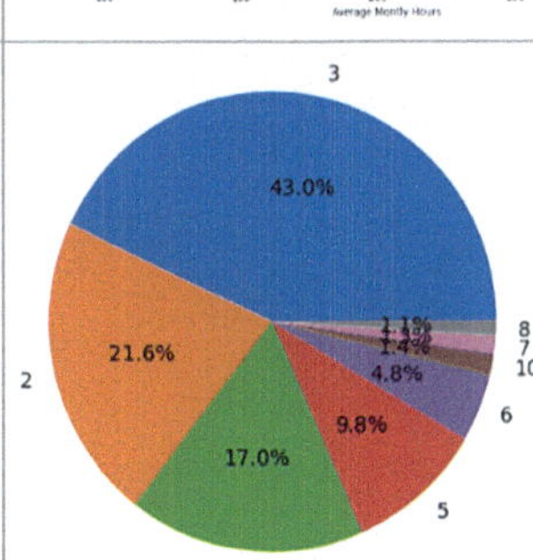

(continued)

Table 3 (continued)

Variable	Description	Type	Distribution
Work Accident	Whether an employee has had a work accident or not. 14.5% of employees had an accident	Boolean	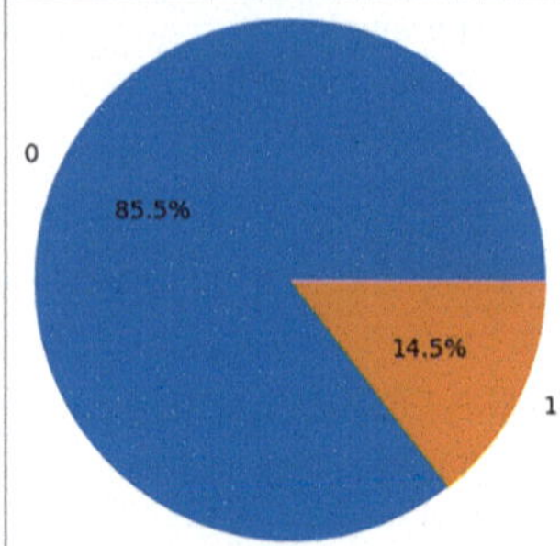
Promotion Last 5 years	Whether an employee has had a promotion in the last 5 years or not. Only 2.1% of employees had a promotion	Boolean	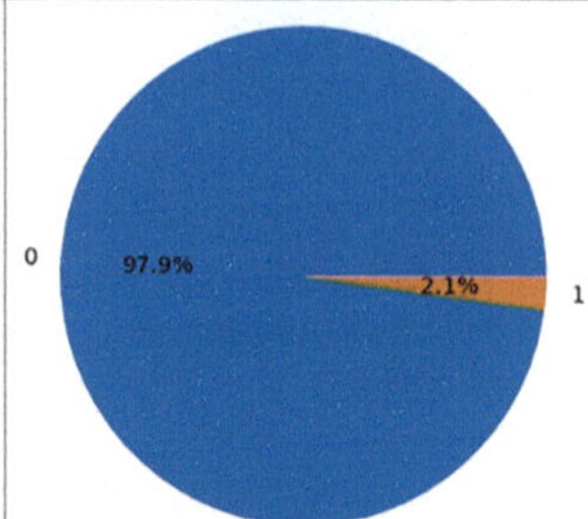
Department	Employee's working department/division	Categorial	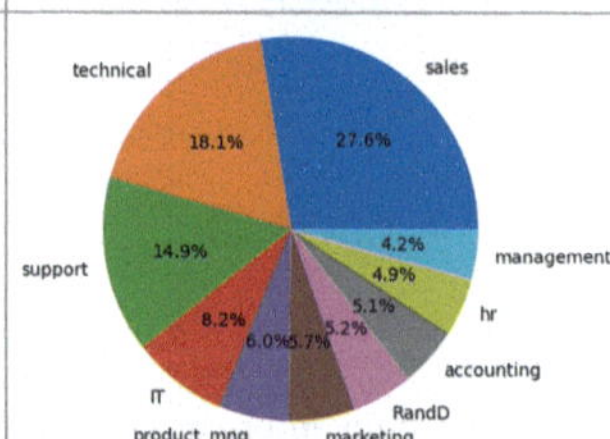
Salary	Salary level of the employee	Categorial	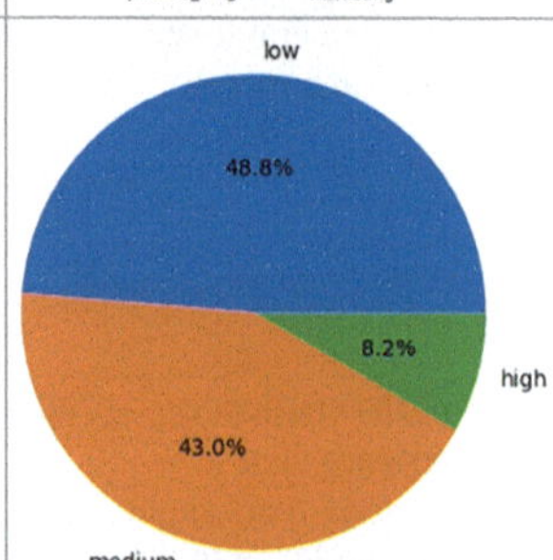
Left	Whether the employee has left the company (23.8%) or not	Boolean	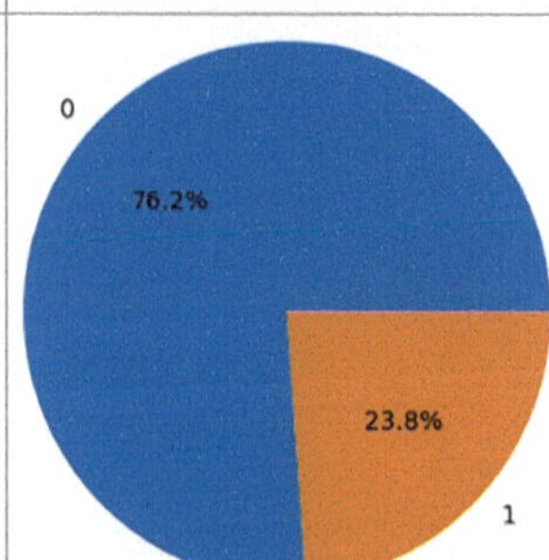

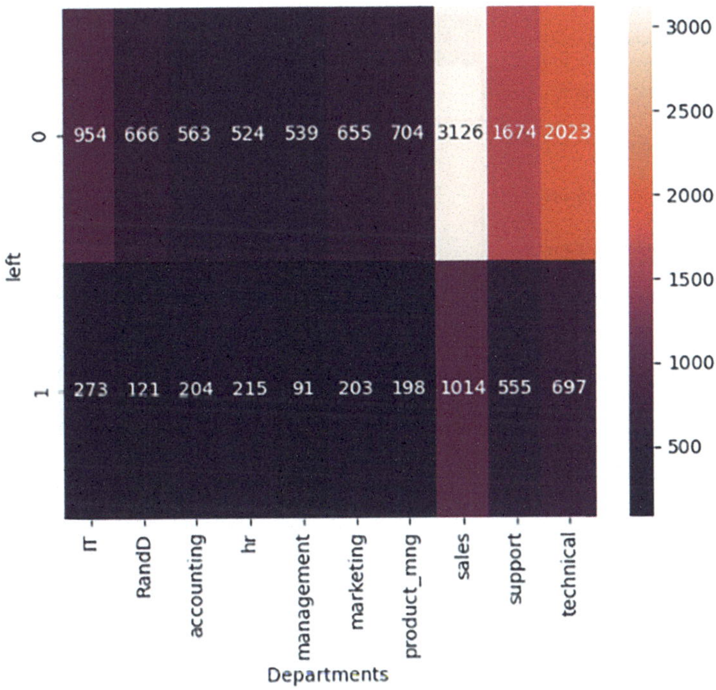

Fig. 2 Departments and target

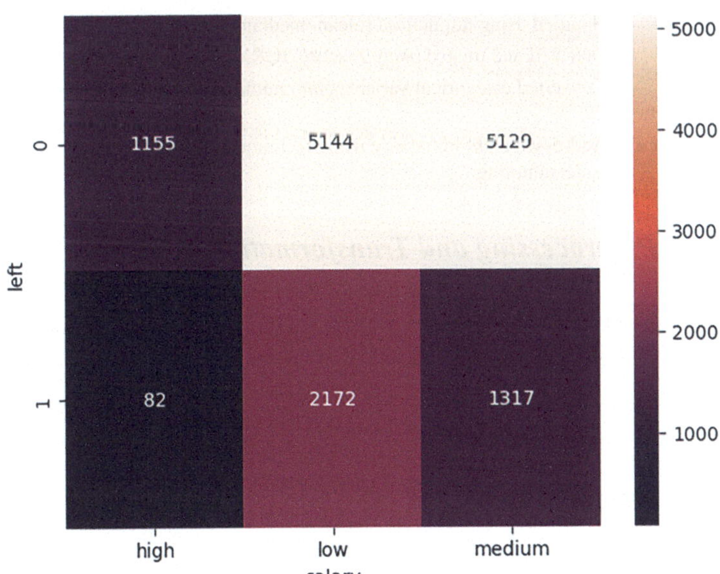

Fig. 3 Salary and target

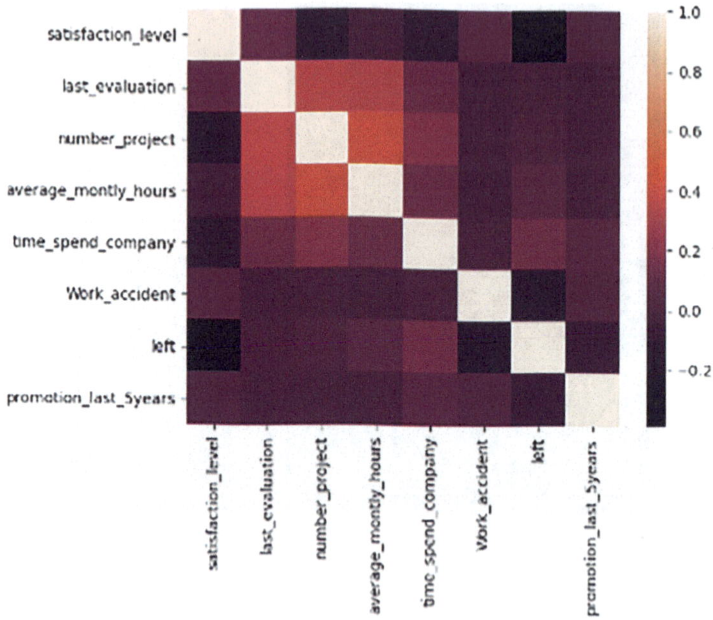

Fig. 4 Correlation matrix

Table 4 Data preprocessing and transformation

Step	Description
Missing Values	Handled using imputation (mean, median, or mode) or removal of records
Outliers	Detected and treated using Z-score, IQR, or visual inspection with box plots
Encoding Categorical Data	Converted categorical variables into numerical form using one-hot encoding
Data Balancing	Addressed class imbalance using techniques like SMOTE or random undersampling

3.2 Data Preprocessing and Transformation

Data preprocessing and transformation involves addressing missing values, outliers, and encoding categorical data [28]. Table 4 presents the steps and the appropriate descriptions.

3.3 Data Mining

Data mining involves applying algorithms to extract patterns and insights from the data, followed by their evaluation [24].

3.3.1 Classification Models

The models used in this study belong to the machine learning classifiers; Radom Forest (RF) [29], Decision Tree (DT) [30], Support Vector Machine (SVM) [31], K Nearest Neighbour (KNN) [32], Logistic Regression (LR) [33], and Gaussian Naïve Bayes (GNB) [34].

Predicting employee turnover is crucial for organizations to maintain a stable and productive workforce. Before applying the XAI algorithms, an evaluation of seven different models, RF, DT, SVM, KNN, LR, and GNB was made.

3.3.2 Models' Evaluation

The evaluation of the performance for classification models could be done based on various metrics [35]. In this paper, five metrics are used: accuracy which is considered as the most commonly used metric to evaluate classification performance [36], confusion matrix, precision, recall, and F1-Score [37]. AUC/ROC Curves (Area Under the Receiver Operating Characteristic Curve) are visual tools for assessing the predictive performance of classification models. These curves plot the true positive rate against the false positive rate. The area under the curve (AUC) offers a quantitative measure of the model's quality [24].

According to the performance metrics displayed in Fig. 5 the best model in this case is the RF with a precision of 0.99. Figures 6 and 7 present other metrics to validate that the RF model is good according to most used metrics.

To enhance the model's performance and avoid overfitting, hyperparameter tuning was conducted for the Random Forest Classifier. The initial accuracy of the model was 0.9887. After tuning, the best parameters identified were n estimators set to 200, min_samples_split to 2, min_samples_leaf to 1, max_features to 'auto',

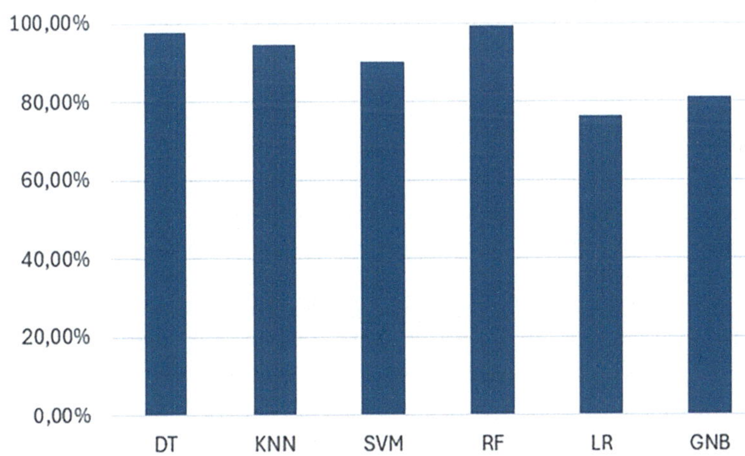

Fig. 5 Models precision

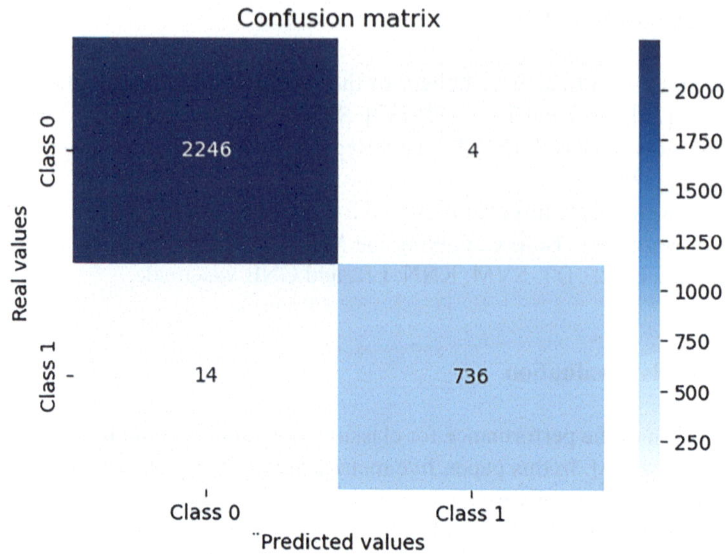

Fig. 6 RF confusion matrix

	precision	recall	f1-score	support
0	0.99	1.00	1.00	2250
1	0.99	0.98	0.99	750
accuracy			0.99	3000
macro avg	0.99	0.99	0.99	3000
weighted avg	0.99	0.99	0.99	3000

Fig. 7 Evaluation metrics

```
Model: RandomForestClassifier
Initial accuracy: 0.9887
Best parameters: {'n_estimators': 200, 'min_samples_split': 2, 'min_samples_leaf': 1, 'max_features': 'auto', 'max_depth': 40,
'criterion': 'gini'}
Final accuracy: 0.9883
```

Fig. 8 RF hyperparameters

max_depth to 40, and criterion to 'gini'. This optimization resulted in a slightly lower final accuracy of 0.9883, as shown in Fig. 8, indicating a more generalized model with reduced risk of overfitting.

The Receiver Operating Characteristic (ROC) curve for RF model demonstrates exceptional classification performance, with an Area Under the Curve (AUC) of 0.99 (Fig. 9). This high AUC value shows that the model has a strong ability to distinguish between the positive and negative classes. The curve is nearly touching the top left corner of the plot, signifying a high true positive rate and a low false positive rate across various threshold settings.

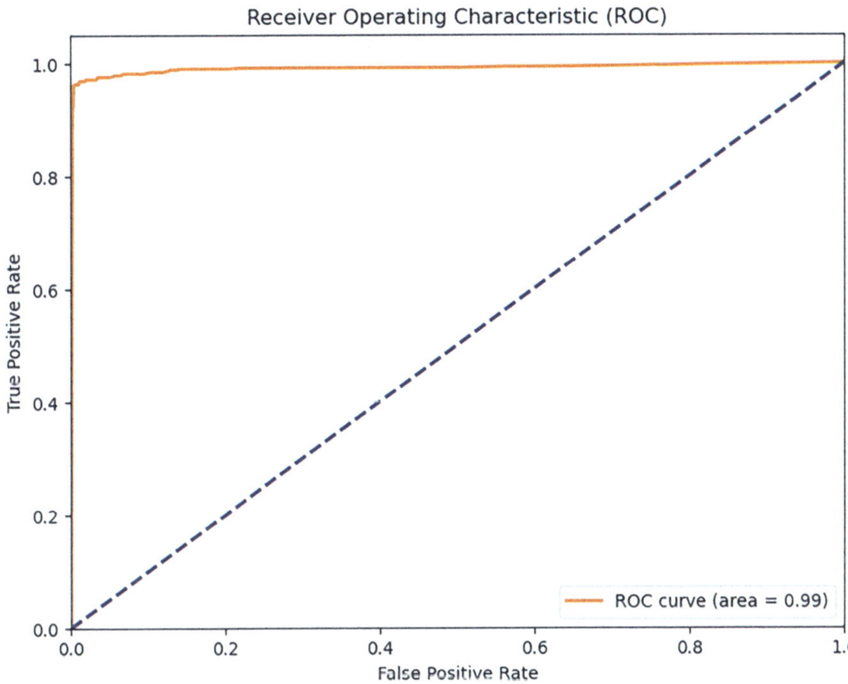

Fig. 9 RF ROC/AUC with balanced data

3.4 *Explainability and Interpretability*

Lately, there's a rising interest in integrating XAI algorithms into real-life applications [38]. Earlier, machines and algorithms didn't face trust issues because they executed precisely what they were programmed to do, embodying a form of absolute trust. A machine lacking autonomy posed no threat, making it easier to trust. Consequently, people often found machine outputs more reliable than human judgments [39]. Nowadays, the question is about why a model results should be trusted [40], so the call of quantifying XAI models' effectiveness is explored from a technical and non-technical point of view [41]. Explanation, meaningful, explanation accuracy and knowledge limits are the most important aspects to understand the models prediction [41].

3.4.1 Partial Dependency Plots (PDPs)

Visualization serves as a powerful interpretive tool for functions, effectively illustrating how their values relate to input variables. While straightforward for low-dimensional cases like single real-valued variables in graphs or categorical variables

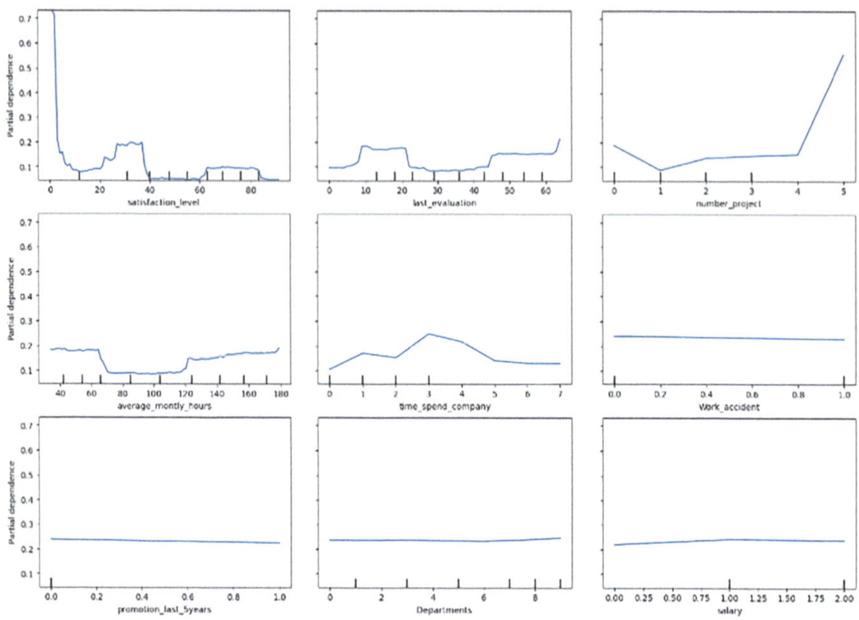

Fig. 10 PDP inputs vs output

in bar plots, and for two real-valued variables in contour or mesh plots, higher-dimensional functions pose challenges. Partial dependence plots address this by averaging function values over non-selected variables, offering insights into the influence of specific subsets of inputs. This method is crucial for understanding complex models such as neural networks or support vector machines, providing clarity on how predictor variables contribute to model outcomes. Particularly useful for regression trees, partial dependence plots simplify interpretation by directly visualizing the impact of chosen variable subsets. They serve as a key tool in interpreting models and understanding their behaviour across different scenarios, despite the complexities introduced by high-dimensional data [41, 42].

The partial dependence plots in Fig. 10 reveal that satisfaction level is the most influential factor in predicting employee turnover, with lower satisfaction markedly increasing the likelihood of leaving. Both extremes in last evaluation scores show a slight uptick in turnover probability, while the number of projects and average monthly hours exhibit U-shaped relationships, indicating that both underworking and overworking contribute to higher turnover rates. Interestingly, while salary shows a slight positive correlation with turnover, factors like work accidents, recent promotions, and department placement have minimal impact on employees' likelihood of leaving the company.

3.4.2 Local Interpretable Model-Agnostic Explanations (LIME)

The robustness of interpretability methods, particularly LIME [40], is a central concern in this study, focusing on how explanations vary with slight changes in input that do not significantly alter model predictions. This instability prompts a deeper investigation into robustness, aiming to quantify it objectively using tools like local Lipschitz continuity. Unlike global criteria, which are impractical for varied input explanations, local Lipschitz continuity measures stability in proximity to specific anchor points, providing a method to estimate the consistency of explanations locally. This approach, though computationally challenging, offers insights into the reliability of interpretability methods under different input conditions, crucial for enhancing their applicability in practical settings [42, 43].

LIME allows the understanding of predictions on a case-by-case basis, offering insights into why a particular decision was made by the model for a single instance. The local interpretability of individual predictions using LIME was applied for the most performant model, the random forest model. In Fig. 11, the prediction probabilities are 34% for Class 0 and 66% for Class 1.

The most influential features leading to the prediction probabilities of class 0 are "satisfaction_level" and "promotion_last_5years", respectively "time_spend_company" and "average_montly_hours" for class 1.

3.4.3 Feature Importance Using (ELI5)

Feature importance involves taking a trained predictive model and a dataset to evaluate the model's performance by computing a reference score, accuracy for a classifier. Then, for each feature in the dataset, it repeatedly shuffles its values to create multiple "corrupted" versions of the dataset. For each shuffled version, it re-evaluates the model's performance. By comparing these new scores to the original

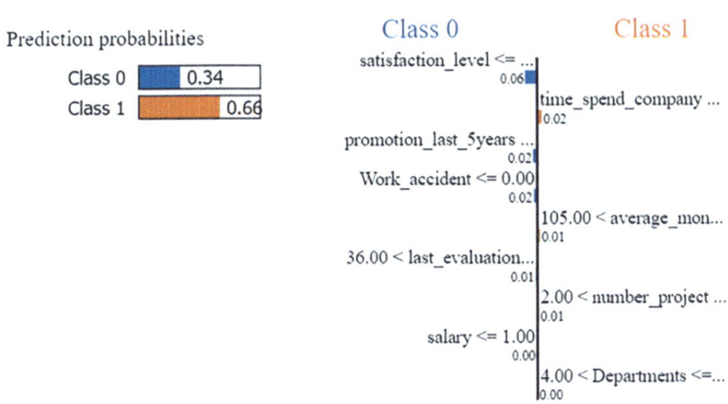

Fig. 11 LIME interpretability graph using RF

Table 5 Feature importance (ELI5)

Weight	Feature
0.2034 ± 0.0099	Satisfaction Level
0.1321 ± 0.0101	Number Project
0.1276 ± 0.0031	Average Monthly Hours
0.1214 ± 0.0082	Last Evaluation
0.0958 ± 0.0057	Time Spend Company
0.0064 ± 0.0017	Departments
0.0027 ± 0.0004	Salary
0.0006 ± 0.0008	Work Accident
0.0003 ± 0.0000	Promotion Last Five Years

reference score, the importance of each feature is determined based on how much the model's performance drops when that feature's values are randomized. The greater the performance drop, the more important the feature is to the model [44].

Feature importance is a way of measuring how important each feature is for a machine learning model. It's like a score that tells you how much each feature contributes to the model's ability to make accurate predictions. The results of feature importance calculations are typically presented as a list of features, each with a corresponding score. The higher the score, the more important the feature is. Interpreting the feature importance results from Table 5, which presents scores for a random forest classifier predicting employee turnover, reveals that weight emerges as the most critical predictor. A high weight indicates a strong likelihood of leaving the company. Following weight, satisfaction_level also plays a significant role, suggesting that job satisfaction lowers turnover risk. Number_project and average_monthly_hours are moderately important, indicating higher workload correlates with increased turnover. Last_evaluation holds less importance, implying good performance reduces turnover likelihood. Time_spend_company, departments, salary, work_accident, and promotion_last_5years show minimal to no impact on turnover, suggesting these factors have little influence on employees' likelihood to leave the company.

3.4.4 Shapley Additive Explanations (SHAP)

Visualizing feature impact in tree ensemble models usually involves bar charts for global importance or partial dependence plots for single feature effects. However, SHAP values provide individualized feature attributions, allowing for richer visualizations. SHAP summary plots, replacing traditional bar charts, illustrate the range and distribution of feature impacts and their relationship to feature values, while SHAP dependence plots offer a detailed alternative to partial dependence plots by capturing interaction effects [45].

SHAP: Global Interpretability

In Fig. 12, the SHAP summary plot elucidates the directional impact of key features on the Random Forest model's turnover predictions. Higher employee satisfaction levels are predominantly associated with negative SHAP values, suggesting a decreasing probability of turnover as satisfaction rises. Conversely, an increasing number of projects initially contributes negatively but then shifts to a positive impact on turnover likelihood, indicating a potential threshold effect. Longer tenure within the company tends to show a positive SHAP value, meaning more time spent at the company increases turnover risk. For last evaluations, higher scores have a mix but lean towards positive SHAP values, hinting that employees with better evaluations may be more inclined to leave. Also, higher average monthly hours worked also tend towards positive SHAP values, implying that greater work hours may lead to a higher chance of turnover. On the other hand, Fig. 12 summarizes the average impact of each feature on the model's output magnitude. Consistent with the summary plot, satisfaction_level stands out as the feature with the highest average SHAP value, followed closely by number_project, confirming their strong influence on the model's decisions.

Local Features Interpretability

Pink arrows signify the extent of impact of input variables that elevate turnover occurrences, while blue arrows denote the influence of input variables that mitigate turnover. The magnitude of the area occupied by the variables within each arrow denotes the degree of influence exerted by the variable [37]. Figure 13 presents the SHAP plot for random instance that were selected randomly.

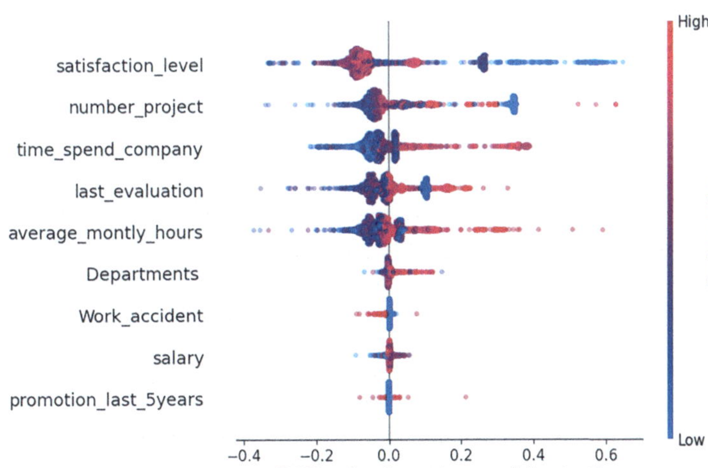

Fig. 12 Global interpretability of the RF model

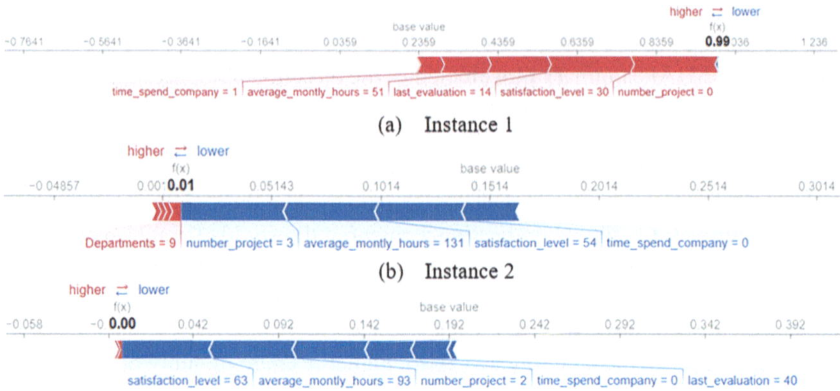

(a) Instance 1

(b) Instance 2

(c) Instance 3

Fig. 13 Explanation of evaluation results based on SHAP of three different instances

Figure 13a illustrates the highest risk of turnover among the three instances, indicating a strong likelihood of the event of interest occurring, such as an employee leaving. The features contributing to this prediction are "number_project", "satisfaction_level", and "last_evaluation".

In Fig. 13b, the risk of turnover is lower than Fig. 13a. Both "number_project" and "satisfaction_level" are higher than the first case.

Figure 13c illustrates the lowest risk of turnover among the three chosen instances, the feature "satisfaction_level" had the highest value when the "number_project" had a higher value than instance than instance 1 and a smaller value than instance 2.

The feature "satisfaction_level" seems to have pushed the prediction significantly for the three instances that were selected randomly.

3.4.5 Individual Conditional Expectation (ICE)

Goldstein et al. [46] introduced Individual Conditional Expectation (ICE) plots as an enhancement of Partial Dependence Plots (PDPs). While PDPs depict the average effect of a feature on the predicted outcome, ICE plots disaggregate this effect to show the relationship between the feature and the predicted response for each individual observation. Using the same definitions as for PDPs, ICE plots generate estimated partial dependence curves for each instance, offering greater interpretability. Each line in an ICE plot represents the prediction for one data point as the feature of interest varies, enabling the detection of heterogeneous relationships that may be obscured by the averaging in PDPs [47].

The ICE plot (Fig. 14) illustrates how the probability of leaving the company varies with the level of satisfaction for different observations. Generally, a low satisfaction level is associated with a high probability of leaving, while high

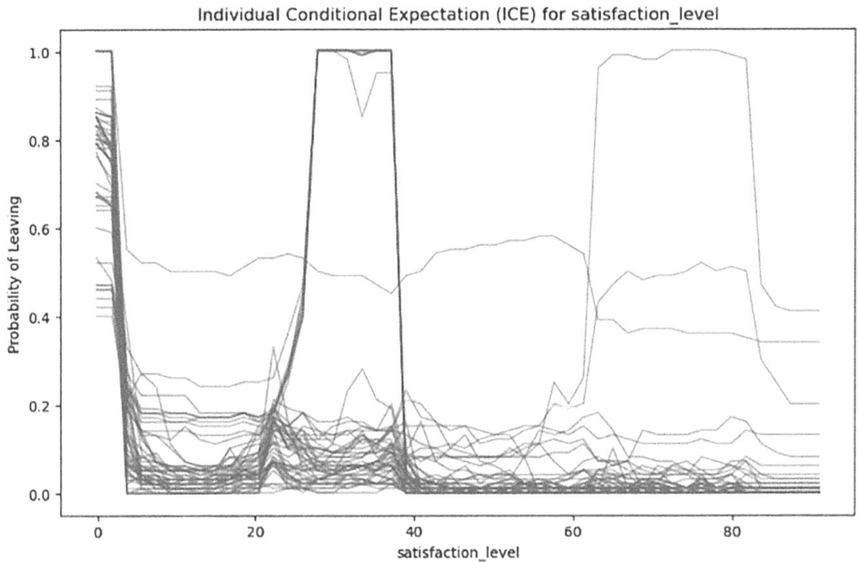

Fig. 14 ICE of satisfaction level

satisfaction corresponds to a low probability of departure. The relationship is non-linear, with sudden transitions and peaks between satisfaction levels of 20 and 40, and more stable variations beyond 60. Some lines show atypical behaviours, suggesting that other factors also influence the decision to leave. In summary, employee satisfaction is a crucial indicator of their likelihood of leaving, but it should be considered alongside other variables for a comprehensive understanding.

3.4.6 Anchored Local Explanation (ALE)

Anchors are rules designed to provide precise local explanations for complex models (black box) by focusing on individual predictions. Unlike other perturbation methods, anchors use an interpretable representation of the data, making it easier for humans to understand. A rule is considered an anchor if, when it applies to a given instance, it ensures with high probability that similar perturbations of this instance will be classified in the same way by the model [48].

The ALE curve (Fig. 15) elaborates the "satisfaction_level" feature plot using the Random Forest classifier. The data includes various employee attributes and a target variable indicating whether an employee left the company. The ALE values are computed by iteratively setting the "satisfaction_level" to a range of values while keeping other features constant and predicting the probability of the target outcome. These ALE values, which indicate the impact of different satisfaction levels on the model's predictions, are then plotted, showing that very low satisfaction has a strong

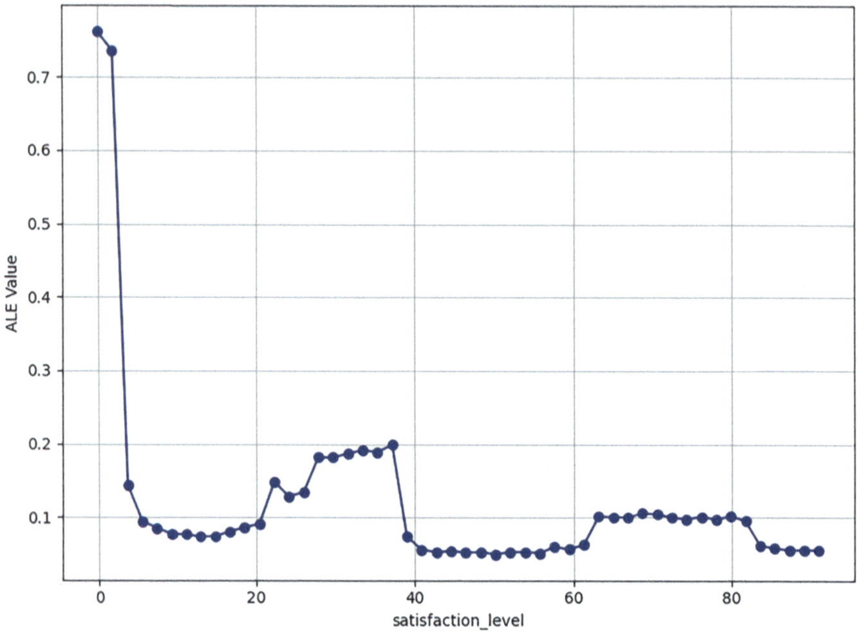

Fig. 15 ALE of satisfaction level

positive effect on the likelihood of an employee leaving, while higher satisfaction levels have a more moderate and stable impact.

4 Results and Discussion

This section presents the results of the model implementation and discusses the implications of the findings. The results are supported by figures and tables to enhance clarity and provide actionable insights.

4.1 Key Findings

The study identifies employee satisfaction as the most significant predictor of turnover, with lower satisfaction levels markedly increasing the likelihood of employees leaving. Additionally, workload dynamics, such as the number of projects and average monthly working hours, exhibit a U-shaped relationship with turnover, indicating that both underworking and overworking contribute to higher turnover rates. Interestingly, factors like salary, promotions, and work accidents have minimal

Table 6 Key findings summary

Predictor	Relationship with turnover
Satisfaction Level	Inversely correlated
Number of Projects	U-shaped relationship
Average Monthly Hours	U-shaped relationship
Salary	Minimal impact
Predictor	Relationship with Turnover

impact on turnover predictions. Table 6 shows the key findings of the study, summarizing the relationships between various predictors and turnover.

4.2 Discussion

The identification of employee satisfaction as the primary driver of turnover highlights the critical need for organizations to prioritize fostering a positive work environment. The U-shaped relationship observed between workload dynamics, specifically the number of projects and average monthly working hours, and turnover further emphasizes the importance of balancing employee workloads to mitigate the risks of both underworking and overworking. Interestingly, factors traditionally assumed to have a strong impact, such as salary, promotions, and work accidents, were found to exert minimal influence on turnover predictions, challenging conventional assumptions and shifting the focus towards intrinsic job satisfaction and workload management.

These findings align with existing literature, which has consistently highlighted employee satisfaction as a key predictor of turnover, but this study goes a step further by leveraging XAI techniques like LIME and SHAP to provide transparent and actionable insights. The transparency offered by these techniques not only enhances trust in AI-driven decisions but also enables HR managers to tailor retention strategies based on interpretable data insights. By demonstrating the efficacy of XAI in bridging the gap between predictive accuracy and interpretability, this research makes a significant theoretical contribution to the field of HRM.

4.3 Discussion of Biases and Limitations

The study acknowledges limitations such as the dataset imbalance, with only 23.8% of employees having left the company, which may skew predictions toward the majority class, and the dynamic nature of satisfaction levels, which can rapidly change, introducing unpredictability into turnover predictions.

Future research should address these limitations by validating the findings across diverse and dynamic datasets, exploring advanced XAI frameworks to capture complex workforce dynamics, and implementing robust bias mitigation strategies to ensure more accurate and generalizable insights. From a practical perspective, HR managers can benefit immensely from the actionable insights provided by this research, such as developing personalized intervention strategies to enhance employee satisfaction and optimize workload distribution, thereby mitigating turnover risks and contributing to sustainable workforce management in smart cities.

5 Conclusion

This study highlights the importance of integrating Explainable AI (XAI) methodologies into workforce management to enhance employee retention strategies. By identifying key predictors of turnover, such as employee satisfaction, workload dynamics, and time spent in the company, this research provides actionable insights for human resource managers to develop targeted retention strategies. The transparency offered by XAI techniques, such as LIME and SHAP, not only enhances trust in AI-driven decisions but also facilitates the adoption of these technologies in workforce management.

The findings underscore the critical role of employee satisfaction as the primary driver of turnover, with lower satisfaction levels significantly increasing the likelihood of employees leaving. Additionally, the U-shaped relationship between workload and turnover suggests that both underworking and overworking contribute to higher turnover rates, highlighting the need for balanced workload management. Interestingly, factors like salary, promotions, and work accidents were found to have minimal impact on turnover predictions, challenging conventional assumptions about their influence.

Future research should focus on validating these findings across diverse datasets to ensure generalizability, developing advanced XAI frameworks to enhance interpretability in capturing complex workforce dynamics, and implementing robust bias mitigation strategies to address inherent biases in workforce datasets. By leveraging these approaches, organizations can align retention efforts with employee needs, contributing to sustainable workforce management and mitigating the adverse effects of high turnover rates.

Acknowledgments This research is supported by the Ministry of Higher Education, Scientific Research and Innovation, the Digital Development Agency (DDA) and the National Center for Scientific and Technical Research (CNRST) of Morocco (Smart DLSP Project—AL KHAWARIZMI IA-PROGRAM).

Declaration of Conflicting Interests The author(s) declared no potential conflicts of interest with respect to the research, authorship, and/or publication of this article.

Funding The author(s) received no financial support for the research, authorship, and/or publication of this article.

References

1. Yorks L, Rotatori D, Sung SY, Justice S (2020) Workplace reflection in the age of AI: materiality, technology, and machines. Adv Dev Hum Resour 22(3):308–319. https://doi.org/10.1177/1523422320927299
2. Garibay O, Ozlem BW, Andolina S, Antona M, Bodenschatz A, Coursaris C, Falco G et al (2023) Six human-centered artificial intelligence grand challenges. Int J Hum Comput Interact 39(3):391–437. https://doi.org/10.1080/10447318.2022.2153320
3. Manganelli L, Thibault-Landry A, Forest J, Carpentier J (2018) Self-determination theory can help you generate performance and well-being in the workplace: a review of the literature. Adv Dev Hum Resour 20(2):227–240. https://doi.org/10.1177/1523422318757210
4. Akdemir, Alper M, Sağbaş M, Sürücü L (2022) Job stress and turnover intention in the hotel employees. J Tourism Gastron Stud. https://doi.org/10.21325/jotags.2022.973
5. Bontrager M, Suzanne Clinton M, Tyner L (2021) Flexible work arrangements: a human resource development tool to reduce turnover. Adv Dev Hum Resour 23(2):124–141. https://doi.org/10.1177/1523422320982930
6. Chaudhary M, Gaur L, Jhanjhi N, Masud M, Aljahdali S (2022) Envisaging employee churn using MCDM and machine learning. Intell Automat Soft Comput 33(2):1009–1024. https://doi.org/10.32604/iasc.2022.023417
7. Kiruthika M, Moorthi K, Anousouya Devi M, Abijah Roseline S (2024) Role of XAI in building a super smart society 5.0. In: XAI Based Intelligent Systems for Society 5.0. Elsevier, pp 295–326. https://doi.org/10.1016/B978-0-323-95315-3.00013-9
8. Klajkó D, Andrea C, Péter R, Szabó ZP, Kázmér-Maye S (2022) Predictors of turnover intention: generational perspective
9. Smokrović E, Kizivat T, Bajan A, Šolić K, Gvozdanović Z, Farčić N, Žvanut B (2022) A conceptual model of nurses' turnover intention. Int J Environ Res Public Health 19(13):8205. https://doi.org/10.3390/ijerph19138205
10. Singh J, Sekhon, Harsh S (2021) Prediction of employee satisfaction level using optimized neural network. https://doi.org/10.34218/IJARET.12.1.2021.011
11. Bandyopadhyay N, Jadhav A (2021) Churn prediction of employees using machine learning techniques. Tehnički Glasnik 15(1):51–59. https://doi.org/10.31803/tg-20210204181812
12. Bhardwaj A, Mishra S, Jain TK (2021) An analysis to understanding the job satisfaction of employees in banking industry. Mater Today: Proc 37:170–174. https://doi.org/10.1016/j.matpr.2020.04.783
13. Paais M, Pattiruhu JR (2020) Effect of motivation, leadership, and organizational culture on satisfaction and employee performance. J Asian Financ Econ Bus 7(8):577–588. https://doi.org/10.13106/JAFEB.2020.VOL7.NO8.577
14. Davidescu AAM, Apostu S-A, Paul A, Casuneanu I (2020) Work flexibility, job satisfaction, and job performance among Romanian employees—implications for sustainable human resource management. Sustainability 12(15):6086. https://doi.org/10.3390/su12156086
15. Moulana M, Priyanka BS, Mohammed Ismail B, Priyanka AM (2020) Early prediction of employee churn
16. Zhao Y, Hryniewicki MK, Cheng F, Boyang F, Zhu X (2019) Employee turnover prediction with machine learning: a reliable approach. In: Arai K, Kapoor S, Bhatia R (eds) Intelligent Systems and Applications, Advances in Intelligent Systems and Computing, vol 869. Springer, Cham, pp 737–758. https://doi.org/10.1007/978-3-030-01057-7_56
17. Zhang H, Lexi X, Cheng X, Chao K, Zhao X (2018) Analysis and prediction of employee turnover characteristics based on machine learning. In: 2018 18th International Symposium on Communications and Information Technologies (ISCIT). IEEE, Bangkok, pp 371–376. https://doi.org/10.1109/ISCIT.2018.8587962
18. Ndayirata S, Belghiti-Mahut S, Briole A (2018) L'influence Des Perceptions Des Pratiques de GRH et de Facteurs de Contexte Organisationnel Sur l'intention de Départ Volontaire: Cas

Des Médecins Du Secteur Public Au Burundi. Journal de Gestion et d'économie Médicales 36(1):61–83. https://doi.org/10.3917/jgem.181.0061
19. Ping DS, Selina, Mohanachandran DK (2015). Understanding and addressing staff turnover in FPSB
20. Sekaran K, Shanmugam S (2022) Interpreting the factors of employee attrition using explainable AI. In: 2022 International Conference on Decision Aid Sciences and Applications (DASA). IEEE, Chiangrai, Thailand, pp 932–936. https://doi.org/10.1109/DASA54658.2022.9765067
21. Das S, Chakraborty S, Sajjan G, Majumder S, Dey N, Tavares JMRS (2023) Explainable AI for predictive analytics on employee attrition. In: Patel KK, Santosh KC, Patel A, Ghosh A (eds) Soft Computing and Its Engineering Applications, Communications in Computer and Information Science, vol 1788. Springer, Cham, pp 147–157. https://doi.org/10.1007/978-3-031-27609-5_12
22. Mohiuddin K, Alam MA, Alam MM, Welke P, Martin M, Lehmann J, Vahdati S (2023) Retention is all you need. In: Proceedings of the 32nd ACM International Conference on Information and Knowledge Management. ACM, Birmingham, UK, pp 4752–4758. https://doi.org/10.1145/3583780.3615497
23. Chowdhury S, Joel-Edgar S, Dey PK, Bhattacharya S, Kharlamov A (2023) Embedding transparency in artificial intelligence machine learning models: managerial implications on predicting and explaining employee turnover. Int J Hum Resour Manag 34(14):2732–2764. https://doi.org/10.1080/09585192.2022.2066981
24. Díaz M, Gabriel JJ, Hernández G, Salvador JLG (2023) Analyzing employee attrition using explainable AI for strategic HR decision-making. Mathematics 11(22):4677. https://doi.org/10.3390/math11224677
25. Heidemann A, Hülter SM, Tekieli M (2024) Machine learning with real-world HR data: mitigating the trade-off between predictive performance and transparency. Int J Hum Resour Manag:1–24. https://doi.org/10.1080/09585192.2024.2335515
26. Shafique U, Qaiser H (2014) A comparative study of data mining process models (KDD, CRISP-DM and SEMMA). 12(1)
27. Kaufman S, Rosset S, Perlich C, Stitelman O (2012) Leakage in data mining: formulation, detection, and avoidance. ACM Trans Knowl Discov Data 6(4):1–21. https://doi.org/10.1145/2382577.2382579
28. Albahra S, Gorbett T, Robertson S, D'Aleo G, Kumar SVS, Ockunzzi S, Lallo D, Hu B, Rashidi HH (2023) Artificial intelligence and machine learning overview in pathology & laboratory medicine: a general review of data preprocessing and basic supervised concepts. Semin Diagn Pathol 40(2):71–87. https://doi.org/10.1053/j.semdp.2023.02.002
29. Ibraheem MK, Ibraheem AA, Al-Abadi J, Mohamed MB, Fakhfakh A (2024) A security-enhanced energy conservation with enhanced random forest classifier for low execution time framework (S-2EC-ERF) for wireless sensor networks. Appl Sci 14(6):2244. https://doi.org/10.3390/app14062244
30. Charbuty B, Abdulazeez A (2021) Classification based on decision tree algorithm for machine learning. J Appl Sci Technol Trends 2(01):20–28. https://doi.org/10.38094/jastt20165
31. Roy A, Chakraborty S (2023) Support vector machine in structural reliability analysis: a review. Reliab Eng Syst Saf 233(May):109126. https://doi.org/10.1016/j.ress.2023.109126
32. Razavi-Termeh SV, Sadeghi-Niaraki A, Razavi S, Choi S-M (2024) Enhancing flood-prone area mapping: fine-tuning the K-Nearest Neighbors (KNN) algorithm for spatial modelling. Int J Digit Earth 17(1):2311325. https://doi.org/10.1080/17538947.2024.2311325
33. Syahab AS, Pamungkas GL, Akmal S (2024) The classification of ultraviolet index using logistic regression and random forest methods for predicting extreme conditions. In: Suryotomo AP, Rustamaji HC (eds) Proceedings of the 2023 1st International Conference on Advanced Informatics and Intelligent Information Systems (ICAI3S 2023), Advances in Intelligent Systems Research, vol 181. Atlantis Press International BV, Dordrecht, pp 4–14. https://doi.org/10.2991/978-94-6463-366-5_2

34. Hartana RD, Sela EI (2024) Nutritional status classification of stunting in toddlers using naive bayes classifier method. J Technol Inf Eng 3(1):01–10. https://doi.org/10.51903/jtie.v3i1.154
35. Li Z (2022) Extracting spatial effects from machine learning model using local interpretation method: an example of SHAP and XGBoost. Comput Environ Urban Syst 96:101845. https://doi.org/10.1016/j.compenvurbsys.2022.101845
36. Maeda-Gutiérrez V, Galván-Tejada CE, Zanella-Calzada LA, Celaya-Padilla JM, Galván-Tejada JI, Gamboa-Rosales H, Luna-García H, Magallanes-Quintanar R, Guerrero CA, Méndez A, Olvera-Olvera CA (2020) Comparison of convolutional neural network architectures for classification of tomato plant diseases. Appl Sci 10(4):1245. https://doi.org/10.3390/app10041245
37. Kim Y (2020) A grounded theory study of burnout among youth workers in Korean Immigrant Churches. Christ Educ J 17(2):373–391. https://doi.org/10.1177/0739891320926512
38. Chuang Y-N, Wang G, Yang F, Liu Z, Cai X, Du M, Hu X (2023) Efficient XAI techniques: a taxonomic survey. arXiv preprint. http://arxiv.org/abs/2302.03225
39. Chamola V, Vikas Hassija A, Sulthana R, Ghosh D, Dhingra D, Sikdar B (2023) A review of trustworthy and Explainable Artificial Intelligence (XAI). IEEE Access 11:78994–78915. https://doi.org/10.1109/ACCESS.2023.3294569
40. Ribeiro MT, Singh S, Guestrin C (2016) Why Should I Trust You?: explaining the predictions of any classifier. In: Proceedings of the 22nd ACM SIGKDD International Conference on Knowledge Discovery and Data Mining. ACM, San Francisco, CA, pp 1135–1144. https://doi.org/10.1145/2939672.2939778
41. Sisk M, Majlis M, Page C, Yazdinejad A (2022). Analyzing XAI metrics: summary of the literature review. arXiv preprint. 10.36227/techrxiv.21262041.v1
42. Friedman JH (2001) Greedy function approximation: a gradient boosting machine. Ann Stat 29(5):1189–1232
43. Alvarez-Melis D, Jaakkola TS (2018) On the robustness of interpretability methods. arXiv preprint. http://arxiv.org/abs/1806.08049
44. Kawakura S, Hirafuji M, Ninomiya S, Shibasaki R (2022) Adaptations of Explainable Artificial Intelligence (XAI) to agricultural data models with ELI5, PDPbox, and Skater using diverse agricultural worker data. Eur J Artif Intell Mach Learn 1(3):27–34. https://doi.org/10.24018/ejai.2022.1.3.14
45. Lundberg SM, Erion GG, Lee S-I (2019) Consistent individualized feature attribution for tree ensembles. arXiv prerpint. http://arxiv.org/abs/1802.03888
46. Goldstein A, Kapelner A, Bleich J, Pitkin E (2014) Peeking inside the black box: visualizing statistical learning with plots of individual conditional expectation'. arXiv preprint. http://arxiv.org/abs/1309.6392
47. Gerelt-Od, Anu-Ujin, Lee Kho, Anhthy Ngo, Andrew Yeh. 2020. 'Generating feature impact from individual conditional expectation plots'
48. Ribeiro MT, Singh S, Guestrin C (2018) Anchors: high-precision model-agnostic explanations. Proc AAAI Conf Artif Intell 32(1). https://doi.org/10.1609/aaai.v32i1.11491

Technologies and Governance of the Adoption of Smart Healthcare Solutions for Urban Health Development

Intissar Haddiya and Sara Ramdani

Abstract The rapid urbanisation of global populations presents complex challenges for healthcare systems, particularly in resource-limited and densely populated areas. Urbanisation amplifies existing health disparities, with informal settlements often lacking access to basic sanitation, healthcare, and infrastructure, thereby increasing the burden of preventable diseases. Recognising these challenges, technological advancements have emerged as pivotal solutions to bridge gaps in healthcare delivery and promote equitable health outcomes. Tools such as mobile health (mHealth) applications, wearable devices, and telemedicine platforms are increasingly employed to enhance disease surveillance, chronic condition management, and patient engagement, especially in underserved urban settings. Studies highlight the potential of the Internet of Medical Things (IoMT) in tracking vital signs and collecting real-time health data, enabling early intervention and personalised care. Moreover, frameworks like the Sustainable Development Goals (SDGs) emphasise the integration of health-focused technologies with broader urban planning efforts to address the social determinants of health and foster inclusive, sustainable urban environments. This study explores the integration of advanced technologies, including mobile health (mHealth), telemedicine, artificial intelligence (AI), and the Internet of Medical Things (IoMT), to address critical urban health needs. These technologies enhance healthcare delivery by enabling real-time monitoring, chronic disease management, and personalised care while bridging geographical and socioeconomic disparities.

This chapter highlights key barriers to implementing smart healthcare solutions, such as data privacy concerns, interoperability issues, and the digital divide, which hinder equitable adoption. It emphasises the importance of sustainable urban plan-

I. Haddiya (✉)
Department of Nephrology, Mohammed VI University Hospital, Faculty of Medicine and Pharmacy of Oujda, Mohammed First University, Oujda, Morocco

S. Ramdani
Laboratory of Epidemiology, Clinical Research and Public Health, Faculty of Medicine and Pharmacy of Oujda, University Mohammed First, Oujda, Morocco

H. Kaddouri et al. (eds.), *Smart Cities for Sustainable and Urban Development*, Green Energy and Technology, https://doi.org/10.1007/978-3-031-99001-4_32

ning aligned with frameworks like the Sustainable Development Goals (SDGs) and demonstrates through case studies how digital tools and community-based approaches have improved health outcomes. Interdisciplinary collaboration, equitable policies, and robust governance are identified as essential to fostering resilient urban healthcare systems. By addressing technological, infrastructural, and social determinants of health, smart healthcare innovations can transform urban public health, promoting efficiency, equity, and sustainability in a digital world.

Keywords Urban health · Smart healthcare · Mobile health · Artificial intelligence · Healthcare disparities

1 Introduction

Healthcare delivery is drastically changing in the last decade, driven by rapid advancements in technology, particularly in mobile connectivity and an expanding array of platforms, applications, sensors, devices, and equipment. These technological innovations are reshaping healthcare by enhancing access, quality, and patient engagement. The development and proliferation of mobile health (mHealth) applications have created new opportunities for patient monitoring, chronic disease management, and telemedicine, improving healthcare access for remote or disadvantaged populations. Studies indicate that mobile connectivity and health data integration significantly improve early detection and intervention, particularly for chronic conditions like diabetes and hypertension, as patients and providers can access real-time data through user-friendly applications on smartphones and personal devices. Such developments have introduced new telecommunication technologies, facilitating data exchange across healthcare services [1, 2].

As mHealth and similar applications grow, network infrastructure must adapt to the Internet of Medical Things (IoMT), a system of interconnected devices designed for healthcare. IoMT tracks vital signs, monitors medical devices, and collects data for clinical research. A study by Acharya and Patil highlights that connected medical devices allow for more proactive health management by enabling continuous monitoring of patient health, particularly in critical care settings [3]. However, IoMT integration requires secure data channels and robust cloud computing frameworks to ensure seamless operation and data security. This modernisation necessitates the adoption of advanced technologies for effective telemedicine delivery, including audio-visual tools that facilitate remote healthcare interactions [4].

Telemedicine, encompassing videoconferencing, e-health services, and remote monitoring of vital signs, has expanded significantly with advancements in high-definition cameras, encryption software, and broadband internet [4]. It transcends geographical limitations, providing cost-effective, high-quality services. Research shows telemedicine improves patient outcomes, reduces hospital readmissions, and alleviates burdens on healthcare facilities. During the COVID-19 pandemic, demand surged, emphasising the need for resilient networks capable of handling large data volumes [5].

In addition to connectivity, healthcare has integrated various information technologies (IT) to enhance patient outcomes and improve care quality. The integration of IT tools has redefined care standards by streamlining workflows, improving communication, and offering data-driven insights that enhance clinical and operational efficiency. Tools such as electronic health records (EHRs), clinical information systems, and remote monitoring systems facilitate efficient healthcare delivery by centralising patient data, offering real-time access to comprehensive information. Research by Kruse et al. highlights that EHRs have improved continuity of care by reducing medical test redundancy and improving provider coordination, particularly in chronic disease management [6]. These systems improve clinical decision-making, reduce medical errors, and increase patient safety, especially for underserved populations [7]. EHRs also increase diagnostic accuracy and support treatment plan formulation through clinical decision support systems that guide providers based on current guidelines.

Additionally, mHealth and wearable devices like fitness trackers empower patients by allowing them to monitor heart rate, blood pressure, and physical activity [8]. Wearable devices like fitness trackers and smartwatches allow patients to monitor metrics such as heart rate, blood pressure, and physical activity, fostering a proactive approach to health management. Research by Piwek et al. indicates that wearable devices enhance patient engagement by providing insights into health patterns, promoting adherence to treatment plans, and enabling early intervention [9]. Furthermore, mHealth applications have proven beneficial in chronic disease management by facilitating continuous monitoring and timely communication with healthcare providers, reducing the likelihood of complications.

Artificial intelligence (AI), including machine learning (ML), revolutionises healthcare with predictive diagnostics, pattern recognition, and treatment optimisation [10]. AI analyses large datasets with speed and accuracy beyond human capacity, identifying risk factors and predicting disease progression. Studies by Topol emphasise that Artificial Intelligence (AI) has advanced radiology, dermatology, and pathology by automating diagnostic processes and identifying patterns in medical imaging that might be missed by human eyes [11]. In oncology, AI algorithms have proven invaluable in early cancer detection by analysing tissue samples with remarkable precision. Similarly, ML models predict hospital readmissions and patient deterioration, allowing for timely interventions and improved outcomes.

However, these advancements are accompanied by significant challenges. Integrating IT into healthcare also presents significant challenges, including privacy concerns, interoperability issues, and the social and ethical implications of AI and ML, such as bias and discrimination [12]. Privacy concerns are acute due to the sensitive nature of health data and the rise of cyberattacks. Interoperability barriers arise as systems across organisations lack standardisation, hindering coordinated care. Ethical concerns include biases in historical data used to train models, leading to potential discrimination. Obermeyer et al. emphasise the need for ethical oversight to address these biases and ensure equitable AI deployment [13].

As urbanisation accelerates globally, particularly in regions like Asia and Africa, these challenges intersect with pressing urban health issues. Resource-constrained and densely populated areas increasingly rely on smart healthcare solutions like IoT, big data analytics, and wearable devices to improve accessibility, optimise resource allocation, and alleviate burdens on healthcare facilities, contributing to sustainable urban development [14, 15]. However, gaps remain in integrating these technologies into urban planning, particularly telecare and Urban Digital Twins, to support aging populations and enhance urban health infrastructure [16, 17]. Addressing these gaps requires interdisciplinary collaboration, comprehensive standards, and joint efforts among industry, policymakers, and academics to create sustainable, inclusive cities [18, 19].

Amid global urbanisation, particularly in rapidly growing regions, unregulated urban expansion exacerbates public health challenges, creating informal settlements with inadequate sanitation and healthcare. These environments, often plagued by pollution, overcrowding, and limited green spaces, further complicate health outcomes and amplify disease transmission risks [20, 21]. These interconnected issues underscore the urgent need for innovative, technology-driven healthcare solutions integrated within sustainable urban development frameworks.

In response to these challenges, the World Health Organisation (WHO) and the United Nations have called for urban planning aligned with the Sustainable Development Goals (SDGs). Particularly relevant is SDG 3, which advocates for health and well-being for all ages [22]. SDG 11, which emphasises making cities inclusive, safe, resilient, and sustainable, also plays a key role in creating healthier urban environments. This approach is supported by frameworks such as the Conceptual Framework for Urban Health, which highlights the interplay between social determinants, physical infrastructure, and health outcomes, emphasising a comprehensive view of the urban health ecosystem [23]. The Health Map of Barton and Grant illustrates how factors such as education, housing, transportation, and the built environment collectively influence health at the urban level [24]. These frameworks underscore the importance of sustainable urban planning and advocate for health impact assessments to mitigate negative health outcomes.

Successful examples of urban health interventions further reinforce the importance of this integrated approach. Green infrastructure and urban parks have been linked to improved mental health, reduced obesity rates, and lower risks of respiratory ailments. Cities prioritising public transportation and active commuting options, such as cycling lanes and pedestrian pathways, have reduced air pollution and traffic-related injuries, contributing to enhanced overall health [25]. Furthermore, digital health technologies and mHealth platforms are increasingly employed to address healthcare accessibility challenges in densely populated areas by facilitating disease tracking, patient monitoring, and preventive health services, particularly in low-resource settings.

Urban public health, as an interdisciplinary field, explores how urban structures influence health outcomes. This comprehensive approach considers the spatial distribution of health-promoting resources and risk factors within urban environments [23, 24, 26–31]. The built environment, including housing, transportation, and public spaces, plays a significant role in health outcomes by promoting or hindering access to essential resources [25]. Social determinants such as socioeconomic status, education, and employment also impact urban health, shaping access to healthcare and health-promoting resources [32].

Digital tools such as EHRs, telemedicine, and mobile health applications offers new opportunities to improve healthcare delivery in urban areas, especially underserved ones. These technologies enable better disease surveillance, chronic condition management, and greater patient engagement, addressing health disparities in urban populations [33].

Complementing these advancements, the integration of AI and ML further strengthens healthcare systems by enabling predictive analytics, early detection, optimised resource allocation, and improved decision-making. Together, these innovations pave the way for more effective and equitable healthcare solutions [10].

However, technological advancements also present challenges, including data privacy, cybersecurity, and the digital divide, which hinder the widespread adoption of healthcare IT solutions. While technology can improve healthcare access and quality, marginalised urban populations may face difficulties accessing digital tools due to economic or technological barriers [34]. The rapid pace of urbanisation and evolving healthcare needs require continuous innovation in healthcare infrastructure, including the integration of smart city technologies and sustainable healthcare systems [35]. These developments must be strategically planned to ensure resilient healthcare systems capable of responding to both current and future challenges.

This chapter aims to highlight research gaps, inform policy, and identify strategies for advancing healthcare delivery and urban public health amidst global urbanisation. By investigating the utility of smart healthcare technologies in fostering sustainable urban development, it underscores the importance of innovation, policy integration, and interdisciplinary approaches to address the multifaceted challenges faced by urban populations. Through the analysis of current research, the chapter emphasises the critical role of equitable policies and technological innovation in addressing the complexities of urban public health in an increasingly digital and interconnected world.

2 Key Challenges in Urban Health

2.1 Overcrowding and Healthcare Infrastructure

Research suggests that high population density can strain healthcare systems and impact health outcomes. Higher population density is associated with increased healthcare costs, hospital bed utilisation, and accident rates [36, 37]. A 1% increase in population density correlates with a 0.4% rise in accident risk, necessitating strategic placement of emergency services in densely populated areas [37]. While urban density can facilitate healthcare delivery, with a one-unit increase in density score linked to a 0.2% increase in maternal health coverage rates, it may also negatively impact mortality [38]. A study in the Netherlands found that higher population density was modestly related to increased mortality, despite promoting active transport [39]. These findings highlight the complex relationship between population density and healthcare systems, suggesting the need for tailored approaches to address the challenges posed by high-density urban environments.

Adding to these pressures, the growing demand for healthcare services further compounds the challenges faced by providers and systems. This rise in demand is attributed to various factors, including changes in health patterns, population aging, and the COVID-19 pandemic [40]. Healthcare providers face difficulties in addressing staffing shortages and ensuring service access. To manage this increasing demand, various strategies have been proposed, including financial incentives, enhancing patient transfer, and community involvement. However, implementing these strategies effectively requires addressing contextual challenges such as decision-making norms and designing appropriate communication strategies [41]. Managing demand is not just about reduction but also about shaping it to best serve health needs with available resources. This requires a comprehensive understanding of how demand is initiated and expressed to better manage the entire healthcare system [42].

2.2 *Environmental and Pollution-Related Risks*

Urban air and water pollution pose significant public health challenges, particularly in developing countries experiencing rapid urbanisation [43]. Air pollution, primarily from fine particulate matter, contributes to premature mortality and respiratory issues [43, 44]. The World Health Organisation estimates that air pollution causes up to seven million deaths annually, mostly in urban areas [44]. Water pollution, often due to inadequate sanitation infrastructure, disproportionately affects the poor [43]. Sources of urban air pollution include increased population, economic activities, and consumption patterns [45]. Despite previous improvements, many urban populations still breathe air that fails to meet health-based guidelines [46]. Addressing these issues requires integrated urban management strategies, including efficient environmental regulation, improved infrastructure, and innovative approaches to minimise pollution [43, 44]. Implementing bold and effective policies based on scientific evidence is crucial for reducing air pollution's impact on public health [46].

2.3 Chronic Diseases and Health Inequities

Lifestyle-related diseases are increasingly prevalent worldwide, posing significant health challenges. In the Philippines, a 2008 survey revealed high rates of hypertension (24.6%), diabetes (5.2%), and dyslipidaemia (72.0%) among adults [47]. Similarly, Saudi Arabia faces a growing burden of NCD, with obesity rates rising from 10.8% in 1975 to 35.4% in 2016 [48]. Among university students globally, overweight/obesity has the highest prevalence at 45.6%, followed by prehypertension at 32.0% [49]. Common risk factors for these diseases include poor diet, physical inactivity, smoking, and alcohol consumption [50]. To address this growing concern, experts recommend implementing comprehensive health education programs, regular screenings, and early interventions to promote healthier lifestyles and reduce the burden of lifestyle-related diseases [48, 49]. However, tackling lifestyle-related diseases cannot be fully achieved without addressing the underlying social determinants of health, which significantly influence health outcomes and exacerbate disparities. Social determinants of health, including poverty, education, housing, and access to healthcare, significantly impact health outcomes and contribute to persistent disparities [51, 52]. These factors influence disease development, healthcare utilisation, and health outcomes across various populations [51]. Addressing social determinants is crucial for reducing health inequalities and improving overall population health [53]. Evidence supports interventions targeting education, urban planning, housing, income supplements, and employment to reduce disparities. While these interventions can lead to long-term societal savings, they require sustained financing and implementation at local, state, and national levels [52]. A comprehensive approach that considers socioeconomic status, race, gender, and insurance status is necessary to promote health equity in healthcare, including specialised fields like otolaryngology [51]. Addressing social determinants through well-designed economic and social policies is essential for achieving health equality [53].

2.4 Access to Healthcare

Barriers to accessing healthcare services are multifaceted and affect various populations. For people with Parkinson's disease, both person-level (e.g., skills, ability to engage, costs) and system-level barriers (e.g., availability of resources) hinder access [54]. Aboriginal people in Australia face challenges related to service coordination, communication, trust, cultural safety, and distance [55]. In developing countries, people with disabilities encounter demand-side barriers (lack of information, costs, limited mobility, stigmatisation) and supply-side barriers (staff attitudes, communication issues, inaccessible facilities) [56]. Economic factors play a significant role, particularly for chronic patients during times of austerity. Unemployed, low-income, and low-educated individuals are more likely to face economic barriers, while women and those with lower health status are more prone to geographical barriers [57]. These barriers can lead to increased inequalities and deteriorated health outcomes, necessitating targeted health policy measures. These access issues are further compounded by regional disparities in healthcare quality and availability, which persist in both developing and developed countries. In India, rural and disadvantaged groups face inequitable access to healthcare, with northern and western regions of West Bengal being particularly underdeveloped [58]. Similarly, Iran exhibits unequal distribution of healthcare facilities across states, with factors such as child mortality rate and number of rural public health centres explaining much of the variance [59]. In Thailand's border regions, over 253,000 people live more than 30 min away from hospitals, highlighting inadequate healthcare accessibility [60]. The United States also experiences geographic variation in healthcare quality, particularly in stroke care. Non-metropolitan hospitals perform worse than metropolitan ones, especially in administering alteplase. Improving stroke centre certification and access to neurologists could help bridge this quality gap [61]. These studies emphasise the need for targeted interventions to address healthcare disparities in underserved areas.

3 Smart Healthcare Technologies

3.1 Telemedicine and Remote Patient Monitoring

Telemedicine has emerged as a transformative tool in expanding healthcare access, particularly for chronic disease management. It improves patient outcomes, enhances access to care, and increases patient engagement, especially in diabetes management [62]. Community pharmacists, as the third largest healthcare professional group globally, are increasingly recognised for their potential in chronic disease management, medication adherence, and reducing adverse events [63]. In low- and middle-income countries, telemedicine has shown promise in maintaining essential health services during the COVID-19 pandemic and reducing socioeconomic disparities in healthcare access [64]. Community pharmacists have demonstrated effectiveness in managing various chronic conditions, including asthma, cardiovascular diseases (CVDs), and diabetes, either independently or as part of disease management teams [65]. However, challenges such as insufficient integration into healthcare pathways and inadequate compensation need to be addressed to fully realise their potential in chronic disease management.

Building upon these innovations, remote medical consultations and monitoring have significantly evolved, leveraging wireless technologies and web-based applications to enhance telemedicine's effectiveness. These technologies allow real-time monitoring of vital signs, interactive consultations, and long-term patient health tracking [66, 67]. Telemedicine encompasses various applications, from interactive video consultations to store-and-forward technologies for transmitting medical data

[68]. Recent advancements include wearable devices for arrhythmia detection and chronic disease monitoring, as well as telemonitoring strategies for more efficient and accurate patient care [69]. The COVID-19 pandemic has accelerated the adoption of telemedicine in healthcare systems [66]. However, challenges remain, such as integrating new technologies into clinical practice, addressing the digital divide, and improving legal regulations [66, 69]. Despite these challenges, telemedicine is becoming increasingly routine for both doctors and patients.

3.2 AI in Urban Health

AI is revolutionising healthcare through improved diagnostics, personalised treatments, and predictive analytics. AI systems, utilising ML and deep learning algorithms, have demonstrated remarkable accuracy in analysing medical images for disease detection and classification [70, 71]. These technologies integrate diverse data sources, including EHRs, genetic information, and lifestyle factors, to provide tailored diagnostic recommendations and treatment plans [70, 72]. AI also accelerates drug discovery processes and optimises clinical trial outcomes [70]. While AI shows promise in enhancing healthcare efficiency and accessibility, challenges remain, including data privacy concerns, ethical considerations, and the need for robust regulatory frameworks [73–77]. Addressing these issues is crucial for the successful integration of AI in healthcare systems, ultimately leading to improved patient outcomes and more effective, personalised care [76].

3.3 Big Data Analytics

Predictive modelling and data-driven approaches are increasingly used to enhance urban health management and address health disparities. Artificial intelligence has shown promise in early detection of disease outbreaks, as demonstrated by the COVID-19 case in China [78]. ML techniques can predict neighbourhood-scale health outcomes, with sociodemographic variables being strong predictors [79]. Predictive analytics have been applied successfully in chronic disease management in the USA and infectious disease control in Africa [80–84]. Models predicting COVID-19-related healthcare resource utilisation across statewide populations have shown high accuracy, although performance variations exist across demographic subgroups [85]. These data-driven approaches can inform policy decisions and improve healthcare delivery. However, challenges remain, including ethical considerations, technological barriers, and the need for enhanced data sharing protocols to ensure privacy and prevent oversight [78].

3.4 *Internet of Things (IoT)*

The integration of connected devices in real-time health monitoring and smart home solutions is rapidly evolving healthcare delivery. IoT-enabled smart devices allow for continuous patient data collection and transfer, enabling proactive health management and personalised care [86]. These systems incorporate wearable devices, ambient sensors, and ML algorithms to detect early signs of health deterioration and predict health-related events [87]. The combination of IoT with cloud computing and big data technologies has led to the concept of "smart health," offering on-demand medical services and empowering patients to take control of their health [86, 88]. However, challenges such as data security, privacy, and interoperability standards must be addressed [87, 88]. Connected health technologies aim to facilitate communication and coordination between patients and healthcare professionals, supporting new models for chronic disease management [89]. These advancements have the potential to improve health outcomes and reduce healthcare costs [82].

3.5 *Blockchain Technology*

Blockchain technology offers promising solutions for secure healthcare data exchange, addressing challenges of interoperability, privacy, and regulatory compliance [90–92]. It enhances data security, integrity, and transparency while complying with HIPAA and GDPR regulations [90]. Innovative approaches combine blockchain with IoT devices, incorporating privacy-preserving techniques like differential privacy and homomorphic encryption [93, 94]. These systems demonstrate improved data storage, retrieval performance, and high success rates in access control and privacy protection [95]. However, blockchain applications face challenges in achieving full GDPR compliance, requiring careful design choices to increase real-world feasibility [96]. Despite technical complexities, regulatory hurdles, and scalability issues, blockchain technology shows potential to revolutionise healthcare data management by balancing accessibility, privacy, and security [90].

4 Smart Healthcare Infrastructure for Urban Settings

4.1 *Smart Hospitals*

Recent technological innovations have significantly impacted healthcare delivery, enhancing patient care and operational efficiency. EHRs and telemedicine have improved access to care, particularly in rural areas, while reducing costs and streamlining workflows [6, 97]. Artificial intelligence and robotics are being explored for precision surgery, rehabilitation, and targeted drug delivery [98]. However, barriers to adoption persist, including technically challenged staff, resistance to change, and cost concerns [6]. Privacy, security, and infrastructure requirements also pose challenges [99]. Despite these obstacles, the integration of technologies like blockchain, AI, and the Internet of Things has shown promise in improving healthcare quality and system management [98, 99]. Ongoing research is necessary to navigate the complexities of technology adoption and optimise healthcare practices effectively [99].

4.2 Digital Health Platforms

Population health management and patient engagement are evolving through various technological tools and strategies. EHRs, health information exchanges, patient portals, and mobile health apps are enhancing population health management capabilities [100]. These technologies enable proactive care management, targeted interventions, and improved health outcomes for entire communities [101]. However, challenges persist, particularly among older adults who may have lower eHealth literacy and concerns about data security [102]. To address these issues, healthcare organisations can implement strategies such as screening for eHealth literacy, providing training, and promoting proxy users. Additionally, leveraging predictive analytics for risk stratification, using population registries to identify care gaps, and automating patient outreach can improve population health management effectiveness [103]. As technology continues to advance, there are opportunities to expand the reach and impact of population health through patient engagement, analytics, and interventions addressing social determinants of health [101].

4.3 Urban Planning and Healthcare

Geographic Information Systems (GIS) have become increasingly important in healthcare planning and public health decision-making. GIS tools can address various healthcare planning issues, including epidemiological surveillance, evaluating spatial inequality of access, and resource allocation [104]. These systems enable the creation of choropleth maps, kernel density models, and distance analyses to support health service planning in urban areas [105]. GIS applications in public health research cover themes such as health-related behaviour across geographic settings, healthcare accessibility, neighbourhood effects on health outcomes, and the construction of appropriate geographic units for analysis [106]. Furthermore, GIS-based decision support systems (DSS) can improve collaborative health planning by increasing stakeholder engagement, enhancing decision-making processes, and improving data availability for health decision-makers [107]. These tools are crucial for addressing challenges in healthcare planning and promoting evidence-driven, localised decision-making in the context of healthy cities.

4.4 Mobile Health Solutions

Community health worker (CHW) interventions have shown promise in improving health outcomes and healthcare utilisation for underserved populations, though more research is needed to establish cost-effectiveness [108]. mHealth technologies offer opportunities to reach minority populations with health information and interventions [109]. Effective cancer prevention programs for African Americans should address barriers such as inadequate access, lack of knowledge, and mistrust, while incorporating community participation, cultural competence, and innovative outreach [110]. Strategies to increase breast and cervical cancer screening among underserved women include management systems for patients and providers, community-based outreach, and integration of preventive services at primary care sites. Mass media campaigns have been successful for Hispanic women but not for African American women. Mobile units can effectively reach elderly women [111]. These approaches demonstrate the importance of tailoring interventions to specific underserved populations.

5 Policy and Governance in Smart Healthcare

5.1 Regulatory Framework

The regulation of AI in healthcare is a complex and evolving field, with various challenges and considerations. Current policies focus on data privacy, patient confidentiality, and algorithmic transparency [112]. Governance mechanisms for AI in healthcare are still developing, with no internationally recognised standards. Key areas for international collaboration include leadership, a whole-systems approach, regulatory processes, and stakeholder engagement [113]. Ethical concerns encompass fairness, transparency, and accountability. Policymakers face challenges in addressing workforce disruption, adapting educational training, and determining insurance coverage for AI technologies [114]. Recommendations include adopting harmonised standards under the World Health Organisation, using European Union law as a model, and reforming the International Health Regulations to better regulate AI in healthcare [115]. International cooperation is crucial for developing coherent solutions and overcoming governance barriers [113].

5.2 Public-Private Partnerships

Collaborative innovation in healthcare requires effective partnerships between governments, private sectors, and academic institutions to address challenges and drive progress. Successful models involve open innovation, co-creation, and public-private partnerships (PPPs) [116–118]. Key factors for fostering innovation include trust-building, user involvement, regulatory support, and incentives [118]. Case studies demonstrate successful outcomes, such as the UK's NHS Innovation Accelerator and the Medicines for Malaria Venture in sub-Saharan Africa [116]. However, challenges persist, including limited resources, divergent priorities, and complex regulatory frameworks [116, 119]. To overcome these obstacles, organisations should focus on developing a culture of respect, investing in individuals, allowing time for interaction, and nurturing kind and empowering leadership [119]. Transparent communication, stakeholder engagement, and sustained political commitment are crucial for successful collaboration and innovation in healthcare [116].

5.3 Integration into National Systems

To ensure equitable access to smart healthcare solutions, several strategies have been proposed. These include co-creation and digitalisation approaches to promote equitable distribution of technological benefits [120]. User-friendly designs, supportive infrastructure, and educational support are key themes identified for advancing digital health equity [121]. The Digital Health Equity Framework can be used to consider factors such as poverty, access, engagement, and digital health literacy barriers [122]. Recommendations include adopting collaborative working approaches, raising awareness of available support, and incorporating digital health equity into healthcare provider training [121, 122]. Community-based approaches involve outreach programs and co-design strategies to address barriers [121]. Policies should focus on inclusive digital health strategies and measurement-based approaches to digital health equity [121, 122].

Effective regulation, collaborative partnerships, and integration into national systems are crucial for advancing AI-driven healthcare, addressing challenges of equity, transparency, and governance while fostering innovation and ensuring equitable access through community-focused and inclusive strategies (Fig. 1).

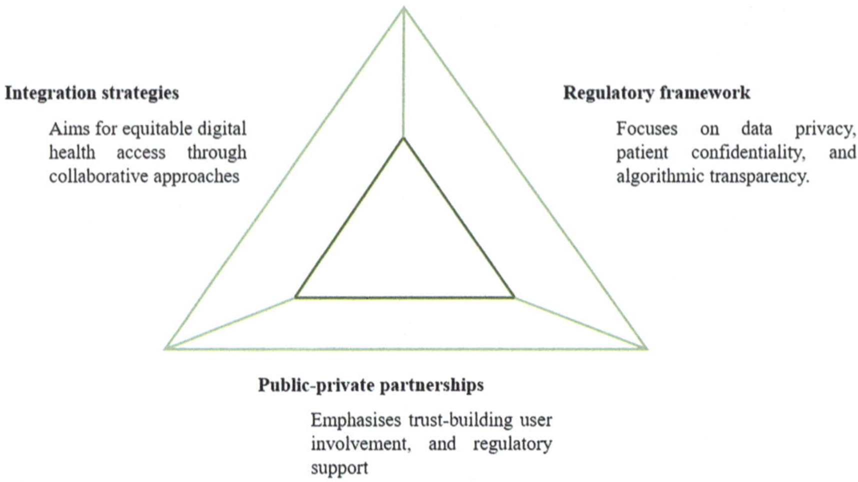

Fig. 1 Advancing healthcare innovation through collaboration and regulation strategies

6 Challenges and Barriers

6.1 Technological and Infrastructure Limitations

The digital divide poses significant challenges to implementing smart healthcare solutions, particularly for vulnerable populations like people with disabilities [123]. Key barriers include inequity in connectivity, affordability, and accessibility of devices and internet services [124]. However, the potential for consumer information technologies to advance health equity is growing as the digital divide narrows. Successful interventions using mHealth, telehealth, and social media have shown promise in addressing health disparities, though challenges in design, implementation, and evaluation remain [125]. To overcome these barriers, strategies include securing political commitment, intersectoral collaboration, economic investments, standardised ethical and regulatory frameworks, and health workforce capacity building [126]. Policy recommendations focus on making internet access a basic right and developing more affordable and accessible technologies without compromising utility [124].

6.2 Data Privacy Concerns

Smart healthcare systems in urban environments face significant data breach risks, with the healthcare sector experiencing frequent security breaches [127]. These breaches can result in substantial financial damage and compromise personal information [128]. To mitigate risks and ensure compliance with regulations like HIPAA and HITECH, organisations should implement robust cybersecurity measures and risk management practices [127, 128]. Best practices include adopting minimum security requirements, encrypting sensitive data, and regularly auditing systems for vulnerabilities [128, 129]. Additionally, organisations should focus on employee training, establishing tighter regulations, and enforcing compliance through civil fines and audits [128]. The integration of IoT devices in healthcare systems presents new challenges, requiring innovative approaches to threat detection and prevention while maintaining data accuracy and efficiency [129–132].

6.3 *Cultural and Behavioural Factors*

The adoption of smart healthcare technologies faces several challenges, including resistance from healthcare professionals, poor infrastructure, and low technical expertise. Key factors influencing adoption include perceived usefulness, belief, willingness, and attitude of healthcare professionals [133, 134]. Human factors play a significant role, with concerns about trust, privacy, and potential job displacement affecting acceptance [135]. Successful implementation strategies involve addressing technological, organisational, and environmental challenges [136]. Patients generally show more enthusiasm for new technologies than medical professionals, valuing increased autonomy in healthcare decisions. To improve acceptance, it is crucial to involve all stakeholders in the development process and address anxieties and feelings of insecurity [137]. Effective change management strategies and a differentiated approach to policies and practices are recommended to scale up adoption and reduce failure rates of smart health projects [136].

6.4 *Financial Constraints*

Implementing smart healthcare solutions in urban development faces significant funding and sustainability challenges. Key issues include limited healthcare resources, rising costs, and the need for long-term financing [138, 139]. Successful strategies involve adopting win-win approaches, such as platform-based collaboration and involving private sector stakeholders in design and implementation. Outcome-based funding models and continuous monitoring can help address financial constraints [138]. To ensure sustainability, smart healthcare systems should integrate with other urban subsystems and consider factors like information flow, material cycles, and lifestyle [139]. Cloud-enabled and data-driven systems can provide cost-effective solutions with reduced IT setup costs [140]. However, challenges persist in legislation, policies, and infrastructure development [141]. Addressing these issues requires a comprehensive framework that considers both technological and organisational aspects of smart healthcare implementation (Fig. 2).

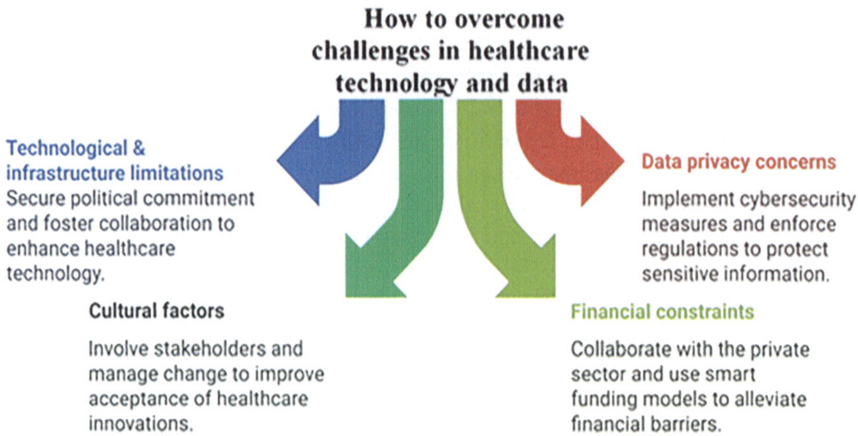

Fig. 2 Overcoming challenges in healthcare technology and data management

7 Case Studies of Smart Healthcare Solutions

7.1 Developed Cities

Smart cities are leveraging technology to revolutionise healthcare delivery and public health initiatives, creating a more responsive, efficient, and patient-centred healthcare ecosystem. These innovations encompass a broad range of applications, including population surveillance to track health trends, active aging support systems for elderly populations, emergency response optimisation, and the reorganisation of care services to better meet community needs [142]. For instance, population surveillance technologies are being employed to monitor disease outbreaks in real-time, enabling faster containment efforts and improved public health decision-making.

Key technologies driving these innovations include IoT sensors, data analytics, mobile applications, and ML algorithms. IoT-enabled wearable devices are increasingly used to collect continuous health metrics such as heart rate, glucose levels, and physical activity. These data points feed into integrated platforms that utilise ML algorithms to predict health outcomes, suggest preventive measures, and provide alerts for potential medical emergencies [15, 143].

Smart city infrastructure plays a critical role in integrating mobile networks, ambient sensor systems, and ML technologies, fostering strategic healthcare management through data-driven approaches [144]. Ambient sensors embedded in public spaces monitor environmental conditions such as air quality, temperature, and noise levels, providing actionable insights for mitigating health risks related to pollution or extreme weather. Similarly, applications focusing on healthy lifestyle promotion use behavioural analytics to encourage exercise, better nutrition, and stress management, often through gamified mobile platforms. Cities like Copenhagen and

Amsterdam have integrated smart technologies into cycling infrastructure to promote active lifestyles, though this has sparked debates around techno-solutionism and pro-innovation bias [145].

For instance, ambient sensors embedded in public spaces can monitor environmental conditions such as air quality, temperature, and noise levels, providing actionable insights for mitigating health risks related to pollution or extreme weather. Similarly, applications focusing on healthy lifestyle promotion use behavioural analytics to encourage exercise, better nutrition, and stress management, often through gamified mobile platforms.

Emergency response systems have seen significant advancements through IoT-connected ambulances equipped with real-time data-sharing capabilities. These systems enable paramedics to access patient histories en route and transmit critical information to hospitals ahead of arrival [146]. Smart health concepts are being integrated into emergency response systems, bridging healthcare services with smart city infrastructures to improve response times and patient outcomes [147]. Additionally, IoT-based systems support aging populations by enhancing elderly care through features such as fall detection, remote monitoring, and voice commands [148, 149].

Smart healthcare initiatives in cities like New York, Barcelona, Seoul, Dubai, and Hong Kong demonstrate the transformative potential of these technologies. Dubai's smart elderly housing incorporates IoT devices for safety and health management, while Hong Kong uses city-wide IoT sensing infrastructures for elderly monitoring and quality-of-life improvements [148, 150]. In developing nations, telemedicine services powered by smart city infrastructure address doctor shortages in rural areas by connecting patients to specialists through video consultations and remote diagnostics, enhancing accessibility and affordability [15].

However, challenges persist in scaling these technologies for widespread adoption. Infrastructure limitations, data privacy concerns, and the need for robust regulatory frameworks are critical barriers [151]. High deployment costs for IoT and ML systems pose obstacles for resource-constrained regions, necessitating innovative funding models and international collaboration to bridge the digital divide [144]. Security, interoperability, and data governance issues further complicate implementation [151].

Smart cities also leverage information and communication technologies to enhance urban services, including healthcare management. Integrated command centres, AI, and innovative applications have proven effective in managing health crises, as demonstrated during the COVID-19 pandemic in India [152]. Cloud-based solutions with real-time data processing and intelligent decision-making capabilities provide a common platform for data analytics in healthcare [153]. Blockchain technology is emerging as a promising solution for secure storage and management of health information in smart city healthcare systems [154].

While many of these advancements are in early developmental stages, their potential to improve accessibility, efficiency, and quality of care is evident. Case studies from cities like Philadelphia, Quebec City, and Seattle highlight efforts to create unified service access points, improving population health and reducing burdens on healthcare professionals [142, 155]. As smart cities continue to evolve, these technologies will play a crucial role in transforming healthcare systems and promoting sustainability.

7.2 Developing Cities

Healthcare initiatives in Nairobi face significant challenges due to the complex dynamics of urban slums. These areas are characterised by substandard living conditions, including overcrowded housing, poor sanitation, and inadequate waste management, all of which contribute to heightened health risks and the prevalence of communicable diseases. The situation is exacerbated by limited access to formal healthcare services, forcing many residents to rely on informal providers who often lack proper training or regulatory oversight. This dependency on informal healthcare systems further entrenches systemic inequities and reduces the quality of care available to vulnerable populations [156]. Efforts to address these issues include the implementation of smart healthcare initiatives targeting specific challenges such as CVD and inefficient medical record systems. In Nairobi's Kibera slum, the introduction of an Electronic Medical Record System aimed to improve healthcare resource utilisation [157]. Similarly, the SCALE-UP study developed a community-centred CVD prevention model that combined public health and private sector efforts to enhance awareness, provide home-based screening, and promote treatment adherence [158]. This model's adaptability was later demonstrated in its application to African migrants in the Netherlands [159]. Other research has focused on age-friendly slum frameworks to address the specific health needs of older residents in Nairobi [160].

Migrants in Nairobi's urban slums face unique and multifaceted barriers to accessing healthcare. These include economic challenges, such as the high cost of medical services and medication, which many migrants struggle to afford due to their precarious employment situations. Furthermore, social barriers, such as harassment, discrimination, and the constant threat of eviction or legal repercussions due to unclear legal status, further deter migrants from seeking care. These issues are compounded by systemic disparities in healthcare delivery, where migrants often face discrepancies in the quality of services compared to local residents [161]. Similar challenges are observed in cities like Mumbai, Dhaka, and Lagos, where urban slums deal with fragmented healthcare systems, financial hurdles, and a lack of adequate infrastructure. For example, in Mumbai, slum residents favour private or mixed public-private providers for healthcare, while in Dhaka, rapid urbanisation necessitates better mapping for health policy planning [162, 163].

One such initiative is the removal of user fees in public health facilities. This policy was designed to reduce financial barriers, particularly for economically disadvantaged populations. While it has improved access for many, the abolition of fees has led to challenges such as overcrowding in public hospitals and inconsistencies in the availability of services, limiting its overall effectiveness.

The Kenya Essential Package for Health (KEPH) is another significant initiative aimed at improving healthcare equity. KEPH outlines a range of essential health services that should be universally accessible. These services include maternal and child health care, disease prevention, and health promotion activities. By defining and standardising these services, KEPH seeks to ensure a minimum standard of healthcare is available to all citizens, particularly those in underserved regions [164]. Technological advancements in eHealth and mHealth have shown promise in urban healthcare, particularly through data analytics, IoT sensors, and mobile applications. These technologies enable real-time health monitoring and personalised treatment [15]. Despite challenges like financial feasibility, infrastructure limitations, and security concerns, increased internet access among healthcare workers has facilitated communication and access to health information [165].

Community-based health strategies have also played a critical role in improving healthcare accessibility. In Nairobi and other cities, the Community Health Strategy (CHS) leverages community health volunteers (CHVs) to connect households with formal healthcare systems, providing preventive services such as immunisations and malaria prevention [164]. CHW programs have demonstrated success in addressing maternal and child health, TB/HIV, and non-communicable diseases, with adaptations made for urban settings [166].

Cities like Addis Ababa and Kinshasa offer lessons in urban health interventions. In Addis Ababa, the Urban Community Health Information System improved data quality and service targeting, while pilot initiatives promoted team-based care despite logistical challenges [167]. In Kinshasa, community-oriented interventions highlighted the importance of effective communication to counter misinformation [168].

Scaling successful pilot programs has been pivotal in expanding urban health interventions. Ghana's CHPS initiative and Bangladesh's BRAC Manoshi Project are examples of community-based models adapted for urban contexts to enhance healthcare accessibility and outcomes [169, 170]. Digital health technologies have further improved maternal and child health (MCH) through mobile apps and SMS-based interventions, although challenges related to accessibility and cultural relevance remain [171, 172].

Effective urban health policies require comprehensive, multisectoral approaches addressing structural barriers and promoting community engagement. Strategies should prioritise underserved populations, incorporate evidence-based planning, and adapt successful models to the unique dynamics of urban slums. Long-term research, such as that conducted in Nairobi's slums, underscores the importance of integrating community participation, addressing root causes of health inequities, and fostering equitable healthcare delivery [173].

7.3 Innovative Pilot Programs

Small-scale urban health projects can provide valuable insights for addressing health challenges in urban settings. Community-based approaches have shown promise, but rural models often require adaptation for urban contexts. In Ghana, the Community-Based Health Planning and Services (CHPS) model was modified for urban areas, incorporating night and weekend home visits to accommodate urban residents' schedules [174]. In India, innovative approaches focused on strengthening community-provider linkages and improving maternal, infant, and adolescent health. Building social capital within communities and encouraging coordination among local stakeholders have been effective strategies for enhancing service utilisation and health outcomes in urban slums [175]. While small, well-implemented programs can provide useful insights, scaling up requires careful testing, learning, and adaptation to achieve sustainable impact at the national level [176].

7.4 Policy Compliance and Addressing Challenges in Urban Healthcare Systems

Cities ranging from developed hubs like Dubai, Hong Kong, and New York to developing centres like Nairobi, Mumbai, and Kinshasa—demonstrate varying levels of policy compliance and innovative approaches to overcoming healthcare challenges. By integrating technology and community-based initiatives, these cities have attempted to meet the complex needs of their populations, though barriers like financial constraints, infrastructural limitations, and systemic inequities persist (Table 1).

Developed cities such as Dubai, Hong Kong, New York, Barcelona, and Seoul have exhibited strong adherence to policies promoting technology-driven healthcare. Dubai's smart elderly housing and Hong Kong's city-wide IoT sensing infrastructures align with goals of improving public health and enhancing quality of life for vulnerable populations [148, 150]. Similarly, New York and Seoul have adhered to policies integrating IoT into emergency response systems, optimising patient care and reducing response times [146, 147].

In Nairobi, the KEPH and the Community Health Strategy (CHS) reflect compliance with policies aimed at improving healthcare equity. These initiatives standardise essential health services and connect households with formal healthcare systems, enhancing preventive care for underserved populations [164]. Similarly, Addis Ababa's Urban Community Health Information System and Kinshasa's community-oriented interventions demonstrate a commitment to improving urban health policy implementation, with a focus on data quality and public health awareness [167, 168].

Despite policy compliance, challenges like financial constraints, infrastructural limitations, and systemic inequities remain critical barriers across these cities.

Table 1 Global smart healthcare initiatives: technologies, challenges, and policy compliance

City	Smart healthcare initiative	Key technologies	Challenges	Policy compliance	How challenges were addressed
Dubai	Smart elderly housing with IoT devices for safety and health management	IoT devices	High deployment costs, data privacy concerns	In alignment with policies for elderly care and IoT integration	Innovative funding models and international collaborations to manage high costs
Hong Kong	City-wide IoT sensing infrastructures for elderly monitoring and quality of life improvements	IoT sensors	Data privacy concerns, regulatory issues	In adherence with policies for elderly monitoring and health metrics	Implementation of regulatory frameworks to address privacy and data governance issues
New York	IoT-connected ambulances for real-time data sharing and optimised emergency responses	IoT-connected ambulances	Infrastructure limitations	In compliance with emergency response optimisation policies	Investment in scalable infrastructure and interoperable IoT systems
Barcelona	Smart health concepts integrated into emergency response systems; robust regulatory frameworks for privacy	IoT, regulatory frameworks	Data governance, scalability	In adherence to privacy frameworks and emergency response standards	Adoption of robust data governance frameworks and blockchain technologies
Seoul	IoT-connected ambulances for emergency response optimisation	IoT-connected ambulances	Scalability, interoperability	In alignment with emergency response optimisation policies	Focus on interoperability and scalable IoT solutions
Copenhagen	Integration of smart cycling infrastructure to promote active lifestyles	Behavioural analytics, gamified platforms	Techno-solutionism debates	Yes, promoting active lifestyles aligns with public health policies	Behavioural analytics and community engagement to counter techno-solutionism critiques
Amsterdam	Integration of smart cycling infrastructure to promote active lifestyles	Behavioural analytics, gamified platforms	Techno-solutionism debates	Yes, promoting active lifestyles aligns with public health policies	Behavioural analytics and community engagement to counter techno-solutionism critiques

(continued)

Table 1 (continued)

City	Smart healthcare initiative	Key technologies	Challenges	Policy compliance	How challenges were addressed
Philadelphia	Creation of unified service access points to improve population health	Unified service platforms	Service integration challenges	Yes, unified access points align with healthcare integration policies	Development of unified platforms to streamline service access and reduce burden on healthcare professionals
Quebec City	Creation of unified service access points to improve population health	Unified service platforms	Service integration challenges	Yes, unified access points align with healthcare integration policies	Development of unified platforms to streamline service access and reduce burden on healthcare professionals
Seattle	Creation of unified service access points to improve population health	Unified service platforms	Service integration challenges	Unified access points align with healthcare integration policies	Development of unified platforms to streamline service access and reduce burden on healthcare professionals
Nairobi	Electronic Medical Record System in Kibera; SCALE-UP CVD prevention model; KEPH and CHS programs	eHealth, mHealth, IoT, community health volunteers	Economic barriers, overcrowding, data privacy concerns	Yes, through KEPH, CHS, and community-based programs	Use of community health volunteers, increased internet access, and eHealth tools
Mumbai	Reliance on private or mixed public-private providers for slum healthcare	Private healthcare providers	Financial hurdles, reliance on private providers	Partial compliance due to reliance on mixed systems	Preference for private or mixed providers to address system gaps
Dhaka	Rapid urbanisation and better health policy planning	Health policy mapping and planning	Infrastructural challenges due to rapid urbanisation	Focused on urban planning for healthcare needs	Implementation of urban health mapping and planning

Lagos	Addressing fragmented healthcare systems in urban slums	Fragmented systems support	Fragmented systems, inadequate infrastructure	Partial compliance; fragmented systems present barriers	Emphasis on public-private partnerships to address system gaps
Addis Ababa	Urban Community Health Information System improving data quality and targeting services	Data analytics, team-based care models	Logistical barriers in scaling pilot programs	In alignment with urban health policy interventions	Use of data-driven decision-making and pilot care models
Kinshasa	Community-oriented interventions to counter misinformation	Community engagement, effective communication	Misinformation and communication gaps	Community engagement aligns with health equity goals	Community-focused communication strategies
Ghana (National)	CHPS initiative adapting community-based models to urban contexts	Community health volunteers, service adaptation	Scaling and funding of community models	In compliance through community-centred health services	Expansion of scalable community health programs
Bangladesh (National)	BRAC Manoshi Project improving maternal and child health outcomes in urban areas	Mobile apps, SMS-based interventions	Accessibility and cultural relevance issues	Yes, compliance through maternal and child health initiatives	Mobile-based interventions and SMS tools for MCH improvements

Developed cities have relied on advanced technologies and innovative funding models to address these barriers. For instance, Barcelona and Amsterdam have implemented regulatory frameworks to address privacy and data security concerns, fostering public trust in smart healthcare technologies [151]. Emerging technologies such as blockchain and cloud-based solutions have further enhanced data security and crisis management, particularly during the COVID-19 pandemic [152, 154].

In Nairobi's Kibera slum, the adoption of an Electronic Medical Record System demonstrated the potential of digital solutions to streamline healthcare resource utilisation. Additionally, the SCALE-UP initiative combined public health and private sector efforts to develop a community-centred cardiovascular disease prevention model, showcasing scalability and adaptability in addressing urban health challenges [157, 159]. Similarly, cities like Mumbai and Dhaka have employed mapping and planning strategies to address fragmented healthcare systems, while Ghana's CHPS initiative and Bangladesh's BRAC Manoshi Project have successfully adapted community-based models to urban contexts [162, 169, 170].

Technological advancements have been pivotal in overcoming systemic barriers. Cities like Copenhagen and Amsterdam have used behavioural analytics and gamified platforms to promote active lifestyles, addressing preventive healthcare needs [145]. Nairobi has leveraged mHealth and eHealth tools, facilitated by increased internet access among healthcare workers, to bridge information gaps and improve real-time health monitoring [165]. However, challenges related to cultural relevance and accessibility persist in the deployment of mobile health applications in cities like Nairobi and Dhaka [171, 172].

Both developed and developing cities have demonstrated compliance with healthcare policies through innovative approaches and the integration of technology. While developed cities like Dubai, Hong Kong, and New York focus on advanced IoT systems and regulatory frameworks, cities like Nairobi, Mumbai, and Dhaka emphasise community-based models and digital health technologies to address systemic inequities. Despite their unique challenges, these cities share a commitment to improving healthcare accessibility, efficiency, and equity, underscoring the transformative potential of smart city initiatives and community-driven healthcare reforms. Long-term success will depend on adapting these models to local contexts, prioritising underserved populations, and fostering multisectoral collaboration to bridge existing gaps.

8 Future Directions in Smart Urban Healthcare

8.1 Emerging AI and ML Technologies

Recent innovations in personalised medicine and smart healthcare are transforming urban health development. EHRs, telemedicine, and wearable devices are improving healthcare delivery and patient outcomes [177]. ML algorithms and optimisation techniques are enhancing disease diagnosis for conditions like CVDs and diabetes mellitus [35]. Advances in genomics and computational biology are enabling the development of targeted therapies and companion diagnostics for biomarker-defined subpopulations [178]. Smart cities are leveraging data analytics, IoT sensors, and mobile applications to facilitate real-time health monitoring and early disease detection [15]. However, challenges persist, including data privacy concerns, interoperability issues, and patient acceptance of technology [177]. Implementation of smart healthcare systems in developing nations can improve accessibility and affordability by optimising resource allocation and streamlining patient care processes [15].

8.2 Genomics and Precision Medicine

Genomic data integration into healthcare systems offers significant potential for personalised medicine and urban health development. Standardisation efforts, like SMART on FHIR Genomics, facilitate the incorporation of genomic data into EHRs, enabling clinico-genomic applications [179]. Implementation science can address disparities in genomic medicine, particularly for high-risk populations in areas such as CVDs and cancer [180]. Artificial intelligence, including ML and natural language processing, plays a crucial role in translating cancer genomics into precision medicine, though challenges in data requirements and algorithmic transparency persist [181]. Several countries have invested billions in national genomic-medicine initiatives, driving transformative change in healthcare systems while addressing implementation barriers. International collaboration and data sharing are essential for accelerating the integration of genomics into healthcare globally [182].

8.3 Integrated Smart Cities

The integration of smart healthcare within urban infrastructure presents both opportunities and challenges for achieving interoperability. Urbanisation-based approaches can facilitate seamless integration and interoperability in heterogeneous health information systems, as demonstrated by the Yaoundé Urban Information Systems Integration and Interoperability Platform [183]. Key challenges include privacy concerns, organisational barriers, and technical limitations [184]. Successful integration strategies involve leveraging IoT for remote patient monitoring and personalised treatments, while addressing security and data governance issues [151]. A comprehensive approach considers the interaction between smart healthcare and other urban subsystems, such as smart homes and intelligent transportation [139]. To ensure seamless data exchange and coordinated services, strategies include adopting interoperability standards, implementing secure IoT frameworks, and fostering collaboration among stakeholders [151, 184].

8.4 Global Collaboration

International standardisation efforts in smart healthcare and urban health development have been progressing for over a decade, aiming to facilitate health information exchange and improve patient safety [185]. The COVID-19 pandemic has highlighted the need for enhanced data sharing protocols in smart city networks to better manage global health emergencies [78]. Standardisation frameworks for smart cities are being developed by international organisations like ITU, focusing on urban infrastructure interoperability and ICT requirements [186]. The Smart Mature Resilience project demonstrates the importance of integrating stakeholders through standardisation, resulting in the development of city resilience tools and the CWA 17300 standards series [187]. These initiatives contribute to advancing global health innovation by promoting knowledge exchange, improving urban health monitoring, and enhancing city resilience against crises like pandemics.

9 Conclusion

Smart healthcare technologies, including AI, IoT and telemedicine, are transforming urban healthcare by addressing the challenges of rapid urbanisation. These innovations enhance healthcare accessibility by bridging geographical barriers and expanding chronic disease management. Telemedicine enables underserved populations to receive essential health services, while IoT devices facilitate real-time health monitoring, supporting early diagnoses and proactive care. Together, these technologies close gaps in healthcare access, creating more inclusive urban health systems.

Improved health outcomes are another hallmark of smart healthcare. AI advances diagnostics and treatment planning through sophisticated data analytics, enabling personalised care and predictive insights, such as outbreak forecasting. Big Data analytics further refines urban healthcare management by optimising resource allocation and informing policy decisions. These tools are instrumental in addressing complex urban health challenges with precision and efficiency.

In addition to accessibility and outcomes, these technologies mitigate systemic inefficiencies. Blockchain solutions enhance healthcare data management by ensuring secure, transparent, and efficient information exchange. Digital platforms facilitate population health management and patient engagement, while smart hospitals integrate advanced technologies to streamline operations and elevate care quality. Collectively, these innovations reduce redundancies and inefficiencies, fostering resilience and equity in urban healthcare ecosystems.

To sustain this progress, collaboration among policymakers, healthcare providers, technologists, and urban planners is essential. Continued investment in emerging technologies such as genomics, ML, and integrated smart city systems will drive innovation. Equitable policies are needed to ensure all populations benefit from these advancements, particularly by addressing the digital divide. Strengthening governance with robust regulatory frameworks for AI, telemedicine, and data security will enhance trust and adoption. Finally, global collaboration through knowledge sharing and standardised protocols will accelerate progress and scalability. By prioritising innovation and inclusivity, urban healthcare systems can become not only efficient but also equitable and resilient for future generations.

References

1. Khujamatov K, Reypnazarov E, Akhmedov N, Khasanov D (2020) Blockchain for 5g healthcare architecture. Paper presented at the 2020 International Conference on Information Science and Communications Technologies (ICISCT), 4–6 November 2020
2. Khujamatov K, Ahmad K, Reypnazarov E, Khasanov D (2020) Markov chain based modeling bandwith states of the wireless sensor networks of monitoring system. Int J Adv Sci Technol 29:4889–4903
3. Acharya AD, Patil SN (2020) Iot based health care monitoring kit. Paper presented at the 2020 Fourth International Conference on Computing Methodologies and Communication (ICCMC), 11–13 March 2020
4. Coombes CE, Gregory ME (2019) The current and future use of telemedicine in infectious diseases practice. Curr Infect Dis Rep 21:41
5. Sosnowski R, Kamecki H, Joniau S, Walz J, Klaassen Z, Palou J (2020) Introduction of telemedicine during the Covid-19 pandemic: a challenge for now, an opportunity for the future. Eur Urol 78:820–821
6. Kruse CS, Kothman K, Anerobi K, Abanaka L (2016) Adoption factors of the electronic health record: a systematic review. JMIR Med Inform 4:e19
7. Levin DC, Rao VM (2004) Turf wars in radiology: introduction. J Am Coll Radiol 1:23–25
8. Free C, Phillips G, Galli L, Watson L, Felix L, Edwards P, Patel V, Haines A (2013) The effectiveness of mobile-health technology based health behaviour change or disease management interventions for health care consumers: a systematic review. PLoS Med 10:e1001362
9. Piwek L, Ellis DA, Andrews S, Joinson A (2016) The rise of consumer health wearables: promises and barriers. PLoS Med 13:e1001953
10. Obermeyer Z, Emanuel EJ (2016) Predicting the future—big data, machine learning, and clinical medicine. N Engl J Med 375:1216–1219
11. Topol EJ (2019) High-performance medicine: the convergence of human and artificial intelligence. Nat Med 25:44–56
12. Price WN 2nd, Cohen IG (2019) Privacy in the age of medical big data. Nat Med 25:37–43
13. Obermeyer Z, Powers B, Vogeli C, Mullainathan S (2019) Dissecting racial bias in an algorithm used to manage the health of populations. Science 366:447–453
14. Ghazal TM, Hasan MK, Alshurideh MT, Alzoubi HM, Ahmad M, Akbar SS, Al Kurdi B, Akour IA (2021) Iot for smart cities: machine learning approaches in smart healthcare—a review. Future Internet 13:218
15. Mohammadzadeh Z, Saeidnia HR, Lotfata A, Hassanzadeh M, Ghiasi N (2023) Smart city healthcare delivery innovations: a systematic review of essential technologies and indicators for developing nations. BMC Health Serv Res 23:1180
16. Fahlevi M, Dandi M, Asyrof RA, Dahlan OP, Asetya DR, Dahlan SP (2024) Elevating healthcare through telemedicine: a Pathto Smart City success in Ikn (Indonesia's Newcapital City). Paper presented at the 2024 International Conference on ICT for Smart Society (ICISS)
17. Mazzetto S (2024) A review of urban digital twins integration, challenges, and future directions in smart city development. Sustainability 16:8337

18. Drobež E, Kavšek M, Bogataj M, Drobez M (2022) Ambient intelligence supporting health and social care services in smart care settings: literature review and research agenda. IFAC-PapersOnLine 55:934–939
19. Khlie K, Benmamoun Z (2024) Towards smarter and greener cities: harnessing Ai and green technology for urban sustainability. J Infrastruct Policy Dev 8:6300
20. Habitat, UN_ (2016) Urbanization and development: emerging futures. World cities Report 3:4–51
21. Okkels N, Kristiansen CB, Munk-Jrgensen P, Goldberg D, Thornicroft G (2017) Mental health and illness in the city. Springer
22. Stuart L (2017) Transforming Our World: The 2030 Agenda for Sustainable Development a/Res/70/1: Theme: Statement, 'Do Not Leave Indigenous Australians Behind'. Paper presented at the Session of the Permanent Forum of Indigenous Issues (PFII): Tenth Anniversary of the United Nations Declaration on the Rights of Indigenous Peoples: measures taken to implement the Declaration
23. Galea S, Freudenberg N, Vlahov D (2005) Cities and population health. Soc Sci Med 60:1017–1033
24. Barton H, Grant M (2006) A health map for the local human habitat. J R Soc Promot Heal 126:252–253
25. Giles-Corti B, Vernez-Moudon A, Reis R, Turrell G, Dannenberg AL, Badland H, Foster S, Lowe M, Sallis JF, Stevenson M, Owen N (2016) City planning and population health: a global challenge. Lancet 388:2912–2924
26. Fudge C, Grant M, Wallbaum H (2020) Transforming cities and health: policy, action, and meaning. Cities Health 4:135–151
27. Galea S, Vlahov D (2005) Urban health: evidence, challenges, and directions. Annu Rev Public Health 26:341–365
28. Grant M, Brown C, Caiaffa WT, Capon A, Corburn J, Coutts C, Crespo CJ, Ellis G, Ferguson G, Fudge C (2017) Cities and health: an evolving global conversation, vol 1. Taylor & Francis, pp 1–9
29. Lovasi GS, Diez AV, Roux, and Jennifer Kolker. (2020) Urban public health: a research toolkit for practice and impact. Oxford University Press
30. Rydin Y, Bleahu A, Davies M, Dávila JD, Friel S, De Grandis G, Groce N, Hallal PC, Hamilton I, Howden-Chapman P (2012) Shaping cities for health: complexity and the planning of urban environments in the 21st century. Lancet 379:2079–2108
31. Vlahov D, Freudenberg N, Proietti F, Ompad D, Quinn A, Nandi V, Galea S (2007) Urban as a determinant of health. J Urban Health 84:16–26
32. Vlahov D, Galea S (2002) Urbanization, urbanicity, and health. J Urban Health 79:S1–S12
33. Sieck CJ, Sheon A, Ancker JS, Castek J, Callahan B, Siefer A (2021) Digital inclusion as a social determinant of health. NPJ Digit Med 4:52
34. Sun N, Esom K, Dhaliwal M, Amon JJ (2020) Human rights and digital health technologies. Health Hum Rights 22:21
35. Chui KT, Alhalabi W, Pang SSH, Ordóñez P, de Pablos R, Liu W, Zhao M (2017) Disease diagnosis in smart healthcare: innovation, technologies and applications. Sustainability 9:2309
36. Jones R (2013) Population density and healthcare costs. Br J Healthc Manag 19:44–45
37. Nieminen S, Lehtonen O-P, Linna M (2002) Population density and occurrence of accidents in Finland. Prehosp Disaster Med 17:206–208
38. Hanlon M, Burstein R, Masters SH, Zhang R (2012) Exploring the relationship between population density and maternal health coverage. BMC Health Serv Res 12:1–7
39. Beenackers MA, Groeniger JO, Kamphuis CBM, Van Lenthe FJ (2018) Urban population density and mortality in a compact Dutch City: 23-year follow-up of the Dutch globe study. Health Place 53:79–85
40. Cruz-Gomes S, Amorim-Lopes M, Almada-Lobo B (2019) The demand for healthcare services and resources: patterns, trends and challenges in healthcare delivery. Operational Research: IO 2018, Aveiro, Portugal, 5–7 September 19, pp 91–106

41. Elmusharaf K, Byrne E, O'Donovan D (2015) Strategies to increase demand for maternal health services in resource-limited settings: challenges to be addressed. BMC Public Health 15:1–10
42. Pencheon D (1998) Managing demand: matching demand and supply fairly and efficiently. BMJ 316:1665–1667
43. Brandon C (1998) Cities and health. In: Environment matters. The World Bank, Washington, DC
44. Samet JM (2019) Pollution. In: Galea S, Ettman CK, Vlahov D (eds) Urban health. Oxford University Press
45. Kura B, Verma S, Ajdari E, Iyer A (2013) Growing public health concerns from poor urban air quality: strategies for sustainable urban living. Comput Water Energy Environ Eng 2:1
46. KATZ MORRIS (1956) Atmospheric pollution and public health. Can J Public Health/Revue Canadienne de Sante'e Publique 47:395–400
47. Sy RG, Morales DD, Dans AL, Paz-Pacheco E, Punzalan FE, Abelardo NS, Duante CA (2012) Prevalence of atherosclerosis-related risk factors and diseases in The Philippines. J Epidemiol 22:440–447
48. Mahmood FM (2018) Prevalence and prevention of lifestyle-related diseases in Saudi Arabia. Int J Health Sci 12:1
49. Nagheer D, Irvin R, Younger-Coleman N (2017) Overview of the prevalence and associated risk, factors of lifestyle diseases in university students. Int J Clin Med 8:344
50. Ng R, Sutradhar R, Yao Z, Wodchis WP, Rosella LC (2020) Smoking, drinking, diet and physical activity-modifiable lifestyle risk factors and their associations with age to first chronic disease. Int J Epidemiol 49:113–130
51. Bergmark RW, Sedaghat AR (2017) Disparities in health in the United States: an overview of the social determinants of health for otolaryngologists. Laryngoscope Investig Otolaryngol 2:187–193
52. Thornton RLJ, Glover CM, Cené CW, Glik D, Henderson JA, Williams DR (2016) Determinants of health evaluating strategies for reducing health disparities by addressing the social.
53. Chelak K, Chakole S (2023) The role of social determinants of health in promoting health equality: a narrative review. Cureus 15
54. Zaman MS, Ghahari S, McColl MA (2021) Barriers to accessing healthcare services for people with Parkinson's disease: a scoping review. J Parkinsons Dis 11:1537–1553
55. Nolan-Isles D, Macniven R, Hunter K, Gwynn J, Lincoln M, Moir R, Dimitropoulos Y, Taylor D, Agius T, Finlayson H (2021) Enablers and barriers to accessing healthcare services for aboriginal people in New South Wales, Australia. Int J Environ Res Public Health 18:3014
56. Baart J, Taaka F (2017) Barriers to healthcare services for people with disabilities in developing countries: a literature review. Disabil CBR Incl Dev 28:26–40
57. Kyriopoulos D II, Zavras A, Skroumpelos K, Mylona KA, Kyriopoulos J (2014) Barriers in access to healthcare services for chronic patients in times of austerity: an empirical approach in Greece. Int J Equity Health 13:54
58. Majumder S, Roy S, Bose A, Chowdhury IR (2023) Understanding regional disparities in healthcare quality and accessibility in West Bengal, India: a multivariate analysis. Reg Sci Policy Pract 15:1086–1113
59. Sepehrdoust H (2009) Health care analysis and regional disparities in different provinces of Iran. Iran Econ Rev 14:113–134
60. Sritart H, Tuntiwong K, Miyazaki H, Taertulakarn S (2021) Disparities in healthcare services and spatial assessments of mobile health clinics in the border regions of Thailand. Int J Environ Res Public Health 18:10782
61. Seabury S, Bognar K, Yaping X, Caroline Huber S, Commerford R, Tayama D (2017) Regional disparities in the quality of stroke care. Am J Emerg Med 35:1234–1239

62. Ezeamii VC, Okobi OE, Wambai-Sani H, Perera GS, Zaynieva S, Okonkwo CC, Ohaiba MM, William-Enemali PC, Obodo OR, Obiefuna NG (2024) Revolutionizing healthcare: how telemedicine is improving patient outcomes and expanding access to care. Cureus 16:e63881
63. Mossialos E, Courtin E, Naci H, Benrimoj S, Bouvy M, Farris K, Noyce P, Sketris I (2015) From "retailers" to health care providers: transforming the role of community pharmacists in chronic disease management. Health Policy 119:628–639
64. Hoffer-Hawlik MA, Moran AE, Burka D, Kaur P, Cai J, Frieden TR, Gupta R (2020) Leveraging telemedicine for chronic disease management in low-and middle-income countries during Covid-19. Glob Heart 15
65. George J, McNamara K, Stewart K (2011) The roles of community pharmacists in cardiovascular disease prevention and management. Australas Med J 4:266–272
66. Shaderkin IA, Shaderkina VA (2021) Remote medical consultations of patients: what has changed in Russia in 20 years. Russi J Telemed e-Health 7:7–17
67. Yuan B, Herbert J (2011) Web-based real-time remote monitoring for pervasive healthcare. Paper presented at the 2011 IEEE International Conference on Pervasive Computing and Communications Workshops (PERCOM Workshops)
68. Field MJ, Grigsby J (2002) Telemedicine and remote patient monitoring. JAMA 288:423–425
69. Rubio LF, Lozano-Granero C, Vidal-Pérez R, Barrios V, Freixa-Pamias R (2023) New technologies for the diagnosis, treatment, and monitoring of cardiovascular diseases. Revista Española de Cardiología (English Edition)
70. Chowdhury RH (2024) Intelligent systems for healthcare diagnostics and treatment. World J Adv Res Rev 23:007–015
71. Devi K, Alghamdi W, Divya N, Alkhayyat A, Sayyora A, Sathish T (2023) Artificial intelligence in healthcare: diagnosis, treatment, and prediction. E3S Web of Conferences 399
72. Sitapati A, Kim H, Berkovich B, Marmor R, Singh S, El-Kareh R, Clay B, Ohno-Machado L (2017) Integrated precision medicine: the role of electronic health Records in delivering personalized treatment. Wiley Interdiscip Rev Syst Biol Med 9:e1378
73. Bartoletti I (2019) Ai in healthcare: ethical and privacy challenges. Paper presented at the Artificial Intelligence in Medicine, Cham, 2019
74. Elendu C, Amaechi DC, Elendu TC, Jingwa KA, Okoye OK, John Okah M, Ladele JA, Farah AH, Alimi HA (2023) Ethical implications of Ai and robotics in healthcare: a review. Medicine (Baltimore) 102:e36671
75. Eshwar M (2023) Exploring the potential of artificial intelligence in healthcare: possibilities and challenges. Int Sci J Eng Manag 02
76. Kewlani R, Rajoriya S, Saxena B (2020) Ai in healthcare for diagnosis and treatment. Int J Psychosoc Rehabil:10118–10122
77. Winter JS (2020) Ai in healthcare: data governance challenges. J Hosp Manag Health Policy 5
78. Allam Z, Jones DS (2020) On the coronavirus (Covid-19) outbreak and the smart city network: universal data sharing standards coupled with artificial intelligence (Ai) to benefit urban health monitoring and management. Paper presented at the Healthcare
79. Feng C, Jiao J (2021) Predicting and mapping neighborhood-scale health outcomes: a machine learning approach. Comput Environ Urban Syst 85:101562
80. Ameh S, Gómez-Olivé FX, Kahn K, Tollman S, Klipstein-Grobusch K (2020) Multilevel predictors of controlled Cd4 count and blood pressure in an integrated chronic disease management model in rural South Africa: a panel study. BMJ Open 10:e037580
81. Battineni G, Sagaro GG, Chinatalapudi N, Amenta F (2020) Applications of machine learning predictive models in the chronic disease diagnosis. J Pers Med 10
82. Bhardwaj N, Wodajo B, Spano A, Neal S, Coustasse A (2018) The impact of big data on chronic disease management. Health Care Manag (Frederick) 37:90–98
83. Carter SE, Ahuka-Mundeke S, Pfaffmann Zambruni J, Navarro Colorado C, van Kleef E, Lissouba P, Meakin S, le Polain de Waroux O, Jombart T, Mossoko M, Nkakirande DB, Esmail M, Earle-Richardson G, Degail MA, Umutoni C, Anoko JN, Gobat N (2021) How

to improve outbreak response: a case study of integrated outbreak analytics from Ebola in Eastern Democratic Republic of the Congo. BMJ Glob Health 6
84. Standley CJ, Graeden E, Kerr J, Sorrell EM, Katz R (2018) Decision support for evidence-based integration of disease control: a proof of concept for malaria and schistosomiasis. PLoS Negl Trop Dis 12:e0006328
85. Kasturi SN, Park J, Wild D, Khan B, Haggstrom DA, Grannis S (2021) Predicting Covid-19-related health care resource utilization across a statewide patient population: model development study. J Med Internet Res 23:e31337
86. Kang M, Park E, Cho BH, Lee K-S (2018) Recent patient health monitoring platforms incorporating internet of things-enabled smart devices. Int Neurourol J 22:S76
87. Haghi M, Spicher N, Wang J, Deserno TM (2022) Integrated sensing devices for disease prevention and health alerts in smart homes. In: Accident and emergency informatics. IOS Press, pp 39–61
88. Linkous L, Zohrabi N, Abdelwahed S (2019) Health monitoring in smart homes utilizing internet of things. Paper presented at the 2019 IEEE/ACM International Conference on Connected Health: Applications, Systems and Engineering Technologies (CHASE)
89. Chouvarda IG, Goulis DG, Lambrinoudaki I, Maglaveras N (2015) Connected health and integrated care: toward new models for chronic disease management. Maturitas 82:22–27
90. Atadoga A, Elufioye OA, Omaghomi TT, Akomolafe O, Odilibe IP, Owolabi OR (2024) Blockchain in healthcare: a comprehensive review of applications and security concerns. Int J Sci Res Arch 11:1605–1613
91. Dagher GG, Mohler J, Milojkovic M, Marella PB (2018) Ancile: privacy-preserving framework for access control and interoperability of electronic health records using blockchain technology. Sustain Cities Soc 39:283–297
92. Hylock RH, Zeng X (2019) A blockchain framework for patient-centered health records and exchange (Healthchain): evaluation and proof-of-concept study. J Med Internet Res 21:e13592
93. Ali A, Pasha MF, Ali J, Fang OH, Masud M, Jurcut AD, Alzain MA (2022) Deep learning based homomorphic secure search-able encryption for keyword search in blockchain healthcare system: a novel approach to cryptography. Sensors 22:528
94. Jia B, Zhang X, Liu J, Zhang Y, Huang K, Liang Y (2022) Blockchain-enabled federated learning data protection aggregation scheme with differential privacy and homomorphic encryption in Iiot. IEEE Trans Industr Inform 18:4049–4058
95. Liu Y, Zhang J, Zhan J (2021) Privacy protection for fog computing and the internet of things data based on blockchain. Clust Comput 24:1331–1345
96. Hasselgren, Anton, Paul Wan, Margareth Horn, Katina Kralevska, and Danilo Gligoroski. 2020. Gdpr compliance for blockchain applications in healthcare
97. Haleem A, Javaid M, Singh RP, Suman R (2021) Telemedicine for healthcare: capabilities, features, barriers, and applications. Sens Int 2:100117
98. Awad A, Trenfield SJ, Pollard TD, Ong JJ, Elbadawi M, McCoubrey LE, Goyanes A, Gaisford S, Basit AW (2021) Connected healthcare: improving patient care using digital health technologies. Adv Drug Deliv Rev 178:113958
99. Alhur A (2024) Redefining healthcare with artificial intelligence (Ai): the contributions of Chatgpt, Gemini, and Co-Pilot. Cureus 16
100. Tierney WM, Rousseau JF, Khurshid A (2018) Measuring and managing population health. Tex Med 114:42–51
101. Mohanty N, Salgado C, Rapoport A, Baxi RS, Panzer J, Simon J (2021) The people, process, and technology for population management in community health. Fam Syst Health 39:112
102. Price-Haywood EG, Harden-Barrios J, Ulep R, Luo Q (2017) Ehealth literacy: patient engagement in identifying strategies to encourage use of patient portals among older adults. Popul Health Manag 20:486–494
103. Handmaker K, Hart J (2015) 9 Steps to effective population health management: the success of an accountable care organization ultimately depends on the extent to which it intelligently

deploys technology solutions for data analytics, care team workflows, and patient engagement to improve population health while reducing costs. Healthc Financ Manage 69:70–77
104. Khashoggi BF, Murad A (2020) Issues of healthcare planning and Gis: a review. ISPRS Int J Geo Inf 9:352
105. Murad A (2018) Planning and location of health care services in Jeddah City, Saudi Arabia: discussion of the constructive use of geographical information systems. Geospat Health 13
106. Wang F (2020) Why public health needs Gis: a methodological overview. Ann GIS 26:1–12
107. Gudes O, Pathak V, Kendall E, Yigitcanlar T (2011) Thinking spatially, acting collaboratively: a Gis-based health decision support system for improving the collaborative health-planning practice. Paper presented at the Proceedings of the International Conference on Health Informatics
108. Viswanathan M, Kraschnewski JL, Nishikawa B, Morgan LC, Honeycutt AA, Thieda P, Lohr KN, Jonas DE (2010) Outcomes and costs of community health worker interventions: a systematic review. Med Care 48:792–808
109. Anderson-Lewis C, Darville G, Mercado RE, Howell S, Di Maggio S (2018) Mhealth technology use and implications in historically underserved and minority populations in the United States: systematic literature review. JMIR Mhealth Uhealth 6:e128
110. Wolff M, Bates T, Beck B, Young S, Ahmed SM, Maurana C (2003) Cancer prevention in underserved African American communities: barriers and effective strategies-a review of the literature. WMJ-MADISON 102:36–40
111. Vellozzi CJ, Romans M, Rothenberg RB (1996) Delivering breast and cervical cancer screening services to underserved women: part I. Literature review and telephone survey. Women's Health Issues 6:65–73
112. Jeyaraman M, Balaji S, Jeyaraman N, Yadav S (2023) Unraveling the ethical enigma: artificial intelligence in healthcare. Cureus 15:e43262
113. Morley J, Murphy L, Mishra A, Joshi I, Karpathakis K (2022) Governing data and artificial intelligence for health care: developing an international understanding. JMIR Form Res 6:e31623
114. Aggarwal M, Gingras C, Deber R (2021) Artificial intelligence in healthcare from a policy perspective. In: Multiple perspectives on artificial intelligence in healthcare: opportunities and challenges. Springer, pp 53–64
115. Bouderhem R (2024) Shaping the future of Ai in healthcare through ethics and governance. Humanit Soc Sci Commun 11:1–12
116. Ebulue CC, Ebulue OR, Ekesiobi CS (2024) Public-private partnerships in health sector innovation: lessons from around the world. Int Med Sci Res J 4:484–499
117. Hyrkäs P, Haukipuro L, Väinämö S, Iivari M, Sachinopoulou A, Majava J (2020) Collaborative innovation in healthcare: a case study of hospitals as innovation platforms. Int J Value Chain Manag 11:24–41
118. Laziz J, Rachid O (2024) Investigating the influence of stakeholders' collaboration on open innovation in healthcare. Int J Multidiscip Res
119. Day-Duro E, Lubitsh G, Smith G (2020) Understanding and investing in healthcare innovation and collaboration. J Health Organ Manag 34:469–487
120. Nambisan S, Nambisan P (2017) How should organizations promote equitable distribution of benefits from technological innovation in health care? AMA J Ethics 19:1106–1115
121. Wilson S, Tolley C, Ardle RM, Lawson L, Beswick E, Hassan N, Slight R, Slight S (2024) Recommendations to advance digital health equity: a systematic review of qualitative studies. NPJ Digit Med 7:173
122. Crawford A, Serhal E (2020) Digital health equity and Covid-19: the innovation curve cannot reinforce the social gradient of health. J Med Internet Res 22:e19361
123. Kolotouchkina O, Barroso CL, Sánchez JLM (2022) Smart cities, the digital divide, and people with disabilities. Cities 123:103613
124. Reddy H, Joshi S, Joshi A, Wagh V (2022) A critical review of global digital divide and the role of technology in healthcare. Cureus 14

125. Bakken S, Sue Marden S, Arteaga S, Grossman L, Keselman A, Le P-T, Creber RM, Powell-Wiley TM, Schnall R, Tabor D (2019) Behavioral interventions using consumer information technology as tools to advance health equity. Am J Public Health 109:S79–S85
126. Iyamu I, Gómez-Ramírez O, Xu AXT, Chang H-J, Watt S, Mckee G, Gilbert M (2022) Challenges in the development of digital public health interventions and mapped solutions: findings from a scoping review. Digit Health 8:20552076221102255
127. Dias FM, Martens ML, de Paula Monken SF, da Silva LF, Santibanez-Gonzalez EDR (2021) Risk management focusing on the best practices of data security Systems for Healthcare. Int J Innov 9:45–78
128. Seh AH, Zarour M, Alenezi M, Sarkar AK, Agrawal A, Kumar R, Khan RA (2020) Healthcare data breaches: insights and implications. Healthcare (Basel) 8
129. Ambarkar SS, Shekokar N (2020) Toward smart and secure Iot based healthcare system. In: Internet of things, smart computing and technology: a roadmap ahead, pp 283–303
130. ElSayed Z, Elsayed N, Bay S (2024) A novel zero-trust machine learning green architecture for healthcare Iot cybersecurity: review, analysis, and implementation. SoutheastCon 2024:686–692
131. Mishra A (2023) Analysis of approaches for identifying threats in the internet of medical things (Iomt) and exploration of research Frontiers. Paper presented at the 2023 IEEE International Conference on ICT in Business Industry & Government (ICTBIG)
132. Williams PAH, McCauley V (2016) Always connected: the security challenges of the healthcare internet of things. Paper presented at the 2016 IEEE 3rd World Forum on Internet of Things (WF-IoT)
133. Pan J, Ding S, Desheng W, Yang S, Yang J (2019) Exploring behavioural intentions toward smart healthcare services among medical practitioners: a technology transfer perspective. Int J Prod Res 57:5801–5820
134. Zayyad MA, Toycan M (2018) Factors affecting sustainable adoption of E-health technology in developing countries: an exploratory survey of Nigerian hospitals from the perspective of healthcare professionals. PeerJ 6:e4436
135. Al-Jaroodi J, Mohamed N, Kesserwan N, Jawhar I (2023) Human factors affecting the adoption of healthcare 4.0. Paper presented at the 2023 IEEE International Systems Conference (SysCon)
136. Renukappa S, Mudiyi P, Suresh S, Abdalla W, Subbarao C (2022) Evaluation of challenges for adoption of smart healthcare strategies. Smart Health 26:100330
137. Safi S, Thiessen T, Schmailzl KJG (2018) Acceptance and resistance of new digital technologies in medicine: qualitative study. JMIR Res Protoc 7:e11072
138. De Rosis S, Nuti S (2018) Public strategies for improving Ehealth integration and long-term sustainability in public health care systems: findings from an Italian case study. Int J Health Plann Manag 33:e131–ee52
139. Zhang L, Ren J, Yuan H, Yang Z, Wang W, Guo M, Cheng G, Zhou L, Tao S, Zhang L (2020) Evaluation of smart healthcare systems and novel Uv-oriented solution for integration, resilience, inclusiveness and sustainability. Paper presented at the 2020 5th International Conference on Universal Village (UV)
140. Demirkan H (2013) A smart healthcare systems framework. IT Prof 15:38–45
141. Khan HH, Malik MN, Zafar R, Goni FA, Chofreh AG, Klemeš JJ, Alotaibi Y (2020) Challenges for sustainable smart city development: a conceptual framework. Sustain Dev 28:1507–1518
142. Rocha P, Nelson AD, Santinha G, Rodrigues M, Queirós A, Rodrigues C (2019) Smart cities and healthcare: a systematic review. Technologies 7:58
143. Alshamrani M (2022) Iot and artificial intelligence implementations for remote healthcare monitoring systems: a survey. J King Saud Univ-Comput Inform Sci 34:4687–4701
144. Cook DJ, Duncan G, Sprint G, Fritz RL (2018) Using smart city technology to make healthcare smarter. Proc IEEE 106:708–722

145. Nikolaeva A (2024) Smart cities and (smart) cycling: exploring the synergies in Copenhagen and Amsterdam. J Urban Technol:1–21
146. Poulton M, Roussos G (2013) Towards smarter metropolitan emergency response. Paper presented at the 2013 IEEE 24th Annual International Symposium on Personal, Indoor, and Mobile Radio Communications (PIMRC)
147. Patsakis C, Papageorgiou A, Falcone F, Solanas A (2015) S-health as a driver towards better emergency response systems in urban environments. Paper presented at the 2015 IEEE International Symposium on Medical Measurements and Applications (MeMeA) Proceedings
148. Jung C, Awad J, Chohan A (2021) The planning of smart elderly housing in Dubai with Iot technologies. Open House Int 46:668–681
149. Saraubon K, Anurugsa K, Kongsakpaibul A (2018) A smart system for elderly care using Iot and mobile technologies. Paper presented at the Proceedings of the 2018 2nd International Conference on Software and e-Business
150. Mulero R, Almeida A, Azkune G, Mainetti L, Mighali V, Patrono L, Rametta P, Sergi I (2017) An Aal system based on Iot technologies and linked open data for elderly monitoring in smart cities. Paper presented at the 2017 2nd international multidisciplinary conference on computer and energy science (SpliTech)
151. Nabi AUL, Snoussi T, Abdalkareem ZA (2023) A review of Iot convergence in healthcare and smart cities: challenges, innovations, and future perspectives. Babylon J Internet Things 2023:23–30
152. Renukappa S, Suresh S, Shetty N, Gandhi L, Abdalla W, Yabbati N, Hiremath R (2024) The role of smart cities in managing the Covid-19 outbreak in India. Smart Sustain Built Environ
153. Wang T, Bhuiyan MZA, Guojun Wang M, Rahman A, Jie W, Cao J (2018) Big data reduction for a smart city's critical infrastructural health monitoring. IEEE Commun Mag 56:128–133
154. Qiu J, Liang X, Shetty S, Bowden D (2018) Towards secure and smart healthcare in smart cities using Blockchain. Paper presented at the 2018 IEEE international smart cities conference (ISC2)
155. Aldama-Nalda A, Chourabi H, Pardo TA, Gil-Garcia JR, Mellouli S, Scholl HJ, Alawadhi S, Nam T, Walker S. 2012. Smart cities and service integration initiatives in North American cities: a status report. Paper presented at the Proceedings of the 13th annual international conference on digital government research
156. Zulu EM, Beguy D, Ezeh AC, Bocquier P, Madise NJ, Cleland J, Falkingham J (2011) Overview of migration, poverty and health dynamics in Nairobi city's slum settlements. J Urban Health 88:185–199
157. Jawhari B, Berenger B, Saleh A (2015) Kibera medical record initiative: barriers of implementing an electronic medical record system. In: Digital healthcare empowering Europeans. IOS Press, pp 1031–1032
158. van de Vijver S, Oti S, Tervaert TC, Hankins C, Kyobutungi C, Gomez GB, Brewster L, Agyemang C, Lange J (2013) Introducing a model of cardiovascular prevention in Nairobi's slums by integrating a public health and privatesector approach: the SCALE-UP study. Glob Health Action. 21;6:22510. https://doi.org/10.3402/gha.v6i0.22510. PMID: 24149078; PMCID: PMC3805842.
159. van de Vijver S, Oti S, Moll E, van Charante S, Allender CF, Lange J, Oldenburg B, Kyobutungi C, Agyemang C (2015) Cardiovascular prevention model from Kenyan slums to migrants in the Netherlands. Glob Health 11:1–6
160. Aboderin I, Kano M, Owii HA (2017) Toward "age-friendly slums"? Health challenges of older slum dwellers in Nairobi and the applicability of the age-friendly city approach. Int J Environ Res Public Health 14:1259
161. Arnold C, Theede J, Gagnon A (2014) A qualitative exploration of access to urban migrant healthcare in Nairobi, Kenya. Soc Sci Med 110:1–9
162. Gruebner O, Sachs J, Nockert A, Frings M, Khan MMH, Lakes T, Hostert P (2014) Mapping the slums of Dhaka from 2006 to 2010. Dataset Papers in Science 2014

163. Naydenova E, Raghu A, Ernst J, Sahariah SA, Gandhi M, Murphy G (2018) Healthcare choices in Mumbai slums: a cross-sectional study. Wellcome Open Res 2:115
164. Brault MA, Ngure K, Haley CA, Kabaka S, Sergon K, Desta T, Mwinga K, Vermund SH, Kipp AM (2017) The introduction of new policies and strategies to reduce inequities and improve child health in Kenya: a country case study on progress in child survival, 2000–2013. PLoS One 12:e0181777
165. Bukachi F, Pakenham-Walsh N (2007) Information technology for health in developing countries. Chest 132:1624–1630
166. Wahl B, Lehtimaki S, Germann S, Schwalbe N (2020) Expanding the use of community health workers in urban settings: a potential strategy for progress towards universal health coverage. Health Policy Plan 35:91–101
167. Ejeta LT, Leta Y, Abuye M, Yasin C, Tebekaw Y, Giday T, Tilahun HA (2020) Implementing the urban community health information system in Ethiopia: lessons from the pilot-tests in Addis Ababa, Bishoftu and Hawassa. Ethiop J Health Dev 34
168. Maketa V, Vuna M, Baloji S, Lubanza S, Hendrickx D, Inocêncio RA, da Luz M, Boelaert, and Pascal Lutumba. (2013) Perceptions of health, health care and community-oriented health interventions in poor urban communities of Kinshasa, Democratic Republic of Congo. PLoS One 8:e84314
169. Marcil L, Afsana K, Perry HB (2016) First steps in initiating an effective maternal, neonatal, and child health program in urban slums: the Brac Manoshi project's experience with community engagement, social mapping, and census taking in Bangladesh. J Urban Health 93:6–18
170. Nyonator FK, Akosa AB, Koku Awoonor-Williams J, Phillips JF, Jones TC (2007) Scaling up experimental project success with the community-based health planning and services initiative in Ghana. In: Scaling up health service delivery: from pilot innovations to policies and programmes, pp 89–112
171. Konduri N, Aboagye-Nyame F, Mabirizi D, Hoppenworth K, Kibria MG, Doumbia S, Williams L, Mazibuko G (2018) Digital health technologies to support access to medicines and pharmaceutical services in the achievement of sustainable development goals. Digit Health 4:2055207618771407
172. Till S, Mkhize M, Farao J, Shandu LD, Muthelo L, Coleman TL, Mbombi M, Bopape M, Klingberg S, van Heerden A (2023) Digital health technologies for maternal and child health in Africa and other low-and middle-income countries: cross-disciplinary scoping review with stakeholder consultation. J Med Internet Res 25:e42161
173. Ezeh A, Mberu B (2019) Case studies in urban health. In: Urban health. Oxford University Press
174. Adongo PB, Phillips JF, Aikins M, Arhin DA, Schmitt M, Nwameme AU, Tabong PT-N, Binka FN (2014) Does the design and implementation of proven innovations for delivering basic primary health care services in rural communities fit the urban setting: the case of Ghana's Community-Based Health Planning and Services (Chps). Health Res Policy Syst 12:1–10
175. Agarwal S, Satyavada A, Patra P, Kumar R (2008) Strengthening functional community–provider linkages: lessons from the Indore urban health programme. Glob Public Health 3:308–325
176. Hodgins S (2020) Learning from community health worker programs, big and small. Global Health Sci Pract 8:147–149
177. Debnath S (2023) Integrating information technology in healthcare: recent developments, challenges, and future prospects for urban and regional health. arXiv preprint arXiv:2307.16296
178. Tsourounis M, Stuart J, Pignato W, Toscani M, Barone J (2015) Current trends in personalized medicine and companion diagnostics: a summary from the Dia meeting on personalized medicine and companion diagnostics. Ther Innov Regul Sci 49:530–543
179. Alterovitz G, Warner J, Zhang P, Chen Y, Ullman-Cullere M, Kreda D, Kohane IS (2015) Smart on Fhir genomics: facilitating standardized clinico-genomic apps. J Am Med Inform Assoc 22:1173–1178

180. Roberts MC, Mensah GA, Khoury MJ (2019) Leveraging implementation science to address health disparities in genomic medicine: examples from the field. Ethn Dis 29:187
181. Xu J, Yang P, Xue S, Sharma B, Sanchez-Martin M, Wang F, Beaty KA, Dehan E, Parikh B (2019) Translating cancer genomics into precision medicine with artificial intelligence: applications, challenges and future perspectives. Hum Genet 138:109–124
182. Stark Z, Dolman L, Manolio TA, Ozenberger B, Hill SL, Caulfied MJ, Levy Y, Glazer D, Wilson J, Lawler M (2019) Integrating genomics into healthcare: a global responsibility. Am J Hum Genet 104:13–20
183. Ngossaha M, Justin YA, Bamana AB, Rodrigue D, Samuel B, Etémé A (2024) Towards a flexible urbanization based approach for integration and interoperability in heterogeneous health information systems: case of Cameroon. In: Lecture notes of the Institute for Computer Sciences, vol 566, pp 258–275
184. Seneviratne O (2023) Enabling data interoperability for decentralized, smart, and connected health applications. Paper presented at the 2023 IEEE/ACM Conference on Connected Health: Applications, Systems and Engineering Technologies (CHASE)
185. Klein GO (2002) Standardization of health informatics-results and challenges. Yearb Med Inform 11:103–114
186. Wang J, Liu C, Zhou L, Jiangpei X, Wang J, Sang Z (2022) Progress of standardization of urban infrastructure in smart city. Standards 2:417–429
187. Lindner R, Jaca C, Hernantes J (2021) A good practice for integrating stakeholders through standardization—the case of the smart mature resilience project. Sustainability 13:9000

Smart Healthcare Solutions for Urban Health Development

Mohamed Eddabbah, CharafEddine Aitzaouiat, Nadia Hachoumi, Kamal Takhdat, and Azidine Guezzaz

Abstract Urbanization presents both opportunities and significant challenges to healthcare systems globally. Rapid urban growth increases the demand for accessible, efficient, and high-quality healthcare, straining traditional models. This chapter employs a PESTEL analysis—evaluating Political, Economic, Social, Technological, Environmental, and Legal factors—to examine how external forces shape smart healthcare solutions. By leveraging advanced technologies such as Artificial Intelligence (AI), the Internet of Everything (IoE), telemedicine, Big Data, and cloud computing, urban healthcare systems can enable real-time monitoring, personalized care, and data-driven decision-making, enhancing patient outcomes and streamlining delivery. Integrating wearables, mobile apps, and remote diagnostics supports early interventions and precision medicine, improving responsiveness to individual and public health needs. However, challenges persist, including patient concerns about usability and privacy, resistance among healthcare professionals requiring retraining, and infrastructure limitations like poor connectivity. Cybersecurity risks, including data breaches, further complicate adoption. Strategies such as encryption, secure cloud storage, and multi-factor authentication are crucial to mitigating these risks. Case studies illustrate successful implementations, emphasizing how smart healthcare optimizes resource use, reduces disparities, and expands care access in urban settings. The chapter concludes with recommendations for policymakers and developers to create sustainable and equitable healthcare ecosystems tailored to urban demands.

M. Eddabbah (✉) · A. Guezzaz
LaRTID laboratory, Higher School of Technology of Essaouira, Cadi Ayyad University, Marrakesh, Morocco
e-mail: m.eddabbah@uca.ac.ma

C. Aitzaouiat
Polydisciplinary Faculty of Sidi Bennour, Chouaîb Doukkali University, El Jadida, Morocco

N. Hachoumi · K. Takhdat
High Institute of Nursing Professions and Health Techniques, Ministry of Health, Marrakesh, Morocco

H. Kaddouri et al. (eds.), *Smart Cities for Sustainable and Urban Development*, Green Energy and Technology, https://doi.org/10.1007/978-3-031-99001-4_33

Keywords Urbanization · E-healthcare · Public health management · Big data · Artificial intelligence · Predictive analytics · Precision medicine · Emergency disaster and pandemic management · Digital maturity · Cybersecurity · Blockchain

1 Introduction

1.1 Urbanization and Healthcare Challenges

Rapid urbanization is a key concern of this decade, bringing forth unique challenges for healthcare systems (Fig. 1), particularly in highly populated urban areas [1]. This growth is largely fuelled by rural-to-urban migration, natural population increase, and the transformation of rural regions into urban zones, all of which place substantial pressure on healthcare infrastructure and services. This accelerated urban growth is also fuelled by the promise of economic opportunities, better access to education, and the perception of improved quality of life in cities [2]. As urban populations swell, traditional healthcare models are increasingly strained, struggling to meet the growing demand for medical services. This demographic shift is driving an urgent need for healthcare solutions that are more accessible, efficient, and of high quality. Urbanization has negative consequences on health due mainly to pollution and overcrowded living conditions. These environmental factors exacerbate existing health issues and create new ones, further burdening urban healthcare systems [3]. Additionally, new lifestyles in cities significantly affect health outcomes. Urban dwellers often engage in less physical activity due to sedentary

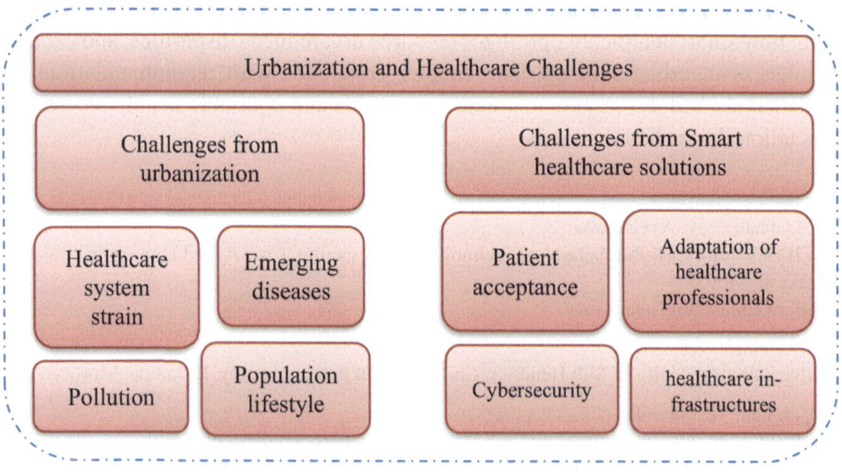

Fig. 1 Urbanization and healthcare challenges

jobs, reliance on transportation, and limited access to safe outdoor spaces. Nutrition habits also change, with increased consumption of processed and fast foods, leading to diet-related health issues. Importantly, the fast-paced urban life, coupled with work pressures, social isolation, and environmental stressors, contributes to significantly higher levels of stress among city residents. This chronic stress has far-reaching implications for both physical and mental health. Moreover, the apparition of new diseases due to urbanization and industrial activities adds another layer of complexity to urban health challenges [4]. Urban healthcare providers must navigate the complexities of serving larger, more diverse populations while contending with limited resources and infrastructure, as well as addressing the unique health challenges posed by urban environments, lifestyle changes, stress-related disorders, and emerging illnesses. The intersection of these factors—rapid urban growth and its underlying causes, population increase, outdated healthcare models, the demand for improved care, the health impacts of urbanization itself, lifestyle-related health issues, chronic stress, and the emergence of new diseases—creates a pressing dilemma for policymakers, healthcare professionals, and urban planners alike [5]. Addressing these challenges requires innovative approaches to healthcare delivery, potentially including Smart Healthcare Solutions, community-based care initiatives, stress management programs, and the strategic use of technology to enhance both access and efficiency in urban healthcare settings, while also tackling the root causes of urban health issues such as air pollution, inadequate housing, sedentary lifestyles, poor nutrition, chronic stress, and the health risks associated with industrial activities.

1.2 The Role of Smart Healthcare Solutions

Smart healthcare solutions are emerging as crucial tools in addressing the complex health challenges posed by rapid urbanization. These technologies, ranging from telemedicine and AI-assisted diagnostics to wearable devices and big data analytics, are transforming the landscape of urban healthcare delivery [6]. By improving access to care, enhancing efficiency, and enabling more personalized and preventive approaches, smart solutions help alleviate the strain on overburdened urban health systems. Importantly, these innovations play a significant role in reducing healthcare costs, both for providers and patients. Telemedicine reduces the need for physical infrastructure and staff in certain scenarios, while AI and automation streamline operations and minimize costly errors [7]. Preventive care enabled by wearable devices and early intervention systems can significantly lower long-term health expenses by catching issues before they become severe [8]. These smart solutions offer innovative ways to monitor and manage public health in densely populated areas, from tracking disease outbreaks to monitoring environmental health risks, often at a fraction of the cost of traditional methods. Moreover, these technologies empower patients with greater control over their health information and

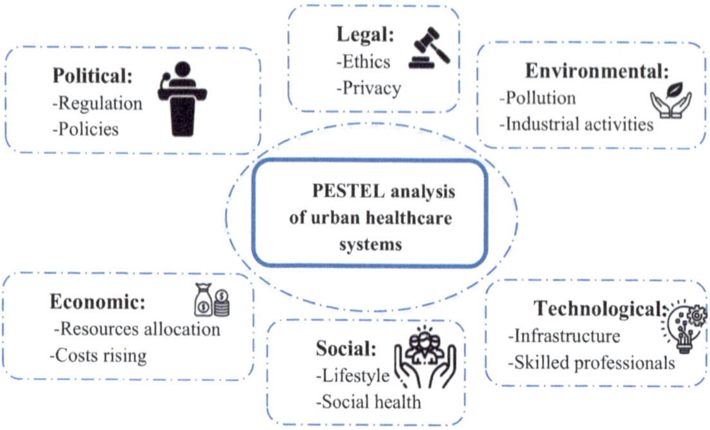

Fig. 2 PESTEL analysis of urban healthcare systems

decision-making, potentially leading to more cost-effective health choices. However, the implementation of smart healthcare solutions in urban settings also raises important considerations about data privacy, equitable access, and the balance between technological efficiency and the human touch in healthcare. As cities continue to grow and evolve, the integration of these smart technologies will play a pivotal role in shaping the future of urban health, potentially leading to more resilient, responsive, cost-effective, and patient-centered healthcare systems.

To comprehensively understand the multifaceted issues surrounding urban health, we employed the PESTEL analysis framework (Fig. 2). This approach examines Political, Economic, Social, Technological, Environmental, and Legal factors influencing urban healthcare systems, offering a structured perspective to identify key challenges and potential solutions. Below is a detailed breakdown of these factors as they pertain to the text.

1.2.1 Political

Urbanization often places governments under immense pressure to implement policies capable of addressing overburdened healthcare systems [9]. Regulatory frameworks must prioritize the expansion of access to healthcare services in urban environments while integrating innovative solutions such as Smart Healthcare to ensure these systems can meet growing demand. Political stability and a strong governmental commitment to healthcare reforms are essential for effectively addressing the complex health challenges posed by rapid urbanization. Governments need to enact targeted policies to tackle critical issues such as air pollution, inadequate housing, and the health risks associated with industrial activities [10]. These measures are fundamental to mitigating the adverse health impacts of urban living and fostering a healthier urban population.

1.2.2 Economic

Urban growth, often driven by the promise of economic opportunities, places significant strain on the financial resources allocated to healthcare infrastructure [1]. As cities grow, investment in healthcare systems must keep pace with population expansion to avoid service degradation and ensure equitable access to care. The economic dynamics of urban life also contribute to healthcare challenges. Sedentary urban jobs and reliance on industries that exacerbate pollution drive to rising costs associated with treating chronic diseases and stress-related illnesses [11]. Addressing healthcare disparities in underserved urban areas will require substantial funding for initiatives such as subsidized care programs, the adoption of digital healthcare technologies, and comprehensive community outreach efforts.

1.2.3 Social

The urban lifestyle drives significant changes in health behaviors, including decreased physical activity due to sedentary jobs, reliance on transportation, and limited access to safe outdoor spaces. These changes often lead to higher rates of obesity, cardiovascular diseases, and other chronic conditions [11]. Overcrowding and social isolation in urban environments further compound health challenges by increasing stress levels, which negatively affect both mental and physical health. Rapid population growth in cities exacerbates healthcare inequities, particularly for marginalized groups. Promoting community-based care initiatives and raising public awareness of preventive health measures are vital to mitigating these social health issues and improving urban health outcomes.

1.2.4 Technological

Smart healthcare solutions, such as telemedicine, wearable devices, and data-driven healthcare planning, are pivotal in efficiently addressing urban health challenges. These technologies optimize resource utilization, enhance service delivery, and facilitate early disease detection, making healthcare systems more responsive to urban needs [12]. However, outdated healthcare models and infrastructure present significant barriers to the adoption of these innovations. Upgrading systems to incorporate technologies like AI and IoT requires not only significant investment but also the availability of skilled professionals who can effectively operate and maintain these advanced systems.

1.2.5 Environmental

Urbanization exacerbates environmental health issues, including air pollution, overcrowding, and limited access to green spaces, all of which directly impact public health. These environmental factors often intensify pre-existing health problems while also contributing to the emergence of new diseases [13]. Industrial activities in urban areas further complicate health challenges by introducing pollutants and creating additional health risks. To create sustainable urban healthcare systems, efforts should focus on mitigating environmental stressors and adopting eco-friendly practices that address the root causes of health issues stemming from urbanization.

1.2.6 Legal

Rapid urban growth necessitates the development of clear regulatory frameworks to ensure the ethical and secure integration of new healthcare technologies. Policies must emphasize patient data privacy, robust security protocols, and equitable access to care to maintain trust in modern healthcare systems [14]. Legal measures should also address broader urban health concerns, such as regulating industrial emissions and improving urban planning. These measures are essential for reducing pollution-related health risks. Furthermore, ensuring compliance with health and safety standards in urban industries can significantly lower the prevalence of occupational diseases and injuries.

This chapter explores the transformative potential of smart healthcare technologies in addressing the unique challenges posed by rapid urbanization (Fig. 3), by

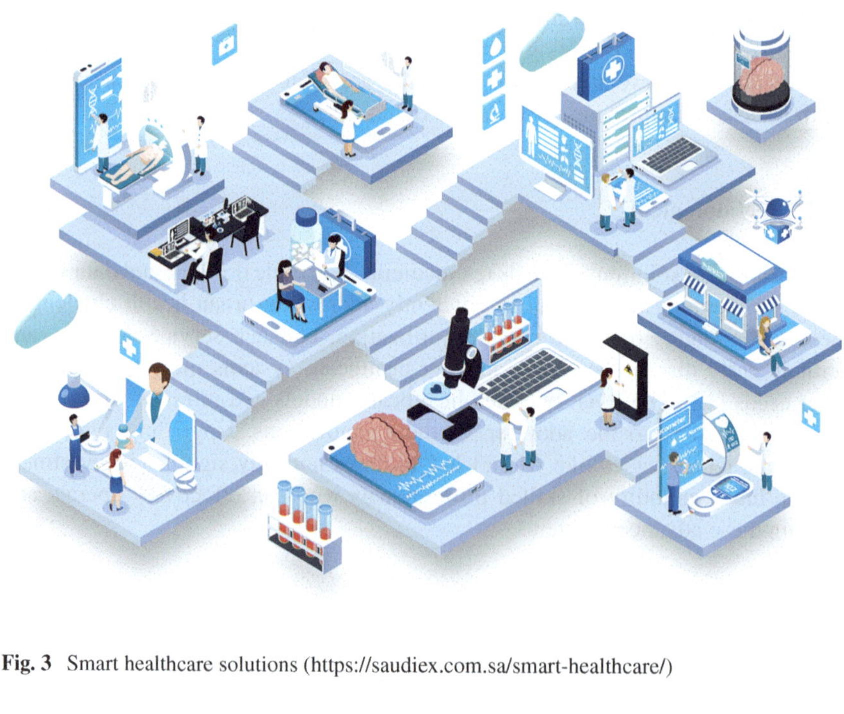

Fig. 3 Smart healthcare solutions (https://saudiex.com.sa/smart-healthcare/)

outlining the challenges of adoption, including infrastructure limitations, patient and professional readiness, and the importance of cybersecurity in safeguarding healthcare data. It provides also an in-depth examination of key technologies such as AI, Internet of Everything (IoE), wearable devices, telemedicine, blockchain, and big data analytics, among others, and their applications in urban healthcare settings. The role of these innovations in improving access to care, optimizing resource allocation, enhancing operational efficiency in hospitals, and ensuring data security through advanced cybersecurity measures is also discussed. Additionally, the chapter addresses the critical need for robust digital infrastructure, the digital maturity of patients and healthcare professionals, and the importance of public-private partnerships in fostering the adoption of these technologies. Case studies and real-world examples are provided to highlight successful implementations in both developed and developing urban environments, offering strategic recommendations for the effective integration of smart healthcare systems in urban setting.

2 Challenges and Obstacles of Urban Healthcare Development

2.1 Digital Maturity of the Healthcare Ecosystem

Digital maturity in the urban healthcare ecosystem can be defined as the comprehensive adoption and integration of technology across all levels of healthcare operations, so to enable more efficient and effective service delivery [15]. This maturity is seen in the transformation of processes and management methods, where administrative tasks, such as patient record management, appointment scheduling, and billing, are automated and digitized, reducing human error and improving efficiency. In terms of logistics, digital maturity enhances the entire logistics chain, from procurement to the distribution of medical supplies, ensuring real-time tracking of resources, better inventory management, and faster response times during critical shortages or emergencies. This integration of technology extends across the healthcare organization's hierarchy, from senior management to frontline healthcare providers. Leadership is responsible for driving strategic adoption of technology, while clinicians and support staff must adapt to new digital tools and workflows. This digital transformation enables seamless communication between departments, better coordination of care, and more informed decision-making, ultimately leading to higher quality patient care. A digitally mature healthcare ecosystem is better equipped to respond to the complexities of urban health demands, ensuring scalability, efficiency, and resilience in the face of growing challenges [16].

2.2 Patient Acceptance of New Technologies and Healthcare Professionals' Adaptation

The integration of new technologies in healthcare often encounters significant barriers, particularly regarding patient acceptance and the adaptation of healthcare professionals (Fig. 4) [17]. Trust plays a pivotal role in this acceptance; patients may hesitate to embrace technologies such as telemedicine or wearable health devices due to concerns about privacy, data security, and the impersonal nature of virtual interactions. They often fear that digital solutions might compromise the quality of care or lead to misunderstandings, which can further erode their confidence in these innovations [18]. For healthcare professionals, the challenge lies in adapting to unfamiliar technologies. Many practitioners may resist changes to their established workflows, fearing that new systems will disrupt their routines or diminish their role in patient care. To combat this resistance, comprehensive training and re-skilling initiatives are essential. These programs can equip healthcare workers with the necessary skills to navigate new technologies confidently and effectively. There is a critical need to balance human expertise with automation [19]. While technologies such as artificial intelligence can enhance diagnostics and treatment planning, they cannot replace the nuanced understanding and empathy that healthcare professionals provide [20]. Thus, integrating technology should aim to augment human capabilities rather than replace them, fostering a collaborative environment where both human and digital resources contribute to improved patient outcomes. This holistic approach can lead to greater acceptance among patients and professionals alike, ultimately facilitating the successful implementation of smart healthcare solutions. The acceptance of new technologies in healthcare involves complex dynamics from both patients and healthcare professionals. Transitioning to digital health solutions like telemedicine, electronic health records (EHRs), and wearable devices requires addressing challenges related to trust, usability, and adoption. Trust issues often arise due to concerns about data privacy, security breaches, and the misuse of

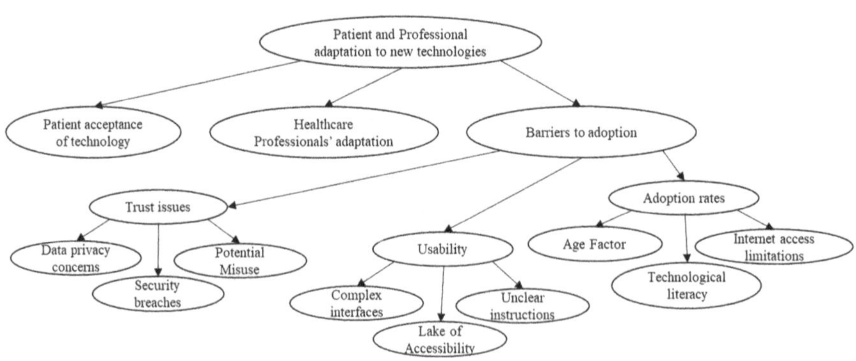

Fig. 4 Barriers to patient and professionals' engagement with new technologies

personal health information, with reported breaches heightening skepticism. Usability is another critical factor, as patients are less likely to engage with digital health platforms that are difficult to navigate, lack accessibility features, or have unclear instructions. Additionally, adoption rates vary based on factors like age, technological literacy, and prior experience with technology. Older adults and individuals with limited access to the internet or smartphones may face significant barriers to adoption, emphasizing the need for inclusive, user-friendly designs and robust privacy protections to foster widespread acceptance.

Addressing privacy, data security, and human interaction concerns is essential for fostering trust and acceptance in digital health. Robust security measures like encryption and multi-factor authentication, coupled with transparent communication about data use, are critical to alleviating patient fears [21]. To counter the perception of impersonal care, providers can emphasize empathy in virtual interactions, ensuring patients feel valued. Resistance to change among healthcare professionals often stems from unfamiliarity with technologies and workload concerns, necessitating comprehensive training programs, peer support, and hands-on experience [22]. Balancing human expertise with automation requires designing tools that augment rather than replace professional skills, fostering interdisciplinary collaboration, and maintaining patient-centric approaches. By addressing these challenges, healthcare systems can seamlessly integrate new technologies to enhance care and efficiency.

2.3 Infrastructure Limitations

The integration of smart healthcare technologies in urban areas relies heavily on robust digital infrastructure, yet various limitations pose significant challenges to their implementation. Connectivity issues are particularly acute in densely populated urban environments, where network congestion often leads to slow internet speeds and unreliable connections [23] These disruptions hinder real-time applications like telemedicine, remote monitoring, and emergency response systems. For example, wearable devices tracking patients' vital signs may fail to transmit critical data promptly, delaying essential care. Furthermore, weak cellular signals or inadequate broadband access in some urban pockets exacerbate disparities in healthcare delivery. Beyond connectivity, many healthcare facilities grapple with outdated digital infrastructure and equipment shortages. Legacy systems, often incompatible with modern technologies, impede the adoption of advanced solutions like AI-driven analytics and cloud-based platforms. For instance, hospitals reliant on outdated EHR systems may struggle to integrate predictive analytics, limiting their capacity to optimize patient outcomes. The unavailability of essential tools, such as telehealth kiosks or remote monitoring devices, further restricts the deployment of smart healthcare solutions, particularly in underfunded areas. These limitations severely impact the scalability and effectiveness of smart

Table 1 Infrastructure limitations and their impact on smart healthcare integration

Infrastructure limitation	Issue	Solution	Impact
Connectivity issues in dense urban environments	High population density can lead to network congestion, resulting in slower internet speeds and unreliable connections, which is problematic for real-time applications like telemedicine and emergency response.	Wearable devices tracking vital signs may fail to communicate in real-time; areas with weak cellular signals face adoption hurdles	Delays in critical care, limits on real-time monitoring, and exacerbation of health disparities in underserved areas
Lack of necessary digital infrastructure and equipment availability	Many urban healthcare facilities lack adequate infrastructure, including outdated hardware, insufficient software, and systems that cannot integrate with new technologies like AI and analytics. Limited availability of essential medical equipment in lower-income areas also restricts technology use.	Legacy EHR systems incompatible with AI platforms; limited telehealth kiosks in lower-income neighborhoods	Inability to leverage data for predictive analytics, limited access to advanced care, and inefficiency in resource management
Impact on the scalability of smart healthcare solutions	Without a solid digital foundation, the scalability of smart healthcare is compromised, limiting the reach and effectiveness of programs like telemedicine. Infrastructure limitations create data silos that hinder collaboration, leading to duplicated tests, inconsistent treatments, and higher costs.	Telemedicine program reach limited by weak internet; inconsistent data flow affects remote monitoring	Lower adoption rates, restricted scalability of healthcare solutions, and compromised quality and coordination of care due to data fragmentation

healthcare systems (Table 1). A telemedicine program, for instance, cannot reach its full potential without reliable internet infrastructure [24], while inconsistent data transmission undermines the reliability of remote patient monitoring. Such gaps also hinder collaboration among healthcare providers, leading to data silos that compromise care quality. Fragmented systems result in duplicated tests, inconsistent treatments, and increased costs, obstructing the creation of a connected healthcare ecosystem. Addressing these barriers requires substantial investment in digital infrastructure, including expanding broadband access in underserved areas and upgrading medical equipment and IT systems. Promoting interoperability between disparate health information systems is also critical to enabling seamless data exchange. By fostering a robust digital foundation, urban healthcare systems can support innovative technologies, enhance care delivery, and achieve better patient outcomes.

2.4 Cybersecurity Risks

As healthcare systems increasingly embrace digital technologies and interconnected networks, the risk of cybersecurity threats has become a significant concern. Protecting sensitive patient data and ensuring the integrity of healthcare operations are paramount to maintaining trust and delivering quality care. Healthcare organizations face various types of cyberattacks that expose patient safety, disrupt operations, and incur significant financial and reputational losses (Fig. 5). One of the most prevalent threats is ransomware attacks, where cybercriminals encrypt critical data and demand a ransom for its release. These attacks can paralyze hospital systems, delay patient care, and result in steep costs. A notable example is the 2020 ransomware attack on Universal Health Services (UHS), which caused widespread system outages across its facilities [25].

In addition to paying ransoms, organizations often face regulatory fines and legal liabilities for compromised data. Phishing attacks are another major concern, involving deceptive emails or messages designed to trick recipients into divulging sensitive information like passwords or financial details. These attacks leverage social engineering tactics, often using personalized information to enhance their credibility and gain unauthorized access to healthcare systems. Similarly, Distributed Denial of Service (DDoS) attacks overwhelm healthcare networks with excessive traffic, rendering systems inaccessible and potentially disrupting patient care. DDoS attacks can also serve as diversions, enabling hackers to steal data unnoticed during the chaos. Data breaches remain a significant challenge, arising from hacking, social engineering, or insider threats. These breaches expose confidential patient information, risking identity theft, fraud, and a loss of patient trust. Organizations suffer reputational damage and legal consequences, including class-action lawsuits. Meanwhile, man-in-the-middle attacks intercept and alter communications between

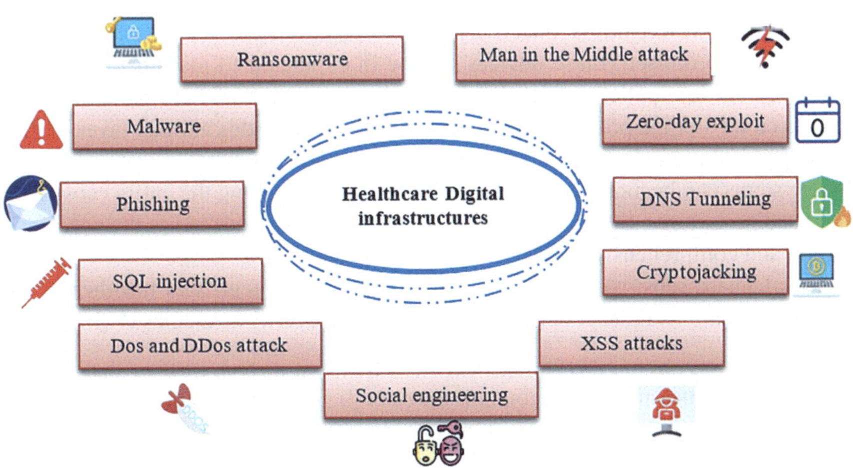

Fig. 5 Cyber attacks in healthcare digital infrastructures

parties, leading to dire consequences, such as tampered medical prescriptions or manipulated patient records that could endanger lives. Malware attacks involve malicious software, such as viruses or spyware, that infiltrate healthcare systems, corrupt data, and compromise functionality [26]. The rise of IoT devices in healthcare increases the risk of malware infections, particularly when these devices lack robust security measures. Social engineering is another significant risk comes from insider threats, where employees or contractors with system access may either intentionally exploit their privileges or unintentionally cause harm through negligence. Insider threats are further amplified by the prevalence of remote work, as employees may access sensitive data from insecure networks [27]. Addressing these threats requires healthcare organizations to adopt comprehensive cybersecurity strategies [28]. These should include regular employee training, advanced threat detection systems, robust encryption protocols, and stringent access controls to safeguard sensitive data and ensure operational resilience.

In addition, the digitization of healthcare systems, including the widespread adoption of EHRs and advanced technologies, has exponentially increased the risk to data privacy and system integrity [29]. Healthcare organizations manage vast repositories of sensitive data, including personal identification, medical histories, and financial details, making them prime targets for cybercriminals. A successful data breach can result in identity theft, insurance fraud, and unauthorized access to medical records, with severe consequences for both patients and healthcare providers. Beyond privacy concerns, system integrity is also at risk. For example, if a malicious actor infiltrates a hospital's EHR system and alters a patient's medication records, it could lead to incorrect treatment, endangering the patient's life. Such tampering not only compromises patient safety but also damages the credibility and operational efficiency of healthcare institutions. The growing interconnectivity of healthcare ecosystems, fueled by the IoE and smart healthcare devices, introduces multiple vulnerabilities [30]. Many modern medical devices, such as insulin pumps, pacemakers, and monitoring equipment, rely on internet connectivity to function efficiently. However, this connectivity creates potential entry points for cyberattacks. If compromised, these devices can be manipulated to malfunction, posing significant risks to patient safety. Similarly, the integration of third-party services, such as cloud storage providers and telehealth platforms, adds another layer of vulnerability. If these external partners fail to adhere to robust cybersecurity practices, their systems can become weak links in the network, exposing the entire ecosystem to cascading failures. Such interconnected vulnerabilities underscore the need for a unified and comprehensive approach to cybersecurity.

To safeguard sensitive patient data and maintain the integrity of healthcare systems, organizations must adopt multi-layered cybersecurity strategies (Fig. 6) [31]:

- Robust Encryption: Encrypting data in transit and at rest ensures that even if intercepted, the information remains unreadable without proper decryption keys. This is particularly critical for protecting data transmitted over IoT devices and cloud platforms.

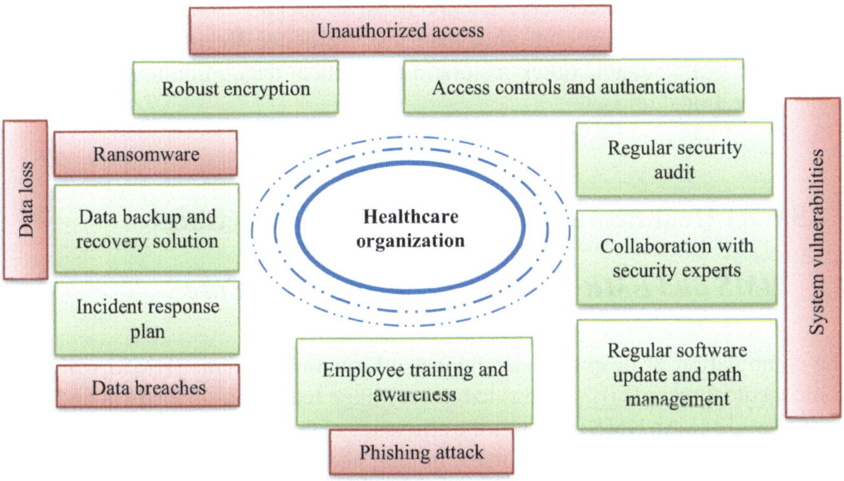

Fig. 6 Cybersecurity risk mitigation in healthcare

- Regular Security Audits: Routine audits and vulnerability assessments help identify weaknesses in healthcare systems. By staying proactive, organizations can reinforce their defenses against emerging threats and address system flaws before they are exploited.
- Employee Training and Awareness: Human error is a leading cause of data breaches. Regular training programs to educate healthcare staff on recognizing phishing attempts, adhering to strong password policies, and handling sensitive data securely are essential. Simulation exercises, such as mock phishing campaigns, can improve employee preparedness.
- Incident Response Plans: A robust incident response plan is critical for minimizing the impact of a breach. These plans should include protocols for detection, containment, eradication, and recovery, ensuring that healthcare services can resume quickly with minimal disruption.
- Access Controls and Multi-Factor Authentication (MFA): Implementing strict access controls ensures that only authorized personnel can access sensitive systems and data. MFA adds an additional layer of protection, while role-based access control (RBAC) limits data exposure based on job responsibilities.
- Collaboration with Cybersecurity Experts: Engaging cybersecurity consultants and joining information-sharing networks can help healthcare organizations stay updated on evolving threats and best practices. Such collaborations enhance an organization's ability to preempt and respond to cyber threats effectively.
- Regular Updates and Patch Management: Keeping all software, systems, and devices up to date is essential for addressing known vulnerabilities. A structured patch management schedule reduces the risk of exploitation by ensuring timely application of security updates.

- Data Backup and Recovery: Comprehensive backup strategies mitigate the effects of ransomware and other data loss incidents. Regular backups stored securely offline enable quick restoration of critical data, minimizing downtime and operational impact.

3 Smart Healthcare Technologies in Urban Settings

3.1 HMS and EMR Integration for Improved Care

The combination of hospital management information systems (HMS) with electronic medical records (EMR) has led to very huge reductions in medical and treatment errors [32]. An HMS improves the efficiency of the hospital through automating programming, allocation of resources, billing, and other administrative tasks. The EMR acts as a database of patient records that is digitally retrievable by all clinicians to provide current and precise information for making decisions about patients. During, for example, real-time access to patient history, allergies, medications, and lab results, a healthcare provider is much more likely to avoid treatment that presents potential harm due to drug interaction (Fig. 7) [33]. Hence, such integration optimizes the workflow of the clinic, reduces time to make a diagnosis, and leads to a much more personalized care plan and faster intervention for the patient, which results in better outcomes. Creating a better and more efficient health care environment would center on reduced administrative burdens and increasing much more accurate and easily accessible patient data, and therefore, the HMS-EMR synergism becomes apparent.

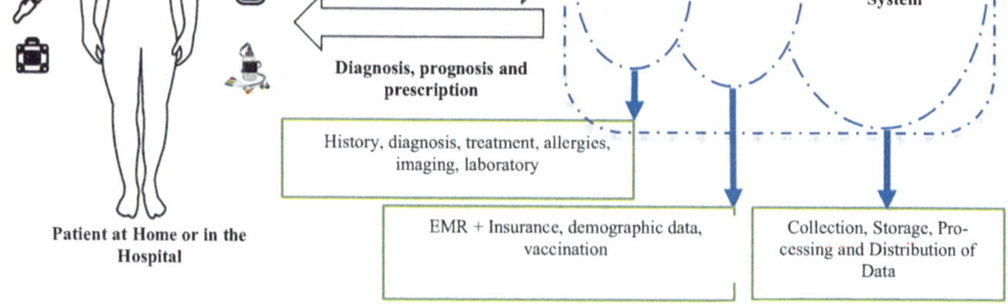

Fig. 7 Hospital information system

3.1.1 Electronic Medical Records (EMR)

EMR represent a significant shift from traditional paper-based patient records to digital formats, offering numerous advantages in terms of accuracy, efficiency, and accessibility. The transition from paper to digital records allows for the streamlined storage and retrieval of patient information, reducing errors that arise from illegible handwriting or misplaced files. EMRs improve patient care by providing healthcare providers with real-time access to comprehensive patient histories, including diagnoses, medications, allergies, and lab results, which are crucial for informed decision-making and personalized treatments [34, 35]. Also, EMRs enhance record accuracy by minimizing manual data entry errors and ensuring that all patient information is consistently updated and stored in a standardized format. This promotes better coordination among healthcare professionals, leading to improved continuity of care across different providers and facilities. EMRs can be integrated with advanced technologies such as AI, big data analytics, and cloud systems to further enhance patient care. AI can analyze EMR data to assist in diagnostics, predictive modeling, and treatment recommendations, while big data analytics can identify trends for public health planning and individualized care. Cloud integration ensures secure, accessible, and scalable storage of patient data, allowing for seamless collaboration between healthcare providers and faster retrieval of records when needed [36]. This comprehensive approach to EMRs is transforming the healthcare landscape, driving efficiency and improving the overall quality of patient care.

3.1.2 Hospital Management Systems (HMS)

HMS play a pivotal role in streamlining hospital operations by automating administrative processes and improving overall efficiency. These systems integrate key functions such as scheduling, billing, patient care, and resource management into a unified platform. Through automation, HMS reduces the burden of manual administrative tasks, enabling healthcare professionals to focus more on patient care and less on paperwork. The integration of scheduling tools within HMS allows for the efficient management of staff rosters, patient appointments, and surgical procedures, minimizing delays and ensuring optimal use of hospital resources. Automated billing systems simplify financial operations by processing payments, generating invoices, and managing insurance claims, all while reducing the risk of errors commonly associated with manual data entry [37]. HMS also enhances patient care by providing real-time access to patient records, treatment histories, and medical data, enabling healthcare providers to make informed decisions more quickly. In terms of resource management, HMS monitors the availability of critical supplies, equipment, and hospital beds, ensuring that these resources are allocated efficiently and avoiding shortages or delays in care.

3.1.3 Cloud Computing and Healthcare Data

Cloud computing has become a vital component in modern healthcare systems, offering transformative solutions for data storage, accessibility, and sharing. In healthcare, vast amounts of patient data are generated daily, ranging from medical histories and treatment plans to diagnostic images and lab results. Storing this data securely and ensuring it is readily accessible when needed is crucial for both healthcare providers and patients [38]. Cloud computing provides layers of healthcare solutions through Software as a Service (SaaS), Platform as a Service (PaaS), and Infrastructure as a Service (IaaS), each addressing distinct needs within healthcare operations (Fig. 8) [39]:

IaaS, healthcare organizations can rely on scalable and cost-efficient cloud infrastructure to store enormous volumes of patient records and medical data on remote servers, rather than maintaining costly physical storage systems. This enhances data security through advanced encryption, disaster recovery capabilities, and data management protocols while reducing the operational burden of managing on-site infrastructure [40].

PaaS offers healthcare developers a platform to build and deploy custom healthcare applications, such as patient management systems or telemedicine platforms. It enables the integration of advanced technologies like AI for diagnostics, IoT for real-time patient monitoring [41], and Big Data analytics for healthcare research.

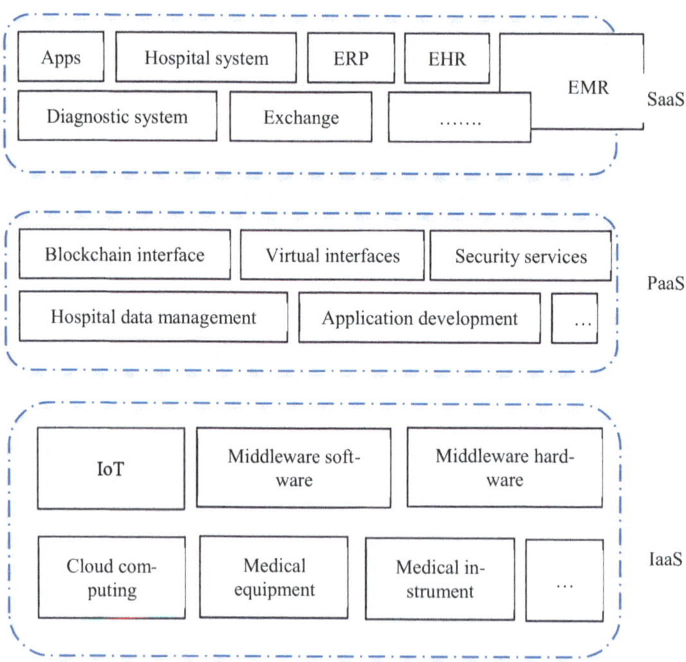

Fig. 8 Cloud Health Resource layers

By leveraging PaaS, healthcare organizations can rapidly innovate and deliver tailored solutions to meet their unique operational needs.

SaaS enables healthcare providers to use ready-made applications hosted in the cloud, such as EHR systems, telemedicine tools, and patient portals. These applications are accessible from any authorized device, allowing continuous access to up-to-date patient information, improving patient care during emergencies or remote consultations [42]. SaaS solutions also facilitate seamless collaboration between healthcare providers, specialists, and patients by enabling secure and efficient data sharing across various platforms.

The ability to share healthcare data seamlessly between different providers, specialists, and patients through cloud computing enhances continuity of care. Secure cloud platforms allow medical professionals to collaborate effectively, reducing delays in diagnosis and treatment by providing immediate access to shared records [43]. Whether the patient is consulting with a primary care physician, a specialist, or receiving treatment in another city or country, cloud-based solutions ensure consistency and accessibility in patient care.

3.1.4 Blockchain Technology in Healthcare

Blockchain technology is revolutionizing healthcare by providing secure, transparent, and tamper-proof solutions for managing medical records [44]. At the core of its application is the decentralized nature of blockchain, allows patient health information to be stored across distributed networks, ensuring that no single entity controls the data. This decentralized approach enhances data security, as tampering with or altering records becomes nearly impossible without detection. Each update to a medical record is added as a block to the chain, creating a permanent, auditable trail that preserves the integrity of the data [45, 46]. One of the primary benefits of blockchain in healthcare is its ability to ensure patient data integrity and security across platforms. Traditional centralized systems are predisposed to breaches and data manipulation, whereas blockchain's decentralized ledger systems provide a more resilient architecture against cyberattacks (Fig. 9). By reducing dependency on centralized databases, blockchain mitigates the risks associated with single points of failure, thereby bolstering the security of sensitive health information. Blockchain also plays a crucial role in fostering trust and interoperability in healthcare data sharing. Seamless sharing of patient data across different healthcare providers can improve care coordination, reduce duplication of services, and accelerate treatment decisions. With blockchain, patients and providers can trust that the shared data is accurate, untampered, and securely managed. Moreover, blockchain's transparency can reduce healthcare fraud by creating immutable records that track every transaction, enhancing accountability throughout the system. Blockchain offers a transformative solution for overcoming security and interoperability challenges in healthcare. By providing a secure, decentralized framework for managing medical records, it ensures data protection while enabling healthcare systems to

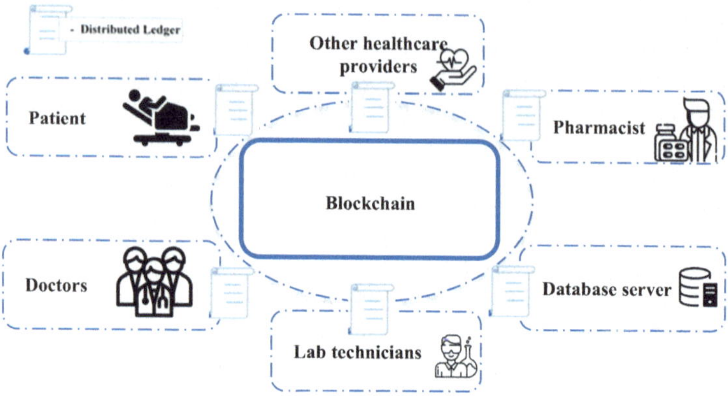

Fig. 9 Blockchain utilization in healthcare: key requirements and challenges

share information seamlessly, ultimately improving patient trust and the efficiency of healthcare delivery.

3.2 Big Data Analytics in Public Health

Big Data analytics in public health is transforming the way healthcare systems collect, analyze, and use data to improve population health outcomes. By leveraging data from diverse sources such as healthcare systems, wearable devices, and mobile health applications, public health authorities can gain deeper insights into the health status of urban populations. These data sources provide real-time information on patient behaviors, chronic disease management, and emerging health trends, which helps in identifying potential health risks early on and formulating targeted interventions [47]. One of the most powerful applications of Big Data in public health is predictive analytics. This technology enables health systems to analyze vast datasets to forecast disease outbreaks, anticipate healthcare demand, and allocate resources more efficiently. For instance, during a disease outbreak, predictive models can analyze patterns of infection spread, patient symptoms, and healthcare usage to predict where and when the next clusters may occur. This allows for faster response times and more targeted interventions, helping to mitigate the impact of the outbreak. Big Data analytics can optimize resource allocation in healthcare settings. By predicting future demand based on historical data and trends, health systems can better manage hospital capacity, staffing, medical supplies, and other critical resources [48]. This data-driven approach ensures that resources are deployed where they are needed most, improving overall healthcare efficiency and reducing the burden on overwhelmed facilities. Big Data analytics is playing an increasingly critical role in public health by enhancing the ability to predict disease outbreaks and optimize resource distribution, thereby enabling more proactive, efficient, and informed public health management.

Fig. 10 The 5Vs of Bigdata

Big data is defined by five key dimensions [49]: Volume, Velocity, Variety, Veracity, and Value (Fig. 10). In healthcare "Volume" refers to the massive amounts of health-related data generated daily, including patient records, imaging, and wearable device outputs. "Velocity" highlights the speed at which this data is created and needs to be processed, especially in real-time scenarios like monitoring vital signs or emergency alerts. "Variety" encompasses the diverse types of data, such as structured EHRs, unstructured physician notes, genomic data, and imaging results. "Veracity" focuses on the reliability and accuracy of the data, which is critical for making informed clinical decisions. Finally, "Value" represents the actionable insights derived from analyzing big data, such as improving patient outcomes, reducing costs, and enabling predictive analytics for better healthcare delivery. Together, these dimensions underpin the transformative impact of big data on modern healthcare systems.

3.3 *Artificial Intelligence (AI) in Healthcare*

Artificial intelligence is radically changing the healthcare system with solutions providing for diagnosis, cure and administration. Application of AI in the diagnosis area makes the procedure faster and more effective while identifying disease entities through different means-analyzing medical images, lab results and patients' data [50]. One example: AI algorithms in radiology can identify cancer or heart diseases in very early stages with better precision than regular methods; it flags abnormalities that might be missed even by human eyes [51]. This promotes early intervention and improves patients' well-being. In treatment, AI plays a role in personalized medicine by analyzing the patient's information, which recommends an appropriate intervention to every single patient. AI-operated systems can evaluate the genetic composition, medical history and lifestyle aspects of a patient and then recommend the therapies it thinks have the best evidence for the treatment results. This also means better therapeutic efficacy and less chance of adverse reactions, ensuring better health care [52]. AI is changing the healthcare industry in every way: increasing efficiency and automation of routine work, streamlining workflows, and managing hospital resources. From optimizing staff scheduling to automating patient

Fig. 11 Applications of AI in healthcare

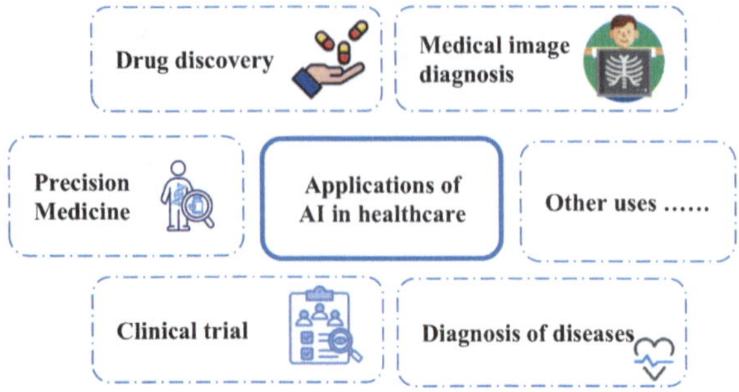

registration and billing, AI frees healthcare professionals to focus on their main concern-patient care. AI can also efficiently process huge amounts of data to support clinical decision-making; hence, it helps in making quick evidence-based decisions (Fig. 11). With improvements to decision making in diagnosis, treatment, and administration, AI is increasing efficiency, accuracy, and patient outcomes in healthcare and driving the future of medicine towards a more intelligent data-oriented processing approach.

3.3.1 Precision Medicine

Precision medicine is transforming urban healthcare by offering individualized treatments tailored to the unique needs of patients. This innovative approach integrates genetic, environmental, and lifestyle data to deliver targeted therapies that address the specific challenges posed by diverse urban populations. The complexities of urban health, shaped by demographic diversity, environmental influences, and varying socio-economic factors, underscore the critical need for precision medicine in these settings. Urban healthcare benefits immensely from precision medicine's ability to craft personalized treatment plans. By utilizing data from EHRs, wearable technologies, and other digital health tools, healthcare providers gain a comprehensive understanding of each patient's health profile [34, 35]. This includes medical history, treatment efficacy, and potential risk factors. For instance, individuals with chronic diseases like hypertension or diabetes can receive management plans that are fine-tuned to their genetic predispositions, daily habits, and environmental exposures. Such customization not only enhances the effectiveness of interventions but also fosters better patient adherence and satisfaction. Over time, this approach contributes to improved health outcomes and reduced healthcare disparities. Genomic research and AI are central to the success of precision medicine, especially in complex urban environments [53]. Genomic sequencing allows for the identification of genetic mutations that influence responses to specific drugs or treatments. This capability is particularly impactful in oncology, where therapies can be designed to target genetic mutations in tumors, significantly improving survival rates and minimizing side effects. AI amplifies these benefits by processing large datasets to uncover patterns and correlations beyond human observation. Machine learning models, for example, can analyze a combination of genomic information, clinical records, and environmental data to predict patient responses to treatments. In urban healthcare systems characterized by genetic diversity, these insights enable equitable and effective care for populations with varied health risks and needs [54]. The intricacies of urban health demand the implementation of precision medicine as a cornerstone of modern healthcare systems. Urban centers face unique challenges such as higher rates of chronic illnesses, greater environmental health risks, and pronounced healthcare inequities. Precision medicine provides a pathway to address these issues comprehensively and sustainably [55]:

- Meeting the Needs of Diverse Populations: Urban environments are a melting pot of cultures, ethnicities, and genetic backgrounds. Precision medicine ensures that treatments are not only biologically relevant but also culturally and socially attuned to diverse patient groups, improving outcomes and fostering trust in healthcare systems.
- Tackling Chronic Diseases: Lifestyle and environmental factors prevalent in urban settings often contribute to chronic conditions such as obesity, cardiovascular disease, and respiratory ailments. Precision medicine leverages individual health data to design prevention strategies and treatment plans tailored to each patient's specific risk profile, promoting long-term wellness.
- Bridging Health Disparities: Marginalized communities in urban areas frequently encounter barriers to quality care. By delivering data-driven, individualized interventions, precision medicine has the potential to reduce preventable illnesses and create a more equitable healthcare landscape.
- Optimizing Treatment Efficacy: Traditional treatments often fail to account for individual variations in drug response. Precision medicine employs biomarkers, genetic data, and lifestyle factors to refine treatments, minimizing adverse effects while maximizing therapeutic benefits. This approach is particularly vital in urban systems, where the volume and diversity of patients demand high efficiency.
- Capitalizing on Advanced Technologies: Urban healthcare systems increasingly adopt tools like AI, big data analytics, and IoE-connected devices. Precision medicine integrates seamlessly with these technologies, enabling the real-time analysis of patient data to refine clinical decisions and improve service delivery.
- Promoting Proactive Health Management: In densely populated urban areas, the strain on healthcare resources can be alleviated by shifting from reactive to proactive care models. Predictive analytics and genomic insights allow healthcare providers to preemptively address potential health issues, reducing the burden of acute care needs.

3.3.2 Accurate Diagnosis with AI-Driven Decision Support

AI-driven decision support systems play a transformative role in improving diagnostic accuracy and efficiency in healthcare. These systems analyze complex datasets, including patient histories, lab results, and imaging studies, to assist clinicians in making informed decisions (Fig. 12). Using machine learning algorithms, these

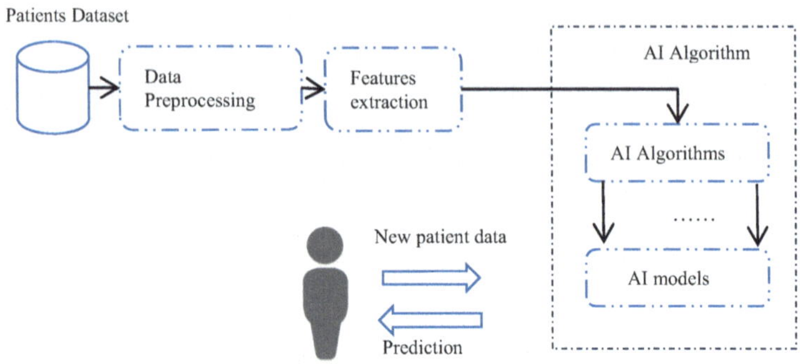

Fig. 12 General AI-Driven medical diagnosis concept

tools can identify patterns and correlations that might be difficult for humans to detect, enabling early identification of conditions such as cancers, cardiovascular diseases, and rare disorders. By providing evidence-based recommendations and flagging potential diagnoses, AI systems reduce diagnostic errors and support more personalized treatment plans [56]. For instance, an AI-powered decision support tool might suggest specific tests or highlight red flags in patient data, ensuring critical conditions are not overlooked. Additionally, these tools are particularly valuable in urban healthcare settings, where the high patient volume can strain specialists. By automating routine diagnostic tasks and prioritizing cases requiring urgent attention, AI enhances the overall efficiency and quality of care delivery. The integration of AI in clinical workflows fosters a collaborative approach to healthcare, where human expertise is augmented by machine precision, ultimately leading to better health outcomes.

3.3.3 Image Recognition in Medical Imaging

AI-driven image recognition is revolutionizing medical imaging by significantly improving diagnostic precision and efficiency. Leveraging advanced algorithms and deep learning, AI tools analyze medical images such as X-rays, CT scans, MRIs, and ultrasounds to detect abnormalities with remarkable accuracy [57]. These tools can identify early signs of diseases, including cancerous tumors, fractures, and pulmonary conditions, often surpassing human capabilities in spotting subtle or complex patterns. For example, AI applications in radiology can swiftly highlight areas of concern on chest X-rays, aiding in the early diagnosis of pneumonia or lung cancer.

The integration of AI in medical imaging enhances workflow efficiency by reducing the time required for interpretation, allowing radiologists to focus on complex cases or consult on high-risk patients (Fig. 13). Additionally, these systems improve accessibility in resource-constrained settings, enabling remote diagnostics and second opinions. By combining AI with telemedicine, healthcare providers can extend their reach to underserved areas, democratizing access to high-quality

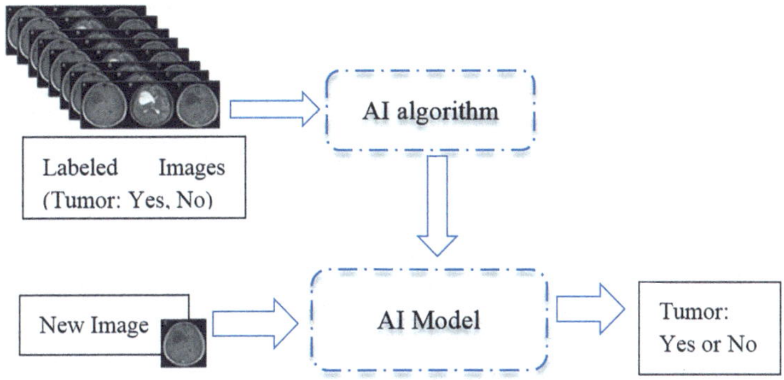

Fig. 13 General workflow of medical imaging processing with AI

medical imaging services. This transformative technology is paving the way for more precise, timely, and equitable healthcare delivery.

3.4 Internet of Everything (IoE) and Connected Devices

IoE is transforming healthcare by integrating various connected devices and systems, enabling seamless communication and data sharing between patients, healthcare providers, and medical technologies. Through the IoE, healthcare devices such as wearable sensors, biosensors, and fitness trackers are connected to a network, allowing continuous monitoring of patient health [58]. This integration enhances real-time data collection and analysis, improving overall patient care and facilitating early interventions. Wearable devices, like fitness trackers and biosensors, are at the forefront of this revolution, offering continuous health monitoring, particularly for chronic diseases such as diabetes and hypertension.

Using several communication protocols (Fig. 14), these devices capture vital signs such as heart rate, glucose levels, and blood pressure, providing real-time insights into a patient's health. By continuously tracking these parameters, patients and healthcare providers can detect early signs of health deterioration, ensuring timely interventions (Fig. 15). The data collected from wearable devices is integrated directly into clinical workflows, enabling healthcare professionals to make informed decisions based on real-time data [59]. This integration helps personalize treatment plans, monitor the effectiveness of interventions, and reduce hospital readmissions. For instance, a patient with hypertension can have their blood pressure monitored remotely, with alerts sent to their physician if critical levels are reached, allowing for immediate action. By combining IoE and wearable devices, healthcare is becoming more proactive and data-driven, ensuring continuous patient engagement, better chronic disease management, and improved healthcare outcomes.

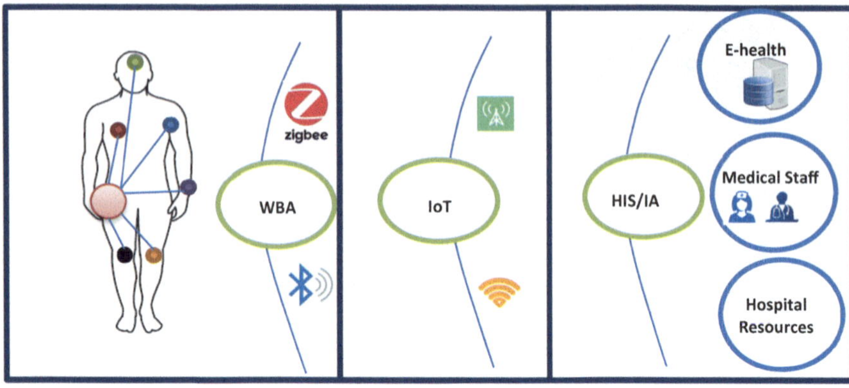

Fig. 14 Health monitoring with wearable devices

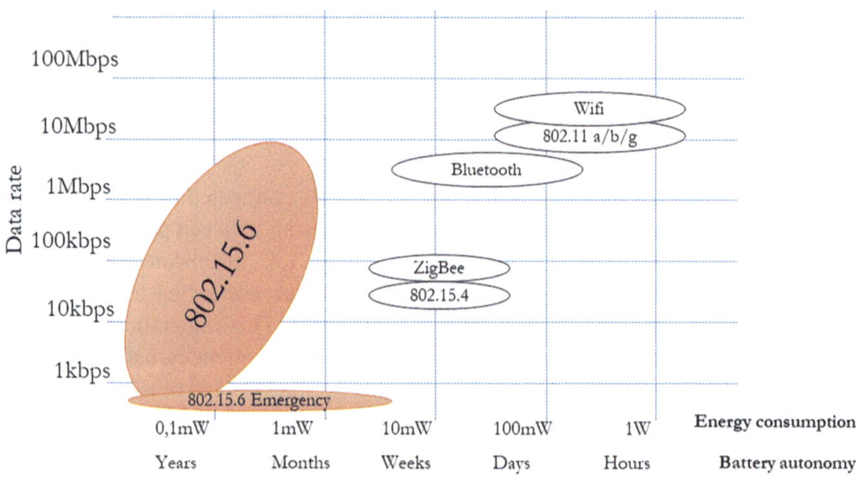

Fig. 15 Devices communication protocols (data rate and energy consumption)

3.5 *Mobile Health Applications*

Mobile health applications (mHealth apps) are playing a transformative role in modern healthcare by enhancing patient engagement, enabling teleconsultations, and streamlining follow-ups. These apps provide patients with easy access to healthcare services and information through their smartphones, empowering them to take an active role in managing their health [60]. Features such as appointment scheduling, medication reminders, and secure messaging with healthcare providers make it more convenient for patients to engage with their care teams. One of the key benefits of mHealth apps is their ability to facilitate teleconsultations, allowing patients to connect with doctors and specialists remotely [61]. This is particularly valuable in urban settings where time constraints or transportation issues can limit access to in-person healthcare visits. With video or chat-based consultations, patients can receive medical advice, diagnoses, and treatment plans without having to visit a clinic, making healthcare more accessible and reducing the burden on hospital systems. mHealth apps also play a significant role in follow-up care, ensuring that patients adhere to their prescribed treatment plans and recover smoothly after procedures or illnesses. Automated reminders for medication, follow-up appointments, and health checks help patients stay on track with their care, reducing the risk of complications and readmissions. Moreover, mHealth apps contribute to early interventions and preventive care by allowing patients to monitor their health metrics, track symptoms, and report any concerning changes to their healthcare providers. This real-time exchange of information facilitates early detection of potential health issues, enabling timely interventions that can prevent conditions from worsening. By promoting continuous patient engagement and preventive measures, mobile health applications are helping to improve healthcare outcomes and reduce the overall strain on urban healthcare systems.

3.6 Remote Diagnostics and Telemedicine

Remote diagnostics and telemedicine have become integral components of modern healthcare, significantly transforming how medical services are delivered. Virtual consultations allow patients to connect with healthcare professionals through video calls, online chats, or phone calls, enabling them to receive diagnoses and medical advice from the comfort of their own homes [62]. This not only enhances accessibility, particularly for those in urban areas with busy schedules or limited mobility, but also ensures that patients can seek care without the need for time-consuming travel or hospital visits. By leveraging telemedicine, healthcare providers can remotely assess patient symptoms, review medical histories, and prescribe treatments or further diagnostic tests. This reduces the burden on healthcare facilities and allows hospitals and clinics to focus on patients requiring in-person care. Remote diagnostics, often supported by wearable devices and mobile health applications, provide real-time health monitoring for chronic diseases such as diabetes or cardiovascular conditions [63]. These devices transmit vital health data to physicians, enabling early detection of issues and timely interventions. For urban populations, where healthcare systems can become overwhelmed due to high demand, telemedicine offers a practical solution to streamline care delivery. It reduces the strain on emergency rooms and outpatient departments by offering an alternative for non-critical conditions, routine check-ups, or follow-up consultations. This ultimately improves healthcare efficiency, reduces wait times, and provides patients with quicker access to professional medical advice, enhancing overall health outcomes.

3.7 From Hospital 4.0 to Hospital X.0

The evolution from Hospital 4.0 to Hospital X.0 represents a transformative journey in the healthcare sector, emphasizing the integration of advanced technologies, patient-centered care, and data-driven decision-making. Hospital 4.0, often characterized by the adoption of Industry 4.0 principles, focuses on automation, interconnected systems, and the use of big data analytics to enhance operational efficiency [64]. This phase incorporates technologies like the IoT, AI, and cloud computing to streamline hospital workflows, improve patient care, and optimize resource management. As healthcare continues to evolve, the concept of Hospital X.0 emerges, envisioning a future where hospitals not only embrace technological advancements but also prioritize holistic patient experiences, personalized medicine, and real-time data utilization. Hospital X.0 emphasizes a seamless integration of health services across various platforms, ensuring interoperability among systems and fostering collaboration between healthcare providers, patients, and technology. It aims to leverage predictive analytics and artificial intelligence to anticipate patient needs, improve health outcomes, and enhance operational efficiencies. Additionally, Hospital X.0 focuses on empowering patients through digital health solutions, allowing them to actively participate in their care processes. With tools like mobile health applications and telemedicine, patients can access their health information, communicate with providers, and receive timely interventions. This shift towards a more patient-centric approach enhances engagement, improves satisfaction, and supports better health management [65]. Ultimately, the transition from Hospital 4.0 to Hospital X.0 marks a significant step toward a more resilient, adaptable, and technologically advanced healthcare ecosystem that prioritizes both operational excellence and patient empowerment in the pursuit of optimal health outcomes.

3.8 Adaptive E-Learning Healthcare Professionals' Lifelong Learning

In the rapidly evolution of healthcare, the concept of lifelong learning has become essential for healthcare professionals [66]. As medical knowledge expands and technologies advance, practitioners must continuously update their skills and knowledge to provide high-quality care (Fig. 16). Lifelong learning empowers healthcare professionals to remain current with the latest research, treatment modalities, and innovations, ultimately enhancing patient outcomes and safety [66]. Adaptive e-learning plays a pivotal role in facilitating this lifelong learning journey. Unlike traditional educational models, adaptive e-learning tailors the educational experience to the individual learner's needs, preferences, and pace. Utilizing data analytics and artificial intelligence, adaptive e-learning platforms assess a learner's existing knowledge and skills, enabling personalized learning pathways that focus on areas requiring improvement. This approach not only enhances engagement but

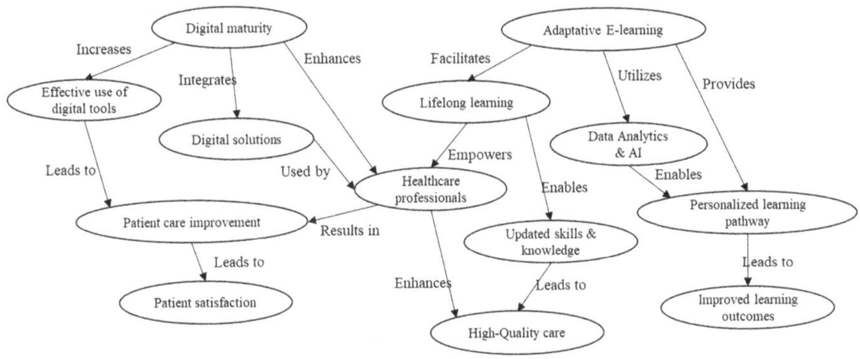

Fig. 16 E-Learning in Healthcare professionals' lifelong learning

also optimizes the learning experience by delivering relevant content that aligns with the learner's professional goals [67]. The digital maturity of healthcare professionals is integral to their effectiveness in utilizing emerging technologies and improving patient care. Digital maturity refers to the ability of healthcare professionals to effectively adopt and integrate digital tools into their practice. As healthcare systems increasingly rely on digital solutions, such as electronic medical records, telemedicine platforms, and big data analytics, professionals must be equipped with the skills and confidence to navigate these technologies. By fostering a culture of digital literacy and competency, healthcare organizations can ensure that their workforce is prepared to leverage digital tools effectively. This involves not only technical training but also cultivating a mindset that embraces innovation and continuous improvement. Adaptive e-learning, combined with initiatives aimed at enhancing digital maturity, provides healthcare professionals with the resources they need to thrive in a technology-driven environment. By offering flexible, tailored learning experiences that build both clinical expertise and digital competencies, organizations can better equip their staff to meet the demands of modern healthcare. This holistic approach not only enhances the competencies of healthcare professionals but also contributes to a more knowledgeable and skilled workforce, ultimately leading to improved patient care and satisfaction in an increasingly complex healthcare system.

4 Case Studies and Real-World Applications

4.1 Smart Healthcare in Developed Urban Areas

Smart healthcare in developed urban areas is driving a technological revolution by integrating advanced systems to improve patient outcomes, optimize resource allocation, and enhance operational efficiency. Cities like Singapore, Seoul, and Boston are leading the way with innovations such as electronic health records, wearable devices, and telemedicine platforms. These technologies facilitate early detection, proactive interventions, and greater access to healthcare services. In Singapore, the National Electronic Health Record (NEHR) system enables seamless patient data sharing, improving care coordination [68]. Seoul's integration of wearable devices allows for continuous monitoring and timely medical interventions [69]. Boston's Massachusetts General Hospital has pioneered telemedicine, offering remote consultations, which have proven essential, especially in rural and underserved areas [70]. Artificial intelligence and wearable devices further enhance diagnosis, with AI revolutionizing medical imaging and wearables enabling real-time monitoring of vital signs. Telemedicine and personalized care improve patient experiences by providing better access and tailored. Predictive analytics and smart inventory systems enable better resource allocation, reducing wait times and ensuring supply availability. Despite these advancements, challenges remain in ensuring equitable access to these technologies, particularly in underserved areas. Critics argue that while these solutions benefit many, they often require significant infrastructure and financial investment, creating a gap between well-resourced urban and rural or low-income communities. Additionally, privacy concerns surrounding patient data and the dependence on technology in healthcare settings pose ongoing issues. Nevertheless, smart technologies have the potential to revolutionize healthcare delivery, making it more efficient and responsive, but careful consideration of equity and accessibility is essential for broader success (Table 2).

4.2 Smart Healthcare in Developing Urban Environments

Smart healthcare technologies, such as mHealth and remote diagnostics, offer significant promise for overcoming healthcare inequities in developing urban environments. These technologies can help address issues like inadequate infrastructure, economic disparities, and limited resources, particularly in underserved areas. Telemedicine facilitates remote consultations, reducing travel and increasing accessibility for patients with chronic conditions [7]. Remote monitoring using wearable devices enables continuous tracking of vital signs, allowing for early intervention and preventing complications [75]. mHealth platforms also play a crucial role in delivering health education, empowering individuals to make informed decisions about disease prevention and management [76]. Moreover, affordable mobile devices and offline capabilities make these solutions accessible even in areas with limited internet connectivity, ensuring healthcare remains available even in resource-constrained environments [77]. Community-based approaches, such as engaging local health workers, further enhance adoption and trust in these technologies [78]. Examples from countries like India, Kenya, and Uganda illustrate the transformative impact of these technologies. In India, the National Health Mission's mHealth initiatives have improved maternal and child health outcomes [79]. In Kenya, mobile money services like M-PESA have facilitated healthcare payments,

Table 2 Applications of emerging technologies in urban healthcare: advantages, limitations, and case studies

Topic	Technology	Advantages	Limitations	City and country	Reference
Smart nation initiative	National Electronic Health Record (NEHR)	– Enhances healthcare coordination. – Reduces medical errors by providing accurate patient data. – Improves communication between healthcare providers.	– Privacy concerns regarding patient data. – Requires high initial investment.	Singapore, Singapore	[68]
Wearable devices	Wearable health monitoring devices (smartwatches, health bands)	– Continuous monitoring of vital signs. – Enables early detection of health issues (arrhythmia). – Improves chronic disease management.	– Data accuracy can vary with device quality. – Requires user compliance for accurate data.	Seoul, South Korea	[69]
Telemedicine	Telemedicine platforms (video consultations)	– Increases access to healthcare, especially in rural areas. – Reduces travel and wait times. – Supports continuous care for chronic conditions.	– Limited physical examination. – Potential technological barriers (poor internet access).	Boston, USA	[70]
AI in medical imaging	AI algorithms for analyzing medical images (X-rays, MRIs)	– Early detection of diseases like cancer. – Reduces human error in image interpretation. – Saves time in diagnosis.	– High dependency on the quality of input data. – Need for trained professionals to interpret AI results.	Germany	[71]
Wearable devices for real-time monitoring	Smartwatches and fitness trackers	– Real-time health tracking of vital signs (heart rate, blood oxygen). – Prevents sudden medical emergencies by alerting to abnormal readings.	– Limited by battery life and data storage. – Privacy issues with collected health data.	USA, Europe	[72]
Big Data in healthcare	Big Data Analytics for patient health data (EHR)	– Identifies trends and patterns in patient health. – Supports evidence-based treatments. – Aids in the development of personalized healthcare strategies.	– Data privacy and security concerns. – Requires massive computational power.	Poland, Mazowieckie	[49]

(continued)

Table 2 (continued)

Topic	Technology	Advantages	Limitations	City and country	Reference
Predictive analytics	Predictive models based on historical patient data	– Optimizes resource allocation (staffing, equipment). – Reduces wait times and improves service efficiency.	– Accuracy depends on the quality of historical data. – May face issues in adapting to sudden changes in patient behavior.	California USA	[73]
Telemedicine	Video consultation platforms	– Increases access for patients in remote locations. – Reduces overhead costs of physical facilities. – Provides quick follow-up care for patients.	– Technical issues like poor internet can disrupt consultations. – Requires a reliable communication infrastructure.	New York USA	[74]
Smart inventory management	RFID, IoT sensors for real-time inventory tracking	– Minimizes waste by managing supply levels accurately. – Reduces the risk of stockouts for critical medical supplies. – Enhances operational efficiency in healthcare facilities.	– Requires significant investment in infrastructure. – Vulnerable to cybersecurity risks.	Texas USA	[34, 35]

reducing financial barriers [80]. Uganda's mobile health platform has improved vaccination rates and health awareness in rural communities [81]. However, despite these successes, challenges remain. The implementation of smart healthcare technologies requires addressing issues of digital literacy and equitable access to technology. In some areas, the cost of smartphones and mobile data can still be prohibitive. Additionally, while these innovations can greatly improve healthcare delivery, their sustainability in the long term is uncertain without continued investment and support. Therefore, while mobile health and remote diagnostics have the potential to bridge healthcare gaps, careful attention must be given to addressing these barriers to ensure their broad and lasting impact (Table 3).

4.3 Public Health Initiatives with Big Data, AI, and Blockchain

The intersection of public health and technology, particularly through big data, AI, and blockchain, is becoming increasingly crucial in managing crises and ensuring transparency in health reporting. These technologies enable advanced data-driven decision-making, resource optimization, and enhanced disease tracking. AI's predictive analytics have proven valuable in early warning systems, as seen with HealthMap, which provides real-time alerts for infectious disease outbreaks [83]. Additionally, machine learning aids in optimizing resource allocation during crises by forecasting hospital needs and targeting vulnerable populations to mitigate risks [84, 85]. Blockchain, on the other hand, ensures data integrity, combats counterfeit medical supplies, and facilitates secure data sharing across stakeholders, as evidenced during the COVID-19 pandemic [45, 46]. Despite these benefits, critics argue that the application of these technologies may face challenges, including ethical concerns, data privacy issues, and accessibility disparities. For instance, while AI-powered surveillance systems enhance outbreak monitoring, their reliance on EHRs and social media data raises questions about privacy and consent [58]. Moreover, blockchain's decentralized structure, though secure, could complicate integration with existing health systems. Addressing these challenges requires robust regulatory frameworks and equitable access to technological infrastructure. Overall, the integration of big data, AI, and blockchain holds immense potential for public health, but careful consideration of its limitations and ethical implications is necessary to maximize its (Table 4).

Table 3 Impact of e-health technologies on healthcare accessibility: case studies and analysis

Topic	Technology	Advantages	Limitations	Reference
Telemedicine	Mobile applications for remote consultations	Increases accessibility by reducing the need for travel, provides timely medical advice, and is particularly useful for chronic diseases and urgent health issues.	Limited by internet access and technological infrastructure in remote areas.	[7]
Remote monitoring	Wearable devices, mobile apps	Enables real-time monitoring of vital signs, allows early detection of health problems, and facilitates timely interventions to prevent hospitalizations and complications.	Requires reliable connectivity for real-time data transmission; device dependency may limit effectiveness in low-resource settings.	[75]
Health education	mHealth platforms	Delivers targeted educational content on disease prevention, healthy lifestyles, and medication adherence, empowering individuals to take control of their health.	Content may be limited in scope and accessibility in areas with low technological penetration.	[76]
Low-cost devices	Smartphones, tablets	Affordable and accessible devices for hosting health apps, enabling healthcare delivery in resource-limited settings.	May not be sustainable long-term if devices or applications are not properly maintained or if there is insufficient training in their use.	[77]
Offline capabilities	Offline mobile health apps	Provides access to healthcare services in areas with limited or no internet connectivity.	Reduced functionality and real-time data transmission is not possible.	[82]
Community-based approaches	Community health workers, local organizations	Ensures that healthcare technologies meet local needs, fosters trust, and encourages the adoption of smart healthcare technologies.	Relies on local capacity and may face resistance from communities or a lack of trained personnel.	[78]
Case study: India	Teleconsultation, mHealth clinics	mHealth initiatives improve maternal and child health outcomes by providing remote consultations and reaching underserved areas.	Limited by local infrastructure, and the effectiveness is dependent on the availability of reliable technology and internet access.	[79]
Case study: Kenya	M-PESA (mobile money system)	Improves access to healthcare for low-income populations by enabling cashless payments for medical services, reducing financial barriers.	Relies on mobile money infrastructure and may not be available or feasible in all rural regions.	[80]
Case study: Uganda	mHealth platform	Tracks immunization records and provides health education, improving vaccination rates and health awareness in rural communities.	May face limitations in rural connectivity and internet access, reducing the effectiveness of the platform.	[81]

Table 4 Leveraging advanced technologies for health management: applications and case studies

Category	Subcategory	Applications	Tools technologies	Examples	References
Predictive analytics	Early warning systems	Identifying emerging health threats and predicting disease spread	AI algorithms, HealthMap	Real-time alerts for infectious disease outbreaks	[83]
	Resource allocation	Optimizing resource distribution during emergencies	Machine learning models	Forecasting hospital patient volumes	[84]
	Risk assessment	Identifying vulnerable populations and transmission hotspots	AI models	Assessing mobility and social interaction patterns	[85]
	Data integrity	Ensuring data authenticity and tamper resistance	Blockchain technology	Reliable storage of public health data	[45, 46]
	Traceability	Tracking medical supplies and vaccine movements	Blockchain systems	Combating counterfeit vaccines during COVID-19	
	Data sharing	Secure sharing of health data among stakeholders	Blockchain platforms	Facilitating research collaboration and public health responses	
Disease tracking & prevention	Surveillance systems	Monitoring real-time data to detect outbreaks	AI-powered systems	CDC's influenza monitoring	[58]
	Contact tracing	Efficiently identifying and isolating potential cases	Blockchain-based systems	Managing COVID-19 outbreaks while ensuring privacy	[45, 46]
	Vaccine management	Tracking vaccine distribution and administration	Blockchain technology	Ensuring safety and efficacy of vaccines	[45, 46]
	COVID-19 response	Managing resource allocation, contact tracing, and vaccination record integrity	AI tools, Blockchain	Crisis management using AI and blockchain	[45, 46]
	Disease surveillance	Monitoring infectious diseases like influenza and measles	AI-based systems	Enhanced rapid responses by health authorities	[84]

4.4 HMS and EMR Implementation in Smart Healthcare Systems

HMS and EMR are pivotal in the transformation of modern healthcare, particularly within the framework of smart healthcare systems (Table 5). Their integration has redefined healthcare delivery by streamlining administrative processes, improving patient outcomes, and significantly reducing medical errors. For instance, Singapore's National Electronic Health Record (NEHR) system exemplifies the seamless sharing of patient data among healthcare providers, which minimizes test duplications and promotes coordinated care [68]. Similarly, the UK's National Health Service (NHS) employs systems like the Summary Care Record (SCR) and the National Programme for IT (NPfIT) to enhance patient safety and facilitate efficient information [86]. In the U.S., EMR systems such as Epic and Cerner have demonstrated substantial improvements in operational efficiency and reductions in errors. A significant advantage of EMR systems is their role in minimizing medication errors by issuing real-time alerts for potential drug interactions and allergies, a critical feature for safe prescribing practices. These systems also foster better communication among healthcare providers, thereby reducing risks of misdiagnosis and inappropriate treatment plans. Furthermore, EMRs facilitate the creation of personalized treatment plans, drawing from comprehensive patient histories to support tailored care strategies [73]. Remote accessibility of patient records ensures continuity of care across multiple providers, crucial in today's interconnected healthcare environment. However, despite these benefits, challenges persist, such as

Table 5 Impact of Electronic Medical Records (EMR) on healthcare: global implementations and benefits

Category	Details	References
Global implementations	– Singapore: National Electronic Health Record (NEHR) enables sharing of patient information, reducing duplication and promoting coordinated care.	[68] [86] [73]
	– **UK (NHS)**: Systems like SCR improve access and safety.	
	– **US**: Widespread EMR systems like Epic and Cerner enhance care and reduce errors.	
Patient safety improvements	– **Medication errors**: EMR reduces errors with real-time alerts for drug interactions and allergies.	[89]
	– **Provider communication**: Enhanced communication reduces misdiagnosis and treatment errors.	
Personalized treatment	EMR systems enable tailored treatment plans based on patient history and health status.	[90]
Access to care	EMR facilitates remote access to patient records, supporting continuity of care across multiple providers.	
Operational efficiency	Automation of administrative tasks allows providers to focus more on patient care, reducing costs and errors.	[87]
Data-driven decision-making	Enables analysis of large datasets for evidence-based practices and population health management.	[88]

interoperability issues, privacy concerns, and the steep learning curve for healthcare personnel during system transitions. While the automation of administrative tasks through HMS and EMRs enhances efficiency and reduces costs, these systems require robust data management and ongoing updates to remain effective. Additionally, their role in enabling evidence-based decision-making underscores the importance of integrating analytical tools for population health management [87, 88].

5 Strategic Recommendations for Implementing Smart Healthcare in Urban Setting

5.1 Policy and Regulatory Support

The effective implementation of smart healthcare technologies in urban environments is exemplified by policies and regulations adopted in various countries, showcasing how supportive governance can drive innovation while addressing public interests. For instance, Singapore's Smart Nation Initiative includes healthcare as a key pillar, with government-backed funding supporting the development of telemedicine platforms and the integration of EMRs like the National Electronic Health Record (NEHR) [68]. This system ensures seamless data sharing among healthcare providers, improving care coordination and operational efficiency. In Estonia, widely regarded as a leader in digital health, the government has implemented robust policies that combine EMRs with blockchain technology, ensuring secure and transparent patient data handling. Estonia's digital infrastructure allows citizens to access their medical records and share them with healthcare providers as needed, promoting trust and engagement [91]. Similarly, the United States provides a regulatory framework that balances innovation and privacy. Programs like the HITECH Act have incentivized healthcare providers to adopt EMR systems through financial grants, while legislation such as HIPAA (Health Insurance Portability and Accountability Act) ensures the protection of sensitive health data [14]. The United Kingdom's National Health Service (NHS) has also demonstrated leadership by integrating urban health systems with smart technologies. Initiatives like the NHS Digital program leverage AI and data analytics to improve patient outcomes, while the General Data Protection Regulation (GDPR) ensures strict compliance with privacy standards [92]. In Morocco, the adoption of smart healthcare technologies is supported by the work of the National Commission of Personal data Protection, which ensures robust data privacy in the digital health sector. The CNDP regulates the use and sharing of patient data in compliance with Morocco's Law 09-08, which aligns with international standards for data protection. This framework enables the secure deployment of electronic health systems and telemedicine platforms, particularly in urban centers like Casablanca and Rabat. Additionally, Morocco's focus on digital transformation through the Maroc Digital 2030 strategy has paved the way for integrating advanced technologies such as AI in diagnostics and telehealth services, enhancing access to healthcare in densely populated areas [93]. In India, the government has launched the National Digital Health Mission (NDHM), which focuses on creating a unified health ID for citizens and integrating urban healthcare systems to enhance accessibility and efficiency. This initiative also addresses urban challenges by promoting telemedicine and digital consultations, particularly beneficial in overcrowded cities [94]. Finally, South Korea combines advanced technologies like AI-driven diagnostics and robotics in urban hospitals, backed by government policies encouraging innovation. Programs such as the Korean New Deal provide funding for digital infrastructure in healthcare, while privacy regulations like the Personal Information Protection Act (PIPA) safeguard patient data [95].

The successful implementation of smart healthcare systems in urban environments requires well-crafted policies and regulations that foster innovation while safeguarding public interests. Governments should prioritize creating a robust digital infrastructure by investing in advanced technologies such as AI, telemedicine, and blockchain while encouraging public-private partnerships to drive research and development tailored to urban healthcare challenges. Providing financial incentives like grants, subsidies, and tax benefits can further motivate healthcare providers to adopt smart solutions. Equally important is the establishment of comprehensive regulatory frameworks to address data privacy, security, and ethical concerns. Regulations should define clear standards for data collection, sharing, and storage, while mandating the use of secure technologies to protect sensitive patient information. Oversight mechanisms, such as independent regulatory bodies, should be implemented to monitor compliance and ensure accountability. To address the unique demands of urban healthcare, policymakers should also focus on interoperability by setting standards for seamless integration across diverse healthcare systems and technologies. Furthermore, policies must promote transparency and equitable access to smart healthcare services, ensuring that all urban residents, regardless of socioeconomic status, can benefit from technological advancements. Governments should engage stakeholders, including healthcare providers, technology developers, and patients, in the policymaking process to create balanced regulations that align with the needs and aspirations of urban populations. By adopting these recommendations, policymakers can create an environment conducive to innovation while ensuring that smart urban healthcare systems remain ethical, inclusive, and effective in improving health outcomes.

5.2 Investment in Digital Infrastructure

The potential of smart healthcare technologies in urban environments is deeply rooted in the development of robust digital infrastructure. This infrastructure encompasses high-speed internet networks, cloud computing resources, and interoperable health information systems that allow seamless data sharing among healthcare providers [96]. Such investments are vital for enabling real-time data exchange, remote monitoring, telemedicine, and mobile health applications, all of which play a significant role in improving patient outcomes and streamlining healthcare delivery.

Countries around the world have increasingly recognized the importance of investing in digital infrastructure for healthcare, particularly in urban settings. For instance, Singapore has been a leader in integrating digital solutions into healthcare. Its NEHR system is a prime example of how interoperable health information systems can provide seamless data sharing among healthcare providers. By ensuring that patient data is readily accessible across hospitals, clinics, and other healthcare settings, the NEHR has reduced duplication of tests, minimized errors, and improved the coordination of care. In addition, Singapore's Smart Nation initiative promotes the use of technology in healthcare by ensuring that all urban residents have access to digital health tools, including mobile health apps and telemedicine platforms. This comprehensive digital infrastructure is complemented by the country's high-speed internet networks and the provision of cloud services for managing health data [97]. In Estonia, a pioneer in digital healthcare, the country's eHealth system connects all citizens with electronic medical records and provides an online portal for accessing health services. This system is integrated with Estonia's digital ID system, enabling secure access to health data for both patients and healthcare providers. The country has made significant strides in promoting digital literacy and providing equitable access to healthcare technologies, even in rural areas [91]. Morocco, recognizing the need for digital infrastructure in healthcare, has made several strides toward modernizing its urban healthcare systems. Through the National Health Information System, Morocco has aimed to enhance the exchange of health data across healthcare providers and ensure that patients receive timely care, particularly in urban areas where healthcare demand is high. The Moroccan Ministry of Health has been working on improving connectivity in healthcare institutions, especially hospitals, by integrating health IT systems, electronic medical records, and telemedicine services. The Moroccan National Digital Strategy (2025–2030) includes specific initiatives aimed at expanding broadband connectivity across the country, focusing on both urban and rural areas. This strategy aligns with efforts to enhance eHealth services, support telemedicine, and ensure that digital health solutions are accessible to all citizens. For example, in urban areas like Casablanca and Rabat, there has been significant investment in upgrading healthcare facilities to handle digital health tools, including mobile applications and online health portals, which help patients access healthcare services remotely [93].

To truly unlock the potential of smart healthcare in urban settings, governments and healthcare providers must invest heavily in digital infrastructure. This includes [96]:

- Enhancing Connectivity: Governments should prioritize the rollout of high-speed internet across urban areas, especially in underserved neighborhoods. This can be achieved through collaborations with private sector technology firms and telecom companies.

- Cloud Computing and Data Security: Investing in cloud computing services will not only facilitate efficient management of healthcare data but also provide secure, remote access to patient records. Countries like Singapore and Estonia demonstrate how secure data-sharing systems can drive improvements in patient care.
- Public-Private Partnerships: As demonstrated in Singapore and Morocco, collaboration between the government and private tech companies can facilitate the development of smart healthcare ecosystems. These partnerships should focus on providing digital health tools, ensuring equitable access, and promoting technological innovation tailored to urban healthcare challenges.
- Promoting Digital Literacy: It is critical to invest in digital literacy programs aimed at improving residents' ability to engage with digital health technologies. These programs should target both healthcare professionals and the general public, particularly in urban areas with lower technological engagement.
- Fostering Inclusivity: Investment efforts must ensure that all urban residents, including those from lower-income backgrounds, have access to the benefits of smart healthcare. This could involve subsidizing digital health tools or providing low-cost access to broadband services in urban slums and underserved areas.

The future of healthcare in urban environments hinges on substantial investment in digital infrastructure. By ensuring high-speed connectivity, secure data sharing, and equal access to digital tools, cities worldwide can foster better healthcare delivery, improved patient engagement, and more efficient systems. Countries like Singapore and Estonia have set global benchmarks, and Morocco's ongoing efforts, coupled with future investment in digital infrastructure, could significantly enhance the healthcare experience for all urban residents, reducing disparities and improving overall public health outcomes.

5.3 Workforce Training and Professional Development

In today's rapidly evolving healthcare landscape, especially within urban environments, the need for continuous workforce training and professional development is paramount [98]. As advanced technologies such as AI, telemedicine, EMR, and data analytics become integral parts of healthcare delivery, it is crucial that healthcare professionals are equipped with the skills and knowledge to navigate and utilize these technologies effectively. Training programs should be comprehensive, addressing both the technical aspects of these innovations and their practical applications in improving patient care, enhancing operational efficiency, and optimizing overall healthcare outcomes [99]. Healthcare workers must not only be adept at using new technologies but also understand how to integrate them into their workflow to enhance care delivery. Collaboration between technology developers and healthcare providers is essential in designing effective training programs. By working together, they can create training modules that are not only technically sound but also practically relevant to the challenges healthcare professionals face daily in urban healthcare settings. These partnerships can result in tailored training initiatives that focus on critical areas such as telehealth practices, data management, and cybersecurity protocols. For example, telemedicine training might cover not only the technical aspects of virtual consultations but also how to maintain patient trust and ensure confidentiality in a digital setting. Additionally, data management training should emphasize the importance of accurate data entry and the implications of data security within healthcare systems. Incorporating regular feedback from healthcare professionals is another key element of an effective training process. This feedback allows for real-time adjustments to training programs, ensuring they remain relevant and practical as healthcare environments evolve. As healthcare professionals engage with new technologies, their firsthand experiences can provide valuable insights into the challenges they face and the areas where further training is needed. This ongoing feedback loop also fosters a culture of continuous learning, encouraging healthcare workers to seek knowledge, enhance their skills, and stay current with new advancements in their field. Furthermore, fostering a culture of lifelong learning within healthcare organizations can lead to a more resilient workforce. This culture encourages healthcare professionals to not only embrace new technologies but also seek opportunities to enhance their competencies over time. Organizations can support this by providing accessible training resources, offering career development opportunities, and recognizing the importance of professional growth. Empowering healthcare professionals with the necessary skills to navigate the complexities of modern healthcare systems can lead to improved patient care, increased job satisfaction, and greater organizational efficiency. Ultimately, investing in workforce training and professional development is essential for the successful integration of smart healthcare technologies. It is not just about adopting new tools; it is about ensuring that healthcare professionals have the competence and confidence to use these technologies to enhance their practice and deliver high-quality care. As urban healthcare systems continue to grow and evolve, providing healthcare workers with the resources and opportunities for continuous learning will be key to building resilient teams capable of meeting the demands of modern healthcare environments.

As shown in Figure 17 Improving Workforce Training and Professional Development is possible by adopting these recommendations [66]:

- Develop Tailored Training Programs: Collaborate with technology developers to create practical, context-specific training modules that address the unique challenges of urban healthcare, focusing on telemedicine, data management, and cybersecurity.
- Promote Lifelong Learning: Establish a culture of continuous learning within healthcare organizations by providing ongoing professional development opportunities, certifications, and career growth initiatives.
- Foster Partnerships for Training Development: Encourage collaboration between healthcare providers, technology companies, and educational institutions to ensure that training programs are aligned with real-world healthcare needs and technological advancements.

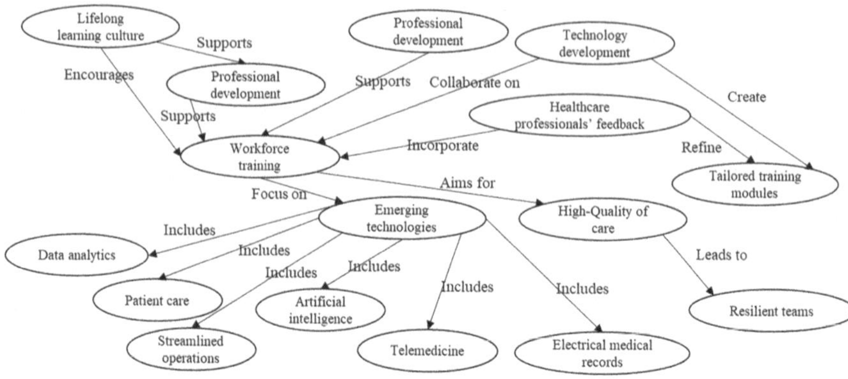

Fig. 17 Workforce training and professional development

- Incorporate Feedback from Healthcare Professionals: Regularly update training materials based on feedback from healthcare workers to ensure that programs remain relevant and effective in meeting the needs of the workforce.
- Support Digital Literacy: Provide access to digital literacy programs to help healthcare professionals effectively navigate and integrate digital tools into their practice.
- Invest in Leadership Training: Equip healthcare leaders with the skills to manage digital transformation, including training on leading change, fostering innovation, and supporting their teams in adapting to new technologies.

5.4 *Public Engagement and Awareness*

Public engagement and awareness are critical to ensuring the successful integration of smart healthcare technologies in urban environments [100]. As the healthcare landscape continues to evolve with the introduction of digital tools and innovations, educating patients about these technologies' benefits and safety is essential. Public education initiatives can help demystify smart healthcare solutions, explaining their functionality and potential to improve health outcomes, accessibility, and patient experience. Informational campaigns that target various demographics through workshops, social media, community outreach, and healthcare provider communications can ensure that all members of the urban population are informed about these advancements. By focusing on clear, accessible communication, these campaigns can break down barriers to understanding and help patients feel more confident in engaging with digital healthcare solutions. Trust and transparency are foundational in establishing a positive relationship between patients and healthcare providers, particularly when it comes to new technologies. Healthcare providers must adopt patient-centered design principles when implementing these smart technologies to ensure that they are user-friendly, accessible, and responsive to individual needs. This approach improves the user experience and encourages patients to take an active role in their healthcare decisions. The more users feel that their needs are considered in the design and application of technologies, the more likely they are to embrace them and participate in their healthcare management. Equally important is transparent communication about the use of patient data, including how it will be collected, stored, and shared. Data privacy concerns often arise when introducing new technologies in healthcare, so it is crucial for healthcare providers to be open about the privacy measures in place and how patient data will be safeguarded. Clear, accessible information about these practices can alleviate fears and build trust in the system. When patients understand that their data is handled securely and ethically, their willingness to engage with digital tools increases, fostering an environment of trust and cooperation. Additionally, involving patients directly in the development, evaluation, and continuous improvement of smart healthcare technologies can provide valuable insights. This collaborative approach not only ensures that the solutions meet the real needs of the community but also fosters a sense of ownership among patients. When patients feel that their input directly influences the technology they use, they are more likely to engage with it and trust its benefits.

Improving public engagement and awareness is possible by adopting these recommendations [101]:

- Implement Public Education Campaigns: Use multiple communication platforms—such as community workshops, social media, and healthcare provider outreach—to inform patients about the benefits, safety, and functionality of smart healthcare technologies. Tailor campaigns to reach different demographics, ensuring accessibility to all community members.
- Adopt Patient-Centered Design: Ensure that smart healthcare technologies are designed with the user in mind, prioritizing accessibility, ease of use, and responsiveness to patient needs. Involve patients in the design process to create solutions that are truly aligned with their preferences and challenges.
- Promote Transparency Around Data Privacy: Communicate openly with patients about data collection, storage, and sharing practices. Provide clear information on privacy protections and security measures to reassure patients about the safety of their personal health data.
- Involve Patients in Technology Development: Engage patients in the co-design and evaluation of smart healthcare technologies to ensure that the solutions meet their needs and provide a sense of ownership. This collaborative approach can improve both the effectiveness and adoption of these tools.
- Encourage Active Participation: Empower patients to take an active role in their healthcare management by demonstrating how digital tools can help them manage their health, access services, and make informed decisions.
- Utilize Multi-Channel Communication: Expand outreach efforts through diverse channels, including social media, websites, and offline community interactions, to ensure that all segments of the population, including vulnerable groups, have access to information about smart healthcare technologies.

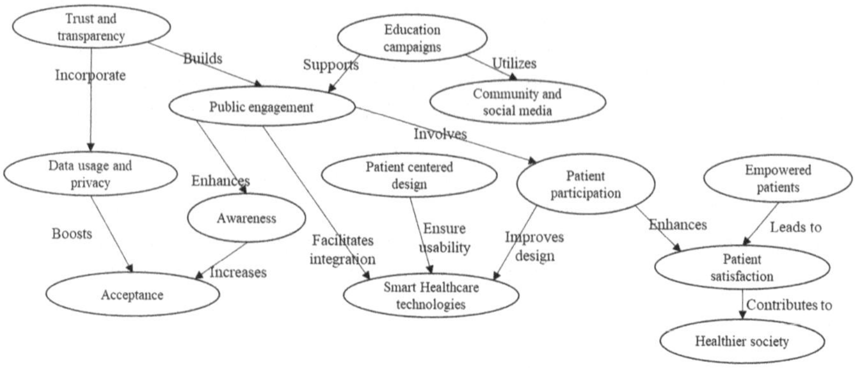

Fig. 18 Public engagement and awareness and smart healthcare technologies

- Foster Trust Through Continuous Engagement: Establish long-term, ongoing communication strategies to maintain transparency and build sustained trust with the public. Offer regular updates, seek feedback, and ensure continuous dialogue between patients and healthcare providers (Fig. 18).

6 Conclusion

The integration of smart healthcare technologies holds immense potential to revolutionize urban health landscapes, offering transformative benefits that include improving access to care, enhancing patient outcomes, and optimizing healthcare delivery processes. The advent of technologies such as AI, the IoE, telemedicine, and big data analytics presents an unprecedented opportunity to address the unique challenges posed by rapid urbanization. Urban centers are often plagued by overcrowded healthcare facilities, an increasing demand for services, and rising healthcare costs, which strain existing systems and resources. By leveraging these advanced technologies, many of these challenges can be mitigated. For instance, AI can assist in diagnostics, treatment recommendations, and predictive analytics, while telemedicine can provide remote care options that alleviate pressure on physical healthcare facilities. Furthermore, IoE and big data analytics can enable real-time monitoring of health data, providing actionable insights that optimize resource allocation and improve the efficiency of healthcare delivery. However, despite these promising advantages, realizing the full potential of smart healthcare technologies is not without significant challenges. One of the most pressing issues is infrastructure limitations, especially in underserved areas. The lack of adequate digital connectivity, internet access, and technological infrastructure in many urban districts can create a digital divide that impedes the widespread adoption of smart healthcare solutions. In these areas, efforts must be made to ensure equitable access to the necessary digital tools and resources, such as high-speed internet, cloud computing capabilities, and devices for telehealth. Without such investments, the promise of smart healthcare may not be accessible to all populations, exacerbating existing health disparities rather than alleviating them. Cybersecurity risks also represent a major concern in the deployment of smart healthcare technologies. The interconnected nature of modern healthcare systems—where health data flows across various platforms and devices—makes these systems highly vulnerable to cyberattacks and data breaches. The privacy and security of patient data are paramount, and any breach could not only compromise individual privacy but also erode trust in digital health solutions. Therefore, it is crucial that robust cybersecurity measures are implemented, such as encryption, secure data storage, and regular system updates. Furthermore, healthcare organizations must invest in creating a culture of cybersecurity awareness, ensuring that healthcare workers understand the importance of protecting patient data and are trained to recognize potential threats. Public engagement and digital literacy are also critical components for the successful adoption of smart healthcare technologies. Even the most sophisticated technologies will fail if patients do not trust or understand them. Thus, educating the public about the benefits, risks, and safety protocols of smart healthcare is essential to fostering an environment of trust and encouraging widespread adoption. This can be achieved through comprehensive public outreach campaigns, community workshops, and clear communication from healthcare providers regarding the safety and efficacy of these technologies. Additionally, it is important to address concerns around data privacy and the ethical use of healthcare data. Transparent communication regarding how data will be handled and protected is vital to alleviating patient concerns and building trust in these technologies. Moreover, continuous training and professional development for healthcare workers are indispensable. As the integration of smart healthcare technologies evolves, healthcare professionals must be equipped with the necessary skills to utilize these tools effectively. Training programs should not only cover the technical aspects of emerging technologies but also emphasize their application in enhancing patient care, improving operational efficiencies, and fostering better health outcomes. Healthcare workers must be prepared to engage with AI-driven diagnostic tools, telemedicine platforms, and data management systems, ensuring that they are comfortable with these tools and capable of maximizing their benefits. Looking ahead, the emerging role of blockchain technology in healthcare is particularly promising. Blockchain offers the potential to enhance data security, interoperability, and transparency, making it easier for healthcare providers to share patient information securely. With blockchain, patient data can be encrypted and stored in a decentralized manner, making it tamper-proof and more resistant to cyber threats. This technology could also streamline the process of transferring medical records between providers, reducing administrative burden and improving care coordination. The use of blockchain could foster greater trust between patients and healthcare providers by ensuring that patient data is handled securely, accurately, and with integrity. Sustainable solutions will be crucial in creating resilient, equitable, and scalable healthcare systems capable of meeting the dynamic needs of urban populations. Investments in digital infrastructure are essential, but so too are collaborative frameworks that bring together technology providers, healthcare organizations, and policymakers. These partnerships can ensure that technological advancements are applied in ways that prioritize patient needs and improve health outcomes. Policymakers must create an environment that supports innovation while also protecting the public interest, ensuring that the benefits of smart healthcare technologies are distributed equitably across different socioeconomic groups. Ultimately, the journey toward a healthier, more connected urban future depends on our ability to effectively leverage the full capabilities of smart healthcare technologies. This will require overcoming existing challenges, such as infrastructure gaps, cybersecurity risks, and public skepticism, while also fostering collaboration across sectors. By addressing these challenges and working together, we can create a healthcare ecosystem that is not only more efficient and effective but also more inclusive and accessible. The integration of smart healthcare technologies holds the potential to improve the quality of care, reduce health disparities, and create more resilient healthcare systems that can better respond to the evolving needs of urban populations. Therefore, the commitment to these technological advancements is not just a technological challenge but a societal imperative to improve the well-being of all citizens.

Competing Interests The authors declare no competing financial or professional interests that could have influenced the content or direction of this chapter.

Ethics Approval This chapter does not involve primary research requiring ethical approval.

References

1. Sukanya R, Tantia V (2023) Urbanization and the impact on economic development. In: New perspectives and possibilities in strategic management in the 21st century: between tradition and modernity. IGI Global, pp 369–408
2. Zhu H, Shen L, Ren Y (2022) How can smart city shape a happier life? The mechanism for developing a Happiness Driven Smart City. Sustain Cities Soc 80:103791
3. Hussain S, Reza M (2023) Environmental damage and global health: understanding the impacts and proposing mitigation strategies. J Big-Data Anal Cloud Comput 8(2):1–21
4. Lawrence RJ (2022) Co-benefits of transdisciplinary planning for healthy cities. Urban Plan 7(4):61–74
5. Cairney P, St Denny E, Mitchell H (2021) The future of public health policymaking after COVID-19: a qualitative systematic review of lessons from Health in All Policies. Open Res Eur 1:23
6. Upadhyay GM, Taneja A, Sharma P et al (2024) Smart cities and healthcare revolution: bridging the future through AI, IoT, and Blockchain. 29 July 2024
7. Haleem A, Javaid M, Singh RP et al (2021) Telemedicine for healthcare: capabilities, features, barriers, and applications. Sensors Int 2:100117
8. Xie Y, Lu L, Gao F et al (2021) Integration of artificial intelligence, blockchain, and wearable technology for chronic disease management: a new paradigm in smart healthcare. Curr Med Sci 41(6):1123–1133

9. Alajlan SA (2024) Governmental policies and healthcare system strengthening in low-income countries. In: Policies, initiatives, and innovations for global health, p 321
10. World Health Organization et al (2021) Policies, regulations and legislation promoting healthy housing: a review
11. Rajagopalan S, Vergara-Martel A, Zhong J et al (2024) The urban environment and cardiometabolic health. Circulation 149(16):1298–1314
12. Olatunji AO, Olaboye JA, Maha CC et al (2024) Revolutionizing infectious disease management in low-resource settings: the impact of rapid diagnostic technologies and portable devices. Int J Appl Res Soc Sci 6(7):1417–1432
13. Murphy L, Markey K, O'Donnell C et al (2021) The impact of the COVID-19 pandemic and its related restrictions on people with pre-existent mental health conditions: a scoping review. Arch Psychiatr Nurs 35(4):375–394
14. Williamson SM, Prybutok V (2024) Balancing privacy and progress: a review of privacy challenges, systemic oversight, and patient perceptions in AI-driven healthcare. Appl Sci 14(2):675
15. Usak M, Kubiatko M, Shabbir MS et al (2020) Health care service delivery based on the Internet of things: a systematic and comprehensive study. Int J Commun Syst 33(2):e4179
16. Liu J, Gatzweiler FW, Kumar M (2021) An evolutionary complex systems perspective on urban health. Socio Econ Plan Sci 75:100815
17. Alsafi Z, Abbas A-R, Hassan A et al (2020) The coronavirus (COVID-19) pandemic: adaptations in medical education. Int J Surg (London, England) 78:64
18. Adeyemi C, Adegboye BO, Odugbose T (2024) The impact of healthcare information technology on reducing medication errors: a review of recent advances. Int J Front Med Surg Res [Online] 5(2):20–29
19. Parker SK, Grote G (2022) Automation, algorithms, and beyond: why work design matters more than ever in a digital world. Appl Psychol 71(4):1171–1204
20. Altamini I, Altamini A, Alhumimidi AS et al (2023) Artificial intelligence (AI) chatbots in medicine: a supplement, not a substitute. Cureus 15(6):e40922
21. Uriawan, W., Adriansyah, S., Maulidiyah, S. J., et al. (2024). Challenges and opportunities: improve patient data security and privacy in distributed systems
22. Shahu A, Kinzer K, Michelles F (2023) Enhancing professional training with single-user virtual reality: unveiling challenges, opportunities, and design guidelines. In: Proceedings of the 22nd international conference on mobile and ubiquitous multimedia, pp 244–256
23. Senk S, Ulbricht M, Tsokalo I, Rischke J, Li SC, Speidel S et al (2022) Healing hands: the tactile internet in future tele-healthcare. Sensors 22(4):1404
24. Omboni S, Padwal RS, Alessa T et al (2022) The worldwide impact of telemedicine during COVID-19: current evidence and recommendations for the future. Connect Health 1:7
25. Thamer N, Alubady R (2021) A survey of ransomware attacks for healthcare systems: risks, challenges, solutions and opportunity of research. In: 2021 1st Babylon International Conference on Information Technology and Science (BICITS). IEEE, pp 210–216
26. Jimmy FNU (2024) Cyber security vulnerabilities and remediation through cloud security tools. J Artif Intell Gen Sci (JAIGS) 2(1):129–171
27. Javaid M, Haleem A, Singh RP, Suman R (2023) Towards insighting cybersecurity for healthcare domains: a comprehensive review of recent practices and trends. Cyber Secur Appl 1:100016
28. Alade OM, Amusan EA, Ojo OJ (2024) Strategic assessment of intricacies in healthcare cyber security: analyzing distinctive challenges, evaluating their ramifications on healthcare delivery, and proposing advanced mitigation strategies. Asian J Res Comput Sci 17(5):238–248
29. Nankya M, Mugisa A, Usman Y et al (2024) Security and privacy in E-health systems: a review of AI and machine learning techniques. IEEE Access, VOLUME 12, 2024, 148796–148816
30. Rafique W, Qaddir J (2024) Internet of everything meets the metaverse: bridging physical and virtual worlds with blockchain. Comput Sci Rev 54:100678

31. Jawad LA (2024) Security and privacy in digital healthcare systems: challenges and mitigation strategies. Abhigyan 42(1):23–31
32. Almalki, Abdullah Hussein, et al. "Revolutionizing Healthcare: The Power of Health Information Systems: Medical records, health information." Journal of International Crisis and Risk Communication Research 7.S10 (2024): 1492.
33. Abrishami R, Aghili SH, Ranjbar MF et al (2024) Assessment of transcription errors and their effects on medical approach, hospitalization duration, and financial costs of traumatic patients: how can we prevent them? Epidemiol Health Syst J 11(3):121–127
34. Khan, Maham, et al. "The impact of artificial intelligence on allergy diagnosis and treatment." Current allergy and asthma reports 24.7 (2024): 361–372.
35. Khan SR, Bernassau AL, & Desmulliez MP (2024) Passive and battery-free RFID-based wireless healthcare and medical devices: A review. IEEE Journal of Radio Frequency Identification.
36. Ştefan A-M, Rusu N-R, Ovreiu E et al (2024) Empowering healthcare: a comprehensive guide to implementing a robust medical information system—components, benefits, objectives, evaluation criteria, and seamless deployment strategies. Appl Syst Innov 7(3):51
37. Prabhod KJ (2024) The role of artificial intelligence in reducing healthcare costs and improving operational efficiency. Q J Emerg Technol Innov 9(2):47–59
38. Chirra DR (2024) Secure data sharing in multi-cloud environments: a cryptographic framework for healthcare systems. Revista de Inteligencia Artificial en Medicina 15(1):821–843
39. Thakur S, Sharma S, Sharma AK et al (2024) Impact of cloud computing in digital healthcare. In: 2024 IEEE International Conference on Computing, Power and Communication Technologies (IC2PCT). IEEE, pp 566–571
40. Arogundade OR (2023) Cloud vs traditional disaster recovery techniques: a comparative analysis. Int Adv Res J Sci Eng Technol 10(4):186–195. https://doi.org/10.17148/IARJSET.2023.10430
41. Rath KC, Khang A, Rath SK et al (2024) Artificial intelligence (AI)-enabled technology in medicine-advancing holistic healthcare monitoring and control systems. In: Computer vision and AI-integrated IoT technologies in the medical ecosystem. CRC Press, pp 87–108
42. Tonetto LM, Saurin TA, Fogliatto FS et al (2023) Information and communication technologies in emergency care services for patients with COVID-19: a multi-national study. Int J Prod Res 61(24):8384–8400
43. Boda, Vishnu Vardhan Reddy. "Edge Computing in Healthcare: What It Is and Why It Matters." International Journal of Emerging Trends in Computer Science and Information Technology 5.4 (2024): 61–70.
44. Thangamani R, Kamalam GK, Vimaladevi M (2024) Revolutionizing healthcare processes: the dynamic role of Blockchain innovation. In: Blockchain for biomedical research and healthcare: concept, trends, and future implications. Springer Nature Singapore, Singapore, pp 229–267
45. Karthikeyan V, Kirubakaran G, Gopalakrishnan K et al (2025) Creative strategies to protect patients' health records and confidentiality using Blockchain technology. In: Blockchain-enabled solutions for the pharmaceutical industry, pp 275–318
46. Jafari, Abdul Muqsit Haji, Ravi Kumar Patchmuthu, and Sharul Tazrajiman Haji Tajuddin. "Immutable COVID-19 vaccination certificate using blockchain." Procedia Computer Science 233 (2024): 194–203.
47. Nwosu NT, Babatunde SO, Ijomah T (2024) Enhancing customer experience and market penetration through advanced data analytics in the health industry. World J Adv Res Rev 22(3):1157–1170
48. Butler CR, Webster LB, Diekema DS (2024) Staffing crisis capacity: a different approach to healthcare resource allocation for a different type of scarce resource. J Med Ethics 50(9):647–649
49. Batko K, Ślęzak A (2022) The use of big data analytics in healthcare. J Big Data 9(1):3

50. Adje KDC, Letaifa AB, Haddad M et al (2023) Smart city based on open data: a survey. IEEE Access 11:56726–56748
51. Ghaffar Nia N, Kaplanoglu E, Nasab A (2023) Evaluation of artificial intelligence techniques in disease diagnosis and prediction. Discov Artif Intell 3(1):5
52. Lustberg MB, Kuderer NM, Desai A et al (2023) Mitigating long-term and delayed adverse events associated with cancer treatment: implications for survivorship. Nat Rev Clin Oncol 20(8):527–542
53. Misra S, Aguilar-Salinas CA, Chikowore T et al (2023) The case for precision medicine in the prevention, diagnosis, and treatment of cardiometabolic diseases in low-income and middle-income countries. Lancet Diabetes Endocrinol 11(11):836–847
54. Ventriglio A, Torales J, Castaldelli-Maia JM, De Berardis D, Bhugra D (2021) Urbanization and emerging mental health issues. CNS Spectr 26(1):43–50
55. Johnson KB, Wei WQ, Weeraratne D, Frisse ME, Misulis K, Rhee K et al (2021) Precision medicine, AI, and the future of personalized health care. Clin Transl Sci 14(1):86–93
56. Gupta P, Manoj KP (2024) "Role of AI for smart health diagnosis and treatment." Smart medical imaging for diagnosis and treatment planning. Chapman and Hall/CRC, 2024. 23–45.
57. KHALIFA, Mohamed et ALBADAWY, Mona (2024) AI in diagnostic imaging: revolutionising accuracy and efficiency. Computer Methods and programs in biomedicine update, 2024, vol. 5, p. 100146.
58. Putra KT, Arrayyan AZ, Hayati N, Damarjati C, Bakar A, Chen HC (2024) A review on the application of internet of medical things in wearable personal health monitoring: a cloud-edge artificial intelligence approach. IEEE Access 12:21437–21452
59. Adeghe EP, Okolo CA, Ojeyinka OT (2024) A review of wearable technology in healthcare: monitoring patient health and enhancing outcomes. OARJ Multidiscip Stud 7(01):142–148
60. Reppou S, Tsoupouroglou I, Pandria N, Bamidis PD (2024, June) Enhancing transitional care for geriatric patients and caregivers: a study on mobile application for disease self-management. In: Proceedings of the 17th international conference on PErvasive technologies related to assistive environments, pp 542–547
61. Bhat KS, Jain M, Kumar N (2021) Infrastructuring telehealth in (in) formal patient-doctor contexts. Proc ACM Human-Comput Interact 5(CSCW2):1–28
62. Painter J, Turner J, Procter PM (2021) Understanding and accommodating patient and staff choice when implementing video consultations in mental health services. CIN: Comput Inform Nurs 39(10):578–583
63. Castillo-Valdez PF, Rodriguez-Salvador M, Ho YS (2024) Research landscape of diabetes mHealth technologies. Foresight STI Govern 18(1):19–32
64. Nikzadipanah A, Rahdar M (2024) Industry 4.0 in smart hospital: a Scientometric study. Hosp Pract Res 9(2):437–447
65. Ansari S, et al. (2025) "Healthcare Marketing and Patient Engagement Strategies." Improving Healthcare Quality and Patient Engagement: Management and Technology Insights. IGI Global Scientific Publishing, 2025, 305–330.
66. Hachoumi N, Eddabbah M, El Adib AR (2023) Health sciences lifelong learning and professional development in the era of artificial intelligence. Int J Med Inform 178:105171
67. Shankar PR (2022) Artificial intelligence in health professions education. Arch Med Health Sci 10(2):256–261
68. Luk CY (2021) The impact of digital health on traditional healthcare systems and doctor-patient relationships: the case study of Singapore. In: Research anthology on telemedicine efficacy, adoption, and impact on healthcare delivery. IGI Global, pp 155–179
69. Son YS, Kwon KH (2023) Utilization of smart devices and the evolution of customized healthcare services focusing on big data: a systematic review. Mhealth 10:7. Korea
70. McCormick R, Estrada J, Whitney C, Hinrichsen M, Lee PT, Cohen AB et al (2021) Teleneurology comprehensive inpatient consultations expedite access to care and decreases hospital length of stay. Neurohospitalist 11(3):229–234. Bosten

71. Maassen O, Fritsch S, Palm J, Deffge S, Kunze J, Marx G et al (2021) Future medical artificial intelligence application requirements and expectations of physicians in German university hospitals: web-based survey. J Med Internet Res 23(3):e26646
72. Dang A (2023) Real-world evidence: a primer. Pharm Med 37(1):25–36
73. Jung K, Kashyap S, Avati A, Harman S, Shaw H, Li R et al (2021) A framework for making predictive models useful in practice. J Am Med Inform Assoc 28(6):1149–1158
74. Larson DN, Schneider RB, Simuni T (2021) A new era: the growth of video-based visits for remote management of persons with Parkinson's disease. J Parkinsons Dis 11(s1):S27–S34
75. Prieto-Avalos G, Cruz-Ramos NA, Alor-Hernandez G, Sánchez-Cervantes JL, Rodriguez-Mazahua L, Guarneros-Nolasco LR (2022) Wearable devices for physical monitoring of heart: a review. Biosensors 12(5):292
76. Umaefulam V, Premkumar K, Koole M (2022) Perceptions on mobile health use for health education in an indigenous population. Digit Health 8:20552076221092537
77. Schacfer MA, Nelson HN, Butrum JL, Gronseth JR, Hines JH (2023) A low-cost smartphone fluorescence microscope for research, life science education, and STEM outreach. Sci Rep 13(1):2722
78. Agonafer EP, Carson SL, Nunez V, Poole K, Hong CS, Morales M et al (2021) Community-based organizations' perspectives on improving health and social service integration. BMC Public Health 21:1–12
79. Verma N, Buch B, Taralekar R, Acharya S (2023) Diagnostic concordance of telemedicine as compared with face-to-face care in primary health care clinics in rural India: randomized crossover trial. JMIR Form Res 7(1):e42775
80. Al Dahdah M (2022) Digital markets and the commercialization of healthcare in Africa: the case of Kenya. Globalizations:1–13
81. Gupta AJ, Turimumahoro P, Ochom E, Ggita JM, Babirye D, Ayakaka I, Mark D, Okello DA, Cattamanchi A, Dowdy DW, Haberer JE, Armstrong-Hough M, Katamba A, Davis JL. mHealth to improve implementation of TB contact investigation: a case study from Uganda. Implement Sci Commun. 2023 Jun 20;4(1):71. https://doi.org/10.1186/s43058-023-00448-w. PMID: 37340456; PMCID: PMC10280918.
82. Woulfe F, Fadahunsi KP, Smith S, Chirambo GB, Larsson E, Henn P et al (2021) Identification and evaluation of methodologies to assess the quality of mobile health apps in high-, low-, and middle-income countries: rapid review. JMIR Mhealth Uhealth 9(10):e28384
83. Bauskar SR, Madhavaram CR, Galla EP, Sunkara JR, Gollangi HK (2022) Predicting disease outbreaks using AI and Big Data: a new frontier in healthcare analytics. Eur Chem Bull 11:4926–4939
84. Tluli R, Badawy A, Salem S, Barhamgi M, Mohamed A (2024) A survey of machine learning innovations in ambulance services: allocation, routing, and demand estimation. IEEE Open J Intell Transp Syst 5:842–872
85. Gianquintieri L, Brovelli MA, Pagliosa A, Dassi G, Brambilla PM, Bonora R et al (2022) Generating high-granularity COVID-19 territorial early alerts using emergency medical services and machine learning. Int J Environ Res Public Health 19(15):9012
86. Pires C (2022) Medical records in community pharmacies: the cases of UK and Australia. Foundations 2(2):399–408
87. Nwosu NT (2024) Reducing operational costs in healthcare through advanced BI tools and data integration. World J Adv Res Rev 22(3):1144–1156
88. Badshah A, et al. "Big data applications: overview, challenges and future." Artificial Intelligence Review 57.11 (2024): 290.
89. Wheeler DA (2021) Using Electronic Health Records (EHR) to reduce documentation and medication errors in ABC Private Duty Nursing Agency. Doctoral dissertation, Trident University International
90. Alam S (2023) PMTRS: a personalized multimodal treatment response system framework for personalized healthcare. Int J Appl Health Care Anal 8(6):18–28

91. SOARES B, FERREIRA A, Veiga PM (2023) The benefits and challenges of blockchain technology and ehealth implementation in Estonia-a literature review. Applied Medical Informatics, 2023 45(4)
92. Johnson T, Kollnig K, Dewitte P (2022) Towards responsible, lawful and ethical data processing: patient data in the UK. Internet Policy Rev 11(1):1–24
93. Bennacer SA, Aaroud A, Sabiri K, Rguibi MA, Cherradi B (2022) Design and implementation of a new Blockchain-based digital health passport: a Moroccan case study. Inform Med Unlocked 35:101125
94. Ghosh R, Kumar S (2021) Mobile health applications during epidemic management in India: a review. EAI Endorsed Trans Smart Cities 5(13):e1–e1
95. Jung CY, Joo HK (2023) Post-'Lee-Luda' personal information protection in Korea: developer responsibility and autonomous AI governance. Int Data Priv Law 13(2):154–167
96. Mohammadzadeh Z, Saeidnia HR, Lotfata A, Hassanzadeh M, Ghiasi N (2023) Smart city healthcare delivery innovations: a systematic review of essential technologies and indicators for developing nations. BMC Health Serv Res 23(1):1180
97. Kwon H, An S, Lee HY, Cha WC, Kim S, Cho M, Kong HJ (2022) Review of smart hospital services in real healthcare environments. Healthc Inform Res 28(1):3–15
98. Tabish SA (2024) Health care management: principles and practice. Springer Books
99. Takhdat K, Eddabbah M, El Hamzaoui H, Lamtali S, El Adib AR (2022) High-fidelity simulation effects on cardiopulmonary resuscitation self-efficacy and knowledge retention in undergraduate nursing students: a two-group, experimental, longitudinal pilot study. Nurs Educ Perspect 43(6):E118–E120
100. Szilvassy P, Širok K (2022) Importance of work engagement in primary healthcare. BMC Health Serv Res 22(1):1044
101. Jiang S, Wu Z, Zhang X, Ji Y, Xu J, Liu P et al (2024) How does patient-centered communication influence patient trust?: the roles of patient participation and patient preference. Patient Educ Couns 122:108161

Overview of Innovations and Challenges of the Smart Healthcare Solutions in the Urban Environments

Wesam Nabil Eid, Fahd Mohammed Aldosari, Amar Yusof Jaffar, and Kanakaprabha Selvakumari

Abstract Urban health development is being influenced more and more by the incorporation of intelligent healthcare solutions that make use of cutting-edge technology like big data analytics, IoT, and artificial intelligence. The effective management of healthcare resources, real-time public health monitoring, and individualized care delivery are some of the special problems faced by densely populated metropolitan environments that these solutions aim to solve. The quality, accessibility, and sustainability of healthcare services in urban areas can all be improved with smart healthcare, as this chapter examines. It sheds light on how cities might use these technologies to address pressing problems like health inequality and the rising prevalence of chronic illnesses while fostering healthier, more resilient communities through the analysis of case studies and new trends. The main objective of chapter is smart medical devices can be influenced by metropolitan health outcomes. Furthermore, it aims to focused on the patient's effective healthcare system that can be enhance in smart cities.

Keywords Artificial intelligence · Big data analytics · Chronic · Healthcare · Metropolitan real-time

W. N. Eid
Department of Cybersecurity, College of Computing, Umm Al-Qura University, Mecca, Saudi Arabia

F. M. Aldosari · A. Y. Jaffar
Department of Computer and Network Engineering College of Computing, Umm Al-Qura University, Makkah, Saudi Arabia

K. Selvakumari (✉)
Department of Artificial Intelligence and Machine Learning, Malla Reddy College of Engineering, Hyderabad, India

H. Kaddouri et al. (eds.), *Smart Cities for Sustainable and Urban Development*, Green Energy and Technology, https://doi.org/10.1007/978-3-031-99001-4_34

1 Introduction

Health is one of the important parts of human well-being. It includes many factors such as quality of life, longevity, and other humanity development. The advanced cases medical science is more improves healthcare, personalized treatments, effective and accessible. However, the cases cities are more urbanized, it creates a lot of health problems such as Asthma, Allergies, and other chronic diseases. There is an urgent need for the effective healthcare solutions to the highly populated areas. The urbanization is facing for both solutions and drawbacks for healthcare. On the one side modern treatments, easy to access the medical facilities, and specialist care unit. On the other side urban regions people are severe cause of health problems particularly like air, noise, sound pollutions, lifestyle of aliments and healthcare facilities. Because of these complications, there is an effective treatment for the smart healthcare solutions in mayor populations. As we know the smart health care solutions means using Artificial Intelligence, Internet of Things, and Wearable devices. The technologies are addressing the complicated health problems in urban areas. It is also enabled for the real time health monitoring system and identify the accurate diseases correctly, also improve the entire health care system. The technologies used to develop these smart systems are described, along with the different approaches used in smart healthcare, such as deep learning, cloud-based IoT applications, fog computing, and challenges and issues faced by smart healthcare. Its scope is broader because it is not designed for a specific application, such as patient monitoring, disease detection, and diagnosis. The quality of life is bestowed by smart health Saba [1]. Modern telemedicine requires the development of wireless body area sensor networks. However, internet criminals and hackers are becoming aware of how they can hack telemedicine systems, and secure health information is now the most expensive on the black market [2]. mHealth, a popular part of telemedicine design, adapts to different sensing technologies and is essential to patient security and privacy. The goal of this study is to improve sensor-based protection and attack mechanisms to safeguard patient privacy on the client side of telemedicine [3].

1.1 Overview of Healthcare Challenges in Urban Settings

The urban areas are facing major obstacles in the entire healthcare system hectic life for standard living and congested areas of the population. Another major reason is highly concentrated on the health care sector is urban areas is overcrowded in malls, hospital, shopping street and hotels etc. And also, longer wait time for even shops and hospitals. Moreover, urban area is facing severe rates of chronic disease such as respiratory issues, diabetes, and cardiovascular disease due to unhealthy foods and contamination. In the metropolitan cities, income is equally divided into the groups of the high quality of the healthcare is major issue. The individuals can be dividing the afford of standard of living and healthcare cost with cover the health insurance. Many of the public health concerns outbreak of the infectious disease is quick and more effective of the system. Another drawback for the healthcare system in urban areas is environmental problems such as noise, air pollution. The populations of the risk are more affected by elderly and children. Meanwhile, mental health issues are most common for metropolitan areas like stress, hypertension, anxiety, and another mental crisis. The different needs for the city's life are increasing accessibility, cost, and better quality of life.

1.2 Urban Health Challenges

The numerous health problems are caused by the urban areas on the health problems. The most common problems are facing the urban areas such as overcrowded hospitals, malls, shopping streets and unequal access of the medical services, Meanwhile, unhealthy habits, more pollution in most of the areas, insufficient nutrition, and people living in cities at higher chance of getting various chronic disease like diabetes, cardiovascular disease, various other respiratory problems. The health inequities are very worsened in the socioeconomic disparities which exposed to the health outcomes. The various other infectious aliments are increased by the more population, necessitating of the public health management and various techniques.

1.2.1 Growing Population and Demand for Healthcare Services

The cities are more population growth as the result of the world's accelerating urbanization of more enormous pressure on the recent healthcare systems. The increasing wait times of people who waited in the hospitals and overcrowded in the populations, this also a heavy burden of the medical professions. The metropolitan areas should be equipped with recent technologies to meet the health services and primarily in the emergency treatment and specialized treatment for the medical procedures. The patients who need for the medical attention, which causes the treatment delays for the insufficient care and other services In addition, urban healthcare system has many kinds of the meet in addition to these operational difficulties. Many of the numerous of demographic groups such as the children and aged people is economically disadvantaged of the groups in the cities. The effective strategies are including the adoption of many healthcare technologies and the address of these challenges in the healthcare delivery in the urban settings.

1.2.2 Environmental Factors (Pollution, Noise, Etc.)

The wellness of the population is greatly affected by the severe levels of pollutions in noise and air which is common seen in the throughout areas. The major caused by the air pollutions such as vehicle emissions, industrial smog from various factories, and the large building construction which leads to various respiratory diseases such as asthma, chronic obstructive pulmonary disease (COPD) and bronchitis etc. This extended prolonged to exposure with air quality can also be the result in cardiovascular issues, raised the mortality rates and especially in groups at risk like children, the old people with prior medical conditions. The contamination levels can be increased by the health concerns in the urban regions, are placing with immense pressure on the healthcare systems. Another critical environmental issue has an impact of noise pollution in urban areas. This is pollution from high level noise from heavy vehicles, industrial noise from large machines, traffic noise can also lead to stress and a higher risk of getting heart disease and hypertension. The environmental variables such as noise and poor air quality is linked to mental health problems like depression and anxiety. Resolving these issues with the environment concerns involves a various strategy is connecting to the public health policy in the effects of pollutions.

1.2.3 Lifestyle-Related Diseases (Diabetes, Hypertension, Etc.)

The urban area is living frequently promotes the lifestyle of the choices in the fuel of the growth in non-communicable various illnesses like overweight, hypertension and diabetes. A recent increase in illnesses is primarily is being into the prolonged behaviour in the bad diets in high level of the stress and all of which are common in the most of the urban areas. The junk food consumption is lack of the physical exercise in the lengthy workdays, and the more convenience of the transportation in cities and all the factors in unhealthy lifestyle choices. The many of the disease like type 2 diabetes and heart disease is more widespread of the heavy burden on the healthcare systems. Many of the lifestyle-related disease in the urban stresses such as traffic congestion, work pressure and lack of less greenery areas this leads to improper food habits and other physical activity.

The long lasting of the consequences from the various aliments is not only the shorten life expectancy of the need for ongoing medical care and management. The urban areas are greater chance of the public awareness campaigns, and preventive healthcare measures and the use of smart healthcare technologies like in various wearable devices and telemedicine platforms can be more helpful in improve the how people can monitor and identify their health. Fig. 1 shows the life style of disease in the health conditions and is influenced in the unhealthy habits and behaviours. This life style disease, including the cancer, diabetes, hypertension, and cardiovascular disease of healthy practices.

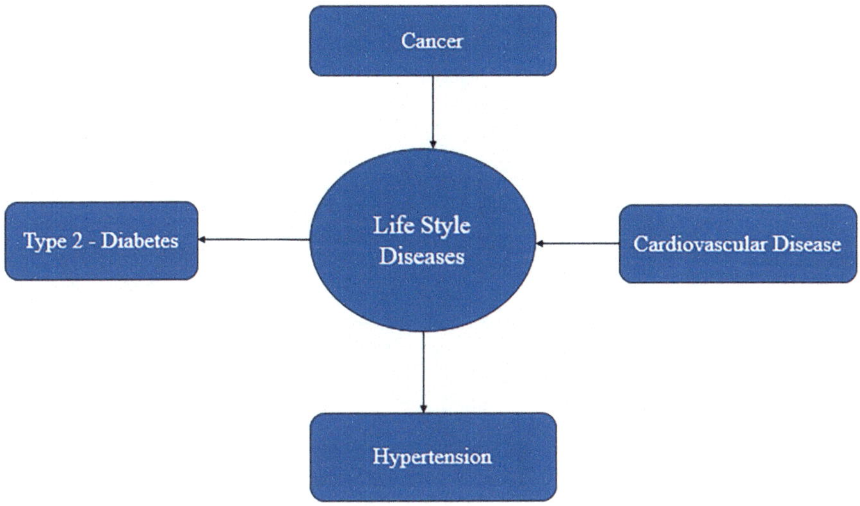

Fig. 1 Life style disease

1.3 Role of Smart Healthcare Solutions in Addressing these Challenges

The smart healthcare solutions can be changing the way of urban healthcare concerns. The solutions can be rectified by Artificial Intelligence, Internet of Things and Big Data. In the heavily populated in urban areas, where the traditional healthcare system is too difficult to manage all the demands.AI—driven technologies for diagnosis is more efficient, quickly identify the disease and also accurately which allows for the medical experts. This is more useful for the chronic disease and life style of related any disease can be recognized easily. The IoT devices are easier to monitor the real-time health monitoring system, which leads to earlier signs of any disease symptoms, ongoing care for those with long term illnesses. This is more useful for reduce the disease rates, particularly beneficial for hospital admissions and easily the burden of overcrowded patients in hospital.

Additionally, telemedicine and mHealth platforms are close gap that can be exist in the urban areas and other health care providers. These technologies can be making more useful remote areas can be accessed medical consultations, more convenient, reduce the traveling cost, especially people who lives in far from urban areas. Smart health app and wearable devices are charging the health track by the fitness levels, tracking by the persistent illnesses and personalized suggestions from smart health apps. When it involves of identifying the disease and handling public health emergencies in metropolitan areas, data analytics is very useful to predicting the disease outbreaks. To control and the spread of more infectious diseases, authorities can be identified by the disease. Everything is considered, smart health care solutions is promoting good healthcare, accessibility, and quality of urban healthcare

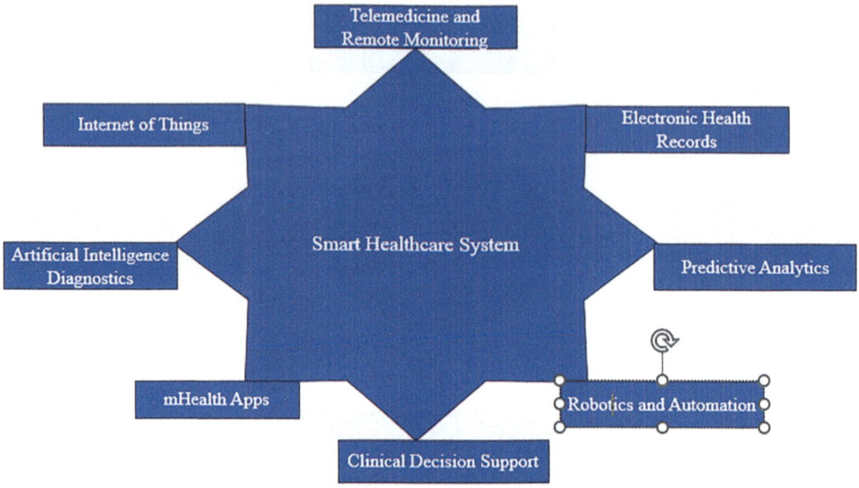

Fig. 2 Smart healthcare system

systems. Fig. 2 illustrates that the smart healthcare system in the various mhealth apps, Clinical decision support, Electronic health records and predictive analytics in the importance of smart healthcare system.

1.4 Objectives of the Chapter

The main objective of the chapter is to explore how efficient of smart health care system is addressing the issues in the urban areas. By rectifying the role of advanced technologies such as Artificial Intelligence, Internet of Things and mHealth have to strengthen the quality and accessibility of healthcare in urban areas. Furthermore, the chapter is also highlighting real time applications of these technologies, how the smart healthcare system can be integrated into urban areas. And also identify the drawbacks and opportunities of with adaption of environment. This chapter is mainly adopted into the insights of how creative approaches can be influenced the future healthcare in urban areas.

2 Smart Healthcare Technologies

The examples of smart healthcare technologies are Artificial Intelligence, telemedicine, and wearable medical devices etc. The IoT enabled devices is supply of real time patient monitoring, particularly for the patient with long term illnesses.

AI-driven solutions can be better diagnosis with specific therapies. The wearable medical technology is allowing people to take their health by tracking their vital signs and promoting proactive healthy lifestyle choices. Telemedicine is also providing the gap between in healthcare to accessible for the remote areas. These technologies allow to the urban healthcare systems meet the increasing demands of a population of the increasing efficiency. One of the most significant sectors that stands to gain from artificial intelligence's (AI) potential to improve numerous areas of human life is healthcare. With problems including data organization, diagnostic accuracy, information retrieval, and cost reduction, medicine continues to be one of the most difficult, costly, and significant industries. AI is well adapted to tackle these issues, which is improving patients' quality of life and decreasing healthcare costs worldwide.

2.1 Telemedicine and Remote Monitoring

The word "telemedicine" is used to describe a broad range of remote medical services. A single description of telemedicine is the provision of medical care through the use of telecommunication technologies. The main objective of telemedicine is the electronic transmission of data between locations. Medical professionals have used electronic communication to improve patient care since the telephone became accessible [4]. The development and broad application of artificial intelligence (AI) in many domains, medical professionals are starting to leverage AI to the fullest extent possible for managing and interpreting data for telemedicine. Telemedicine platforms offer innovative and distinctive options for outpatients who need medical care but cannot afford it or have restricted access to it. The telemedicine's challenges, its capabilities and adaptability must be developed in order to solve telemedicine's issues. The global health sector, particularly telemedicine, has increasingly substantial opportunities because to the Internet of Things (IoT), a continual advancement in technology that enables the extension of contacts across the internet. Additionally, the development of information management capacity has also contributed to this growth. There were countless opportunities for growth when telemedicine and AI were combined, and this paper reviewed the research that has already been done in the field. The four most recent advancements in this subject are patient monitoring, healthcare information technology, intelligent aid diagnosis, and information analysis collaboration [5]. Telemedicine, often known as telehealth, is the application of electronic communication to transfer medical data between sites in order to enhance the health of patients [6]. Telemedicine is emerging as the next big thing in electronic health care. The health care sector is becoming more pliable and adaptable thanks to the new telemedicine stages. It involves arming the patient with the necessary equipment so they may communicate with the healthcare professionals from the comfort of their own homes [7].

2.2 *Wearable Health Devices*

Electronic components or computers intended for wear on the human body are referred to as wearables. Every day, there are more and more health problems, including numerous chronic illnesses that require ongoing care. For this reason, a device that can continuously monitor a person is needed. System that has been developed will continuously monitor the individual without professional assistance. Furthermore, measured parameters will be sent to the server so that they can be monitored remotely by anybody. Utilizing some sort of connectivity capabilities, wearable devices will let users access information instantly. Mobile data acquisition (DAQ) solution based on Android will gather end-user-specific physiological signal data that is obtained by sensors attached to the body, store, analyze, and display it on the smartphone, and then transmit it to a server for additional processing. Because a system needs to be power-efficient for continuous monitoring, Bluetooth low energy (BLE) devices are utilized instead of traditional Bluetooth ones. As a result, the gadget will function more efficiently, and since continuous bodily monitoring is possible, it will also reduce the chance of death by helping to prevent future health problems as well as aid in the treatment of many chronic conditions [8]. The wearable health technology community is highly engaged in wearable health monitoring systems. The smart wristband is one of the wearable health device categories in which the market has made the greatest financial investments. A variety of products are available targeting specific target markets, such as individuals with cardiovascular problems [9]. Wearable technology is one development that is encouraging this advancement and has the potential to be a creative response to problems facing the healthcare industry because it may be used for patient management and sickness prevention and maintenance, including physical monitoring [10]. Numerous wearable gadgets are available to assist people in tracking their vital body indicators, which provides a general sense of their health. The development of high-performance wearable gadgets has been spurred by improvements in the effectiveness of healthcare systems. Portable healthcare solutions have the ability to significantly reduce healthcare expenditures and enable important patients to have ongoing health monitoring from remote areas. The creation of a reliable medical device that is safe, efficient, and trustworthy is the most urgent requirement in this field. It should be able to employ flexible sensors to monitor vital signs from a variety of human organs or the environment within or outside the body [11].

2.3 *Artificial Intelligence in Diagnosis and Treatment*

A major shift in many industries, particularly the medical and healthcare sectors, is being brought about by artificial intelligence. It is observing and becoming increasingly important, something that only seemed unthinkable a few years ago. AI is an enormous benefit to the medical industry and is becoming more and more crucial for physicians, radiologists, nurses, and other healthcare workers. In the modern period, healthcare has grown to be both a major industry and a basic human need. AI has the potential to enhance medical decision-making and address mistakes like incorrect diagnoses and needless, drawn-out procedures that are part of healthcare systems. The many technologies that have been developed or are currently being developed in the context of healthcare include the Internet of Things (IoT), cloud computing, edge computing, artificial intelligence, and big data analysis. The healthcare sector supplies a vast amount of patient health data that machine learning algorithms may utilize to identify issues early on. Better health monitoring and management systems that operate more efficiently and within time constraints are what patients, physicians, and hospitals anticipate [12]. AI has actively evolved to easily and easily integrate into remarkable, interdisciplinary domains such as computer science, education, engineering, and medicine. The purpose of this survey is to present and emphasize the value of AI research in the fields of biomedicine and medical informatics. In this vast field of medical science, we have studied the most recent AI research, paying particular emphasis to medical diagnostics [13].

Figure 3 shows the Artificial Intelligence in Diagnosis and Treatment is transformative role in the healthcare, enhancing the better accuracy of the diagnosis and treatment. The procedure of diagnosing anaemia is time-consuming and costly for the medical community because it involves numerous tests before a definitive diagnosis can be made. The process of classifying is much more intricate and frequently takes years to finish, which deteriorates the prognosis and delays appropriate treatment [14]. Generative Adversarial Networks, Extreme Learning Machines, and Long-Short Term Memory are a few Deep Learning approaches that can do this. Medical practitioners and academics can easily navigate and use platforms created by combining informatics techniques from several structured and unstructured data sources. Identifying and treating COVID-19 cases more quickly is the main advantage of these AI-powered platforms [15]. The primary domains of artificial intelligence application in medicine and showcase some exciting international

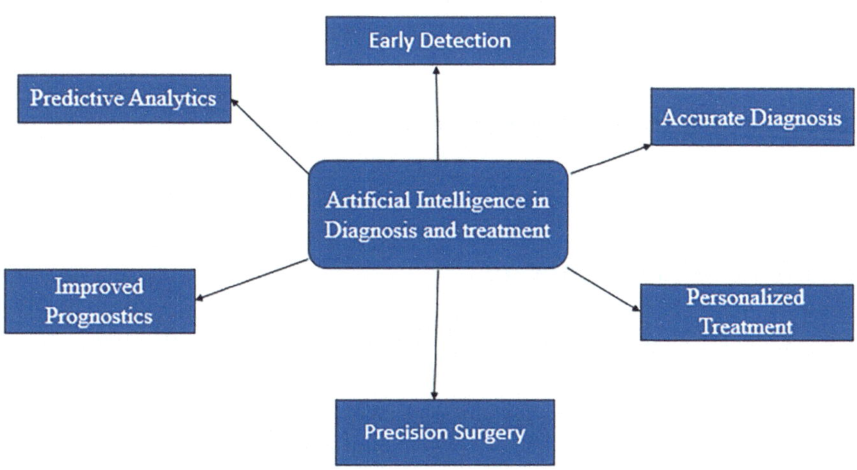

Fig. 3 Artificial intelligence in diagnosis and treatment

advancements in specific fields, including telemedicine, surgical robots, diagnostics, cardiology, and novel medication discovery. The effectiveness of teamwork amongst experts from many domains is critical to the successful implementation of artificial intelligence technologies. Artificial intelligence is being used in the medical and healthcare fields to detect unhealthy processes early on and reduce patient treatment times [16].

2.4 Robotics in Healthcare: Automated Surgery and Hospital Management Systems

Information and communication technology (ICT) and mechatronics are critical fields that include medical robots and computer-assisted therapy. ICT technology has actually had a major impact on the healthcare sector by bringing new approaches to preserving a patient's therapy. Medical robotics is a development of professional and exam robots together with other technologies, as surgical navigation has become more common, especially in minimally invasive surgery [17]. Medical robots are classified into the following groups: robots for neurosurgery, orthopedics, endoscopic surgery, robotics for instructional purposes, robots for rehabilitation assistance, and robots for capsules. In addition to studying medical 3D image modeling, virtual surgery simulation, open-source surgical systems, and remote surgery in difficult situations including medical robots, a quick overview is given of the research and application status of conventional surgical robots both domestically and internationally [18] The increasing usage of telerobotic devices in surgery, patients can expect more dependable results and faster recovery times. By effectively providing physical and occupational therapy, robotic rehabilitation systems allow for more intensive care that is constantly tailored to the demands of the patient. The discovery of novel treatments for a wide range of illnesses and disorders could be sparked by technological advancements in robotics, which could also raise patient outcomes by raising the bar for care coordination and accessibility. The purpose of this article is to suggest a few of the most significant medical and health-care robotics technological advancements and skills that are required to enhance human health and wellbeing [19].

3 IoT in Urban Healthcare

3.1 Overview of the Internet of Things (IoT) in Healthcare

The Internet of Things (IoT) has enhanced patient care in the healthcare industry by enabling remote monitoring, telemonitoring, and expanding the use of telemedicine. Because it offers improved medical facilities for patients, doctors, and hospitals alike, IoT plays a significant role in the medical industry. The five main parts of the proposed system are diagnostics, hospital patient care, pharmacist patient care, primary report preparation, and patient data collecting. The goal of the suggested system is to offer top-notch patient care even in isolated locations. It may even be intelligent enough to evaluate the information gathered by wearable Internet of Things sensors and suggest a checkup [20]. The seamless platform that the Internet of Things (IoT) provides to connect people and things to one another can improve and simplify our lives. The help of this vision, from compute-based, centralized schemes to a more distributed environment with a wide range of applications, including smart cities, smart wearables, smart homes, and smart mobility. Massive data management, including the management of individual person photos and reports, increases security concerns and human labor [21]. IoT devices, software, UI, and communication networks are all used to improve the functionality of smart cities. Long wait times, high costs, manual payment procedures, travel distances to medical facilities, and a need for frequent doctor visits are some of the inherent shortcomings of the traditional healthcare system that are criticized. All the necessary components for smart healthcare systems to supplant traditional healthcare systems are available, and the Internet of Things (IoT) is regarded as the best option for low-cost, 24/7 patient monitoring [22]. A nation's level of development is significantly influenced by its healthcare system. The factors influencing the quality of healthcare are the cost, nature, and availability of medical treatments. They provide primary patient monitoring and care assistance to improve our nation's health care system from rural to urban sectors by keeping an eye on health metrics including blood pressure, heart rate, temperature, ECG, and so on. The kit may be used around-the-clock since real-time data can be stored and retrieved from the cloud at any time or location. The 1:10,000 doctor-to-patient ratio, which is an important issue in rural India, is now addressed by this kit since professionals across the country can be the data on their timeline [23]. The most important aspects of the human telemedicine idea are the IoT based smart health monitoring system. Modern information and communication technologies have made it possible to create the Internet of Things (IoT), which has numerous practical uses. The storage, display, and communication of information are the fundamental functions of the Internet of Things. As a result, the IoT-based e-Health monitoring system is customized for ongoing remote patient monitoring and collects, aggregates, and analyzes data. The area of managing esmart health for patients in cities or rural areas [24].

3.2 *IoT-Enabled Healthcare Devices and Systems for Real-Time Health Monitoring*

The Internet of Things of today is altering the way typical gadget users think. Our lives comfortable and hassle-free due to the ongoing advancements in the field of the internet of things, which are causing an uprising in the machinery of the present.

Risking one's health is not a restricted specification in the modern era as compared to all age groups, owing to the hectic lifestyle and inappropriate behaviors. The Internet of Things has enabled a far simpler solution to remote real-time health monitoring of victims from the home and hospital in these harmful settings [25] real-time patient health monitoring system in an Internet of Things setting that makes use of vital signs like body temperature and pulse rate. The liquid crystal display unit which is connected with a mobile application or device software displays the results [26]. An Internet of Things (IoT)-based smart health monitoring system that continuously checks a person's body temperature, heart rate, and oxygen saturation to provide medication on schedule. To put the suggested system into practice, a system with a temperature sensor, heartbeat sensor, and SpO2 sensor based on the Arduino platform and cloud Sheets API has been built [27]. The Internet of Things, or IoT, is a quickly expanding network of linked objects that includes wearables, smart appliances, actuators, sensors, and actuators. IoT development systems are becoming more sophisticated as a result of the spread of these connected devices, which produces enormous volumes of data and makes a wide range of applications and services possible. Scalability, flexibility, and maintainability of IoT systems require careful consideration of design patterns and modeling methodologies in order to handle this complexity. IoT-based technologies are being developed in the healthcare sector to manage costs and enhance patient care and outcomes [28]. IoT use in healthcare has increased significantly for a variety of specialized medical applications. As a result, in the end, it improves the diagnosis of significant medical anomalies compared to earlier technologies. Thus, it increases the system's effectiveness and efficiency to the point where it may be used as a portable device. In order to diagnose serious and essential cardiac conditions, medical practitioners still favor the patient's ECG data. The electrocardiogram (ECG) provides important insights into the many functions of the heart across time [29].

3.3 Smart Hospitals and Urban Health Systems Integration

Smart Hospitals is combination of advanced technology such as Artificial Intelligence, Internet of Things, and data analytics to produce an effective, patient-centered care settings. This is future for the healthcare industry. The increasing medical requirements is significant populations of the smart hospitals in the urban settings must be connected to form a larger health system. These can be reduced by the AI-driven diagnostic tools for the correct medial choices, real-time patient monitoring system, and computerized technologies to speed up hospital operations from the patient admission to the discharge. The patients can be receiving continuous, better medical treatment in the smart hospital and urban health systems are combined. This promotes the better care of coordination across the several healthcare facilities. For example, Electronic medical records can be safely throughout the clinics, hospitals, and specialists, increasing information exchange and cutting down on repetition. In Additional, telemedicine platform and health apps enhance remote consultations in highly crowded locations, It have access to timely in healthcare. It is integration reduces wait times and both efficient and urban health needs.

4 Data-Driven Approaches for Urban Health Development

The data-driven methods are transformed in the growth of metropolitical areas is identified by the huge amount of medical data to improve the patient data and other decision-making. The emergence of the wearables and electronic health records and Internet of Things. These data can be identified into the insights of the disease outbreaks and other public health trends, and other requirements of the specific patients. Data Analytics and Machine Learning algorithms are used for the healthcare providers are able to identify treatments plans for the urban areas. The data-driven techniques can be helpful for solve problems like hospitals and equal access to care in the metropolitan areas. The predictive analytics can be identifying the patient inflow patterns, hospitals can be better control of employees' levels and other allocations. The public health authorities can also be used in data for identifying risk locations for the transmission of diseases or chronic illnesses. The medical professional can be identified more greater awareness of understanding of urban health issues and efficient healthcare solutions.

4.1 Big Data and Predictive Analytics for Personalized Healthcare

Big data and predictive analytics can be customized treatment for the strategies on each patient of the health. The more use of the big data is allowing for the collecting and analysis of large amount of patient's information such as medical history, lifestyle of factors, real-time patient monitoring system in the metropolitical areas. The advanced algorithms in machine learning models are used to identify data in order to find patterns, predict the health risks and specific treatment recommendations. The chronic disorders such as diabetes, hypertension, and cancer, in the early management is vital, predictive analytics is improving early diagnosis of diseases. The healthcare practitioners are mitigating the need for the emergency care and hospitalizations in identifying in potential risks of health and implementing preventive measures by the examining trends in patient data. In addition, customized treatment is made possible by the prediction of models, which treatment is more successful and unique demands of every patient. The individualized treatment is more effective and the sustainable for the city healthcare systems, and also improving the patient outcomes.

4.2 Use of Electronic Health Records (EHR) in Urban Healthcare Systems

The modern urban health care systems are significant on the use of digital health records, which improve stability, accessibility, and efficiency of the service. The healthcare providers of may easily and quick and safely patient data across multiple levels of electronic health records, which digitize patient information such as treatment plans, previous history of medical treatment and medication records. The digital records can be frequently care from various providers. And the chance errors unnecessary tests and enhancing the quality of care. Additionally, the digital health records to data-driven city healthcare administration.

Figure 4 depicts the Electronic Health Records is enable to healthcare providers to store, access, and share patient data. The digital health records authorities and healthcare practitioners of identify epidemics of illness, detect the gaps in care for marginalized populations, and monitor populations health patterns by collecting patient data. The decision making can be improved and public health emergencies is more effectively due to real-time data sharing and analysis. The digital health records are combining with predictive analytics, which makes it possible to create more accurate treatment plans for each patient based on their whole medical history. The digital healthcare systems are improving in both more general public health initiatives and the outcomes for the specific patients.

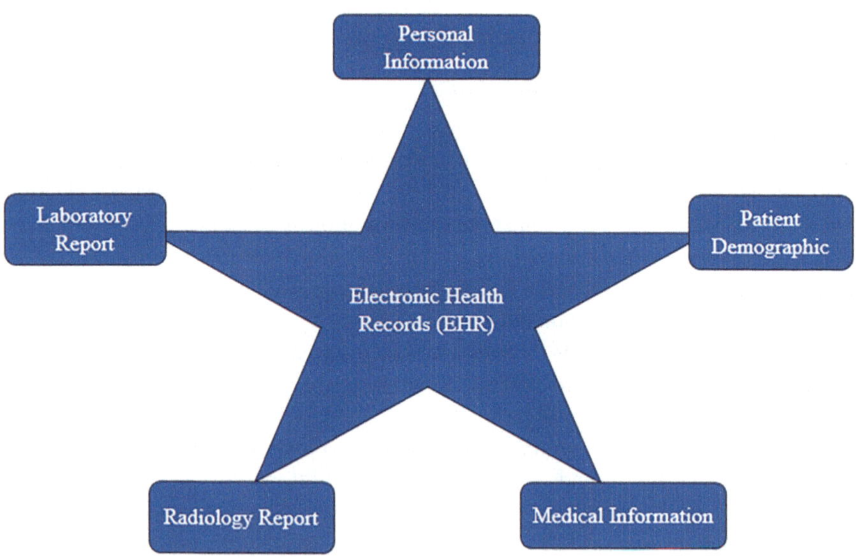

Fig. 4 Electronic Health Records (EHR) in urban healthcare system

5 Mobile Health (mHealth) Applications

The applications for the mHealth have transformed in urbanized healthcare through improving the efficiency and accessible of healthcare services. These applications are more used smartphones and other devices to securely get medical guidance, manage chronic illnesses, and other goals to monitor their health. Healthcare security value and healthcare confidence have had a beneficial effect on the way users interact with mHealth app functions, information, and doctors to increase user satisfaction with public health governance. Additionally, the parasocial relationships between doctors and mHealth app users have a positive effect on the way users interact with doctors to improve user fulfillment with public health governance [30].

Figure 5 shows the Mobile Health Applications are the digital versions of patient's and real time collection of health Information. And also, for the crucial for efficient healthcare delivery, ensuring continuity of care and data decision-making. Furthermore, mHealth solutions allow patients to reduce the need for in-person visits by arranging appointments, prescription of reminders and specialists via telemedicine. The overcoming gaps in conventional healthcare delivery, and other apps improve patient engagement and contribute to other outcomes in fast-paced other contexts by promoting proactive healthcare delivery.

5.1 *Role of mHealth Apps in Urban Areas*

The mHealth apps are essential to metropolitical area healthcare because is they provide easy, quick access to health management resources and other services. mHealth apps to provide patients a convenient in highly places of the medical facilities is overcrowded and appointments in the long time. These apps, living in the cities on their vital signs and manage long term conditions like diabetes and

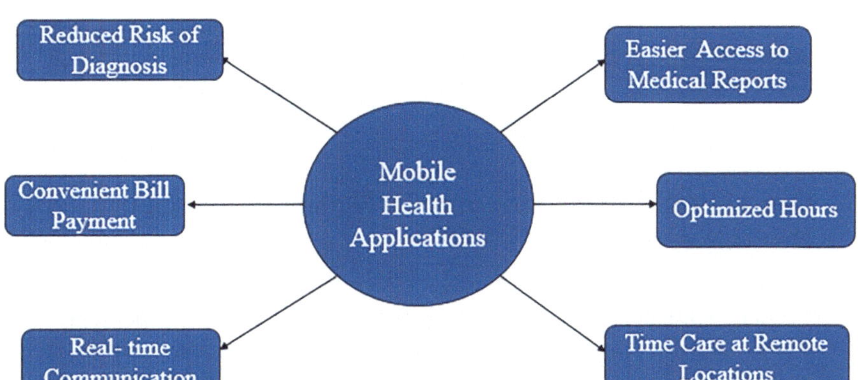

Fig. 5 Mobile health applications

hypertension and medical advice. The telemedicine functions in mobile health apps for the online consultations with medical specialists, guidance without requiring in-person meetings. The Managing personal health, mHealth apps support preventative care and other awareness, is improves urban health. They include any mental health support, vaccine and other screening reminders and educational resources. The health outcomes are provided by mHealth apps, which also lessen the congested healthcare systems through improving and the tools need to take charge of their health. The increasing popularity of mHealth apps as essential resources for managing health and avoiding chronic illnesses. This study focuses on mobile health apps that help people control their diabetes autonomously [31]. Personalization of mHealth applications and enabling users to identify new personal health aims were widely reported, showing that mHealth may be able to promote the implementation of personalized healthcare [32].

5.2 *Disease Management and Health Education Through Mobile Apps*

The mHealth applications are increasing in importance as resources for the illness management, education, and health, especially in metropolitan areas to healthcare can be limited. These apps can give users in real-time in the health tracking, prescription reminders and personalized health analytics to help manage in chronic disease such as diabetes, hypertension, and asthma. The mHealth apps enable patients to take control of their health by providing tools like blood pressure monitoring, and symptom management logs. This encourages patients to follow treatment more closely and stays in hospital. Disease management are important for mHealth apps as well as health education. They can access to reliable information about a variety of health-related subjects, such as exercise, mental health, nutrition, and preventive medicine. These apps can make in making education decisions about their health and adopting better habits. Urban population is public health results are improved by mHealth apps interactive elements, which included video, articles, audio, and quizzes. These features promote lifelong learning and increase public awareness of new health.

6 Overview of Innovations and Challenges of the Smart Healthcare Solutions

This section mentioned the key contributions of each paper. Each paper represents the various smart healthcare solutions of the technology used and the applications to improving the healthcare sectors. The innovation of the technology such as wearable sensors, remote monitoring tools and AI-powered diagnostic tools for the efficient data streaming, improving the healthcare outcomes. This unique technology is used for the predictive healthcare models to real-time health monitoring systems. In overview of the smart healthcare systems can be identified by the way of services. These advanced technologies are offered by the treatment plans, continuous health monitoring systems faster efficiency of the healthcare outcomes. These technologies (Table 1) are widely used in urban solutions.

This research highlights in IoT, Deep Learning, Blockchain, and the other areas. The different aspects of the smart healthcare systems and the different ways of the significant among the healthcare solutions in these similarities. This chapter mainly aims to holistic view of the smart healthcare solutions in urban areas.

- Integrating the latest and most modern advances in technology.
- Address challenges related to connectivity, safety, and transparency.
- Highlighting the practical applications like IoT, AI, and blockchain.
- Combining the gaps between the city planning and medical services

Table 1 Literature survey

Title	Summary	Technology discussed
Kumari et al. [33]	The aim of this effort offers a standard online platform that physicians and parents may use to save and retrieve children's medical records, including recommendations for vaccinations.	Web and mobile-based technologies
Safavi et al. [34]	The security and privacy issues in smart homes and healthcare are covered in this analysis, with a focus on the importance of safe methods to protect user data from online threats.	Smart health technologies
Safari et al. [35]	For smart care facilities, this research proposes a deep learning-based clinical decision management and patient monitoring system which includes visual data for quick responses.	Deep learning
Khan et al. [36]	This research demonstrates the use of a speech and video-based patient monitoring system in smart city contexts during the COVID-19 pandemic.	Smart health applications
Delsi and Amutha [37]	The integration of wearable technology with real-time data transmission is covered in this paper, which analyzes IoT technologies and communication frameworks in the health care sector.	IoT technologies
Chaudhary et al. [38]	The edge of things-based secure system for smart healthcare is put out, which enables immediate access and real-time health monitoring.	Real-time analysis
Zaw et al. [39]	The paper explores the role of ICT in smart healthcare and emphasizes the need for a comprehensive definition of smart healthcare in the context of smart cities.	ICT advancements

(continued)

Table 1 (continued)

Title	Summary	Technology discussed
Hassan et al. [40]	Blockchain and Machine Learning be combined to improve security and efficiency in the smart healthcare industry while emphasizing patient autonomy and rights.	Machine learning
Zhang et al. [41]	This study highlights the importance of information flow for effective healthcare delivery through investigating the interactions between smart healthcare systems and other smart city subsystems.	Smart city subsystems
Chandni et al. [42]	The health hazards associated with pollution in urban settings, the article addresses real-time air quality monitoring utilizing internet of things devices.	IoT
Elhosseini et al. [43]	The promise of AI and big data analytics to improve operational effectiveness and innovative governance is emphasized in this in-depth investigation of the fields in healthcare.	AI and big data
Vidap et al. [44]	A blockchain with permissions system for electronic medical records is proposed in the report, ensuring patient control over access and data protection.	Blockchain
Al-Jaroodi et al. [45]	This study highlights the significance of technology in addressing healthcare issues by looking at the human factors affecting stakeholders' acceptance of Healthcare 4.0 technologies.	Healthcare 4.0
Pradhan et al. [46]	The influence of 5G technology on smart healthcare systems is addressed in the study, along with a revolutionary architecture that reduces costs and improves efficiency.	5G technology
El Majdoubi et al. [47]	To ensure stakeholder satisfaction, a privacy ontology for S-healthcare is recommended in order to align safeguards with patient preferences and legal constraints.	Privacy in healthcare
Shaikh et al. [48]	The article discusses Internet of Things (IoT)-based healthcare technologies, outlining issues and offering solutions for enhancing smart health systems.	IoT
Yamanoor et al. [49]	To promote affordability and accessibility, this paper offers methods for creating a Personal Health Dashboard for patients using wearable technology.	Wearable technologies
Haryadi et al. [50]	To improve security and traceability in medication workflows, an e-prescription management system with integration with the NFT is recommended.	NFTs and blockchain
Volkov et al. [51]	The establishment of an extensible platform for healthcare services reduces the costs related to creating new products and permits flexibility in service delivery.	Healthcare platforms
Alghanim et al. [52]	To detect privacy gaps and provide engineering solutions, this piece suggests a privacy structure for medical care in smart cities.	Privacy engineering

7 Conclusion and Perspectives

The smart healthcare solutions, which provide cutting-edge instruments and other technologies is improve performance, accessibility, and patient care are transforming the way other health concerns are identified. The modern healthcare systems are more effectively in the rising needs of other growing populations, tackle the spike in chronic disease, an lessen the negative effects of other external factors on health by through the use of techniques like mHealth applications, data-driven techniques, and IoT-enabled devices. The urban healthcare facilities require the combination of smart hospitals, personalized healthcare, and predictive analytics. The future healthcare will more depend on technologies of each solution that can be encourage into preventive medicine, better health literacy, and ensure access to everyone as the cities continue to change. An urban health system that can be serve as all kinds of populations and enhance public health outcomes overall, other businesses, and healthcare of professionals they must continue to collaborate. Smart healthcare is a significant advancement in addressing health problems that urban environments in currently and future generations.

References

1. Raoof SS, Durai MAS (2022) A comprehensive review on smart healthcare: applications, paradigms, and challenges with case studies. Contrast Media Mol Imaging 2022:4822235. https://doi.org/10.1155/2022/4822235
2. Mohsin AH, Zaidan AA, Zaidan BB, Albahri AS, Albahri OS, Alsalem MA, Mohammed KI (2018) Real-time remote health monitoring systems using body sensor information and finger vein biometric verification: a multi-layer systematic review. J Med Syst 42(12):238. https://doi.org/10.1007/s10916-018-1104-5
3. Shuwandy ML, Zaidan BB, Zaidan AA, Albahri AS (2019) Sensor-based mHealth authentication for real-time remote healthcare monitoring system: a multilayer systematic review. J Med Syst 43(2):33. https://doi.org/10.1007/s10916-018-1149-5
4. Moghadas A, Jamshidi M, Shaderam M (2008) Telemedicine in healthcare system. In: 2008 World automation congress, Waikoloa, HI, USA. IEEE
5. Kadu A, Singh M (2021) Comparative analysis of e-health care telemedicine system based on Internet of Medical Things and Artificial Intelligence. In: 2021 2nd International conference on smart electronics and communication (ICOSEC), Trichy, India. https://doi.org/10.1109/ICOSEC51865.2021.9591941
6. Park H-A (2018) Secure telemedicine system. In: 2018 International conference on computational science and computational intelligence (CSCI), Las Vegas, NV, USA. https://doi.org/10.1109/CSCI46756.2018.00146
7. Riffat SA, Harun F, Hassan T (2020) An interactive tele-medicine system via Android application. In: 2020 Advanced computing and communication Technologies for High-Performance Applications (ACCTHPA), Cochin, India. https://doi.org/10.1109/ACCTHPA49271.2020.9213200
8. Wanjari ND, Patil SC (2016) Wearable devices. In: 2016 IEEE international conference on advances in electronics, communication and computer technology (ICAECCT), Pune, India. https://doi.org/10.1109/ICAECCT.2016.7942600

9. Fei H, Ur-Rehman M (2020) A wearable health monitoring system. In: 2020 International conference on UK-China emerging technologies (UCET), Glasgow, UK. https://doi.org/10.1109/UCET51115.2020.9205468
10. Caixeiro T, Čale D, Coutinho C (2022) Wearable devices for health remote monitoring system. In: 2022 International symposium on sensing and instrumentation in 5G and IoT. https://doi.org/10.1109/ISSI55442.2022.9963458
11. Verma K, Preity P, Ranjan R (2023) An insight into wearable devices for smart healthcare technologies. In: 2023 13th International conference on cloud computing, Data Science & Engineering (confluence), Noida, India. https://doi.org/10.1109/Confluence56041.2023.10048809
12. Deeba F, Patil SR (2021) Implementation of artificial intelligence in disease prediction and healthcare system—a survey. In: 2021 Innovations in power and advanced computing technologies (i-PACT), Kuala Lumpur, Malaysia. https://doi.org/10.1109/i-PACT52855.2021.9696698
13. Farrugia A, Al-Jumeily D, Al-Jumaily M, Hussain A, Lamb D (2013) Medical diagnosis: are artificial intelligence systems able to diagnose the underlying causes of specific headaches? In: 2013 Sixth international conference on developments in eSystems engineering (DeSE), Abu Dhabi, United Arab Emirates. https://doi.org/10.1109/DeSE.2013.72
14. Kovačević A, Lakota A, Kuka L, Bečić E, Smajović A, Pokvić LG (2022) Application of artificial intelligence in diagnosis and classification of anemia. In: 2022 11th Mediterranean conference on embedded computing (MECO), Budva, Montenegro. https://doi.org/10.1109/MECO55406.2022.9797180
15. Kaur S, Kaur M (2022) Using deep learning techniques for diagnosis & treatment with COVID-19 & AI technology. In: 2022 11th International conference on System Modeling & Advancement in research trends (SMART), Moradabad, India. https://doi.org/10.1109/SMART55829.2022.10047343
16. Semenov VP, Baranova LY, Yagya TS (2022) Application of artificial intelligence in medicine. In: 2022 XXV International conference on soft computing and measurements (SCM), Saint Petersburg, Russian Federation. https://doi.org/10.1109/SCM55405.2022.9794891
17. Alseddiqi M, AlMannaei B, Najam O, Al-Mofleh A (2022) The importance of medical robots in improving healthcare services. In: 2022 International conference on automation, computing and renewable systems (ICACRS), Pudukkottai, India. https://doi.org/10.1109/ICACRS55517.2022.10029105
18. Zhang X, Ma X, Zhou J, Zhou Q (2018) Summary of medical robot technology development. In: IEEE international conference on mechatronics and automation (ICMA), Changchun, China, pp 443–448. https://doi.org/10.1109/ICMA.2018.8484458
19. Okamura AM, Matarić MJ, Christensen HI (2010) Medical and healthcare robotics. IEEE Robot Autom Mag 17(3):26–37. https://doi.org/10.1109/MRA.2010.937861
20. Gera S, Mridul M, Sharma S (2021) IoT-based automated healthcare monitoring system for smart city. In: International conference on computing methodologies and communication (ICCMC), Erode, India, pp 364–368. https://doi.org/10.1109/ICCMC51019.2021.9418487
21. Jassim DA, Hreshee SS (2022) Internet of things in health care system: cloud computing review. In: International conference on engineering technology and its applications (IICETA), Al-Najaf, Iraq, pp 348–354. https://doi.org/10.1109/IICETA54559.2022.9888733
22. Umar U, Khan MA, Irfan R, Ahmad J (2021) IoT-based cardiac healthcare system for ubiquitous healthcare service. In: International congress of advanced technology and engineering (ICOTEN), Taiz, Yemen, pp 1–6. https://doi.org/10.1109/ICOTEN52080.2021.9493478
23. Bhutada S, Singh A, Upadhyaya K, Gaikar P (2021) Ru-Urb IoT-AI powered healthcare kit. In: International conference on intelligent computing and control systems (ICICCS), Madurai, India, pp 417–422. https://doi.org/10.1109/ICICCS51141.2021.9432257
24. Swamy TJ, Murthy TN (2019) eSmart: an IoT-based intelligent health monitoring and management system for mankind. In: International conference on computer communication and informatics (ICCCI), Coimbatore, India, pp 1–5. https://doi.org/10.1109/ICCCI.2019.8821845
25. Sainadh AVMS, Mohanty JS, B. P. S., G. V. Teja, and R. K. Bhogal. (2021) IoT-enabled real-time remote health monitoring system. In: International conference on intelligent com-

puting and control systems (ICICCS), Madurai, India, pp 428–433. https://doi.org/10.1109/ICICCS51141.2021.9432103
26. Dwivedi S, Bairwa B, Shriharinayaka P (2024) Design and development of IoT-enabled health monitoring system using node MCU. In: International conference on intelligent technologies (CONIT), Bangalore, India, pp 1–6. https://doi.org/10.1109/CONIT61985.2024.10626290
27. Khaparkar S, Mishra A (2023) A smart tele-healthcare system for real-time health monitoring and remote consultation. In: International conference on innovations in high-speed communication and signal processing (IHCSP), Bhopal, India, pp 57–62. https://doi.org/10.1109/IHCSP56702.2023.10127169
28. Hassine MB, Tounsi I, Kacem MH (2023) Developing IoT-based smart health monitoring systems using design patterns. In: ACS/IEEE international conference on computer systems and applications (AICCSA), Giza, Egypt, pp 1–8. https://doi.org/10.1109/AICCSA59173.2023.10479310
29. Deshmukh GR, Chaskar UM (2018) IoT-enabled system design for real-time monitoring of ECG signals using TIVA C-series microcontroller. In: International conference on intelligent computing and control systems (ICICCS), Madurai, India, pp 976–979. https://doi.org/10.1109/ICCONS.2018.8663001
30. Cao J, Zhang G, Liu D (2022) The impact of using mHealth apps on improving public health satisfaction during the COVID-19 pandemic: a digital content value chain perspective. Healthcare 10(3):479. https://doi.org/10.3390/healthcare10030479
31. Shen Y, Wang S, Shen Y, Tan S, Dong Y, Qin W, Zhuang Y (2024) Evaluating the usability of mHealth apps: an evaluation model based on task analysis methods and eye movement data. Health 12(13):1310. https://doi.org/10.3390/healthcare12131310
32. Arafa A, Mostafa ZM, Sheerah HA, Alzahrani F, Almuzaini Y, Senosy S, Hassan RIA (2022) mHealth app barriers, usability, and personalization: a cross-sectional study from Egypt and Saudi Arabia. J Pers Med 12(12):2038. https://doi.org/10.3390/jpm12122038
33. Kumari S, Haripriya A, Aruna A, Vidya DS, Nithy MN (2017) Immunize—baby steps for smart healthcare: smart solutions to child vaccination. In: 2017 International conference on innovations in green energy and healthcare technologies (IGEHT), Coimbatore, India. https://doi.org/10.1109/IGEHT.2017.8094048
34. Safavi S, Meer AM, Joel Melanie EK, Shukur Z (2018) Cyber vulnerabilities on smart healthcare, review and solutions. In: 2018 cyber resilience conference (CRC), Putrajaya, Malaysia. https://doi.org/10.1109/CR.2018.8626826
35. Safari G, Majidi B, Khanzadi P, Manzuri MT (2018) Cross-platform e-management for smart care facilities using deep interpretation of patient surveillance videos. In: 2018 25th National and 3rd international Iranian conference on biomedical engineering (ICBME), Qom, Iran. https://doi.org/10.1109/ICBME.2018.8703545
36. Khan UT, Zia MF (2021) Smart city technologies, key components, and its aspects. In: 2021 International conference on innovative computing (ICIC), Lahore, Pakistan. https://doi.org/10.1109/ICIC53490.2021.9692989
37. Delsi RS, Amutha B (2023) IoT revolutionizing healthcare: a survey of smart healthcare system architectures. In: 2023 International conference on research methodologies in knowledge management, artificial intelligence and telecommunication engineering (RMKMATE), Chennai, India. https://doi.org/10.1109/RMKMATE59243.2023.10369980
38. Chaudhary RRK, Chatterjee K, Singh A (2021) A health monitoring framework based on edge of things for smart healthcare system. In: 2021 IEEE 6th international conference on computing, communication and automation (ICCCA), Arad, Romania. https://doi.org/10.1109/ICCCA52192.2021.9666367
39. Zaw TO, Kyaw S, Muthaiyah A, Jasbi A (2021) Contextualization of smart healthcare: a systematic review. In: 2021 7th International conference on research and innovation in information systems (ICRIIS), Johor Bahru, Malaysia. https://doi.org/10.1109/ICRIIS53035.2021.9617060
40. Hassan M, Chen J, Zhu C, Zukaib U (2022) Adoption of blockchain-based artificial intelligence in healthcare. In: 2022 5th International conference on artificial intelligence and big data (ICAIBD), Chengdu, China. https://doi.org/10.1109/ICAIBD55127.2022.9820137

41. Zhang L et al (2020) Evaluation of smart healthcare systems and novel UV-oriented solution for integration, resilience, inclusiveness, and sustainability. In: 2020 5th International conference on Universal Village (UV), Boston, MA, USA. https://doi.org/10.1109/UV50937.2020.9426210
42. Chandni NS, Ismail M, Muzahidul Islam AKM (2022) IoT-driven solution for indoor air quality monitoring system to develop a smart healthcare environment: A review-based study. In: 2022 International conference on decision aid sciences and applications (DASA), Chiangra, Thailandi. https://doi.org/10.1109/DASA54658.2022.9765098
43. Elhosseini MA, Gharaibeh NK, Abu-Ain WA (2023) Trends in smart healthcare systems for smart cities applications. In: 2023 1st International conference on advanced innovations in smart cities (ICAISC), Jeddah, Saudi Arabia. https://doi.org/10.1109/ICAISC56366.2023.10085212
44. Vidap P, Bhargav A, Paswan R, Jewalikar A (2023) Blockchain solution to electronic healthcare records. In: 2023 International conference on intelligent and innovative technologies in computing, electrical and electronics (IITCEE), Bengaluru, India. https://doi.org/10.1109/IITCEE57236.2023.10090970
45. Al-Jaroodi J, Mohamed N, Kesserwan N, Jawhar I (2023) Human factors affecting the adoption of healthcare 4.0. In: 2023 IEEE international systems conference (SysCon), Vancouver, BC, Canada. https://doi.org/10.1109/SysCon53073.2023.10131064
46. Pradhan B, Das S, Roy DS, Routray S, Benedetto F, Jhaveri RH (2023) An AI-assisted smart healthcare system using 5G communication. IEEE Access 11:108339–108355. https://doi.org/10.1109/ACCESS.2023.3317174
47. El Majdoubi D, El Bakkali H, Sadki S, Leghmid A, Maqour Z (2021) HOPPy: holistic ontology for privacy-preserving in smart healthcare environment. In: 2021 Fifth world conference on smart trends in systems security and sustainability (WorldS4), London, United Kingdom. https://doi.org/10.1109/WorldS451998.2021.9514051
48. Shaikh Y, Parvati VK, Biradar SR (2018) Survey of smart healthcare systems using internet of things (IoT). In: 2018 International conference on communication, computing and internet of things (IC3IoT), Chennai, India. https://doi.org/10.1109/IC3IoT.2018.8668128
49. Yamanoor S, Yamanoor NS (2021) Position paper: low-cost solutions for home-based healthcare. In: 2021 International conference on communication systems & networks (COMSNETS), Bangalore, India. https://doi.org/10.1109/COMSNETS51098.2021.9352859
50. Haryadi GA, Allwinnaldo J, Lee M, Kim D-S (2023) Enhanced healthcare system with NFT-prescription. In: 2023 14th International conference on information and communication technology convergence (ICTC), Jeju Island, Republic of Korea. https://doi.org/10.1109/ICTC58733.2023.10392924
51. Volkov I, Radchenko G (2021) Architecture of mHealth platform for storing, exchanging, and processing of medical data in smart healthcare. In: 2021 Ural symposium on biomedical engineering, radio electronics and information technology (USBEREIT), Yekaterinburg, Russia. https://doi.org/10.1109/USBEREIT51232.2021.9455081
52. Alghanim AA, Rahman SMM, Hossain MA (2017) Privacy analysis of smart city healthcare services. In: 2017 IEEE international symposium on multimedia (ISM), Taichung, Taiwan. https://doi.org/10.1109/ISM.2017.79

A Domain Specific Language-Based Chatbot Development for Smart Tourism: Applications in Smart Cities

Lamya Benaddi, Adnane Souha, Charaf Ouaddi, Abdeslam Jakimi, and Brahim Ouchao

Abstract With the rise of smart cities, the tourism sector is adopting smart solutions to enhance visitor experiences by leveraging urban infrastructures and technologies for personalized and sustainable services. Among these solutions, Artificial Intelligence (AI) and NLP-based smart applications are pivotal in improving tourist interactions throughout their journey. However, integrating these advanced functionalities into smart applications remains challenging, driving the development of chatbots. These automate interactions to provide fast, personalized assistance, with platforms like Watson, Dialogflow, Bot Framework, and Amazon Lex simplifying chatbot creation. Despite this, the process remains complex, requiring expertise in software development, AI, and financial resources for NLP services. To address these challenges, this study proposes designing and implementing a Domain-Specific Language (DSL) tailored for chatbot development in smart tourism applications. The DSL aims to simplify and standardize the development of chatbots, facilitating their integration with urban services such as public transport, local events, and tourism flow management. This research highlights the challenges and opportunities associated with the use of DSL-based chatbot development in the tourism sector. Thus, it offers promising prospects for improving the ecosystem of smart cities, enhancing their attractiveness and ability to effectively manage tourism infrastructures. In addition, this paper provides an analysis of DSL implementation possibilities, supported by a comparative study of existing approaches.

Keywords Smart cities · Smart tourism · Artificial intelligence (AI) · Domain specific language (DSL) · Chatbot

L. Benaddi · A. Souha · C. Ouaddi · A. Jakimi (✉) · B. Ouchao
Software Engineering and Information Systems Engineering Team, Faculty of Sciences and Techniques of Errachidia, UMI-Meknes, Meknes, Morocco

H. Kaddouri et al. (eds.), *Smart Cities for Sustainable and Urban Development*, Green Energy and Technology, https://doi.org/10.1007/978-3-031-99001-4_35

1 Introduction

Smart cities are urban spaces that integrate advanced digital technologies to efficiently manage their infrastructures and improve the quality of life of their residents [15]. These connected cities exploit sensor networks, the Internet of Things (IoT), Big Data systems and AI to optimize key areas such as transport, energy, waste management and security. In this context, smart tourism has become an important pillar for attracting tourists while ensuring sustainable and efficient management of passenger flows [28]. Indeed, smart tourism, also developed on a digitized basis, can offer tourists a personalized, fluid and immersive experience, while integrating sustainable practices that meet the needs of modern cities [6].

However, managing tourist flows in smart cities presents numerous challenges. It involves not only controlling visitor movement to prevent overcrowding and conserve resources but also delivering information and services that meet visitors' expectations in real time. As data volumes continue to grow and visitor satisfaction requirements increase, cities need to adopt innovative solutions to manage these flows effectively. Intelligent chatbots are needed to support and enhance the visitor experience in these cities. Designed to provide instant access to information, personalized recommendations, and real-time assistance, chatbots are a promising solution to the challenges of smart tourism. By integrating with smart city infrastructures, these chatbots can not only answer visitors' questions, but also tailor responses to personal preferences and needs, creating a richer and more interactive travel experience [9].

The development of chatbots has evolved significantly in recent years, from the simple automated response tools to sophisticated conversational agents capable of understanding users' needs in natural language and responding to them in a contextualized mode. This progress is based on advances in AI, and NLP. Creating effective chatbots for specific sectors, such as tourism, education, or healthcare, requires specialized technical skills, including building complex conversations and managing a variety of scenarios. Consequently, there is a growing need for DSLs to simplify and structure chatbot development. DSL provides syntax and concepts tailored to the specific requirements of each domain, making it easy to create customized dialogs, precise interactions and business logic without the need for in-depth expertise in general programming. By providing an appropriate framework, DSL accelerates chatbot development and improves chatbot accuracy, enabling developers and designers to focus on user experience and interaction efficiency.

In this chapter, we present the theoretical foundations of an MDE (Model Driven Engineering) approach aimed at accelerating the creation of chatbots dedicated to smart tourism in the context of smart cities. Our approach proposes a DSL and a metamodel enabling users with limited programming skills to design chatbots in the tourism domain, while liberating them from the technical complexities associated with frameworks and programming languages. In addition, we explore the possibilities of DSL implementation through a comparative study of existing approaches, in order to highlight the advantages and limitations of these solutions.

The remainder of this chapter is structured into four main sections. First, Sect. 2, we introduce the key concepts needed to understand our approach, exploring the notions of smart tourism and smart cities, chatbots, as well as the fundamentals of MDE and DSL, while highlighting their role in developing solutions tailored to smart tourism. Second, Sect. 3, we provide a literature review of DSLs for chatbot development. Then, in Sect. 4, we present a research gap in DSLs and the objective of this chapter. In Sect. 5 we detail our proposed approach by presenting an overview of DSL, and the associated metamodel, as well as an analysis of implementation possibilities, supported by a comparative study of existing approaches. Section 6 gives the challenges and limitations of our research. Finally, Sect. 7 concludes this work with a summary of the contributions made and a reflection on prospects for improvement, accompanied by a discussion of future work to extend and validate this approach.

2 Key Concepts

2.1 *Smart Cities*

Smart cities are urban environments that utilize advanced technologies to enhance the quality of life for residents and improve the experiences of visitors [15]. They incorporate digital solutions such as the Internet of Things (IoT), Artificial Intelligence (AI), and 5G connectivity to create interconnected systems capable of collecting, analyzing, and acting on data in real-time. These technologies support efficient monitoring and management of urban services, optimizing city functionality and promoting sustainable development.

One of the key aspects of smart cities is their intelligent infrastructure, which forms the foundation for a range of applications, including smart tourism. These infrastructures enable cities to offer tailored, efficient, and engaging experiences to visitors by seamlessly integrating tourism into the broader urban ecosystem.

2.2 *Smart Tourism*

Smart tourism builds on the principles of smart cities by leveraging advanced technologies such as recommender systems and chatbots to revolutionize the travel experience. Through tools like mobile applications and booking platforms, destinations can provide real-time information, personalized recommendations, and seamless interactions with visitors. Features such as automatic translation, real-time updates, and services for people with reduced mobility ensure accessibility for all,

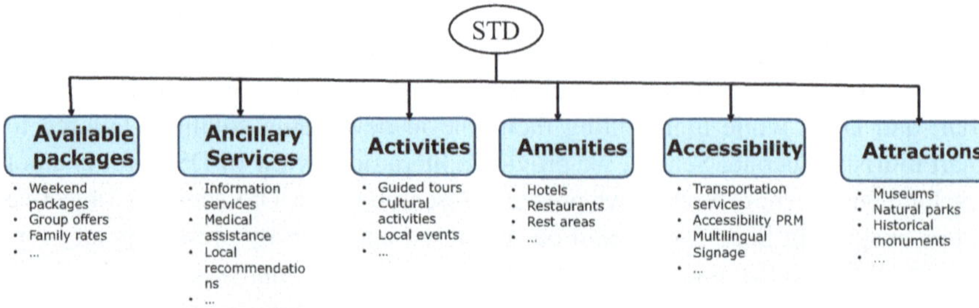

Fig. 1 Key components of the 6A Framework for STD analysis

while applications like Visit a City[1] and TripAdvisor[2] offer tailored itineraries and reviews that enhance engagement and satisfaction.

The integration of these smart technologies in tourist destinations has given rise to the concept of Smart Tourism Destinations (STDs) [8]. To illustrate this concept, some destinations are leveraging mobile applications that align with the six core components of STDs identified by Buhalis et al.: Attractions, Accessibility, Amenities, Activities, Available Packages, and Ancillary Services. These applications highlight the crucial role of mobile technology in creating personalized and intelligent tourism experiences that enhance visitor satisfaction and engagement.

To explore the STD concept, Fig. 1 shows the six fundamental components identified by Buhalis and Amaranggana Buhalis and Amaranggana [8]. These components form the basis for creating a fluid, rich travel experience. Each component plays a specific role in the functionality of an STD. For example, Attraction includes cultural sites such as museums and historical monuments that attract tourists. Accessibility highlights transportation services, multilingual signage and facilities for persons with reduced mobility. Amenities focus on basic services such as hotels and restaurants, while Activities emphasize interactive opportunities such as guided tours and cultural events. Available Packages offer customized travel deals, while Ancillary Services provide support services such as medical assistance and local advice. By combining these components with advanced technologies, destinations can create interconnected ecosystems that stimulate innovation, increase visitor satisfaction and support sustainable growth in the tourism industry.

A variety of emerging technologies are being used to optimize each component of Smart Tourism Destinations (STDs). Table 1 illustrates how tools such as AI, NLP, IoT, recommender systems, and many others are transforming these components into truly intelligent, interconnected ecosystems. These innovations not only improve the quality of services but also offer personalized experiences that meet the specific needs of visitors, while boosting the overall efficiency of STDs. For example, attractions benefit from interacting with chatbots to provide information about tourist sites, while accessibility is improved by harnessing Big Data and smart

[1] https://www.visitacity.com/.

[2] https://www.tripadvisor.com/.

Table 1 Integration of emerging technologies across the 6A components of STD

Technologies	6A					
	Attractions	Accessibility	Amenities	Activities	Available Packages	Ancillary Services
AI	Personalized recommendations for visitors	Traffic flow analysis	Optimization of hospitality services	Personalized experiences	Creation of customized packages	Optimization of ancillary services
NLP (chatbot)	Interaction via chatbots to provide information	Voice and chatbot assistance for visitors	Request management via chatbots	Chatbots for organizing activities	Reservation assistance via chatbots	Real-time customer support via chatbots
IoT	Attraction flow management	Real-time transport tracking	Equipment automation (connected rooms)	Data collection for interactive experiences	Reservation monitoring	Management of local services (shuttles)
Big data	Analysis of visitor preferences	Itinerary optimization	Data analysis to anticipate needs	Activity recommendations	Creation of customized packages	Service improvement
Recomender system	Attraction suggestions based on tastes	Transportation and itinerary recommendation	Hotel and guesthouse recommendation	Activity recommendations	Personalized packages	Travel agency recommendations, tour guide recommendation
AR (augmented reality)	Interactive visits	Site navigation assistance	Visualization of amenities	Immersive experiences	–	–
VR (virtual reality)	Virtual visit simulations	–	–	Immersive games, educational virtual tours	–	–
Geolocation	Local news in real time	Optimized browsing	–	Locate nearby activities	–	–

sensors. Similarly, hotel services and tourism experiences are enriched by personalized technologies and mobile apps, contributing to a more fluid and engaging overall experience.

2.3 Chatbot in Tourism

Chatbots, also known as conversational agents [20] or bots [13] or virtual assistants [27], are software applications designed to facilitate human-computer interactions through natural language. These agents can interpret user inputs, process them, take relevant actions, and generate responses [1]. In recent years, chatbots have demonstrated their value across numerous fields, streamlining tasks and enhancing user experience in areas such as automated customer service, flight booking, and e-commerce [2]. Additionally, they offer convenient access to a wide range of services through websites, mobile apps, and social media platforms like Twitter, Telegram, and Messenger [26]. This allows users to engage with services without the need to install additional applications. By supporting text and voice-based conversations in natural language, chatbots simplify the interaction process and make accessing services more intuitive [12].

Tourism dedicated chatbots are virtual assistants designed to help travelers throughout their journey, from trip planning to return. Their main advantage is their 24/7 availability, enabling travelers to receive information, advice and instant assistance without time or language restrictions. In addition to rapidity, chatbots can also help reduce the workload of human agents by handling repetitive and routinized requests such as bookings, information on local attractions or the provision of personalized itineraries [29].

However, chatbots in the tourism sector still face some challenges. On the one hand, they require a detailed understanding of the user's intent to provide relevant answers, which poses difficulties for natural language processing, particularly for regional expressions or uncommon languages. Moreover, chatbots need to be constantly updated according to local information and changing conditions, such as temporary closures or new recommendations. Security and confidentiality aspects are also essential, as users often share personal data and payment information when interacting with these services.

Tourism chatbots are increasingly used to enhance the user experience by providing personalized information and services. Notable examples include the VisitOrlando[3] chatbot, which assists travelers in planning their stay in Orlando, the Booking.com Bot[4] for seamless hotel bookings, and the Expedia Bot[5] for managing reservations and providing personalized recommendations.

[3] https://fr.visitorlando.com/.

[4] https://www.booking.com/.

[5] https://www.chatbotguide.org/expedia-bot.

2.4 Model Driven Engineering (MDE)

MDE is an approach to software development that places models at the core of all stages, from business analysis to code generation [7]. Unlike traditional, source-code-driven methods, MDE favors abstract, formalized representations of systems. This general approach aims to improve software quality, productivity and maintainability by structuring development around models.

Model Driven Architecture (MDA) [17], standardized by the Object Management Group (OMG) [18], is a specific application of MDE. It is based on the use of models as the main development artifacts for separating business and technological preoccupations. This approach enables the creation of flexible, scalable systems, particularly suited to multi-platform projects. MDA relies on three main types of models: PIM, PSM and PDM (Fig. 2):

- PIM (Platform-Independent Model): An abstract model that focuses on the business logic and core functionality of a system, without reference to a specific technology platform. For example, a PIM for a chatbot would define user interactions and data flows without specifying the programming language or implementation platform.
- PSM (Platform-Specific Model): A model derived from PIM, completed with technical details specific to a particular platform (e.g., Android, iOS, AWS, etc.). For example, a PSM for a chatbot would include configurations specific to an NLP platform such as Dialogflow or Rasa.
- PDM (Platform Description Model): A model that describes the technical characteristics and constraints of a given platform. This model serves as a guide for transforming PIM into PSM.
- Model transformations: MDA is based on automated processes for converting PIM into PSM, and then into executable code. These transformations are based on standards such as QVT (Query/View/Transformation) to ensure consistency and efficiency.

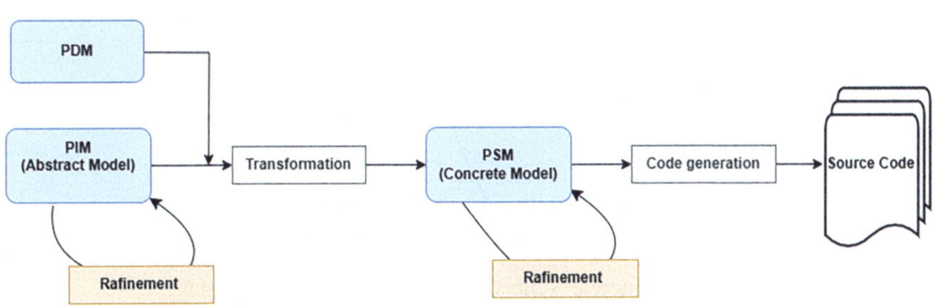

Fig. 2 Model transformation process in MDA

2.5 Domain Specific Language (DSL)

DSL is a language designed specifically to respond to the needs of a specific application domain [11], as opposed to a general-purpose language (GPL) such as Java or Python, designed for more general uses. DSLs enable concepts and solutions to be expressed in a language closely related to the business domain, providing abstractions that facilitate communication between business experts and developers. By focusing on specific domains, DSLs reduce complexity and increase development efficiency.

DSLs are classified into two major categories: textual DSLs and graphical DSLs (or visual DSL). Textual DSLs are languages used to define scripts, configurations or queries, such as SQL, which enables interaction with databases, or other specific languages designed for automation tools. Graphical DSLs, on the other hand, are used to represent graphical models or diagrams, such as UML for software modeling, or tools dedicated to the design of visual workflows. These two types of DSL offer solutions tailored to specific needs, facilitating modeling and automation in a variety of fields.

A DSL is composed of four main elements (Fig. 3):

- **Abstract Syntax**: Defines the logical structure of the DSL, focusing on its underlying concepts rather than its representation.
- **Concrete Syntax**: Specifies how the constructs of the DSL are presented, whether in textual or graphical form.
- **Semantics:** Describes the meaning and behavior of the constructs, including how they are interpreted and executed.
- **Serializable Syntax**: Ensures that the constructs of the DSL can be reliably saved and exchanged in a consistent format.

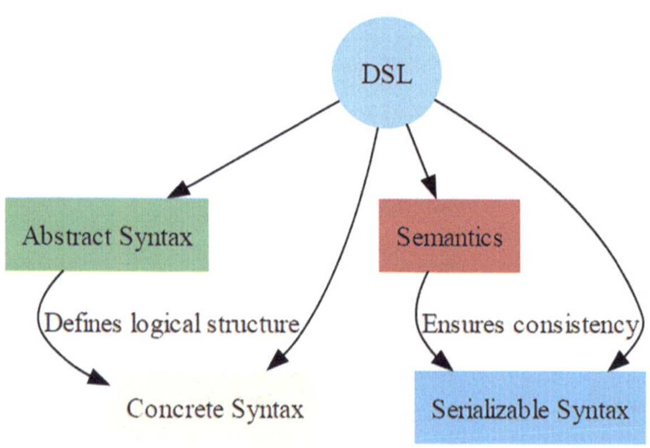

Fig. 3 Key components of a DSL

3 Literature Review of DSLs

To simplify and accelerate this process, researchers have introduced DSLs that offer specialized abstractions and enable chatbot models to be structured efficiently. For example, Conga [25], Xatkit [10], dFlow [14], and the studies [3, 21] indicated in facilitate the design and implementation of essential chatbot components, while offering the possibility of generating code adapted to different platforms like rasa and Dialogflow.

In parallel, other research has focused on the definition of metamodels to structure DSLs. These works, notably those related to Bot Composer [5], the Rasa framework [22], and Dialogflow [24], provide conceptual frameworks for organizing the elements of conversational systems. These contributions make it possible to standardize the structure of chatbots while simplifying their customization [4]. Furthermore, the systematic literature reviews [16, 19] provide insights related to chatbot development using MDE. They have thoroughly examined existing DSLs, as well as the approaches and technologies employed in their development.

4 Research Gap of DSLs and Objectives

DSLs have emerged as promising tools to simplify the development of chatbots, particularly by addressing the complexities associated with NLP, conversational modeling, and integration with various platforms. Several research efforts have attempted to advance the use of DSLs in this domain. For example, Xatkit [10] offers a metamodel-driven DSL focusing on three key aspects: natural language understanding (NLU) or intent metamodel, execution metamodel, and platform metamodel. This enables developers to create chatbots tailored to specific tasks.

Another significant contribution is Conga cited in [25], which facilitates the creation of conceptual models for chatbots. It supports the generation of chatbot implementations on platforms such as Dialogflow and Rasa, bridging the gap between high-level design and practical deployment. Similarly, Ouaddi et al. [21] propose a graphical DSL that simplifies the modeling of conversational scenarios and generates the necessary project files for Rasa-based chatbots. This approach offers a visual interface for defining interactions, making it accessible to users with minimal technical expertise.

Despite these advancements, existing DSLs for chatbot development face notable limitations. First, they do not comprehensively address the diverse range of chatbot development platforms and frameworks currently available. For instance, many frameworks, such as Amazon Lex, Microsoft Bot Framework, and others, are not fully integrated into these solutions, leaving developers to manually adapt their designs for unsupported platforms. Second, the reliance on APIs with high associated costs poses a significant barrier to adoption, particularly for small businesses and developers working within limited budgets. Moreover, existing DSLs often focus on isolated aspects of chatbot development, such as conversation modeling or NLU, without offering end-to-end support for the entire lifecycle, including deployment, and maintenance. This fragmented approach limits their applicability in real-world scenarios, where interoperability and scalability are critical.

This chapter addresses these gaps by exploring the theoretical foundations of a Model Driven Engineering (MDE) approach aimed at accelerating the creation of chatbots dedicated to smart tourism in the context of smart cities. Our approach proposes a DSL and a metamodel enabling users with limited programming skills to design chatbots in the tourism domain while liberating them from the technical complexities associated with frameworks and programming languages. In addition, we explore the possibilities of DSL implementation through a comparative study of existing approaches, in order to highlight the advantages and limitations of these solutions.

5 Proposed Approach

Our approach is based on the use of a DSL specifically designed for developing intelligent chatbots in the field of smart tourism.

5.1 *Principle of the Proposed Approach*

Our approach proposes a conceptual framework based on the Model-Driven Architecture (MDA) paradigm to accelerate the development of chatbots dedicated to the tourism sector. It focuses on creating chatbots that are flexible, scalable, and adapted to specific user needs, taking advantage of the separation of business and technological concerns. Based on models at different abstraction levels (PIM, PDM, and PSM), this framework relies on automatic model transformation, guaranteeing a seamless transition from business specifications to executable code. Figure 4 provides a detailed illustration of this approach.

The metamodel defines key concepts for chatbot design, including the types of user interactions (e.g., requesting information about attractions or booking services), tourism entities such as hotels, museums, restaurants, or local events, as well as the data flows needed to personalize responses, such as user preferences or location.

The approach involves creating PIM models, transitioning to PSM, and generating the source code of chatbots through successive transformations between these models.

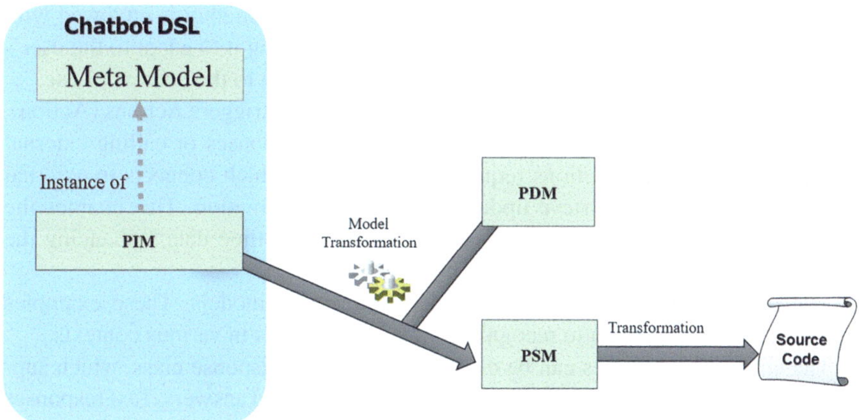

Fig. 4 DSL and MDE-based approach for developing smart tourism chatbots

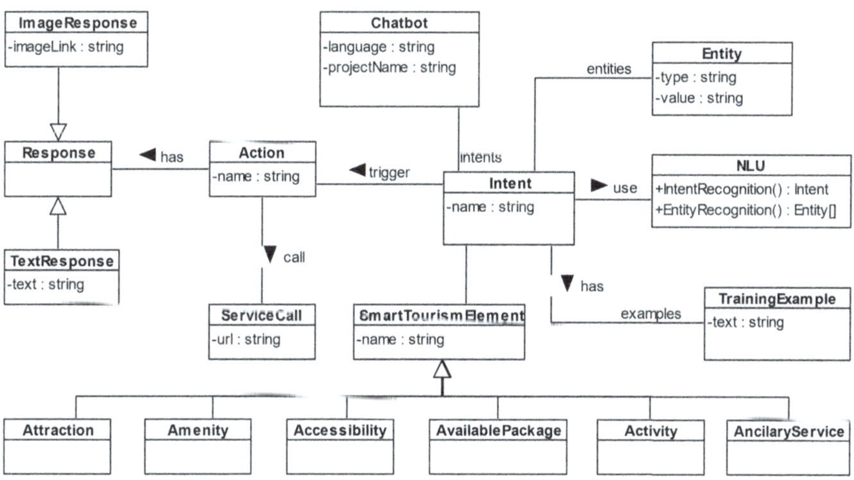

Fig. 5 Chatbot metamodel for smart tourism

5.2 *Metamodel of DSL*

In the context of smart tourism, the development of chatbots requires specific tools [23] that can meet personalization and interoperability requirements. To respond to these needs, a dedicated DSL is provided. This language enables chatbots to be modeled independently of any platform while facilitating their transformation into concrete implementations adapted to the various technologies available.

The proposed chatbot metamodel (Fig. 5) is based on an architecture composed of several key elements, specially designed to meet the specific requirements of the smart tourism field. At the core of this metamodel is the main element, the chatbot, responsible for managing interactions with the user. This chatbot is endowed with key attributes, such as language and project name, enabling it to adapt to the user's linguistic preferences, while being personalized according to the context of use.

To accomplish tasks based on user intents, the chatbot triggers Actions (Action), which represent specific operations such as sending responses or calling external services. Some of these actions require a ServiceCall, which connects to external services via a URL to retrieve updated or specific information. This enables the chatbot to leverage external resources by retrieving real-time data, enhancing the completeness and relevance of its responses.

TrainingExamples are sentences used to train NLU models. These examples enable the chatbot to learn to recognize intents and entities in various contexts.

The chatbot's responses can be customized using a Response class, which supports various formats, including text-based and image-based answers. Text responses provide information in written form, while image responses transmit information visually, such as a photo of a requested attraction. This flexibility enhances the user experience by diversifying response formats to better meet user needs.

Finally, to adapt to the specificities of smart tourism, the model introduces the SmartTourismElement class. This component groups together the 6As of the STDs: Attraction, Accessibility, Activity, Amenity, AncillaryService and AvailablePackages. This approach ensures that the diverse needs of visitors are fully and consistently considered.

5.3 DSL Implementation Approaches

The design of a DSL is based on various ecosystems and frameworks that provide the essential tools for its creation, transformation, and management. These ecosystems are key to the modeling, code generation and integration of DSLs within complex software environments. Table 2 provides an overview of the main elements and characteristics of these ecosystems and frameworks.

The table above (Table 2) compares various ecosystems and frameworks for creating DSLs, including the Eclipse ecosystem, the Microsoft ecosystem, and frameworks such as ANTLR, Spoofax, JetBrains MPS, MetaEdit+ and others.

The Eclipse ecosystem offers a complete set of open-source tools covering all stages from modeling to code generation: Xtext for textual DSLs, Sirius for graphical DSLs, EMF for metamodel-based modeling, and Acceleo for code generation. It is particularly well suited to projects using the MDE approach, offering great flexibility and solid support for model management. However, these tools can be complex to use, and the Eclipse IDE is often considered resource intensive.

The Microsoft ecosystem is mainly centered around Visual Studio (VS), supplemented by tools such as DSL Tools for graphical DSLs and Roslyn for internal DSLs. It integrates seamlessly with C# and .NET development and provides solid documentation and an intuitive user experience. However, major limitations include dependency on VS and limited adaptability to external DSLs.

Table 2 Comparison of ecosystems and frameworks for designing and managing DSLs

Criteria	Eclipse ecosystem[a]	Microsoft ecosystem[b]	Other frameworks/ approach
License	Open source	Proprietary	– Open source (ANTLR, Spoofax) – Commercial (MetaEdit+)
Main objective	Creation of textual, graphical and model-based DSLs.	DSL design mainly in visual studio (**VS**) and related tools.	Flexibility to create external, internal or graphical DSLs.
Supported DSL types	– Textual – Graphical – Metamodel-based (EMF)	– Textual and graphical – Internal	– Textual (ANTLR, Spoofax) – Graphical (MetaEdit+). – Internal (JetBrains MPS)
Main tools	– **EMF**: Metamodel modeling – **Xtext**: Textual DSLs – **Sirius**: Graphical DSLs – **Acceleo**: Code generation	– **DSL tools**: Design in VS – **Roslyn**: Syntax analysis and creation of internal DSLs – **Power platform**: Specific workflows	– **JetBrains MPS**: Model-based DSLs – **ANTLR**: External DSLs – **Spoofax**: Rapid DSL development.
Host language/ base	Java	C#, .NET	Java, Scala, python, etc.
Community	Strong open-source community (XText, EMF, etc.)	Microsoft community focused on .NET.	Varied: Depending on the tool (Spoofax, ANTLR, etc.)
Advantages	– Wide range of free tools – Flexibility for graphical and textual DSLs – Strong support for MDE	– Integrated with VS. – Good documentation for C# developers	– Versatility for specific cases – Modern tools like MPS
Disadvantages	– EMF/Sirius complexity – Eclipse IDE is resource intensive	– Less suitable for external DSLs – Strong dependence on VS	– Often requires advanced expertise – Some tools are commercial (MetaEdit+).

[a]https://eclipse.dev/modeling/emf/.
[b]https://learn.microsoft.com/en-us/visualstudio/modeling/.

Finally, other frameworks such as JetBrains MPS, ANTLR and Spoofax provide various tools for specific needs. These methods are distinguished by their modernity and versatility, but may require advanced expertise and, in some cases, financial investment (e.g., MetaEdit+). These tools are ideal for projects requiring a customized solution or a highly specific DSL.

6 Challenges and Limitations of Developing DSL for Chatbots in Smart Tourism

6.1 Challenges of Developing DSL

The development of a DSL for chatbots in the smart tourism sector presents several significant challenges. Firstly, the availability of data for training NLU models is a pressing issue. The tourism domain requires high-quality, domain-specific datasets that capture diverse user intents, regional language variations, and real-time contextual information. Acquiring such datasets is complicated by the dynamic and heterogeneous nature of tourism-related data. Secondly, there is a strong dependency on proprietary development tools. Many chatbot development platforms are provided by large technology companies, leading to potential vendor lock-in. This dependency restricts flexibility, increases costs, and limits adaptability, making it challenging to transition to other tools or platforms as requirements evolve. Thirdly, research in this field remains minimal. The domain of DSL-based chatbot development for tourism lacks substantial studies and established frameworks. This limitation hampers innovation and slows the adoption of robust solutions tailored to the specific needs of the tourism sector. Finally, the implementation tools available for DSLs are not widely adopted within the developer community. Frameworks such as those in the Eclipse or Microsoft ecosystems have steep learning curves and are limited in their ability to foster collaboration and the sharing of best practices. This lack of widespread adoption restricts the development of a strong, supportive ecosystem around these tools.

6.2 Limitations of Developing DSL

The proposed approach to developing a DSL for chatbots in the smart tourism sector has several limitations. Firstly, the availability of domain-specific data remains a significant barrier. Tourism data is often fragmented, inconsistent, or not readily available in machine-readable formats. Secondly, there are constraints related to the tools used for implementing the DSL. Many of these tools, such as those in the Eclipse or Microsoft ecosystems may not be user-friendly for non-expert developers. This limitation can hinder the accessibility of DSL to a broader audience and reduce its practical applicability. Finally, the lack of interoperability poses a challenge. The current DSL-based solutions may not easily integrate with existing chatbot frameworks, diverse technologies, or platforms. This limitation restricts the versatility and usability of the developed chatbots across various systems.

7 Conclusion

In this work, we have presented an MDE-based approach to simplify and accelerate the creation of chatbots dedicated to smart tourism in smart cities. This contribution is based on the design of a dedicated metamodel, which forms the basis for the future development of a DSL. This DSL will enable users, without programming expertise, to easily design chatbots while reducing the technical complexities associated with traditional tools. This approach aims to offer an accessible, high-performance solution, aligned entirely with the needs of the smart tourism sector.

At the same time, we analyzed the 6A model of STDs and explored the impact of emerging technologies on their various components, highlighting their key role in the evolution of STDs.

In addition, a comparative analysis of ecosystems and frameworks dedicated to DSLs enabled us to identify solutions adapted to various needs: Eclipse for open-source projects, Microsoft for enterprise environments, MetaEdit+ for creating graphical DSLs rapidly, and Spoofax for designing advanced textual DSLs.

In the future, we plan to implement our metamodel with the Eclipse ecosystem to develop a modeling tool and code generator capable of automatically producing chatbots for smart tourism. This solution will considerably reduce the time and effort required for development while promoting the adoption of this technology in the context of smart tourism.

Acknowledgments This work was supported by the Ministry of Higher Education, Scientific Research and Innovation, the Digital Development Agency (DDA) and the CNRST of Morocco (Alkhawarizmi/2020/32).

Competing Interests The authors declare that they have no competing interests.

References

1. Adamopoulou E, Moussiades L (2020) An overview of chatbot technology. In: Artificial intelligence applications and innovations. Springer, Cham
2. Benaddi L, Ouaddi C, Jakimi A, Ouchao B (2024) A systematic review of chatbots: classification, development, and their impact on tourism. IEEE Access 12:78799–78810
3. Benaddi L, Ouaddi C, Souha A, Jakimi A, Ouchao B (2024) Code generation templates for accelerating chatbot development in smart tourism. In: 2024 7th international conference on advanced communication technologies and networking (CommNet)
4. Benaddi L, Ouaddi C, Souha A, Jakimi A, Ouchao B (2024) Model-driven engineering to develop chatbots for smart tourism. In: 2024 international conference on global aeronautical engineering and satellite technology (GAST)
5. Benaddi L, Souha A, Ouaddi C, Jakimi A, Ouchao B (2024) Towards a unified metamodel for developing the conversational agents for smart tourism. Procedia Comput Sci 236:241–247
6. Bhuiyan KH, Jahan I, Zayed NM, Islam KMA, Suyaiya S, Tkachenko O, Nitsenko V (2022) Smart tourism ecosystem: a new dimension toward sustainable value co-creation. Sustainability 14(22):15043
7. Brambilla M, Cabot J, Wimmer M (2017) Model-driven software engineering in practice. Morgan & Claypool Publishers

8. Buhalis D, Amaranggana A. (2014) Smart Tourism Destinations. In Z. Xiang & I. Tussyadiah (Eds.), Information and Communication Technologies in Tourism 2014 (pp. 553–564). Springer, Cham. https://doi.org/10.1007/978-3-319-03973-2_40
9. Calvaresi D, Ibrahim A, Calbimonte J-P, Schegg R, Fragnière E, Schumacher M (2021) The evolution of chatbots in tourism: A systematic literature review. In: Information and communication technologies in tourism 2021: proceedings of the ENTER 2021 eTourism conference, January 19–22, 2021
10. Daniel G, Cabot J, Deruelle L, Derras M (2020) Xatkit: a multimodal low-code chatbot development framework. IEEE Access 8:15332–15346
11. Hermans F, Pinzger M, Van Deursen A (2009) Domain-specific languages in practice: a user study on the success factors. In: Model Driven Engineering Languages and Systems: 12th International Conference, MODELS 2009, Denver, CO, USA, October 4–9, 2009, proceedings 12
12. Jain A, Bhati P (2022) Comparative analysis and development of voice-based chatbot system for differently-abled. J Phys Conf Ser
13. Lebeuf C, Storey M-A, Zagalsky A (2017) Software bots. IEEE Softw 35(1):18–23
14. Malamas N, Panayiotou K, Symeonidis AL (2023) dFlow: a domain specific language for the rapid development of open-source virtual assistants. arXiv preprint arXiv:231002102
15. Margherita EG, Esposito G, Escobar SD, Crutzen N (2021) Exploring the smart city adoption process: evidence from the Belgian urban context. arXiv preprint arXiv:210105670
16. Martínez-Gárate ÁA, Aguilar-Calderón JA, Tripp-Barba C, Zaldívar-Colado A (2023) Model-driven approaches for conversational agents development: a systematic mapping study. IEEE Access 11:73088–73103
17. Miller J, Mukerji J (2003) MDA Guide Version 1.0.1. OMG, Framingham
18. Omg. (2012) OMG systems modeling language (OMG SysML), version 1.3. OMG
19. Ouaddi C, Benaddi L, Bouziane EM, Jakimi A, Chehri A, Saadane R (2024) DSL-driven approaches and metamodels for chatbot development: A systematic literature review. Expert Systems, 41(5), e13787. https://doi.org/10.1111/exsy.13787
20. Ouaddi C, Benaddi L, Jakimi A (2024) Architecture, tools, and DSLS for developing conversational agents: an overview. Procedia Comput Sci 231:293–298
21. Ouaddi C, Benaddi L, Naimi L, Jakimi A (2024) A graphical DSL for accelerating chatbot development. In: 2024 7th international conference on advanced communication technologies and networking (CommNet)
22. Ouaddi C, Benaddi L, Souha A, Jakimi A (2024) Towards a Metamodel for proactive Chatbots. In: 2024 ASU international conference in emerging Technologies for Sustainability and Intelligent Systems (ICETSIS)
23. Ouaddi C, Benaddi L, Souha A, Jakimi A (2024) A comparative and analysis study for recommending a chatbot development tool. In: 2024 international conference on global aeronautical engineering and satellite technology (GAST)
24. Pérez-Soler S, González-Jiménez M, Guerra E, de Lara J (2019) Towards conversational syntax for domain-specific languages using Chatbots. J Object Technol 18(2) 5:1–21
25. Pérez-Soler S, Guerra E, de Lara J (2021) Creating and migrating chatbots with conga. In: 2021 IEEE/ACM 43rd international conference on software engineering: companion proceedings (ICSE-companion)
26. Revang, M., Baker, V., Manusama, B., & Mullen, A. (2018). Market guide for conversational platforms
27. Sarikaya R (2017) The technology behind personal digital assistants: an overview of the system architecture and key components. IEEE Signal Process Mag 34(1):67–81
28. Simon Elias B, Alahi A, Ayyoob S (2023) Environmentally sustainable smart cities and their converging AI, IoT, and big data technologies and solutions: an integrated approach to an extensive literature review. Energy Inform 6:9
29. Ukpabi DC, Aslam B, Karjaluoto H (2019) Chatbot adoption in tourism services: a conceptual exploration. In: Robots, artificial intelligence, and service automation in travel, tourism and hospitality. Emerald Publishing Limited, pp 105–121

Energy Conservation: A Case Study on Electrical Audits for Sustainable Smart Cities

L. Ramesh, P. Kanakaraj, K. Kalyan Sagar, P. Aruna Jeyanthy, and R. Narayanamoorthi

Abstract Energy auditing plays a vital role in an academic institution toward energy consumption, which manages the energy crisis through conservation schemes. Energy audit research studies are carried out worldwide towards implementation in industry, commercial and residential organizations, with significantly less preference for academic institutions. This work satisfies the above issue by executing an electrical energy audit at dental college with the proposed methodology and suggested the best recommendation for implementation. The wastage audit, preliminary audit, and detailed audit are conducted at the institution with the help of a power quality meter and the planned methodology. The required recommendations are suggested for energy saving at lighting system, fans, air conditioner and compressor and sterilizer with the permissible payback period. The challenges on the local power distribution system are analyzed, and the implementation is inducted to restructure the overall design. The post-audit was conducted and measured the percentage of energy saving as compared with recommendations.

Keywords Distribution system · Energy audit · Energy efficiency · Power quality meter

---

L. Ramesh
Department of EEE, Dr. M.G.R. Educational and Research Institute, Chennai, Tamil Nadu, India

P. Kanakaraj (✉) · R. Narayanamoorthi
Department of EEE, S.R.M. Institute of Science and Technology, Kattankulathur, Chengalpattu, Tamil Nadu, India

K. K. Sagar
Department of EEE, Shri Vishnu Engineering College for Women, Bhimavaram, Andhra Pradesh, India

P. A. Jeyanthy
Department of EEE, Kalasalingam Academy of Research and Education, Virudhunagar, Tamil Nadu, India

H. Kaddouri et al. (eds.), *Smart Cities for Sustainable and Urban Development*, Green Energy and Technology, https://doi.org/10.1007/978-3-031-99001-4_36

1 Introduction

The operating cost and environmental impact associated with the utilization of equipment or system can be decreased with the help of energy conservation and energy efficiency [1]. Energy efficiency is started in few countries due to its perspective and existed for all energy sectors and transportation. The biggest challenges are to improve the efficiency of production and distribution, along with the reduction in cost and lower the environmental impacts. Therefore, the need to improve energy efficiency as suggested by UN SDG Goals 7 can have helpful influences on the energy intensity and green environment. Energy Audits (EA) are the initial step to improve buildings, institutions, and industrial facilities. There are several steps to carry out the energy audit, and efforts depend on the audit. However, a universal process is presently adopted for most facilities to reduce energy consumption. A recent study resulted that predicted power requirement will be doubled in India by implementing electric transport. But the Predicted generation is deficient. So, all the public needs to conserve energy in all the energy application organizations. The literature review, energy conservation opportunity, and projected work are found out in chapter two in line with this. This work focuses on the energy audit study at an academic institution through the methodological proposal and suggested recommendations.

This paper discussed the energy audit review and identified opportunities for energy conservation in Sect. 2. Section 3 explained the data collection process involved in the energy audit. Critical issues and recommendations were explored in Sect. 4. The results and discussion were detailed in Sect. 5. Finally, the paper concluded in Sect. 6.

2 Energy Audit Review and Energy Conservation Opportunity

2.1 Energy Audit Literature Study

The energy audit analysis is done for the house, commercial building, university by the different authors discussed in this section. In Lin and Tsai [2], an advanced HEMS facilitated by a non-intrusive load monitoring (NILM) technique with an automated non dominated sorting genetic algorithm-II (NSGA-II)-based multi-objective in-home power scheduling mechanism is discussed. The authors suggested five solutions to reinforce the observation and management of energy flow and reduce the energy waste in optimal conclusion. In Qian et al. [3], the simulation-based nationwide energy savings analysis of different energy conservation measures (ECMs) enhances building energy performance. The taxonomy of DL algorithms available in the literature applied to various problems in EUI is presented in Mishra et al. [4]. The main objective of this survey is to provide a comprehensive idea to the researcher/utility engineer about the applications and future research scope of DL methods for power systems studies.

In Mahmoudzadeh et al. [5] observed the critical situation after assessing the daylight in the classroom and suggested a practical solution by redesigning the shading device. In Cunha and Oliveira [6] presented characteristics of energy use and analysis of nZEB hotel buildings. The survey has been carried for 45 hotels, and based on the survey author concludes with five findings to reduce energy saving. In Ferreira et al. [7] developed a tool to stimulate energy performance in a building. The software tested by implementing multi-owner building energy service companies helped maintain or improve energy efficiency in the long term. In Ruggeri et al. [8], a decision support system is developed to plan and manage energy retrofit campaigns tailored for cultural heritage. Costs and energy uses are assessed, and the compatibility of interventions and their impact on indoor environmental quality. In Barja-Martinez et al. [9], a multi-objective hybrid energy management system minimizes both the electricity expenses and the household greenhouse gas emissions released due to consumption, considering the entire life cycle of the generation assets used to provide energy. In Martinez et al. [10], an energy audit is carried out at academic institutions and suggested various conservative measures to make energy consumption more economical. The detailed study carried out resulted from cost mitigation, reduced environmental pollution, and a demand-supply gap. Evgenia [11] established an energy audit survey that facilitated irrational energy consumption with rank distribution and resulted in the demonstration in the MathCAD package with the example of the infrastructure of Neryungrinsky Region educational establishments. The study conducted [10] on a commercial building and restructured existing lighting fixtures' performance and resulted in 63% of total power consumption, followed by lights with 23% and plug-in loads with 14%. The results are presented in Table 1 as a numerical comparison of existing research articles.

To support energy research initiative towards United Nations Sustainable Development Goal 07, Dr. MGR Educational and Research Institute have taken the initiative in the year 2017. Inaugurated, The Institution of Green Engineers—Energy Saving and Awareness' Club named "IGEN GREEN9 MGR ENSAV CLUB. The

Table 1 Comparison table for existing energy audit articles

Aspect	Keerthi Jain et al. [12]	Kumar et al. [13]	Priya et al. [14]	Janakiraman et al. [15]	Keerthi Jain et al. [16]	Arya et al. [17]
Energy savings (%)	15–20	50	25	30	25	29.54
Cost reduction (%)	20	50	25	35	30	30
Payback period (years)	1.5	1	1.2	2	1.8	0.5
Software used	–	ETAP	ETAP	ETAP	ETAP	–

executed research works under the completed projects were published in Scopus Journals and Conferences [18–24], and the reports are published in IGEN—GREEN9 (Energy Efficiency Research Group) digital library. This project represents the collaborative research in IGEN-GREEN9 with global researchers and executed detailed Electrical Energy Audit and recommendation at Dental collage. The wastage audit, preliminary audit, and thorough audit are conducted and concluded with the proposals for the existing issues. The recommendations are suggested, implemented in five stages by the head of the institution in 6 months. The post-audit was conducted and measured the improvement in the electrical energy efficiency.

2.2 Opportunities in Energy Conservation

- **Low Consumption Energy Lamps** consume one-fifth of all electricity production. Low energy lamps in commercial buildings use 30% of the total amount.
- **Intelligent Monitoring and Targeting System** is the most efficient part of energy management, which will get the fastest payback. Variable speed drives controlled the voltage and frequency of the motor, expressively dropping energy consumption.
- **High-Efficiency Motors:** In worldwide, 65% of total industrial energy is consumed by electrical motor industries. Averagely, a single 75 kW motor nearly consumes power with a cost equivalent to hundred times of its purchase value in its lifetime. In simple words, its purchase value is low as compared to its energy consumption in 2 months.
- **Building Controls:** Generally, 40% of energy is used for buildings, and it can be reduced to 30% by adopting effective building control strategies. In line with that, the outcome of the expert's review can also provide the building controls.
- **Intelligent Lighting Controls:** By connecting intelligent lighting systems to house electricity, 40% of energy consumption is controlled. The sunlight level sensors and position detection controls can be implemented to benefit natural daylight savings and maximize off periods, which helps to reduce maintenance costs.
- **Combined Heat and Power (CHP)** is a mean feature/part of energy efficiency. CHP plants outcome with a greater degree of energy efficiency when compared to conventional methods. It can realize good returns on investment, especially when maximum use is made of the heat generated.
- **Supply Voltage Optimization:** The standard operating voltages of electrical apparatus are designed to work at 400 V/230 V. However, most equipment suffers from persistent over-voltages or under voltage, resulting in efficiency, equipment issue, and maintenance. This can be resolved by maintaining the supply voltage between 220 and 225 V, which can help to save 5%–15% in electricity consumption.

- **Power Management Solutions:** The intelligent power monitoring and load management systems can significantly reduce the tariff structures and incentive payments. This will also help to reduce maintenance and improve the lifetime of the apparatus.
- **Increased Factory or Process Automation:** Implementing advanced automation can have potential savings of 60% with improved productivity, reduction in downtime, and reduced maintenance requirements.

3 Data Acquisition

This work was executed at the leading academic institution, which educates the medical community. The electrical quantity such as voltage, current and physical phenomenon like sound, pressure, or temperature is measured with the help of an electrical sensor, hardware, and programmable software by the process of data acquisition. The real-time data acquired through Power Quality Analyzer for a specified period. The college power system is provided with six electricity services from the Indian electricity board, catering to the complete power requirement. The existing loads like indoor and outdoor lightings, ACs, sterilizers, fans are operated without a control mechanism to reduce energy consumption. Air conditioning systems, lighting, steam system and boilers, electric motors, compressed air systems, hot water systems, and special purpose apparatus are considered. The power distribution single line diagram is represented in Fig. 1.

Fig. 1 represents six service connections presently available to cater to all the needs, and all are LT services. The line incoming from the distribution transformer with 500 A capacity connected with series and passed through nearby college by 200 A. The power utilization with the equipment is represented in Fig. 2. It means the primary utilization by the air-conditioning of 35% and 24% by the lighting instruments. Different tariff graphs are also analyzed, and a significant issue in the meter is presented. Fig. 3 shows the recorded and sanctioned demand; The recorded demand shows a maximum value of 16 kVA, and a minimum value of 9 kVA. The sanctioned demand is 4.8 kVA for service connection no. 531. Figure 4 indicates that no action was taken for demand management. In the years 2014 and 2015, the college paid a penalty of Rs. 9738/month for exceeding the sanctioned demand.

The selected days are identified, and the daily load curve for all the blocks is tabulated and recommended. With all the observed readings, the main block daily load curve is represented in Fig. 4 and shows unit consumption in kWh. The maximum unit consumption of 40.68 units/day is noted from Fig. 4. It is observed that lighting plays the maximum power consumption in the main block. It resulted in maximum energy saving possible if the lightings are converted to LED. As mentioned, a power quality meter helps us measure real-time measurements of voltage and current in respective buses and lines. Fig. 5 represents the three-phase voltage measurements at the sample bus, and it was noted under voltage up to 195 volts and shown in red. Fig. 6 illustrates the 3 phase current measurements at a sample bus. It

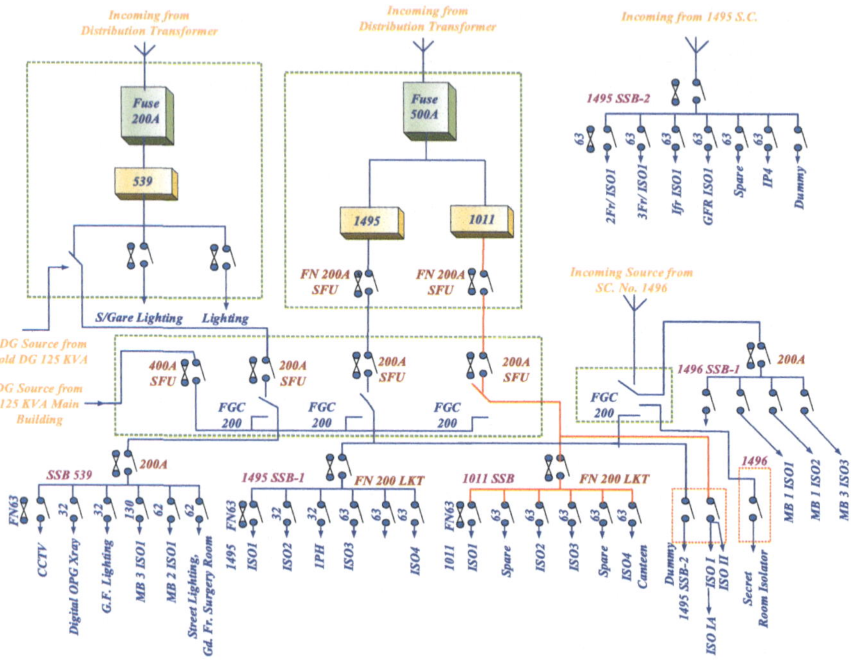

Fig. 1 Dental college power distribution layout

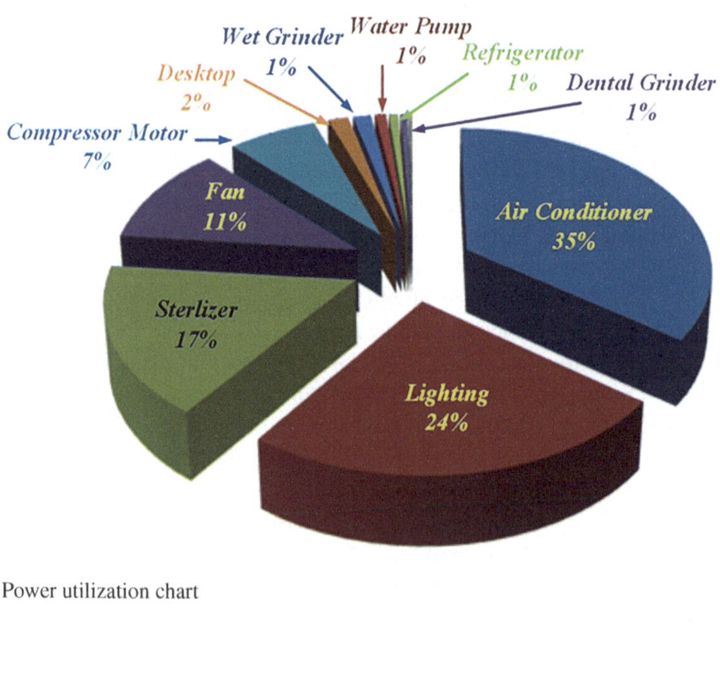

Fig. 2 Power utilization chart

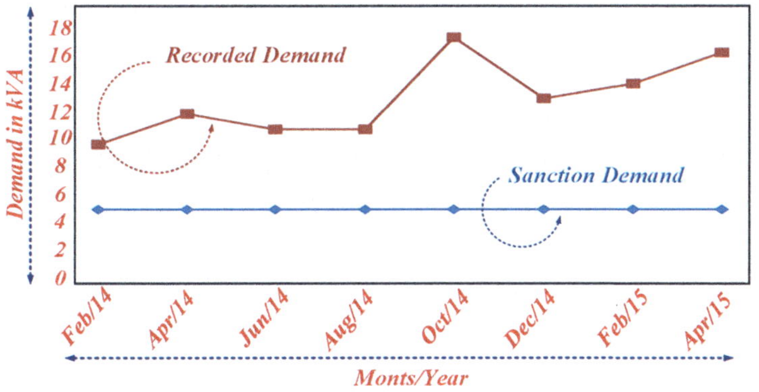

Fig. 3 Sanctioned and recorded demand

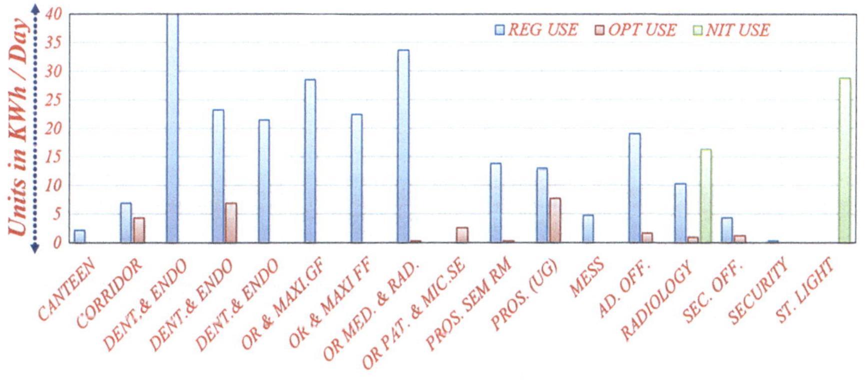

Fig. 4 Sample daily load curve

was pointed out that they currently vary from 10 to 70 amps; they are a vast over current issue up to 110 amps during normal operations.

The critical parameter which plays a significant role in the quality of power was measured for a particular period and resulted in Fig. 7. The monitored power factor touched a maximum of 0.95 and a minimum of 0.83. Significant recommendations to be reported for improvement of power factor. The power quality monitoring is estimated through the impression of harmonic measurement and represented in Fig. 8. The recorded power factor leads to significantly less, in which colossal current is drawn, leading to overheating of cable, which affects breaker trip. To avoid this, it is suggested to connect the required ratings of the capacitor in parallel to the loads.

The level of the harmonics is recorded as A3 THD (B-Phase) 6.45, 6.69, 11.51. The accepted two categories of voltage and current harmonics are less than 5 and

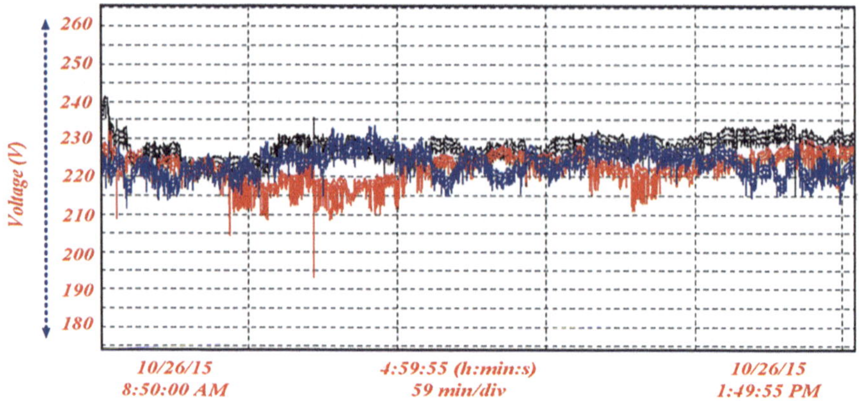

Fig. 5 Recorded voltage at sample bus

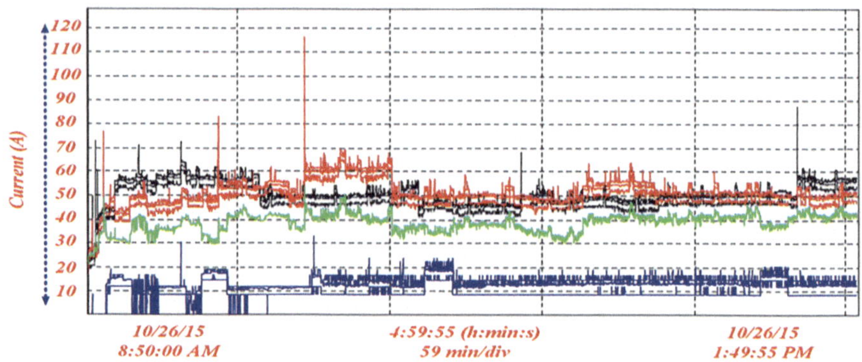

Fig. 6 Recorded current at sample bus

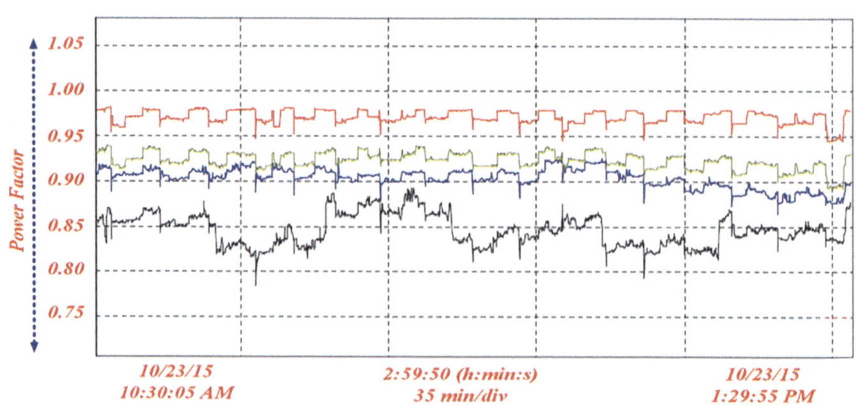

Fig. 7 Recorded power factor at sample bus

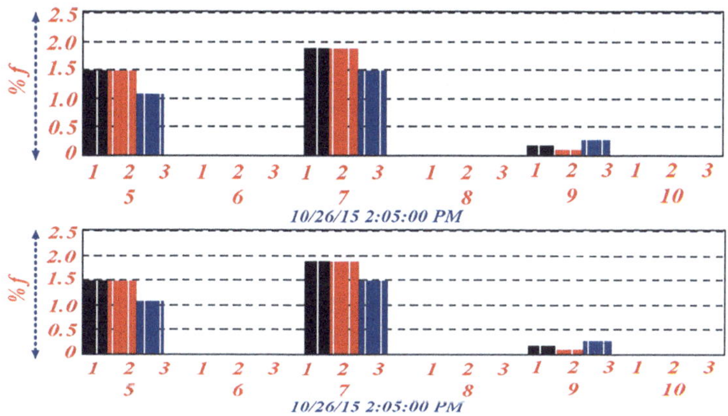

Fig. 8 Recorded voltage harmonics

less than 8. The recorded high-level harmonic, 11.5%, leads to failure of electronics card, overload of the transformer, increase in power loss and heating of cable, which will create regular breakdown and increase the operating cost of maintenance. To avoid that, with huge investment, a harmonic distortion panel is recommended in parallel with the load.

4 Critical Issues and Recommendations

The proposed audit process methodology is carried out at the dental college. The audit team identified all the load details and the energy-consuming loads; the report stated that all kinds of loads are available like lighting, air conditioner, fan, air compressor, sterilizer, etc. The identified loads are connected through different power distribution systems towards several energy meters. Some service connections exceeded the allotted demand and paid the penalty, and one service connection with minimum load and one with no load. It takes to pay a minimum charge for two energy meters. The various critical issues and recommendations are discussed below.

4.1 Critical Issues

Issues with TNEB Connections—six LT meter services are available in the college. After the study, it is found that some non-performance recorded, like low power factor, over demand resulted in paying the penalty. The low power factor will converge to voltage drop and consume high current; this will reduce equipment performance and lead to machines getting the breakdown. The increase in current will heat the cable, which leads to fire accidents for men and materials.

Issues with Lighting Connections: The audit result indicates that most of the lightings through fluorescent tubes and CFL in indoor and outdoor areas focuses on high-pressure mercury and sodium vapor street lighting. All the indoor lighting is fitted with 3 × 36 W fluorescent tube lights with a consumption of 108 watts. It is observed that, due to minimum switch operation, all the lights are in a continuous process and not switched off when there is no need for lighting. After the audit of the room, the excess lighting capacity is represented here. The block, third floor, the department of orthodontics connected with 50 fluorescent fittings, each of 3 × 36 W with consumption 5400 watts. This lighting is not essential for this room, concerning the room illumination level.

Another example is the department in B Block, fitted with 47 nos of fluorescent fittings, each of 3 × 36 watts, with consumption 5076 watts in a single room that is not essential as per the report of room illumination level. The street lightings also did not match the illumination level and found that 60% of saving opportunities provided good recommendations. Excellent energy-saving opportunities existed in lighting, recommended in the next session.

Issues with Air Conditioning: The auditorium connected with low air conditioning capacity, rated 2 × 2 ton and 4 × 1.5 ton air conditioner, which is not sufficient concerning air conditioning's planned capacity. This will lead to run air conditioning in full action, due to large spacious room, which will cause more power consumption. All other ten numbers of air conditioning are old type and filled R22 refrigerant gas; this affects the environment by air pollution. It is suggested to replace 80% of air conditioning with star ratings and offers an optimized number of air conditioners as per standards.

Issues with Fan: The big foremost problem is 75% of the fan operated without speed control, which leads to an increase in energy consumption. The cleaning maintenance of the fan did not satisfactory, resulting in an increase in the load of leaf, which reduces the fan's speed. Identified 15 fans are not in operation and few fans operation too slow. Al the fan available in the building are not rated quality. One hundred seven fans operated, in which 50% working for maximum time to be provided recommendations.

Issues with Sterilizer and Compressor: The used sterilizers are mostly old types without temperature controllers, mainly for power saving. Twenty-five percentage sterilizers with steam leakage and constant steam leak due to poor seal, five sterilizers are faulty, and heating is not proper due to the heating element's damage. Periodic maintenance needs for fan blades for the improved cleaning due to more dust.

No Proper Monitoring of Air outlet: The following issues are noted, (1) no proper temperature and pressure gauge in inlet and outlet lines (2) three compressor motors in operation continuously with low requirement of air pressure (3) two compressors are operated in the air blocked room (4) No required dampers provided to two compressors (5) ten dental chairs in the position of air leakage due to damaged valves and hose.

The issues are represented in this session. The respective energy-saving recommendations are suggested in the next session, resulting in a monthly reduction in the energy tariff and improved power quality and rectified all the local electrical faults.

4.2 Recommendations

The primary objective of energy audit and management is to deliver the best solution to the organization in reducing energy costs, finding the solution for implementation, and suggesting new opportunities in energy equipment, meaning the methodology to decrease energy necessities presented through recommendations. The recommendations for the service connection, lighting, air conditioner, fan, sterilizer, and power quality meter report are represented with no investment option, minimum investment, and maximum investment option.

Rearrangement of Service Connections: The recommendation starts with the no investment option. With no load at 213 service connection, the college paid a minimum monthly charge; thus, they paid Rs. 4154 as a penalty within 16 months. It is recommended to surrender the service connection. The 531 service connection sanctioned demand was 4.8 kVA, and the current maximum demanded is 16 kVA. It is suggested to submit 531 and shift all loads to service no. 1496. With the above action, Rs. 7300/annum can be saved. The recommendation with investment option is represented below with savings opportunity. The power factor recorded at 531 connections is 0.77 pf to 0.81 pf and 0.4 pf to 0.8 pf lag for service no. 1496; the team recommended parallel capacitors implementation to improve power factors. The demand exceeds 55 kVA for service no. 1495 compared with the sanctioned demand of 46 kVA; this result paid Rs. 25,258 as penalty within 13 months. The load sharing is suggested to avoid penalty; 100 kVA can be done from service no. 1495 to 539 with an estimated approximate cost is Rs. 50,000/– for cable and labor fees. Another opportunity to save consumption is by load share in service no. 1011 by transferring 250 kVA to another service no. 1496. This will reduce the penalty of Rs. 41,164 with an estimated investment cost of Rs. 45,000. To avoid penalty, 18.6 kVA sanctioned demand of service no. 1011 is to be increased to 46 KVA. Surrendering the LT connection is the final suggested idea and to go for HT connection. The cost/unit is the same, so there is no unit price variation, but it is a proposed investment for upgrading—the cost of Rs. 280,000 for sanctioned demand of 200 kVA.

Optimal Lighting Proposal: The recommendation without investment is presented below with a savings graph. It is suggested to create awareness across the college to switch off the lights when not in use. The other proposed proposal is to maintain the lighting effectively by cleaning interior walls, periodic cleaning, painting with a brighter color for better reflection, control and maintenance of lamp fixtures, etc. The recommendation with low investment is presented below with a savings graph as shown in Fig. 9. It is suggested to implement pull cord switches, which will help each doctor operate the light. The pull cord switch

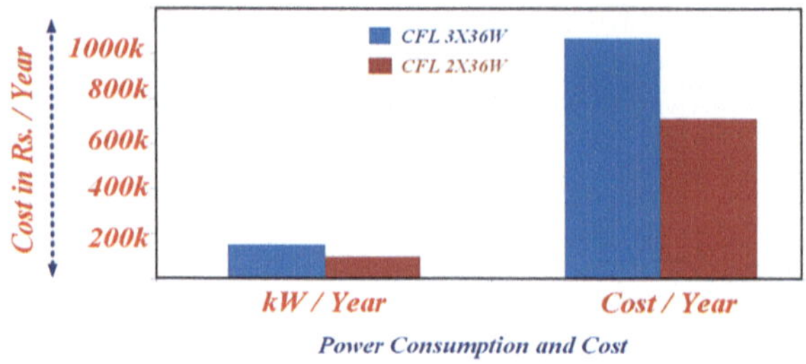

Fig. 9 Power saving through lighting

Table 2 Pull cord switch implementation analysis

Sr. No.		1	
Description		Pull guard switch provide for light	
Invest	**Lakhs**	43,200	
Payback period monthly		0.3	
Status		Before	After
No		1728	1728
Hrs.		8	7
Power consume	**Watts**	36	36
Total consumption	**KW**	498	435
	Month	12,939	11,322
	Year	1,55,271	1,35,862
Power cost	**Month**	1,03,514	90,575
	Year	12,42,169	10,86,898
Saving (yearly)		–	1,55,271
Payback period (month)		–	0.3

recommendations are presented in the Table 2. The recommended medium investment and solar implementation is shown below with a savings graph. The study of the daily usage chart of all the rooms and halls suggested that the maximum usage for lighting is represented in the administrative block. The conversion of lighting to LED is offered for the executive office, as illustrated in Table 3. It is recommended, 2 × 36 W LED instead of 3 × 36 W with a maximum cost saving of Rs. 27,256/year, it is shown in Fig. 10. It is suggested to implement solar power implementation with Rs. 1,30,000, without backup storage for the administrative office. The total predicted cost saving is Rs. 27,260, and the payback period is 3.4 year.

Table 2 represents the estimated cost saving by implementing 566 pull chord switches of 38,653.8 kw/year and Rs. 2, 67,345/year. The estimated cost of Rs. 22,440 spend to implement pull chord switch. The power-saving options in street lighting are represented in Table 4 by replacing 150 W HPMV with 80 W LED. The suggested energy-saving is 4475 kW/year, and cost reduction is appropriate Rs. 30,623/year. Implemented 14 numbers LED lamps with a reasonable cost of 9400/

Table 3 LED recommendation analysis

Sr. No.		1	
Description		3*10 W LED replaced	
Invest	**Lakhs**	0.3	
Payback period monthly		0.023	
Status		Before	After
No		20	20
Hrs.		8	8
Power consume	**Watts**	108	30
Total consumption	**KW**	17.3	4.8
	Month	449.3	124.8
	Year	5391.40	1497.6
Power cost	**Month**	3145	874
	Year	37,740	10,488
Saving (yearly)			0.27
Payback period (month)			13

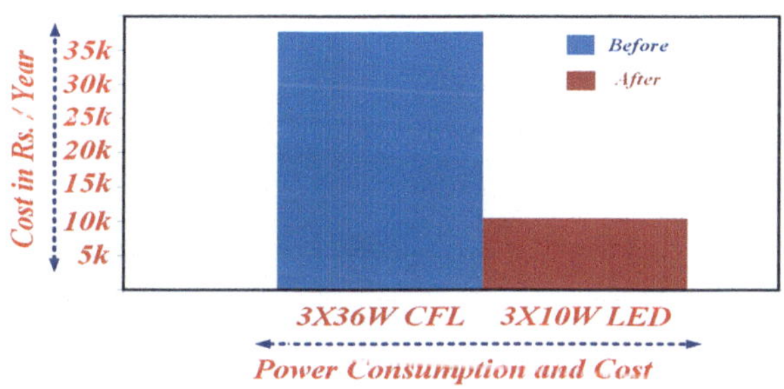

Fig. 10 Savings analysis with lighting

each, an investment of Rs. 1,27,000, and the payback period is 4.1 years. The overall lighting energy saving graph is represented in Fig. 11.

4.2.1 Air Conditioning Savings Opportunity

Concerning the issues in the air conditioner, the following action taken for implementation.

- Optimal placement of air conditioners is identified by using the author's meter placement program. It has resulted in four number of air conditioner can fulfill the capacity of the conference hall.
- According to the star-rated specification, it is suggested to service air conditioner once in 6 months to reduce current consumption.
- Replace non-rated ac with 5 stars ac can reduce Rs. 28,500 per year.

Table 4 Street light LED conversion

Sr. No.		1	
Description		80 W LED lamp replaced	
Invest	**Lakhs**	1.35	
Payback period monthly		0.021	
Status		Before	After
No		14	14
Hrs.		12	12
Power consume	**Watts**	150	80
Total consumption	**KW**	25.2	13.4
	Month	655.2	349
	Year	7862	4188
Power cost	**Month**	4914	2443
	Year	55,034	29,316
Saving (yearly)			0.26
Payback period (month)			63

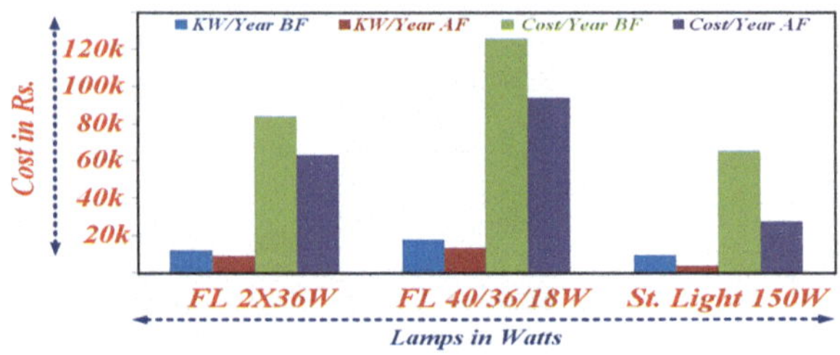

Fig. 11 Cost analysis after and before audit

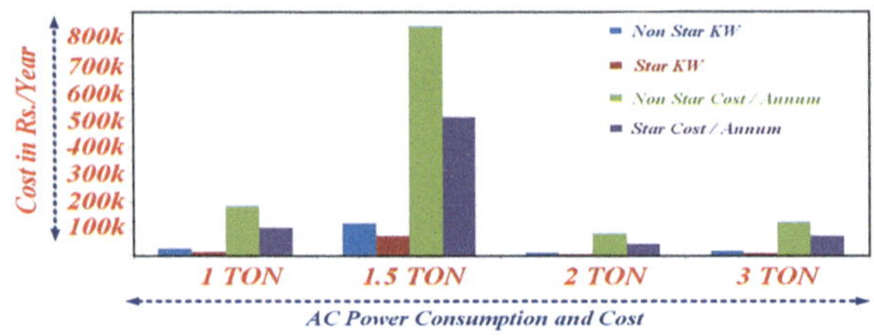

Fig. 12 Cost analysis with various ton AC

Figure 12 shows the power cost in comparison with many air conditioners. It is observed that energy consumption through air conditioner in the college is the peak, because of non-rated and old model air conditioners. The team recommended five stars rated air conditioner with 30% energy saving or an inverter type air conditioner with 40% energy saving.

4.2.2 Recommendation for Fan Cooling

In summer as a protection from heat, fan cost is 30 paisa per unit, which is very low compared to a charge of air conditioner cost of 10 rupees per unit. The recommendation without investment is mentioned below. The 50% heat comes to the house from the terrace, 30% from the house door and windows, and remaining through walls. It is suggested to keep the sun-oriented warmth entering through the patio. It can be through financially savvy non-contact asbestos sheets & many courses exist to diminish. There are multiple approaches to lessen the warm entrance on your open patio. The alternate ways to save energy by using step-type regulators instead of knob type consume 30% more energy. The recommendation with the replacement option is presented below with a savings graph. The star decides the appliance's performance; the high ratings mean high performance and low energy consumption.

The sample calculation for two department room fan replacements is calculated. The cost difference between 75 W fan and 50 W fan is shown in Fig. 13. By replacing five-star rated 70 fans, we can save power 3500 kw/year and cost-saving around 26,811/year.

4.2.3 Recommendation for Sterilizer and Air Compressor

A steam at 2700 F temp. is used to sterilize the dental equipment in all the laboratories. It is advised to follow the below action periodically: (a) clean the blocks in the pressure regulating valve, safety valve (b) clean leaky valve and hole, replace if defective, clean leaking seal or gasket (c) replace if broken; Remove any dust/dirt with a damp cloth and dry off; and remove water and waste matter from inside. It also suggested improving the performance by replacing the present damaged heating element and controlling accessories. Fixing the air compressor at the perfect place will allow getting more fresh air through the inlet, increasing the life span. Every 3 °C reduction will save 1% compressor energy use [25]. The identified 10 leaks are rectified with a proper solution. The solution on it is to use an ultrasonic

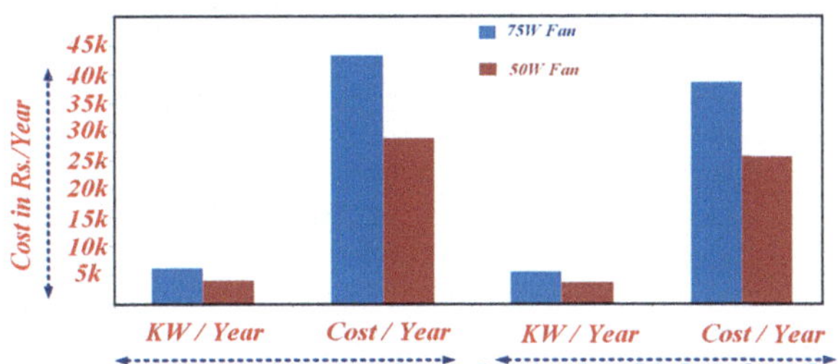

Fig. 13 Cost analysis with various fans

acoustic detector. In a rotary compressor, an adjustable drive system will save 15% of the annual energy consumption.

5 Results and Discussion

The life span of electrical equipment depends on the quality of the power delivered. The electrical health monitoring of the Dental College power distribution system and all the required real-time data are collected through a power quality meter. The issues are identified, and recommendations are suggested to improve the power quality and conservation of energy performance. Table 5 represents the audit summary and the expected power saving under the category of zero investment, low investment, and high investment.

The complete recommendation ended with the total estimated savings of 1,51,138 Kw/Year with cost savings is Rs. 98,680/month. The calculated investment cost is Rs. 1,477,240 with a payback period of 2.6 years. Figure 14 shows the comparison of connected load when calculated before and after the audit. It is an outcome that best possibility of saving in lighting and air conditioner recorded.

5.1 *Post Audit Outcome*

The energy audit report with recommendations submitted to the head of the institution and given strong advice to implement in a continuous phase of action. The institution started implementing the suggested request in five stages of activity. The post-audit was conducted at the institution after 6 months of the implementation stage—the post-audit result is described below.

Table 5 Summary of audit power saving

Sr. No.	Investment categories	Description	Saving KW/ year	Invest in Rs.	Cost-saving/ month (Rs.)	Cost-saving/ year	Payback period in the year
1	Simple measured/no investment	Simple actions and no/less investment activity	89,003	22,640	53,027	633,436	–
2	Low cost	Low investment activity	3894	172,000	10,578	126,948	1.5 Avg
3	Reconstruction action	High investment activity	58,241	1,282,500	35,075	420,910	3.7 Avg
Total			151,138	1,477,140	98,680	1,181,294	2.6 Avg

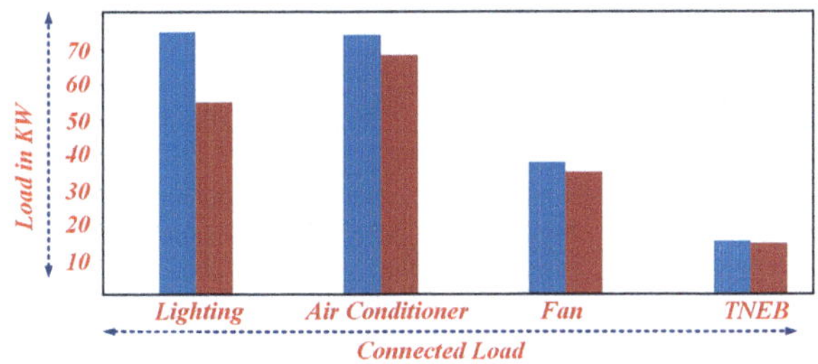

Fig. 14 Merged connected load (before and after audit)

- All the recommendations suggested in the rearrangement of service connections were implemented and achieved a 100% outcome.
- The optimal lighting proposal, as suggested by the audit team, recorded effective implementation with energy-saving outcomes. It is observed that 75% of the lighting changed to LED and achieved 40% energy saving. Another observation that solar generation was implemented for 25% lighting and achieved 25% savings in energy.
- As suggested, all the air conditioning, not replaced to star rated one, presently 50% of the air conditioning replace with BEE rated AC and achieved 60% saving in the monthly tariff.
- The maintenance supervisor and two-member team allotted and recorded all the fans are in good maintenance. The institution implemented 50% of fans with 35 W, Superfan, and 40% fan with 45 W energy saver fan. Its outcome 15% energy saving in electricity bill.
- The team was delighted with the post-audit implementation and appreciated the institution's head.

6 Conclusions

UN SDG goal 7 was started and implemented successfully through the proposed work on electrical energy audit using six recommended recommendations to save energy in the dental college in Chennai. The recommendations are suggested to reduce energy demand with a permissible payback period. The audit states that total cost savings are 98,780 rupees per month, with an average payback period of 2.6 years. The post-audit outcome concluded a 25% reduction in connected load and achieved a 30% reduction in monthly tariff. The authors suggested implementing the same action towards all the academic institutions across the country, which will decrease a massive drop in energy demand.

References

1. Krarti M (2020) Introduction to energy audit. Energy Aud Build Syst:1–24. https://doi.org/10.1201/9781003011613-1
2. Lin Y-H, Tsai M-S (2015) An advanced home energy management system facilitated by non-intrusive load monitoring with automated multiobjective power scheduling. IEEE Trans Smart Grid 6(4):1839–1851. https://doi.org/10.1109/tsg.2015.2388492
3. Qian D, Li Y, Niu F, O'Neill Z (2019) Nationwide savings analysis of energy conservation measures in buildings. Energy Convers Manag 188:1–18. https://doi.org/10.1016/j.enconman.2019.03.035
4. Mishra M, Nayak J, Naik B, Abraham A (2020) Deep learning in electrical utility industry: a comprehensive review of a decade of research. Eng Appl Artif Intell 96:104000. https://doi.org/10.1016/j.engappai.2020.104000
5. Mahmoudzadeh P, Afacan Y, Adi MN (2021) Analyzing occupants' control over lighting systems in office settings using immersive virtual environments. Build Environ 196:107823. https://doi.org/10.1016/j.buildenv.2021.107823
6. Cunha FO, Oliveira AC (2020) Benchmarking for realistic NZEB hotel buildings. J Build Eng 30:101298. https://doi.org/10.1016/j.jobe.2020.101298
7. Ferreira M, Almeida M, Rodrigues A (2016) Cost-optimal energy efficiency levels are the first step in achieving cost effective renovation in residential buildings with a nearly-zero energy target. Energ Buildings 133:724–737. https://doi.org/10.1016/j.enbuild.2016.10.017
8. Ruggeri AG, Calzolari M, Scarpa M, Gabrielli L, Davoli P (2020) Planning energy retrofit on historic building stocks: a score-driven decision support system. Energ Buildings 224:110066. https://doi.org/10.1016/j.enbuild.2020.110066
9. Barja-Martinez S, Rucker F, Aragues-Penalba M, Villafafila-Robles R, Munne-Collado I, Lloret-Gallego P (2021) A novel hybrid home energy management system considering electricity cost and greenhouse gas emissions minimization. IEEE Trans Ind Appl 57(3):2782–2790. https://doi.org/10.1109/tia.2021.3057014
10. Martinez J, Ireta F, Hernández MA, Lozano JM, Gutiérrez H (2017) A modified analysis of electrical energy consumption in university buildings. IEEE Lat Am Trans 15(3):408–414. https://doi.org/10.1109/tla.2017.7867169
11. Al-Shemmeri T, Naylor L (2017) Energy saving in UK FE colleges: the relative importance of the socio-economic groups and environmental attitudes of employees. Renew Sust Energ Rev 68:1130–1143. https://doi.org/10.1016/j.rser.2016.08.004
12. Keerthi Jain K, Kishore Kumar N, Murali Krishnan S, Ramesh L (2015) Comparative analysis of residential houses for effective reduction in power demand. Int J Appl Eng Res 10(6):5489–5494
13. Kumar A, Shashi Ranjan M, Singh BK, Kumari P, Ramesh L (2015) Electrical energy audit in residential house. Procedia Technol 21:625–630. https://doi.org/10.1016/j.protcy.2015.10.074
14. Priya TM, Sanjana V, Gohila B, Lavanya R, Anbazhagan A, Veerasundaram M, Ramesh L (2015) Design and analysis of a sustainable LV residential microgrid. Procedia Technol 21(January 2015):139–146. https://doi.org/10.1016/j.protcy.2015.10.081
15. Janakiraman D, Hariharan J, Umamaheswaran N, Ramesh L, Velrajkumar P (2021) My optimal residential house implementation through power audit. In: 2021 2nd International conference for emerging technology, INCET 2021, pp 1–6. https://doi.org/10.1109/INCET51464.2021.9456207
16. Keerthi Jain K, Kishore Kumar N, Muralikrishanan S, Ramesh L (2016) Residential house electrical energy audit case study with ETAP. Adv Intell Syst Comput 394:1181–1194. https://doi.org/10.1007/978-81-322-2656-7_108
17. Arya A, Jyoti P, Arunachalam NB, Awanish Kumar L, Ramesh VG, Egbert H (2017) Review on industrial audit and energy saving recommendation in aluminium industry. In: 2016

International conference on control instrumentation communication and computational technologies, ICCICCT 2016, pp 758–764. https://doi.org/10.1109/ICCICCT.2016.7988054

18. Adika CO, Wang L (2014) Autonomous appliance scheduling for household energy management. IEEE Trans Smart Grid 5(2):673–682. https://doi.org/10.1109/tsg.2013.2271427
19. Ozturk Y, Senthilkumar D, Kumar S, Lee G (2013) An intelligent home energy management system to improve demand response. IEEE Trans Smart Grid 4(2):694–701. https://doi.org/10.1109/tsg.2012.2235088
20. Mammoli A, Robinson M, Ayon V, Martínez-Ramón M, Chen C-f, Abreu JM (2019) A behavior-centered framework for real-time control and load-shedding using aggregated residential energy resources in distribution microgrids. Energ Buildings 198:275–290. https://doi.org/10.1016/j.enbuild.2019.06.021
21. Pedrasa MAA, Spooner TD, MacGill IF (2010) Coordinated scheduling of residential distributed energy resources to optimize smart home energy services. IEEE Trans Smart Grid 1(2):134–143. https://doi.org/10.1109/tsg.2010.2053053
22. Zhao Z, Lee WC, Shin Y, Song K-B (2013) An optimal power scheduling method for demand response in home energy management system. IEEE Trans Smart Grid 4(3):1391–1400. https://doi.org/10.1109/tsg.2013.2251018
23. Abdelaziz EA, Saidur R, Mekhilef S (2011) A review on energy saving strategies in industrial sector. Renew Sust Energ Rev 15(1):150–168. https://doi.org/10.1016/j.rser.2010.09.003
24. Andersson E, Arfwidsson O, Thollander P (2018) Benchmarking energy performance of industrial small and medium-sized enterprises using an energy efficiency index: results based on an energy audit policy program. J Clean Prod 182:883–895. https://doi.org/10.1016/j.jclepro.2018.02.027
25. CADDET Renewable Energy Newsletter, March 1997. / NREL. 32 p. 1997.

Underwater Image Enhancement Using Recurrent Neural Network for Disaster Recovery and Emergency Response Analysis in Smart Cities

Deluxni Natarajan and Pradeep Sudhakaran

Abstract After any tsunami, hurricane, flood, or any other natural disasters, immediate assessment and response are important for the minimization of damage and protection of human lives. Underwater image processing has emerged as one such crucial tool in disaster recovery and emergency response situations. It is used for the assessment of damage caused to underwater and coastal infrastructure, tracking environmental changes, and supporting rescue operations. However, underwater imaging is very often adverse conditions of low visibility, turbidity, and low light, which extremely hampers the acquisition of sharp, actionable images during such events. This chapter introduces an Underwater Image Enhancement Approach Using Recurrent-based Convolutional Neural Networks (RCNN) to enhance the image clarity and supply accurate real-time data for disaster recovery in smart cities. The RCNN model improves underwater image quality by eliminating effects such as blurring, noise, and contrast decay, which are often seen in similar images. This will help make assessments concerning the degree of damage sustained by various underwater structures like bridges, dams, and pipelines more effective. The improvement provided will also help the rescuers and recovery teams in real-time imagery of submerged areas in order to navigate safely during rescue missions and identify survivors or obstacles. Hence, the enhanced underwater images provide good information for post-disaster planning by gaining further insight into modifications in terrain below water, displacement of sediments, and erosion of the coastline. With smart city infrastructure allied with advanced image processing techniques, the approach provides an effective response to a disaster and the planning period for long-term resiliency so that greater restoration and safety are achieved within a shorter timeframe after the occurrence of a natural disaster.

D. Natarajan · P. Sudhakaran (✉)
Department of Computing Technologies, School of Computing, SRM Institute of Science and Technology, Kattankulathur, Chennai, India
e-mail: pradeeps1@srmist.edu.in

H. Kaddouri et al. (eds.), *Smart Cities for Sustainable and Urban Development*, Green Energy and Technology, https://doi.org/10.1007/978-3-031-99001-4_37

Keywords Recurrent convolutional neural networks (RCNN) · Convolutional neural networks (CNN) · Remote operated vehicle (ROV) · Very deep super-resolution network (VDSR) · Border collie flamingo optimization (BCFO) · Cross-scale feature differential module (CSFDM) · Adaptive histogram equalization (AHE)

1 Introduction

Underwater environments often present challenges for clear imaging due to factors like water turbidity, low lighting, color distortion, and scattering. In scenarios like disaster recovery and emergency response, especially within smart cities, obtaining clear underwater images is crucial for tasks like infrastructure assessment, search and rescue, and environmental monitoring. In these contexts, underwater images can help assess structural damage, locate missing persons, and identify hazardous areas. However, conventional imaging systems struggle with the unique characteristics of underwater conditions, leading to the need for enhancement techniques that can improve image clarity and detail. This has driven researchers to explore advanced machine learning and computer vision methods for underwater image enhancement.

Underwater image enhancement, RCNNs allow us to highlight and improve specific sections of an image without wasting computational resources on unimportant areas. This region-focused approach is more efficient and helps in rendering clearer, more informative visuals of underwater scenes, even in challenging visibility conditions. Underwater imaging of conventional imaging method in underwater environments. Motivation of using RCNN in underwater image enhancement to detection the localized feature of suffer from issues like poor visibility, color distortion, and low contrast, which are not uniform across the entire image. Some region may be more affected than others. Real time application of underwater, such as autonomous underwater vehicles or remotely operated vehicles require fast and efficient processing of image data. To diverse underwater condition in adaptability vary greatly depending on factors like depth water turbidity and salinity, which affect image quality in different ways. By isolating regions, RCNN can identify areas with more noise and apply targeted noise reduction techniques. This selective noise reduction ensures that finer details are not lost during the process. The rationale for the recovery strategy in order to assist a client in obtaining data from any other server should that server lose information and is unable to deliver the information to the client. Regarding the achievement for that purpose, many various procedures have been suggested in the way. In Flood, Fire, seismic tremors or any equipment glitch or any accidental deletion of information may never again remain accessible incapable to provide information to the client [13]. On the way to accomplish the reason, numerous diverse procedures have been proposed. In circumstances like Flood, Fire, seismic tremors or any equipment glitch or any accidental deletion of information may never again remain accessible. These contradictions by engaging scholars who highlight that procedural justice cannot wholly be achieved and that distributive justice especially in grassroots processes using the politics of neo-liberalism and the ontology of possibilities [9]. In this paper, the multi-sited instrumental case study methodology is employed to identify and describe the community experiences, and the factors prolonging the structural inequality & injustice as well as the vision of the justice and their claims within the disaster recovery environment. Application of enhance underwater imaging strategies for disaster recovery, remote sensing, autonomous systems or underwater robotics are increasingly applied to other applications like environmental monitoring, underwater archeology or marine research. In monitoring environment, they measure climate change, emissions, and ecology allowing instant information for conservation. Submerged environments are advantaged by autonomous vehicles that provide a survey of the sites and photographic documentation of objects and structures without disturbing the ecosystems. In marine research these systems study deep sea environment, learn patterns flow of water current, track marine lives, etc. These technologies therefore offer a flexibility that is revolutionizing the way we explore and conserve some of the world's most inhospitable terrains.

2 Background and Related Work

Convolutional Neural Networks (CNNs) are a form of Deep Learning algorithms intended for handling data that is in a grid like structure such as images. CNNs have almost changed the scene of image processing tasks and are now the central components of various computer vision systems. In convolutional layer, filters (or kernels) move across the image to get some features like edges, textures and other patterns. Every filter convolves pixels by their weights and then add them up providing a feature map as the result. These feature maps refer to specific pattern in the image such as vertical or horizontal edges, corners or something more complicated. Pooling layer decreases size of feature maps, so the model is not very large and it is less sensitive to translations or distortions of input image. In order to reduce the dimensionality of the feature maps these are flattened and then put into Convolution and pooling layers. These layers serve higher order purpose such as deciding the type of the image that has been classified. CNNs are highly effective at learning hierarchical features from images: The first layers identify low-level features such as outlines, and specific surface structures. Intermediate layers discern more complicated patterns; curves, corner. Lower layers hold information on objects structure and other high semantic knowledge. Extending the RCNN work in deep learning models has provided a region proposal from the input image. Recurrent proposal are candidate bounding boxes that may contain object of interest. The advantage of RCNN over traditional CNNs is their capability to designate a specific region in an image. This ability is very important in environments where up to objects may be present within the image and only a specific region should be targeted for detection or even enhancement.

RCNN employs the use of recurrent proposal to crop out potential object in an image irrespective of the background or many objects in the scene. These proposals assist to allow the model to concentrate on the relevant aspect of the image and not the entire image at one time. Underwater deep learning dataset is collected from an underwater video which is captured by sea cucumber fishing ROV (Remote Operated Vehicle). This work based on the network structure and improvements are derived from the Faster RCNN and Hyper-net method, for the underwater dataset, the method proposed in this paper have a well performance in recall and object detection accuracy [6]. To surpass the deficiency in train data, transfer introduction of learning is made to network training [14]. The fine-tuning process makes the model to adapt with the requirement gradually one at a time, underwater image enhancement. The multi-scale CNN, we suggest a transfer learning of getting rid of the bubbles and the projected-patterns on the object and for the system designed for underwater environment, several points should be compensated such as the interference of fluctuation and bubbles [7]. The simple solution is to infuse convolutional neural network for stereo to neutralize the impact of bubbles and/or water displacement. Since the difficulty in training CNN involves small size of database with large variation, we design a special bubble generation device for generating real bubble databases of multiple size and density. Traditional approaches toward image optimization submerged environments utilizes processes of histogram equalization, dehazing and contrast stretching for purposes of correcting color shift, improving light and reducing light scattering due to water content. However these methods are not able to provide consistent results in different kinds of underwater environments. Perception model for detecting objects submerged in water based on high quality underwater images and video using deep learning approach. For enhancing the underwater image quality, Very Deep Super-Resolution Network (VDSR), that has a higher visuals, is used, hence it is adequate for object detection [12]. Next, the object is recognized from the improved underwater image using the developed Border Collie Flamingo optimization-based deep CNN classifier (BCFO-based deep CNN). The highlight of the developed research is a BCFO-based algorithm for tuning the hyper-parameter of the classifier. Apply data augmentation three various ways to be able to detect a human in any situation where the camera is blurry or in any environmental conditions. During the training phase, the Faster R-CNN (Object Detection) [3] algorithm should be employed. Subsequently, the trained model must be tested to assess its accuracy and effectiveness. Regarding human detection, it is appropriate to assess the model using certain PASCAL VOC indicators. To incorporate local information, there is developed a novel local transformer model. This model employs the concept of PSNR attention as well as linear computations. Nevertheless, in order to build the convolutional neural network, the global features must be extracted. In order to combine the temporal and frequency domain data. Color corrected and color distorted underwater photograph are subjected to both qualitative and quantitative analysis. This is done so as to show how effective the algorithm is in practice. In order to capture local information, a unique transformer model is presented. This model makes use of peak-signal-to-noise ratio (PSNR) [11] attention and linear operations. In order to construct the convolutional neural network, it is necessary to extract global features, and incorporating both temporal and frequency domain data is also necessary. Both the high frequency and the low frequency information of the image are exploited in order to combine various aspects [5]. Both qualitative and quantitative assessments are performed on underwater photographs that have varying degrees of color distortion. This is done in order to illustrate the usefulness of the algorithm. To this end, an FPN-motivated network structure is developed, and both a CNN and a Transformer are employed in tandem so that both multiscale features can be learned and long-range cross-pixel dependencies can be captured [4]. An additional structure called cross-scale feature differential module (CSFDM) is then developed to fulfill the redundancy of information between features of distinct scales [8]. A series of experiments are carried out to demonstrate that our presented FBDPN can achieve better UOD performance and lower computational cost than other benchmark algorithms. Also discuss and compare a part-based classification approach with segmented classification using small-scaled CNN classifiers (with less than 20,000 hyper-parameters and 3M unity vector operations) together with a mean and expandable bounding box approach to re-label the described ROIs iteratively [2]. The iterative bounding box algorithm in conjunction with bounding box intersection check eliminated wrong erroneous segment classification as well as offered the optima and accurate matching ROI for a specific classification label such as surfaces with pockmark coverage. To support disaster and crisis management, a warm optimization algorithm (SOA) [1] is developed. In this system, the UAV search and rescue team adopted the delay tolerant network strategy that can explore. The particular mentioned strategy proved to be an effective to find the global optimum in the given space without the step to a suboptimum. Conventional methods for underwater image enhancement rely on techniques like histogram equalization, dehazing, and contrast stretching to correct color distortion, improve visibility, and reduce scattering caused by water's absorption of light [15]. Recurrent Convolutional Neural Networks (RCNN) have recently established themselves as powerful tools because they merge spatial and sequential information processing thus becoming ideal for underwater disaster detection work in smart cities. Underwater surveillance systems continue to escalate in use so scientists are applying RCNNs across applications that include flood monitoring as well as underwater structural damage assessment and tsunami prediction and submerged object detection. On our work of research Gap underwater depth not found the waste and disaster, so paper rectified that wastage and fish wastage from the different depth how long it store the in-depth and to removing with clear view of water. Several improvements need to be applied to the real-time performance of RCNN when operating in underwater conditions. Models become deployable on IoT devices with low-power hardware through efficient compression tools such as pruning and knowledge distillation and quantization. The feature extraction efficiency gains from incorporating MobileNet and EfficientNet alongside recurrent attention works to maintain accuracy levels. A combination of these processing methods makes RCNN perform efficiently for underwater images which improves disaster detection and performance in marine environments.

RCNN models have been adapted for disaster recovery and emergency response by enabling rapid and accurate detection of critical objects, such as damaged infrastructure, vehicles, and human survivors, in aerial or satellite images. In non-underwater scenarios, RCNNs assist in post-disaster assessments by analyzing large-scale images to prioritize rescue efforts and allocate resources efficiently. Within smart city frameworks, RCNNs can enhance disaster preparedness by integrating with surveillance systems to monitor potential hazards, detect early signs of structural damage, and support autonomous systems in recovery operations, improving overall urban resilience.

3 Underwater Image Enhancement

The nature of the underwater image can be enhanced to obtain information for further study. An image's quality can be raised by enhancement or restoration. Image enhancement aims to improve an image's attractiveness from a human perspective. For marine photos, numerous researchers have created a variety of techniques and algorithms. Corrections and extending are part of an underwater image improvement system that blends dual intensity photos with Rayleigh stretching. Conventional picture enhancement techniques mostly rely on processing in the frequency and spatial domains. Distortion of color, wavelength at which light is absorbed by water varies. The majority of underwater photos are dominated by blue or green tones due to the rapid absorption of red and green light. Low Contrast: A hazy look and less contrast are the results of scattering processes brought on by particles and the water itself. Blurriness is caused by turbulence and scattering, which weaken an object's clarity. Noise: Because of the bad lighting, underwater photos frequently have a lot of noise, especially in deeper areas. Enhancement has two different domain phase are enhancement process spatial domain and Transform domain. By extending the grayscales, the spatial domain of enhancement techniques separates the redistribution's histogram intensity. Using mathematical functions like the Fourier Transform (FT), Discrete Cosine Transform (DCT), or Discrete Wavelet Transform (DWT), one can enhance an image by converting it into the frequency domain. After entering the frequency domain, picture processing is carried out by taking use of its unique features. Following processing, the ultimate output of the image is a conversion back to its original spatial domain.

3.1 Techniques of Enhancement

Method directly aim to improve the perceptual quality of image without modeling the specific degradation process. For all techniques has important roles with have Histogram Equalization, CLAHE (Contrast Limited Adaptive Histogram Equalization) and White Balance adjustment.

3.1.1 Histogram Equalization

An image processing technique called Histogram Equalization (HE) is used to increase an image's contrast. Redistributing pixel intensity levels is intended to improve feature visibility. The distribution of pixel intensity values in an image is represented by a histogram, which displays the number of pixels with a given intensity value (in an 8-bit grayscale image, this range is 0–255). Pixel intensities in brighter photos will be concentrated toward the higher end of the histogram, while those in darker images will be concentrated toward the lower end.

The original intensity values are changed via Histogram Equalization to make them more uniformly distributed throughout the whole range of potential intensities (0–255 for an 8-bit image). Different of histogram equalization of compute the histogram which is calculate the histogram of the sea image, which gave the frequency of each pixel intensity. And it cumulative distribution function (CDF) to computed the summing the frequency counts of the intensity values. The help to maps the original intensity values to new equalized values. The CDF is used for transforming the intensity of each pixel. Pixel intensities are distributed more uniformly throughout the image thanks to the new intensity, which is proportional to the cumulative distribution. In Image of HE example of very dark with pixel intensities most between 0 and 50. Advantage to enhances the visibility of details in both bright and dark region of image. Simple and effective is easy to execute and performs admirably for a wide range of image types. Various of HE are AHE (Adaptive Histogram Equalization) while apply the entire sea image it have small region (titles) of water images. To developing image local contrast but can amplify local and sea noise in some depth cases and CLAHE (contrast Limited Adaptive Histogram Equalization) version of AHE the contrast amplification in areas with little variation to avoid noise amplification. A histogram Equalizing modifies the distribution of intensity in an image to improve contrast. Overall contrast is improved as a result of stretching the range of pixel intensities.

$$s_k = \left(L - 1 | M \times N\right) \sum_{i=0}^{k} n_i \tag{1}$$

Where, s_k is the new pixel intensity of image, L total number of intensity level, $M \times N$ total number of pixels, n_i number of pixel with intensity i. Figure 1 from the below image of histogram equalization of enhancement process histogram of original image and equalized image process graph.

3.1.2 White Balance Adjustment

A method for resolving color imbalances in photos brought on by different lighting conditions is called white balance adjustment. Underwater photos exhibit a bluish or greenish color cast due to the substantial color shift caused by light absorption by

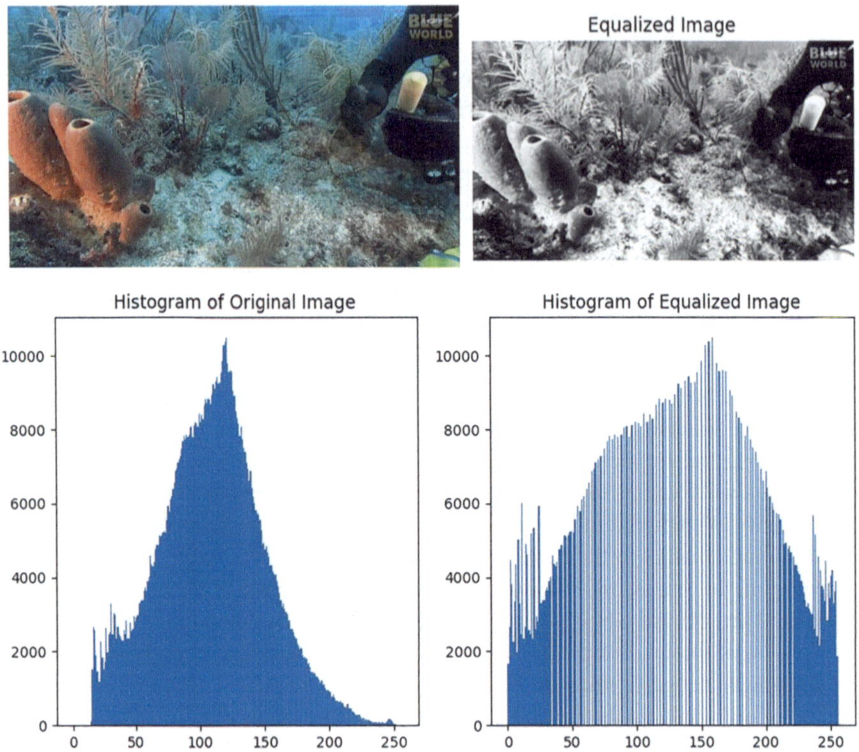

Fig. 1 Different for original image of underwater scene histogram equalization graph and equalized image and histogram process graph

the water, particularly the loss of red tones. Principal of white balance of the overall color tone of an image may change due to variations in the color temperatures of various light sources. By adjusting the white balance, colors are guaranteed to appear as they would in neutral lighting. In usually accomplished by modifying the image's red, green, and blue (RGB) channel gains. In order to compensate for the loss of red light in underwater photos, the red channel is typically boosted. Process of reference point and adjust RGB channel and transformation formula, picture, a reference white or neutral gray point is chosen. This is an example of a surface that, in neutral lighting, ought to look white. To regulate the color quantity and scaling factors are used to change each of the color channels (red, green, and blue) separately. Figure 2 shows by equation process of white balance adjustment process with it find the remaining region of red and blue and green color process and white balanced image. Restoring the red channel is prioritized in underwater photos. Basic equation for white balance adjustment is:

$$I_{wb}(x) = I(x).\frac{1}{g_r}\,|\,\frac{1}{g_g}\,|\,\frac{1}{g_b} \tag{2}$$

Fig. 2 Different region and depth of underwater image to improve the white balanced image

Where, $I_{wb}(x)$ is white balanced pixel value position x, $I(x)$ original pixel value at x, g_r, g_g, g_b gain applied to red, green and blue channel respectively. Once gain has apply the underwater image appears more of naturally colored with corrected white areas and improved overall improved color balance.

3.1.3 Gamma Correction

Gamma correction is a nonlinear operation that modifies pixel values in accordance with a power-law relationship to change an image's brightness. The gamma adjustment can be especially helpful in submarine picture processing to make darker parts light without overexposing the bright areas, which are frequently impacted by inadequate lighting and absorbed light in the water. Gamma correction is important of undersea images of light absorption in water, depending on its wavelength, light is absorbed at varying rates as it passes through water, particularly in deeper regions. Images become darker, less apparent, and have less color intensity as a result. Nonlinear Brightness Perception: The way that brightness is detected by the human brain is nonlinear. Gamma correction improves details in darker regions and makes the image appear more naturally illuminated by adjusting the pixel intensities to match human visual perception. The equation for gamma correction is

$$I_{\text{corrected}} = I_{\text{input}}^{\gamma} \tag{3}$$

where, $I_{\text{corrected}}$ gamma-corrected pixel intensity, I_{input} original pixel intensity to normalized between 0 and 1, γ is gamma value which control the type of correction applied. $\gamma < 1$ (Low Gamma): Makes details in shade more apparent by darkening the bright parts and enhancing the amount of contrast in the darker areas. This is particularly helpful for photos taken underwater that seem too gloomy. $\gamma > 1$ (High Gamma): Enhances the image's brightness by making the lighter areas more intense. However, it should be utilized carefully because it might cause details to be lost in extremely bright places. Raw image of water image process with Eq. (3) to converted the gamma corrected image (gamma = 1.5) Fig. 3. $\gamma = 1$: does not make any changes to the image. Some process correction of gamma with normalization of pixel values and gamma equation, re scaling. In order to normalize pixel intensities to a range between 0 and 1, each pixel value is usually divided by the maximum pixel intensity (255 for 8-bit pictures). The original intensity range—for example, 0–255 for 8-bit image is then restored using the rectified pixel values.

3.1.4 Retinex Based Enhancement

By simulating how the human visual system interprets color in different lighting circumstances, retinex-based enhancement is a useful method for enhancing underwater photos. Retinex-based improvement attempts to rectify the color imbalances and reconstruct the actual reflectance of objects in order to address these problems. Logarithmic Transformation: The image undergoes an exponential change in order to distinguish between sunlight and reflectance. Simplifies the process of estimating and manipulating components. Several filtering techniques, such as Gaussian filtering or bilateral filtering, are used to estimate the lighting component $L(x)$. Smoothing out illumination changes while maintaining the image's crucial structural elements

Fig. 3 Raw image of water scene is gamma with normalization process of gamma corrected Image (gamma = 1.5)

is the aim. The estimated illumination is subtracted from the logarithmic image to get the reflectance (x), which depicts the true colors and textures:

$$\log(R(x)) = \log(I(x)) - \log(L(x)) \tag{4}$$

Following reflectance recovery, Eq. (4) the logarithmic transformation is reversed using the exponential transformation:

$$R(x) = \exp(\log(R(x))) \tag{5}$$

Each designed for specific enhancement Purposes, SSR (single scale retinex) and MSR (Multi scale retinex) and MSRCR (Multi scale retinex with color restoration) Eq. (5). The illumination is estimated by SSR using a single filter, often Gaussian. It can produce halo effects or strange hues, but it also improves local contrast. By using multiple Gaussian filter scales to estimate illumination at varying levels of detail, MSR outperforms SSR. As a result, the overall balance between local and global contrast is improved and the improvement becomes more natural. The MSR formula is

$$R(x) = \sum_{s=1}^{n} w_s . \log(I(x)) - \log(I(x) * G_s(x)) \tag{6}$$

Fig. 4 Undersea image improved image following by equation process of retinex enhanced image

Weight for each scale w_s, $G_s(x)$ is the Gaussian filter at scale s, * represents the convolution operation. Equation (6) MSRCR is especially helpful for underwater photos since it addresses color distortion by combining MSR with color restoration. The color restoration feature improves detail visibility by correcting the color shift caused by uneven illumination.

$$R_{MSRCR}(x) = C(x).R_{MSR}(x) \tag{7}$$

Where $C(x)$ is the color restoration function, Eq. (7) which compensate for the loss of color due to the illumination imbalance (Fig. 4).

4 Technical Approach and System Architecture

The proposed RCNN-based underwater image enhancement system is designed to improve image quality by detecting and enhancing regions of interest in underwater environments. The high-level architecture begins with the input of raw underwater images, which are first processed through a recurrent proposal network to identify key areas affected by issues such as blurriness, color distortion, or noise. These regions are then passed through convolutional layers to extract features and classify the image quality issues. Finally, the system applies enhancement techniques such as contrast adjustment, deburring, and color correction to produce a refined, clearer output, enabling better visibility and object recognition in underwater conditions.

4.1 Pre-Processing Module

The pre-processing module of the RCNN-based underwater image enhancement system involves techniques such as color correction, noise reduction, and contrast enhancement to prepare raw images for further analysis. Color correction adjusts

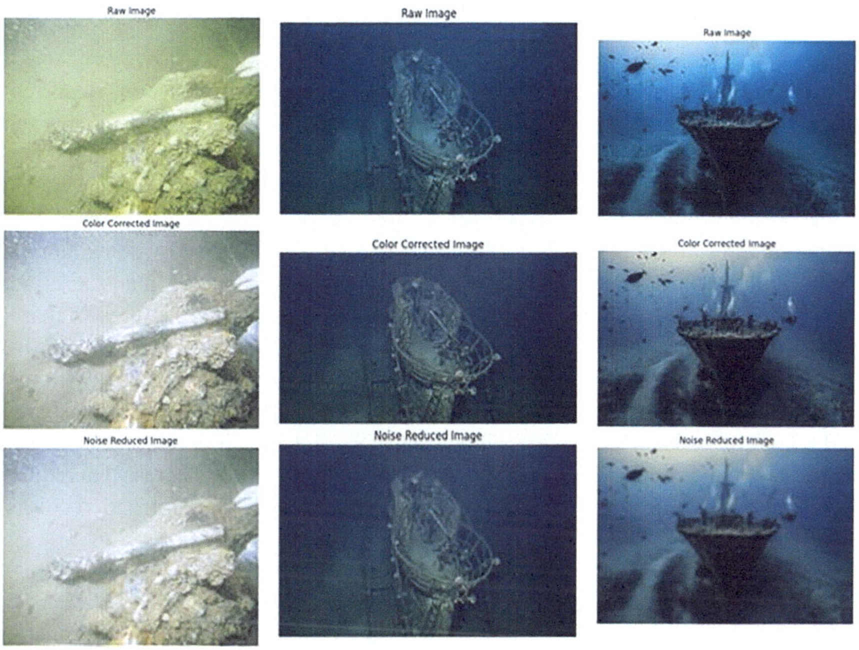

Fig. 5 Undersea image pre-processing of raw image, color correction, noise reduced image

the distorted hues caused by the water's absorption of light, restoring natural tones. Noise reduction filters out sensor or environmental noise, ensuring a clearer image. Contrast enhancement improves the visibility of objects by adjusting brightness levels and boosting details in low-contrast areas. These pre-processing steps ensure that the RCNN receives cleaner, more structured images, allowing it to accurately detect and enhance regions of interest, leading to more effective underwater image processing. In edge enhancement, within this step Sobel edge detection is used to enhance the structural edges within the image. Since RCNN only uses edges good for object detection and segmentation this step enhances edges that are important to the model. Image Resizing and Normalization: The image is resized back to the general input dimensions of most RCNN models, which is 224*224 in this instance. The pixel values are scaled to the range between 0 and 1 to bring all of the inputs to the same scale which benefits the model. Pre-processed Image Output: Finally, the pre-processed image is shown in Fig. 5. After that, the data is ready to be proceeded to the RCNN model for disaster analysis such as disaster classification, damage detection and so on.

4.2 Region Proposal and RCNN Processing Module

Disaster recovery and emergency response are heavily reliant on Recurrent Convolutional Neural Networks (RCNNs) improved ability to analyze images and extract temporal features. It was found out that with the help of convolutional layers, RCNNs can detect several spatial features from aerial or satellite imagery such as the presence of a distorted and broken skyline and buildings after a calamity. Key spatial features of actual aerial or satellite imagery are determined by convolutional layers where such elements as damaged buildings and changed landscapes are revealed by RCNNs. The recurrent layers introduce time-related features so that it is possible to monitor changes and updates about disaster effects. RCNNs improve the quality of low-resolution images, which can be essential in conditions with low visibility, enabling clearer, more precise assessments. This combination of spatial and temporal analysis capabilities allows RCNNs to detect disaster types, forecast progression, and support emergency response teams in prioritizing areas that require immediate attention, ultimately streamlining disaster management and recovery efforts (Fig. 6).

4.2.1 Feature Extraction of Underwater Image Disaster Analysis

Using RCNNs, key features can be identified and changes monitored, and semantically assisted the emergency response teams on the assessment of the effects of disasters and the formulation of response strategies.

- Data Collection

Imaging data collection for analysis of disasters involves taking satellite or drone imagery over different time instance. This also gives an RCNN (recurrent-based Convolutional Neural Network) the ability to watch and analyze the effects of the disaster. Disaster classification identifying the feature extracted from feature extraction of RCNN.

Feature extraction for RCNN identifies key visual elements in the image, such as water accumulation: indicates a flood or tsunami which charred or burnt areas of wildfire of sign and debris and structural damage. Most likely to occur in place experiencing an earthquake, a tornado or even a hurricane. Landslides with characterized by moved topographical feature or broken terrains. Thus, the presence of such features, extracted and analyzed separately by the RCNN, enables it to identify the disaster type, even if several occurrences related to different types of disaster happen at the same time—as, for instance, an earthquake followed by a landslide.

- Severity Assessment

Some of the key findings indicates that RCNN can assess the degree of destruction in order the directing emergency responders to the most critical areas. What it provides is specific quantitative and visual data, with the form of maps pointing out the extent and intensity of the occurrence. Damage Metrics of proposed work with

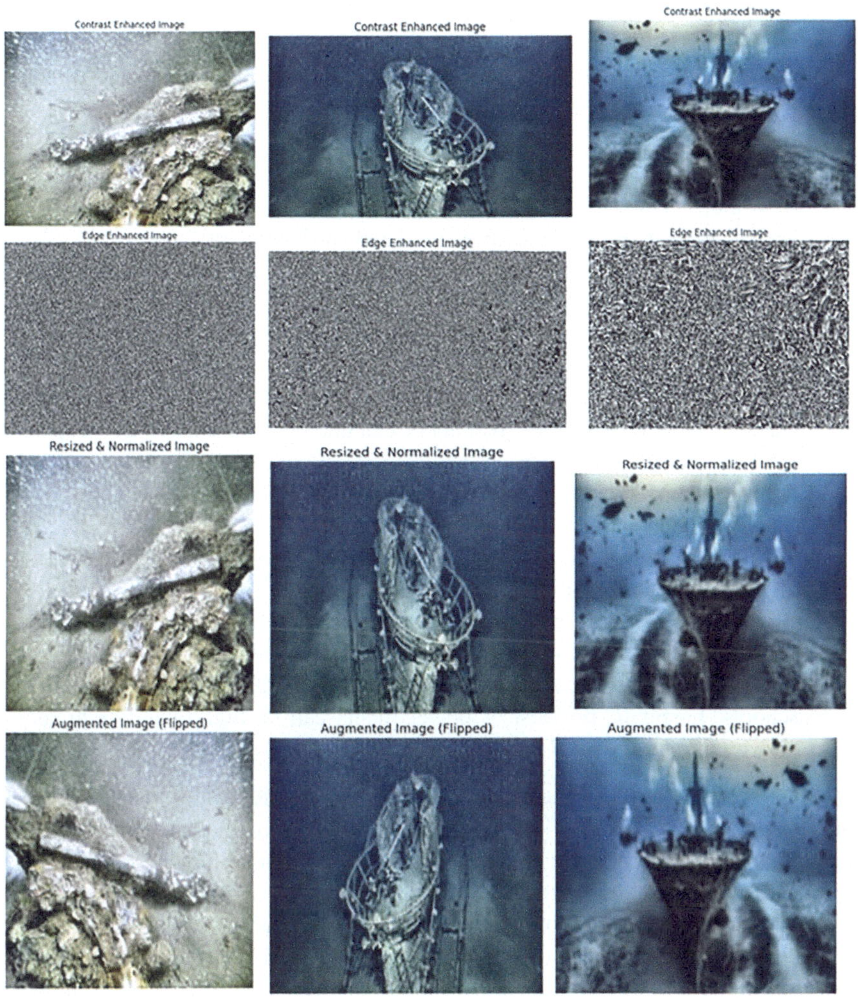

Fig. 6 Undersea image pre-processing edge image, resized and normalized image and augmented image (flipped)

percentage of Area Affected the RCNN can determine what percent of the total landscape has been affected. For example, it may forecast how much of a city or a forest has been flooded or burned. Structural Damage The model is also capable of identifying and counting the extent of damages or loss of buildings or infrastructures (roads and bridges etc.).

This includes:

- Minor Damage: Discoloration or other breach of continuity or broken edges anywhere, in walls or roofs.
- Major Damage: Destruction of partial or full parts of various buildings.

- Total Destruction: Crumbling from the inside or incredibly poor and unnoticed structures.
- Object Detection: The RCNN can point to a particular object such as vehicle, a fallen tree or a utility pole that requires to be cleared or rebuild.

• Components of Visual Damage Maps

  - Heat-maps: These are graphical representations in that the shade of red, yellow or green is representative of the level of damage encountered. Cells in deeper color areas were significantly damaged compared to cells in brighter colors or pastels. They are usually overlaid on the satellite or drone image which provides the subject with a clear visual interest map showing areas of highest intensity.
  - Damage Overlays: An original satellite or drone image is still shown below the overlay layer so that the responder can get a general idea of geographical location and find such items as streets, landmarks damaged in the certain area.
  - Point-based Markers: In some instances, the RCNN may also detect of interest points such as
  - Damaged buildings: Marked with different colored markers where structures that need inspection or repair are pointed out.
  - Blocked roads: Signs that signal that a road cannot be traveled on because of debris or water, allowing the logistics of supplies to be redirected.
  - Critical infrastructure: Blown roofs, caved in ceilings, or crumbling walls, in schools, hospitals, bridges, or power plants can be highlighted as areas of priority attention (Fig. 7).

The RCNN processing module function ad key part of underwater image enhancement. To identify regions of interest (ROIs) by sliding a small network over the image feature map, generating potential bounding boxes that indicates areas likely needing enhancement. They undergo feature extracting using convolutional layers, which capture essential visual details as edges, textures and patterns. Each extracted region is then passed through classifier that categorizes the detected issues, such as low contrast, blurriness or color distortion. To enhance these ROI, several techniques are applied: contrast adjustment improve visibility by making areas darker and brighter areas lighter. Clarity enhancement involves techniques like de-blurring to refine edges and reduces image fuzziness. Details enhancements sharpens fine textures and edges, bringing out intricate details that might have been lost due to underwater conditions. Figure 7 raw image sea work in damage map and simulated heat-Map with infrastructure and it flow of disaster recovery.

The post processing and generation modules is the final stage of the RCNN based underwater image enhancement system, aimed at fine-tuned the imaging quality before delivering the output. Post processing begins with brightness adjustment, ensuring uniform illumination across the image, correcting area that are too dark or too bright. Color balance is then optimized, addressing any residual distortions by restoring natural color tones affected by underwater conditions. This step is

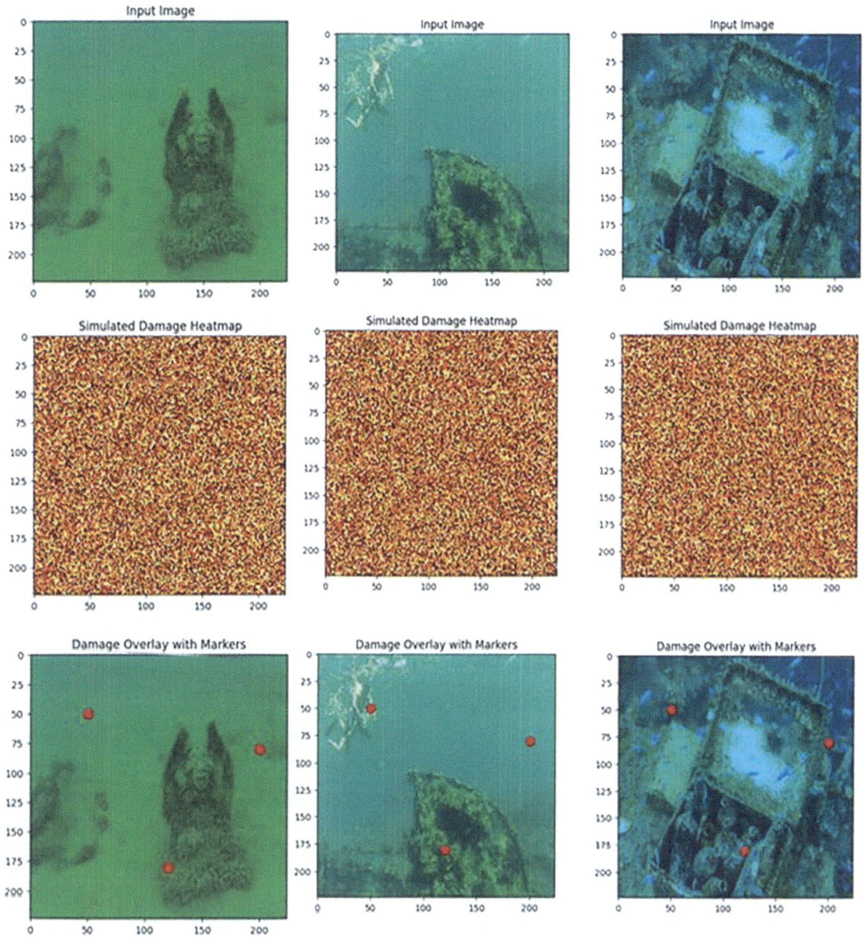

Fig. 7 Original image it simulated damage heat-map and it damage overlay with markers

crucial for making the image visually realistic and easier to interpret. Clarity adjustment further refine the images by enhancing sharpness and reducing any remaining noise or artifacts introduced during earlier stages. These adjustments ensure that the final images are crisp and detailed, particularly in regions of interest (ROIs), which are crucial for precise decision-making. For emergency response usage, the final output is formatted to be compatible with real-time decision-making tools. Images are optimized for quick loading, compressed for efficient transmission across networks, and annotated or marked for ease of interpretation by rescue teams. This ensures that the enhanced images can be rapidly deployed and integrated into disaster recovery systems, aiding responders in accurately assessing underwater conditions and locating key objects or survivors.

5 System Integration in Smart Cities for Disaster Recovery

Integrating the RCNN-based underwater image enhancement system within smart city infrastructures offers a transformative approach to disaster recovery.

5.1 Integration into Smart City Infrastructure

Image analysis employing the system developed by the RCNN enables authorities in smart city systems to identify the losses, find people in need of help as well as distribute the resources responsibly within the shortest time possible. Remote controlled aerial vehicles with advanced stabilized cameras can hence obtain better and clearer images of affected regions as preprocessing improves the captured images' quality. The improved images are analyzed by RCNN algorithms to identify those structures that have collapsed, areas that are flooded, or roads that are blocked so that first responders can knowing where to locate and target their search and rescues. China-Arab States Consortium on Disaster Resilient ICT Development—Conference Proceedings tying in with the on-going smart city technologies, this system uses the existing architecture of cameras and sensors for perpetual surveillance, analysis, and quick reactions hence making smart cities sustainable to both the dynamic urban issues besides major catastrophes.

The RCNN system can be seamlessly integrated into existing smart city frameworks by connecting it with IoT devices and various monitoring systems. For instance, control centers can utilize enhanced underwater imagery for real-time assessments of water quality and safety in aquatic environments, aiding in urban planning and disaster management. Additionally, environmental sensors, such as those measuring water clarity or pollution levels, can feed data into the RCNN system, allowing for dynamic adjustments in image processing based on current conditions. This integration enhances situational awareness and enables proactive management of water bodies in urban areas, ensuring that stakeholders have timely and accurate information for effective decision-making. For RCNN systems to connect with smart city environments, architectures, IoT gadgets, and control systems, there are several essential parts that require connection to facilitate and support the transfer of data, real-time processing, and resulting automated actions. It is the IoT gadgets like Fig. 8 the street cameras, traffic signals, environmental sensors, and drones who together constitute sensors, constantly feeding the city with visuals. These devices can be made conducive for edge computing in which they are able to perform basic data conditioning such as noise removal or contrast enhancement at the point of origination. This helps to partially unload the overload of centralized servers while guaranteeing the receipt by the RCNN system of high-quality preprocessed data.

Working Process of IOT device connection to Pre-Processing from initial image processing to find the noise reduction and color correction are applied at source.

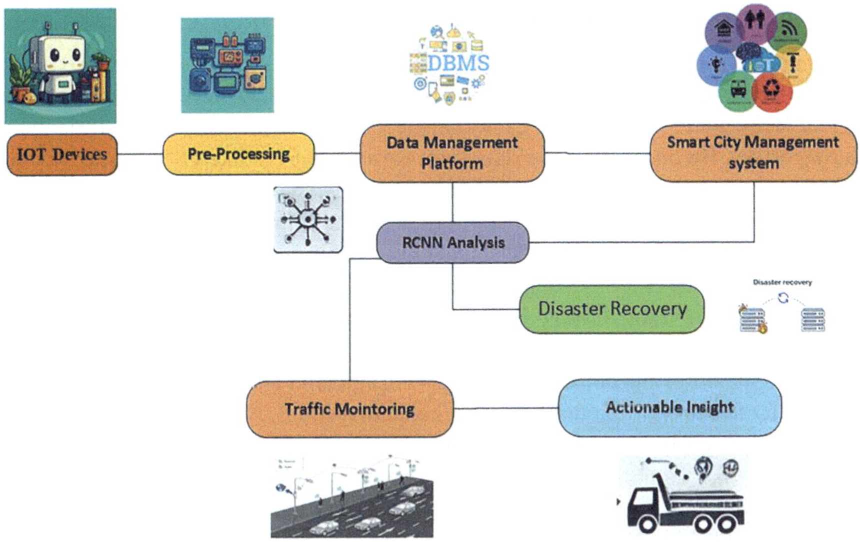

Fig. 8 Smart city infrastructure work on Iot work with disaster recovery with continue traffic monitoring with finally actionable insight

Data Management platform collected preprocessed data from IoT devices. RCNN analysis to detect and classify objects. The RCNN system conveys the images to perform identification of the objects such as vehicles, building, and the damaged area, as well as the region of interest. It receives analyzed data and gives timely decisions to address urban infrastructure to the organization. The insights obtained are employed to modify the traffic signals, inform the relevant control agents or warrant response actions. Disaster recovery and traffic management genres are unique instances where it is possible to use the specialized type of RCNN analysis, in order to enhance safety and organization of the cities. It entails development of a middle ware for data management that links all IoT devices to the RCNN framework. This platform takes in visual data feeds and identifies objects (like cars, pedestrians, or destroyed buildings as an example) using RCNN algorithms before feeding that information to the city's management systems. For example, in traffic supervision, the data from street cameras in real time can be processed by RCNNs to identify traffic jams or accidents, and violations, and traffic signals will be adjusted or authorities will be notified. In the same way, the information gathered by drones or sensors in the disaster can be used to find the place that needs urgent response—for example, the buildings that have been collapsed or the dangerous areas. The integration of RCNN with IoT and monitoring system means that we can have a dynamic urban environment to manage, getting better response time during incidences as well as promoting public safety for smart cities to be more effective and efficient.

5.2 Real-Time Processing for Emergency Response

Real-time processing for emergency response leverages advanced data analytics, IoT devices, and AI to rapidly gather, analyze, and act on incoming information during crises. By integrating live data from various sources like GPS, social media, weather sensors, and surveillance systems, emergency response teams can make swift, informed decisions. This enables quicker coordination of resources, timely communication with first responders, and effective evacuation plans. Real-time processing minimizes response delays, optimizes resource allocation, and ultimately saves lives by providing critical insights and automated alerts during high-pressure situations.

Data acquisition and communication during and after an event or a crisis are vital to the effectiveness of the disaster management systems. Such information can be delivered in real time from numerous sources including airborne systems, wearable technology like cameras, portable handheld devices, sensors etc., enabling the emergency teams capture real information of the situation on the ground. The resulting continuous data flow helps the responders adapt their tactics on the fly so that the situation is always analyzed timely for optimal management. And the ability to transmit information seamlessly also helps other agency, including the Police, Fire, medical services, and Disaster Management agencies to work jointly for a common cause. In availability of detailed communication technologies such as fifth-generation or 5G and satellite network means that real-time data can be passed across regions or during disaster situations that affect the normal functioning of other networks. This guarantees that the response teams are always on the same network irrespective of where they are and that important information which includes location tracking, assessment of damage, and amount of resource required is passed quickly. Hence these technologies do not only aid decision making during crises but also assist in averting risks, preventing the loss of many people, and assist in the building of more infrastructures that can hold out during disasters, mishaps or other calamities.

The enhancement by the RCNN improves the decision-making process in the disaster situations as graduated retention of the imagery provided the capacity for rapid and accurate image analysis based on the deep learning algorithms. Recurrent Convolutional Neural Networks (RCNNs) are very effective for object detection and segmentation for purposes such as recognizing debris, collapsed structures and trapped persons which is very important to any emergency service. The RCNN enriched image recognition can further faster zoo min to interesting regions, set priority over critical areas and even identify patterns which another human eye can do not. This hastens the determination of losses and more importantly assists the response team in correct resource calibrating (Fig. 9).

The automation of large volumes of visual data by RCNN technology saves the amount of time that is taken when the data is analyzed manually, enabling decision makers to act as and when within a complex environment. This capability is very important especially when few minutes make a lot of difference like after any

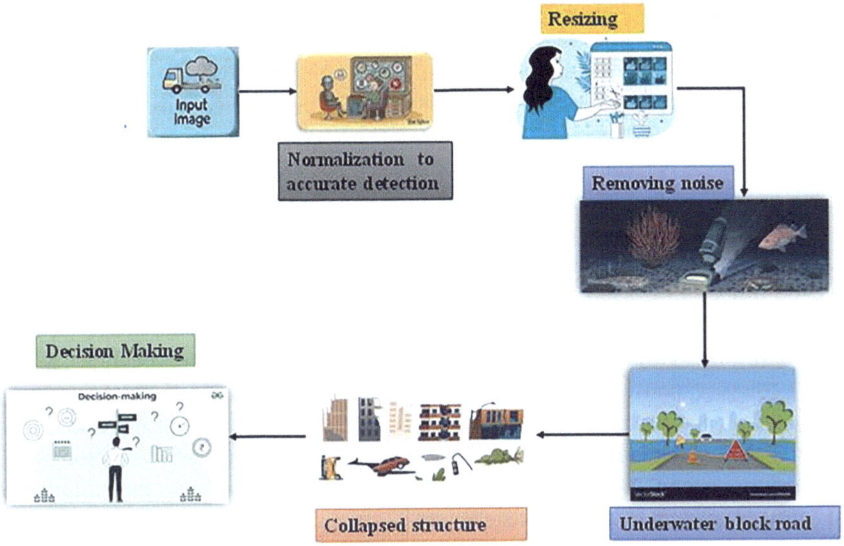

Fig. 9 Emergency response with normalization to accurate detection and removing noise of sea work and block road, collapsed structure with proposed process making finally decision

disaster like twister, flood or fire. The RCNNs allow getting a detailed understanding of the situation when emergency teams know exactly where to focus and which needs should be considered as priorities in order to efficiently allocate the resources and reduce response time. In general, by utilizing the RCNN-enhanced images, first responders and other members of the disaster management teams get better and faster decisions toward the emergency incidents and eventually, better outcomes for the incidents. In real world process of input image with Normalization to accurate detection with region of interest with image. With the connection of resizing the removing noise with real application disaster and underwater block road and to identifies key of areas with underwater block road and collection of collapsed structure process and provide critical insight each region represents. The extracted information is used assist in decision.

5.3 *Automated Alerts and Analysis*

An important use of automated alerts and analysis is to act as notification systems for real time information and subsequent analysis to supplemented image data in case of disasters. Using top-notch image recognition systems like RCNN and link them with the automatic alert system instantly, key detections like hazards, damaged infrastructure or persons in peril can be easily highlighted and passed to the concerned personnel. The first one involves real-time image processing in which images as well as video streams generated by drones, or even surveillance cameras or satellite cams are fed into an array of AI models. RCNN model identifies and segment features of interest such as, collapsed buildings, a blocked road or any other structure. These results are immediately translated by the system. Once a major event or threat occurs, the mechanism of sending notification creates notification depending on the specified parameters, for example with respect to level of risk, type of loss or presence of critical situations. Through short message services, email, or emergency-specific mobile applications, these warnings are distributed to the relevant recipients, emergency services, local authorities, or even the general public. The alerts can consist of quintessential data obtained from Image analysis like the venue of the event, the extent of destruction or the resources needed for the response. For instance, if the system identifies a trapped man in the image the alert may return geographical coordinates and recommend the approaching rescue squads. Slow response rates are eliminated, early and necessary actions are promptly commenced, and there are little calls for human interjection during important incidents. This mechanism involving the use of automated image analysis along with smart alert systems promotes better control of situation awareness thus improving on the management of disasters and emergencies from the emergency response system.

The inclusion of alerts as part of the operational procedures in an emergency response system is a revolution in disaster response in the aspect of time and efficiency. Such information is particularly valuable when image-based alerts become integrated into existing response workflows, ensuring emergency teams can receive clear and tangible real-time information directly in their working environments in order to respond more effectively and accurately to the ongoing developments [10]. On the basis of RCNN-augmented image data, automated notification can be delivered directly into the incident management systems or command centers that are utilized by first responders. These systems can be either pre-programmed to give priority to certain events to other based on level of priority of the event say or on the basis of location to tend to critical situations such as instances of building collapse or any person in perilous conditions. With this integration, the response teams can respond very fast, and determine which areas need what, and thus respond in the right manner. Further, the combined use of automated alerts enhances the management of resources. Since deep analysis of the images gives specific data like number of injured people, the degree of damage, or availability of the access ways, coordinators are able to send the right amount of people with necessary tools without excessive time.

To further enhance emergency response capabilities, the RCNN system incorporates a mechanism for sending automated alerts and notifications based on the analysis of enhanced image data. When specific conditions are detected such as the identification of objects, potential hazards, or significant changes in underwater environment the system can generate alerts that are automatically disseminated to relevant authorities and emergency responders. These alerts can be integrated into existing emergency response workflows, ensuring that teams receive real-time updates and insights as they occur. This integration streamlines communication and

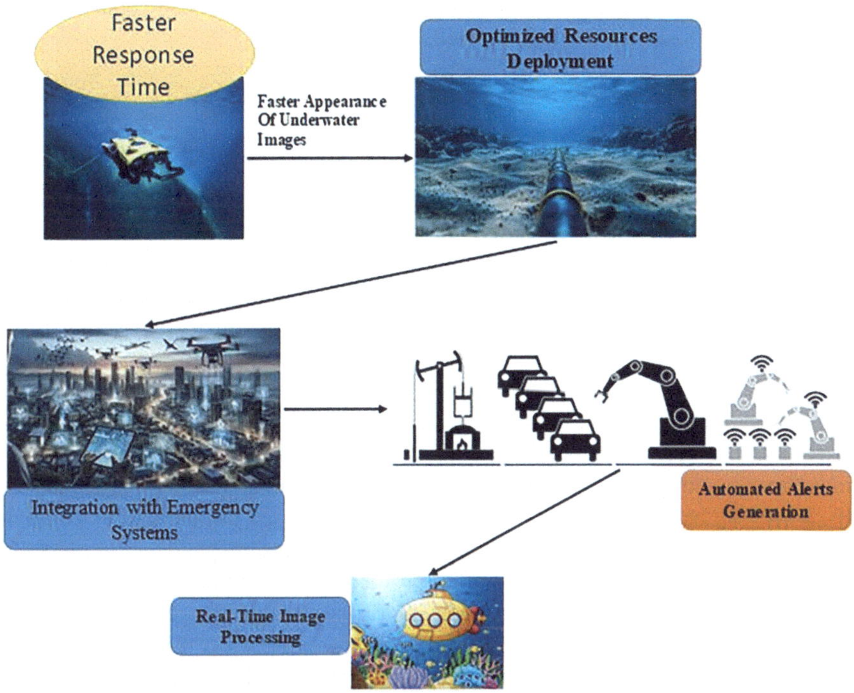

Fig. 10 Faster response of time is faster appearance of underwater image to optimized deployment and automated alerts generation in real time image processing

coordination among response teams, significantly improving response times and enhancing the overall effectiveness of disaster management efforts.

Alerts can also be combined with communication systems to make sure that response teams are getting the alert right on their corresponding devices like Smartphones or handheld radios all in real time. This helps in elimination of time waiting for manual update or directions that would be expensive to await in the market. Finally, the application of the technology improves the speed of decision making and communication, the efficiency of resulting actions and activities, and, thus, the overall prognosis and mitigation of the loss of life and property in case of emergencies (Fig. 10).

Enhancing the quality and visibility of images in underwater conditions is a pressing issue in disciplines including marine science, underwater archeology and environment surveillance. Difficulties that can be faced include low light, color shift and scattering in water even in clear water such as in in turbid waters are compensated by methods such as image enhancement algorithms, dehazing, and light compensation. Real-life samples like the ones applying adaptive filtering and deep learning models significantly improve the results in paradoxically fuzzy waters and are capable of reconstructing the details. For instance, projects covering the coral reefs have cleaned up their image and the ability to assess marine diversity regardless the level of sedimentation is restored.

6 Efficient Resource Utilization

The research demonstrates the benefits of RCNN for disaster management because it enables precise feature detection along with object recognition capabilities and predictive data analysis. To implement flood detection through combination of sensor the IOT alongside to sensing the remote of data to identifies area at immediate flood risk. The Coldwater focus optimization technique operates as an example to enhance Region of Interest (ROI) processing by dedicating unique resources solely toward flood vulnerable areas. Different approach to efficient and sustainable disaster response to decrease processing every while supporting city wide implementation. Drones controlled by RCNNs assess structural damage following earthquakes by identifying building fractures, debris, and compromised infrastructure. Through focused analysis, critical infrastructure evaluation helps responders successfully manage repair operations and evacuation protocols. Through targeted computation the technology reduces expenses by focusing on essential structures which leads to reduced energy consumption and cost of operations. Underwater disaster areas present substantial operational challenges because submersed elements become particularly hard to see through dark surroundings and harsh sea environments. AUVs are using integrated RCNN technology for underwater real-time object identification tasks. The technology utilizes ROI-based processing for cost-effective computation which enables purposeful examination of underwater debris objects and underwater dangers. This method cuts down extra data processing which allows for efficient disaster recovery operations within challenging deep-sea wreckage areas together with flooded urban settings.

The example scenarios demonstrate that RCNN functions efficiently alongside its capacity to adapt during disaster situations. The advanced technology performs exceptionally well in resource management tasks because it eliminates wasteful data processing operations while preserving computational efficiency in time-sensitive and energy-limited operational circumstances. RCNN shows its value in diverse disaster management applications through its combination capabilities with edge computing and IoT technology alongside UAVs and AUVs. Through ROI processing in its operations RCNN directs emergency responders to distribute their resources effectively toward important assessment regions. The targeted application of RCNN results in reduced costs to allocate resources properly and produces superior decision-making capabilities which generate better outcomes for disaster preparedness in smart cities.

6.1 Enhanced Integration with IoT and Edge Computing

Underwater image enhancement along with IoT devices and the edge computing is a promising future work. This is because edge computing, which is the method of performing data analyzing on the device and not a server that can be located miles away, has the ability to reduce latency to mere seconds where the need for enhanced and analyzed images is needed. Through employing a model of IoT sensors and edge computing together with other high-end imaging technologies, it will be rather easy to establish distributed intelligent and independent systems that willingly process and analyses visible data in the wide, big underwater regions. This could help better execute disaster recovery scenarios, where quick decision making is important, and allow autonomous transport or sensors to adapt instantly to the operating environment without relying solely on cloud services. Together, it would promote quicker, more readily scalable and reliable quantitative underwater evaluations to further augment awareness and effectiveness in operations.

RCNN known as Recurrent Convolutional Neural Networks have been demonstrated in several use cases for smart city disaster detection. In flood prediction, RCNNs work from one image to another satellite and other relevant meteorological data to produce effective predictions that allow for intervention. For earthquake damage estimation, RCNNs examine the rates of earthquakes and geographical images of cities for estimating risky zones and efficient rescuing tasks. However, the RCNNs also outcompete their counterparts in multi-hazard detection since they are capable of utilizing multiple data streams, for instance, IoT sensor data flow and satellite imagery to enable detection of multiple disasters at once. Besides, it operationalizes handling spatial temporal data with real-time monitoring and decision in disaster situations. Based on the findings of efficient resource utilization, the RCNNs manage and improve the computation resources efficiency, decrease the response time and scale up the system. Deploying and Interoperating RCNNs with IoT and edge computing enhance the reliability and adaptability of disaster management thereby fostering better safe city development.

6.2 Case Studies on the Use of RCNN in Disaster Detection in Smart Cities

To understand the effectiveness of RCNN in disaster detection and emergency response, this section presents various case studies where RCNN-based underwater image enhancement played a key role in disaster recovery operations. These case studies highlight the usage of RCNN in addressing real-time decision-making, visibility challenges, and efficient resource utilization in smart cities.

6.2.1 Case Study 1: Assessment of Flooded Urban Infrastructure

In a metropolitan area prone to monsoon floods, authorities deploy underwater drones, or underwater robotics are equipped with RCNN-enhanced imaging systems to assess critical infrastructure and submerged transportation networks. Traditional image processing techniques are failed due to sediment interference, low visibility, and varying light conditions. However, the RCNN model successfully reconstructed high-quality underwater images, supporting the disaster response teams to accurately prioritize recovery efforts and evaluate structural integrity.

The real-time image enhancement provided by RCNN model reduces the need for multiple manual inspections, optimizing the use of technical and human resources. Additionally, by enabling precise damage assessment, emergency teams could allocate manpower and funding more effectively.

6.2.2 Case Study 2: Search and Rescue Operations After Coastal Storms

Following a severe coastal storm, the rescue teams uses to face difficulties in locating potential survivors and submerged debris in low-visibility waters. RCNN-based enhancement techniques significantly improved underwater images captured by remotely operated vehicles (ROVs) and support the rescue team. The enhanced images facilitated better detection of hazardous obstructions and trapped individuals, leading to a 30% improvement in rescue efficiency compared to conventional imaging methods.

By improving object detection accuracy and reducing search time, the RCNN framework ensures that rescue operations will use the available resources such as diving teams, fuel and communication systems more efficiently. This technological advantage enhanced response times and minimized operational delays in disaster scenarios.

6.2.3 Case Study 3: Submerged Infrastructure Monitoring Post-Earthquake

After an earthquake-induced tsunami, critical bridges and underwater pipelines needed immediate inspection to prevent secondary disasters such as structural collapses and gas leaks. RCNN-enhanced imaging allowed authorities to detect underwater damage and micro-cracks that were otherwise imperceptible using traditional vision systems.

The precise identification of structural vulnerabilities helped prioritize urgent repairs, preventing resource wastage and redundant inspections. The use of AI-driven enhancement ensuring safety, reduced reliance on divers and efficient time management.

6.3 *Findings and Utility of RCNN in Disaster Detection*

Based on the above analyzed case studies, the following key findings justify the utility of RCNN in disaster detection operations:

- The RCNN aids emergency responders in making informed decisions with minimal uncertainty by significantly improving underwater image clarity.
- The application of RCNN ensures that financial, human and technical re-sources are deployed effectively, preventing inefficiencies in disaster management.
- The use of RCNN which brings improved visibility of underwater environments accelerates the recovery and assessment process, reducing downtime for affected infrastructure.
- The RCNN models can be integrated into broader disaster management frameworks, which could support in real-time monitoring across urban landscapes.

Overall, by integrating RCNN-enhanced underwater imaging into smart city disaster monitoring and response systems, the authorities can ensure safer, faster and more cost-effective emergency interventions.

7 Conclusion

Enhancement of underwater images using RCNN is a revolutionary technology that has a powerful impact on disaster-related activities since accurate view in underwater conditions can be created. Due to the implementation of deep learning methods, RCNN facilitate a more efficient evaluation of the submerged infrastructures and risks in smart cities in the short time required to make the required decisions during disaster responses. Combining all these models with IoTs and edge computing guarantees real-time analysis and immediate autonomy due to the low latency involved. In the future, improving multi-hazard detection, incorporation with the autonomous system, and multimodal data fusion would augment the RCNN application. Further, fine tuning these models can improve their future applicability to environmental purposes and regions that are less resource endowed making for smarter, sustainable disaster management. Main perspective through swarm intelligence, real-time detection systems driven by RCNN collaborate more effectively with unmanned surface vehicles (USVs) and autonomous underwater vehicles (AUVs) to direct coordinated exploration during disaster response. By improving its multi-hazard detection capabilities, the RCNN will become better equipped to identify and handle intricate underwater disaster situations. Enhancing comprehension through the incorporation of Explainable AI (XAI) will boost confidence in RCNN-based emergency response decision-making. Developments in neural computation and innovative artificial intelligence will enable real-time processing on underwater devices with limited resources. Model durability will be strengthened further by growing open-source underwater datasets, promising adaptability in a variety of aquatic situations.

References

1. Alawad W, Halima NB, Aziz L (2023) An unmanned aerial vehicle (UAV) system for disaster and crisis Management in Smart Cities. Electronics (Switzerland) 12(4). https://doi.org/10.3390/electronics12041051
2. Bosse S, Kasundra P (2022) Robust underwater image classification using image segmentation, CNN, and dynamic ROI approximation. Eng Proc 27(1). https://doi.org/10.3390/ecsa-9-13218
3. Dulhare UN, Ali MH (2023) Underwater human detection using faster R-CNN with data augmentation. Mater Today Proc 80:1940–1945. https://doi.org/10.1016/j.matpr.2021.05.653
4. Fedele R, Merenda M (2020) An Iot system for social distancing and emergency management in smart cities using multi-sensor data. Algorithms 13(10):1–24. https://doi.org/10.3390/a13100254
5. Gosavi A, Fraioli G, Sneed LH, Tasker N (2020) Discrete-event-based simulation model for performance evaluation of post-earthquake restoration in a smart city. IEEE Trans Eng Manag 67(3):582–592. https://doi.org/10.1109/TEM.2019.2927318
6. Han F, Yao J, Zhu H, Wang C (2020) Marine organism detection and classification from underwater vision based on the deep CNN method. Math Probl Eng 2020. https://doi.org/10.1155/2020/3937580
7. Ichimaru K, Furukawa R, Kawasaki H (2019) CNN based dense underwater 3D scene reconstruction by transfer learning using bubble database. In: Proceedings—2019 IEEE winter conference on applications of computer vision, WACV 2019, pp 1543–1552. https://doi.org/10.1109/WACV.2019.00169
8. Ji X et al (2024) FBDPN: CNN-transformer hybrid feature boosting and differential pyramid network for underwater object detection. Expert Syst Appl 256. https://doi.org/10.1016/j.eswa.2024.124978
9. Joseph J, Mohammed Irshad S, Alex AM (2021) Disaster recovery and structural inequalities: a case study of community assertion for justice. Int J Disast Risk Reduct 66(December 2020):102555. https://doi.org/10.1016/j.ijdrr.2021.102555
10. Kumar A, Singh JP (2022) Randomized convolutional neural network architecture for eyewitness tweet identification during disaster. J Grid Comput 20:20
11. Liu X et al (2024) Enhancement of underwater images through parallel fusion of transformer and CNN. J Mar Sci Eng 12(9):1467. https://doi.org/10.3390/jmse12091467
12. Lyernisha SR, Seldev Christopher C, Fernisha SR (2023) Object recognition from enhanced underwater image using optimized deep-CNN. Int J Wavelets Multiresolution Inf Process 21. https://www.worldscientific.com/doi/abs/10.1142/S0219691323500078
13. Ravindranath K, Raghupriya N, Krishna Vamsi P, Sharath Kumar D (2018) Study on disaster recovery in cloud environment. Int J Eng Technol 7(2):100–103. https://doi.org/10.14419/ijet.v7i2.32.13537
14. Sun X, Liu L, Dong J (2018) Underwater image enhancement with encoding-decoding deep CNN networks. In: 2017 IEEE SmartWorld ubiquitous intelligence and computing, advanced and trusted computed, scalable computing and communications, cloud and big data computing, internet of people and smart city innovation, SmartWorld/SCALCOM/UIC/ATC/CBDCom/IOP/SCI 2017, pp 1–6. https://doi.org/10.1109/UIC-ATC.2017.8397462
15. Zeng F, Pang C, Tang H (2023) Sensors on the internet of things systems for urban disaster management: a systematic literature review. Sensors 23(17). https://doi.org/10.3390/s23177475

Optimizing Urban Systems Through Digital Twin Technology for a Smart City, Insights from a Real Case Study

El Mahdi Bouyahrouzi, Ali El Kihel, Mouna Berquedich, and Soufiane Embarki

Abstract With the continuous rise in urban populations, cities are under increasing pressure to develop and optimize infrastructure and services to meet the evolving needs of their residents. This rapid urbanization presents significant challenges in efficiently managing resources, mobility, and public services. In response, digital twin technology has emerged as a promising solution for the development of smart cities. By enabling the collection and analysis of real-time data, this technology enhances urban system efficiency and allows local authorities to make more proactive and predictive decisions. This study introduces a novel approach for creating an immersive virtual environment that replicates the functionalities of a smart city. Through this methodology, critical challenges in urban management, such as infrastructure optimization, resource allocation, and mobility improvement, are addressed. The project also leverages the Unity platform to develop interactive models that simulate various urban dynamics, including waste management, transportation, and environmental monitoring, with the goal of providing a deeper understanding of urban challenges and supporting more effective urban planning. A case study conducted in the El Qods neighborhood, located in the eastern part of Oujda, Morocco, demonstrates the effectiveness of the proposed solution in optimizing energy use, managing infrastructure, and enhancing urban planning, including the allocation of green spaces.

Keywords Smart city · Digital twin · Internet of things · Urban planning · Urban policy · Optimization · Sustainable development

---

E. M. Bouyahrouzi (✉) · A. El Kihel · S. Embarki
Industrial Engineering Laboratory, ENSA, UMP, Oujda, Morocco
e-mail: elmahdi.bouyahrouzi@ump.ac.ma

M. Berquedich
Green Tech Institute—Mohammed VI Polytechnique University, Benguerir, Morocco

H. Kaddouri et al. (eds.), *Smart Cities for Sustainable and Urban Development*, Green Energy and Technology, https://doi.org/10.1007/978-3-031-99001-4_38

1 Introduction

With the rapid development of computing power and communication technologies, the Fourth Industrial Revolution is underway, marking a profound transformation of society. In the field of urban planning, this revolution manifests through the concept of the "smart city". This innovative solution aims to improve the quality of life for urban residents by making cities more adaptable and responsive to their inhabitants' needs, while integrating groundbreaking solutions to address major challenges such as environmental sustainability, infrastructure optimization, and decision-making efficiency [13, 17, 22]. Smart cities rely on advanced digital infrastructure, leveraging Industry 4.0 technologies to connect and synchronize various urban systems [15].

A key pillar of this solution is the connectivity of infrastructures to cloud servers via high-speed communication networks. This connectivity enables the entire urban ecosystem-streets, buildings, transport infrastructures, energy networks, and more to exchange real-time data among its components. Moreover, this connectivity allows the city to monitor and respond swiftly to changes in urban conditions. For instance, collected data may include air quality, building energy consumption, or vehicle traffic flow, enabling centralized and rapid decision-making as well as intelligent resource management. This data, analyzed in real time with the support of artificial intelligence, transforms raw data into actionable insights, enhancing the city's resilience to complex urban challenges [14].

However, a major breakthrough in smart cities lies in the integration of the digital twin concept, marking a significant step toward connected and proactive cities [8]. A digital twin involves creating a detailed and dynamic virtual replica of the physical city in a digital environment [1]. This virtual model, known as the urban digital twin, is designed to remain continuously connected to the physical city, allowing real-time data to feed the virtual model [24]. Consequently, decision-makers can visualize and analyze the current state of infrastructures, anticipate the impacts of development projects, and simulate the city's response to unforeseen events.

Since its emergence in the early 2000s, the applications of digital twins have diversified. Initially developed for industrial use, digital twins have gradually expanded to various fields such as aerospace, automotive, and now, smart cities. Exemplary projects like Virtual Singapore [11] and the S-Map virtual city [19] illustrate the impact of urban digital twins in optimizing urban systems and supporting proactive infrastructure management. In this regard, [9] consider a digital twin as an "advanced virtual representation" capable of simulating the real-time performance of urban systems, enabling visualization, prediction, and improvement of operations across different levels of a complex environment. They affirm that a digital twin, when applied in a smart city, becomes a fundamental tool for anticipating future needs, optimizing energy management, and increasing resilience to urban challenges. Similarly, other researchers argue that by integrating physical infrastructures, real-time data, and simulation capabilities, digital twins play a crucial role in creating "intelligent and responsive" cities on a global scale.

The integration of urban digital twins into smart cities thus becomes a central tool for strategic planning and management, offering engineers, decision-makers, and municipal officials the ability to test various development scenarios in a secure environment before their implementation in the real world [5, 14]. By reducing uncertainties, optimizing investments, and anticipating responses to change, the digital twin supports a sustainable and resilient vision of urban management. In this context, the urban digital twin or "digital twin city" is considered one of the fundamental components of an autonomous and intelligent city, capable of continually adapting to technological developments and the needs of its population [6, 24].

Despite the progress made, limited research has been conducted to develop an ecosystem that merges smart cities with digital twins. In [8], the authors examine different ways digital twins can be integrated into smart cities to optimize urban management. They highlight the importance of digital twins in providing a real-time replica of the urban environment, enabling continuous monitoring and service adjustments based on current conditions and needs. However, this requires significant computing resources and raises challenges in terms of data storage and security. In Mandl and Kirschner [16], the authors classify smart city components into several segments: smart energy, smart buildings, smart mobility, smart technology, smart healthcare, smart infrastructure, smart governance, smart security, and smart citizens. A clear classification of components is essential for the successful implementation of digital twins in cities, enabling coherent integration and smooth operations.

This chapter proposes an innovative approach to integrating digital twins into smart cities, focusing on an optimized method for managing urban systems. The core innovation lies in using a digital twin model capable of synchronizing real-time data collected from various urban sectors through a centralized platform. Unlike traditional approaches to smart city integration [10], which often rely on segmented data and domain-specific management, this methodology unifies data and processes in a shared digital space, offering a consistent and real-time view of the city.

The remaining sections of this document are divided into six main parts: Sect. 2 reviews existing work on the use of digital twins across various smart city domains. Section 3 introduces the smart city concept utilized in this study and details the methodology for developing an "Urban Digital Twin". Section 4 applies this methodology to a real-world case study, examining the practical aspects of its implementation. Section 5 summarizes the results achieved, evaluating the methodology's effectiveness in enhancing resource management and service responsiveness. Finally, Sect. 6 concludes by summarizing the study's main contributions and emphasizing the methodology's significance for improving smart city efficiency. It also outlines future research directions to refine the methodology further and expand the application of digital twins to address evolving urban challenges.

2 State of the Art

Urban Digital Twins represent urban hubs that embrace digital and sustainable transformation to tackle contemporary challenges. These centers leverage technological advancements from Industry 4.0, particularly in information and communication technologies, to enhance citizens' quality of life. By integrating connected infrastructures such as IoT devices, communication networks, and analytics platforms, an Urban Digital Twin model collects and analyzes real-time, continuous data. This enables proactive resource management, quicker and more efficient responses to citizens' needs, and more informed decision-making for municipal officials. Additionally, Urban Digital Twins aim to create urban environments that are more sustainable, safe, resilient, and inclusive while promoting economic development and enhancing social well-being.

In this context, a study was conducted to analyze the evolution of publication frequency on the themes of "Smart Cities" and the concept of "Urban Digital Twin". This reflects the integration of digital twins into smart cities. The data was collected using the international research platform Scopus, covering a 10-year period from 2015 to 2024. The analysis of publication frequency on the themes of Smart Cities and Urban Digital Twins from 2015 to 2024 reveals significant trends, as illustrated in Fig. 1. These trends highlight the growing academic and professional interest in these two key concepts of urban transformation.

The graph in Fig. 1a shows a steady increase in the number of publications related to Smart Cities and their synonyms over the past decade, with a marked upward trend through to the present day. The analysis can be divided into two phases: Phase I between 2015 and 2017: The number of publications grew from 1341 in 2015 to 4522 in 2017, indicating rapid growth. This reflects the rising awareness of the benefits offered by Smart Cities in addressing urban challenges such as sustainability, resource management, and quality of life improvement. This period represents a phase of discovery and conceptualization, during which researchers actively explored how technology could transform urban planning. Phase II between 2018 and 2023: The number of publications stabilized at around 6000 articles per year, with a notable peak of 6503 publications in 2019. This stabilization may indicate that Smart Cities research has entered a maturation phase, where fundamental concepts are well-established, and recent research likely focuses on optimizing and implementing specific solutions in real-world urban contexts.

Figure 1b illustrates the evolution of publications on the concept of Urban Digital Twins, which started more slowly but has shown a steady increase in interest over the years. The first publications on Urban Digital Twins appeared in 2017, with only a few articles. This initial scarcity can be attributed to the novelty of the concept, which was primarily applied to specific and isolated use cases at the time. Starting in 2018, interest in integrating digital twins into smart cities began to grow. This moderate increase in the early stages reflects the gradual adoption of IoT technologies and high-speed connectivity, enabling more advanced applications of digital twins in urban environments. Furthermore, the number of publications on Urban Digital Twins surged significantly in 2023. This rapid growth demonstrates that

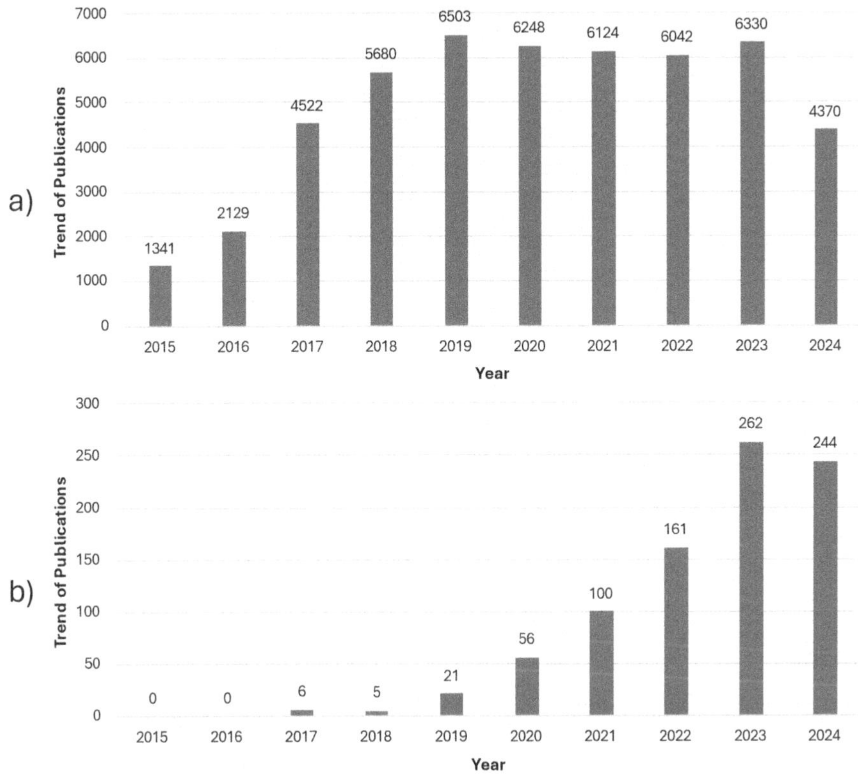

Fig. 1 Evolution of the number of publications per year for the period 2015–2024: (**a**) Number of publications on smart cities and their synonyms, (**b**) number of publications on urban digital twins and their synonyms

Urban Digital Twins are increasingly seen as a strategic tool for urban management and planning. This can be interpreted as a positive sign of vitality and engagement in this field of research. Researchers are actively exploring how this technology can enhance urban resilience, sustainability, and efficiency by providing virtual models that simulate various scenarios and improve decision-making processes.

In the Fig. 2, a complementary study was conducted to highlight the growing interest in Industry 4.0 technologies in the context of Urban Digital Twins and to identify the most utilized technologies between 2015 and 2024. This study demonstrates how technological advancements, such as the IoT, AI, cloud computing, and 5G connectivity, play a crucial role in the development and optimization of Urban Digital Twins. These technologies enable real-time data collection, processing, and analysis, which are essential for creating dynamic virtual replicas of physical cities. This integration supports better resource management, more effective responses to citizen needs, and strategic urban planning. Additionally, Urban Digital Twins serves as a testbed for simulating scenarios, improving decision-making, and fostering sustainable urban development.

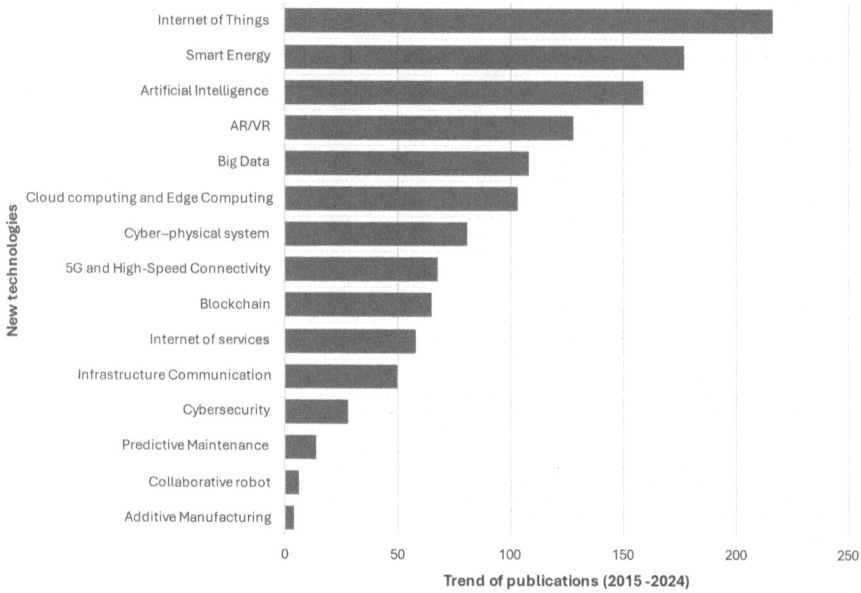

Fig. 2 Number of publications on I4.0 technologies applied to urban digital twins between 2015 and 2024

The results of this study demonstrate the increasing adoption of Industry 4.0 technologies, reflecting their ability to transform traditional cities into interconnected and smart cities. Over the study period, certain technologies, e.g., IoT, Smart Energy, and AI, have emerged as pillars in the creation of urban digital twins due to their capabilities in providing real-time data, automating processes, and facilitating predictive, data-driven decision-making. Moreover, by analyzing publications from 2015 to 2024, the study highlights trends and priorities in technological integration within smart cities, focusing on areas such as smart energy, security, infrastructure management, and urban mobility. This analysis provides insights into the technological directions embraced by researchers and decision-makers to build more sustainable, resilient, and citizen-centric cities.

To better understand the significance of each key Industry 4.0 technology in the context of Urban Digital Twins, a complementary study was conducted. The results of this study are summarized in Table 1.

3 Methodology

The proposed methodology introduces a comprehensive and innovative approach to integrate digital twins into smart city management, focusing on leveraging Industry 4.0 technologies to enhance urban systems' efficiency, adaptability, and resilience. It is divided into four main components, as illustrated in Fig. 3.

Table 1 Overview of Industry 4.0 technologies and their applications in urban smart cities

Technology	Definition and role of the technology	Application of the technology in urban smart city
Internet of things (IoT)	IoT lies at the core of digital twins, enabling real-time data collection via a network of sensors and connected devices. In a smart city, these sensors are deployed across sectors such as buildings, transport infrastructure, and energy networks [23].	Monitoring energy use, traffic management, air quality tracking, waste management, and public safety. IoT enables the creation of a living digital replica of the city, responsive to current conditions and capable of informing real-time decisions.
Smart energy	Smart energy management uses digital twins to optimize the production, distribution, and consumption of energy within the city [20].	Energy demand forecasting, renewable energy integration optimization, monitoring, and predictive maintenance of electrical grids. Digital twins balance energy supply and demand in real-time, reducing losses and increasing efficiency.
Artificial intelligence	AI is essential for analyzing and interpreting the vast amounts of data generated by digital twins. It aids in automating urban management processes and anticipating issues [3].	Traffic flow prediction, public transport system optimization, predictive maintenance, resource management. AI transforms collected data into actionable insights, facilitating proactive decision-making across diverse domains.
AR/VR	Augmented and virtual reality allow immersive visualization and interaction with the digital twin, making them valuable for urban planning and maintenance training [25].	Modeling and simulating urban development scenarios, evaluating the impact of infrastructure projects, and training maintenance teams in simulated environments. These technologies enhance decision-making by providing realistic visual representations.
Big data	Big Data enables the storage of vast amounts of real-time data generated by IoT sensors and other city devices [2].	Trend analysis of consumption patterns, anomaly detection in infrastructure, urban service optimization. Big data technologies support predictive and analytical capabilities of digital twins, identifying patterns and optimizing resources.
Cloud computing and edge computing	Cloud computing provides scalable storage and processing for digital twin data, while edge computing improves response times by processing data near its source [18].	Real-time data storage and analysis, large-scale urban operations optimization, latency reduction for critical applications. These technologies offer a flexible and scalable infrastructure to support city-scale digital twins.
Cyber–physical systems	CPS combine physical-world data with computational capabilities to efficiently control urban infrastructures [12].	Integration of intelligent transport systems, management of critical infrastructure, automated energy network control. CPS provides a layer of intelligent and automated management, capable of responding to real-time city events.

(continued)

Table 1 (continued)

Technology	Definition and role of the technology	Application of the technology in urban smart city
5G and high-speed connectivity	5G networks provide the fast and reliable connectivity necessary for real-time digital twin applications in smart cities [26].	Real-time traffic management, augmented/virtual reality applications, emergency response management. 5G ensures smooth and rapid data transmission, making urban systems more interconnected and responsive.
Blockchain	Blockchain secures data exchanges and enhances transparency in urban information management [4].	Identity management systems, IoT data security, urban supply chain tracking. Blockchain ensures data integrity and traceability within urban ecosystems.
Internet of services	The Internet of Services delivers personalized digital services to residents, based on data collected by the digital twin [21].	Digital public services, emergency services management, automation of municipal services. This technology contributes to a more interactive and citizen-centric city.

3.1 Detailed City Analysis

This step ensures an accurate representation of the city in the digital twin by identifying key sectors and assessing digital maturity [7]:

- **Identification Layer:** The first layer focuses on identifying key sectors. It is essential to pinpoint the strategic areas of the city that would benefit most from integration into the digital twin, such as transport infrastructure, energy networks, buildings, and environmental systems.
- **Maturity Layer:** The second layer involves evaluating Digital Maturity. An analysis of the digital maturity of each sector and the city is carried out using a questionnaire. The questionnaire is designed to gather precise and detailed data across various domains: Policy and Governance, Organization and Human Resources, Transportation, Services and Quality of Life and Environment and Sustainability. A formula calculates the maturity level based on sector-specific weighting factors, highlighting strengths and areas needing improvement.

$$M_d = \sum_{i=1}^{n} \frac{M_{Dli} \times g_{Dli}}{g_{Dli}} \quad (1)$$

With:

M_d: Maturity level for each domain.

M_{Dli}: Response provided to the questionnaire.

g_{Dli}: Weighting factor, representing the importance assigned by decision-makers and municipal officials to each elementary component of the associated dimension.

$M_{Dli} \times g_{Dli}$: Product of the elementary components (M_{Dli}) and their weighting factors (g_{Dli}).

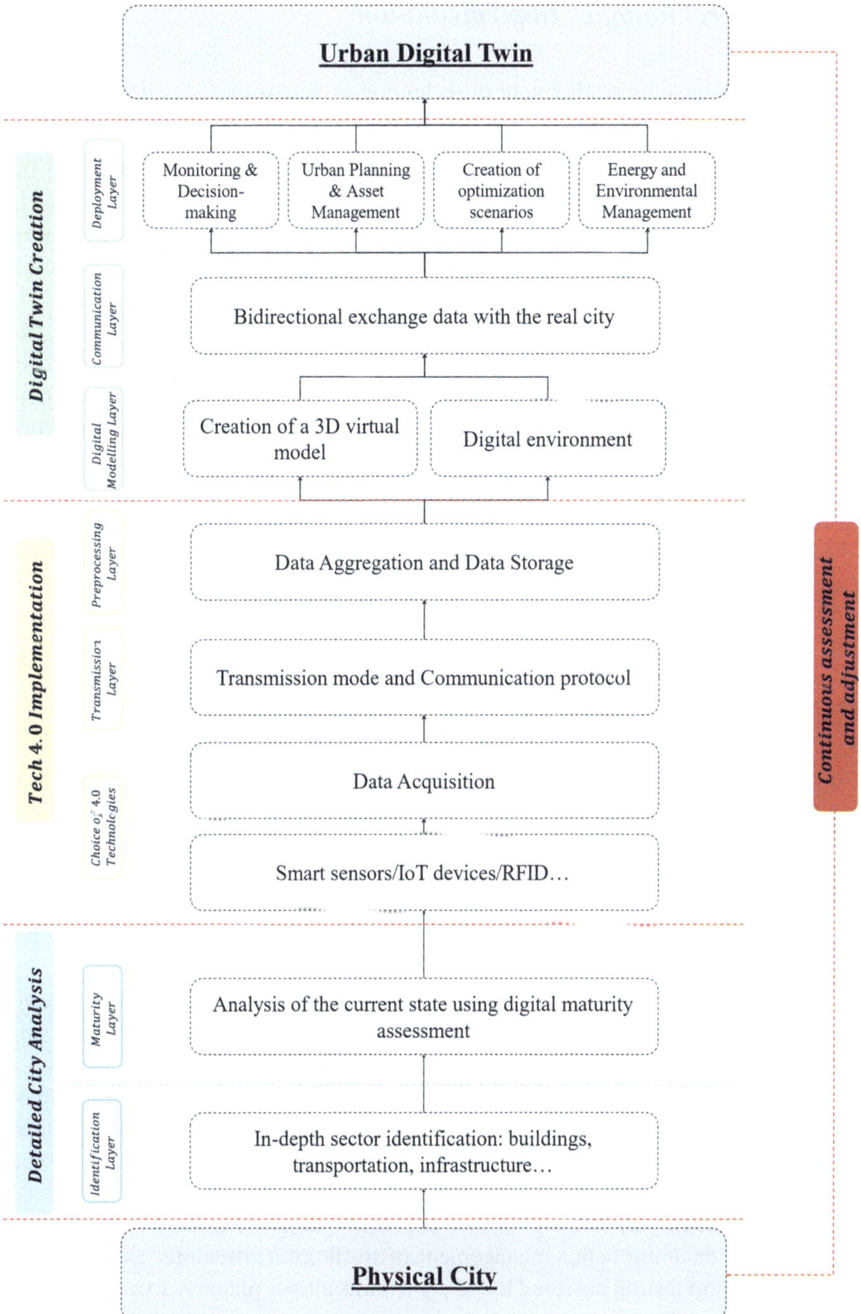

Fig. 3 Methodological framework for integrating I4.0 technologies and digital twin in smart city management

3.2 I4.0 Technologies Implementation

This phase ensures the deployment of technologies necessary for collecting, transmitting, and processing data efficiently and securely. It consists of three layers:

- **Choice of I4.0 Technologies:** IoT sensors and advanced devices such as RFID, drones, and smart cameras are deployed to monitor urban conditions. These technologies collect data on infrastructure, traffic, environmental factors, and more. The diversity of devices ensures comprehensive monitoring across all critical city sectors.
- **Data Transmission Layer:** Robust communication protocols like Wi-Fi, 5G, NB-IoT, and LPWAN are selected to transmit data efficiently based on the needs of each sector. For example, critical incidents such as security alerts use real-time transmission, while less urgent metrics like energy usage can be transmitted in batches. This tailored approach optimizes bandwidth and energy consumption while ensuring the availability of critical information.
- **Data Preprocessing Layer:** Collected data is aggregated, stored, and secured in centralized cloud systems or Big Data platforms. Key features include scalability to accommodate growing datasets and advanced security measures like encryption and authentication to protect sensitive information.

3.3 Digital Twin Creation

The heart of the methodology lies in creating a dynamic digital replica of the city, capable of interacting with physical systems in real-time. This step includes three layers:

- **Digital Modeling Layer:** A 3D model of the city is created, incorporating buildings, roads, parks, and other infrastructure. The digital twin evolves dynamically, reflecting real-world changes and enabling simulations of various scenarios (e.g., traffic flow optimization, energy consumption reduction). This capability supports informed urban planning and decision-making.
- **Communication Layer:** Continuous synchronization ensures the digital twin remains updated with real-time data from sensors and IoT devices. For instance, in the event of an infrastructure failure, the digital twin can visualize the issue, recommend solutions, and guide emergency responses.
- **Deployment Layer:** The digital twin becomes a decision-making tool for urban managers, enabling better management of traffic, infrastructure, and emergencies. Scenario testing powered by AI algorithms allows planners to anticipate the impacts of policy changes, such as implementing energy-saving measures or enhancing mobility. Additionally, it supports sustainable urban management by monitoring environmental factors like air quality and recommending actions for improvement.

3.4 Continuous Assessment and Adjustment

This phase ensures that the digital twin remains accurate, relevant, and aligned with the city's evolving needs. It emphasizes:

- **Performance Monitoring:** Key performance indicators (KPIs), such as energy efficiency, traffic flow, and air quality, are tracked to evaluate the twin's effectiveness in achieving urban management goals.
- **Anomaly Detection:** AI-powered tools identify deviations from normal patterns, such as sudden increases in water consumption or air pollution, enabling quick responses. These insights help improve the city's resilience and responsiveness to challenges.
- **Parameter Adjustment:** Feedback loops refine the twin's algorithms and parameters based on performance evaluations. This ensures the digital twin adapts to changes in urban conditions, continuously improving its predictive and decision-making capabilities.
- **Update of I4.0 Technologies:** The model is updated with emerging Industry 4.0 innovations, such as more advanced sensors or enhanced communication protocols, ensuring it remains at the forefront of technological capabilities.

4 Case Study

The case study applied the methodology to a prototype of a real neighborhood instead of an entire city. This smaller-scale approach demonstrated how the methodology could be deployed in a localized, practical urban environment. Data for the neighborhood's digital twin was gathered through sensors, IoT devices, and geospatial sources, covering infrastructure, traffic, energy use, and environ-mental parameters. Focusing on the neighborhood allowed for precise analysis of resource management needs, energy efficiency, and environmental monitoring. This prototype validated the methodology's feasibility, relevance, and effectiveness. Additionally, it provided an opportunity to refine processes before scaling up to a city-wide application. The iterative approach ensures the model is adaptable to real urban conditions and transferable to other similar contexts.

4.1 Case Study Description

Hay Nakhil, a neighborhood in Oujda, Morocco, faces significant urban challenges due to rapid population growth and urbanization, including issues with traffic, energy efficiency, air quality, and waste management. To address these, a digital twin project was initiated to optimize urban management by providing real-time

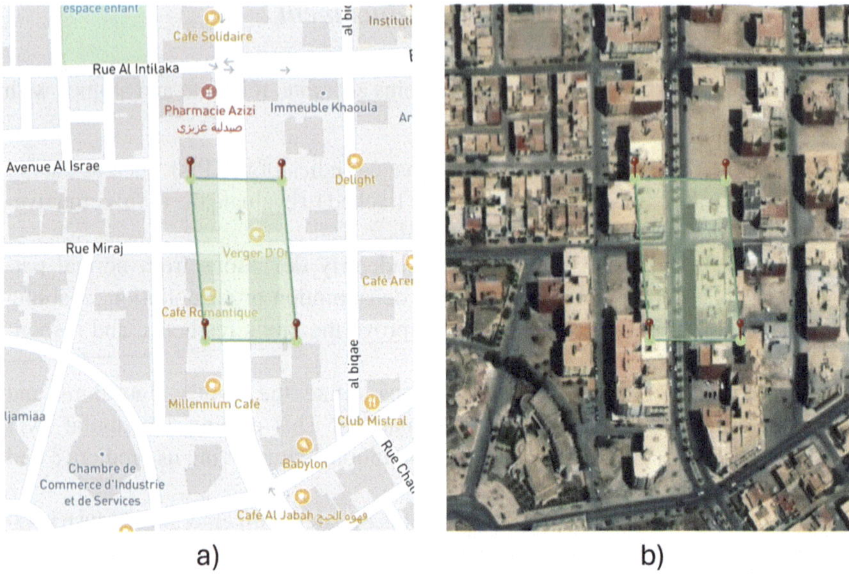

Fig. 4 Extension of the modeled Florence area. In green the area covered by our Digital Twin. This map is shown in the GEO Field Area coordinate system

insights, anticipating future needs, and simulating intervention impacts. The project leverages Industry 4.0 Technologies like IoT, smart sensors, and AI to enable proactive management and sustainable development.

The methodology, applied to a 7207 m^2 area in Hay Nakhil as shown in Fig. 4, followed four stages: analysis, implementation, creation, and continuous evaluation. This localized case study validated the digital twin's efficiency and its potential to solve pressing urban challenges, such as resource optimization and environmental monitoring. The project demonstrates how digital twins can transform urban management in developing regions, laying the groundwork for expansion across Oujda and beyond. This initiative highlights the role of digital twins in driving sustainable development and improving quality of life in fast-growing urban areas.

4.2 Application of the Method Proposed

4.2.1 Detailed City Analysis

Following the proposed methodology, the first phase focuses on an in-depth analysis of the Hay Nakhil neighborhood to identify key sectors for inclusion in the digital twin model and evaluate their digital maturity. This phase highlights the neighborhood's pressing challenges through field analysis and feedback from residents and local managers. The identified issues include:

Traffic Management: Frequent congestion due to an inadequate road network and poor traffic flow coordination.

- **Energy Efficiency:** High energy consumption in public and residential buildings, linked to a lack of real-time monitoring and optimization.
- **Air Quality:** Elevated pollution levels caused primarily by vehicle emissions.
- **Waste Management:** Inefficient waste collection and sorting practices, along with illegal dumping sites impacting cleanliness and hygiene.
- **Loack of Digital Infrastructure:** Insufficient infrastructure for real-time monitoring, limiting proactive urban management capabilities.

To address these challenges, a structured diagnostic process was implemented using a questionnaire to evaluate the neighborhood's current digital maturity. The questionnaire covers five critical domains: Governance and Policy, Human Resources Transportation, Services and Quality of Life and Environment and Sustainability.

Each domain is assessed using a Likert scale (1–5) to ensure consistent and objective evaluation. The responses are analyzed individually, creating a detailed profile of the neighborhood's strengths and areas for improvement. This profile provides a clear understanding of the current state of urban systems and digital readiness, forming the foundation for targeted interventions within the digital twin framework.

This comprehensive analysis highlights both the neighborhood's challenges and opportunities for improvement. It establishes a baseline for developing tailored solutions, such as real-time traffic optimization, energy-efficient systems, enhanced environmental monitoring, and improved waste management. The results will guide the subsequent stages of digital twin integration, ensuring interventions are aligned with the neighborhood's specific needs and sustainability goals.

The results of the questionnaire are summarized in a radar chart, shown in Fig. 5. This visualization allows for a clear comparison of the maturity levels across the five key domains, helping stakeholders identify priority areas for development and action.

4.2.2 I4.0 Technologies Implementation

The first step in Industry 4.0 technologies implementation involves carefully selecting and deploying solutions tailored to address the neighborhood's specific challenges. The chosen technologies are designed to provide comprehensive coverage across various urban management domains, ensuring a detailed and dynamic understanding of the environment. Table 2 summarizes the impact of each technology on urban management across various domains.

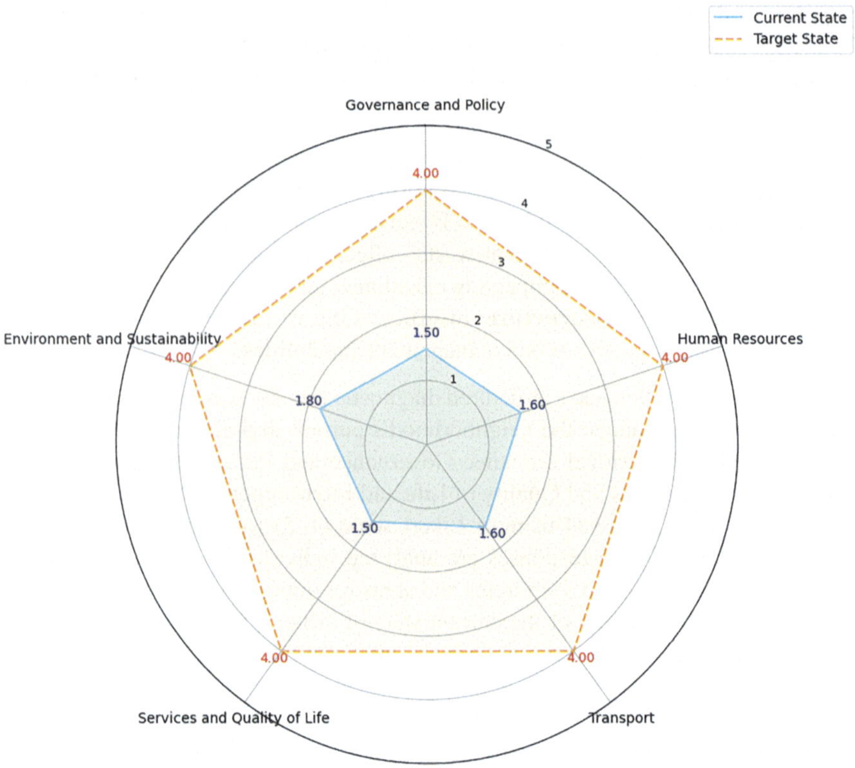

Fig. 5 Difference between the state before integration of the proposed solution and the desired state

4.2.3 Transmission Layer and Preprocessing Layer

The transmission and preprocessing layers play a critical role in ensuring the seamless flow and preparation of data collected by sensors, IoT devices, and other urban sources. These layers act as a bridge between raw data acquisition and its integration into the digital twin framework, providing the foundation for real-time insights and actionable analysis.

Communication Protocols

- **MQTT (Message Queuing Telemetry Transport):** A lightweight and reliable protocol ideal for IoT systems, enabling real-time data exchange with minimal bandwidth usage. Its publish/subscribe architecture ensures fast and efficient communication.
- **JSON (JavaScript Object Notation):** A simple, flexible data formatting standard for organizing sensor data. Its lightweight nature supports rapid data transmission and easy integration.

Table 2 Technologies relevant to specific urban functions and their sub-functions, aligning with the goals of smart city transformation

Technologies		*IoT*	*Smart Energy*	*AI*	*AR/ VR*	*Big Data*	*CC and EC*	*CPS*	*5G and High-speed connectivity*	*Blockchain*	*Internet of Services*	*Infrastructure communication*	*Cybersecurity*	*Predictive maintenance*	*Additive manufacturing*
Functions	Sub-function														
Urban Management	Monitoring	X		X	X	X	X	X	X			X	X	X	
	Real-time traffic monitoring	X		X		X	X	X	X	X		X	X		
	Predictive maintenance for infrastructure	X		X										X	
	Resource optimization	X	X	X		X	X	X							
Public services	Energy grid optimization	X	X	X		X	X								
	Emergency response systems	X		X		X	X	X	X			X	X		
	Smart waste management	X		X		X									
	Citizen-centric service portals	X		X			X			X	X				

(continued)

Table 2 (continued)

Technologies		*IoT*	*Smart Energy*	*AI*	*AR/VR*	*Big Data*	*CC and EC*	*CPS*	*5G and High-speed connectivity*	*Blockchain*	*Internet of Services*	*Infrastructure communication*	*Cybersecurity*	*Predictive maintenance*	*Additive manufacturing*
Functions	Sub-function														
Mobility	Smart traffic systems	X		X	X	X			X	X		X	X		
	Autonomous vehicles	X		X				X	X				X		
	Smart parking solutions	X		X											
	Mobility analytics and planning	X		X		X	X								
Environment	Air quality monitoring	X		X	X	X	X		X						
	Smart water management	X		X											X
	Renewable energy integration	X	X	X			X								
	Waste management and recycling	X		X		X									X

Transmission Modes

- **Wi-Fi:** Ideal for environments with clustered sensors near network access points.
- **4G/5G:** Suited for dispersed sensor networks, ensuring connectivity in wider urban areas. These wireless technologies provide comprehensive coverage and real-time data transmission in diverse urban settings.

Data Preprocessing

- **Data Filtering:** Eliminates noisy or erroneous data to maintain high data quality.
- **Data Aggregation:** Combines data from multiple sensors, reducing redundancy and optimizing system performance.
- **Data Normalization:** Standardizes data formats for seamless integration into the digital twin model, enabling consistent analysis and decision-making.

This structured approach to transmission and preprocessing guarantees that the data entering the digital twin system is reliable, consolidated, and standardized. It forms the foundation for creating an accurate digital twin, capable of reflecting real-world urban conditions and supporting effective urban management.

4.2.4 Digital Twin Model

The digital twin creation phase involves constructing a realistic and functional virtual model of the Hay Nakhil neighborhood using advanced modeling tools and real-world data. This phase includes the following components:

Digital Modeling Layer

The software used in our case study is "Unity3D". The advanced modeling software "Unity3D" enabled the creation of a realistic and functional digital twin (as shown in Fig. 6). For the 3D model of the neighborhood, we included three main layers: the terrain, buildings, and infrastructure. These were developed using two methods:

Fig. 6 Several 3D objects are detailed in the model. (**a**) 3D objects representing buildings. (**b**) 3D objects representing vehicles. (**c**) 3D objects representing avatars

modeling in Blender and then importing files in FBX format and directly importing assets from the Unity Store. Moreover, to enhance the utility of the digital twin, urban mobility models were integrated, allowing the simulation of pedestrian and vehicle movements within the neighborhood. These models include different types of agents, such as transport vehicles (Fig. 6b) and pedestrians (Fig. 6c), whose behaviors were calibrated using experimental data collected on-site. These simulations aim to evaluate traffic flows, manage crowds, and anticipate the impacts of infrastructural changes.

Communication Layer

This phase is considered a stage during which the 3D simulation model created in the design phase is deployed in the real world. The objective of this module is to establish bidirectional communication, enabling real-time synchronization of data between the physical neighborhood and its digital twin. Real-time information and changes are used as input data for the developed model. In other words, instead of running the simulation and design model based on historical data and physical properties, the model operates using real-time data and changes collected by sensors and IoT devices, as illustrated in Fig. 7. The historical database is automatically updated based on continuous data collection, meaning the behavior of the virtual system can evolve in response to the data collected from the physical system.

The collected data is then integrated into the Unity engine. Furthermore, the top layer ensures data visualization, allowing for user interactivity within the platform.

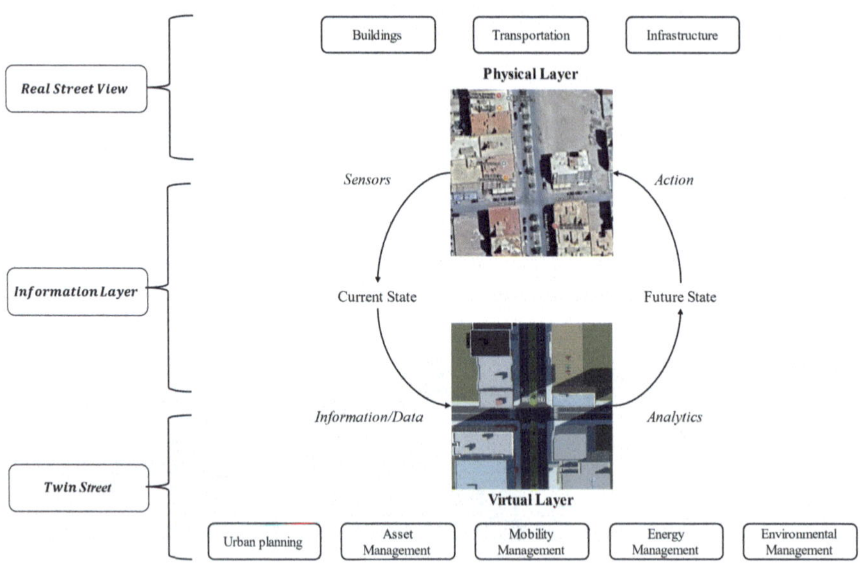

Fig. 7 Our concept of digital twin

The primary objective was to synchronize the information gathered by IoT sensors, e.g., traffic flows, energy consumption, and environmental parameters, with the digital model.

5 Results and Discussion

In this section, we conduct a series of tests on the proposed digital twin model to evaluate its performance in various dynamic urban scenarios. These scenarios include sudden variations in traffic flows, abrupt changes in energy consumption, or unexpected alterations in environmental parameters.

5.1 *Deployment Layer*

During the deployment phase of the digital twin model for the Hay Nakhil neighborhood, various scenarios were simulated to test the model's functionality in a real urban environment. A bidirectional communication system was established between the virtual model and physical infrastructure, ensuring real-time data exchange.

The platform visualized the neighborhood in 3D and displayed real-time data collected from installed sensors. For example, real-time vehicle counts were visualized using data from CCTV cameras at key intersections. These cameras employed video analysis to categorize and count vehicles by type, including cars, motorcycles, buses, trucks, and bicycles. This setup demonstrated the digital twin's ability to provide actionable insights for urban management (see Fig. 8). The digital twin

Fig. 8 The different views of the 3D city developed using unity. (**a**) Detailed aerial 3d view; (**b**) 3d view with object; (**c**) simplified view

platform offers three types of 3D visualizations tailored to various analytical and operational needs:

- **Simplified View:** Focuses on roads and intersections, omitting details of buildings or the environment. Best suited for analyzing traffic flows, planning routes, and studying road infrastructure. Supports proactive and optimized traffic management.
- **3D View with Objects:** Provides an immersive, ground-level perspective, replicating the experience of being within the urban model. Useful for examining architectural details and simulating human interactions. Allows isolation and analysis of specific elements without broader context distractions.
- **Detailed Aerial 3D View:** Offers a comprehensive, elevated view of the city, showing buildings, roads, and infrastructure. Ideal for visualizing the urban layout and analyzing the spatial arrangement of infrastructure. Enables precise localization of urban elements for accurate navigation and planning.

The platform is designed for continuous development, integrating real-time data from city-installed sensors. It provides a powerful tool for government authorities and users to monitor the neighborhood's state remotely, eliminating the need for physical site presence.

5.2 Continuous Assessment and Adjustment

The Continuous Assessment and Adjustment phase ensures the digital twin remains dynamic, accurate, and aligned with the evolving needs of the neighborhood. This phase emphasizes a feedback loop for continuous improvement, optimizing the twin's performance to meet strategic objectives.

- **Real-Time Monitoring:** Monitoring systems play a critical role in real-time urban management: (1) Traffic Flow Monitoring: Sensors identify congestion hotspots during peak hours, enabling real-time traffic route adjustments and reducing delays. (2) Air Quality Measurement: Environmental sensors detect pollution spikes linked to peak traffic times, guiding efforts to reduce congestion and improve sustainability. 3) Energy Consumption Monitoring: Smart devices track energy usage in buildings, highlighting peaks and suggesting optimizations like energy-efficient equipment adoption.
- **Simulation and Optimization:** Data synchronized with the digital twin facilitates simulations to enhance urban management: 1) Traffic Optimization: Testing scenarios improves signal synchronization and suggests alternative routes for better flow. 2) Proactive Infrastructure Management: Sensors detect road anomalies, enabling preventive maintenance that minimizes costs and disruptions. 3) Waste Management: Data analysis optimizes collection routes, reduces illegal dumping, and improves city cleanliness.

- **Feedback Mechanism:** Validated virtual modifications are implemented in the real world, maintaining a seamless synchronization between the digital model and reality. This feedback loop drives continuous refinement and adapts the twin to evolving conditions.

Finally, part of the continuous evaluation involves reassessing the digital maturity defined during the first phase. This re-assessment ensures that the solutions proposed and implemented through the digital twin deliver tangible added value to the urban management of the neighborhood, with notable improvements in several domains.

The diagram in Fig. 9 visually illustrates these advancements and objectives, comparing the state before integration, the state after integration, and the targeted state across various key dimensions. Furthermore, it clearly highlights the gap between the initial objectives and the state achieved after implementing the proposed approach, which is primarily based on the integration of I4.0 technologies and the digital twin.

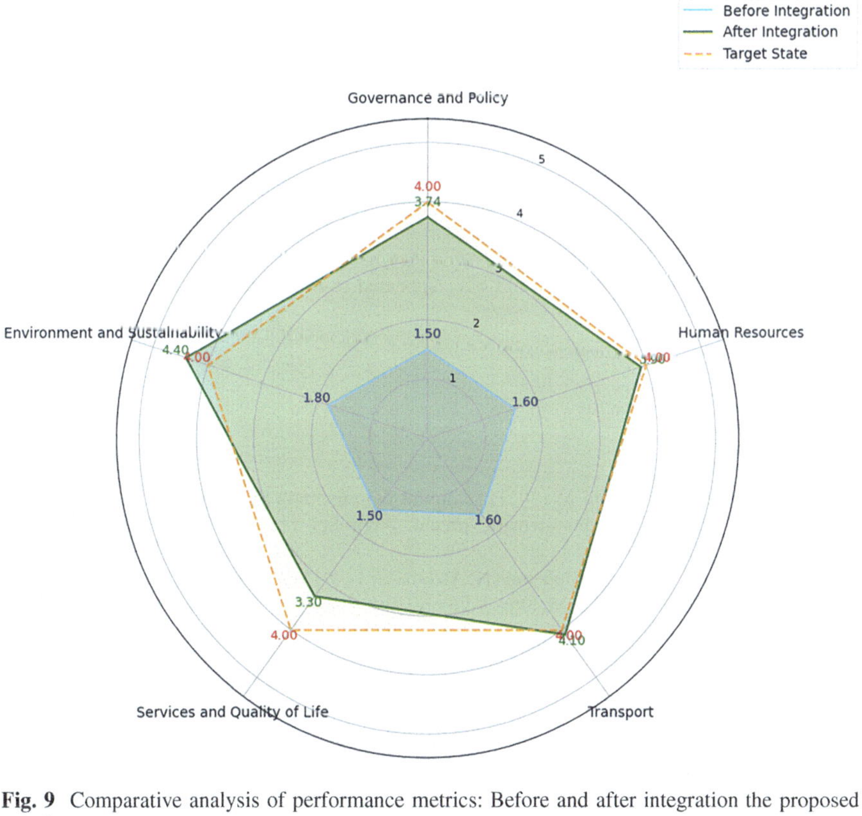

Fig. 9 Comparative analysis of performance metrics: Before and after integration the proposed solution

6 Conclusion

The digital transformation of cities, embodied by the concept of the urban digital twin, marks a significant advancement in the proactive and sustainable management of urban environments. Through this article, we have demonstrated how integrating 4.0 technologies and a digital twin model can optimize resource management, enhance urban resilience, and significantly improve the quality of life for citizens. The case study on the Hay Nakhil neighborhood illustrates the tangible impact of these approaches, with concrete results such as a 15% improvement in traffic flow, a notable reduction in energy consumption, and greater efficiency in waste management.

The simulations conducted using the digital twin also highlight its potential to inform urban decision-making, actively involving citizens in the planning process. Interactive mechanisms, such as object tagging or 3D scenario simulations, enable citizens to report issues or suggest improvements, thereby fostering participatory urban governance. These tools anticipate the impacts of public policies and test solutions for rare or critical situations, such as flooding or crowd movements.

Despite these promising results, several challenges remain. Issues related to data security, the rapid evolution of technologies, and accessibility in less-developed areas pose significant obstacles. However, these limitations pave the way for future research aimed at improving digital twin models, integrating more emerging technologies, and developing inclusive solutions tailored to a wide variety of urban contexts.

Competing Interests The authors declare that they have no known competing financial interests or personal relationships that could have appeared to influence the work reported in this chapter.

References

1. Agnusdei G, Elia V, Gnoni M (2021) Is digital twin technology supporting safety management? A bibliometric and systematic review. Appl Sci 11:2767. https://doi.org/10.3390/app11062767
2. Al Nuaimi E, Al Neyadi H, Mohamed N, Al-Jaroodi J (2015) Applications of big data to smart cities. J Internet Serv Appl 6:25. https://doi.org/10.1186/s13174-015-0041-5
3. Allam Z, Dhunny ZA (2019) On big data, artificial intelligence and smart cities. Cities 89:80–91. https://doi.org/10.1016/j.cities.2019.01.032
4. Bhushan B, Khamparia A, Sagayam KM, Sharma SK, Ahad MA, Debnath NC (2020) Blockchain for smart cities: a review of architectures, integration trends and future research directions. Sustain Cities Soc 61:102360. https://doi.org/10.1016/j.scs.2020.102360
5. Dani AAH, Supangkat SH, Lubis FF, Nugraha IGBB, Kinanda R, Rizkia I (2023) Development of a smart city platform based on digital twin technology for monitoring and supporting decision-making. Sustainability 15:14002. https://doi.org/10.3390/su151814002

6. Dembski F, Wössner U, Letzgus M, Ruddat M, van der Laag Yamu C (2020) Urban digital twins for smart cities and citizens: the case study of Herrenberg, Germany. Sustainability 12, 17 pp. https://doi.org/10.3390/su12062307
7. Facchini F, Oleśków-Szłapka J, Ranieri L, Urbinati A (2020) A maturity model for logistics 4.0: an empirical analysis and a roadmap for future research. Sustainability 12:86. https://doi.org/10.3390/su12010086
8. Gade D, Aithal S (2021) Smart cities development during and post COVID-19 pandemic—a predictive analysis. Int J Manag Technol Soc Sci 189–202. https://doi.org/10.47992/IJMTS.2581.6012.0140
9. Grieves M, Vickers J (2017) Digital twin: mitigating unpredictable, undesirable emergent behavior in complex systems. In: Transdisciplinary perspectives on complex systems. Springer, Cham, pp 85–113. https://doi.org/10.1007/978-3-319-38756-7_4
10. Halúsková B (2023) Digital twin in smart city. Transp Res Procedia, TRANSCOM 2023: 15th international scientific conference on sustainable, Modern and Safe Transport 74:1471–1478. https://doi.org/10.1016/j.trpro.2023.11.308
11. Ignatius M, Wong NH, Martin M, Chen S (2019) Virtual Singapore integration with energy simulation and canopy modelling for climate assessment. IOP Conf Ser Earth Environ Sci 294:012018. https://doi.org/10.1088/1755-1315/294/1/012018
12. Jabbar MA, Samreen S, Aluvalu R, Reddy KK (2018) Cyber physical Systems for Smart Cities Development. Int J Eng Technol 7:36. https://doi.org/10.14419/ijet.v7i4.6.20229
13. Kim HM, Sabri S, Kent A (2021) Chapter 2—smart cities as a platform for technological and social innovation in productivity, sustainability, and livability: a conceptual framework. In: Kim HM, Sabri S, Kent A (eds) Smart cities for technological and social innovation. Academic, pp 9–28. https://doi.org/10.1016/B978-0-12-818886-6.00002-2
14. Lim Y, Edelenbos J, Gianoli A (2024) What is the impact of smart city development? Empirical evidence from a smart city impact index. Urban Gov 4:47–55. https://doi.org/10.1016/j.ugj.2023.11.003
15. Lom M, Pribyl O, Svitek M (2016) Industry 4.0 as a part of smart cities. In: 2016 smart cities symposium Prague (SCSP), pp 1–6. https://doi.org/10.1109/SCSP.2016.7501015
16. Mandl C, Kirschner T (2014) European smart city initiative assessment of best practices to stimulate market-/demand-pull. https://doi.org/10.13140/2.1.4559.7449
17. Manoharan G, Durai S, Rajesh GA, Razak A, Rao CBS, Ashtikar SP (2023) An investigation into the effectiveness of smart city projects by identifying the framework for measuring performance. In: Artificial intelligence and machine learning in smart city planning. Elsevier, pp 71–84. https://doi.org/10.1016/B978-0-323-99503-0.00004-1
18. Massobrio R, Nesmachnow S, Tchernykh A, Avetisyan A, Radchenko G (2018) Towards a cloud computing paradigm for big data analysis in smart cities. Program Comput Softw 44:181–189. https://doi.org/10.1134/S0361768818030052
19. Morton P, Thompson E, Dalton R, 2013. Virtual city models: a global perspective
20. O'Dwyer E, Pan I, Acha S, Shah N (2019) Smart energy systems for sustainable smart cities: current developments, trends and future directions. Appl Energy 237:581–597. https://doi.org/10.1016/j.apenergy.2019.01.024
21. Paskaleva K, Cooper I (2018) Open innovation and the evaluation of internet-enabled public services in smart cities. Technovation 78:4–14. https://doi.org/10.1016/j.technovation.2018.07.003
22. Rehman U, Faria P, Gomes L, Vale Z (2023) Future of energy management systems in smart cities: a systematic literature review. Sustain Cities Soc 96:104720. https://doi.org/10.1016/j.scs.2023.104720
23. Syed AS, Sierra-Sosa D, Kumar A, Elmaghraby A (2021) IoT in smart cities: a survey of technologies, practices and challenges. Smart Cities 4:429–475. https://doi.org/10.3390/smartcities4020024
24. Weil C, Bibri SE, Longchamp R, Golay F, Alahi A (2023) Urban digital twin challenges: a systematic review and perspectives for sustainable smart cities. Sustain Cities Soc 99:104862. https://doi.org/10.1016/j.scs.2023.104862

25. Xiang M, Wang R-H, Zhong Z, Hong CY, Yi YW, Xiang SW, Roy D (2023) Empowering society 5.0 through AR-VR applications: studying the potential in Aizuwakamatsu smart city. AIP Conf Proc 2909:070007. https://doi.org/10.1063/5.0182561
26. Xu W, Zhou J, Yang L, Li L (2018) The implications of high-speed rail for Chinese cities: connectivity and accessibility. Transp Res A Policy Pract 116:308–326. https://doi.org/10.1016/j.tra.2018.06.023

Smart Container Terminal Using Metaheuristic: Genetic Algorithm and Tabu Search

Sanaa Aidi and Mohamed Mazouzi

Abstract The Quay Crane Scheduling Problem (QCSP) is essential for optimizing smart port operations, significantly impacting sustainable urban logistics. Effective scheduling enhances cargo handling efficiency, reduces vessel turnaround times, and minimizes port congestion, thus improving the overall productivity of urban transport networks. The QCSP aims to determine the optimal sequence of crane operations to reduce container handling time while addressing constraints such as crane interference and safety distances. This paper presents a genetic algorithm (GA) approach to solve the QCSP, utilizing natural selection-inspired metaheuristic techniques. The GA evolves an initial population of crane schedules through iterative refinement based on a fitness function that evaluates total handling time and critical constraints. A simulation demonstrates the GA's effectiveness across various scenarios, revealing significant reductions in handling time and showcasing its scalability and adaptability to real-world conditions. The findings suggest broader applications of GA-based scheduling for other urban logistics challenges, contributing to the development of resilient and sustainable smart cities.

Keywords Smart cities · Sustainable development · Container terminal · Genetic algorithm · Tabu search · Scheduling · Optimization

---

S. Aidi (✉) · M. Mazouzi
Department of Mechanics, Engineering and Innovation, National Higher School of Electricity and Mechanics, University Hassan 2, Casablanca, Morocco
e-mail: sanaa.aidi-etu@etu.univh2c.ma

H. Kaddouri et al. (eds.), *Smart Cities for Sustainable and Urban Development*, Green Energy and Technology, https://doi.org/10.1007/978-3-031-99001-4_39

1 Introduction and Literature Review

1.1 Introduction

Container terminals are essential nodes within global supply chains, functioning as strategic hubs for transferring cargo between ships, trucks, and railways [2]. The efficiency of these terminals directly influences the speed and cost of shipping, thereby impacting both operational expenses and the competitiveness of logistics companies [24]. Central to terminal efficiency is the optimized handling of containers, which involves tasks like unloading containers from vessels, transferring them to storage yards, and loading them onto various modes of transport [7]. A critical element in these operations is the coordination of quay cranes large cranes that manage container loading and unloading from ships.

The Quay Crane Scheduling Problem (QCSP) is a fundamental challenge in optimizing terminal operations, as it dictates the sequencing and task allocation for multiple quay cranes along a berth. The primary objective of the QCSP [22] is to minimize the total handling time, or makespan, which represents the duration from the start of the first crane's operation to the completion of the last. Efficient crane scheduling reduces the makespan, enhances ship turnaround time [23], lowers vessel waiting times, and decreases costs related to crane idle time and container handling.

However, the QCSP is highly complex and falls into the category of combinatorial optimization problems. It is constrained by factors such as crane interference [14], safety distances between cranes, task precedence requirements, and variability in container handling times. As the number of cranes and tasks increases, the solution space grows exponentially, rendering exact optimization methods impractical for large-scale problems. Consequently, [6] heuristic and metaheuristic approaches, particularly genetic algorithms (GA) [1], have gained traction as effective solutions for QCSP [15], capable of navigating extensive solution spaces and yielding high-quality results within a manageable timeframe ideal for real-time scheduling in dynamic terminal environments.

Genetic algorithms, known for their adaptability to complex constraints and rapidly changing conditions, have shown significant promise in solving the QCSP [25]. By simulating the process of natural selection, GAs generate and evolve a population of solutions through selection, crossover, and mutation, making it possible to discover optimal or near-optimal crane schedules. This paper aims to demonstrate the effectiveness of the GA approach for QCSP [21], focusing on minimizing handling times while accommodating realistic operational constraints.

This chapter is structured to comprehensively explore the Quay Crane Scheduling Problem (QCSP) and the proposed innovative solutions. It begins with an introduction, which sets the stage by highlighting the importance of QCSP in container terminal optimization and its relevance to modern logistics and smart city initiatives. This section also outlines the research problem, objectives, and contributions of the study. The literature review follows, presenting a detailed analysis of existing studies on QCSP, including heuristic and metaheuristic approaches such as Genetic Algorithms (GA) and Tabu Search (TS). This review identifies the limitations in current methodologies and establishes the need for the proposed hybrid approach.

Next, the chapter delves into the methodologies for modeling and solving QCSP, where the problem formulation is detailed, including constraints and objectives. This section introduces the hybrid GA-TS framework, explaining how these techniques are integrated to address the limitations of individual methods. The focus is on the adaptive mechanisms and algorithmic enhancements that improve solution quality and computational efficiency.

The subsequent section, discussion of results, presents a thorough analysis of the performance of the proposed approach. Comparative studies with existing methods are included, supported by numerical experiments and visualizations, to demonstrate the advantages in terms of solution quality, computational time, and practical applicability.

Finally, the conclusion and perspectives summarize the key findings, emphasizing the contributions of the hybrid approach to QCSP optimization. This section also discusses the implications for container terminal operations and provides insights into future research directions, including the potential for integrating emerging technologies such as artificial intelligence and real-time data analytics to further enhance scheduling efficiency.

1.2 Literature Review

Here is an overview of key literature related to the Quay Crane Scheduling Problem (QCSP) [23], container terminal optimization, and genetic algorithms for complex scheduling [19], which supports the context and objectives outlined in the introduction:

- Container Terminal Operations and Optimization

  Container terminals play a pivotal role in the global supply chain [1] by facilitating the efficient transfer of containers between various modes of transportation. Several studies have highlighted the criticality of optimizing terminal operations to enhance [25] overall supply chain efficiency. For instance, extensively review operations research approaches in container terminals, covering aspects like berth allocation, yard management, and equipment scheduling, including quay cranes [26]. Their work underscores the interconnected nature of terminal processes and the need for integrated approaches to optimize multiple subsystems simultaneously (Fig. 1).

- Quay Crane Scheduling Problem (QCSP)
  QCSP specifically addresses quay crane tasks sequencing and allocation [22] to minimize makespan or total handling time. Meisel and Bierwirth [6] examine the QCSP in depth, identifying the complexities and constraints involved, such as

Fig. 1 Quay crane in container terminal

crane interference, safety distances [7], and task precedence. Their work categorizes QCSP [3] as a combinatorial optimization problem and evaluates various solution methodologies. The authors highlight that [4], as terminal sizes and throughput demands grow, efficient QCSP solutions become essential for minimizing operational delays and enhancing port productivity.

- Constraints in QCSP
  The complexities of QCSP arise from its numerous constraints, including crane [27] interference and safety distance requirements. Bierwirth and Meisel [7] expand on this by detailing how QCSP solutions must address physical and operational limitations, which affect the feasible movements and positioning of cranes along the berth. They explore methods for managing these constraints, which are essential for real-world applications. The study indicates that constraint-handling mechanisms are a core component in the development of viable scheduling algorithms for container terminals.
- Genetic Algorithms (GA) for Scheduling in Container Terminals
  Genetic algorithms (GA) have gained attention in QCSP solutions due to their effectiveness in exploring large and complex solution spaces. Were among the first to apply GAs to QCSP, demonstrating that GAs can generate near-optimal solutions within reasonable time frames by iteratively improving crane schedules. Their findings underscore the adaptability [9] of GAs to dynamic terminal conditions, as well as their flexibility in handling multiple constraints like crane interference and task sequencing. Following this, researchers such as Lee et al. (2010) have further refined GA-based approaches [5], introducing multi-objective fitness functions [21] that balance between minimizing makespan and reducing operational costs, which align well with the goals of sustainable smart port operations.

- Heuristic and Metaheuristic Approaches for QCSP
  Heuristic and metaheuristic approaches, particularly those based on evolutionary principles, have become popular for solving large-scale QCSP due to their computational efficiency [19]. For example, Cordeau et al. (2005) explore various heuristic methods, including simulated annealing and tabu search, for quay crane scheduling and show [8] their comparative performance against GAs in real-time scheduling scenarios [6]. These methods are found to be beneficial for dynamic environments, where new tasks and constraints frequently arise.
- Applications of QCSP in Smart Cities and Sustainable Development
  Recent studies have also examined the relevance of efficient QCSP [13] solutions to the broader goals of smart cities and sustainable urban logistics [10], discussing how optimizing container terminal operations contributes to the sustainability of urban transport networks by reducing port congestion, emissions, and energy consumption [23]. By implementing advanced scheduling algorithms like GA [16], ports can improve throughput while contributing to sustainable urban growth, aligning with smart city goals of resource efficiency and reduced environmental impact.

These studies provide a foundation for understanding the critical role of QCSP [8] in container terminal operations and highlight the importance of advanced algorithms like GA in addressing the problem's inherent complexities [20]. The literature also reflects the relevance of optimized QCSP solutions within the context of sustainable urban development, where enhanced logistics and transport operations are essential for building smart, resource-efficient cities.

The extant research on the Quay Crane Scheduling Problem (QCSP) sheds light on container terminal optimization and the application of heuristic and metaheuristic methodologies such as Genetic Algorithms (GA). However, it has numerous limitations that provide potential for improvement. While GA-based approaches are flexible and adaptable in solving QCSP, they frequently encounter the premature convergence problem, limiting their capacity to explore varied portions of the solution space. Similarly, studies that use Tabu Search (TS) [11] effectively escape local optima but can suffer from computational inefficiencies when faced with large-scale or highly dynamic problems. Additionally, the reviewed literature rarely combines multiple metaheuristic techniques to leverage their complementary strengths, particularly in addressing the multi-objective nature of QCSP, such as minimizing makespan while balancing resource utilization and operational costs.

Our approach addresses these gaps by proposing a hybrid metaheuristic framework that integrates GA and TS to optimize QCSP. The synergy between these algorithms enhances the exploration and exploitation phases of the solution process, ensuring robust and near-optimal scheduling even under dynamic and constrained environments. This hybrid method introduces adaptive mechanisms for constraint handling and dynamic fitness evaluation, aligning well with the complexity of real-world terminal operations. Furthermore, by focusing on sustainability metrics such as energy efficiency and reduced emissions, this approach contributes to the broader goals of smart city development and sustainable urban logistics. Thus, the proposed framework not only advances computational efficiency and solution quality but also provides practical, scalable strategies for modern container terminal challenges.

2 Research Method

2.1 Problem Definition

The QCSP can be framed as a scheduling problem in the context of smart cities and sustainable development, where a set of quay cranes must service specified container bays on a vessel [18]. The goal is to minimize the total makespan, which represents the time from the start to the completion of all crane tasks. Achieving efficient crane scheduling enhances port operations and supports sustainable development by reducing resource consumption and improving logistics efficiency, essential for smart city infrastructure [13]. Key constraints in this problem include

- Crane Interference: Quay cranes must not collide or interfere with each other.
- Safety Distance: A minimum distance must be maintained between adjacent quay cranes.
- Task Precedence: Some tasks may need to be completed before others.

2.2 The Mathematical Formulation of the Quay Crane Scheduling Problem (QCSP)

The mathematical formulation of the Quay Crane Scheduling Problem [7] (QCSP) can be expressed as an optimization problem to minimize the total handling time (makespan) while adhering to constraints such as crane interference and safety distances [17]. The formulation involves defining decision variables, objective functions, and constraints.

2.2.1 Decision Variables

Decision variables for QCSP modeling:

- T_i is the start time of the operation at container bay i, where $i = 1, 2, \ldots, N$ (with N is the total number of bays).
- C_k is the quay crane k, where $k = 1, 2, \ldots, M$ (with M is the total number of cranes).
- X_{ik} is the binary variable verifying:

$$\begin{cases} X_{ik} = 1 \text{ ,if crane } k \text{ is asigned to bay } i \\ \quad 0 \text{ , otherwise} \end{cases}$$

- S_{ij} is a binary variable indicating the sequence between tasks *i* and *j*, where:

$$\begin{cases} S_{ji} = 1, \text{if task } j \text{ starts after task } i \\ \quad 0 \text{ , otherwise} \end{cases}$$

2.2.2 Objective Function

The objective is to minimize the makespan, which is the total handling time of all tasks. The makespan can be expressed as:

$$Minimize\, C_{\max} = \max\left(T_i + P_i\right) i \in \{1,\ldots,N\} \quad (1)$$

Where P_i represents the processing time of bay *i*.

2.3 Assignment Constraint

Each task *i* must be assigned to exactly one crane:

$$\sum X_{ik} = 1, \forall i \in \{1,\ldots,N\} \quad (2)$$

2.3.1 Crane Non-Interference Constraint

Two cranes *k* and *l* working on tasks *i* and *j*, respectively, should not interfere if they are assigned to adjacent or nearby bays. This is enforced by:

$$Ti - Tj \geq \delta\left(1 - S_{ij} - S_{ji}\right) \qquad \forall i \neq j, \forall k, \quad (3)$$

Where δ is the minimum safety distance (in terms of time) between crane operations to avoid interference.

2.3.2 Precedence Constraint

If task *i* precedes task *j*, then:

$$T_i + P_i \le T_j + M\left(1 - S_{ij}\right), \qquad \forall i, j \tag{4}$$

Where M is a large positive constant.

2.3.3 Safety Distance Constraint

To maintain a minimum safety distance between two adjacent quay cranes k and l, ensure:

$$i - j \ge \text{Distance}_{\min} . X_{ik} . X_{jl} \qquad \forall i, j, k, l \tag{5}$$

Where $\text{Distance}_{\min}$ is the required minimum number of bays separating two cranes.

2.4 Handling Time Constraints

For each crane k, the total handling time should not exceed the maximum time:

$$T_i + P_i \le C_{\max} \qquad \forall i \tag{6}$$

This formulation captures the QCSP's essential features, providing a framework for optimizing the allocation and scheduling of quay cranes while considering key operational constraints. Due to its combinatorial nature and large solution space, the problem is typically solved using heuristic methods like genetic algorithms.

3 Results and Discussion

3.1 Genetic Algorithm Approach

The genetic algorithm (GA) is a metaheuristic inspired by natural selection [2]. It is suitable for solving complex optimization problems like the QCSP [22]. The main components of the GA are:

- **Encoding**: Each chromosome represents a potential crane scheduling solution, where genes correspond to the sequence of tasks assigned to each crane.
- **Initialization**: A population of chromosomes is randomly generated.
- **Fitness Function**: The fitness of each chromosome is evaluated based on the total handling time, penalizing crane interference, and violations of safety distances.
- **Selection**: The tournament selection method is used to choose parents for reproduction based on their fitness.

- **Crossover and Mutation**: Genetic operators are applied to generate new offspring, ensuring diversity in the population. Crossover exchanges segments of parent chromosomes, while mutation introduces small random changes.
- **Termination**: The algorithm stops when a predetermined number of generations is reached, or when the improvement in the solution becomes negligible.

3.1.1 Example Simulation

An example QCSP simulation is provided to demonstrate the effectiveness of the genetic algorithm. Suppose there are four quay cranes and ten container bays to be serviced [12]. The cranes must complete the tasks while avoiding interference and maintaining a minimum safety distance of two bays (Table 1).

The genetic algorithm is run for 100 generations with a population size of 50. The crossover rate is set to 0.8, and the mutation rate is 0.1 (Table 2).

3.2 Discussion

The application of a genetic algorithm (GA) to the Quay Crane Scheduling Problem (QCSP) demonstrates substantial potential for addressing complex scheduling challenges in container terminals, which is essential for advancing smart cities and sustainable development. The GA efficiently generates feasible schedules that satisfy all necessary constraints, such as crane interference, safety distances, and task

Table 1 Initial container bays and crane assignment

Bay number	Container task	Initial crane assignment
1	Unloading	Crane 1
2	Loading	Crane 2
3	Unloading	Crane 3
4	Loading	Crane 4
5	Unloading	Crane 1
6	Loading	Crane 2
7	Unloading	Crane 3
8	Loading	Crane 4
9	Unloading	Crane 1
10	Loading	Crane 2

Table 2 Sample solution after 100 generations

Crane	Sequence of tasks	Total handling time (minutes)
1	1 → 5 → 9	120
2	2 → 6 → 10	130
3	3 → 7	110
4	4 → 8	115
Total	–	130

precedence. By minimizing makespan, the GA not only reduces ship turnaround times but also enhances terminal productivity, aligning with sustainable development goals by optimizing resource use and reducing operational inefficiencies.

One major advantage of the GA in solving the QCSP lies in its ability to explore extensive solution spaces and escape local optima an area where traditional optimization methods often struggle. Approaches like mixed-integer linear programming may falter with large-scale QCSP instances due to exponential solution space growth. In contrast, the GA leverages stochastic operators, including crossover and mutation, to diversify the solution pool and refine promising solutions through successive generations. This adaptability positions GAs as particularly suitable for real-time scheduling in dynamic environments, where conditions and requirements may evolve rapidly, a critical feature for smart city logistics.

Additionally, the GA framework's flexibility allows for easy adaptation to various QCSP scenarios. It can accommodate additional constraints, such as time windows for crane operations or variable processing times depending on container weight and location, thereby enhancing sustainable practices by accounting for energy efficiency and minimizing idle times. The GA can also handle multiple objectives, like workload balancing or minimizing energy consumption, through multi-objective optimization, making it well-suited for sustainable and efficient terminal management.

Despite its strengths, the GA has limitations, particularly concerning parameter settings like population size, crossover rate, and mutation rate. Suboptimal settings can hinder performance, and dynamic adjustments during the search process may improve outcomes. Future enhancements could involve hybridizing the GA with other heuristic methods, such as local search or simulated annealing, to capitalize on the strengths of multiple techniques and enhance convergence speed and solution quality.

In conclusion, the genetic algorithm provides a robust, adaptable solution to QCSP, well-suited for large-scale, complex scheduling issues in container terminals. Its capability to handle diverse constraints and adapt to evolving operational needs highlights its potential in optimizing quay crane scheduling, thus contributing to smarter, more sustainable container terminal operations. Future research could focus on hybrid models, adaptive parameter tuning, and multi-objective optimizations to further maximize GA benefits, supporting smart city logistics and sustainable development objectives.

4 Conclusion and Future Research

In conclusion, the genetic algorithm (GA) approach for solving the Quay Crane Scheduling Problem (QCSP) in container terminals proves to be a highly effective and versatile method for optimizing crane operations. The GA consistently produces feasible, high-quality scheduling solutions by successfully minimizing the makespan while meeting essential operational constraints such as crane interference, safety distances, and task precedence. Its capacity to explore large and complex solution spaces makes it particularly suitable for large-scale scheduling problems, where traditional exact methods struggle due to computational limitations. Additionally, the GA's flexibility allows it to incorporate various problem variations and additional objectives, providing a robust solution adaptable to real-world demands in dynamic terminal environments.

Future research directions could further refine this approach by integrating hybrid methods, adaptive parameter tuning, and multi-objective optimization techniques to enhance scheduling quality and efficiency. Overall, the GA-based approach makes a substantial contribution to the optimization of container terminal operations, reducing ship turnaround times and significantly boosting port productivity, thus supporting the advancement of smart, sustainable logistics within modern urban infrastructures.

Competing Interests The authors would like to express their sincere gratitude to the management and staff of Port of MARSA MAROCCO whose insights and operational data greatly contributed to the understanding of the Quay Crane Scheduling Problem in a practical context.

Lastly, the authors would like to thank University Hassan 2 for providing access to computational facilities and research materials that enabled the successful completion of this study.

References

1. Agra A, Oliveira M (2018) MIP approaches for the integrated berth allocation and quay crane assignment and scheduling problem. Eur J Oper Res 264(1):138–148. https://doi.org/10.1016/j.ejor.2017.05.040
2. Aidi S, Mazouzi M (2023) Optimization approach for yard crane scheduling problem using genetic algorithm in container terminals. In: Mejdoub Y, Elamri A (eds) ITM web of conferences, vol vol. 52: 02002. https://doi.org/10.1051/itmconf/20235202002
3. Aidi S, Torbi I, Mazouzi M, Belassiria I (2021) Planning of one-way quay cranes using the genetic algorithm. In: Moumen A, Mejjad N, Slimani H (eds) SHS Web of Conferences, vol vol. 119: 07005. https://doi.org/10.1051/shsconf/202111907005
4. Amani H, Bouyaya L, Chaib R, Djekrif FZ, Aizi M (2023) Optimisation of quay crane scheduling problem at the port of Algeria. In: Kabashkin I, Yatskiv I, Prentkovskis O (eds) Reliability and statistics in transportation and communication, Lecture notes in networks and systems. Springer International Publishing, Cham, pp 232–241. https://doi.org/10.1007/978-3-031-26655-3_21
5. Azizi A, Zhang Z, Rui Z, Yunxi Li C, Igathinathane PF, Mathew J et al (2024) Comprehensive wheat lodging detection after initial lodging using UAV RGB images. Expert Syst Appl 238:121788. https://doi.org/10.1016/j.eswa.2023.121788
6. Bierwirth C, Meisel F (2009) A fast heuristic for quay crane scheduling with interference constraints. J Sched 12(4):345–360. https://doi.org/10.1007/s10951-009-0105-0
7. Bierwirth C, Meisel F (2010) A survey of berth allocation and quay crane scheduling problems in container terminals. Eur J Oper Res 202(3):615–627. https://doi.org/10.1016/j.ejor.2009.05.031
8. Bouklata A, Bensfia C (2020) Digitalization of port passage procedures: focus on the transit time of goods. In: 2020 IEEE 13th international colloquium of logistics and supply chain management (LOGISTIQUA), Fez, Morocco. IEEE, pp 1–6. https://doi.org/10.1109/LOGISTIQUA49782.2020.9353931

9. Chen JH, Lee D-H, Goh M (2014) An effective mathematical formulation for the unidirectional cluster-based quay crane scheduling problem. Eur J Oper Res 232(1):198–208. https://doi.org/10.1016/j.ejor.2013.06.051
10. Feng X, He Y, Kim K-H (2022) Space planning considering congestion in container terminal yards. Transp Res B Methodol 158:52–77. https://doi.org/10.1016/j.trb.2022.02.012
11. Gulić M, Maglić L, Valčić S (2018) Nature inspired metaheuristics for optimizing problems at a container terminal. Pomorstvo 32(1):10–20. https://doi.org/10.31217/p.32.1.16
12. Guo P, Cheng W, Wang Y, Boysen N (2018) Gantry crane scheduling in intermodal rail-road container terminals. Int J Prod Res 56(16):5419–5436. https://doi.org/10.1080/00207543.2018.1444812
13. Haghani A, Jung S (2005) A dynamic vehicle routing problem with time-dependent travel times. Comput Oper Res 32(11):2959–2986. https://doi.org/10.1016/j.cor.2004.04.013
14. Hop DC, Van Hop N, Anh TTM (2021) Adaptive particle swarm optimization for integrated quay crane and yard truck scheduling problem. Comput Ind Eng 153:107075. https://doi.org/10.1016/j.cie.2020.107075
15. Lee D-H, Wang HQ, Miao L (2008) Quay crane scheduling with non-interference constraints in port container terminals. Transp Res E Logist Transp Rev 44(1):124–135. https://doi.org/10.1016/j.tre.2006.08.001
16. Legato P, Mazza RM (2018) A decision support system for integrated container handling in a transshipment hub. Decis Support Syst 108:45–56. https://doi.org/10.1016/j.dss.2018.02.004
17. Li N, Chen G, Govindan K, Jin Z (2018) Disruption management for truck appointment system at a container terminal: a green initiative. Transp Res Part D: Transp Environ 61:261–273. https://doi.org/10.1016/j.trd.2015.12.014
18. Li W, Xiaoning Z, Zhengyu X (2020) Efficient container stacking approach to improve handling: efficiency in Chinese rail–truck transshipment terminals. SIMULATION 96(1):3–15. https://doi.org/10.1177/0037549719843347
19. Ma S (2020) Series: Routledge maritime masters; 7. In: Economics of maritime business, 1st edn, New York, Routledge. https://doi.org/10.4324/9781315658124
20. Ng WC, Mak KL (2006) Quay crane scheduling in container terminals. Eng Optim 38(6):723–737. https://doi.org/10.1080/03052150600691038
21. Sanaa A, Imane T, Mohamed M (2025) Quay crane scheduling in container terminals using a hybrid genetic algorithm. J Comput Sci
22. Skaf A, Lamrous S, Hammoudan Z, Manier MA (2019) Single quay crane and multiple yard trucks scheduling problem with integration of reach-stacker cranes at port of Tripoli-Lebanon. In: 2019 IEEE international conference on systems, man and cybernetics (SMC), Bari, Italy. IEEE, pp 852–857. https://doi.org/10.1109/SMC.2019.8914667
23. Vernadat, François, Anis Chelbi, Pierre Baptiste, Ali Siadat, Lahcen Mifdal, and Ilias Majdouline. (2021), 13th International conference on MOdeling, optimization and SIMulation: "new advances and challenges for sustainable and smart industries"
24. Wang W, Peng Y, Xinglu X, Li X, Liu H, Liu S, Yan X (2024) Smart container port development: recent technologies and research advances. Int Transp Infrastruct 3:liad022. https://doi.org/10.1093/iti/liad022
25. Yue L-J, Fan H-M, Fan H (2023) Blocks allocation and handling equipment scheduling in automatic container terminals. Transp Res C Emerg Technol 153:104228. https://doi.org/10.1016/j.trc.2023.104228
26. Zhang Z, Liu M, Lee C-Y, Wang J (2018) The quay crane scheduling problem with stability constraints. IEEE Trans Autom Sci Eng 15(3):1399–1412. https://doi.org/10.1109/TASE.2018.2795254
27. Zhao N, Lodewijks G, Zhuorui F, Sun Y, Sun Y (2022) Trajectory predictions with details in a robotic twin-crane system. Complex Syst Model Simul 2(1):1–17. https://doi.org/10.23919/CSMS.2021.0028

AI-Based Chabot for Enhancing Urban Service Delivery in Smart Cities

Charaf Ouaddi, Lamya Benaddi, Adnane Souha, Abdeslam Jakimi, and Rachid Saadane

Abstract With the rapid growth of urban populations and the increasing complexity of urban environments, cities are adopting innovative technologies to enhance service delivery and improve residents' quality of life. The concept of smart cities has emerged as a strategic framework to address issues such as traffic congestion and inadequate public services. In this context, AI chatbots have become essential tools, providing efficient, real-time, and personalized interactions between city authorities and their constituents. By leveraging advanced technologies, these chatbots can facilitate access to information and services, thereby enhancing citizen engagement and responsiveness. Despite the significant advantages offered by AI chatbots, their integration into urban management is not without its challenges. Data privacy concerns, the digital divide that limits equitable access, and the requirement for reliable infrastructure can impede the effective deployment of these technologies. Moreover, there is a lack of comprehensive research exploring the multifaceted impact of chatbots on urban service delivery. This paper aims to investigate the transformative role of AI-based chatbots in enhancing urban service delivery within smart cities. Specifically, it analyzes how chatbots can improve communication, enhance access to essential services, and foster active citizen engagement. Through an in-depth examination of various case studies, this research highlights successful implementations of chatbots in smart cities. It identifies key challenges that need to be addressed for their effective integration. Ultimately, this work contributes to a deeper understanding of how AI chatbots can shape the future

C. Ouaddi (✉) · L. Benaddi · A. Souha · A. Jakimi
Software Engineering and Information Systems Engineering Team, FST Errachidia, University of Moulay Ismail, Meknes, Morocco
e-mail: c.ouaddi@edu.umi.ac.ma

R. Saadane
Electrical Engineering Department, Hassania School of Public Works, Casablanca, Morocco

H. Kaddouri et al. (eds.), *Smart Cities for Sustainable and Urban Development*, Green Energy and Technology, https://doi.org/10.1007/978-3-031-99001-4_40

of urban management, emphasizing the importance of collaboration among stakeholders to leverage the benefits of these technologies.

Keywords Smart cities · Chatbots · Artificial intelligence · Urban service delivery

1 Introduction

Urban populations are expanding rapidly, and cities face increasing challenges due to the growing complexity of urban environments. To address these issues, innovative technologies are being adopted to enhance service delivery and improve residents' quality of life [9]. The concept of smart cities [10] has emerged as a strategic approach to tackling urban challenges such as traffic congestion and inadequate public services. Within this context, AI chatbots have become vital tools [1], enabling real-time, efficient, and personalized interactions between city authorities and citizens. Despite the significant advantages offered by AI chatbots, their integration into urban management is not without its challenges. Data privacy concerns, the digital divide that limits equitable access, and the requirement for reliable infrastructure can impede the effective deployment of these technologies. Moreover, there is a lack of comprehensive research exploring the multifaceted impact of chatbots on urban service delivery.

This chapter aims to investigate the transformative role of AI-based chatbots in enhancing urban service delivery within smart cities. Specifically, it analyzes how chatbots can improve communication, enhance access to essential services, and foster active citizen engagement. Through an in-depth examination of various case studies, this research highlights successful implementations of chatbots in smart cities. It identifies key challenges that need to be addressed for their effective integration. Ultimately, this work contributes to a deeper understanding of how AI chatbots can shape the future of urban management, emphasizing the importance of collaboration among stakeholders to leverage the benefits of these technologies.

The remainder of the paper is organized as follows: Sect. 2 delves into the theoretical background of smart cities, exploring their key concepts and challenges. Section 3 focuses on AI chatbots, discussing their technologies and theoretical foundations, including an overview, architecture, and current tools for development. Section 4 outlines the methodology adopted for integrating chatbots into smart city services. Section 5 presents case studies of AI chatbot applications in smart cities, highlighting a conversational scenario and its implementation. Section 6 addresses the challenges encountered in deploying chatbots and suggests future directions for overcoming these obstacles. Finally, Sect. 7 concludes the chapter.

2 Smart Cities: Theoretical Background

Smart cities have emerged in recent years as a response to the challenges of rapid urbanization and the need for sustainable development. A smart city is a concept without a universally agreed-upon definition [9], despite significant global interest and investment in the field. Generally, smart cities aim to improve the quality of life for residents by leveraging technology, particularly through initiatives focused on sustainability, efficiency and enhanced urban services. However, definitions of smart cities vary widely, often substituting terms like "intelligent" or "digital" within the concept of smart cities [11]. Generally, the smart city represents an urban environment that integrates technology and innovation to address modern urban challenges, enhance citizen engagement, and ensure long-term sustainability.

Figure 1 shows the significant increase in publications on smart cities in Scopus over recent years. The real change began in 2012, coinciding with the emergence of big data and cloud computing technologies, which provided the foundation for many smart city innovations. These technologies enabled cities to collect, store, and analyze vast amounts of data, facilitating data-driven decision-making and resource optimization. The surge in publications from 2012 onward reflects the rapid integration of these technological advancements into urban management and planning, driving increased interest and research in smart cities.

Smart cities rely on a set of advanced technologies and innovative techniques that aim to improve urban life, optimize resources, and promote sustainability. These technologies are interconnected, forming a digital ecosystem that facilitates real-time monitoring, analysis, and control of various urban functions.

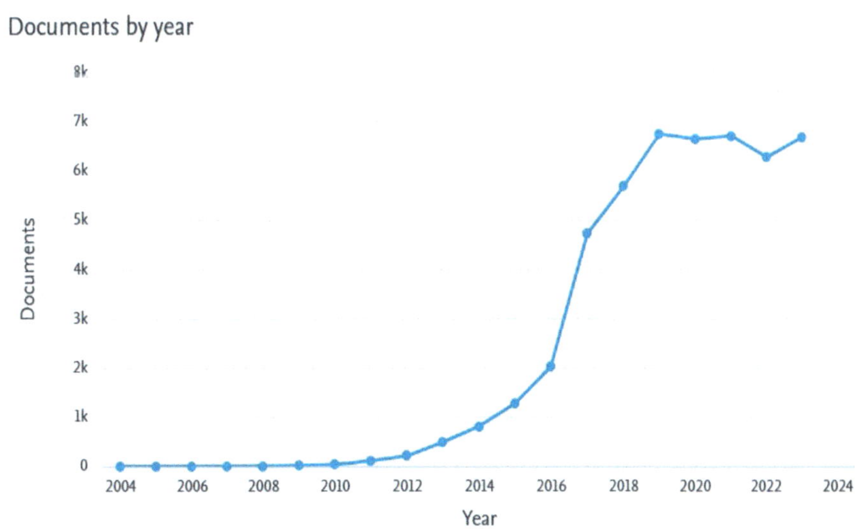

Fig. 1 Key components of the 6A framework for STD analysis

The Internet of Things (IoT) is one of the cornerstones of smart cities. Through connected sensors and devices, IoT enables continuous data collection and exchange across the city. For example, IoT sensors installed in public lighting systems automatically adjust brightness based on ambient conditions or pedestrian presence, resulting in significant energy savings. In waste management, smart bins equipped with sensors can signal when they are full, optimizing collection routes and reducing operational costs [27].

Big Data and data analytics also play a crucial role in smart cities. Big data refers to the massive amounts of data generated by sensors, citizens, and digital services. By processing this data, cities can obtain valuable insights that facilitate decision-making. For example, data analysis allows cities to monitor traffic flows, predict energy demands, or adjust services based on citizen habits. This data-driven decision-making helps urban managers address issues more accurately and efficiently, while enhancing service quality [3].

Artificial Intelligence (AI) and Its Subsets (Machine Learning and Deep Learning) are essential for interpreting complex datasets and for automating processes intelligently. AI can recognize patterns in data and make predictions. For example, in public safety, AI can analyze surveillance footage to detect suspicious behaviors, enhancing security in public spaces. In transportation, machine learning algorithms optimize traffic flows and reduce congestion in real time, improving mobility for citizens [2].

Cloud computing is another key technology that supports smart cities by enabling remote data storage and processing. This infrastructure allows centralized information management and reduces costs associated with physical servers. For instance, data collected from sensors can be stored in the cloud, where it is accessible to various municipal agencies to improve coordination and response speed [23].

Blockchain is increasingly used in smart cities to ensure data transparency and security. Due to its decentralized structure, blockchain allows secure transactions and enhances trust in digital public services, such as electronic voting systems or digital identity management [26].

Internet networks provide fast and reliable connectivity, which is crucial for the proper functioning of connected devices and real-time applications. Reliable internet improves the performance of IoT devices and enables applications such as autonomous vehicles, real-time traffic monitoring, and smart healthcare solutions.

In the next section, we will discuss chatbots as a smart solution integrated into smart cities. Powered by artificial intelligence, chatbots improve interactions between citizens and urban services as they automate responses to common questions, guide users to the appropriate services, and gather useful data for better public service management.

3 AI Chatbots: Technologies and Theoretical Foundations

3.1 Overview of Chatbots

Chatbots, also called conversational agents [15], bots [12], or virtual assistants [22], have emerged as a powerful technological solution in various sectors, such as tourism Benaddi, Ouaddi, Souha, Jakimi, and Ouchao [5], customer service in businesses, and particularly in smart cities, as they enable automated and efficient interactions between citizens and urban services. A chatbot is a software application that simulates human-like conversations, either through text or voice, and uses technologies such as natural language processing (NLP) and machine learning [8]. They enhance service delivery as they provide instant responses, reduce waiting times, and ensure 24/7 availability. Chatbots are broadly categorized into two types: rule-based (intent-based) chatbots and AI-powered chatbots. Rule-based chatbots operate on pre-defined rules and respond to specific keywords or phrases, while AI-powered chatbots use advanced AI and NLP to handle complex and dynamic conversations.

The adaptability of AI-powered chatbots makes them ideal for smart city applications such as public inquiries, real-time traffic updates, and personalized citizen support. Chatbots enhance communication and automate repetitive tasks, which makes cities more efficient and citizen centric.

The history of chatbots dates back to the 1960s with the famous chatbot ELIZA [25], developed by Joseph Weizenbaum at the Massachusetts Institute of Technology (MIT). ELIZA was one of the first programs capable of simulating human-like conversations by using pattern matching to respond to user inputs. It gained popularity for mimicking a Rogerian psychotherapist, engaging users in simple and structured dialogs.

Since ELIZA, chatbot technology has evolved significantly. In the 1990s, advancements led to the development of ALICE, which integrated natural language processing (NLP) capabilities to enable more complex interactions. Later, chatbots like Siri, Alexa, and Google Assistant introduced voice recognition and machine learning, making conversational systems more interactive and intuitive.

The release of ChatGPT by OpenAI in 2022 marked a new milestone in the evolution of chatbots. ChatGPT employs advanced natural language processing and machine learning techniques, built on the GPT (Generative Pre-trained Transformer) architecture [24]. Unlike its predecessors, ChatGPT can manage complex conversations, generate human-like responses, and adapt to a wide range of topics, which makes it one of the most powerful chatbots today. Similarly, the launch of Gemini in 2023 brought further advancements in chatbot technology, incorporating enhanced capabilities for understanding and interacting with users more effectively.

Fig. 2 The timeline of chatbot evolution over time. (Adapted from Ouaddi et al. [19] and Benaddi et al. [4])

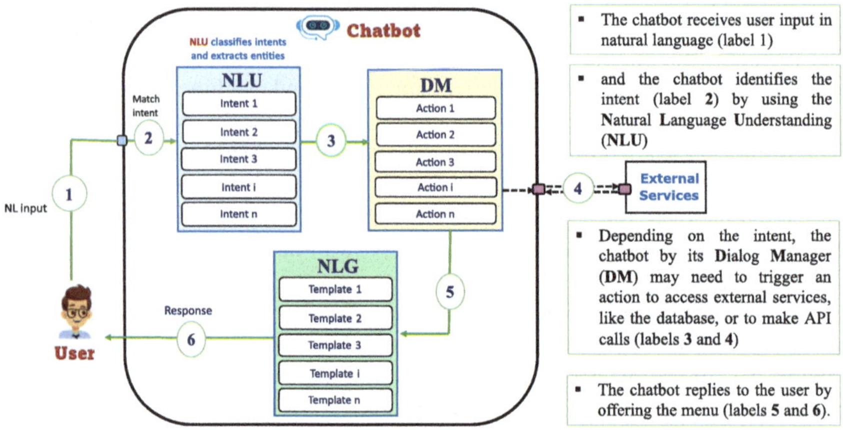

Fig. 3 The conceptual architecture of chatbots. (Adopted from Benaddi et al. [6])

This evolution, shown in Fig. 2, has allowed chatbots to better understand the context of conversations, provide more accurate responses, and engage in more sophisticated dialogs with users.

3.2 *Architecture of Chatbots*

The conceptual architecture of chatbots, as illustrated in Fig. 3, typically consists of three main components: Natural Language Understanding (NLU), Dialog Management (DM), and Natural Language Generation (NLG). This architecture is designed around a set of intents that users may want to accomplish. For instance, given a user input like "What is the weather in Tanger?" (the first arrow in Fig. 3), the chatbot tries to identify the corresponding intent. If the chatbot does not find any matching intent, some tools allow for a default fallback intent [15, 18].

After matching an intent at the NLU level, the chatbot extracts the parameters from the utterance, such as the location and date (the second arrow). The chatbot, facilitated by its DM component, can execute diverse actions based on user intent, including interfacing with external services to retrieve data. Finally, the NLG component receives data from the DM and uses predefined templates to generate natural language responses and reply to the user.

3.3 Current Tools for Developing Chatbots

The development of chatbots relies on a wide range of tools [17] that address various aspects of their creation, functionality, and deployment. These tools include platforms, which provide ready-to-use environments for designing and deploying chatbots easily using graphical forms, such as Dialogflow, Amazon Lex, and IBM Watson Assistant. Frameworks, like Rasa and Xatkit, allow developers to build custom chatbots with greater control over functionality through programming. Libraries, such as TensorFlow and PyTorch, support the integration of machine learning capabilities for natural language processing. Services, including cloud-based options like AWS Lex or Azure Bot Services, simplify hosting and scalability. Additionally, Domain-Specific Languages (DSLs) are emerging in some research projects, offering approaches that facilitate chatbot development independently of any platform. Examples of such DSLs include Conga [20], Xflow [13], Xatkit [7], and graphical DSL for developing rasa framework chatbots [16].

Table 1 provides an overview of chatbot development tools, showcasing their examples, advantages, and limitations. It highlights key features to help developers understand their functionalities.

Table 1 Advantages and limitations of chatbot development tools

Tool type	Examples	Advantages	Limitations
Platforms [21]	• Watson assistant • Dialogflow • Amazon Lex • Wit.ai	• Easy to use with graphical interfaces • Provides pre-built NLP capabilities	• Can be locked • Require significant costs
Frameworks	• Bot framework • Rasa framework • Xatkit	• Offers high flexibility for custom implementations • Supports local and cloud deployment • Open access	• Requires programming expertise
Services	• LUIS	• Cloud-based solutions • No access to source code	• Can be locked • Require significant costs
Libraries	• Chatterbot	• Simple implementation • Open access	• Does not support NLP services • Used to create simple bots with code blocks
DSLs [14]	• Conga	• Tailored for specific use cases • Platform-independent	• Limited scope of application • Smaller developer community

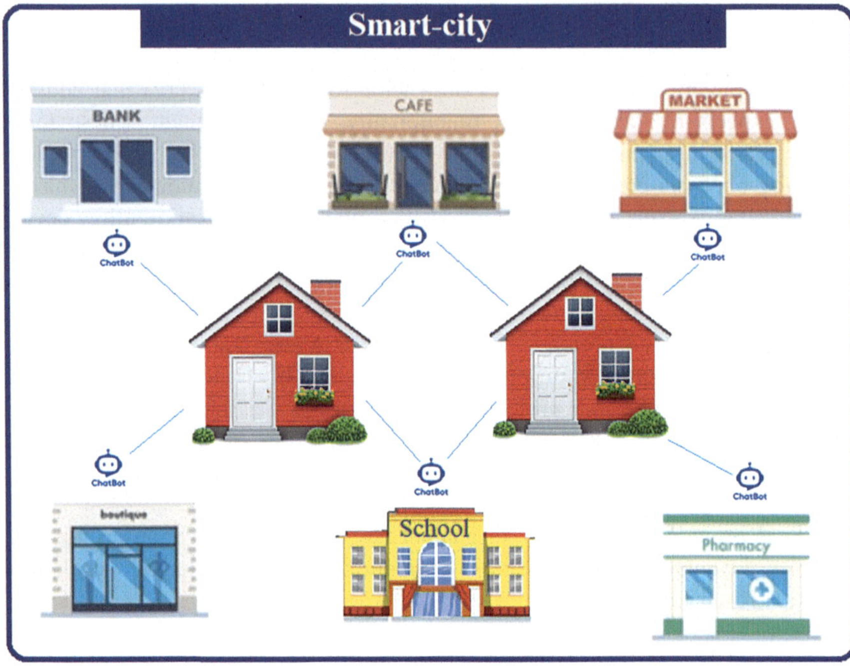

Fig. 4 Methodology for integrating chatbots into smart city ecosystems

4 Methodology

The methodology for integrating chatbots into smart city services involves enabling each service to offer its dedicated chatbot, allowing users to access these services through these virtual assistants. As indicated in Fig. 4, chatbots are associated with key service points such as banks, schools, markets, and pharmacies. These chatbots facilitate communication by instantly responding to citizen inquiries and automating service requests. Each chatbot is tailored to the specific needs of its service, leveraging AI and NLP to enhance user interactions and service accessibility. This approach ensures that citizens can interact efficiently with various urban services through dedicated chatbots.

5 Case Study

5.1 Conversational Scenario

This case study explores the implementation of an AI chatbot designed to provide restaurant services in a smart city environment. The chatbot assists users in booking rooms, checking availability, requesting amenities, and managing reservations.

Table 2 A conversational scenario of a chatbot in a smart restaurant

User	Chatbot
Hello	Hi, I'm the restaurant bot. How can I assist you today?
I'd like to see the menu.	Sure! Here is the menu with the details of our available services.
I'll have the pasta and a side salad.	Great choice! You've selected pasta and a side salad. Do you confirm your order?
Yes.	Perfect! Your order has been confirmed. The total amount to pay is 15 dollars.
I'd like to make a complaint.	I'm sorry to hear that! Could you please provide me with the details of your complaint?
The food I received was cold.	I apologize for the inconvenience. We will take your complaint into account and improve our service.
Thank you. Goodbye!	You're welcome! Have a great day. Goodbye!

Following the conversation scenario outlined in Table 2, the chatbot interacts with customers using natural language processing to understand and respond to their requests.

The system ensures personalized communication, enhancing user experience. By integrating such a chatbot into the smart city's service ecosystem, restaurants can optimize operations and deliver better customer satisfaction, aligning with the goals of smart urban living.

5.2 Implementation

To implement this scenario and make the restaurant smarter, we utilized the Rasa Framework, a powerful open-source tool for building AI-driven conversational agents. Rasa provides robust capabilities for natural language understanding (Rasa NLU) and dialog management (Rasa Core). It enables the creation of highly interactive and personalized chatbots by using Python. It supports intent recognition, entity extraction, and managing flows of conversations.

The framework allows integration with external APIs and databases to enhance the functionality of chatbots. Figure 5 illustrates an example of interaction between the chatbot and the user.

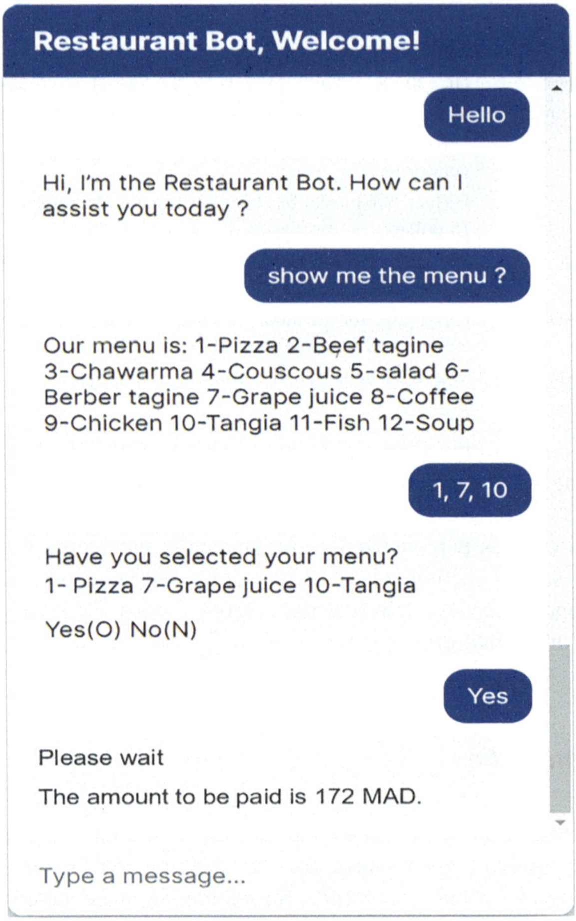

Fig. 5 Sample chatbot interaction in a smart restaurant

6 Challenges and Future Directions

Developing chatbots for smart cities presents several challenges. Firstly, training the natural language understanding (NLU) components of chatbots requires extensive datasets tailored to specific services such as banks, schools, and public utilities. Secondly, the diversity and complexity of development tools often result in high costs and reliance on proprietary systems, leading to potential vendor lock-in. Thirdly, ensuring the availability, security, and privacy of the required data introduces significant obstacles, especially given the sensitive nature of urban applications. Additionally, the underlying infrastructure, such as the Internet of Things (IoT) and computing power, plays a critical role. Limited or inadequate IoT connectivity and insufficient computational resources can further hinder the effective deployment and performance of chatbots in smart cities.

To address these challenges, future efforts should focus on preparing comprehensive and domain-specific datasets to support chatbot development for each service. This will enable chatbots to deliver more accurate and context-aware responses. Additionally, creating domain-specific languages (DSLs) can simplify and accelerate chatbot development by providing reusable components and reducing technical barriers [14]. These directions aim to make chatbot development more efficient and scalable while addressing the unique demands of smart cities.

7 Conclusion

In this work, we explored the technologies used in smart cities as well as those related to chatbots. We analyzed the transformative role of AI-based chatbots in enhancing urban management and citizen services. Through case studies, we demonstrated how these tools can facilitate access to information, strengthen user engagement, and optimize the delivery of essential services.

Finally, we identified key challenges, such as data privacy concerns and access disparities, while proposing future directions for the successful integration of chatbots into smart city ecosystems.

Acknowledgments This work was supported by the Ministry of Higher Education, Scientific Research and Innovation, the Digital Development Agency (DDA) and the CNRST of Morocco (Alkhawarizmi/2020/32).

Competing Interests The authors declare that they have no competing interests.

References

1. Adamopoulou E, Moussiades L (2020) An overview of chatbot technology. In: Artificial intelligence applications and innovations. Springer, Cham
2. Allam Z, Dhunny ZA (2019) On big data, artificial intelligence and smart cities. Cities 89:80–91
3. Bassoo V, Ramnarain-Seetohul V, Hurbungs V, Fowdur T, Beeharry Y (2018) Big data analytics for smart cities. In: Internet of things and big data analytics toward next-generation intelligence. Springer, pp 359–379
4. Benaddi L, Ouaddi C, Jakimi A, Ouchao B (2024) A systematic review of chatbots: classification, development, and their impact on tourism. IEEE Access
5. Benaddi L, Ouaddi C, Souha A, Jakimi A, Ouchao B (2024) Chatbots in tourism sector: classification, evolution, and functionalities. In: IEEE 12th international symposium on signal, image, video and communications (ISIVC). IEEE
6. Benaddi L, Ouaddi C, Souha A, Jakimi A, Ouchao B (2024) Model-driven engineering to develop chatbots for smart tourism. In: 2024 International conference on global aeronautical engineering and satellite technology (GAST). IEEE
7. Daniel G, Cabot J, Deruelle L, Derras M (2020) Xatkit: a multimodal low-code chatbot development framework. IEEE Access 8:15332–15346

8. El Alaoui H, El Aouene Z, Cavalli-Sforza V (2023) Building intelligent chatbots: tools, technologies, and approaches. In: 2023 3rd international conference on innovative research in applied science, engineering and technology (IRASET). IEEE
9. Gracias JS, Parnell GS, Specking E, Pohl EA, Buchanan R (2023) Smart cities—a structured literature review. Smart Cities 6(4):1719–1743
10. Gretzel U, Zhong L, Koo C (2016) Application of smart tourism to cities. Int J Tourism Cities 2(2)
11. Ismagilova E, Hughes L, Dwivedi YK, Raman KR (2019) Smart cities: advances in research—an information systems perspective. Int J Inf Manag 47:88–100
12. Lebeuf C, Storey M-A, Zagalsky A (2017) Software bots. IEEE Softw 35(1):18–23
13. Malamas N, Panayiotou K, Symeonidis AL (2023) dFlow: a domain specific language for the rapid development of open-source virtual assistants. arXiv preprint arXiv:231002102
14. Ouaddi C, Benaddi L, Bouziane EM, Jakimi A, Chehri A, Saadane R (2024) DSL-driven approaches and Metamodels for Chatbot development: a systematic literature review. Expert Syst:e13787
15. Ouaddi C, Benaddi L, Jakimi A (2024) Architecture, tools, and DSLS for developing conversational agents: an overview. Procedia Comp Sci 231:293–298
16. Ouaddi C, Benaddi L, Naimi L, Jakimi A (2024) A graphical DSL for accelerating chatbot development. In: 2024 7th international conference on advanced communication technologies and networking (CommNet). IEEE
17. Ouaddi C, Benaddi L, Souha A, Jakimi A (2024) A comparative and analysis study for recommending a chatbot development tool. In: 2024 International conference on global aeronautical engineering and satellite technology (GAST). IEEE
18. Ouaddi C, Benaddi L, Souha A, Jakimi A, Ouchao B (2024) A sketch of an approach for discovering UML use-case diagrams from textual specifications of systems using a chatbot. In: 2024 IEEE 12th international symposium on signal, image, video and communications (ISIVC). IEEE
19. Ouaddi C, Benaddi L, Souha A, Jakimi A, Ouchao B, Saadane R (2024) Exploring and analyzing the impact of chatbots in tourism industry. In: Proceedings of the 7th international conference on networking, intelligent systems and security. Association for Computing Machinery
20. Pérez-Soler S, Guerra E, de Lara J (2021) Creating and migrating chatbots with conga. In: 2021 IEEE/ACM 43rd international conference on software engineering: companion proceedings (ICSE-companion). IEEE
21. Revang M, Baker V, Manusama B, Mullen A (2018) Market guide for conversational platforms. Gartner
22. Sarikaya R (2017) The technology behind personal digital assistants: an overview of the system architecture and key components. IEEE Signal Process Mag 34(1):67–81
23. Suciu G, Vulpe A, Halunga S, Fratu O, Todoran G, Suciu V (2013) Smart cities built on resilient cloud computing and secure internet of things. In: 2013 19th international conference on control systems and computer science. IEEE
24. Vaswani A, Shazeer N, Parmar N, Uszkoreit J, Jones L, Gomez AN et al (2017) Attention is all you need. In: Proceedings of the 31st international conference on neural information processing systems, Long Beach, California, USA. Curran Associates, Inc
25. Weizenbaum J (1966) ELIZA—a computer program for the study of natural language communication between man and machine. Commun ACM 9(1):36–45
26. Xie J, Tang H, Huang T, Yu FR, Xie R, Liu J, Liu Y (2019) A survey of blockchain technology applied to smart cities: research issues and challenges. IEEE Commun Surv Tutor 21(3):2794–2830
27. Zanella A, Bui N, Castellani A, Vangelista L, Zorzi M (2014) Internet of things for smart cities. IEEE Internet Things J 1(1):22–32

Optimizing Urban Management through the Internet of Things (IoT): Sustainable Solutions and Challenges for Smart Cities

Mounir Boumhamdi, Aziz Atmani, Yamna Azizi, and Marouane Ikira

Abstract The increasing urbanization worldwide has imposed significant pressures on urban management systems, necessitating innovative solutions for sustainable development. The Internet of Things (IoT) has emerged as a transformative technology to optimize energy, water, and waste management systems in smart cities. This study employs a mixed-method approach to investigate the role of IoT in enhancing urban management in three leading cities: Singapore, Copenhagen, and Barcelona. These cities were selected due to their advanced adoption of IoT technologies and commitment to sustainability. Data were cleaned and analyzed using thematic and statistical methods, including regression modeling. The findings demonstrate that IoT significantly enhances urban sustainability by reducing inefficiencies, improving resource utilization, and promoting green initiatives. However, challenges such as high investment costs, data privacy concerns, and the digital divide persist. Recommendations include integrated policy frameworks, public-private partnerships, and community engagement to address these barriers and maximize IoT's potential in urban management.

Keywords Smart cities · Sustainability · Digital technologies · Urban development · Public-private partnerships · Urban resilience

---

M. Boumhamdi (✉) · A. Atmani
Laboratoire de Recherche Business Intelligence, Gouvernance des Organisations, Finance et Criminalité Financière, Faculté des Sciences Juridiques Économiques et Sociales de Casablanca, Université Hassan II, Casablanca, Morocco

Y. Azizi
Laboratoire du Droit Public et Sciences Politiques, Faculté des Sciences Juridiques Économiques et Sociales de Casablanca, Casablanca, Morocco

M. Ikira
Laboratoire de Recherches en Gestion, Économie et Sciences Sociales, Faculté des Sciences Juridiques, Économiques et Sociales d'El Jadida, Université Chouaib Doukkali, El Jadida, Morocco

H. Kaddouri et al. (eds.), *Smart Cities for Sustainable and Urban Development*, Green Energy and Technology, https://doi.org/10.1007/978-3-031-99001-4_41

1 Introduction

The increasing complexity of urban management, driven by rapid urbanization, resource constraints, and environmental challenges, necessitates innovative approaches to ensure sustainable development. According to the United Nations [20], by 2050, more than 68% of the world's population is projected to live in metropolitan areas, creating immense pressure on existing urban infrastructures. These challenges demand smarter and more efficient solutions to optimize resource utilization and improve urban livability. One promising avenue lies in the integration of smart city technologies powered by the Internet of Things (IoT), enabling real-time data collection, analysis, and automation to address these multifaceted issues effectively.

1.1 Background and Motivation

The Internet of Things (IoT) has emerged as a transformative tool for enhancing urban management. As cities expand, traditional systems for managing energy, water, and waste often prove inadequate, leading to inefficiencies, rising costs, and environmental degradation [20]. IoT offers innovative solutions, leveraging sensors and automation to optimize operations, improve resource efficiency, and support sustainability [2]. For instance, IoT has been successfully applied in energy management through smart grids, which enhance energy distribution and reduce waste [19]. Similarly, IoT-driven water management systems, such as Tokyo's predictive maintenance technologies, have minimized water loss [21]. In waste management, IoT-enabled smart bins and optimized collection routes have significantly improved efficiency, as evidenced by Barcelona's initiatives [8]. Furthermore, Singapore's IoT-integrated traffic and air quality management systems demonstrate how IoT can contribute to improved urban mobility and environmental health [7].

Despite these advancements, several challenges hinder IoT adoption. Economic barriers, such as the high cost of deploying and maintaining IoT infrastructure, remain a significant concern, particularly for cities in developing countries [12]. Privacy concerns associated with extensive data collection, including potential misuse and surveillance risks, further complicate widespread IoT implementation [14]. Additionally, interoperability issues stemming from the lack of standardized IoT protocols limit the scalability and integration of these systems [20]. Finally, the digital divide exacerbates inequalities, as marginalized communities often lack access to IoT technologies, reducing the inclusivity of smart city benefits [19].

While existing studies highlight IoT's transformative potential, critical gaps persist. Many investigations prioritize technical solutions without addressing the socio-economic and policy dimensions that influence successful adoption. For example, research focusing solely on technological efficiency overlooks challenges such as public trust and regulatory frameworks essential for long-term sustainability (Smith and Dawson 2021). Furthermore, the majority of case studies explore advanced cities like Copenhagen and Tokyo, leaving behind insights on how emerging cities can adopt similar technologies under constrained resources [7]. Addressing these gaps requires a comprehensive approach that integrates technical, economic, and social perspectives to ensure equitable and sustainable urban transformation.

This chapter explores the transformative potential of IoT technologies in urban management by analyzing their applications in energy, water, and waste systems. Through a mixed-methods approach, it combines qualitative and quantitative analyzes of case studies from Copenhagen, Tokyo, and Barcelona. The findings aim to uncover actionable insights, addressing both the opportunities and barriers associated with IoT adoption, and providing policy recommendations to guide future implementations. By bridging the gaps in existing literature, this work contributes to a more inclusive and practical understanding of IoT's role in achieving sustainable urban development.

1.2 Preview of Related Work

The integration of the Internet of Things (IoT) into urban environments has been a focus of extensive research, underscoring its transformative potential in optimizing critical city functions. IoT technologies enable cities to collect and analyze real time data, providing actionable insights for managing complex urban systems. This capability has been particularly evident in energy, water, and waste management sectors, where IoT has demonstrated its ability to enhance efficiency, reduce wastage, and improve sustainability.

In energy management, IoT-enabled smart grids have redefined the operational landscape by facilitating real-time monitoring, predictive analytics, and efficient distribution of electricity. Angelidou [1] highlights the pivotal role of IoT in reducing energy wastage and seamlessly integrating renewable energy sources into existing grids, thereby aligning urban energy systems with sustainability goals. Similarly, Batty et al. [4] emphasize the role of IoT in predictive maintenance, where sensor networks continuously monitor the health of energy infrastructure to preempt disruptions and optimize performance. These applications not only improve energy efficiency but also reduce costs and enhance system reliability.

Water resource management has also significantly benefitted from IoT technologies. Smart water systems leverage sensors for leak detection, quality monitoring, and distribution optimization, enabling cities to conserve resources and maintain high standards of service delivery. For instance, Barcelona's IoT-driven water management systems have been instrumental in reducing water losses and improving operational efficiency, serving as a model for other cities worldwide [3]. Similarly, Tokyo's IoT-enabled water infrastructure has demonstrated how real-time monitoring and automated systems can achieve a 15% reduction in water losses, highlighting the potential of IoT in addressing urban water challenges [19].n waste management, IoT technologies have introduced innovative solutions to longstanding inefficiencies. Smart bins equipped with IoT sensors detect fill levels and notify authorities when collection is required, enabling dynamic routing that minimizes operational costs and reduces environmental impacts. Singapore's urban initiatives, as documented by Hollands [13], exemplify the successful implementation of such systems, which have streamlined waste collection and enhanced recycling efforts. Barcelona has also achieved a 12% reduction in waste-related inefficiencies by leveraging IoT to optimize collection processes and promote sustainability [8].

Despite these successes, critical challenges persist in the widespread adoption and scalability of IoT technologies. Zuboff [21] critiques IoT systems for enabling what she terms "surveillance capitalism," wherein personal data is commodified, raising concerns about citizen autonomy and privacy. Goodman and Powles [10] further argue for the necessity of robust cybersecurity frameworks and transparent governance to address the ethical and privacy risks associated with IoT. Additionally, Mora et al. [16] identify interoperability issues as a major barrier to IoT integration, as proprietary systems often lack the compatibility required for large-scale deployment. Such fragmentation limits the ability of cities to create cohesive and efficient IoT ecosystems.

The digital divide further exacerbates inequities in IoT adoption, as marginalized populations frequently lack the infrastructure, digital literacy, and financial resources needed to benefit from IoT advancements. Deursen et al. [9] stress that this disparity undermines the inclusivity of smart city initiatives, leaving vulnerable communities unable to access the benefits of IoT-enabled systems.

While these studies highlight the significant opportunities and challenges associated with IoT integration, they often analyze individual domains—such as energy, water, or waste management—in isolation, neglecting their systemic interdependencies. Furthermore, the social and ethical dimensions of IoT implementation are frequently addressed as secondary considerations, despite their profound implications for governance, equity, and public trust. The interconnected nature of these challenges calls for a more holistic approach to IoT research and policy.

This study seeks to address these gaps by providing a comprehensive analysis of IoT applications, barriers, and strategies for fostering sustainable and inclusive smart city development. By exploring the synergies between different urban systems and considering the social and ethical dimensions of IoT adoption, this research aims to offer actionable insights for policymakers, urban planners, and stakeholders to leverage IoT technologies for building resilient, equitable, and sustainable urban environments.

1.3 Research Gap

Although the Internet of Things (IoT) has demonstrated transformative potential in urban management, several critical research gaps remain insufficiently addressed. First, much of the existing literature adopts a siloed approach, focusing predominantly on specific domains such as energy, water, or waste management, without adequately examining the interconnectedness of urban systems. Urban environments are inherently complex, with energy systems influencing water distribution, waste management impacting energy efficiency, and water systems shaping urban resilience. However, this systemic interdependence is rarely considered, limiting the scope of current research and practical implementations.

Second, the interplay between economic, ethical, and technical challenges remains underexplored, despite its significant impact on the scalability and inclusivity of IoT solutions. Privacy concerns, for instance, are often treated as isolated issues without considering their intersection with other barriers, such as interoperability or regulatory inadequacies. Proprietary IoT systems often lack standardization, creating integration challenges that prevent cities from adopting cohesive and scalable solutions [16]. Moreover, the digital divide exacerbates these issues, as marginalized communities frequently lack access to IoT technologies, perpetuating existing inequities and limiting the societal benefits of smart city initiatives.

Third, existing frameworks for IoT implementation often fail to prioritize inclusivity, resulting in solutions that do not adequately address the needs of marginalized populations. Vulnerable groups, particularly those in low-income or underserved communities, are frequently excluded from IoT-enabled urban improvements due to financial, infrastructural, or technical barriers. This exclusion not only undermines the social equity goals of smart city initiatives but also limits the potential of IoT to create truly sustainable and resilient urban environments.

This research seeks to bridge these gaps by adopting a holistic approach that examines IoT applications across interconnected domains of urban management—namely, energy, water, and waste. In addition to analyzing the technological benefits, this study will delve into the challenges that hinder IoT implementation, including economic, ethical, and social considerations. By addressing the systemic nature of urban challenges and prioritizing inclusivity, this research aims to provide actionable insights for developing sustainable and equitable IoT solutions tailored to the needs of diverse urban populations.

1.4 Focus of this Paper

This paper explores the key ethical and social challenges associated with IoT integration in smart cities, with a particular focus on applications in energy, water, and waste management. It examines economic, privacy, interoperability, and digital inclusion barriers that impede IoT adoption and proposes strategies to overcome these obstacles. By drawing insights from case studies in Barcelona, Tokyo, and Singapore, this study highlights the importance of balancing technological innovation with social equity and ethical considerations, offering a roadmap for sustainable and inclusive urban development.

2 IoT Applications in Urban Management

The integration of the Internet of Things (IoT) into urban management has transformed the way cities optimize their resources, enhance operational efficiency, and deliver improved services. By leveraging real-time data collection, advanced analytics, and automation, IoT technologies offer cities the ability to adapt dynamically to the demands of urbanization while reducing inefficiencies and promoting sustainability [2]. This transformation has proven essential as urban populations grow rapidly, putting increasing pressure on infrastructure and resources [20].

IoT applications in urban management can be broadly categorized into three key domains: energy, water, and waste management. Each of these sectors presents unique opportunities for innovation through IoT. Smart grids in energy systems have redefined power distribution, enabling cities to reduce energy wastage and integrate renewable energy sources seamlessly [1]. Similarly, IoT-driven water management systems leverage smart sensors and predictive analytics to optimize water usage and improve service delivery, as demonstrated in Tokyo's IoT-enabled water infrastructure [19]. In waste management, IoT technologies such as smart bins and route optimization have enhanced operational efficiency, minimized environmental impact, and contributed to better resource recovery [8, 13].

Despite these advancements, the integration of IoT is not without challenges. High implementation costs often deter resource-constrained cities from adopting these technologies [12]. Furthermore, interoperability issues arising from the lack of standardized IoT protocols hinder the seamless integration of devices and systems [16]. Privacy concerns and the digital divide exacerbate these issues, creating barriers to equitable IoT adoption, particularly in marginalized communities [9].

2.1 Energy Management

IoT has redefined energy systems through the introduction of smart grids and renewable energy optimization. By integrating sensors, IoT-enabled grids provide real-time data that enhance energy distribution, reduce waste, and improve overall efficiency [16].

Case Study: In Copenhagen, IoT-based smart grids have reduced energy consumption by 10%, demonstrating the role of demand-side management in improving sustainability. These grids also support renewable energy integration, balancing supply and demand dynamically [22].

Despite these benefits, high implementation costs and technological fragmentation hinder widespread adoption, particularly in cities with limited financial resources [20].

2.2 *Water Resource Management*

IoT technologies in water management focus on reducing inefficiencies and conserving resources. Smart meters, leak detection systems, and predictive analytics have been instrumental in modernizing urban water distribution networks [20].

Case Study: Tokyo's IoT-driven water management system has minimized water loss by 15%, highlighting the effectiveness of real-time monitoring and automated maintenance systems [21]. Additionally, IoT-based irrigation systems have optimized water use in urban parks, reducing waste and promoting green urban spaces [19].

However, challenges such as data privacy concerns and the high cost of IoT infrastructure upgrades remain barriers to adoption [14].

2.3 *Waste Management*

The application of IoT in waste management has improved efficiency, reduced operational costs, and supported sustainability. Smart bins equipped with IoT sensors detect fill levels, enabling optimized collection schedules and routes [12].

Case Study: Barcelona's IoT-enabled waste collection system has significantly reduced greenhouse gas emissions by streamlining vehicle routes. Real-time data analysis has also improved recycling initiatives, enhancing overall waste management efficiency [8].

Despite these advancements, IoT adoption in waste management is hindered by socioeconomic disparities, which limit access to smart technologies in underserved communities [19].

2.4 *Challenges*

While IoT has demonstrated significant potential in urban resource management, its widespread adoption is constrained by several challenges:

- **Economic Barriers:** High installation and maintenance costs often deter low-income cities from adopting IoT solutions [12].
- **Privacy and Security Concerns:** The extensive data collected by IoT systems raises ethical questions about surveillance and the misuse of sensitive information [14].
- **Interoperability Issues:** The lack of standardized protocols across IoT platforms limits integration and scalability, creating inefficiencies in system deployment [20].
- **Digital Divide:** Unequal access to IoT technologies exacerbates urban inequalities, leaving vulnerable communities without the benefits of smart systems [19].

The literature and case studies reviewed highlight the transformative potential of IoT in addressing urban resource management challenges. Successes in cities such as Copenhagen, Tokyo, and Barcelona demonstrate the ability of IoT to enhance energy efficiency, conserve water, and optimize waste management. However, the challenges of cost, privacy, and technological fragmentation must be addressed to achieve equitable and sustainable urban systems.

Addressing these barriers requires collaborative efforts among governments, private sectors, and communities. By establishing robust regulatory frameworks, fostering public-private partnerships, and promoting technological inclusivity, IoT can drive significant progress toward sustainable urban development.

3 Methodology

This study employs a qualitative approach to explore the transformative role of IoT in urban management, focusing on its applications in energy, water, and waste systems. The analysis combines an extensive literature review with case studies of three cities—Barcelona, Tokyo, and Singapore—recognized for their advanced IoT implementations. These cities were selected for their innovative projects, measurable impacts, and the availability of reliable data.

The qualitative approach is complemented by a comparative framework that highlights both the shared benefits and the unique challenges encountered in each city. This ensures a comprehensive understanding of how IoT technologies can optimize urban systems while addressing economic, technical, and social barriers.

3.1 Data Collection and Cleaning

The methodology incorporates data from diverse sources, including academic articles, government reports, case studies, and industry documentation, to ensure a comprehensive understanding of IoT applications in urban management. The data collection process targeted information directly relevant to energy, water, and waste management systems across the selected case studies: Barcelona, Tokyo, and Singapore.

3.1.1 Dataset Overview

- **Initial Dataset**: 150 documents were initially gathered, comprising peer-reviewed journal articles, white papers, government policy documents, and technical reports. These sources were chosen based on their relevance to IoT implementation, governance models, and measurable outcomes in urban contexts.

- **After Data Cleaning**: Following a rigorous cleaning process, the dataset was refined to 102 high-quality sources. This ensured the inclusion of only credible, relevant, and non-redundant materials, eliminating noise and improving the reliability of the analysis.

3.1.2 Data Cleaning Techniques

The cleaning process involved multiple steps:

- **Relevance Filtering**: Sources were evaluated to ensure alignment with the study's focus on IoT applications in energy, water, and waste management.
- **Duplicate Removal**: Redundant entries, such as repeated findings or overlapping case studies, were eliminated to streamline the dataset.
- **Quality Assessment**: Each source was assessed for credibility, prioritizing peer-reviewed journals, official government reports, and well-documented industry publications.

3.1.3 Rationale for Cleaning

This refinement process enhanced the reliability and focus of the analysis by:

- Ensuring that only high-quality data relevant to the study's objectives were included.
- Reducing biases introduced by extraneous or non-credible sources.
- Improving the clarity of insights by retaining context-specific information from the selected cities.

3.1.4 Summary

- **Initial Dataset**: 150 documents.
- **Final Dataset**: 102 high-quality, relevant, and credible sources focusing on IoT's transformative role in urban management.

3.2 Analytical Framework

The analysis was guided by three core dimensions:

- **Efficiency Gains:** Measuring how IoT technologies improved resource utilization, reduced costs, and minimized waste.
- **Sustainability Contributions:** Assessing the impact of IoT on environmental goals, including emissions reduction and conservation.

- **Implementation Barriers:** Identifying challenges such as cost, privacy concerns, and technological fragmentation.

These dimensions were applied across the selected case studies, enabling a comparative analysis to identify best practices and lessons learned.

3.3 Ethical Considerations

The study adhered to ethical standards by ensuring transparency in source attribution and avoiding any form of bias. However, some limitations were identified:

- **Data Availability:** Variations in the depth and quality of available data across the case studies.
- **Focus on Advanced Cities:** The findings may not fully apply to cities with differing socio-economic contexts.

The methodology integrates a robust research design with detailed data refinement and analysis. By focusing on three exemplary cities and applying a structured analytical framework, this study aims to uncover actionable insights into the role of IoT in sustainable urban management.

4 Results and Discussions

The findings from this study emphasize the significant efficiency gains achieved through IoT adoption in urban management while also identifying persistent challenges that limit its full implementation. The results are presented through a combined analysis of efficiency gains, challenges encountered, and visual insights provided in Table 1, Table 2, Fig. 1, and Fig. 2.

IoT technologies have substantially improved energy, water, and waste management systems across Copenhagen, Tokyo, and Barcelona. Table 1 summarizes these efficiency gains, highlighting reductions in energy consumption, water losses, and waste inefficiencies. In energy management, Copenhagen achieved a 10% reduction in energy consumption by deploying IoT-enabled smart grids that dynamically optimize energy distribution and integrate renewable energy sources. Tokyo and Barcelona also recorded improvements of 7% and 9%, respectively, demonstrating the versatility of IoT in improving energy efficiency across different urban contexts.

Table 1 Efficiency gains across IoT applications

City	Energy efficiency gains (%)	Water savings (%)	Waste reduction (%)
Copenhagen	10	5	8
Tokyo	7	15	6
Barcelona	9	4	12

Table 2 Challenges in IoT adoption

Challenge	Description	Cities impacted
High costs	The financial burden of deploying and maintaining IoT infrastructure.	Copenhagen, Tokyo, Barcelona
Privacy concerns	Issues related to data collection, surveillance, and potential misuse of sensitive information.	Copenhagen, Tokyo, Barcelona
Interoperability issues	Lack of standardization and compatibility between IoT platforms, hindering integration.	Copenhagen, Barcelona
Digital divide	Inequitable access to IoT technologies, particularly affecting marginalized communities.	Barcelona

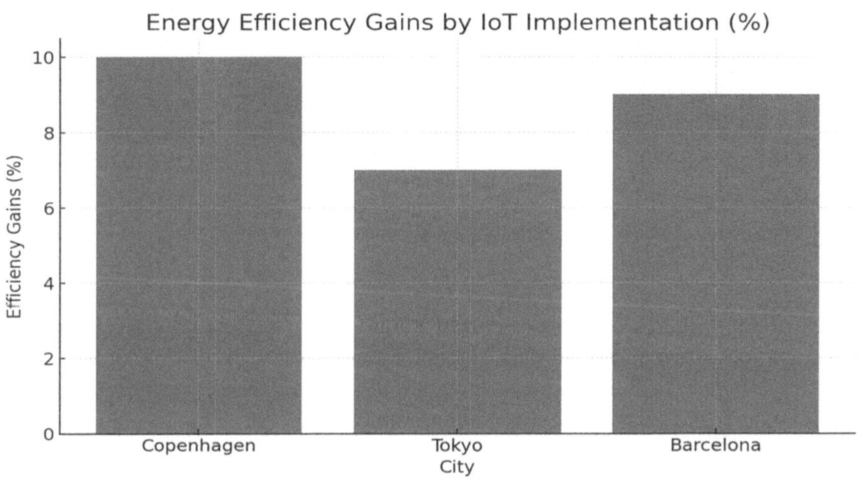

Fig. 1 Illustration of the percentage of energy efficiency gains achieved through IoT implementation in Copenhagen, Tokyo, and Barcelona

In water management, Tokyo achieved a 15% reduction in water loss through IoT systems that enable real-time leak detection and predictive maintenance. Copenhagen and Barcelona reported smaller but notable gains of 5% and 4%, respectively, underscoring IoT's potential to conserve water resources and enhance distribution networks. Similarly, waste management systems in Barcelona showcased a 12% reduction in inefficiencies through the deployment of smart bins and optimized collection routes. These improvements significantly reduced greenhouse gas emissions and operational costs, while Copenhagen and Tokyo achieved reductions of 8% and 6%, respectively, in waste inefficiencies. Despite these advancements, the study also revealed critical challenges, summarized in Table 2, which categorizes barriers into four main dimensions: economic, privacy, interoperability, and digital divide. High costs were consistently identified as a primary obstacle, affecting all three cities. The financial burden of deploying and maintaining IoT infrastructure often limits adoption, particularly in cities with constrained resources. Privacy concerns were equally pervasive, with the vast amount of data collected by IoT systems raising

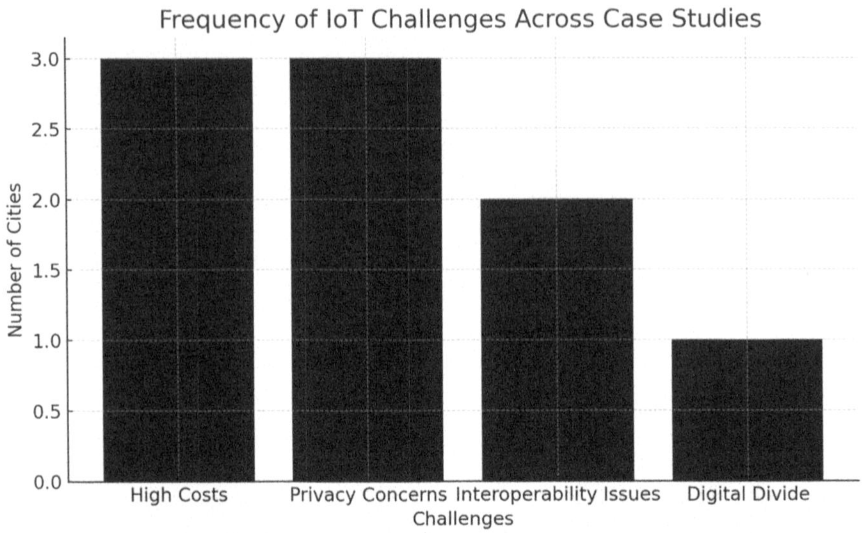

Fig. 2 Illustration of the prevalence of challenges encountered during IoT implementation across Copenhagen, Tokyo, and Barcelona

issues of surveillance, data misuse, and public trust. Interoperability challenges were reported in Copenhagen and Barcelona due to the lack of standardized IoT protocols, which hinder seamless integration and scalability of these systems. Additionally, the digital divide was most pronounced in Barcelona, where marginalized communities lacked access to IoT technologies, exacerbating existing inequalities and limiting the inclusivity of smart city initiatives.

These findings are further contextualized through Fig. 1 and Fig. 2, which visually represent the efficiency gains and the prevalence of challenges across the three cities. Figure 1 highlights the comparative efficiency improvements, with Copenhagen leading in energy efficiency (10%), Tokyo excelling in water conservation (15%), and Barcelona showcasing significant gains in waste management (12%). These variations reflect the differing priorities, strategies, and governance models of each city. On the other hand, Fig. 2 illustrates the frequency and severity of challenges encountered during IoT implementation. High costs and privacy concerns were the most universally reported barriers, while interoperability issues were more context-specific, particularly affecting Copenhagen and Barcelona. The digital divide in Barcelona further emphasizes the need for inclusive policies to ensure equitable access to IoT benefits.

The results underscore the transformative potential of IoT technologies in urban management. Copenhagen's advancements in energy management demonstrate how IoT-enabled systems can dynamically optimize resource utilization and reduce emissions. Tokyo's achievements in water management highlight the value of IoT in conserving vital resources and ensuring infrastructure reliability. Barcelona's success in waste management showcases the role of IoT in enhancing operational efficiency, promoting recycling, and reducing environmental impact. However, the challenges identified—such as high costs, privacy concerns, interoperability issues, and the digital divide—highlight the complexities of scaling IoT solutions in diverse urban contexts.

To overcome these barriers, strategic interventions are required. Governments must prioritize creating supportive regulatory environments, fostering public-private partnerships, and investing in IoT solutions that address both technical and social challenges. Public-private collaboration can help mitigate financial constraints by leveraging shared investments, while inclusive policies and global standardization efforts can address interoperability and accessibility issues. Enhancing data governance frameworks will also be critical in addressing privacy concerns and building public trust.

In conclusion, IoT technologies hold immense potential to drive sustainable urban development, optimize resource use, and create resilient and inclusive cities. By addressing the challenges identified in this study, stakeholders can align IoT innovation with broader societal goals, ensuring its transformative impact is realized to its fullest extent.

4.1 *Efficiency Gains*

The implementation of IoT technologies has resulted in measurable improvements across energy, water, and waste systems in Copenhagen, Tokyo, and Barcelona:

- **Energy**: Copenhagen achieved a 10% reduction in energy consumption through smart grids that optimize distribution and integrate renewable sources. Tokyo and Barcelona also recorded improvements of 7% and 9%, respectively, in energy efficiency.
- **Water**: Tokyo demonstrated a 15% reduction in water loss using IoT-driven leak detection and predictive maintenance. Copenhagen and Barcelona reported smaller but notable gains of 5% and 4%.
- **Waste**: Barcelona leads with a 12% reduction in waste inefficiencies, leveraging IoT-enabled smart bins to optimize routes and improve recycling. Copenhagen and Tokyo achieved reductions of 8% and 6%, respectively.

This table summarizes the efficiency gains achieved through IoT implementation in energy, water, and waste management systems across Copenhagen, Tokyo, and Barcelona. Copenhagen leads in energy efficiency (10%), Tokyo excels in water savings (15%), and Barcelona demonstrates significant waste reduction (12%).

4.2 Challenges in IoT Adoption

The study identified recurring challenges that hinder the effective implementation of IoT technologies:

- **High Costs**: All three cities face significant financial barriers due to the high cost of IoT deployment and maintenance.
- **Privacy Concerns**: Concerns about data collection, surveillance, and misuse are prevalent across all cities.
- **Interoperability Issues**: Copenhagen and Barcelona reported difficulties in integrating diverse IoT systems due to a lack of standardization.
- **Digital Divide**: In Barcelona, inequitable access to IoT technologies limits the inclusivity of its benefits.

This table outlines the key challenges faced by Copenhagen, Tokyo, and Barcelona in implementing IoT technologies for urban resource management. High costs and privacy concerns are universal challenges, while interoperability issues primarily affect Copenhagen and Barcelona. The digital divide is highlighted as a significant concern in Barcelona, reflecting unequal access to IoT technologies.

4.3 Visualizing Key Insights

The results are further illustrated through the following figures:

- **Energy Efficiency Gains**: Fig. 1 highlights the variation in energy efficiency improvements among the three cities, showcasing Copenhagen's leadership in this domain.
- **Challenges Frequency**: Fig. 2 depicts the frequency of challenges faced during IoT implementation, with high costs and privacy concerns being the most pervasive.

The following figure illustrates the percentage of energy efficiency gains achieved through IoT implementation in Copenhagen, Tokyo, and Barcelona. Copenhagen leads with a 10% reduction in energy consumption, followed by Tokyo (7%) and Barcelona (9%), showcasing the transformative potential of IoT in optimizing energy systems.

The following figure highlights the prevalence of challenges encountered during IoT implementation across Copenhagen, Tokyo, and Barcelona. High costs and privacy concerns are the most frequently reported issues, affecting all three cities, followed by interoperability issues (reported by two cities) and the digital divide (reported by one city).

The findings emphasize the transformative potential of IoT technologies in enhancing urban resource management while acknowledging the challenges associated with their adoption. IoT has demonstrated its ability to drive efficiency in energy, water, and waste management systems, offering actionable insights for sustainable urban development.

The efficiency gains achieved highlight IoT's critical role in addressing urban challenges. Copenhagen's 10% reduction in energy consumption showcases the power of IoT-enabled smart grids in optimizing distribution and integrating renewable energy sources. Tokyo's 15% reduction in water loss underscores the value of IoT technologies in improving infrastructure through leak detection and predictive maintenance. Similarly, Barcelona's 12% improvement in waste reduction highlights the ability of IoT-enabled systems to streamline collection processes, optimize routes, and promote recycling. These results collectively validate IoT as a vital tool for achieving sustainable urban development goals.

Despite these successes, the study identifies persistent barriers that hinder widespread IoT adoption. High costs remain a universal concern, requiring innovative financing mechanisms, such as public-private partnerships, to mitigate financial burdens. Privacy concerns are prevalent due to the extensive data collection required by IoT systems. These issues call for robust regulatory frameworks and transparent governance practices to ensure ethical data use. Interoperability issues, particularly in Copenhagen and Barcelona, reflect the lack of standardization among IoT platforms, complicating integration and scalability. The digital divide, exemplified by challenges in Barcelona, highlights the inequitable access to IoT technologies, necessitating inclusive policies to bridge gaps and extend benefits to marginalized communities.

The implications of these findings are significant. Governments must prioritize creating supportive regulatory environments that promote IoT adoption while safeguarding privacy and equity. Urban planners should integrate IoT into long-term strategies, leveraging its capabilities to address resource challenges and align with global sustainability goals. The private sector has a role to play in developing cost-effective, interoperable, and user-friendly IoT solutions, reducing barriers to adoption and ensuring broader accessibility.

To overcome these challenges, several recommendations emerge. Strengthening public-private partnerships can lower costs, facilitate technology transfer, and accelerate IoT deployment. Inclusive policies are critical to bridging the digital divide and ensuring equitable access to IoT benefits. Promoting global standardization efforts will enhance compatibility across IoT platforms, reducing fragmentation and fostering scalability. Enhancing data governance frameworks is equally essential to build public trust and address privacy concerns.

The findings underscore both the opportunities and complexities of IoT adoption in urban resource management. By addressing the identified barriers and leveraging the demonstrated efficiency gains, cities can harness IoT technologies to drive sustainable development, optimize resource use, and create more resilient and inclusive urban environments. These insights provide a clear pathway for stakeholders to align IoT innovation with broader societal goals, ensuring its transformative potential is realized to its fullest extent.

5 Conclusion

The integration of Internet of Things (IoT) technologies into urban management represents a pivotal shift in the way cities address resource optimization, operational efficiency, and sustainability. This study explored the transformative potential of IoT through case studies in Copenhagen, Tokyo, and Barcelona, each showcasing measurable efficiency gains in energy, water, and waste management. Copenhagen's smart grids reduced energy consumption by 10%, Tokyo's IoT-enabled water systems minimized water losses by 15%, and Barcelona's smart waste management solutions achieved a 12% reduction in inefficiencies. These achievements underscore IoT's ability to deliver tangible benefits across diverse urban contexts.

However, the study also identified critical challenges that constrain IoT's scalability and inclusivity. High implementation costs remain a universal barrier, often deterring cities with limited resources from adopting these technologies. Privacy concerns, stemming from extensive data collection, raise ethical questions about surveillance, data misuse, and public trust. Interoperability issues, due to the lack of standardized protocols, hinder seamless integration of IoT platforms, while the digital divide exacerbates inequities by limiting access to smart technologies for marginalized populations.

To unlock the full potential of IoT, a holistic and inclusive approach is necessary. Policymakers must prioritize creating supportive regulatory environments that address privacy concerns and promote global standardization efforts. Public-private partnerships can alleviate financial burdens and accelerate the deployment of cost-effective IoT solutions. Moreover, inclusive policies are essential to bridge the digital divide, ensuring that all communities, including underserved populations, benefit from IoT innovations.

The findings from this study highlight the need for tailored strategies that balance technological advancements with social equity and environmental sustainability. Cities must adopt long-term visions that integrate IoT into broader urban development plans, leveraging these technologies to align with global sustainability goals. Collaboration among governments, industry stakeholders, and communities will be key to overcoming the technical, economic, and social challenges identified in this research.

In conclusion, IoT technologies hold immense potential to redefine urban management by enabling smarter, more efficient, and sustainable cities. By addressing the barriers to implementation, cities can harness IoT to create resilient and inclusive urban environments. This study offers a roadmap for policymakers and urban planners to leverage IoT's transformative power while ensuring equitable access to its benefits, paving the way for a sustainable and innovative future for urban systems.

Competing Interests The authors declare that they have no competing interests in relation to this work. All efforts have been made to ensure that the analysis, findings, and recommendations presented in this chapter are impartial and based solely on academic research and objective analysis.

Ethics Approval This study did not involve any experiments with human participants or animals. All data used in the study were derived from publicly available sources, and no personal or sensitive information was collected or used. As such, formal ethics approval was not required for this research. The authors adhered to ethical guidelines in conducting and reporting the study.

References

1. Angelidou M (2015) Smart cities: a conjuncture of four forces. Cities 47:95–106. https://doi.org/10.1016/j.cities.2015.02.001
2. Atzori L, Iera A, Morabito G (2017) Understanding the internet of things: definition, potentials, and societal role of a fast-evolving paradigm. Ad Hoc Netw 56:122–140
3. Barns S (2020) Platform urbanism: negotiating platform ecosystems in connected cities. Urban Geogr 41(3):453–472. https://doi.org/10.1080/02723638.2020.1730632
4. Batty M, Axhausen KW, Giannotti F et al (2012) Smart cities of the future. Eur Phys J Spec Top 214(1):481–518. https://doi.org/10.1140/epjst/e2012-01703-3
5. Chen X, Huang Z, Li Y (2021) IoT innovations in urban sustainability: the Singapore perspective. Smart Cities 3(1):55–72. https://doi.org/10.3390/smartcities3010008
6. Ajuntament de Barcelona. (n.d.) Smart waste management in Barcelona. Retrieved from https://ajuntament.barcelona.cat
7. Deursen V, Alexander JAM, Van Dijk JAGM (2019) The first-level digital divide shifts to differences in usage: a five-year panel study. New Media Soc 21(2):354–375. https://doi.org/10.1177/1461444818797082
8. Goodman B, Powles H (2019) Transparency dilemmas in smart cities. Big Data Soc 6(1):1–12. https://doi.org/10.1177/2053951719858748
9. Greenfield A (2013) Against the smart city. Do Projects
10. Hashem IA, Targio VC, Anuar NB et al (2016) The role of big data in smart city development. Inf Syst 59:69–83. https://doi.org/10.1016/j.is.2016.01.001
11. Hollands RG (2008) Will the real smart city please stand up? Intelligent, progressive or entrepreneurial? City 12(3):303–320. https://doi.org/10.1080/13604810802479126
12. Kitchin R (2016) The ethics of smart cities and urban science. Philos Trans R Soc A Math Phys Eng Sci 374(2083):20160115. https://doi.org/10.1098/rsta.2016.0115
13. Leszczynski A (2016) Speculative futures: data, algorithms, and urban space. Environ Plann D Soc Space 34(5):816–835. https://doi.org/10.1177/0263775815622978
14. Mora L, Deakin M, Reid A (2019) Smart-city development paths: insights from the first two decades of research. J Urban Technol 26(2):1–23. https://doi.org/10.1080/10630732.2018.1552791
15. Smith A, Stirling A (2018) Innovation, sustainability, and democracy: an analysis of grassroots contributions. J Environ Innov Soc Trans 27:54–67. https://doi.org/10.1016/j.eist.2017.10.004
16. Wu C, Shen Q, Cheng H (2022) Green urban planning and the role of IoT technologies. Sustain Cities Soc 78:103545. https://doi.org/10.1016/j.scs.2022.103545
17. Zhao X, Zuo J, Wu G (2018) The intersection of smart cities and environmental management. Energy Policy 39(5):1210–1220. https://doi.org/10.1016/j.enpol.2018.02.001
18. Zuboff S (2019) The age of surveillance capitalism: the fight for a human future at the new frontier of power. PublicAffairs, New York
19. Boumhamdi, M., Atmani, A., & Ikira, M. (2025) Smart Cities and Urban Green Initiatives: Integrating Technology and Sustainability, https://link.springer.com/chapter/10.1007/978-3-031-88653-9_59, https://doi.org/10.1007/978-3-031-88653-9_59
20. United Nations (2019) World urbanization prospects: the 2018 revision. Department of Economic and Social Affairs, Population Division, United Nations, New York.

21. Tokyo Metropolitan Government n.d. Smart water management initiatives in Tokyo: leakage prevention and smart metering projects. Official information published by the Tokyo Metropolitan Government, Bureau of Waterworks. Available at: https://www.waterworks.metro.tokyo.lg.jp/eng/efforts/smart_water.html
22. City of Copenhagen (n.d.) Smart city and energy management initiatives in Copenhagen. Technical and Environmental Administration, City of Copenhagen. Available at: https://international.kk.dk

Addressing Hidden Urbanization: Challenges and Strategic Approaches for Sustainable Urban Growth in Sri Lanka

Kuda Singankutti Kankanamalage Nathasha Janani Kudasinghe, Jagath Munasinghe, Weerakoon Dasanayakalage Bimsara Jayanath Jeewananda, and Attygalage Ranjith Rupasinghe

Abstract Sri Lanka's urban classification system, based on local government types, categorizes Municipal Councils (MCs) and Urban Councils (UCs) as urban areas, while Pradeshiya Sabhas are considered rural. However, this classification fails to capture the dynamic nature of urban growth, resulting in challenges for policy-making, urban investment, and long-term planning. The phenomenon of "hidden urbanization," marked by urban sprawl and ribbon development, continues despite a lag in urban population growth. This creates difficulties for planners addressing the expanding informal settlements, such as slums, which increase social vulnerability and inequality. This study explores Sri Lanka's urbanization process through a literature review and user perception survey, highlighting the impact of unplanned development. It emphasizes the need for integrating Geographic Information Systems (GIS) and Remote Sensing (RS) technologies to monitor growth patterns in underserved areas. These tools, though underutilized, can provide real-time data, supporting better urban management. The paper advocates for their use in improving land-use planning, monitoring informal settlements, and creating smarter, more connected towns. The study also emphasizes the importance of decentralizing urban governance and empowering local governments to address region-specific challenges. Local authorities, closer to their communities, are better equipped to implement tailored solutions and strategies that meet local needs. In the context of smart cities, decentralization is essential for building inclusive, resilient urban environments that serve diverse residents, especially those in informal settle-

Kuda Singankutti Kankanamalage Nathasha Janani Kudasinghe (✉) · W. D. B. J. Jeewananda · A. R. Rupasinghe
General Sir John Kotelawala Defence University, Colombo, Sri Lanka
e-mail: ksknjkudasinghe@kdu.ac.lk

J. Munasinghe
University of Moratuwa, Moratuwa, Sri Lanka

H. Kaddouri et al. (eds.), *Smart Cities for Sustainable and Urban Development*, Green Energy and Technology, https://doi.org/10.1007/978-3-031-99001-4_42

ments. In conclusion, an integrated approach combining technology, decentralized governance, and community-focused planning is key to managing Sri Lanka's urbanization challenges. Such an approach will contribute to the development of sustainable, equitable cities capable of supporting future growth and improving the quality of life for all citizens.

Keywords Hidden urbanization · Urban sprawl · Slums and shanties · GIS · Local governments

1 Introduction

Urban development plays a crucial role in driving Sri Lanka's economic progress. However, the accelerated pace of urbanization has introduced several challenges for urban planners and policymakers. While urbanization in Sri Lanka is a significant phenomenon, it is often characterized by "hidden urbanization," which complicates efforts to plan and manage development effectively. The World Bank [57] highlights that the country's urban expansion has surpassed the growth rate of its urban population. This imbalance is evident in the prevalence of urban sprawl and ribbon development, which dominate Sri Lanka's urban landscape. These unregulated and poorly planned expansion patterns hinder efforts to create sustainable, well-connected urban areas [36, 48].

A critical consequence of this uncontrolled urban expansion is the proliferation of underserved settlements, commonly referred to as "slums" or "shanties". These informal settlements lack essential infrastructure and services, emerging as a direct outcome of uncoordinated urbanization. Such developments intensify social inequality and marginalization, further dividing urban spaces [6, 13].

The traditional approach to classifying urban areas in Sri Lanka, which relies on local government categories, compounds the issue. Under this system, Municipal Councils (MCs) and Urban Councils (UCs) are designated as urban, while regions governed by Pradeshiya Sabhas are considered rural. However, this rigid administrative classification fails to reflect the dynamic and multifaceted nature of urbanization. Consequently, it leads to inefficiencies in policy and planning [52]. This mismatch between the expansion of urban areas and population growth underscores the need for more nuanced and comprehensive urban classification systems to inform future planning strategies.

Addressing the challenges posed by hidden urbanization within the framework of smart cities necessitates innovative solutions. Geographic Information Systems (GIS) and Remote Sensing (RS) technologies emerge as critical tools for understanding and managing urban growth. These technologies enable urban planners to analyze land-use changes and monitor the proliferation of informal settlements in real time [31]. By leveraging GIS and RS, policymakers can develop a clearer understanding of urban growth patterns and implement data-driven strategies to address them effectively.

Additionally, decentralizing decision-making processes in urban governance is vital for fostering inclusive and adaptive planning. Local authorities, being more attuned to the needs of their communities, are well-positioned to address the specific challenges posed by informal settlements. Empowering these authorities with greater autonomy and resources would pave the way for more localized and effective solutions, including the integration of smart city initiatives [25].

This research aims to explore the complexities of hidden urbanization in Sri Lanka by examining its implications, particularly in terms of urban sprawl, informal settlements, and the limitations of existing classification systems. By integrating findings from literature reviews and user perception studies, this study provides actionable insights into strategies for sustainable urban growth in Sri Lanka.

The paper begins by exploring the extent of hidden urbanization in Sri Lanka, starting with its definitions and core characteristics. The second section delves into the drivers and patterns of hidden urbanization, focusing on urban sprawl and the emergence of informal settlements. The third section analyzes the shortcomings of Sri Lanka's current urban classification system and its impact on urban planning. The fourth section highlights the potential of smart city frameworks, particularly the use of GIS and RS technologies, in addressing these issues. The final section proposes recommendations for decentralized governance and sustainable urban planning. Together, these sections aim to provide a holistic understanding of hidden urbanization and outline practical approaches to enhance urban management in Sri Lanka.

2 Methodology

This study aims to explore the phenomenon of "hidden urbanization" in Sri Lanka, examining its impact on urban growth patterns, informal settlements, and the implementation of smart city solutions. The research methodology is structured around a combination of qualitative and quantitative methods to provide a comprehensive understanding of the urbanization process. The approach includes a literature review, a user perception survey.

A thorough literature review was conducted to explore the existing body of knowledge on urbanization, urban sprawl, informal settlements, and smart city development in Sri Lanka. The review focused on academic articles, government reports, urban planning documents, and case studies related to urban growth patterns in Sri Lanka and other developing countries with similar challenges. This review provides a theoretical foundation for understanding the dynamics of hidden urbanization, the role of informal settlements in urban development, and the potential of GIS and RS technologies for urban planning.

To complement the theoretical insights gained from the literature review, a user perception study was conducted to understand the experiences and perspectives of urban residents in both formal and informal settlements. A policy analysis was conducted to evaluate the current urban development policies in Sri Lanka, particularly those addressing informal settlements and urban sprawl. Key documents from urban planning authorities.

3 Results and Discussion

3.1 Results from Literature Review

3.1.1 Urbanization on a Global Context

Urbanization has long been understood as the migration of people from rural areas to urban centers. In contemporary discussions, it is often framed through the rise of megacities—urban regions with populations surpassing ten million. Over the past decade, the number of megacities has grown from 23 to 31, and projections indicate that this figure could reach 41 by 2030 [47, 50]. Alongside the growth of megacities, many smaller urban centers are emerging to accommodate expanding populations. However, focusing only on these large urban hubs and relying on socio-demographic data may not fully reflect the complexity of urbanization processes.

Typically, population data used to assess urbanization comes from periodic surveys conducted every 2 to 10 years, depending on each country's census schedule [9, 15]. Most nations rely on traditional methods that focus on residential population distribution as the primary factor for defining urban areas and measuring urbanization patterns [18]. These approaches, however, come with notable limitations. They are time-consuming, costly, and labor-intensive, which means updates are infrequent. Additionally, urban dynamics are shaped by more than just residential populations—commuter flows also play a significant role. Defining urban areas is further complicated by the absence of a universally accepted definition and standardized methodologies. Moreover, conventional statistics may fail to capture "hidden urbanization," a phenomenon driven by growing mobility, technological advancements, and shifting aspirations, which reshape lifestyles regardless of where people live [8]. Thus, urbanization should be viewed not just as a population shift, but as a broader transformation in how people live and interact with their environment [56].

This evolving pattern of urbanization will significantly influence the spatial structure of cities in the future [28]. As urban areas expand, local governments are faced with increasingly complex challenges in managing growth and maintaining quality of life for residents [32]. Meanwhile, national and provincial governments must work to ensure the equitable distribution of essential resources, which requires a comprehensive understanding of urban areas, particularly in the context of sustainable development [23, 59].

As highlighted earlier, relying solely on surveys and census data to define urbanization is insufficient. These datasets provide only snapshots of data at specific intervals, which can miss important trends [10, 11, 20, 60]. Accurately identifying urban areas is essential for effective governance, planning, and resource allocation. This identification forms the basis for spatial policies at both national and regional levels, which guide the future development of urban areas [1]. Furthermore, understanding the complex and often hidden dimensions of urbanization in low- and middle-income countries is critical for devising strategies that ensure sustainable urban growth in these regions [19].

The concept of "urban" lacks a universally accepted definition, resulting in different criteria used across countries to define "urban areas" [47]. Without a global standard, each country has developed its own criteria to classify urban areas and determine the proportion of the population living in urban spaces. These criteria generally fall into four broad categories:

- Population-based
- Administrative boundary-based
- Land use-based
- Multi-criteria based

Among these, demographic characteristics and land use are the most commonly employed to define urban areas [30]. However, urbanization is much more intricate than what these indicators alone can convey [27]. It encompasses a range of economic, social, and technological factors that collectively push a larger share of the population to adopt urban lifestyles [53]. As Wirth [56] aptly noted, "urbanism is a way of life," highlighting that urbanization is not merely about the physical relocation of people from rural to urban spaces, but also about profound changes in social norms, lifestyles, and societal organization as urban living becomes the norm for more individuals.

3.1.2 Urbanization in Sri Lanka

In Sri Lanka, urban areas are defined based on local government institutions, with municipal councils (MCs) and urban councils (UCs) classified as urban, while other areas are categorized as "rural" [53]. In contrast, countries like India use stricter criteria, such as a population exceeding 5000, a population density of over 400 people per square kilometer, and 75% of the male workforce engaged in non-agricultural activities [18]. Bangladesh, on the other hand, adopts a broader definition that considers areas urban if they are densely populated, well-developed around a central hub, and dominated by non-agricultural employment [30].

However, these definitions often rely heavily on statistical measures like population density and economic activity, which can be too simplistic to capture the complex, evolving nature of urban areas and their functions. They also tend to depend on administrative boundaries, which don't always reflect how people live and interact within urban spaces. This calls for more nuanced methods that consider lifestyles, social behaviors, and spatial practices driving urban dynamics, rather than relying solely on static elements like population counts and administrative boundaries.

Over the past decades, Sri Lanka's urbanization pattern has displayed an unexpected trend. Official data show a decline in the urban population from 21.5% in 1981 to 14.6% by 2001, with a slight increase to 18.2% by 2012. This sharp decrease, which seemed inconsistent with observable urban growth on the ground, was largely due to changes in how the Department of Census and Statistics (DCS) defined "urban areas." Before 1987, towns and city councils were classified as urban areas, but in that year, town councils were abolished and merged with village councils, making them "rural" areas [19].

The local government-based classification approach has proven misleading in Sri Lanka, leading to inaccuracies in policy decisions, urban investment priorities, and development planning. To better understand and address these challenges, more sophisticated and descriptive methods are needed to accurately capture urbanization patterns and the true urban footprint.

Recent studies have found minimal urban sprawl in satellite towns specifically designed to manage urban growth, while urban sprawl is more pronounced in monocentric cities where development is concentrated around a central point [45]. This highlights the importance of assessing the unique needs of each town and creating policies tailored to local conditions. By doing so, planners can manage growth effectively and promote long-term urban viability.

Sri Lanka is currently experiencing an urban paradox, as evidenced by the 2012 census. According to the data, only 18.2% of the population lived in 64 Municipal Council areas, including nine provincial capitals. However, international estimates suggest the urban population could be between 35% and 45% of the total population [46]. Government documents estimate an even higher urban population of 50%. This discrepancy is further supported by analyzes of nighttime lights and other data sources.

Colombo, the capital and economic hub of the country, illustrates this urban paradox starkly. The official census recorded a population of 561,314 for Colombo in 2012. However, when considering the entire Western Region Megapolis (WRM), including all of the Western Province, the population reached 5.8 million—ten times the official figure and larger than the total urban population of the entire country. This emphasizes the rapid spatial growth of Colombo and the WRM over the past decade.

Effective urban planning in Sri Lanka requires evidence-based policies that can respond to the challenges and opportunities of urbanization. The large gap between official urban population data and other estimates reveals a need for clearer definitions and consistent methodologies for defining urban areas. Ambiguity in this area can create significant obstacles to understanding urban systems, which in turn impacts effective policy-making.

The Colombo Metropolitan Region Special Plan, introduced in 1998, and the more recent Western Region Megapolis Plan of 2016, developed by the Ministry of Megapolis and Western Development, have been key initiatives in managing urban expansion within this region. These plans aim to coordinate development efforts across municipalities in the Western Province, including Colombo. The WRM has become a powerful economic engine, contributing around 40% of the national GDP and hosting a range of high-growth industries.

Globally, two primary drivers of urban expansion are population growth and economic development [7]. However, in Sri Lanka, population growth is not as significant a factor driving urbanization. Instead, rural-to-urban migration is a major contributor to population increases in provincial capitals. In many cities worldwide, urban expansion is closely linked to economic development rather than population growth [7]. This trend is evident in Sri Lanka, where large-scale redevelopment projects have positioned Colombo as a regional trade and investment hub.

Traditional global measurements of urban growth, based purely on population figures, have limitations, particularly in areas that are administratively classified as rural but exhibit urban characteristics [46]. To better understand urban dynamics, spatial analysis tools are used to map and analyze urban areas over time. These methods offer a more accurate picture of urban expansion and change [2, 3].

Preliminary analyzes have shown that urban and semi-urban areas extend beyond official Municipal Council boundaries in all nine provincial capitals of Sri Lanka. To better understand urban expansion, planners defined a "fringe area"—an area with continuous urban characteristics—that extends beyond these official boundaries. Spatial analysis revealed that built-up areas in and around these provincial capitals have been growing at an average annual rate of 6.42%, driven by both economic development and population growth. This rate of urbanization surpasses that observed in European, North American, African, and Indian cities, though it is slightly lower than that observed in Chinese cities [40].

3.1.3 Urban Sprawl and Informal Settlements: The Consequences of Unplanned Urbanization in Sri Lanka

Urban sprawl describes a pattern of urban expansion that is marked by low-density, dispersed growth, often jumping over developed areas and leading to disconnected and isolated land uses. This leapfrogging pattern often results in fragmented urban spaces that are spread out, creating irregular shapes and a lack of cohesive land use [21, 42]. Sprawl is commonly characterized by single-use zoning [22] (Wei and Ewing 2018), low-density development [3, 45], and outward growth that stretches away from city centers [3, 51]. It often leads to the transformation of rural areas into urban ones, reducing available agricultural and natural land [24].

Unlike more compact forms of urban development, sprawl is typically a result of unregulated and rapid urban growth. This uncontrolled expansion can create significant challenges for cities, especially in terms of sustainability. The environmental, social, and economic impacts of sprawl make it critical for urban areas to address the factors driving this growth [33]. Effective management is needed to ensure that cities develop in ways that balance population growth with environmental preservation and social equity. The causes of sprawl vary between developed and developing countries, meaning that tailored solutions are required to address specific urban issues. Urban planners must fully grasp the extent and the drivers behind urban sprawl to implement effective urban policies [5]. This unplanned expansion often leads to the development of informal settlements, leaving many communities living in vulnerable conditions.

In spatial planning, one of the primary challenges is mapping and quantifying urban sprawl. This is complicated by the varied characteristics of sprawl and the lack of universally accepted definitions [55]. As Bhatta et al. [14] point out, sprawl is a multi-dimensional issue that needs a combination of scales and parameters for accurate analysis. The process of measuring sprawl requires multiple approaches, both relative and absolute. For example, Galster et al. [22] propose eight dimensions to evaluate sprawl: density, continuity, concentration, clustering, centrality, nuclearity, mixed uses, and proximity. These dimensions reveal the complex nature of sprawl, where factors like population growth [43], built-up density, and urban infrastructure play significant roles in shaping urban landscapes [14, 21, 44].

The challenge lies in accurately measuring these diverse factors. Relying on just one or two indicators could lead to incomplete conclusions [14]. To get a clearer understanding of sprawl, urban planners need to examine how these dimensions interact and influence each other. By recognizing these nuances, planners can develop more effective strategies for managing urban growth.

Technological advancements, particularly Geographic Information Systems (GIS) and Remote Sensing (RS), have become indispensable in contemporary urban planning. These tools allow for real-time monitoring and analysis of urban expansion, enabling planners to assess growth patterns and land use changes with high precision [29]. GIS and RS have revolutionized the way cities are planned, as they integrate spatial and temporal data that provide deep insights into urban dynamics [37]. Various methods, such as Shannon's entropy and the Urban Expansion Intensity Index, offer quantitative ways to measure urban sprawl and track its evolution over time [2, 38]. These tools enable urban planners to make more informed decisions based on data-driven insights.

The concept of urban form is essential in understanding the impact of sprawl. Urban form refers to the arrangement and physical structure of urban spaces, which in turn affects land use patterns. There are different types of urban forms, including mono-centric, poly-centric, linear, and satellite towns [4]. Understanding how these forms shape the urbn environment can help planners determine which patterns are best suited for sustainable development.

For example, a mono-centric urban form is one where development clusters around a central point, often creating a ring-like structureof activity. In contrast, poly-centric urban areas have multiple centers of development, creating clusters of activity spread out over a larger area [58]. A linear urban form arranges activities along a single axis [54], while a satellite town involves a main urban center supported by smaller, secondary centers that are more self-sufficient.

Remote sensing imagery is invaluable for monitoring urban growth and land use changes. These images provide up-to-date views of land cover, helping urban planners track changes in the built-up environment. GIS and remote sensing technologies are critical in decision-making processes, as they provide the necessary data to assess and guide urban development [26].

In the context of modern urban planning, GIS and remote sensing are not just tools for measurement—they also enable urban planners to analyze and predict the impact of urban sprawl. With the help of statistical methods such as Shannon's entropy and the Urban Expansion Intensity Index, planners can gain insights into how fast a city is expanding and how this growth is affecting urban structure. These analyzes are essential for managing the long-term consequences of sprawl, ensuring that cities grow in ways that are both sustainable and equitable.

Landscape metrics also play a crucial role in assessing urban sprawl, providing a way to evaluate the spatial distribution of urban areas. These metrics help to measure the degree of fragmentation, aggregation, and connectivity within urban landscapes, which are all essential for creating cohesive, sustainable urban environments.

3.1.4 Slums and Shanty Settlements: Understanding Informal Urban Living

The term "slum" is typically used to describe areas where people live in substandard conditions, often associated with poverty and neglect. According to Cities Alliance [17], slums are "neglected parts of cities where housing and living conditions are appallingly poor." The terms "slum" and "informal settlements" are often used interchangeably in the literature, though this may lead to confusion as they both represent underdeveloped and disadvantaged communities that continue to shape the physical and economic landscape of cities [17].

A key challenge in addressing slums lies in defining them clearly. There is no universally agreed-upon definition of a slum, and various perspectives exist depending on the context, location, and issues under consideration. For example, Smith [41] identifies that the Affordable Housing Institute uses as many as 12 different definitions of slums. Definitions can vary depending on factors like land legality, temporary nature, or public health considerations. In some places, like Nairobi (Kenya) and Mexico City (Mexico), slums are not even formally defined, further complicating efforts to classify and address them [49].

Shantytowns, often viewed as a specific form of slum, are generally characterized by high-density, low-income residents living in uninhabitable, informal housing. These areas represent a widespread global phenomenon where the living conditions are far from ideal, and the residents are often left outside the formal planning and development processes.

While slums and shantytowns are often seen as burdens on urban development, their residents contribute significantly to the local and national economy. The informal economy that thrives within slums plays a crucial role in supporting the formal sector. Slum dwellers often provide cheap labor that subsidizes the middle class and business sectors, often accepting jobs that others are unwilling to take due to their low pay and lack of formal education [15, 35]. These informal jobs are closely linked to the needs of surrounding formal communities [34].

In some cases, the informal economy in slums has a measurable impact on the national economy. Dharavi, India, for example, employs up to 70% of its residents, with an estimated economic output of $700 million annually [16]. Similarly, other major Indian cities such as Delhi, Bangalore, and Kolkata rely heavily on informal sectors for manufacturing activities [12]. Some researchers suggest that if the economic contributions of slum dwellers were considered, areas like Dharavi would no longer be considered slums [39].

3.1.5 Government Approaches to Slum and Shanty Development

In Colombo, Sri Lanka's largest urban center, urban renewal initiatives such as the Sustainable Township Development Programme (STP) and the more recent Urban Regeneration Project of Colombo City (URPCC) have been implemented to address issues related to slums and shanty settlements. Despite these efforts, critics argue that such programs are unsustainable and short-term in nature. The government tends to switch between programs without establishing a long-term strategy to provide secure housing for the urban poor.

The STP, for instance, focuses on clearing informal settlements (Under-served Settlements or USS) and relocating residents to high-rise complexes on the urban periphery. However, there are two main perspectives on the effectiveness of this approach. One supports the relevance of the STP as a step toward sustainable urban development in Colombo, while the other argues that such relocations violate the rights of slum dwellers. Critics claim that the program's emphasis on "rationalizing land use" has not been effective in addressing the root causes of unplanned urbanization, nor does it provide viable long-term solutions for slum residents.

The development of slums and shanty towns is often driven by the efforts of the residents themselves, who build their homes with limited resources and amidst harsh conditions. Despite this, these communities are frequently marginalized, labeled as unlawful, and stigmatized. In many cities, informal settlements are socially and spatially excluded, as their residents are often seen as "other" or different from those in formal neighborhoods. This exclusion is a major barrier to addressing the challenges of informal housing and slums, despite decades of research and policy efforts aimed at improving conditions in these areas.

One factor that complicates the problem is the discourse surrounding informality. The words, meanings, and images used to describe informal settlements often reinforce negative stereotypes and misconceptions. These "discourses" may contribute to the ongoing challenges faced by slum dwellers, as they shape the way we view and approach informal urban living.

3.2 Results from the User Perception Study

This study focused on Wanathamulla, one of the largest informal settlements in Sri Lanka, and provided valuable insights into the living conditions of the residents as well as their involvement in the planning process.

3.2.1 Design Planning and Housing Improvement

One of the main findings was the lack of user involvement during the planning phase. To remedy this, it is suggested that a housing advisory service be introduced within the settlement. Such a service would offer residents practical advice on improving their homes and provide guidance on sustainable building techniques.

According to the survey, the most pressing concerns for the community were the poor quality of housing structures and the insufficient number of toilets. Over 60% of households reported spending money on maintaining and improving their homes within the last year. Despite these efforts, many homes still lacked basic planning features, such as organized room layouts and efficient use of space. The structures were built with low-cost local materials, which often failed to provide adequate protection from the elements.

3.2.2 Community Development and Community Participation

Currently, there is no organized community involvement in addressing the settlement's challenges. This lack of engagement is particularly apparent when it comes to housing quality and infrastructure issues. Homes are poorly insulated and vulnerable to temperature extremes, while the lack of clean water and sanitation facilities leaves residents at greater risk of waterborne diseases.

The study suggests that upgrading the settlement to improve housing and infrastructure must go hand in hand with preserving the community's way of life and making the area more disaster-resilient. Establishing community development councils based on the physical and social characteristics of the settlement could be an effective solution. These councils would provide a platform for residents to participate in the planning and decision-making processes. By involving them in discussions about the design of new residential units, choice of building materials, and the overall settlement layout, these councils would ensure that the upgrades meet the specific needs and desires of the residents.

Moreover, the study advocates for using community labor in the construction process. This approach not only addresses the housing shortage but also provides job opportunities and equips residents with skills that can improve their long-term economic prospects. Collaborating with non-governmental organizations (NGOs) could help to implement this strategy effectively.

3.2.3 Government Involvement and Local Decision Making

The study also identified a significant gap in governance, especially regarding the involvement of local authorities. Insights from policy makers within the Urban Development Authority (UDA), including 10 architects and urban planners, revealed that there are no active development plans for monitoring the growth patterns of informal settlements like slums and shanties. Currently, all decisions are made at the central government level, leaving local authorities with little say in the management of these areas. This centralization has created a disconnect between government policy and the communities it aims to serve.

It is clear from the study that local governments need to be granted more authority to manage and improve informal settlements. Local authorities, who are closer to the communities, have a better understanding of the residents' unique needs and challenges. Decentralizing decision-making would allow for more tailored, responsive solutions that align with the needs of the people.

Another key recommendation from the study is the integration of Geographic Information Systems (GIS) and Remote Sensing (RS) technologies into urban planning efforts. The results of the perception study indicate that the UDA currently lacks a comprehensive development plan to monitor slum growth patterns. The study highlights the necessity of creating a regularization plan for these informal settlements, which would mark out plot boundaries and provide structure for future development.

By incorporating GIS and RS, urban planners could track land-use changes in real time, providing them with the data necessary to monitor the growth of informal settlements more effectively. These tools would help policymakers gain a deeper understanding of urban growth dynamics, leading to more informed, data-driven solutions for managing and developing these areas.

4 Conclusion

Sri Lanka is facing a growing challenge in how its cities are developing. While urban areas are expanding, the population growth in these areas is not keeping pace, leading to a phenomenon known as "hidden urbanization." This results in sprawling neighborhoods and informal settlements, such as slums and shanties, which are growing rapidly but without the proper infrastructure or planning to support them. As a result, many residents live in inadequate conditions, with limited access to clean water, sanitation, and basic services, making them vulnerable to health risks and social inequality.

The paper highlights the need for a more inclusive approach to urban development. Residents in these informal settlements must be given a voice in the planning process, as they are the ones who understand the challenges they face every day. Involving them through community councils, for example, would help ensure that any improvements made to their living conditions truly meet their needs. Similarly, empowering local governments, who are closer to these communities, can lead to more responsive and effective solutions, as opposed to centralized decision-making that often overlooks local realities.

The study also reveals the potential of modern technologies, such as Geographic Information Systems (GIS) and Remote Sensing (RS), to help manage urban growth. These tools provide urban planners with valuable real-time data, making it easier to track the development of informal settlements and plan more effectively for the future. By using this data, policymakers can better understand the dynamics of urban growth and create more sustainable and inclusive cities.

Further studies need to be carried on how policy interventions at the national level can be better aligned with local urban dynamics, ensuring that initiatives such as affordable housing, sanitation, and transportation systems are not just implemented but are effective in meeting the actual needs of communities. Moreover, exploring the role of digital technologies and smart city concepts in improving urban governance could offer new avenues for enhancing the responsiveness and effectiveness of local governments in addressing the needs of marginalized urban populations. By fostering an inclusive, data-driven approach to urban development, Sri Lanka can pave the way for more sustainable, equitable, and prosperous cities.

Ultimately, Sri Lanka's urban development challenges require a comprehensive approach that brings together technology, community involvement, and local governance. If done right, this approach can help create cities that are not only better planned but also more resilient, inclusive, and livable for everyone.

References

1. Abeysinghe U, Saparamadu S, Hewawasam C (2023) Effective integration of walkability-enhancing neighborhood environment attributes into urban transport planning in Sri Lanka. University of Moratuwa
2. Aburas MM, Abdullah SH, Ramli MF, Zulfa H, Asha'ari. (2017) Land suitability analysis of urban growth in Seremban Malaysia, using GIS-based analytical hierarchy process. Procedia Eng 198:1128–1136
3. Aguilera PA, Fernández A, Fernández R, Rumí J, Salmerón A (2011) Bayesian networks in environmental modelling. Environ Model Softw 26(12):1376–1388
4. Alpkokin P, Black JA, Hayashi Y (2010) A comparative analysis of metropolitan poly-centric employment growth and its impacts on commuter travel patterns. In: Paper presented at the 12th world conference on transport research, Lisbon, July 11–15
5. Al-sharif AA, Pradhan B (2016) Spatio-temporal prediction of urban expansion using bivariate statistical models: assessment of the efficacy of evidential belief functions and frequency ratio models. Appl Spat Anal Policy 9(2):213–231
6. Amarasinghe U, Marasinghe R (2020) Urban inequality and informal settlements in Sri Lanka: challenges for sustainable development. J Urban Stud 14(3):45–60
7. Angel S, Blei AM, Civco DL, Parent J (2012) Atlas of urban expansion. Lincoln Institute of Land Policy, Cambridge, MA
8. Anwar B, Xiao Z, Akter S, Rehman RU (2017) Sustainable urbanization and development goals strategy through public–private partnerships in a south-Asian metropolis. Sustainability 9(11):1940
9. Baffour B, Brown JJ, Smith PW (2013) An investigation of triple system estimators in censuses. Stat J IAOS
10. Balk D, Leyk S, Montgomery MR, Engin H (2021a) Global harmonization of urbanization measures: proceed with care. Remote Sens 13(24):4973
11. Balk D et al (2021b) Urbanization and the global environment. Glob Environ Chang 68:102258
12. Benjamin S (2004) Urban land transformation for pro-poor economies. Geoforum 35(2):177–187
13. Bexell M, Jönsson K (2017) Responsibility and the United Nations' sustainable development goals. In: Forum for development studies, vol 44(1). Routledge, pp 13–29

14. Bhatta B, Saraswati S, Bandyopadhyay D (2010) Urban sprawl measurement from remote sensing data. Appl Geogr 30(4):731–740
15. Brugmann J (2012) Financing the resilient city. Environ Urban 24(1):215–232
16. Chege PW, Mwisukha A (2013) Benefits of slum tourism in Kibera slum in Nairobi, Kenya. Int J Arts Commerce 2(4):94–102
17. Cities Alliance (2006) Analytical perspective of pro-poor slum upgrading frameworks. UN-HABITAT
18. Coelho K, Sood A (2022) Urban studies in India across the millennial turn: histories and futures. Urban Stud 59(13)
19. Ellis P, Roberts M (2016) Leveraging urbanization in South Asia: managing spatial transformation for prosperity and livability. World Bank Publications, Washington, DC
20. Firman T (2015) Urbanization in Indonesia: a critical assessment of the current situation. Urban Stud 52(1):67–88
21. Frenkel A, Ashkenazi M (2008) Measuring urban sprawl: how can we deal with it? Environ Plann B Plann Des 35(1):56–79
22. Galster G, Hanson R, Ratcliffe MR, Wolman H, Coleman S, Freihage J (2001) Wrestling sprawl to the ground: defining and measuring an elusive concept. Hous Policy Debate 12(4):681–717
23. He W, Li X, Yang J, Ni H, Sang X (2022) How land use functions evolve in the process of rapid urbanization: evidence from Jiangsu Province, China. J Clean Prod 380:134877
24. Heidarinejad N (2017) The effects of urban expansion on spatial and socioeconomic patterns of the Peri-urban areas: a case study of Isfahan City, Iran. TU Dortmund University
25. Jain S, Raj M (2022) Decentralized governance for urban development: lessons from South Asia. Urban Gov Pol Rev 10(2):112–130
26. Laatikainen T, Tenkanen H, Kyttä M, Toivonen T (2015) Comparing conventional and PPGIS approaches in measuring equality of access to urban aquatic environments. Landsc Urban Plann 144:22–33
27. Li J, Lei J, Li S, Yang Z, Tong Y, Zhang S, Duan Z (2022) Spatiotemporal analysis of the relationship between urbanization and the eco-environment in the Kashgar metropolitan area, China. Ecol Indic 135:108524
28. Liao H, Xie D (2020) A review on urbanization and its impacts on urban spatial structure. ResearchGate
29. Liu T, Yang X (2015) Monitoring land changes in an urban area using satellite imagery, GIS and landscape metrics. Appl Geogr 56:42–54
30. McGranahan G, Satterthwaite D (2014) Urbanisation concepts and trends, vol vol. 220. International Institute for Environment and Development, London
31. Miller A, Small C (2021) Using GIS and remote sensing for urban planning: innovations and applications. Urban Plan Technol J 8(4):200–215
32. Morad H, Zeng Y (2020) Urbanization and cities of the future. ResearchGate. Accessed January 26, 2025. https://www.researchgate.net/publication/339881580_Urbanization_and_cities_of_the_future
33. OECD (2012) Compact City policies: a comparative assessment. OECD green growth studies. OECD Publishing, Paris
34. Opeyemi AY (2012) Empirical analysis of resource curse in Nigeria. Econ Manag Sci 1(6):19–25
35. Pawar DH, Mane VD (2013) Socio-economic status of slum dwellers with special reference to women: geographical investigation of Kolhapur Slum. Res Front 1(1):69–72
36. Perera M, Wijesinghe L, Gamage S (2022) Urban sprawl and its impacts on Sri Lankan cities: a critical review. Sust Urb Stud 9(1):34–52
37. Ramachandra TV, Aithal BH, Sowmyashree MV (2014) Urban structure in Kolkata: metrics and modelling through geo-informatics. Appl Geomat 6(4):229–244
38. Ren Y, Li H, Shen L, Zhang Y, Chen Y, Wang J (2018) What is the efficiency of fast urbanization? A China study. Sustainability 10(9):3180
39. Saglio-Yatzimirsky MC (2014) Megacity slums: social exclusion, space and urban policies in Brazil and India, vol 1. World Scientific

40. Seto KC, Dhakal S, Bigio A, Blanco H, Delgado GC, Dewar D, Zwickel T (2014) Human settlements, infrastructure, and spatial planning. In: Edenhofer O, Pichs-Madruga R, Sokona Y (eds) Climate change 2014: mitigation of climate change. Cambridge University Press, Cambridge, pp 923–1002
41. Smith SJ (2013) Crisis and innovation in the housing economy: a tale of three markets. In: Financial innovation: too much or too little, vol 71, p 73
42. Suárez M, Benayas J, Justel A, Sisto R, Montes C, Sanz-Casado E (2024) A holistic index-based framework to assess urban resilience: application to the Madrid Region, Spain. Ecol Indic 166:112293
43. Sudhira HS, Ramachandra TV, Jagadish KS (2004) Urban sprawl: metrics, dynamics and modelling using GIS. Int J Appl Earth Obs Geoinf 5(1):29–39
44. Torrens PM, Alberti M (2000) Measuring sprawl
45. Tsai Y-H (2005) Quantifying urban form: compactness versus sprawl. Urban Stud 42(1):141–161
46. Uchida H, Nelson A (2010) Agglomeration index: towards a new measure of urban concentration (No. 2010/29). WIDER Working Paper
47. UNDESA, U, UNECLAC, U (2015) Water for a sustainable world. World Water Development
48. UN-Habitat (2021) World cities report: envisioning the future of cities. United Nations Human Settlements Programme
49. UN-Habitat (2004) The challenge of slums: global report on human settlements 2003. Manag Environ Qual Int J 15(3):337–338
50. United Nations (2014) World urbanization prospects: the 2014 revision. United Nations, New York
51. Verbeek T, Boussauw K, Pisman A (2014) Presence and trends of linear sprawl: explaining ribbon development in the north of Belgium. Landsc Urban Plann 128:48–59
52. Weerakoon T, Samarawickrama R (2019) Challenges of urban classification in Sri Lanka: a policy perspective. Sri J Urban Dev 7(2):78–92
53. Weeraratne B (2016) Re-defining urban areas in Sri Lanka, ResearchGate. Accessed January 26, 2025. https://www.researchgate.net/publication/303292752_Re-defining_urban_areas_in_Sri_Lanka
54. Wijesundara J, Wijekoon N (2017) Transitory urban interventions for effective place making in shopping streets: case of Pettah in Sri Lanka. CPP Int J Urban Environ 2(1):10–25
55. Wilson EH, Hurd JD, Civco DL, Prisloe MP, Arnold C (2003) Development of a geospatial model to quantify, describe and map urban growth. Remote Sens Environ 86(3):275–285
56. Wirth L (1938) Urbanism as a way of life. Am J Sociol 44(1):1–24
57. World Bank (2015) Sri Lanka: leveraging urbanization for economic growth. World Bank Group, Washington, DC
58. Yasin MY, Yusoff MM, Abdullah J, Noor NM, Noor NM (2021) Urban sprawl literature review: definition and driving force. Geografia 17(2)
59. Yigitcanlar T (2010) Making space and place for the knowledge economy: knowledge-based development of Australian cities. Eur Plan Stud 18(11):1769–1786
60. Zang J et al (2022) The dynamics of urbanization in China and its impact on spatial patterns. J Urban Plann 49(4):537–556

Inclusive Smart Cities: Ensuring Accessibility for Disabled Children in Urban Spaces

Shashwata Sahu and Navonita Mallick

Abstract Smart cities have been launched in response to the problems caused by urbanization due to the increased use of non-renewable resources. Even though we explore how technology can lead to better living conditions, it seems that many smart city initiatives ignore the needs of children with disabilities. In cities, accessibility is more than just a convenient solution. It's a basic human right. This can only materialize in its total capacity if the smart cities also incorporate inclusive infrastructure, services and digital solutions addressing all residents, particularly vulnerable groups such as disabled children. This chapter addresses the accessibility of smart cities to disabled children and legal frameworks through technology, aiming at equal access to urban resources. It does this by utilizing doctrinal research to analyze international and national laws on urban accessibility and the effect of digital technology on social inclusion. Although improvements have been realized, this research has identified inadequacies in legal frameworks and inclusivity goals, with a large proportion of disabled children under-served. It argues that smart city development must prioritize sustainability by integrating digital innovation, legal support, and socially inclusive urban design centered on disabled children.

Keywords Smart cities · Disability rights · Urban accessibility · Inclusive infrastructure · Disabled children

1 Introduction

It is no secret that urbanization worldwide has been advancing faster than ever, radically altering how cities form and are governed. As per the United Nations, more than half of all inhabitants in the world (55%) live in urban areas at this point; by 2050, it will increase to two-thirds (68%) [95]. As cities continue to boom, the task

S. Sahu (✉) · N. Mallick
School of Law, KIIT-DU, Bhubaneswar, India

H. Kaddouri et al. (eds.), *Smart Cities for Sustainable and Urban Development*, Green Energy and Technology, https://doi.org/10.1007/978-3-031-99001-4_43

of handling large populations in ever more intricate urban landscapes is growing accordingly. Resource management, housing, transportation infrastructure, and health care, among other issues, present real challenges. To solve this issue, governments and urban planners must seek the ideal of smart cities designed with technology that can improve the quality of citizens' lives and optimize resource use [85]. Through new information and communication technologies (ICT) and artificial intelligence (AI) combined with IoT, we establish smart cities and interconnected urban spaces that use this network, providing comfort for citizens and public administration. Smart cities offer opportunities for better urban living through improved traffic control, energy efficiency and access to public services [82]. According to reports by global agencies such as the World Economic Forum and International Telecommunication Union, smart city projects are proliferating rapidly in most developing countries across North America, Asia, and Asia-Pacific regions, with a projection of over 500 new cities expected to be developed by 2025 [39]. These cities aim to address various urban issues, such as pollution, overpopulation and finite resources, and optimize living conditions for the people within them.

Yet, discussions about smart cities often emphasize efficiency, economic growth and technology at the expense of social inclusion and human rights. This leads to one of the biggest concerns raised in this context: are smart cities actually for everyone, especially those who must rely on accessible public transportation and services, such as children living with disabilities? There is nothing social or ethical about inclusiveness in urban planning. It is a human right. Those urban spaces, though, can be a challenge, especially for children with disabilities. With transportation difficulties, access to only the most basic education and healthcare needs unmet in traditionally built cities but even further left wanting under newer smart city structures [76]. The "World Health Organisation (WHO)" shows that about 15% of the world's population suffers from some form of disability and a significant part are children [94]. Still, the needs of children and adolescents are often unmet in urban planning, leading to isolation or a lower quality of life. Access to light is still seen as a luxury in many urban areas, where physical barriers are often built by the architecture of public spaces and digital barriers emerge from non-inclusive technologies.

Smart cities could serve to widen these chasms without a focus on inclusivity. However, as technology becomes the backbone of urban development if smart city initiatives are not inclusive for all children, cities may exacerbate existing inequalities and further marginalize disabled young people. Smart Cities for All? Bridging Digital Divides for Socially Sustainable and Inclusive Cities. For instance, AI-powered public transport utilizing IoT could provide faster and more efficient routes. Still, it may not be accessible if accessibility features like wheelchair ramps, audiovisual work aids or digital design are generally inadequate for children with mobility and vision or hearing impairments [62]. On a usability level, if mobile apps and online platforms in smart healthcare solutions do not include children with disabilities at the design stage, they will surely fail to meet their needs. This urban planning and smart city development gap reinforces the immediacy of an integrated human rights approach to city building, one not limited to technology or efficiency, as reiterated in my blog on this issue but firmly anchored within fundamental human rights for all, including children with disabilities [75]. The "United Nations Convention on the Rights of Persons with Disabilities (UNCRPD) states that all people, including children living with disabilities, have a right to participate and live independently within communities, two elements directly linked to access spaces, urban and services [84]. If smart cities disregard these global human rights standards, then disabled children risk being excluded from the digital transformation wave.

However, it must also be accompanied by sustainability inclusivity. While sustainability in urban contexts used to be synonymous with minimizing carbon footprints and saving natural resources, walking lightly on the earth is one way it is described. Being sustainable also means creating social environments that support all city residents at an equitable level for wide-ranging physical or cognitive abilities [19]. A city that can only cater to the needs of less than a quarter of its population is hardly sustainable. But urban sustainability in the long term can also be undermined when cities fail to include vulnerable groups like disabled children, who are then perpetually shut out of society. In sharp contrast, cities that emphasize inclusive design in their smart city agendas tend to reap more benefits on the social front, such as higher levels of interpersonal trust and well-being, as all citizens can actively participate and thereby benefit from urban resources. The purpose of this chapter is to identify existing gaps in terms of smart city planning that address the special needs for transport accessibility among disabled children. The chapter first assesses the existing national and international legal frameworks that oblige accessibility in urban spaces for the disabled population. Secondly, it will highlight how digital technologies and the urban infrastructure within smart cities can promote accessibility for disabled children, then provide solutions to improve urban mobility, health and education for children with a disability using smart city technologies and finally, will also recommend policy changes to make Smart city planning and governance Inclusive.

2 Research Methodology

The chapter uses a qualitative research method, and it adopts a doctrinal approach to examine laws, regulations, and policies related not only to urban planning but also to smart cities and disability rights. The implied effect of an accessibility and inclusivity perspective will be comparatively analyzed by analyzing various legal instruments at both the national and international levels. The chapter will also draw upon secondary sources such as academic literature, case studies and policy reports to explore how the needs of disabled children are being met or not in existing smart cities (Fig. 1).

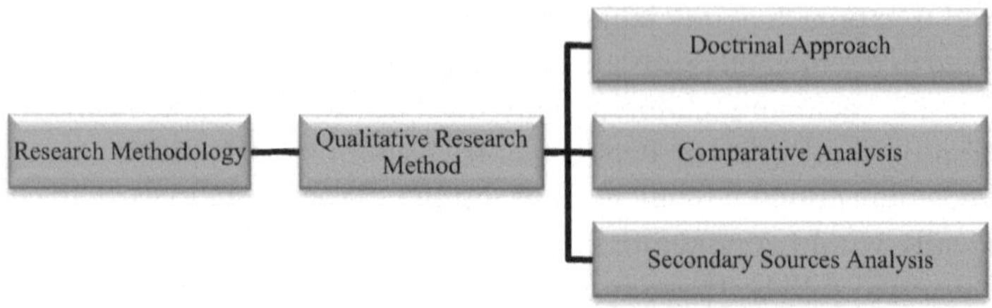

Fig. 1 Illustration of the research methodology

3 Conceptualization and Theoretical Framework for Smart Cities

These cities intend to redefine urban living by enhancing the ideas of governance, infrastructure and public services through digital technologies. All these sustainability and efficiency features form integrated parts of the smart cities leveraging tech for truly positive repercussions so that our civilization lasts longer on this planet. Yet access remains the paramount problem, especially for the most marginalized among us, children with disabilities [53]. Inclusive urbanism aims to address those traditional accessibility problems of cities by the urgent necessity not only for the disabled population but even for their potential exclusion within smart city designs. The social model of disability and human rights theories argue for a society designed without barriers, where smart cities provide physical infrastructures suitable for everyone.

3.1 Conceptualization of Smart Cities

A smart city is an urban area that has employed information and communication technology to gain sustainability control on critical infrastructure components, hence optimizing efficiency during its daily operations by taking input from the devices. This idea is based on the development of smarter cities that are more sustainable, efficient, and better at looking after their citizens through high-tech intervention [42]. At the heart of this stake is a belief that urban challenges, traffic, water use or waste disposal in desert cities and concerns about air quality or climate-proofing European infrastructure for powerful storms made more frequent as our planet heats can all be better tackled with data-driven solutions, IoT answer [71] (Fig. 2).

Good governance in smart cities is one in which the government and people actively participate. It also involves establishing clear, responsible mechanisms for decision-making using digital platforms on which residents can participate in the planning of cities. Driving smart cities is a set of digital technologies, including the IoT and AI-based big data analytics as key components. These technologies allow

Fig. 2 Key aspects of city planning in smart cities

the collection and processing of huge data records, which are then used for real-time decision-making to optimize urban operations. Smart cities need contemporary infrastructure for implementing digital mechanisms. Examples include smart grids for energy distribution, next-gen transportation systems and intelligent public utilities that can eliminate waste and inefficiencies. In smart cities, public services become more competent by using technology to improve operations and leverage customized service delivery for its citizens in areas like healthcare, education or waste management. However, the potential of smart cities for creating a high-quality environment and building urban sustainability can be realized only if they are inclusive. In other words, when city life is supported by everyone from children with disabilities to anyone else living in our society.

For decades, urban areas have been a difficult place to navigate for those who are disabled. Ancient and more modern cities are largely unsuitable for most people, people with mobility issues and sense- or brain-based limits. This is even more of a struggle for disabled children, who require access to transport, healthcare and education services as well as public spaces that enable them to lead rich lives [14]. With the infusion of digital and technological solutions, the move to smart cities may serve the positive or the opposite side. From one perspective, smart cities can contribute to universal urban designs that are user-centered for all citizens, for example, through implementations of digital solutions such as intelligent transportation systems that give live data about accessible routes or IoT-enabled buildings having disabled-friendly access options with automated features [36]. Conversely, new walls can be erected if smart city technologies are not designed to include all citizens from the start. For example, looking ahead, a city service increasingly delivered through exclusive and inaccessible digital interfaces may deny children with visual or cognitive disabilities access to the services. So, a smart city, in my belief, should go beyond technological innovations and be more of a universal design.

3.2 Theoretical Framework

International conventions, such as the UNCRPD, are helping to frame a rights-based approach at the heart of inclusive smart city design. The UNCRPD focuses on reinforcing the rights of persons with disabilities to participate in everyday life freely, including public spaces and services [83]. Smart city technologies need to be implemented using the principles of UNCRPD, not just for efficiency and effectiveness but also for equitable access for all, including disabled children. It highlights the need for digital applications, whether in healthcare, transportation or education, to ensure they are based on accessibility and inclusive design [25]. Building human rights into smart cities means disabled kids are not only part of development but also get support from the places they live.

The social model of disability stands in contrast to the medical model, which sees a disabled body as an individual problem that needs fixing. Instead, the social model claims that disability is a result of societal barriers, not individual impairments [18]. Physical, communication, and attitudinal barriers that people with disabilities face include not just the structural properties of physical places like buildings without ramps and doors too narrow for a wheelchair to pass through unaided but also obstacles such as websites designed in ways screen readers cannot read or prejudice on the part of others [73] (Fig. 3).

The Social Model of Disability is an approach which moves the focus from individual impairments to barriers created by society. The social model asserts that people are not disabled as a result of their impairments but because the infrastructure, both physical and systemic, around them does not accommodate what they require [20]. Lack of ramps, lifts and stairs and poorly designed buildings in poor conditions with walls collapsing on top of students or houses that otherwise trap people inside also make subjects in the built environment disabled. Smart cities, not developed inclusively, can more likely prove such strong barriers [55]. This model, therefore, makes clear the need for smart city technology and infrastructure to eliminate this social disablement by accommodating not only disabled children but all people, as Full participation would represent. For smart cities, the social model can describe how urban infrastructure and technology generate or eliminate obstacles for disabled persons. If we have a city based on the social model, it would put these societal barriers first and eliminate them to make an environment in which all can thrive much more accessible, with priorities placed on places where disabled children are reliant, from producing transportation systems that children with mobility

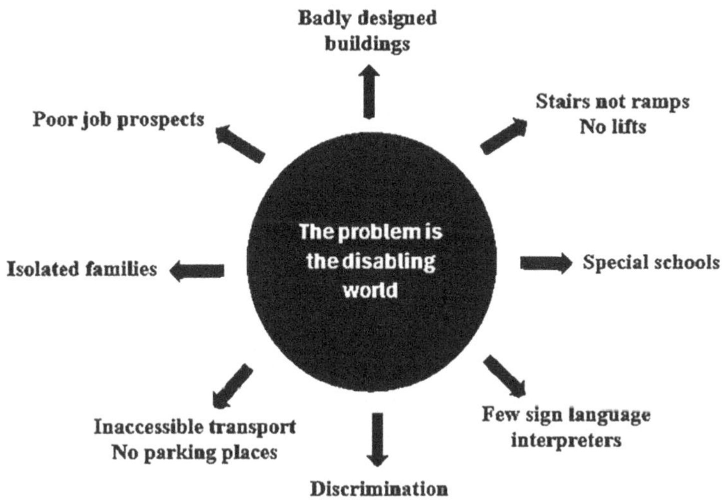

Fig. 3 The social model of disability and its impact on accessibility in urban environments

issues can easily navigate to providing public spaces and facilities outfitted by structures able to cater to multiple disabilities or developing digital platforms reachable for the sensory impaired-to-cognitive disabled child [65]. Thus, by emancipating itself from the medical model, smart cities can finally go beyond symbolism in promoting an inclusive world and start building that environment with very young disabled citizens. In this context, technology becomes a tool to overcome barriers rather than unintentionally put them up.

The TAM examines how new technology can be adopted and integrated into individuals' lives. The model states that the perceived ease of use and usefulness are crucial determinants to rationalize why an individual will accept or reject a technology. The technology acceptance model can be used to evaluate the adoption rate of digital solutions added for disabled children and families by smart cities [9]. In a smart city, an IoT transportation system uses information to know where things are located and sends updates on available paths. But suppose the experience with those products is not tailored for disabled children, especially child-friendly interfaces, settings configurations responsive to different disabilities, and instructions that are straightforward enough on all accessibility levels. In that case, it may pass unused by families. This ease of use is even more important for children who have disabilities and may or may not be able to interact with a digital platform the same way an average child would. The application of the TAM to smart city technologies, on the other hand, is pivotal for successful implementation and utilization by the communities they serve.

Inclusive urban design theory focuses on the need to provide public spaces that are accessible and usable by everyone, irrespective of physical impairment. It is based on the idea that cities should be planned for everyone, including people with special needs, such as disabled ones, senior citizens and children, or anybody who may need help [70]. In the arena of smart cities, inclusive design refers to infrastructure and services that cater to a range of different requirements of members of society. It features physical changes such as ramps, tactile paths, optional public transport services, and virtual adaptations using voice command systems or signs for the visually disabled, who can get instructions in Braille format. Visible displays to make UX more accessible for children with cognitive impairments [28]. A prime example of inclusive design is the application for a top-down implementation, where cities have successfully integrated smart technologies in public transportation systems to ensure access and availability of data reflecting disability, such as Barcelona [29]. The buses in the city support blind and deaf people with audio means; they also show their location through digital platforms to inform about accessible routes. Singapore is another such example, where urban planners have integrated smart technology to transform public spaces into being inclusive, right from parks with wheelchair-friendly tracks to sensor-based systems helping disabled children navigate through cityscapes. Therefore, inclusive urban design strategies will allow smart city-oriented development to collabosrate with planners for disability eradication. It is also an important pillar in terms of sustainability and social equity, as cities designed for all are better poised to serve diverse populations over time by promoting urban resilience.

4 Legal Frameworks and Policy Analysis

Cities globally are adopting technologies to create smart cities, which have the potential to be more effective cities with accessibility in mind for everyone. Indeed, accessibility in smart city initiatives is not only a nice thing to do ethically speaking, which is required by law [82]. The international and national legal instruments that reflect accessible urban Disabled will further be discussed in this section, focusing on the scenario of India.

4.1 International Legal Frameworks

The UNCRPD is one of the most important international legal frameworks for promoting, protecting and ensuring that persons with disabilities have rights, especially those children. It came into force in 2006 and was a ground-breaking treaty that introduced the social model of disability to this international context [57]. The UNCRPD does not see disability only as the physical or mental state in which we need to intervene but also stresses that those with disabilities are excluded because our societies make it difficult for them to participate more fully. The Convention lists several human rights concerning which state parties agree to promote, protect and ensure the full and equal enjoyment of all human rights by persons with disabilities [50]. Article 9 is a central provision of the UNCRPD that explicitly covers accessibility. It requires state parties to adopt measures needed so that persons with autism can readily access the physical environment, transportation, information and communication technologies (ICT), and other public services. The importance of people with disabilities having equal access to public spaces, for example, in urban areas, is acknowledged by Article 9 [31]. Part of smart cities, when it comes to Article 9, UNCRPD obligated countries to make sure that technology does not leave behind people with disabilities. This concept means designing products, environments and services so that they can be used by everyone, regardless of their age or physical abilities [44]. For example, suppose smart city applications like IoT and AI-based public services do not offer suitable means for disabled people to interact with them as required by the Convention. In that case, it cannot be called an inclusive technology solution. This may include voice recognition for individuals with visual disabilities or updates on public transport in real time that is assistive to those who need support due to a mobility impairment [88]. The UNCRPD also underlines the importance of inclusive education (Article 24) and access to healthcare (article 25), which are central components in urban development put forward by smart cities. Accessible schooling in healthcare facilities is also crucial to the well-being and development of children with disabilities [33]. Therefore, smart cities should have these principles in the design and governance of their infrastructure to meet this inclusiveness mandated by UNCRPD.

From Table 1, international legal frameworks, such as UNCRPD and CRC, play an important role in ensuring accessibility to smart cities. These frameworks point to the rights of disabled children and their need for cities, services and technology. By aligning smart city projects to these global standards, planners can open an integrated, inclusive environment for disabled children and help enable them full participation in all facets of urban life.

4.2 National Laws and Policies Specific to India

The "Rights of Persons with Disabilities (RPwD) Act, 2016" is the main legislative framework in India on accessibility rights to disabled populaces and was framed following principles enshrined under UNCRPD. RPwD Act is a full-fledged law that protects the rights of persons with disabilities, including in areas like employment and education [79]. Section 40 of the RPwD Act requires that public buildings, transport facilities and information and communication services be accessible to persons with disabilities [44]. According to the Act, public infrastructure of any sort must be created, taking into consideration that it is used not only by regular people but also by those with disabilities. Of particular note for smart cities are their provisions related to the impact of rapid urbanization and increasing technology integration in how our modern urban areas are being shaped [90]. Indian laws say that every smart city mandatorily incorporates certain basic features such as ramps,

Table 1 Comparative analysis of international legal frameworks on accessibility for disabled populations

International legal framework	Key provisions	Relevance to smart city planning for disabled children
United Nations Convention on the Rights of Persons with Disabilities (UNCRPD), 2006	– Article 9 mandates accessibility to the physical environment, transportation, information, communication technologies (ICT), and public services. – Emphasizes full participation in society for persons with disabilities. – Promotes universal design for accessibility in public infrastructure.	– Smart cities must ensure that all infrastructure, services, and technologies are universally designed and accessible to disabled children, complying with international human rights standards.
Sustainable Development Goals (SDGs), 2015	– Goal 11 aims to make cities and human settlements inclusive, safe, resilient, and sustainable. – Focuses on accessible urban infrastructure and inclusive public spaces for all, including vulnerable populations.	– Smart city initiatives must align with the goal of creating accessible urban spaces, ensuring disabled children have access to safe transportation, education, and healthcare.
World Health Organization (WHO) Global Disability Action Plan 2014–2021	– Aims to improve health outcomes for disabled populations. – Stresses the importance of removing barriers to healthcare, mobility, and social inclusion. – Promotes the use of assistive technologies for persons with disabilities.	– Smart healthcare services in cities must integrate assistive technologies (e.g., telemedicine platforms, IoT-based health monitoring) for disabled children. – Promotes inclusive urban mobility and social services for better health outcomes.
Universal Declaration of Human Rights (UDHR), 1948	– Article 25 states the right to a standard of living adequate for health and well-being, including medical care and necessary social services. – Article 21 emphasizes equal participation in public affairs and services.	– Smart cities need to ensure equal access to public services, healthcare, and infrastructure for disabled children to uphold their human rights.
Convention on the Rights of the Child (CRC), 1989	– Article 23 specifically protects the rights of children with disabilities. – Calls for appropriate assistance and services to ensure access to education, healthcare, and recreation for disabled children.	– Smart cities must prioritize accessible education, healthcare, and recreational services for disabled children, ensuring their full inclusion in urban life.

Table 1 (continued)

International legal framework	Key provisions	Relevance to smart city planning for disabled children
International Labour Organization (ILO) Vocational Rehabilitation and Employment (Disabled Persons) Convention (No. 159), 1983	– Promotes access to vocational training and employment for persons with disabilities. – Encourages inclusive policies for public services that enhance the employability of disabled persons.	– Smart cities need to incorporate inclusive digital education and training platforms to prepare disabled children for future employment, particularly in tech-driven environments.
European Accessibility Act, 2019	– Sets standards for accessible products and services in the European Union. – Focuses on ICT, public transportation, and urban services to be accessible to all, including persons with disabilities.	– Smart cities in the EU must ensure that digital services, transportation, and public infrastructure are fully accessible to disabled children, complying with the Act's accessibility standards.

elevators, tactile paving in public spaces, and digital platforms that should be accessible to people with disabilities.

The third promise that most angers disability rights activists concerns the Smart Cities Mission, an initiative launched by the Government of India in 2015 to make urban areas accessible and inclusive for citizens, credited as one step toward transforming Indian cities into 'smart' sustainable ones. They aim to provide technology-based initiatives to enhance urban living through inclusive development [96]. The Smart Cities Guidelines from the Ministry of Urban Development reiterate that cities must be more inclusive for all sections of society, including disabled people. This includes making the physical world, like public spaces, transportation systems and the digital one, accessible to all. For instance, cities such as Pune and Ahmedabad, under India's Smart Cities Mission, have started integrating accessible features into their transportation system with low-floor buses similar to that of other metros and giving real-time updates for disabled passengers [56]. These efforts are in keeping with mandates under both the RPwD Act and those of the UNCRPD, showing that India intends to make its smart cities inclusive. The problem is getting these measures to happen in some standard way across all Smart City projects.

The Table 2 highlights the provisions of accessible laws with different national coverage for Orang Peka in various jurisdictions. They require accessible infrastructure, transportation, and digital services, including India's RPwD Act or the U.S. ADA. We believe that by including disabled children in the urban transformation, better mobility, healthcare and education can be achieved by integrating these legal mandates into smart city planning. This is a way to create truly inclusive cities for all kinds of citizens and on which laws are based.

Table 2 Comparative analysis of national laws on accessibility for disabled populations in smart cities

Country	Relevant national law	Key provisions	Relevance to smart city planning for disabled children
India	Rights of Persons with Disabilities (RPwD) Act, 2016	– Mandates accessible infrastructure in public buildings, transportation, and digital services. – Introduces the concept of universal design in public spaces. – Emphasizes equal access to education, healthcare, and mobility.	– Smart cities must ensure that all public spaces, transport systems, and digital platforms are designed with accessibility in mind. – Strong legal backing for incorporating inclusive urban designs and IoT-enabled services for disabled children.
United States	Americans with Disabilities Act (ADA), 1990	– Prohibits discrimination based on disability. – Requires public and private sectors to provide accessible public services and infrastructure. – Includes specific provisions for accessible transportation and digital technologies.	– Smart city technologies (e.g., public transportation systems, healthcare, and education platforms) must be accessible to disabled children under ADA compliance.
United Kingdom	Equality Act, 2010	– Outlaws discrimination against disabled people in public services, education, and transportation. – Mandates reasonable adjustments in urban design and service delivery. – Covers physical and digital accessibility.	– Smart cities are required to ensure that digital and physical infrastructure, such as public spaces, transportation, and educational systems, are accessible to disabled children.
Australia	Disability Discrimination Act (DDA), 1992	– Prohibits discrimination in access to public places, services, and education. – Requires accessible public buildings, spaces, and transport. – Includes provisions for digital accessibility, ensuring services and websites are usable by disabled individuals.	– Australian smart cities need to integrate accessibility features, such as ramps, tactile paths, and IoT-enabled public services, that cater to disabled children's needs.
European Union	European Accessibility Act, 2019	– Sets accessibility standards for products and services across the EU. – Includes requirements for ICT accessibility and urban infrastructure. – Mandates that public transportation and digital services are accessible to all.	– The Act obliges EU smart cities to ensure accessible ICT solutions, such as public transport apps and real-time data systems, are inclusive for disabled children.

Table 2 (continued)

Country	Relevant national law	Key provisions	Relevance to smart city planning for disabled children
Canada	Accessible Canada Act (ACA), 2019	– Aims to create a barrier-free Canada by 2040. – Includes requirements for accessible transportation, public spaces, and digital services. – Enforces accessibility in federally regulated sectors.	– Canadian smart cities must prioritize the inclusion of accessible infrastructure and technology (e.g., real-time accessible public transportation apps) to serve disabled children.

4.3 Gaps Between Policy Formulation and Implementation

Though there are comprehensive legal provisions on the international level, like those outlined by UNCRPD and national laws through the RPwD Act ensuring accessibility in urban spaces, we find, among other things, a large gap that exists between policy-making and the actual execution of smart city projects [4]. One of the biggest hurdles in bridging this gap is uneven application across cities or regions. Smart city projects across India have demonstrated considerable progress in making accessible infrastructure part of their offerings. Still, some cities fall behind because they haven't prioritized it or do not have resources [15]. For example, many public buildings and transport systems in smaller towns and rural areas continue to be unpractical for people with a disability, even as per the mandates of the RPwD Act. This inconsistency may sabotage the goal of building urban environments that are truly inclusive because, all too often, accessibility is treated as an afterthought rather than a fundamental feature of city planning [93]. Another related problem arises from a lack of coordination between government bodies and other stakeholders. Smart city projects generally are multi-actor, where the actor mixes with local governments, private companies and international organizations. There is no way to enforce it and ensure accessibility will be prioritized throughout the stages of your projects [69]. The results of the interviews suggest that many urban planners may not have sufficient technical knowledge and resources to apply accessible technologies properly (e.g., infrastructure design concepts) toward satisfying the needs for access among disabled populations.

There is also a financial cost to setting up accessible technologies, which affects the outcome of policies fairly heavily. Smart city technologies could help cities be more accessible or at least drive efficiency even when not the outright beneficiary. Still, they often come with expensive price tags. Broke local governments will likely do cheaper, less inclusive public infrastructure design, resulting in zero tactile paving, no ramps at corners or even accessible digital interfaces, technology and inclusivity are not realized [5]. Smart city implementations rightly focus on increasing efficiency and quality of life, but these efforts may leave citizens behind. To illustrate, while IoT technologies in public transport may make services faster and more dependable, they might lack dedicated features for users with disabilities, such as real-time audio announcements, a boon to the visually impaired [30]. This reflects

Gap Between Policy and Execution	Uneven Application Across Regions
	Accessibility as an Afterthought
	Lack of Coordination Among Stakeholders
	Insufficient Technical Knowledge
	Financial Constraints
	Lack of Inclusive Smart Technologies
	Delayed Integration of Inclusivity
	Weak Enforcement of Legal Frameworks
	Absence of Regular Audits & Accountability Measures

Fig. 4 Gap between policy and execution

the tech industry, where inclusivity falls through to later steps in a process instead of being accounted for as an essential part of design. This would mean that the approach to smart city design taken by both governments and urban planners needs to shift from merely focusing on technology implementation to combining accessibility in a specific way at the start of project planning rather than performing it as an add-on. Further, robust enforcement mechanisms are necessary to guarantee the implementation of legal frameworks such as CRPD and RPWD Act. Regular audits on public infrastructure and punitive measures for non-compliance could close the gap between policy, theory & practice (Fig. 4).

5 Literature Survey on Disability of Children in Smart Cities

The integration of smart city initiatives can significantly raise the quality of life for a child with a disability through the extensive use of developed technologies to offer more accessible, friendly, and inclusive urban areas. A complete literature survey, in this connection, reveals quite a few approaches and considerations related to this very domain. A systematic literature review by Kolotouchkina et al. calls up the need for creating a barrier-free digital urban environment because leadership in this area can inspire transformative change. The need for cities to prioritize creating a digitally inclusive city, which guarantees equal access for all citizens, including children with impairments, is therefore of great essence [48]. Based on their research, Soares de Oliveira Neto introduces a concept known as Inclusive Smart Cities, utilizing technology to further improve the life experiences of citizens with disabilities. This approach champions the design and implementation of city technologies that ensure the diverse needs of all children are met through accessibility and inclusive features for kids with disabilities [63]. In addition, Al-Sayed et al. demonstrate ways in which smart city initiatives might facilitate the quality of life for children with disabilities by using technology to create an accessible, connected, and participating environment. Among other beneficial contributions of smart infrastructure, this research indicates that sensor-equipped traffic lights and accessible transportation facilitate the successful inclusion of children with disabilities into an urban setting [13]. Despite these developments, challenges still exist. According to a systematic literature review, Bricout et al. pointed to a pertinent gap in research concerning the applications of smart cities in disadvantaged neighborhoods. This means more inclusive planning and implementation strategies need to account for unique needs experienced by children with disabilities in various urban contexts [46]. Zhou et al. [98] have performed a systematic literature review to illustrate how smart cities need effective collaboration from various stakeholders in the sustainable and inclusive environment of the cities. It highlights that the development, implementation, and usage of smart technologies for special needs in children with disabilities are very important and need to be addressed by interdisciplinary research [98]. Pearson and Shafik explore the challenges that face people with disabilities in urban areas and how smart technology can address these challenges. Their research will discuss strategies to use smart technology to create a more inclusive environment that enhances accessibility, mobility, and quality of life for people with disabilities [68]. Salha et al. come up with an integrated approach ensuring that no people with disabilities will be left out of smart cities. They also advise using checklists on design as guidelines in initial architectural ideas and smart city design to aid in reasonable access to buildings, public and private. This means cities are getting smarter and also inclusive to everyone, particularly persons with disabilities [76]. Despite these strides, challenges exist. While promising routes for improving the lives of children with disabilities come with smart city initiatives, the design and implementation of such initiatives must go through inclusive and participatory approaches. Through the consideration of accessibility in a process actively engaging stakeholders, cities can offer conditions that support the well-being and development of children alike, regardless of their abilities.

6 Accessibility in Smart Cities: Current Gaps

Smart cities are heralded as innovative solutions that will change how we live in our urban environments by improving efficiency, sustainability and quality of life through technology integration. Unfortunately, many of these cities concentrate on technological advancements and infrastructure improvements at the cost of accessibility needs, especially when it comes to disabled populations such as children. Ironically, the most vulnerable members of society, those who stand to benefit greatly from those technologies, are often excluded because they cannot use them in their present form [34]. It will illustrate cases in which smart city projects have not included disabled children effectively, present incomplete or failed pilot studies and show the difficulties of introducing inclusive technologies into an already designed infrastructure for this type of urban project.

6.1 Real-World Examples of Neglect

Along with the environmental impact, one of the most crucial problems in many smart city projects comes from their concentration on urban efficiency and technological innovation rather than ensuring those improvements make it to every citizen. Take Barcelona, one of the most internationally renowned smart cities that has become synonymous with technological innovation in city life [32]. Although the city has been seeing progress in becoming a more pedestrian-friendly urban place, efforts to make these areas accessible for people with disabilities have come in short supply. Barcelona has integrated many aspects of smart city technologies, such as free wifi and sensor-powered traffic control. Still, disabled children in Barcelona have been neglected in terms of these benefits [58]. There is still an uneven distribution of accessible elevators and ramps throughout the city. At the same time, many public spaces that are being architecturally modernized do not take into account children with physical or cognitive disabilities. Songdo, a South Korean smart city, is often heralded as a model for future urban development. The city was criticized for not dealing with connectivity despite having high tech to monitor traffic, waste and energy [67]. A common criticism is that more focus has been devoted to building a futuristic playground for companies and wealthy denizens rather than prioritizing solutions that would address the needs of susceptible groups like disabled children. While smart systems for public transport offer real-time updates and AI-assisted navigation, they are not always available to children with visual impairments or mobility challenges. Moreover, many digitally integrated public spaces do not have feature sets that include senses other than touch or audition and thereby risk excluding large segments of society.

6.2 Case Studies of Incomplete Accessibility Implementations

Urban mobility, in particular, has been an area where smart cities have faced difficulties. One of the best guinea pigs for all this is London, a city that frequently gets credit for being at the forefront of smart transportation, including its use of digital payment systems, real-time tracking in public transit, and heavy penetration with contactless fare technology [21]. However, as we have already discovered in a previous post, the city has often been criticized for not taking into account disabled children while planning its smart city. Despite some advancements in London, buses are largely accessible, and metered black priority taxis are available to assist passengers with physical disabilities. Still, most tube stations remain a fortress for wheelchair users [17]. Furthermore, many transportation-related digital interfaces in the city were hard to use for children with cognitive disabilities due to the absence of poor-quality guidance tools or alternative communication methods.

The UAE's Smart Dubai initiative, which aspires to deliver exceptional standards of living for Emiratis and city residents, including those with special needs, also continues to confront issues surrounding technological advancement [6]. It is a problem, however, even in an AI-powered service like the smart healthcare platform or digital customer service built on this data as we've previously designed solutions for with Stackla and Terem Technologies, again created with Accessibility goals at their core, what is not guaranteed to deliver, without special consideration of accessibility when analyzed jointly [35]. For example, Dubai's smart healthcare system offers virtual consultations and telehealth solutions for patients in remote areas, which is good progress. Still, these services may not serve the special needs of children with disabilities who need personal assistance or alternative communication methods [38]. The high-tech structure of the city has not aided access to tech-reliant healthcare, as children with developmental difficulties or, for example, those who struggle using digital interfaces are often left out. Accessibility to education is also a problem many smart city initiatives may have failed to address. In the example of Singapore, a city worth cutting dividends to pay for, particularly in education, many schools use AI and Internet of Things devices. Despite this, children with disabilities are less likely to access these new technologies, particularly those with learning or cognitive needs. Smart education tools create new efficiency and personalization for mainstream students, but in doing so, they have added barriers for some kids with disabilities [37]. Furthermore, the teaching tools used to emphasize this disadvantage are also out of reach for many, if not most, disabled children. This means these students end up receiving an inferior form of education compared to their able-bodied peers.

Similarly, public spaces in smart cities are also exclusionary. Cities like Toronto aim to revolutionize public spaces by weaving digital components into the physical environment of places everyone can use and enjoy, primarily advancing esthetics or services that support diverse members of the general population [49]. Digital kiosks and smart benches provide Wi-Fi, for example, and may be useful in the lives of most citizens, but they exclude many children with disabilities. Things like the digital kiosks rarely include touch-based interfaces for visually impaired users or most smart benches, which largely disregard wheelchair access and children with mobility issues. Consequently, places like parks are meant to be available for everyone and become exclusionary.

6.3 Challenges in Adopting Inclusive Technologies within Smart City Frameworks

A major obstacle in incorporating accessible technologies into smart city models is that it represents an expensive and ambitious task to implement universal design principles across all areas of urban planning. Incorporating some inclusive technology resources may involve specialized hardware and software that might be deemed cost-prohibitive, particularly within municipalities with limited budgets. In infrastructure systems such as public transport, it is technically possible to incorporate tactile paving or auditory guides that would enable access for some individuals [64]. Still, these features are often not considered part of the essential smart city infrastructure and are viewed more like a set of non-obligatory add-ons.

Urban planners and technologists similarly do not understand or appreciate the special needs of disabled children. Few people who design the new technologies that go into these smart cities are trained in accessibility or are familiar with disability rights [59]. Consequently, these technologies tend to reflect average user needs and not the diverse requirements of different groups of individuals. One example of this is that the digital offerings in smart cities will typically be built to cater for users who can see or hear well, leaving out blind and deaf children or those with cognitive difficulties. Equally, the breakneck speed with which tech in smart cities can become outdated presents a problem for inclusivity [78]. In the smart city gold rush, municipalities rarely have time to develop ways of providing access to these AI, IoT or data analytics technologies. Accessibility features, in many cases, are bolted on instead of being considered early during the process. In other words, while smart city tech appears to be futuristic and innovative, what we discover time and again is that it does nothing for kids with disabilities [97]. However, smart city projects are often fragmented and ensuring universal access throughout different sectors can be challenging. It is an open system based on collaboration among federal agencies, private industry and international institutions. Since there are no set rules of what accessibility entails, it is often overlooked and does not get enforced [60]. For instance, one section of a smart city may concentrate on accessible transport while neglecting to encompass access features in public spaces or digital services (Fig. 5).

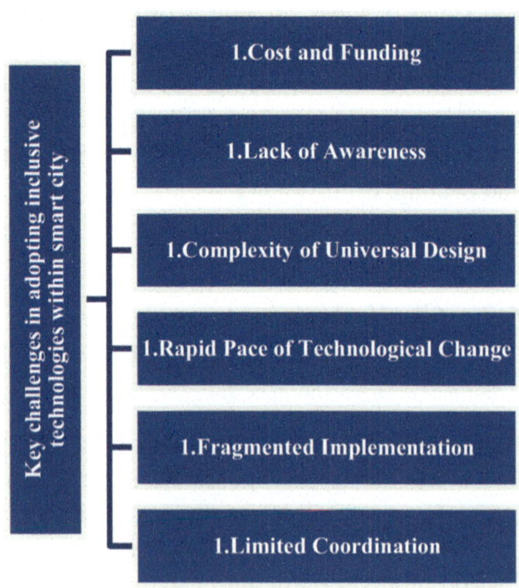

Fig. 5 Key challenges in adopting inclusive technologies within smart city

7 The Role of Technology in Ensuring Accessibility

The digitization wave is sweeping over the city as beneficiation from intelligent technology helps solve myriad problems that range from traffic congestion to ecological imbalance. Yet, at the same time, technology increasingly plays a vital role in urban spaces, one of its most crucial functions being that it should help ensure those environments are as inclusive and accessible to all bodies with an emphasis on particularly vulnerable populations like disabled children, making public spaces more democratic [61]. This in-depth part investigates how these are implementable for creating an accessible city. It provides some successful examples of cities that have been implemented, their compatibility with global accessibility standards, and their scalability (Fig. 6).

That is why the Internet of Things (IoT) has been identified as an essential part of building upon existing smart city infrastructure systems that are more integrated and can collect real-time data. Additionally, smart sensors could be placed along pedestrian paths in common public areas. As wheelchair users get closer to them, the system will guide them through which path is most accessible for disabled passengers. These innovations help improve safety and maintain autonomy for disabled children when moving through urban environments. Similarly, AI-powered technologies could be a core part of ensuring availability across all these urban systems [40]. For example, with AI manually integrated into public transportation management systems, we can see updates in real-time about whether the bus or train will arrive on time for disabled children and make sure there are wheelchair-accessible

Fig. 6 Roles of technology in ensuring accessibility

vehicles present at all times and predict traffic obstacles. AI can also be used in the education systems of smart cities so that children with learning disabilities can have personalized lessons [43]. Personalizing education with ever-evolving AI algorithms for the individual and doing so at scale, easily densifying educational content asynchronously presents an opportunity to meet each student where they are as disparate human minds.

In doing so, augmented reality gives them an extra layer of access through visual and auditory cues in the built environment. For example, city streets could be adorned with real-time overlays through AR apps that help visually impaired children reach their destinations. Those overlays can contain audio directions that help visually impaired children navigate spaces in a building or public space [51]. Moreover, with AR, we can make museums, natural parks, etc., all more accessible to disabled children, providing a real immersive experience that fits the different needs of each child. The multimedia content of AR provides contextual information in audio, text, or visual form so that all children can easily and highly comprehend their environment regardless of physical disabilities [26]. We have also seen wearable technological devices for disabled children in cities to help them navigate. Pairing those with IoT and AI, they provide immediate feedback on traverse pathways available for walk or roll to appraisable alert potential hazards or direct the way out to assist in case of emergency. Wearable devices with haptic feedback, such as the subtle vibrations that guide a child through city streets or warn them of an obstruction from which they are about to trip [89]. Although these technologies hold many promises, it is important to think critically about their implementation in practice to not reinforce an existing context.

AR further enhances accessibility by providing visual and audio aids to people navigating complex urban spaces. For instance, AR apps could make real-time overlays directing blind kids via city streets. These overlays might offer audio instructions for a child who is blind to navigate public spaces or buildings [86]. At cultural sites and parks, AR can provide additional layers of information that include braille text for the visually impaired or sound recordings to make the experience more immersive. AR provides contextual information in multiple modes without discrimination so that all children can interact properly with their surroundings, regardless of their abilities due to physical constraints or cognitive faculties. Wearable mobility devices that help steer disabled children through the city [23]. These devices, often connected to the IoT and AI, can provide immediate feedback on navigational routes, even alert an individual of possible danger, and open a conduit between that person and emergency services if they run into trouble. Similarly, a wearable device with haptic feedback that offers subtle vibrations can assist children in navigating streets in the city and notify them of potential hazards. The potential these technologies offer is huge, but ensuring they are genuinely inclusive demands the same rigor with which we deploy them [91]. Toronto is another example of successful implementation in the public sphere and AR technology. They enable disabled children to move around in their environment while listening and viewing the augmented reality devices placed at checkpoints throughout a public park [22]. The AR systems come with IoT devices connected to them, which in turn tell the child's location and provide real-time feedback on walkable paths that help children experience their independence safely outdoors. AI is also being deployed in schools to create personalized learning platforms for students with cognitive disabilities to enjoy the same type of education as their classmates. AI meets wearable technology in London has employed a combination of artificial intelligence and smart wearables to help disabled children better explore the metropolis. Smartphone applications use artificial intelligence to deliver instant information about accessible routes and modes of transportation [11]. At the same time, wearables vibrate for blind children so they can pass over streets or public places. These inventions have dramatically improved the lives of children with mobility barriers, giving them more freedom to traverse urban environments.

7.1 Cost, Infrastructure and Policy Support

In addition to these technical challenges, this chapter outlines in detail a number of scalability factors that not only refer to its role in existing cloud computing infrastructures such as Amazon web services. This work presents how IoT and AI systems could be implemented citywide at a great investment cost, such as for smart sensors, data networks and digital platforms [2]. The salary required for setting down such systems can be prohibitive in smaller cities or developing regions. On the other hand, as technology progresses, we can expect that IoT sensors, AI systems and AR applications will also decrease in price over time, which might make these solutions available to a much wider range of cities. Besides, the data aspect of these AI systems in urban landscapes is fundamental to their scalability [52]. Predictive and actionable AI, regarding accessibility, needs large data sets. To do this, cities must determine how to collect reliable and accurate data that considers the whole disabled population when making decisions with other real-time AI systems. Furthermore, developing a single urban environment that combines several technologies, IoT, AI, and AR, requires close interaction between market participants: government agencies preparing legislative acts and incentives for residents, equipment manufacturers and developers of solutions based on it [41]. However, the scalability of these technologies will be constrained unless there are strong policy frameworks and cooperation. Concerning the legal alignment, these technologies must be implemented in accordance with international standards for accessibility as defined by UNCRPD. Under the UNCRPD, public spaces and services, including digital ones, are supposed to be open to facilitate access for persons with disabilities on an equal basis.

8 Proposed Solutions for Inclusive Smart Cities

This is the future of urban development, where technology and data are used to uplift the quality of life for all residents. Still, for any of these cities to truly thrive, they must be places where the full range of residents, especially vulnerable children with disabilities, can live and play. This kind of inclusivity will depend on innovations in technology and new models and methods for planning cities, strengthened infrastructure, or well-enforced policy environments that guarantee access as a basic human right [27]. It will deliver recommendations for the systematic integration of inclusive urban planning, infrastructure and digital solutions across smart city design phases; suggest tech-driven interventions to cater to disabilities faced by children in cities; and provide policy pointers toward sustainable and enforceable accessibility measures within future-proofed urban development plans.

8.1 Recommendations for the Integration of Inclusive Urban Planning, Infrastructure, and Digital Solutions in Smart Cities

To build genuinely inclusive smart cities, the first step must always be cemented in making intelligent urban planning truly inclusive. This means ensuring that public spaces are designed to be universally accessible, meaning usable by all people, to the greatest extent possible, without requiring specific adaptation. Suppose a smart cities initiative wants to plan public transportation systems, for instance [12]. In that case, it is not enough to integrate high-tech solutions such as IoT-capable buses or AI-driven traffic management. There is also a need for children with disabilities to be taken into account, as these systems must provide specific resources [80]. These solutions can range from wheelchair-accessible bus stops and audible intersection signals for visually impaired children to easy and customizable digital interfaces for children with cognitive disabilities. One mistake smart cities make that is costly and less effective than incorporating accessibility from the very beginning.

Building accessible public buildings and spaces should be a high priority for smart cities. This involves ensuring that schools, hospitals and recreational areas are adequately facilitated with ramps, lifts, and fixtures such as tactile paving, etc., enabling disabled children to access all necessary facilities. Moreover, digital infrastructure, everything from public Wi-Fi networks to interactive kiosks and city apps, must be accessible by design so that disabled children can engage with these systems just as effortlessly as their non-disabled peers [87]. Digital should also be an integrated part of creating inclusive smart cities. Smart city platforms need to be designed to consider what disabled children require. This could be in the form of voice assistants powered by AI to guide children through city services, IoT devices capable of monitoring and improving accessibility on the go or AR apps that give visual or auditory cues for kids with sensory impairments. More broadly, integrating these digital solutions with the very substance of cities will ensure that all disabled children are not left behind in this journey toward smarter and better-connected urban futures.

8.2 *Proposals for Specific Technological Innovations*

Thoughtless of whether disabled children were to benefit from technology in urban spaces before, the seamless blending of state-of-the-art tech with the living environment creates a new vision for what can be done differently. One of the most interesting spaces for innovation addresses smart urban mobility systems. This is an even greater ordeal for disabled children in traditional public transportation systems, which lack physical accessibility and are difficult to navigate due to numerous routes and schedules. IoT-powered solutions allow cities to develop transportation systems that offer up-to-date information on the location of accessible vehicle routes and stops. For example, a mobile application could inform a wheelchair user that the following bus has an elevation or inform a blind child through some voice alert system about when it will reach their stop. Smart traffic signals with sensors may also be able to detect an approaching disabled pedestrian and automatically extend crossing times; this is one of the tools that might help make urban mobility safer for all without discrimination [74].

IOT-enabled health solutions in Healthcare Another opportunity for improvement is in healthcare; IOT brings many possibilities from the perspective of intelligent cities. Regular access to healthcare is also critical for children with disabilities, including those living with chronic conditions. IoT devices can improve healthcare in a smart city by helping health systems become more responsive and efficient with the available resources [1]. Long-term, wearable health monitors might measure vital signs and automatically send the numbers to healthcare providers so they could keep track of remote and timely interventions made. AI-based telemedicine platforms could also provide consultations designed for children with disabilities, reducing the need to be physically present in healthcare centers and limiting their contact with healthcare professionals for children with mobility issues. The ability to avoid a live office visit can be advantageous.

Smart systems for education will make technology play a major role in the lives of disabled kids. Being self-reliant and living full of liberty is another vital sexagenarian field where technology can solve practical problems related to smart cities, particularly educating children. By nature, AI-driven education can also tailor specific learning paths and programs, such as customizing the lesson plan to fit the needs of each child with cognitive disability. Again, consider that a child who is autistic might learn best through pictures and visual stimuli versus a child with dyslexia who may need auditory tools [72]. Using speech-to-text and text-to-speech technologies in these platforms can further enable children with hearing or speech impairment to be part of classroom activities. On the other hand, designing classroom electronic products as an IoT, like interactive whiteboards and smart desks, are, for example, home automation devices that can be customized to adjust the environment according to all kinds of disabilities no matter what level they reach; this ensures its usage between different age groups effectively in response care among disabled children. Apart from these innovations mentioned above, specifically regarding assistive technology solutions that cater to children with disabilities, there is a whole new world of use cases around immersive learning and entertainment for disabled kids using augmented reality (AR) and virtual reality (VR) [77]. For visually or auditorily impaired kids, AR applications can overlay digital information on top of a physical space to help them navigate urban scenarios. In contrast, VR can help to recreate the outside world in a safe setting where children with disabilities might be able to simulate walking around freely and gain some independence, like learning how to take public transport.

8.3 Policy Recommendations for Long-Term Sustainability and Enforceability of Accessibility Measures

Policies that can bring about systematic change are equally important for building smart, accessible cities, technological innovations and inclusive master planning. A key obstacle to the effective implementation of accessibility is that it tends to be perceived as something optional or an add-on rather than a core aspect of urban development. This requires all levels of government to build smart city accessibility features and enforce these standards. These regulations must be consistent with global accessibility legislation, such as the UNCRPD's principle of ensuring that public spaces, services and technologies are accessible to people with disabilities. The legislation added that national regulators should secure the necessary powers for a status of institutional independence, binding all smart city initiatives to accessibility standards from their planning up until the final roll-out phases [10]. Enforcement bodies should also be able to find any project that does not follow the standard to encourage financial compliance. Another key policy recommendation is the creation of funding mechanisms for accessibility alongside that specific to smart city projects. There is a cost associated with integrating accessibility features to urbanism that adds up at the initial phases of development, causing high costs for developers, from offering a grant/subsidy to developers who design with accessibility in mind to providing tax breaks for tech companies constructing solutions that target children affected by special needs and disabilities (Table 3).

To maintain these measures for long-term success, smart city policies also need provisions for maintenance and updates. In light of the rapidly changing nature of technology and population needs, they may become outdated. This is why smart cities need to plan to regularly update their digital infrastructure to stay open and accessible for everyone who lives there. Such monitoring could involve routine auditing of public spaces and services for their accessibility, as well as systematic reviews concerning the application of new technologies in transportation, e.g., legal

Table 3 Proposed solutions for inclusive smart cities

Category	Recommendations
Inclusive urban planning and infrastructure	Ensure universal accessibility in public spaces, including wheelchair-accessible bus stops, audible intersection signals, and digital interfaces for children with cognitive disabilities.
	Prioritize accessible public buildings (schools, hospitals, recreational areas) with ramps, lifts, and tactile paving.
	Develop smart city platforms with AI voice assistants, IoT monitoring for accessibility, and AR apps for sensory-impaired children.
Technological innovations	Implement IoT-powered smart mobility solutions, such as real-time accessibility updates for public transport.
	Deploy AI-based telemedicine for remote healthcare access and wearable health monitors for continuous health tracking.
	Introduce AI-driven personalized education platforms with speech-to-text, text-to-speech, and smart classroom technology.
	Develop AR/VR applications to enhance navigation and simulate independent movement for disabled children.
Policy recommendations	Enforce legal frameworks aligning with UNCRPD to ensure mandatory accessibility in urban development.
	Establish funding mechanisms like grants, subsidies, and tax breaks for accessibility-driven smart city projects.
	Mandate regular audits of public spaces and digital infrastructure to ensure ongoing compliance with accessibility standards.
	Foster community engagement through co-design practices, public consultations, and collaboration with disability advocates.

regulations regarding driverless cars, health care or electronic environments used by disabled children regularly. Last, public participation must play a crucial role in smart city policy-making. They argue that the needs of disabled children, their families and carers must be placed centrally within planning for smart city projects. This requires community consultation, co-design practices, and constant conversations between elected officials, tech teams and disability advocates.

9 General Discussions

This chapter revealed major shortcomings in smart city planning when it comes to making sure disabled children have access and includes examples of the virtual tours offered during lockdowns. However, smart city initiatives have tended not to mainstream inclusive design principles that ensure disabled children can access public spaces, facilities and technologies. IoT, AI, and augmented reality can transform cities into more inclusive communities. Still, in practice, even cutting-edge smart city deployments serve general populations rather than catering specifically to disabilities affordances of children [81]. Most often, the fact that accessibility is an afterthought in smart city designs, and developers believe efficiency, sustainability, or innovation cannot be available. Smart public transportation systems, for example, provide real-time tracking and IoT-powered convenience features that cater to most citizens but may not be available or are prohibitively expensive if you have a cognitive disability as simple as ADHD [45]. Online platforms that are being utilized in the fields of education, health, and public services to educate or serve differently-abled children do not provide any special settings customized for them; therefore, disabled kids face difficulties if they want to use these online systems.

Still, the research also finds that technology can be used to narrow these gaps with dexterous application. The same IoT sensors, AI-driven systems, and AR applications which make our cities smart can also be employed to ensure that the urban spaces are both productive and accessible for all. For example, if IoT-enabled devices help provide real-time data for accessible routes, AI platforms may offer personalized education to children with cognitive disabilities. The main point of this work is that it changes the paradigm in which solutions are wallowed because technology itself or complex digital innovations aren't a bottleneck, but how they have been implemented and contextualized into the city fails to deliver health outcomes accessible for disabled children [7]. Legal frameworks have a crucial role in efforts to close the accessibility gap; Smart cities with inclusivity are built from the ground up. The law must require these high standards; they cannot be an afterthought. International agreements such as the UNCRPD lay a solid legal foundation that enables cities to guarantee full accessibility of public places and services for children with disabilities [13]. The obligation to provide inclusive infrastructure and services is then solidified through national laws like the RPwD Act in India. Nonetheless, these legal frameworks will only be successful if they are effectively operationalized. The results indicate that although a large number of countries, including India, have framed legislation for safeguarding the rights of disabled persons, an 'implementation gap' exists between policy and practice [16]. Moreover, these legal mandates are often not possible to be well accounted for in the implementation of smart city initiatives, with no clear enforcing mechanisms to ensure compliance, e.g. through auditing or fiscal sanctions, too little funding and lackluster planning. In the future, the city will be smarter, not just built on dumb legal frameworks but adapted and integrated into every walk of our urban development from inception to the implementation period. There needs to be a partnership between government regulators and policymakers, with innovative urban planners and technologists alongside the disabled advocates [24]. Therefore, through the availability of strong legal frameworks to measure and verify openness and accountability, developers can adopt a more inclusive approach in any section involving smart city. Governments must adhere to these standards so that cities are both inclusive and comply with their obligations under international laws and national legislative frameworks. One of the most critical insights from this research is the need for a seamless alignment between policy, technology, and urban planning to ensure accessibility for disabled children in smart cities [47]. The success of inclusive smart cities hinges on the ability of policymakers to work closely with urban planners and technologists to integrate accessibility into the core of smart city designs. Technology alone cannot solve accessibility challenges; it must be guided by strong policy frameworks and supported by an urban infrastructure that reflects the population's diverse needs.

There is a disconnect between these three elements in many smart city projects. Technological solutions are often developed in isolation from the policy frameworks that mandate accessibility, and urban planners may lack the resources or expertise to implement these technologies in a way that serves disabled children [8]. A more collaborative approach is necessary to address this, where policy, technology, and planning are aligned from the beginning of a smart city project. For example, when developing IoT-powered transportation systems, policymakers must ensure that accessibility standards are embedded into the system's design, such as by requiring real-time updates on the availability of accessible vehicles or designing apps with customizable features for disabled users [66]. Similarly, urban planners must collaborate with technologists to ensure that digital infrastructure, such as public Wi-Fi networks or digital kiosks, is accessible to children with sensory or cognitive disabilities.

10 Conclusion and Future Directions

While every city continues to bring out the smart, tech-driven urban landscape in different shades of execution and innovation, there is also a reminder that not much can be done without inclusion, at least for children with disabilities. While smart cities promise greater efficiency, sustainability and improved quality of life for all residents, those benefits will not be fully realized if every citizen cannot participate in urban living. Human rights and the basic rights of children to public services, transportation, healthcare, education and access to all places within their community are not dispensable in service of smart city initiatives [99]. This research has demonstrated the potential for transforming accessible settings through cross-reference and interaction between legal frameworks, urban planning provisions, and technology in smart cities, especially regarding children with disabilities. Yet, solid legal underpinning, which the UNCRPD and national laws such as India's RPwD Act ought to be, can provide a sound basis for protecting rights. However, for such legal frameworks to lead to critical change, they need effective implementation and enforcement [3]. Urban planning determines how these legal requirements are embodied in physical and digital spaces, making them accessible. Innovations such as IoT, AI, and augmented reality are powerful means of further enhancing accessibility [54]. This can range anywhere from smart transportation systems that provide real-time accessibility routes to AI-based educational platforms which generate a personalized learning experience. This three-pronged combination of legal backstopping, urban planning, and technology could forge smart cities that are not merely smart but, above all else, inclusive.

Although great advancements have been achieved in technology to increase accessibility within smart cities, the territory is still very new. Emerging technologies are being introduced daily to integrate into some form of other. The chapter concludes by suggesting the research and application agenda, including AI, ML blockchain, and 5G networks, to boost urban inclusivity further. For instance, AI and ML could help predict the accessibility needs of these children at a given time to offer personalized solutions as they move through the city [92]. Similarly, blockchain technology might be a way to develop openly available pathways for managing accessibility-related data that comprehensively documents the extent of smart city efforts that are or are not inclusive of all residents. Second, broader research is needed to understand how 5G networks and, hence, IoT or AR applications based on them can better cover more advanced various types of everyday environments toward addressing accessibility in the world around us. We are already seeing this, as Singapore is racing toward becoming the world's first smart nation, so more research has to be done on how these emerging technologies can integrate into existing smart city frameworks for an even wider-reaching urban development next-generation that will include all.

References

1. Abdulmalek S, Nasir A, Jabbar WA, Almuhaya MAM, Bairagi AK, Al-Masrur Khan M, Kee S-H (2022) IoT-based healthcare-monitoring system towards improving quality of life: a review. Healthcare 10(10):1993. https://doi.org/10.3390/healthcare10101993
2. Abdulsalam YS, Hedabou M (2021) Security and privacy in cloud computing: technical review. Future Internet 14(1):11. https://doi.org/10.3390/fi14010011
3. Abhiyan, Wada Na Todo. 2017. Sustainable development goals: agenda 2030
4. Adlakha R, Guha SK (2012) Protecting the disabled persons under the human rights regime—the shift from welfare to rights. J Ind Law Soc 3:67
5. Alahi MEE, Sukkuea A, Tina FW, Nag A, Kurdthongmee W, Suwannarat K, Mukhopadhyay SC (2023) Integration of IoT-enabled technologies and artificial intelligence (AI) for Smart City scenario: recent advancements and future trends. Sensors 23(11):5206. https://doi.org/10.3390/s23115206
6. AL-Dabbagh R (2022) Dubai, the sustainable, Smart City. Renew Energy Environ Sustain 7:3. https://doi.org/10.1051/rees/2021049
7. Aledhari M, Razzak R, Qolomany B, Al-Fuqaha A, Saeed F (2022) Biomedical IoT: enabling technologies, architectural elements, challenges, and future directions. IEEE Access 10:31306. https://doi.org/10.1109/ACCESS.2022.3159235
8. Ali SA, Elsaid SA, Ateya AA, ElAffendi M, Abd El-Latif AA (2023) Enabling technologies for next-generation smart cities: a comprehensive review and research directions. Future Internet 15(12):398. https://doi.org/10.3390/fi15120398
9. Alsyouf A, Lutfi A, Alsubahi N, Alhazmi FN, Al-Mugheed K, Anshasi RJ, Alharbi NI, Albugami M (2023) The use of a technology acceptance model (TAM) to predict patients' usage of a personal health record system: the role of security, privacy, and usability. Int J Environ Res Public Health 20(2):1347. https://doi.org/10.3390/ijerph20021347
10. Altaf M, Mostafa S, Rodney A. Stewart. (2025) Enhancing accessibility in smart Cities: AI-based autonomous inspection and certification framework. Dig Eng 5:100034. https://doi.org/10.1016/j.dte.2024.100034
11. Araya ED (2023) Augmented education in the global age; artificial intelligence and the future of learning and work. Routledge
12. Asian Development Bank (2022) Inclusive Cities: Urban Area Guidelines. Asian Development Bank. https://doi.org/10.22617/TIM200037-2

13. Attar RW, Habes M, Almusharraf A, Alhazmi AH, Attar RW (2024) Exploring the impact of smart Cities on improving the quality of life for people with disabilities in Saudi Arabia. Front Built Environ 10:1398425. https://doi.org/10.3389/fbuil.2024.1398425
14. Babik I, Gardner ES (2021) Factors affecting the perception of disability: a developmental perspective. Front Psychol 12:702166. https://doi.org/10.3389/fpsyg.2021.702166
15. Bajpai N, Biberman J (2021) India's smart city program: challenges and opportunities. Center for Sustainable Development, Earth Institute, Columbia University
16. Moumita Barman. 2023. "Towards inclusivity: assessing the rights of persons with disabilities act, 2016.". https://www.cdpp.co.in/articles/towards-inclusivity-assessing-the-rights-of-persons-with-disabilities-act-2016 (October 26, 2024)
17. Belli L, Cilfone A, Davoli L, Ferrari G, Adorni P, Di Nocera F, Dall'Olio A et al (2020) IoT-enabled smart sustainable cities: challenges and approaches. Smart Cities 3(3):1039–1071. https://doi.org/10.3390/smartcities3030052
18. Berghs M, Atkin K, Graham H, Hatton C, Thomas C (2016) Scoping models and theories of disability. In: Implications for public health research of models and theories of disability: a scoping study and evidence synthesis. NIHR Journals Library. https://www.ncbi.nlm.nih.gov/books/NBK378951/ (October 25, 2024)
19. Bibri SE, Krogstie J, Kärrholm M (2020) Compact city planning and development: emerging practices and strategies for achieving the goals of sustainability. Dev Built Environ 4:100021. https://doi.org/10.1016/j.dibe.2020.100021
20. Buder S, Perry R (2024) The social model of disability explained. Social Creatures. https://www.thesocialcreatures.org/thecreaturetimes/the-social-model-of-disability (December 15, 2024)
21. Campisi T, Severino A, Al-Rashid MA, Pau G (2021) The development of the smart cities in the connected and autonomous vehicles (CAVs) era: from mobility patterns to scaling in Cities. Infrastructures 6(7):100. https://doi.org/10.3390/infrastructures6070100
22. Carnevale MJ (2017) Toronto augmented reality map: enhancing citizen engagement with open government data using contemporary media platforms (masters thesis). OCAD University
23. Chen Y, Wang X, Le B, Wang L (2024) Why people use augmented reality in heritage museums: a socio-technical perspective. Herit Sci 12(1):108. https://doi.org/10.1186/s40494-024-01217 1
24. Christofi A (2023) Smart cities and cumulative effects on fundamental rights. Internet Policy Rev 12(1) https://policyreview.info/articles/analysis/smart-cities-and-cumulative-effects-on-fundamental rights (October 26, 2024)
25. Colding J, Nilsson C, Sjöberg S (2024) Smart cities for all? Bridging digital divides for socially sustainable and inclusive cities. Smart Cities 7(3):1044–1059. https://doi.org/10.3390/smartcities7030044
26. Chris Creed, Maadh Al-Kalbani, Arthur Theil, Sayan Sarcar, and Ian Williams. 2023. "Full article: inclusive augmented and virtual reality: a research agenda.". https://www.tandfonline.com/doi/full/10.1080/10447318.2023.2247614 (November 18, 2024)
27. Disability Inclusive and Accessible Urban Development Network. 2017. "The inclusion imperative: towards disability-Inclusive and accessible urban development"
28. Dolphin S, Downing M, Cirrincione M, Samuta A, Leite K, Noble K, Walsh B (2024) Information accessibility in the form of braille. IEEE Open J Eng Med Biol 5:205–209. https://doi.org/10.1109/OJEMB.2024.3364065
29. Dudley J, Yin L, Garaj V, Kristensson PO (2023) Inclusive immersion: a review of efforts to improve accessibility in virtual reality, augmented reality and the Metaverse. Virtual Reality 27(4):2989–3020. https://doi.org/10.1007/s10055-023-00850-8
30. Elassy M, Al-Hattab M, Takruri M, Badawi S (2024) Intelligent transportation systems for sustainable smart cities. Transp Eng Aust 16:100252. https://doi.org/10.1016/j.treng.2024.100252
31. Fernández-Batanero JM, Montenegro-Rueda M, Fernández-Cerero J, García-Martínez I (2022) Assistive technology for the inclusion of students with disabilities: a systematic review. Educ Technol Res Dev 70(5):1911–1930. https://doi.org/10.1007/s11423-022-10127-7

32. Ferrer J-R (2017) Barcelona's smart city vision: an opportunity for transformation. Field Actions Sci Rep (Special Issue 16):70–75. https://journals.openedition.org/factsreports/4367 (October 25, 2024)
33. Ferri D (2017) Unveiling the challenges in the implementation of Article 24 CRPD on the right to inclusive education. A case-study from Italy. Laws 7(1):1. https://doi.org/10.3390/laws7010001
34. Field MJ, Jette AM, Institute of Medicine (US) Committee on Disability in America (2007) Transportation patterns and problems of people with disabilities. In: The future of disability in America. National Academies Press, US. https://www.ncbi.nlm.nih.gov/books/NBK11420/ (October 25, 2024)
35. Gao X, He P, Zhou Y, Qin X (2024) Artificial intelligence applications in smart healthcare: a survey. Future Internet 16(9):308. https://doi.org/10.3390/fi16090308
36. Garcia-Font V (2021) Conceptual technological framework for smart cities to move towards decentralized and user-centric architectures using DLT. Smart Cities 4(2):728–745. https://doi.org/10.3390/smartcities4020037
37. GEM Report UNESCO (2023) Technology in education: a case study on Singapore. GEM Report UNESCO. https://doi.org/10.54676/HOOV5879
38. Haleem A, Javaid M, Singh RP, Suman R (2021) Telemedicine for healthcare: capabilities, features, barriers, and applications. Sens Int 2:100117. https://doi.org/10.1016/j.sintl.2021.100117
39. Herweck P (2023) Why developing nations are defining the smart cities of the future. World Economic Forum. https://www.weforum.org/stories/2023/01/smart-cities-developing-nations-davos23/ (January 16, 2025)
40. Hou L, Latif J, Mehryar P, Withers S, Plastropoulos A, Shen L, Ali Z (2024) An autonomous wheelchair with health monitoring system based on internet of thing. Sci Rep 14(1):5878. https://doi.org/10.1038/s41598-024-56357-y
41. Hsu Y-C, Huang T, Verma H, Mauri A, Nourbakhsh I, Bozzon A (2022) Empowering local communities using artificial intelligence. Patterns 3(3):100449. https://doi.org/10.1016/j.patter.2022.100449
42. Hui CX, Dan G, Alamri S, Toghraie D (2023) Greening smart cities: an investigation of the integration of urban natural resources and smart city technologies for promoting environmental sustainability. Sustain Cities Soc 99:104985. https://doi.org/10.1016/j.scs.2023.104985
43. Iyer LS (2021) AI enabled applications towards intelligent transportation. Transp Eng Aust 5:100083. https://doi.org/10.1016/j.treng.2021.100083
44. Jain S, Jain M (2024) Revisiting the conceptual terrains of the right to accessibility in India: the role of judicial enforcement. Laws 13(4):54. https://doi.org/10.3390/laws13040054
45. Johri A, Sayal A, Chaithra N, Jha J, Aggarwal N, Pawar D, Gupta V, Gupta A (2024) Crafting the techno-functional blocks for Metaverse—a review and research agenda. Int J Inf Manag Data Insights 4(1):100213. https://doi.org/10.1016/j.jjimei.2024.100213
46. Kaharevic A, Wihlborg E (2024) Whose future is smart?: A systematic literature review of smart Cities and disadvantaged Neighborhoods. In: Proceedings of the 17th international conference on theory and practice of electronic governance, Pretoria South Africa. ACM, pp 318–326. https://doi.org/10.1145/3680127.3680220
47. Kaiser ZRM, Abdullah. (2024) Smart governance for smart cities and nations. J Econ Technol 2:216–234. https://doi.org/10.1016/j.ject.2024.07.003
48. Kolotouchkina O, Barroso CL, Sánchez JLM (2022) Smart cities, the digital divide, and people with disabilities. Cities 123:103613. https://doi.org/10.1016/j.cities.2022.103613
49. Languillon-Aussel R (2021) Digitalization of public spaces. The great urban change? Artic J Urban Res 22. https://doi.org/10.4000/articulo.4859
50. Lawson A, Beckett AE (2021) The social and human rights models of disability: towards a complementarity thesis. Int J Hum Rights 25(2):348–379. https://doi.org/10.1080/13642987.2020.1783533
51. Li J (2024) Beyond sight: enhancing augmented reality interactivity with audio-based and non-visual interfaces. Appl Sci 14(11):4881. https://doi.org/10.3390/app14114881

52. Mahdi MN, Ahmad AR, Qassim QS, Natiq H, Subhi MA, Mahmoud M (2021) From 5G to 6G technology: meets energy, internet-of-things and machine learning: a survey. Appl Sci 11(17):8117. https://doi.org/10.3390/app11178117
53. Makkonen T, Inkinen T (2024) Inclusive smart Cities? Technology-driven urban development and disabilities. Cities 154:105334. https://doi.org/10.1016/j.cities.2024.105334
54. Math SB, Gowda GS, Basavaraju V, Manjunatha N, Kumar CN, Philip S, Gowda M (2019) The rights of persons with disability act, 2016: challenges and opportunities. Indian J Psychiatry 61(Suppl 4):S809. https://doi.org/10.4103/psychiatry.IndianJPsychiatry_105_19
55. Mattie JL, Borisoff JF, Wong AS, Miller WC (2015) User perceptions of existing home access solutions and a novel home access device. Disabil Rehabil Assist Technol 11(8):668. https://doi.org/10.3109/17483107.2015.1027302
56. Mindell JS, Amin S, Mackett RL, Taylor J, Yaffe S (2024) Chapter two—disability and travel. In: Mindell JS, Watkins SJ (eds) Advances in transport policy and planning, health on the move 3: the reviews. Academic, pp 47–87. https://doi.org/10.1016/bs.atpp.2023.11.009
57. Mladenov T (2013) The UN convention on the rights of persons with disabilities and its interpretation. Alternatives 7(1):69–82. https://doi.org/10.1016/j.alter.2012.08.010
58. Mone G (2015) The new smart cities. Commun ACM 58(7):20–21. https://doi.org/10.1145/2771297
59. Moon NW, Baker PMA, Goughnour K (2019) Designing wearable technologies for users with disabilities: accessibility, usability, and connectivity factors. J Rehabil Assist Technol Eng 6:2055668319862137. https://doi.org/10.1177/2055668319862137
60. Mora L, Gerli P, Ardito L, Petruzzelli AM (2023) Smart city governance from an innovation management perspective: theoretical framing, review of current practices, and future research agenda. Technovation 123:102717. https://doi.org/10.1016/j.technovation.2023.102717
61. Musa AA, Malami SI, Alanazi F, Ounaies W, Alshammari M, Haruna SI (2023) Sustainable traffic management for smart cities using internet-of-things-oriented intelligent transportation systems (ITS): challenges and recommendations. Sustainability 15(13):9859. https://doi.org/10.3390/su15139859
62. Nasser, Nidal, Asmaa Ali, Lutful Karim, and AbdulAziz Al-Helali. 2024. "Enhancing mobility for the visually impaired with AI and IoT-enabled mobile applications." ScienceOpen https://doi.org/10.14293/PR2199.000775.v1
63. Neto O, De JS (2019) Inclusive smart cities: theory and tools to improve the experience of people with disabilities in urban spaces (Doutorado em Sistemas Eletrônicos). Universidade de São Paulo. https://doi.org/10.11606/T.3.2019.tde-30012019-090025
64. Office of Technology Assessment (1995) Wireless technologies and the national information infrastructure. U.S. Government Printing Office
65. Okonta DE, Vukovic V (2024) Smart cities software applications for sustainability and resilience. Heliyon 10(12):e32654. https://doi.org/10.1016/j.heliyon.2024.e32654
66. Oladimeji D, Gupta K, Kose NA, Gundogan K, Ge L, Liang F (2023) Smart transportation: an overview of technologies and applications. Sensors 23(8):3880. https://doi.org/10.3390/s23083880
67. Overstreet K (2021) Building a city from scratch: the story of Songdo, Korea. ArchDaily. https://www.archdaily.com/962924/building-a-city-from-scratch-the-story-of-songdo-korea (October 25, 2024)
68. Pearson SR, Shafik M (2024) Smart cities enhancing the lives of people with disabilities. In: Qin Y, Zhou X, Yang S, Zhao J, Chen B, Al-Ahmad M (eds) MATEC Web of Conferences, vol vol 401. 07004. EDP Sciences. https://doi.org/10.1051/matecconf/202440107004
69. Pereira V, Gabriela MA, Cunha TJ, Lampoltshammer PP, Testa MG (2017) Increasing collaboration and participation in smart city governance: a cross-case analysis of smart city initiatives. Inf Technol Dev 23(3):526–553. https://doi.org/10.1080/02681102.2017.1353946
70. Pineda VS (2022) What is inclusive and accessible public space? J Public Space 7(2):5–8. https://doi.org/10.32891/jps.v7i2.1607

71. Piracha A, Chaudhary MT (2022) Urban air pollution, urban heat island and human health: a review of the literature. Sustainability 14(15):9234. https://doi.org/10.3390/su14159234
72. Poli M (2021) Smart technologies and the case of people with disabilities: a preliminary overview. IOS Press. https://doi.org/10.3233/FAIA210096
73. Reber L, Kreschmer JM, James TG, Junior JD, DeShong GL, Parker S, Meade MA (2022) Ableism and contours of the attitudinal environment as identified by adults with long-term physical disabilities: A qualitative study. Int J Environ Res Public Health 19(12):7469. https://doi.org/10.3390/ijerph19127469
74. Redij S, Rao A, Kumar SA, Jadhav A (2022) Smart bus tracking for blind and visually impaired. In: Patil MD, Vyawahare VA (eds) ITM Web of Conferences, vol vol 44: 02001. EDP Sciences. https://doi.org/10.1051/itmconf/20224402001
75. Rezaee R, Ghaffari M, Rabiei R, Kavousi A, Rakhshanderou S (2024) Design and usability evaluation of a mobile application for self-care among Iranian adolescents. BMC Public Health 24(1):892. https://doi.org/10.1186/s12889-024-18341-z
76. Salha RA, Jawabrah MQ, Badawy UI, Jarada A, Alastal AI (2020) Towards smart, sustainable, accessible and inclusive city for persons with disability by taking into account checklists tools. J Geogr Inf Syst 12(04):348–371. https://doi.org/10.4236/jgis.2020.124022
77. Semary HE, Al-karawi KA, Abdelwahab MM, Elshabrawy AM (2024) A review on internet of things (IoT)-related disabilities and their implications. J Disabil Res 3:20240012. https://doi.org/10.57197/JDR-2024-0012
78. Shin S-Y, Kim D, Chun SA (2021) Digital divide in advanced smart city innovations. Sustainability 13(7):4076. https://doi.org/10.3390/su13074076
79. Singh S (2019) Medical council of India's new guidelines on admission of persons with specified disabilities: unfair, discriminatory and unlawful. Indian J Med Ethics IV(1):29–34. https://doi.org/10.20529/IJME.2018.064
80. Singh P, Elmi Z, Meriga VK, Pasha J, Dulebenets MA (2022) Internet of things for sustainable railway transportation: past, present, and future. Clean Logist Supply Chain 4:100065. https://doi.org/10.1016/j.clscn.2022.100065
81. Szaszák G, Kecskés T (2020) Universal open space design to inform digital technologies for a disability-inclusive place-making on the example of Hungary. Smart Cities 3(4):1293–1333. https://doi.org/10.3390/smartcities3040063
82. Ullah A, Anwar SM, Li J, Nadeem L, Mahmood T, Rehman A, Saba T (2024) Smart cities: the role of internet of things and machine learning in realizing a data-centric smart environment. Complex Intell Syst 10(1):1607–1637. https://doi.org/10.1007/s40747-023-01175-4
83. UNCRPD (2022) United Nations convention on the rights of persons with disabilities. National Disability Authority. https://nda.ie/disability-policy/uncrpd (October 25, 2024)
84. UNHR (2006) Convention on the rights of persons with disabilities. OHCHR. https://www.ohchr.org/en/instruments-mechanisms/instruments/convention-rights-persons-disabilities (October 25, 2024)
85. Upreti M, Saikia P, Shilky PL, Kumar A (2024) Major challenges in the urbanizing world and role of earth observations for livable cities. In: Earth observation in urban monitoring. Elsevier, pp 23–52. https://doi.org/10.1016/B978-0-323-99164-3.00002-1
86. Valvo L, Alice DC, Garlisi D, Giuliano F, Giarré L, Tinnirello I (2021) A navigation and augmented reality system for visually impaired people. Sensors 21(9):3061. https://doi.org/10.3390/s21093061
87. van Hoof J, Marston HR, Kazak JK, Buffel T (2021) Ten questions concerning age-friendly cities and communities and the built environment. Build Environ 199:107922. https://doi.org/10.1016/j.buildenv.2021.107922
88. van Twist A, Ruijer E, Meijer A (2023) Smart cities & citizen discontent: a systematic review of the literature. Gov Inf Q 40(2):101799. https://doi.org/10.1016/j.giq.2022.101799
89. Wang W-H, Hsu W-S (2023) Integrating artificial intelligence and wearable IoT system in long-term care environments. Sensors 23(13):5913. https://doi.org/10.3390/s23135913

90. Wang CH, Steinfeld E, Maisel JL, Kang B (2021) Is your smart city inclusive? Evaluating proposals from the U.S. Department of Transportation's Smart City challenge. Sustain Cities Soc 74:103148. https://doi.org/10.1016/j.scs.2021.103148
91. Wegen V, Myla JL, Herder RA, Pastore-Wapp M, Van Wegen EEH, Bohlhalter S, Nef T, Krack P, Vanbellingen T (2023) An overview of wearable haptic technologies and their performance in virtual object exploration. Sensors 23(3):1563. https://doi.org/10.3390/s23031563
92. Whaiduzzaman M, Barros A, Chanda M, Barman S, Tania Sultana M, Rahman S, Roy S, Fidge C (2022) A review of emerging technologies for IoT-based smart cities. Sensors 22(23):9271. https://doi.org/10.3390/s22239271
93. Whaley BA, Martinis JG, Pagano GF, Barthol S, Senzer J, Williamson PR, Blanck PD (2024) The Americans with disabilities act and equal access to public spaces. Laws 13(1). https://doi.org/10.3390/laws13010005
94. WHO (2011) World report on disability. WHO. https://www.who.int/teams/noncommunicable-diseases/sensory-functions-disability-and-rehabilitation/world-report-on-disability (October 15, 2024)
95. WHO (2024) Urban Health. WHO. https://www.who.int/health-topics/urban-health (January 16, 2025)
96. Wilding D. 2007. The impact of digital platforms on news and journalistic content
97. Wolniak R, Stecuła K (2024) Artificial intelligence in smart cities—applications, barriers, and future directions: a review. Smart Cities 7:1346–1389. https://doi.org/10.3390/smartcities7030057
98. Zhou S, Loiacono ET, Kordzadeh N (2024) Smart cities for people with disabilities: a systematic literature review and future research directions. Eur J Inf Syst 33(6):845–862. https://doi.org/10.1080/0960085X.2023.2297974
99. Zhu H, Shen L, Ren Y (2022) How can smart city shape a happier life? The mechanism for developing a happiness driven smart city. Sustain Cities Soc 80:103791. https://doi.org/10.1016/j.scs.2022.103791

Index

H. Kaddouri et al. (eds.), *Smart Cities for Sustainable and Urban Development*, Green Energy and Technology, https://doi.org/10.1007/978-3-031-99001-4

The manufacturer's authorised representative in the EU is Springer Nature Customer Service Centre GmbH, Europaplatz 3, 69115 Heidelberg, Germany. If you have any concerns regarding our products, please contact ProductSafety@springernature.com

Printed and bound by CPI Group (UK) Ltd, Croydon, CR0 4YY
21/09/2026
02220016-0001